QUALITY INSPIRED MANAGEMENT

The Key to Sustainability

C. Harold Aikens
The University of Tennessee

Prentice Hall
Boston Columbus Indianapolis New York San Francisco Upper Saddle River
Amsterdam Cape Town Dubai London Madrid Milan Munich Paris Montreal Toronto
Delhi Mexico City Sao Paulo Sydney Hong Kong Seoul Singapore Taipei Tokyo

Vice President and Editor in Chief: Vernon R. Anthony
Acquisitions Editor: David Ploskonka
Editorial Assistant: Nancy Kesterson
Director of Marketing: David Gesell
Executive Marketing Manager: Derril Trakalo
Senior Marketing Coordinator: Alicia Wozniak
Marketing Assistant: Les Roberts
Project Manager: Maren L. Miller
Senior Managing Editor: JoEllen Gohr
Associate Managing Editor: Alexandrina Benedicto Wolf
Senior Operations Supervisor: Pat Tonneman
Operations Specialist: Laura Weaver
Art Director: Jayne Conte
Cover Designer: Suzanne Duda
Cover Image: Istockphoto.com
AV Project Manager: Janet Portisch
Full-Service Project Management: Smitha Pillai/ S4Carlisle Publishing Services
Composition: S4Carlisle Publishing Services
Printer/Binder: Edwards Brothers
Cover Printer: Coral Graphics
Text Font: Minion

Credits and acknowledgments borrowed from other sources and reproduced, with permission, in this textbook appear on the appropriate page within the text. Unless otherwise stated, all figures and tables have been provided by the author.

Library of Congress Cataloging-in-Publication Data

Aikens, C. Harold.
Quality inspired management : the key to sustainability/C. Harold Aikens.
p. cm.
Includes bibliographical references and index.
ISBN 978-0-13-119756-5 (alk. paper)
1. Total quality management. 2. Sustainable development. I. Title.
HD62.15.A3953 2011
658.4'013—dc22

2009039241

10 9 8 7 6 5 4 3 2 1

Prentice Hall
is an imprint of

PEARSON

www.pearsonhighered.com

ISBN 10: 0-13-119756-8
ISBN 13: 978-0-13-119756-5

To Helen, without whose love, support, and inspiration this book would not have been possible.

FOREWORD

It is a privilege and a very pleasant surprise to be invited to write the Foreword for this latest work of Hal Aikens. It is a surprise because I have been (and remain) a trenchant critic of quality management or rather the various and varied management models that were constructed (especially those of Crosby, Deming, and Juran and their followers) on the quality ideology first formalized by Walter Shewhart—and practiced by the Moravian shoemaker, Thomas Bat'a, in the early twentieth century in a way seldom surpassed to this day. It is my contention that without very careful modification to suit the stage of development of an organization and its unique market context, advice that is drawn from a model that has no explicit theoretical foundation (and as a consequence almost as many descriptions as promoters), little empirical support for its assumptions and "principles," and sees the customer as omni-important is more likely to lead to failure than success for the twenty-first-century multistakeholder organization. Professor Aikens has observed that "the dawn of the twenty-first century ushered in a new era." What he has done in this most welcome and long overdue book is demonstrate that this new era has also ushered in a new, reinvigorated quality management.

It is a privilege to have this opportunity to be associated, albeit tenuously, with what is a refreshing and refreshingly honest appraisal and application of the quality ideology. The title is most apt and the methodology is sufficiently clever to evince envy—I do so wish I had thought of looking at quality management from four different levels. Had I done so many of the difficulties I (and no doubt many others) have faced with understanding and using quality management could have been avoided.

Professor Aikens has provided a masterful description, analysis, and application of quality management. The work is comprehensive, free of cant and idolatry, and most importantly is set in an organizational context that is of unprecedented complexity and volatility, and where the principal resource is knowledge. At the earliest point (in the first paragraph of the preface) the reader is left in no doubt that the world has changed; the knowledge economy is now a reality and knowledge has characteristics that are fundamentally different from the physical resource that fueled the manufacturing economy in which quality management was first developed. A rapidly growing number of organizations (most notably in information technology, e.g., Microsoft) find that in contrast to what happens with physical resources the continued application of knowledge yields increasing returns—value increases rather than decreases with use.

The shift from a manufacturing to a knowledge economy has changed the way in which organizations behave, and perforce changes the context in which management models such as those of quality management must be assessed. That shift also changes the assessment criteria. By placing quality management in the knowledge economy, emphasizing its strategic role and documenting its methodologies and extraordinary array of tools, Professor Aikens shows that quality management has moved into what he describes as its third generation. In this new form, quality as a whole-of-organization guiding strategy (which is NOT applicable to *all* organizations at *all* stages of their development) is able to be distinguished from other strategies and separated from the quality methodologies (PDCA, the process approach) and tools (statistical process control, Taguchi loss function, Poka-Yoke), which ARE relevant to all organizations because they reduce cost. Of course it is the magic of the quality methodologies and tools that, if used to improve quality, concomitantly reduce costs and, if used to reduce costs, concomitantly improve quality. Third-generation quality management also identifies the customer as one of several (often many) stakeholders, any one of which has the capacity to impede the organizational aim if their wants and expectations are not met. For all of those reasons *Quality Inspired Management* carries a warning: Managers should treat with extreme caution advice derived from a management model that was written in (and for) a very different era, and fails to recognize that customers are not the only stakeholders that can cause an organization to fail if their wants and expectations are not met.

Whether it is used as a whole-of-organization strategy utilizing its proven cost-reducing/quality-improving methodologies and tools, or a *coherent* set of methodologies and tools used to achieve *other strategies* at least cost (and with improved quality), quality management is a *sine qua non* of organization success and demands the attention of all who seek that goal.

Although this outstanding book will be read with profit by a great many audiences from the quality expert to the CEO, it is likely to be most appreciated by students and their instructors. Thus far instructors have labored without appropriate tools to convince their students that quality management is not a management fad and to demonstrate its relevance to the contemporary (business and nonbusiness) multistakeholder organization that so obviously functions in an environment where rapid, often frame-breaking change is the dominant feature and knowledge is the dominant resource.

Kevin Foley, BComm (Hons) (UNSW),
MComm (UNSW), Ph.D. (ANU)
Patron, Fellow, and Member of the Board of Directors,
Australian Organisation for Quality; Adjunct Professor,
Faculty of Business, University of Technology, Sydney;
Adjunct Professor, Centre for Management Quality
Research, Royal Melbourne Institute of Technology;
Visiting Professor, Business School,
University of Versailles.

PREFACE

The dawn of the twenty-first century ushered in a new era—one characterized by ubiquitous, affordable, and portable telecommunications; the collapse of the Berlin Wall signaling the end of the Cold War; the development of and easy access to the World Wide Web; and the widespread practice by the private sector of trying to penetrate markets on a global scale using strategies such as supply chaining, outsourcing, offshoring, and insourcing. These factors have dramatically changed the nature of industrial warfare, and the total annihilation of barriers to international competition has radically changed the rules of engagement. In short, during the last decade the world has become a flatter place[1] in the sense that the impediments that once existed to the free exchange of labor, materials, information, and capital between the industrialized developed nations and the emerging economies of the world (places like India and China) have all but disappeared. For the most part, corporations have the advantage today of conducting business on a worldwide level playing field.

The business implications are staggering. Viability and sustainability are no longer regional issues and consequently can no longer be resolved from a local perspective. Success and growth depend on visions and strategies that cut across political, cultural, geographical, and religious boundaries. And if the complexity thus created were not enough, the contemporary manager is confronted with continual and fast-moving change. Driven by spiraling technological developments, upward trends in customer expectations, societal demands for greater accountability in resource utilization, better stewardship in conserving the planet, and growing concerns for the human condition with specific reference to the quality of work life, the senior leaders in today's industrial organization must make decisions that are multifaceted, involve complex relationships that are often unknown and in some cases unknowable, are unstructured, require information that is incomplete or unreliable, and have fuzzy solution sets. These factors do not necessarily represent recent phenomena. Nevertheless, their impacts are exacerbated by pressures imposed by the system to respond more effectively and quickly to stakeholder demands in an environment where the dynamics can change in a heartbeat. Many managers are ill-equipped or unwilling to rise to these challenges and, while well-intended, the decisions they make are often not based on a systems view or holistic approach. Decision approaches tend to accommodate the reactive instantaneous response to crises. Once fires are fought and damage control contains a problem, little if anything is done to leverage the experience and turn it into a learning opportunity to correct the system and prevent a future problem from recurring. That takes time—a luxury unavailable in a culture where, upon containing one fire management's reward is an urgent call to extinguish another. The system can control management rather than the other way around and as a result nothing ever really improves.

The fix lies in a smarter management paradigm—one that embodies sound principles—those that have formed the common bases for the many sundry philosophies each of which was debuted in its time as the *solution suprême*, the evolution of a body of standards to govern quality control in international commerce, and the emergence of business excellence models that provide rewards and recognition for superior performance. In the early days the most widely accepted approaches were those that were popularized through the association and promotion of their respective guru creators, and were given names that include statistical process control (SPC), total quality management/control (TQM/TQC), business process reengineering (BPR), theory of constraints (TOC), Six Sigma (SS), design for Six Sigma (DFSS), and lean production/Six Sigma. Each of these at the time of its introduction professed to be the paragon of contemporary management practice and masqueraded as something innovative and distinctive from the rest. On the standards front, drawing what Britain and America had used during World War II, and later broadened to include management and strategic concepts, the International Organization for Standardization (ISO) established a set of international standards that have gone through numerous revisions. To incentivize quality initiatives, many countries adopted award-based programs to serve as a set of guidelines for excellence—the most notable of these being the Malcolm Baldrige National Quality Award and the European Foundation for Quality Management business excellence model.

Ironically, and fortunately, the ideas underlying all of these approaches, developed at different times by different people, in different parts of the world, can be found to be rooted in a common set of values and principles. The reason why there continues to be a market for new names for old ideas, albeit each time delivered with some novel twists, may be management's failure to properly comprehend and apply the basic principles that have been available to them for so long. A typical and flawed implementation strategy, particularly in the early days, has been to cherry-pick those components of any of the professed philosophies that are comfortable, convenient, politically correct, or economically viable, and to ignore or defer the rest. This is in stark contrast

[1]Friedman, Thomas L. *The World Is Flat: A Brief History of the Twenty-first Century.* New York: Farrar, Straus and Giroux, 2005.

to the political will and intestinal fortitude required to develop a culture that supports a meta-management perspective.[2] Meta-management takes a holistic view of quality, recognizing its legitimacy as a cohesive management theory that focuses on stakeholder interests. By contrast, traditional models drive decision-making from a competitor/investor perspective, and this can and often does counter the organization's proclaimed commitment to value the needs of *all* stakeholders.

Such observations as these motivated me to embark on this book project.

THE VISION

My vision to write a book that would comprehensively treat the subject began eight years ago. Initially my purpose was to provide a resource that students and practitioners alike could use to cognitively integrate the foundation principles of quality thereby comprehending the underlying theory of management. It is my belief that through an elevated level of understanding, management practices can be vastly improved and resources more wisely invested.

The reviewers of the initial prospectus advised that achieving my goals within a single book was too ambitious and would be best served through a two-volume series. The pervasive theme of the first book in the series emphasized that quality is central to all business decisions at all levels—and that quality principles must be engrained in the culture and become the most significant driver in all activities, from formulating strategy in the boardroom to determining transportation modes on the shipping dock. In essence, quality should be viewed as a *corporate force*.[3] As the first book dealt heavily with high-level management themes such as strategy, accountability, leadership, decision making, and human resources, its intended audiences are those who are foremost concerned with the management issues and not particularly interested in the nuances of details that are relevant to the contributions made at the operations and engineering levels. Thus, there is a strong emphasis on the case method of teaching. Notwithstanding, *Quality: A Corporate Force* does provide a liberal treatment of statistical process control and Six Sigma.

This is the second book of the series and is intended to fill in the gaps left by the first. *Quality: A Corporate Force* takes a systems approach and views the organization from a top-level management perspective. This book completes the story by building on the systems view and filling in some of the technical blanks. In the first book I coined the term *quality inspired management* (QIM) as a meta-management tagline that conveys the idea that quality is (or should be) the driving force behind management. QIM represents the development and practice of an explicit theory of quality management that goes beyond mere definitions and sets of principles or rules. QIM defines a paradigm that draws from a coherent body of cross-disciplinary knowledge where quality emerges as a strategic concern.

Use of a fresh term also helps to eliminate any preconceived biases so students can objectively evaluate the theory of quality on its own merits. As students and practitioners read the book it is my belief that substituting QIM for other taglines (such as TQM) facilitates comprehension and reduces the liabilities associated with any prior misconceptions. Rather than an approach separate and distinct from those that went before, QIM should be interpreted to be an umbrella that covers the positive aspects of the many and varied methodologies that were the forebears of our current thinking—and furthermore, when considered in the multidisciplinary and dynamic context of working organizations, provides coherence and relevance.

Central to the QIM meta-management culture is the ability to take an open systems view of the enterprise and its supply chain extensions, and to elicit appropriate responses to feedback that is relevant to the organization's mission. Within the system the contributions of engineering and operations toward the achievement of purpose are inextricably linked to the firm's strategic positioning and the interests of all stakeholders. My premise is that with the proper focus on stakeholders the ultimate sustainability of the organization will be ensured. Consequently, the title for this second book of the series, *Quality Inspired Management: The Key to Sustainability*, was thoughtfully selected.

ORGANIZATION

The book is organized in four major parts: *Quality from 30,000 Feet*, *Quality from 5,000 Feet*, *Quality at Sea Level*, and *Quality Below the Surface*.

Part I, *Quality from 30,000 Feet*, is a macroview of quality from top management's point of view. There is some unavoidable overlap with the content of *Quality: A Corporate Force*; however, the information in common has been significantly condensed in this book. In *Chapter 1* we begin the study by setting the stage for all that follows. We first beg the question *Why study quality?* followed by an analysis of the job market for quality professionals. We then introduce the concept of a stakeholder model as superior to the traditional view of customer-focused approaches. We round out the chapter by providing a historical perspective showing how quality thinking has evolved through the ages and its impact on an organization's readiness to progress on the quality front. We show how quality is a journey over which an organization's quality capabilities evolve. This evolutionary process can take many years and an organization's quality maturity can be defined in terms of three distinct generations. We close the chapter and pave the way for Chapter 2 by noting that to be successful, quality must be strategic and driven from the top. *Chapter 2* continues this theme and investigates the role that quality (QIM) principles play in developing corporate strategy. We discuss the open systems theory

---

[2]Foley, Kevin J. *Meta Management: A Stakeholder/Quality Management Approach to Whole-of-Enterprise Management*. Sydney: Standards Australia, 2005.

[3]Aikens, C. Harold. *Quality: A Corporate Force—Managing for Excellence*. Upper Saddle River, NJ: Prentice Hall, 2005.

and the importance of bringing in fresh energy, ideas, and resources from the environment to renew the organization and reverse the destructive effects of the second law of thermodynamics, which states that all systems if not reenergized will eventually unravel and become dysfunctional. We then discuss organizational supporting structures, leadership, and accountability. The balanced scorecard approach is suggested as a tool to ensure that an acceptable balance is maintained among all stakeholder interests. We then turn to the subject of information engineering and discuss the important and growing role that information (and knowledge) plays in the control of quality at all levels. We take a hard look at the decision-making process and knowledge management systems. *Chapter 3* is about the human element and begins with the introduction of an open systems model that describes a generic human resource system. The chapter then goes on to examine the work environment and its relationship to an organization's quality objectives. The characteristics of high-performance work organizations (HPWO), workforce empowerment, and the building of a supportive culture are discussed. Next, the chapter examines work design and motivation and finishes with a section on team building.

Part II, *Quality from 5,000 Feet*, examines those topics that management should consider in advance when planning for production, such as product and process design, and maintenance and reliability. *Chapter 4* treats the requirements side of quality with a review of specifications and tolerancing by introducing dimensional, geometrical, and statistical tolerancing concepts. Included is how to read a shop drawing and how to develop composite specifications that make sense from a statistics viewpoint. Also covered is a review of dimensional metrology including an introduction to some of the commonly used gauges and measuring devices. The sections that follow discuss the meaning of *conformance to specifications* and the chapter concludes with a brief discussion of how the concept of requirements, expressed as a specified tolerance, applies to the service sector. *Chapter 5* covers both product and process design and begins with the illustration of some improvement methodologies including the engineering method, the Plan-Do-Check-Act (PDCA) cycle, and the Define-Measure-Analyze-Improve-Control (DMAIC) tactics. We examine the principles of value engineering and Taguchi's design principles before moving on to describe the various generic production flow configurations that will influence process designs, ranging from line flow designs (high-volume, low-variety), through intermediate flow patterns (low-volume, family processing, cellular designs), to flexible flow designs (low-volume, high-variety, customized). Each flow design is evaluated with respect to how it can best support an organization's goal to create value for stakeholders. Under product design some integrated product development approaches are discussed including an examination of quality function deployment (QFD). Next, we explore a number of product design issues that deal with product life-cycle considerations and are presented under the broad heading *Design for Supportability*. *Chapter 6* is on maintenance and reliability and takes the student through the hierarchy of maintenance philosophies, including run-to-failure (RTF), preventive maintenance (PM), predictive maintenance (PdM), proactive maintenance (PaM), and reliability-centered maintenance (RCM). Methodologies are then presented for prioritizing and measuring maintenance work. The chapter next focuses attention on the link between quality and the maintenance function with detailed discussions of total productive maintenance (TPM) and lean maintenance. The next section discusses how various maintenance practices impact the workforce. Concluding and rounding out the maintenance and reliability themes are discussions on root cause and prevention methodologies, including failure mode effects analysis (FMEA) and Global 8D problem solving.

Part III, *Quality at Sea Level*, consists of six chapters that deal with the application of quality principles where most of the action takes place—at the work interface. *Chapter 7* is an exposition of a large collection of the "soft" tools of quality—those that can be applied to new-idea creation, planning, process analysis, and problem solving. *Chapters 8–12* cover methodologies that are classified more on the "hard" side of the spectrum, as their applications are generally based on the availability of process data. Specifically, *Chapter 8* introduces the concept of statistical control and the construction and use of variables control charts, including $\overline{X}$ and R charts, $\overline{X}$ and s charts, and X and MR charts. Transformations of nonnormal data are also discussed. *Chapter 9* covers a variety of issues relating to control chart use including measurement processes, sampling strategies, and process capability. This chapter also demonstrates how the control chart can be a useful tool for identifying improvement opportunities by isolating and separating the major sources of process variability. *Chapter 10* turns attention from variables to attributes measurement with a description of *p/np* and *c/u* charts. The omnipresent issues of inspector consistency and sample size determination are examined and some guidelines are provided on how to select the type of control chart best suited for a particular application. This is followed by a brief introduction to some commonly used attributes acceptance sampling plans. *Chapter 11* focuses on those characteristics that are pertinent to lean environments and how they impact and are influenced by QIM principles. We discuss the principles that underlie lean thinking and present Little's law to explain the relationship between work-in-process inventory, cycle time, and flow time. The concepts of pull scheduling, kanban control, and one-piece flow are presented as important components of a lean production system. The chapter then discusses the application of lean thinking to the service sector (lean service) and to the consumption cycle (lean consumption). The next two sections deal respectively with visual controls, poka-yoke, and autonomation. Specifically, Chapter 11 drives home the point that lean environments lead to short production runs and associated quality challenges. Some tools to address these issues include mixed model scheduling approaches, changeover reduction strategies, and short-run control chart procedures. For effective short-run control the chapter examines the use of deviation from nominal (DNOM), deviation from target (DTAR), stabilized, and Z charts. The chapter next examines various inspection methods for defect prevention, including successive check, self-check, and source inspections. Chapter 11 concludes with a discussion of kaizen blitz—the team-based radical improvement methodology. *Chapter 12* emphasizes the application of quality in the service sector, focusing on those characteristics that in general differentiate the provision of services from goods manufacturing. In this chapter we discuss customer satisfaction and the challenges of how to capture reliable data in order to measure it. In this regard we discuss customer survey methodologies, including the structured SERVQUAL approach. We conclude the chapter and Part III with a discussion of how control charts can be applied in a services environment, the problem with autocorrelated data and how to overcome it, and the role services play as secondary value streams supporting the primary value streams of manufacturing systems.

Part IV, *Quality Below the Surface*, constitutes the final two chapters of the book and covers those advance tools necessary to break processes down—that is, to perform "open heart surgery"—in an attempt to determine not only how the processes behave but why they behave the way they do. Part IV is about discovery—developing new insights and comprehensions that can be used to re-engineer systems and processes to make them better. *Chapter 13* covers analysis of variance (ANOVA) and design of experiments (DOE). This chapter deals with topics ranging from correlation to one-way ANOVA studies to screening designs to fractional two-level designs. It is not our intention to create DOE experts, as the material in *Chapter 13* does not go into sufficient depth to achieve that goal. However, students should learn enough to participate constructively in experimentation, and be able to design and conduct small experiments on their own. The DOE material is followed by a discussion of gauge repeatability and reproducibility (R&R) studies. *Chapter 14* is devoted to continuous improvement and Six Sigma. We begin this chapter with an investigation into the intriguing question of what it means to continuously improve. We then discuss the theory of constraints (TOC) philosophy for driving improvement efforts. This is followed by a discussion of how successful Six Sigma programs are designed and implemented. We then conclude the chapter with an overview of design for Six Sigma methodologies and a discussion of how organizations leverage the compatibility of lean thinking with Six Sigma into simultaneous implementations called Lean Six Sigma.

PEDAGOGICAL FEATURES

Quality Maxims

Each of the 14 chapters of the book has an overall theme that is related to quality inspired management. These themes have collectively been encapsulated in a compact set of 14 quality maxims. Each chapter begins with a statement of the maxim that is relevant to that chapter's theme. Collectively the maxims represent a cohesive set of quality principles that span the theory of quality management.

As an example, Chapter 1 is devoted to the development of quality thinking through the ages and the difficulty in coming up with a single monolithic definition of quality that everyone can agree on. Recognizing this one is quick to conclude that quality is more about perception than reality. So, the following was selected as the first quality maxim: *Quality is a perception of how well the balanced needs of all stakeholders have been met or exceeded.*

Vignettes

Following each maxim, students are presented with a short vignette, a chapter lead-in that is designed to tie some quality principle to a well-known song; and although the lyricist had a totally different context in mind, the words (or message) can be spun in a way that supports a strong quality theme. For example, the song title for Chapter 11 (lean production) is "Lean on Me," about a renegade high school principal who against all obstacles used radical methods to turn a crime-ridden high school around. Through the lyrics the students at this particular high school are reminded that he (and probably he alone) truly cares for them and that they can depend (lean) on him for support. The vignette methophorically compares this situation to the environmental factors in industry that militate against successful lean implementations and reminds the reader that (quality) tools are available to help them survive the tough times. It is our intent that these vignettes will help stimulate interest as well as serving as memory devices to enhance student learning.

Learning Objectives and Key Outcomes

Following the vignette, each chapter commences with a list of learning objectives and key outcomes, all of which are stated in behavioral action–verb language completing the sentence "After careful study of the material in Chapter (blank), you should be able to . . ." These objectives should enable professors and students alike to clearly focus on the desired outcomes for each chapter. With these outcomes in mind instructors should be able to develop relevant assessment plans that are fair and do not *shock and awe* their students.

Key Terms

Throughout each chapter, key terms have been set in bold type in the body text. Students and teachers will be able to find definitions for each of these terms by consulting the glossary at the back of the book.

Key Concepts

Throughout the body of each chapter, key concepts or phrases have been lifted from the main body of the text and placed in larger type font as framed banners across the page. The purpose is to continually remind students of the key message *as they read* and to provide a mechanism for later being able to quickly scan chapter material for key points.

References and Web Links

Throughout the book, references are included as footnotes and many of these have links to relevant Web sites where students can access additional information.

Showcase of Excellence

Starting in Chapter 2, at the end of each chapter just before the *Summary of Key Ideas* section students will find a boxed feature that showcases five organizations that have won the Malcolm Baldrige National Quality Award. These are provided with the organization's name, the award category (e.g., *small business, manufacturing, service, etc.*) and a link to the Baldrige and company Web sites where students and teachers can go to obtain more information. This showcase feature is designed to emphasize three points: (1) the diverse nature of successful quality implementations (industry-type, geography, public/private, size, etc.); (2) the degree to which quality has become a core value (ingrained in the culture) of superior organizations; and (3) the benefits that superior organizations claim have resulted from their quality efforts.

Summary of Key Ideas

Following the *Showcase of Excellence* box, students will find a *Summary of Key Ideas* section that states the key ideas drawn from the chapter and followed by a short paragraph summarizing the idea and its salient points. It is intended that this summary will facilitate learning and help students with study and review.

Mind Expanders to Test Your Understanding

The final section in each chapter is titled *Mind Expanders to Test Your Understanding* which provides some thought-provoking questions and exercises or problems designed so that students can assess their individual mastery of major concepts. In most cases the questions and problems are not trivial. Questions are designed with the intent of forcing thought processes to a *depth* requiring synthesis and integration. The problems have been devised to simulate as clearly and practicable as possible some of the issues and data complexities that students are likely to encounter in practice.

Minitab

Wherever possible the statistical illustrations throughout the book have been developed using Minitab software Release 15. The worked examples have also used Minitab for the statistical analyses and graphical displays. Although numerous quality management software packages are commercially available to support quality management, Minitab has been widely adopted by industry due to its user-friendliness. Files can easily be exported to and from Excel, and the reports generated are of exceptional quality.

Appendices

Materials have been included in several appendices to supplement and support the main text. Appendix A contains Deming's celebrated 14 points and 7 deadly diseases and Crosby's 14-step approach. Appendix B contains statistical tables that students will need with data analysis. This includes a table specially designed for directly obtaining an estimate of defects per million opportunities for Six Sigma applications. Appendix C includes the bodies of knowledge for Six Sigma black belts and green belts. Appendices D and E illustrate a detailed concept map of industrial engineering, and a map providing statistical information concerning important discrete probability distributions, respectively. Finally, Appendix F provides tables that can help with design of experiments applications. As mentioned, a complete glossary defining key terms used through the textbook is included at the back of the book.

SUPPLEMENTS

In addition to the text, there is a Companion Website (www.pearsonhighered.com/aikens) for student and instructor use. For instructors, there is an Instructor's Manual and PowerPoint slides. To access supplementary materials online, instructors need to request an instructor access code. Go to **www.pearsonhighered.com/irc**, where you can register for an instructor access code. Within 48 hours after registering, you will receive a confirming e-mail, including an instructor access code. Once you have received your code, go to the site and log on for full instructions on downloading the materials you wish to use.

THE INTENDED AUDIENCE

Quality Inspired Management: The Key to Sustainability has been written to provide the skills needed by practicing engineering and operations personnel, and should suit most introductory baccalaureate-level quality courses in engineering, business, technology, or other programs where the curriculum incorporates a body of knowledge in quality principles. This book would also be suitable for graduate studies in business and for engineering or engineering management graduate programs requiring a course in contemporary quality management and control. Additionally, the book can play a role as a resource for industrial training courses, as much of the content is compatible with the green and black belt bodies of knowledge.

In writing this book the aim was to provide a scope of material that could be covered in a one-semester three-credit-hour college or university course. Given the wide variation of needs by instructors and students, the actual coverage will depend on the background and prior preparation of students enrolling in a particular course and the specific objectives of the course. The following provide some broad guidelines of how the material could be distributed to meet various needs.

- A three-credit-hour core course in an undergraduate degree program in engineering. With engineering students, instructors will undoubtedly want to heavily emphasize the material found in Chapters 8–14. It is recommended that one week (2–3 lectures) be spent on Chapter 1, four weeks (8–12 lectures) on Chapters 2–7 carefully selecting those topics students have not been introduced to in other subjects, and eight weeks (16–24 lectures) on Chapters 8–14. This will leave approximately two weeks (4–6 lectures) for testing and review.
- A three-credit-hour undergraduate course in a degree program in business, strategy, or operations/production management; or an elective unit supporting any undergraduate business major. It is recommended that roughly half of the semester be devoted to Chapters 1–7 (14–21 lectures). More or less time can be spent on each topic depending on the class and the curriculum. Approximately three weeks each should be set aside for Chapters 8 and 10 (a total of 6–9 lectures), and two weeks (4–6 lectures) should be devoted to Chapters 12 and 14. This will leave approximately two weeks (4–6 lectures) for testing and review. Depending on students' interests and course objectives, selected topics from Chapters 9 (advanced SPC), 11 (lean production), and 13 (advanced statistical methods) could be introduced as appropriate.
- A three-credit-hour graduate course in a MBA program, other business-related masters program (e.g., accounting, strategy, marketing, or finance), or in an engineering management masters-level curriculum. In general, graduate courses should emphasize the integrative nature of quality management across the entire business system. It is recommended that three weeks (6–9 lectures) be spent on Chapters 1–3, one week (2–3 lectures) on Chapter 4 placing emphasis on statistical tolerancing and cost of quality, two weeks (4–6 lectures) on Chapters 5–6, two weeks (4–6 lectures) on Chapter 7, four weeks (8–12 lectures) on Chapters 8–12 carefully selecting those topics appropriate to the student audience, and one week (2–3 lectures) on Chapter 14. This will leave approximately two weeks (4–6 lectures) for testing and review.
- Industrial, noncredit short courses, and reference.
  - Champions/Senior Management Level Awareness Training. It is recommended that senior managers read all of Chapters 1, 3, 7, and 14, with selected topics of interest from Chapters 2, 5, 6, and 8–13.
  - White Belt: All of Chapter 1 with selected topics from Chapters 3, 7, and 14.
  - Green Belt: All of Chapters 1, 7, 8, 10, and 14 with selected topics from Chapters 3–6, Chapter 9, and Chapters 11–13.
  - Black Belt: All of Chapters 1, 3, 7, 8–14, with selected topics from Chapters 2, 4–6, and 9.
  - Master Black Belt: All of Chapters 1, 4–14 with special emphasis on Chapter 13, and selected topics from Chapter 3.

QUALITY CERTIFICATIONS

This text covers the body of knowledge for the American Society of Quality (ASQ) Certified Six Sigma green belt (CSSGB) in its entirety and approximately 95% of the body of knowledge for the ASQ Certified Six Sigma black belt (CSSBB). With the growing interest in Six Sigma and Lean Sigma programs the ability to produce graduates who are ASQ certified has wide appeal and would constitute a competitive advantage for academic institutions that can provide this opportunity for its students. Appendix C contains tables that list each topic that appears on the relevant certification exams, the level tested according to Bloom's taxonomy, and a cross-reference to the chapter(s) in the book where the topic is covered.

ACKNOWLEDGMENTS

I would like to express my appreciation to those who have helped make this book a reality. A special thanks to all the companies that have permitted me to observe and participate in their quest for excellence. I shall forever be indebted to you for affording me the opportunity as it has been a pleasure to work with such dedicated and passionate people. I would also like to acknowledge the contributions of my students who continue to provide inspiration and optimism for a brighter future with an improved quality of life for all.

As this is the second in a series, those individuals and institutions acknowledged in *Quality: A Corporate Force—Managing for Excellence*, though not mentioned again by name, also deserve recognition here.

I would like to acknowledge my reviewers, Frank J. Falcione, Lorain County Community College; Jonathan Greer III, Grand Rapid Community College; Marilyn M. Helms, Dalton State College; David Leonard, John Wood Community College; Patrick L. McCormick, Ivy Tech Community College, Fort Wayne; and Harold D. Zarr Jr., Des Moines Area Community College; they made suggestions that helped shape the content to its intended purpose. Thanks also to Minitab, Inc. which kindly provided me with a copy of Minitab Release 15 for use during the development of this project.

Finally, I would like to give special recognition to my family, who have been patient and sacrificed much so that this project could reach a satisfactory conclusion. I am especially eternally grateful to my wife Helen, who is my rock of Gibraltar and who has patiently endured many a lonely evening and continual separations as the book project seemed to erode the sanctity of every free moment. I am also indebted to my daughter Jocelyn, an MBA qualified health care professional, who has spent countless hours meticulously proofreading my manuscript and making constructive suggestions for improvement. To both of you, I thank you for your unconditional love and support.

—C. Harold Aikens

CONTENTS

PHOTO CREDITS

CHAPTER 1

p. 4: Pearson Learning Photo Studio
p. 4: Pearson Learning Photo Studio, photo by David Mager
p. 6: Pearson Education/PH College, photo by Irene Springer
p. 7: Pearson Education/PH College, photo by George Dodson
p. 7: Pearson Learning Photo Studio, photo by David Mager
p. 11: © Dorling Kindersley
p. 24: U.S. Air Force
p. 24: U.S. Air Force, photo by Master Sergeant Joe Cupido
p. 25: Pearson Education Corporate Digital Archive

CHAPTER 2

p. 30: Merrill Education, photo by Liz Moore
p. 31: Pearson Education/PH College, photo by Vincent P. Walter
p. 31: Pearson Education/PH College, photo by Laimute Druskis

CHAPTER 3

p. 60: Pearson Education/PH College, photo by Brady

CHAPTER 4

p. 84: Getty Images, Inc.—Photodisc, photo by Lawrence Lawry
p. 98: Pearson Education Custom Publishing
p. 99: Andy Crawford © Dorling Kindersley
p. 99: Pearson Education/PH College
p. 100: Pearson Education/PH College, photo by Frank LaBua

CHAPTER 5

p. 117: Pearson Education/PH College
p. 132: Pearson Education/PH College

CHAPTER 6

p. 156: Pearson Education/PH College, photo by Laima Druskis
p. 156: Pearson Education/PH College, photo by Michal Heron

CHAPTER 7

p. 190: Merrill Education, photo by Krista Greco

CHAPTER 8

p. 247: Pearson Education/PH College, photo by Rhoda Sidney

CHAPTER 9

p. 292: Pearson Education/PH College, photo by Frank LaBua
p. 322: Pearson Education/PH College, photo by Laima Druskis

CHAPTER 10

p. 362: Pearson Education/PH College, photo by Vincent P. Walter

CHAPTER 11

p. 415: Pearson Education/PH College, photo by Annette Brieger/Goldpitt

CHAPTER 12

p. 452: Pearson Education/PH College, photo by Charles Gatewood

CHAPTER 13

p. 484: Pearson Education/PH College, photo by Peter Buckley

CHAPTER 14

p. 530: Pearson Education/PH College, photo by Stephen Agricola

PART ONE

QUALITY FROM 30,000 FEET

CHAPTER ONE

QUALITY OVERVIEW

QUALITY MAXIM #1: *Quality is a perception of how well the balanced needs of all stakeholders have been met or exceeded.*

Where Has All the Quality Gone?

"Where Have All the Flowers Gone?"[1]—a popular folk song of the '60s—depicts a poignant cycle that begins in a field where young girls pick flowers and in time marry young men, who are then taken by the military and sent off to war where they are killed and end up in graveyards, which in due course become fields of flowers, thus completing a cycle of life.[2] Sometimes it seems like that ballad, written a half century ago, is a metaphor for the evolution of quality management. Throughout the '80s and '90s the United States and indeed all industrialized nations placed great emphasis on quality—building mainly on the teachings and inspiration of several notable individuals credited with helping rebuild the Japanese economy after World War II. Enormous resources, especially in the private sector, were directed toward the quality movement (more like a crusade) as personnel in vast numbers were trained, expanding the circle of participation and forever changing work cultures, environments, and methodologies. Overnight, a new megadollar quality services industry was born. Consultants and academics in growing numbers offered various brands of quality elixirs, promising cures to all of industry's woes. However, with the passage of time the process went full circle. Failing to see satisfactory results, disciples gradually became cynics and, faced with seemingly new panaceas, many were quick to defect. Now, three decades later, as the old folk song prophesied, the world may seem to have rewound the clock. While many products of today are superior to those of the '70s, product unreliability and shoddy service are still all too common. The skeptic may claim that much of the investment in quality in the last quarter century was largely ineffective, and what was learned was soon forgotten. Nevertheless some organizations have been able to get it right—and it is these that inspire others to continue the quest, understanding that quality is not a set of rules but a *theory*, the application of which can be the key to sustainability in a highly competitive world.

[1]Written by Pete Seeger and Joe Hickerson between 1956 and 1960, "Where Have All the Flowers Gone?" became a hit folk song in 1962 when the Kingston Trio recorded it as a single (advancing to #21 on the charts), and in the same year Peter, Paul, and Mary included it on their first album, which was listed on *Billboard Magazine*'s top-10 list for 10 months.

[2]The music and lyrics for "Where Have All the Flowers Gone?" can be accessed at http://kids.niehs.nih.gov/lyrics/flowrsgone.htm.

LEARNING OBJECTIVES AND KEY OUTCOMES

After careful study of the material in Chapter 1, you should be able to:

1. Articulate an understanding of why the study of quality is important.
2. Have an appreciation for the job market and employment opportunities for quality professionals.
3. Describe in general how the challenges in quality management differ in manufacturing, government, health care, and education.
4. Demonstrate an understanding of the quality movement from a historical perspective.
5. Describe how the quality movement has evolved through three generations of development.
6. Identify some of the notable early pioneers of the quality movement and describe their contributions.
7. Define quality from numerous perspectives.
8. Define the term *operational definition* and develop an operational definition of quality that suits your particular work environment.
9. Define the fundamental principles underlying *quality inspired management* as a meta-management philosophy for improved competitiveness.

1.1 QUALITY—WHY SHOULD YOU STUDY IT?

Together we are about to embark on a journey into the fabulous world of quality management. Before we begin, it is appropriate to take leave and ask some simple but very important questions: *Why study this subject? What is in it for me? How can the study of the foundations of quality help me in my future job? How do these principles relate to today's business world?* These are all legitimate and relevant questions, which we will begin to address by posing a series of additional questions.

Think back to when you last rode a bus. Was the bus driver kind and pleasant or grim and grumpy? Think about the last time you shopped for clothes—perhaps to select an outfit for some special occasion. Did you readily find what you wanted? How many stores did you have to visit before you found what you were looking for, gave up, or settled for something less? Were you happy with your final choice? Did the outfit you selected fit properly when you got it home? If you had to return it to the store how were you treated by the salesperson? Did the salesperson make you feel good or guilty about the return? Have you ever ordered a meal at a restaurant and been unhappy with how it turned out—either the way the meal was cooked or the quality of the food itself (flavor, temperature, texture, etc.)? How did this experience make you feel? Have you ever bought an item that broke down after only a short time in use? Do you ever feel as if you have been deceived or cheated by a service provider or the manufacturer of a product? Do you ever feel that you got less than you bargained for? Have you ever asked yourself, or heard others remark, "What is happening to the quality in this country?" Or, you may have found yourself wondering "Why can't I find anything that is made in America anymore?" As a member of society are you concerned about rising prices, shoddy workmanship, and sloppy service? If your answer was yes to any of these questions this course is certainly worthwhile and worthy of your time. All of these sentiments are from a consumer's perspective. Let's consider the question of quality from a broader viewpoint.

Do you have a bank account? How do you know your bank manager is looking after your interests to make sure you get the best return and that your money is safe? Are you concerned that customers who have more money in the bank might receive preferential treatment at your expense? When you turn on a faucet at home how certain are you that the water is safe to drink? Or, when you purchase meat at the local grocer that it is safe to eat? How confident are you that the pristine stream running through your grandmother's property will not be polluted by upstream manufacturing plants?

Quality is everyone's concern—it is a very real part of our daily lives. Leading scholars have recognized the inextricable synergy between organizational *excellence* and *quality*; it is our position therefore that these two key terms are synonymous. *Excellence* has been defined as "the overall way of working that balances stakeholder concerns and increases the probability of long-term organizational success through operational, customer-related, financial, and marketplace performance excellence."[3] Almost everyone would understand the concept of excellence—that is, the merits of striving to be the best—to excel in all walks of life.

Excellence is characteristic of a work style that balances stakeholder concerns.

Unfortunately, when it comes to the provision of goods and services, the quest for profits has too often trumped quality. The mantra of many a boardroom is profits, profits, profits—and the relentless quest for profits is a road paved with bad decisions leading to cost-cutting, downsizing, and sacrificing the interests of stakeholders. Not only are stakeholders often not considered

Job Opening

Lean Six Sigma Black Belt

- Manufacturer of Flow Control Products
- Salary Range: $95,000–$115,000
- Locations: Alabama, Pennsylvania, Illinois, Tennessee

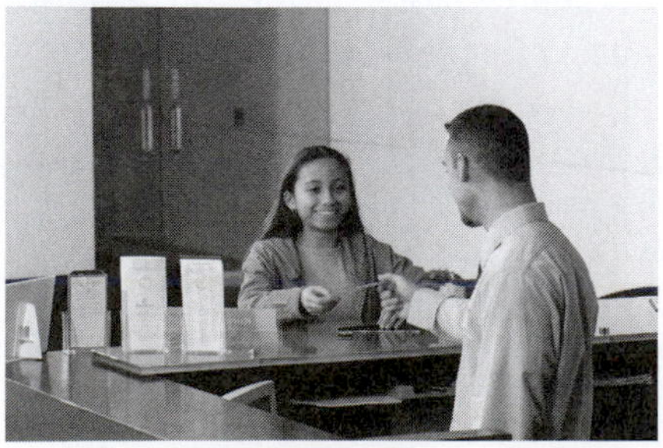

A quality moment of truth—the way a bank teller treats a customer shapes that customer's opinion of the entire banking organization.

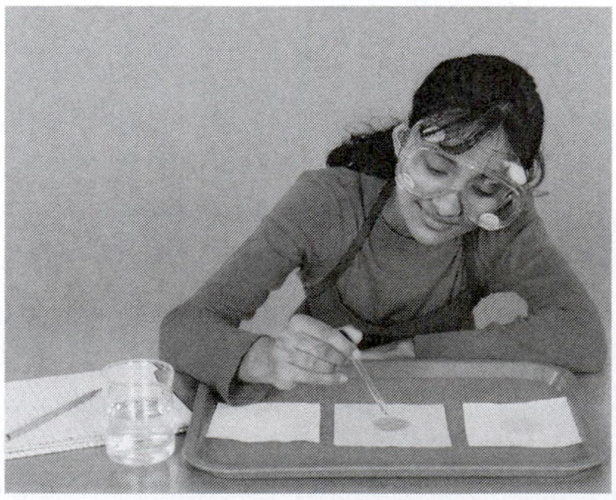

Quality behind the scenes—a quality professional performs a laboratory test to ensure product compliance with government health regulations.

[3]Edgeman, R. L., S. M. P. Dahlgaard, J. J. Dahlgaard, and F. Scherer. "Leadership, Business Excellence Models and Core Value Deployment." *Quality Progress* 32, no. 10 (1999): 49–64.

in such corporate decisions, but typically are hard to find in corporate plans and objectives. It does not have to be that way. With smarter management practices organizations can have it both ways: high bottom-line impacts and happy stakeholders, an ideal formula for growth and prosperity. The knowledge and technology are available to ensure performance excellence. The key to organizational excellence is through improved knowledge acquisition and sharing, a greater capacity to think systemically, and the political will to excel. This text is about all those things, in addition to providing a solid grounding in the technical skills necessary to achieve them.

The keys to satisfying stakeholders while maximizing the bottom line lie in improved knowledge sharing, a greater capacity to think systemically, and the political will to excel.

The need is apparent. Satisfying that need is perhaps the most compelling reason why you should study this subject. The market for qualified quality professionals is growing and the applicable compensation and benefit packages can secure a comfortable lifestyle. Since quality is about excellence and excellence is about balancing the needs of all stakeholders, we shall pause here to discuss how a stakeholder perspective differs from the traditional customer-focus view, and then we shall take a closer look at the quality management job market.

1.2 THE IMPORTANCE OF ORGANIZATION STAKEHOLDERS

All organizations, whether large or small, public or private, for profit or not-for-profit, can be defined as an open system, continually contributing to and taking sustenance from the environment in which it operates. Its sustainability depends on how the organizational system defines itself relative to its environment on the one hand and its strategic response to this self-actualization on the other. We will discuss the open systems view in further detail in Chapter 2. For now, let us create a mental picture of some generic business enterprise (this could be any organization). The purpose of this enterprise is to provide some product or service, or a mix of products and services, to other organizations or individuals (called markets) that reside within the environment. The environment also provides refuge for other parties that have an interest in or are affected by the operations of this enterprise. Examples of these include competitors, suppliers, regulators, investors, governments, employees and their families, members of society, and so on. By its mere existence this organization directly affects the lives of many and its survival is of interest to many more. Within this cosmos of interested and affected parties there exists a subset of parties who possess *power* over the destiny of the organization. These parties with power have needs and if those needs are not met, can induce failure, inflict damage, incur excessive and undesirable costs, or act in other ways that inhibit the organization's ability to achieve its goals. These interested and affected parties who possess such power are the organization's *stakeholders*.[4]

It is important that we make a distinction between an organization that has a stakeholder perspective and one with a customer perspective. Much of the work of the early quality pioneers, discussed below and leading to a movement called total quality management, placed the customer at the center of an organization's universe, establishing the customer's supreme sovereignty, and mandating that all that really mattered was satisfying customers' needs. This philosophy went further by taking the position that competitive advantage could almost always be achieved by creating the "WOW" factor—that is, pleasantly surprising customers with service or products that clearly exceeded expectations. In theory this view makes customers the most important, if not the only, stakeholder group, and under the customer perspective corporate strategies are more interested in mounting defensive plays and counter measures against competitors than identifying and satisfying stakeholders. However, this view leads to an inevitable conflict concerning how the organization measures its success. While satisfying customers may be the corporate mantra, performance metrics that satisfy the needs of shareholders and other investors may be emphasized the most and have little to do with how happy customers are.

Sustainability ultimately depends on how well an organization's senior leadership is able to identify and respond to all stakeholders. Some interested and affected parties, such as shareholders, will always be stakeholders. Others, such as customers, in some cases may not possess the power to inflict negative consequences on the firm and therefore do not fit the stakeholder definition. Consider, for example, who the stakeholders would be for a large oil company. At the consumer level the transaction is the only relationship the person who drives into a service station has with the company. The individual customer who needs fuel has no power over prices, service, or quality because at that level no action on the customer's part could inhibit the oil company's achievement of any of its objectives.

Where customers have been identified as stakeholders, the stakeholder view does not diminish the importance of satisfying their needs and indeed creating the WOW factor is an important corporate goal. Nevertheless, it is equally important to impress all stakeholder groups and many of the tools presented throughout this text can be applied to stakeholders who are not customers.

1.3 THE JOB MARKET FOR QUALITY PROFESSIONALS

Sustainability is the number one issue for the twenty-first-century enterprise. The workforce will need a special set of skills to solve the unstructured, complex, and multifaceted problems that are encountered in a market-driven,

[4]Foley, Kevin J., Douglas A. Hensler, and Jan Jonker. "Quality Matters." In *Quality Management and Organizational Excellence: Oxymorons, Empty Boxes, or Significant Contributions to Management Thought and Practice*, edited by Kevin J. Foley, Douglas A. Hensler, and Jan Jonker, 1–38. Sydney, Australia: SAI Global, 2007.

technology-dominated, socially conscious, politically correct, information rich, and global environment. The skills sets that are developed through the study of quality management fundamentals are suitable for resolving some of the important issues facing many of the organizational sectors, including government, health care, manufacturing, education, and service. While many hard economic times and fierce competition have meant that some industrial sectors have had to downsize, eliminating jobs in the process—especially in manufacturing—companies are increasingly turning to quality professionals to provide the answers. In health care, government, and education the future market for quality specialists seems promising.

The quality management skill set can help organizations solve many of the complex problems they face in a global economy.

1.3.1 Quality in Government

Organizations in the public sector face some unique challenges. First, there is the issue of improvement incentives and the availability of capital. Government organizations are funded by their citizen constituencies through various methods of taxation. There is no profit motive and the performance of each department is budget driven. Cost cutting usually only becomes a priority when budgets are cut—and producing under budget is not regarded as a virtue since to do so does not mean that a department can keep the excess and roll it over to the subsequent year. Not only does the relevant department lose the savings, it will more than likely receive less during the next budget cycle. In addition, there are no dollars to fund improvement projects unless such projects have previously been built into the budget.

Job Opening

Design for Six Sigma Black Belt/Master Black Belt

- Fortune 1000 Company
- Salary Range: $100,000–$150,000
- Location: Massachusetts

There are also some notable differences between private and public sector work cultures. A supportive culture is essential to the success of any continuous improvement program and it may be more difficult within the bureaucratic structure of governmental organizations to establish the infrastructure, measurement processes, and management practices needed to drive quality. A supportive culture would be one in which information is freely and widely shared, problems are publicly aired and analyzed, and there is total transparency with respect to reporting and accountability. These concepts are foreign to many government agencies. Bureaucracies tend to be more rigid in their processes and practices and can be slow to embrace change and inflexible to shifting requirements.

Quality management in bureaucratic not-for-profit organizations must overcome some unique challenges with resources, commitment, and culture.

Quality without profit motive: government departments offer boundless opportunities for doing more with less.

Relationships between the organization and its stakeholders also differ. In the government, investors and customers are members of the same group—the taxpayers. Government officials must be accountable to the taxpayers who, due to the lack of competition, have little control over the quality or scope of services provided. The only recourse available in a democracy is through the ballot box, and elected officials who fail to perform or respond to the unique needs of their constituencies can periodically be voted out of office. The electorate unfortunately has little control over the performance of nonelected government officials.

1.3.2 Quality in Health Care

Quality and efficiency in the provision of health care in the United States have been on the national radar screen for some time, and were a major issue during the 2008 U.S. presidential campaign. Politicians and those with vested interests in the existing system boast that the United States has the world's best health care system. Whether that is true or not is a matter for informed debate. Notwithstanding, every day millions of Americans receive a high quality of medical services that

Job Opening

Master Black Belt

- Hospital Health Care
- Salary Range: $100,000–$145,000
- Location: Maryland

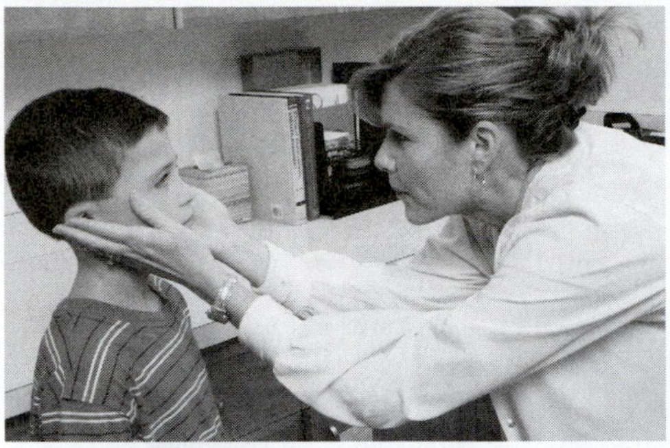

A doctor examines a juvenile patient—quality healthcare professionals strive to improve care while reducing costs and eliminating waste.

enables them to maintain or restore the level of health and well-being necessary to function adequately in society. Still, many Americans who need medical attention do not receive it—due in part to lack of access, but mostly due to prohibitive costs. And the spiraling cost of health care continues to place quality care beyond the reach of increasing numbers of people. Quality problems in health care can generally be grouped according to the following four categories:[5]

1. **Variation in the use of services**—There is a pattern of wide variation in the practice of medicine and the provision of health care services across the United States. Differences in the quality of health care are evident across regions, cities, socioeconomic divisions, rural-versus-urban, and so forth.
2. **Underuse of some services**—Inefficiencies in the current system deny many people necessary intervention treatments that could prevent needless additional medical complications that later add costs and reduce overall system productivity.
3. **Overuse of some services**—On the other side of the issue, many Americans receive treatment regimens that are unnecessary, increase costs, and ironically can even endanger their health—ultimately leading to increased future costs. Some of the treatments that are unnecessary or redundant are due to excessive precautions by practitioners who believe that by erring on the conservative side of patient care they can protect themselves against future lawsuits. Another source of overuse stems from customers demanding prescriptions for the brand-name drugs they see advertised on TV, when cheaper generic forms of the same drug would be equally effective, or non-prescription remedies would work just as well.
4. **Misuse of some services and an unacceptable level of errors**—There is an increasing trend in the number of people who are injured, permanently disabled, or die prematurely as a result of medical treatments in the United States each year. Some of these are due to errors in prescribed treatments and others are due to negligence.

Quality in health care must address variation in use, underuse of some services and overuse of others, and the commission of errors.

Quality tools can play an important role in the design and improvement of health care systems, and employment of quality professionals in the health care industry is growing steadily.

1.3.3 Quality in Education

The No Child Left Behind Act of 2001[6] set federal quality standards concerning teachers, curriculum, and student performance, creating a public awareness of the importance of providing the resources to deliver the knowledge and skills that students need, and to design educational systems that can respond as the needs change. Quality management principles have increasingly been applied to the design of improved educational systems from kindergarten through university studies.[7]

In education, the three key drivers for quality and sustainability all begin with the letter *A*. These are accountability, alignment, and assessment. Proper accountability, which includes systemic procedures for ensuring that all educational institutions are meeting their goals and producing desired results, helps focus attention on improvement opportunities.

Quality management in education must deal with the three A's: accountability, alignment, and assessment.

Alignment is a process issue. In the delivery of quality education it is imperative that curricula and support programs match federal- and state-mandated standards and assessment requirements. The assessment procedures must be designed to reliably measure what students are learning relative to what they should be learning. In keeping with the open systems view described in Chapter 2, effective assessment programs are based on the extensive use of feedback from all stakeholder groups, including teachers, administrators, parents, students, community citizens, legislators, and other interested parties.

In the private sector, where for-profit corporations compete to fulfill the needs of industry or consumers, the focus typically emphasizes the importance of satisfying

Education is New Frontier for Quality Professionals.

[5]*Improving Health Care Quality: Fact Sheet,* Agency for Healthcare Research and Quality (AHRQ), U.S. Department of Health and Human Services. http://www.ahrq.gov/news/qualfact.htm.

[6]Public Law 107-110, signed into law by President George W. Bush on January 8, 2002.

[7]The Malcolm Baldrige National Quality Award has a unique set of criteria for performance excellence for educational institutions and makes annual awards in this category.

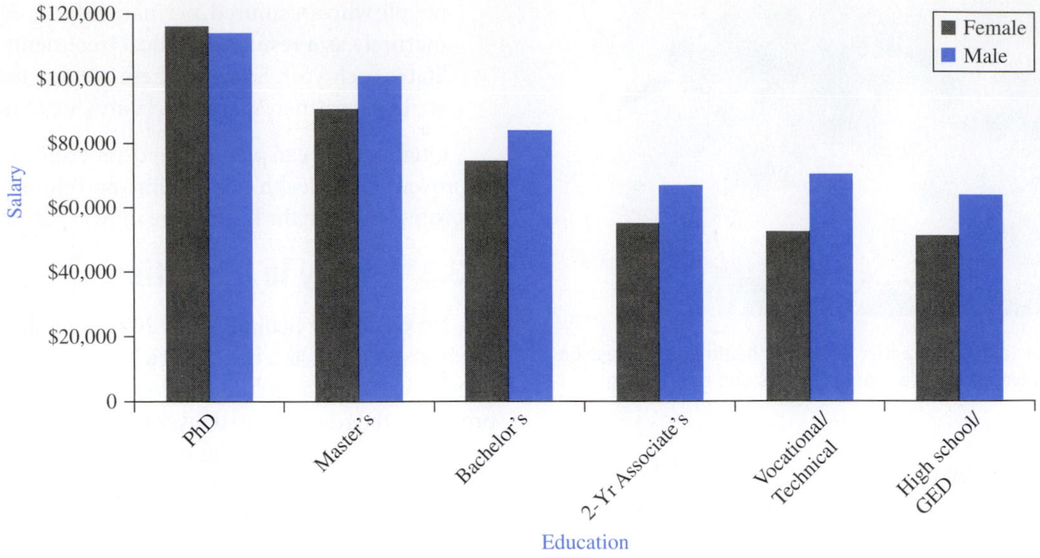

FIGURE 1-1. U.S. Salaries for Quality Professionals by Education Level

Source: "2008 Salary Survey: Our Annual Look at Who Makes What in the Quality Industry." *Quality Digest*, August 2008. http://www.qualitydigest.com/pdfs/SalarySurvey2008.pdf.

stakeholders who are *customers*, even in extreme cases declaring customers to be of sovereign status. By comparison, in education the needs of a multiplicity of stakeholders must be carefully balanced. One important stakeholder group includes *students*. In an education system the role of the students provokes an interesting dichotomy. On one level students are like customers—they pay, or someone pays on their behalf—an amount for a service (i.e., education) with some well-defined expectations of what constitutes desirable outcomes (e.g., diploma, degree, good job, etc.). On another level, students are more like raw materials entering the educational system without knowledge, going through a transformation process that takes them down an educational value chain, until they depart the system transformed into productive educated citizens. This dichotomy places unique challenges on educational systems committed to the principles of excellence.

1.3.4 Salaries for Quality Professionals

The job market and salaries for quality professionals are excellent and have good future prospects, as organizations increasingly turn to quality experts to help them find improvements in cost, process, products, and service.

As a quality professional one can expect a level of compensation that will ensure a comfortable quality of life. As one might expect, and as shown in Figures 1-1 and 1-3, salaries increase with education and experience, and as Figures 1-2 and 1-4 illustrate, salaries also vary respectively by geographical region and industry type.

Figures 1-1 through 1-3 also indicate a gender gap in pay. Nevertheless, it is not clear whether the disparities reported on surveys are not more related to industry type or title than to gender.[8] It is quite likely that more women are employed in lower-than-average-paying jobs or industries than men.

Quality professionals hold many and varied job titles such as manager, specialist, coordinator, engineer, consultant, analyst, and auditor. About 45% of those responding to a recent survey had formal American Society of Quality (ASQ) certifications and 27% had earned a Six Sigma belt.[9,10] Even though some universities offer programs of study to prepare students for careers in quality, those who hold quality management and support positions come from a variety of backgrounds. One discipline that provides an excellent foundation for a quality career is industrial engineering (IE), as a typical IE curriculum blends technical, mathematical, and business content together with an emphasis on data analysis and complex decision making. As Figure 1-5 illustrates, the market for industrial engineers in the United States has grown steadily over the last five years and the future prospects look bright.[11] Figure 1-6 provides information on the mean and median salaries for IEs over the last decade.[12] Despite a decline in

---

[8] "2008 Salary Survey: Our Annual Look at Who Makes What in the Quality Industry." *Quality Digest*, August 2008. http://www.qualitydigest.com/pdfs/SalarySurvey2008.pdf.

[9] Ibid., Figures 4 and 5.

[10] ASQ has established criteria that lead to certifications as Quality Auditor, Quality Engineer, Quality Manager, Six Sigma Black Belt, Six Sigma Green Belt, Quality Technician, Mechanical Inspector, Quality Improvement Associate, Certified Calibration Technician, Reliability Engineer, Software Quality Engineer, Quality Auditor—Biomedical, Quality Auditor—Hazard Analysis and Critical Control Point (HACCP), Manager of Quality/Organizational Excellence, Quality Process Analyst, and Pharmaceutical Good Manufacturing Practice (GMP) Professional. Many quality professionals hold more than one certification.

[11] From *Occupational Employment Statistics (OES) Survey,* Bureau of Labor Statistics, Department of Labor. http://stats.bls.gov/oes/.

[12] Ibid.

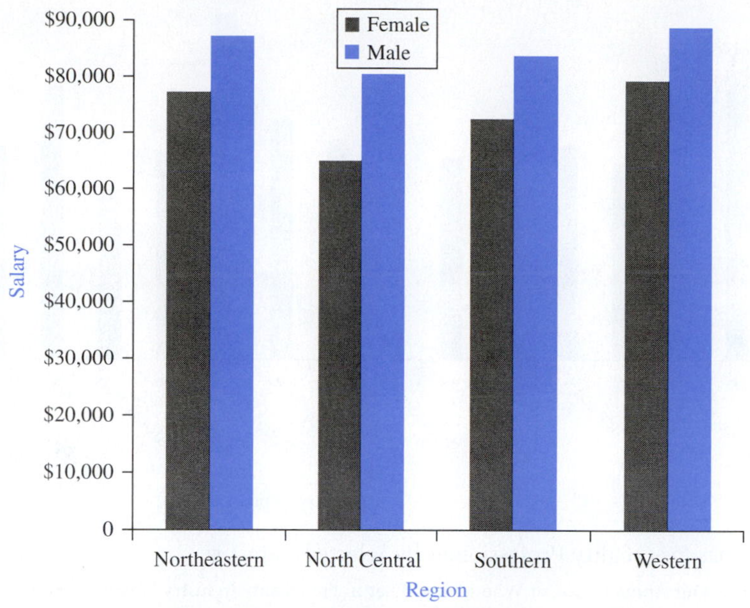

FIGURE 1–2. U.S. Salaries for Quality Professionals by Region

Source: "2008 Salary Survey: Our Annual Look at Who Makes What in the Quality Industry." *Quality Digest*, August 2008. http://www.qualitydigest.com/pdfs/SalarySurvey2008.pdf.

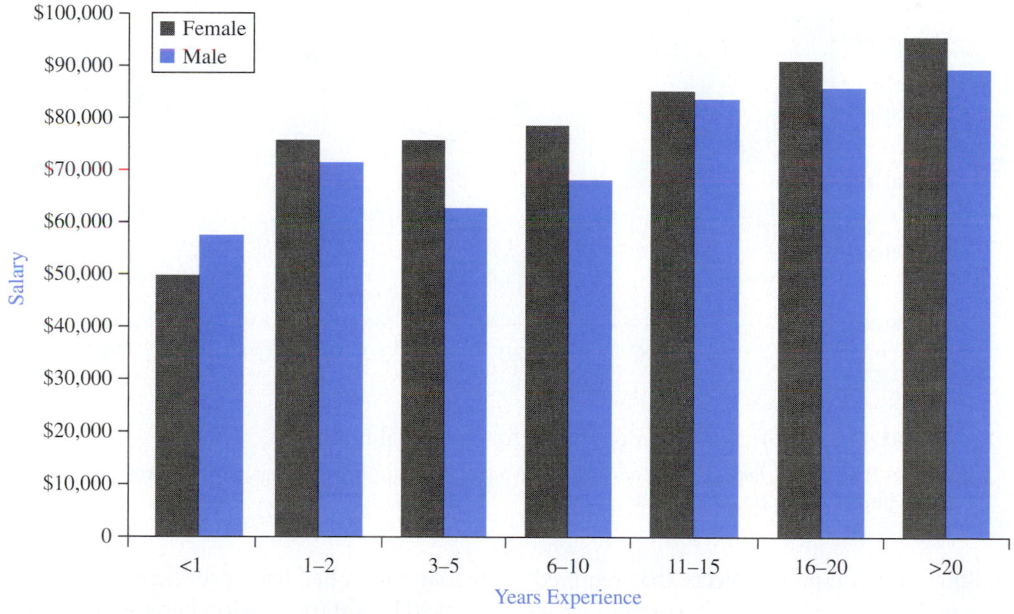

FIGURE 1–3. U.S. Salaries for Quality Professionals by Experience

Source: "2008 Salary Survey: Our Annual Look at Who Makes What in the Quality Industry." *Quality Digest*, August 2008. http:// www.qualitydigest.com/pdfs/SalarySurvey2008.pdf.

some industries, notably manufacturing, the demand for IEs is projected to grow by 20% by the year 2016, as companies increasingly turn to the IE profession to help them find new ways to trim waste and increase productivity.[13]

[13]"Job Outlook," *Occupational Outlook Handbook*, 2008–2009 Edition, Bureau of Labor Statistics. http://www.bls.gov/oco/ocos027.htm#outlook.

1.4 THE HISTORY OF QUALITY

1.4.1 Quality in Antiquity

We shall begin this journey by climbing into a time capsule and taking a trip back through history. Applying quality standards to determine how *good* something is dates back to early civilizations. The temples, monuments, coliseums, roads, bridges, and pyramids of the ancient cities of Rome, Athens, and Alexandria

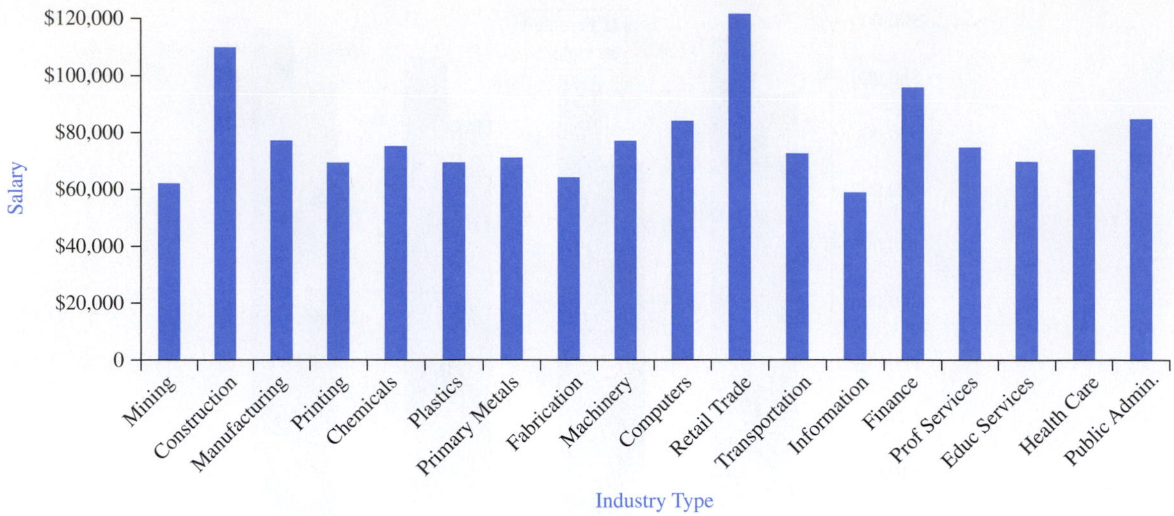

FIGURE 1–4. U.S. Salaries for Quality Professionals by Selected Industry

Source: "2008 Salary Survey: Our Annual Look at Who Makes What in the Quality Industry." *Quality Digest*, August 2008. http://www.qualitydigest.com/pdfs/SalarySurvey2008.pdf.

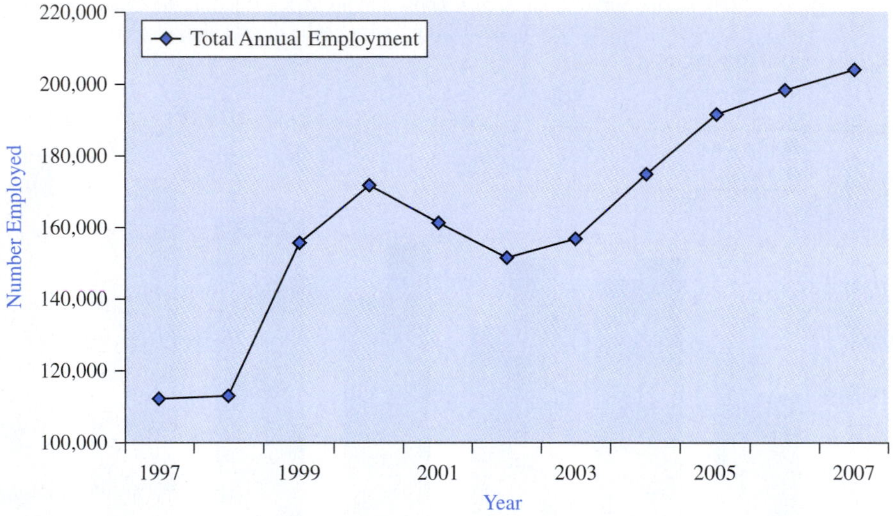

FIGURE 1–5. U.S. Trends in Employment for Industrial Engineers

Source: Occupational Employment Statistics (OES) Survey, Bureau of Labor Statistics, Department of Labor. http://stats.bls.gov/oes/.

represented monolithic and complex projects that required meticulous attention to detail. In antiquity a high premium was placed on good quality, which typically ranked higher in importance than time, cost, or even human safety. However, quality during these early periods was often built on the backs of slaves and the products were one-of-a-kind. The concepts of mass production, replication, and efficiency were still centuries away.

In antiquity quality was controlled through pride and close supervision.

The first account of any organized approach to the production of goods and services was in the Middle Ages, a period that lasted roughly a millennium—dating from the fall of the Western Roman Empire in the fifth century to the middle of the Renaissance in the sixteenth century. The feudal system had been prevalent in the early part of this era marked by a stark division between noblemen and the common people. The noblemen were the landowners and consisted of lords and vassals. Vassals owed their allegiance to their lords who in return had typically granted them fiefs which were normally parcels of land. A vassal could then become a lord by further dividing his land into fiefs which he would distribute to his vassals in return for their loyalty. So at the top of the social ladder there was this hierarchy of noblemen, each owing his allegiance to a stronger, more powerful lord than himself. At the bottom were the common people representing about 90% of the population. The lower class consisted of the peasants who did most of the work. The peasant class was divided into two categories: serfs and freemen. Most were serfs who were indentured to the land

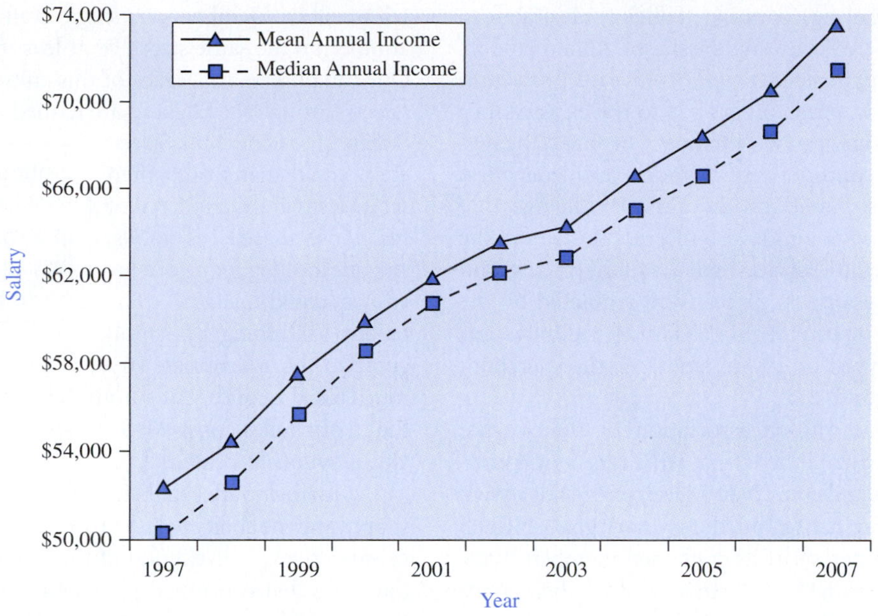

FIGURE 1–6. U.S. Salary Trends for Industrial Engineers

Source: Occupational Employment Statistics (OES) Survey, Bureau of Labor Statistics, Department of Labor. http://stats.bls.gov/oes/.

Job Opening

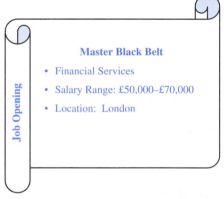

Master Black Belt

- Financial Services
- Salary Range: £50,000–£70,000
- Location: London

they tended. They were not free to leave but they also could not be removed from their land by their lord. If the land was sold, the serfs were sold along with it and consequently fell under the jurisdiction of the new landowner. Serfs were usually granted a small strip of land for their own use, which could be passed on to heirs as an inheritance.

The life of a peasant family was hard. Multiple families typically lived together in a single one- to two-bedroom house with no amenities, few cooking utensils, and one or two tables and chairs. Peasants had little variety in their diet which mainly consisted of pottage (a thick soup), bread, and berries. In addition to their farm chores, peasants were also responsible for the overall caretaking of their lord's manor. In return for taking care of their lord, and in addition to food and shelter, the lords provided the serfs with protection against invading armies—which were a medieval fact of life.

Some peasants were freemen and could move from place to place if they were unsatisfied with their lord or situation. Freemen typically lived in towns near the protection of castles and manors and ran their own independent businesses. Nearly all free peasants were involved in manufacturing, contributing to their communities in skilled trades such as carpenters, blacksmiths, weavers, bakers, masons, painters, cobblers, tanners, candle makers, apothecaries, and so on. The craftsmen paid their respective lord (who owned the land where the town was located) rent for using their small plots to conduct their respective businesses. However, in addition to the rent the lords had the power to tax these small business owners and in time the taxes became excessive. It was impossible for a single person to object to these taxes so the idea of the tradesmen banding together into guilds was born. The word *guild* comes from the Saxon word *gilden* meaning "to pay" and refers to the dues required for a member to join a guild.

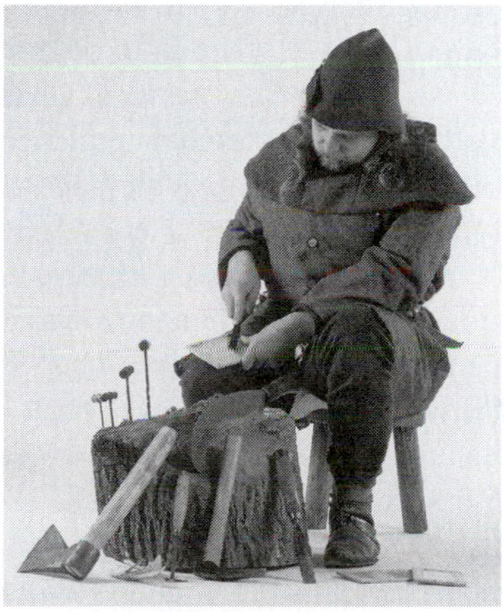

A medieval carpenter at work—quality of his craft was determined by individual skill and pride of workmanship.

A guild in the Middle Ages had many characteristics in common with modern labor unions. Benefits to guild membership included protection from excessive taxation, regulation of competition (achieved by fixing prices, and prohibiting advertising and discounting), banning of illicit trade practices, protection of trade secrets, limiting the size of guild membership to ensure sufficient business for all, health care, protection for goods and horses when members had to travel, assistance with funeral expenses and care for members' orphans, the regulation of working conditions and hours, and education (guilds funded the first public schools of the Middle Ages).

The establishment of guilds also offered benefits to the consumers of goods and services, such as fair pricing and quality. Under this system all prices were regulated by the guilds and rigorous inspection and training regimens ensured quality. Two types of guilds emerged: the merchant guilds and the craft guilds.

A merchant guild was an association of traders that banned together to regulate the way in which trade was conducted in entire towns and cities. Nearly every citizen of a town would belong to its merchant guild, the primary goal of which was to look after the interests of its craftsmen members. Outsiders were allowed to come into the town and sell their goods only if they paid a toll, and only then if the goods and services sold would not interfere with the sales of guild members. Merchant guilds became the principal negotiators with the lords in determining fair tax rates and levies. Over time this placed the officers of the merchant guilds in very influential positions within a community. With their vested power the guild officers typically became the civic leaders and members of the elite. These were the wealthiest citizens and the principal source of money for loans. When the town or any of its residents needed money they would approach the guild first. Merchant guilds would often even make deals with the king, collecting his required taxes in return for complete freedom to continue to raise additional taxes as the guild saw fit.

Guilds were formed to provide tradesmen unity against excessive taxation by feudal lords.

Guild officers also had absolute control over quality, requiring members to submit their goods for guild inspection before any items could be sold. The inspection process covered quality of workmanship, as well as weights and measures. If problems were found the craftsman involved was required to correct the problem and provide assurances that such problems would not recur. Once the guild had passed inspection on a manufactured good, declaring that the item was well made and complied with guild standards, the craftsman was free to sell the item. Members of a single craft usually lived near one another on the same street, or at least the same area of town. One can observe the legacy of this culture even today as many streets throughout England are named after trades (e.g., Paper Street, Shoe Lane, Milk Street, etc.). A craftsman would often set up a small shop in front of his house to sell his goods. Anyone looking for a particular item could engage in comparative shopping by simply going to the appropriate street or district for the purchase—much like the concept of automobile marketing of today where the motor miles consists of numerous competing dealerships all contiguously located.

Guild officers had absolute control over quality, requiring guild inspection before goods could be sold.

With their trade regulations, the power to control prices and supply, and the ability to levy taxes, the merchant guilds began to threaten the very livelihood of the individual tradesmen. Also, the issues that concerned plasterers were different from those that concerned weavers, and different again from those that concerned candle makers, different still again for tailors, and so on. In response, those in the crafts and trades established their own guilds—the **craft guilds**. Craft guilds were formed when a group of tradesmen or craftsmen engaged in the same occupation joined together as a formal organization. There were guilds for every craft or trade performed within a city or town (e.g., a masons' guild, a carpenters' guild, a bakers' guild, a weavers' guild, etc.). The craft guilds provided both protection and mutual aid for their members. No one could practice a craft or trade without belonging to the appropriate craft guild, and each guild established rules that defined obligations and benefits similar to those of the merchant guilds. The craft guilds had two primary objectives. The first was to ensure that there was enough work to keep each member of the guild busy. This was accomplished by placing quotas on the number of new people who could enter each trade (as apprentices), thus creating monopolies and barring external competition. The second objective was to ensure that all work produced by guild members was of high quality.

Quality was totally controlled within the guilds, which set high standards and carefully monitored compliance. An important part of the quality control system was the rigorous requirements for guild membership. To become a member of a craft guild a man (women were not allowed to participate) had to advance through three stages of development—apprentice, journeyman, and master.

Generally, in his early teens a young man could become an apprentice by being sent to work under the close supervision and mentorship of a master in a specific trade. The apprenticeship lasted between five and nine years, depending on the craft, and during this period the apprentice was not permitted to marry, was provided room and board, and received no pay. Once the apprenticeship period was served the trainee achieved the status of journeyman.

As a journeyman a young craftsman was paid for his labor but his tenure in the journeyman status was open-ended. Some journeyman remained in that status forever as achieving the rank of master was difficult and uncertain. To advance a journeyman had to create in his own time a *masterpiece* that he

Job Opening

Director, Lean Six Sigma

- Fortune 1000 Company
- Salary Range: $100,000–$150,000
- Location: New Jersey

Job Opening

Senior/Master Black Belt

- Pharmaceutical/Biotechnology
- Salary Range: €80,000–€100,000
- Location: Ireland

could present to the relevant craft guild as evidence of his craftsmanship skills. This process had some political ramifications and a journeyman's chance of gaining the guild's acceptance depended as much on his relationship and reputation with the top guild members as on the quality of the masterpiece submitted.

Once accepted into guild membership a craft guild master could set up his own business and begin to train his own apprentices. Within the master's workshop the craftsman would typically perform all of the tasks necessary to complete a particular job—from raw materials to a finished product. There was no specialization of labor or attempts to achieve economies of scale. Products were often one-of-a-kind produced one-at-a-time.

The craft guilds played a dominant role in quality control. Craftsmen were obliged to submit items to guild officers for inspection both during the making and when finished. The guilds strictly forbade work on Saturday afternoons, Sundays, religious holidays, and nights. The idea behind forbidding evening work was that, considering the poor lighting of the time, working at night would result in low productivity, reduced attention to detail, and poor quality.

To become a member of a craft guild a man had to advance through three stages—apprentice, journeyman, master.

1.4.2 The Evolution of the Quality Movement

Quality as we know it today, addressing all the complex business issues that must be balanced and fine-tuned if the business is to succeed in an increasingly competitive and challenging environment, has been called a *movement* by many—a movement with roots that can be traced back only as far as the beginning of the Industrial Revolution in the late eighteenth and early nineteenth centuries, a time when major changes in agriculture, manufacturing, and transportation changed the cultural and socioeconomic landscapes of the Western world. During this period the invention and refinement of the *steam engine* sparked a successive chain of technological advancements that enabled human labor to be replaced by machines, thus sharply increasing manufacturing productivity. The concept of mass production was spawned by the invention of the *printing press* and, with the invention of the *clock,* came the application of timing and time-keeping to the measurement of production and efficiencies. During the twentieth century, inspired mainly by the need for improved quality during wartime, much of the seminal work that provides the theories and bases for many of the methods in use today were conducted by early pioneers of the movement—some of whom were popularized and elevated to *guru* status when interest in the movement became widespread in the latter part of the twentieth century. Including this groundbreaking work by the early pioneers, the quality movement has experienced three distinct stages (or generations) of development.[14]

The modern quality movement was spawned by three significant inventions: the steam engine, the printing press, and the clock.

1.4.2.1 First Generation: Pre-1980—Measurement and Control. First-generation quality was a microview that was product or service oriented and based largely on **measurements, controls,** and **detection.** The work involved periodically taking scientific measures from critical process stages and applying statistical tools to predict process outputs. The primary focus of these early QC departments was to keep poor quality out of the hands of customers by sorting out the good quality from the bad, and the emphasis was on inspection of output rather than on process. Improvement was achieved mostly through remedial actions whenever the quality of outputs failed to meet minimum requirements. First-generation programs were internally focused and product-oriented—and failed to recognize the importance of building stakeholder relationships. The absence of a systems view generated little or no interest in cross-functional integration, supplier networking, or engaging the workforce.

First-generation quality failed to recognize the importance of building relationships.

Problem-solving efforts were limited to production lines or workstations with known difficulties, and the front-line workers were blamed when things did not go as planned. Management thought that improved quality could be achieved through **incentives** and controls. Formal efforts at controlling quality were mainly aimed at manufacturing enterprises, with little or no application to the service sector.

By the early part of the twentieth century the wide use of mass production had created the need for quality **prediction**—using a small fraction of the total process output as an estimator of the quality of a much larger production batch. A new set of analytical tools were required. In the 1930s Walter A. Shewhart,[15] a physicist at Bell Telephone Laboratories, pioneered the use of statistical methods in what later became formally known as quality control (QC)[16]—a term used to

---

[14]Foster, David, and Jan Jonker. "Third Generation Quality Management: The Role of Stakeholders in Integrating Business in Society." *Managerial Auditing Journal* 18, no. 4 (2003): 327.

[15]For a photo and short biography of Walter Shewhart, consult http://www.iso-9000.ro/eng/9001/gurus.htm.

[16]Shewhart, W. A. *Statistical Method from the Viewpoint of Quality Control.* Washington, DC: Graduate School, Department of Agriculture, 1939.

describe systems that are dedicated to ensuring that products or services are designed, produced, and delivered in a manner that meets or exceeds stakeholder expectations. Quality engineering (QE) is a term that is also used when referring to these systems. Shewhart's book, *Economic Control of Quality in Manufacturing Product*, published in 1931, introduced the world to **control charts.** In investigating how to reduce the frequency of failures of telephone hardware in 1924, Shewhart coined the terms "assignable" and "chance" causes of variation,[17] and developed the control chart as a statistical tool for distinguishing between the two.

During this same era two of Shewhart's colleagues at Bell Labs, Harold F. Dodge[18] and his assistant Harry Romig,[19] developed some statistical sampling methods that could be used for accepting or rejecting entire production lots. The war effort created the critical need for better methods of quality control, and among the first application for the 1940 Dodge-Romig Sampling Inspection Tables was the acceptance of rifle bullets. However, formal quality tools were scarce in the 1940s and the quality advocates of the day fought an uphill battle for recognition. The top priority during World War II was on meeting schedules even if quality suffered.

The quality movement got its first organized and government-supported boost during the postwar reconstruction of Japan under General Douglas MacArthur. It was during this time that an unsung hero of the movement, Homer Sarasohn,[20] had a significant impact. Sarasohn was a physicist who had designed radio transmitters before the war. When the United States entered the war he joined the army and served as a paratrooper. Then, following his release from active duty, Sarasohn went to work at the Massachusetts Institute of Technology Radiation Laboratory (called the Rad Lab), which was the major radar development center for the United States. At the Rad Lab his responsibility was to convert prototypes of new designs into products that could be mass produced. He must have been very good at his job because, at age 29, he was handpicked by General MacArthur to head up an effort to rebuild the communications infrastructure in Japan. This involved building transmitters and hundreds of thousands of receivers throughout the occupied country. What Sarasohn found was a country that was almost completely devastated by Allied bombing. Refuges were everywhere and all factories and factors of production had been destroyed. Any production equipment that had been spared was hard to locate and it was still more difficult to find people with the skills to operate them. The quickest path to restoring manufacturing capacity was to resurrect some of the large companies that had been operating before and during the war—members of the *zaibatsu*.

Job Opening

Director, Process Improvement & Lean Six Sigma

- Internet Marketing
- Salary Range: $110,000–$150,000
- Location: Nebraska

During the nineteenth century, as part of the industrialization of Japan, business groups called *zaibatsu* were formed. Each *zaibatsu* was composed of several large corporations, typically representing one of each major industry. For example a *zaibatsu* might be made up of a major bank, a major trading company, a major metals company, a major chemicals company, and so on. An entire *zaibatsu* was under the control of the family running the parent holding company. Some of the large *zaibatsu* carry names that are still prominent today, such as Mitsui, Mitsubishi, Sumitomo, Yasuda, Nissan, Nomura, and Furukawa (Fujitsu, Ltd.). These combines formed an interlocking structure where each corporation owned shares in the others and exercised favoritism among its members in purchasing and sales.[21] The major firms within a *zaibatsu* offered lifelong employment, compensation systems based on seniority, and company-sponsored unions to create stable working conditions and attract the best workforce. Price discrimination against companies outside the *zaibatsu* and payments to parties and politicians for political influence were common practices. Small corporations were less stable and at the mercy of the *zaibatsu* parent corporations and less attractive to employees. The small companies consequently hired the largest percentage of female, foreign, and/or unskilled workers. Until the end of the Second World War the *zaibatsu* controlled the Japanese economy, dominating price structures, supply chains, and employment.

One of MacArthur's first acts as Supreme Commander of the Allied Powers (SCAP) was to abolish the *zaibatsu*, believing that through the manipulation of supplies and prices these enormous conglomerates had largely been responsible for the war.[22] MacArthur banned any participation by previous *zaibatsu* top managers as their companies were revitalized.[23] The lack of top corporate leadership

Until the end of World War II, commerce in Japan was controlled by the *zaibatsu*.

[17]These terms were later changed by Dr. W. Edwards Deming respectively to "special" and "common" causes.

[18]For a photo and short biography of Harold Dodge consult http://www.asq.org/about-asq/who-we-are/bio_dodge.html.

[19]For a photo and short biography of Harry Romig consult http://www.asq.org/about-asq/who-we-are/bio_romig.html.

[20]For a photo tribute to Homer Sarasohn consult http://honoringhomer.net.

[21]Beer, Michael, and Bert Spector. *Readings in Human Resource Management*. New York: Free Press, 1985, 273.

[22]Cringely, Robert X. "Stranger in a Strange Land." *The Pulpit*, PBS, May 25, 2000. http://www.pbs.org/cringely/pulpit/2000/pulpit_20000525_000408.html.

[23]These leaders were able to return when the U.S. occupation ended and their combined expertise was a major factor in Japan's economic resurgence during the decade following.

presented a challenge for Sarasohn, who had no choice but to promote inexperienced middle managers into top management positions. Within a year some of the factories had been restored to the point where they were capable of producing some, albeit primitive, output. However, the quality was shocking with yields less than 10% in some cases. As Sarasohn was to learn, such poor quality was not extraordinary or a consequence of the war. Under the *zaibatsu* quality and productivity had never been a concern and was not a part of the work culture. With few exceptions designs were bad and many of the manufacturing techniques dated back to the mid-nineteenth century. Factories were dirty and disorganized. There was no sense of quality on the part of workers[24] and mass production resulted in shoddy output. It was difficult for Sarasohn to even communicate with the workers on improvement. He was trying to instill the principles of quality through a translator to an audience that had no understanding of the concepts. To complicate matters, the translators used were Hawaiian Japanese who did not understand the technical terms they were asked to translate, and who were distrusted by the native Japanese who viewed them as disloyal traitors.

Sarasohn decided that if he were to be effective he needed to learn the Japanese language himself. So he moved in with a Japanese family and lived as one of the locals, learning the language and the culture until he left Japan in 1950.

As new production techniques were introduced production yields began to climb, reaching 75%–85% in the various vacuum tube plants. Nevertheless, these gains were due more to method than attention to quality. Factories across the country were still poorly supervised, disorganized, and cluttered. Furthermore, quality had not yet become ingrained in the workforce culture. By 1949, a war between North and South Korea[25] seemed inevitable and as a consequence the U.S. occupation of

Job Opening

Process Improvement Manager

- Fortune 100 Insurance Company
- Salary Range: $100,000–$130,000
- Location: New Hampshire

Japan began to wind down. Sarasohn needed to quickly find a way to sustain and build on all the progress that had been made. His first action was to set up a national Electrical Test Laboratory (ETL) which served as an approval authority for new radio and electrical equipment prototypes. Once ETL approval was granted a new design could go into full-scale production and would be subjected to random sampling by the ETL during production to ensure quality. If quality problems were found the ETL had the power to shut down a production process and the responsible managers had to satisfactorily demonstrate that the quality problems had been fixed before production was allowed to resume. This clearly made quality the responsibility of management and could not be delegated.

Sarasohn's next project was to create a quality management course tailored to the needs of Japanese plant managers. He teamed up with Charles Protzman, a Western Electric engineer who had been brought to Japan to manage the telecommunications system. Together they wrote a textbook and designed a seminar syllabus that was compulsory for senior electronics executives. The seminar—called the Civil Communication Section Management Seminar, conducted eight hours per day, four days a week, over an eight-week period—covered the philosophies and techniques of quality control and management practices.[26] Many of the graduates of Sarasohn's seminars went on to build very successful companies, such as Sony, Matsushita, and Mitsubishi.

At the end of 1950 the time finally arrived for Sarasohn and Protzman to return to the United States. It was their job to nominate a successor who could carry on the work they had started. They preferred an American since in their view the leading Japanese statisticians were too steeped in tradition to effectively build high-performance mass production systems.[27] Sarasohn turned to Walter Shewhart, inventor of the control chart and singly responsible for much of the theory and techniques of statistical quality control. Unfortunately, at the young age of 59, Shewhart was in poor health and had to decline Sarasohn's invitation. Sarasohn then turned to W. Edwards Deming,[28] one of Shewhart's students who had already visited Japan in 1946 and again in 1948 to assist with postwar census activities.

To understand Deming and the contribution he was to make to the quality movement, we shall again enter our time machine and momentarily go back a quarter of a century to the year 1925 when he was a graduate student at Yale. It was during the summer of 1925, and again in the summer of 1926, that Deming was employed at the Hawthorne Works of

[24] Workers were typically untrained or poorly trained and poorly supervised.

[25] In 1948 the United Nations sanctioned general elections as a move to unify North and South Korea. These elections were held in the South resulting in the establishment of the Republic of Korea (ROK). The Soviets prevented the U.N. elections from going forward in the North, conducting their own independent elections, which established the Democratic People's Republic of Korea (DPRK). The U.N. promptly recognized the ROK as Korea's legitimate government, while the Soviet Union and the entire communist bloc countries recognized the DPRK as the only legitimate sovereign Korean nation. These elections resulted in a permanent division between the two countries at the 38th parallel and set the stage for a civil war.

[26] The CCS Management Seminar would be conducted in Japan over the next 25 years by the Japan Management Association. Sarasohn and Protzman's textbook, *Fundamentals of Industrial Management*, is still in print.

[27] The Japanese tradition stressed the artistry of individual craftsmanship, typically striving for perfection with little regard for time or cost. Sarasohn perceived that the native Japanese statisticians of the day might become more focused on an appreciation for the art and lose sight of the main objective, which was the ability to replicate quality in large numbers.

[28] For a gallery of photos of W. Edwards Deming consult http://www.iso-9000.ro/eng/9001/gurus.htm or http://deming.org/index.cfm?content=67.

Western Electric.[29] It was here that Deming first became acquainted with Shewhart's work and was so impressed that he decided to pursue a career path in statistics. When he earned his doctorate from Yale in 1928, Deming accepted a research job with the U.S. Department of Agriculture. He remained in government service and eventually became the head mathematician at the U.S. Census Bureau. In this role he was able to indulge in his love for statistics and made many contributions to the discipline, particularly in the area of sampling theory. Ironically it was the Second World War that gave Deming's career its big break. When the United States entered the war at the end of 1941 production capacities across the country were diverted to the war effort as factories converted from consumerism to the needs of the military. Widespread concerns arose over quality, and the solution seemed to lie in a broader application of statistical methods to production processes. American statisticians with the necessary skills to meet this growing need were scarce. Deming readily recognized the opportunity and capitalized on it.

Beginning in 1942, with the support of Stanford University and the War Production Board, Deming organized and conducted a series of short courses to teach Shewhart's principles to managers and engineers working in munitions factories. These courses proved to be extremely successful and were expanded to other industrial audiences. By the end of the war it is estimated that over 10,000 personnel had been through one of these seminars.[30]

The war effort of WWII made quality a national priority and created a demand for statisticians with quality management experience.

In 1946 Deming joined the faculty at New York University's Graduate School of Business Administration and began the task of marketing himself to industry as a statistical consultant. Deming's expectations that the firms that he had worked with during the war would beat a path to his door went unrealized. He was largely spurned by American industry that, in satisfying the tremendous pent-up demand for goods and services, put quality issues aside in favor of higher production and inflated profits. When Deming went to Japan in 1950 he had a significant advantage over Sarasohn. Sarasohn had been MacArthur's man and therefore was identified with the occupation. Deming, on the other hand, was postoccupation and entered Japan under the sponsorship of the Japanese Union of Scientists and Engineers (JUSE). This gave him instant credibility and was largely instrumental in ensuring him a revered place in Japanese history. Dr. Deming conducted executive training in statistical methods and consulted with Japanese industry leaders. His popularity increased, reaching national folk hero status, and JUSE eventually established the most highly coveted quality prize bearing Deming's name. The Deming Prize is awarded each year to the Japanese company that is judged to have made the most advanced use of statistical methods to improve quality.

Dr. Deming's focus was on **management principles,** which called for analyzing the production **system** first, and then using statistical tools as a means for gaining the necessary knowledge to improve overall performance. Deming was unrelenting in his insistence that quality is the responsibility of management, and all problems are primarily due to faulty management practices or systems' deficiencies. Deming's logic was that, since management owns the systems, ultimately management is to blame when things go wrong. It was Deming's mantra that workers are not responsible for the sins of their bosses, and his celebrated 14 points and 7 deadly diseases[31] later became a rallying point for second-generation quality implementations.

During the 1950s Dr. Deming served as a consultant to a number of foreign governments, including India, West Germany, Mexico, and Turkey.[32] Even though Deming failed to reach the notoriety at home that he might have expected, he did maintain a viable consulting practice over the next quarter of a century until, at the age of 80, his life changed dramatically. I will explain how this happened a little later.

Let us again use our time machine to wind the clock back a couple of decades to the year 1934. Despite the popular belief that the concepts of quality management were solely introduced into Japan by Americans, the idea of applying statistics to ensure that production conforms to customer requirements was formally proposed by Kiribuchi Kanzō in a 1931 monograph.[33] Even earlier, some Japanese companies had used control charts on a limited basis. With the outbreak of war came a heightened interest in quality control, particularly in those industries supporting the military. In 1942 a university professor, Kitagawa Toshio, published a Japanese translation of Pearson's classic monograph[34] that improved upon and applied Shewhart's methods to production processes with a specific emphasis on improving conformance. A year later a research "think-tank" consisting of Japanese engineers

[29]The Hawthorne plant located in Cicero, Illinois—a suburb of Chicago—was the site of the famous Hawthorne studies, the seminal body of research that laid the groundwork for the application of sociology, social psychology, and anthropology to the industrial workplace. The Hawthorne studies were conducted between 1927 and 1932, and the original purpose was to test the effect of illumination on productivity. Instead of proving a connection with illumination, some unexpected results were achieved: the experimenter effect and the social effect. It was found that short-term improvements in productivity occur when workers think management cares for them or is giving them some sort of special treatment (experimenter effect) and that workers who are isolated and placed in a group bond together and form a special kind of camaraderie that also increases productivity (social effect).

[30]Gabor, Andrea. "Deming Demystifies the 'Black Art' of Statistics." *Quality Progress* 24, no. 12 (December 1991): 26–28; Grant, Eugene L. (as told to Theodore Lang). "Statistical Quality Control in the World War II Years." *Quality Progress* 24, no. 12 (December 1991): 31–36.

[31]Deming, W. Edwards. *Out of the Crisis.* Cambridge, MA: MIT Center for Advanced Engineering Study, 1982. Deming's 14 points and 7 deadly diseases are listed in Appendix A.

[32]Tsutsui, William M. "W. Edwards Deming and the Origins of Quality Control in Japan." *Journal of Japanese Studies* 22, no. 2 (Summer 1996): 295–325.

[33]Ibid.

[34]Pearson, E. S. *The Application of Statistical Methods to Industrial Standardization and Quality Control.* London: British Standards Institution, Publication Department, 1935.

and statisticians was formed to study mathematical approaches to controlling mass production.[35] Such efforts proved to be more theoretical than practical and, apart from the application of statistics to some inspection processes, not much progress was made in the systematic application of quality principals to working production systems. It was perhaps no surprise, therefore, that Sarasohn found such pervasive low levels of quality and productivity throughout Japanese factories.

The real genesis for the quality movement in Japan was primarily due to the efforts of JUSE, the organization that sponsored Dr. Deming in 1950. JUSE had its beginnings in 1944 as the Greater Japan Technological Association[36] and was created through the merger of three prominent scientific professional societies.[37] In 1949 JUSE received a grant from the Economic Stabilization Board to report on technologies abroad that would be relevant to Japan's economic recovery. JUSE seized this as an opportunity to gain prestige and, after evaluating numerous possibilities, including atomic energy and high-frequency communications, statistical quality control was selected as the single new technology that offered the most promise and relevancy for the future.[38] Contrary to what has been widely assumed, the content of Deming's lecture series proved to be a disappointment to the JUSE scholars who were already well read on Western quality control literature and were hoping to learn some novel and innovative approaches. Prior to Deming's arrival a core group of quality activists, familiar with Western methods, had assembled under the organizational structure of JUSE, had begun to forge industrial relationships, and started a process of knowledge sharing across the engineering and management communities.

Nevertheless the Japanese found Deming to be an exceptional and engaging teacher, who could explain difficult statistical concepts in clear and simple terms, and his message was enthusiastically received. Even though Deming's seminars contributed little new in the way of intellectual merit, his presence certainly reinforced and provided credibility to the stated mission of JUSE, and his value lay more in his presence than in his knowledge.[39] Deming had a commanding presence that could be overpowering. He was tall, well-muscled, and had a military demeanor that could humble a chief executive officer with a few words.[40] Towering over the average Japanese citizen, he was a fascination and authority figure. Deming became a brand and the extensive media attention his visits received provided valuable free advertising for the quality movement.

From a mundane standpoint, even though Deming's work with the Japanese received much pomp and ceremony, and helped create momentum for the cause, practitioners generally found Deming's techniques too sophisticated and too mathematically rigorous for typical shop-floor applications. As Sarasohn had discovered, the factories of postwar Japan were using production methods that were primitive compared to those of Western nations where advanced quality management tools were being successfully applied. There were exceptions as quality control methods were well-diffused in some of the large firms with mass production technologies. However, a 1954 survey revealed that only 13% of 46,000 factories were employing modern quality control methods.[41] The perception was that the teachings of Deming, though interesting, were too abstract to be applied to existing systems. By the early 1950s JUSE was on the horns of a dilemma as the enthusiasm and forward progress for the quality movement appeared to be losing momentum and support.

Meanwhile, back in the United States, a quality revolution was brewing. In 1946, 253 professionals representing 17 quality-related disciplines had come together to form the American Society for Quality Control (ASQC). While the formation of ASQC provided credence to the quality movement, this action also reinforced the notion that quality control was highly specialized and not a diffuse component of management practice. A pivotal event was the publication of an article written by Ralph Wareman, one of the founding members of ASQC. Wareman's article[42] appeared in *Fortune* magazine and for the first time explained quality principles in a language managers understood, published in a magazine highly respected by them, and linked anticipated results directly to the bottom line. This article piqued the interest of a vast audience of general managers, and QC departments, staffed with highly skilled analysts, sprang up across the industrial landscape, particularly in manufacturing organizations. Still, progress would be slow and it would be several

[35]This "think-tank" of researchers was formed under the auspices of the Technology Agency (Gijutsu-in).

[36]The association was formally renamed the Japanese Union of Scientists and Engineers in May 1946 with the expressed mission of creating opportunities for scientists and engineers to make contributions toward Japan's economic recovery.

[37]The three societies were the Industrial Policy Association, founded in 1918; Japan Technological Association, founded in 1935; and the All-Japan Federation of Science and Technology, founded in 1940.

[38]Tsutsui, "W. Edwards Deming."

[39]Ibid.

[40]Hunter, J. Stuart. "Obituary: W. Edwards Deming (1900–93)." *Journal of the Royal Statistical Society, Series A (Statistics in Society)* 157, no. 3 (1994): 497–99.

[41]Tsutsui, "W. Edwards Deming."

[42]Wareman, Ralph. "Statistical Quality Control Is Among the Sharpest Management Tools Developed in Half a Century." *Fortune,* December 1949.

Job Opening

Six Sigma Black Belt

- Global Communications
- Salary Range: €50,000–€75,000
- Location: Belgium

decades before quality was taken seriously, as the postwar emphasis would be to satisfy the monumental pent-up demand for consumer goods, giving little attention to whether products were of high quality or met customer needs. From the early 1950s and into the 1960s, markets were flush with products and demand was high. Customers' expectations were low and there was little incentive for U.S. manufacturers to install tighter quality controls. Because the United States was the only industrialized super power to come through the war with manufacturing capabilities in tact, there was little threat from foreign competition. It is no wonder that few business leaders during this period took note of the teachings of the early pioneers of the quality movement, such as Dr. Deming.

Back in Japan the revitalization of the quality movement was due to the intervention of another American—Joseph Juran,[43] a mechanical engineer who began his career as an inspector at Western Electric's Hawthorne plant in 1924, where he met Shewhart and Deming. After leaving Western Electric Juran established a lucrative consulting practice and developed international prominence as a QC expert, spending the next half of a century researching, lecturing, consulting, and writing. His work in quality management led to an invitation by JUSE in 1954 to lecture to the CEOs of the 140 largest companies in Japan. While in Japan Juran toured factories, delivered lectures, and evaluated the state of the quality movement. Unlike Deming, Juran took a less technical view of quality control and, based on his observations, he pronounced that the greatest inhibitor to further progress with the movement was that quality had been too narrowly defined and was too mathematical. He stressed the importance of pragmatism over theory, the need for organization-wide participation, and use of QC as a management tool as opposed to the domain of specialists. Juran's message resonated well with an audience that was searching for a cure that would help reevaluate, reengineer, and restart the Japanese movement. Perhaps his key contribution was to shift the focus from the perfection of mathematical technique to a broader issue—the achievement of corporate goals. Juran's approach stressed the need for hands-on management, better training, and continuous process improvement. His famous trilogy called for the management integration of **planning, control,** and **improvement.**[44] To Juran these elements were the keys to success and a sequence that management could follow that was analogous to financial budgeting, cost control, and profit improvement.

Juran advocated hands-on management, better training, and continuous improvement.

From 1955 to 1965 the Japanese quality movement underwent a major rebirth as it reassessed, retooled, and refashioned itself. Sweeping reforms were introduced as the movement experienced a transformation from a narrow science-driven specialty into a comprehensive systematic and scientific framework for manufacturing management and control. A new synthesis emerged, called total quality control (TQC) that ironically was to become a benchmark that industrial leaders worldwide would in time strive to emulate.

Armand V. Feigenbaum,[45] also an engineer, was the first to actually use the term *TQC*, and intended to convey the idea that quality must be so embedded in the culture that it becomes a way of life. Many of his ideas correspond to the teachings of Juran, and his book,[46] printed in more than 20 languages including Chinese, Japanese, French, and Spanish, has become a global foundation reference for the practice of quality management. Feigenbaum maintained that all areas of a company needed to be involved in the quality effort and that the quality professional has an opportunity to make wide-sweeping contributions that go beyond those of a functional specialist by providing information and direction. His philosophy encompassed the following nine principles:

- TQC is company-wide, including all functions at all levels.
- Quality is what the customer says it is.
- Quality and production costs are complementary—higher quality results in lower costs.
- Individual and team zeal are required.
- Quality is a management style, requiring leadership that continuously and relentlessly emphasizes quality principles.
- Quality and innovation together are required for the development of new products.
- Quality must form a part of corporate ethics, involving all of management, and should not be delegated to specialists.

[43]For photos of Joseph Juran consult http://www.iso-9000.ro/eng/9001/gurus.htm or http://www.jmjuran.com/photoLibrary/jmjLibrary1.htm.

[44]Juran, Joseph M. *Juran's Quality Control Handbook*. 3rd ed. New York: McGraw-Hill, 1974.

[45]For a short biography and photo of Armand Feigenbaum consult http://www.iso-9000.ro/eng/9001/gurus.htm or http://www.asq.org/about-asq/who-we-are/bio_feigen.html.

[46]Feigenbaum, Armand V. *Total Quality Control*. 3rd ed., revised. New York: McGraw-Hill, 1991. The first edition was published in 1951 under the title *Quality Control: Principles, Practice, and Administration.*

- New and existing technologies must be deployed continuously to the improvement of processes.
- Quality is the most cost-effective and least capital intensive route to productivity, and should be implemented in conjunction with suppliers and customers.

Feigenbaum's work provided the foundations for what became total quality management (TQM) in the 1980s.

Armand Feigenbaum provided the foundations for total quality management.

In 1979 Philip Crosby[47] introduced the **cost of quality (COQ)**[48] concept, which places the spotlight on the real costs of doing things poorly. Crosby was an advocate of **zero defects** (ZD), a theme that is consistent with the modern thinking behind **Six Sigma** (SS).[49] Crosby asserted that coming *close* to meeting requirements was simply not good enough, and that zero defects can be achieved only if management establishes the leadership and systems that enable success. Philip Crosby, one of America's first corporate vice presidents of quality (at ITT), believed that senior managers must take charge of quality and make COQ a part of financial accounting systems. This set him apart from other quality pioneers of the day who were mainly academicians, and top managers across the country flocked to the quality college he established in Winter Park, Florida. Ironically, like Deming in his courses, Crosby outlined a 14-step approach for achieving quality improvement.[50] Crosby's most significant contributions to the quality movement were his *four absolutes* which he argued are basic requirements for understanding the purpose of an organization's quality program.

Philip Crosby introduced the concepts of cost of quality (COQ) and zero defects (ZD).

Absolute 1: Quality means conformance to requirements, and requirements are what the customer says they are. Workers at all levels should be admonished to "do it right the first time every time."

Absolute 2: Quality comes from prevention, and prevention is the result of finding and correcting problems with the system. Opportunities for improvement exist in all systems.

Absolute 3: "Zero defects" is the only suitable quality performance standard; otherwise shipping nonconforming product to customers would be acceptable. When a nonconformance occurs, action should be taken to identify and eliminate the cause thereby preventing its recurrence in the future.

Absolute 4: Quality measurement is the price of nonconformance, and is essential to garner management attention, to prioritize and correct problems, and to monitor progress.

By the 1970s the creation of formal **quality assurance** (QA) programs (in contrast to *quality control)* began to divert attention from detection to prevention, but the interest was not widespread.[51] By this time some remarkable changes, which had been largely transparent in the West, were beginning to occur in Japan as production capabilities began to reap the benefits of its quality transformation.

Quality requires doing it right the first time every time.

Like a Phoenix rising from the ashes, Japan was silently reinventing itself and its manufacturing infrastructure, giving birth to a new postwar empire. The Western world got its first glimpse of the transformation when a new breed of automobiles made their debut on foreign shores. Not only did customers notice superior quality, the Japanese vehicles were more fuel-efficient, a criterion of great importance during the energy crisis of the mid-1970s. Still, this did not arouse the interest of the U.S. automobile executives who refused to believe that consumer loyalty would ever abandon the Western (predominantly American) market.

The real attention-getter came in 1980 with the airing of a TV documentary titled *If Japan Can, Why Can't We?* Through this program the National Broadcasting Corporation showcased the work of Dr. Deming, crediting the successes in Japan to his teachings and positioning Japan as a major competitive threat to the U.S. economy. This was a wakeup call and a catalyst that propelled the country into a second generation of quality, and made Dr. Deming an overnight superstar. This milestone event also marked the end of the first generation of the movement and the beginning of the second, in which quality gradually moved from a micro product view to a macro organizational view.

In 1980, Western industrialized nations became aware of the magnitude of the impending threat from Japan.

1.4.2.2 Second Generation: Post-1980—The Emergence of Quality as a Universal Mantra.

In addition to pioneering new concepts of quality management, the Japanese introduced innovative work practices to the world, and laid the groundwork for the **high-performing work organizations** (HPWOs) of today. Skilled Japanese workforces, loyal to their companies through guaranteed lifetime employment, used just-in-time production and knowledge-based quality control systems to drive levels of productivity far

[47]For a biography and photo of Philip Crosby consult http://www.iso-9000.ro/eng/9001/gurus.htm or http://www.wppl.org/wphistory/PhilipCrosby/index.html.

[48]Crosby, Philip B. *Quality Is Free: The Art of Making Quality Certain.* New York: McGraw-Hill, 1979.

[49]COQ and Six Sigma are discussed in more detail in Chapter 14.

[50]Crosby's 14-step approach is included in Appendix A.

[51]The basic and perhaps subtle difference between QC and QA was that the former focused on the products and services produced while the latter was more concerned with the processes that produced them.

beyond those attainable by their Western counterparts. Meanwhile, changes were on the way that would irrevocably affect the business climate. The world was beginning to shrink. Spawned by jet air travel, computer technology, and satellite communications, domestic markets were rapidly becoming global ones. Led by corporate giants—companies like General Motors, Ford, GE, Motorola, and Procter & Gamble—quality management began to be promoted as more than a mere set of tools to be used at the discretion of management.

By the late 1980s and early 1990s quality advocates saw the need to establish standards that could be used strategically as weapons of choice on competitive battlegrounds. The problem with this idea was the lack of a unified international effort to develop quality standards capable of garnering universal unanimity and support. During this period three sets of documents were developed, each independently crafted, by different groups on different continents using disparate languages and terminologies, and employing divergent approaches.[52] These document sets, each in its own domain, gained wide recognition and attracted a covey of loyal followers who were willing to readily subscribe to the teaching embodied by their "new standard" of choice. The three standards were:

1. **Books authored by quality pioneers**—During the decade of the 1980s the work of three of the quality pioneers previously discussed gained widespread prominence. They were W. Edwards Deming,[53] Joseph Juran,[54] and Philip Crosby.[55] Each of these "gurus" capitalized on the groundswell of public interest by publishing and promoting his own individual set of ideals that were neatly and succinctly laid out as a prescriptive set of points, steps, or principles that, if followed closely, would be the *garanti* recipe for success.[56] Each of the books and seminars, workshops, consulting projects, and the like were primarily aimed at the middle to top levels of management in U.S. organizations.
2. **ISO9000 standards**—The Geneva-based International Organization for Standardization (ISO) formed a committee, with representation from over 100 companies, for the purpose of developing a consensus document that could be implemented as a set of international standards. These standards were given the title ISO9000 and were targeted primarily at the middle level of management. The content of the first version, published in 1987, was heavily influenced by the British quality standards (BS5750) used during World War II, and by the U.S. Department of Defense standards (MIL STDS). ISO9000 was particularly well suited for manufacturing organizations and was readily accepted as the basis for accreditation and certification. Compliance with this set of internationally agreed-upon standards formed the only auditable and credible evidence that organizations could use to profess themselves "quality capable."[57]

The 1987 version was actually a family of three standards, each of which covered a different scope of activities. ISO9001 (*Model for quality assurance in design, development, production, installation, and servicing*) was for organizations whose activities included not only the production and delivery of existing services but also the creation of new products and services; ISO9002 (*Model for quality assurance in production, installation, and servicing*) was for organizations engaged in the production and delivery of products and services, but not involved in new product development; and ISO9003 (*Model for quality assurance in final inspection and test*) was for organizations that were involved in only the final inspection of a finished product, with no involvement or concern for how the product was produced.

Over the years ISO9000 has gone through several major reforms. The first reform, in 1994, addressed an

---

[52]Foley et al., "Quality Matters."

[53]During the 1980s and into the 1990s (until his death in 1993) Dr. Deming was in great demand as a consultant and teacher. He took his message to the world traveling to Japan, England, Europe, Australia, New Zealand, Canada, and other countries. He worked with large corporations including Ford Motor Company, General Motors, Dow Chemical, and Procter and Gamble. It has been estimated that through his seminars and work with industry Dr. Deming touched the lives of tens of thousands—approximately 100,000 according to Boardman, Thomas J. "The Statistician Who Changed the World." *The American Statistician* 48, no. 3 (1994).

[54]In 1979 Joseph Juran established the Juran Institute to provide training and consulting services in quality improvement and management. Headquartered in Southbury, Connecticut, the institute expanded rapidly and branch offices were opened in China, Egypt, Germany, the Netherlands, South Africa, South Korea, Spain, the United Arab Emirates, and the United Kingdom. Juran's consulting clients included large U.S. corporations such as Armour and Company, Dennison Manufacturing Company, Merck, Sharp & Dohme, Otis Elevator Company, Xerox, and the United States Navy Fleet Ballistic Missile System.

[55]In 1979 Crosby established Philip Crosby Associates to provide consulting and training services in quality management. Headquartered in Winter Park, Florida, the firm helps organizations wordwide with offices in Latin America, Australia, New Zealand, Canada, China, India, and Southeast Asia (Malaysia, Singapore, Thailand, and Brunei). http://en.wikipedia.org/wiki/ISO_9001#History_of_ISO_9000, http://www.jstor.org.proxy.lib.utk.edu:90/stable/2684713?seq=2, http://en.wikipedia.org/wiki/Joseph_M._Juran, http://findarticles.com/p/articles/mi_qa3618/is_200507/ai_n14799744/pg_4?tag=artBody;col1.

[56]Deming, W. Edwards, *Quality, Productivity, and Competitive Position* (1982) and *Out of the Crisis* (1986). Cambridge, MA: MIT Center for Advanced Engineering Study; Juran, Joseph M. *Juran's Quality Control Handbook*, 5th ed. with Blanton A. Godfrey (1999), 4th ed. with Frank M. Gryna (1988), 3rd ed. with Frank M. Gryna and Richard S. Bingham (1974), 2nd ed. with Richard S. Bingham (1962), 1st ed. (1951). New York: McGraw-Hill; Crosby, Philip B. *Quality Is Free: The Art of Making Quality Certain* (1979), *Quality without Tears: The Art of Hassle-Free Management* (1984, 1995), and *Let's Talk Quality: 96 Questions You Always Wanted to Ask Phil Crosby* (1989). New York: McGraw-Hill. *Completeness: Quality for the 21st Century.* New York: Plume, 1992; *Philip Crosby's Reflections on Quality: 295 Inspirations from the World's Foremost Quality Guru.* McGraw-Hill, 1996; and *Quality Is Still Free: Making Quality Certain in Uncertain Times.* New York: McGraw-Hill, 1996.

[57]The timing of the introduction of the ISO 9000 standards coincided with the burgeoning and global interest in the need for improved quality. Almost overnight a multibillion-dollar industry was born based on the use of structured nonfinancial audits to improve competitiveness.

important shortcoming. Up to that point being ISO9000 certified did not mean that an organization was producing quality outputs at a high level; the certification simply meant that the company was following the appropriate documentation requirements. In 1994 the standards were revised to place increased emphasis on quality assurance through prevention as opposed to stressing documentation and final inspection. The most radical reform came in 2000 when ISO9001, ISO9002, and ISO9003 were all combined into a single standard called IS09001:2000 in which continuous improvement and senior management involvement became key requirements for certification.

The most recent change by the ISO committee resulted in ISO9001:2008. In the most current standard the requirements have not changed significantly from the 2000 version; however, in the current document more emphasis has been placed on quality as an intrinsic component of a total management (as opposed to a quality management) system and the *condition requise* for sustainability.

3. **National quality awards or business excellence models**—During the 1980s and 1990s a number of countries established national quality awards or business excellence models typically written by a relatively small group of senior managers from industry to provide a federally endorsed set of quality standards and principles. The first of these was the establishment of the Deming Prize[58] mentioned previously.

When Dr. Deming visited Japan in 1950 he delivered a series of eight-day and one-day seminars. The content of these lectures was compiled into a book and sold. Dr. Deming donated his royalties from the book sales to JUSE whichin turn used them to establish the annual award. The Deming Prize, which is still administered by JUSE and is completely funded through company memberships, is awarded in three categories:

During the period from 1980 to 1990 three sets of independent standards emerged: (1) books from prominent authors, (2) ISO9000, and (3) national quality awards.

a. **Individuals or groups**—to recognize those individuals or groups of individuals who have made outstanding contributions to the study, application, or dissemination of quality principles.

b. **Organizations or divisions**—to recognize distinctive and superior performance improvement through the application of quality principles in a designated year.

c. **Operations business units that reside within a larger corporate structure**—to recognized achieved levels of distinctive and superior performance improvement through the application of quality principles in a designated year.

[58]http://www.juse.or.jp/e/deming/index.html.

It would be nearly four decades before the idea of business excellence models resonated with the rest of the Western world. The most prominent of the more recent models, clearly targeted at top management, are the Malcolm Baldrige National Quality Award (MBNQA)[59] in the United States and the European Foundation for Quality Management (EFQM).[60] Though very similar to the EFQM and other excellence models in content and purpose, the MBNQA differs in that it was created by an act of Congress in 1987 and signed into law by President Ronald Reagan. The MBNQA was established as an attempt by Congress to jumpstart a flagging quality movement in the United States by promoting quality awareness, recognizing notable achievements, and publicizing successes. The Baldrige process evaluates organizations against seven major criteria—leadership; strategic planning; customer and market focus (students, stakeholders, and market forces for education institutions; patients, other customers and markets for health care organizations); measurement, analysis, and knowledge management; workforce focus; process management; and results. Awards are presented annually by the U.S. president to those organizations that have demonstrated excellence in manufacturing, service, small business, education, health care, and not-for-profit. Since the first awards were made in 1988, the MBNQA program has played a significant role in helping U.S. organizations improve their ability to deliver sustained value to their stakeholders.

The EFQM *Excellence Model*, promoted in the United Kingdom as *Business Excellence* by the British Quality Foundation (BQF), was introduced in 1992 and has grown to become the most widely used framework for quality standards throughout Europe and forms the basis for many national and regional quality awards. The EFQM Excellence Model is nonprescriptive and has been used by organizations for self-assessment, as the basis for benchmarking, as a guide for improvement, as a tool to effect change and influence internal thinking, and as a structure for management practices. The model employs nine criteria. Five of these are enablers—leadership, policy and strategy, people, partnerships and resources, and processes. Enablers cover what an organization *does*. The other four criteria are results—customer results, people results, society results, and key performance results. Results cover what an organization *achieves*.

Each of the three groups described quality management differently. The gurus saw quality as a progression of quality control, then quality management, advancing finally to a state of **total quality management** (TQM). The ISO language stressed quality assurance whereas the quality awards groups presented quality management as essential to business excellence.

[59]http://www.quality.nist.gov/.

[60]http://www.efqm.org/Default.aspx?tabid=154.

The existence of these disparate approaches fueled the movement, but also detracted from it, since it was difficult for any one program to gain traction when seemingly conflicting philosophies were available and there appeared to be no cohesive theory of quality management to use as the basis for systems design. This may have been a contributing factor to the failure of many programs and the abandonment of others around the turn of the century.

The beginning of the second generation probably occurred around the mid-1980s when industry began to realize that quality can be a major factor in gaining a competitive edge. As a strategic priority quality was treated as a major driver, influencing all important business decisions, and a minor industry was born. The 1980s saw the emergence of a growing army of consultants, academics, and a new breed of professionals who marched into battle armed with statistical tools, training aids, and implementation strategies. Company after company was willing to commit significant resources to the cause, believing that in doing so they could beat or at least keep abreast of their competition. In many cases companies had no choice as formal statistically based quality programs became a customer requirement or a prerequisite for supplier certification. Quality programs ranked high on the list of corporate goals. Business glossaries began to include some new three-letter buzzwords, such as **statistical process control** (**SPC**) and TQM.

During the second generation management accepted responsibility for poor quality and emphasized systems, participation, and assessment.

Companies invested heavily in wholesale training until all employees were thoroughly indoctrinated in the teachings of the quality guru of choice. The attention turned from product orientation and quality of output to the overall management system responsible for producing the output. Following the teachings of the gurus such as Shewhart, Deming, Juran, and their contemporaries, industry leaders began to accept the idea that the root causes for poor quality were largely systemic. Because management owned the systems, the idea that inferior quality was management's responsibility and *not the worker's fault* gained widespread acceptance.

With an expanded role, managers began to understand the holistic nature of their responsibilities and the need to reach out beyond organizational boundaries. Leadership teams were formed. The emphasis shifted from measurement to **assessment** and from control to employee **self-management.** Quality control turned from detection to prevention and problem solving from reactive to proactive. These concepts were often formally written into job descriptions and team charters.

Companies that joined the new movement were often encouraged by dramatic short-term results. Continuous improvement became the cornerstone of cultural change. Companies soon learned that there were no quick fixes and that total commitment to quality meant to be in it for the long haul. The mantra *quality is a journey, not a destination*[61] could often be found in training manuals, management briefs, and shop-floor banners. The quest for improved quality, including acquiring process knowledge and learning how to effectively respond to it, became a process in itself.

Six Sigma was developed at Motorola as a disciplined approach for implementing TQM.

During this period Bill Smith, an engineer at Motorola, coined the term *Six Sigma* (a registered trademark of Motorola) to represent a statistical approach he had developed for reducing defects. The idea behind Six Sigma is to achieve a process capability where production is nearly perfect. The name originates from the statistical idea that almost all of a normal distribution will comply with requirements if the process is well-centered and the specification limits are more than six standard deviations apart. To drive home the importance of zero defects, Six Sigma programs count the number of defects per million opportunities. The goal is to produce less than 10 defects per million, a condition that for all practical purposes is defect-free.

Dr. Mikel Harry, a senior staff engineer at Motorola, and Richard Schroeder, an ex-Motorola executive, were responsible for transforming Smith's statistical measurement tools into a TQM spin-off philosophy. Their integrated change and data-driven management paradigm won Motorola the MBNQA in 1988. Seeking an alternative to what they perceived to be some deficiencies in the TQM approach, high-profile business leaders of the day including Bob Galvin, CEO of Motorola; Larry Bossidy, CEO of Allied Signal (now Honeywell); and Jack Welch, CEO of General Electric, threw their support behind the development of the Six Sigma movement.

The appeal of Six Sigma is in its prescriptive approach to implementation and cultural change. The emphasis is on selecting high-leverage projects that offer clearly defined benefits tied to the bottom line—a concept that is sure to be a rallying point for top managers and shareholders. During the execution of each project the program offers a clear delineation of responsibilities: top to bottom, as well as the articulation of training requirements, including a formal certification plan for credentialing all participants. The Six Sigma methodology is covered in further detail in Chapter 14.

In the mid-1980s Kaoru Ishikawa[62] created a Japanese version of quality management[63] that stressed the benefits of broad employee participation, permitting bottom-up as well as top-down involvement. To differentiate between the Japanese-style and Western versions of total quality control, Ishikawa called the Japanese programs **company-wide quality control** (**CWQC**). CWQC systems used employee teams,

---

[61]Rhinehart, Emily. "Quality Management Is a Journey, Not Just a Destination." *Managed Healthcare* 10, no. 5 (2000): 52–55.

[62]For a biography and photo of Kaoru Ishikawa and other quality pioneers consult http://www.iso-9000.ro/eng/9001/gurus.htm.

[63]Ishikawa, Kaoru. *What Is Total Quality Control? The Japanese Way*. Upper Saddle River, NJ: Prentice Hall, 1985.

which he called **quality circles,** as the vehicle to draw on the talents and expertise of front-line employees. In his writings Ishikawa claimed that he originated the concept of internal customers, where each operator views the next operation in line as a valued customer. This thinking underscored the importance of breaking down barriers between departments. Ishikawa is best known for the fishbone diagram, which are also called Ishikawa diagrams in his honor. The Ishikawa diagram is a powerful tool that can be taught to anyone and can be effectively used by nonspecialists to identify root causes.

In 1986 Masaaki Imai[64] introduced to the Western world a philosophy called **Kaizen,**[65] which loosely translated means *ongoing improvement involving everybody, without requiring much money.* Kaizen places attention on incremental and continuous improvement and, as a philosophy, advocates that every aspect of daily life (personal and business) can be and should be improved. The Kaizen method was founded on five elements: (1) teamwork, (2) discipline, (3) improved morale, (4) quality circles, and (5) suggestions for improvement.

Ishikawa contributed the idea of using worker teams, called quality circles, to help solve process problems.

By the mid- to late 1990s most industries faced fierce global competition, complex management issues, and despite their best efforts in TQM, many corporations could not forestall declining performance.

During the second generation the ASQC grew in stature and launched a certification program for quality professionals. The name of the professional society was shortened in 1997 to the American Society for Quality (ASQ).[66]

Imai is responsible for introducing Kaizen to the world.

1.4.2.3 Third Generation: Quality as a Strategic Driver in Whole-of-Enterprise Management. By the late 1990s, inspired by the quality standards exemplified by prestigious business excellence models such as the Deming Prize, MBNQA, and the EFQM, and a worldwide acceptance of the ISO9000 family of standards, many organizations began to reach a level of maturity that clearly set them apart from their industrial counterparts. These organizations became trailblazers, often cited in the literature and widely used as benchmarks for those wishing to emulate success. The Six Sigma philosophy gained popularity as ASQ assumed the role of administering tests leading to the various levels of Six Sigma certification. In the third generation, quality is not seen as a means to an end, but as a *contributor to a strategic commitment to create sustained customer value.* Third-generation quality is an advanced stage of quality deployment and stresses consensus, engagement, accountability, relationships, and response. We will discuss third-generation quality in further detail in Chapter 2.

Third-generation quality recognizes the need to make quality diffuse throughout the strategic and operational planning processes.

[64]For a short biography and photo of Masaaki Imai consult http://www.qualitydigest.com/june97/html/imai.html.

[65]Imai, Masaaki. *Kaizen: The Key to Japan's Competitive Success.* New York: McGraw-Hill/Irwin, 1986.

[66]ASQ's full-range of services can be accessed from its Web site at http://asq.org.

1.5 THE MANY FACES OF QUALITY

In the third generation, as the fulcrum for forward planning is the satisfaction of stakeholder interests, quality is a **corporate force**[67] and is at the heart of an organization's ability to compete. Quality principles are the foundation blocks of corporate sustainability, and the centerpiece of strategy, literally impacting every important decision that a company makes—from the boardroom to the shipping dock. However, quality is not defined the same way by everyone, and achieving a consensual view of quality can be one of the most significant steps any organization will take in its quality journey.

All quality philosophies share one common goal: quality is a core value—a corporate force.

1.5.1 Various Perspectives

Throughout the literature various terms have been used to define differing sets of principles aimed at improved performance, better competitive capabilities, and higher customer satisfaction. A few of the better known ones have been *quality assurance (QA),*[68] *total quality control (TQC),*[69] *customer-driven quality (CDQ),*[70] *continuous process improvement (CPI),*[71] *zero defects (ZD),*[72] *total quality management (TQM),*[73] *business process reengineering (BPR),*[74] *lean production (LP),*[75] *Six Sigma (SS),*[76] and *Lean Six Sigma (LSS).*[77] Such programs vary widely in scope and implementation, but they all have one goal in common: the centricity of quality as a corporate core value.

[67]Aikens, C. Harold. *Quality: A Corporate Force—Managing for Excellence.* Upper Saddle River, NJ: Prentice Hall, 2006.

[68]http://www.qaiworldwide.org/.

[69]Feigenbaum, "Total Quality Control."

[70]http://www.ccdq.com/.

[71]Robson, George D. *Continuous Process Improvement.* New York: Free Press, 1991.

[72]http://www.daimlerchrysler.com/dccom/0,,0-5-7182-1-465545-1-0-0-0-0-0-243-7165-0-0-0-0-0-0-0,00.html.

[73]http://www.tandf.co.uk/journals/titles/14783363.asp.

[74]Landon, C., and G. Miller. *Business Process Re-engineering: A Management Handbook.* 3rd ed. Redmond, WA: Vertical Systems, 2002.

[75]Womack, James P., Daniel T. Jones, and Daniel Roos. *The Machine That Changed the World: The Story of Lean Production.* New York: First Harper Perennial, 1991.

[76]Pande, Peter S., Robert P. Neuman, and Roland R. Cavanagh. *The Six Sigma Way: How GE, Motorola, and Other Top Companies Are Honing Their Performance.* New York: McGraw-Hill, 2000.

[77]George, Michael L. *Lean Six Sigma: Combining Six Sigma Quality with Lean Production Speed.* New York: McGraw-Hill, 2002.

Quality can mean good timing—a flawed or late supply drop or airborne refueling could result in an aborted mission. Getting the right materials in the right quantities at the right location and at the right time can be critical.

Contemporary quality management must be the embodiment of many and varied perspectives—folded into a committed corporate culture with a strong knowledge-driven leadership. Many failed implementations can be attributed to too great a reliance in a single guru or philosophy and in so doing, inadvertently neglecting the holistic approach necessary for a complete cultural transformation. To eliminate some of the baggage and predispositions associated with previous programs some authors are beginning to adopt new terms (for example, **quality inspired management** [QIM][78] and **meta-management**[79]) to describe a paradigm of management that is holistically committed to principles that value stakeholder needs, strive for perfection, and embrace cultures that reflect participation, equity, teamwork, and learning. Throughout this book, *quality inspired management (QIM)* has been used as an umbrella term that encompasses the principles and theory of this holistic management paradigm.

***QIM* is an umbrella term used to refer to any management paradigm that is holistically committed to stakeholder needs, human value, and the pursuit of excellence.**

Different perspectives of quality are evidenced by the numerous definitions of "quality" used to inspire and champion the cause. With respect to product quality one of the earliest recorded definitions asserts that quality can be assessed on three levels: how well a product fits its intended function, how well it withstands the rigors of its operating environment (durability and reliability), and to what extent the product delights its user (aesthetics, pleasure, beauty, intangibles, etc.).[80] In the modern era leaders in the quality movement have chosen to place their particular brand on one or more of these dimensional definitions—for example, fitness for use,[81] durability, and unexpected operating performance.[82]

There may be instances where, though the beneficiaries of quality are unable to clearly *articulate* what it is, they are certain that *they know it when they see it.*[83] Hence, quality is largely perceptual, reflecting personal feelings or attitudes at a moment in time. Quality is an individual value judgment, and at the end of the day, quality is simply what the recipient says it is. If a stakeholder ever perceives that a better deal is available elsewhere, loyalty yields to the stronger force of self-gratification. Whether in truth an alternative opportunity proves to be as utopian as the stakeholder perceives may be irrelevant, because *subjective perception will always trump objective reality!*

Quality is a value judgment and perception is more important than reality.

Perceptions can be shaped through a transcendental comparison between what the stakeholder expected and what was actually received. On the one hand, quality may be rated poorly if certain features or attributes, known as static quality factors, are missing. On the other hand, certain unanticipated features, known as dynamic quality factors, can intensify positive perceptions. The *principal of neutral awareness* states that the perception of quality is affected only by the absence of static factors or the presence of dynamic factors.[84]

In the case of a service, good will is often made or lost at the point of sale based on the spontaneous encounter a customer has with the employee of a service provider (e.g., store clerk, bellhop, ticket agent, etc.). These *moments of truth*[85] can more often than not mean the difference between positive or

[78]Aikens, *Quality: A Corporate Force,* 3

[79]Foley, Kevin. *Meta Management: A Stakeholder/Quality Management Approach to Whole-of-Enterprise Management.* Sydney: Standards Australia: 2005.

[80]Pollio, Marcus Vitruvius. "The Ten Books of Architecture," translated by Morris Hicky Morgan. New York: Dover, 1960. Originally published by Harvard University Press in 1914.

[81]Juran, *Juran's Quality Control Handbook,* 1974.

[82]Leffler, Keith B. "Ambiguous Changes in Product Quality." *American Economic Review* 72, December 1982.

[83]Persig, Robert M. *Zen and the Art of Motorcycle Maintenance: An Inquiry into Values.* New York: William Morrow, 1974, 185–213.

[84]Cusins, Peter. "Understanding Quality through Systems Thinking." *The TQM Magazine* 2 (1994): 19–27.

[85]Carlzon, Jan. *Moments of Truth.* Reprint ed. New York: HarperCollins, 1989, 1–2.

Service quality offers unique challenges—what is perceived to be enjoyable and fulfilling by some can be perceived by others to be stressful and traumatic.

The principle of neutral awareness states that the perception of quality increases if dynamic factors are present and decreases if static factors are absent.

negative perceptions of quality. Anyone who has had to deal directly with service representatives manning 1-800 complaint hot-lines, personnel working behind merchandise return counters, contact personnel at lost baggage counters, or ticket agents trying to sort out the itineraries of angry and weary travelers on an overbooked aircraft know full well how such moments of truth can influence one's attitude toward the entire company.

Service quality is not limited to intangibles. A product is often involved, and the delivery service can be an important factor. Both of these (that is, the product and the service) have quality characteristics that can readily be measured. For example, guests of a hotel expect to receive tangible products (e.g., room supplies, food, beverages) together with the service (comfortable room consistent with price), delivered in a timely and efficient manner (responsiveness of front desk, bellhop, concierge, housekeeping to needs or requests). Most services can be decomposed along four dimensions: physical product, service product, service environment, and service delivery.[86] A physical product is defined as any tangible entity that is exchanged as part of a transaction. A service product is defined as a bundle of tasks that are offered to a customer to satisfy specific needs, and the service environment is the place where the service experience takes place. Finally, the service delivery is the method and mechanisms through which the service provider interacts with the customer throughout the transaction.

This decomposition can be illustrated further through two examples. In the automobile industry a car of a particular make and model is a physical product; service products include various pricing policies, leasing arrangements, and extended warranty options available either through a dealer or direct from the manufacturer; a dealer's showroom or service facility are examples of service environments; and a sales pitch, test drive, and incentives are all part of service delivery. At a university a diploma is a physical product; choices in majors or elective courses are service products; classrooms, laboratories, library facilities, and student centers are service environments; and pedagogies, assessment instruments, and teaching methods are all examples of service delivery.

Moments of truth are instantaneous contacts with customers.

As stakeholders, customers make choices. This is the essence of the free enterprise system and strong competition drives continuous and break-through improvements. In making choices customers may claim that they are applying specific criteria (e.g., features, functions, price, aesthetics, etc.). In the final analysis customer decisions are based on value judgments, as customers weigh (consciously or subliminally) the difference between expectations and what the customer perceives as reality. This phenomenon is especially evident in service industries where a customer's satisfaction is often based on such intangible factors as assurance, empathy, reliability, and responsiveness.[87]

Job Opening

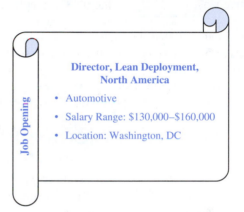

Director, Lean Deployment, North America

- Automotive
- Salary Range: $130,000–$160,000
- Location: Washington, DC

1.5.2 An Operational Definition of Quality

In the world of quality management operational definitions are essential in order to minimize the variability that arises from inconsistent behaviors. Such inconsistencies can arise due to misunderstandings and lack of effective communications. An **operational definition** is a clear concise communicable written statement that will be understood by everyone the same way. Our previous discussion concerning the various perspectives on quality begs the question of whether it is possible to craft an operational (working) definition of quality.

Viewing quality as a stakeholder value judgment leads to the following insights. Quality as perceived by a stakeholder is an assessment of how far actual results either exceed or fall short of what the stakeholder expected in the first place. If expectations are exceeded the result will be a positive perception. In this case one would say that the quality is *good* and someone (presumably the stakeholder) derived pleasure from the experience (and hopefully if given the opportunity would opt to return in the future for a repeat performance or

[86] Rust, Roland T., Anthony J. Zahorik, and Timothy L. Keiningham. *Return on Quality: Measuring the Financial Impact of Your Company's Quest for Quality*. Chicago: Probus, 1994.

[87] Parasuraman, A., Leonard L. Berry, and Valarie A. Zeithaml. "Refinement and Reassessment of the SERVQUAL Scale." *Journal of Retailing* 27, no. 4 (Winter 1991): 420–50.

Quality is a two-dimensional gap measure: (1) the difference between the outcomes a stakeholder expected and what the stakeholder perceives were the actual outcomes delivered; and (2) the difference between what a stakeholder had to sacrifice to achieve the outcomes (e.g., dollars, time inconvenience, talents, foregone opportunities, etc.) and perceived benefits derived from the actual outcomes delivered.

FIGURE 1–7. Operational Definition of Quality

recommend the organization to a friend or colleague). If the experience does not meet expectations, the perception will be negative. In this case one would rate quality as *poor*, engendering negative feelings that could range anywhere from disappointment to anger. Under such circumstances stakeholders are likely to defect (that is, withdraw support, seeking alternative affiliations). Even when expectations are met *exactly*, negative perceptions could be garnered if stakeholders are not convinced that alternative relationships offer the prospects of better results. This rationale leads to the operational definition shown in Figure 1–7.

An operational definition is a clear, concise statement that is understood by everyone the same way.

These comparisons then form the basis for making personal judgments as to whether an experience is worth repeating.

1.6 QUALITY AND CORPORATE STRATEGY

As Will Rogers once said, "*even if you are on the right track, you'll get run over if you just sit there.*" Competitiveness in the global information economy requires adaptability to changing conditions. Maintaining the status quo is not an option; companies either move forward or decline, but of one thing they can be certain—they will not stand still. Quality is central to good strategic planning. Companies must know who their stakeholders are, what those stakeholders need, how the company's core competencies can be used to meet and exceed stakeholder expectations, and the dynamics of the environment in which they operate. Over time strategy *emerges* as the consequence of strong leadership, vision, and systems knowledge. To set the stage for excellence, an investigation of the strategic planning process is appropriate.[88] This topic is discussed further in Chapter 2.

1.7 SUMMARY OF KEY IDEAS

Key Idea: *An organization that has achieved excellence will have succeeded in identifying its stakeholders, responded to their needs in a balanced fashion, and produced exceptional outcomes in operations, customer relations, financials, and the marketplace.*

Excellence and *quality* are synonymous terms. Quality is not only about satisfying the needs of customers; quality is about balancing the needs of all stakeholders. In some cases customers may not even be stakeholders if the customers' only relationship with the organization is through the transaction. With smart management practices an organization can experience sustained growth and profitability by focusing on its stakeholders rather than markets, competitors, and customers.

Key Idea: *Quality principles are rooted in history, paralleling developments in industrial practice, organizational theory, social engineering, and technology.*

The concept of quality has been around since the time that one person relied upon another to provide a needed product or service. In antiquity, quality was controlled by craft guilds through pride of workmanship. Master tradesmen would closely supervise the work of apprentices who learned their trades over a period of many years. Quality as well as working conditions were tightly controlled by the guilds. Production was slow and skills were not specialized, as an individual tradesman usually performed all the tasks necessary to complete the job.

The twentieth century introduced mass production techniques, standardization, specialization of labor, and tools for predicting process output. By the middle of the century statistical tools had been applied to acceptance sampling and process control. By the late 1990s many organizations had begun to realize that by making quality a core value, and the basis for corporate strategy, quality could become a competitive weapon. Quality as a core value had dramatic implications. Workplace cultures changed, and relationships up and down the value stream were affected. Management valued their employees in new ways, empowering them with greater responsibilities and access to decision making. The emphasis began to shift from productivity to value creation. Technological developments launched the information age and with it the concept of the knowledge worker. The decision-making process became fact-based, increasingly using multidisciplinary teams to solve problems and improve operations.

Key Idea: *The maturity of quality thought and practice has been gradual and the quality movement has evolved through three distinctive generations of development.*

The evolution of the quality movement has been characterized by three identifiable generations. The first generation, which lasted until about 1980, was concerned primarily with measurements, controls, and detection. The belief was that quality control could be achieved primarily through effective detection—that is, effective inspection methods could detect defective product early enough to avoid that product reaching the customer. Many of the statistical principles that form the basis of modern-day statistical process control were developed during this period, pioneered by the efforts of Shewhart, Sarasohn, Dodge, Romig, Juran, and Deming.

---

[88] Strategy is the topic of Chapter 2.

The second generation, which started around 1980 and lasted two decades, marked an era during which attention began to focus on prevention and the use of quality as a standard. The problem was that three different document sets, independently developed, were available fueling confusion on how best to proceed. Multitudes of organizations subscribed to one of the many doctrines advanced by a quality pioneer—their "guru" of choice. Others took a more structured approach, seeking certification under one of the ISO9000 standards. Some subscribed to both approaches. A third set of guidelines were the criteria stipulated in national quality awards and business excellence models, the predominant ones being the Deming Prize, the MBNQA, and the EFQM. Each of these families of standards singularly and in combination provided the foundation for the implementation of TQM, in which quality became the cornerstone of corporate governance and a zero defects mentality became ingrained in organizational cultures.

In time, the interest in TQM and zero defects waned as companies turned to what they perceived as alternative strategies with greater appeal. Programs like lean production, business process reengineering, high quality-work organizations, and Six Sigma appeared on the landscape. Through the efforts of Smith, Harry, and Schroeder, and with the support of the CEOs of some key companies, Six Sigma gained in popularity.

At the dawn of the new millennium, a third generation became evident—an era that placed more emphasis on structure, accountability, and management participation. Six Sigma has achieved widespread use because it has satisfactorily addressed many of the shortcomings of failed TQM programs. In the third generation, the focus is on continual improvement, resource leverage, management commitment from the top, and creation of a culture that can sustain customer value.

Key Idea: *Quality inspired management (QIM) is a management paradigm holistically committed to stakeholder needs, human value, and the pursuit of excellence.*

Quality is fundamental to all major business decisions, including structures, policies, strategies, decisions, planning, and operations. Organizations exist because they have stakeholders who depend on them for various outcomes. For example, as stakeholders customers expect certain products or services; investors expect adequate returns and responsible fiscal management; employees expect safe working conditions, fair treatment, and job security; and so on. Stakeholders will alliances or transact business with the organization that gives them the greatest perceived value, defined as the difference between benefits and sacrifices. For a company to sustain its position in the market, value must play a central role in all of its major decisions and quality must be accepted as a core value.

Key Idea: *Quality can be defined from many different perspectives.*

Product quality has been defined as function in use, firmness, durability, and delight—all as perceived by customers. Quality is also an instantaneous value judgment made during a moment of truth, which may have more to do with the psychology of the experience rather than the substance of the service. Quality can be the absence of static factors or the presence of dynamic factors. Quality is personal and in the end is what the recipient says it is. Quality is a judgment call, made at the boundary of a system, and is based on perceived value and whether the perceived benefits meet or exceed expectations.

As a core value, quality is inculcated into an organization's culture and transforms the management paradigm, which includes strategy, structure, problem solving and decision making, internal and external relationships, production philosophy and configuration, and personnel policies. Quality inspired management (QIM) is the embodiment of an integrated management paradigm with a committed culture and a strong knowledge-driven leadership.

Key Idea: *An operational definition is a clear, concise statement that is understood by everyone the same way.*

Quality depends on clear and concise communications to all stakeholders on expectations, measures, and procedures, so that all personal efforts will be in alignment with quality goals. Such consistency requires that operational definitions be developed for all key requirements and measures. It is essential that each organization also develop a working definition of quality. The definition should be communicable and understood the same way by everyone. I have suggested that a working definition might treat quality as a two-dimensional gap assessment. The first gap is the perceived difference between what a customer receives and what that customer expected. The second gap is a measure of perceived value (i.e., the difference between the benefits received and what the customer had to sacrifice in order to get them).

Key Idea: *Quality inspired management requires a culture built on seeking and acting on the truth, which is derived from the application of appropriate tools of analysis to relevant and timely process data.*

Information technologies have provided decision makers with access to large databases and the capability to manage and act on knowledge rather than supposition. As a result a new breed of worker—the knowledge worker—has emerged. These employees are valued not only for job-specific skill sets, but also for their abilities to think, reason, innovate, solve problems, and collaborate with associates.

Quality inspired management requires developing a culture built on seeking and acting on the truth, a management mindset that bases decisions and policies on knowledge, and that relentlessly pursues a profound understanding of process behaviors, relationships, and cause and effects. This culture is characterized by the unwavering dedication to continual learning, individually and holistically, and the ability for personnel at all levels to think statistically and effectively manage variability.

1.8 MIND EXPANDERS TO TEST YOUR UNDERSTANDING

1-1. For each of the following organizations define the different stakeholder groups. Suggest actions that each group could take that inflict negative consequences on the organization concerned.

a. A large public university
b. A privately owned restaurant
c. A carpet-cleaning franchise
d. A municipal bus service
e. A department store that is part of a national chain
f. The federal postal service
g. A dental practice

1-2. How would you define quality for each of the stakeholders identified in questions 1a–g?

1-3. Discuss some of the challenges that must be overcome to apply quality management principles in the following industries.

a. Government
b. Health Care
c. Education

1-4. Describe how the craft guild system of the Middle Ages helped ensure quality. How do you think quality was defined during this era?

1-5. In a sentence or two describe the contributions to the quality movement made by each of the following pioneers.

a. Shewhart
b. Dodge and Romig
c. Sarasohn
d. Deming
e. Juran
f. Wareman
g. Feigenbaum
h. Crosby
i. Smith
j. Harry and Schroeder
k. Ishikawa
l. Imai

1-6. Describe the three generations that characterize the evolution of the quality movement. Explain how a progression through the three generations impacts the following:

a. Management practices
b. Management–labor relations
c. Stakeholder relationships
d. Relationships with suppliers
e. Relationships with competitors

1-7. Write an operational definition for the following:

a. A quality set of tires for the vehicle you use to commute to school or work
b. Being on time for a meeting
c. Good service at a restaurant
d. A good return on an investment
e. An A grade on a research paper or report

CHAPTER TWO

QUALITY AS A STRATEGIC IMPERATIVE IN THE INFORMATION AGE

QUALITY MAXIM #2: *Management is getting things done, leadership is doing the right things, and strategy is knowing what to do.*

What's Strategy Got to Do with It?

A 1993 movie *What's Love Got to Do With It?*[1,2] chronicled the life of rock singer Tina Turner and her efforts to break free from the domination of an abusive husband. Tina was a young singer in a small rural community in western Tennessee when she was discovered by Ike, an accomplished songwriter, performer, and record producer. Ike went on to make Tina a star, but ironically as her fame grew, the story is that he became jealous and abusive. She found herself caught up in a vicious circle that not only threatened her sustained success with a mounting vengeance, but also seemed at times to be inescapable. Tina's story is a metaphor for the corporate world. How many failed companies started out strong—with good ideas, capable leadership, and solid financial backing—only to discover several years down the track that their markets have evaporated, or their investors have withdrawn support, or they can no longer honor labor agreements? Organizations can become so caught up in their glitter of success that they forget (if they ever knew) the core values that made them great in the first place. When corporate eyes are more focused on self than service, organizations can spiral into a destructive freefall that is often not recognized until it is too late to save the company. Toward the end of 2008 the CEOs of the Detroit three[3] U.S. automobile manufacturers came to Washington to collectively petition the U.S. Congress for a bailout of their beleaguered companies. The trio claimed that billions ($25 billion initially) were needed to forestall the imminent collapse of these stalwarts—the pillars of the American economy. Their supporters argued that the Detroit three were too large to fail, and without government intervention thousands of jobs would be lost. The critics countered that the companies had already failed and bankruptcy was the only option that would ensure the swift organizational reengineering necessary. Where and with whom does the failure lie? Where did the process break down? Why have some foreign automakers been profitable, even expanding their operations on U.S. soil, while the Detroit three were on the verge of collapse? Are not all auto companies competing in the same global markets and pledged to serve the same, or similar, stakeholder groups? What role does information play? Do the successful and unsuccessful collect the same sort of data, from similar sources, and transform the data into usable information, and the information into knowledge using comparable methodologies? How do the strategic questions and their associated knowledge-transfer mechanisms compare between successful and failing enterprises? In the final analysis *what does strategy have to do with it?* Can an organization that fails, particularly in the company of others in the industry with healthy balance sheets, argue convincingly that its strategy was not flawed, and that it simply fell victim to hard economic times and conditions beyond its control or ability to predict? Good strategy is long term, visionary, and equips an organization to weather destructive forces and be able to land on both feet. That is sustainability!

[1]Directed by Brian Gibson and based on a book by Tina Turner, the movie was released on June 9, 1993. Angela Bassett played Tina Turner and Laurence Fishburn played the part of Ike Turner. The movie cost around $15 million to produce and grossed over $80 million worldwide. Both Bassett (Best Actress in a Leading Role) and Fishburn (Best Actor in a Leading Role) were nominated for Oscars based on their performances.

[2]The music and lyrics for the title song "What's Love Got to Do With It?" can be accessed at http://new.music.yahoo.com/tina-turner/tracks/whats-love-got-to-do-with-it—861061#lyrics.

[3]Ford Motor Company, General Motors, and Chrysler Corporation.

LEARNING OBJECTIVES AND KEY OUTCOMES

After careful study of the material in Chapter 2, you should be able to:

1. **Describe the significance of open systems thinking to the successful implementation of QIM principles.**
2. **Differentiate between an output and an outcome, and discuss how undesirable outputs or outcomes are forms of waste.**
3. **Define and discuss the major subcomponents of a value stream.**

(continued)

4. Differentiate between deviation-amplifying and deviation-correcting forms of feedback, and describe their respective roles in the achievement of quality goals.
5. Describe and discuss the four steps of the strategic planning cycle, with particular emphasis on the importance of open systems thinking.
6. Briefly describe benchmarking, its objectives, and the different types of benchmarking.
7. Differentiate between mechanistic and organic organization, define Theories X, Y, and Z management styles, and discuss the relationship between structure and management practice.
8. Describe the qualities of an effective leader and discuss those attributes that are consistent with good leadership.
9. Demonstrate an understanding of the balanced scorecard as an integrated performance metric and how to apply its principles to develop some realistic measures for an industrial organization.
10. Describe the evolution of the quality movement in terms of three generations of development.
11. Articulate some of the attributes that have led to successful quality programs.
12. Demonstrate an understanding of the difference between analytic and naturalistic decision making, how decisions are made, and what constitutes a good (quality) decision.
13. Describe how data, information, knowledge, and wisdom are related and contribute to the achievement of an organization's quality goals.
14. Appreciate the various sources of information and information repositories that exist in organizations.
15. Describe some differences between statistical process control (SPC) and intelligent process control (IPC) in the creation and maintenance of high-quality process output.
16. Appreciate the value of effective technical communications to the achievement of quality goals.

2.1 OPEN SYSTEMS THINKING

Quality is about the performance of systems, and an understanding of systems theory is a necessary building block for learning how to design effective quality programs.[4] A dictionary definition describes a system as a group or set of related or associated material or immaterial things forming a unity or complex whole.[5] The word **system** is encountered frequently in many contexts and varies in scope and purpose. Most of us have studied (and even marveled at) the intricacies of our solar system, navigated the nation's highway network that comprises an elaborate system of freeways, derived pleasure (or annoyance) from high-tech sound systems, experienced the unpleasantness associated with a blocked sewage system, reaped the benefits of educational systems, placed faith in the infallibility of a just judicial system, and enjoyed a standard of living that would not be possible without a complex network of efficient production systems. All of these systems share some common characteristics. Each has component parts that must work together synergistically to achieve some intended *function* or *purpose*. The intricate and sometimes mysterious and complex interactions that occur between its components are a system's defining features.

Those who hold management positions, especially if their education or experiential backgrounds are in operations or engineering, may be ill-equipped to deal with the complex nature of many systems. Engineers and technologists are trained to apply the scientific method to problem solving—a structured methodology that advocates a form of analysis that breaks down a system into its component parts and then studies each part independently. The rationale is that it is much easier to analyze a small piece in detail rather than the behavior of the entire system. The scientific approach worked well in the past during a historical period known as the *machine age* when business systems were less complex and could be decomposed, studied, and improved like machines—that is, by first disassembling, then diagnosing, and finally modifying or repairing. The origin of the machine age can be traced to the birth of scientific management and the work of pioneers such as Frederick W. Taylor[6]

A chef confers with the restauranteur concerning a quality problem with his ovens. Improvement requires a systems approach.

A system is a group or set of related or associated material or immaterial things forming a unity or complex whole.

[4]Harrington, H. James, Joseph J. Carr, and Robert P. Reid. "What's This 'Systems' Stuff, Anyhow?" *The TQM Magazine* 11, no. 1 (1999): 54–57.

[5]*Shorter Oxford English Dictionary*. 5th ed. Oxford, England: Clarendon Press, 2002.

[6]Frederick Winslow Taylor, who lived from 1856 to 1915, was an American mechanical engineer and is generally regarded as the father of scientific management. Applying a new philosophy on how work should be organized, studied, and improved at Midvale Steel Works, he was able to achieve some dramatic improvements in productivity.

Quality begins with strategy—an organization's leaders must identify who their stakeholders are and how they can best serve stakeholder interests.

A car salesman serves a prospective buyer. A successful transaction depends on understanding the customer's needs and two-way communications.

and the Gilbreths,[7] who taught that the key to efficiency was through standardization and tight central controls. Manufacturing systems during this era were designed with an emphasis on specialized equipment and labor, standardization, and economies of scale. Industrialists like Henry Ford[8] perfected the model and it became possible to produce high volumes at low cost, placing many products for the first time within the reach of the middle to lower socioeconomic classes. This sparked a new wave of consumerism. During the machine age organizations were typically characterized by closed systems thinking and managed using practices and procedures that were inflexible. The need to collect and act on environmental **feedback** was not recognized. It was not until the coming of the information age that organizations began to appreciate that an open systems approach is crucial to sustainability.

2.1.1 The Enterprise as an Open System

Information technologies and market forces have made it virtually impossible for any organization to operate without continual interaction with (and nourishment from) its environment. In the contemporary world survival depends on an open systems view, and *open systems theory* provides a convenient model for explaining how modern organizations function.[9] The open systems view not only supports the stakeholder model of quality (the majority of stakeholder groups reside in the environment), but it also gives special emphasis to the establishment of communal relationships necessary for sustainability. This principle has its roots in the biological sciences where biological systems are dynamic and engage on a personal level with coinhabitants of a shared environment.

In a system all component parts must work together synergistically to achieve an intended purpose.

The open systems view recognizes the need for an organization to commune with its environment.

Some important parallels between industrial and biosystems can be observed. For example, both types of systems depend on feedback from their respective environments for sustainability and growth. Industrial organizations use this feedback to validate behaviors, provide a sense of self-worth, and guide future activities. The second law of thermodynamics, from the engineering sciences, foretells the consequences of a closed system approach. Specifically the second law states that after any processing step a closed system always finishes with less total energy than was present at the beginning. That is, energy is always dissipated in the form of waste. Think of a business as a ball that has been set in motion and is rolling toward some intended destination (goal). Eventually, if additional energy is not introduced, the ball will cease its forward motion and ultimately come to rest, due to physical forces such as friction and gravity. As the dissipated (wasted) energy, called entropy, builds up the ball slows down. When the stored entropy reaches a sufficient level the system shuts down entirely, halting any further forward progress by the ball. However, if the ball receives a push (new energy) from an external source, it can again be set into motion.

[7]Frank (1868–1924) and Lillian (1878–1972) Gilbreth were one of the great husband and wife teams of science and engineering. They made many contributions to the design of work and how work was measured, including the use of micro-motion analysis that breaks tasks down into basic elements called therbligs (*Gilbreth* spelled backward).

[8]Henry Ford (1863–1947) was founder of the Ford Motor Company and is credited with assembly-line techniques that made large numbers of automobiles affordable through mass production.

[9]Scott, Richard W. *Organizations: Rational, Natural, and Open Systems.* Upper Saddle River, NJ: Prentice Hall, 1997.

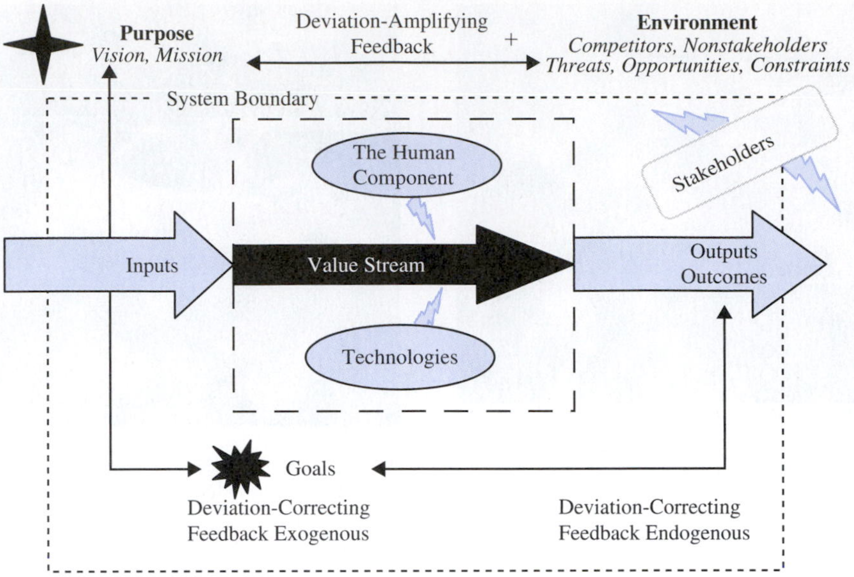

FIGURE 2-1. The Business Organization as an Open System

Entropy sucks the life and vitality from any system. In the case of biosystems entropy buildup leads to aging, loss of function, and eventual death. Similar results can occur in the business world. Examples abound of startups, or even long-established concerns, where the leadership turned a blind eye and deaf ear to its environment ensuring its doom and sealing its fate.

Nevertheless business failure is not inevitable because, unlike biological systems, *businesses have the ability to renew themselves*![10] Renewal can take place as long as external activation energies breathe enough new life into the organization to sustain it. Similar to providing fresh momentum to a rolling ball that has slowed or stopped, activation energies are "boosts" to a system aimed at obstructing the second law. In the management of systems, activation energies can come in many disguises—such as reorganization, business process reengineering, leadership turnover, external consultancies, Six Sigma, employee retreats, industry conferences, and so forth.

2.1.2 The Components of the Open Systems Business Model

Figure 2-1 illustrates the open systems business model, and highlights the following important elements.

- **Boundaries. System boundaries** separate what belongs to an organization from those things that are external. The system interfaces at its boundaries should be carefully managed, because it is across these boundaries that the organization is able to interact with its environment.
- **Purpose and Goals.** An organization's **purpose** is its justification to exist, and can be thought of as a contract between the organization and its varied constituencies. **Goals** are internally set by an organization's leadership and define how the organization intends to achieve its purpose. In a QIM culture, a commonality of purpose is inculcated throughout the structure, top to bottom, and is reflected in performance goals established at each level.
- **Inputs and Outputs or Outcomes.** Organizations depend on many types of **inputs** from the environment, including personnel, information, energy, capital, and materials; and also certain constraining forces that arise from legal, political, ethical, and bureaucratic controls. Since quality of output depends on quality of input, managing the supplier network and particularly each vendor's critical-to-quality (CTx) characteristics.[11] CTx attributes are those that are key to internal process performance and that directly influence value stream outputs. Outputs and outcomes are different, and can be desirable or undesirable. An **output** is a system deliverable—as seen from the value stream—looking outward toward the customers' systems, and represents the cumulative effect of all transformations that occur up to the point where a unit of output is measured. Outputs can be measured from within a system, usually at the system boundary at the point where the product or service is delivered and exits the system. By contrast, an **outcome** is defined as the effect, either immediate or ultimate, that an output has on a subsequent system or systems. Outcomes can be beneficial or adverse and can occur several tiers removed from the original system

[10]Hanna, David P. *Designing Organizations for High Performance.* Reading, MA: Addison-Wesley, 1988.

[11]In defining CTx characteristics those performance factors that are important to customers are inserted in place of the x factor. The most common factors are CTQ—critical to quality, CTD—critical to delivery, CTC—critical to cost, CTP—critical to process, and CTS—critical to safety.

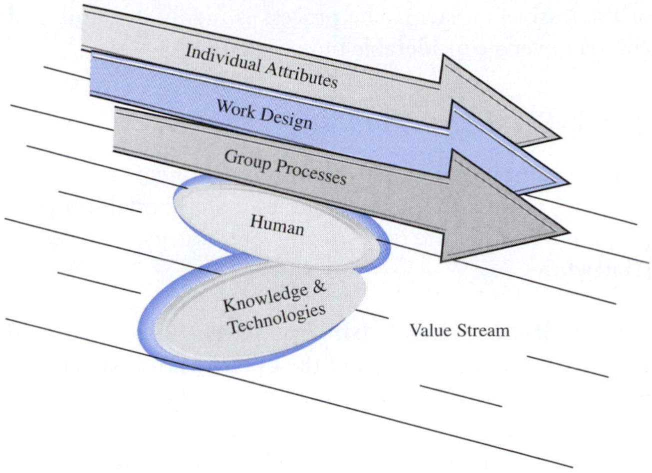

FIGURE 2-2. Value Stream Components

(e.g., several stages down the supply chain). Although it is usually possible to observe outputs immediately, outcomes may take considerable time to become apparent, if ever.

- **The Value Stream.** The **value stream**, a term first coined by Michael Porter,[12] is the system's transformation process, and includes all the tasks, skills, and knowledge required to convert a system's inputs into desirable outputs. When the value stream is expanded to include external supplier systems upstream and the customers' supply and distribution points downstream, the resulting network is called the **supply chain.** As Figure 2-2 shows, there are five critical subelements that work together to create the value stream.[13]

  Individual attributes are the mix of skills and human traits that contribute to making each member of the workforce unique, and include such factors as education, experience, attitudes, physical capabilities, preferences, wit, and intelligence. The **group process** represents the human ability to work together—a propensity to cooperate, coordinate, and delegate. Teaming, social, and communications skills are important to the group process, whether or not formal team structures are used. The **work design** defines the collective elements that prescribe the task and work sequence that must be followed to create a product or service. Where work design and group processes intersect, the value stream depends on teaming or interpersonal relations to get work accomplished.

  All system transformations no matter how primitive utilize **technologies** to get things done. The word *technology* is defined in a broad sense, ranging from an abstract "soft" side (paradigms, procedures, and computer software) to a more complex and rigorous "hard" side (innovation, experimentation, and computer hardware). Incremental continuous improvements and quantum breakthroughs (a key to competitive superiority) can impact either the soft or hard side or both. The **knowledge base** is the organization's collective wisdom concerning how processes behave and why they behave the way they do, including the complex interrelationships that exist between the technologies and work design. A QIM organization requires that senior management know profoundly *what it is doing and why*, and consequently base its decisions on information and fact.
- **Environment.** The **environment** comprises everything that is not contained within a system's boundaries. This includes other systems, owned by competitors and stakeholder and nonstakeholder groups. The environment also contains any opportunities, threats, and constraints that can have positive or negative impacts on sustainability.
- **Feedback.** Feedback distinguishes a closed system from an open one. A closed system has no feedback loops, and, once a process starts, it continues unaltered to completion. Feedback provides information on outputs or outcomes that occur at some particular process step, and is then fed back as input to a previous process step to modify system behavior if necessary. External, or exogenous, feedback captures information about system outcomes, and internal, or endogenous, feedback is concerned with outputs. Feedback can be positive or negative.

  Positive feedback loops amplify divergence, modifying the context of an action in such a way that the conditions favoring that action are reinforced. The consequence is that the action continues to be exerted, and possibly with increasing fervor. The feedback then indicates even more favorable conditions, so the action is amplified repeatedly, sometimes leading to explosive and destructive consequences. An example of positive feedback, which is called **deviation-amplifying,**[14] arises when an organization's leadership continues to answer the following question in the affirmative: "*Is the purpose on target with the needs and expectations of existing and potential stakeholders, giving due consideration to the threats, constraints, and opportunities that exist at this point in time?*" This type of positive feedback is what some call the **voice of the customer** (VOC) or more appropriately the **voice of the stakeholder** (VOS). Positive feedback provides a form of vindication and reinforcement of management decisions and actions. Confident that its policies and strategies are working, managers may resist those changes that could better satisfy existing stakeholders, attract new stakeholders, and prepare for impending contingencies, such as downturns in the economy and new government regulations.

---

[12] Porter, Michael E. *Competitive Advantage.* New York: Free Press, 1985. Porter actually refers to this as the "value chain."

[13] Hanna, *Designing Organizations.*

[14] Ibid.

The VOC/VOS is an important component in integrated process and product development methodologies such as quality function deployment (QFD).[15] An effective VOC/VOS process requires a multifaceted integrated approach.[16] However, some organizations make the mistake of acting on data that appear to support their preconceived perceptions of the market, rather than listening to and carefully considering any data that may provide contrarian views. If positive feedback drives reinforcing behaviors that are increasingly beneficial, the process is called a **virtuous cycle.** If the behaviors are increasingly destructive the process is called a **vicious cycle.**

Positive feedback cannot last indefinitely. An organization will eventually seek a state of equilibrium and will survive in the long term only if it can adapt to change. This is the role of negative feedback, which strives to balance a system by returning it to some stable state. Negative feedback, called **deviation-correcting,**[17] is characterized by a controller mechanism that continually measures the gap between existing and desired states, transmitting instructions for corrective action to close the gap. A good illustration of a negative feedback system is the operation of a thermostat on a home's heating/air conditioning system. When the internal temperature of the home reaches some preset target the system shuts off; when the temperature rises above or drops below the target the system turns on to cool or heat as necessary until the target is again achieved.

In an open system negative feedback exists at two levels. At the endogenous level the feedback is wholly contained within system boundaries and attempts to answer the following questions: "*Did the organization do what was intended? How did the outputs compare to the internal goals?*" At the exogenous level, negative feedback is obtained by looking outward from the system—across its boundaries and into the environment to inquire, "*How well do the organization's goals align with its purpose?*" Purpose and mission should be the litmus tests used to validate each organizational goal. Forward planners should ask, "*How does the achievement of each goal support mission and purpose, and ultimately help move the organization closer to achieving its vision?*"

2.2 QUALITY AND STRATEGIC PLANNING

An effective strategic plan requires knowing how and when to adapt to change, and is not something that can be relegated to an individual or even a group. A strategy is not a task nor is it an end product. A good business strategy is the product of a strong leadership with vision and a solid understanding of the business system. The process is highly dynamic and emerges over a considerable time period.

2.2.1 Strategic Planning Process

Most organizations employ some variation of a strategic planning process that invokes the four steps: establish purpose, address interface issues, formulate strategy, and execute strategy.

2.2.1.1 Step 1: Establish Purpose. Refer again to Figure 2-1. All components of the open business system are driven by purpose. The business strategy is the blueprint that takes an organization from the present into the future, so it is essential that the organization's leaders understand its environment and markets, the role it plays (i.e., its niche), and those affected/interested parties with power (i.e., its stakeholders). Establishing purpose usually begins with some variation of a process known as a **SWOT analysis.**[18] This is a four-step process, and the effort required for each step is not necessarily the same. In practice it is usually somewhat easier to gain information on strengths and weaknesses (internal issues) than on opportunities and threats which require compiling reliable information on the environment from external sources. However, new information technologies have broadened access, enabling organizations to gain useful competitive intelligence, often without cost, to modify or validate purpose.[19,20]

Strategy requires strong leadership, vision, knowledge, commitment, and relationships.

A SWOT analysis helps an organization understand its uniqueness relative to the competition, and its purpose is clarified when all intelligence concerning markets, competitors, stakeholders, and internal capabilities are assimilated and integrated to determine how it should position itself in the market. Purpose can change as a result of planning activities further down the chain. For example, competitive intelligence could reveal new information on emerging threats or opportunities, or alternatively, improvement initiatives such as business process reengineering, lean production, or Six Sigma could result in a reevaluation of core competencies.

Vision and mission are closely tied to purpose. A vision is a projection of realistic aspirations matched to core strengths and new opportunities. The mission is a succinct statement of purpose. Strategy and internal capabilities are mutually supportive. The SWOT analysis can be used to identify core competencies, and this knowledge can then be used to exploit the firm's strategic uniqueness. Conversely, as strategy emerges, management will establish priorities and

---

[15]This topic is discussed in further detail in Chapter 5.

[16]The integrated approach considers input from sources such as competitive intelligence, gemba visits, market research, industry surveys, etc.

[17]Ibid.

[18]*SWOT* is an acronym that stands for strengths, weaknesses, opportunities, and threats.

[19]Lang, Eva M. "Using Competitive Intelligence as a Strategic Planning Tool." *CPA Consultant* 15, no. 4 (August/September, 2001): 5–7.

[20]Curtis, James. "Behind Enemy Lines." *Marketing,* May 24, 2001, 28–29.

set goals that may lead to the creation of additional capacity or the development of new competencies.

Establishing purpose requires individuals who are possibility thinkers, can recognize new opportunities, and can spot potential. Such visionaries are internal strategists who can guide an organization through a process that helps them think about the world and their place in it in dramatically new and different ways.[21]

2.2.1.2 Step 2: Address Interface Issues. The second step of the strategic planning process is to address interface issues—that is, to assess how the organization interacts (communes, transacts, markets, barters, etc.) with all its many constituencies, stakeholders and nonstakeholders. Managing these external relationships is crucial to the creation and delivery of quality products and services and requires an in-depth understanding of stakeholders (including stakeholder customers), competitors, markets, the economy, business constraints, and in particular stakeholder expectations. The open systems planning (OSP) model[22] employs a sequence of four steps to help organizations address strategic interface issues.

- Environmental scan—the process of gathering intelligence from the environment to help the organization determine where it should position itself and how it will compete.
- Predicting the future without intervention—projecting the present into the future, assuming that all trends and current environmental conditions will continue without change.
- Projecting an ideal future with intervention—creating an ideal future scenario that could be made possible through appropriate interventions to alter the course of nature.
- Establishing action plans—deciding on specific actions, aimed at creating the ideal future that will balance the perceived needs of the system with those of the environment, and will help the organization determine how system interfaces will be managed.[23]

The mission is a succinct statement of purpose.

Strategic visioning permits managers at the top to see their company differently.

2.2.1.3 Step 3: Formulate Strategy. The formulation of strategy should involve broad and diverse participation throughout the organization, encouraging innovative thinking at every level and stressing the importance of new *voices, conversations, passions, perspectives*, and *experiments*.[24] The corporate strategy should be focused, divergent (identifying how the organization differentiates itself from its competitors), and should have a compelling tagline (a slogan that captures the essence of how the strategy wins against the competition).[25] Strategy should capitalize on those core competencies that the organization identified in Step 1, and should attempt to match competencies to stakeholder needs. To do this the organization should ask itself,

- *Who are our stakeholders?*
- *What can we provide that each of our stakeholders values?*
- *Are there any conflicts of interest that could lead to satisfying one stakeholder only at the expense of another? How can these conflicts be resolved and can the needs of all stakeholder groups be satisfactorily balanced?*
- *Why do (or would) a particular stakeholder choose to affiliate with our organization? For each stakeholder group are we the preferred provider of their particular needs?*
- *How well do we understand the options available to each of our stakeholders and the process that each stakeholder follows in forming alliances?*

The answers to these questions help an organization design its business system around stakeholder needs. This includes the determination of how the organization intends to compete through the establishment of competitive priorities[26,27] and the design of a well-matched production system. For example, if *cost* was selected as a competitive priority, the production system would be designed to efficiently generate outputs that can be sold at a price below what competitors charge. On the other hand, if *quality* is a competitive priority, the production system would be designed for outputs with superior features, such as close tolerances; durability; maintainability; available, courteous and knowledgeable service; convenient location; or safety.

Increasingly in some markets *time* (for example, customers requiring *just-in-time* delivery service or strategies emphasizing *time-to-market*) and *flexibility* (for customers demanding greater customization or the need to change order specifications or quantities on short notice) are emerging as important competitive priorities.

Establishing competitive priorities can help an organization determine market strategies based on timing (that is, whether the intent is to lead the market or delay entrance until the market is mature) and to determine those product features that are order qualifiers (those which meet minimum customer expectations)

Great strategies are focused, divergent, and have compelling taglines.

---

[21]LaBarre, Polly. "The New Strategic Paradigm: Getting to the Future First by Getting Different." *IndustryWeek* 243, no. 21 (November 21, 1994): 30.

[22]Jayaram, G. K. "Open Systems Planning." In *The Planning of Change*, 3rd ed., edited by W. G. Bennis, K. D. Benne, R. Chin, and K. E. Corey. New York: Holt, Rinehart and Winston, 1976.

[23]Hanna, *Designing Organizations.*

[24]Hamel, Gary. "Strategy Emergence." *Executive Excellence* 15, no. 12 (December 1998).

[25]Kim, W. Chan, and Renee Mauborgne. "Charting Your Company's Future." *Harvard Business Review*, June 2002.

[26]Hayes, Robert H., and Steven C. Wheelwright. *Restoring Our Competitive Edge: Competing Through Manufacturing.* New York: Wiley, 1984.

[27]Foo, G., and D. J. Friedman. "Variability and Capability: The Foundation of Competitive Operations Performance." *AT&T Technical Journal*, July/August 1992, 2–9.

and order winners (those which will woo the business away from the competition).[28]

2.2.1.4 Step 4: Execute Strategy. The final step is to execute the emerged strategy. A tool, used widely by world-class organizations for executing strategy, originated in Japan and is called **hoshin kanri** (or policy deployment).[29] **Policy deployment** emphasizes participation by management in a linked team structure that vertically spans the organizational structure, calls for periodic systematic reviews, and facilitates learning through feedback.

2.3 BENCHMARKING

Benchmarking, sometimes referred to as *best practices,* is a tool that is important to strategic planning and to the development of quality management capabilities. *Benchmarking* is the term used to describe a continuous and formal process for defining the best practices—usually but not necessarily within the same industry sector, which can then form the basis for productive change without having to reinvent what has already been tried and tested by others. Benchmarking can help break down internal resistance to change that typically comes with new initiatives that challenge existing paradigms.

2.3.1 Types of Benchmarking

Benchmarking can be used to generate intelligence in a number of areas that have strategic importance. There are six major benchmarking types.

- **Process benchmarking**—the reengineering of primary value streams, or core business processes, can be facilitated by identifying and observing how other organizations have designed and are operating similar systems. A possible outcome of process benchmarking is the internal reevaluation of core competencies leading to decisions to outsource certain products or services.
- **Financial benchmarking**—long-term sustainability, financial stakeholder returns, and cost competitiveness can be assessed by comparing an organization's financial performance with industry norms and reported performance by major competitors.
- **Performance benchmarking**—an organization's competitive position in the marketplace can be evaluated by comparing its products and services with those provided by major competitors.
- **Product benchmarking**—an organization's ability to design new products or services can be evaluated against competitors' products (sometimes by reverse engineering) to determine strengths, weaknesses, and innovative approaches.
- **Strategic benchmarking**—an organization's approach to developing strategy, and determining how to compete within a defined market, can be improved by observing how other firms, even in different industry sectors, have identified their keys to success.
- **Functional benchmarking**—an organization can improve a single function, particularly a support function such as human resources, marketing, purchasing, accounting, information and technology, distribution, and so on by comparing practices within and across industries.

2.3.2 Benchmarking Steps

Figure 2-3 illustrates the six generic steps that most organizations will follow in the benchmarking process. These steps are further explained below.

Step 1: Identify benchmark type. Selection of the type depends on your objective. For example, benchmarking can be done to gather competitive intelligence for strategic planning, or benchmarking can be a tool employed by a Six Sigma process improvement team. This step includes the establishment of baseline information that will provide the basis for making meaningful comparisons with the benchmark organizations.

The types of benchmarking are process, financial, performance, project, strategic, and functional.

Step 2: Identify benchmark targets. This step requires the identification of other organizations that have similar processes or issues to the specific area of interest. For example, if the interest is in improving the quality of service in a chain of restaurants, only those organizations engaged in food service operations and serving a similar clientele in similar markets would be eligible to be included on this list.

Step 3: Identify leaders in the benchmark area. This step can be challenging as it requires the identification of those organizations that are known to excel in the specific area of interest. For example, if one is interested in improving the human resource management system, the companies for study and comparative analyses would be those organizations that are widely recognized as being leaders in managing their human assets.

Step 4: Identify best practices and measures. Detailed surveys are typically used to flesh out those practices that work, those that should be avoided, and the measures that are the most effective in gauging performance. Surveys are normally masked to protect the confidentiality of data and to facilitate cooperation by the companies surveyed.

Step 5: Conduct site visit. Cutting-edge practices can often be fully understood only through personal observation. This is an important step to bring about an effective knowledge transfer from the best practices company to

[28]Hill, Terry. *Manufacturing Strategy: Text and Cases.* 3rd ed. Boston: Irwin/McGraw-Hill, 2000.

[29]Norton, Brian R. "Breaking the Constraints to World Class Performance." *Accountancy SA*, June 2000, 5–7.

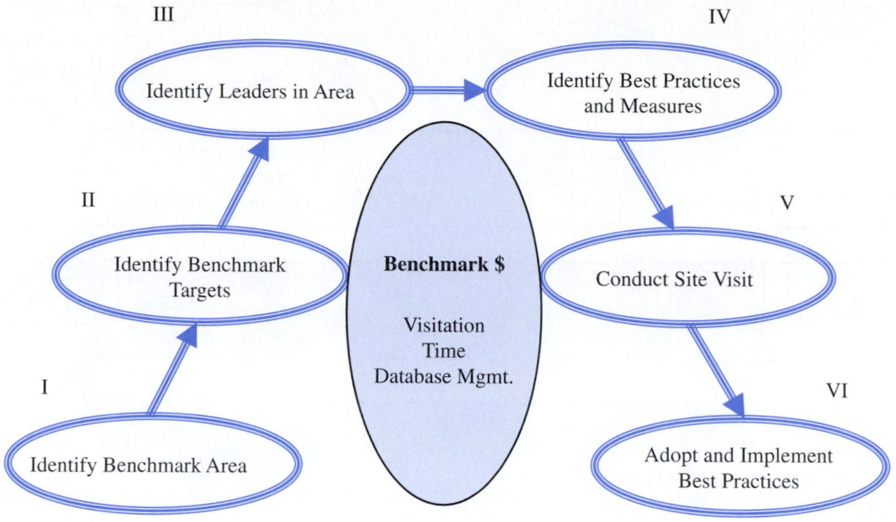

FIGURE 2-3. Benchmarking Process

the benchmark organization; however, it may not always be possible to conduct, particularly if the firm being benchmarked is a competitor.

Step 6: Adopt and implement best practices. During this step the organization acts on what it has learned during the previous five steps. Implementation plans are developed to operationalize targeted leading-edge practices. These plans include identifying specific improvement opportunities, finding the necessary resources to fund related improvement projects, and achieving organization-wide buy-in on the value of the proposed changes.

The four steps of the benchmarking process are to identify the type, the targets, the leaders, and the best practices, then conduct site visits, and, finally, adopt best practices.

2.3.3 Benchmarking Costs

Considerable resources can be consumed through benchmarking and it is important to ensure that the value outweighs its cost. There are three main areas where costs can accrue: visitation, time, and database management.

Visitation costs include all travel expenses necessary for making personal contact with companies on the benchmark list, such as accommodation, transportation, meals, gifts, and lost time on the job by participating staff.

Time costs cover the total impact on payroll as a result of personnel who must be diverted from their regular tasks to engage in benchmarking activities, and then following through on implementation strategies.

The database management costs accrue as a result of the development and maintenance of a comprehensive benchmark knowledge base.

2.4 QUALITY AND ORGANIZATION STRUCTURES

2.4.1 Organization Structures

An organization's structure is the means through which work gets done and information is exchanged. Redesigning the structure can be the means for achieving a better response to stakeholder needs and *superior quality*. For example, flatter structures with fewer management layers can often foster improvements in cross-functional relationships and greater flexibility in the distribution of resources.

Figure 2-4 illustrates the two fundamental types of organizational structure: *mechanistic* and *organic*.[30]

2.4.1.1 Mechanistic Structures. Mechanistic structures are formal, generally nonparticipative, hierarchical, tightly controlled, and inflexible. **Mechanistic** designs, also called *silo* or *bureaucratic* structures, were typical of the machine age and have their origins in the works of Frederick W. Taylor[31] and Max Weber.[32] The earliest mechanistic models relied on standardization with highly defined tasks that were broken down into specialized work elements. The work environment was governed by numerous rules and a strict hierarchy of authority and central control, an authoritarian

---

[30]Zaltman, Gerald, Robert Duncan, and Jonny Holbeck. *Innovations and Organizations.* New York: Wiley, 1973.

[31]Taylor, Frederick W. *The Principles of Scientific Management.* New York: Harper Brothers, 1911.

[32]Weber, Max. *Wirtschaft und Gesellschaft, 1922: Economy and Society: An Outline of Interpretative Sociology, a New and Complete Translation of Wirtschaft und Gesellschaft,* edited by Guenther Roth and Claus Wittich. Berkeley: University of California Press, 1978.

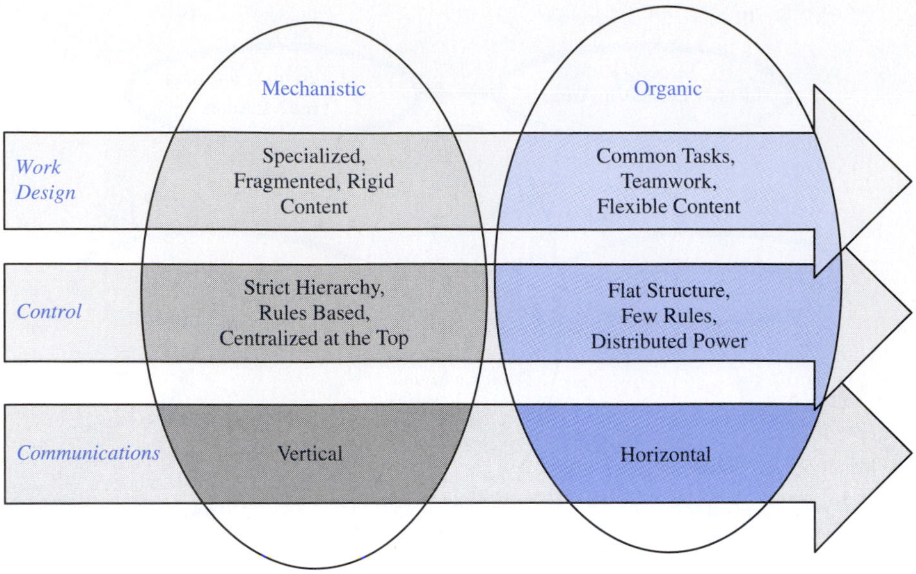

FIGURE 2-4. Types of Organizational Structures

management style that came to be known as *Theory X*.[33] Theory X is based on the premise that the average worker dislikes and will avoid work, and that in order to get things done workers must be forced with threats of reprisal. Further, the Theory X manager assumes that workers lack ambition, desire security above all else, and prefer direction to autonomy. Mechanistic structures and Theory X are still practiced in some modern organizations, and appear to be best suited to certain types of production technologies, particularly in mass flow and process settings, or where production is organized in small batch quantities or made to order.[34,35]

Nevertheless mechanistic structures can act as inhibitors to quality. When a business is organized with a clearly defined vertical chain of command, partitioned along functional lines, a fundamental contradiction can arise between the *structure of work* and the *execution of work*. Internal value streams are defined by the organization's cross-functional linkages and, as work is handed off from one workstation to another, it must cross the boundaries of numerous functional subsystems. This is illustrated in Figure 2-5 and represents the way work typically gets done. While mechanistic structures provide excellent within-function control and communications, none of the value stream participants has an overall view of the entire process. It is difficult to build-in quality when no single person has responsibility (ownership) of the entire system.

A second problem with mechanistic structures is that they are intrinsically inflexible. An organization that aspires to respond to stakeholder needs must be capable of readily adapting to change. By contrast, vertical organizations partitioned by function can sometimes develop a myopic view of what is required, and change is approached solely from a disciplinary perspective. The merits of change are often based on the size of the threat rather than the magnitude of the benefit. In a mechanistic structure, it is difficult at best and impossible at worst to stimulate the need for change, let alone elicit a timely and effective response. When coupled with a Theory X management style, it becomes all the more difficult for workers to buy into the need for change and support management-led change initiatives.

Mechanistic structures support Theory X management styles, and are formal, nonparticipative, hierarchical, tightly controlled, and inflexible.

2.4.1.2 Organic Structures.

One of the first proponents of an **organic** structure was Herbert A. Simon[36] who advocated more flexibility and broader participation in decision making, aligning the organization design to a management style called *Theory Y*.[37] On the opposite side of the spectrum from Theory X, Theory Y assumes that workers treat their jobs as a natural part of life and sincerely desire to put forth the required effort to achieve excellence, without the necessity of threats or coercions on the part of management. In addition Theory Y assumes that workers will actively seek and readily accept responsibility, and that they will

---

[33]McGregor, Douglas. *The Human Side of Enterprise.* New York: McGraw-Hill, 1960.

[34]Clegg, Stewart R. *Modern Organizations: Organization Studies in the Postmodern World.* London: Sage, 1990.

[35]Mintzberg, H. *The Structuring of Organizations: A Synthesis of the Research.* Upper Saddle River, NJ: Prentice Hall, 1979.

[36]Simon, Herbert A. *Administrative Behavior.* 3rd ed. New York: Free Press, 1976.

[37]McGregor, *The Human Side.*

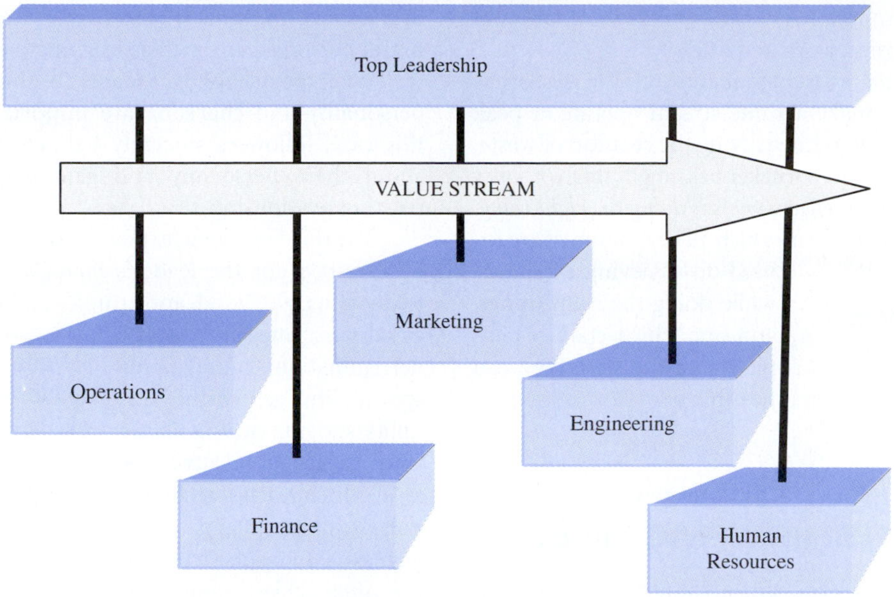

FIGURE 2-5. Cross-Functional Nature of Work

achieve a high degree of job satisfaction if they are provided with rewards commensurate with high-order psychological needs as posited by Maslow's hierarchy of needs.[38,39]

Though supportive of the Theory Y philosophy, Simon's model viewed the organization as a closed system in which all crucial issues were handled internally. In contrast, modern organic structures are flexible, have few rules, vest control where the work is performed, emphasize the use of teams, stress open communications, and use an open systems view to enable change.[40,41] However, organizing or reorganizing for the sake of improved quality comes with a caveat. Some organizations with an organic-type structure can still be hampered by mechanistic thinking when it comes to managing uncertainties. It is human nature to look for panaceas and invest heavily in programs that promise flexible and right-sized organizations—and to do so faster than the competition. However, if changes are introduced for the wrong reasons valuable resources can be wasted and nothing actually improves. Sometimes conditions get worse.

Organic structures are most effective when organizational designs align with a management style that has been coined *Theory Z*.[42] Theory Z fundamentally has its roots in Deming's 14 points and emphasizes the value of the employee. The assumption is that employee loyalty will be achieved by promising lifelong employment and stressing the well-being of all workers, both on and off the job. Typical of Theory Z organic structures are goal-oriented self-managed work teams, a commitment to high levels of training and human development, and employee rewards for self-development in addition to work performance.[43]

Organic structures are flexible, have few rules, are team oriented and participative, and can easily adapt to change.

2.4.2 Contingency Theory

Contingency theory (CT), a term originally coined by Lawrence and Lorsch,[44] broadens the scope of structural design. Since business organizations are open systems and intrinsically responsive,[45] CT acknowledges that decisions are based on an organization's dependencies with its environment. As challenges and opportunities arise, an organization draws on a survival instinct to make whatever changes are necessary to differentiate itself from its competitors. This requires access to a wide range of options, including the use of mechanistic designs when appropriate.

Information technology (IT) has also had a significant impact, as the information age has created a new breed of knowledge worker. Consequently structures have emerged

---

[38]Maslow, A. H. "A Theory of Human Motivation," *Psychological Review* 50, no. 4 (1943): 370–96.

[39]According to Maslow's theory human beings are motivated by unsatisfied needs, and human needs form a hierarchy in which lower-level needs must be satisfied before higher-level ones. At the lowest level are the basic physiological needs, followed by safety, then love, then esteem, and finally self-actualization.

[40]Khandwalla, P. N. *The Design of Organizations.* New York: Harcourt Brace Jovanovich, 1977.

[41]Randolph, W. A., H. J. Sapienza, and M. A. Watson. *Psychometric Theory.* New York: McGraw-Hill, 1991.

[42]Ouchi, William G. *Theory Z: How American Business Can Meet the Japanese Challenge.* Reading, MA: Addison-Wesley, 1981.

[43]Aughton, Peter, and Neville Brien. *Applying Open Systems Theory for Dramatic Improvements in Business Performance.* The Sixth European Ecology of Work Conference, Bonn, May 1999.

[44]Lawrence, Paul R., and Jay William Lorsch, *Organization and Environment: Managing Differentiation and Integration.* Homewood, IL: Richard D. Irwin, 1969.

[45]Scott, Richard W. *Organizations: Rational, Natural, and Open Systems.* Upper Saddle River, NJ: Prentice Hall, 1987.

that facilitate the assimilation, analysis, and dissemination of knowledge in performing work activities.[46]

Structure *does not guarantee quality.* If the system is ill-designed, trying to make a value stream operate at peak efficiency may only result in intensifying the creation of waste. The management guru Peter Drucker has taught that *efficiency is doing the thing right, but effectiveness is doing the right thing.* A well-designed system is one in which all key resources are focused on achieving excellence while doing the right things, and one which enables people to be the best they can possibly be.[47]

Organizations adapt their structures to respond to the challenges and opportunities around them.

2.5 MANAGEMENT AND LEADERSHIP PRACTICES

It has been said that poor leadership accounts for the vast majority of an organization's problems.[48,49] An organization's leadership sets the tone, establishes the priorities, creates the work culture, and is responsible for employee morale. A good leader is able to influence positive feelings, provoke creativity, arouse enthusiasm, build consensus, strike compromise, negotiate buy-in, communicate effectively, and create a team spirit. There is a difference between commitment and leadership, both of which are important to the success of QIM. Leaders should be united in their commitment to common goals, but not every committed manager will be an effective leader. To further our understanding of leadership, it is helpful to investigate why people follow a leader in the first place.

2.5.1 Why People Follow a Leader

Research studies suggest that all human motivation is based on a quest for outcomes that are *good*, and that dimensions of leadership arise from three different definitions of that word—useful, pleasant, or moral.[50]

Defining "good" as *useful*, the allegiance of some people to a particular leader is because they acknowledge that the leader possesses skills and knowledge superior to their own—that is, they choose to follow because they trust the leader's technical competence. The follower need not admire, or even like the leader, who may be seen as the source for useful personal gain such as learning or rewards.

When people follow a leader for the purpose of *pleasure*, personality and charisma are important determinants. In this case, followers sincerely believe that the leader cares about them personally and genuinely wants to build a trusting relationship.

The third reason why people follow someone is that they have respect for the leader's *moral authority.* Morality has to do with right, good, and ethics—and includes such virtues as fairness, integrity, loyalty, determination, courage, responsibility, generosity, humility, and humor. A manager who has recognized attributes at all three levels—technical skills, charisma, and moral virtues—can inspire trustworthiness and loyalty in others and can often achieve remarkable improvements in productivity with personnel that have failed under the direction of others.

People will follow a leader who has superior skills, is charismatic and genuinely cares for them, or has moral authority.

2.5.2 Leadership Skills

Whereas only certain people are "born leaders," leadership skills can be taught to anyone with proper training and experience.[51] These skills fall into five categories.

- Having a sense of timing—knowing when and how to decisively intervene in a problem situation.
- Possessing an ability to enable—being able to get the best out of people.
- Being an advocate for change—being able to listen, seek advice, and be receptive to new ideas.
- Having a vision—being able to recognize potential, spot opportunities, and champion change.
- Possessing empathy—being sensitive to opposing viewpoints, and able to build consensus and resolve conflicts.

2.6 ACCOUNTABILITY AND IMPLEMENTATION ISSUES

2.6.1 Performance Measurement and the Balanced Scorecard

It is common for top managers to equate bottom line, or financial measures, with corporate health. Measures such as return-on-assets and earnings-per-share worked well under machine theory where the primary focus was profit maximization and cost containment. While it is true that long-term survival depends on financial performance and that success will be reflected in such measures, financial metrics are merely indicators of the consequences of a system

[46] Aikens, C. Harold. *Quality: A Corporate Force—Managing for Excellence.* Upper Saddle River, NJ: Prentice Hall, 2006.

[47] DePree, Max. *Leadership Is an Art.* New York: Dell, 1989.

[48] Scholtes, Peter. *The Leader's Handbook.* New York: McGraw-Hill, 1997. Scholtes asserts that 95% of an organization's problems are systemic and that 95% of organizational change does not lead to improvement; the fault must be borne by the leadership who own the systems and establish priorities.

[49] Deming, W. Edwards. *Out of the Crisis,* MIT Center for Advanced Engineering Study, 1986, Cambridge, MA., page 315. Deming states that 94% of the improvement opportunities are the responsibility of management and 6% can be dealt with through operator intervention.

[50] Guillen, Manuel, and Tomas F. Gonzalez. "The Ethical Dimension of Managerial Leadership: Two Illustrative Case Studies in TQM." *Journal of Business Ethics* 34, no. 3/4 (December 2001): 75–189.

[51] Smith, August W. "Leadership Is a Living System: Learning Leaders and Organizations." *Human Systems Management* 16, no. 4 (1997): 277–84.

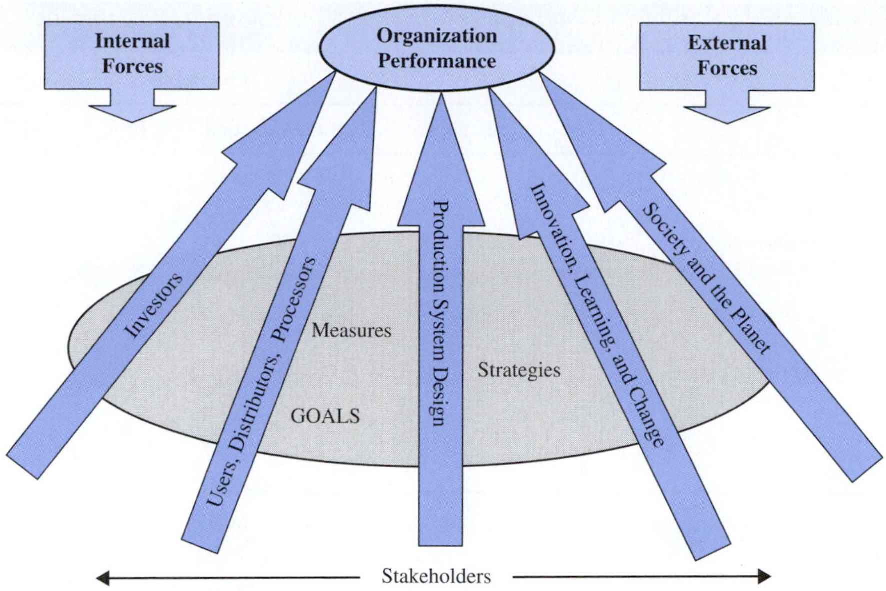

FIGURE 2-6. Balanced Scorecard Performance Metrics

working well and cannot pinpoint problems arising along the value stream or improvement opportunities. Furthermore, financial metrics measure the outcomes that are important to only a subset of the organization's stakeholders. It is equally vital to an organization's sustainability that those metrics of interest to nonfinancial stakeholders be carefully managed. The **balanced scorecard**[52] is an approach that can effectively direct management attention to those key areas that contribute to the creation of quality for *all* stakeholders. Illustrated in Figure 2-6, the balanced scorecard emphasizes those operational perspectives that are critical to stakeholder needs. Figure 2-6 shows five possible categories: *investors; users, distributors, and processors of products or services (customers); production system design; innovation, learning, and change; and society and the planet.* The scorecard helps management achieve superior performance by managing the tradeoffs that occur across the various relevant performance categories. Sometimes improvement in one area can be achieved only at the expense of another; therefore keeping all perspectives in a state of balance is an ongoing challenge.

2.6.1.1 User Perspective. This perspective covers any identified stakeholder group that is a direct user of a product (or service), or distributes the product to some lower-tier user, or processes the product into a product mix of some other form. Under this category measures are developed that relate directly to those things that mean the most to users. It is imperative that the organization try to see the world through user stakeholders' eyes, ascertaining whether these customers can trust that the organization will deliver on its promises. Stakeholder satisfaction is based on how the recipients of the products or services delivered perceive the quality they got compared to what they expected to get. If quality levels exceed expectations, then the perception will be favorable; if they fall short, an unfavorable perception will result. It is vital therefore that the measures employed reflect quality as defined by the stakeholders themselves, which may differ from the definitions used by the service or product provider. As Figure 2-6 illustrates, the balanced scorecard process first sets goals, then identifies specific attributes relative to each goal, and finally establishes measures for each attribute.

2.6.1.2 Internal Production System Perspective. The internal production metric is concerned with performance in the core competency areas. Here, it is important to establish measures that differentiate the company from its competitors—factors in which the organization knows that it must excel if it is to achieve and sustain competitive advantage for all stakeholder groups that benefit directly from value stream outputs and outcomes. The internal production system perspective is concerned with systemic factors that govern how well quality is created at the source. *Quality at the source* is quality of process, and it is imperative that the relationship between process variables and the measures listed under the users' perspective be well understood. Process measures include such factors as output quality, cycle times and yields, employee skills and productivity, machine changeover and down times, inventory turns and levels, process throughput (flow) times, production costs, and the like.

2.6.1.3 Innovation, Learning, and Change Perspective. The innovation, learning, and change perspective is concerned with the organization's ability to continuously improve, and addresses those specific factors that have been identified as essential to the continuous creation of value for

[52]Kaplan, Robert S., and David P. Norton. "The Balanced Scorecard—Measures That Drive Performance." *Harvard Business Review,* January/February 1998.

TABLE 2-1 Comparison of Various Quality Paradigms

	Philosophy						
Theme	**TQM**	**Six Sigma**	**Lean Production**	**BPR**	**TOC**	**SCM**	**HIO**
Systems Approach	X	X	X	X	X	X	
Process Improvement	X	X	X	X	X	X	
Employee Involvement and Empowerment	X	X	X		X		X
Teamwork	X	X	X				X
Customer Focus	X	X	X	X		X	
Supply Chain Integration	X	X	X			X	
Emphasis on Performance Measurement	X	X	X	X	X	X	X
Cycle Time Reduction	X	X	X	X	X	X	
Value Enhancement	X	X	X	X	X	X	
Waste Reduction	X	X	X	X			

TQM = total quality management; BPR = business process reengineering; TOC = theory of constraints; SCM = supply chain management; HIO = high-involvement organization.

all stakeholders. This is also where an organization addresses those things that are necessary to improve its ability to learn and become more knowledge driven. Metrics in this category focus on such factors as new product and process introductions, improvement initiatives, the implementation of change, and the holistic development of personnel.

Innovation, learning, and change measures the organization's ability to continuously improve.

2.6.1.4 Investor Perspective. The metrics in the investors' category take a critical look at a firm's financial performance from the perspective of those shareholders that have invested capital and expect a competitive rate of return. The measures developed should address whether creditors and investors consider the organization low risk and whether returns are attractive relative to alternative investment options that may be available from time to time. In this area it is common to include conventional financial ratios, such as return on equity, debt to equity, and total profits, as well as new nontraditional ones such as profit to total investment or profit to equity investment[53] and economic value added (EVA).[54,55]

2.6.1.5 Society and the Planet Perspective. An organization may have stakeholders whose needs are not satisfied through the metrics defined by any of the previous perspectives: users/customers, internal production system, learning, or investors. Such stakeholders reside in the environment and their power derives from the ability to shape the organization's future through regulation, public opinion, political persuasion, incentives, and the like. This category might include groups such as federal regulators, industry inspectors, environmental activists, and so on. The balanced scorecard should include a set of metrics that appropriately addresses the needs of each of these *stakeholders-at-large*. For example, a health inspector may be identified as a stakeholder of a restaurant if he or she has the power to close down the establishment. In this case the balanced scorecard should include measures that coincide with those used in the inspection report.

2.6.2 Implementing Second Generation Quality

During the 1980s and 1990s organizations grappled with how they could best move from first- to second-generation quality as global competition intensified. This proved to be a period of learning and development of new quality principles that have become the foundations for the modern era of management practice. By the year 2000 it appeared to many that these second-generation programs, which had been broadly implemented under the catch phrase **total quality management (TQM)**, had become passé. Attention began to turn to more trendy and popular concepts such as lean production, theory of constraints (TOC), business process reengineering (BPR), Six Sigma (SS), supply chain management (SCM), and high-involvement organizations (HIOs). The proponents of each of these alternative approaches had deliberately distanced themselves from TQM in an attempt to emphasize the uniqueness of what they had professed to be new thinking. However, as Table 2-1 shows, these methodologies all share some common themes that have evolved along with the quality movement and are derivatives of a shared body of knowledge. TQM did not invent any of these themes, but it can be argued that TQM legitimized their usage and had a major influence on the transition of management paradigms from machine age practices to open systems thinking.

Research studies of companies with mature quality programs have found that successful implementations are evolutionary and occur in four distinct phases, each of which

[53]Gold, Bela, and Ralph M. Kraus. "Integrating Physical and Financial Measures for Managerial Controls." *Academy of Management Journal 7*, no. 2 (June 1964): 109–27.

[54]Stern, Joel M., and John S. Shiely, with Irwin Ross. *The EVA Challenge: Implementing Value Added Change in an Organization.* New York: Wiley, 2001.

[55]Stewart, G. Bennett III. *The Quest for Value: A Guide for Senior Managers.* New York: HarperCollins, 1991.

takes years to complete.[56] The first three of these phases takes an organization through the second generation of quality and the fourth phase marks a firm's entry into the third generation.

The first phase is a period of predeployment, during which the organization must come to terms with basic QIM principles, develop insights into the significant differences between the operation of a QIM and non-QIM enterprise, and convince all employees throughout that there is a genuine need for change on a grand scale. This is a period characterized by setting goals, learning new concepts, seeking advice from external experts, and benchmarking best practices.

Two phases follow predeployment and represent the organization's individual evolution through the knowledge and practice of QIM. During the first phase the focus is mainly on internal issues, stressing the company-wide participation of employees through teams, training, and focused problem solving. During phase one, the company starts pressuring suppliers for improved quality, paying increased attention to measurements and analysis.

When the focus broadens from individual processes to a systems view, the organization enters the second phase. During this period fundamental process changes are evident, such as the introduction of cellular manufacturing and the creation of self-managed teams. As management becomes more aware of its responsibility for quality and begins to take a more active role, employee involvement is expanded and teams receive training in more rigorous analytical tools.

Successful implementations take time.

2.6.3 Advancing to the Third Generation

An organization has advanced to the third generation of quality when managing critical processes using cross-functional teams is the norm. At this level, all employees have a stakeholder mindset and the principles of QIM are fully operational at all levels. Internal quality tools are routinely used to monitor and improve all primary and support value streams, and a mature quality-based strategic planning process formally collects and acts on feedback from reliable sources on how well the organization's performance is meeting the needs of all stakeholders. Strategic decisions are based on metrics that anticipate and respond to future stakeholder requirements. Feedback is used to develop new products, and new product development has become a core business process. The supplier management system has reached a level of sophistication with few suppliers and an integrated supply chain. All employees understand and support a common set of goals and teams are proficient in problem solving and data analysis.

Ironically, in phase three, organizations discover that the relentless pursuit of continuous improvement is not always the best course. Foley et al. argue that "there may be times in the life of an enterprise where, to satisfy its survival criteria, it might be necessary to discontinue or slow down the rate of quality improvement activity."[57] In third-generation QIM the key idea is to create *value to the total enterprise*, giving due consideration to the needs of all stakeholder groups. This insight is part of the magic of Six Sigma, in which projects are selected based on their potential value to the business.

In phase three, decisions are all based on how stakeholder needs will be impacted.

2.6.4 Keys to Success

What does it take to succeed with quality management? It would appear that successful implementations are not either/or propositions—the *degree* to which the factors have been integrated into the organization's culture matters.[58] Rhetoric cannot create quality. QIM initiatives that are simply mandated are doomed from the start. Implementations must be led and resourced from the top. Not surprisingly, where quality has become an integral part of an organization's standard procedures, programs tend to be more successful and financial performance stronger. There is also some compelling evidence to suggest that a synergistic effect exists between control and learning. For QIM to add value, innovations that will better serve stakeholders must come from knowledge and learning. Studies show that 10 factors have generally been present in companies that claim successful implementations.[59,60]

1. Top management commitment
2. A clear and concise policy that focuses on the customer (more appropriately, the stakeholders)
3. Creation of a culture committed to quality
4. A steering committee that assumes the dual role of quality leadership and quality guardianship
5. A committed and empowered workforce
6. Emphasis on consistent, relevant, and application-based training with a common language set of tools
7. Continuous improvement of processes
8. Effective two-way communications
9. Management-by-fact, evaluating performance, and making decisions based on hard data
10. Adequate rewards and recognitions

Successful organizations emphasize learning, stakeholders, a supportive culture, the value of the workforce, communications, continuous improvement, and management-by-fact.

---

[56]Easton, George S., and Sherry L. Jarrell. "Patterns in the Deployment of Total Quality Management: An Analysis of 44 Leading Companies." In *The Quality Movement and Organization Theory,* edited by Robert E. Cole and W. Richard Scott. Thousand Oaks, CA: Sage, 2000.

[57]Foley, Kevin, Richard Barton, Kerry Busteed, John Hulbert, and John Sprouster. *Quality Productivity and Competitiveness: The Role of Quality in Australia's Social and Economic Development.* Sydney: Standards Australia.

[58]Douglas, Thomas J., and William Q. Judge Jr. "Total Quality Management Implementation and Competitive Advantage: The Role of Structural Control and Exploration." *Academy of Management Journal* 44, no. 1 (February 2001): 158–69.

[59]Ibid.

[60]Gurnani, Haresh. "Pitfalls in Total Quality Management Implementation: The Case of a Hong Kong Company." *Total Quality Management* 10, no. 2 (March 1999): 209–28.

2.7 INFORMATION ENGINEERING AND DECISION MAKING

2.7.1 Quality Decision Making

Hundreds of decisions are made each day as workers up and down the value stream carry out assigned tasks, monitor routine operations, identify and resolve problems, plan for the future, and work together to improve key processes. Figure 2-7 highlights the importance of knowledge management (KM) to value creation in the open systems model. A QIM-led organization has a leadership that stresses learning and knowledge acquisition. Knowledge workers (KWs) at all levels use facts (retrieved from the knowledge base) to design, manage, and deliver goods and services. The knowledge base is continually updated as learning takes place based on improved process knowledge, experience, and feedback (endogenous and exogenous, positive and negative). When data are converted to information, and information into knowledge, it becomes possible for management to base its decisions on hard data and facts. Facts are used to assess performance, to validate improvements, and in strategic planning. The word *fact* is a misnomer because by implication one would assume that the information stored in the knowledge base reflects *truth*. In reality knowledge is always imperfect and the systems designed to acquire, validate, and analyze raw data, and convert the data into useful and retrievable information that can be easily accessed by those who need it, and in a timely fashion, can always be improved. When validated, consolidated, and archived, knowledge becomes part of the collective corporate wisdom. The design, maintenance, operation, and improvement of these hierarchical systems are the concerns of a discipline known as **information engineering.** Information engineering performs a crucial function in modern organizations to ensure that information is reliable, relevant, and timely, and that data conflicts are effectively resolved. It is also important that data be cleansed of redundancies, outliers, and other sources of error.

Decision-making capabilities can be developed with practice.

Decision making is a process by which individuals withdraw relevant information from the knowledge base to address some issue at hand. The decision-making process is somewhat less predictable than production processes, since the way in which humans make decisions is cognitive and not well understood. Human behavioral outcomes in any setting are difficult to measure and are affected by factors that are complex and highly variable. Also, unlike its production counterparts, decision making has a built-in improvement mechanism called *learning*. As the decision maker gains in experience, learning becomes expertise, which in turn improves decision making capability. Therefore the quality of decision making improves with practice.

2.7.2 Analytic versus Naturalistic Models

Decisions occur at two levels: analytical and naturalistic. At the analytical level, decisions are based on deductive reasoning, probability theory, and the application of formulae. Grounded in solid mathematical and statistical theories, analytical modeling is an essential tool for refining the firm's knowledge base by converting raw data into relevant information. Analytical decision tools are available from the

Decision making has a built-in improvement mechanism called learning.

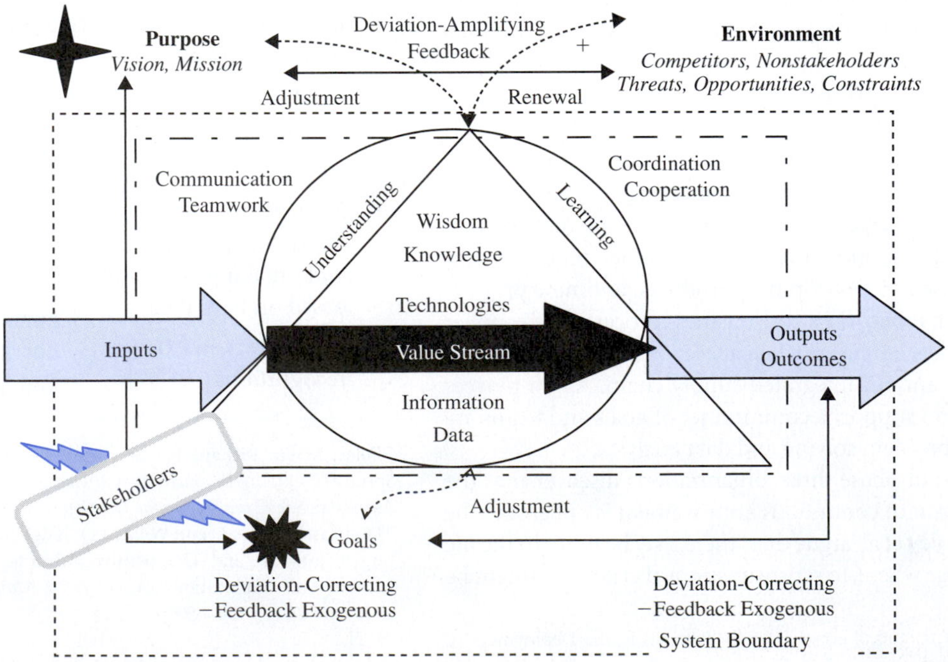

FIGURE 2-7. The Role of Knowledge in Quality Management

decision and statistical sciences and operations research, and include such methodologies as simulation, queuing theory, linear programming, Markov processes, analysis of variance, and design of experiments. Such tools can help the decision maker better understand the decision environment, to analyze hypothesized scenarios, perform "what if?" analyses, and aid in the design of optimum policies.

Most of the important decisions in quality management are naturalistic, and though analytic models can provide valuable insights, those tools alone are incapable of dealing with the uncertainties and complexities of real-life decision scenarios. With naturalistic decision making, the object is to reach a decision that is good enough as opposed to one that is optimal. This is called **satisficing**. Even when deterministic outcomes result from the use of analytic tools, converting the information into effective actions often requires some decisions in the naturalistic domain. While a mathematical or statistical model can be invaluable for working *on* a process, it cannot identify *which* process to study, *how* to study it, or by *whom*, or even how to interpret information in a *changing context* of the work environment. The analytical model cannot provide guidance on how to select from a number of alternatives or how to evaluate the trade-offs. Making decisions in complex, unstructured, or pressure situations requires **naturalistic** capabilities, such as perception, judgment, and intuition. There are two types of naturalistic decisions—programmed and nonprogrammed.

Programmed decisions are actions that are predetermined.

2.7.3 Programmed Decisions

Programmed decisions are usually made subconsciously and occur in circumstances where the outcomes have been predetermined. As an example, an operator on an assembly line may not have to consciously decide whether a red or blue wire should be connected to "terminal A" on a particular circuit. After thousands of repetitions such decisions take place *naturally* and without any conscious thought on the part of the worker. A human is programmed to make such routine decisions through repetition, training, familiarity, and experience. A person need not be conscious of all the decisions necessary to get home (e.g., turn right here, left there, go three blocks, etc.) but would need to carefully follow instructions or a map to find a friend's house for the first time.

Naturalistic decision making strives for good decision, not optimal ones.

2.7.4 Nonprogrammed Decisions

Those decisions that are nonroutine, or require new information, are nonprogrammed. While it is possible to build skills and experience in nonprogrammed decision making, each scenario presents a uniqueness such that the *best* decision in one instance will not necessarily be the best in the next albeit similar instance. In addition, different people bring their individuality to a problem situation and may not approach the problem the same way, reach similar conclusions from a common data set, or agree on the best solution. Research has identified the following nine decision skills that nonprogrammed decision makers use[61] and many of these capabilities can be taught.

1. **Intuition**—the ability to recognize patterns without being able to explain how.
2. **Abstract mental simulation**—the ability to take a set of outcomes and then imagine a plausible process that could have created them.
3. **Leverage**—the ability to exploit a situation, capitalize on its strengths, turn problems into opportunities, and spot weaknesses.
4. **Metacognition**—the ability to look beyond the obvious, size up a situation, and see things others have missed.
5. **Story telling**—the ability to return, and vicariously take others, to a problem situation in order to analyze a decision that was made and its subsequent effects. When a computer is programmed to use past decisions as a basis for learning and improving future decision making, the technique is called *case-based reasoning*.
6. **Metaphors** and **analogues**—the ability to mentally create situations different from (metaphors) and similar to (analogues) the decision scenario for the purpose of generating improved decision options.
7. **Mind reading**—the ability to listen and go beyond the written or spoken word, and to understand the intent behind such communications.
8. **Rational analysis**—the ability to apply the scientific method to develop workable decision options.
9. **Team mind**—the ability to collaborate with others where a group decision is required, and to act and think as one.

2.7.5 What Is a Good (Quality) Decision?

A good decision does not necessarily lead to good results. Decisions are always based on the information available at the time, and less than favorable consequences can occur even if the decision itself is of high quality. One can draw from quality management terminology and define a poor decision as a defective one. *A decision is defective if what is learned as a consequence of the decision would result in a different decision in a future similar situation.* Phrased differently, a decision is the output of the decision process, and negative corrective feedback in the form of experiential learning is an outcome. If feedback results in a different decision (output) in the future, then the decision is a poor one. As in any process, if one can identify and remove the cause of

---

[61] Klein, Gary. *Sources of Power: How People Make Decisions.* Cambridge, MA: MIT Press, 1998.

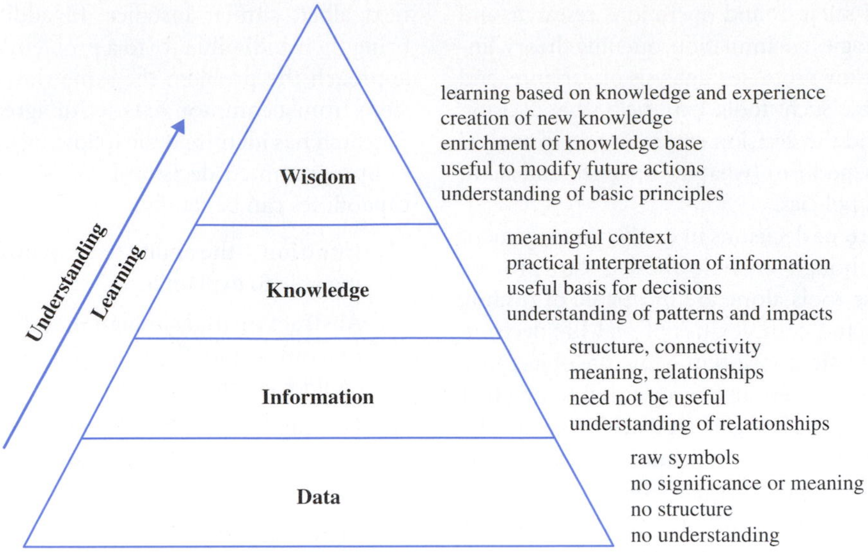

FIGURE 2-8. DIKW Information Hierarchy

poor decisions, the process will be improved and higher-quality decisions can be expected in the future.

Poor decisions may be due to biases in either the way the decision maker thinks or the way the available information is processed. It has been shown that decision makers are apt to approach a new problem situation, starting with what *they think they know* and then making adjustments, rather than approaching the problem objectively, genuinely *seeking the truth*.[62]

When confronted with a problem it is all too typical for the decision maker to formulate a hypothesis (that is, form an opinion) and then try to collect data that will support that position. It is often the case that more could be learned about the problem if the aim was to compile evidence that would refute the á priori assumptions. That is, it may be better to try and collect data that would refute the original hypothesis than try and build a case supporting it.

Capabilities can be learned to help make better nonprogrammed decisions.

A second source of poor decision making is referred to as systemic **latent pathogens**.[63] These are inherent flaws in the design of the system that make it impossible to avoid errors. A properly executed causal analysis can often help flesh out and solve problems with the system that could be improved by either permanently removing the cause or nullifying its effect on future decisions.

A defective (poor) decision is one that would not be repeated in similar situations in the future.

[62]Kahneman, D., P. Slovic, and A. Tversky (eds.). *Judgment Under Uncertainty: Heuristics and Biases.* Cambridge, England: Cambridge University Press, 1982.

[63]Reason, J. *Human Error.* Cambridge, England: Cambridge University Press, 1990.

2.8 THE INFORMATION ENVIRONMENT AND KNOWLEDGE-BASED MANAGEMENT PRACTICES

2.8.1 The Information Hierarchy

The extent to which decision making and knowledge-based management is effective and leads to *good* decisions depends on the collective content of the human minds of those who participate in the process and their ability to efficiently process information in a way that induces continual learning. It has been suggested that the intellectual capital of humans can be classified using the information hierarchy[64] model depicted in Figure 2-8.[65,66] The components shown, which help to describe the evolution in human understanding and learning, are defined as follows:

- **Data**—raw facts, measurements, or observations with no established relationships to each other, or to some external problem, issue, or objective.
- **Information**—data that have been structured by creating relationships and connectivity between the various data points. Unlike data, information can be applied to a "who/what/where/when" line of questioning relevant to a specific problem, issue, or objective.

[64]Reference to the DIKW hierarchy in the literature goes by several different names depending on the discipline. In most knowledge management literature the hierarchy is called the *knowledge* hierarchy or *knowledge* pyramid. In the information science profession DIKW is more often referred to as the *information* hierarchy or *information* pyramid. In this context the distinction between the words *information* and *knowledge* is based on the specific orientation of a particular profession or discipline.

[65]Bellinger, Gene, Durval Castro, and Anthony Mills. *Data, Information, Knowledge, and Wisdom.* http://www.systems-thinking.org/dikw/dikw.htm.

[66]Ackoff, R. L. "From Data to Wisdom." *Journal of Applied Systems Analysis* 16 (1989): 3–9.

- **Knowledge**—information in action. With knowledge, not only the "who/what/where/when" of an issue can be addressed, but also the "how." Knowledge puts information into a meaningful context with a practicable interpretation to guide decisions and actions.
- **Wisdom**—learning based on knowledge. Wisdom is created when knowledge is widely and freely shared and when those who have acquired knowledge are able to reflect on and learn from the consequences of their actions. Wisdom can add the question "why" to "who/what/where/when/how." When past actions were based on that prior knowledge, and new wisdom creates new knowledge based on that experience, future behaviors and actions will be modified whenever similar instances arise.

As Figure 2-8 suggests, the DIKW hierarchy defines an evolutionary process of understanding and learning. At the first level, understanding is limited to the symbols used to document and communicate the raw data that have been captured by some data acquisition schema. There is little that can be understood beyond that since raw data lack structure and context. Data have no value until they are transformed into information. At the information level, data have not only been given structure and context, but relationships within the data set and its connectivity with and applicability to specific problem areas have been established. Understanding has also progressed to include relational concepts and contextual application. Information has the ability to address the "who/what/where/when" of a problem, but may not be that useful in the context of the problem or issue at hand. To be useful, information must provide the basis for improved decision making.

The DIKW hierarchy helps us understand the evolution of human understanding and learning.

Knowledge is created when information takes on a contextual meaning and forms the basis for problem solving and decision making. At the knowledge level, learning has evolved to a point where value can readily be extracted from the information, relevancy to specific applications can be established, patterns can be recognized, and impacts can be anticipated and evaluated. The three tiers—data, information, and knowledge—are cross-linked with feedback loops that act as drivers to move learning and understanding forward. Knowledge is enhanced through experience and fresh infusions of new information. At the level of being able to understand how information relates to the issues at hand, new questions inevitably arise that require new data, which are subsequently transformed into information, and when analyzed in the context of the targeted questions, become new knowledge. This cycle of knowledge renewal forms a continuum that drives the level of cognition to the highest tier—wisdom. Wisdom is the ultimate level of understanding.

Decisions are made at several levels in an organization and range from short-term operational issues to the development of long-range strategic initiatives. The first three tiers of the DIKW hierarchy deal with the past (i.e., what is known).[67] Only wisdom can shape a desired future based on understanding that evolves with time and experience. Hence, wisdom typically supports decision making that resides in the strategic domain. Here, visionaries can create new systems and reengineer existing ones by applying wisdom to "what if?" scenarios. Day-to-day operational issues must be dealt with in the context of existing systems and therefore are normally supported by the information and knowledge tiers. The relationship between the DIKW hierarchy and organizational decision making is illustrated in Figure 2-9.

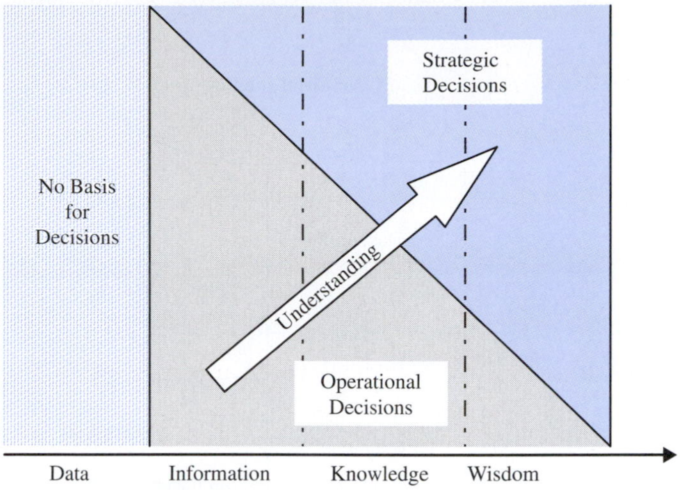

FIGURE 2-9. How DIKW Supports Decision Making

2.8.2 Information[68] Sources

In a knowledge-rich environment information can come from many sources—ranging from manual and crude methods to highly sophisticated information technologies. Effective quality management programs rely on relevant, timely, and reliable information, and quality professionals will seek to learn where knowledge throughout the organization is kept, how it is retrieved, the raw data sources, how data are transformed into information, how information is converted into knowledge, and how knowledge becomes wisdom—while striving to fully understand the underlying measurement systems. Some quality award programs have included in their evaluation criteria how well the knowledge transfer mechanisms work. For example, the Tennessee Center for Performance Excellence evaluates organizations on how the "quality and availability of needed data, information, software, and hardware for (their) workforce, suppliers and partners, collaborators, and customers are ensured."[69] These criteria are

[67]Ackoff, "From Data to Wisdom."

[68]The term *information* is used here to include the first two tiers of the DIKW that form the platform for an organization's knowledge base and for creating corporate wisdom.

[69]Excerpt from the 2008 *Criteria for Performance Excellence: Travelling the Road to Success,* Tennessee Center for Performance Excellence, p. 41. http://www.tncpe.org/downloads/2860_TNCPE_Lores.pdf.

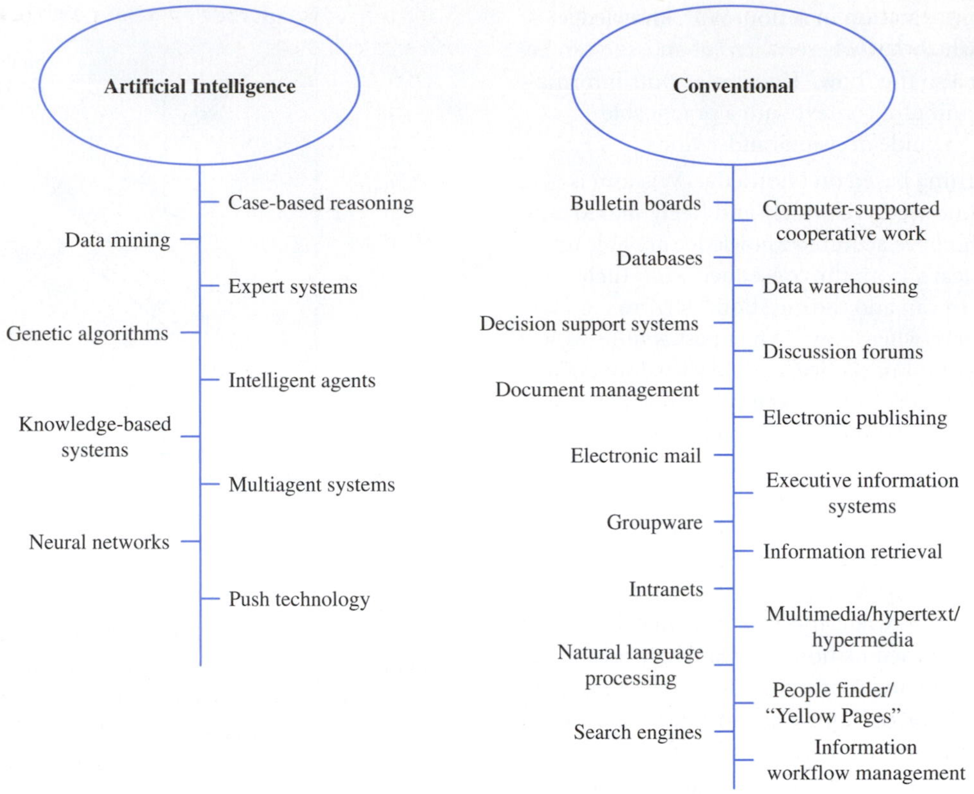

FIGURE 2-10. Sources of Quality Management Information

based on Criteria 4 of the Malcolm Baldrige National Quality Award which measures how well organizations are able to ensure data accuracy, integrity and reliability; timeliness; and security and confidentiality. Organizations are also asked to demonstrate accessibility to data and information, and to show how they manage their knowledge systems to accomplish effective transfer throughout the workforce and from and to all stakeholders.[70]

The numerous and various measurement processes employed in a typical business are capable of generating mountains of data that can be unstructured and scattered. Depending on the level of technological sophistication of a particular measurement process, we can place the information source in either of two categories: artificial intelligence (AI) based or conventional.[71] Some commons sources in each of these categories are shown in Figure 2-10. Brief definitions of each of these can be found in the Glossary at the back of the textbook.

2.8.3 Information Repositories

Once defined, information will be placed in central repositories from which it can be widely accessed as required. Repositories generally fall into one of the following three groups:[72]

- **Competitive intelligence**—information concerning the external environment in which the organization operates. This includes intelligence concerning competitors, stakeholders, suppliers, regulatory constraints, political factors, economic conditions, market trends, and international developments.
- **Structured internal decision support**—information that is used internally as inputs to primary and secondary value streams. This includes blueprints, specifications, bills of materials, production routings, operating procedures, workplace discipline, and quality standards.
- **Informal internal human knowledge and collective wisdom**—this repository may be incomplete or not formally captured, and can easily be lost through personnel turnover. The informal repository includes know-how,[73]

---

[70]2009–2010 *Criteria for Performance Excellence,* Baldrige National Quality Program, p. 17. http://www.baldrige.nist.gov/PDF_files/2009_2010_Business_Nonprofit_Criteria.pdf.

[71]Edwards, John S., Duncan Shaw, and Paul M. Collier. "Knowledge Management Systems: Finding a Way with Technology." *Journal of Knowledge Management* 9, no. 1 (2005): 113–22.

[72]Lau, H. C. W., A. Ning, K. Fl Pun, K. S. Chin, and W. H. Ip. "A Knowledge-Based System to Support Procurement Decision." *Journal of Knowledge Management* 9, no. 1 (2005): 87–100.

[73]Davenport, T. H., and L. Prusak. *Working Knowledge.* Boston: Harvard Business School Press, 2000.

experience, insights, communication and social skills, analytical and deductive capabilities, and the ability to work with others.

The information engineer is concerned with maximizing the effectiveness of each repository with respect to three primary functions—knowledge creation, knowledge sharing, and knowledge utilization.

2.8.4 Knowledge Creation

Knowledge creation can occur in a number of ways. For example, prior knowledge can be combined or placed in a new contextual framework or new analytical tools can be applied. A problem situation can be viewed from fresh perspectives or knowledge can be created through a socialization process as communication and teaming skills are used to facilitate a learning environment.

Innovative organizations have good knowledge creation capabilities.

It is perhaps not surprising that organizations committed to innovation generally have good knowledge creation capabilities which have been shown to positively contribute to stakeholder satisfaction.[74] QIM organizations, which are committed to continual learning and the relentless pursuit of improvement, will have in place programs to perfect their internal knowledge creation capabilities.

2.8.5 Knowledge Sharing

Knowledge sharing occurs between individuals, within groups, and across departments or other subsystem interfaces. This is adversely affected when uncertainties, mistrust, and insecurities pervade the workforce culture. QIM is based on the empowerment and participation of workers at all levels. This has the intrinsic effect of creating a sense of ownership and positive outlooks. It has been shown that the internal transfer of knowledge is facilitated when cultures are stable, performance is strong, and the organization is growing.[75]

2.8.6 Knowledge Utilization

While the ability to change enhances knowledge creation, it is ironic that **knowledge utilization** seems to be stronger in organizations that have been slow to innovate.[76] This is perhaps due to the fact that knowledge workers become more confident in the application of old established technologies and more fearful of readily adopting new practices. On the other hand, the utilization of knowledge is facilitated in organizations that incorporate knowledge transfer in their performance metrics (for example, as part of the balanced scorecard).[77] In addition, studies have shown that the more diversified the company and the more complex the decision making, the more likely will be found a climate that encourages multifunctional knowledge sharing.[78]

Knowledge utilization is facilitated in cultures receptive to change.

2.9 INTELLIGENT PROCESS CONTROL

Statistical process control (SPC), a topic that we cover in detail in Part III, is used to regulate product quality by monitoring quality attributes during each processing step. SPC uses a tool called a control chart to track the measurements of these attributes, and relative to each attribute, to determine whether or not the process is in control (i.e., stable and can be predicted). Process data are collected by taking small samples at predetermined intervals. The items selected are then measured and the measurements analyzed statistically to determine if the process as a whole is producing as required—that is, whether process outputs are on target and within a tolerance range that the customer is willing to accept. If the process outputs appear to be satisfactory, no action is taken. If, on the other hand, there is evidence that the process is not performing as it should, an operator intervenes and makes adjustments. Future samples will help determine whether the adjustments worked or further action is necessary.

SPC has some shortcomings. Although computationally simple, control charts can require special skills to correctly interpret data patterns. The recognition of unnatural data patterns and taking appropriate action in a timely manner are crucial. Statistical sampling cannot guarantee 100% perfection due to random variation and process dynamics. In addition, for high-volume processes such as flow shops and continuous production, process changes can come and go and never be detected by SPC, particularly if the changes are small or short-lived. The top diagram in Figure 2-11 describes a typical SPC quality control system.

Another less transparent problem with SPC is its limited application. SPC is generally used only to track those variables thought to be of interest. But who decides and on what basis? Many of the variables tracked are thought to be important because they have always been the ones measured or they relate directly to a specified quality attribute. In the latter case such variables are nearly always dependent, or output, variables and are not process drivers, or independent variables. While the attention is on these measures there is no way to directly tie the monitored variables to the process drivers that are the true determinants of quality. SPC can be effective in monitoring specific product features one-at-a-time. However, there may be numerous variables that affect quality

---

[74]Huang, Hua-Wei, Hong-Yu Shih, Hsiao-Wen Huang, and Che-Hung Liu. "Can Knowledge Management Create Firm Value? Empirical Evidence from the United States and Taiwan." *The Business Review* 5, no. 1 (September 2006): 178–83.

[75]Ibid.

[76]Ibid.

[77]Ibid.

[78]Sabherwal, Rajiv, and Sanjiv Sabherwal. "Knowledge Management Using Information Technology: Determinants of Short-Term Impact on Firm Value." *Decision Sciences* 36, no. 4 (2005).

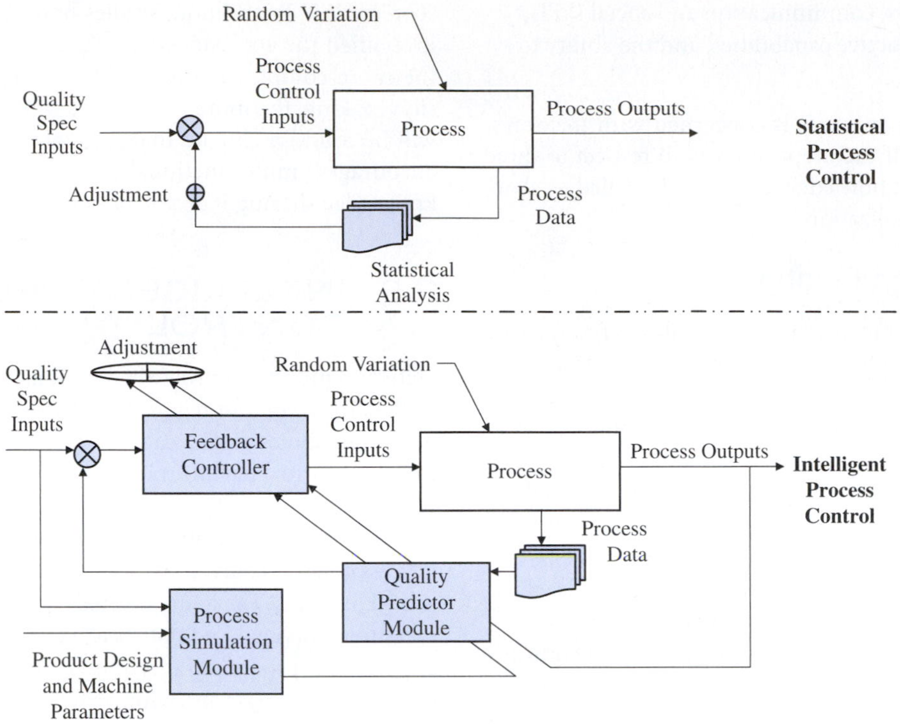

FIGURE 2-11. Comparison of SPC with IPC

and are not specifically SPC-monitored—variables such as humidity, tool wear, raw material variability, and housekeeping are examples. Although these variables can be known to have an impact, can be tracked independently, and accounted for and accounted for in process settings, the SPC tools themselves fall short of being able to define the strength of the correlation or interaction between such variables and some important variables may be ignored entirely. SPC can also lead to a condition called *overcontrol* in which an operator, reacting to data patterns on a control chart, overadjusts and, in so doing, unwittingly introduces variability into the process.

Statistical sampling cannot guarantee perfection due to sampling error and process dynamics.

Intelligent process control (IPC), often used with control charts, has the ability to address many of the shortcomings of SPC in a high-volume fast-moving and automated manufacturing environment. A typical IPC system is illustrated in the lower diagram of Figure 2-11. As a starting point this system uses a simulation process model to conduct an inline experimental design to determine how process variables are related to critical output measures. The output from the simulator—which is ongoing with process data inputs and operator feedback—together with actual signals from the process are fed into a quality predictor that represents continual process learning. The predictor is linked to a feedback controller that automatically adjusts the process to close any gaps between outputs and requirements based on the current state of process knowledge.

A number of different models have been used to process signals from output data to predict quality. For example, applications of artificial neural networks,[79] fuzzy logic,[80] expert systems,[81] and regression[82] have been reported in the literature. Specifically, artificial neural networks have achieved high accuracy in control chart pattern recognition in some applications.[83] It has been shown, for example, that a neural network can detect small to moderate shifts in a process mean better than conventional control chart techniques. In addition, neural networks have been trained to recognize out-of-control patterns, such as trends, sudden shifts, cycles, and data stratification.[84]

An artificial neural network (ANN) is a system of parallel distributed processors that are designed to model the way the human brain processes information, makes decisions, and learns. It is beyond the scope of this text to delve too deeply into the subject of the theory and methodology of building and

[79]Woll, Suzanne L. B., Douglas J. Cooper, and Blair V. Souder. "Online Pattern-Based Part Quality Monitoring of the Injection Molding Process." *Polymer Engineering and Science* 36, no. 11 (1996): 1477–88.

[80]Zadeh, L. A. "Fuzzy Sets." *Information and Control* 8, no. 3 (1965): 338–53.

[81]Kameoka, Seiji, Nobuhiro Haramoto, and Tadamoto Sakai. "Development of an Expert System for Injection Molding Operations." *Advances in Polymer Technology* 12, no. 4 (1993): 403–18.

[82]Kleinbaum, David G., Lawrence L. Kupper, Keith E. Muller, and Azhar A. Nizam. *Applied Regression Analysis and Other Multivariable Methods.* 3rd ed. Pacific Grove, CA: Brooks/Cole, 1998.

[83]Barghash, Mahmoud A., and Nader S. Santarisi. "Pattern Recognition of Control Charts Using Artificial Neural Networks—Analyzing the Effect of the Training Parameters." *Journal of Intelligent Manufacturing* 15 (2004): 635–44.

[84]Guh, R. S., and J. D. T. Tannok. "Recognition of Control Charts Concurrent Patterns Using Neural Network Approach." *International Journal of Production Research* 37, no. 8 (1999): 1743–65.

SHOWCASE of EXCELLENCE **Winners of the Malcolm Baldrige National Quality Award Manufacturing Category**

- 2008: Cargill Corn Milling North America, a manufacturer of corn and sugar-based products serving food, feed, and fermentation markets. *http://www.nist.gov/public_affairs/releases/cargill_profile.html, http://www.cargill.com.*
- 2005: Sunny Fresh Foods, Inc., a manufacturer of egg-based food products serving quick-service restaurants, business and institutional food services, schools, and the military. *http://www.quality.nist.gov/PDF_files/Sunny_Fresh_Foods_Profile.pdf, http://sunnyfreshfoods.com.*
- 2004: The Bama Companies, Inc., a manufacturer of frozen ready-to-use food products served worldwide by quick service and casual dining restaurant chains. *http://www.nist.gov/public_affairs/BamaPDFfinal.pdf, http://www.bama.com.*
- 2003: Medrad, Inc., a worldwide leader in developing, manufacturing, marketing, and servicing medical devices that enhance imaging of the human body and enable therapy. *http://www.nist.gov/public_affairs/baldrige2003/Medrad_3.3.04.pdf, http://www.medrad.com.*
- 2002: Motorola Inc. Commercial, Government and Industrial Solutions Sector, the leading worldwide supplier of two-way radio communications including radio networks, systems, products and services, and integrated communications and information technology solutions. *http://www.nist.gov/public_affairs/releases/cgiss.htm, http://www.motorola.com/cgiss/malcolmbaldrige.*

applying an ANN. For our purposes, we will simply describe the concept by saying that an ANN uses a model to simulate multiple layers of data processing elements called neurons. Each neuron is then linked to neighboring neurons with linkage coefficients that describe the relative strength of the connectivity. The network acquires new knowledge through a training algorithm that learns the solution to a particular problem by adjusting the connectivity coefficients. Once an ANN has gone through a learning cycle, the new knowledge is validated through a testing phase during which new input patterns (with known outputs) are introduced. By holding the coefficients fixed, the actual system outputs can be compared to the desired (known) ones, thereby measuring learning effectiveness.

2.10 TECHNICAL COMMUNICATIONS

Technical communications has been defined as the process of conveying usable information about a specific technological subject to an intended audience for the purpose of performing an action or making a decision based on its contents.[85] The technology can range from low-end business processes and practices to high-tech automation and computing. The word *technical* can refer to any specialized body of knowledge that is not widely known outside the domain of a small community of subject matter experts. The success of QIM implementations can rise or fall on the capability that quality professionals have in communicating effectively to a general nonspecialist audience. That is, among other essential skill sets, the quality professional must master the art of technical communications.

To be effective, technical communications must connect with the intended audience—the receivers of the information. Communications must be delivered in a manner that meets the needs of readers, listeners, or viewers, at their level of understanding, in language and terminology appropriate to their backgrounds—including an appreciation for diverse cultures, experiences, formal education, ethnicities, genders, ages, and social status. The key is to be able to translate technical concepts into a lay vocabulary and phraseology suitable to a general audience. This is especially critical in a QIM culture where diverse multicultural teams are required to work together toward common quality goals and open communications are stressed.

In quality programs, technical communicators work with a variety of media, including paper (e.g., reports, manuals, specifications, proposals, memoranda), video (e.g. video slide presentations, promotional video clips, DVDs), sound (e.g., sound bites, CDs), local area networks (LANs) and the Internet (e.g., e-mails, Web sites, blogs).

The key to effective technical communications is the ability to translate technical concepts into language that is suitable to a general audience.

Technical documentation is critical to the success of QIM, and is the term used to describe technical communications when the delivery medium is paper. This can include various forms of formal documents such as user manuals, equipment descriptions, spare parts lists, standard operating procedures, quality testing procedures, marketing and promotion materials, and product specifications. These documents can be for internal use or for the benefit of stakeholders, suppliers, or other interested parties.

2.11 SUMMARY OF KEY IDEAS

Key Idea: *The open systems view perceives the organization as an entity that communes with the environment that sustains it, with specific emphasis on the necessity of obtaining continual feedback that can be used to adapt to changing requirements.*

To be competitive in today's marketplace an organization must understand its environment and actively interact

[85] Johnson-Sheehan, Richard. Technical Communication Today. New York: Pearson/Longman, 2006.

with it. A competitive strategy requires a good knowledge of stakeholders, competitors, the supply chain, and all environmental forces that shape decisions and business strategy. Under open systems theory, feedback is used extensively to guide a change management process that will ensure the sustained creation of value for all stakeholders.

Key Idea: *Outputs, which are the immediate deliverables from internal value streams, should be differentiated from outcomes, which are the consequential effects that outputs have on subsequent systems. Outputs can be defined at the system level or, more narrowly, at an individual processing step along a primary or supporting value stream.*

An output is measured from within a system, either at the end of an entire value stream, or at the exit point of an intermediate processing step. An outcome is an effect that an output has on another system or downstream process, and may take some period of time to show up. Outputs and outcomes can be desirable or undesirable. Any output that cannot be used by the customer or outcome that impacts negatively on the achievement of a customer's mission is called waste. Waste is toxic if it has a damaging effect on the system where it collects.

Key Idea: *All of the elements required to transform a production system's collection of inputs into a mix of desirable outputs is called the primary value stream. All interconnected activities required to provide material, financial, human resource, maintenance, and administrative support for primary value streams are called support, or secondary, value streams.*

The primary value stream includes all the tasks, skills, and knowledge required to convert inputs into outputs. When the value stream is expanded to include external supplier systems upstream and the customers' supply and distribution points downstream, the resulting network is called the *supply chain.* At the human–work interface the value stream is supported by three components—individual attributes that include the mix of skills and human traits that define individual differences; group processes that represent the collective teaming skills that enable people to work together to achieve common goals; and the design of the task and work sequence necessary to complete the various value-adding processes.

Key Idea: *Positive and negative feedback are both critical to the effective management of a QIM organization.*

Positive feedback amplifies either good or bad behavior. The consequence is that, as a result of the feedback, the behavior is continued but with greater enthusiasm. In the open business system positive feedback, referred to as deviation-amplifying, is used by organizations to learn what they are doing right (in the eyes of their stakeholders) and to continue to do those things—perhaps even better in the future—by focusing greater attention and directing more resources to those areas. By comparison, negative feedback has a built-in adjustment mechanism that balances a system in order to return it to some desired and stable state. In QIM organizations negative feedback, called deviation-correcting, is used to evaluate whether goals and objectives have been achieved and, if not, to take appropriate corrective actions.

Key Idea: *QIM organizations have the ability to arrest the destructive effects of entropy through learning, self-renewal, and the continual infusion of new energies from the environment.*

The second law of thermodynamics states that systems will always move from a state of organization to a state of chaos as wasted energy, called entropy, accumulates. Entropy will never decrease and will increase whenever possible, unless there is some form of external intervention to arrest and reverse this natural process. With an open systems approach organizations recognize that sustainability requires the continual communion with the external environment, which is seen as the source for new ideas, energies, and self-renewal. Unlike a biological system, an industrial organization has the ability to reinvent itself and adapt—even if that means a fundamental change in direction, purpose, relationships, membership, and structure.

Key Idea: *A strategy defines an organization's intended relationships with all interested and affected parties, including stakeholders and nonstakeholders, with a particular emphasis on balancing the needs of all identified stakeholders.*

Sound business plans are built around a good strategy. Strategy evolves as an organization learns more about itself and its markets—and all the relevant players. Strategic planning helps sharpen purpose and better match core competencies with clearly defined product and service niches in the marketplace. A good strategy is focused, divergent, and has a compelling tagline. It is capable of articulating the market, the organization's position within it, what stakeholders value, the relative strengths and weaknesses of major competitors, and how the organization intends to compete.

Key Idea: *There is no optimal organizational structure that can guarantee quality. A QIM organization will adopt a suitable structure to support its purpose and competitive priorities, and to respond to its unique set of challenges and opportunities.*

Organic structures that are flexible, participative, and team oriented appear to be more consistent with the open systems view than mechanistic structures that are rules based with highly defined standardized tasks, a strict hierarchy, and tight central controls. However, contingency theory suggests that because organizations have an inherent will to survive, when faced with challenging situations an organization will adapt to whatever structure works best at the time.

Key Idea: *A good strategy identifies and capitalizes on core competencies as the means to identify how an organization can deliver sustained value to its stakeholders and do this better than competitors.*

Core competencies include strengths such as unique manufacturing technologies, human resource development, marketing, and innovation. Such factors are market differentiators because they are difficult for competitors to imitate. Strategy should capitalize on these factors by assessing how stakeholder value is achieved relative to these strengths, how this can be used to establish marketing goals and competitive priorities, and what production system design would support this strategy.

Key Idea: *Strong leadership is essential to good quality—many leadership qualities can be learned.*

A good leader is someone who is an enabler, a change advocate, a visionary, is empathetic, and has a good sense of timing. Leadership skills are defined by the motivations of those who choose to follow. People generally follow a leader because they perceive that person has superior technical skills, and is therefore someone from whom they can learn; they trust that the leader genuinely cares for them and is secure that the person will look out for their best interests, and/or they recognize and respect the leader's moral authority.

Key Idea: *The balanced scorecard approach places an appropriate emphasis on those things that matter most for all stakeholders of an organization, and especially focuses attention on process rather than outcomes, and on how the diversified concerns of stakeholders are related and integrated.*

The balanced scorecard is an approach that gives appropriate performance weightings to a number of factors that are critical to success. Rather than basing business performance just on financials, the balance scorecard requires the establishment of measures in five perspective areas: investors; users/customers; production system; innovation, learning, and change; and society and the planet. Under each of the five areas, a company sets specific goals that are consistent with its business plan. Attributes that are critical to the achievement of each goal are established, and specific measures are developed for each goal. In doing so, conflicts—such as the achievement of financial goals that may be in conflict with sustained customer value—are identified and compromises made. Use of the balanced scorecard approach helps an organization better understand the various causal systems that affect the performance measures its stakeholders care about most. Armed with this knowledge the organization can set priorities and allocate resources in a manner that attempts to simultaneously satisfy the diverse needs of all stakeholders.

Key Idea: *A successful implementation of a QIM program is evolutionary and involves a journey through three distinctive phases (generations) of development. Starting from scratch an organization can expect this process to take a long time—usually many years—to complete.*

To be successful, QIM must be an evolutionary process and can take years to complete as an organization evolves through three distinct generations of development. When a company embarks on the quality journey, it begins a transformation from first- to second-generation quality. Research has demonstrated that this process requires a company to go through four distinct phases. During the first three, a company will be in the second generation. Phase three will take it into the third generation. Phase one is a predeployment phase during which the organization goes through a period of self-actualization, learning about itself, its operations, and its people. It may run pilot programs during this time, benchmark others, conduct training, use consultants, and start applying some statistic tools on a limited basis. During phase two the organization begins to make quality a priority, establishing teams, proliferating training, and stressing improvement. Most of the activity is to achieve incremental improvements on existing systems. In this phase, the scope broadens to include a systems approach. Quality is extended to the design function and to include supply chain relationships. Accountability and measurement have become the norm in evaluating and improving performance.

In phase three, the organization has entered the third generation and quality is central to business strategy. A third-generation culture is characterized by a committed and visionary leadership, trusting relationships with its stakeholders, high-performance work systems, an empowered workforce who value continued learning, well-designed and managed processes, decisions that are data driven, and strong financial and market outcomes. Management decisions are fact based, the culture is united in a common purpose, the supply chain is integrated, and all participants up and down the chain are striving for the same goals. Stakeholder value is the primary driver, and the organization grows by learning.

Key Idea: *QIM implementations must be led and resourced from the top, quality must become engrained in the culture, and synergy must exist between learning and an organization's internal and external relationships.*

Successful organizations emphasize learning, stakeholders, a supportive culture, the value of the workforce, communications, continuous improvement, and management by fact. Change cannot be mandated from the top—it must be led. Management must carefully prepare the culture for QIM readiness and actively engage stakeholders at all levels, inviting and welcoming their active participation in the change process.

Key Idea: *Human decision making is a cognitive process that includes an inherent improvement mechanism called learning.*

Decision making is a process in which humans use their cognitive abilities to apply knowledge to some problem situation. Human behavior in any given set of circumstances is variable and unpredictable and is difficult to measure. In addition, the ability to learn imbues the process with an automatic improvement mechanism that achieves better-quality through experience and practice.

Key Idea: *Decisions occur at two levels: analytical and naturalistic.*

Analytical tools are based on deductive reasoning, probability theory, and the application of mathematical formulae. Given a set of inputs, analytical models convert those inputs into a structured set of outputs or conclusions that generally indicate a decision that best meets the decision maker's objectives. Naturalistic decisions deal with the complexities, uncertainties, and lack of structure characteristic of real-world problems. This requires perception, judgment, and intuition, as well as the ability to analyze, interpret, and properly apply new information to the problem at hand.

Key Idea: *A good decision does not necessarily lead to good results.*

A good decision is one that, given the same set of circumstances and available information, would be repeated in the future. An outcome of the decision process is the experience and learning that results from the observed consequences of the decision made. If the consequences of a decision were undesirable and, in retrospect the decision maker has learned how to avoid those consequences in the future by making a different decision, then the original decision can be judged to have been a poor one.

Key Idea: *The information (DIKW) hierarchy defines the process by which data are transformed into information, information into knowledge, and knowledge into wisdom to advance the state of understanding and learning within an organization.*

Raw data lack structure and meaning. Meaning comes only when the data are structured and relationships within the data set, and in the context of a problem situation, are established. As raw data become contextually relevant information is created and a certain level of understanding is achieved. Information becomes knowledge when the information can be put into action and provide the basis for decision making through pattern recognition and impact assessment. The creation of knowledge takes understanding to a practical level. Knowledge becomes wisdom when a deep understanding of basic principles and cause and effects, derived from experience and expertise, can be used to enrich the database and guide overall policy making and strategic planning.

Key Idea: *Organizational knowledge can be accessed from numerous and disparate sources, ranging in technological sophistication based on artificial intelligence to relatively crude conventional systems based on computerized or manual techniques.*

Sophisticated information technology systems use artificial intelligence methodologies such as neural networks, fuzzy logic, genetic algorithms, and expert systems. These systems have the capability to convert raw data into information, and then process this information continuously to *learn* how process behavior is affected by controllable and uncontrollable factors and to consequently positively influence future behavior through improved decision making (e.g., process adjustments). The more common sources of information tend to be scattered, lack integration, and are of varying levels of sophistication. Collectively, these sources form the backbone of an organization's communication system and include such things as electronic mail and bulletin boards, discussion forums, decision support systems, groupware, and databases.

Key Idea: *Organizations that place a high premium on innovation will have good knowledge creation capabilities.*

When a company is willing to commit significant resources to research and new product development, and places a high value on innovation, risk taking, and new ideas, the creation of knowledge is a core value and central to the execution of strategy. QIM organizations, striving to continually improve and learn, emphasize the importance of strengthening their internal capabilities to create knowledge.

Key Idea: *For high-volume automated manufacturing environments intelligent process control (IPC) can improve quality by overcoming some of the shortcomings of statistical process control (SPC).*

A typical IPC system is able to continuously monitor process output and can, through the use of a predictor module that contains a dynamic model of the specific operation or system, predict quality. If the predicted quality is in variance with the requirements, a process controller automatically makes adjustments. For high-volume automated processes an IPC system can be more effective at creating quality than SPC since an IPC system has the ability to learn how a process behaves, the critical parameters that have the greatest impact on quality, and how to best respond when requirements are not met.

Key Idea: *The key to effective technical communications is to be able to translate a specialized technical subject into language suitable to a general audience.*

Information that is critical to decision making is often technical in nature and the translation of such information into usable knowledge requires effective technical communications. This means that the information must be delivered to the intended receivers in a manner and language that can be understood by all—taking into account diverse cultures, experiences, formal education, ethnicities, genders, ages, and social status.

2.12 MIND EXPANDERS TO TEST YOUR UNDERSTANDING

2-1. Molly and Henry, a wife and husband team, suffered layoffs from jobs they had held in the automotive industry for over 20 years. In their mid-40s they now face the daunting task of how to satisfactorily cope with a major career change and what options are available. They both like to cook and Molly is extremely talented with cakes, pies, and fancy pastries which she loves to prepare for dinner parties and as gifts to friends and family. Henry is also good at baking and has a passion for experimenting with different bread recipes and organizing menus around his *boulangerie* creations. They decide to try and turn their avocation into a vocation by opening a small bakery/café where the locals could congregate for coffee, sandwiches, and desserts or take home loaves of "homemade" bread, pies, or cakes. They name their new venture (by combining both of their names) the Mollenry Bread Shoppe. Henry sells his Harley and, together with some personal savings and a second mortgage on their home, the couple is able to put up enough equity to obtain financial backing from their local bank. Both Molly and Henry believe that there is a good market for their concept but are not really sure.

a. Who are the stakeholders of the Mollenry Bread Shoppe? Give some examples of affected/interested parties that may not be stakeholders?

b. How can Molly and Henry take these stakeholder needs into account as they start their new business venture and develop a strategy to guide their business decisions?

c. How can Molly and Henry use the balanced scorecard approach to ensure that all stakeholder needs are met? Specifically, can you suggest the relevant dimensions (main categories) of the scorecard and some measures that might be appropriate to each?

d. How would Molly and Henry's approach to developing a business plan for the Mollenry Bread Shoppe differ under closed and open systems views?

e. Give an example of deviation-amplifying feedback in the case of this company and how this positive form of feedback could lead to a vicious circle. How could this feedback lead to a virtuous circle?

f. Give an example of endogenous deviation-correcting feedback in the case of this company. Give an example of exogenous deviation-correcting feedback. How could the use of these negative forms of feedback help Mollenry better satisfy stakeholder needs?

g. How should Molly and Henry define quality? Can you suggest how they can ensure that quality goals are being met?

h. Since both Molly and Henry expected to be busy baking and handling the administrative aspects of the business (such as purchasing and accounts) they decided to hire someone to operate the front counter, greeting the public, taking orders, and operating the cash register. After sifting through a mountain of applications they decided to hire Sally, a recent college graduate who had majored in political science and was having difficulty finding a job. They had many applicants that seemed well qualified but they settled on Sally because they wanted someone who was smart and who could take responsibility with little supervision and direction. However, their faith in Sally was ill-placed. Sally was clearly "overqualified" for the job and felt that it was beneath her. She saw Mollenry as a way of paying her rent until something more suitable came along. Therefore she spent every spare moment while on the job reading job ads, searching the Net, and brushing up her resume. At times she was quite irritated when a customer entered the shop and interrupted her job-seeking activities, and her body language and tone reflected her feelings. Given this consequence, do you believe Molly and Henry's decision to hire Sally was a poor one? Why or why not?

2-2. The Mollenry Bread Shoppe became an overnight success. Molly and Henry's success story was written up in several trade journals, and they were the guests on several TV talk shows. They were even featured in a *New York Times* article advising travelers of the best places to eat. They began to open up additional shops—first across town, then neighboring towns, and finally in different states. After five years, the baking duo had a network of 32 bread shops in 11 states, with 150 employees.

a. What environmental factors does the company strategy have to consider now that were different when the company was simply a mom and pop single-shop operation? Do you think that the stakeholders have changed from the original definition?

b. Has the definition of quality changed as a result of the organization's growth? How can demographic and cultural differences be accommodated in the various locations while maintaining control of quality and ensuring that stakeholders needs are satisfied?

c. With so many employees in several geographical areas, how would Molly and Henry's leadership styles differ under Theory X, Y, and Z? How might their organization structures differ under each leadership style? What are the strengths and weaknesses of each style?

d. How would you define the primary and secondary value streams for this business system?

e. From a quality management perspective, how do you think the Mollenry Bread Shoppe chain differs from a large regional commercial bakery that produces large batches of bread, cakes, and pastries for distribution through supermarkets?

2-3. In December 2008 the CEOs of the three major U.S. automobile companies—General Motors, Chrysler, and Ford—appeared before a congressional committee seeking government assistance to prevent the collapse of the U.S. automotive industry. The reasons given for needing a government "bailout" were many and complex. For example, each CEO testified that the company's cars could not be sold at competitive prices due to the high labor costs imposed by union contracts. These officials also blamed the collapse of the financial system that led to the freezing of credit on a global scale. On the congressional side there was the issue of fuel efficiency. The government had previously given the industry a $25 billion grant to produce fuel-efficient vehicles, yet by the end of 2008 dealer lots across the country were filled with gas-guzzling sports utility vehicles, trucks, vans, and sedans. Those opposed to a bailout argued that the "big three" should be allowed to proceed into bankruptcy so that some needed restructuring could take place. They pointed to the many Asian and European automobile companies with plants located in the United States, which in spite of the same challenges and constraints were doing very well—some actually building new plants. The opponents thought the car manufacturers should have done a better job of reading the market forces and preparing for contingencies. For example, with the escalating cost of gasoline, they asked why the three car giants had not been more aggressive in investing in new fuel-saving technologies. Opinions on quality were divided. Some felt that public perceptions favored imports over domestic quality. Others believed that the quality of the American-made products met or exceeded world standards, and the lack of competitiveness was primarily a cost issue. The proponents of the bailout countered that this was not a quality issue and that failing to help the ailing companies would result in massive layoffs—warning that assembly plant closures would trigger a domino effect across the industry as the impacts were felt further up the supply chain.

a. Who do you think are the stakeholders of the big three? Do you think that the plight of these automobile companies was unavoidable and that they properly identified all stakeholders together with strategies designed to meet stakeholder needs? If yes, what factors led to their demise that were not stakeholder related? If no, which stakeholders' needs were not properly accounted for as these companies formulated strategy, set goals, and established performance metrics over the past decade?
b. What information sources should automobile companies consult in order to determine whether they are meeting stakeholder needs? Do you think that any of these sources was ignored or not taken seriously in the case of the big three?
c. How do you think automobile companies should conduct the first two steps in strategic planning—that is, establishing purpose and addressing interface issues? What would be some of the relevant issues that companies in this industry would want to address relative to their various stakeholder groups? Do you believe that anything could have been done by any of the big three, during their respective strategic planning processes, that could have averted their impending collapse?
d. What role do you think leadership could have played in the ultimate need for each of the big three automobile companies to ask for a federal bailout?

2-4. List an output for each of the following systems. Now, suggest a possible outcome for each of these systems.

a. An automobile engine
b. A university statistics course
c. A process that produces a chemical that is used as a calcium additive by orange juice manufacturing plants
d. A process that produces tires for SUVs

2-5. Describe in your own words the concept of a value stream. How would one differentiate between a primary and secondary value stream? Why does quality depend on managing the entire value stream rather than focusing on its component parts? Specifically, how could an improvement of part of a value stream, without considering the whole, result in an overall reduction in quality? Is it possible for an organization to have a core competency that relates to a support value stream? Explain.

2-6. In the following leadership situations, describe what you think are the most likely reasons why the people involved follow the lead of the person mentioned.

a. After a devastating first half, one of the team's members rallied his teammates in the locker room, imploring each of them to put forth extra effort and win a victory for their coach and alma mater. The team experienced an astonishing comeback during the second half.
b. The astronomy professor, an internationally recognized expert in her field, inspired her students to put forth effort that was far in excess of a normal class load expectation, in order to take full advantage of the learning opportunity.
c. A cult leader convinces young people to leave their homes and jobs, join his movement, and live in a commune in South America.

2-7. The following situations require strong leadership. What are some of the qualities and skills you would look for if it were your responsibility to recruit someone to fill these positions?

a. President of the United States
b. Plant manager of an automobile assembly plant
c. Team leader of a process improvement team
d. Engineering department head of a large public university
e. Coach of a high school football team
f. CEO of a small regional airline company

2-8. Give an example of how an academic department in a large university might obtain and use the following types of feedback.

a. Deviation-amplifying
b. Endogenous deviation-correcting
c. Exogenous deviation-correcting

2-9. Place yourself in the position of the president of a large public research university. Suggest some of the attributes that would be important under each of the five segments of a balanced scorecard. How do you think that each of these attributes might be measured? How can the balanced scorecard help the university formulate strategy and improve performance?

2-10. Assume that you have been given the task to evaluate a large public electric utility and to make suggestions for improvement. One of the factors you want to look at is how performance in the production and distribution of electric power is and should be measured. You are particularly intrigued at the prospect of introducing the balanced scorecard approach to public utilities. List two goals and at least one corresponding measure for each of the following scorecard perspectives.

a. Investors' perspective
b. Users' perspective
c. Internal production system perspective
d. Innovation, learning, and change perspective
e. Society and the planet perspective

2-11. Why is QIM an evolutionary process? Is it necessary for the implementation of quality principles to pass the test of time before taking an organization to a higher level? If a company that is at the phase two or three level merges with a company that is at the predeployment level (or below), how do you think the comingling of cultures should be handled?

2-12. Why is fact-based decision making essentially good strategy? What does it mean to base strategic planning on knowledge? Is an open systems view important to management by fact? Explain. How does fact-based management support organizational learning? Explain.

2-13. Classify each of the following decisions. Are they analytical, programmed naturalistic, or nonprogrammed naturalistic? Explain.

a. Dialing your best friend's cellular phone
b. Using a mathematical formula to determine the minimum cost reorder point for an item in inventory
c. Inspecting the output of an assembly process for defects
d. Purchasing a new photocopying machine
e. Selecting the members for a quality improvement team
f. Selecting a university to attend
g. Driving to your grandmother's house
h. Performing the repetitive task of soldering two wires on a circuit board in a factory that manufactures TV sets
i. Calling an offensive play in a football game when it is fourth down with one yard to go for a first down

2-14. Mattie is planning an automobile trip to attend her cousin's wedding. She has identified four criteria for determining the route she will take from her hometown, Cypress Grove, to where the wedding will take place in Central Metropolis. The criteria are time, distance, cost, and stress. She has selected two routes. She can take Route 86—a heavily traveled limited-access four-lane highway that will take her through several large cities, some of which are not on a direct path to Central Metropolis. She also must pay tolls on several segments along the route. Route 42 is a more direct albeit two-lane route and is somewhat shorter, though access is not limited and it does travel through several small towns which has the potential of slowing down the trip. Nevertheless, Route 42 is preferred choice of Mapquest and Yahoo. Mattie also considered that Route 42 offered a more scenic trip, with less traffic and therefore less stress. Taking this into account, the distance differential and the cost of tolls, Mattie opted for Route 42. When she was a hundred miles into the trip she encountered a bridge that had collapsed where the state patrol diverted all traffic onto a detour route around the bridge site. The detour turned into a nightmare—a small, winding road through the mountains with bumper-to-bumper traffic. Mattie's car overheated, broke down, and she had to call a tow truck which took 2 hours to arrive. Mattie did reach the wedding eventually, but tired, frazzled, and 12 hours later than planned—not to mention the $500 she had to shell out for the tow truck and mechanic.

a. Was Mattie's decision to use Route 42 in lieu of Route 86 a poor one? Explain.
b. Was Mattie's decision an analytical or naturalistic process? Was it programmed or nonprogrammed?
c. What could Mattie have possibly learned from this experience that could help improve her decision-making quality in a similar future situation?

2-15. For each of the following sets of raw data, provide examples of how the data could be transformed into information, knowledge, and wisdom, respectively.

a. Number of people attending a rock concert
b. Diameters of 1,000 holes, measured on a coordinate measuring machine on randomly selected parts that were produced over a 10-day period
c. List of candidate causes for misapplied labels, generated from group brainstorming

d. Customer complaints that were lodged at a 1-800 hot-line and a company's Web site over a one-year period
e. Number of students admitted to a particular university with an SAT score above 1250
f. Number of voters supporting a particular referendum
g. Total number of gold medalists at the Olympic games in Athens

2-16. Place each of the following information sources in its proper classification: AI based or conventional.

a. A genetic algorithm used to solve a large-scale constrained integer programming problem to determine a factory's optimal product mix
b. A company's published annual report that is made available to all employees over a local area network
c. A notice from the president e-mailed to all employees
d. Six Sigma improvement team collaboration using Lotus Notes
e. Computerized storage of an organization's personnel files
f. Use of a neural network to qualify bank customers for loans in order to minimize the risk of defaults
g. Use of a CAD/CAM system to generate engineering drawings of product designs, manufacturing routings, bills of materials, and specifications for tooling

2-17. Do you think that an intelligent process control (IPC) system can be effectively applied to a manufacturing system where production is scheduled in short runs of small batch quantities? Explain. *Note:* When the production mix represents a family of similar product designs and the batches are produced within a manufacturing cell, this is referred to as *cellular manufacturing.* A cell is a self-contained mini-factory where all operations necessary to complete a product are performed by employees who have multiple skills and can rapidly change over from one product to the next.

CHAPTER THREE

PEOPLE—THE MOST VALUABLE ASSET

QUALITY MAXIM #3: *The genius of quality lies in the informal, impromptu and inspired ways that real people solve real problems in ways that formal processes cannot anticipate.*

Just a Step on the Boss-Man's Ladder?

What is a job? Is it simply a paycheck to pay the bills or is it a reason for getting up in the morning? Dolly Parton's lyrics from her hit song "Nine to Five"[1] echo the sentiments of many workers and their relationships with their bosses as she laments that she works nine to five, with full service and devotion, but she never seems to get the credit for all her hard work. The fruits of her labors just make her boss look good and help him advance while she is held back—merely a step on his career ladder.[2] While Dolly's song was topping the charts a bankrupt tire plant on the west coast of the United States closed down, laying off hundreds of workers. Six months after the closure the company reopened under Japanese ownership, re-calling many of the former workers back to their jobs. Within weeks plant production was up to full capacity and productivity had reached levels much higher than had ever been achieved before. Long-term sustainability no longer seemed to be an issue. A television reporter interviewed some of the workers and asked them how the work environment had changed to bring about such a radical turn-around. One of the tread line operators appeared on camera and said that the biggest difference was in how the employees were treated. The old management expected the workers to "park their brains at the gate when clocking in," whereas the new leaders actively sought and sincerely listened to employee input. "They may not always respond in the way we would like, but we do feel that our ideas are seriously considered and that they [the management] are sincere in wanting to know what we think." Quality inspired managers recognize that their most valuable asset is the human resource pool—this irreplaceable repository of knowledge and wisdom that the people who create the value, and produce the goods and services, carry around in their heads, and their ability to work together and share what they know. This is a departure from the conventional view that employees are a cost center, and when cost reductions are mandated, labor—a dominant line item in the annual budget—is one of the first areas targeted for cuts. On the other hand, viewing workers as assets underscores the connection between people and an organization's profitability. The mindset shifts from trying to eliminate to striving to cultivate. Enlightened quality managers typically employ some variation of the *Theory Z* style of leadership,[3] knowing that by helping people be the best they can be, showing concern for the whole person, concentrating on the development of people as productive workers and citizens, and acknowledging that everyone has a contribution to make, the achievement of corporate goals are facilitated. In essence, QIM organizations will almost always identify the workforce (commonly given inclusive titles such as associates or partners) as valued stakeholders. When this is done policy decisions concerning change, compensation and benefits, and workforce reductions are handled very differently than in environments where employees are treated as expendable resources.

[1]"Nine to Five" was written and performed by Dolly Parton for the 1980 film comedy of the same title. This was the film debut for Parton and costarred Jane Fonda and Lily Tomlin. The title song was nominated for an Academy Award and received four Grammy nominations, winning Parton awards for Best Country Song and Best Country Vocal Performance.

[2]The music and lyrics for "Nine to Five" can be accessed at http://www.metrolyrics.com/nine-to-five-lyrics-dolly-parton.html.

[3]Theory Z, a style of leadership discussed in this chapter, places a high premium on the human component, offering in return for loyalty job security, career development opportunities, and a commitment to honoring employee needs.

LEARNING OBJECTIVES AND KEY OUTCOMES

After careful study of the material in Chapter 3, you should be able to:

1. Describe human resources as an open system and explain how the HR system supports an organization's purpose and quality goals.
2. Describe the characteristics of a high-performance work organization (HPWO) and how it differs from a traditional one.
3. Define the term *high-performing work practice (HPWP)* and discuss how HPWP supports the creation of an HPWO and QIM.
4. Discuss examples of some rewards that have been used to create within the workforce a sense of shared ownership, and how each is administered and supports the implementation of HPWPs.

(continued)

5. Discuss the difference between participatory empowerment and self-management empowerment and how these two approaches support the establishment of an HPWO.
6. Discuss the link between organizational culture and QIM and some of the major factors that a company must address prior to bringing about significant cultural change.
7. Compare the advantages and disadvantages of job rotation, job enlargement, and job enrichment with respect to the contributions each approach can make to a high-performing work environment.
8. Define the five factors that determine the intrinsic motivating potential of a job, how these factors are used to compute a job's motivating potential score (MPS), and how the MPS can be used in making personnel assignments.
9. Understand the fundamentals of effective team building and discuss the criteria that can be applied to evaluate team effectiveness.
10. Define the attributes of teaming, and explain how team decision-making differs from individual decision-making.

3.1 THE OPEN SYSTEM OF HUMAN RESOURCES MANAGEMENT

Figures 3-1 and 3-2 depict the human resource management (HRM) function as an open subsystem of the total enterprise model. The more comprehensive business system that was introduced in Figure 2-1 constitutes the environment within which HRM operates and its components include technologies, culture, constraints, knowledge, problem-solving tools, and leadership. Inputs to the HRM system include those attributes that differentiate one person from another, such as knowledge, skills, experience, attitudes, beliefs, and aptitudes. Outputs can be classified in one of two ways: those that directly impact organizational performance (such as quality and productivity) and those that directly influence human behavior (what the psychologists call the **affective domain**), such as pride and loyalty. From the open systems perspective it should be obvious that HRM spans the entire organization and cannot be relegated to a single department.

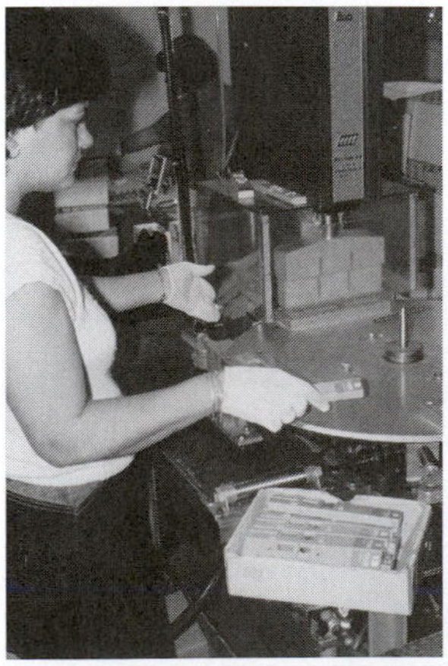

A factory worker assembles parts in a manufacturing cell. Productivity and quality depend on how management treats its workforce.

In the HR system, the transformation process is concerned with aligning individual and group behaviors with job-related tasks in a way that supports the organization's strategy and is consistent with the work design. Incentives, rewards, and training should be carefully determined so that mixed signals and conflicts are avoided and personnel are unified behind a common purpose. An emphasis that integrates the DIKW hierarchy and knowledge management with the human condition is especially important. Guidelines from a company's top leadership must be clearly understood by all concerned, and conversely, senior management should actively seek input down the chain to guard against misinterpretations and noncompliance. Decisions at all levels should be based on information retrieved from a central knowledge base that is timely and utilizes rules that can adapt to changing conditions and accommodate complex, fuzzy, and unstructured problems.

The primary objective of HRM is to improve the quality of outputs through productive behaviors.

HRM has two primary objectives: **competence management** and **behavior management**.[4] The former is concerned with enriching the quality of system inputs (i.e., building a competent workforce) while the latter focuses on the quality of process (i.e., transforming the mix of inputs into desirable behaviors). Together, both objectives are aimed at enriching the quality of system outputs.

3.2 HIGH-PERFORMANCE WORK ORGANIZATIONS

During the last decade dramatic changes have occurred worldwide in the way organizations operate and are managed. Such practices as out-sourcing; downsizing (or right-sizing); consolidations through mergers, acquisitions, and takeovers; organization flattening; broadened spans of control; high management turnover rates; and globalization have become common trends. In a climate of unprecedented technological change pressures to better protect the environment, demands for improved quality of life for employees, the desire to achieve greater returns for financial stakeholders, and pressures for greater accountability to all

[4]Wright, P. M, and S. A. Snell. "Toward an Integrative View of Strategic Human Resource Management." *Human Resource Management Review* 1 (1991): 203–25.

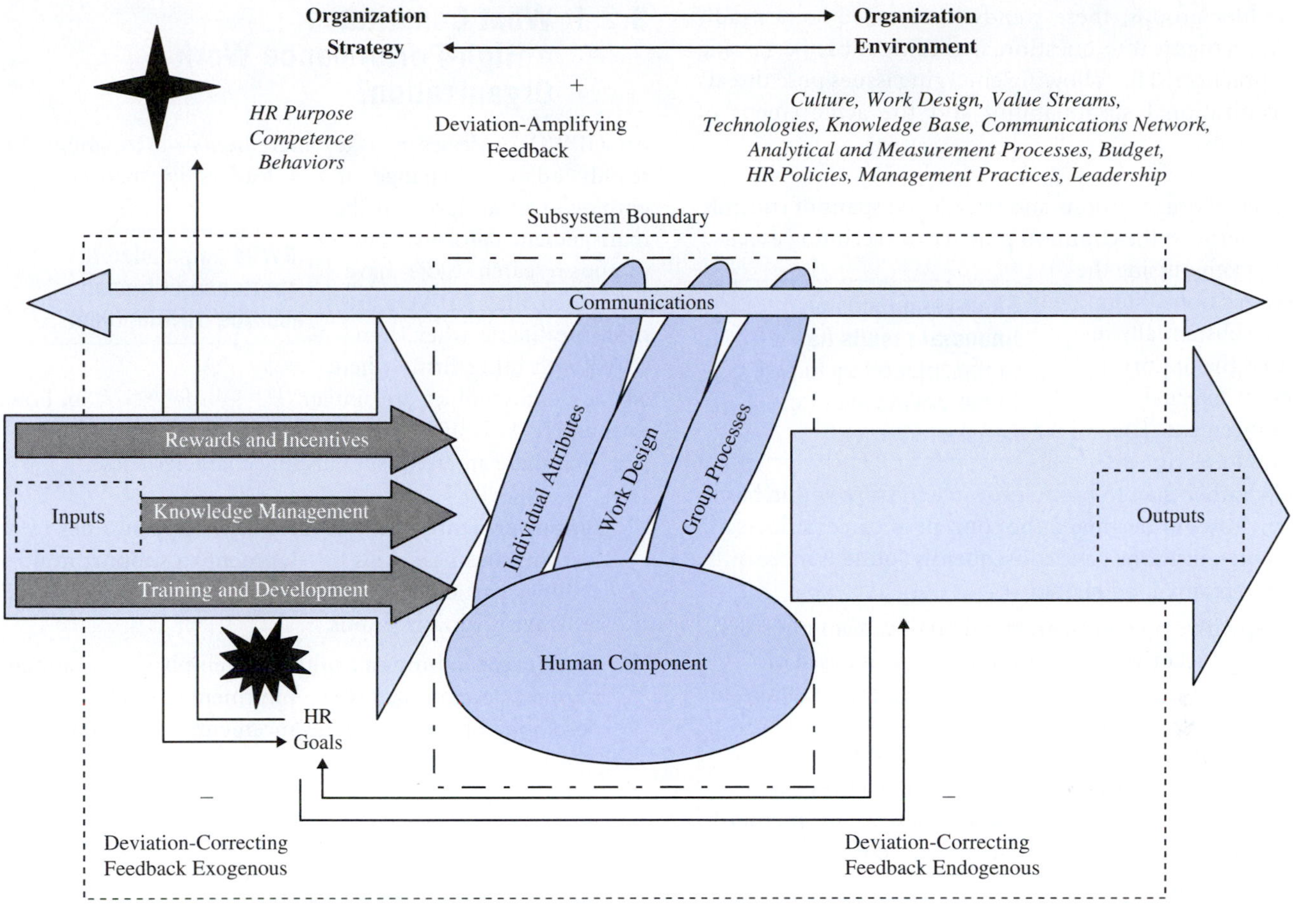

FIGURE 3-1. Human Resources System

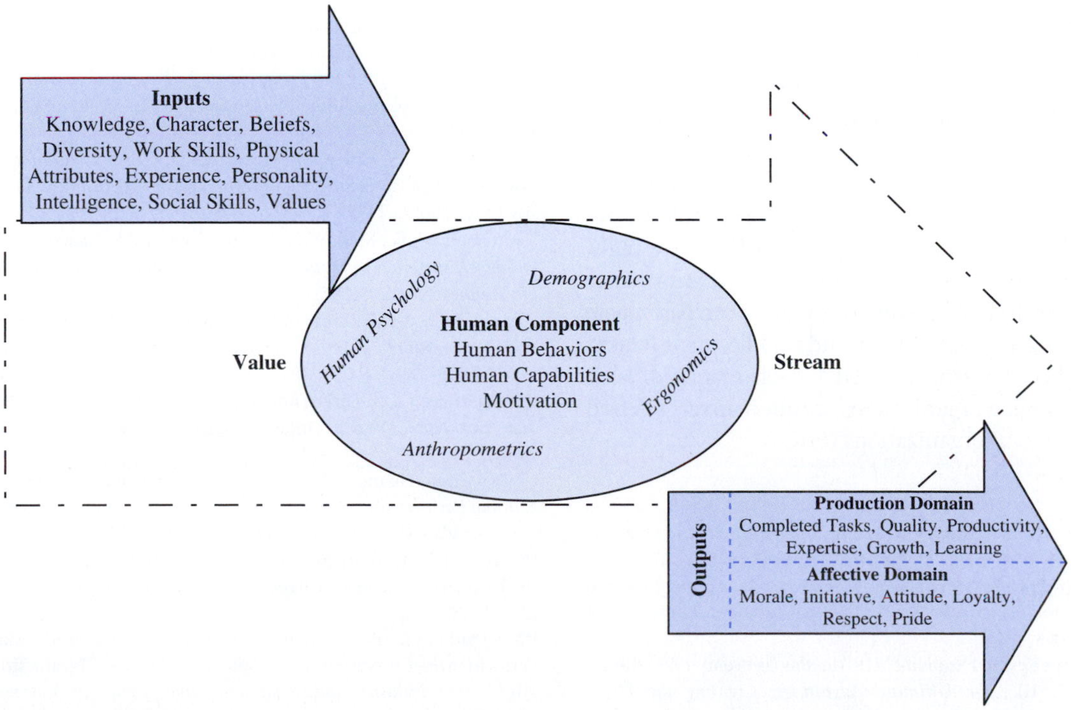

FIGURE 3-2. Human Component of Value Stream

stakeholder groups, these trends have called some traditional paradigms into question, including those pertaining to HR practices. The following emerging issues pose threats to organizational sustainability and the achievement of quality goals.

- Flattened organizations and broadened spans of control, together with the common practice of recruiting executives from outside the organization system, have substantially reduced the opportunities for internal advancement. The trend to recruit externally rather than to home-grow talent from within has in many cases undermined the concept of career ladders, and organizations have consequently found it more difficult to retain good people.

Short-term focus on financial results has led to the reluctance to invest in the development of people.

- Competitive pressures, greater demands from investors, and top executive compensation packages based on short-term financial results have forced organizations to cut spending, focus on cost containment, and divert resources away from long-term investments including the development of people.

The trend of external recruiting has reduced opportunities for internal advancement.

- Technological change, lean production methods, the growth of e-business, the use of the Internet, and globalization strategies have fundamentally changed the way that work is performed.[5]

The traditional HR functions of staffing, recruiting, compensation, and benefits are becoming decentralized and replaced by progressive approaches including high-involvement work practices, training, strategic workforce planning, and organizational continuous improvement programs.[6] QIM organizations realize that people are the key to competitiveness, and the HRM system governs how well people are deployed and utilized to achieve company goals. Increasingly, emphasis has been placed on the need for high-performance work organizations (HPWOs) where employees can participate in decision-making and supervisory and middle-management positions have been largely replaced by self-managed work teams (SMWTs).[7] Such organizations are alternatively referred to as high-involvement organizations (HIOs).[8]

[5]Lester, R. K. *The Productive Edge: How U.S. Industries Are Pointing the Way to a New Era of Economic Growth*. New York: Norton, 1998, 191–257.
[6]Morgan, John P. *HR Practices for High-Performance Organizations*. Research paper, Foundation for Sustainable Economic Development, University of Melbourne, April 2001.
[7]Gollan, P. "Finding the Keys to Flexibility." *HR Monthly* (February 1998): 10–11.
[8]Lawler, Edward E. III. *The Ultimate Advantage: Creating the High-Involvement Organization*. San Francisco: Jossey-Bass, 1992.

3.2.1 What Constitutes a High-Performance Work Organization?

An HPWO possesses two key ingredients—a capability to readily adapt to change and a leadership that engages employees at all levels in the management process. Numerous research studies have confirmed that HPWOs are more profitable when compared with other firms where all the demographics are similar.[9,10,11,12,13,14,15,16,17,18] So, how does an HPWO differ from other organizations? It would appear that there are five distinguishing characteristics.

HPWOs emphasize the importance of human resource development.

1. **Empowerment**—the creation of opportunities for learning and new skills development to support multi-skilling, participative decision-making, and the use of self-managed work teams.
2. **Investment in human capital**—an emphasis on human resource development is a commitment to learning, skills development, and career advancement; and the physical,

[9]Kling, J. "High Performance Work Systems and Firm Performance." *Monthly Labor Review* (May 1995): 29–36.
[10]Ichniowski, C. K. Shaw, and G. Prenushi. "The Effects of Human Resource Practices on Productivity: A Study of Steel Finishing Lines." *American Economic Review* 87, no. 3 (1997): 291–313.
[11]Becker, B. E., M. Huselid, P. S. Pickus, and M. Spratt. "High Performance Work Systems and Firm Performance: A Synthesis of Research and Managerial Implications." In *Research in Personnel and Human Resource Management*, vol. 16, edited by Gerald R. Ferris, 53–101. Stamford, CT: JAI Press, 1998.
[12]Applebaum, E. T. Bailey, P. Berg, and A. L. Kalleberg. *Manufacturing Advantage: Why High Performance Work Systems Pay Off*. London: Cornell University Press, 2000.
[13]Patterson, M. G., M. A. West, R. Lawthorn, and S. Nickells. "The Impact of People Management Practices on Business Performance." *IPD Issues in People Management* 22 (1997).
[14]Wood, S., L. de Menezes, and A. Lasaosa. *High Involvement Management and Performance*, paper delivered at CLMS, University of Leicester, May 2001.
[15]Betcherman, G., N. Leckie, and K. McMullen. *Developing Skills in the Canadian Workplace*. Ottawa, Ontario, Canada: Canadian Policy Research Networks, Inc., 1997.
[16]Tung-Chun, Huang. "The Effect of Participative Management on Organizational Performance: The Case of Taiwan." *International Journal of Human Resource Management* 8, no. 5 (October 1997): 677–89.
[17]Guthrie, J. P. "High Involvement Work Practices, Turnover, and Productivity: Evidence from New Zealand." *Academy of Management Journal* 44, no. 1 (2001): 180–90.
[18]MacDuffie, J. P. "Human Resource Bundles and Manufacturing Performance: Organizational Logic and Flexible Production Systems." *World Auto Industry, Industrial and Labor Relations Review* 48 (1995): 197–221.

TABLE 3.1 High Performance Work Practices

HPWO Attribute	High Performance Work Practices					
Empowerment	Sense of employment security	Self-managed work teams	Off-line cross-disciplinary and multi-level participation in problem-solving	Group participation in process improvement	Job enrichment and cellular designs places decision-making authority at operations level	Use of genuine suggestion programs
HR Development	Psychometric tests for selection in hiring and promotion	Well-developed induction training	Extensive and broad training opportunities for experienced employees	Job rotation	De-centralization of quality efforts	Commitment to families and support for child-care and special needs of working parents
Human Performance	Regular performance appraisals	Employee mentoring	On-line work teams	Good selection procedures for hiring and promoting	Performance feedback from numerous sources	Flexible job descriptions
Equitable Rewards	Comparatively high compensation	Objectives-based and profit-related incentives	Individual performace-based pay	Minimal status distinctions	No compulsory redundancies and avoidance of voluntary redundancies	Harmonized holiday entitlement
Knowledge Sharing and Learning	Broad dissemination of business plan	Company-wide understanding of performance metrics and targets	Full disclosure of financial information	Knowledge base learns quickly and is adaptive to changing conditions	Decision rules are based on information retrieved from knowledge base	Knowledge base self-organizing and maintaining

mental, intellectual, and psychological health (and improvement) of people.

3. **Systems that enable excellence in human performance**—work practices aimed at improved performance through continuous learning, with built-in mechanisms that help people perform at their maximum potentials. These systems incorporate such features as formal goal-driven appraisals, mentoring and coaching, educational incentives and assistance, standardized on-the-job training, and train-the-trainer programs.
4. **Equitable reward systems**—policies and procedures that are designed to motivate employees and acknowledge the value of individual and group contributions.
5. **Culture that values learning and knowledge sharing**—systems that facilitate knowledge transfer and learning; communicate relevant and timely information to all employees; and ensure that feedback reaches those responsible for formulating an executing strategy.

HPWOs enable people to perform at their maximum potentials.

3.2.2 High-Performing Work Practices

The term *high-performing work practice (HPWP)* has been coined to describe any management initiative that promotes an organization's transition to an HPWO. Table 3-1 shows how various HPWPs support those attributes commonly found in HPWOs, and is based on the findings from some research conducted in the United States and the United Kingdom.[19,20,21]

While there is some controversy on a precise definition of HPWP, it has been generally accepted that the design of an HPWP will include three elements: high employee involvement, aggressive human resource management involvement, and shared ownership based on an egalitarian reward

[19]Guest, D. E. "HR and the Bottom line: Has the Penny Dropped?" *People Management* 20 (July 2000): 26–31.

[20]Pfeffer, J., *The Human Equation: Building Profits by Putting People First*, Boston, Harvard Business School Press, 1998.

[21]Pil, F. K., and J. P. MacDuffie. "The Adoption of High-Involvement Work Practices." In *The American Workplace: Skills, Compensation and Employee Involvement*, edited by C. Ichniowski, T. A. Kochan, D. I. Levine, C. Olson, and G. Strauss, 1–37. Cambridge, England: Cambridge University Press, 2000.

system. Research has consistently shown that high levels of productivity, financial performance, strong employee commitment, sustained competitiveness, and innovation are observed when HPWPs are present.[22,23,24]

3.2.2.1 High Employee Involvement. The objective of high involvement is to build trust and imbue employee loyalty by creating a sense of **shared ownership**. This idea is not new and was a concept first introduced by Peter Drucker who referred to this new management–workforce relationship as management by objectives (MBO).[25] The principle behind MBO is to make certain that everyone in the organization has a clear understanding of where the organization is heading, how it intends to get there, and how they can help. This is accomplished by permitting employees to play an expanded role in planning and operating decisions; and to develop full and open communications up and down the chain of command. This is usually accompanied by empowering the workforce to utilize a high degree of discretion in performing their daily tasks. The HRM system supports high employee involvement with practices that facilitate communications, promote transparency, and encourage participation. In developing HPWPs an organization should consider the following fundamental building blocks.

- **Transparency**—the full disclosure of organizational financial information, performance, and strategy. This includes providing the entire workforce with unedited versions of the business plan, goals, objectives, and operating targets. Employees need to believe that the leadership is truthful and that there are no "hidden agendas." If each person is aware of where the organization is headed, his or her individual role in helping it get there, and all relevant issues (e.g., market pressures, budgetary constraints, impending threats and opportunities, etc.), a shared sense of "we are all in this together" can be created. Employees also need to view the organization from systems and stakeholder perspectives, and learn who has been identified as stakeholders, the needs of each stakeholder group, and how the organizational objectives support those needs. As employees gain a greater strategic awareness of the business, they begin to develop empathy for those issues that the senior leadership grapple with continuously, and can better understand the decisions that are made. When they are also given the opportunity to participate in decisions, the workforce moves from the level of understanding to ownership. Nothing can equal *pride of ownership* in its ability to pique enthusiasm and evoke loyalty.

An HPWP requires transparency and free and open communications.

- **Collegiality**—the creation of workforce unanimity through association. A staff association's membership that is inclusive of all organizational levels can be an effective vehicle for breaking down barriers. The staff association can be an effective conduit for issues, concerns, ideas, and information to be exchanged. It provides a convenient forum for management to genuinely seek input from the workforce and for the workforce to gain a legitimate hearing of grievances. As a formal organization the association can help to build trust, bridge gaps, and create a companywide community spirit.
- **Feedback**—the recognition of employees as stakeholders and a genuine desire to respond to their needs. High involvement means recognizing the importance of the human resource component, a process that starts by identifying an organization's personnel as a key stakeholder group. As with any stakeholder group a QIM company will strive to continually keep abreast of employee needs and actively solicit feedback on how well those needs are being met. Internal surveys, such as *360-degree feedback*, can be a valuable tool in collecting this information. The 360-degree feedback, also known as multirater or multisource feedback, was originally used successfully in the military and gets its name from the fact that the feedback comes from all around the employee—analogous to the 360 degrees that create a perfect circle. The feedback comes from subordinates, peers, managers, self-assessment, and if appropriate from external sources such as customers, suppliers, or other stakeholders. The results from a 360-degree feedback can be used by the employee to plan his or her own training and development, and are used by some organizations in making promotional or merit pay raise decisions.

HPWPs make it easy for employees to make improvement suggestions.

- **Employee input**—the use of formal suggestion schemes. Some of the best ideas for improvement can originate with the people who have the greatest technical knowledge. HPWPs recognize this and not only encourage fresh ideas, but also invent ways to make it easy for employees to make suggestions. For any suggestion scheme to be effective workers must believe that their ideas will be evaluated with integrity and that they will receive proper recognition if any of their ideas are adopted.
- **Teaming**—leveraging intellectual capital through collective wisdom. Teams are more than just a collection

---

[22]MacDuffie, John Paul. "Human Resource Bundles and Manufacturing Performance: Organizational Logic and Flexible Production Systems in the World Auto Industry." *Industrial and Labor Relations Review* 48 (January 1995).

[23]Ashton, D., and J. Sung. *Supporting Workplace Learning for High Performance Working.* Geneva, Switzerland: ILO.

[24]Blysshe, Simon, and Wendy Hirsh. "Career Interventions in the Workplace." *Briefing Papers, Summary of Thematic Literature Review for the Consultative Workshop,* University of Derby, March 2006: Joint Sponsors—National Institute for Careers Education and Counseling; Centre for Guidance Studies.

[25]Drucker, Peter F. *The Practice of Management.* New York: Harper and Row, 1954.

of people placed in a group and told to work together. A team is established for a specific purpose and the members of the team are expected to work together to achieve a common set of goals. In the early days of the quality movement quality improvement teams, called **quality circles**, were expected to meet periodically—usually weekly—and collectively come up with ideas to improve the quality of their work. Quality teams can still be found as an integral component of formal quality programs (e.g., Six Sigma improvement efforts) and are an essential component in HPWPs. With the maturity of quality programs (late second generation to third generation) organizations often adopt flat organic structures that support the creation of self-managed (or self-directed) teams having discretionary powers, cross-functional teams formed to solve problems affecting the performance of primary or support value streams, and "kaizen blitz" teams that are established for the ad hoc purpose of identifying opportunities for quick improvements. We discuss team building in further detail in Section 3.6.

3.2.2.2 Aggressive Human Resource Management Involvement. In a high-performing QIM organization there is an important principle that guides policies. To maximize the value of human capital HR policies must recognize that the source of improved productivity and innovation resides within the creative spirit of the people who populate the system. The key to unlocking human potential begins with the recognition that the workforce represents a pool of valuable assets, and the return on those assets is directly proportional to how much the organization is willing to invest to develop new skills; stimulate brains and thought processes; promote learning; and ensure the safety, well-being, and quality of work life for all employees. It is interesting to contrast this perspective with the traditional school of thought that treats the labor force as a cost center that should be reduced at worst and minimized at best. A typical HPWP work environment will showcase four important features. It will be goal-driven, embrace progressive recruitment practices, engage in comprehensive training, and be committed to continuous improvement.

> **High-performing organizations treat human resources as a profit center, not a cost center.**

- **Performance reviews and goal-setting.** Feedback is critical to improvement. Just as this principle applies to the value stream it is also true for human performance. Understanding job expectations and performance metrics is fundamental. Periodically each employee should have an opportunity to meet with his or her supervisor to review performance. In today's business environment it is desirable that such meetings take place monthly or quarterly so that problems can be dealt with as they occur and employees have the opportunity to take corrective action before unsatisfactory behaviors are repeated. If such frequency is not practicable, then reviews should be conducted at least annually. During these review sessions the worker and manager can take the opportunity to discuss and revise goals and, if appropriate, discuss feedback from peer workers, customers, and other relevant stakeholders. The review sessions should be conducted in the spirit of genuinely trying to correct weaknesses to help the employee improve. Above all, they should never be punitive. The guiding principle is to identify and praise those things the employee has done very well (stellar performance) and to be slow to criticize those areas that have been below par. The object and tone of these sessions should be on improvement—developing the employees so that they can achieve their fullest potential—be the best they can be. To be effective these sessions should not be promoted as "appraisals" or "evaluations." Terms such as *meetings, sessions,* or *consultations* are preferred, as such descriptions convey a positive message. In addition to discussing performance issues relative to the current job, some organizations also use these meetings to discuss career goals and the alignment of those goals with opportunities for advancement within the company. Such discussions can help identify relevant training or education needs.
- **Progressive recruitment practices.** Filling vacant positions should be consistent with the organization's needs as specified by business strategy. In evaluating candidates in a management-by-fact culture (e.g., QIM) formal assessment tools are applied to ensure that competencies match needs, consistent and equitable differentiation between applicants, and to maintain a balanced and diversified workforce.
- **Comprehensive training programs.** High-performing organizations require a well-trained workforce that can flexibly adapt to changing requirements. The modern trend has been away from specialization (which characterized the machine age) and toward multiskilling (providing workers with the skills to perform numerous tasks). Multiskilling is especially important in lean factories where self-contained manufacturing cells must manage the production for an entire family of products produced in small batch quantities. In addition to breadth of skills, in-house training programs focus on technical depth and updating. Continual training is an important component of process improvement, and group training is used to reduce variability between workers and identify best methods. In addition, the practice of assigning a worker to train a new employee has been abandoned in favor of structured induction training, which is consistently applied to all new hires. Many high-performing organizations have added an important nontraditional component to in-house training—mentoring. On technical matters, subject matter experts are relied upon to coach less experienced personnel. In nontechnical matters, such as leadership and teaming, an employee can be assigned a mentor from anywhere in the company or in some cases from an external source.

> **HPWPs are based on comprehensive training programs.**

- **Redesigning work for improved performance.** Self-managed or self-directed work teams are not only responsible for operational expediency but are also empowered to implement changes for continuous improvement. Therefore employees are able to make changes in work methods to improve productivity and efficiency. In addition, QIM organizations have structured procedures for achieving quantum breakthrough improvements and for introducing new technologies. As major changes are anticipated, all personnel affected are brought into the process at the earliest practicable opportunity. That way, workers are fully involved in the decisions relative to the transition. To guide the design of work, quality standards are used as benchmarks. These include such widely accepted guidelines as the ISO 9000 and the business excellence models, such as the Malcolm Baldrige National Quality Award and the European Foundation for Quality Management.

One way to create a sense of employee ownership is to allow workers to become investors—financial stakeholders in the business.

3.2.2.3 Shared Ownership.

Most organizations are in business to make a profit. As a tool for creating a strong sense of belonging and commitment, nothing can compare to the power derived from allowing employees to become financial stakeholders. Financial rewards such as profit sharing, stock options, and other performance-related schemes give explicit recognition to the contributions that individuals and teams make toward the achievement of organizational objectives. It is crucial that all rewards be perceived to be egalitarian and rewards need not be financial. Other benefits, often referred to as "perks," convey an important message: *The organization cares about its people and values what they do.*

Nonfinancial benefits include such things as the free use of facilities, family-friendly services, and eligibility for certain privileges (e.g., reserved parking spots). Some benefits can be extended to spouses and children. Benefits may be permanent or be granted for a limited period (e.g., employee of the month). The total benefit package can help employees better identify with the organization and with each other, and to feel like they are part of a community and belong to something worthwhile. In short, the rewards structure heightens the importance of the work that people do and increases their self-esteem. Combined, these positive feelings translate into improved quality and better overall organizational outcomes. The following practices are examples of the types of rewards that support HPWOs.

- **Performance-based pay (for some or all employees).** The best form of compensation is performance based, provided the system is designed around meaningful and measurable objectives. Setting performance targets is an idea that is consistent with the empowered and knowledge worker concepts. As a caveat, systems perceived to be the most equitable will have built-in safeguards to protect against the human element. When performance rewards are left to the judgment of supervisors personal biases can potentially favor some and short-change others. To prevent this, the best plans are those with the most quantifiable criteria—for example, meeting productivity, quality, delivery, and cost goals can normally be measured with ease and may be less controversial than the subjective determination that employee A should get twice the raise of employee B. We learned in Chapter 2 that strategy flows from the top and is translated into goals, then objectives, and finally targets with increasing specificity the closer one gets to the actual human–work interface. If all goals, objectives, and targets are aligned, then it makes sense to base at least some portion of each employee's compensation on achievement. This should apply to everyone, at every level—no matter what the job. However, it is important that all goals-objectives-targets be achievable; otherwise the incentive is lost. Individuals need to believe they can succeed or they will stop trying and what was designed to be a motivation scheme can then have exactly the opposite effect. Performance-based pay should also include team efforts, which can take into account broader goals and focus on the importance of a team's collective contributions to the success of the entire enterprise. A variation of profit sharing is called **gainsharing**, which is usually awarded to a team based on increased productivity or decreased costs. With gainsharing there is no direct cost in implementing the plan. If costs are reduced a share of that reduction is passed on to the employees responsible. If team productivity improves, a portion of the resulting profits are passed on to the members of the team. If there are no documented improvements the team receives nothing from the gainsharing allocation.

Goals, objectives, and targets must be achievable; otherwise employees will stop trying and the incentive is lost.

- **Stock options (for some or all employees).** Stock options provide employees with the right to purchase company stock at a given price during a certain period of time in the future. Usually the price of the option, called the grant price, is set to the market price of the stock at the time the option is issued and normally lasts for 10 years. As the market price increases above the grant price, the option becomes more valuable. If the market price drops below the grant price, the option is worthless. As an encouragement to retain employees, some companies will establish a 4- or 5-year required vesting period before personnel can execute their options. **Stock options**, more than profit sharing, create a strong emotional tie between employees and a company. There is no more powerful way to create a sense of ownership than to permit employees to actually *own* part of the company by becoming shareholders. Stock options can effectively align the interests of employees with those of the company.
- **Flexible job descriptions.** A job description contains a complete description of a job or position, including the tasks required, prescribed methods, the purpose and responsibilities of the job, how the job relates to other jobs and the value stream, the qualifications needed for the job,

and any other expectations. It is important to keep job descriptions current and flexible. An inflexible job description can inhibit innovation and experimentation. An HPWP job description (the formal written word) will empower workers so they can do their best and encourage them to test new ideas (the informal spoken word). The reality is that the workplace is dynamic and jobs are constantly changing due to personal growth, organizational reengineering, or the introduction of new technologies. Flexible job descriptions enable employees to spread their wings, grow their positions, and learn how to contribute more substantially.

An HPWP job description is flexible, empowering, and encourages fresh ideas.

- **Flexible working arrangements.** Many HPWPs accommodate flexible working arrangements in the HRM system. For example, a working mother may need to drop small children at school before coming to work, so her employer allows her to *flex* her eight-hour workday and work the hours from 9 a.m. to 6 p.m. instead of having to rigidly comply with the normal operating hours of 8 a.m. to 5 p.m. With the challenges presented by two-working-parent households this practice, called *flex-time*, is increasing in popularity. Similarly, some companies can permit employees to choose a work location of choice—even to include their own homes. With the ready access to the Internet, if a person's primary work responsibilities involve the computer, it may make little difference where the computer is situated. Working from a computer terminal at a remote location is called **telecommuting**. There is also a trend that has proven to be attractive for people who, due to personal circumstances, cannot work full time. This is called **job share** and involves an employer hiring two people to fill a single vacancy. The downside to this approach is finding two compatible employees to share the job; when a good match is made, however, the concept has worked well. Employees have reported improved job satisfaction and more positive attitudes toward life in general when flexible working arrangements are available.[26]
- **Job rotation, job enrichment, and job enlargement.** Routine and boring work can be the source of declining morale. As SMWTs are given the responsibility for managing their respective operations and the members of these teams are trained in multiple skills, job rotation, job enrichment, or job enlargement may be used to make work more interesting. These practices are described more fully in Section 3.6.
- **Family-friendly policies and practices.** HPWPs are most effective when they take into account the issues that are the most important in the work and private lives of the people who do the work. This is accomplished through policies and practices termed **family-friendly**. These are intended to help working parents balance personal and family responsibilities with work. In addition to the flexible work arrangements discussed previously, family-friendly policies include maternity (and paternity) leave benefits, allowances for personal and sick time, breastfeeding arrangements, on-site child care, and counseling referral services.

Fringe benefits can make the difference between places where people want to work and those that are undesirable.

- **Nonpay (fringe) benefits.** The provision of nonpay compensation, sometimes called **fringe benefits**, can make the difference between creating a desirable place where people would want to work (and work hard) and an undesirable one. Fringe benefits are normally not taxed. Examples are contributions paid by the employer to retirement, such as pension or 401(k) plans; employer-paid health insurance; low-priced or free meals in a company canteen; on-site sports or exercise facilities; counseling services; child care; company-operated buses or public transport subsidies to get employees to and from work; reserved and convenient parking; ability to take computer, photographic, or other equipment home for private use; housing assistance through low-cost mortgage loans; noncash gifts for reasons not connected to the job (e.g., Christmas, retirement, new baby, wedding, etc.); and provisions for employees who are disabled. One employer in California, for example, provides a pickup and delivery service for employees when their cars need servicing—the company takes care of making the appointment, taking the car to the service facility, and returning it to the company parking lot at the end of the day.

Some workers do not necessarily respond positively to empowerment.

3.3 WORKFORCE EMPOWERMENT

Achieving a high-performing culture is essential to the principles of QIM in which the work environment must act as an enabler—helping workers do their best and learn new skills by tapping human tacit knowledge, leveraging emotional capital, and channeling new energies into enhanced performance.[27] We have seen that in a high-performing QIM company the intellectual content of job responsibilities expands and workers routinely exercise discretion, make decisions, and solve problems as they go about their daily tasks.

[26]"What Is a Family-Friendly Company?" *Mothering* 128 (January/February 2005). http://www.mothering.com/articles/growing_child/family_society/family-friendly-company.html.

[27]Detert, J. R., R. G. Scroeder, and J. J. Mauriel. "A Framework for Linking Culture and Improvement Initiatives in Organizations." *Academy of Management Review* 25, no. 4 (2000): 850–63.

We can also note that, while there are countless and proven advantages to the HPWO culture, there is also a downside. Some workers do not respond well to the job-related stress levels imposed by the new expanded set of responsibilities and accountabilities. Jobs that could once be performed on an individual basis, and in isolation, now require interaction with coworkers and extensive communications. Some workers simply prefer not to be empowered, and that is why some companies have resorted to psychological testing to help match personality types to job responsibilities. We discuss this in further detail in Section 3.6.

There are two approaches for empowering a workforce. A **participatory empowerment** strategy gives employees partial decision-making authority while management maintains veto powers. This means that management retains the right to judge the relative merits of any employee inputs—including the rejection of ideas, the refusal of requests, or the countermanding of decisions made. Formal suggestion schemes and the use of quality circles are examples of participatory approaches.

The second is a **self-management empowerment** strategy and is the approach favored by QIM. Autonomy is given to front-line employees using management structures such as SMWTs, enabling them to follow through on their recommendations up to and including implementing changes.[28] Transferring control to the level where the work is actually performed integrates all three subprocesses of the HR transformation—individual attributes, work design, and group processes.

Workforce empowerment represents a fundamental cultural transformation in most organizations and requires dramatic changes in the way that managers and workers interact. Management has to be genuinely prepared to accept the consequences of an empowered workforce, and employees must be prepared to assume the new intellectual dimensions that will be added to their job responsibilities. Top management commitment to empowerment is not sufficient to bring about the transformation. If ordinary employees are to become knowledge workers, a culture that facilitates learning is essential.

3.4 CULTURE AND THE VALUE STREAM

Organizational culture is holistic, historically determined, and socially constructed; involves beliefs and behaviors; exists at a variety of levels; and manifests itself in a variety of ways that affect the work environment.[29] Culture is not something that an organization *has*; rather, it is the essence of what the organization *is*,[30] and has been defined as "*that bundle of signs, symbols, beliefs, traditions, myths, ways of thinking, speaking and doing which characterize the life and behavior of a given group of people.*"[31] A culture that supports QIM must be defined by and support the constant attainment of stakeholder satisfaction through an integrated system of tools, techniques, and training.[32]

Since an organization's culture is woven around its people, reengineering the human resource system will lead the cultural transformation. Workers who are assigned to job improvement and problem-solving teams will discover the need to learn new methods of job analysis, assessment, recruitment, and socialization.[33] A new paradigm for human resource management—one based on quality consciousness, HPWPs, and the alignment of culture with quality objectives—will emerge.

Cultural change is led through the redesign of the human resource system.

Cultural change is a slow, painstaking process that evolves over a long protracted period, and will take shape as people are gradually won over to a new way of thinking and working, and begin to accept a new sense of what is normal. Old habits and conventions die hard, and cultures are particularly resistant to changes imposed on them from the outside. It is especially important that employees at all levels participate in the change management process, as culture directly impacts the value stream. Motivation, orientation to work, and control (all of which are cultural dimensions) affect the quality of working life, work design, and productivity. The human response to HRM policies will differ by individual.

Nevertheless, in the information age HRM is only partially accountable for good or bad work practices. Individual values and attitudes can be profoundly influenced by how information is communicated and interpreted, and the system by which knowledge is captured, archived, and transferred. A culture is based on universally accepted interpretations, understandings, values, and behavioral norms.[34]

In Chapter 2 we learned that the second law of thermodynamics teaches that sustainability depends on how well a system can adapt to changes in its environment. If it cannot adapt fast enough it will eventually self-destruct and die. With respect to HRM this phenomenon says that in order to

---

[28]Dimitriades, Zoe S. "Total Involvement in Quality Management." *Team Performance Management* 6, no. 7/8 (2000): 117–21.

[29]Hofstede, G., B. Neuijen, D. D. Ohayv, and G. Sanders. "Measuring Organizational Cultures: A Quantitative Study Across Twenty Cases," *Administrative Science Quarterly* 35, no. 2 (1990): 286–316.

[30]Hawkins, P. "Organizational Culture: Sailing between Evangelism and Complexity." As cited in "A Framework for Linking Culture and Improvement Initiatives in Organizations," by J. R. Detert, R. G. Scroeder, and J. J. Mauriel. *Academy of Management Review* 25, no. 4 (2000): 850–63.

[31]NAGCELL (National Advisory Group for Continuing Education and Lifelong Learning). *Creating Learning Cultures: Next Steps in Achieving the Learning Age*, Second Report, 1999. As cited in *Building a Learning and Training Culture: The Experience of Five OECD Countries*, by Peter Kearns and George Papadopoulos. Kensington Park, SA: Australian National Training Authority, 2000.

[32]Cameron, K. S., and D. A. Whetten. "Organizational Effectiveness and Quality: The Second Generation." In *Higher Education: Handbook of Theory and Research*, edited by J. R. Samart. New York: Agathon Press, 1996.

[33]Klimoski, R. J., and R. G. Jones. "Suppose We Took Staffing for Effective Group Decision-making Seriously?" In *Team Decision-Making Effectiveness in Organizations*, edited by R. A. Guzzo and E. Salas. San Francisco: Jossey-Bass, 1994.

[34]Jackson, Susan E., and Randall S. Schuler. "Understanding Human Resource Management in the Context of Organizations and Their Environments." *Annual Review of Psychology* 46 (1995): 237–71.

survive, it is necessary that the system keep abreast of changing expectations and roles of all its players. Even the compatibility requirements for membership will be shaped as the work culture evolves, as it is essential that the culture and its individual members remain a good fit. Employees who feel like misfits are likely to be unproductive, and such disgruntled individuals can also tear down the morale of those around them. In a high-performing QIM organization recruiting practices balance the importance of technical skills with a candidate's potential to produce, learn, and grow in the relevant work and organizational cultures.

3.5 MOTIVATION AND WORK DESIGN

It has been said that if an organization wants its workers to do a good job, then its management has to give them a good job to do. On the surface this seems like sage advice; however, what constitutes a *good* job can be controversial. Underlying best quality practices has been the assumption that, given the option, humans will generally prefer jobs that provide greater autonomy, the use of intellectual capital, and increased responsibility for outcomes. That is, it has been widely accepted that people are more motivated by jobs that allow them to have greater self-control, more decision-making authority, and imbue a personal ownership for results. However, there is some contradictory evidence that suggests this assumption may be a myth, possibly contributing to some of the failed TQM implementations during the1990s,[35] and that some workers regard repetitive, routine work to be less stressful and more desirable than a broader scope of responsibilities that require decision-making and creativity.[36] Notwithstanding, three approaches have been used to motivate workers by trying to make work more interesting and challenging. These are job rotation, job enlargement, and job enrichment.

3.5.1 Job Rotation

Figure 3-3 illustrates the concept of **job rotation**—a practice that rotates personnel through a number of related tasks. This technique not only helps avoid the monotony of repetitive work, but it also provides management with flexible capacity. For job rotation to work, each employee within a production cell must learn how to perform every job required. As a consequence schedules can easily be flexed to accommodate cases of illness, vacations, or production fluctuations. Figures 3-4 and 3-5 illustrate how job rotation might be applied in a medical doctor's office and in a restaurant, respectively.

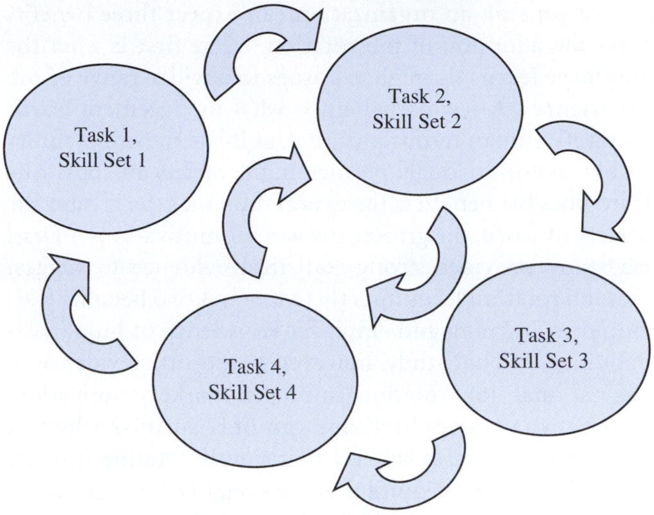

FIGURE 3-3. Job Rotation

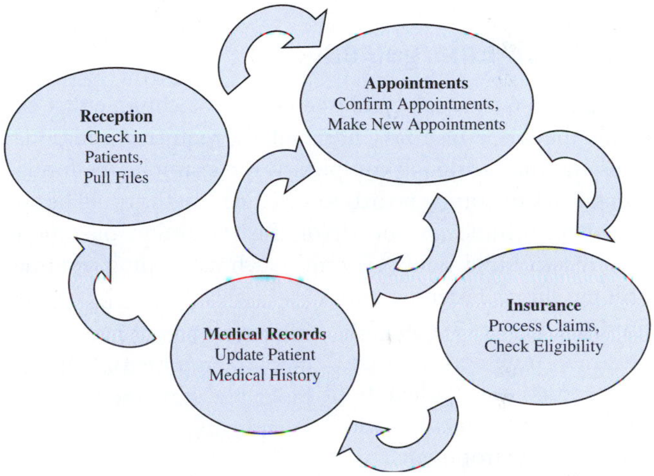

FIGURE 3-4. Job Rotation in a Doctor's Office

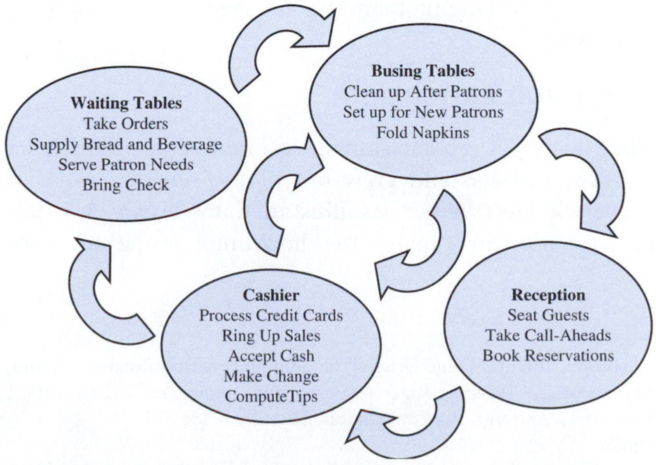

FIGURE 3-5. Job Rotation in a Restaurant

---

[35]Hartman, Sandra J., Anthony L. Patti, and Joseph R. Razek. "Human Factors Affecting the Acceptance of Total Quality Management." *International Journal of Quality and Reliability Management* 17, no. 7 (2000): 714–29.

[36]Juran, J. M., and Frank M. Gryna. *Quality Planning and Analysis (From Product Development Through Use)*. 3rd ed. New York: McGraw-Hill, 1993.

In general, an organization can expect three benefits from the adoption of job rotation.[37] The first is what the employee learns through exposure to a wider range of job experiences. A second benefit is what management learns about its human resource capital, as it has the opportunity to observe individuals performing a variety of jobs. The third possible benefit is the expectation that the greater the variety of work, the greater the worker motivation. A Danish study provides strong statistical evidence to suggest that job rotation leads directly to the first two benefits (i.e., employee learning and employer knowledge of human capabilities).[38] That study, however, found little evidence to suggest that job rotation improves worker motivation. Without further research one can only surmise why this might be the case. It is likely that simply rotating workers through jobs that require the same relative level of skill—even if the skills required are different—does not necessarily make the work more interesting. The other two work design tools—job enlargement and job enrichment—have a much better chance of increasing the motivating potential of jobs.

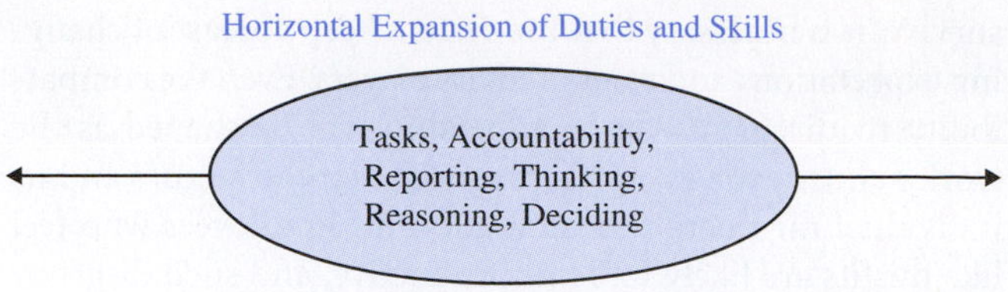

FIGURE 3-6. Job Enlargement

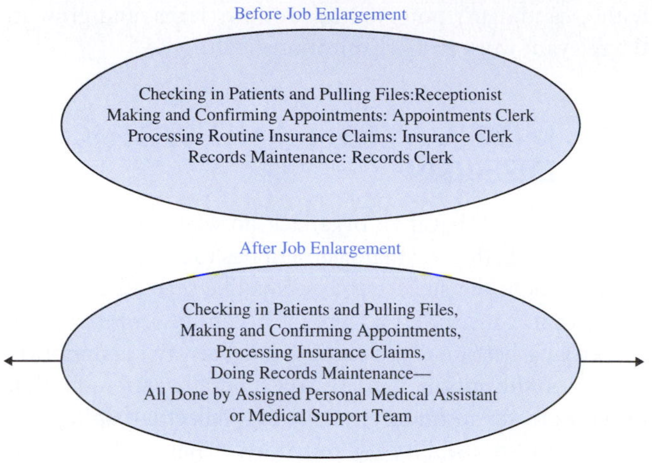

FIGURE 3-7. Job Enlargement in a Doctor's Office

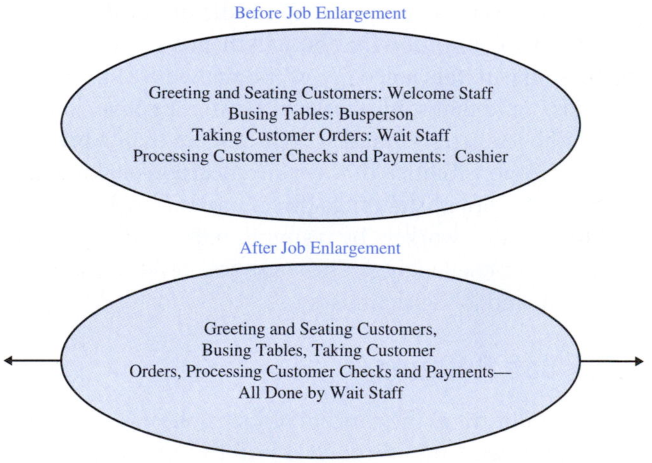

FIGURE 3-8. Job Enlargement in a Restaurant

3.5.2 Job Enlargement

As Figure 3-6 shows, **job enlargement** is a technique that expands the scope of a task horizontally, requiring additional skills and increasing job complexity. For example, the fundamental task of cutting boards to specified lengths could be enlarged to include any or all of the remaining operations required to build bookcase units, such as sanding, routing, drilling, gluing, fastening, and finishing. Figures 3-7 and 3-8 illustrate how job enlargement could be applied respectively to the medical and restaurant environments.

Job enlargement has motivating potential since tasks are increased in complexity.

Worker motivation through job enlargement is possible. Representing the opposite end of the spectrum from specialization which creates repetitive work, job enlargement expands the job. Workers are able to perform a number of tasks, sometimes being able to actually experience the pride in observing firsthand the finished part, component, or product.

3.5.3 Job Enrichment

The concept of **job enrichment** as a motivator originated a half century ago and grew out of the research work of Frederick Hertzberg.[39] As illustrated by Figure 3-9, job enrichment can expand the horizontal scope of work content; however, its primary emphasis is on expanding the vertical, or intellectual, requirements. Thus, job enrichment will make an individual worker responsible for making decisions, solving problems, and using judgment without the need for direct supervision. Typically, the worker is accountable for the quality of outcomes, responsible for communicating production status, and for alerting management when problems arise that require higher-level attention. Figures 3-10 and 3-11 illustrate how the medical and restaurant jobs have vertically expanded to require greater intellectual capacity.

---

[37]Eriksson, Tor, and Jaime Ortega. "The Adoption of Job Rotation: Testing the Theories," *Working Paper 04-3*. Aarhus, Denmark: Department of Economics, Aarhus School of Business, 2004 (ISSN 1397-4831).

[38]Ibid.

[39]Herzberg, F. "One More Time: How Do You Motivate Employees?" *Harvard Business Review* 46, no. 1 (1968): 53–62.

What is the process for enriching a job? How would one go about improving the motivating potential of work? Hertzberg suggested that job designers follow seven guiding principles.

1. Remove controls.
2. Increase personal accountability.
3. Assign work in complete units with clear starting and ending points.
4. Grant authority and freedom.
5. Provide periodic reports directly to workers—not only to supervisors.
6. Introduce new and difficult tasks.
7. Assign specialized tasks—thus encouraging the development of expertise.

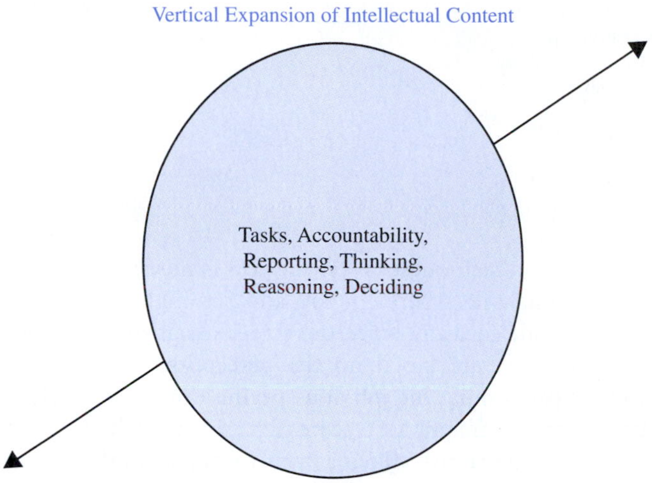

FIGURE 3-9. Job Enrichment

Although Hertzberg did not use the term *empowerment* the concept is implicit in his fourth principle, and job enrichment has become an important tool for empowerment. It is therefore essential that the organizational culture be ready and that management is attuned to the specific needs of individual employees. An ill-conceived or executed job enrichment program can unwittingly have the opposite effect of what was intended. Rather than motivating the worker, additional duties (that were previously the domain of managers and supervisors) can intimidate employees who are ill-prepared or unwilling to embrace new responsibilities or accountabilities. These are issues that should be kept in mind when recruiting new personnel, designing training programs, and inculcating values into the culture's consciousness.

A *good* job has been defined as one where the level of enrichment has, for a particular employee, achieved an operational match between the job's motivating potential score (MPS) and the individual's growth needs strength (GNS).[40] The MPS is a measure of the intrinsic motivating potential of work and is calculated using the following equation.

Motivating Potential Score (MPS)

$$MPS = \frac{SV + TI + TS}{3} \times A \times F \qquad \textbf{3-1}$$

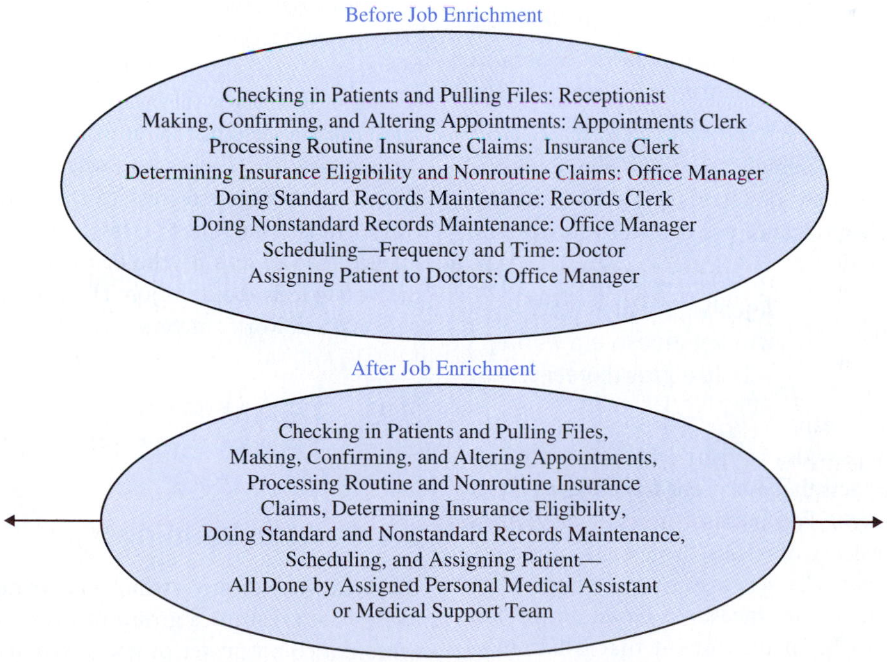

Before Job Enrichment
Checking in Patients and Pulling Files: Receptionist
Making, Confirming, and Altering Appointments: Appointments Clerk
Processing Routine Insurance Claims: Insurance Clerk
Determining Insurance Eligibility and Nonroutine Claims: Office Manager
Doing Standard Records Maintenance: Records Clerk
Doing Nonstandard Records Maintenance: Office Manager
Scheduling—Frequency and Time: Doctor
Assigning Patient to Doctors: Office Manager
After Job Enrichment
Checking in Patients and Pulling Files,
Making, Confirming, and Altering Appointments,
Processing Routine and Nonroutine Insurance
Claims, Determining Insurance Eligibility,
Doing Standard and Nonstandard Records Maintenance,
Scheduling, and Assigning Patient—
All Done by Assigned Personal Medical Assistant
or Medical Support Team

FIGURE 3-10. Job Enrichment in a Doctor's Office

[40]Hackman, J. R., and G. R. Oldham. *Work Redesign*. Reading, MA: Addison-Wesley, 1980.

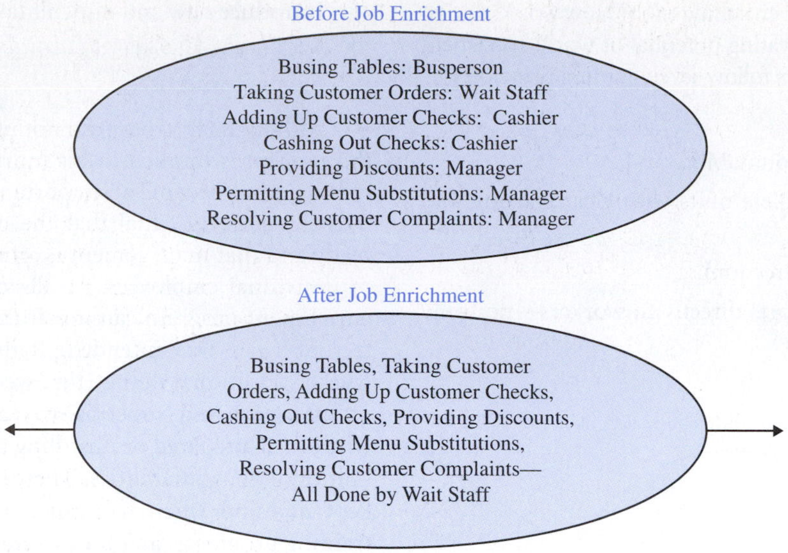

FIGURE 3-11. Job Enrichment in a Restaurant

where,

- SV = Skill Variety: the degree to which a job requires different skills, abilities, or talents.
- TI = Task identity: the extent to which a job requires completion of a whole and identifiable unit of output.
- TS = Task Significance: the degree to which the job impacts the lives of other people, the organization, and the environment.
- A = Autonomy: the freedom the worker has to determine pace, sequencing, and the work methods.
- F = Feedback: the degree to which the worker obtains information on performance, from personal observation, supervisors, and other sources.

Jobs with a high MPS generally result in higher job satisfaction and productivity than jobs scoring low. **Skill variety**, **task identity**, and **task significance** are related to the specific task and address the questions, *Why am I doing this job, is it important, and does it allow me to utilize my talents?* Taken together they provide the worker with a sense of "meaningfulness." **Autonomy** provides workers with a sense of "responsibility"—an awareness that *others are depending on them, so they had better do it right!* **Feedback** satisfies the individual's personal need for "knowledge of results," begging the question, *Did I get it right, and if not, what do I need to do differently?*

A job with a high MPS will not motivate a worker with low growth needs.

The five characteristics are measured based on the Job Diagnostic Survey (JDS)[41], an instrument that is administered to workers who are performing a particular job. The response for each question on the JDS is measured using a seven-point scale, where 1 = low and 7 = high. The MPS is only a crude measure since the scores assigned to the five characteristics are based on the perception of a specific worker performing the job at a specific time. Two workers can generate different MPS scores for the same job. It should also be noted that the MPS is a measure of *potential* and does not mean that every job with a high MPS will motivate every worker performing it. To get the best results a job's MPS should be matched with a worker's growth needs.

Industrial studies have found that workers with high growth needs respond to jobs with a high MPS and those with low growth needs do not.[42] Growth needs are determined by having an employee fill out a growth needs strength (GNS) questionnaire designed to measure individual preferences concerning likes and dislikes, and what constitutes the "ideal" job. When a person is matched to a job, the job characteristics (MPS) and the employee's growth needs (GNS) combine to influence the worker's psychological state, and specifically, how the worker feels about the job. These feelings then translate into personal and work outcomes. This is illustrated in Figure 3-12.

3.6 TEAM BUILDING AND TEAM-BASED DECISION MAKING

3.6.1 Team Building

Unlike a work group—which is a group of people working together—a team is a group of people working together to achieve a common set of goals. When the group can, at its discretion, develop and execute its own methods to achieve the common goals, the team is a **self-managed work team**

[41]Hackman, J. R., and G. R. Oldham. *The Job Diagnostic Survey: An Instrument for the Diagnosis of Jobs and the Evaluation of Job Redesign Projects.* Technical Report No. 4. New Haven, CT: Yale University Press, 1974.

[42]Hackman and Oldham, *The Job Diagnostic Survey.*

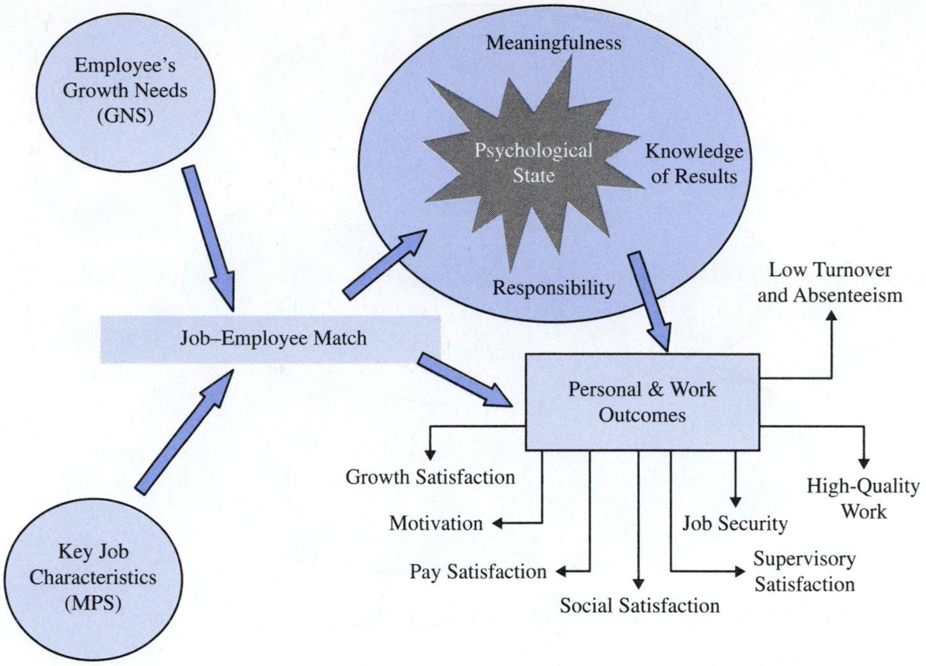

FIGURE 3-12. Job Motivation

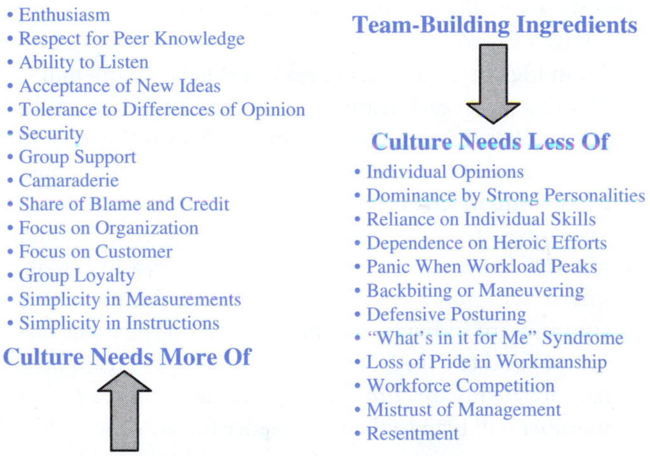

FIGURE 3-13. Changing the Culture to Support Team Building

(SMWT). The team goals are established outside the team (e.g., by top management) and to achieve those goals the team typically has full autonomy to perform such management functions as establishing schedules, performing training, distributing rewards, evaluating performance, recognizing exceptional contributions, and approving vacation leave.

When a team defines its own set of goals the self-managed team becomes a self-directed one. The **self-directed work team (SDWT)** operates as a profit center and a microcosm of the larger system of which it is a part. It performs all of the functions of an SMWT but also has the authority to shape its own future by making decisions on compensation, discipline, hiring and firing, and supplier certifications. While the SDWT may be the ideal to support the goals of an HPWO, senior management may be reluctant to relinquish total control to any work group.

Nevertheless, the trend toward the use of teams to make decisions, manage operations, and effect improvements has become a means for broadening participation and empowering the workforce. Simply placing people in a group and calling it a team does not ensure it will work as one, however. Teamwork is a building process that needs to become ingrained in an organization's work culture. This requires a gradual transformation that includes eliminating fear, building trust, providing adequate training, and implementing egalitarian rewards. Teaming skills do not come naturally; they need to be taught. However, training alone is not sufficient. Team building is the process of *enabling* the group to effectively and efficiently reach a common goal. This is a management issue and the most effective environment for team building is one in which management plays an active role—taking a genuine interest in and guiding the team-building activities. Applying this principle has proven to be one of the keys to success in Six Sigma companies. Figure 3-13 illustrates some important attributes that a culture needs more of and also those it needs to give up to be conducive to effective team building.

A self-directed work team has the autonomy to set and achieve its own goals.

Figure 3-14 depicts three dimensions of team development that derive from child developmental psychology: *identity, conceptual skills*, and *self-monitoring*.[43] This process

[43]Klein, G. A., C. E. Zsambok, and M. L. Thordsen. "Team Decision Training: Five Myths and a Model." *Military Review* (April 1993): 36–42.

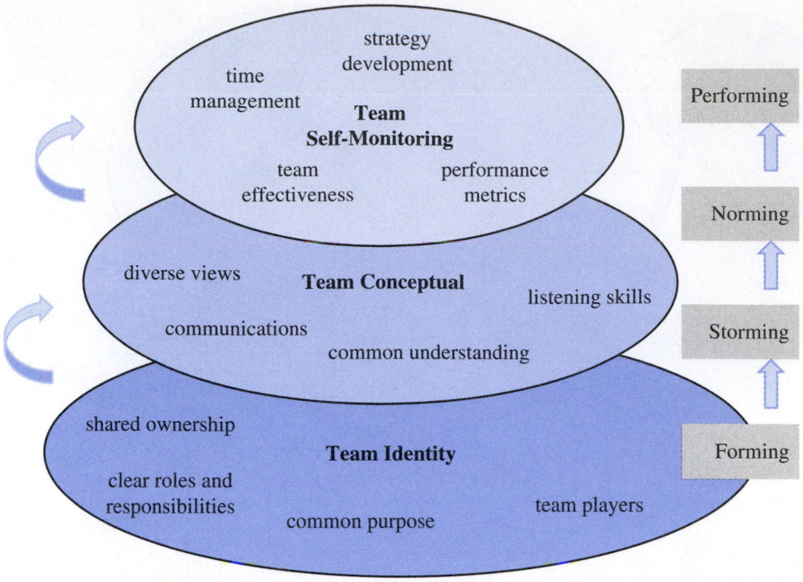

FIGURE 3-14. Team-building Development Model

has also been characterized in terms of four developmental stages through which teams evolve as they mature and coalesce as an effective unit. The stages are *forming, storming, norming, and performing.*[44]

- **Team identity.** Inexperienced teams have fragmentary identities with each team member focusing on individual tasks rather than team requirements. This is the development stage referred to as *forming* during which a group of individuals first become defined as an entity and jointly begin to face the process of finding their common identity. Expectations and roles are unclear, and personal interrelationships are superficial. During this phase it is common for individual team members to "push the envelope" and test the boundaries of acceptable team behavior. During this time of inexperience each team member will turn to the team leader for guidance. This leads into *storming*, the next and possibly most difficult stage in a team's road to development, which is characterized by conflict and resistance. This is a period when authority is challenged and the team tries to sort out personality and cultural differences. Anxiety levels, hurt feelings, blame, and divided loyalties are all typical when a team is in the storming phase. Such turbulence is necessary as the team moves toward an integrated identity when team members can identify themselves holistically in relationship to the team. In an effective team everyone shares ownership, all opinions are welcome and respected, no one dominates, and there are clearly defined roles and responsibilities. All members are united in a clear and mutually understood sense of purpose.
- **Team conceptual skills.** Teams must rely on shared practices and routines. As teams mature they develop competencies in the ability to work as one and the performance of basic tasks becomes an automatic process. When a team begins to coalesce and work together toward common goals, it has reached the *norming* stage. When a team emerges from the storming stage, conflicts are rare and the atmosphere is calm, steady, and relaxed. Team members can agree on and abide by group norms—the team's code of conduct. A sense of cooperation and collaboration makes it possible to begin to tackle problems, collect data, and reach consensus on process and solutions. During the process conceptual skills are developed at four levels. First, the team develops the ability to define clear goals and a unified sense of purpose. Second, the team begins to share a common understanding of a particular problem situation or data interpretation. This requires good listening, communications, and interpersonal skills. Third, the team is able to focus and collectively look ahead, anticipating problems, obstacles, and other contingencies. Finally, the team develops the ability to manage uncertainty, resolve conflicts, and reconcile opposing viewpoints.
- **Team self-monitoring.** The ability to manage the flow of ideas, sometimes referred to as *metacognition*, is a characteristic that differentiates a mature team from an immature one. When a team reaches this capability level it is said to have entered the *performing* stage—the pinnacle state where a team can effectively act as a single unit. At this level team members have developed relationships based on mutual respect and empathy. All are perceived as peers and together the group has established a workable structure and a unified purpose. A team does not know in advance who is going to show up at a

[44]Tuckman, B. W. "Development Sequence in Small Groups." *Psychological Bulletin* 63 (1965): 284–399.

1. **Team Goals.** Are goals developed through a group process that ensures broad interaction and buy-in from each team member?
2. **Broad Participation.** Do all team members actively participate in the accomplishment of tasks and are roles within the team equitably shared? Do all members actively participate in team decision making?
3. **Feedback.** Is feedback frequently asked for and freely given in the spirit of evaluating the team's performance and clarifying individual feelings and interests?
4. **Leadership.** Is leadership shared among team members or do a few strong personalities dominate?
5. **Enthusiasm.** Do members willingly and enthusiastically contribute to team requirements?
6. **Problem Solving.** Are members encouraged to solve problems, critique the team's effectiveness, and freely discuss team issues?
7. **Conflict Resolution.** How does the team deal with conflict? Are members encouraged to express negative feelings and confrontation within the group? Are opposing views welcome and dealt with in a sincere and respectful manner?
8. **Exploitation of Individual Potential.** Has the team fully recognized and utilized all the resources that each member is capable of contributing to team goals? These include talents, skills, knowledge, and experiences.
9. **Risk Taking and Creativity.** Does the team encourage risk taking and creativity? When mistakes are made, are they treated as a source of learning or reprimand?

FIGURE 3-15. Guidelines for Team Improvement

meeting, who will be alert, what ideas will be presented, and how the ideas will coalesce, yet such contingencies can be dealt with as the team has created a team mind in which the collective acts as one. The team mind makes the whole greater than the sum of its parts, creates new and unexpected outcomes from all individual contributions, manages the flow, and keeps the discussion relevant. A performing team works smoothly and on-purpose most of the time. However, since a team is comprised of humans, personal clashes that arose during the storming stage can recur. When conflicts do arise a performing team has the maturity to resolve such problems swiftly and not let such problems distract it from the achievement of its goals.

It is important that during the building process, teams be periodically evaluated to determine how well they are working together and to identify areas needing improvement. Figure 3-15 lists nine guidelines for team improvement.

3.6.2 Team-Building Culture

Team building strives to create a climate that encourages and values the contributions of all members. The combined energies of individual members are directed toward problem solving, task effectiveness, and the achievement of the team's goals. As the team progresses through the development process previously described, it begins to recognize that the performance of one individual member cannot be separated from the performance of others. To maximize effectiveness the team must strive to meet certain conditions.[8] For example, teamwork must be highly interdependent. Members must be committed to their goals and recognize that the only way to achieve them is by working together. Next, the team leader must have good people skills and be committed to the group. Team management should be a shared function, and each member should have the opportunity to exercise leadership when the needs of the group require his or her particular skills or experiences. As for the members, it is essential that each be committed to the team and willing to contribute.

To be effective a climate must be created where team members feel relaxed and can participate in open and direct communications. In addition, team members must all share a mutual respect for one another, be prepared to take risks, and be permitted to develop individual skills and abilities.

A further condition is the creation of a clear set of goals that are understood by everyone and that set ambitious but achievable performance targets. On a personal level the roles of individual members must be well defined and a subculture must be cultivated in which members can examine group and individual weaknesses and errors without engaging in personal attacks. This enables the team to learn from its experiences. To improve, the team must be willing to frequently and critically evaluate its own performance.

Effective teams work on their capacity to create new ideas, which can incubate from within the group through interaction or from external inputs. Good ideas should be followed up, and creativity and risk taking appropriately rewarded. Finally, each team member should genuinely believe that he or she can influence the team's agenda. The feeling of trust and equality facilitates a culture of open and honest communications.

Team building occurs naturally when members work together on a task that all agree is important (e.g., when a crisis occurs). When a group organizes to achieve a common purpose working relationships emerge. An effective team will develop either formal or informal operating guidelines that will govern how the team will function, serve as a benchmark for evaluating performance, and facilitate individual cooperation and productivity. The team will learn how to manage conflict, critique itself without personal attack or anyone taking offense, and provide feedback that encourages individual commitment and participation.

3.6.3 Team Decisions

Everyone makes personal decisions as they go about their daily lives. When these decisions impact only themselves they can be made in isolation and the decision maker is the only one who is harmed by a bad decision. However, in a lean environment decisions that are made affect a number of people; even a team decision usually impacts people who are not members of the team. When a team is empowered to make a decision it assumes an awesome responsibility to use its power wisely and make certain that the interests of *all*

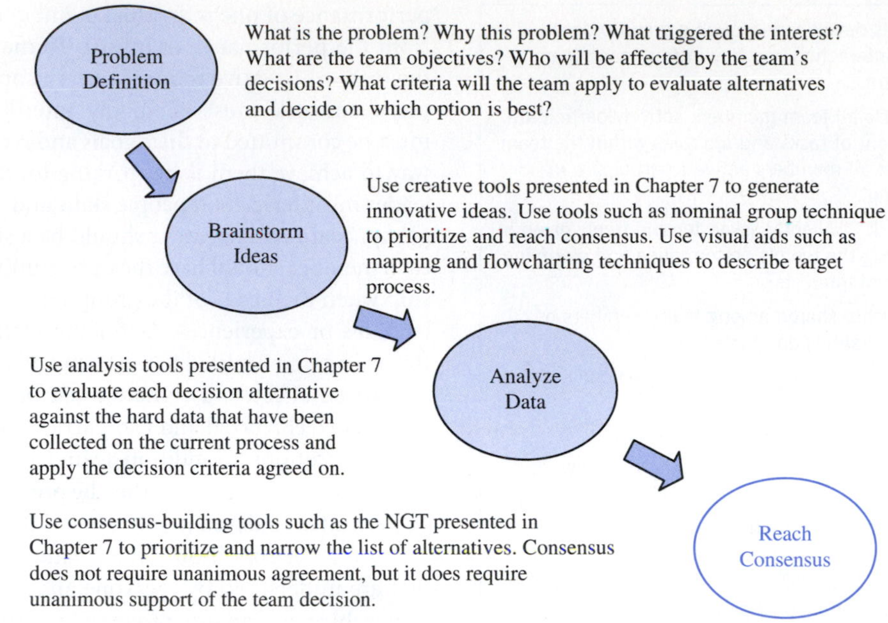

Problem Definition
What is the problem? Why this problem? What triggered the interest? What are the team objectives? Who will be affected by the team's decisions? What criteria will the team apply to evaluate alternatives and decide on which option is best?
Brainstorm Ideas
Use creative tools presented in Chapter 7 to generate innovative ideas. Use tools such as nominal group technique to prioritize and reach consensus. Use visual aids such as mapping and flowcharting techniques to describe target process.
Use analysis tools presented in Chapter 7 to evaluate each decision alternative against the hard data that have been collected on the current process and apply the decision criteria agreed on.
Analyze Data
Use consensus-building tools such as the NGT presented in Chapter 7 to prioritize and narrow the list of alternatives. Consensus does not require unanimous agreement, but it does require unanimous support of the team decision.
Reach Consensus

FIGURE 3-16. Team Decision Process

SHOWCASE OF EXCELLENCE **Winners of the Malcolm Baldrige National Quality Award Healthcare Category**

- 2008: Poudre Valley Health System, a private not-for-profit health care organization serving residents of northern Colorado, western Nebraska, and southern Wyoming. http://www.pvhs.org, http://www.nist.gov/public_affairs/releases/poudre_profile.html.
- 2007: Mercy Health System, a fully integrated health care system with three hospitals and a network of 64 facilities consisting of 39 multispecialty outpatient centers throughout northern Illinois and southern Wisconsin. http://www.mercy healthsystem.org, http://www.quality.nist.gov/PDF_files/Mercy_Health_System_Profile.pdf.
- 2007: Sharp HealthCare, San Diego County's largest integrated health care delivery system in southern California. http://www.sharp.com, http://www.quality.nist.gov/PDF_files/Sharp_HealthCare_Profile.pdf.
- 2006: North Mississippi Medical Center, a not-for-profit health care delivery system serving 24 rural counties in northeast Mississippi and northwest Alabama. http://www.nmhs.net/bpress, http://www.quality.nist.gov/PDF_files/NMMC_Profile.pdf.
- 2005: Bronson Methodist Hospital, a state-of-the-art all-private-room facility in southwest Michigan designed as a peaceful, healing environment with features such as an indoor garden atrium, complete with lush trees, plants, and bubbling water. http://www.bronsonhealth.com, http://www.quality.nist.gov/PDF_files/Bronson_Profile.pdf.

people who will be impacted by the decision are properly represented. This can be accomplished two ways. First, management should try and get the broadest representation possible in selecting the composition for team membership. Second, the team should aggressively seek input from stakeholders outside the team on relevant issues. These strategies help facilitate support and buy-in from a large and diverse constituency.

> **When a team makes a decision it has a responsibility to make certain that the interests of all those who will be affected are properly represented.**

Figure 3-16 depicts the four major phases of team decision-making. The first step is for the team to reach agreement on a problem statement and to set some goals. Next, the team needs to generate some ideas using brainstorming techniques. These ideas can range from how the team should proceed to critical issues that need to be resolved to data requirements to alternative solutions. The next step is to use tools of analysis to collect and evaluate hard data, to map brainstormed ideas onto the knowledge base generated from data analysis, and to produce some options that can be implemented. The final step is to prioritize and narrow the list of alternatives, possibly testing them for viability, and to reach team consensus on final recommendations.

3.7 SUMMARY OF KEY IDEAS

Key Idea: *Human resource management is a complex multifaceted open system that spans the entire organization.*

Human resource management is an open system that operates within the context of the larger business system. Its inputs are the attributes that distinguish one person from another, and outputs are in the form of performance measures (such as quality) and behavioral attributes (such as feelings and emotions). The transformation process is concerned with optimizing the deployment of the human factor within the value steams to produce the best results. Communications and rewards are key determinants to optimum results. The primary purpose of the HR system is to manage inputs, through improved workforce competence, and to improve quality through productive behaviors.

Key Idea: *An HPWO has two distinguishing features: It can readily adapt to change and it has a management style that fully engages with the workforce, permitting broad participation in decision-making at every level.*

HPWOs can readily adapt to change because the people make the difference. These organizations are high performing because the senior leaders understand the importance of developing its human resource capabilities and then giving its employees freedom to fully deploy their skills and utilize their talents. In an HPWO the workers are empowered to participate in the decision-making process, perform their jobs in a climate of HPWPs, and are confident that their efforts will be appropriately recognized and rewarded.

Key Idea: *A high-performing work practice (HPWP) is a work system that supports the goals of an HPWO and includes high employee involvement, aggressive HRM involvement, and shared ownership.*

Employee involvement begins with full disclosure by management of all information relative to business conditions, strategy, goals, objectives, and targets. Next, a community of common interest needs to be created. One way to accomplish this is to form a staff association where issues, concerns, ideas, and information can be freely exchanged. Finally, management needs to devise systems that can genuinely engage with the workforce. Surveys and suggestion schemes, when properly implemented, can be effective in tapping the "voice of the employee." However, the most productive way to actively involve the workforce is through the use of teams. Teams can be permanent or formed for an ad hoc purpose. The common objective can be quality improvement or problem solving, and teams are often cross-functional and empowered to make limited decisions.

Aggressive HRM involvement means that the HRM system must fully support the establishment of HPWPs. This involves frequent feedback and guidance through performance reviews, progressive recruitment practices to ensure a good match between new hires and the strategic needs of the business, comprehensive training programs, and freedom for worker groups to experiment and change their work methods to improve productivity and efficiency.

The best way to create a sense of shared ownership is to allow employees to become financial stakeholders in the company. Financial rewards such as profit sharing, stock options, and other performance-related schemes give explicit recognition to the contributions that individuals and teams make toward the achievement of organizational objectives. However, all rewards need not be financial. Nonfinancial benefits include such things as the free use of facilities, family-friendly services, and eligibility for certain privileges. Some of these benefits can be extended to spouses and children.

Key Idea: *Workforce empowerment represents a fundamental cultural transformation in most organizations and requires dramatic changes in the way that managers and workers interact.*

Before management can proceed with a strategy to empower its workforce it should realize that not all workers will necessarily respond positively to the idea of increased intellectual responsibilities and performance accountabilities. Therefore, many companies resort to psychological testing to determine personality types which are used to match employees with changing job requirements. There are two approaches to empowerment: participatory and self-management. The greatest benefit is likely to be derived from self-management because workers are in truth granted discretionary powers over their respective spheres of operational influence. However, this approach has a much greater cultural impact than the participatory approach under which management retains veto rights over all low-level decisions. In either approach there is a potential for backlash if employee fears are heightened and empowerment is thrust upon an ill-prepared workforce.

Key Idea: *Cultural change is a slow, painstaking process that takes shape gradually as people begin to accept a new sense of what is normal.*

Culture is the single defining characteristic of an organization, and quality can be achieved only when the culture is unified in its desire to satisfy customers. To transform the culture, the human resource system needs to be reengineered to bring it in line with corporate quality goals. A cultural change will have taken place when people accept a new sense of what is normal. In recruiting new personnel it is more important that they be selected on their potential to fit into the new culture than on the strength of their technical skills.

Key Idea: *Job rotation expands the skill base of individuals and provides management with improved knowledge of human resource capabilities by being able to observe employees in a variety of job situations; this practice leads to increased operational flexibility.*

Job rotation is a practice that moves employees from job to job and in so doing offers the prospect of three advantages. The first advantage is that rotating a worker through a number of jobs contributes to the worker's learning and development with an expanded skills base. The second advantage is that job rotation gives management the opportunity to observe workers performing in a variety of jobs. This improved knowledge of human resource capabilities provides management with more flexibility in making personnel assignments. The third advantage is that job rotation has the potential of motivating employees by making work more interesting. However, research studies have not generally supported the motivation theory.

Key Idea: *Job enlargement is a horizontal expansion of work and increases the complexity of the tasks involved and the number of skills required to perform them.*

Job enlargement is the opposite of specialization, and can motivate workers by removing boredom and tedium that can result from highly repetitious work narrow in scope and requiring few skills. Apart from eliminating monotony, job enlargement can provide the worker with a sense of pride of accomplishment as he or she is able to see a more complete component of the end product.

Key Idea: *Job enrichment is an HPWP that can potentially motivate workers by expanding job responsibilities both horizontally and vertically.*

Job enrichment increases the intellectual content of a job, providing the worker with greater freedoms, less controls, greater authority to make decisions, and greater job complexity. Nevertheless, job enrichment will not always lead to increased motivation. If an employee has low growth needs he or she will respond negatively to a job that has a high intrinsic motivating potential.

Key Idea: *Each job carries with it an intrinsic motivating potential that is determined by five factors: skill variety, task identity, task significance, autonomy, and feedback. These factors combine to motivate employees with high growth needs.*

Intrinsic motivating potential is determined by skill variety, task identity, task significance, autonomy, and feedback. People appear to be more motivated if they can see how their contribution impacts something larger, if they think their work is important, if they have a say in how they do their work, and if they are told that they have done a good job. The five factors can be used to compute a motivating potential score (MPS) for each job. A job with a high MPS will not necessarily motivate a worker who has low growth needs. By asking personnel to complete a growth needs strength (GNS) survey, those with high growth needs can be matched to jobs with high MPS and vice versa.

Key Idea: *The idea of forming teams and using those teams to make important operational decisions is fundamental to the principles of QIM.*

The use of teams to make decisions, manage operations, and effect improvements is a fundamental mechanism for broadening participation and empowering the workforce in lean production environments. However, simply placing people in a group and calling it a team does not ensure it will work as one. Teamwork is a building process. People need to be trained on how to productively participate on a team and teaming has to become an integral part of the workforce culture.

Key Idea: *Teamwork is a building process that needs to become an integral part of the culture, and requires a gradual transformation that includes eliminating fear, building trust, providing adequate training, and implementing egalitarian rewards.*

Team building is the process of enabling a group of people to work effectively to achieve a common goal. An effective team has to come together on four levels: competence, identity, cognition, and metacognition. Team competence measures how well its members can learn to act and think as one. Team identity is a measure of how well each member can set aside personal interests and relate to the team and comembers. Team cognition refers to how well the team as a unit has developed conceptual skills, and team metacognition is a measure of how effective the team can manage the flow of ideas and develop a team mind in analyzing data and making decisions.

3.8 MIND EXPANDERS TO TEST YOUR UNDERSTANDING

3-1. Harriett Harris has worked for the same company for 10 years. Due to the early and untimely death of her mother, Harriett did not have the opportunity to attend college, although she could have easily handled the demands of academe. Instead, she opted to go to work to support her two younger siblings. The local plant of Aztec Electronics readily employed her and trained her for their injector line, a critical process in the manufacture of electronic semiconductors. Harriett's job was to operate equipment that performs an important function in the manufacture of printed circuit boards, an operation where quality and yields are watched closely and play an important role in corporate profitability. Harriett takes pride in her work, always striving to do a good job. She has, over the years, become one of the most highly skilled and productive workers. Her job is highly specialized and requires only a few tasks, but each must be performed with great precision and close attention to detail.

Harriett's world has just been turned upside down. As a result of customer pressures on Aztec for just-in-time deliveries, the company has recently hired a consultant to help transform the manufacturing plant into one that supports cellular manufacturing. New social work groups have emerged. Harriett has been relocated from her line position, next to her friend Cathy, and placed in a work cell with people she does not know very well. She misses the daily chats with Cathy as they worked side-by-side on the line. She is now in this new group and has been told that her job responsibilities have changed in breadth and depth. She will have to undergo some intensive training and her performance will be based not only on her individual achievements but also on the collective performance of her work team. Harriett is frightened and has suddenly become overwhelmed by a sense of insecurity. Like an anchor she feels that the tide of anxiety and frustration is slowly pulling her under. "Coming to work used to be such fun—now I dread it and look forward to the weekends."

a. How could Harriet's management have anticipated what impact the change in Harriet's job responsibilities might have on her as a person and the quality of her future contributions to Aztec Electronics?
b. If the employees, including Harriet, are defined as a valued stakeholder group, what could Aztec management do differently in managing change so that the needs of these stakeholders are considered?
c. Assuming that Aztec is able to identify good employees such as Harriett who do not wish to have their jobs enriched, and who do not wish to be empowered, how should they resolve this issue if their intentions are to achieve greater decentralization, a flatter organizational structure, and more autonomy at the work interface?

3-2. Dennis Mitchell is the plant manager of a company in northern Alabama that produces interiors for luxury yachts. The ethnicities of his workforce are mixed: about 40% are white, 25% African American, 15% Hispanic, 10% Libyan, and 10% Asian (Korean, Chinese, and Vietnamese). As for gender breakdown, 60% are women. Twenty percent either do not speak English at all or they have very poor English conversational skills. His departmental supervisors report that they have grave difficulty in giving orders and having those orders carried out.

Lately the business has been in a downturn. A soft economy together with growing foreign competition has had a serious effect on sales. As a result, last week Dennis reluctantly had to cancel one of the three production shifts and lay off 70 workers. The plant is one of the primary employers for a small town of 15,000 and the layed-off workers have little prospects of gaining work locally. That incident only exacerbated an already volatile workplace environment. Just last month, when Dennis expanded the duties of the sewing room foreperson to include the cut-out department, the workers downed their tools and walked off the job. The threatening language they used at the time gave Dennis cause to believe some intended to physically damage the property so he summoned the local sheriff's department for protection. That inflamed the group more and it took some skillful negotiating on the part of an independent mediator to calm things down and get the workers to return.

Responding to pressures from stockholders the company CEO is concerned about sustainability, and is now pressuring Dennis to turn things around. Dennis would like to adopt a more participatory style of management, but is not certain how to go about it, particular since he perceives that there is a lack of trust between management and labor.

a. What are some of the principal issues that need to be articulated before this company can begin to build a new workplace culture? That is, if you were summoned to this plant as a consultant, what are the most important questions that you would need answered intially?
b. In building a supportive culture how can this company overcome the language barrier? How can it begin to build trust between management and the employees?
c. How could the team structure be used to advantage in an environment like this?
d. How could the management of this company use its human resources to help it gain competitive advantage?
e. If the company does find it necessary to downsize, how do you think its approach would differ under these two perspectives:
    i. Losing personnel is like writing off assets on the company balance sheet.
    ii. Losing personnel is like shedding costs from the company profit and loss statement.

3-3. Define some of the outcomes of HRM as an open system. How can feedback be used to improve these outcomes? How would these outcomes be measured?

3-4. Place yourself in the position of someone who has just been hired to help an organization move toward an HPWO. You find that the workforce is diverse, and, after testing the employees, you find that many of them have low growth needs. You also find that they mistrust management. What would you do?

3-5. Imagine that you are working in a small factory that makes high-end cabinets, built-in bookcases, and mantels. The work has been organized so that a team of workers performs all of the operations, such as cutting, milling, sanding, gluing, and staining on a family of similar products. Although similar, all products are distinct in that they are custom made for customers and will typically be installed in residential homes. The workers within the team all have multiple skills in all of the operations necessary. How would a team such as this operate if it were an SMWT? How would it differ as an SDWT? Explain.

3-6. Place yourself in a top leadership position in the following organizations. What steps could you take to create a sense of shared ownership on the part of your employees?

a. Large discount retailer, like Target or Walmart
b. Large, publicly traded company that manufactures computers
c. Small, privately owned bake shop
d. Hotel chain
e. Government agency
f. Airline
g. School system

3-7. Consider each of the organizations listed in problem 3-6. To what extent do you believe the employees can be empowered? As a senior leader, how would you empower them?

3-8. How does a team differ from a club (e.g., a service club like Rotary)? How does a team differ from a committee? Explain.

3-9. For each of the following teams, indicate whether the team is self-managed, self-directed, neither self-managed nor self-directed, or not a team at all.

a. College football team
b. School debating team
c. First-response emergency team
d. U.S. diplomatic negotiating team
e. International consulting group
f. Surgical team
g. U.S. Army brigade infantry forward combat team
h. Cockpit crew of a Boeing 747 aircraft
i. NASCAR pit crew

PART TWO

QUALITY FROM 5,000 FEET

CHAPTER FOUR

SPECIFICATIONS AND PRODUCTION IMPLICATIONS

QUALITY MAXIM #4: *Excellence is no distinction among the excellent.*[1] *Competitiveness is not defined by how often a process can deliver within a specification range, but how close the process can consistently deliver around a target value.*

Specifications: What Customers Need versus What Can Be Made

The lyrics of a 1960s rhythm and blues song "Ain't No Mountain High Enough"[2] proudly proclaim that no mountain can be too high, no valley too wide, or no river too deep to keep the singer from his or her ultimate goal (passion)[3]—a message that resonates well with those who are committed to the relentless pursuit of excellence. A fundamental principle of quality is to always strive to meet customers' requirements and to never stop trying to identify and eliminate barriers to perfection. With tangible products customer requirements are often quantifiable and expressed as a range of acceptable values—referred to as the product specifications (specs). The lower specification limit is the smallest acceptable value of the quality attribute and the upper specification limit represents the largest. When, due to process variability, the process output results in a range of part values that is larger than the spec range, the process is incapable of meeting the specs and defective parts are inevitable. Even if all parts are within the spec range, output quality can still be questioned if many of the parts are larger or smaller than the target value—typically the midpoint of the range. The quality challenge is a two-step process, both of which are aimed at improving manufacturing capability. First, through reduced process variability and better targeting, all parts produced can be brought into compliance with the specs. Second, through the continued reduction of process variability more parts can be produced closer to the target. This cycle gets harder and more challenging with each successful improvement and requires commitment and passion to achieve high standards of excellence. The goal is to hit the target precisely every time, and as the song suggests, no obstacle—no matter how ominous—is a legitimate excuse for failing to do so.

LEARNING OBJECTIVES AND KEY OUTCOMES

After careful study of the material in Chapter 4, you should be able to:

1. **Define the differences between specifications that are critical to function, critical to manufacture, and critical to assembly.**
2. **Differentiate between dimensional, geometric, and statistical tolerancing.**
3. **Analyze and develop statistical tolerances for composite parts.**
4. **Calculate the proportion of a composite part that is outside specifications and by how much the total variability would have to be reduced in order to bring the total composite distribution within the allowable specification range.**
5. **Identify and describe the functions of a number of gauges used for dimensional metrology.**
6. **Discuss the concept of conformance to specifications, including some of the related sampling issues involved.**
7. **Discuss some of the issues involved in establishing specifications for the provision of services.**

4.1 SPECIFICATIONS AND TOLERANCES

4.1.1 Introduction

The creative genius of a product's design assumes that the dimensions and geometry (shape) of its component parts will be without error, and where important, all surfaces will be smooth. Preproduction prototypes can come close to this ideal. However, when parts are produced in production quantities, variability (which can come from numerous sources) can cause each part to differ from technical drawings and from other parts of the same design. Distortions in shape, variability in dimensions, and some degree of roughness can be

[1] Serbo-Croation proverb.

[2] Written by Nickolas Ashford and Valerie Simpson in 1966, this song was a hit single recorded by Marvin Gaye and Tammi Terrel in 1967. The song became a hit again in 1970 when it was recorded by Diana Ross, who was also nominated for a Grammy for this performance.

[3] The music and lyrics for "Ain't No Mountain High Enough" can be accessed at http://kids.niehs.nih.gov/lyrics/aintno.htm.

Two quality engineers take time out to discuss some industry trends. An good understanding of statistical principles is essential to establish competitive design specifications for assembled products.

expected. If the design parameters are properly controlled within prescribed limits, the part will be regarded as acceptable and passed on to the next processing step in the value chain.

Product specifications can be grouped into three major categories (dimensions, geometry, and surface), and are covered by an international set of standard requirements called the **geometrical product specifications** (GPS). These are illustrated in Figure 4-1.

Superior design practices give consideration to the criticality of each specification, as follows.

- **Critical to function** (CTF)—*Is the specification critical to the product doing what the customer expects and what the manufacturer promises?* Failing to meet CTF specifications can jeopardize a product's intended function-in-use. *Function* is defined to include use, but also safety and dependability. For example, the customer of a power saw expects the tool not only to provide a true and smooth cut through a variety of materials, but also to be properly guarded, electrically grounded, and to have a long working life. The customer may also expect the designer to consider certain ergonomic factors, such as ease of handling, grip, and weight.
- **Critical to manufacture** (CTM)—*Is the specification critical to the efficient manufacture of the product in the quantities required?* CTM specifications give consideration to manufacturing capabilities and can reflect the capability of a particular processing center to consistently meet the specification, or the specification can be a key to controlling variability in downstream operations. Some processes, especially automated ones, can experience downtime if the variability of input is excessive for critical parameters.
- **Critical to assembly** (CTA)—*Is the specification critical to all parts fitting together properly in a final assembly?* A CTA classification is assigned to the specification of a component part that is key to the successful mating with another part, or component, in a downstream assembly operation. For example, a certain assembly operation might require the insertion of a shaft through a hole. The diameter of the shaft must be less than the diameter of the hole or the two parts will not fit. In addition, the clearance between the shaft and hole cannot exceed a specified tolerance or excessive vibration would cause the assembly to fly apart during use.

A product designer may intentionally fail to indicate a tolerance range on the shop drawings if a dimension is not thought to be critical to function, manufacture, or assembly. In these cases the drawing will reflect only the nominal dimension.

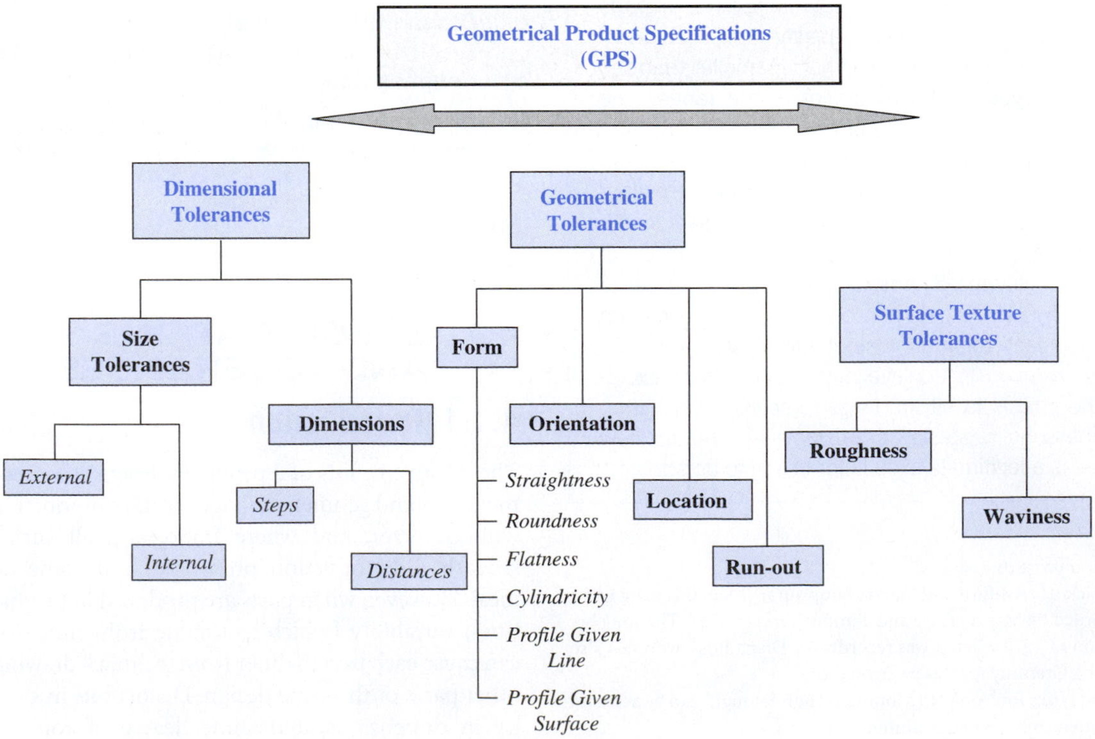

FIGURE 4-1. Types of Product Specifications Covered by GPS

However, in the absence of a stated specification range and, given the reality of process variability, it is impossible to determine whether a specific part is acceptable or not. In these cases **general tolerances** apply. A general tolerance is determined by customary workshop accuracies that have been defined in ISO standards for various manufacturing processes. As an example, Table 4-1 shows the general tolerances for machining operations (material removal) as specified in ISO Standard 2768.

> **A general tolerance is used when a specification is not critical to function, manufacture, or assembly.**

4.1.2 Dimensional Tolerancing

Dimensional tolerances can be categorized as either *external*, *internal*, *step sizes*, or *distances* as illustrated in Figure 4-2.

- **External Sizes.** An external size is a dimension measured between two extreme boundaries in a particular

TABLE 4-1 General Tolerances Specified in ISO 2768

Tolerances for Linear Dimensions (*mm*)

Tolerance Class	Range of Lengths (*mm*): 0.5 up to 3.0	Over 3.0 up to 6.0	Over 6.0 up to 30.0	Over 30.0 up to 120.0	Over 120.0 up to 400.0	Over 400.0 up to 1000.0
fine (f)	0.1	0.1	0.2	0.3	0.4	0.6
medium (m)	0.2	0.2	0.4	0.6	1.0	1.6
course (c)	0.4	0.6	1.0	1.6	2.4	4.0
very coarse (v)	-	1.0	2.0	3.0	5.0	8.0

Deviations in Angular Dimensions (degrees and minutes)

Tolerance Class	Range of Lengths (*mm*): Up to 10.0	Over 10.0 up to 50.0	Over 50.0 up to 120.0	Over 120.0 up to 400.0	Over 400.0
fine (f) medium (m)	± 1°	± 30′	± 20′	± 10′	± 5′
course (c)	± 1°30′	± 1°	± 30′	± 15′	± 10′
very coarse (v)	± 3°	± 2°	± 1°	± 30′	± 20′

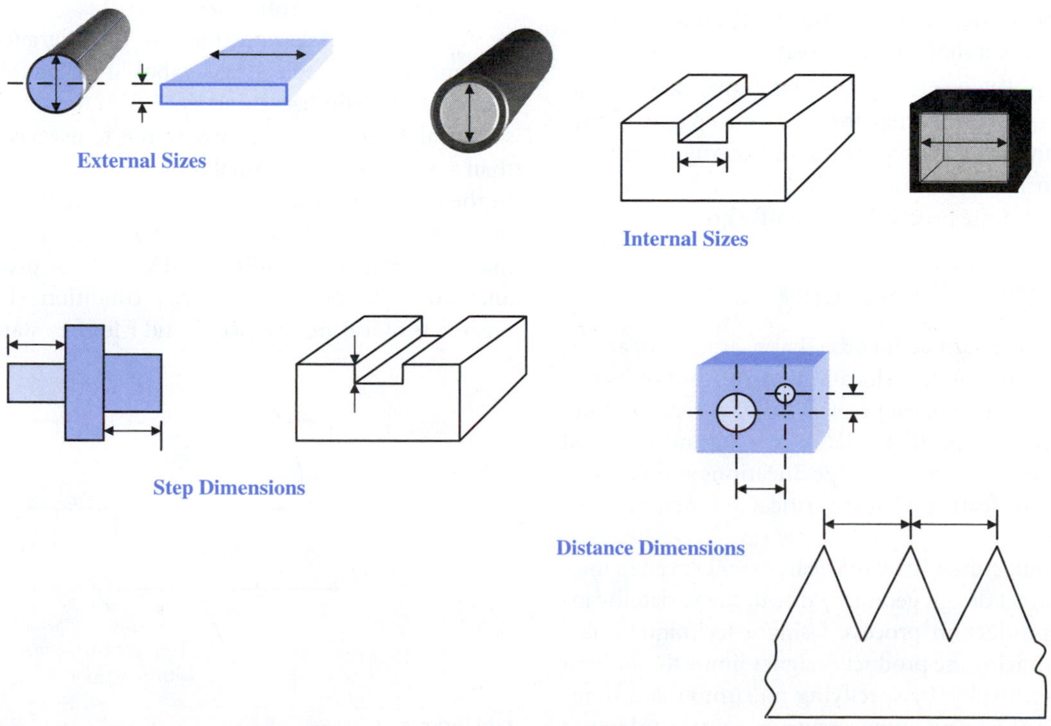

FIGURE 4-2. Dimensional Tolerances on Size

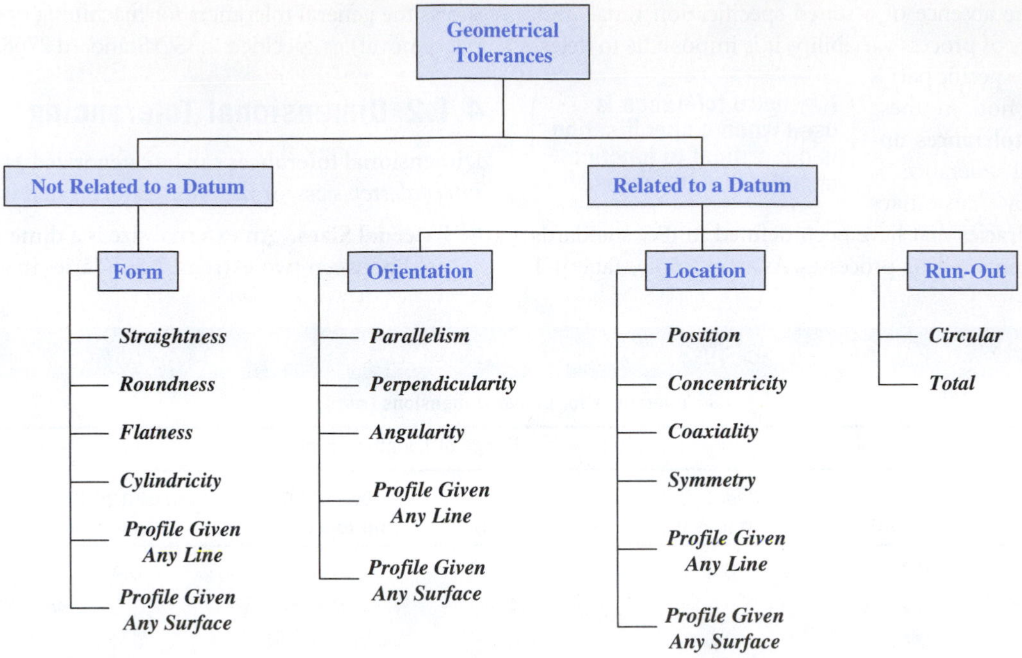

FIGURE 4-3. Classification of Geometrical Tolerances

plane relative to a part's feature. Examples are the diameter of a shaft, the thickness of a board, or the length of a mounting plate.

- **Internal Sizes.** An internal size is a dimension that is internal to a part, measured between two points that reside on internal part surfaces. Examples are the inside diameter of cup, the inside dimension of a box, and the width of a machined keyway.
- **Step Sizes.** A step size is a dimension between two points, only one of which resides on a surface of the part. Examples are the depth of a groove and the distance between the end of a shaft and the position of a coupling.
- **Distances.** A distance is any dimension that does not fit the definition of an external, internal, or step size. Examples of distances are the separation of the origins of two holes, the position of a hole relative to the edge of a metal plate, and the precise location of a groove.

4.1.3 Geometric Tolerancing

The geometry of a design defines the shape, or *form*, of an object as well as the orientation, location, or run-out of its features. This is illustrated in Figure 4-3. Orientation, location, and run-out must be specified relative to a datum (i.e., fixed points or planes). Product design specifications stipulate target values for those features that are critical to function, manufacture, or assembly.

When manufactured, an actual feature will deviate somewhat from its target design geometry due to the variability inherent in the production process. Using a technique called **geometric tolerancing** the product designer limits the amount of deviation permissible by specifying maximum and minimum values on the engineering drawings using a **tolerance frame.** As Figure 4-4 illustrates, the tolerance frame is constructed using a rectangle containing two or more compartments. Those compartments contain, from left to right, the symbol for the geometric characteristic, the tolerance (expressed in millimeters with modifiers as appropriate), and the letter (or letters) identifying the datum system (points or planes). Table 4-2 shows the standard symbols that are used for each tolerance characteristic.

> **A manufactured part will deviate from its design geometry due to process variability.**

In the second compartment of the tolerance frame, the tolerance is preceded by the symbol "ø" if the tolerance zone is circular or cylindrical, and by "S ø" if the tolerance zone is spherical. Note that the convention is to use a comma rather than a period in the decimal representation of the tolerance. In the circle following the stated tolerance value any appropriate modifiers will be inserted. For example, M is used for **maximum material condition** (MMC); P for **projected tolerance zone**; L for **least material condition** (LMC); S for **regardless of feature size** (RFS); and F for **free state condition.**

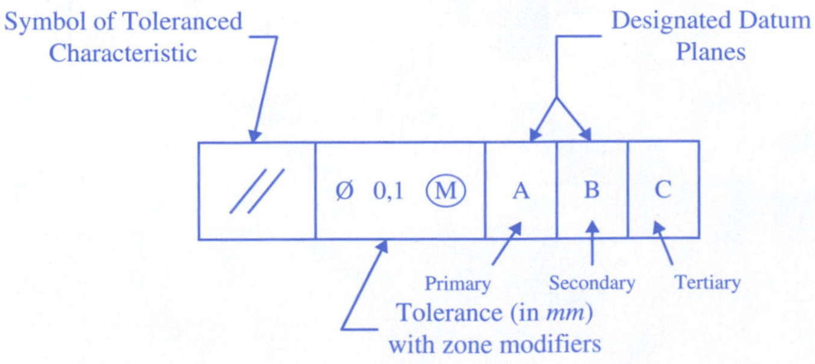

FIGURE 4-4. Tolerance Frame Contents

TABLE 4.2 Geometrical Tolerances, Symbols, and Datum Requirements

Tolerances	Characteristics	Symbol	Datum Needed
Form	Straightness	⏤	no
	Flatness	▱	no
	Roundness	○	no
	Cylindricity	⌭	no
	Profile any Line	⌒	no
	Profile any Surface	⌓	no
Orientation	Parallelism	∥	yes
	Perpendicularity	⊥	yes
	Angularity	∠	yes
	Profile any Line	⌒	yes
	Profile any Surface	⌓	yes
Location	Position	⌖	yes or no
	Concentricity (for center points)	◎	yes
	Coaxiality (for axes)	◎	yes
	Symmetry	⌯	yes
	Profile any Line	⌒	yes
	Profile any Surface	⌓	yes
Runout	Circular	↗	yes
	Total	⌰	yes

A definition of each of these modifiers can be found in the Glossary.

In Figure 4-4 the tolerance frame specifies a 0.1 *mm* round or cylindrical tolerance zone as the maximum material condition (e.g., minimum hole or maximum shaft) relative to planes A (primary datum), B (secondary datum), and C (tertiary datum).

- **Tolerances of Form. Tolerances of form** state how far features of form (flatness, straightness, roundness, cylindricity, surface profile, and line profile) are permitted to vary from the target (nominal) value stipulated on the drawings. ISO standards govern the terms and measurement methods used for specifying and determining compliance with tolerances. The basis for evaluating form is a *local deviation*, which is defined as the distance from each measuring point to a referenced feature. The convention used to determine compliance with specifications is to compute the sum of the largest local positive deviation from the referenced feature to the absolute value of the largest local negative deviation. The symbols given for the measured difference of the maximum deviations are STRt (straightness), RONt (roundness), PLNt (flatness), and CYNt (cylindricity). The ISO standards

TABLE 4-3 Abbreviations Used in GPS Standards for Reference Features in Tolerances of Form

Evaluation Method	Form Tolerance			
	Straightness	Flatness	Roundness	Cylindricity
Minimum-zone feature	MZLI	MZPL	MZCI	MZCY
Least-squares reference feature	LSLI	LSPL	LSCI	LSCY
Upper reference feature	OPLI	–	–	–
Lower reference feature	LOLI	–	–	–
Minimum circumscribed reference feature	–	–	MCCI	MCCY
Maximum inscribed reference feature	–	–	MICI	MICY

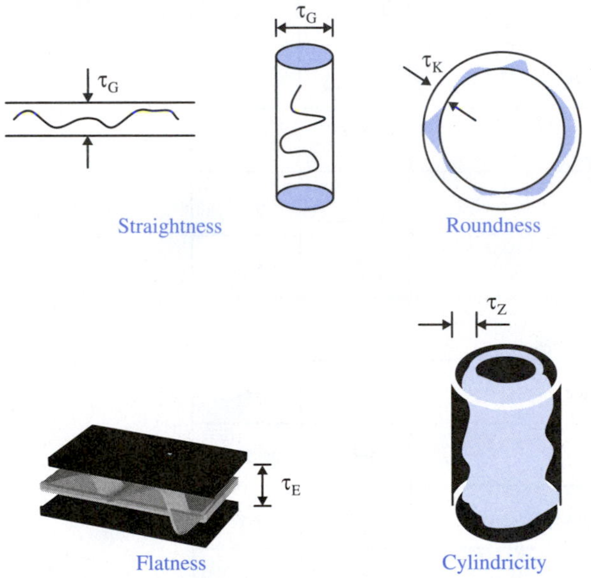

FIGURE 4-5. Zone Measures for Tolerances of Form

specify six different methods for evaluating form deviations, and are shown in Table 4-3.

Figure 4-5 illustrates the minimum zone method for determining tolerance limits for form.

Straightness. A **straightness** tolerance is defined as a line that must lie between two parallel lines or surfaces that are τ distance apart, or alternatively, that must lie inside a cylinder of diameter τ. The two parallel lines and cylinder define what is called the minimum zone for the line (or MZLI). Whenever STRt (MZLI) $\leq \tau_G$, a part is said to comply with the specifications.

Roundness. The tolerance zone for **roundness** is the difference τ_K between the radii of two concentric circles that together define the maximum and minimum circumferential profiles of each cross section of the part. The part complies with specifications when RONt (MZCI) $\leq \tau_K$.

Flatness. The tolerance zone used to determine **flatness** is defined by two parallel planes separated by a distance τ_E. When all points on the surface of a part lie between the two planes—that is when PLNt $\leq \tau_E$—the part is considered to be in compliance. Figure 4-6 illustrates a drawing specifying a flatness tolerance zone $\tau_E = 0.01$ *mm.*

Cylindricity. For **cylindricity,** the tolerance zone is defined by two coaxial cylinders that are a distance τ_Z apart. To be in compliance the entire surface of the cylindrical part cannot exceed the separation of the two coaxial cylinders—that is, CYLt (MZCY) $\leq \tau_Z$.

- **Tolerances of Orientation. Tolerances of orientation** include parallelism, perpendicularity, and angularity. There are five different situations in which tolerances of orientation can be required. In each of these cases the tolerance zone is defined by two planes that are separated by the prescribed tolerance τ, and is parallel to, perpendicular to, or at a prescribed angle relative to a datum plane, a datum straight line, or a datum system (consisting of two planes) as appropriate. Five possible scenarios can arise.

1. **Parallelism, perpendicularity,** or **angularity** tolerance of a straight line related to a datum system. In the case of parallelism, the datum system can be a fixed reference straight line and plane. In all three cases the datum system can consist of two planes. Figure 4-7 is an example of an angularity tolerance of a straight line related to a datum plane. In this case the straight line is the axis of a hole that is to be drilled in a plate at a 45° angle. The plane of the surface of the plate (shown as plane A) is the datum, and the tolerance zone is defined by two parallel planes (planes B and C) that are a distance of 0.05 *mm* apart and inclined at the specified angle of 45° relative to the datum.
2. Parallelism, perpendicularity, or angularity tolerance of a straight line related to a datum straight line.
3. Parallelism, perpendicularity, or angularity tolerance of a straight line related to a datum plane. Figure 4-8 shows an excerpt from a drawing specifying a perpendicularity zone of 0.5 *mm* of a part component relative to two planes.
4. Parallelism, perpendicularity, or angularity tolerance of a plane related to a datum straight line.
5. Parallelism, perpendicularity, or angularity tolerance of a plane related to a datum plane.

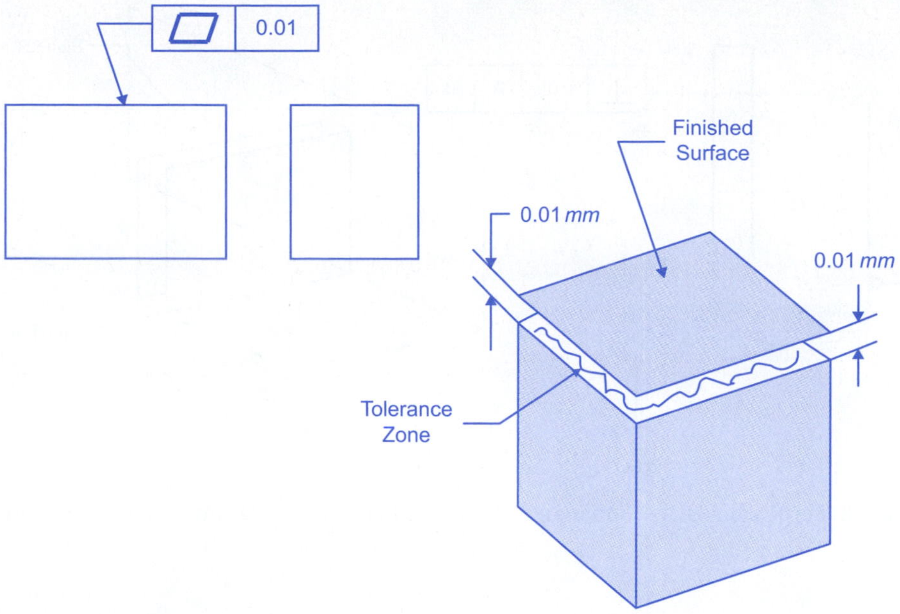

FIGURE 4-6. Flatness of a Finished Surface

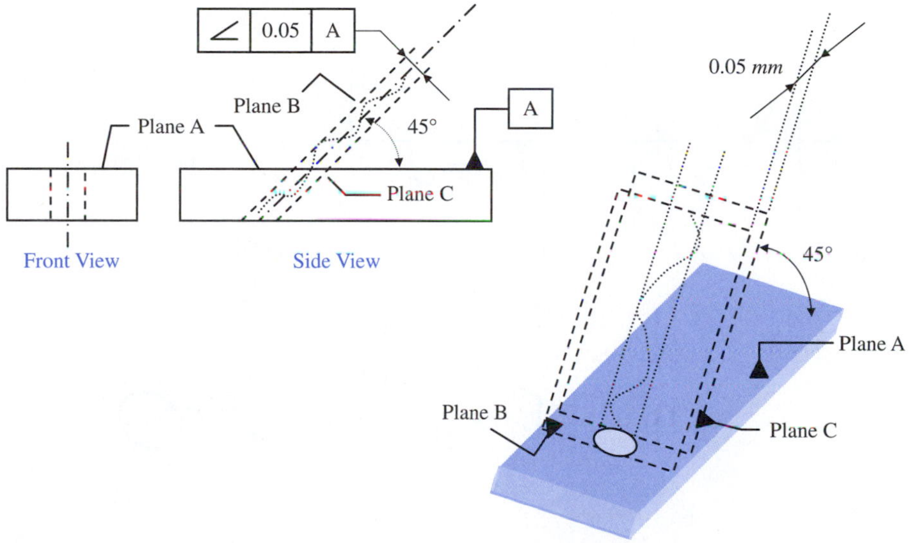

FIGURE 4-7. Angularity Tolerance of a Straight Line Related to a Datum Plane

- **Tolerances of Location. Tolerances of location** are concerned with **position, concentricity, coaxiality,** and **symmetry.** Position is relative to a specified datum (point or plane) and can refer to a point, line, plane, or surface. Figure 4-9 depicts the positioning of the axis of a hole within a cylindrical tolerance zone of 0.08 *mm* and dimensioned relative to three planes: A, B, and C.

  Concentricity is a condition in which two or more part features (cylinders, cones, spheres, etc.) in any combination share a common center, coaxiality is a condition in which two or more part features share a common axis, and symmetry is a condition in which a feature or features is symmetrically disposed about the centerline of a datum. Figure 4-10 is an example of a symmetry tolerance zone (0.7 *mm*) of a slot relative to a plane of two outer surfaces, and Figure 4-11 is an example of a concentricity tolerance zone (0.1 *mm*) of a circle center point on a work piece related to another point in the same plane.
- **Run-Out Tolerances. Run-out** is a measure associated with a work piece surface that possesses some circular symmetry. Run-out is defined as the amount by which the geometric form deviates from an exact circle, cylinder, or sphere. A run-out tolerance can be specified as either circular or total. Circular run-out is subclassified as either radial or axial depending on the relationship between the direction of measure and the relevant

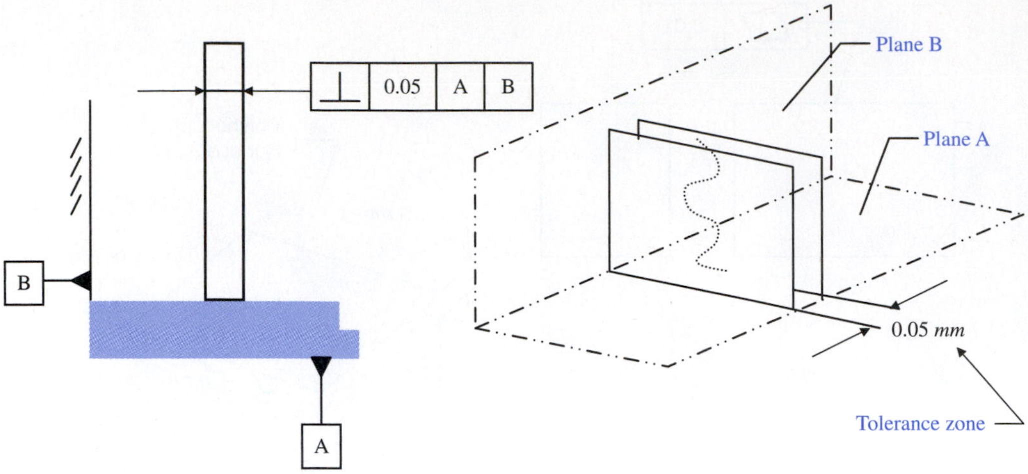

FIGURE 4-8. Perpendicularity (Squareness) of a Line Related to a Datum System of Two Planes

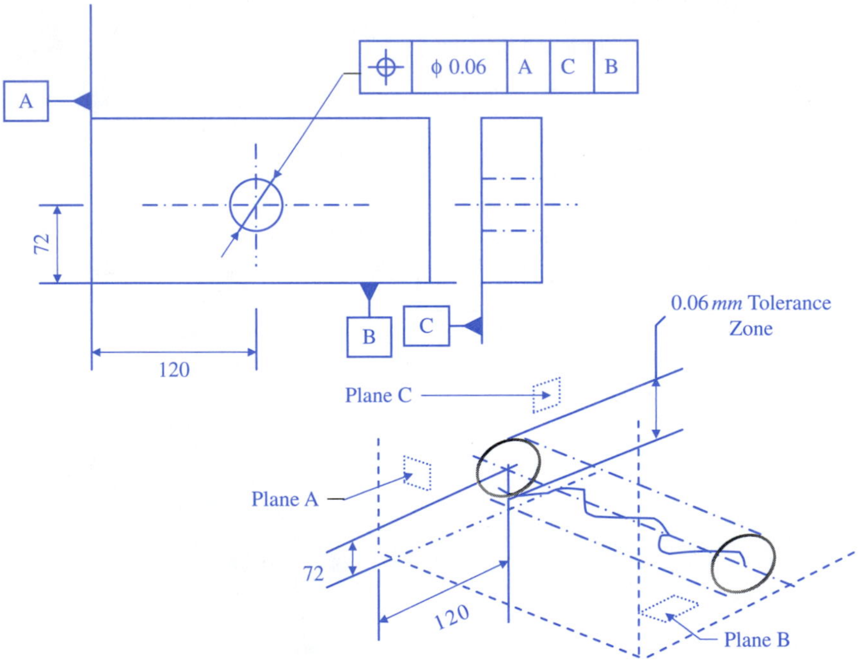

FIGURE 4-9. Position Tolerance of a Straight Cylinder Related to Three Datum Planes

axis of rotation. Figure 4-12 illustrates a circular run-out tolerance zone of 0.05 *mm* specified in both the radial and axial directions, relative to a rotational axis shown as datum plane A. In the former (radial direction) case, the largest permissible deviation between the maximum and minimum radial measurement around the axis of rotation is 0.05 *mm*. In the latter (axial direction) case, acceptable parts cannot exhibit a discernable difference in radii from the axis of rotation that exceeds the specified tolerance zone of 0.05 *mm* in the axial direction. The surface profile of the part, for every radius, must fit within the cylindrical zone defined by the tolerance.

Measurements of circular run-out are independent. Gauges placed at different positions along the surface will likely have different zero points and therefore are not directly comparable. In the case of total run-out all measuring positions are strictly related and have a common zero point. The specified tolerance is over the entire surface of the work piece. Figure 4-13 illustrates total run-out in both the radial and axial directions. Where the radial direction is specified, the tolerance zone is the volume between two concentric cylinders from the common axis A-B defined by the datum systems A and B. Acceptable work pieces will be those whose actual surfaces lie between the two cylinders with radii that differ by 0.05 *mm*.

Geometric tolerancing is concerned with whether individual parts comply with specs.

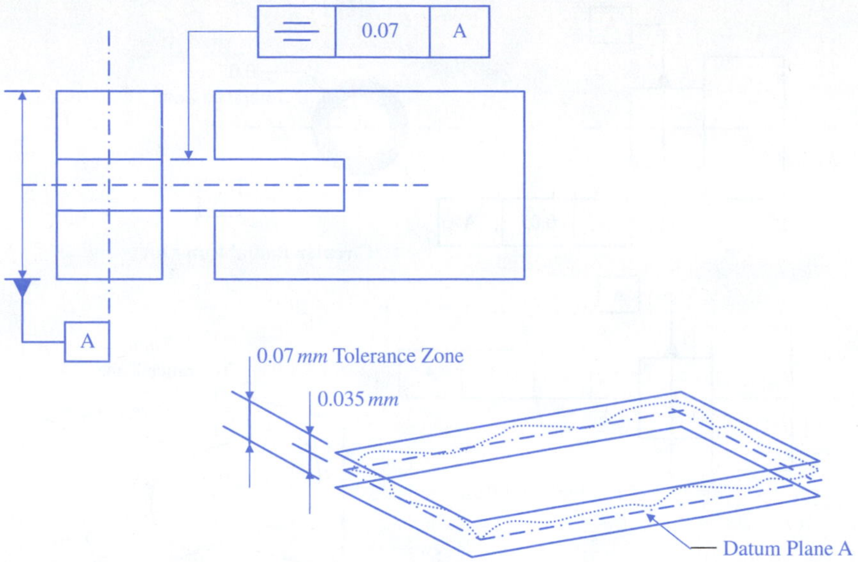

FIGURE 4-10. Symmetry Tolerance of a Slot Related to a Plane of Outer Surfaces

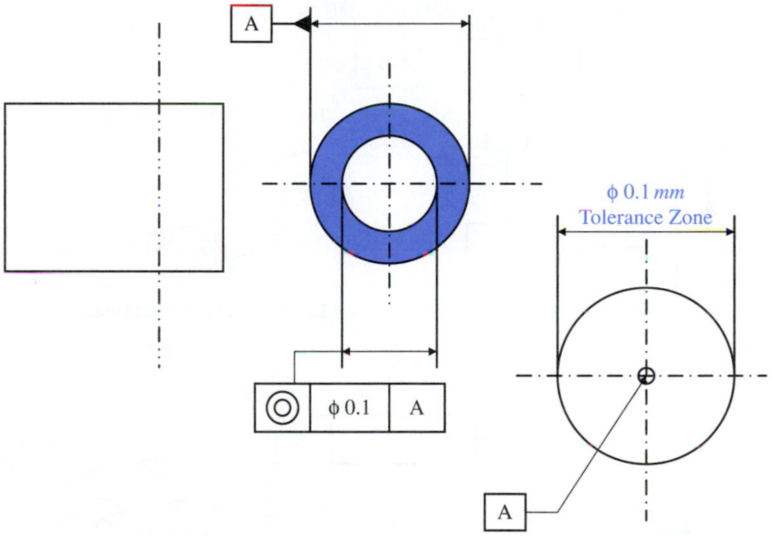

FIGURE 4-11. Concentricity Tolerance of a Circle Center Point Related to Another Point in the Same Plane

When the direction of measure is axial, the tolerance zone is defined by two parallel planes that are perpendicular to the datum axis, A, and are separated by the tolerance $\tau = 0.05$ *mm*. The difference between the axial total run-out and the circular total run-out is that the positions along the axis for the cylindrical sections (of width $\tau = 0.5$) for total run-out must be the same for every section.

4.1.4 Statistical Tolerancing

Geometric tolerancing assigns specifications to part parameters based on *worst-case* scenarios. As discussed previously, the factors used to determine the worst-case tolerances can be based on function, manufacturability, or ease of assembly. Parts that comply with worst-case geometric specifications should be completely interchangeable.

There are some cases, however, where the interchangeability of parts is not as important as a "derived" parameter such as clearance or a composite dimension, which results when two or more component parts are joined in an assembly. Refer to Figure 4-14. With such composite dimensions it is possible to assembly two parts that are out of spec individually (e.g., an oversized and an undersized) with a resultant assembly that meets specifications. The ability to use out-of-specification components enables the product designer to widen the component specifications and potentially lower

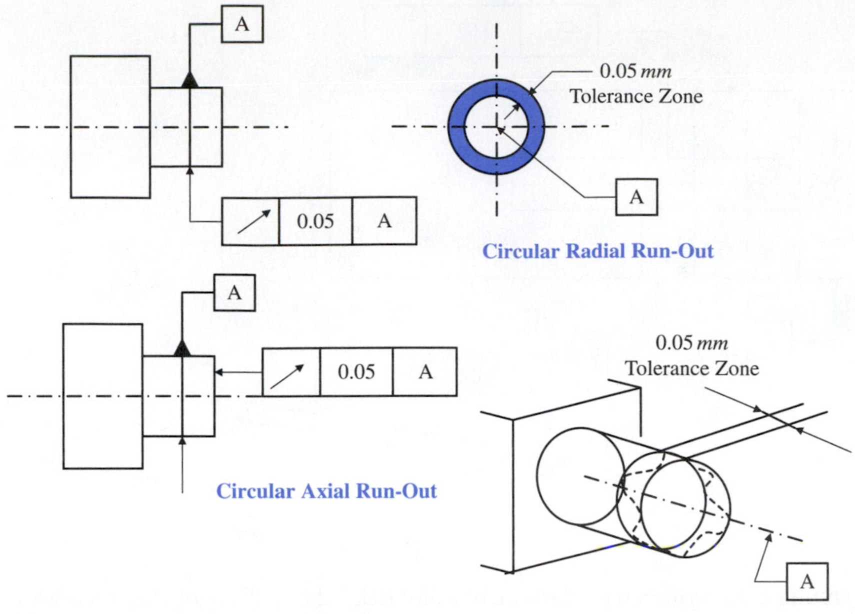

FIGURE 4-12. Circular Run-out Tolerances Related to a Datum Plane

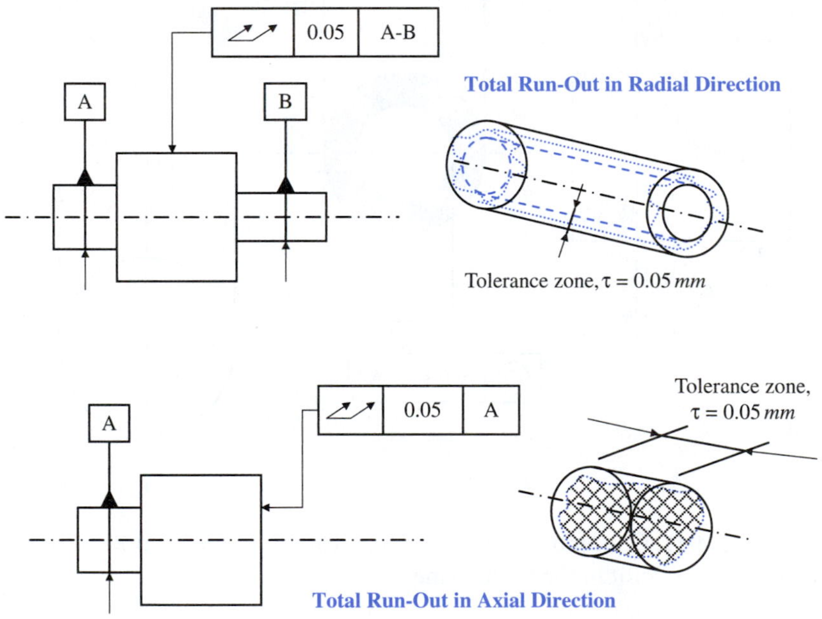

FIGURE 4-13. Total Run-out Tolerances Related to a Datum Plane

the cost of scrap and rework without adversely affecting the quality of the final assembly.

While geometric tolerancing is concerned with whether an *individual* part is acceptable, **statistical tolerancing** focuses on the acceptability of an entire *population* of parts. The notion of a population implies that, due to variability in the production process, individual parts will differ in their geometric characteristics. A population can be a specific production batch or parts produced over some specified period of time. When the focus is on populations (e.g., a production batch) instead of individual parts, two observations can be made concerning tolerances.

1. A population with parts that are on the average out of the tolerance range may contain some parts that are within specification.
2. A population with parts that are on the average within the tolerance range may contain some parts that are outside the specification limits.

Whether or not the manufacture of out-of-specification parts is inevitable is a capability issue, a topic which will be discussed in some depth in Chapter 9.

4.1.4.1 Fundamental Statistical Concepts. Statistical tolerancing requires a fundamental understanding of probability

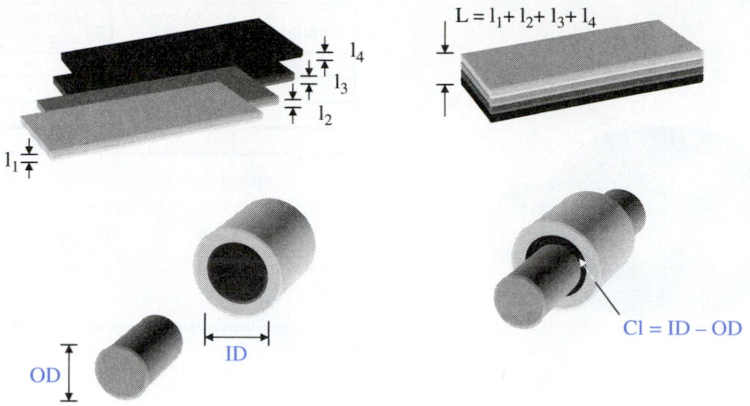

FIGURE 4-14. Composite and Clearance Dimensions

theory. When any geometric characteristic of an individual part is measured, the measurement (x) is a random variable that belongs to the population of all parts produced. The set of all possible values in the population is called the distribution of the random variable and can be described in terms of a probability density function $f(x)$. When sample measurements are taken from a specific process, this raw data can be summarized in a bar chart called a histogram. The histogram is an approximation of the underlying density function $f(x)$ and can provide valuable insights into important process parameters such as the mean, spread, and shape of the distribution. If $f(x)$ is known, the following equations can be used to calculate the mean and variance of the underlying distribution.

Statistical tolerancing is concerned with the acceptability of an entire population of parts.

For discrete distributions:

$$\text{Mean}(x) = E(x) = \mu = \sum_i x_i f(x_i) \quad \text{4-1}$$

$$\text{Var}(x) = \sigma^2 = E(x^2) - \mu^2 = \sum_i (x_i - \mu)^2 f(x_i) \quad \text{4-2}$$

For continuous distributions:

$$Mean(x) = E(x) = \mu = \int_{-\infty}^{\infty} x f(x)dx \quad \text{4-3}$$

$$Var(x) = \sigma^2 = E(x^2) - \mu^2 = \int_{-\infty}^{\infty} (x - \mu)^2 f(x)dx \quad \text{4-4}$$

A parameter such as a composite dimension or clearance can be represented by a sum or difference of individual part dimensions.

Composite

$$y = x_1 + x_2 + \cdots + x_n \quad \text{4-5}$$

$$\mu_y = \mu_{x_1} + \mu_{x_2} + \cdots + \mu_{x_n} \quad \text{4-6}$$

$$\sigma_y^2 = \sigma_{x_1}^2 + \sigma_{x_2}^2 + \cdots + \sigma_{x_n}^2 \quad \text{4-7}$$

Two-component Clearance

$$y = x_1 - x_2 \quad \text{4-8}$$

$$\mu_y = \mu_{x_1} - \mu_{x_2} \quad \text{4-9}$$

$$\sigma_y^2 = \sigma_{x_1}^2 + \sigma_{x_2}^2 \quad \text{4-10}$$

Mixed

$$y = x_1 - x_2 + x_3 \cdots + x_n \quad \text{4-11}$$

$$\mu_y = \mu_{x_1} - \mu_{x_2} + \mu_{x_3} \cdots + \mu_{x_n} \quad \text{4-12}$$

$$\sigma_y^2 = \sigma_{x_1}^2 + \sigma_{x_2}^2 + \sigma_{x_3}^2 \cdots + \sigma_{x_n}^2 \quad \text{4-13}$$

4.1.4.2 Statistical Assumptions. The validity of Equations 4-7, 4-10, and 4-13 require **independence** of all component dimensions. On the other hand, Equations 4-6, 4-9, and 4-12 will hold even if all components are not independent. The distributions of individual components need not be normal for Equations 4-6, 4-7, 4-9, 4-10, 4-12, and 4-13 to be valid. However, if each distribution $f(x_i)$ is normally distributed, then $f(y)$ will also follow a normal distribution. If any of the $f(x_i)$ are not normal, then $f(y)$ will still approach normality as the number of components increases. The validity of Equations 4-6, 4-7, 4-9, 4-10, 4-12, and 4-13 depend on an important final assumption: **randomness.** It is important that parts be randomly mated during the assembly process.

4.1.4.3 Statistical Tolerancing Implications. Consider the part shown in Figure 4-15. The overall length, L, is determined by summing the lengths of each of the components A, B, and C, respectively. The *worst case* approach would determine the specifications of L based on combining the shortest possible lengths of all components to obtain the smallest L, and the longest possible lengths of all components to obtain the largest L. This method is referred to as **stacking tolerances** and may be valid if the component dimensions are not independent or if the mating process is nonrandom.

However, in most industrial applications, components are manufactured under conditions where the assumption of independence seems reasonable and parts are sufficiently

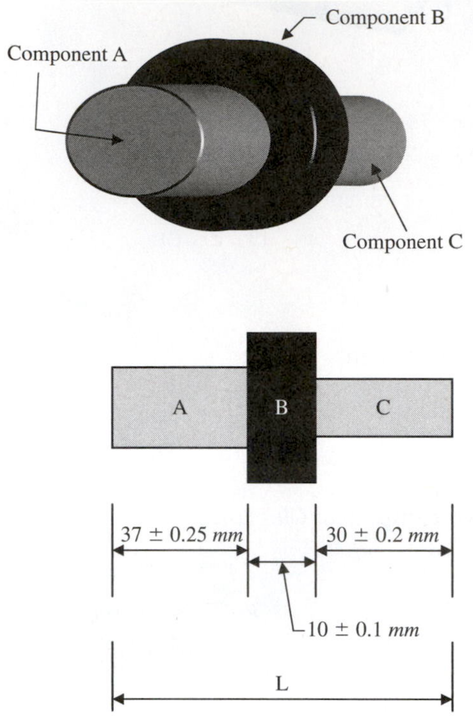

Component	Average Length	Shortest Length	Longest Length
A	37	36.75	37.25
B	10	9.9	10.1
C	30	29.8	30.2
L	77	76.45	77.55

Specifications for L Based on Stacking Individual Component Tolerances: 77 ± 0.55 *mm*

$$\mu_A = 37 \quad \sigma_A = \frac{37.25 - 36.75}{6} = \frac{0.50}{6} = 0.0833$$
$$\mu_B = 10 \quad \sigma_B = \frac{10.1 - 9.9}{6} = \frac{0.20}{6} = 0.0333$$
$$\mu_C = 30 \quad \sigma_C = \frac{30.2 - 29.8}{6} = \frac{0.40}{6} = 0.0667$$
$$\mu_A + \mu_B + \mu_C = 37 + 10 + 30 = 77$$
$$\sigma_L^2 = \sigma_A^2 + \sigma_B^2 + \sigma_C^2 = (0.0833)^2 + (0.0333)^2 + (0.0667)^2$$
$$\sigma_L^2 = 0.0069 + 0.0011 + 0.0044 = 0.0125$$
$$\sigma_L = \sqrt{0.0125} = 0.1118$$
$$3\sigma_L = 3(0.1118) = 0.335$$

Overall Lengths for L Actually Produced in Assembly: 77 ± 0.335 *mm*

FIGURE 4-15. Stack Up Tolerancing Example

mixed prior to assembly, so that one can also assume a process of random mating. Under these conditions the probability of actually achieving worst case scenarios is minuscule, so for practical purposes we can disregard the possibility of such cases.

In the foregoing example, let us assume that all components are produced within their respective specification ranges—that is, nearly all of the process output is within the stipulated specification interval. We know from our study of statistics that if a process is stable[4] almost all of its output will be contained within an interval that is six standard deviations wide.[5] Hence we can estimate the value of the standard deviation for each component as one-sixth its respective tolerance interval.

$$\sigma_A = \frac{37.25 - 36.75}{6} = 0.0833$$
$$\sigma_B = \frac{10.1 - 9.9}{6} = 0.0333$$
$$\sigma_C = \frac{30.2 - 29.8}{6} = 0.0667$$

[4]A stable process is one that is in a state of statistical control, a concept that is covered in detail in Chapter 8.

[5]If a distribution is stable and normal, a Six Sigma interval will capture 99.73% of its members; if the distribution is stable and non-normal then a Six Sigma interval will capture between 95% and 99% of its members.

The average of the distribution of overall lengths L will be the sum of the averages of each of the component distributions and the variance will be the sum of the variances. Then, the resulting spread of values of L actually produced can be estimated by six standard deviations of the overall lengths.

$$\mu_A + \mu_B + \mu_C = 37 + 10 + 30 = 77$$
$$\sigma_L^2 = \sigma_A^2 + \sigma_B^2 + \sigma_C^2 = (0.0833)^2 + (0.0333)^2 + (0.0667)^2$$
$$\sigma_L^2 = 0.0069 + 0.0011 + 0.0044 = 0.0125$$
$$\sigma_L = \sqrt{0.0125} = 0.1118$$
$$3\sigma_L = 3(0.1118) = 0.335$$

As we can see the actual range of the final composite dimensions, L, will be only one-third the engineering tolerance range (0.335 *mm* compared to 0.55 *mm*).

From a quality perspective there appears to be an inherent inconsistency between the specifications of the components and the specification of the assembly. Either the specifications on L are unnecessarily wide or the specifications on the components are too tight.

What are some of the implications? Referring again to Figure 4-15 let us assume that, compared to components A and C, the component B dimension is difficult to control. If components A and C are maintained within the tolerance range, it might be of interest to determine how far component B can be permitted to drift out of its individual tolerance range before effecting the acceptable tolerance range of the overall length L.

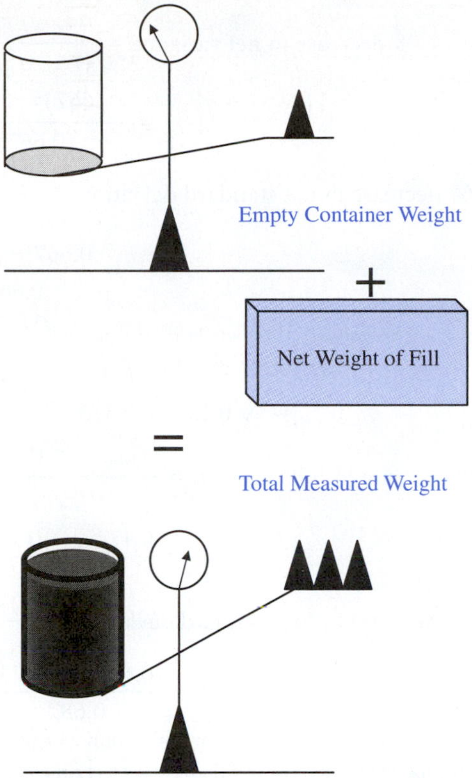

x_{empty} = weight of empty container

x_{full} = weight of filled container

$$x_{net} = x_{full} - x_{empty}\,;\ \mu_{net} = \mu_{full} - \mu_{empty}$$

$$\sigma^2_{net} = \sigma^2_{full} + \sigma^2_{empty}$$

Label Weight = 454 *gm*

Specifications on Net Weight: 454 *gm* ± 2 *gm*

Specifications on Empty Containers: 20 *gm* ± 0.5 *gm*

What specifications must production meet on filled container weights?

$$\mu_{net} = 454 = \mu_{full} - \mu_{empty} = \mu_{full} - 20$$

$$\mu_{full} = 454 + 20 = 474$$

$$\sigma^2_{full} = \sigma^2_{net} + \sigma^2_{empty} = \left(\frac{2gm}{3}\right)^2 + \left(\frac{0.5gm}{3}\right)^2 = 0.472$$

$$\sigma_{full} = \sqrt{0.472} = 0.687$$

Specifications on Filled Containers:

$$474gm \pm 3\,(0.687) = 474 \pm 2.061gm$$

FIGURE 4-16. Derived Weight/Net Weight from Filling Operation

We shall start by again assuming that the standard deviation for the overall length L is equal to one-sixth the allowed tolerance.

$$\sigma_L = \frac{2(0.55)}{6} = 0.1833$$

The variance of the overall length L is the square of the standard deviation.

$$\sigma_L^2 = (0.1833)^2 = 0.0336$$

The variance of the overall length L is the sum of the variances for each of the components that combine to create L.

$$\sigma_L^2 = \sigma_A^2 + \sigma_B^2 + \sigma_C^2 = 0.0069 + \sigma_B^2 + 0.0044 = 0.0336$$

From this expression we can calculate the variance, standard deviation, and tolerance interval for component B.

$$\sigma_B^2 = 0.0223$$

$$\sigma_B = \sqrt{0.0223} = 0.149$$

$$6\sigma_B = 0.896$$

Note that this represents a 1927% increase in the variance required to exactly meet the specs. The longest and shortest component lengths can now be computed.

Longest expected length = 10 + 3(0.149) = 10.447

Shortest expected length = 10 − 3(0.149) = 9.553

Clearly this permits some of component B parts to be out of spec since the specs call for parts no longer than 10.1 *mm* or shorter than 9.9 *mm*. This means that component B can drift out of spec to a certain extent without generating out-of-spec assemblies. If we assume normality we can consult the normal tables found in Table II of Appendix B to estimate what percentage out of spec can be tolerated.

$$z_u = \frac{10.1 - 10}{0.149} = \frac{0.1}{0.149} = 0.671$$

$$\Phi(0.671) = 0.221 \Rightarrow 22.1\%$$

$$z_l = \frac{9.9 - 10}{0.149} = \frac{-0.1}{0.149} = -0.67$$

$$\Phi(-0.671) = 0.221 \Rightarrow 22.1\%$$

Hence one can produce a total of 44.2% of component B parts outside the specification limits, provided that the other two components are within spec.

4.1.4.4 Net Weight Example. Let us consider a filling operation where specifications have been placed on the weight of the contents added to an empty container as illustrated in Figure 4-16. The variables that can be directly measured are the total weight of the filled containers and the weight of the empty containers. The net weight—the parameter of interest—has to be indirectly computed using statistical tolerancing methods.

We can see that with label weight specifications of 454 *gm* ± 2 *gm* and container specs of 20 *gm* ± 0.5 *gm*, the production process must comply with filled container specs of 474 *gm* ± 2.061 *gm*. This assumes that the process producing the empty containers (which is typically outsourced) and the filling process are in statistical control[6] and on target.

[6]The concept of statistical control is discussed at length in Chapters 8–12.

An alternative scenario is one where the product label weight represents a specified *minimum*. If this is the case, then it is not possible to target the process on 454, because to do so would result in at least 50% of the product underweight (assuming a normal-shaped distribution). Therefore, overfills (that is, fills that exceed the label weight) are inevitable as shown in Figure 4-17.

When the label weight represents an absolute minimum, there is an incentive for the producer to reduce the standard deviation of the fill distribution by as much as possible. As shown by the dotted distribution at the bottom of Figure 4-17 this will make it possible to reduce the average fills while complying with the minimum fill condition. Suppose the production manager of this process wants to know by how much the standard deviation of fill weights would have to be reduced in order to cut the average overfill in half—that is, from 2 grams to 1 gram. As the following calculations show, to reduce the average overfill by 50% requires a 75% reduction in the net fill variance, a 50% reduction in the net fill standard deviation, a 70.6% reduction in the full fill variance, and a 45.7% reduction in the full fill standard deviation.

$$\text{Average Overfill} = \mu_{net\,(new)} - 454\ gm = 1\ gm$$

$$\mu_{net\,(new)} = 454\ gm + 1\ gm = 455\ gm$$

$$\mu_{net\,(new)} = 454\ gm + 3\sigma_{net} = 455\ gm$$

$$\sigma_{net\,(new)} = \frac{455\ gm - 454\ gm}{3} = 0.333\ gm$$

$$\sigma^2_{net\,(new)} = 0.111$$

$$\%\ \text{decrease in net variance} = \frac{\sigma^2_{net\,(old)} - \sigma^2_{net\,(new)}}{\sigma^2_{net\,(old)}} = \frac{(0.667)^2 - 0.111}{(0.667)^2} = 75\%$$

$$\%\ \text{decrease in net standard deviation} = \frac{\sigma_{net\,(old)} - \sigma_{net\,(new)}}{\sigma_{net\,(old)}} = \frac{0.667 - 0.333}{0.667} = 50\%$$

$$\sigma^2_{full\,(new)} = \sigma^2_{net\,(new)} + \sigma^2_{empty} = (^{1\ gm}/_{3})^2 + (^{0.5\ gm}/_{3})^2 = 0.139$$

$$\sigma_{full\,(new)} = \sqrt{0.139} = 0.373$$

$$\%\ \text{decrease in full variance} = \frac{\sigma^2_{full\,(old)} - \sigma^2_{full\,(new)}}{\sigma^2_{full\,(old)}} = \frac{0.472 - 0.139}{0.472} = 70.6\%$$

$$\%\ \text{decrease in full standard deviation} = \frac{\sigma_{full(old)} - \sigma_{full(new)}}{\sigma_{full(old)}} = \frac{0.687 - 0.373}{0.687} = 45.7\%$$

For a manufacturing process such as this there is a significant incentive to get the fill volumes to consistently be as close to the target as practicable. As previously noted, if the average is exactly on target a symmetrical distribution of fill volumes (such as a normal distribution) would result in half (50%) of the fills less than target. The only way to counteract that is to deliberately

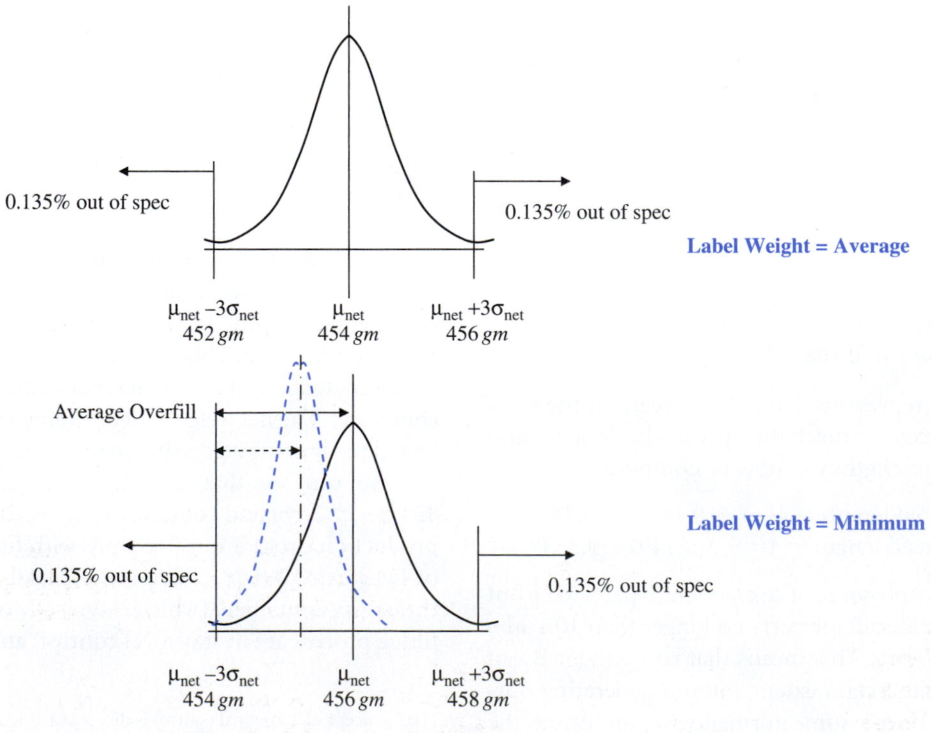

FIGURE 4-17. Impact of Fill Variance on Average Overfill

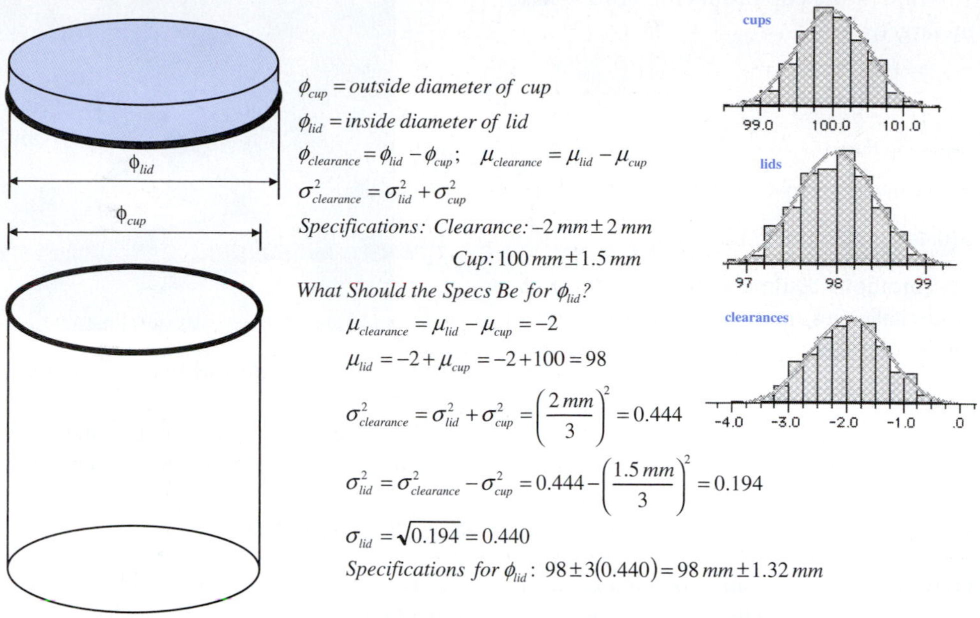

ϕ_{lid}
ϕ_{cup}
ϕ_{cup} = outside diameter of cup
ϕ_{lid} = inside diameter of lid
$\phi_{clearance} = \phi_{lid} - \phi_{cup}$; $\mu_{clearance} = \mu_{lid} - \mu_{cup}$
$\sigma^2_{clearance} = \sigma^2_{lid} + \sigma^2_{cup}$
Specifications: Clearance: −2 mm ± 2 mm
Cup: 100 mm ± 1.5 mm
What Should the Specs Be for ϕ_{lid}?
$\mu_{clearance} = \mu_{lid} - \mu_{cup} = -2$
$\mu_{lid} = -2 + \mu_{cup} = -2 + 100 = 98$
$\sigma^2_{clearance} = \sigma^2_{lid} + \sigma^2_{cup} = \left(\frac{2\ mm}{3}\right)^2 = 0.444$
$\sigma^2_{lid} = \sigma^2_{clearance} - \sigma^2_{cup} = 0.444 - \left(\frac{1.5\ mm}{3}\right)^2 = 0.194$
$\sigma_{lid} = \sqrt{0.194} = 0.440$
Specifications for ϕ_{lid}: $98 \pm 3(0.440) = 98\ mm \pm 1.32\ mm$
cups
99.0 100.0 101.0
lids
97 98 99
clearances
-4.0 -3.0 -2.0 -1.0 .0

FIGURE 4-18. Lid Fit Example

overfill. If overfills are excessive the extra material is "free" to the customer and profits are reduced. Too excessive an overfill, or too great a variability in fills, can cause line spillage and the necessity for messy and expensive cleanups. If containers are underfilled by a discernable amount, customers may feel cheated or, worse still, may fear that product tampering has occurred between the points of production and consumption. The answer to all of these issues lies in the ability to consistently control production targeting (maintain constant average fills) and reduce the variability in fills (achieve and sustain a relatively small error, plus and minus, in individual fill volumes).

4.1.4.5 Lid Fit Example. Figure 4-18 illustrates an example of statistical tolerancing applied to a drinking cup with a lid that is supposed to be watertight. You can imagine that this product is similar to the paper drinking cups given out at fast-food establishments. The clearance specifications ensure an *interference* fit so that the lids will not only prevent spillage as required but also are not so loose that they fall off. In this case the clearance specifications should be established based on functionality. If the clearance is too large, the watertight integrity will be compromised and possibly the lids will fail to stay on. If the clearance is too small, the lids will not fit on as intended and will also fall off. Given a tolerance range on the cups, the problem is to compute the required specification range for the lids.

The critical dimensions that govern cap fit are the outside diameter of the cup, ϕ_{cup}, and the inside diameter of the lid, ϕ_{lid}. As Figure 4-18 shows, the specified clearance for cap fit, $\phi_{clearance}$, is $-2\ mm \pm 2\ mm$ which would yield a maximum clearance of 0 *mm* and a minimum clearance of −4 mm. The production specifications for ϕ_{cup} are $100\ mm \pm 1.5\ mm$. The upper specification limit of the cups (USL_{cup}) is equal to 101.5 *mm*, and the lower specification limit (LSL_{cup}) is equal to 98.5 *mm*. Given this information, the design engineer might wish to establish the appropriate production specifications for ϕ_{lid} to ensure compliance with the $\phi_{clearance}$ specifications.

The statistical mating of parts is governed by two parameters: the means of the distributions of the components to be mated and their respective standard deviations. In establishing the specification range for the lids we shall assume that the process producing the cups can be controlled statistically to meet the target (nominal) dimension on the average, and will produce the individual diameters within the specified tolerance range. That is

$$\mu_{cup} = 100\ mm \text{ and } \sigma_{cup} = \frac{USL_{cup} - \mu_{cup}}{3} = \frac{101.5\ mm - 100\ mm}{3} = 0.5\ mm$$

When the variances of the cup and lid combine, they must not exceed the variance allowed by the clearance specification:

$$\sigma_{clearance} = \frac{USL_{clearance} - \mu_{clearance}}{3} = \frac{0\ mm - (-2\ mm)}{3} = 0.667\ mm$$

Note that statistically it is the variances, not the standard deviations, that combine. Hence,

$$\sigma^2_{clearance} = (0.667)^2 = 0.444 = \sigma^2_{lid} + \sigma^2_{cup}$$

Then,

$$\sigma^2_{lid} = \sigma^2_{clearance} - \sigma^2_{cup} = 0.444 - (0.5)^2 = 0.194;$$
$$\text{and } \sigma_{lid} = \sqrt{0.194} = 0.440$$

Since statistically the means of the component distributions are additive,

$$\mu_{clearance} = -2 = \mu_{lid} - \mu_{cup}$$

Assuming the distribution of the cup producing process has an average that is on target,

$$\mu_{cup} = 100$$

Then,

$$\mu_{lid} = \mu_{clearance} + \mu_{cup} = -2 + 100 = 98$$

Therefore, the appropriate design specifications for the lid are

$$98 \pm 3(0.440) = 98\ mm \pm 1.32\ mm.$$

These design specifications assume statistical control of both components and challenges manufacturing to strive to consistently produce the cups and lids within their respective specification ranges on a sustained basis.

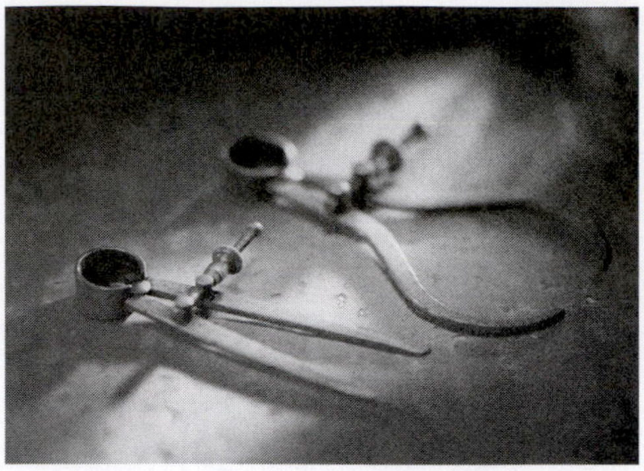

FIGURE 4–19. Outside and Inside Calipers.

4.2 DIMENSIONAL METROLOGY

4.2.1 Introduction to Metrology

The word **metrology** is derived from the Greek word *metron*—for measure, and has been defined by the International Bureau of Weights and Measures as "the science of measurement, embracing both experiment and theoretical determinations at any level of uncertainty in any field of Science and Technology."[7] An important concept in metrology is **traceability,** which establishes the degree to which measurements can be compared—for example, one measurement compared to the previous one, or to one taken last month or last year, or one taken in any facility anywhere in the world. Traceability is achieved through a process called **calibration,** which establishes the relationship between the output of a measurement process and the *known* value of a recognized measurement standard. Calibration is used whenever possible to check the accuracy of industrial measurement processes.

In establishing metrological traceability, measurement error must also be evaluated since any physical measurement consists of two components: an estimate of the true value of the measured attribute and the uncertainty of this estimate. The evaluation of measurement uncertainty provides insights into the **precision** of a measurement process—that is, how repeatable and reproducible are the measurement outputs over an extended period of time. These topics are covered in greater depth in Chapter 9.

In practice, metrology covers a wide range of measurement needs such as using scales to calculate bag weights, evaluating the particle size distribution in a dry chemical batch using a particle size analyzer, using a spectrophotometer to determine the level of whiteness in a crop of mushrooms, or using a dynamometer to measure the torque produced by an engine. Some measurement processes are simple and crude while others are high tech and require great skill. Some measures can be obtained on-line by process personnel whereas others result from the use of specialized equipment by specially trained personnel in a laboratory environment.

It is beyond the scope of this text to present a comprehensive treatment of metrology and its variety of associated processes. We will limit our discussion to an overview of some of the tools commonly used for dimensional measurements.

4.2.2 Caliper

A **caliper** is a device that is used to measure the distance between two points with great precision.[8] This measuring tool has tips that are carefully adjusted to fit between the points to be measured and then the distance is read on an accompanying scale. Depending on the intended use various designs are available. These include inside, outside, divider, oddleg, vernier, digital, and dial calipers. Figure 4-19 illustrates inside calipers (lower) and outside calipers (upper).

Inside calipers are used to measure dimensions that are internal on a work piece and can be designed to be manually adjusted or can be fitted with an adjusting screw. Outside calipers are used to measure dimensions that are on external surfaces. With proper skill outside calipers can provide good accuracy and repeatability, and can be successfully applied to the measurement of relatively large dimensions such as huge diameter pipes.

A caliper is used to measure the distance between two points with great precision.

Divider calipers have sharp points on each tip enabling the measurement between two points on a surface, such as on maps or scale drawings. The two caliper points are placed between the respective points to be measured and then transferred to a ruler to translate the distance into the actual measurement. **Oddleg calipers,** also called **hermaphrodite calipers** or **oddleg jennys,** are a special type of caliper used to scribe a line on a work piece a set distance from its edge. A "bent" leg tracks along the edge of the work piece while the sharp point of the scriber makes its mark, ensuring a scribed line that is parallel to the edge.

Vernier calipers, illustrated in the lower portion of Figure 4-20, offer an improvement to the inside and outside calipers, since they are capable of measuring both internal dimensions (using a set of upper jaws) and external dimensions (using a set of lower jaws). Some versions also provide a depth probe that is attached to the movable head and slides along

[7]http://www.bipm.org/en/convention/wmd/2004/.

[8]More information on calipers can be obtained by consulting http://en.wikipedia.org/wiki/Caliper.

[9]Note that the vernier scale in Figure 4-19 is calibrated to read only to the nearest one-hundredth of a centimeter. Most vernier scales in practice read to the nearest one-hundredth of a millimeter.

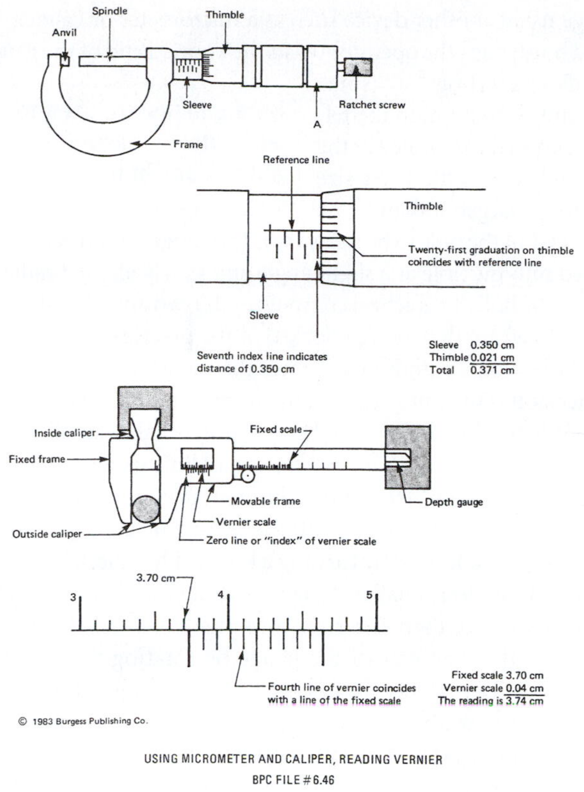

FIGURE 4–20. Diagrams of a Micrometer and Vernier Caliper.

the center of the body of the device, enabling it to reach deep grooves in a work piece.

The vernier caliper consists of a fixed scale, movable frame, and a zero index point. The nearest tenth of a millimeter is read directly from the fixed scale under the zero point on the movable frame. Then the nearest hundredth of a millimeter is read from the vernier scale where markings on the fixed and movable scales line up.[9] Some vernier scales include measurement readings in both metric and imperial units, providing precision to either one-hundredth of a millimeter, or one-thousandth of an inch. If better precision than this is required a micrometer, discussed below, is the device to use. Figure 4–21 shows a quality control technician using a vernier caliper to measure the thickness of an automobile clutch fricition plate.

Dial and **digital calipers** are refinements to vernier calipers that enhance user-friendliness and help reduce human error. In the case of a dial caliper, a gear rack drives a pointer on a circular dial that rotates once every inch, tenth of an inch, or 10 millimeters, allowing the operator to observe the reading directly without having to interpolate values on a vernier scale. However, the operator still has to add basic inches or tens of millimeters read from the slide on the caliper. The device is typically designed with a rotatable dial beneath the pointer, enabling differential measurements between two objects or the comparison of an object against some standard. On some models the caliper setting can be locked permitting consecutive and simple go/no go measurements.

A digital caliper offers a further refinement by replacing the analog dial with an electronic digital display. This design permits the operator to read the measurement directly, and some models can be switched directly between metric and imperial units. These devices also provide for zeroing a display at any point along the slide, providing the same differential measurement capability of the dial calipers, but without the need to read numbers that are often upside down. Some digital calipers also offer a feature that permits the freezing of a reading so that dimensions can be read and recorded even in awkward locations where the display cannot be directly seen by the operator.

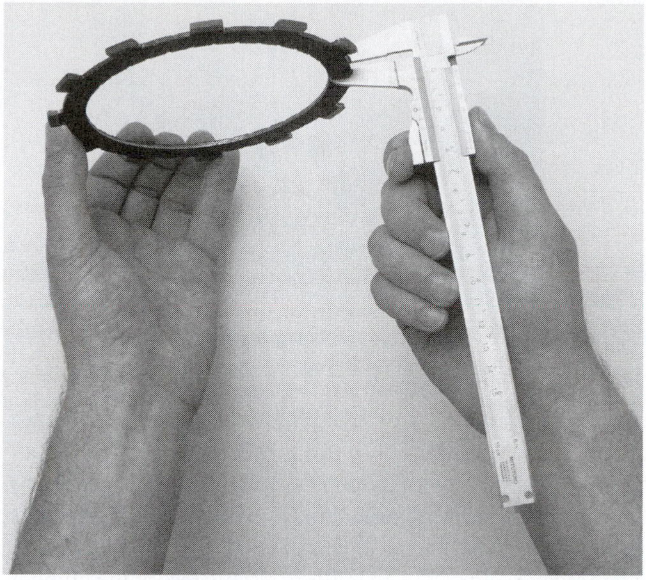

FIGURE 4–21. Calipers Can Be Used for Precise Point-to-Point Measurements.

Many digital calipers have the capability of providing a serial data output that can be interfaced with personal computers. This permits measurements to be taken and instantly stored in a spreadsheet, thus significantly reducing the time required to take and record data.

4.2.3 Micrometer

A **micrometer** is a device used for obtaining precise measurements of thicknesses, outer and inside diameters of shafts, and depths of slots.[10] This device offers some advantages over calipers—they are easy to use, provide high precision, and produce consistent readouts.

> **A micrometer is used for obtaining precise measurements of thicknesses, diameters, and depths.**

There are three common types of micrometers: external, internal, and depth. The upper portion of Figure 4-20 shows the various components of an external micrometer. Figure 4-22 illustrates its use, which is typically applied to the measure of diameters of wires, spheres, and shafts; and external surface dimensions. Internal micrometers are applied to the diameter of holes, and a depth micrometer is used to measure slots or steps.

When using these devices, precision is achieved through a fine-pitched screw mechanism. For example an inch-system micrometer has a spindle with 40 threads per inch, and one

[10]More information on micrometers can be obtained by consulting http://en.wikipedia.org/wiki/Micrometer.

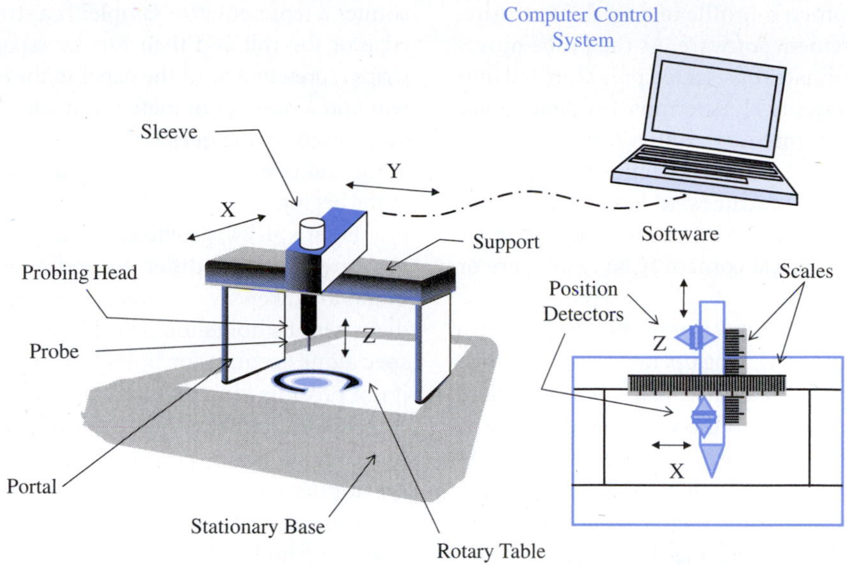

FIGURE 4-23. Coordinate Measuring Machine

to using these tools. Little skill is involved in using the gauges and interpreting the results, and the cost of gauge maintenance is relatively low compared to other measuring devices.

Go/no go gauges that are used to check the diameter of holes are referred to as plug gauges. The "go" end is used to test whether a bore diameter is too small—if it is smaller than the minimum specified, the gauge cannot be inserted, and the part is rejected. The "no go" end of the gauge tests whether the bore diameter exceeds the maximum specified. If the diameter is within the specification range the gauge cannot be inserted. If the diameter is too large, the gauge can be inserted, the test fails, and the part is rejected. If the part passes both tests—that is, the "go" end can be inserted and the "no go" end cannot be inserted—the part is accepted.

A go/no go gauge is used to collect data that signifies whether a part is acceptable or unacceptable.

A **thread gauge** is a special type of plug gauge designed to check threaded parts.[13] With this device the "go" end should screw fully into the work piece; the "no go" end should not. The "go" end of a gauge is often marked green and the "no go" end colored red to facilitate operator use.

Another variation of a go/no go gauge is a **gap gauge,** which typically has two pairs of jaws—the first (outermost) pair is set at the upper specification limit of the dimension of interest and the second (inner) pair is set at the lower specification limit. An acceptable part will pass through the first pair of jaws and stop at the second. Therefore, unlike the plug gauges both specification limits are checked on a single pass. Gap gauges can be used to measure outside distances such as the outside diameter of a pipe, or specialized versions can be used to measure dimensions of a threaded part. There are numerous other go/no go gauge designs in use that are applied to the inspection of lengths, heights, internal grooves, keyway dimensions, and splines. The designs of individual gauges are often customized to the specific application.

4.2.6 Other Gauges

Other types of gauges are available to measure various geometric dimensional tolerances. A **feeler gauge** consists of small blades of steel of different thicknesses used to measure clearances.[14] The respective thicknesses are marked on each blade and several can be combined to measure intermediate values. Feeler gauges are available in both metric and imperial units.

A **fishtail gauge** and a **center gauge** are used for checking angles during metal grinding processes.[15] As the name implies a **radius gauge** is used to measure radii.[16] The gauge is placed against the edge to be checked in front of a bright light source. Any light leakage between the blade of the gauge and the edge of the part indicates noncompliance. These gauges are available in both concave and convex profiles.

A feeler gauge is used to measure small clearances.

A **goniometer,** which gets its name from two Greek words, *gonia* (angle) and *metron* (measure), is an instrument that measures angles.[17] There are numerous design variations, each specialized to a particular application.

4.2.7 Coordinate Measuring Machine

A **coordinate measuring machine** (CMM), as the diagram in Figure 4-23 shows, is a high-tech device used for measuring geometric dimensions and consists of a measuring probe that

[13]More information on thread gauges, including illustrations, can be found at http://en.wikipedia.org/wiki/Thread_pitch_gauge.

[14]More information on feeler gauges, including illustrations, can be found at http://en.wikipedia.org/wiki/Feeler_gauge.

[15]More information on fishtail and center gauges, including illustrations, can be found at http://en.wikipedia.org/wiki/Center_gauge.

[16]More information on radius gauges, including illustrations, can be found at http://en.wikipedia.org/wiki/Radius_gauge.

[17]More information on goniometers, including illustrations, can be found at http://en.wikipedia.org/wiki/Goniometer.

is guided around the geometric profile of a work piece using a computer and measurement software. As the probe moves over the surface the coordinates of selected points are fed into the computer and used to accurately determine key dimensions. CMMs are available in numerous configurations of size, design, and probe technologies. Size variations range from freestanding floor-mounted machines, to benchtop models, to handheld portable designs. A CMM can be operated manually or by computer numerical control (CNC) software or personal computer (PC).

Probes can be mechanical, optical, or lasers. The latter two types are noncontact methods. During operation the probe detects a beginning point on the material from a chosen direction, normally a vector perpendicular to the work piece surface. A mechanical probe touches the material surface using a small ball which contacts the material with a measured force, causing a deflection that can be measured and used to compare with acceptable touch points within the tolerance limits specified.

Optical probes are moved over the surface similar to mechanical ones but do not make direct contact. A continuous picture is taken of the surface and edges and key points can be identified by contrasting black and white zones in the image. Laser probes detect the distance between reference points using deflections of light and what is called a half beam shadowing principle.

A coordinate measuring machine is a computer-driven machine that uses sensitive probes to take highly accurate measurements.

4.3 WHAT DOES CONFORMANCE MEAN?

What does it mean to conform to specifications? On the surface this seems like a simple question. One measures a part dimension, checks the dimension against the specification limits shown on the part drawing, and makes a decision accordingly. However, this issue is not always so clear cut. Consider a large roll of paper emerging from a paper-making process. The specifications require the paper product (e.g., 60-pound basis weight) to have a certain nominal tensile strength (e.g., 44 pounds in the machine direction and 25 pounds in the cross-web direction) with the stipulation of maximum and minimum allowable tensile strengths. These are important requirements to ensure that when the parent rolls are converted into individual consumer products, the material in those products will not break down or tear apart in use. The roll of paper represents the output of a continuous manufacturing process and presents us with a legitimate quality question. *Just what does conformance to the strength specifications mean?* Does conformance dictate that *any* sample piece taken from the roll from *any* location selected at random comply with the stated requirements? The obvious answer is yes, as subsequent converting operations have no way of discriminating the conforming from the nonconforming product. How then will overall product quality be tested? As the test for tensile strength is destructive, some form of sampling inspection is imperative. But what constitutes a representative sample? Is a strip cut from the leading edge of the roll and then further separated into smaller test strips representative of the paper in the entire roll? Is it better to wind off a number of meters and select small test areas at random? Even so, the question will remain as to whether product at the end of a roll is equivalent to product that was produced at the beginning of the roll or midway through it.

Figure 4-24 provides some examples showing how measurements can differ depending on where the measurements are taken. For example, if a specification states that the diameter of a hole should be $20\ mm \pm 0.5\ mm$, how does this spec alone account for holes that are not perfectly round? Is it not possible for a diameter measured at one position to be within spec and at another out of spec? What does conformance mean then? It cannot be left to luck as to where an inspector decides to position the measuring device.

Consider another example where a granular product is stored in cylindrical bins. Will a sample taken at the top, middle, bottom, sides, and so on all yield the same particle size distribution? Not likely. A certain plant produces a mineral product that is ground into a fine palletized form and the "whiteness" checked using a lab spectrophotometer. A small sample of product is placed in a bottle and shaken. A measurement is then taken. If the inspector shakes the sample again and remeasures, a different reading can result. If only a single test is performed, the difference between acceptance and rejection can be determined on particle size distribution and light reflectivity characteristics of those product pellets that happen to reside at the top of the test vessel.

Similar examples can be found in continuous processes where inspectors are physically limited to the extraction of samples from sampling ports that are strategically positioned along the value stream. What confidence is there that product samples taken from these ports is representative of the entire production lot? Is there a false belief in the capability of a particular process to produce a homogeneous output steam? *For quality to have any meaning conformance to specifications must be operationally defined in the context of sampling schemes, testing procedures, and customer expectations.*

Production variability can come from many sources. It can be due to *special* or *common* causes.[18] If special causes are present (e.g., a bad run of raw material), statistical process control techniques can detect and remove defective product before it gets to customers. Common causes, on the other hand, are systemic and can be positional within a production unit, short term across production units within a common production run, or longer term run-to-run differences. In determining conformance to specifications it is important that managers, engineers, and operations personnel understand variation and be able to describe the parameters of the distributions of products coming off production lines. Measuring processes must be aligned with requirements and target the right things. The relative size of measurement variation must also be well understood.[19] It is important that operational definitions exist that

[18]Special and common causes are defined in the Glossary and are also discussed in Chapter 7.

[19]Refer to the material concerning measurement studies covered in Chapter 9.

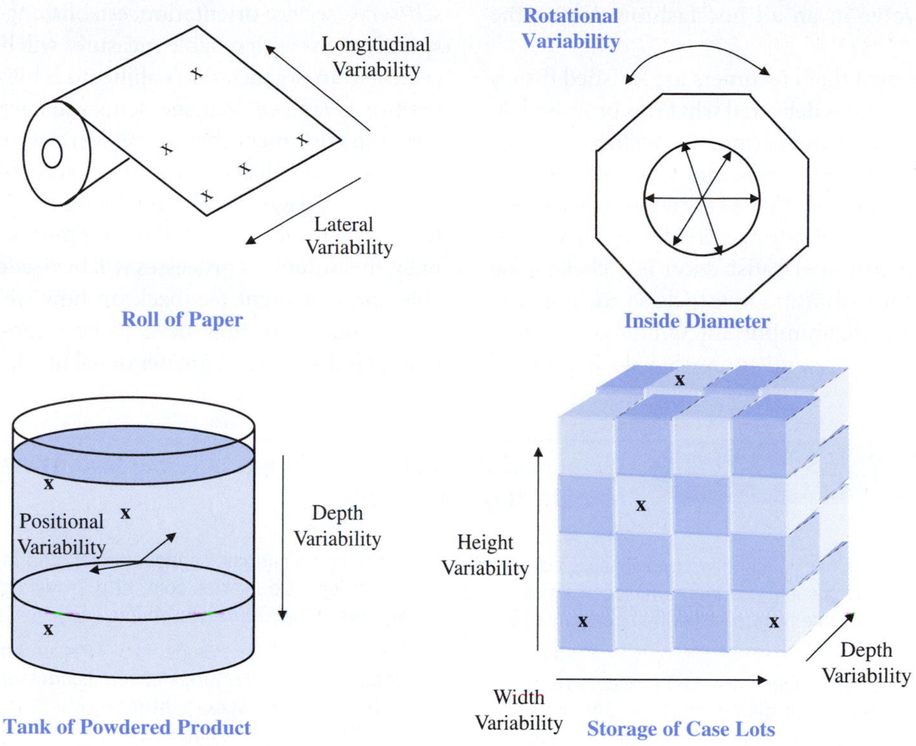

FIGURE 4-24. Variability Due to Position Measured

provide clear and concise instructions on what measurements to take, where and how to take them, how many units should be measured, how they should be selected, and the frequency of the measurement.

Operational definitions for how measures will be taken and how items should be selected for testing are essential.

Data samples are a microcosm of the production universe and, even though they are intended to be representative, they are still artifactual. At the end of the day customers are satisfied or dissatisfied based on the individual units of output they receive, not on summary production statistics that convey distributions and capability performance based on measurement processes that may be flawed.

4.4 SPECIFICATIONS IN THE SERVICE SECTOR

The concept of establishing specifications in the service sector is much more elusive than in the case where a tangible product is fabricated from a set of engineering drawings. The quality of a service outcome is difficult to measure and, unlike product geometry, it is difficult to achieve consistency in service delivery across customers. This is because quality of service is largely about perceptions—what one customer perceives to be excellence differs from one person to the next. Therefore service quality is about the ability of the service provider to create a sense, on the part of all customers, that value has been created, and to work on customer loyalty. This means that relationships must be built based on trust and goodwill.

Often service specifications are vague, using terms such as *customer satisfaction* to describe requirements. In some instances what it takes to satisfy (or even delight) customers can be reduced to quantifiable measures. For example, a certain fast-food chain advertised that if a customer's order is not filled in two minutes there will be no charge. Quality in this case can be easily measured by timing how long it takes to complete a service. However, such measures are rare and even when they exist it may be difficult to prove a direct correlation between the measure and customer satisfaction. For example, it may be possible to fill a customer's order well under the two-minute standard, but if in so doing the customer does not receive what was ordered, or the quality is off, the customer may be far from satisfied.

Service specifications are often vague as they relate to something that is intangible, variable, inseparable from the provider, and perishable.

Unlike most products, services are intangible and cannot be inspected in advance of actually making the purchase. Another important difference is that the customer typically comes in direct contact with the provider during the transaction, and the production and consumption processes may occur simultaneously or in very close proximity. A third aspect to setting service specifications is the inherent variability that does not exist at the same level for tangible products. The provision of a service is often a personalized process and the nature and quality is determined by who provides it, when it is provided, and where. As services commonly involve an interpersonal interaction between provider and customer, the determination of service requirements can evolve in an ad hoc fashion during the service transaction.

Research has revealed that customers are satisfied if they perceive that the provider has delivered what was promised, in a manner that was personal and caring, was willing to go the extra mile to meet customers' needs, and was responsive in handling problems and queries.[20] These are all somewhat nebulous and highly variable and setting performance specifications that will ensure customer satisfaction is a challenging reality. That does not mean that attempts to do so are futile and that goals and measures are unimportant. On the contrary, as the economy gravitates more and more to a do-it-yourself, self-serve, service orientation, establishing meaningful service specifications with reliable measures will become increasingly critical to an organization's ability to achieve and sustain competitive advantage. Managements will have to be more in tune with the customers they serve, learn what makes them tick, their likes and dislikes, and their needs. Processes will then need to be designed and staffed with trained personnel whose focus is on those things that matter most to customers. And finally, measurement processes will be needed that provide reliable and consistent feedback on how well the processes are doing and where they need to be improved. Service sector quality is discussed in greater detail in Chapter 12.

SHOWCASE of EXCELLENCE **Winners of the Malcolm Baldrige National Quality Award Manufacturing Category**

- 2001: Clarke American Checks, a printer of personalized checks and provider of checking-account and bill-paying accessories. http://www.clareamerican.com, http://www.nist.gov/public_affairs/clarke.htm.
- 2000: Dana Corporation, Spicer Driveshaft Division, North America's largest independent manufacturer of automotive driveshafts and related components, supplying customers in the United States and worldwide. http://www.spicerdriveshaft.com, http://www.nist.gov/public_affairs/baldrige00/Dana.htm.
- 2000: KARLEE Corp., a contract manufacturer of precision sheet metal and machined components for customers in the telecommunications, semiconductor, and medical-equipment industries. http://karlee.com, http://www.nist.gov/public_affairs/baldrige00/Karlee.htm.
- 1999: STMicroelectronics, Inc.—Region Americas, one of the world's top manufacturers of semiconductor integrated circuits. http://www.st.com, http://www.nist.gov/public_affairs/bald99/stmicroelectronics.html.
- 1998: Boeing Airlift and Tanker Programs, designer, developer, and manufacturer of the C-17 Globemaster 111 airlifter. http://www.quality.nist.gov/Boeing_98.htm.

4.6 SUMMARY OF KEY IDEAS

Key Idea: *Superior designs designate tolerances as critical to function, critical to manufacture, or critical to assembly.*

If quality is to be built at the source it is important that the design engineers communicate the importance of the tolerances they specify on product drawings. There are three ways that a specified tolerance can be critical. Critical to function (CTF) ties the tolerance directly to customer needs. Critical to manufacture (CTM) means that the tolerance impacts the parts manufacturability and is set based on the capabilities of manufacturing processes. Critical to assembly (CTA) means that a tolerance is important to the successful mating with other parts or subassemblies.

Key Idea: *Tolerances can be either dimensional or geometrical.*

Dimensional tolerances refer to measurements between two points or planes located on external surfaces, internal to the parts geometry, step sizes, or other critical distances. Geometric tolerancing defines the required form, orientation, location, or run-out of product features. Form tolerances include straightness, roundness, flatness, and cylindricity. Tolerances of orientation include parallelism, perpendicularity, and angularity; and tolerances of location are concerned with position concentricity, coaxiality, and symmetry. Finally, run-out is a measure of out-of-roundness and is applied to circular, cylindrical, and spherical forms.

Key Idea: *Geometric tolerancing is concerned with whether individual parts comply with specifications; statistical tolerancing is concerned with the acceptability of a population of parts.*

Geometric tolerances apply to critical dimensions on individual parts or product components. When parts are brought together (mated) into an assembly, the individual part dimensions are not as important as the ability for the parts to fit together and function as a composite unit. It is therefore possible to have an assembly meet a set of composite specifications with components that fail to meet their individual specifications. This is due to the way variables combine when parts are randomly mated. A composite set of specifications, based on stacking the tolerances of its individual components, which is a worst case parts-mating scenario, will typically be too loose. The probability of worst case mating of parts is practically zero. A better alternative, which will lead to tighter composite specifications and improved quality, is to establish composite specifications based on a statistical tolerancing approach.

Key Idea: *Metrology is the science of measurement; the application of metrology to the selection of the proper measurement device, and an understanding of the capabilities of that device, are critical to the proper collection and interpretation of process data.*

[20]Johnston, Robert. "Towards a Better Understanding of Service Excellence." *Managing Service Quality* 14, no. 2/3 (2004): 129–33.

A measurement process involves a person using a device, such as a gauge, to take a measurement that is hopefully close to, if not equal to, the true value of the dimension being measured. In reality there will always be measurement error that can arise due to bias, lack of precision, or resolution. Selection and maintenance of measuring devices are important, as well as consistent and thorough training. In evaluating measurement error the ability to effectively compare two measures is important. This is called metrological traceability.

There are many different devices available for taking data. In this chapter I introduced some of the more common tools used for measuring geometric dimensions. Included were calipers (used to measure the distance between two points), micrometers (used to measure thicknesses, diameters, and depths), bore gauges (used to measure the dimensions of holes), go/no go gauges (used to sort out parts that fail to meet specifications), feeler gauges (used to measure small clearances), fishtail and center gauges (used to check angles), radius gauges (used to measure the radii of circular geometries), and goniometers (used to measure angles).

On the high-tech end of the spectrum coordinate measuring machines (CMMs) use computer software to scan a part surface and, using probes, take accurate measurements of numerous key dimensions.

Key Idea: *Conformance to specifications must be operationally defined in the context of sampling schemes, testing procedures, and customer expectations.*

We have seen that specifications can be critical to function, manufacturability, or assembly. Nevertheless, conformance is sometimes not clear-cut, as a conclusion on compliance can depend on what is tested, how it is selected for testing, how it is tested, or who does the testing. If a process or storage facility is likely to possess some degree of is "homogeneity" the intended term? could it be "heterogeneity"? or if a measurement process is nonrepeatable, then the definition of conformance becomes an issue and should be operationally defined.

Key Idea: *Setting specifications in the service sector is different from a manufactured product; service requirements need to account for those factors that can lead to a satisfactory experience that is mutual between the customer and the service provider.*

Service specifications are often vague because they relate to something that is intangible, variable, inseparable from the provider, and perishable. The provision of a service can be a personalized process where the nature and quality is determined by who, when, and where a service is provided. It is also possible for the determination of service requirements to evolve in an ad hoc fashion during the service transaction itself. Customers are usually satisfied if four criteria are met: delivering on promises, delivering in a style that projects a personal touch and caring, being willing to go above and beyond the minimum requirements to please, and being responsive to problem solving and queries.

4.6 MIND EXPANDERS TO TEST YOUR UNDERSTANDING

4-1. Interpret each of the following tolerance frames.

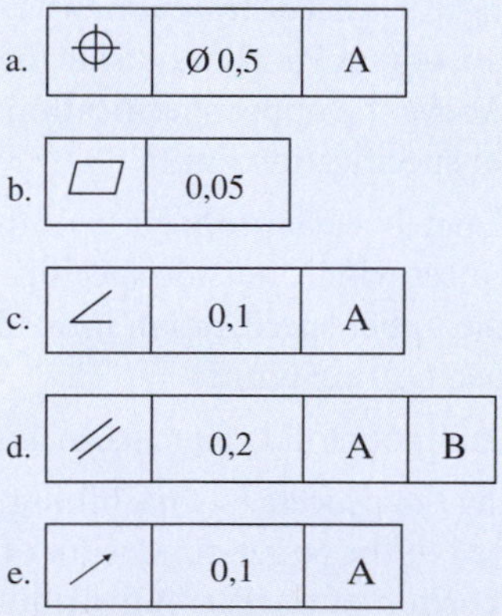

4-2. A certain production process produces two parts: a bearing and a shaft. In the final assembly the shaft and bearing are randomly mated and a critical specification is the clearance between the two parts. The distribution of the inside diameters of the bearing is normally distributed with a mean equal to 5.40 *mm* with a standard deviation of 0.735 *mm*. The distribution of the outside diameters of the shaft is normally distributed with a mean equal to 3.93 *mm* with a standard deviation of 0.413 *mm*.

a. Estimate the mean and variance of the distribution of clearances for the bearing-shaft assemblies.

b. Assuming that the mating of bearings and shafts is random estimate the proportion of assemblies that will not fit together. (*Hint:* The assemblies will not fit together if the clearance is less than or equal to zero.)

c. By how much would the variance in clearances have to be reduced so that the proportion of defective assemblies is less than 1%?

4-3. The research and development team of a medical device manufacturer is designing a new diagnostic test strip to detect the breath alcohol level. The materials are assembled as shown in the following figure.

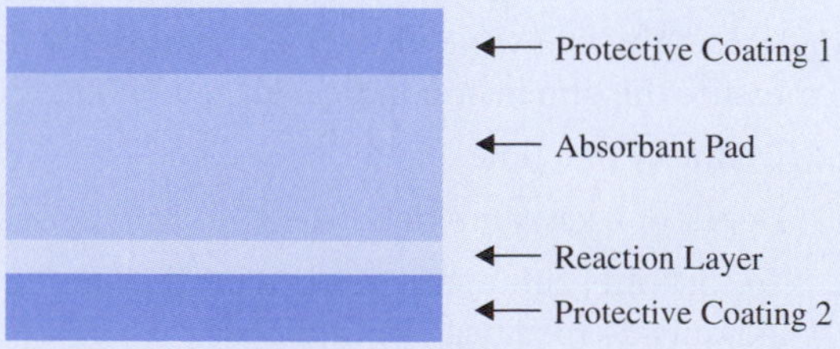

Production data for each of the four components is summarized in the following table. All measurements are in millimeters.

	Materials			
	Protective Coating 1	**Absorbant Pad**	**Reaction Layer**	**Protective Coating 2**
Mean	9.972	49.775	4.979	7.964
Standard Deviation	2.007	9.692	0.988	0.958
Distribution Shape	Normal	Normal	Normal	Normal

a. Compute an estimate of the mean and variance of the thickness of assembled strips.

b. Assuming random assembly, what is the probability that the thickness of a strip, selected at random, will exceed 75 *mm*?

4-4. A food processor manufactures a certain canned vegetable that has a printed label weight of 305 *gm*. The specifications for the net contents in each can are 305 *gm* ± 10 *gm*. Individual cans are randomly selected from the end of the filling line and weighed after sealing, and weighed. These are gross weights and reflect the weight of the empty cans, the lids, and the processed vegetable contents. The gross weights are in a state of statistical control with an average of 324.1 *gm* and a standard deviation of 4.5 *gm*. The cans and lids come from a supplier who provides data that shows that the processes producing these components is in statistical control with a process average and standard deviation for the cans of 15 *gm* and 2 *gm*, respectively; and for the lids the average and standard deviation are 3 *gm* and 0.3 *gm*, respectively.

a. Estimate the average and standard deviation of net weights of this canning operation.

b. Assuming a normal distribution, what percentage of the canning process is producing cans that are overweight (i.e., above the upper specification limit)? What percentage is underweight?

c. Assuming that the average of the canning operation can be brought to and controlled at the target weight of 305 *gm*, by what percentage would the variance of the gross fill weights have to be reduced so that the average fills are 3.5 standard deviations away from each specification limit? Assume that the variances of the cans and lids remain the same.

4-5. For each of the following identify an appropriate device to measure the dimension indicated.

a. Diameter of a hole

b. Depth of a keyway slot

c. To sort out shaft diameters in a production lot that are outside specification limits

d. Distance between the center of a hole and a particular edge of a part

e. Thickness of a silicon wafer

f. Threads on a pipe

g. Small, critical clearance between contacts in an electronic controller

4-6. Two machined parts must be assembled as shown in the diagram below. To function properly a clearance must be maintained between the two parts. The critical dimension on Part A has specifications of 1.8 ± 0.05 and the critical dimension on Part B has specifications of 1.4 ± 0.05 as indicated. The design engineers have specified that the clearance on each side be 0.2 ± 0.05. Production data indicates that for Part A the average dimension is 1.790 with a standard deviation of 0.0189; for Part B the average dimension is 1.406 and a standard deviation of 0.0220.

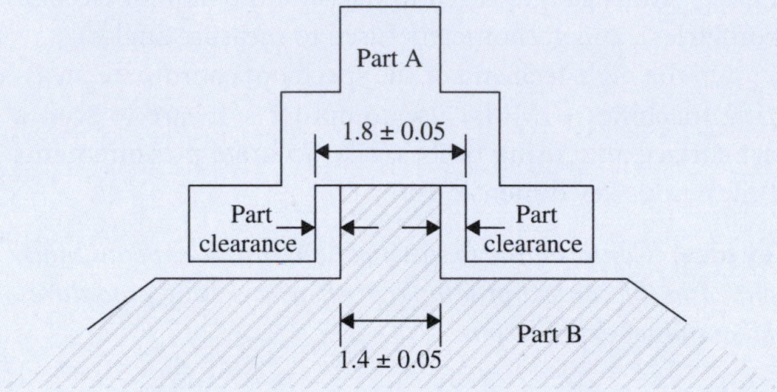

Tolerances for Machined Parts Assembly

a. Assuming a normal distribution, what proportion of the production of Part A is outside the specification limits? Above the upper specification limit? Below the lower specification limit?

b. Assuming a normal distribution, what proportion of the production of Part B is outside the specification limits? Above the upper specification limit? Below the lower specification limit?

c. Approximately what proportion of the distribution of clearances will lie outside specification limits? Above the upper specification limit? Below the lower specification limit?

d. What assumptions did you make in answering part c?

e. Assuming the process for machining Part A can be recentered to the target dimension of 1.8, estimate the proportion of the Part A distribution that would be out of specification on the upper and lower sides respectively.

f. If the process for Part A machining can be recentered to the target dimension of 1.8, what proportion of the clearance distribution will lie outside the specification limits on the high and low sides respectively?

g. If you assume that six standard deviations covers most of the distribution, can you suggest a tighter specification range for the clearance specifications?

4-7. The management of Greenfield Tire and Rubber has asked you to review the gum ply and wire reinforce thickness at one of its tire manufacturing facilities. The rework shop is receiving more than normal jobs that were not meeting the overall CTA (critical to assembly) specification of 0.25 *in* ± 0.1 *in*. Meeting these specifications on each assembly is important so that parts do not exceed the limitations of the curing mold. Each assembly consists of two layers of material: a gum ply layer and a wire reinforce layer. You have been provided with production data for the gum ply and wire reinforce processes respectively. These data are shown in the following two tables.

Sample Number	Gum Ply Layer Thickness (in)			
	1	2	3	4
1	0.1700	0.1087	0.0961	0.0856
2	0.1888	0.1362	0.2005	0.1371
3	0.1559	0.1320	0.1465	0.1610
4	0.1149	0.1566	0.1381	0.1219
5	0.1289	0.0913	0.1586	0.0903
6	0.1768	0.1150	0.1615	0.1278
7	0.1678	0.1357	0.1942	0.0928
8	0.1927	0.0879	0.1197	0.1379
9	0.1130	0.1590	0.1527	0.1363
10	0.1221	0.1814	0.1434	0.0931
11	0.1621	0.1392	0.0877	0.1392
12	0.1248	0.1322	0.1294	0.1434
13	0.1586	0.1777	0.1601	0.1533
14	0.1617	0.1146	0.1659	0.1385
15	0.1668	0.1311	0.1612	0.1847
16	0.1857	0.1360	0.1416	0.1249
17	0.1216	0.1292	0.1060	0.1207
18	0.1171	0.1406	0.1481	0.1827
19	0.1164	0.1064	0.1283	0.1832
20	0.1366	0.1189	0.1427	0.1566
21	0.1980	0.1784	0.1656	0.1574
22	0.1090	0.1083	0.1340	0.0707
23	0.1291	0.1076	0.1424	0.1833
24	0.1248	0.1285	0.1329	0.1407
25	0.0932	0.1474	0.1659	0.1623
26	0.1485	0.1483	0.1159	0.1278
27	0.1261	0.1458	0.0956	0.1718
28	0.1801	0.1698	0.1403	0.1184
29	0.1628	0.1056	0.1511	0.1421
30	0.1543	0.0906	0.1405	0.1685

Sample Number	Wire Reinforce Layer Thickness (in)			
	1	2	3	4
1	0.1118	0.1045	0.0889	0.0700
2	0.0981	0.0586	0.0949	0.0920
3	0.1114	0.0800	0.0990	0.1619
4	0.0834	0.1154	0.0911	0.0780
5	0.0858	0.1000	0.0930	0.0977
6	0.1100	0.1092	0.1197	0.0637
7	0.1085	0.1139	0.0927	0.1145
8	0.1237	0.1173	0.0687	0.0580
9	0.0875	0.1089	0.0482	0.1220
10	0.0795	0.0886	0.1373	0.1029
11	0.0877	0.0765	0.1174	0.0706
12	0.1346	0.1176	0.1059	0.1225
13	0.1095	0.1078	0.1172	0.0938
14	0.1058	0.0819	0.1343	0.1169
15	0.0919	0.1144	0.1256	0.1128
16	0.0878	0.0682	0.0950	0.0898
17	0.1004	0.0923	0.1074	0.0944
18	0.0903	0.1339	0.1195	0.0886
19	0.0969	0.1177	0.0743	0.0864
20	0.0736	0.0914	0.1014	0.0658
21	0.1245	0.1025	0.0774	0.1293
22	0.1181	0.0704	0.1012	0.0984
23	0.1216	0.1003	0.0776	0.1094
24	0.1086	0.0727	0.0844	0.0783
25	0.0855	0.0849	0.0536	0.0916
26	0.0820	0.1075	0.1387	0.1110
27	0.0924	0.1594	0.0867	0.0811
28	0.1142	0.0695	0.1081	0.0759
29	0.0944	0.1082	0.0974	0.0936
30	0.1282	0.1164	0.1130	0.1306

a. Use these 120 points of raw data provided for each of the components to estimate the parameters (μ and σ) of the distribution of the final thicknesses of individual gum ply wire reinforce assemblies. We shall assume that the processes are stable and repeatable.

b. Using the solution you obtained in part a, estimate the percentage of assemblies that will be outside the specification limits. How many parts per million is this?

c. What would you recommend that the management of Greenfield Tire and Rubber do in order to get the gum ply wire reinforce assembly thickness closer to the nominal specification?

d. If management is successful in achieving the target specification, what percentage can be expected to be outside specifications assuming that the process variances for each component remain unchanged? How many parts per million is this?

e. Assuming that both components are on target, by how much would the variance of both components have to be reduced in order to achieve no more than 3.4 parts per million outside either specification limit? Assume that the variances of each component will be in the same proportion of the total after the change as before.

4-8. SunTech Co. is developing a new window film that will prevent windows from shattering upon impact. The specifications on the film thickness are 370 ± 10 μm (micrometers). The film is made up of three layers: a dyed polyester film, metallic lining, and a scratch resistant coating. Data collected from the production of each layer component are presented in the table below. The specifications for each component are as follows (all measurements shown are in μm):

Polyester film	100 ± 5
Metallic lining	120 ± 2
Scratch-resistant coating	150 ± 3

a. Using the 50 data points provided for each component, estimate the mean, variance, and standard deviation for each component and also for the final assembled film thicknesses.

b. For each component calculate the estimated percentage of the relevant production output that will be outside the stipulated specification limits.

c. Estimate the percentage of the assembled films that will be outside the stipulated specification limits.

Polyester Layer

	Observation				
Sample	1	2	3	4	5
1	107.1397	99.2086	100.9042	100.0536	102.4504
2	100.8674	102.3427	100.6385	96.3919	100.1238
3	100.3446	100.3614	102.1023	101.2086	101.2238
4	103.2022	98.9172	104.1252	100.0115	98.3616
5	104.7148	100.7343	101.6746	100.3884	104.4914
6	100.8449	103.0214	102.8825	103.3160	102.7463
7	99.1435	100.3274	100.6272	98.9970	101.8641
8	98.8761	99.3938	99.2531	101.7511	99.3864
9	102.7593	100.6944	97.4749	98.8227	104.2506
10	96.3279	100.3979	101.6386	102.2754	100.7074

Metallic Lining

	Observation				
Sample	1	2	3	4	5
1	118.9965	119.4768	120.1195	121.3607	119.5365
2	119.9275	118.5908	119.0238	120.1219	119.7238
3	121.0552	120.7206	120.5632	120.0192	118.4176
4	120.4201	121.724	118.9609	120.8402	119.2126
5	120.0643	119.5225	120.0839	119.8735	119.9827
6	120.4571	119.2172	119.1524	119.5827	120.6158
7	118.7643	118.5731	119.9608	118.6721	118.0054
8	119.5879	120.8373	121.3822	119.2283	119.0834
9	120.6605	121.2018	118.9951	120.5087	118.8764
10	119.3968	120.8473	120.6957	120.9016	118.3809

Scratch Resistance Coating					
	Observation				
Sample	1	2	3	4	5
1	147.4471	150.0942	150.9554	150.8616	151.3109
2	153.342	154.1938	149.4951	150.4375	150.8502
3	149.8125	151.6689	149.6252	151.7198	149.5837
4	152.2016	150.9156	150.873	152.1419	149.4748
5	149.2191	149.002	151.3166	148.825	150.9926
6	150.5601	150.2813	150.0309	151.1049	150.8316
7	150.5922	151.0496	148.2893	151.0391	148.8795
8	151.8168	152.6681	148.701	149.4111	151.992
9	150.4225	151.3513	147.2515	148.5842	148.5399
10	151.0835	149.8955	153.0184	150.0018	147.9945

Sample	Layer 1				
1	0.0552	0.0565	0.0573	0.0524	
2	0.0537	0.0550	0.0568	0.0557	
3	0.0561	0.0513	0.0519	0.0560	
4	0.0543	0.0570	0.0564	0.0576	
5	0.0534	0.0580	0.0543	0.0551	
6	0.0559	0.0550	0.0555	0.0555	
7	0.0528	0.0576	0.0566	0.0579	
8	0.0561	0.0559	0.0541	0.0537	
9	0.0537	0.0564	0.0549	0.0515	
10	0.0553	0.0555	0.0537	0.0546	
11	0.0574	0.0564	0.0531	0.0538	
12	0.0539	0.0543	0.0558	0.0539	
13	0.0556	0.0559	0.0544	0.0550	
14	0.0559	0.0557	0.0531	0.0540	
15	0.0540	0.0543	0.0550	0.0544	
16	0.0534	0.0560	0.0556	0.0505	
17	0.0564	0.0553	0.0565	0.0553	
18	0.0533	0.0531	0.0516	0.0583	
19	0.0581	0.0556	0.0579	0.0567	
20	0.0559	0.0559	0.0553	0.0544	
21	0.0543	0.0542	0.0549	0.0571	
22	0.0557	0.0530	0.0568	0.0574	
23	0.0567	0.0553	0.0527	0.0549	
24	0.0512	0.0535	0.0552	0.0560	
25	0.0552	0.0563	0.0569	0.0568	
26	0.0552	0.0565	0.0559	0.0562	
27	0.0537	0.0529	0.0523	0.0532	
28	0.0552	0.0552	0.0546	0.0532	
29	0.0569	0.0555	0.0560	0.0556	
30	0.0538	0.0528	0.0541	0.0565	
					0.001634
					0.0550

Sample	Layer 2			
1	0.0259	0.0263	0.0263	0.0281
2	0.0272	0.0271	0.0275	0.0310
3	0.0266	0.0260	0.0257	0.0285
4	0.0289	0.0283	0.0270	0.0268
5	0.0247	0.0267	0.0275	0.0272
6	0.0270	0.0257	0.0267	0.0286
7	0.0294	0.0273	0.0278	0.0253
8	0.0265	0.0279	0.0264	0.0272
9	0.0261	0.0274	0.0275	0.0272
10	0.0296	0.0287	0.0277	0.0278
11	0.0274	0.0285	0.0259	0.0246
12	0.0287	0.0274	0.0257	0.0268
13	0.0284	0.0269	0.0278	0.0275
14	0.0273	0.0281	0.0275	0.0275
15	0.0261	0.0249	0.0290	0.0266
16	0.0280	0.0292	0.0290	0.0269
17	0.0273	0.0273	0.0276	0.0256
18	0.0251	0.0253	0.0259	0.0260
19	0.0259	0.0285	0.0275	0.0278
20	0.0259	0.0260	0.0273	0.0270
21	0.0262	0.0258	0.0261	0.0255
22	0.0276	0.0274	0.0283	0.0268
23	0.0267	0.0247	0.0243	0.0286
24	0.0263	0.0273	0.0258	0.0271
25	0.0260	0.0274	0.0266	0.0265
26	0.0245	0.0260	0.0268	0.0292
27	0.0263	0.0269	0.0259	0.0276
28	0.0265	0.0281	0.0260	0.0275
29	0.0262	0.0221	0.0270	0.0260
30	0.0294	0.0283	0.0274	0.0280

d. Assuming that the production distributions for each of the three components can be centered on the relevant target specifications, recommend a set of revised assembly specifications that will ensure that no more than 1% of the films will be too thick and no more than 1% of the films will be too thin.

e. Assuming that the natural process spread (i.e., 6σ) is exactly equal to the allowed tolerance (i.e., Upper Spec Limit - Lower Spec Limit) for each of the three components, how would your answer to part c change?

4-9. A process at a manufacturing plant produces a two-layer film. Each layer thickness is measured by a properly calibrated measuring device. The first layer has a target mean specification of 0.055 in. and the second layer has a target mean specification of 0.030 in. The combined finished product has specifications of 0.085 in. ± 0.005 in. The following data were obtained from measurements of each layer individually. Four samples were taken each hour for 30 hours.

a. Assume that the processes used to produce each layer component are stable and predictable. Use the raw data (120 data points) for each component layer to estimate the respective layer standard deviation and variances. What is the estimated variance of the resultant product?

b. Estimate the percentage of layered films that will be outside the specification limits.

c. What needs to done to retarget the components so that the composite product (layer film) production is on target?

d. If the target suggested in part c can be achieved, estimate the total percentage of output that would be outside the specification limits.

e. If the suggestions in part c can be achieved, how much would the variance of Layer 1 have to be reduced in order to meet the composite specifications in the plus/minus 3 sigma sense?

CHAPTER
FIVE

QUALITY OF DESIGN

QUALITY MAXIM #5: *A superior product or process requires not only good engineering. Teamwork, open-mindedness, communications, broad perspectives, diversity of input, customer focus, cost-effectiveness, and environmental consciousness are also essential.*

The Miracle of Design: Make It into the Shape You Want!

A John Denver song "The Potter's Wheel"[1] hails the miracle of design, as through the lyrics he describes the creative power that the designer feels as he or she takes something without function and reshapes it, molding it into something of value.[2] The quality of a product or service lies in the genius of its design. Quality of output can be only as good as the process designed to produce it, and in the case of a tangible product, quality is dependent on how well the design of the product conforms to the needs of the customer, its manufacturability, and its environmental impact. When a U.S. automobile assembly plant in Tennessee was constructed, a large area was set aside and devoted to process and product improvement. On one occasion while visiting this plant I discovered a partially disassembled Citroën—a vehicle that was in the process of being reverse engineered so that the American designers could unlock some of the mysteries of French creativity. Of particular interest was the headlight assembly which engineers found to contain only 7 parts. Its U.S. counterpart required 32 to achieve the same function. Learning how to reduce 32 parts to a mere 7 had valuable implications for cost reduction, manufacturing simplicity, better maintainability, and improved reliability. Sustainability and competitiveness are often determined by how well an organization can identify and adopt best practices. Like clay on the potter's wheel, production methods or product designs can be shaped into a wasteful disaster or an artistic model of excellence. The principles guiding this process are to eliminate, combine, simplify, change, and convince. The first improvement step is to question each operation, task, motion, or feature and challenge its necessity—with a view toward eliminating unnecessary or wasteful elements. It is also helpful to consider whether any two operations can be combined either by simplifying the manufacturability of design or by streamlining the production process. Value engineering can be applied to the design through a structured who/what/where/when/why/how line of questioning, referred to in an earlier chapter. This logically leads to innovation as the follow-up questions reveal opportunities for change. Can we change the where? Can we change the when? The who? The how? Finally, any improved methods or product designs must be thoroughly tested and then sold to those who are responsible for a successful implementation. Like the potter's wheel, design is a never-ending cycle as the designer strives to turn, tweak, and revise until he or she creates the shape and outcome desired.

---

[1]Written by Bill Danoff, "The Potter's Wheel" was recorded by Mack Bailey on his 1985 album *On My Way*, and then again by John Denver in 1991 on the album *Different Directions*.

[2]The lyrics for "The Potter's Wheel" can be accessed at http://www.john-denver.org/Default.asp?id=127.

LEARNING OBJECTIVES AND KEY OUTCOMES

After careful study of the material in Chapter 5, you should be able to:

1. **Describe the process of value engineering, how it supports the creation of superior designs, and how it relates to the PDCA process improvement cycle.**
2. **Discuss the Taguchi loss function and describe why this concept is important to the setting of quality targets.**
3. **Define and discuss Taguchi's four principles of design.**
4. **Discuss integrated product and process design (IPPD), including its objectives and the relevant issues that must be addressed in implementing this approach.**
5. **Describe quality function deployment (QFD), including the development of a house of quality (HOQ), and how this approach supports QIM.**
6. **Discuss the concept of design for supportability (DFS) and how some of the product life-cycle issues can be addressed during a product's design and development phase.**
7. **Discuss some design guidelines that, if followed, will facilitate the maintainability and serviceability of a product during its life cycle.**

(continued)

8. Discuss some design guidelines that, if followed, will facilitate product reliability during its life cycle.
9. Discuss the anticipatory failure determination (AFD) approach and how the AFD methodology can lead to more reliable product designs.
10. Discuss some design guidelines that, if followed, will facilitate the testability and inspectability of a product.
11. Define the objectives and benefits of a green products design process and discuss how designing products that are environmentally friendly can support an organization's quality goals.
12. Discuss some design to cost guidelines that, if followed, will achieve some predetermined cost objectives.
13. Briefly describe the concept of target costing.
14. Briefly describe the concept of activity-based costing.
15. Discuss the importance of considering other life-cycle issues in the product design phase including installability, usability/trainability/documentation, and updateability/upgradeability.
16. Describe the major process flow design patterns and the production parameters that best suit each design.
17. Balance the cycle time of a process line to a specified takt time and compute the line efficiency and the efficiencies of each individual workstation in the line.
18. Define *cellular manufacturing,* including the principal objective behind this form of workplace organization, and discuss how it relates to the concept of process flow patterns.
19. Demonstrate an understanding of the three methods employed by group technology (GT) to classify products into family groupings.
20. Demonstrate an understanding of the PERT method of project management including the creation and analysis of a PERT network.

5.1 ENGINEERING IMPROVEMENT METHODOLOGIES

5.1.1 The Engineering and Scientific Methods

An engineer has been defined as someone who uses scientific knowledge to solve practical problems.[3] The emphasis on solving real and acknowledged problems distinguishes *engineers* and the work they do from *scientists*, who possess advanced knowledge in one or more of the natural or physical sciences, and *researchers*, who engage in systematic investigations to establish facts.[4] Engineers, scientists, and researchers share one thing in common—a rational approach to problem solving. The approach used by scientists and researchers is called the **scientific method**, illustrated in Figure 5-1. The **engineering method**, illustrated in Figure 5-2 is similar, differing mainly in objective. The scientific researcher seeks knowledge as an end goal by itself. The engineer is also a seeker of knowledge. However, the engineer unlike the scientist would not place a high premium on new knowledge that did not lead to the solution of some previously identified problem.

As Figures 5-1 and 5-2 illustrate, the scientific method focuses on testing a hypothesis while the engineering method is concerned with the analysis and improvement of designs. The engineer concentrates on developing and validating models that will be tested to achieve and improve functional designs, using a rationale that can be understood by other engineers. The work of scientists and engineers alike involves activities that alternate between the concrete and the abstract, and both groups are subject to peer scrutiny. The scientific researcher normally publishes his or her findings in peer-reviewed academic journals where assumptions, methodology, logic, and conclusions can be openly challenged. Similarly, an engineer's original design, or improvement of an existing design, is often subjected to review by other engineers who are called on to "check" the work for errors in logic, calculations, or assumptions.

Existing knowledge is typically the starting point for either method. The researcher will normally conduct a detailed literature search to benefit from the work of others and provide a frame of reference for advancing the knowledge base. Like the scientist, engineers are creators of new knowledge as every problem has unique factors that have not been encountered before in the same context or configuration. Engineering is a creative activity, and in the evolution of new conceptual designs and in the resolution of complex problems, engineers rely heavily on the collective technical data base (within the specific engineering discipline) on what has worked in the past and what has not worked so well, best practices, pitfalls, and technological capabilities.

The engineering method advances knowledge specific to some identified problem.

The engineering and scientific methods both employ the application (and development) of theory in the abstract domain, followed by experimentation (and testing) in the

[3]http://en.wikipedia.org/wiki/Engineer.
[4]Ibid.

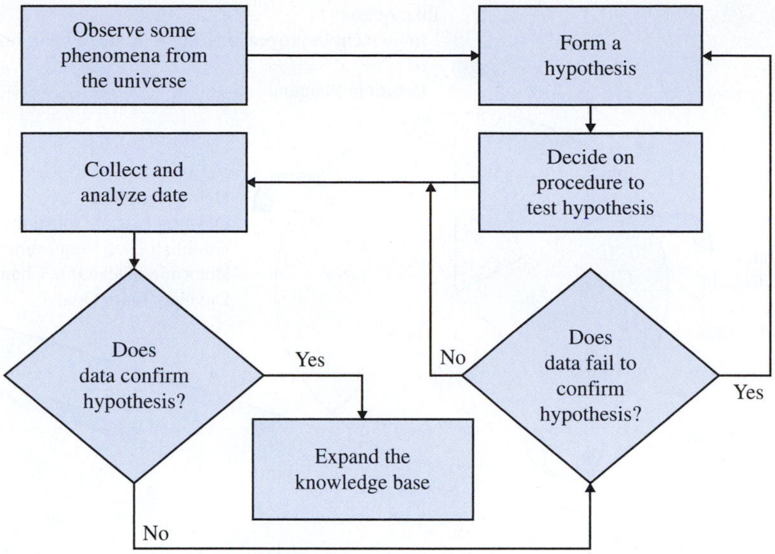

FIGURE 5-1. The Scientific Method

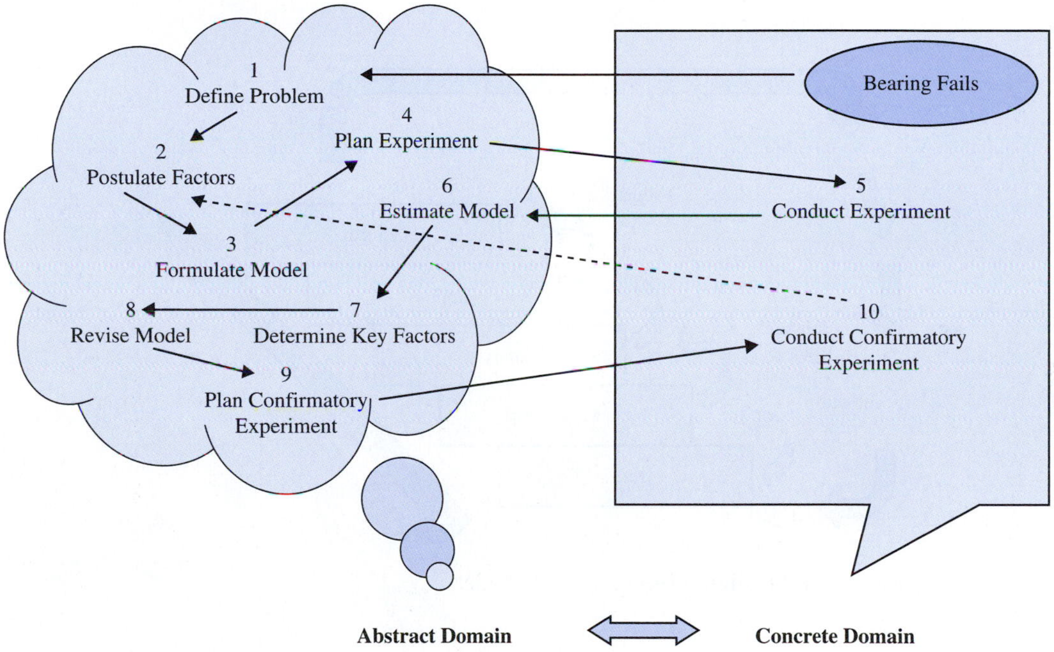

FIGURE 5-2. The Engineering Method

concrete domain. As the engineer develops and refines his or her model the work is similar to the scientific researcher testing a hypothesis. The outcomes are the same. The researcher cannot advance a new theory until the hypothesis has been subjected to rigorous testing and scrutiny. Likewise, the difference between a superior engineering design and an inferior one can often be traced to how carefully and thoroughly the concept was tested under scientifically controlled conditions.

The Plan-Do-Check-Act (PDCA) cycle (Figure 5-3) and the Define-Measure-Analyze-Improve-Control (DMAIC) process of Six Sigma[5] (Figure 5-4) represent applications of the engineering and scientific methods to quality improvement.

5.1.2 The PDCA/PDSA Cycle

The **Plan-Do-Check-Act (PDCA)** cycle has been widely accepted as the basic model for the process of continuous improvement, and has its roots in the work of Walter Shewhart.

[5] Pande, Peter S., Robert P. Neuman, and Roland R. Cavanagh. *The Six Sigma Way: How GE, Motorola, and Other Top Companies Are Honing Their Performance.* New York: McGraw-Hill, 2000.

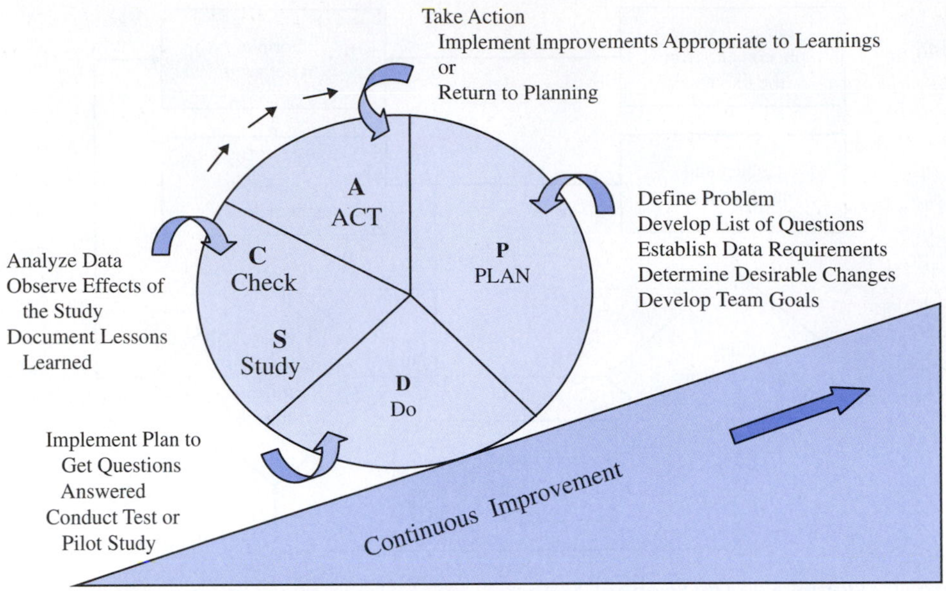

FIGURE 5-3. PDCA/PDSA Cycle

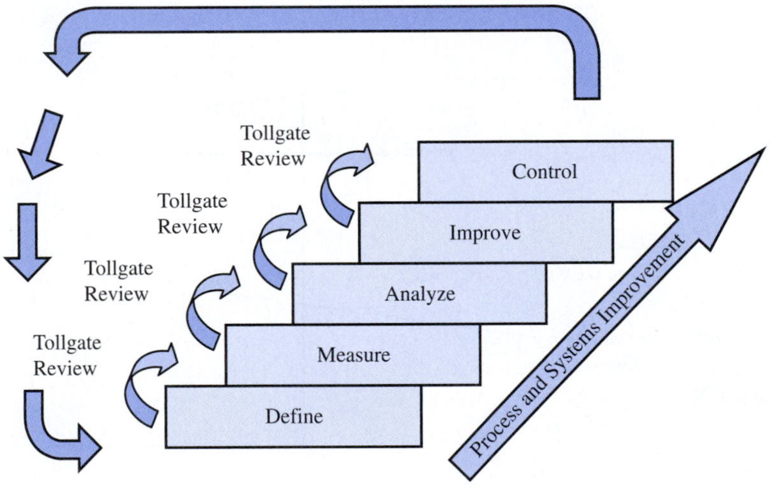

FIGURE 5-4. Six Sigma Improvement Model

In 1950 Deming instructed the Japanese on how to apply the Shewhart product development cycle which involves five steps:[6] design the product, build the product, put the product on the market, use market research to test the product in service, and use customer feedback to redesign the product, and continue the cycle. The Shewhart cycle inspired the Japanese to develop a general problem solving process called the PDCA cycle[7] that became popular due to its simplicity and logic. The process consists of four steps: Plan, Do, Check, and Act.

In the *Plan* step, the problem solvers grapple with issues such as why the particular issue is worthy of consideration, what is known about the problem, how the problem should be defined, what questions need to be answered, the best way to obtain answers to those questions, who will do what tasks, and how will each task be best accomplished.

During the *Do* step problem solvers implement their plan attempting to collect the data that will answer their questions and aid in the resolution of the problem at hand. This step can entail a broad spectrum of activities ranging

---

[6]Kolsar, P. J. "What Deming Told the Japanese in 1950." *Quality Management Journal* 2, no. 1 (1994): 14.

[7]Mizuno, S. *Companywide Total Quality Control.* Tokyo: Japanese Union of Scientists and Engineers, 1984.

from collecting performance data to conducting a designed statistical experiment.

The *Check* step is where the data (or experimental outcomes) are analyzed and transformed into new knowledge aimed at addressing the issues and answering the questions raised during the *Plan* step. The new insights and learnings will either provide the basis for action or generate a new set of issues and questions.

The final step in the cycle is to *Act* on what was learned. Actions can include piloting a new method, implementing new controls, solving the stated problem, or simply recognizing that the team was trying to answer the wrong set of questions or the set was incomplete. In the latter case, the process cycles back to the *Plan* step and the process repeats.

Deming preferred the term PDSA cycle—substituting a *Study* step for the *Check* step[8]—a process that he claimed was more in keeping with Shewhart's intent and that is more supportive of team-oriented problem solving. It was Deming's contention that in a team setting progress is made when three conditions are met.

1. Every team member has the opportunity to contribute ideas that are considered and ultimately become part of a consensus position.
2. Team learning takes place.
3. The team makes periodic fresh starts from a higher level of understanding.

5.1.3 The DMAIC Tactics

The Six Sigma improvement model, called **Define-Measure-Analyze-Improve-Control (DMAIC)** (pronounced Duh-May-Ick), adds an additional step to the PDCA/PDSA cycle. The PDCA and DMAIC models are grounded in a common philosophy—*use data to create knowledge relative to a problem and then act on the knowledge to solve the problem.* The tactics of DMAIC[9] are summarized below.

The *Define* tactic sets the stage and justifies the project. During this phase a clear and communicable problem statement is developed, key processes and customers are identified, roles and responsibilities are assigned, and project benefits are clearly articulated.

DMAIC is the continuous improvement model of Six Sigma.

The *Measure* tactic focuses on the design and implementation of a data collection strategy. The data collected provides a baseline which can be used to evaluate the merits of any future proposed changes.

The objective of the *Analyze* tactic is to isolate those factors that are most likely causing the defined problem. During this phase it is important to identify root causes and sources of variation—specifically the causes of process variation and the sources of lead time variation.

The purpose of the *Improve* tactic is to generate workable solutions that will eliminate the confirmed causes of variability or nullify their effects. This phase starts with creative and innovative thinking, finally narrowing the focus to solutions that are practicable and implementable.

The purpose of the *Control* tactic is to ensure sustainability. Problem solutions are implemented and documented, and workers are trained in the new procedures. Ownership of the new improved system is then passed to those accountable for process performance.

5.1.4 Improvement Model Relationships

A feature that distinguishes DMAIC from other improvement models is the provision for "tollgates" between each of the phases. The tollgates represent milestone events when the improvement team meets formally with senior management and other project participants to review progress and ensure that the project stays on target and in alignment with corporate goals. Figure 5-5 shows how the three improvement models are related.

5.1.5 Enumerative and Analytic Studies

The primary purpose of any statistical analysis is to create sufficient knowledge so that the analyst can take action that is rational and defendable. Dr. Deming was one of the first to articulate a distinction between two fundamental types of studies—enumerative and analytic[10] which differ with respect to the purpose of the investigation and the objects that are targeted for action.

As illustrated in Figure 5-6 an **enumerative study** is concerned with some fixed quantity of material and uses sampling data to estimate the parameters of that finite population.[11] Such estimates can be framed in questions beginning with "what?"—for example, *What is the average of this fixed lot? What is its variance? What percentage fails to meet specifications?* It should be noted that any inferences made from the study cannot be reliably extended to material outside the fixed quantity sampled. Enumerative studies deal with data that are not only finite but also static. These studies are incapable of modeling changes in quality that are time related.

Examples of enumerative studies are (1) checking an incoming lot of material to determine if the supplier has met specifications, or (2) checking lots of raw stocks in a warehouse to determine compatibility for commingling in a future production run.

The aim of an enumerative study is to estimate the parameters of a fixed lot of material by sampling a fraction of it.

---

[8]Deming, W. Edwards. *The New Economics for Industry, Government, Education*, Second Edition, Cambridge, Massachusetts: MIT Press, 2000, 131–133.

[9]DMAIC is covered in greater detail in Chapter 14.

[10]Deming, W. Edwards. "On Probability as a Basis for Action." *The American Statistician* 29 (1975).

[11]A statistical population is discussed in Chapter 9 and defined in the Glossary.

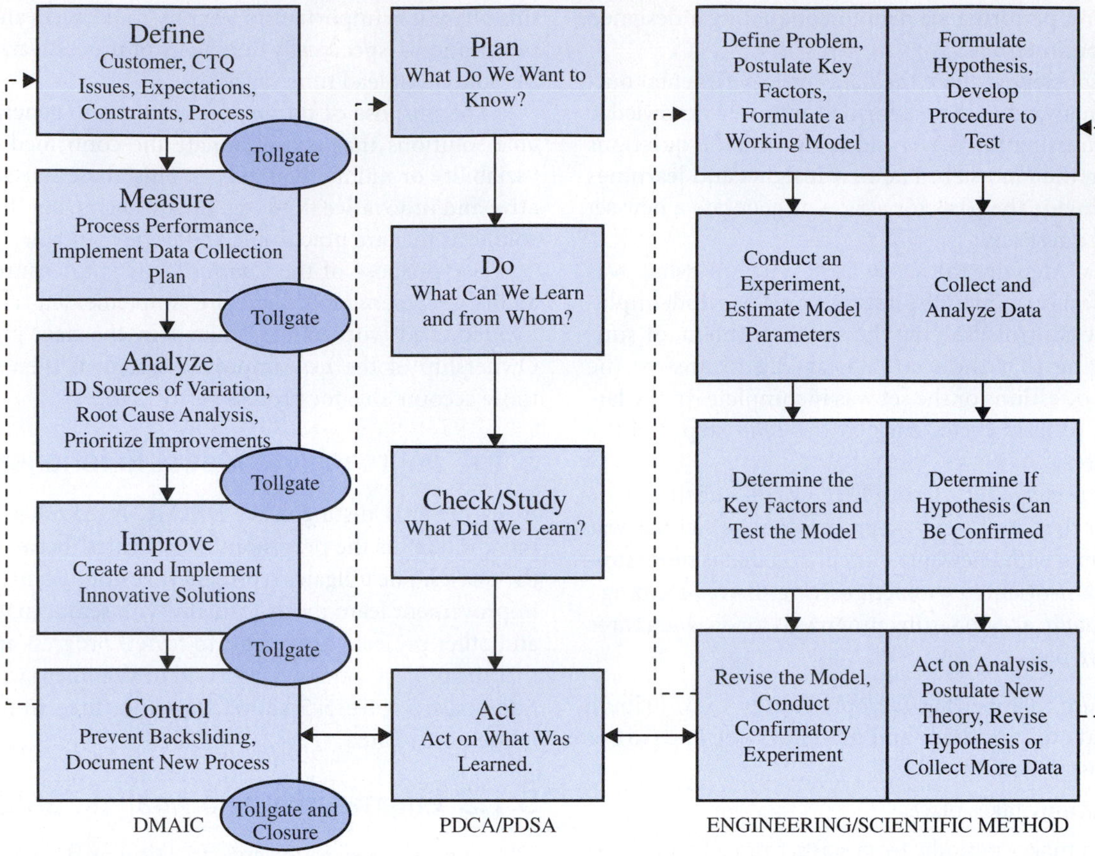

FIGURE 5-5. Improvement Models

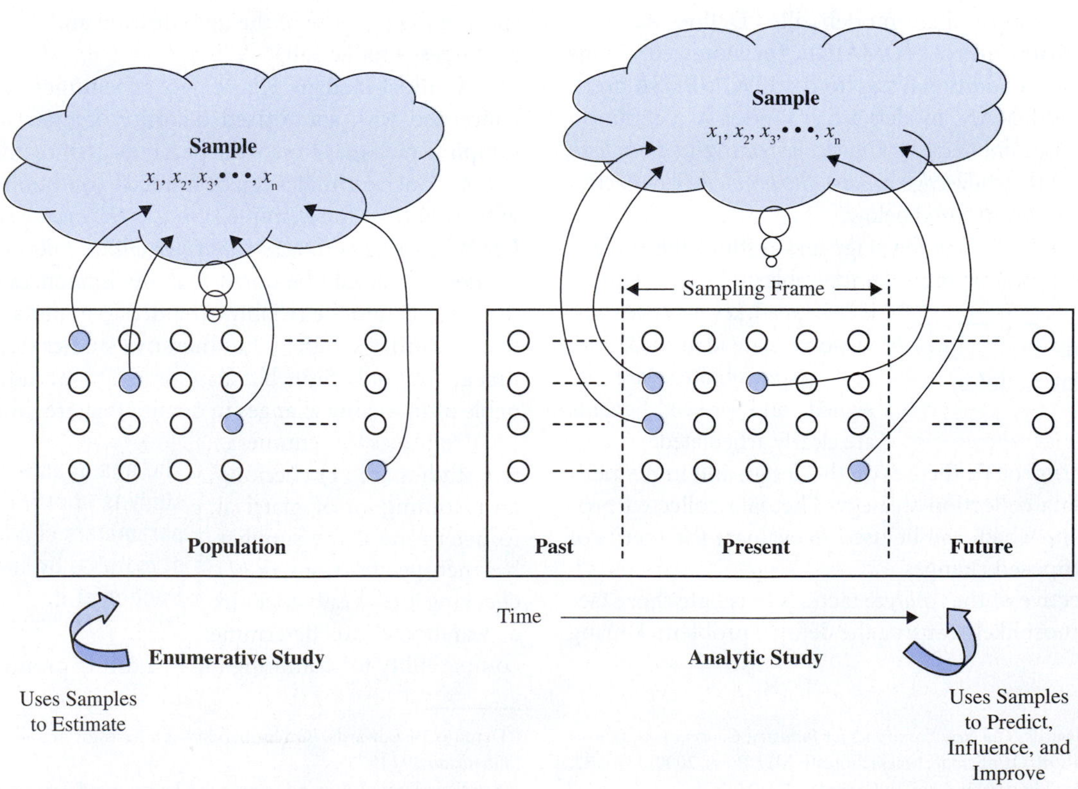

FIGURE 5-6. Enumerative Versus Analytic Studies

By contrast an **analytic study** is concerned with observations of a process over time. In this case the concept of a population is ill-defined. That is, in the case of a dynamic process much of the output was produced in the past and cannot be retrieved for study, and much of the output has yet to be produced in the future. Therefore, statistically the analyst must define a frame that is the set of candidates from which individual sampling units are selected. Defining a frame can be thought of as establishing a time frame during which sampling takes place. The object of the analytic study is not so much the "what?" but the "why?" and the "how?" The aim is to better understand process behavior and to establish cause-and-effect relationships so that intervention strategies can be employed to positively influence future behavior. In short, the object of this type of investigation is process improvement.

Analytic studies deal with data that is infinite and dynamic. Inferences made regarding time-related influences can be very helpful in making predictions with regard to outputs that occur outside the sampling frame. Examples of analytic studies are 1) the use of control charts to monitor process performance, or 2) conducting a design of experiment to determine how to reduce process variation.

The aim of an analytic study is to understand how a process behaves and to define the principal process drivers.

5.2 THE IMPACT OF QUALITY ON PRODUCT AND PROCESS DESIGNS

In the development of a new product about 70% of a product's life-cycle cost is determined at the conceptual (design) stage. There was a time when design engineers represented a select group of employees, endowed with special prophetic gifts of being able to know what customers needed or wanted. In relative solitude, armed with computers, CAD programs, and test facilities, these designers applied their powers of imagination to the creation of product drawings, bills of materials, and manufacturing processes. Designs were validated when prototypes were built and tested by highly skilled technicians. It was only then that the design was handed over to manufacturing personnel, who were expected to figure out how to efficiently produce the product.

The problem with this approach was that invariably sources could not be found to reliably supply some of the specified materials, or manufacturing had to purchase new and costly equipment to meet arguably unimportant characteristics, or manufacturing engineers had to make shop-floor design modifications (called engineering change orders or ECOs) to achieve compatibility between the product drawings and production capabilities. When a design engineer works in isolation, he or she develops a product within the context of existing production capabilities that may be inflexible to major modifications. However, this presents a kind of catch-22 situation because often when the designs are released, manufacturing engineers must redesign processes as well as modify the designs until they reach a point where the two are congruous.

In a QIM environment, design and manufacturing engineers work together from concept to launch—and even afterward—addressing such issues as supply chain deployment, product use and maintenance, and disposal/recycle at the end of product life. The design engineer shares in the ownership of the entire production process and in the product life cycle. When the product design, production, and support processes are rationalized, the designs take into account manufacturability, cost, reliability, and maintainability systemically through integrated approaches that are discussed further in Section 5.6. The principal objective is improved response. Stakeholder needs become the focus of an integrated design process with the aim of meeting those needs faster, with improved quality and less cost. Processes and products are developed concurrently. Processes include those that are part of the primary value streams and those that are supportive and part of secondary value streams such as sales and marketing, materials management, test and verification, maintenance, training and human resource development, distribution, and product disposition or recycling.

In a QIM environment product and process design activities are fully integrated.

Continuous and early life-cycle planning involves an ongoing consideration of how customers, suppliers, and the manufacture-deliver-disposition systems interact throughout a product's entire life cycle so that resources can be better managed and resource constraints better understood. Superior (robust) designs leads to products that are less sensitive to variations in production processes, and the use of quality tools on the manufacturing side lead to processes with reduced variability, less waste, and improved capabilities. The use of continuous (and breakthrough) improvement methodologies offers the propensity for exceeding customers' expectations and ultimately influencing future needs (as customers become more discerning).

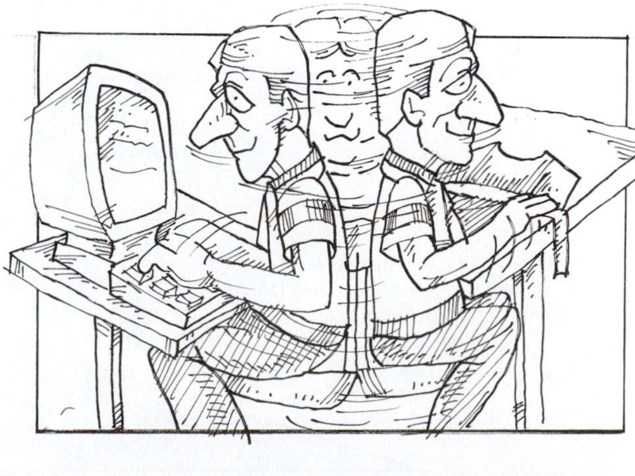

Today's product designers must have the ability to adapt their creative genius to the capabilities of manufacturing processes.

5.3 QUALITY OF PRODUCT OR SERVICE DESIGN

5.3.1 What Constitutes a Quality Design?

What is a quality design? Once a design has been released to production, measurements are taken of randomly selected units to determine how well the key design features comply with what was specified. However, in the eyes of the customer an ill-conceived product will fail to live up to expectations no matter how well it performs during production testing. What satisfies, impresses, or otherwise excites a customer must be well understood by design engineers and incorporated into the product development process. While this is necessary, customer needs alone are not sufficient. A quality (superior) design must also be producible. This means that the design engineer should have a complete understanding of process capabilities so that tolerances can be realistically set and a product is neither overdesigned (manufacturing has to meet tolerances that are too tight) nor underdesigned (resulting in shoddy quality leading to assembly problems or loss of function in use). Either scenario can lead to costly waste.

The design process does not end with the customer and manufacturability. Other considerations also must factor into the final design. Increasingly, public awareness of environmental protection has forced manufacturing concerns to be more socially responsible. Many product designs now reflect concerns of how the environment will be impacted throughout the product's life cycle, including what happens to the product once it has outlived its usefulness. When these factors are incorporated into the product development process, the product is said to be green and the design an ecodesign.

An ill-conceived product design will fail to satisfy customers even if it performs well during production testing.

5.3.2 Value Engineering

Value engineering (**VE**), also called **value analysis** (**VA**), has been around since World War II, and was motivated by the need to develop viable substitutes for production inputs that were in short supply. In modern times, some variations such as lean production and business process reengineering have expanded the scope of VE to cover entire systems and be an integral part of business strategy. As the name implies VE is a methodology aimed at improving the *value* of goods or services. In VE value is defined as the ratio of *function* (i.e., customer benefit) to *cost* (i.e., customer sacrifice). Therefore, VE is concerned with improving value by either increasing function (in search of the *wow* factor) or reducing cost, and doing so without sacrificing on quality.

VE is a systematic process similar to the scientific method, and can be illustrated using Deming's PDCA cycle (refer again to Figure 5-3) as shown in Figure 5-7.

As Figure 5-7 indicates, VE employs an intuitive reasoning approach through an intensive *how/why* questioning regimen that follows a four-step cycle that mirrors PDCA/PDSA. The first (Plan) step is concerned with gathering information concerning the function of an object or product feature of interest, and moreover gaining an understanding of how those functions and features affect customer perceptions. During this functional analysis questions are raised concerning what the object does *(what? how?)*, what it is designed to do (should/could do), what it should not do, and the possibilities

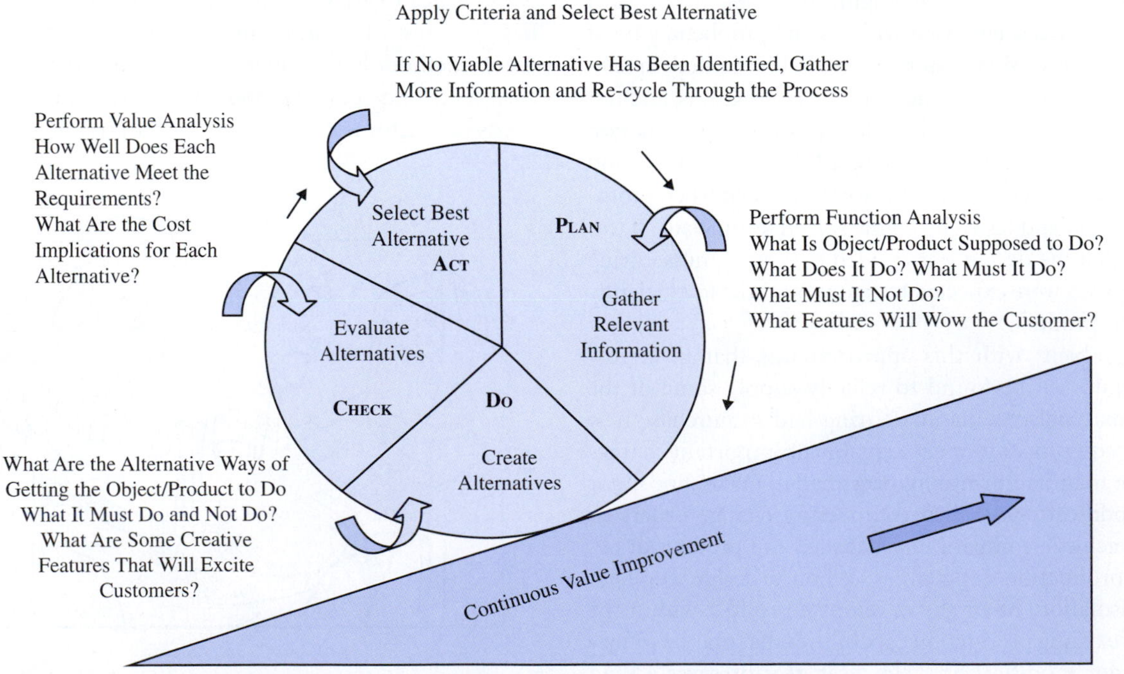

FIGURE 5-7. Value Engineering Version of PDCA Cycle

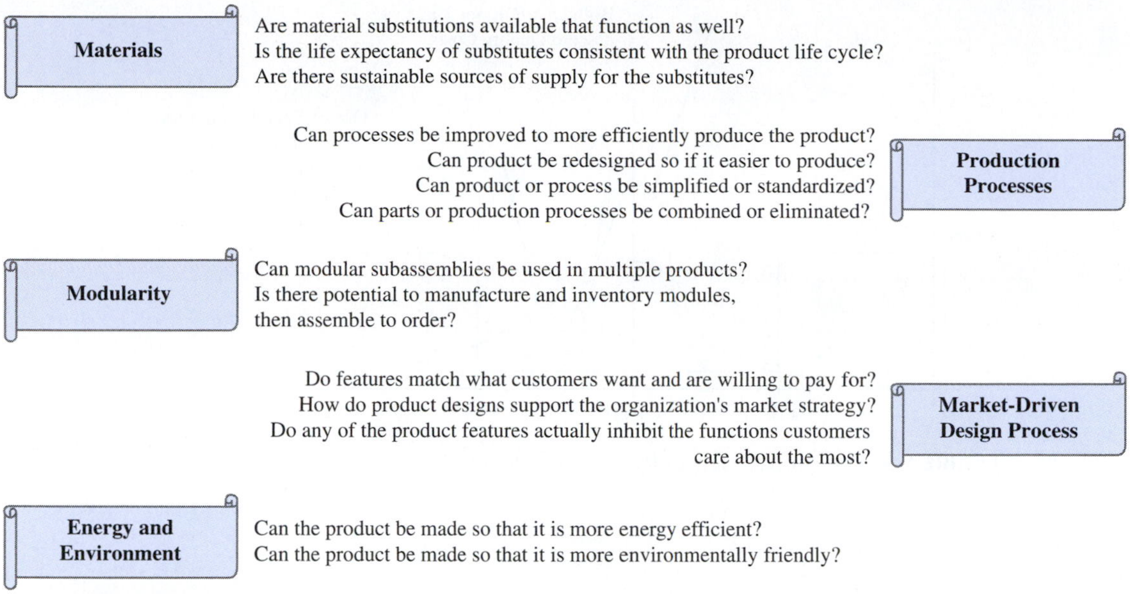
Materials
Are material substitutions available that function as well?
Is the life expectancy of substitutes consistent with the product life cycle?
Are there sustainable sources of supply for the substitutes?
Can processes be improved to more efficiently produce the product?
Can product be redesigned so if it easier to produce?
Can product or process be simplified or standardized?
Can parts or production processes be combined or eliminated?
Production Processes
Modularity
Can modular subassemblies be used in multiple products?
Is there potential to manufacture and inventory modules, then assemble to order?
Do features match what customers want and are willing to pay for?
How do product designs support the organization's market strategy?
Do any of the product features actually inhibit the functions customers care about the most?
Market-Driven Design Process
Energy and Environment
Can the product be made so that it is more energy efficient?
Can the product be made so that it is more environmentally friendly?

FIGURE 5-8. Target Areas of Opportunity for Value Engineering, Lean Production, and Business Process Reengineering

that could be created by reengineering the product. The first three of these questions are followed by the additional question—*why*? The fourth question is followed by *why not?* After a thorough understanding of function and customer needs has been gained, the analysis team is ready to proceed to the second step—creating alternatives to achieve the function.

A number of tools that can be employed to help in the generation of creative ideas are covered in Chapter 7. For example, a VE team might use structured brainstorming to generate ideas, and then employ the nominal group technique (NGT) to reach consensus and rank the candidates. This step, analogous to "Do" in the PDCA/PDSA cycle, investigates alternative ways of fulfilling the object's function. This is the time to challenge any assumptions, abandon convention, and think outside the box. Some key areas that offer possibilities for improvement for VE are shown in Figure 5-8. Once a list of alternatives has been identified and ranked the team moves on to the evaluation phase.

Value engineering is concerned with improving value by increasing function or reducing cost without sacrificing quality.

In the next (Check) step the focus is on evaluating the top-ranked alternatives to see how well each achieves the desired function and improves value. Those that are judged to satisfactorily meet all requirements are then compared using an engineering economic analysis—that is, how much will each alternative cost and how much will each save? The objective of this step is to provide enough information so that the process can proceed to the final step where a decision can be made.

The last (Act) step in the VE cycle is to make a final selection. At this stage the VE team presents and "sells" its ideas to the organization's decision makers who must give consideration to many factors, such as sustainable materials supply, quality, customer acceptability, and cost. Therefore, the decision made may not always favor the alternative offering the greatest cost advantage. As with PDCA/PDSA, there is also the possibility that the cycle fails to produce any acceptable alternatives and the process moves back to step one and repeats the four steps.

5.3.3 Cost of Quality Considerations in Product or Service Design

5.3.3.1 Taguchi Loss Function. Taguchi has defined quality as the "loss imparted to society from the time the product is shipped."[12] This view, referred to as a **Taguchi loss function**, is a departure from conventional wisdom that considers all spec-compliant products to have incurred zero quality (noncompliant) cost, and that it is legitimate to assess an immediate and continuing constant unit cost to products that reach and surpass the tolerance limit threshold. According to this conventional view, illustrated in Figure 5-9, any output that lies between the upper and lower specification limits is considered to be acceptable (with zero quality cost) and is passed on to the next process in the value stream or shipped to the customer.

Quality control tools may be applied to ensure that minimal if any unacceptable output is produced. Any units of output that occur outside the limits, if detected, will be routed back up the value stream for rework or will be

[12] Taguchi, Genichi. *Introduction to Quality Engineering: Designing Quality into Products and Processes.* Japan: Asian Productivity Organization, 1988.

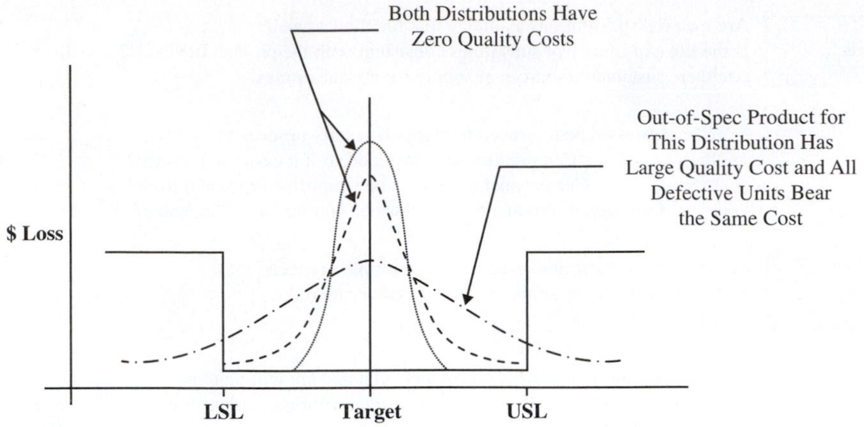

FIGURE 5-9. Traditional View of the Cost of Quality

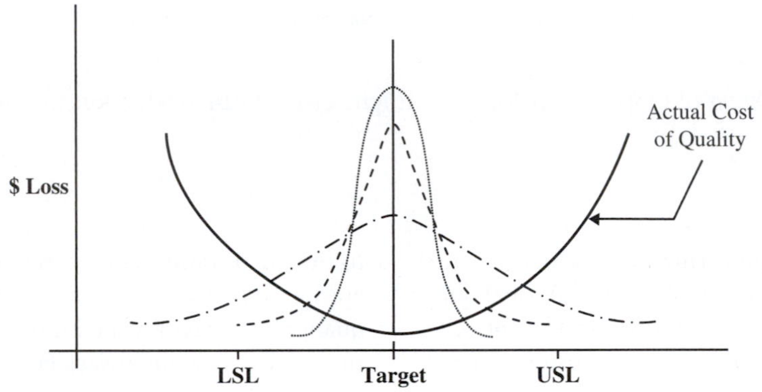

FIGURE 5-10. Taguchi Loss Function

discarded as scrap. The cost of these rejects is considered to be the same regardless of the degree of noncompliance. Specification limits are considered to be sharp cut-off points where quality instantaneously goes from being perfectly acceptable (with a quality cost of zero) to totally unacceptable (with some significant and positive cost).

In reality the idea that such a sharp, clearly defined point of demarcation exists is a myth. Taguchi's philosophy states that the objective of any production process is to achieve the target geometrical dimensions rather than to just hit the tolerance window. During the production process, as actual output begins to differ from the target in either direction, quality costs accrue and increase continuously as the output passes the tolerance limit and beyond. This concept is illustrated in Figure 5-10.

The Taguchi loss function describes the impact on quality costs of not meeting a target requirement.

Taguchi has asserted that quality is best when the desired features are at the nominal and that customers become increasingly dissatisfied as performance drifts further away from the target. The Taguchi loss function defines quality cost as the cost associated with the absolute deviation between actual and nominal performance. Taguchi theorized that the loss increases parabolically as the deviation increases and presented the following quadratic model to describe the cost of quality.

Taguchi Loss Function

$$L(x) = k(x - T)^2 \qquad \textbf{5-1}$$

where,

x = an actual measurement of a quality feature
$L(x)$ = loss as a function of the output
T = target (usually nominal) value of the quality feature
K = a constant, the value of which depends on orders of magnitude, units of measure, and loss function monetary units

The Taguchi quadratic loss function does not necessarily apply in all cases. Sometimes, albeit rarely, a loss function can be created and fit to actual market research data. In the absence of such data, the quadratic loss function is a good option. Even if there is no apparent benefit in generating the equation to describe the cost of quality, the Taguchi loss function emphasizes the importance of trying to minimize variation around a target,

Production processes should be designed to minimize variation.

even for capable processes when most of the output is within the acceptable range. The genesis of the Taguchi view places the spotlight squarely on the customer where it belongs, and derives from four fundamental principles.

5.3.3.2 Taguchi's First Principle: Quality of Design. Quality should be designed into a product. Quality of process, screening, or inspection cannot compensate for a poor design. A good design can render a product robust and immune to many of the common-cause sources of variability that are inherent in the production process. The design emphasis should be on prevention and not rely on the ability to detect errors when they occur downstream in the production process. Good design practices exploit the capabilities of manufacturing processes to minimize cost, variability, scrap, rework, and downtime while facilitating flow.

5.3.3.3 Taguchi's Second Principle: Minimize Variation around the Target. The primary objective of each processing center should be to produce units of output with minimum variation around the target value for each design parameter. Simply conforming to specifications is not enough. The quest should be to continually flesh out and eliminate those sources that are creating variability. This is the primary motivation behind widely practiced quality programs such as Six Sigma. One way to address the problem is through robust designs that are relatively insensitive to certain variability sources, such as workplace environmental factors like noise, dust, humidity, and temperature.

Sources of variation can be split into two main categories—those that cause an individual unit to deviate from its target and those that keep successive units from being identical. Quality programs need to be vigilant to both types, minimizing deviations from target, and at the same time minimizing variability between units of output that are intended to be identical.

5.3.3.4 Taguchi's Third Principle: Quality Should Be Based on Product Capability Rather than Features. Features are related to quality in that options and different configurations can help an organization cultivate markets and use price discrimination as a competitive tool. However, product features should not be used as the basis for quality. Quality should be based on a product's capability which is a design issue. Has the product been designed in a way that recognizes the needs and capabilities of the manufacturing processes? Is the design robust against those factors that cannot be directly controlled by the manufacturing system? Will the design meet customer needs and expectations? Can the design be produced at a competitive cost?

5.3.3.5 Taguchi's Fourth Principle: Quality Costs Should Be Measured over a Product's Entire Life Cycle. Deviations from target should be measured for all critical design parameters and should be captured as a component of overall **product life-cycle costs**, which in addition to costs incurred while the product is in production (such as inspection, scrap, and rework) also takes into account the costs associated with warranty servicing, returns, product replacement, and disposal after the product has outlived its usefulness.

5.3.4 Integrated Approaches for Product Design

5.3.4.1 Integrated Product Development. There was a time in the past when communications between product designers and operations personnel was limited during the product development stages. As a result, designs often failed to take into account production considerations. Even though the designs were well engineered and prototypes worked as intended, the designs could not be efficiently produced in multiple quantities under factory conditions. This often led to a flurry of engineering change orders (ECOs)—as production engineers tried to reach some realistic middle ground with the designers to achieve compatibility between the design concept and production capabilities. Typically, numerous iterations occurred as ECOs traveled back and forth between manufacturing and the product development groups, thus creating the need for new trial runs—and more ECOs—until finally a design was created that could be produced. The process was inefficient, time-consuming, and counterproductive. The final designs were often not optimal as the design engineers were forced to compromise on too many good ideas.

In the competitive environment of today some companies cannot afford the luxury of time in getting new products to market. When corporate strategy is based on being first-to-market—that is, leading markets with new products and having the ability to rapidly adapt to changing customer (and other stakeholder) needs—sustainability depends on shortening the new product development cycle time. This requires an **integrated product development** (**IPD**) approach that considers not only product and process design technologies, but also customer needs, management practices, and production and supply chain capabilities. In short, the product development process must be consistent with an organization's strategy and business plan, and requires an open systems approach. Cross-functional communications and collaboration are essential and the design process must be a team effort with participation across the entire organization. **Integrated product and process design** (**IPPD**), **concurrent engineering** (**CE**), **simultaneous manufacturing** (**SM**), **design for manufacturability** (**DFM**), and **design for Six Sigma** (**DFSS**) are all terms that have been used to describe an organizational strategy that uses cross-functional teams of design engineers, manufacturing engineers, operations personnel, marketing specialists, and accountants to rationalize product designs with production and support processes. This concept is illustrated in the diagram shown in Figure 5-11.

Adopting an integrated product and process development strategy requires a commitment by top management and a culture that embraces organization-wide communication, collaboration, and cooperation. In many instances, this represents a fundamental paradigm change and can therefore take considerable time to put into place.

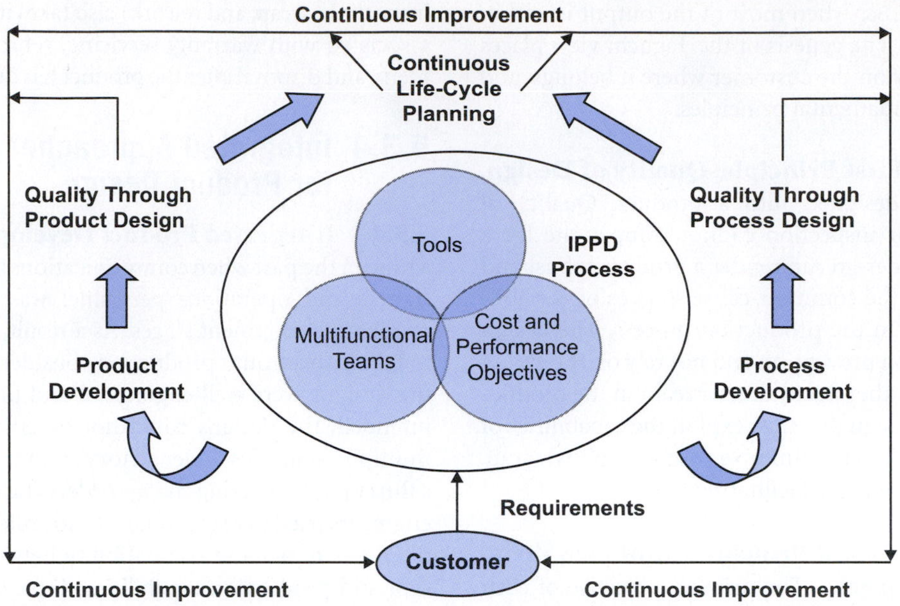

FIGURE 5-11. Integrated Product and Process Design

Along the way, structural changes, a reeducation of the workforce, and a major overhaul of existing product development practices may be needed.

The objective is to create an optimal design that can take advantage of the strengths and overcome the limitations of the expanded make-and-deliver system, which includes the supply chain (suppliers and distributors), production processes, human capabilities, and materials handling equipment. Normally, existing production capabilities are treated as design constraints and are subject to only minimal modifications. However, in the case of a major new product line the design or redesign of production systems can actually be driven by the new product concepts.

An IPPD strategy requires structural changes, a reeducation of the workforce, and a major overhaul of existing practices.

The DFM process has significant and long-term economic impact on profitability. As previously stated, research studies have indicated that up to 70% of a product's life-cycle costs are determined by decisions made during the design period—a mere 20% are due to decisions made during production. Another interesting finding is that most of the initial 70% are due to what takes place during the first 5% of the design cycle when decisions are made that have lasting implications on quality and manufacturability.

Superior designs incorporate two basic principles: standardization and simplification. Whereas the creative genius of the designer can yield a preference for custom parts for any new design, the use of standardized parts results in lower costs. The proliferation of parts that have specialized applications requires larger inventories, smaller replenishment orders, and customized tooling, all of which militate against efficient manufacturing and procurement practices.

Standardization can be facilitated using GT systems that are capable of retrieving descriptions of existing parts with similar characteristics to those required. The design engineer determines those features desired by a part or component, and then accesses the GT system to identify similar parts that match the requirements. If similar parts are identified, there are three possibilities: (1) an existing part is usable unmodified, (2) an existing part may be slightly modified to suit the purpose, or (3) no existing parts are satisfactory. In the latter case, it may still be possible to use the retrieved data to facilitate the design of a new part. Additional efficiencies are also possible if existing tooling and fixtures can be used in the part manufacture. An alternative knowledge base is one commonly referred to as the **component supplier management** (CSM) system—supply chain management software that maintains information on parts and their respective suppliers with the ability to cross-reference information.

The second key principle is simplification—fewer parts in a design mean reduced material costs and fewer manufacturing steps—which in turn create opportunities for cost reduction and quality improvement. Product complexity is a major cause of production variability. If designs can be simplified, the need for sophisticated technologies, skilled labor, and dedicated processes is diminished. An integrated product development program, properly implemented, will result in superior designs consistent with the design principles shown in Figures 5-12 and 5-13.

5.3.4.2 Quality Function Deployment.

Quality function deployment (QFD), originally developed by the Japanese in the 1960s, is a technique that ties product and service designs directly to customer needs. QFD was first introduced in the United States in the 1980s. Differing from other IPPD approaches, QFD uses the **voice of the customer** (VOC) as

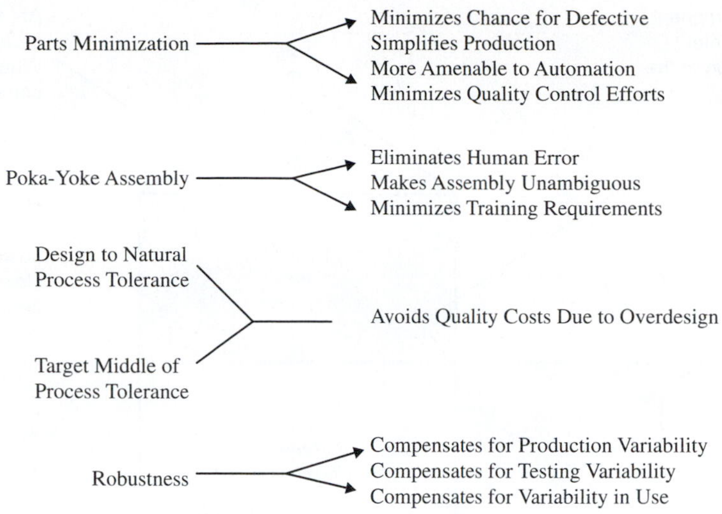

FIGURE 5-12. Principles of Superior Design I

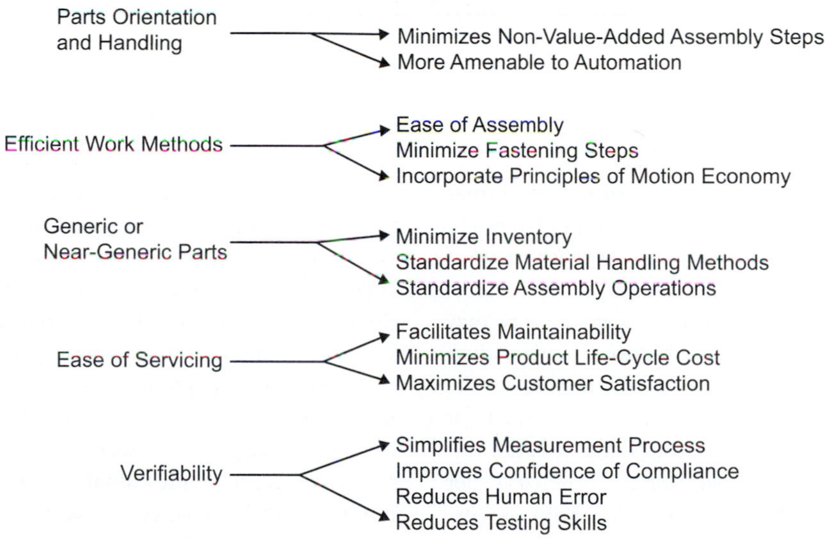

FIGURE 5-13. Principles of Superior Design II

the main driver in product or service design decisions, and is used to derive a series of matrices that document relationships in a framework called the **house of quality (HOQ)**. The HOQ is an integrated set of product planning matrices that can be used to translate customer needs into product or service design parameters. The various components of a typical HOQ are depicted in Figure 5-14.

As Figure 5-14 illustrates, the construction of the HOQ focuses on five major areas:

- The left wing of the house contains the **voice of the customer** (VOC)—a list of customer needs.
- The ceiling of the house contains the **voice of the engineer** (VOE)—the design features and technical requirements.
- The right wing of the house contains a competitive analysis, comparing customer priorities against the performance of competitors.
- The foundation contains a technical evaluation including benchmarks and target values.
- The middle of the house provides a correlation analysis of design parameters.
- The roof of the house contains performance correlations—a matrix that describes the relationships between design features.

The QFD approach supports QIM's goal to place customers at the center of all strategic business decisions. The VOC can be captured in a number of ways—using a variety of

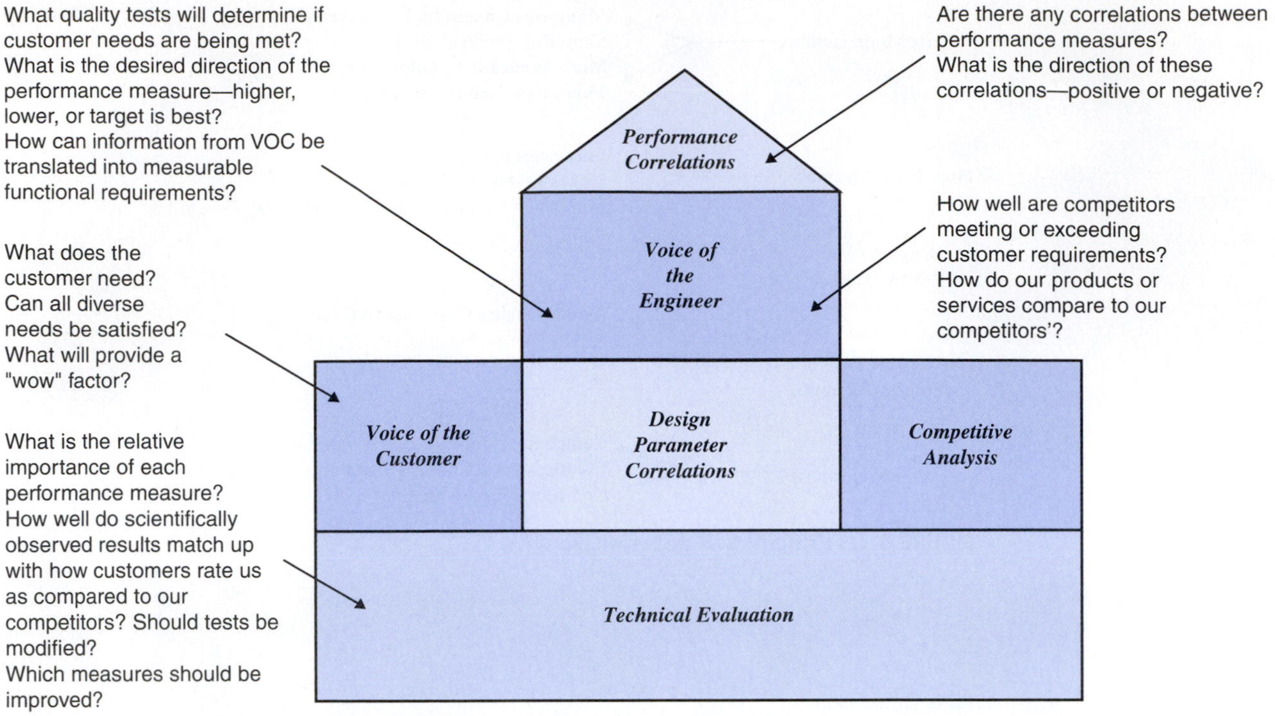

What quality tests will determine if customer needs are being met?
What is the desired direction of the performance measure—higher, lower, or target is best?
How can information from VOC be translated into measurable functional requirements?
Are there any correlations between performance measures?
What is the direction of these correlations—positive or negative?
Performance Correlations
Voice of the Engineer
How well are competitors meeting or exceeding customer requirements?
How do our products or services compare to our competitors'?
What does the customer need?
Can all diverse needs be satisfied?
What will provide a "wow" factor?
Voice of the Customer
Design Parameter Correlations
Competitive Analysis
What is the relative importance of each performance measure?
How well do scientifically observed results match up with how customers rate us as compared to our competitors? Should tests be modified?
Which measures should be improved?
Technical Evaluation

FIGURE 5-14. The House of Quality

tools including interviews, surveys, focus groups, customer specifications, observation, warranty data, or field reports. However, one should be wary of relying too heavily on what customers *state* they need. Surprisingly customers may be unable to articulate their needs in a manner that can be translated into product or process designs. It is the unspoken voice of the customer that must be revealed, and this requires getting inside customers' heads and seeing the world from their perspectives.

Customers may be unable to articulate their needs in a manner that can be translated into product or process designs.

Another complication is the fact that the customer does not speak with a monolithic voice; the marketplace is diverse with multiple segments all acclaiming different needs. All viewpoints must be considered, reconciled, and balanced to develop an overall successful product or service.

Traditionally it was the role of the marketing function to determine customer needs and to translate these needs into design specifications. This practice placed a barrier between product development and the customer, thereby isolating product designers from the ultimate beneficiaries of their designs. Direct involvement by these personnel with the customer can provide firsthand knowledge of and create empathy for customers—revealing their needs, trials and tribulations, aspirations, and those features that can produce a "wow" factor.

In addition to ensuring that product development is involved in defining customer requirements, QFD stresses the need to develop a customer knowledge base through **gemba visits.** *Gemba* is the Japanese word meaning "the place where the real action takes place." A customer's gemba is where a product or service is put to use—for example, if the product is a tractor, the gemba may be Farmer Brown's north-40 cornfield; if the service is hotel accommodation, the gemba is the reception desk or the concierge station. Through gemba visits, a manufacturer or service provider can observe firsthand how a product or service is used and the key factors that influence satisfaction. What may be obvious to customers may not be apparent to service providers or manufacturers. Market research, while valuable, can often result in misleading interpretations or can miss vital information. Customer surveys, interviews, and focus groups are only as good as the questionnaire design, the skill of the interviewers, the ability to obtain participation from a representative cross section of customer groups, and a level of trust that will cause interviewees to provide complete and honest answers. Nothing can substitute for real-life experience; however, well-executed and productive gemba visits are not easy. They need to be well planned and carried out by trained personnel who have excellent observation and listening skills and who can properly interpret and communicate visit outcomes in terms of design specifications and improvement opportunities. Gemba visits are *not* sales calls. It is therefore advisable that the regular sales rep not even participate. The visit is a chance for members of the product development design team to gain an in-depth understanding of the customer and the environment and

Through a gemba visit a service provider can learn firsthand the key factors that influence satisfaction.

conditions under which a product or service is put to use. It is extremely important to the success of a gemba visit that the customer fully understand its purpose.

The building of the HOQ parallels the product development process. This process begins with the VOC, a column vector of customer needs that is positioned on the left "wing" of the HOQ. This vector may be organized by category and generated using a brainstorming tool such as an *affinity diagram*.[13] It is important that customer needs reflect the targeted market segment and address any assumed or "wow" factor features. If the number of customer needs exceeds 25 to 30, the vector should be decomposed into smaller segments or categories which are then analyzed separately. For each product feature, either a customer priority can be assigned using a rating scale of 1 to 5, or a relative percentage can be assigned to each requirement (where all percentages sum to one).

The information gathered on customer needs is then translated into technical performance measures that are tied to each key product feature. These measures are organized into related categories and should be meaningful, measurable, and stated in a way that avoids the implication of a particular technical solution. This is the voice of the engineer (VOE), sometimes called the voice of the company, where internally the organization has to determine how it can or will verify that a product or service is actually doing what the customer wants. This step adds a row vector of process/system capabilities to the HOQ. Together the VOC and VOE vectors form the rows and columns of a matrix the can be used to evaluate how well system capabilities match customer needs.

Phase three of the HOQ is a correlation analysis that takes place at two levels. The first level involves a consideration of whether correlations exist between any of the selected technical performance measures and, for any that are discovered, the direction of the correlation (positive or negative). The second level of analysis is to establish correlations between the technical measures and product features. This component of the HOQ provides the information necessary to complete the relationship matrix defined in phases one and two. The product development team enters a rating score in each cell that represents a collective judgment of how well each technical performance measure will determine the functional performance of the product feature specified.

In phase four of the HOQ the emphasis is on the competition. During this step, the product development team tries to obtain as much information as possible concerning how well competitors are, or are capable of, fulfilling customers' needs. This is where a good competitive intelligence function can pay huge dividends. Prior generation products are evaluated against competitive products–obtaining feedback from surveys, customer meetings or focus groups. The purpose of this step is to assess strengths and weaknesses relative to the competition, thereby gaining valuable insights on what the organization needs to do to achieve competitive advantage. This includes information on how well current product offerings are perceived by customers when compared with similar products or services available from competitors.

During the fifth and final phase a technical evaluation is conducted of prior generation products and competitive products, using all relevant data such as warranty and repair histories and product life-cycle costs. As a result preliminary target values for product features can be established. An important objective during this phase is to validate performance measures (do they measure what customers actually want?) and production capabilities (do we have the capacity and technologies to deliver what the customer wants?), to identify and prioritize improvement targets, and to assess the relative difficulty of moving forward with each improvement option. Any potential positive and negative interactions between requirements and technical performance measures are then determined, and the negative ones become the focus of improvement efforts to overcome the need for compromise.

Next, ratings based on weighted importance and level of difficulty are assigned to each technical performance measure. The difficulty rating can be on a scale of 1–5, 1–10, or any other scale that is convenient (the higher the number the more difficult). Difficulty is based on technology, personnel skills, business risk, manufacturing capabilities, supplier capabilities, costs, and time frames. Figure 5-15 shows an example of a completed HOQ.[14]

HOQ is only the first step in the structured seven-step systems engineering approach of QFD illustrated in Figure 5-16.

The possible benefits of QFD include:

- The creation of a customer-driven culture
- The reduction of the cycle time for new product design and development
- The platform for implementing concurrent engineering methods
- The reduction of product development costs due to fewer engineering design changes
- The facilitation of communications through the use of cross-functional teams
- The creation of new knowledge
- The clarification of design priorities and the improvement of quality

5.3.5 Design for Supportability—Product Life-Cycle Considerations

The integrated approaches discussed in the previous section typically deal with product features, cost, quality, and manufacturability. Limited resources available for product development together with pressures to get new products on the market often militate against an early design consideration for after-sales support. From a QIM perspective this constitutes a missed opportunity to gain competitive advantage.

[13]Affinity diagrams are covered in Chapter 7.

[14]http://en.wikipedia.org/wiki/Image:A1_House_of_Quality.png.

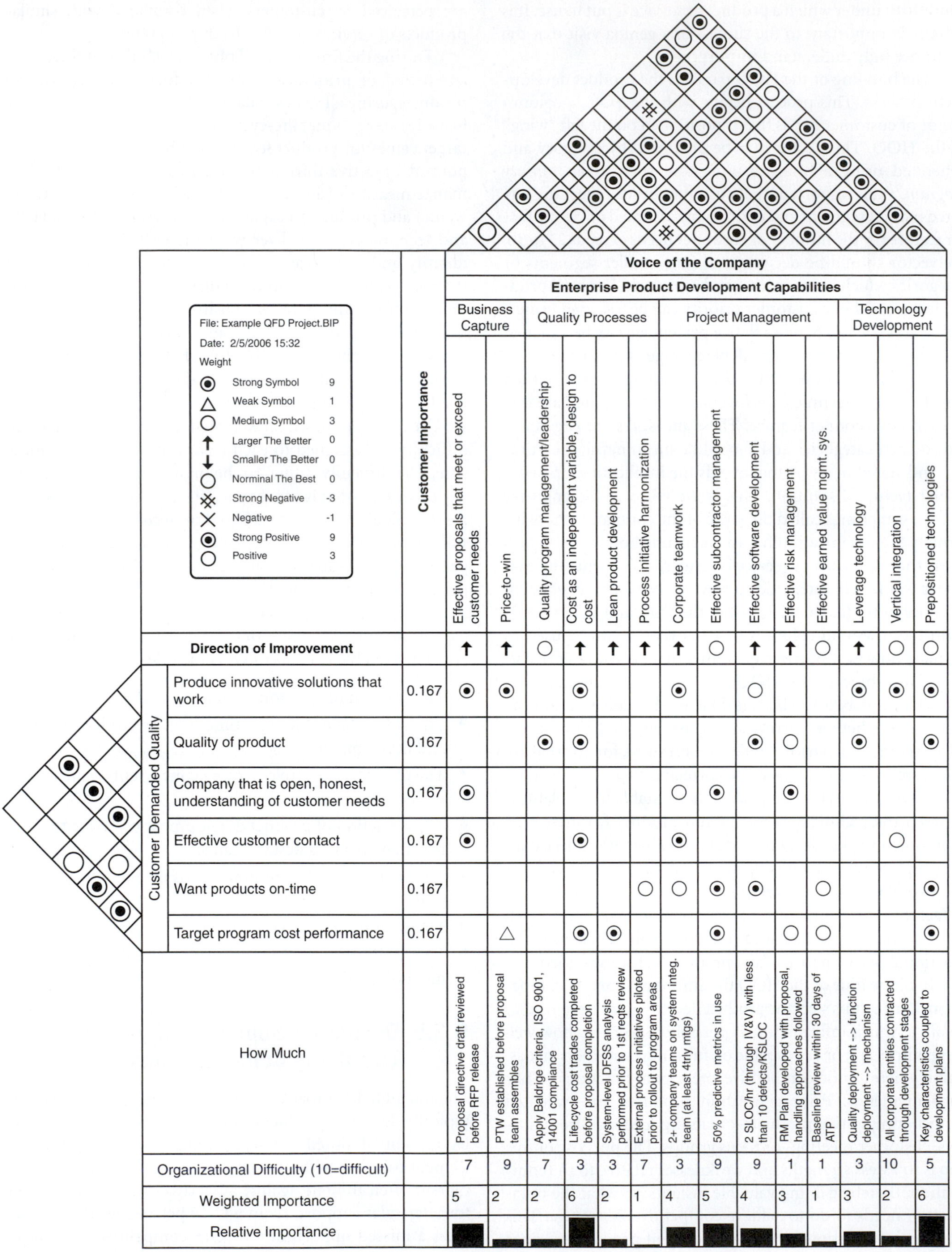

FIGURE 5-15. Example House of Quality

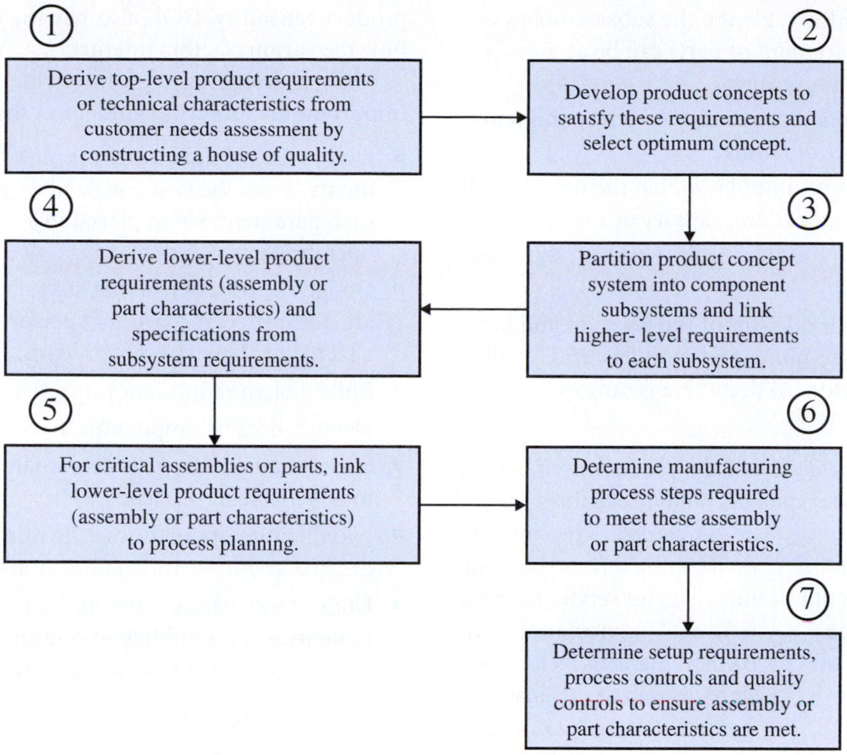

FIGURE 5-16. Systems Engineering Approach of QFD Methodology

In many markets the quality of after-sales support is a major determinant in overall customer satisfaction and customer support can represent a major source of continuing revenue to the organization. **Design for supportability** (DFS) requires design engineers to think well beyond shipment of the product and consider factors that cover a product's entire life cycle. DFS incorporates the following life-cycle considerations.

Design for supportability considers factors that cover a product's entire life cycle.

- Design for maintainability/serviceability
- Design for reliability
- Design for testability/inspectability
- Ecodesign/design for decommissioning and replacement
- Design to cost
- Design for installability
- Design for usability/trainability/documentation—the human factor
- Design for updateability/upgradeability

5.3.5.1 Design for Maintainability/Serviceability.

Maintainability and serviceability may be unimportant for low cost, highly reliable, or consumable products. However, in the case of some durable goods with long life cycles, the cost to maintain the product while in service may be more significant than the initial purchase cost. For example, hours of labor and out-of-service downtime can be saved if the parts most likely to wear out can be readily accessed and easily replaced. **Design For Maintainability/Serviceability** is the process by which features are included in a product design that facilitate the in-service maintenance and serviceability of a product and helps to address such issues as the following: How many automobile designs locate critical v-belts so that half of the engine components must be removed for replacement? Costs and frustrations that result from designs with poor supportability can be minimized if maintenance and service personnel are involved early in the product development process. Their unique perspectives on maintenance and service issues can then be incorporated into an integrated design, leading to lower overall product life-cycle costs.

When designing a product for ease of maintenance the following guidelines should be employed.

- Identify the parts, assemblies, or modules that are subject to wear and have the largest probability of requiring replacement during the life of the product. Design quick fastening mechanisms for items that will likely need to be removed and replaced in service.
- Design quick fastening mechanisms for items that will likely need to be removed and replace in service.
- Design products so that disassembly and reassembly can be accomplished using common shop tools.
- Minimize the number of subassemblies or modules requiring service by placing the parts most likely to fail or wear out together in the minimum number of

subassemblies or modules. Design the subassemblies or modules so that replacement of parts can be accomplished using simple procedures.

- Design built-in sensors that can readily detect and report failures and the source of failure.
- Design subassemblies or modules so that the need for adjustment or maintenance is unnecessary or minimized.
- Design assembles and modules with common, standard, interchangeable parts.
- Use **poka-yoke**[15] to mistake-proof the position and location of parts in subassemblies and modules, and on all fasteners and connectors so that there is only one correct way to reassemble.

Concurrent with the design of the product itself, service and after-the-sale support policies and procedures should be developed. This includes establishing and implementing a plan for service training, writing maintenance and owners' manuals (refer to design for usability/trainability/documentation discussed below), and establishing inventory policies regarding spare parts support.

It is important that product designs take into account how the product will be maintained and serviced during its useful life.

5.3.5.2 Design for Reliability. The process by which features are included in a product design that help to improve a product's reliability over its useful service life is called **Design for Reliability**, and it is important that this be factored into the early concept phase of product development. Involving reliability engineers early in the integrated product design process can help identify reliability issues and enable the assessment of reliability implications as the design takes shape and undergoes modifications. This approach is in contrast to the typical reactive way that organizations have addressed product reliability. Traditionally, reliability has been more of a design afterthought and improved through design modifications that result from rigorous and iterative prototype testing. This is inefficient and can also be limited in its effectiveness because pressures to get a product to market can outweigh the desire for design improvements.

Numerous techniques are available to improve product reliability in the early stages of design—and are far more effective and less costly than building, testing, and modifying prototypes. Tools such as finite element analysis, fluid flow, thermal analysis, and integrated reliability prediction models can be used in computer simulation models to build greater reliability into product designs. **Design of experiments** (DOE)[16] methodologies provide a structured approach that can consider how both product and process parameters affect product reliability. DOE also has the capability of assessing how the various factors interact.

When designing a product where reliability issues are important the following guidelines should be employed.

- Design the product so that it meets customer requirements across the entire expected operating range for each parameter when placed in service in an actual operating environment.
- Design the product so that stresses and thermal loads are minimized and product performance is minimally affected by these stresses or loads.
- Build system redundancy into the design.
- Derate[17] design components to increase the safety margin.
- Specify component parts and materials that have proven to be reliable in service.
- Avoid failure opportunities by minimizing the number of parts and interconnections in the design.
- Understand process capabilities to deliver reliable components and assemblies; and improve processes where they are not capable of meeting reliability standards.

In designing products for reliability several structured tools are available, including the **design failure mode and effects analysis (DFMEA)** and **anticipatory failure determination (AFD). Failure mode and effects analyses** can be applied to the mitigation of failures in both design (DFMEA) and process (**PFMEA**).[18] AFD has a similar objective to the FMEA approach—that is, the identification and removal of the root causes for failure. However, a reverse logic is used in the deployment of AFD. Instead of studying a part or component looking for possible failure modes, the designers invert the problem. Failures are treated as desired outcomes and design engineers try to create designs in which the failure of interest will occur with reliable predictability. AFD offers an advantage over the DFMEA approach. A DFMEA relies on the identification of failures based on the organization's knowledge base—that is, documented failures or failures known to occur through personal experience and technical expertise. This view can restrict creative insights that can accrue from a diversified group openly brainstorming all the things that might potentially go wrong in the functioning of a part, assembly, or module.

AFD strives to create designs that are more reliable by incorporating an understanding of the underlying causes of failure.

AFD employs the following steps.

- Invert the failure problem. Instead of hypothesizing possible root causes for a particular failure, invert the problem to state how the failure can be forced to occur.

[15]Poka-yoke techniques are discussed in some detail in Chapter 7.

[16]Design of experiments is addressed in Chapter 13.

[17]The term *derating* is used particularly in reference to electrical apparatus and means deliberately reducing performance expectations. For example, a 100 ohm resister, for design purposes, might be de-rated to 80% of its maximum load—that is, 80 ohms.

[18]FMEA procedures are discussed in Chapter 6.

- Seek methods or a sequence of events that can produce the failure. The focus of the problem is not on possible things that can happen to induce a failure; rather the focus is on those things that must happen consistently for the failure to reliably occur. This requires the engineering of a sequence of events that will assuredly lead to a failure scenario. AFD is a creative process that helps designers better understand the problems of failure and how product components function and interact to invoke a failure.
- Verify that the appropriate resources are available to cause each potential failure. Resources will fall in one of seven categories: materials, field effects, space, time, structure, system functions, and other system data.

5.3.5.3 Design for Testability/Inspectability. Inspection processes can be time-consuming and costly, particularly if prescribed test procedures require the development or acquisition of sophisticated test equipment. **Design for Testability and Inspectability** includes in the design those features that facilitate final testing, inspection, and acceptance at the end of the manufacturing process. Inspection costs can be minimized if quality engineers are represented on the design team. Certain design choices can mitigate the need to test the product with specialized equipment at various stages of production. Using methods such as QFD a common understanding can be established between the engineering, manufacturing, and quality control functions regarding minimum requirements for product acceptance—both at the end of the manufacturing value stream and at the point of use. A concurrence on what is required can then form the basis for designing in parallel the product and its associated testing procedures. If this is handled in an integrated manner in advance of product prototyping, product development cycle time can be significantly reduced. In designing products for testability/inspectability the following principles should be followed.

- Use geometrical dimensioning and tolerancing[19] to unambiguously represent and communicate design intent.
- Specify process knowledge product tolerances that are within the natural capabilities of the manufacturing process as indicated by capability indices such as C_p and C_{pk}.[20]
- Make provision in the product design for test points, access to test points and connections, and sufficient space to support test points, connections, and built-in test capabilities. Consideration should be given to the ease with which any testing and inspection that will be required in service throughout the life of the product can be performed.
- Employ standard connections and interfaces to facilitate the use of standard test equipment and connectors and minimize the effort required to set up and run test procedures.
- Consider the compatibility with automated test equipment. High volumes and standardized test methods can help support the justification of the development, acquisition, and implementation of automated test equipment.
- Provide built-in test and diagnostic capabilities to self-test and self-diagnose in the factory and in the field.
- Provide physical and electrical partitioning to facilitate testing and to isolate faults and defects.

5.3.5.4 Ecodesign and Green Product Design. Increasingly society and organized environmentalists are pressuring governments and industry to clean up the planet and put into place systems that will conserve Earth's limited resources. An **ecodesign,** or what some refer to as a **green product** design, is one that strives to minimize the environmental impacts throughout a product's entire life cycle. Manufacturers can no longer be solely concerned with producing and distributing a product and are beginning to realize the need to take responsibility for its disposition at the end of its useful life. Take-back, recycle, salvage, and initiatives of similar ilk are **end-of-life management** (**EOLM**) strategies. These activities are referred to by some as *design for decommissioning and replacement (DDR)*.

Ecodesign has become a worldwide movement and continues to spread. Many nations are now encouraging green designs and provide special labeling to identify those products that comply with environmental standards. The number of labels and participating countries continues to grow as labels depict ecodesign in buildings, carbon reduction, electronics, energy, food, forest products, retail goods, textiles, and tourism.[21]

A green design is one that minimizes the environmental impacts throughout a product's entire life cycle.

Because ecodesign is concerned with cradle-to-the-grave, its implementation requires cross-functional involvement up and down the supply chain. As soon as the first design sketches are drawn and a new product concept is born, its environmental impact is largely determined. The ecodesign process, also known as **design for the environment** (**DFE**), is concerned with five areas: raw materials, manufacturing processes, end use of the product, distribution of the product, and end-of-life management (EOLM). Some of the major design issues under each of these headings are shown in Figure 5-17.

The beginning of the product development cycle is the best time to address environmental requirements. These efforts help force a compromise between VE efforts, which are primarily focused on maximizing customer value (creating

[19] These principles were discussed in Chapter 4.

[20] Process capability is discussed in Chapter 9.

[21] Up-to-date information on ecolabels can be found at http://ecolabelling.org/.

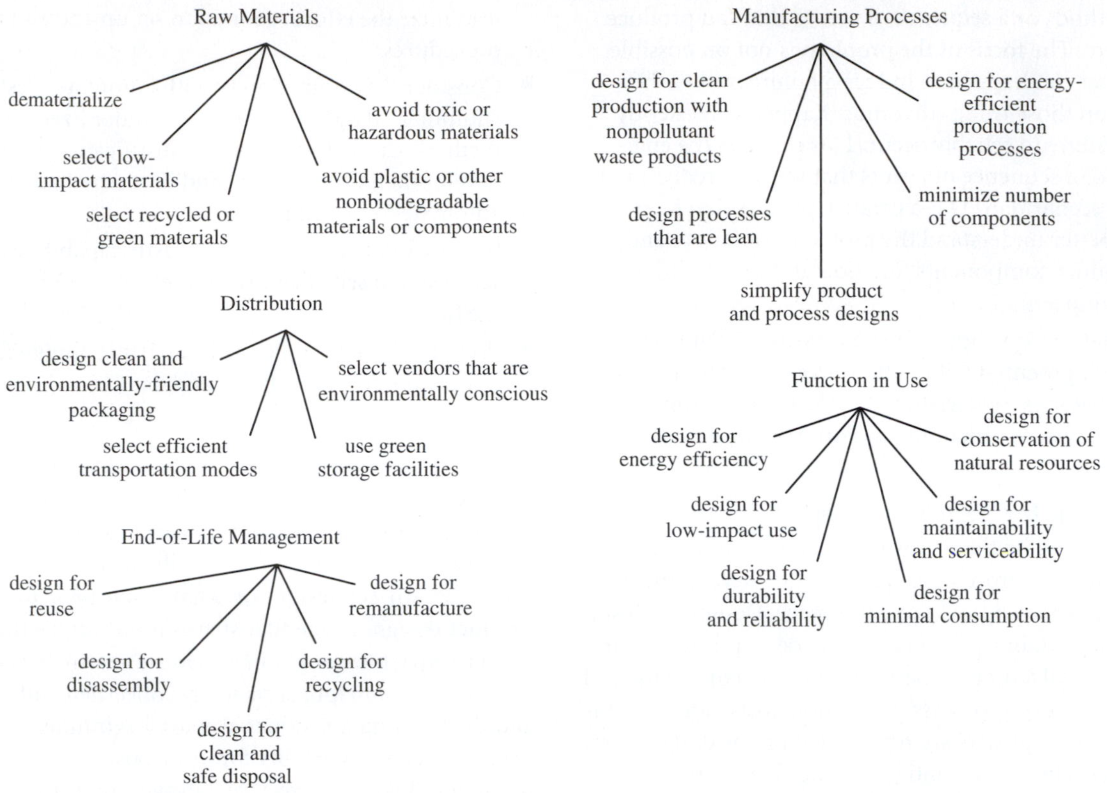

FIGURE 5-17. Ecodesign Considerations

customer delight at minimum cost) and producing a design that is environmentally friendly. Ironically, introducing the concepts of ecodesign to the VE process can actually produce superior outcomes, as this brings a fresh perspective and new way of thinking about integrated product and process development.

Whenever possible, and during the product development cycle, ecodesign engineers strive to identify suitable materials from green sources with low or no toxicity. In addition the design team investigates the energy requirements, identifies alternative energy sources, and selects the most energy efficient methods for manufacturing and design. Finally, as part of EOLM, the engineers address recycling or reuse and design an associated plan for disposing of the product at the end of its useful life.

Green products use less energy and natural resources, do not create unwanted waste or pollutants, and can be recycled or reused.

The design considerations that specifically apply to green products should be balanced with those that relate to quality, economics, manufacturability, and function. Ironically, as the ecofactors track closely with lean and QIM principles, producing green products actually requires minimal sacrifices of other design goals. Ecodesign typically produces products that use less energy and natural resources, do not create unwanted waste or pollutants, and can be recycled or reused. When linked to a lean Six Sigma program, ecodesign can simultaneously improve profitability, customer satisfaction, and harmony with nature.

A Scandinavian study found that a company that successfully integrates ecodesign into its product development process can expect the following benefits, all of which contribute to sustainability.[22]

- Reduced material, component, and manufacturing costs
- Increased market share
- Improved market image
- Increased staff commitment, involvement, and satisfaction
- Improved capability to innovate
- Increased competitiveness and profitability
- Reduced inventory and associated administration costs
- Reduced maintenance and service costs

An Australian appliance manufacturer serves as an example of good ecodesign.[23] Email Major Appliances, formerly Southcorp Whitegoods, used a multidisciplinary team consisting of 15 experts to develop a new line of environmentally friendly dishwashers. The new Dishlex line can wash a

[22]Johannson, Glenn, Johan Widheden, and Carl Gunnar Bergendahl. "Green Is the Color of Money." *Greenpack Report 2001-02.* Molndal, Sweden: IVF Industrial Research and Development Corporation, funded by the Nordic Industrial Fund, August 2001.

[23]http://www.deh.gov.au/settlements/industry/corporate/dfe.html.

full load of dishes with less than 18 liters (4.75 gallons) of water compared to normal dishwashers that require an average of 30–53 liters (8–14 gallons). The weight of each unit on the new line was trimmed by up to 7 kilograms (15.4 pounds) through dematerialization, material substitutions, and parts reduction. Component coding facilitates disassembly and recycling. These new improved dishwashers have also been designed to use enzyme-based detergents that require lower temperatures, and other water-saving features.

Further information on green initiatives and case studies can be found by contacting the University of Tennessee Center for Clean Products and Clean Technologies,[24] the Royal Melbourne Institute of Technology Center for Design,[25] the University College for the Creative Arts Center for Sustainable Design in Surrey, UK,[26] Carnegie Mellon University Green Design Consortium,[27] the United States Environmental Protection Agency Design for the Environment Program,[28] and the O2 Global Network.[29]

5.3.5.5 Design to Cost. Since approximately 70% of the total life-cycle costs of a product are determined during the design phase, efficiency concerns that have traditionally been limited to supply chain issues (e.g., the procurement, manufacturing, replenishment, or customer order cycles) are no longer sufficient. Customer satisfaction is often based on cost performance over a product's entire life cycle. **Design to cost (DTC)** is a process that is rooted in Taguchi's design principles and follows the five general guidelines listed below.

- Concurrently design the product and the process that will be used to produce the product. This approach minimizes costs that accrue as a result of engineering design changes that may be necessary to facilitate product manufacturability, and the waste that occurs as a result of designs that are incompatible with manufacturing capabilities.
- Whenever possible, avoid complexity of design, process, and quality measures. Simple designs are less costly than complex ones.
- Minimize the number of parts per component, subassemblies, and assemblies.
- Minimize the number of processing steps in the production routing required to manufacture the product.
- Minimize the number of material sources that will be necessary to support the manufacturing process.

Design to cost is a methodology that builds into a product design the ability to meet cost performance requirements over a product's entire life cycle.

[24]http://eerc.ra.utk.edu/ccpct/index.html.
[25]http://www.cfd.rmit.edu.au.
[26]http://www.cfsd.org.uk.
[27]http://www.ce.cmu.edu/GreenDesign.
[28]http://www.epa.gov/dfe.
[29]http://www.o2.org/index.php.

Methodologies such as VE, QFD, and DFM can be used to incorporate the concept of life-cycle costing into product designs. Two additional approaches that can be employed are target costing and activity-based costing.

5.3.5.5.1 Target Costing. The Japanese were among the first to discover an important underlying relationship between design and cost. Simply trying to reduce the costs associated with a poor design will not work. Competitiveness depends on the ability to design the cost out of products. **Target costing (TC)**, a term that is often used synonymously with DTC, is a technique that starts with the determination of a target price that needs to be met if the proposed product is to be competitive. This predetermined price is translated into a target cost after deducting desired profit margins. The product is then designed in a way that will not exceed the targeted cost while still meeting other product requirements such as functional performance and aesthetics.

In some cases simultaneously meeting the target cost and other customer requirements may prove to be infeasible. When this is discovered, ideally early in the product development cycle, the decision can be made to revisit and possibly revise the target cost, or to abandon the design and decide not to launch the product.

Target costing is an iterative multidisciplinary process that should include participation from the marketing and accounting functions during every step of the design cycle. During this process estimates are developed for each of the components necessary to produce a finished product. This includes materials, direct and indirect manufacturing costs, packaging, logistics and transportation, and the allocation of indirect costs. Once the overall target cost is determined it is apportioned by the team across the major functional areas of design, manufacturing, distribution, sales, and marketing.

Target costing strives to design a product so that it meets a target cost while satisfying all customer requirements.

It is not uncommon to miss the target on the first pass. This can be due to poor cost estimates, an incorrect distribution of the costs across the areas and components, or an unrealistic target. The design team, in consultation with the marketing and accounting experts, may have to go through numerous iterations to achieve a workable design, or abort the project in the event that no feasible design can be created.

5.3.5.5.2 Activity-Based Costing. Accountants have traditionally allocated indirect costs (also called overhead and burden) to products on the basis of certain direct costs—typically labor. This method does not take into account the fact that different products are not equal in their respective contributions to indirect costs. This can lead to one product *subsidizing* another, since the actual cost of production may not be captured by the cost accounting method used. For example, advanced technologies and productivity improvements have reduced some direct cost centers (e.g., labor) while increasing certain indirect costs (e.g., capital investment).

This phenomenon can be clearly seen in automated environments, where machines have taken on the work that once was relegated to humans. The investment in such automated processes adds to the plant's indirect costs while reducing the direct labor costs. **Activity-based costing (ABC)**[30] attempts to remedy the shortcomings of conventional methods by using cause-and-effect analyses to identify the key activities (called cost drivers) where costs are generated and then use these activities as the basis for allocating costs. Cost drivers can be categorized as either structural or executional.[31] Structural cost drivers derive from an organization's strategy and relate to economic factors such as product complexity, scale and scope of production operations, technology deployment, stakeholder relations, and marketing strategies. Executional cost drivers are rooted in routine business operations and drive indirect costs in areas such as equipment utilization, quality control, inventory control, maintenance, materials handling, and workforce participation. Some examples of indirect costs and possible executable cost drivers are as follows.

- Maintenance costs—possibly driven by total number of machine hours run or the average size of a production batch.
- Materials handling costs—possibly driven by average size of production batch or the total number of orders processed.
- Quality control costs—possibly driven by total number of inspections required or the total number of orders processed.

ABC is a four-step process:[32]

Step 1 Determine the relevant activities.

Step 2 Allocate overhead cost to the relevant activities.

Step 3 Select an appropriate allocation cost driver for each activity.

Step 4 Use the cost drivers to allocate costs to products.

The following example illustrates the mechanics of ABC. Q-One Enterprises makes two products, the Nifty Whiz and the Econo Clean. The monthly production volumes for the two products are 20,000 and 35,000, respectively. Total overhead costs for the Q-One plant facility are $550,000 per month and the plant uses an average of 3,500 direct labor hours per month to produce the Nifty Whiz and 4,500 direct labor hours per month to produce the Econo Clean.

Activity-based costing uses cause-and-effect analysis to identify cost drivers, which are then allocated to products in proportion to their respective use of the cost driver resource.

Conventional accounting methods would allocate the overhead based on the number of direct labor hours expended.

$$\text{Nifty Whiz: } \frac{3{,}500 \text{ direct labor hours expended}}{8{,}000 \text{ total direct labor hours}}$$

$$\times \$500{,}000 = \$218{,}750$$

$$\frac{\$218{,}750}{20{,}000} = \$10.94 \text{ per unit}$$

$$\text{Econo Clean: } \frac{4{,}500 \text{ direct labor hours expended}}{8{,}000 \text{ total direct labor hours}}$$

$$\times \$500{,}000 = \$281{,}250$$

$$\frac{\$281{,}250}{35{,}000} = \$8.04 \text{ per unit}$$

Table 5-1 illustrates the ABC method. Four relevant categories and their respective cost drivers have been identified: materials control (cost driver—number of orders received); operations (cost driver—number of direct labor hours expended); logistics and transportation (cost driver—number of units shipped); and sales (cost driver—number of sales contacts made). For each category, the number of cost driver units for each product and the average unit cost for each cost driver have been determined. The total overhead allocated to each product is the sum of the unit costs across all categories. The total overhead allocated to Nifty Whiz is $233,313 (more than the conventional method) and to Econo Clean $266,687 (less than the conventional method). Converting these totals to unit overhead charges we have,

$$\text{Nifty Whiz: } \frac{\$233{,}313}{20{,}000} = \$11.67 \text{ per unit}$$

$$\text{Econo Clean: } \frac{\$266{,}687}{35{,}000} = \$7.62 \text{ per unit}$$

By allocating indirect costs in a rational manner that considers where the true costs are generated, some products will be revealed as more profitable and others less profitable than under the conventional methods.

A limitation to the ABC method is that not all indirect costs can be directly assigned to specific products. For example,

A supervisor and foreman confer on a production problem, while the operator awaits instructions. Supervisory salaries are overhead costs which should be allocated to products on a rational and equitable basis.

[30]Kaplan, Robert S., and W. Bruns. *Accounting and Management: A Field Study Perspective.* Boston: Harvard Business School Press, 1987.

[31]Shank, John, and Vijay Govindarajan. *Strategic Cost Management: The New Tool for Competitive Advantage.* New York: Free Press, 1993.

[32]Baker, W. M. "Understanding Activity-Based Costing." *Industrial Management* 36 (1994): 28–30.

TABLE 5-1 Illustration of Activity-Based Accounting

Overhead Allocations Based on Activity-Based Accounting

Overhead Category	Materials Control	Operations	Logistics and Transportation	Sales	Totals
Total Overhead Generated	$100,000	$200,000	$50,000	$150,000	$500,000
Cost Driver Units Consumed: Nifty Whiz	2,000 orders	3,500 direct labor hours	20,000 units shipped	650 sales contacts	
Cost Driver Units Consumed: Econo Clean	6,000 orders	4,500 direct labor hours	35,000 units shipped	300 sales contacts	
Cost Driver Unit Cost	$12.50	$25.00	$0.91	$157.89	
O/H Allocation: Nifty Whiz	$25,000	$87,500	$18,182	$102,632	$233,313
O/H Allocation: Econo Clean	$75,000	$112,500	$31,818	$47,368	$266,687

how can one apportion a share of the CEO's salary to any single product or activity? Costs that cannot be directly allocated are called **business sustaining costs** and are lumped together in an unallocated cost category. Management has the prerogative to allocate the business sustaining pool to products, using traditional cost accounting methods, or may choose not to allocate these costs to products at all. In general, business sustaining costs should represent only a small portion of an organization's total overhead.

Business sustaining costs are those that cannot directly be allocated to products or services.

The ABC method determines activity costs using a variety of tools that include activity charts, worksheets, and cost buildup tables. During the design phase, the product design team is often able to identify and highlight those activities where the majority of product costs are incurred. Efforts can then be directed toward design modifications in either product or process that will result in reducing or eliminating costs in these key areas.

5.3.5.6 Other Life-Cycle Design Factors. Other life-cycle design considerations include the following:

- **Design for installability**—Customer satisfaction for products that require installation, activation, or implementation of any kind depends on the ease and effectiveness with which the user can get the product to function as intended and within a reasonable timeframe. Satisfaction can easily turn into frustration if a customer tries to install the product only to discover that the power supply cord is a meter too short, or the plumbing fittings are incompatible with the connections for the main water supply. Designers should anticipate where and how products will be used and installed and try to adapt product designs to facilitate the most likely customer scenarios.
- **Design for usability, trainability, and documentation**—Products should be designed with the human factor in mind. Human–product interfaces should incorporate the principles of sound ergonomic design, taking into account average human capabilities and the ease with which humans will be able to use the product. Designs should also facilitate the ability to clearly document proper operating, care, and maintenance instructions with a view to how users can be trained and those issues that trainers are likely to encounter.
- **Design for updateability and upgradeability**—As technology or customer needs change there may be a need to either update or upgrade product functionality. In the early stages of design, decisions need to be made on whether the design should or will support updateability and/or upgradeability and how.

5.4 QUALITY OF PROCESS DESIGN

5.4.1 Flow Designs

In Chapter 2 we discussed the importance of establishing competitive priorities as part of an organization's corporate strategic planning process. Competitive priorities define how the organization intends to compete and are derived after some internal soul-searching has determined how core competencies can best match stakeholders' needs in multiple areas, including cost, quality, time, flexibility, and responsiveness. The corporate strategy provides the framework for product and process development, and the process design component shapes the organization's operations strategy, which in turn determines the flow strategy used to produce the product. The strategies at all levels are driven by the organizations quality management system (QMS) that consists of all the policies, procedures, controls, standards, and regulations that are relevant to the planning and execution activities described in Chapter 2. Feedback received during the execution phases in turn are used to modify the QMS. The hierarchy of strategies and their relationships are illustrated in Figure 5-18.

The **flow strategy** determines how factors of production will be configured (that is, the transformation subsystem design) in order to support an organization's competitive priorities and marketing strategy. Two principal factors are variety (product mix) and volume (individual batch quantities). As Figure 5-19 shows, high-volume, low-variety combinations generally favor **line flow** designs, and it is not uncommon to find production lines dedicated to running large batch quantities of a single product. In some situations, such as oil refining, chemicals, and paper making the relevant processes run continuously. Here, flow patterns are linear with minimal if any backtracking.

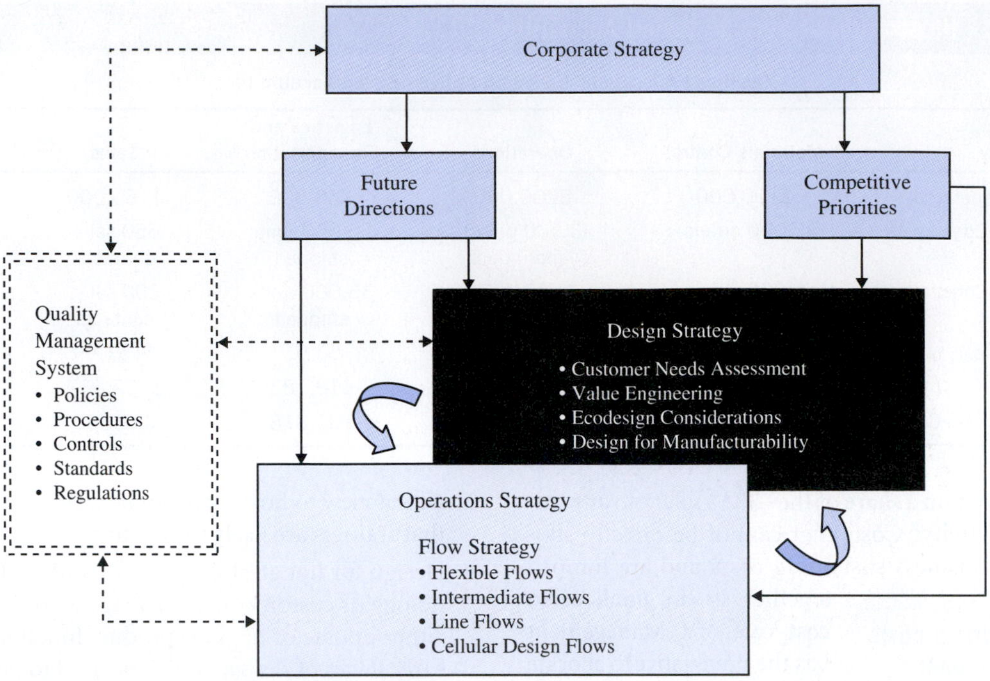

FIGURE 5-18. The Hierarchy of Strategies

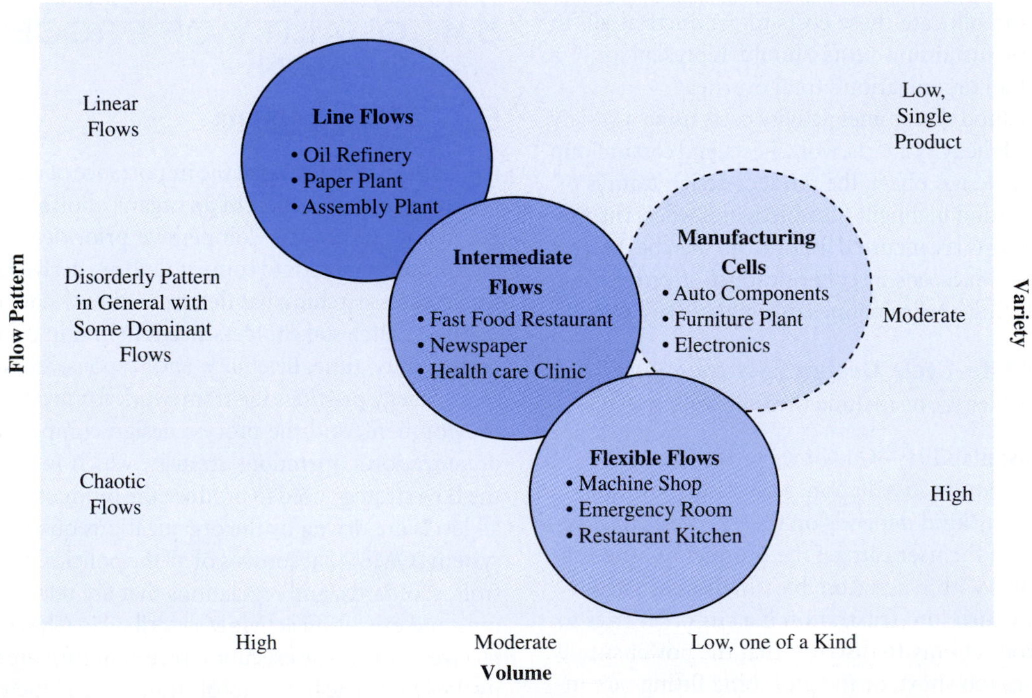

FIGURE 5-19. Flow Strategies

On the opposite end of the spectrum are flexible flows, used when customer satisfaction is achieved through quick response to customized requirements that can be changed on short notice. The extreme case is when each order is unique and requires a custom routing, such as a machine or tool and die shop. With flexible flows, patterns tend to be chaotic with significant crisscrossing and backtracking to accommodate multiple production routings. Where volume and variety are at moderate levels, production processes are designed to accommodate intermediate flows. At this level one can expect to find some disorderly flow patterns, although certain patterns will tend to dominate.

> **A line flow process design suits high-volume, low-variety production with linear flow patterns.**

Figure 5-20 shows how individual process selection fits within the overall flow strategy continuum.

5.4.2 Line Flow Designs

5.4.2.1 Continuous Processes. Line flow designs are suited for continuous processes and flow shop operations. **Continuous processes** are characterized by standardized products produced in large volumes, and typically on a 24-hour, 7-day basis. These processes use automated specialized equipment that can be prohibitively expensive to start, stop, or shut down. The initial setup is complex and time-consuming. The plant layout for a continuous process normally follows the processing stages, with a product-oriented style. Lines are usually dedicated to a single product and include all the processing steps in a prescribed routing sequence. The output rate is controlled by equipment capacity and the input rates of flow mixtures. There is usually one primary input to these processes. Material costs are high relative to labor costs which are low (due to automation). Many of these processes involve the production of commodities, such as oil, sugar, and minerals. Other examples of continuous processes are glass, steel, and paper manufacturing. Figure 5-21 is a diagram of a sugar refinery. Note the smooth linear pattern of the primary flow as material advances along the value stream converting cane into consumable sugar.

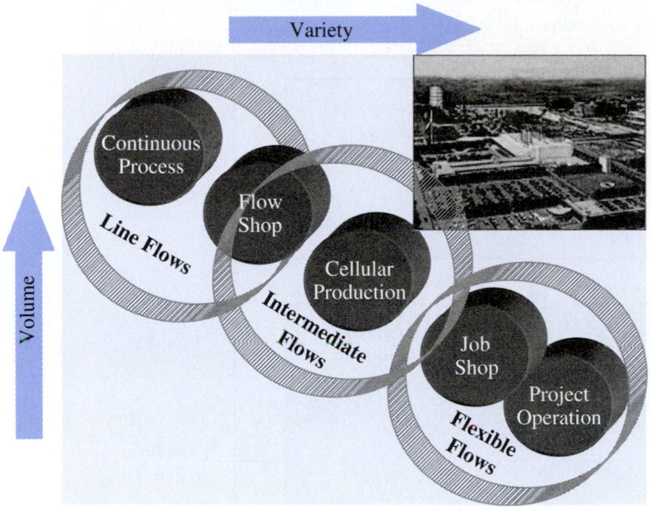

FIGURE 5-20. Process Selection

5.4.2.2 Flow Shops. **Flow shop** designs are similar to continuous processes except in the case of a flow shop a discrete product is concerned. These processes employ specialized automated equipment to produce high volumes of output with a low-variety product mix. Even with automation these processes can in some cases be labor intensive, and direct labor and indirect costs can be dominant. The best way to conceptualize a flow shop is to think of an assembly-line process. The high volumes generate economies of scale

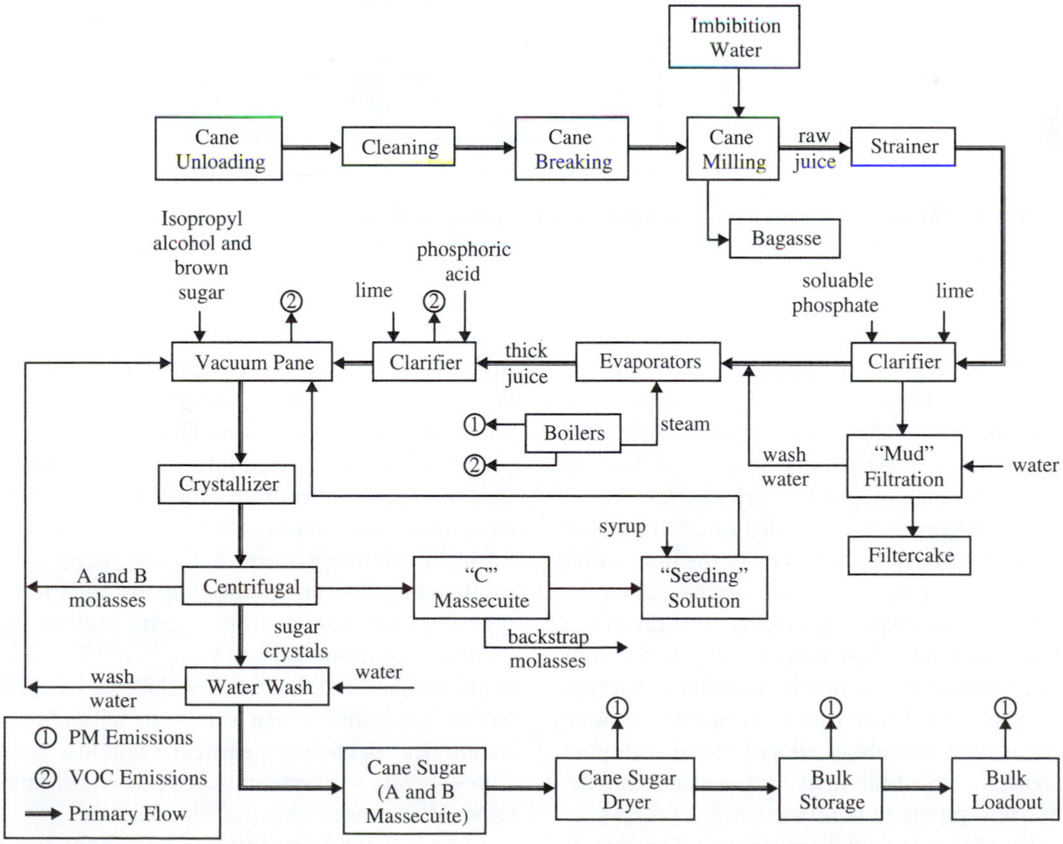

FIGURE 5-21. Diagram of Sugar Refinery

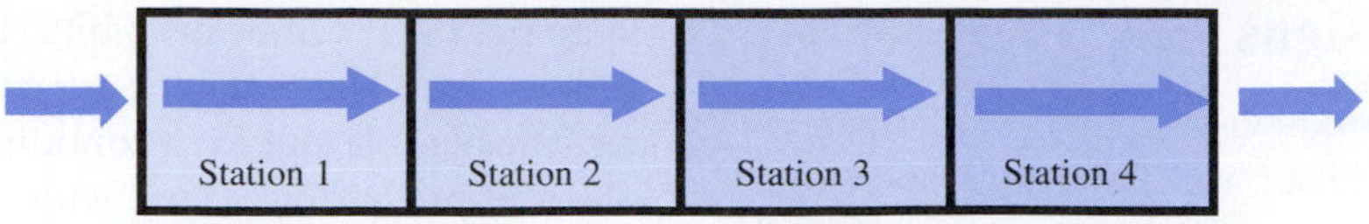

FIGURE 5-22. Schematic of a Flow Shop and Product-Oriented Layout

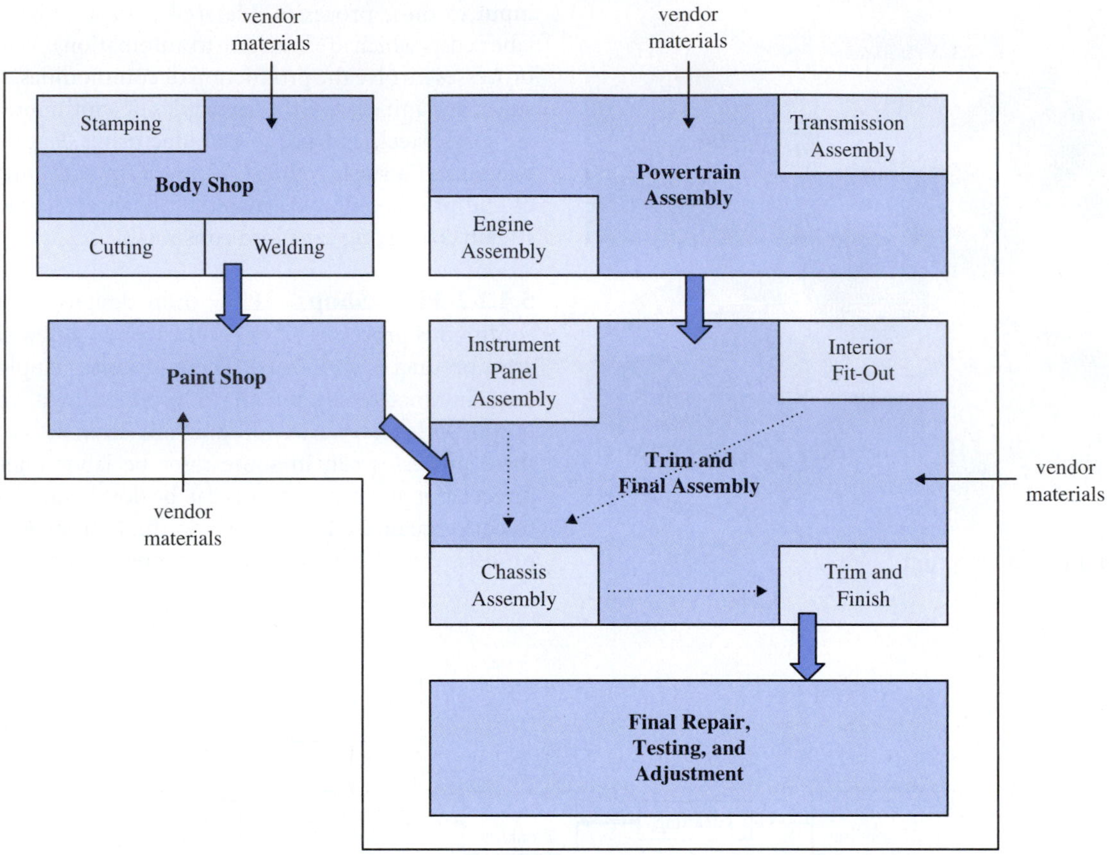

FIGURE 5-23. Simplified Flow Diagram of Automobile Assembly Plant

through lower direct labor costs and volume discounts (achieved with bulk purchasing). These production systems are well suited to just-in-time vendor deliveries, enabling low in-process inventory levels of materials. Due to the high-volume, low-variety nature of flow shops, **product-oriented layouts** are typical where factors of production are dedicated to product groupings, often with entire lines dedicated to the production of a single product. In general, flow shops can be easily controlled with simple managerial procedures. Figure 5-22 shows a conceptualized schematic of a flow shop and Figure 5-23 is a simplified flow diagram of an automobile assembly plant, typifying the flow shop pattern. There is a smooth linear flow along the value stream as parts are fabricated and mated with purchased components, finally resulting in a finished automobile, tested and ready for shipment to the customer.

Flow shops have some distinct disadvantages. The specialized nature of the equipment (sometimes an integrated line) makes changes in rates of output and/or product design difficult. Even minor product design changes can mean substantial equipment modifications. Since a large proportion of the production routing can be interconnected within a single production line, the entire production schedule can be vulnerable to equipment breakdowns. A further disadvantage concerns the human element. In this type of production environment it is common for work to be standardized and worker skills to be specialized. Each worker is responsible for a relatively small and repetitive contribution (e.g., tightening a particular bolt to a specified torque). Consequently, workers can feel unchallenged, and this can lead to boredom, low morale, and absenteeism.

Flow shops are inflexible to changes in production rates and product design.

While direct labor costs for flow shops are generally low, just the reverse is true of indirect costs. Flow shop systems are

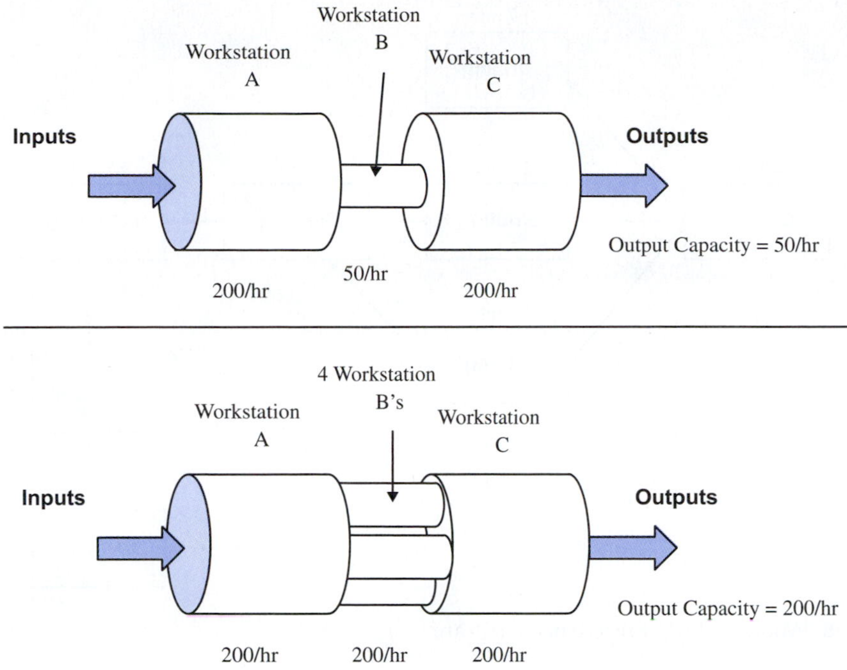

FIGURE 5-24. Effect of Line Balance on Capacity

complex, and their planning, design, purchase, and installation are time-consuming and costly. The capital intensive nature of the highly specialized equipment, coupled with the requirement for a sizable support staff, creates large overheads. This also limits an organization's flexibility because the equipment is difficult to dispose of or to modify for an alternative use.

In a flow shop's product layout, material is moved in one direction along the value stream and in a fixed pattern. Fixed path equipment, such as conveyors and overhead hoists, are common. There is little need for space for work-in-process (WIP) since materials are continually consumed as the product advances down the line. All workstations are sometimes tied together, leaving no area between operations for WIP to accumulate. If products are made-to-stock, substantial warehousing space may be necessary to store finished goods until it can be sold.

A major concern with a product layout is being able to balance the work so that no one workstation becomes a major bottleneck that severely restricts the smooth flow of work further down the line.

5.4.2.3 Line Balancing. As for the rate of output, production lines must be balanced to the slowest least productive workstation. This is called **line balancing.** Refer to Figure 5-24. The flow shop on the top depicts a line consisting of three workstations. This line is ill-balanced and the total system capacity is limited to the lowest capacity in the line—workstation B. Workstation B is called a **bottleneck** and forces workstations A and C to be underutilized. The bottom of Figure 5-24 illustrates the impact of eliminating the bottleneck by adding three more type B workstations. The total line capacity can now be perfectly balanced at 200 units per hour and the output capacity has quadrupled.

An important factor in the design of a flow shop is how much output is needed to satisfy customer demand. Converted to time requirements this can be stated as how much time is available for each unit produced in order to meet demand requirements. This important design parameter is called **takt time** and is defined as follows:

Takt Time

$$\text{Takt time} = \frac{\text{Total time available for production}}{\text{Total number of units to produce}} \quad \text{5-2}$$

For any operating flow shop, the actual production rate is called **cycle time,** defined as follows:

Cycle Time

$$\text{Cycle time} = \frac{\text{Some specified time}}{\text{Total number of units produced during that time}} \quad \text{5-3}$$

A good process design will match the cycle time with the takt time. This can never be accomplished exactly due to the variability that is present in the output of each individual workstation and that accumulates along the value stream. The design challenge is to try and get the actual (cycle) time as close to the required (takt) time as possible.

To illustrate how a flow shop can be designed with a value stream balanced to takt requirements, consider the following simple example. In a woodworking shop, a certain product goes through the eight steps shown on the following page.

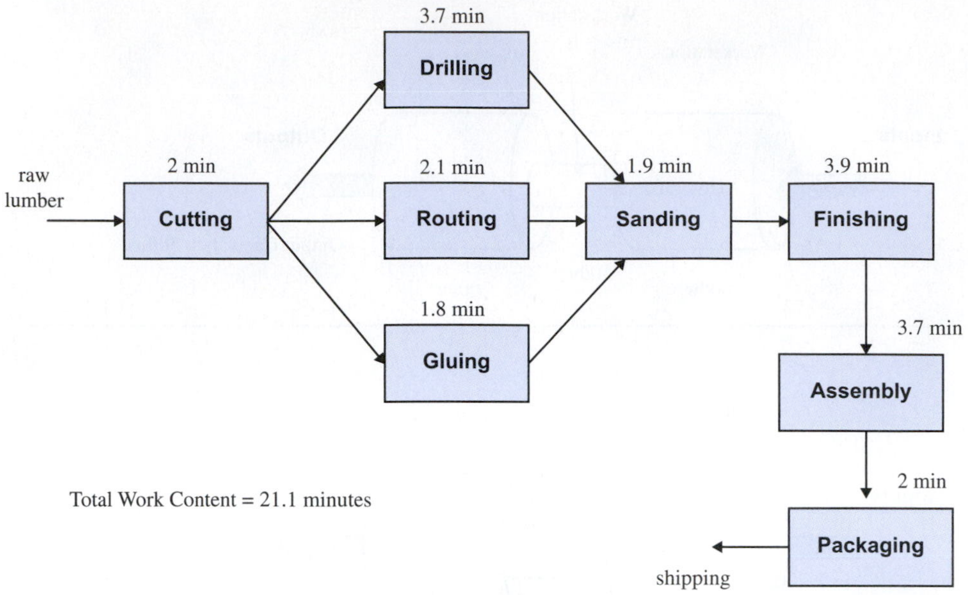

FIGURE 5-25. Woodworking Precedence Diagram

Operational Step	Operation	Standard Time (min)
1	Cutting	2
2	Drilling	3.7
3	Routing	2.1
4	Gluing	1.8
5	Sanding	1.9
6	Finishing	3.9
7	Assembly	3.7
8	Packaging	2

To manufacture this product, the eight steps must be performed in the defined sequence shown in the **precedence diagram** in Figure 5-25.

The total work content (the value-adding time to complete all eight steps) is the sum of the individual activity times, or 21.1 minutes. An order for 525 finished units of this product must be produced within the next five days. The plant operates on a single shift and each shift has seven hours available for productive work. The takt time is equal to 4 minutes per unit, computed as follows:

$$\text{Takt time} = \frac{7\ \text{hrs/shift} \times 5\ \text{shifts} \times 60\ \text{min/hr}}{525\ \text{units required}}$$

$$\text{Takt time} = \frac{2{,}100\ \text{min available}}{525\ \text{units required}} = 4\ \text{min/unit}$$

The first step in designing the line is to set the design cycle time equal to the computed takt time. Then we proceed to assign tasks to workstations, making certain that none of the precedence requirements are violated. We can roughly determine a priori the number of required workstations as follows.

$$\text{Number of workstations} = \frac{21.1\ \text{min}}{4\ \text{min}}$$
$$= 5.275\text{, or 6 workstations}$$

Number of Workstations

$$\#\ \text{of workstations} = \frac{\text{Total work content}}{\text{Takt time}} \qquad \textbf{5-4}$$

There are several methods used for assigning tasks to a workstation. We will use the longest operation time (LOT) method, which gives priority to those tasks with the largest work content. A convenient framework to guide the process is to establish a spreadsheet as follows. For each workstation the time available column will be reinitialized to the design cycle time of 4 minutes. The sequence of line balancing steps is illustrated in Figure 5-26.

Workstation Number	Time Available (min)	Eligible for Assignment	Assignment
1	4	Cutting	Cutting

For the first step, there is clearly only one operation choice: *cutting*. Since the work content for the cutting operation is less than the design cycle time we can assign cutting to workstation 1. That assignment reduces the available time to 2 (4−2) as follows.

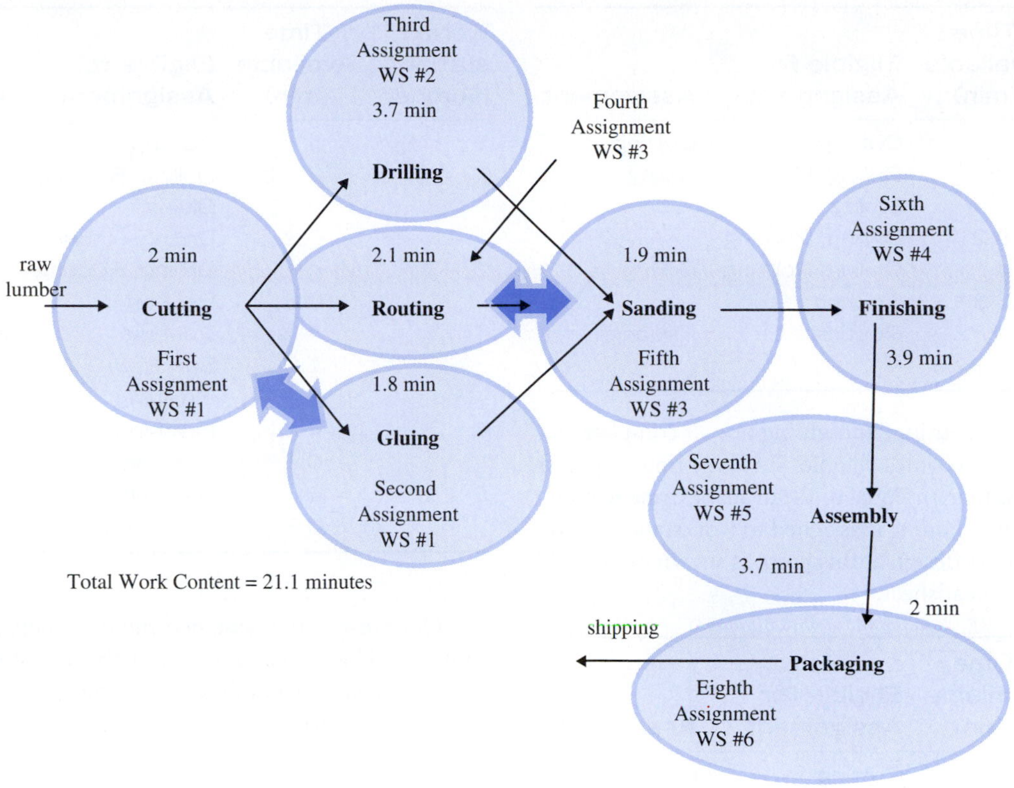

FIGURE 5-26. Balancing Line

Work-station Number	Time Available (min)	Eligible for Assignment	Assignment
1	4	Cutting	Cutting
	2	Drilling, Routing, Gluing	Gluing

Once cutting is complete, according to the precedence diagram, drilling, routing, and gluing can take place. The LOT technique will select the one with the largest work content that can fit into the available time. While LOT would favor drilling, the 3.7 minutes work content is greater than the available time for workstation 1. The only operation that will fit is gluing, so that assignment is made, reducing the available workstation time to 0.2 minute leading to the following updated spreadsheet.

Work-station Number	Time Available (min)	Eligible for Assignment	Assignment
1	4	Cutting	Cutting
	2	Drilling, Routing, Gluing	Gluing
	0.2	Drilling, Routing	—

Since none of the remaining eligible candidates have work content times that are less than or equal to 0.2 minute, we close off workstation 1 and start on workstation 2. Of the two remaining eligible operations the LOT technique favors the one with the largest time—drilling. This assignment is made and the spreadsheet is updated accordingly.

Work-station Number	Time Available (min)	Eligible for Assignment	Assignment
1	4	Cutting	Cutting
	2	Drilling, Routing, Gluing	Gluing
	0.2	Drilling, Routing	—
2	4	Drilling, Routing	Drilling
	0.3	Routing	—

We now move on to workstation 3 where an assignment is made to the last eligible operation, routing. At the end of this step the spreadsheet looks as follows.

Work-station Number	Time Available (min)	Eligible for Assignment	Assignment
1	4	Cutting	Cutting
	2	Drilling, Routing, Gluing	Gluing
	0.2	Drilling, Routing	—
2	4	Drilling, Routing	Drilling
	0.3	Routing	—
3	4	Routing	Routing
	1.9		

Once drilling, routing, and gluing are *all* complete the sanding operation becomes eligible. Since the time required for *sanding* is less than or equal to the available time remaining for assignment, *sanding* is assigned to workstation 3. The status of the process design at this point is shown on the following updated spreadsheet.

Work-station Number	Time Available (min)	Eligible for Assignment	Assignment
1	4	Cutting	Cutting
	2	Drilling, Routing, Gluing	Gluing
	0.2	Drilling, Routing	—
2	4	Drilling, Routing	Drilling
	0.3	Routing	—
3	4	Routing	Routing
	1.9	Sanding	Sanding
	0	Finishing	—

This process is repeated until all assignments are made. Once *sanding* is assigned *finishing* becomes an eligible candidate, and the updated spreadsheet shows the progression to workstation 4.

Work-station Number	Time Available (min)	Eligible for Assignment	Assignment
1	4	Cutting	Cutting
	2	Drilling, Routing, Gluing	Gluing
	0.2	Drilling, Routing	—
2	4	Drilling, Routing	Drilling
	0.3	Routing	—
3	4	Routing	Routing
	1.9	Sanding	Sanding
	0	Finishing	—
4	4	Finishing	Finishing
	0.1	Assembly	—

Once *finishing* is assigned *assembly* becomes an eligible candidate and the updated spreadsheet shows the progression to workstation 5.

Work-station Number	Time Available (min)	Eligible for Assignment	Assignment
1	4	Cutting	Cutting
	2	Drilling, Routing, Gluing	Gluing
	0.2	Drilling, Routing	—
2	4	Drilling, Routing	Drilling
	0.3	Routing	—
3	4	Routing	Routing
	1.9	Sanding	Sanding
	0	Finishing	—
4	4	Finishing	Finishing
	0.1	Assembly	—
5	4	Assembly	Assembly
	0.3	Packaging	—

Once *assembly* is assigned the final step, *packaging*, becomes an eligible candidate and the updated spreadsheet shows the assignment of the packaging operation to the final workstation, workstation 6.

Work-station Number	Time Available (min)	Eligible for Assignment	Assignment
1	4	Cutting	Cutting
	2	Drilling, Routing, Gluing	Gluing
	0.2	Drilling, Routing	—
2	4	Drilling, Routing	Drilling
	0.3	Routing	—
3	4	Routing	Routing
	1.9	Sanding	Sanding
	0	Finishing	—
4	4	Finishing	Finishing
	0.1	Assembly	—
5	4	Assembly	Assembly
	0.3	Packaging	—
6	4	Packaging	Packaging
	2	—	—

Line or individual workstation efficiency can be computed as follows:

Efficiency

$$\text{Efficiency} = \frac{\text{Total work content}}{\text{Takt time}} \qquad \textbf{5-5}$$

Line Efficiency

$$\text{Efficiency} = \frac{\text{Total work content}}{\text{\# of Workstations} \times \text{Cycle time}} \qquad \textbf{5-6}$$

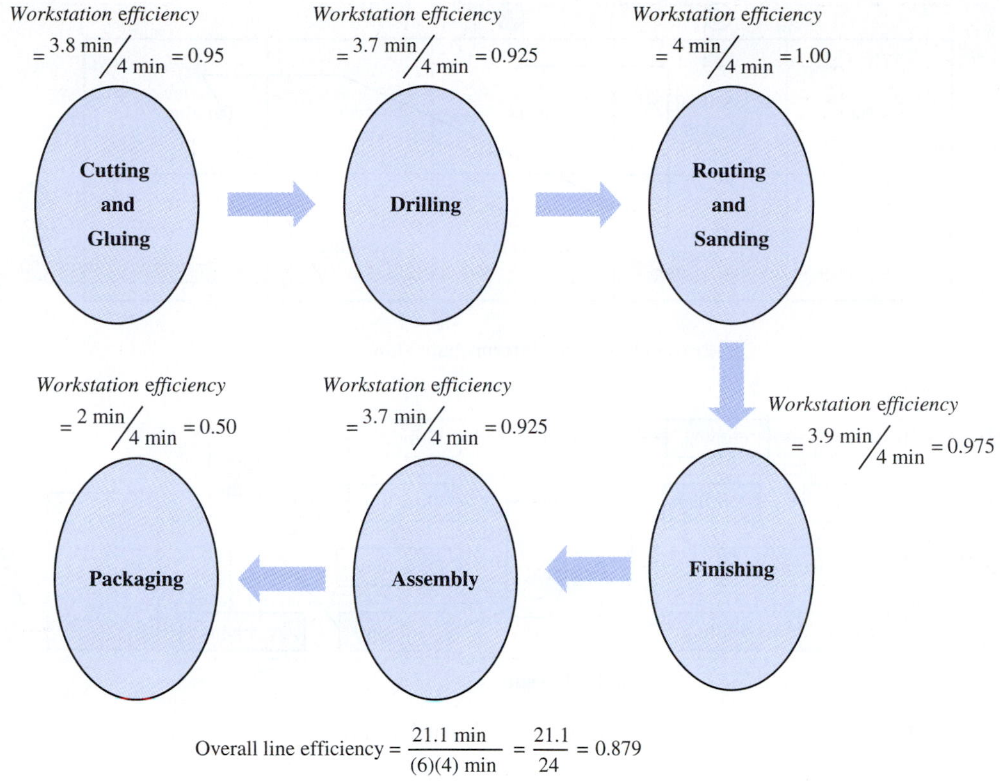

FIGURE 5-27. Balanced Line

The six-station line is shown in Figure 5-27. The projected efficiency of the overall line is 87.9%, computed as follows.

$$\text{Line efficiency} = \frac{21.1}{6 \times 4} = \frac{21.1}{24} = 0.879$$

Individual workstation efficiencies range from 100% (workstation 3) to 50% (workstation 6). It is obvious that for this product line, workstation 6 has an excess capacity. The options are to decrease the capacity or divert the excess to another product line if possible.

It is important to realize that this balanced design is based on two fundamental assumptions: (1) the standard production time estimates do not vary, and 2) each workstation has a yield of 100%. In reality neither of these assumptions will hold. The output of each workstation will be variable and follow a unique distribution that also might change over time. (Refer to problem 5-1 at the end of this chapter for an exercise that will illustrate this concept.) It is also possible that some or all of the workstations will experience scrap or rework that will yield outputs less than 100%.

5.4.2.4 Computerized Line Balancing. Balancing a line using manual calculations can become unwieldy as the problems grow in size. Fortunately some commercial line balancing software packages are available to help. Applied Computer Services, Inc. offers an Excel-based package called Timer Pro Professional,[33] and EASE, Inc. offers a line balance module of its EASEworks[34] software that has the capability to import a solution into a process simulator to observe results. Others that have line balancing modules are the Balancer Domain of AssemblyLinePro[35] by TSF Manufacturing Solutions, and ProBalance Line Balancing[36] software by Proplanner which employs IBM's computer method for sequencing assembly-line operations (COMSOAL). The COMSOAL algorithm was originally developed by Chrysler for assigning tasks to workstations while preserving precedence, codependency, and resource constraints. Most of the commercially available software packages including these do not guarantee an optimal solution. They use heuristics (rules) to balance a line to an acceptable level of efficiency. A popular package is POM-QM for Windows[37] which lets the user choose from a list of five heuristics that specify the order in which tasks are considered for allocation to workstations: ranked positional weight (RPW), longest operation time (LOT), shortest operation time (SOT), most number of following tasks, and least number of following tasks.

[33] http://www.acsco.com/linebalancing.htm.
[34] http://www.easeinc.com/solutions/lineBalancing.php.
[35] http://www.assemblylinepro.com/Walk-thru/Balancer_Domain/BalancerDomain.htm.
[36] http://www.proplanner.com/Product/Details/LineBalancingSA.aspx.
[37] Weiss, Howard J. *POM-QM for Windows, Version 3*. Upper Saddle River, NJ: Prentice Hall, 2006.

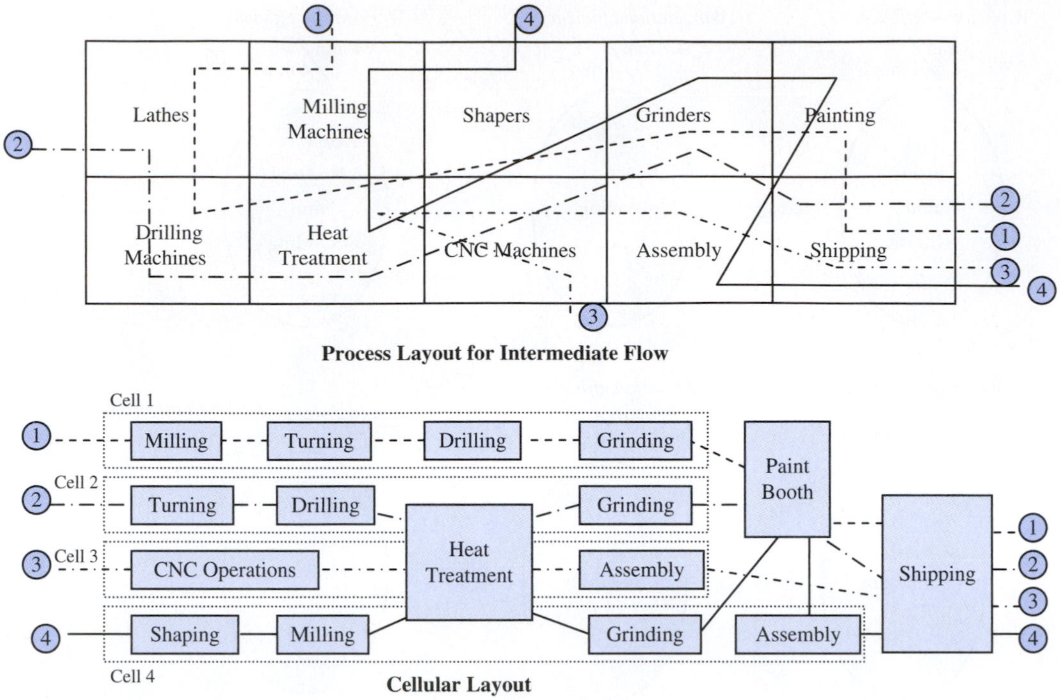

FIGURE 5-28. Comparison of Process Layout with Cellular Design

5.4.3 Intermediate Flow Designs

5.4.3.1 Process Layouts. When individual production runs and product variety are of moderate size a **process-oriented layout** can be used that departmentalizes by grouping together similar functions (i.e., type of machines, tasks, or operations). This is illustrated in the top diagram of Figure 5-28.

Process layouts offer some advantages over product-oriented line flow designs. Since a variety of products are routed through each department, process machines are usually general purpose and less capital intensive than the specialized equipment found in product layouts. In addition, general purpose machines usually require less skill to operate, yielding a lower direct labor cost. These machines also provide the flexibility to implement changes in product mix or the introduction of new products more easily. As the same equipment is used for all products, capacity can be balanced to anticipated demand, thereby enhancing equipment utilization. Process layouts are also called **functional layouts.**

> **A process layout is an intermediate flow design that groups similar functions together within an operating department.**

On the negative side it is difficult to schedule products in a process-oriented environment in a manner that achieves a smooth and continuous flow. Backtracking, disorderly flow patterns, and disruptions are common. Idle time can occur in a department if it has to wait for work to arrive from another department and to avoid this it is common to permit the accumulation of work-in-process (WIP) between departments. The buildup of WIP consumes valuable space and can lead to factory floor inefficiencies as materials handling and operations are hampered by the clutter of stacks of material waiting to be processed.

If products are made-to-order, finished goods inventory can be low as products are shipped to customers soon after they exit the value stream. Process layouts require flexible materials handling equipment such as forklifts that can follow multiple paths, move in any direction, and carry large unit loads. This equipment is typically manned by materials handlers with special skills who must serve the entire manufacturing facility. If multiple demands for material occur simultaneously, disruptions can occur. Inevitably some departments have to wait for materials to be brought in or removed when required.

5.4.3.2 Cellular Manufacturing. Cellular manufacturing is a philosophy that uses a technique called **group technology** (GT) to sort products into families with similar manufacturing characteristics. The principal objective is to achieve some of the benefits associated with line flow designs while preserving the small batch and rapid response capabilities of flexible flow designs. Although not essential, work cells are fundamentally important to the principals of lean production—a topic covered in Chapter 11. The substitution of a process layout with a cellular design has proven to increase productivity and improve quality by streamlining materials flow, empowering workers, enriching jobs, reducing inventories, and simplifying shop-floor control.

Refer to the bottom layout of Figure 5-28. The jumbled flow pattern of the process layout has been streamlined by establishing cells for each of the parts. Note that capital or space intensive operations, such as heat treating and painting, are shared by all cells. The first step in establishing

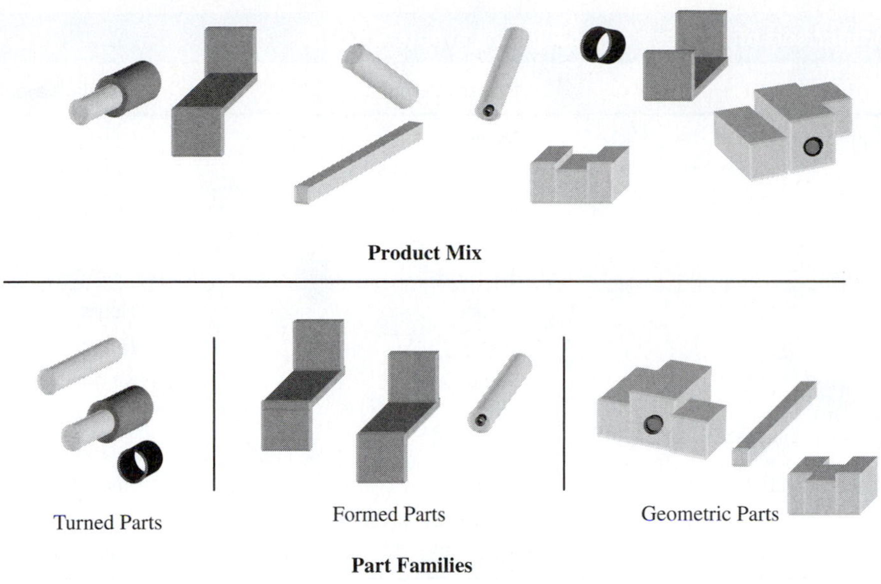

FIGURE 5-29. Example of the Use of Group Technology

a cellular design is to determine part families and assign each part to a family. The cells will then be designed to support families rather than individual parts.

A manufacturing cell is an autonomous collection of workstations that represent a mini-factory. A cell is a microcosm of the larger value stream containing most if not all of the capabilities (equipment and resources) necessary to produce a final product or component. Typically, a cell consists of 2 to 10 multiskilled workers who can operate more than one (sometimes all) of the 3 to 15 workstations contained in the cellular layout.[38] The workers are assigned to a cell function as a team and share decision-making powers and accountabilities for outcomes.

> **Cellular manufacturing achieves some of the benefits of line flow while preserving the small batch and rapid response capabilities of flexible flow.**

In a cellular layout departments are organized around product families, similar to product-oriented layouts with line flow designs. Materials processing is managed in small batches, called kanban quantities, which are usually small enough to enable manual materials handling. Shop floor controls consist of visual and other simple triggering devices. Throughput is fast and efficient, and communications and teamwork are facilitated since all operators contributing to the value stream are in close proximity. Table 5-2 demonstrates the comparative benefits that a cellular design offers over a traditional process layout.

In classifying parts into families and machines into work cells, GT seeks to group products according to similarities in design and manufacture as illustrated in Figure 5-29. By putting products together that can take advantage of a common setup (or major setup), process routing, or similar operator skills, economies of scale can be achieved at levels approaching those of line flows.

Three methods are commonly used to determine machine-part families. The first of these is a visual intuitive method, the second is based on part characteristics, and the third is based on commonality of production processes.[39,40] With the visual intuitive method an experienced person or team examines the parts and assigns them to logical groupings based on technical knowledge of the part geometries and the manufacturing processes required in manufacturing. The visual method is prone to human error as it is based on expertise, experience, and the ability for a group of people to reach consensus. However, this procedure is fast and can be reliable when the number of parts is small (e.g., 20 or less); the technique becomes less effective with a growth in part variety.

The second GT method focuses on part characteristics, and is known as **part coding and classification analysis (PCA)**. With this approach, each part is assigned a unique code that conveys information concerning product features and/or process requirements. The coding system is used to assign numerical weights to product or process characteristics. A classification scheme is then derived by retrieving and grouping code combinations. PCA methods work well when

[38]Lee, Quarterman. *Facilities and Workplace Design: An Illustrated Guide.* Engineering and Management Press, Institute of Industrial Engineers, 1997. Available as free download at http://strategosinc.com/dl0_facdes.htm.

[39]Ibid.

[40]Offodile, O. Felix, Abraham Mehrez, and John Grznar. "Cellular Manufacturing: A Taxonomic Review Framework." *Journal of Manufacturing Systems,* January 1, 1994. http://www.sme.org/cgi-bin/get-item.pl?JM133196&2&SME&.

TABLE 5-2 Comparison of Process and Cellular Layouts

Operations Factor	Process Layout	Cellular Layout
Number of material handling moves	Many	Few
Typical travel distances	500 ft–4,000 ft	100 ft–400 ft
Flow path	Variable/Jumbled	Fixed/Smooth
Work in process inventory levels	Days–Months	Hours
Raw material inventory turns	3–10	15–60
Throughput time	Weeks	Hours
Response time	Weeks	Hours
Production scheduling	Complex	Simple
Supervision	Close	Minimal
Teamwork	Inhibits	Promotes
Quality feedback and corrective action	Hours–Days	Instantaneous
Skill levels	Narrow, Specialized	Broad, Multiple
Equipment utilization	85%–95%	70%–80%

the product mix is large (i.e., tens of thousands) and when products or processes are complex, or product routings are inconsistent. The coding and classification process can be sophisticated, complex, and costly. External consultants are normally required to perform a proper PCA. However, the application of fuzzy logic, combined with the **analytic hierarchy process (AHP)**, has been proposed as a methodology for grouping products based on the perceived weighted importance of product features;[41] and a back propagation neural network has been suggested as a method for classification based on part geometry.[42]

In the third GT method, called **production flow analysis (PFA)**,[43] products with similar production routings are grouped together into families. Using manual or computer route sheets or spreadsheets, relationships between parts and the machines required to produce them are recorded. The next step is to create a production flow analysis chart—a machine-part incidence matrix of 1's and 0's, with a row for each machine and a column for each part. The cell value $x_{ij} = 1$ if machine i processes part j, or 0 otherwise. Table 5-3 illustrates a PFA chart for a factory that has five parts and five machines.

In a PCA each part is assigned a unique code that conveys information concerning features and process requirements.

Next, the PFA chart is rearranged as shown in Table 5-4. It is clear that a logical clustering is to place parts A, C, and E in one family that will be produced in a cell that contains machines 1, 3, and 5. Parts B and D will comprise a second family to be produced in a cell containing machines 2 and 4. PFA is powerful and fast if process routings are consistent, the number of parts are moderate (e.g., less than 100), and the value stream is not complex. Even when these criteria are met, PFA has some limitations. It is a subjective process and can be

TABLE 5-3 PFA Chart

	Part				
Machine	A	B	C	D	E
1	1	0	1	0	1
2	0	1	0	1	0
3	1	0	1	0	1
4	0	1	0	1	0
5	1	0	0	0	1

TABLE 5-4 PFA Chart Showing Natural Groupings

		Part Family				
		I			II	
Cell	Machine	A	C	E	B	D
I	1	1	1	1	0	0
	3	1	1	1	0	0
	5	1	0	1	0	0
II	2	0	0	0	1	1
	4	0	0	0	1	1

[41] Sangwan, Kuldip Singh, and Rambabu Kodali. "Fuzzy Part Family Formation for Cellular Manufacturing Systems." *Production Planning and Control* 15, no. 3 (April 2004): 292–302.

[42] Chung, Yunkung, and Andrew Kusiak. "Grouping Parts with a Neural Network." *Journal of Manufacturing Systems* 13, no. 4 (1994): 262–75.

[43] Burbidge, J. L. "Production Flow Analysis." *The Production Engineer* 50 (1971): 139.

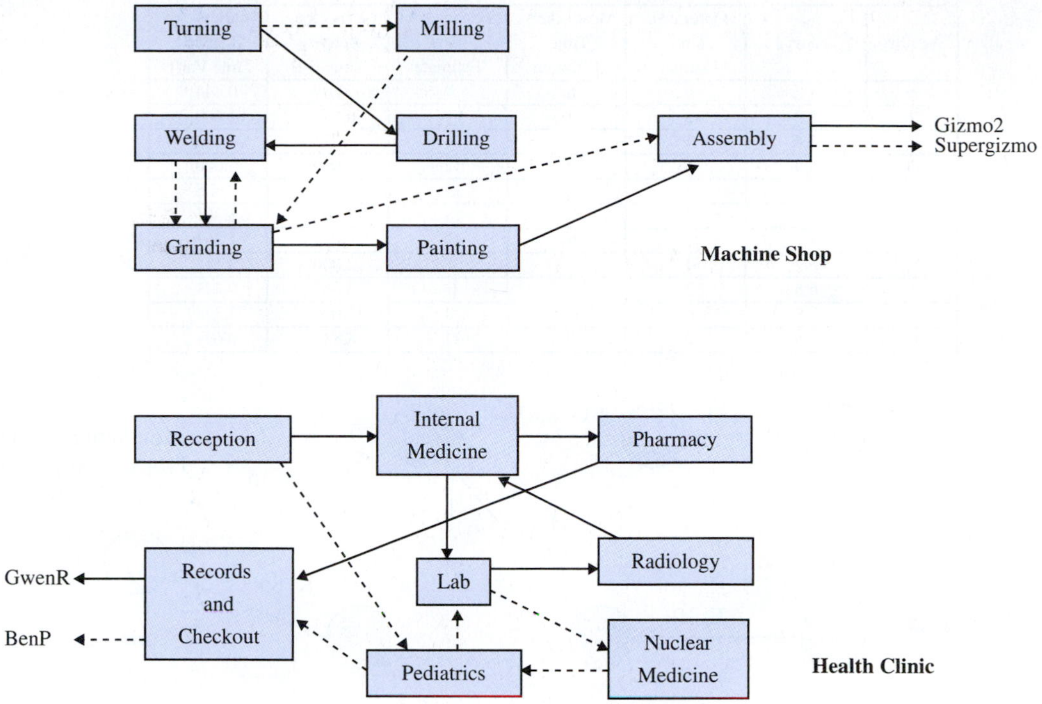

FIGURE 5-30. Examples of Job Shop Flexible Flow Designs

unwieldy for large PFA charts—even a matrix approaching 100 columns is difficult to analyze. Some systematic methods for effectively analyzing and using PFA charts have been suggested in the literature (i.e., similarity coefficient–based, array-based, and mathematical-based methods).[44]

Cellular designs have certain disadvantages. Production volumes are typically too low to justify the installation of efficient high-volume equipment. Overall equipment utilization can be expected to decrease when cells are introduced since the need to make each cell self-sufficient usually requires the purchase of additional machines. Similar to line flows, cell production is vulnerable to equipment breakdowns, and it is sometimes difficult to balance work across cells. Also, after forming cells throughout a plant, it is not uncommon to find that there are tasks that do not fit into any cell. These tasks are all lumped into a **remainder cell** that operates more like a traditional job shop than a work cell.

5.4.4 Flexible Flow Designs

The objective of **flexible flow design** is to satisfy customer demand for fast response and customization. On this end of the spectrum production must be configured to accommodate large variety and low volumes which dictates a functional process layout. Figure 5-30 shows examples of two job shop flexible flow layouts. One shows a typical machine shop and the other is a health clinic, demonstrating the applicability of flexible flow designs to a service application.

A flexible flow process design suits low-volume, high-variety production with chaotic flow patterns and where fast response is required.

5.4.4.1 Job Shop. A **job shop** arrangement offers the ultimate in flexibility. This arrangement provides the capability to respond to individual customized requirements with general purpose equipment that is less expensive to purchase, install, and maintain than equipment needed to support continuous or discrete production lines. In addition, general purpose equipment is easier to modify for changing requirements and less susceptible to obsolescence. Compared to line flows, a job shop has certain additional advantages. Dangerous activities can be segregated from other less dangerous ones. Workers must have higher skill levels, and this leads to greater fulfillment and pride in their work. The work pace is not dictated by a moving line and production rates are less vulnerable to equipment breakdowns.

A job shop design has some inherent disadvantages. The jumbled flow patterns can require excessive backtracking with numerous material moves over long distances. WIP levels can be expected to be high as inventories of materials queue up between departments waiting to be processed. This is driven by a complex production scheduling system that can result in long throughput times and slow response times. Management control, particularly on the shop floor, is difficult. A job shop environment is not conducive to team building, and quality is often achieved only through the close scrutiny of supervisors. And even with close supervision quality can suffer because no one supervisor owns the entire process. Quality information can be delayed and sometimes never get to those who need to

[44]Offodile et al., "Cellular Manufacturing."

Activity	Arc	Optimistic Time Estimate	Most Likely Time Estimate	Pessimistic Time Estimate	Expected Activity Time E[t]	Variance of Activity Time Var[t]
A	1,2	4	6	8	6	0.444
B	1,3	2	4	8	4.333	1.000
C	2,4	1	3	7	3.333	1.000
D	3,4	6	9	12	9.000	1.000
E	3,5	5	10	15	10.000	2.778
F	3,6	7	12	18	12.167	3.361
G	4,7	5	9	12	8.833	1.361
H	5,7	1	2	3	2.000	0.111
I	6,8	2	3	6	3.333	0.444
J	7,9	10	15	20	15.000	2.778
K	8,9	6	9	11	8.833	0.694

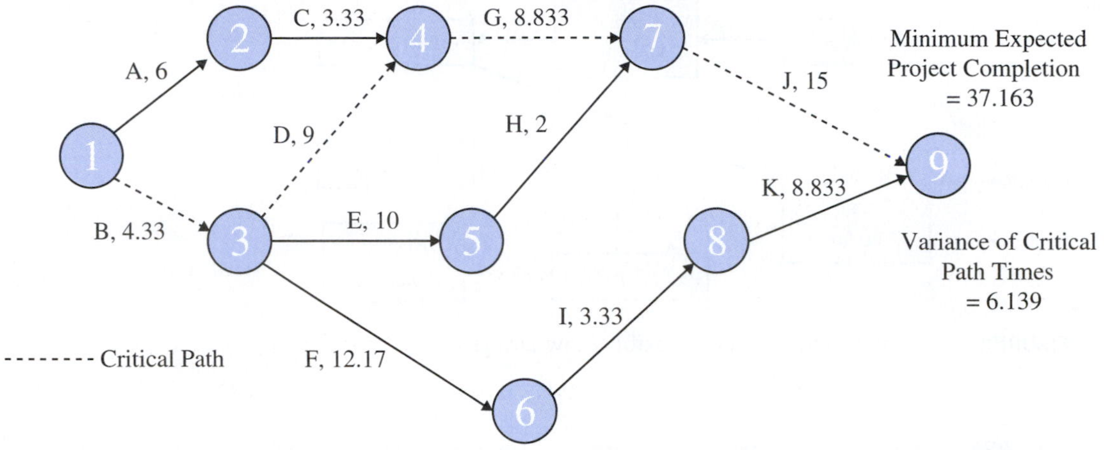

FIGURE 5-31. A PERT Project Network Showing Expected Activity Times

take corrective action. On the positive side, since all parts share common equipment good machine utilizations can be achieved with proper capacity planning.

5.4.4.2 Project Designs.

Contrasted with a job shop, a **project** is temporary and usually consists of a unique collection of tasks undertaken in a defined sequence to create a desired product or service. Projects can be large scale and complex, such as building a skyscraper. Multiple interdependent activities can be involved and there is always a finite time frame. The unique nature of the product or service created makes a project nonrepetitive. Projects are usually conducted under the control of a project manager who can react to changes on short notice as required. The project manager owns the project outcomes and is responsible for ensuring that the project is completed on time and within budget. As for process, the project manager will assemble and integrate all inputs needed to meet project objectives and will carefully monitor progress. It is the project manager who is ultimately accountable for the quality of project outcomes.

There are several techniques for defining and managing project activities and their relationships. One of the most popular is the **program evaluation and review technique (PERT)** that uses activity time distributions to predict expected project completion times. Because projects usually represent one-of-a-kind and never-to-be-repeated endeavor, there is always an element of uncertainty in how long it will take to complete any or all of the activities. The PERT method permits the project manager to factor in possible contingencies and develop activity expected times and time variances based on best case, worst case, and most likely scenarios. A common tool for managing progress is the **Gantt chart** discussed in further detail in Chapter 7.

A project is a temporary unique collection of tasks undertaken in a defined sequence to create a desired product or service.

Figure 5-31 shows a PERT network for a small project involving 11 activities. The network shows the precedence relationships for all project activities, and no activity can commence until all its predecessors have been completed. The longest time path through the network is called the **critical path**. Activities lying on the critical path are similar to the bottlenecks in a value stream. A project can be expedited only by shortening the duration of critical activities, and conversely, any delay in a critical activity will lengthen a project. The project shown in Figure 5-31 has an expected completion time of a little over 37 days with a total variance of the accumulated critical path time equal to 6.139. As shown, the critical path consists of the four activities B, D, G, and J.

Figure 5-32 provides two Gantt charts for this project. The first shows the earliest times that each activity can be

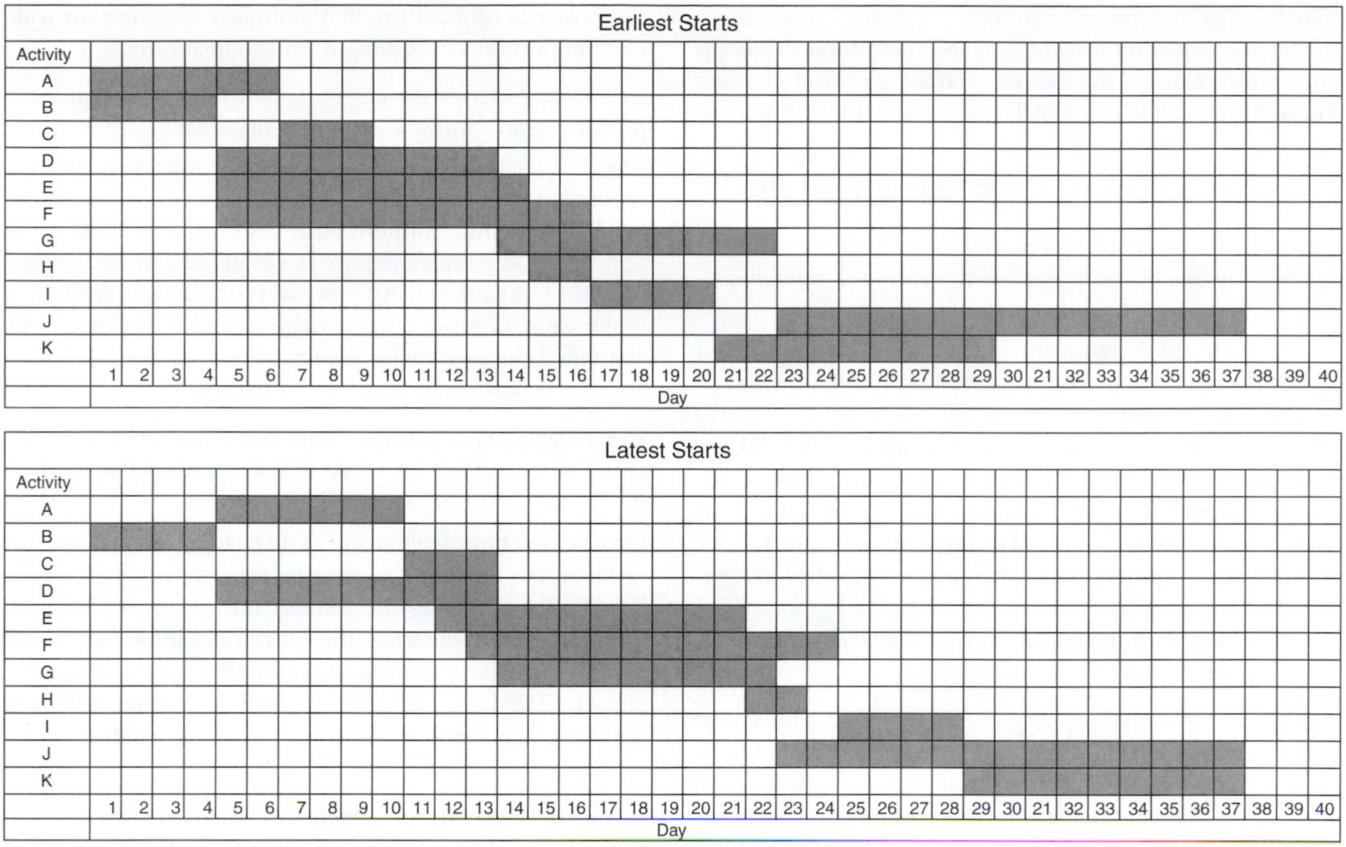

FIGURE 5-32. Gantt Charts for Earliest and Latest Activity Start Times

SHOWCASE of EXCELLENCE — Winners of the Malcolm Baldrige National Quality Award Small Business Category

- 2007: PRO-TEC Coating Company, manufacturer of coated sheet steel used primarily in the U.S. automotive industry for use in cars, trucks, and sport utility vehicles. *http://www.proteccoating.com, http://www.quality.nist.gov/PDF_files/PROTEC.pdf.*
- 2006: MESA Products, Inc., a small privately held business that designs, manufactures, and installs cathodic protection systems that control the corrosion of metal surfaces in underground and submerged structures, such as pipelines and tanks. *http://www.mesaproducts.com, http://www.quality.nist.gov/PDF_files/MESA_Profile.pdf.*
- 2005: Park Place Lexus, a dealership that sells and services new and preowned Lexus vehicles, and sells parts to the wholesale and retail markets. *http://www.parkplacetexas.com, http://www.quality.nist.gov/PDF_files/Park_Place_Lexus_Profile.pdf.*
- 2004 and 1998: Texas Nameplate Company, Inc., a privately held family business that produces custom nameplates in frequent orders for small business in the United States. *http://www.nameplate.com, http://www.nist.gov/public_affairs/Texas_Nameplate_PDF_final.pdf.*
- 2003: Stoner, Inc., a small family-owned business that produces more than 300 specialized cleaners, lubricants, coatings, and car care products. *http://www.stonersolutions.com, http://www.nist.gov/public_affairs/baldrige2003/Stoner_3.3.04.pdf.*

scheduled to start (in order to meet the projected minimum expected completion time of 37), and the second shows the latest times. Comparing the two charts the project manager knows how much scheduling flexibility exists for each activity.

5.5 SUMMARY OF KEY IDEAS

Key Idea: *A superior design is one that meets customer expectations, can be efficiently manufactured, and is environmentally friendly throughout the entire product life cycle.*

An ill-conceived design will not satisfy customers even if it performs well during production testing. Design must start with an understanding of what product features are likely to produce a customer "wow" factor and can be produced at a competitive cost. Superior design processes also consider the immediate and longer-term environmental impacts of the production, consumption, and disposal of the product. Value engineering is a process that, when properly applied, can ensure that a design meets customer needs while minimizing costs. Ecodesign strategies can address the product life-cycle environmental issues.

Key Idea: *A principal goal of quality is to produce a product or deliver a service as close to the target value as possible; the cost of quality is quadratically proportional to the degree with which the actual process output deviates from the target—even if the output is still within the allowable tolerance range.*

The definition of excellence is to hit the target—this is usually the nominal value within a specification range or some stipulated requirement. The Taguchi loss function is a model that describes the impact on quality costs of not meeting the target—and the reality is that the cost of quality varies quadratically to how much the actual output deviates from the target. This model teaches that simply delivering within the allowable specification range is not good enough—that there is still a cost associated with not hitting the target. The Taguchi function demonstrates the importance of focusing quality efforts on variance reduction and is embodied in Taguchi's four quality principles (build quality into the design; minimize variability around the target; define quality on manufacturability, function, and cost—not on features; and apply quality concepts over a product's entire life cycle—not just at the point of sale, but over a product's entire life cycle).

Key Idea: *The competitive environment today requires an integrated approach to product design and development—one that considers product and process design technologies, stakeholder needs, management practices, and production and supply chain capabilities.*

The process that an organization uses to design and develop new products must be consistent with its strategy and business plan, and above all should be based on an open systems approach. Cross-functional communications and collaboration are essential and the design process becomes a team effort with participation across the entire organization. This approach requires a major commitment on the part of top management, and a successful implementation depends on the existence of a culture that embraces organization-wide communication, collaboration, and cooperation.

Key Idea: *Superior designs are based on two fundamental principles: standardization and simplification.*

The use of standardized parts results in lower costs through reduced inventories, larger replenishment orders, and standardized tooling. Simplification also reduces costs since fewer parts in a design require less material and a simplified less costly manufacturing process. Product complexity is a major cause of production variability, and if designs can be simplified, the need for sophisticated technologies, skilled labor, and dedicated processes is diminished.

Key Idea: *Quality function deployment (QFD) is an integrated process and product development approach that is driven by the voice of the customer (VOC).*

Quality function deployment (QFD) supports the goals of QIM as the customer needs are central to all decisions. Specifically, QFD uses the voice of the customer (VOC) to derive a series of matrices that document information and relationships in a framework called the house of quality (HOQ). The HOQ, an integrated set of product planning matrices, then translates customer needs into product or service design parameters.

Key Idea: *Designing for supportability strives to gain competitive advantage by focusing attention, during the design phase, on those support issues that will likely arise during a product's entire life cycle.*

In some markets quality after the sale can be more important than quality at the point of sale. Competitive advantage can be enhanced by considering how well a customer is satisfied over the entire life cycle of a product, not simply how well the product may perform hot off the assembly line. The life-cycle perspective requires that design teams consider issues relevant to the ease with which the product can be serviced and maintained; how reliable the product is when used in typical field environments; the ease with which the product can be inspected or tested for faults; how eco-friendly the product is when the time comes to decommission the part either recycling or disposing of its materials and components; the life-cycle cost performance of the product; the ease with which the product can be installed or removed from service; the ease and simplicity with which personnel can be trained to use the product and simplicity and comprehension level of relevant documentation; and the ease and cost effectiveness of being able to update or upgrade product capabilities to meet expanding needs or technologies.

Key Idea: *Continuous processes are suited for a line flow process design where low variety, often a single product, is produced in high volumes on specialized automated equipment.*

Continuous processes are typically run on automated specialized equipment lines that are prohibitively expensive to start up, stop, or shut down. Therefore, these processes typically are operated around the clock seven days a week. Layouts for continuous processes are normally organized around products and it is common to find lines dedicated to a single product. Continuous processes are generally inflexible; it is difficult to change product designs and to add or reduce capacity. Material costs are high relative to labor costs which are typically low due to automation.

Key Idea: *A flow shop is a discrete manufacturing design that uses highly specialized equipment to produce high volumes of output of a low-variety product mix.*

The high-volume, low-variety production of discrete products suits a manufacturing layout oriented around products using assembly-line methods. These production systems are well suited to pull scheduling methods, with just-in-time deliveries and low work-in-process inventory levels. Simple shop-floor control methods can be employed to ensure that production requirements are met.

Key Idea: *To minimize the disruption of flow and excessive inventory buildups, the activities in a flow shop should be balanced by designing the work content for each activity such that the cycle time matches the takt time as closely as possible.*

A good process design will match the cycle time to the takt time. The takt time is the time allowed for the production of each unit of output and is derived by computing the ratio of the time available to the demand that must be met during that time. A production line that consists of a sequence of workstations is limited in its output by the workstation that has the smallest capacity. This workstation is called the bottleneck operation. By establishing a line cycle time equal to the takt time and then balancing each workstation as close to the design cycle time as possible, the chance of excessive WIP inventory buildup between workstations will be minimized. This can never be accomplished exactly due to the variability of output that will be present in each individual workstation, and the cumulative effect of variation along the value stream.

Key Idea: *The program evaluation and review technique (PERT) is an effective method for organizing and managing large complex projects, and is capable of incorporating uncertainties that are associated with the completion of project activities.*

The PERT technique uses activity time distributions to predict expected project completion times. The expected times and time variances are based on three time estimates for each activity—a pessimistic estimate, an optimistic estimate, and a most likely estimate. By representing a project as a network the relationship between project activities is apparent—especially which activities must be completed before other activities can commence. The longest expected path through the network is called the critical path and represents the shortest expected completion time for the project. Activities on the critical path deserve the closest management attention since any delay in a critical activity will delay the project. Due to uncertainties the critical path can change as the project progresses. A common tool for monitoring progress is the Gantt chart.

5.6 MIND EXPANDERS TO TEST YOUR UNDERSTANDING

5-1. Perform the following experiment. Place 200 pennies in a container (as an alternative, some other item such as poker chips or marbles will do). Line up five Styrofoam cups. The first four cups represent work-in-process between five workstations in a production line. Cup number five represents the output from the entire line that can be shipped to customers. Now let the roll of a single die represent the output for any of the workstations on one production cycle. Roll the die and place the corresponding number of pennies in the first cup. This is the production output for workstation number one on the first cycle. Now roll the die again and, based on the number that comes up, transfer that number of pennies from the first cup to the second cup. If the number on the die is greater than the number of pennies in cup number one, then transfer the entire contents to cup number two. Continue this process until the die has been thrown and pennies transferred for all five workstations. This completes the first cycle. Now, repeat this process for 20 cycles. Then answer the following questions. (*Note:* This experiment is more interesting if five different people participate as operators of the individual workstations and a sixth person keeps track of the cycle count.)

a. What is the expected output for each workstation? (*Hint:* What is the expected outcome from a single throw of a fair [i.e., nonbiased] die?)

b. What is the expected total production output for 20 cycles?

c. From the experiment, how many units did you count in cup number five that could actually be shipped to customers? How do you explain any observed discrepancy between the number observed and the number you expected in part b?

d. What are the implications of this experiment on process variability?

e. What assumptions did we make with this experiment and are these assumptions realistic? (*Hint:* Consider the concepts of line balancing and workstation capacity.)

5-2. How can the application of value engineering (VE) ensure that a product or service meets the three criteria for a superior design: excites customers with a "wow" factor, can be produced efficiently, and is environmentally friendly over the entire life cycle?

5-3. Suggest some of the issues that an ecodesign engineer should consider in developing green designs for the following products.

a. Computer
b. Automobile
c. Sport coat
d. Shoes

5-4. For each of the following processes indicate whether the process flow is continuous, flow shop, process flow, cellular, flexible flow, or project.

a. Bakery that produces commercial loaves of bread
b. Development of a retail shopping mall
c. Shipbuilding
d. Production of roofing shingles
e. Bank
f. Production of kitchen cabinets

5-5. The following machine-incident matrix has been prepared for a manufacturing plant that produces seven parts on six machines. Use the PFA group technology technique to cluster the seven parts in logical family groupings.

	Part						
Machine	**A**	**B**	**C**	**D**	**E**	**F**	**G**
1	0	1	0	0	1	0	1
2	0	1	0	0	1	0	1
3	1	1	0	1	1	1	1
4	1	0	1	0	0	0	0
5	1	0	0	1	0	1	0
6	1	0	1	1	0	1	0

5-6. Horizon Aerospace Industries has received a contract from NASA to assemble a module for an upcoming space mission. A team of Horizon engineers has developed the list of activities, precedence constraints, and time estimates shown below.

Space Module Assembly

			Time Estimates (days)		
Symbol	**Activity**	**Preceding Activities**	**Optimistic**	**Most Likely**	**Pessimistic**
A	Construct shell of module	none	25	30	45
B	Order life support system and scientific experimentation package from supplier	none	10	15	20
C	Order components of control and navigational system	none	20	25	35
D	Wire module	A	3	3	5
E	Assemble control and navigational system	C	5	7	12
F	Perform preliminary test of life support system	B	1	1	1
G	Install life support in module	D, F	4	5	7
H	Install scientific experimentation package in module	D, F	2	2	3
I	Perform preliminary test of control and navigational system	E, F	4	4	6
J	Install control and navigational system in module	H, I	8	10	14
K	Final testing and debugging	G, J	6	8	15

a. Draw a PERT network for the NASA project. Include dummy activities if needed to ensure that each activity has a unique *i,j* designation. (Hint: the expected time for each activity is determined by a weighted average that gives a weight of 1 to the optimistic and pessimistic times and a weight of 4 to the most likely time).

b. For each activity calculate the total float and free float. (Hint: Total float for any activity is defined to be the difference between the earliest event time for the activity's *i* node and the latest event time for the activity's *j* node. Free float is the difference between the earliest event times for the *i* and *j* nodes.)

c. Determine the critical path for the project.

d. What is the probability that the project will be completed in 54 days or less? (Hint: The standard deviation of any activity is estimated to be one-sixth the range (optimistic–pessimistic). The variance of the critical path is the sum of the variances of all critical activities. Assume project completion times are normally distributed).

e. What is the probability that the project will be completed by its earliest expected completion date?

f. What is the probability that the project will take more than 70 days to complete?

5-7. An advertising agency is planning a campaign to launch a new product using three principal media: poster boards, television, and newspapers. The campaign has been planned and the activities, precedence order, and time estimates are shown below.

			Time Estimates (weeks)		
Activity	Description	Pred Act.	o	p	m
A	Plan ad campaign	—	2	2	2
B	Establish official launching date	—	12	25	17
C	Write text for newspapers	A	3	5	4
D	Draw illustrations for newspapers	A	3	5	4
E	Negotiate contract for newspapers	A	2	5	3
F	Design poster art	A	6	12	8
G	Negotiate contract for poster	A	2	5	3
H	Write script for TV	A	4	5	3
I	Negotiate film contract for TV	A	3	6	4
J	Negotiate display contract for TV	A	2	4	3
K	Make blocks for newspapers	C, D	1	3	2
L	Send blocks to newspapers	E, K	1	2	1
M	Print poster boards	F	1	2	1
N	Distribute posters	G, M	1	2	1
O	Make TV film	H, I	4	9	6
P	Send film to program company	J, O	1	2	1
Q	Arrange news conference	P	3	6	4

a. Draw the PERT network for this project.

b. What is the minimum expected time for project completion?

c. Compute the total and free floats for each activity.

d. Which activities are critical and should be carefully managed?

e. How many weeks from now, at the earliest, could we start to make the halftone blocks for the newspaper ads?

f. How many weeks, at the latest, would the agency need to start to make the halftone blocks for the newspaper ads in order not to delay the project?

g. The agency chief wants to roll out the new ad campaign 20 weeks from today. What is the probability of meeting this deadline?

h. List some near-critical activities that should also be carefully managed if the project is to be completed on time.

i. The team member who is responsible for the poster medium has arranged to have the posters printed as soon as they are ready and at the earliest possible opportunity. By how much can the design work on the poster boards be delayed without delaying the printing?

5-8. A company manufactures a consumer product that involves 12 distinct production steps as shown below. The demand for this product is 2,700 per week and the manufacturing plant operates three eight-hour shifts five days per week. Each shift works 450 productive minutes.

a. Compute the takt time necessary to fill the demand of 2,700 within the time allotted.

b. What is the theoretical number of workstations for a balanced line?

c. Using the LOT criteria balance this line to a design cycle time equal to the takt time computed in part a.

d. Compute the projected efficiency of the line you designed in part c.

e. Identify the bottleneck(s).

f. Which station has the lowest efficiency? What could you do to improve this?

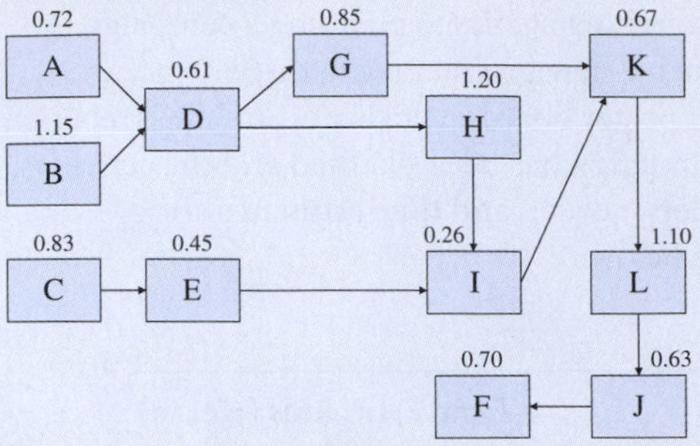

Total work content = 9.17minutes

5-9. Keen Edge, Pty. Ltd. manufactures a range of high-quality knives. The table shown below represents the steps required to produce the QZ106 model pocket knife. The company wishes to set up a production line that can efficiently meet a demand of 250 per week. The production line will be scheduled to work eight hours per day with two 15-minute breaks, five days a week.

a. Draw a precedence diagram for this manufacturing process.

b. Compute the takt time necessary to fill the demand of 250 within the time allotted.

c. What is the theoretical number of workstations for a balanced line?

d. Using the LOT criteria balance this line to a design cycle time equal to the takt time computed in part b.

	Activity	Predecessor Activity	Time Required (min)
A	Cut blades from sheet steel	None	5.2
B	Cut springs from sheet steel	None	6.3
C	Mark, drill, and nail nick blades	A	3.2
D	Mark, drill, and nail nick springs	B	2.7
E	Straighten blades	C	2.5
F	Straighten springs	D	3.8
G	Harden and temper blades	E	5.4
H	Harden and temper springs	F	4.3
I	Grind blades to edge	G	2.7
J	Clean blades	I	3
K	Blank out linings	None	4.6
L	Pierce holes in linings	K	2.6
M	Solder bolsters to linings	L	1.6
N	Assemble blades, springs, bolsters, linings	H, J, M	4.3
O	Crop off assembly	N	1.5
P	Knock up assembly	O	2.8
Q	Glaze	P	2
R	Grind and shape blades	Q	2.7
S	Final polish and sharpen	R	3.5
T	Package	S	3
		Total Time	67.7

e. Compute the projected efficiency of the line you designed in part d.

f. Identify the bottleneck(s).

g. Which station has the lowest efficiency? What could you do to improve this?

h. How would variability in the times to compete each activity impact the line cycle time? How would variability affect the ability to meet the takt time requirement? How would variability impact the projected efficiency of the line?

i. How do quality problems affect the ability to meet takt time? (*Hint:* What if some of the workstations have yields that are less than 100%?)

5-10. For each of the following products or services, indicate the customer's gemba.

a. Pair of running shoes

b. High-definition television set

c. Insurance policy

d. Automobile service and repair

e. College education

f. Air travel

g. Dishwasher

5-11. For each of the following products or services, list some life-cycle supportability issues that are important to customers.

a. Automobile

b. Computer

c. Refrigerator

d. Lawn mower

e. Insurance policy

f. College education

g. Investment counseling

5-12. Continental Ltd. is the manufacturer of three products: the Alpha, the Gamma, and the Omega. Information on overheads is shown in the following table.

Continental Ltd. Overheads

Overhead Cost Categories	Materials Control	Setups	Machine Operations	Logistics and Transportation	Engineering	Totals
Total Overhead Generated	$980,000	$92,000	$1,100,000	$975,000	$750,000	$3,897,000
Cost Driver Units Consumed: Alpha	90,000 Sales units	42 Production runs	198,000 Machine hours	22 Shipments	50 Production orders	
Cost Driver Units Consumed: Gamma	43,000 Sales units	120 Production runs	141,900 Machine hours	78 Shipments	26 Production orders	
Cost Driver Units Consumed: Omega	18,000 Sales units	33 Production runs	12,600 Machine hours	43 Shipments	37 Production orders	

a. Allocate the overhead costs to each of the three products using the conventional accounting method of machine hours.

b. Allocate the overhead costs to each of the three products using activity-based costing.

c. How do your answers in parts a and b compare? What is the advantage of using your answer in part b to establish the price for each of the three products?

5-13. A fundamental idea behind programs such as Six Sigma is to follow Taguchi's second principle and minimize variability around the target for all critical dimensions. While achieving Six Sigma capability achieves good quality, the specifications remain in tact. Using Taguchi's loss function as a guide make an argument for tightening specification limits as variability is reduced as an alternative to aspiring to achieve Six Sigma quality.

CHAPTER SIX

SYSTEM MAINTENANCE AND RELIABILITY

QUALITY MAXIM #6: *Quality-at-the-source begins with processes, equipment, and machines that are well maintained and reliable.*

Quality and Maintenance: A Horse and Carriage Duo!

Over a half century ago, in "Love and Marriage"[1] Frank Sinatra metaphorically compared the union of two people in matrimony to a horse and carriage—indivisible, one necessary to the other. The lyrics caution that no matter how hard a person tries to separate love from marriage one simply cannot be had without the other.[2] Through a fable we will illustrate how the metaphor could just as well apply to quality and maintenance. Hilda and Ingrid owned competing pastry shoppes in a small German village. An annual February festival brought hordes of visitors from neighboring towns and afar. Hilda and Ingrid both recognized this as a great business opportunity. However, each had a different approach to coping with the inevitable surge in demand.

Because of the difficulty and time required for changing from one product (for example bread) to another (such as bagels) Ingrid would estimate her demand for each product a day in advance and produce that quantity in large batches. That way she would have to perform only the cleanups and setups (which were substantial) between each batch. Ingrid's philosophy toward maintenance was "don't fix what's not broken." She firmly believed it was in her best business interest (translation: profits) to repair equipment only when it broke down. Since the demand spike was substantial during the busy season, keeping up often meant running dough mixers, rollers, ovens, and slicers on a nonstop basis. Ingrid would sometimes bake several days ahead and store the product in coolers. A tour through her bakery revealed a casualty of the intensity with which she strived to maintain her production schedules. Flour and dough carpeted the floor, materials were scattered and difficult to locate, and the facility was in a general state of disarray.

Hilga's approach was different. During the slack season she had carefully planned on how she could maximize baking capacity and have the ability to flexibly match any day-to-day demand volatility. She carefully reviewed the steps required to changeover from one product line to another and found that many of the steps could be performed while the previous batch was still in production. As a result she was able to shave 68% off the time previously required for a typical changeover. This she felt would enable her to produce multiple batches if needed during each day to flexibly respond to short-term changes in demand. She also contracted with a village repairperson to check her equipment weekly, performing all necessary routine maintenance and replacing any parts he deemed necessary to keep the machines running at peak efficiency.

As the festival drew near, the population of the village swelled as accommodation became hard to find. Ingrid and Hilda ramped up production to full strength, working long days to keep up with demand. Ingrid would produce large batches of her breads, bagels, pastries, and cakes, and she would place any leftover production in her coolers for the next day and discard what she could not store. During slack times she baked ahead and stored her product in anticipation of demand spikes. When the bearings in her dough mixer started squeaking, she ignored the problem reasoning that she could not afford to halt production to check out the problem and would deal with it later. When one of her ovens overheated she brushed it off saying she would simply shorten the bake cycle and again defer the problem to a more convenient time. In the shoppe over the road Hilda was busy baking her wares in small batches, having available her complete product variety when her shoppe opened each day. Additional batches were then produced during the day as needed. Hilda religiously insisted that a standard of pristine cleanliness be maintained throughout her facility and she also had a strict policy that manufacturers' maintenance guidelines (including periodic lubrication of equipment and replacement of drive belts, heating elements, and other components) be followed. When problems were encountered she would immediately take equipment offline and get the problem fixed, even if it meant having to work through the night to make up for lost time.

Initially, Ingrid and Hilda shared the town's bakery business equally. Then customers began to notice a difference in quality. Ingrid's customers complained that products were often under- or overcooked, or stale. At times, bread loaves were smaller than expected and cakes were not as light and fluffy as Hilda's. The shoppe was often found to be sold out of some items. When the German chancellor came to town the local dignitaries hosted an event that required catering to a substantial number of guests. Due to the size and instantaneous

[1]The lyrics for "Love and Marriage" were written by Sammy Cahn and the music was by Jimmy Van Heusen. The song was first performed by Frank Sinatra in 1955 in the television special *Our Town* and also as a recording on the hit parade.

[2]The music and lyrics for "Love and Marriage" can be accessed at http://www.rhapsody.com/frank-sinatra/the-capitol-years/love-and-marriage/lyrics.html.

nature of the demand, orders were placed with both shoppes. Hilda had little difficulty fitting the extra demand into her schedule and was even able to accommodate a request for custom-decorated cakes and tarts. Ingrid had to work her processes, which were already overtaxed, even harder. As a consequence, a bearing in one of her mixers froze up, cutting her throughput capacity in half, and an oven caught on fire. She had to close her facility for a major overhaul. Luckily, Hilda was able to save the day! She temporarily put on a second shift and delivered all orders on time.

LEARNING OBJECTIVES AND KEY OUTCOMES:

After careful study of the material in Chapter 6, you should be able to:

1. Describe the maintenance and reliability management (MRM) function as an open system operating within the broader context of the overall business system.
2. Discuss and differentiate between the run-to-failure (RTF), preventive maintenance (PM), predictive maintenance (PdM), proactive maintenance (PaM), and reliability-centered maintenance (RCM) philosophies of maintenance management.
3. Apply the analytic hierarchy process (AHP) to develop a priority ranking for proposed maintenance projects.
4. Describe the hierarchy of maintenance performance measures and discuss how this hierarchy can be used to develop a balanced scorecard for the effective evaluation of maintenance services.
5. Define the overall equipment effectiveness (OEE) score, including how it is derived, its use, and its limitations.
6. Define the objectives of a total productive maintenance (TPM) program and describe how it supports the goals of QIM.
7. Describe the 5S workplace discipline and its importance to the objectives of lean maintenance and quality.
8. Describe the seven deadly wastes and their importance to the objectives of lean maintenance and quality.
9. Discuss how kaizen and jidoka can contribute to the objectives of lean maintenance and quality.
10. Describe how the philosophy of just-in-time can be applied to the maintenance function.
11. Discuss how the implementation of lean maintenance is likely to impact the workforce.
12. Describe how an FMEA is conducted and used to identify potential failures and their associated root causes so that processes or products can be designed in a way that these failures can be prevented.
13. Describe how the Global 8D problem-solving process helps identify root causes of failures, after the failures have occurred, and develop, implement, and verify permanent corrective actions.

A shopping mall moments before opening its doors to enthusiastic holiday shoppers. Effective maintenance systems are necessary to ensure the products they seek will be available for purchase and of the quality customers expect.

6.1 THE OPEN SYSTEM OF MAINTENANCE AND RELIABILITY MANAGEMENT

Like other support value streams **maintenance and reliability management** (MRM) can be defined as an open system, as depicted in Figures 6-1 and 6-2. The larger business system (discussed in Chapter 2 and illustrated in Figure 2-1) constitutes the environment for MRM including components such

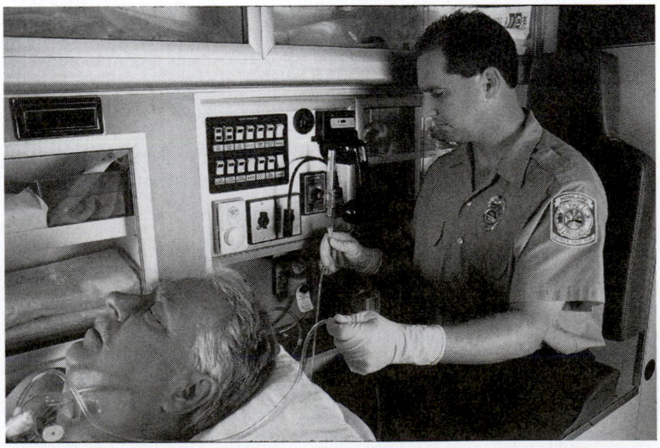

An emergency medical services (EMS) paramedic tends an accident victim enroute to the hospital. Effective maintenance systems that support the manufacture and operation of first responder equipment can literally mean the difference in life or death.

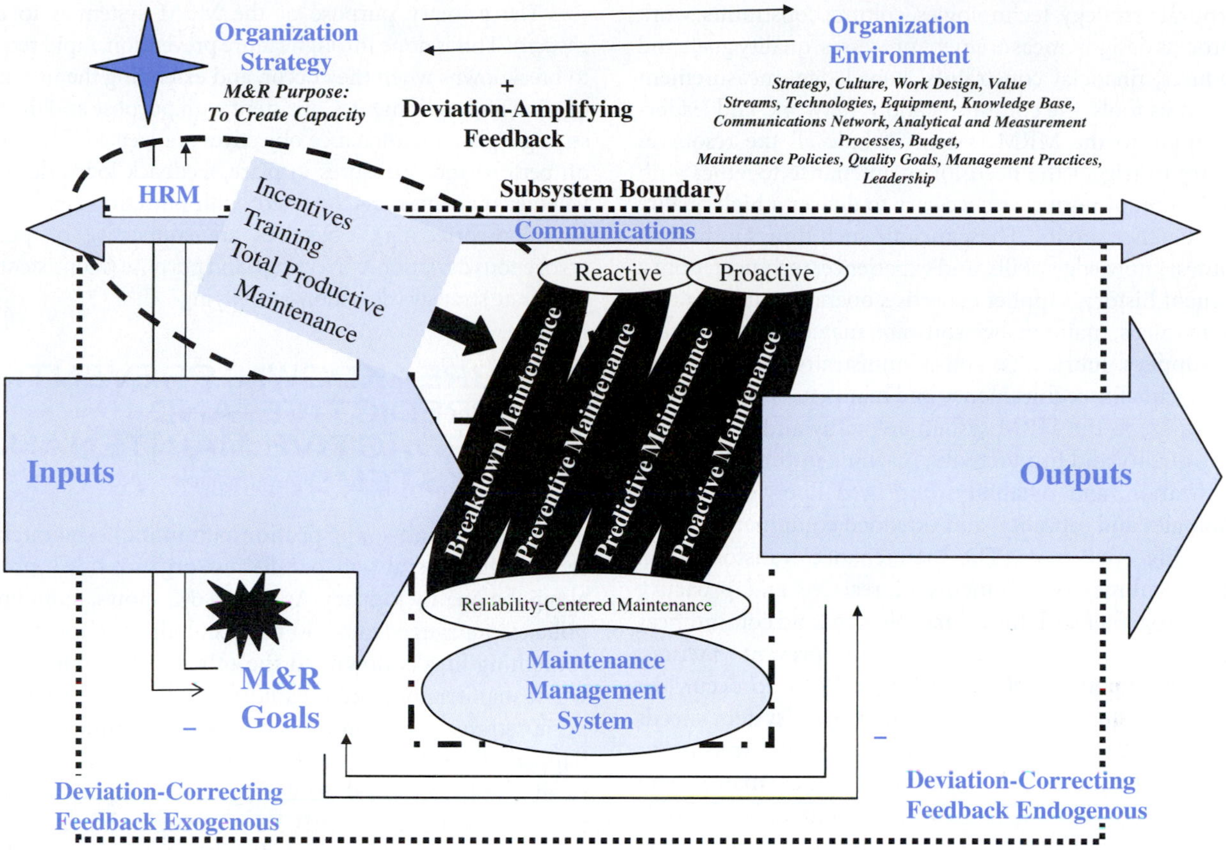

FIGURE 6-1. Maintenance and Reliability Management System

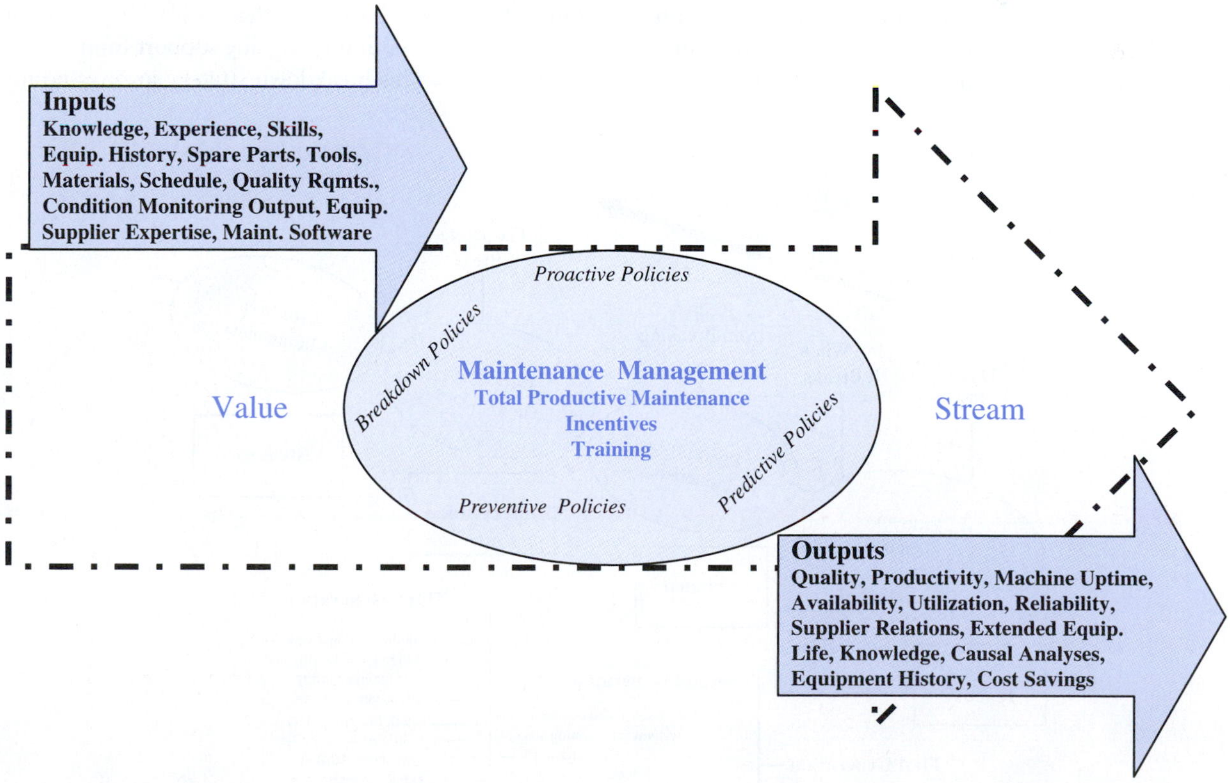

FIGURE 6-2. Maintenance Component of Value Stream

as corporate strategy, technologies, culture, constraints, work and process designs, measurement processes, quality goals and procedures, financial constraints, knowledge, measurement and analysis tools, the communications network, and leadership. Inputs to the MRM system include all the resources necessary to trigger the need for maintenance together with the collection of resources necessary to deliver a high quality of maintenance service. These include such things as human attributes: knowledge, skills, and experience; technical inputs: equipment history, supplier expertise, operating and maintenance manuals, maintenance software; materials: tools, spare parts, supplies, lubricants; and administrative components: schedules, quality requirements, and maintenance policies.

Outputs to the MRM system are many and diverse, including quality and productivity; machine uptime, availability, utilization, and reliability; improved knowledge about technologies and causality; and extended equipment life and reduced life cycle costs. The maintenance transformation process combines the elements of reactive and proactive strategies together with the human element and communications to prevent failures. When failures do occur, the maintenance system needs to respond and correct problems in a timely manner and with a standard of excellence. Figure 6-1 illustrates the interaction that exists between the HRM and the MRM subsystems. Leadership in incentives and training programs normally comes from the HRM group. HRM also provides the momentum for programs that lead to broadened participation and empowerment, such as **total productive maintenance** (TPM)—a strategy that relegates many of the routine maintenance tasks to front-line machine operators.

The primary purpose of the maintenance and reliability management system is to create capacity.

The primary purpose of the MRM system is to *create capacity*. This is done through failure prevention, rapid response to breakdowns when they occur, and extending the useful lives of plant and equipment. Consistent with purpose and the firm's strategic plan, maintenance objectives are created. Then, based on performance measures in place, feedback loops determine whether outcomes are consistent with objectives (endogenous deviation-correcting), objectives are supportive of purpose (exogenous deviation-correcting), and purpose is consistent with corporate strategy (deviation-amplifying).

6.2 BREAKDOWN, PREVENTIVE, PREDICTIVE, AND PROACTIVE MAINTENANCE SYSTEMS

The maturity and sophistication pertaining to maintenance services in general will parallel an organization's progress along the quality journey. As Figure 6-3 shows, maintenance policies can range from "do not fix at all" or "fix only when something breaks down" to the other end of the spectrum where maintenance and reliability is well integrated into corporate strategy, operations, and forward planning. When this happens, the primary concern is not so much rapid response to the breakdown and rendering needed repairs (although this must be accomplished), but viewing each breakdown as an opportunity to identify and eliminate root causes so that the probabilities of recurrence are reduced.

6.2.1 Run-to-Failure

Run-to-failure (RTF) is a policy that is still widely used in industry and requires minimum ongoing support from the maintenance function. This breakdown strategy involves no forward

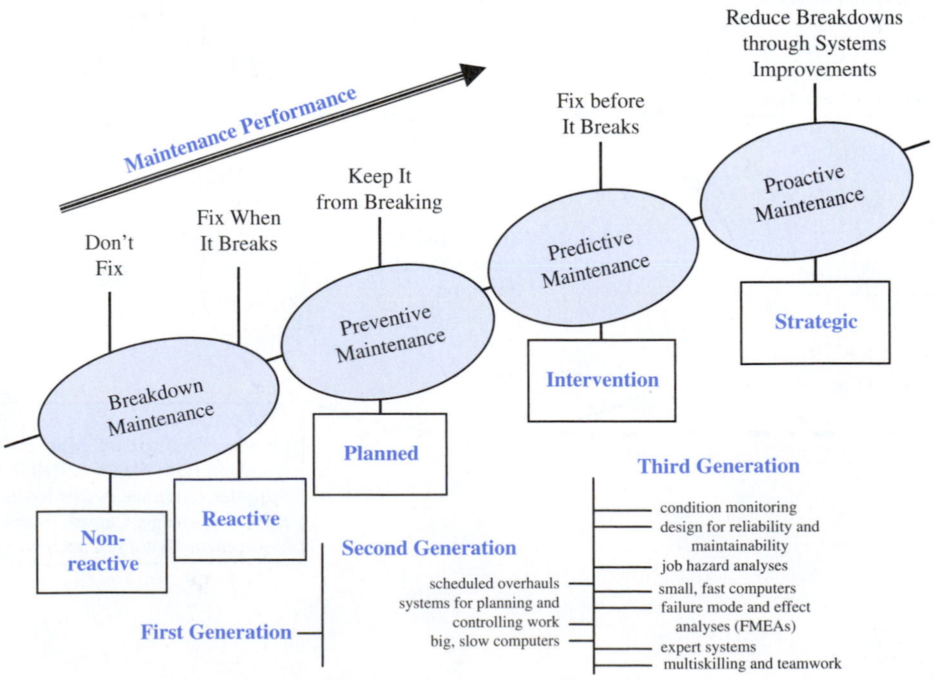

FIGURE 6-3. Hierarchy of Maintenance Service Maturity

planning at all. Under this policy, equipment is run, with little or no maintenance surveillance, and when it can no longer perform its function it is either repaired or replaced. Maintenance costs are accrued only on an "as needed" basis. If something breaks down the decision is made at that time to either repair it or not. In the latter case the decision might be to replace the item rather than fix it. RTF strategies may work well in the case of low-cost equipment, and especially for equipment that is not part of a line flow process, is underutilized and nonbottleneck, and where backup capacity is available. There is always a trade-off between the cost of planned maintenance and the cost of failure induced by a reactive policy.

Run-to-failure can work well if equipment is low cost, underutilized, and nonbottleneck—or if backup capacity is available.

6.2.2 Preventive Maintenance

Preventive maintenance (**PM**) forms the lowest level of planned activities, and when properly implemented can return cost savings in excess of 25%.[3] Under PM, downtime for maintenance is scheduled in advance, thereby avoiding any disruptions to production. The primary goal of PM is to prevent failures before they occur by preserving and enhancing equipment reliability. This is done through the application of systematic procedures to determine if operating conditions and machine wear are within prescribed and acceptable limits. PM activities include going through an equipment check list, conducting machine overhauls (partial or complete), changing oils and lubricants, and replacing any parts or components that show signs of excessive wear and the potential for failure. If a PM inspection indicates that the rate of wear exceeds expectations, then the cause of the accelerated deterioration is investigated and corrected so that the equipment will not fail in service. A common measure for prioritizing PM projects is the **mean time between failures** (**MTBF**) rate accumulated in equipment histories over time. PM is generally better than RTF; however, there is some debate as to whether PM is cost-effective. Critics argue that it costs more to regularly schedule process downtime and to replace parts that may still have useful lives than to go with a breakdown strategy and just repair equipment when it is necessary to do so. They also point out the labor intensity of PM and suggest that this may not be the most efficient use of maintenance personnel. And what is perhaps the major concern, implementation of PM does not warrant against unanticipated abrupt and catastrophic breakdowns.

PM is cost-effective if the failure rate increases with age or if breakdown costs exceed the costs of scheduled maintenance.

The proponents of PM retort that cost comparisons should look long term and consider total equipment life-cycle costs—for example, in addition to the maintenance service–related costs such as spare parts inventories, materials, labor (including interest and tax implications), the costs of production lost time, disrupted schedules, and effective system service life are also relevant. The potential long-term benefits of PM, which should factor into any comparative analysis, are improved system reliability, decreased cost of equipment replacement, decreased system downtime, and improved spare parts inventory management. In general, PM is a cost-effective strategy if the equipment failure rate increases with age—that is, the older the equipment (or part) the more likely it will wear out. If the equipment is found to have a constant failure rate over its life (e.g. exponentially distributed) then PM generally will not pay.

The second factor that needs to be considered is the trade-off between PM and RTF. PM is the best strategy when its cost is less than the total cost of reacting to unscheduled breakdowns. Care should be taken not to underestimate the cost of corrective action when breakdowns occur, which should include such costs as lost production time, lawsuits, and loss of customer goodwill.

6.2.3 Predictive Maintenance

In the last 20 years maintenance costs have risen sharply relative to operating costs, largely as a result of more complex machines and greater sophistication of manufacturing processes. Along with advances in technology has come the need for companies to invest a greater share of operating capital in production assets, and sharp increases in the cost of that capital have fueled interest in maximizing asset life. In addition, with a global movement toward lean production and just-in-time, the cost of downtime has also risen considerably. In lean plants, with minimum inventories on hand, an unscheduled equipment failure can shut down the entire value stream. On the other hand, a rise in automation, computer controls, and process and product complexity present greater opportunities for process failures. With these developments it has become clear that progressive organizations must institute more intelligent practices if machine availability, reliability, and useful life are to be improved.

Equipment suppliers are increasingly incorporating lean principles in their design and manufacturing processes. Consequently, equipment performance requirements can be written into purchase contracts with the stipulation that suppliers provide information on expected reliability-in-service. Nevertheless, how well equipment actually holds up and performs to its design specifications under field conditions cannot be simulated in an equipment supplier's plant. With the free sharing of knowledge between users and suppliers, equipment designs can improve over time.

As the name implies, **predictive maintenance** (**PdM**) uses systems that are designed to detect and correct failures before they occur, thereby preventing equipment breakdowns. Time-based PM tasks are replaced by condition-based tasks (PdM). **Condition monitors** are portable or in-line instruments used to predict the probability of failure

[3]Chalifoux, Alan, Jearldine I. Northrup, and Nina Y. Chan. "Streamlined Reliability Centered Maintenance (RCM)." *CERL Technical Report* 99/50, U.S. Army Corps of Engineers, Construction Engineering Research Laboratory, May 1999.

of critical components. Condition monitoring technologies include, but are not limited to:

- **Vibration measurement and analysis**—monitoring critical rotating components to detect early signs of bearing wear to prevent seizure.
- **Infrared thermography**—using an infrared imaging and measurement camera to see and measure thermal energy emitted from critical process parameters, such as the detection of underperforming heat exchangers; or dysfunctional pumps, gearboxes, bearings, or motors.
- **Oil analysis and tribology**—monitoring the quality of oil and other lubricants to detect deterioration in critical functional characteristics in lubricants, friction buildup, or excessive wear.
- **Ferrography**—identifying particles suspended in lubricating fluids to determine the association between the size, shape, composition, and concentration of particles and the wear characteristics of machine components.
- **Ultrasonics**—using ultrasound to detect air and gas leaks, and also to monitor the operating condition of mechanical components such as bearings and gearboxes.
- **Motor current analysis**—monitoring electric motors to detect faults such as broken or cracked bars; high resistance; faults to ground; and internal faults such as turn-to-turn, coil-to-coil, and phase balance.

Condition monitoring need not be a technological device. Operators on the front line can monitor process condition continuously using any of the five human senses (see, feel, hear, smell, and taste). An experienced operator can often sense when something is not right and a good understanding of process behavior and causality can often lead to early warning of an impending breakdown. Other tools that can contribute to condition monitoring are trend analysis, pattern recognition, data comparison, tests against limits and ranges, correlation of multiple technologies, and statistical process control. The principal benefit of PdM is the early warning that can reduce the incidence rate of breakdowns—signaling impending failures (such as a bearing overheating) or the detection of small failures that could trigger a chain reaction leading to a catastrophic event (e.g., if a bearing seizes many other parts could be damaged). PM and PdM programs are often implemented together.

PdM systems are designed to detect and correct failures before they occur.

6.2.4 Proactive Maintenance

While the objective of PM is to prevent machine failures through periodic maintenance performed according to a set procedure and schedule, and PdM combats breakdowns through early warning detection of emerging signs of impending failure, **proactive maintenance (PaM)** is a discipline that focuses on root causes rather than active symptoms, faults, or machine wear conditions. A PaM program is aimed at extending equipment life and differs fundamentally from PM and PdM in the following ways.

- Under PaM, only necessary repairs are carried out, parts are never replaced prematurely nor are overhauls conducted until required.
- Failure is not treated as a routine and normal part of doing business—the goal is to ultimately eliminate all root causes of failure so that they no longer occur.
- The implementation of PaM avoids the necessity of crisis maintenance management, which requires dropping everything and deploying maintenance crews to the scene of a breakdown to get a critical machine back on line.

A typical PaM project involves three steps: (1) setting a quantifiable target for a root cause parameter under investigation (e.g., target level of cleanliness for a machine lubricant); (2) implementing a program to control the parameter to the target level (e.g., cleansing the lubricant of contaminants); and (3) routine monitoring of the parameter using a reliable measurement process (e.g., particle analysis) to verify the status is on target.

PaM identifies and manages the root causes of failure in order to extend equipment life.

6.2.5 Reliability-Centered Maintenance

Reliability-centered maintenance (RCM) was developed in the early 1960s in the airline industry. Keeping aircraft in the air is maintenance intensive and the airline companies realized that they simply did not have the resources to support PM and, even if they could, PM was an ineffective strategy for controlling aircraft failures. A task force, commissioned by the U.S. government and consisting of representatives from the Federal Aviation Administration and the airline industry, was charged to investigate the strengths and shortcomings of PM. Their findings concluded that PM has little effect on the overall reliability of a complex system (such as an aircraft) unless there is a dominant failure mode, that PM alone cannot guarantee reliability, and revealed that there are some systems for which there is no effective form of PM. The term *reliability-centered maintenance* came from the title of a report issued by United Airlines in the early 1970s.

RCM is a maintenance strategy that combines the benefits of reactive and proactive approaches and utilizes the features of RTF, PM, PdM, and PaM—all of which are integrated into a comprehensive asset management system. The idea is to appropriately deploy all these tools in order to optimize the use of production resources. Each maintenance strategy has an important role to play and suits a particular equipment type or production scenario. Experience has shown that the implementation of RCM can shave as much as 30%–50% off an

[4]Ibid.

organization's annual maintenance budget.[4] An initial investment in training, tools, and in building a knowledge base to provide equipment condition benchmarks is necessary, but recovery of the startup cost should be swift as the call for reactive maintenance measures decreases due to the prevention of failures through PaM, and the replacement of PM with effective condition monitoring. As reactive maintenance costs go down, total maintenance costs are also reduced; energy costs are reduced (through condition monitoring instrumentation); machine availability, reliability, and utilization are improved; capacity expands; and productivity increases. An added expected benefit is that RCM extends the useful lives of production facilities, equipment, and processes.

RCM combines the benefits of reactive and proactive maintenance strategies.

6.3 PRIORITIZING AND EVALUATING MAINTENANCE WORK

6.3.1 Prioritization Methodologies

The maintenance support function, like all other areas of an organization, has limited resources (e.g., human resources, materials, time, space, tools, etc.). Also, like other cost centers of the enterprise, the maintenance department is held accountable for performing its services within a prescribed budget. When equipment breakdowns occur and cause disruptions in primary-value-stream flow, calls for maintenance support usually get high priority, even jumping to the head of the maintenance queue—similar to life-threatening events in a hospital emergency room. Notwithstanding, in a typical factory there are a multitude of not-so-critical calls for repairs or routine maintenance. In the absence of a rational system for prioritizing work the subjective assignment and scheduling of specific maintenance tasks can result in the jobs in greatest need failing to receive the priority they deserve, and some jobs may never advance high enough on the list to ever receive any attention at all.

A rational system for prioritizing maintenance work can ensure that the jobs in greatest need receive the attention they deserve.

Figure 6-4 is a modeling framework for prioritizing maintenance projects and draws on the computerized maintenance management (CMM) system—the organization's knowledge repository of plant and equipment information. The CMM system contains historical information on machines, instrumentation, buildings, equipment, fixtures, and utilities, which is useful for predicting service lives, evaluating risks, estimating life-cycle costs, and assessing conditions. Such information can be used to determine decision criteria within the modeling step labeled *life-cycle and benefit analysis,* which in turn can be applied in the context of strategic, production, budget, and administrative needs in order to develop a final list of priorities.

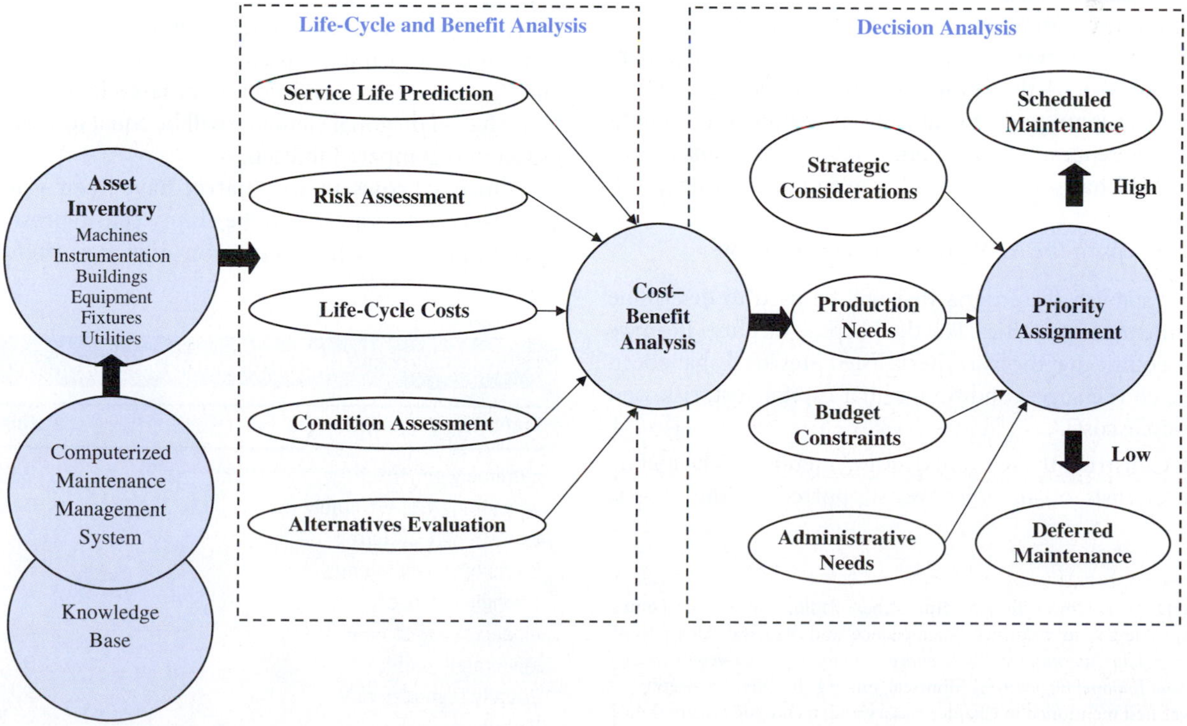

FIGURE 6-4. Decision Model for Maintenance Prioritization

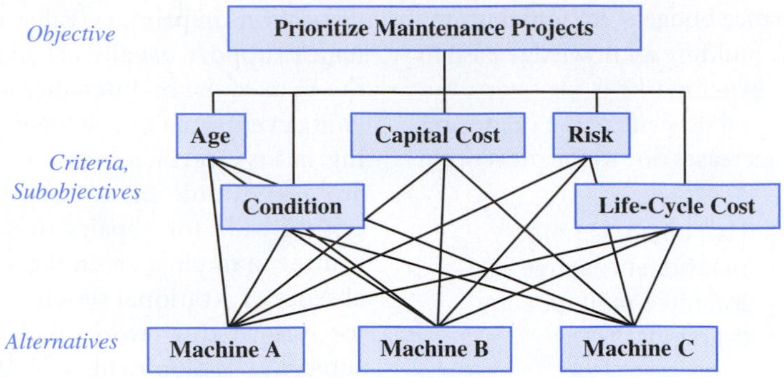

FIGURE 6-5. AHP Procedure for Maintenance Criteria

Using a single criterion, priorities can be based on age, condition, initial capital cost, risk, or life-cycle cost.[5] This can be shown to be ineffective since the projects receiving the worst scores on any of the five criteria (i.e., oldest, worst condition, highest cost, riskiest, or having the highest life-cycle cost) would be selected. That these criteria can be conflicting is obvious. Using the age criteria, one might place a high priority on the oldest machine, which might actually be in the best condition or pose the least risk.

> **Maintenance priorities can be based on age, condition, capital cost, risk, or life-cycle costs.**

The shortcomings of a single criterion approach can be overcome by employing a methodology such as the **analytic hierarchy process** (AHP) that simultaneously considers multiple criteria.[6] Originally proposed by Saaty in 1980,[7] AHP is a technique that can be used by a decision team to assign relative importance scores using subjective and personal preferences. The idea is that humans can express a preference between two items and can subjectively express the strength of that preference on a scale from 1 to 9. When teams are involved, AHP has proven to be effective in facilitating group consensus.

The steps of the AHP procedure are as follows:

Step 1 Establish the criteria that will be used to determine maintenance priorities. For the purpose of illustration we will assume that the five criteria listed previously have been selected (i.e., age, condition, initial capital cost, risk, and life-cycle cost).

Step 2 Construct the relevant decision hierarchy. The hierarchy consists of an objective, supported by the criteria (subobjectives), with various alternatives (maintenance projects) as illustrated in Figure 6-5.

Step 3 Establish priorities.

Step 4 Synthesize judgments.

Step 5 Check consistencies.

Step 6 Develop a prioritized list.

At steps 3–5 the decision team conducts a number of pairwise comparative analyses—one for the selected criteria and one for the maintenance projects against each criterion. If there are n criteria, there will be $n + 1$ pairwise comparisons. This step is then completed as follows. Using Table 6-1 the team determines its preferences between each pair of criteria. This process produces a matrix similar to the one shown in Table 6-2. As an example, in this matrix the team has expressed an equal-to-moderate preference (score $= 2$) for age as compared to life-cycle cost, a strong preference (score $= 5$) for condition over age, a moderate preference (score $= 3$) for condition over capital cost, a strong-to-very-strong preference (score $= 6$) for condition over life-cycle cost, and so on. Note that all diagonal elements will be equal to 1 where each criterion is compared to itself.

Once all rows in the matrix have been visited and preference scores awarded, the empty cells represent those ij combinations where criterion i is not preferred to

TABLE 6-1 AHP Preference Ratings

Preference Judgment	Rating
Extremely preferred	9
Very strongly to extremely	8
Very strongly preferred	7
Strongly to very strongly	6
Strongly preferred	5
Moderately to strongly	4
Moderately preferred	3
Equally to moderately	2
Equally preferred	1

[5] Vanier, Dana, Solomon Tesfamariam, Rehan Sadiq, and Zoubir Lounis. "Decision Models to Prioritize Maintenance and Renewal Alternatives." *Proceedings, Joint International Conference on Computing and Decision Making in Civil and Building Engineering,* Montreal, June 14–16, 2006, 2594–2604.

[6] AHP was first mentioned in Chapter 5 as a group technology methodology for grouping families of products.

[7] Saaty, T. L. *The Analytic Hierarchy Process.* New York: McGraw-Hill, 1980.

TABLE 6-2 Pairwise Comparison of Maintenance Criteria—Selection of Initial Preference Scores

Criteria	Age	Condition	Capital Cost	Risk	Life-Cycle Cost
Age	1				2
Condition	5	1	3		6
Capital Cost	3		1		5
Risk	7	3	5	1	9
Life-Cycle Cost					1

TABLE 6-3 Pairwise Comparison of Maintenance Criteria—Completion of the Matrix

Criteria	Age	Condition	Capital Cost	Risk	Life-Cycle Cost
Age	1	0.2	0.333	0.143	2
Condition	5	1	3	0.333	6
Capital Cost	3	0.333	1	0.2	5
Risk	7	3	5	1	9
Life-Cycle Cost	0.500	0.167	0.2	0.111	1

TABLE 6-4 Normalized Pairwise Comparison of Maintenance Criteria

Criteria	Age	Condition	Capital Cost	Risk	Life-Cycle Cost
Age	1	0.2	0.333	0.143	2
Condition	5	1	3	0.333	6
Capital Cost	3	0.333	1	0.2	5
Risk	7	3	5	1	9
Life-cycle Cost	0.500	0.167	0.2	0.111	1
	16.500	4.700	9.533	1.787	23

⇩

Criteria	Age	Condition	Capital Cost	Risk	Life-Cycle Cost	Row Avg
Age	0.0606	0.0426	0.0350	0.0799	0.0870	0.0610
Condition	0.3030	0.2128	0.3147	0.1865	0.2609	0.2556
Capital Cost	0.1818	0.0709	0.1049	0.1119	0.2174	0.1374
Risk	0.4242	0.6383	0.5245	0.5595	0.3913	0.5076
Life-cycle Cost	0.0303	0.0355	0.0210	0.0622	0.0435	0.0385

criterion *j*. We will place in those cells the reciprocal of the score that was inserted in cell *ij*. That is, if $a_{ij} = k$, then $a_{ji} = 1/k$. Following this procedure, the pairwise comparison matrix for the five criteria is completed as shown in Table 6-3.

The next step is to normalize the pairwise comparison matrix by dividing each element by its respective column total. This is illustrated in Table 6-4. Once the normalized matrix has been developed, the average of each row is computed. The resulting column of row averages is the **priority vector** for the criteria, ω. In this case ω = [0.0610 0.2556 0.1374 0.5076 0.0385]. Note that all columns in the normalized matrix, including the priority vector, will sum to one. The components of the priority vector represent the relative weights for the criteria, and are in a one-to-one correspondence with the order in which the criteria were placed in the pairwise comparison matrix.

The next issue is how consistent was the team's logic in assigning the preference scores. Checking consistency is a four-step procedure.

Step 1 Let A represent the original pairwise comparison matrix (i.e., Table 6-3). Compute $A\omega^T$ as follows:

$$A\omega^T = \begin{bmatrix} 1 & 0.2 & 0.333 & 0.143 & 2 \\ 5 & 1 & 3 & 0.333 & 6 \\ 3 & 0.333 & 1 & 0.2 & 5 \\ 7 & 3 & 5 & 1 & 9 \\ 0.5 & 0.167 & 0.2 & 0.111 & 1 \end{bmatrix} \begin{bmatrix} 0.0610 \\ 0.2556 \\ 0.1374 \\ 0.5076 \\ 0.0385 \end{bmatrix}$$

$$= \begin{bmatrix} 0.3074 \\ 1.3728 \\ 0.6995 \\ 2.7345 \\ 0.1954 \end{bmatrix}$$

Step 2 Compute

$$\lambda_{\max} = \frac{1}{n}\sum_{i=1}^{n} \frac{i^{th} \text{ element of } A\omega^T}{i^{th} \text{ element of } \omega^T}$$

$$\lambda_{max} = \frac{1}{5}\left[\frac{0.3074}{0.0610} + \frac{1.3728}{0.2556} + \frac{0.6995}{0.1374} + \frac{2.7345}{0.5076} + \frac{0.1954}{0.0385}\right] = 5.1937$$

Step 3 Compute the consistency index $(CI) = \dfrac{\lambda_{max} - n}{n - 1}$

$$CI = \frac{5.1937 - 5}{4} = 0.0484$$

Step 4 Compute the consistency ratio $(CR) = \dfrac{CI}{RI}$

TABLE 6-5 Table of Random Indices for AHP Consistency Tests

n	3	4	5	6	7	8	9	10
RI	0.58	0.9	1.12	1.24	1.32	1.41	1.45	1.51

where RI is a random index taken from Table 6-5 where n =the number of rows (categories) in the pairwise comparison matrix.

$$CR = \frac{0.0484}{1.12} = 0.0432$$

The rule of thumb is that consistency is acceptable as long as the $CR \leqslant 0.10$.

A pairwise comparison analysis is next performed on the tentative maintenance projects (e.g., machines), starting with the assignment of preference scores for each pair of machines for each specific criterion. Assume that there are three machines requiring maintenance (Machines A, B, and C). The pairwise comparison analyses against the five criteria are illustrated in Tables 6-6 through 6-10.

Summarizing this analysis, we have the following:

$$\omega_{criteria} = [0.0610\ 0.2556\ 0.1334\ 0.5076\ 0.0385]$$

$$\Omega_{machines} = \begin{bmatrix} \omega_{age} \\ \omega_{condition} \\ \omega_{capital\ cost} \\ \omega_{risk} \\ \omega_{life\text{-}cycle\ cost} \end{bmatrix} = \begin{bmatrix} 0.6333 & 0.2605 & 0.1062 \\ 0.7644 & 0.0698 & 0.1659 \\ 0.0882 & 0.6687 & 0.2431 \\ 0.0833 & 0.1932 & 0.7235 \\ 0.2364 & 0.7013 & 0.0623 \end{bmatrix}$$

TABLE 6-6 Pairwise Comparison of Machines Against Age

Age Criteria	Machine	Machine B	Machine C
Machine A	1	3	5
Machine B	0.333	1	3
Machine C	0.2	0.333	1
	1.533	4.333	9.000

⇩ ⇩

Age Criteria	Machine A	Machine B	Machine C	Row Avg ω^T
Machine A	0.6522	0.6923	0.5556	0.6333
Machine B	0.2174	0.2308	0.3333	0.2605
Machine C	0.1304	0.0769	0.1111	0.1062

$$A_{\omega}{}^T = \begin{matrix} 1.945621 \\ 0.790082 \\ 0.319658 \end{matrix}$$

$\lambda_{max} = 3.038715$

CI = 0.019357

CR = 0.033375

TABLE 6-7 Pairwise Comparison of Machines Against Condition

Condition Criteria	Machine A	Machine B	Machine C
Machine A	1	9	6
Machine B	0.111	1	0.333
Machine C	0.167	3	1
	1.278	13.000	7.333

Condition Criteria	Machine A	Machine B	Machine C	Row Avg ω^T	
Machine A	0.7826	0.6923	0.8182	0.7644	2.387504
Machine B	0.0870	0.0769	0.0455	0.0698	$A_\omega{}^T = 0.209993$
Machine C	0.1304	0.2308	0.1364	0.1659	0.502584

$\lambda_{max} = 3.054399$

$CI = 0.027199$

$CR = 0.046896$

TABLE 6-8 Pairwise Comparison of Machines Against Capital Cost

Capital Cost Criteria	Machine A	Machine B	Machine C
Machine A	1	0.143	0.333
Machine B	7	1	3
Machine C	3	0	1
	11.000	1.476	4.333

Capital Cost Criteria	Machine A	Machine B	Machine C	Row Avg ω^T	
Machine A	0.0909	0.0968	0.0769	0.0882	0.264764
Machine B	0.6364	0.6774	0.6923	0.6687	$A_\omega{}^T = 2.015415$
Machine C	0.2727	0.2258	0.2308	0.2431	0.730606

$\lambda_{max} = 3.00703$

$CI = 0.003515$

$CR = 0.006061$

We can now obtain a vector of machine priorities by computing $\Theta_{\text{priority}} = \omega_{\text{criteria}}\Omega_{\text{machines}}$ as follows.

$$\Theta_{priority} = [0.0610\ 0.2556\ 0.1334\ 0.5076\ 0.0385] \begin{bmatrix} 0.6333 & 0.2605 & 0.1062 \\ 0.7644 & 0.0698 & 0.1659 \\ 0.0882 & 0.6687 & 0.2431 \\ 0.0833 & 0.1932 & 0.7235 \\ 0.2364 & 0.7013 & 0.0623 \end{bmatrix}$$

$$\Theta_{priority} = \begin{bmatrix} \textit{Machine A} & \textit{Machine B} & \textit{Machine C} \\ 0.297 & 0.251 & 0.452 \end{bmatrix}$$

Based on this analysis Machine C would receive the highest priority, followed by Machine A and then Machine B.

For those wishing greater sophistication, other tools that have been proposed to solve the multicriteria optimization problem are the **weighted mean** (WM), **multiobjective optimization** (MOO), and **fuzzy synthetic evaluation** (FSE).[8] A detailed description of these methodologies is beyond the scope of this text. However, no method can ensure that the "best" projects receive the highest priorities. Regardless of what tool is employed

[8]Vanier, et. al., "Decision Models."

TABLE 6-9 Pairwise Comparison of Machines Against Risk

Risk Criteria	Machine A	Machine B	Machine C
Machine A	1	0.333	0.143
Machine B	7	1	0.2
Machine C	3	5	1
	11.000	6.333	1.343

Risk Criteria	Machine A	Machine B	Machine C	Row Avg ω^T	
Machine A	0.0909	0.0526	0.1064	0.0833	0.251061
Machine B	0.2727	0.1579	0.1489	0.1932	$A\omega^T = 2.587811$
Machine C	0.6364	0.7865	0.7447	0.7235	2.272592

$\lambda_{max} = 3.065819$

CI = 0.032909

CR = 0.05674

TABLE 6-10 Pairwise Comparison of Machines against Life-Cycle Costs

Life-cycle Cost Criteria	Machine A	Machine B	Machine C
Machine A	1	0.250	5
Machine B	4	1	9
Machine C	0.2	0.111	1
	5.200	1.361	15.000

Life-cycle Cost Criteria	Machine A	Machine B	Machine C	Row Avg ω^T	
Machine A	0.1923	0.1837	0.3333	0.2364	0.723033
Machine B	0.7692	0.7347	0.6000	0.7013	$A\omega^T = 2.207343$
Machine C	0.0385	0.0816	0.0667	0.0623	0.187464

$\lambda_{max} = 3.072263$

CI = 0.036132

CR = 0.062296

there will always be an element of subjectivism, personal bias, and incomplete information that will influence the results. Excellent opportunities exist for researchers, working with practitioners, to develop new decision tools that use real or simulated data to explore and compare a variety of decision options that will improve the management of maintenance services.

6.3.2 Maintenance Performance

Assessing the quality of any product or service requires reliable measures that are applied to those key characteristics that are the most important to customers. Maintenance performance measures fall broadly into three categories:[9]

- Measures of equipment performance, such as availability, reliability, or overall equipment effectiveness (OEE).
- Measures of cost performance, such as operation and maintenance (O&M), labor, material, and parts.
- Measures of process performance, such as unscheduled downtime, ratio of planned to unplanned maintenance work, schedule compliance, and response time.

[9] Campbell, J. D. *Uptime: Strategies for Excellence in Maintenance Management.* Portland, OR: Productivity Press, 1995.

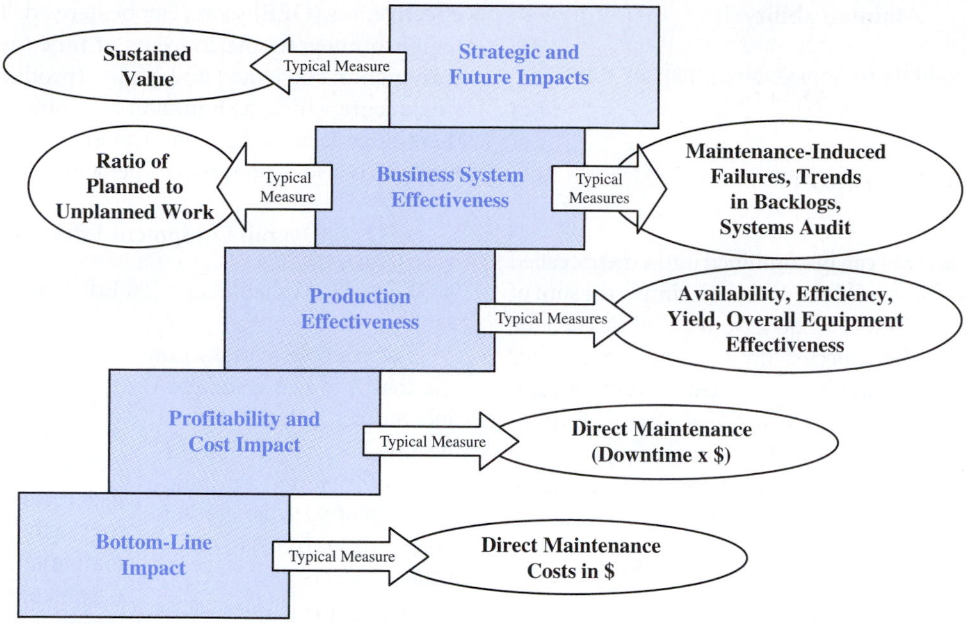

FIGURE 6-6. Hierarchy of Maintenance Performance Measures

However, these categories are somewhat narrow and do not fully take into account the impact that the maintenance function has on the open business system. A broader perspective portrays maintenance performance as a hierarchy that is closely correlated with an organization's quality maturity. As Figure 6-6 depicts, the measure of maintenance performance can be portrayed as a building-block process consisting of five levels.[10]

6.3.2.1 Maintenance Performance Levels 1 and 2. At the first two levels, only the direct costs of delivering maintenance services are captured. The second level is differentiated from the first in its ability to consider the impact that performance has on profit and loss as opposed to simply treating maintenance as a cost center. On these two levels the organization assumes that maintenance activities in the present timeframe have negligible impacts on machine downtime, quality, yields, or future maintenance requirements.

6.3.2.2 Maintenance Performance Level 3. At the third tier, measures are developed to assess how maintenance performance affects the operation and effectiveness of the production system, with particular emphasis on how well maintenance contributes to those operations on primary value streams. An important set of metrics employed at this level focus on whether equipment (a system component) or process (several components) or system (interconnected processes) can be relied on when and where they are needed to support production operations. One of these is **reliability**, a statistical measure of the probability that a particular piece of equipment, operational process, or production system will perform its intended function over its expected operation time or life. Reliability in a particular instance depends on the failure rate. For example, when failures occur randomly, *number of failures* (a random variable) will often fit a Poisson distribution with a mean rate of occurrence equal to λ, expressed as failures per some stipulated period of time t. In this case reliability can be stated as either the *probability of failure* (Equation 6-1) or the *mean time to failure* (Equation 6-2).[11]

Reliability

$$R_{(t)} = \text{Probability of a failure by time } t = 1 - e^{\lambda t} \quad \textbf{6-1}$$

$$\text{Mean Time to Failure (MTTF)} = \frac{1}{\lambda} \quad \textbf{6-2}$$

An alternative view of equipment/process/system performance is to measure *unreliability* – that is, the statistical probability of restoring a component, process, system to full functionality after it fails. The term used to describe this metric is **maintainability** which is defined to be the probability that a successful maintenance action can restore a component/process/system within some specified period of time. If one assumes randomness in repair times such that the *number of repairs* (a random variable) is Poisson distributed with mean μ, expressed as the number of successful repairs per stipulated time period t, maintainability can be computed as either the *probability of a successful repair* (Equation 6-3) or the *mean time to repair* (Equation 6-4).[12]

[10] Tsang, Albert H.C., Andrew K.S. Jardine, and Harvey Kolodny. "Measuring maintenance performance: a holistic approach," *International Journal of Operations and Production Management,* **19** (7), 1999, 691–715.

[11] Equations 6-1 and 6-2 assume that the time between failures can be approximated by an exponential distribution with mean $1/\lambda$.

[12] Equations 6-3 and 6-4 assume that the time to restore service can be approximated by an exponential distribution with mean $1/\mu$.

Maintainability

$$M(t) = \text{Probability of a successful repair by time } t = 1 - e^{\mu t} \quad \textbf{6-3}$$

$$\text{Mean Time to Repair (MTTR)} = \frac{1}{\mu} \quad \textbf{6-4}$$

Equations 6-2 and 6-4 can be combined into a metric called **mean time between failures (MTBF)** which is simply the sum of the MTTF and the MTTR. As Equation 6-5[13] illustrates, the MTBF is the sum of the expected times that a component/process/system is running and being repaired, respectively, and is a concept that is applied to the useful life of a piece of equipment, a process, or a system as applicable. Specifically, premature failures (infant mortality) and end-of-life instances (when repairs are either impracticable or uneconmomical) are excluded.

Mean Time Between Failures (MTBF)

$$MTBF = MTTF + MTTR = \frac{1}{\lambda} + \frac{1}{\mu} = \frac{\mu + \lambda}{\mu\lambda} \quad \textbf{6-5}$$

When the concepts of reliability (the probability that a process will not fail) and maintainability (the probability that a process is successfully restored following a failure) are combined a metric can be created that depicts the probability that a component/process/system is operational at a given time. This metric is **availability** and is defined as the probability that a component/process/system has not failed and is not undergoing a repair action when it needs to be used.

Availability is usually expressed as a percentage – an availability of 99.9% would mean that only in one case out of a thousand when someone needed to use the component/process/system would they find that the component/process/system is either broken down or in the process of being repaired. Equation 6-6 represents the calculation of availability for a single component, taking into account only breakdown repairs (i.e. no scheduled maintenance activities that would take equipment out of service).[14]

Availability

$$Availability = \frac{E[uptime]}{E[uptime] + E[downtime]} = \frac{MTTF}{MTTF + MTTR} = \frac{1/\lambda}{1/\lambda + 1/\mu} = \frac{\mu}{\mu + \lambda} \quad \textbf{6-6}$$

When availability is combined with two other measures a widely used macro-measure, called the **overall equipment effectiveness** (OEE) score, can be derived. The OEE is an attempt to integrate the concepts of time (availability), speed (production efficiency) and quality (production yield) into a single score which, as Equation 6-7 shows, is the product of these three factors. As each of the measures is a percentage, the OEE is also expressed as a percentage.

Overall Equipment Effectiveness

$$\%OEE = (\%\ \text{Avaliability}) \times (\%\ \text{Efficiency}) \times (\%\ \text{Yield}) \quad \textbf{6-7}$$

The measure of maintenance performance occurs at five levels that progressively capture the increasing impact of maintenance on an organization's strategic intent.

For example, a production line that is available for production (i.e., uptime) 92% of the time, has a production efficiency of 95%, and a production yield of 90% has an OEE score = (0.92) (0.95) (0.90) = 0.786 or 78.6%.

There are a number of sources of nonproduction time losses that will impact the OEE score. These include the following.

- Availability Losses
  - Scheduled shutdown losses
  - Periods when no production is planned
  - Planned maintenance that delays production
- Downtime losses
  - Equipment breakdowns
  - Product changeovers
  - Tooling or part changes
  - Job startup and adjustment
- Performance Efficiency Losses
  - Minor line stops, such as breaks, jams, circuit breaker trips, etc.
  - Reduced speed, cycle time, or capacity
- Yield Losses Due to Quality Issues
  - Defects resulting in rework
  - Scrap
  - Yield losses during changeovers and startup

The OEE provides a means for comparing machines and work centers, and helps teams focus on the most problematic areas for improvement. However, it can be somewhat misleading since the OEE calculation assumes that all three factors are of equal importance, which in general is untrue. A 1% loss in equipment downtime does not have the same financial impact on a business as a 1% loss in efficiency; and a 1% efficiency loss is not the same as a 1% loss in yield. An OEE equal to 85% has been widely recognized as the world-class standard and as such has been used to compare equipment and to set goals for individual employees.

Some companies have even adopted an overall score, called the **overall plant effectiveness (OPE)**—an aggregate of all OEEs in the plant. The OPE is computed by taking a

---

[13]Equation 6-5 assumes Poisson failure and repair rates and exponential time between failures and repair times.

[14]Equation 6-6 is valid for a single component. For systems MTBF should be substituted for MTTF.

weighted average of the availability, efficiency, and yield computations respectively for each component/process/system and then applying Equation 6-7 on these aggregate measures. This formula is shown in Equation 6-8.

Overall Plant Effectiveness

$$\%OPE = \left[\frac{\sum_{i=1}^{k}(\text{Availability}_i(\tau_i)}{\sum_{i=1}^{k}\tau_i}\right] \times \left[\frac{\sum_{i=1}^{k}(\text{Efficiency}_i)(\tau_i)}{\sum_{i=1}^{k}\tau_i}\right] \times \left[\frac{\sum_{i=1}^{k}(\text{Yield}_i)(\tau_i)}{\sum_{i=1}^{k}\tau_i}\right] \qquad \textbf{6-8}$$

where,

i = the number of components/processes/systems

τ_i = the scheduled operation time for the i component/process/system

Using an OPE score to conclude anything meaningful about maintenance performance is generally unwise for two reasons. First, it is unfair and unrealistic to compare two or more types of equipment /processes/systems that perform different functions in different environments. And second, the use of OPE as a metric is not actionable. Gathering any information that cannot form the basis for management action is useless and a type of waste.

6.3.2.3 Maintenance Performance Levels 4 and 5. Sadly, many organizations never progress up the hierarchy past the third level, which suffers under the flawed assumption that present conditions will continue indefinitely into the future. At the fourth level of the measurement hierarchy, the attention turns to the open system of the enterprise and how well system goals have been supported by maintenance performance. The fifth level is about sustained value, and measures at this level must be devised to recognize some elusive cost areas, such as depletion of the value of a fixed asset that may depend on future demand, technological changes, and maintenance strategies. Measures at the fifth tier also must flexibly allow for maintenance actions to be assessed relative to the depreciating value of equipment, processes, or products.

The hierarchy of maintenance measurement progresses from measures that are primarily financial to those that are more strategic.

6.3.2.4 Establishing Maintenance Performance Metrics. The lower-level metrics in the hierarchy tend to be financial in nature and the higher-level strategic. Financial measures have a major drawback in that they focus managerial behavior on short-term results. As a result, few managers will make the investment decisions necessary for long-term sustainability. Another shortcoming is that using financial measures is like trying to drive a car by looking in the rearview mirror. Financials are better at gauging the consequences of past decisions than indicating what can be expected in the future. Such indicators are also subject to managerial manipulation—that is, whatever it takes to make the short-term metrics look good.

Leading performance indicators tend to be those that are customer focused such as response time, service, and satisfaction.[15] Widening the perspective from the narrow lagging measures to the forward-looking leading ones requires a balanced approach. In Chapter 2 the balanced scorecard[16] was introduced as a means to focus management attention on a multidimensional performance metric, tying quantifiable measures to strategic objectives. As previously discussed, the balanced scorecard approach defines relevant perspectives for evaluating performance. For any specific case the number of perspectives and their respective descriptions will depend on an organization's stakeholders.Chapter 2 suggested a generic breakdown as follows:

- **Users.** How can an organization reliably determine how well it is doing from the perspective of those stakeholders that directly use or distribute its products and/or services?
- **Internal production systems.** How can the organization assess how well its operations, technologies, capacities, skills, and capabilities work together to achieve strategic goals and create sustained value for all stakeholders?
- **Innovation, learning, and change.** How can the organization measure its capability to improve, grow, and create sustained value over the long term for all stakeholders?
- **Investors.** How can the organization know how well it is performing in the eyes of its shareholders, creditors, and lenders; how do investors assess the level of risk; and what would it take to attract new investors?
- **Society and the Planet.** How can the organization know how well it is complying with regulations, public expectations, societal values, and the preservation of natural resources.

Applied to maintenance performance the following process has been advocated for implementing the balanced scorecard approach.[17]

Step 1 Formulate a strategic plan for maintenance services, including developing and improving maintenance capabilities to include expanded condition-based, predictive, and proactive programs; broadening the skills of the maintenance

[15] Eccles, R. G."The Performance Measurement Manifesto." In *Performance Measurement and Evaluation*, edited by J. Holloway, J. Lewis, and G. Mallory, 5–14. London: Sage.

[16] Kaplan, Robert S., and David P. Norton. "The Balanced Scorecard—Measures That Drive Performance." *Harvard Business Review*, January/February 1998.

[17] Tsang, A. H. C. "A Strategic Approach to Managing Maintenance Performance." *Journal of Quality in Maintenance Engineering* 4, no. 2 (1998): 87–94.

TABLE 6-11 An Example Balanced Scorecard for Maintenance Performance

Balanced Scorecard for Maintenance Performance of Electricity Transmission and Distribution Company

Perspective	Strategic Objectives	Key Performance Indicators (KPIs)
Financial	Reduce operation and maintenance (O&M)	O&M costs per customer
Customer	Increase customer satisfaction	Customer-minute loss Customer satisfaction rating
Internal Processes	Enhance system integrity	% of time voltage exceeds limits Number of contingency plans reviewed
Learning and Growth	Develop a multi-skilled and empowered workforce	% of cross-trained staff Hours of training per employee

Tsang, Albert H. C., Andrew K. S. Jardine, and Harvey Kolodny. "Measuring maintenance performance: a holistic approach." *International Journal of Operations & Production Management* 19(7), 1999, page 706.

workforce and knowledge base; and implementing and/or expanding total productive maintenance practices with broader workforce participation all along the value streams.

Step 2 Operationalize the maintenance strategy by developing long-term supporting objectives for each of the five sectors of the balanced scorecard. Identify relevant key performance indicators (KPIs) and establish performance targets for each. Cascade the KPIs and targets down to teams and workers in the form of individual performance goals.

Step 3 Develop action plans that provide the means for achieving the targets established during step 2.

Step 4 Conduct periodic reviews to assess progress made toward meeting strategic objectives and validating causal relationships between measures.

The balanced scorecard approach can be used to measure maintenance performance.

An example of a balanced scorecard format[18] for an electricity transmission and distribution company[19] is shown in Table 6-11.[20]

6.4 QUALITY IN MAINTENANCE AND RELIABILITY

6.4.1 Total Productive Maintenance

Research has shown that in many factories machines are running only about 60% of the time, and the 40% downtime is equally split between personal breaks and maintenance on the one hand and setup time and machine breakdowns on the other.[21] These numbers are alarming because they suggest a goldmine of untapped capacity. Figure 6-7 illustrates a

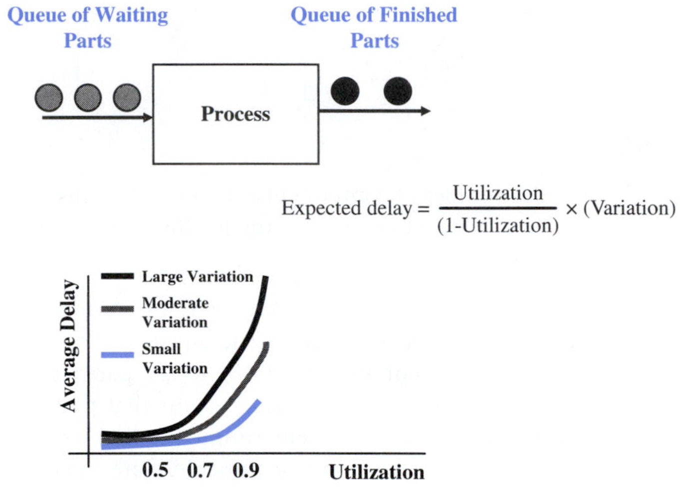

FIGURE 6-7. Effect of Variability on Process Delays

principle drawn from queuing theory that defines the relationship between wait time, utilization, and variation. As machine utilization approaches 100%, the denominator of the first term on the right of the expected delay formula approaches infinity. At that level, any variability at all in workstation processing time can drive the WIP (waiting in front of that workstation), and its corresponding wait time, to astronomical levels!

A second insight from Figure 6-7 is that as processing variability increases, the waiting time climbs exponentially. Even a moderate amount of variability can induce a waiting time that is 9 to 10 times the average service time. High variability can result in substantially longer delays. This is significant because wait time adds cost and no value. By increasing uptime, capacity is increased and the effect of variability on delays is reduced. As an added benefit, the actions that are taken to increase uptime sometimes also reduce the variability in processing times.

TPM permits operators to participate in solving unproductive time problems.

Total productive maintenance (TPM) is a lean tool that uses the operators of machines to help identify and solve

[18]Note that the four perspectives shown in this example coincide with those of Kaplan and Norton, the architects of the balanced scorecard approach.
[19]Tsang et al., "A Strategic Approach," 706.
[20] Note that the management of this utility elected to use only four performance metrics - those proposed by the originators of the concept.
[21]George, Michael L. *Lean Six Sigma: Combining Six Sigma Quality with Lean Speed.* New York: McGraw-Hill, 2002.

similar problems related to equipment downtime. TPM programs track unproductive time in six loss categories.

- Machine downtime
- Machine setups
- Minor production stoppages
- Unscheduled breaks
- Production time spent in rework
- Waste

Data collected in these six categories are tied to the three measures used to compute the OEE (availability, efficiency, and yield). The real value of the OEE as a maintenance metric is to characterize and communicate the sources of equipment losses and develop Pareto charts (refer to Chapter 7) to guide improvement efforts. As procedural changes are implemented, a corresponding improvement in the OEE validates the improvement efforts. Since the sources of nonproductive losses, listed previously, relate to value stream and overall production system performance, OEE is usually incorporated into a TPM program. In TPM improvement efforts are expanded to include the participation of nonmaintenance personnel who are actually operating the equipment and also of those who support them in secondary roles.

OEE is incorporated into TPM so that front-line employees are involved in solving the problems causing nonproductive losses.

TPM has two primary goals—zero breakdowns and zero defects. Simultaneously reducing downtime and production errors will improve the OEE by improving equipment availability, increasing utilization and efficiency, reducing costs, and minimizing inventory costs associated with spare parts. TPM involves a five-prong attack.

1. Track and document losses. Compute and evaluate the OEE for each critical workstation.
2. Involve the machine operators by allowing them to take responsibility for certain maintenance tasks, such as
   a. Performing routine repairs and maintenance
   b. Taking proactive action to prevent problems before they occur
   c. Seeking solutions to problems to prevent their recurrence
3. Take a systematic approach to maintenance where standards are developed and preventive plans are devised. Maintenance specialists are used to oversee maintenance activities, provide assistance in problem diagnostics, and specifically train and supervise operators in carrying out routine maintenance and repair tasks.
4. Provide relevant and continuous training to all personnel in maintenance skills related to their particular work environments.
5. Achieve zero maintenance through maintenance prevention (MP). MP is a proactive approach that tracks failures back to their root causes and the information is then used to eliminate those causes at the earliest possible point in the equipment design phase. Equipment design should not only consider failures but also take into account maintainability during the life of the machine.

TPM is an important component of Lean Six Sigma. It has been estimated that within three years of adopting TPM, most companies can expect between 15% and 25% increases in machine uptime and up to 50% increase in productivity.[22] Lean Six Sigma is a topic that is covered in more detail in Chapter 14.

6.4.2 Lean Maintenance

Lean production, which has been mentioned in previous chapters and is discussed in further detail in Chapter 11, is a business philosophy pioneered by Toyota after World War II. Under the lean philosophy a set of standard tools are applied to the design, organization, and management of primary and support value streams and of the extended enterprise including suppliers, customers, and all intermediaries that comprise the firm's various supply chains. The goal of *lean* is to meet or exceed customer requirements at minimum cost by identifying and eliminating the sources of *muda*, the Japanese word for waste.

The goal of lean maintenance is to eliminate waste while improving throughput, cycle time, and quality.

Lean maintenance is the application of the lean philosophy to the maintenance system, and the same goal applies—that is, identify and eliminate sources of **waste**, or **muda** (wasted time, effort, materials, space) while improving throughput, cycle time, and quality. To provide the highest quality output at the lowest cost in the shortest possible time, an organization must be able to depend on reliable systems and equipment that will be up and running when needed, and will perform according to equipment vendors' specifications.

Numerous tools have been used to implement lean principles to improve manufacturing and service operations. Those that apply to lean maintenance include the **5S workplace discipline**, identification and elimination of the seven deadly wastes, kaizen, jidoka, and just-in-time (JIT).[23]

6.4.2.1 The 5S Workplace Discipline. Workplace efficiency begins with good housekeeping. The Japanese used five words, all beginning with the letter *S*, to define a workplace discipline that facilitates flow. These words are *seiri, seiton, seiso, seiketsu*, and *shitsuke*. Some companies use these original words and train all personnel in their meanings. Others have tried to adopt equivalent English words that also begin with the letter *S*.

---

[22]Nakajima, S. *Introduction to Total Productive Maintenance.* Cambridge, MA: Productivity Press, 1988.

[23]Hawkins, Bruce. "The Many Faces of Lean Maintenance." *Plant Engineering* 59, no. 9 (September 2005): 63–65.

- **Seiri (sift or sort).** The first step in establishing a workplace discipline is to go through the clutter and keep only what is necessary. What is unnecessary should be discarded. If something is not used regularly or there is no anticipated use for it, it is best to remove it from the workplace. This includes tools or materials that have not been used for an extended period and are not likely to be needed in the future. Files and papers that have no use should also be discarded or boxed up for storage (if there is an archival purpose). The idea is to free up the space so that items and materials that are needed on a daily basis can easily be located. Seiri also includes a thorough cleaning of the workplace, eliminating dust, oil, debris and the like. Applied to lean maintenance, everything that does not add value should be removed including parts removed from broken-down machines; equipment or piping systems that have been "abandoned-in-place"; obsolete charts, graphs, drawings, or O&M manuals; and unrepairable assemblies or tools. The 5S discipline demands that anything not needed for maintenance jobs at hand should be discarded.

Workplace clutter must be eliminated, keeping only what is necessary and discarding the rest.

Deciding to dispose of some specific item can be questionable and a difference of opinion within the work group can easily arise. In such cases a tagging system has worked well in removing the controversy. A tag is placed on the item in question with a date—usually about 6 months into the future. If the item is used before the date is reached, the tag is removed. Otherwise, the item is discarded.

- **Seiton (straighten).** An efficient workstation has a place for everything and everything is kept in its rightful place. Human factors engineering can be used to design the work area so that frequently used items are within easy reach to avoid bending, stretching, or even excessive walking. The need for heavy lifting can be eliminated through the use of wheeled containers or conveyors. Tools can be mounted on fixtures that hang from overhead gantries and that enable easy access with little effort. In tool shops shadow boards can be constructed to ensure hand tools are returned to their proper locations. The use of labeling and compartmented drawers can also help personnel find what they need, when they need it, without wasting valuable time. Applied to lean maintenance, equipment and storage facilities should be positioned to eliminate wasted motions or travel time giving due consideration to work flow patterns and directions.

An efficient workstation has a place for everything and everything in its place.

- **Seiso (sweep, scrub, or shine).** Seiso is about maintaining cleanliness and tidiness. Once all the clutter is gone and everything needed has been assigned a proper place, it is important that the workplace stay that way. Daily cleaning routines must become a part of every worker's job. The maintenance of a neat working environment includes keeping equipment clean and in good working order, and ensuring that tools and materials are where they should be. It is a good idea to institute a daily check sheet procedure where a crew member is required to sign off on each item. This procedure provides documentation that seiso is being carried out. Certain tasks may require special skills (e.g., performing routine maintenance) and it is management's responsibility to ensure that workers are properly trained to perform them. Applied to lean maintenance, shop spaces used to rebuild equipment should be *clean* rooms, free of all dust, dirt, and other contaminants. In all operating areas concrete floors should be sealed so that spills can easily be cleaned up. Lighting fixtures should be kept in good repair, blown bulbs replaced expeditiously, and should be sufficient to provide adequate lighting. Painting work spaces with light colors also helps. Maintenance problems are more easily spotted when work spaces are well lit and contaminant free.

Workstations should be kept clean and tidy.

- **Seiketsu(spick-and-span or systemize).** Seiketsu is the continual maintenance of seiri, seiton, and seiso. This step simply says, "maintain the routine that has been established for workplace tidiness and order." The 5S discipline becomes the new company-wide standard and a system should be implemented that ensures the condition of the workplace does not deteriorate. However, it is often difficult to get all employees onboard with this new mindset. Applied to lean maintenance, area visual controls can be used to assist workers with tool selection and use—examples that have been employed in practice are registration marks on fasteners, and color codes to show correct operating ranges on gauges.

Seiketsu can be an indicator of how rapidly an organization's culture is transforming to the new order. Workplace inspections offer the opportunity for positive reinforcements through generous praise, which can be accompanied by tangible recognition, for the deserving; or alternatively, suggestions for improvement can be offered in cases where standards have not been met. In the latter case there are times when management needs to assist a workstation that is having trouble meeting organizational goals.

Workstation discipline should be systematized and a company-wide standard.

- **Shitsuke(sustain).** Shitsuke is the kaizen[24] of the 5S discipline.The first four steps are about establishing and

[24]Kaizen was first introduced in Chapter 1 and is discussed in further detail later in this chapter and again in Chapter 11.

maintaining a neat and organized workplace. Shitsuke is concerned with improvements—*how can the state of cleanliness and order be improved?* This goes beyond simple housekeeping. In this step workers tackle the underlying causes of disorder and grime. For example, spills and leaks can be quickly cleaned, but it is far better to discover and eliminate the causes so that they occur with less frequency, or do not recur. Tools such as fishbone diagrams and Pareto charts[25] can help, as workers strive to identify problems. Under shitsuke, work is no longer viewed as the execution of standard methods. Workers collectively gain an improving team mind that is continually on the lookout for better methods. A process for conducting regular audits or surprise inspections should be considered. When management shows a genuine interest in the cleanliness and orderliness of work spaces, workers are more likely to pay attention to safety, order, and cleanliness.

Workstations must be disciplined to continuously strive for improvement in orderliness and cleanliness.

6.4.2.2 Seven Deadly Wastes.[26] There are seven primary sources of muda—referred to as the **seven deadly wastes**—that lean production targets for elimination. Listed as follows, these are also areas of concern for lean maintenance.

1. **Overproduction.** Producing more than can be sold or is needed by the next workstation along the value stream feeds inventory, which increases costs and reduces efficiency. As with any operation, **overproduction** in maintenance is defined as performing any work that does not add value. Examples of maintenance overproduction include excessive preventive maintenance, rework (having to repeat jobs or return to workstations because work was done incorrectly the first time), misapplying of predictive maintenance technologies (using condition monitoring devices not applicable to equipment being monitored), and any other unnecessary or non-value-added work activities.
2. **Unnecessary Delays.** Disruptions due to **unnecessary delays** constitute a significant source of waste. In production areas delays can occur due to line imbalances, lack of material, machine breakdowns, quality problems, scheduling errors, and the like. In maintenance, delays can be due to personnel waiting to be assigned to a job, queuing at a storeroom parts issue window or tool crib, waiting for consultation with operating or technical personnel, waiting for the answer to a technical question, waiting for equipment history files, or waiting for necessary approvals.
3. **Transportation and Materials Handling. Transportations** occur when people move from one location to another. **Materials handling** is the transportation of materials. Movements of people or materials add cost but no value (except perhaps place utility), and a goldmine of opportunity can be tapped by designing layouts and processes so that the distances traveled and numbers of transportations and materials handlings are minimized. Under lean maintenance, this form of waste can occur through poor planning or procedure—for example, personnel having to travel back and forth between the operating area and the shop to obtain technical information and special tools, or to the storeroom for repair parts. Other sources of waste include poorly designed preventive maintenance sequences and lack of proper coordination of the work activities.
4. **Process Method.** A somewhat elusive form of process waste relates to **process method.** Unnecessary waste occurs whenever inefficient methods are employed that require redundant, unnecessary, or extra steps. Under lean maintenance processing wastes can result from poorly designed work order systems that require multiple entries of the same data, a material requisitioning system that requires redundant approvals, or ineffective job plans that create backtracking or unnecessary steps.
5. **Inventory. Inventory** is treated by the lean philosophy as a major opportunity for improvement since investment in materials ties up working capital and consumes resources to store and manage them. Types of inventories include raw materials, work in process, finished goods, supplies, and repair parts. The latter category is of interest to lean maintenance. Waste in spare parts management include such sources as the inability to identify and dispose of obsolete parts, multiple entries of the same part number in the computerized parts management system, incorrect settings for minimum and maximum inventory levels, and hidden storage areas where parts and supplies get squirreled away such as lockers and toolboxes.
6. **Motion Inefficiencies.** A source of waste in any job pertains to how each individual employee performs assigned tasks. It is important to investigate what movements are used and whether they are ergonomically efficient and safe. Good methods studies always take into account ergonomic design guidelines and the principles of motion economy. **Motion inefficiencies,** or losses, occur in maintenance when shops or storerooms have poor layouts; materials are not (or are improperly) kitted for repairs and overhauls; or operating and maintenance (O&M) manuals cannot be readily accessed through poor storage or cataloging procedures.
7. **Scrap, Rework, and Defects. Scrap, rework, and defects** are the most obvious forms of waste. It is far better to do a job correctly the first time through than to have to produce two in order to get one right. The further the product has traveled down the value stream before it is scrapped out or is recycled for rework the more the

[25]These tools are covered in detail in Chapter 7.

[26]These are not to be confused with Deming's seven deadly diseases found in Appendix A.

value-added cost accrual. Root causes are numerous and typically the object of improvement studies. Under lean maintenance, defects can arise due to poor quality work, which can be the result of inadequate training; ineffective or ambiguous procedures; or the unavailability of proper tools, parts, or supplies to do a particular job. Defects can also be produced during unnecessary intrusive preventive maintenance tasks, inadequate teardown or rebuild methods, or defective repair parts. Defects in spare parts can be vendor induced or can be the result of improper storage and handling techniques.

6.4.2.3 Kaizen. *Kaizen* is the Japanese word meaning *continuous improvement.* **Kaizen** is a key principle employed by QIM to create a supportive business strategy. The lean production philosophy defines a Kaizen event as any focused intensive rapid-improvement effort that uses cross-functional teams to investigate specific identified problems or opportunities. Such events form the basis for improvement projects in Lean Six Sigma programs, and are successful in broadening participation and generating enthusiasm, momentum, and energy. Kaizen events facilitate teambuilding by unifying workers behind a common set of goals in which they have a vested interest.

Under lean maintenance, kaizen events can be used to improve equipment reliability and maintenance service. Any system, equipment, or machine that chronically causes production disruptions or produces excessive scrap, rework, or defects are good candidates for Kaizen studies. A typical Kaizen event will be a team project, and will begin with the construction of a simplified **failure mode and effects analysis (FMEA)**[27] where failure modes are listed and prioritized according to impact (both frequency and duration). The team will then analyze those failure modes that are high priority to determine their root causes. Then, using techniques similar to those presented in Chapter 7 the team will brainstorm (or brainwrite) practical and implementable solutions, and finally reach consensus on the best alternative. Once selected, the solution is implemented and monitored to validate its impact. When results have been validated the solution may be extended to similar systems, equipment, or machines in other areas. Configuration management tasks, such as the updating of drawings, computer files, records, and other documentation, may also be performed as the project is brought to closure.

> **Under lean maintenance kaizen can be used to improve equipment reliability and maintenance service.**

6.4.2.4 Jidoka. *Jidoka* is the Japanese word for autonomation, which roughly translated means *quality at the source.* **Jidoka** is therefore a fundamental tenet of QIM, which subscribes to the philosophy that quality must be built (rather than inspected) into a product or service. On the factory floor a jidoka policy requires operators to stop a production line anytime they see a problem. Jidoka is human intervention in an automated process to avoid the production of unacceptable quality.

The English word, *autonomation*, comes from the two root words *autonomous* and *automation*. The idea is to empower people to autonomously detect problems and stop production. However, **autonomation** is more than simply stopping the line; it also means tracking down the root cause and solving the problem. This means that operators must be trained to spot problems and empowered to stop a line to fix them when they occur. It is much better to halt production than to continue to produce poor quality. Jidoka can be intimidating to some employees who are fearful of having to accept such an awesome responsibility. Halting an entire production line is a big deal! A possible solution to this is a three-color andon system. Andons are visual signaling devices and operate similar to traffic lights with red, yellow, and green. When the light is green, production proceeds smoothly. If something goes wrong, an operator can stop the line outright by activating the red light, or if uncertain can summon a supervisor by turning on the yellow light.

The jidoka production mindset can be extended to the maintenance function when **quality-at-the-source** thinking has become part of the workforce culture. This will result in the following behavioral outcomes.

- Excellence in work quality and job performance becomes an expectation. Anything less than the best will never be an option. It will always be better, and faster, to do a job right than to do it over!
- Training programs will be tied directly to best practices, quality requirements, and skills improvement, so that workers know how to do their jobs correctly and can perform at their fullest potential.
- Detailed plans and specifications together with written operational instructions on quality requirements, testing, and inspection procedures will be provided at all workstations. Quality and production information should be clearly documented, including photographs and samples when appropriate. For the maintenance function, this includes clear instructions on machine breakdown, repair, and rebuild procedures; maintenance routines; machine operating procedures; and proper use of tools, gauges, and test instruments. Memory, perceptions, and poor training breed variability in results.
- Post maintenance audit procedures for critical equipment will be established whereby operators and maintenance personnel will collectively be required to verify that maintenance activities have produced satisfactory outcomes.
- All machine breakdown events will be investigated to determine and eliminate the root cause.
- Root cause analyses will be performed whenever production processes are found to be producing excessive scrap or rework.

6.4.2.5 Just-in-Time. In the quest to become leaner and more efficient, organizations have sought ways to cut costs while improving the capability to respond to changing customer needs on short notice. The **just-in-time (JIT)**

[27]FMEA was first discussed in Chapter 5 as it applies to superior design; FMEA methodology is covered in detail later in this chapter.

philosophy aims to accomplish these goals by requiring that materials are ordered only when and in the exact quantities needed. The implementation of JIT has resulted in substantial reductions in raw materials inventories and freed up valuable floor space that could be reassigned to value-adding operations. JIT and how it relates to Six Sigma®quality management is discussed further in Chapter 11.

Maintenance inventories are those used to support maintenance operations, such as spare parts, and can represent a significant proportion of the total inventory value—up to 20% in some organizations. Reducing these inventory levels is important. Parts to support unscheduled equipment breakdowns need to be stocked, but not in excessive quantities. Materials and parts for planned work (e.g., preventive maintenance) can be purchased as needed. Inventory usage (and therefore inventory levels) can also be minimized by reducing the number and frequencies of unscheduled machine breakdowns, and by increasing the overall reliability of critical processes. This will happen as kaizen events become a normal part of the maintenance culture and improvements begin to have positive impacts on the quality of maintenance services.

JIT can be applied to the maintenance function when maintenance requirements become predictable.

6.5 IMPACTS ON THE MAINTENANCE WORKFORCE

In Chapter 3 we explored how the adoption of QIM impacts an organization's culture and the lives and jobs of the workers who populate it. Employees at all levels are affected by and participant in the change process. Maintenance is no exception. Traditional roles, skills, and organization must give way to innovative approaches that are more responsive to QIM. In pre-third-generation cultures, maintenance is performed by workers who are assigned to a specific support silo (e.g., maintenance, engineering, or physical plant) and work under the direct supervision of master tradespersons (e.g., master mechanic, electrician, pipe fitter, etc.). These masters typically serve as the collective repositories of technical and historical information relative to a plant's operating equipment, and the primary mission of maintenance is seen to be a quick expeditious response to equipment failures. When a machine goes down, performance is measured by how quickly the maintenance crew can get the machine back up and on line.

In the increasingly complex world of today, the foremost maintenance issue is how to anticipate and prepare for the unexpected so that lost production time is minimized. This requires new skills and a different approach. The team-oriented approach of QIM (and third-generation thinking) extends to the maintenance function where teams of workers, representing a crosssection of skills, are empowered to collectively determine the allocation of resources needed to respond to maintenance requirements. These teams share an expanded mission that is three-faceted: to improve operational efficiencies, to improve use of resources and thereby reduce overall maintenance costs, and to preserve the value and extend the useful lives of operational assets.[28]

To accomplish these goals, the maintenance teams have to turn from a *reaction-to-failure* mindset to one that is more proactive and predictive. The focus must broaden from simply trying to quickly repair broken down machines to the long-term improvement of operating capacity. A secondary goal is to achieve this while also reducing costs. A strategy that has helped organizations achieve this expanded mission, and do so without increasing the headcount of the maintenance workforce (and thereby increasing costs), is to transfer some of the burden of routine maintenance tasks to operations personnel through the implementation of TPM programs.

Maintenance personnel need the skills and mindset to anticipate and prepare for the unexpected.

In the wake of such dramatic change, researchers are beginning to develop sophisticated techniques to address the important issues of how to minimize the size of the maintenance workforce without sacrificing on service, and how to optimize the skills mix through cross-training. For example, the problem of designing a workforce structure that is capable of performing a variety of maintenance jobs within a specified time interval can be formulated as a nonlinear mixed integer program. This model can then be solved efficiently using a well-known heuristic that was developed for the generic bin packing problem.[29]

Some examples of analytical models used to address the skills composition issue include the use of queuing theory to determine system impacts of various workforce structures, Markov analysis to obtain optimal assignments of personnel to tasks, and a linear programming algorithm to enable the simultaneous optimization of assignments and overall workforce structure.[30]

As QIM principles become more entrenched in a corporate culture and workforce empowerment becomes the norm, individuals within the workforce will increasingly accept personal responsibility for the quality of those process outcomes within their spheres of influence. Participation of maintenance workers will be broadened to include membership on cross-functional project teams as a consequence of the Lean Six Sigma implementations, which place high value on the technical expertise and knowledge of production technologies that maintenance workers can contribute.

The Lean Six Sigma process has also been applied directly to the proactive improvement of maintenance by increasing equipment reliability and machine uptimes. Dramatic results have been reported. For example, John Deere reduced unscheduled machine downtime by 80% in two years and was able to increase the mean time between failures, prolonged

---

[28]Yolton, John. "Restructured Maintenance Workforce Yields Better Response, Efficiency." *Pulp and Paper,* October 1997.

[29]Chang, Soo Y., Yushin Hong, Joong H. Kim, and Xere Kim. "A Heuristic Algorithm for Minimizing Maintenance Workforce Level." *Production Planning and Control* 10, no. 8 (1999): 778–86.

[30]Dietz, Dennis C., and Matthew Rosenshine. "Optimal Specialization of a Maintenance Workforce." *IIE Transactions* 29 (1997): 423–33.

reliability, and machine uptime and availability by eliminating stressors such as heat, vibration, dirt buildup, oxidation, hydraulic contamination, power surges, and so on.[31]

The nature of maintenance work places the maintenance workforce in a unique position relative to a firm's critical technologies. By having to actively interface with operating technologies on a daily basis and with a depth and breath that exceeds the exposure usually gained by operations personnel, the typical maintenance worker accumulates a knowledge and experience base that can be gained only by working with the equipment in a maintenance context. This knowledge can be harnessed and channeled to guide and influence an organization's internal technological change, which is essential to breakthrough improvements and the sustainability of competitive advantage. A 1997–2000 study of the maintenance departments of five large and medium-sized process/manufacturing companies revealed that the role of maintenance workers in the selection, acquisition, and installation of new technologies can be expanded in two ways.[32] First, as an indirect player, maintenance can be called upon to provide feedback on asset performance as part of an organization's strategic planning process, or as advice to vendors who design, select, or install new technologies. Second, in a direct role the collective wisdom of the maintenance workforce can be tapped to facilitate the introduction of new technologies, helping integrate these technologies with existing processes. Additionally, maintenance expertise can be directed toward making new or existing technologies more compatible to the changing needs of the work environment.

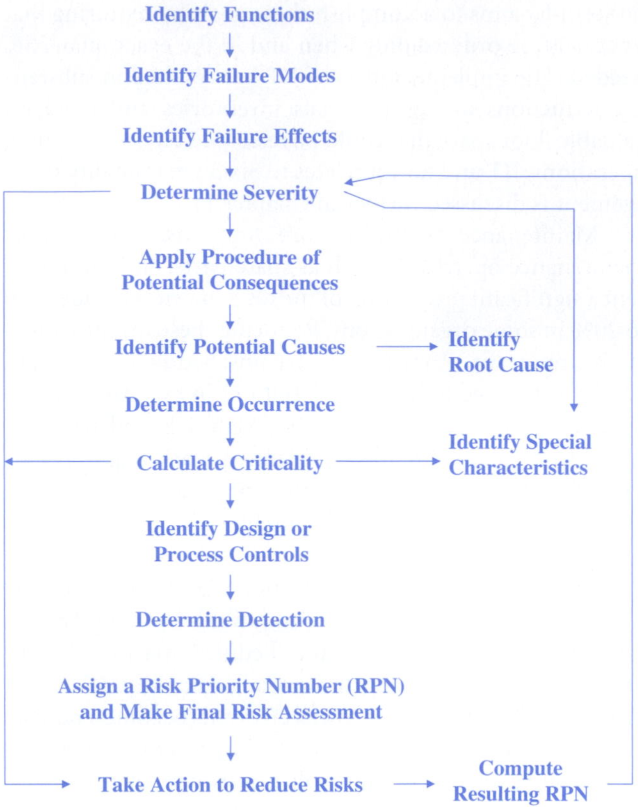

FIGURE 6-8. FMEA Process

6.6 FAILURE MODE AND EFFECTS ANALYSIS

A **failure mode and effects analysis** (**FMEA**) is a systematic procedure for identifying and preventing problems before they have a chance to occur. An FMEA can be applied to a product or to a process and is customarily applied during the product design and development or process design stages. However, the use of FMEAs as a continuous improvement tool can yield big dividends. For example, an automotive supplier formed FMEA teams to analyze both product and process designs. The results of these activities produced a decrease in defectives to 0.2 part per million, equipment uptime increased from 74% to 89%, customer complaints were completely eliminated, and productivity per labor hour increased by 22%.[33]

An FMEA is a systematic procedure for identifying and preventing problems before they have a chance to occur.

A typical FMEA process, shown in Figure 6-8, examines a product or process in an attempt to identify all the ways it can potentially fail. A failure can occur due to a malfunction in operation, a design error, or when a human makes a mistake in either using a product or operating a process. All the ways in which a process or product failure can occur are called *failure modes*. There are two types of FMEAs: a **design failure mode and effects analysis** (**DFMEA**) and a **process failure mode and effects analysis** (**PFMEA**). The primary objective of a DFMEA is to identify potential failures associated with a product's design that would lead to

- Product malfunctions or reduced product reliability
- A shortened product life
- Safety hazards to customers while using the product

A PFMEA is applied for the purpose of discovering any potential failures in a process that could

- Impact product quality or reliability
- Reduce process reliability
- Cause customer dissatisfaction
- Create safety hazards or cause injury to workers
- Cause harm to the environment

Associated with each failure mode is a *failure effect*—the consequences that a failure mode has on the operation, function, or status of a product or process. All effects are not equally likely, so for each failure mode–failure effect combination a relative risk is assigned. The relative risk is expressed as a **risk priority number** (**RPN**) which is the product of three factors: severity, occurrence, and detection. These factors are defined as follows:

- **Severity**—an assessment of how serious the effect of the failure mode is on the customer.

[31]Cooper, Howard. "Lean Maintenance Maximizes Cost Savings." *Manufacturing Engineering* 133, no. 6 (December 2004): 16–17.

[32]Cooke, Fang Lee. "The Important Role of the Maintenance Workforce in Technological Change: A Much Neglected Aspect." *Human Relations* 55, no. 8 (2002): 963–88.

[33]McDermott, Robin E., Raymond J. Mikulak, and Michael R. Beauregard. *The Basics of FMEA*. Portland, OR: Productivity Press, 1996.

- **Occurrence**—an assessment of the likelihood that a particular cause of a failure mode will occur during the life of a product or process.
- **Detection**—an assessment of the likelihood that the failure mode or its cause will be detected before the impact of the effect reaches the customer.

Each of the three risk factors is assigned a number from one (low) to ten (high). Tables 6-12 to 6-14 provide some guidelines for assigning ratings to each factor. An RPN is obtained for each failure mode and effect by multiplying the three factors together (severity *x* occurrence *x* detection). RPNs therefore range from 1 to 1,000 and reflect the relative rankings of the importance of each type of failure for corrective action. Failure modes with the highest RPNs should get management attention first. Nevertheless, any failure mode with a high severity rating (9 or 10) should be investigated regardless of its overall RPN.

When corrective action is taken, a revised RPN, called the resulting RPN, is calculated by updating the severity, occurrence, and detection factors. The FMEA cycle repeats with continuing improvements until the resulting RPNs for all failure modes are considered to be acceptable. Conducting an FMEA is a 10-step process.

The RPN is an index that reflects the relative importance of each failure mode, and is useful in prioritizing improvement efforts.

Step 1 Review the design using blueprints or schematics (DFMEA) or the process using flowcharts (PFMEA) to identify key components.

Step 2 Brainstorm potential failure modes by consulting existing documentation and data, looking for clues.

Step 3 List potential effects of failure, keeping in mind that there may be more than one for each failure mode.

Step 4 Consult the severity rating guidelines and assign severity ratings based on the perceived severity of the consequences of each failure.

Step 5 Consult the occurrence rating guidelines and assign occurrence ratings based on the perception of how frequently each failure is likely to occur.

Step 6 Consult the detection rating guidelines and assign detection ratings based on the chances the failure will be detected prior to a product design being released to production (DFMEA) or a process failure would be detected before its consequences would be felt by a customer (PFMEA).

Step 7 Calculate the risk priority number (RPN) for each failure mode by multiplying the product of the relevant severity, occurrence, and detection ratings.

Step 8 Develop an action plan that defines *who* will do *what* and by *when*.

Step 9 Implement the improvements identified by the DFMEA or PFMEA team.

Step 10 Calculate the resulting RPN by reevaluating each of the potential failures after improvements have been made and have had time to take effect, recalculating the relevant severity, occurrence, and detection ratings, and their corresponding RPNs.

Conducting an FMEA is a team effort. The most effective size for an FMEA is between four to six people, with representation from all relevant areas (e.g., manufacturing, engineering, maintenance, materials, information technologies, and so on),

TABLE 6-12 Severity Rating Scale Guidelines

Severity Rating	Description	Impact on Customer Process or Product
10	Critically dangerous	Failure could be life threatening or cause serious injury to a customer or cause total system breakdown.
9	Extremely dangerous	Failure could do irreparable harm to the environment, cause noncompliance with laws or regulations, or cause major or permanent injury to a customer or serious system disruption.
8	Very dangerous	Failure could cause major harm to the environment or render the process or product inoperable or unfit for its intended use.
7	Dangerous	Failure could cause minor injury, quality problems resulting in a high level of customer dissatisfaction, or major system problems requiring significant repairs or rework.
6	Moderate danger	Failure could cause minor injury or a partial malfunction of a product resulting in some customer dissatisfaction, or it could result in subsystem shutdown.
5	Low danger	Failure could result in problems severe enough to generate some customer complaints, or system problems requiring moderate repairs or rework.
4	Very low danger	Failure could cause minor or no injury but an annoyance to the customer and/or result in minor system problems that can be overcome with minor modifications, but a minor loss in performance will result.
3	Minor danger	Failure could create some minor customer dissatisfaction and/or minor system problems that can be overcome with minor modifications with no noticeable loss in performance.
2	Slight danger	Failure would cause no injury and may not be apparent to the customer, and could have little or no effect on the customer's product or process.
1	No danger	Failure would cause no injury and would not be noticeable to the customer and would have no effect on the customer's process or product.

TABLE 6-13 Occurrence Rating Scale Guidelines

Occurrence Rating	Description	Potential for Failure to Occur
10	Critically High - Probability of failure is certain	More than one occurrence per day or a probability of more than 3 failures in 10 events ($C_{pk} < 0.33$).
9	Extremely High - Failure is almost inevitable	One occurrence every three or four days or a probability of between 1 and 3 occurrences in 10 events ($0.33 \leq C_{pk} < 0.67$).
8	Very High: Failures occur repeatedly	One occurrence per week or a probability of between 2 and 5 occurrences in 100 events ($0.67 \leq C_{pk} < 0.83$).
7	High: Failures occur often	One occurrence per month or a probability of between 4 and 10 occurrences in 1,000 events ($0.83 \leq C_{pk} < 1.00$).
6	Moderately High: Failures occur with some noticeable frequency	One occurrence every three months or a probability of between 6 and 30 occurrences in 10,000 events ($1.00 \leq C_{pk} < 1.17$).
5	Moderate: Failures occur occasionally	One occurrence every six months or a probability of between 7 and 50 occurrences in 100,000 events ($1.17 \leq C_{pk} < 1.33$).
4	Moderately Low: Failures occur infrequently	One occurrence per year or 6 occurrences per 100,000 events ($C_{pk} \cong 1.33$).
3	Low: Relatively few failures	One occurrence every one to three years or six occurrences in 10 million events ($C_{pk} \cong 1.67$).
2	Very Low: Failures are few and far between	One occurrence every three to five years or two occurrences in 1 billion events ($C_{pk} \cong 2.00$).
1	Remote: Failures are unlikely	Less than one occurrence every five years and a probability of less than two occurrences in 1 billion events ($C_{pk} > 2.00$).

TABLE 6-14 Detection Rating Scale Guidelines

Detection Rating	Description	Potential for Failure Detection
10	Absolute uncertainty: there is no chance of detection	The product is not inspected or the process is not monitored; or failures cannot be detected using available methods.
9	Very remote chance of detection	Process monitoring is insufficient to detect failures, or product is sampled, inspected, and released based on acceptance sampling methods.
8	Remote chance of detection	Process monitoring is insufficient to detect all but catastrophic failures, or product is accepted is based on no observed defects in any samples taken.
7	Very low chance of detection	Process monitoring is insufficient to detect all but major failures; product defects can be detected using manual inspection, but detection is left to chance.
6	Low chance of detection	Process monitoring is insufficient to detect all but minor failures; product defects can be detected using attributes measurement tools such as go/no go gauges.
5	Moderate chance of detection	Limited statistical process control is used to detect failures in product and process; system of product inspection exists but is based on samples and is not automated.
4	Moderately high chance of detection	There is continual process monitoring and/or 100% inspection of product but it is not automated; SPC is used with adequate response to out-of-control signals.
3	High chance of detection	The application of SPC is widespread and has effectively achieved process capabilities $C_{pk} > 1.33$ for critical operations; process monitoring and product inspection is not automated.
2	Very high chance of detection	The process is continuously monitored and product is 100% inspected using automated methods.
1	Almost certain chance of detection	The failure in process or product is obvious or there is 100% automatic inspection with periodic preventive maintenance and calibration of the inspection equipment, or there are automatic shutdown mechanisms that prevent failures.

and should include personnel who have familiarity with the product design and/or expertise with the production process. Table 6-15 is an example of a convenient format for a worksheet that can be used to guide the FMEA process and help keep the team focused.

Table 6-16 is an abbreviated illustration of a PFMEA applied to the operation of a pressure washer.[34]

[34]Other FMEA examples can be found at http://www.fmeainfocentre.com/examples.htm.

TABLE 6-15 FMEA Worksheet

Process/Product: ______________

FMEA Team: ______________

Team Leader: ______________

FMEA Number: ______________

FMEA Date: (Original): ______________

(Revised): ______________

Page ______ **of** ______

FMEA Process									Actions and Results							
Component and Function	Potential Failure Mode	Potential Effects of Failure	Severity	Potential Cause of Failure	Occurrence	Current Controls	Detection	RPN	Recommended Action	Responsibility and Target Completion Date	Action Taken	Severity	Occurrence	Detection	RPN
Total RPN Before Action									Total RPN After Action						

TABLE 6-16 PFMEA Example

Process/Product:	**Pressure Washer**	**FMEA Number:**	**7-0090**
FMEA Team:	**PW Alpha Team**	**FMEA Date: (Original):**	**1-Mar**
Team Leader:	**Angela P.**	**(Revised):**	**23-May**
		Page	**1 of 1**

FMEA Process									Actions and Results						
Component and Function	**Potential Failure Mode**	**Potential Effects of Failure**	**Severity**	**Potential Cause of Failure**	**Occurrence**	**Current Controls**	**Detection**	**RPN**	**Recom-mended Action**	**Responsi-bility and Target Comple-tion Date**	**Action Taken**	**Severity**	**Occurrence**	**Detection**	**RPN**
Motor	Motor won't start	Cannot operate	8	Low oil level	6	Preventive Maint Check	3	144	Add check oil to start-up Procedure	ABR 6/15	Added to Start-up	8	1	3	24
		Cannot operate	8	Ignition Switch "off"	3	Operator Start-up Procedure	1	24	None			8	3	1	24
		Cannot operate	8	Excess Pressure Build-up	5	None	4	160	Install pressure gauge	CHL 6/15	Installed gauge	8	5	1	40
Pump	Water or oil leak	Loss of Pressure	6	Gasket seal broken	4	None	5	120	Add to PM procedure	HRC 6/15	Added to PM	6	2	2	24
	Pump not building max pressure	Failure to wash as designed	7	Inadequate water supply	3	Operator Start-up Procedure	4	84	Install pressure regulator valve	CHL 6/30	Valve installed	7	2	1	14
Gun/Nozzle Assembly	Excess press. w/gun closed	Can blow main gasket	6	Suction or discharge valves clogged or worn out	5	Preventive Maint Check	7	210	Replace at each PM	HRC 6/15	Changed PM rules	6	2	3	36
Total RPN Before Action								**742**	**Total RPN After Action**						**162**

6.7 GLOBAL 8D PROBLEM SOLVING

The Global 8D (eight disciplines) problem-solving methodology was originated by the U.S. government following World War II as MIL-STD 1520, and was later adopted and improved by the Ford Motor Company. Where the FMEA approach attempts to anticipate and correct problems before they occur, Global 8D deals with the modes and causes of failure after the fact. Once employed, the results of Global 8D can help improve the quality of a future FMEA analysis of a product, process, or service.

Global 8D is a methodology aimed at identifying and eliminating root causes of failure, and then implementing permanent corrective actions to prevent recurrence. The method is also designed to evaluate the effectiveness of control mechanism failures that allowed the problem to escape undetected in the first place. As Figures 6-9a and 6-9b show, Global 8D employs a gateway step (D0) followed by eight discipline steps (D1–D8).

Global 8D is a problem-solving methodology that strives to identify and eliminate root causes of failure after the fact.

- **Gateway Step Discipline D0—Prepare for Global 8D.** This step confirms that the Global 8D methodology is needed. Symptoms are documented showing clearly that a problem has occurred, and a formal assessment is conducted to justify the deployment of resources needed to conduct a full team-based cross-disciplinary problem-solving effort. Immediate damage control in the form of emergency response actions (ERAs) is taken to prevent further undesirable consequences.

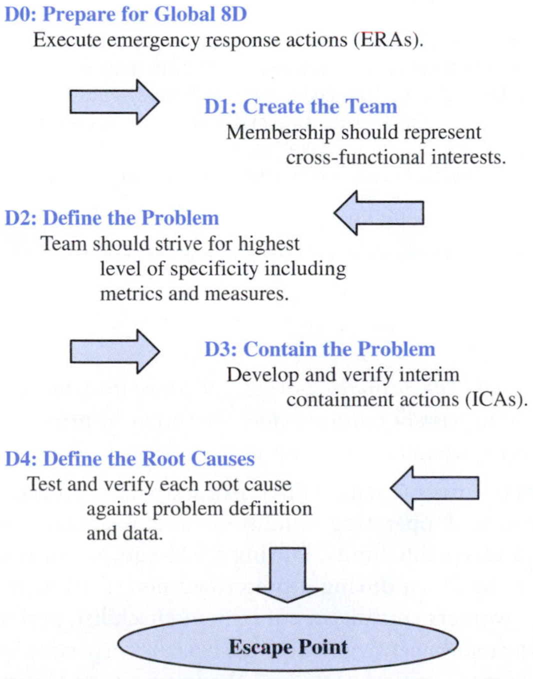
D0: Prepare for Global 8D
Execute emergency response actions (ERAs).
D1: Create the Team
Membership should represent cross-functional interests.
D2: Define the Problem
Team should strive for highest level of specificity including metrics and measures.
D3: Contain the Problem
Develop and verify interim containment actions (ICAs).
D4: Define the Root Causes
Test and verify each root cause against problem definition and data.
Escape Point

FIGURE 6-9a. Global 8D Process: First Five Steps

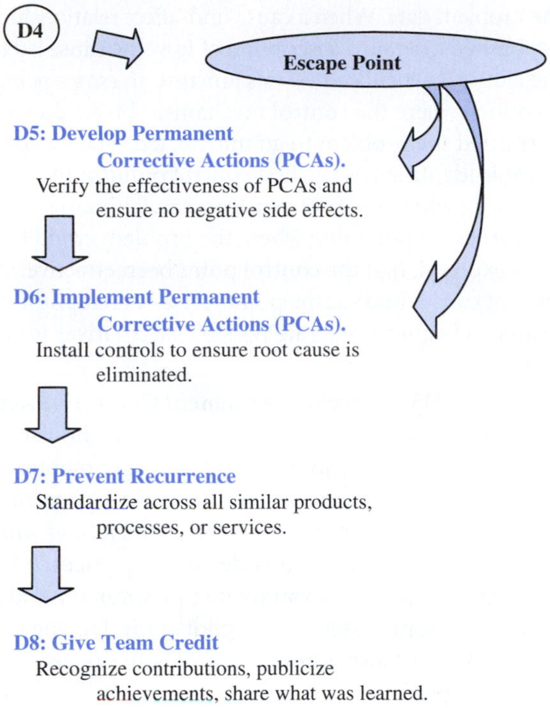
D4
Escape Point
D5: Develop Permanent Corrective Actions (PCAs).
Verify the effectiveness of PCAs and ensure no negative side effects.
D6: Implement Permanent Corrective Actions (PCAs).
Install controls to ensure root cause is eliminated.
D7: Prevent Recurrence
Standardize across all similar products, processes, or services.
D8: Give Team Credit
Recognize contributions, publicize achievements, share what was learned.

FIGURE 6-9b. Global 8D Process: Last Four Steps

- **Discipline D1—Create the Problem-Solving Team.** The problem-solving team should consist of individuals who have the knowledge, time, authority, and skills to solve the problem and to follow through by implementing corrective actions. The team composition should be cross-functional and include representation from all areas that need to be involved in containing, analyzing, correcting, and preventing the problem at hand.
- **Discipline D2—Define the Problem.** In describing the problem the team needs to be as specific as possible, including such details as who suffers if the problem goes unsolved; what is the cost of not solving the problem; when was the problem discovered, how was it discovered, where was it discovered, and by whom; a description of the failure mode and rate; and any metrics and measures relevant to the problem situation.
- **Discipline D3—Contain the Problem.** In this discipline the team places bounds on the problem to contain it. Intermediate containment actions (ICAs) are developed and implemented. The ICAs are to protect customers from the problem until permanent corrective actions can be taken. For example, all production that has been potentially affected by the problem should be isolated for detailed inspection. Those that have been shipped may need to be recalled and those still in the manufacturing facility placed on hold until the problem has been fixed. The team needs to verify the effectiveness of the ICAs.
- **Discipline D4—Identify Root Causes and Escape Point.** During the fourth discipline the problem-solving team identifies all potential causes and gathers as much evidence as possible to reliably test each potential cause against

the problem data. When a cause-and-effect relationship is established a detailed description of how the cause led to the failure is articulated. At this juncture an *escape point* is identified where the control mechanism broke down and permitted the problem to go undetected. The escape point is identified as the first control point in the process following where the root cause occurs. This constitutes the earliest opportunity where the problem could have been exposed, had the control point been effective. Once all root causes and escape points have been identified the team can begin to generate possible alternatives for corrective action.

- **Discipline D5—Develop Permanent Corrective Actions (PCAs).** Discipline D5 is concerned with choosing permanent corrective actions (PCAs) and documenting the rationale for each. The team confirms that the recommended PCAs will in fact solve the problem and will not produce any negative unintended consequences. Implementation of the PCAs can require preliminary evaluation and in some cases a small pilot study. During this step PCAs must also address control issues posed by the escape point.
- **Discipline D6—Implement Permanent Corrective Actions.** In discipline six the team implements the PCAs identified in discipline five. Data are collected showing that the corrective actions put into place are effective in preventing a recurrence of the root cause. This includes a demonstration of how the escape point control mechanism has been improved in its capability for early detection.
- **Discipline D7—Prevent Recurrence.** Preventing recurrence of the problem requires expanding the scope of the PCAs and controls to apply to other similar products, processes, or services. The standardization and deployment of corrective actions across all products or services that might be subject to the same or similar problem leverages the problem-solving effort, which then becomes a preventative and proactive measure across the production facility.
- **Discipline D8—Give Team the Credit.** The last step of the Global 8D process is the formal recognition of the collective efforts of the problem-solving team, and to formally approve its report. Achievements should be widely publicized and the knowledge and learning acquired should be freely shared.

SHOWCASE of EXCELLENCE **Winners of the Malcolm Baldrige National Quality Award Manufacturing Category**

- 1998: Solar Turbines, Inc., the world's largest supplier of midrange industrial gas turbine systems. http://mysolar.cat.com, http://www.quality.nist.gov/Solar_98.htm.
- 1997: 3M Dental Products Division, a Minnesota-based manufacturer of more than 1,300 dental products including restorative materials, crown and bridge materials, dental adhesives, and infection control products. http://solutions.3m.com, http://www.quality.nist.gov/3M_97.htm.
- 1997 and 1991: Solectron Corporation, an electronics design and manufacturing services firm. http://www.solectron.com, http://www.quality.nist.gov/Solectron_97.htm.
- 1996: ADAC Laboratories, a Silicon Valley manufacturer of high-technology healthcare products. http://www.quality.nist.gov/ADAC_96.htm.
- 1995: Armstrong World Industries—Building Products Operations, manufacturer of acoustical ceilings and wall panels. http://www.armstrong.com, http://www.quality.nist.gov/Armstrong_95.htm.
- 1995: Corning Inc.—Telecommunications Products Division, the world's largest manufacturer of optical fibers. http://www.corning.com, http://www.quality.nist.gov/Corning_95.htm.
- 1993: Eastman Chemical Company, manufacturer of over 400 chemicals, fibers and plastics, that show up in hundreds of consumer goods such as beverage bottles, chewing gum, coatings for flooring materials, window cleaners, aspirin, and computer diskettes. http://www.eastman.com, http://www.quality.nist.gov/Eastman_Chemical_Corp.htm.

6.8 SUMMARY OF KEY IDEAS

Key Idea: *The primary purpose of the maintenance and reliability management (MRM) system is to create capacity.*

The purpose of the maintenance and reliability management (MRM) system is to ensure that the organization obtains the maximum value from its plant assets. When equipment runs at less than peak efficiency or breaks down, valuable capacity is lost. The MRM focuses on preventing failures, maximizing the productive efficiencies of plant equipment, and extending the useful lives of all plant physical assets.

Key Idea: *The primary purpose of preventive maintenance (PM) is to prevent failures before they occur by preserving and enhancing equipment reliability.*

Preventive maintenance (PM) uses systematic procedures to determine if operating conditions and machine wear are within acceptable limits. During a PM equipment is scheduled to be down during a prescribed period while maintenance workers consult an equipment check list, performing routine maintenance activities such as the inspection and replacement of critical parts, and changing oils and lubricants. A PM can be used to conduct either a partial or major machine overhaul, which involves tearing down and rebuilding major components or an entire line. There is some debate concerning the cost-effectiveness of this strategy, as PM is labor intensive and may require the unnecessary replacement of parts. An additional concern is that, even with PM, unanticipated catastrophic breakdowns can still occur.

Key Idea: *Predictive maintenance (PdM) is designed to detect and correct potential problems before they occur and result in equipment failures.*

Predictive maintenance (PdM) differs from PM in the fundamental methodology that is used to prevent breakdowns. PM is strictly a time-based approach whereas PdM is condition based. There are a variety of condition monitoring technologies that can be employed, including vibration analysis, infrared thermography, oil analysis and tribology, ferrography, ultrasonics, and motor current analysis. In addition, there are low-tech condition monitoring options, such as continuous operator vigilance that employ the five human senses and statistical tools that can check for trends, patterns, and comparisons to established benchmarks. PM and PdM are not mutually exclusive and can be companion approaches that are implemented together.

Key Idea: *Proactive maintenance (PaM) is designed to prevent failures by identifying and removing the root causes before they have a chance to cause breakdowns.*

As the name implies, proactive maintenance (PaM) takes the lead in eliminating root causes so that failures do not occur, rather than focusing on how the system can effectively respond when a failure does occur. Thus, PaM strives to extend equipment life with the ultimate goal of preventing breakdowns altogether. Under PaM only necessary repairs are carried out and any equipment failure is regarded as unacceptable. PaM is a systems approach to the maintenance function.

Key Idea: *Reliability-centered maintenance (RCM) is an integrated strategy that combines the benefits of all reactive and proactive approaches.*

Reliability-centered maintenance (RCM) is an integrated approach that attempts to optimize maintenance services by drawing on the strengths of RTF, PM, PdM, and PaM programs. The idea is to match each maintenance strategy to the application for which it is best suited, thereby maximizing the deployment of maintenance resources and extending the useful lives of production facilities, equipment, and processes.

Key Idea: The *analytic hierarchy process (AHP) is an effective method for prioritizing a list of pending maintenance projects.*

The analytic hierarchy process (AHP) is a methodology that facilitates team consensus on decisions that involve multiple criteria. The process requires the team members to assign preference scores on a pairwise basis to the criteria as a whole and to repeat the process by assigning preference scores between the pending projects, also on a pairwise basis, against each criterion. AHP has a built-in consistency test that can expose any major errors in logic. The outcomes of the AHP procedure are individual priority vectors that reflect the relative importance of the list of criteria, and a ranking of the projects against each criterion. An overall score vector provides the final prioritization of the list of projects.

Key Idea: *Maintenance performance can be measured at various levels; the level an organization chooses will be closely tied to the maturity of its quality implementation.*

Maintenance performance is a building-block process that has five levels that correspond to an organization's quality maturity. At the first two levels only the direct costs of delivering maintenance services are captured. At the third level, measures are developed to assess how maintenance performance affects the operation and effectiveness of the production system, particularly how well maintenance contributes to operations along the primary value streams. At this level it is common for organizations to use an OEE score for internal comparisons and to compare overall performance with external benchmarks. The fourth level of the hierarchy is system focused and emphasis is placed on how well system goals have been supported by maintenance performance. The fifth and highest level measures sustainability and captures data relative to depreciating values, technological changes, and maintenance strategies.

Key Idea: *Total productive maintenance (TPM) is a lean production tool that uses operators and front-line personnel to identify and solve problems related to equipment downtime.*

Total productive maintenance (TPM) is a team-oriented approach that is part of a participatory style of management. Teams consisting of operations personnel track unproductive time in six categories: downtime, setup time, production stoppages, unscheduled breaks, production time devoted to rework, and waste. These data are then used to compute machine availability, machine efficiency, and production yields, the three factors that are used to compute an OEE score. The two principal goals of TPM, which is integral to quality programs such as Lean Six Sigma, are to drive the defect rate to zero and to eliminate waste.

Key Idea: *Lean maintenance has as its goal the elimination of waste that is related to the maintenance function. Since maintenance is concerned with creating capacity, lean maintenance targets those areas where capacity is lost. This includes throughput, cycle time, and quality—the major factors that impact availability, efficiency, and yield.*

Lean maintenance is the application of the lean philosophy to the maintenance system. To provide the highest quality output at the lowest cost in the shortest possible time, an organization must be able to depend on reliable systems and equipment that will be up and running when needed, and will perform according to equipment vendors' specifications. The lean tools that are widely employed to achieve these aims are the 5S workplace discipline, the seven deadly wastes, kaizen events, jidoka, and just-in-time (JIT).

Key Idea: *To respond to the complex needs of the technological and information age the design of the HRM system must be compatible with the objectives of the MRM system. The culture must embrace a new mindset that is proactive and predictive rather than reactive, and workforce training must build new skills that will enable the workers to accept an expanded set of responsibilities.*

A systems approach to maintenance and reliability management requires new skills and a different approach. A team-oriented approach is employed to improve operational efficiencies and the deployment of resources, thereby reducing costs and extending the useful lives of operational assets. To accomplish these goals, the maintenance teams have to turn from a reaction-to-failure mindset to one that focuses on the long-term improvement of operating capacity.

Key Idea: *A failure mode and effects analysis (FMEA) is a systematic procedure for identifying and preventing problems before they have a chance to occur.*

An FMEA is a structured process for identifying, prioritizing, and correcting all the ways a process or product can potentially fail. All the ways in which a process or product failure can occur are called failure modes, and associated with each failure mode is a failure effect. All effects are not equally likely, so for each failure mode–effect combination a relative risk index, called the risk priority number (RPN), is assigned. The RPN is a composite of three factors: severity, occurrence, and detection. Failure modes with the highest RPNs normally would receive the highest priority, although the severity rating usually trumps the other factors and any failure mode with a high severity rating should receive a high priority regardless of its overall RPN. Following corrective action a revised RPN, which should be less than the original RPN, serves as an indicator that the corrective actions are having a positive impact on reducing the occurrence and severity of failures and/or increasing the ability to detect problems prior to failure.

Key Idea: *Global 8D is a systematic procedure for identifying the root causes after a problem has occurred and developing permanent corrective actions to prevent recurrence of the problem in the future.*

Global 8D is a methodology that is closely related to the engineering (scientific) method and the PDCA improvement cycle. Once a failure has occurred, and the decision is made to allocate resources to preventing its recurrence, a nine-step process attempts to identify and eliminate the root causes and develop permanent corrective actions. The first step, discipline D0, serves as a gateway for the methodology during which the problem symptoms are identified and a case is made to justify the necessary effort to solve the problem. Meanwhile, emergency response actions (ERAs) are taken to immediately curtail the negative consequences associated with the problem. During the next two steps, disciplines D2 and D3, a team is assembled and the problem crisply defined. In discipline D3 the problem is contained by implementing intermediate corrective actions (ICAs) that serve as temporary solutions pending a more permanent fix. The next two disciplines (D4 and D5) identify root causes and the escape point, and then proceed to develop some alternative permanent corrective actions (PCAs) to ensure that the problem does not recur. Discipline D6 is concerned with implementing the permanent solutions, and D7 leverages the improvement effort by expanding the application across the organization to include other similar products, processes, or services. Finally, discipline D8 focuses on an official recognition of the effort and broad publicity of the achievements.

6.9 MIND EXPANDERS TO TEST YOUR UNDERSTANDING

6-1. For each of the following choose the maintenance system (RTF, PM, PdM, PaM, or RCM) that you think is the most appropriate and explain your reasoning.

a. Bicycle
b. Automobile
c. Oil refinery
d. Sawmill
e. Heating and air conditioning system
f. Semiconductor wafer plant
g. Commercial airliner

6-2. A company has three machines that require maintenance attention. Four criteria are used to prioritize these machines to determine the order in which they will be serviced. The criteria are cost of production disruption, condition of the equipment, age of the equipment, and estimated time out of service. The preference scores shown in the following tables have been assigned.

Pairwise Comparison of Maintenance Criteria

Criteria	Cost of Production Disruption	Condition of Equipment	Age of Equipment	Estimated Time out of Service
Cost of Production Disruption	1	5	7	2
Condition of Equipment		1	4	
Age of Equipment			1	
Estimated Time out of Service		3	3	1

Preferences for Machines Against Criteria

Cost of Production Disruption	Machine A	Machine B	Machine C
Machine A	1		
Machine B	6	1	4
Machine C	2		1

Age of Equipment	Machine A	Machine B	Machine C
Machine A	1	3	2
Machine B		1	
Machine C		2	1

Condition of Equipment	Machine A	Machine B	Machine C
Machine A	1	5	3
Machine B		1	
Machine C		2	1

Estimated Time of Service	Machine A	Machine B	Machine C
Machine A	1		
Machine B	5	1	3
Machine C	3		1

a. Compute a priority vector for the criteria as a whole and for the machines against each criterion.

b. Compute the consistency ratio for each of the five pairwise comparison matrices. Comment on your results.

c. Using the priority vectors develop an overall priority list for the three machine projects. Which machine should be serviced first?

6-3. Describe how an FMEA process supports each of the following maintenance approaches.

a. RTF

b. PM

c. PdM

d. PaM

e. RCM

6-4. What role does knowledge play in an FMEA? How is an FMEA related to the PDCA cycle?

6-5. How does FMEA compare to Global 8D? How is Global 8D related to the PDCA cycle?

6-6. For each of the following indicate whether a PFMEA, DFMEA, or Global 8D approach is the most appropriate.

a. A progressive stamping operation is having trouble holding the allowed tolerance for the distance between two holes in the part.

b. Customer feedback has focused attention on the need to improve the reliability of a certain electronic component.

c. Contaminants have been discovered in a critical machine lubricant.

d. The senior management of a large bank wants increased productivity in the processing of loan applications.

e. Competitive pressures suggest that the unit cost of a particular consumer product needs to be reduced by 15%.

6-7. How does lean maintenance differ from the traditional approach under which a maintenance department has the responsibility for all maintenance functions? Which level on the maintenance performance hierarchy would you expect to find a lean maintenance organization? What do you think is the difference between lean maintenance and lean production? Is there a relationship between lean maintenance practices and HPWPs? Explain.

6-8. How does the maintenance performance hierarchy relate to the balanced scorecard approach to measuring organizational effectiveness?

PART THREE

QUALITY AT SEA LEVEL

CHAPTER SEVEN

UNDERSTANDING PROCESS BEHAVIOR—THE PROBLEM-SOLVING TOOLKIT

QUALITY MAXIM #7: *Quality in planning, design, operations, maintenance, and service requires the continuous application of creative problem-solving tools.*

Creativity: Want to Change the World? There's Nothing to It!

A 1971 movie[1] tells the story of Charlie, a poor, kindhearted boy who finds a golden ticket hidden in a chocolate bar thereby winning a tour of the world-renowned candy factory of Mr. Willy Wonka. At the factory Charlie meets up with four other lucky winners and is launched into what proves to be an amazing adventure. As the group starts on the tour, Willy Wonka sets the tone for the day with the song "Pure Imagination."[2] In it he explains that what the group is about to experience will defy explanation, as they are entering a world of pure imagination where, to change the world all one need do is imagine the unthinkable and it will come true.[3] Willy Wonka's mantra is sage advice for quality managers and process improvers. The key to success is the ability to create to innovate—to think beyond the ordinary and imagine the possibilities. This process begins with a profound understanding of *how a process is behaving, why is it behaving that way, how it got that way, how an intervention can modify the behavior, and how it should be modified.* Walt Disney once said, "Disneyland will never be completed. It will continue to grow as long as there is imagination left in the world." The same philosophy can be applied to process improvement, a quest which will never be complete. As long as the power of human minds can collectively generate fresh ideas and invent new methods, those who are united in a common cause will inevitably find a better way to get things done. The process cannot be wasteful or lack purpose—it must be focused and based on knowledge that comes from the application of sound information strategies. When knowledge is combined with the boundless abilities of human thought, seemingly insurmountable problems can be scaled down to manageable proportions.

[1] *Willy Wonka and the Chocolate Factory* is a 1971 film adaptation of the 1964 novel *Charlie and the Chocolate Factory* by Ronald Dahl. The film, directed by Mel Stuart and starring Gene Wilder in the title role, received an Academy Award nomination for Best Original Score.
[2] Written and composed by Anthony Newley and Leslie Bricusse for the film *Willy Wonka and the Chocolate Factory.*
[3] The music and lyrics for "Pure Imagination" can be accessed at http://kids.niehs.nih.gov/lyrics/pureimag.htm.

LEARNING OBJECTIVES AND KEY OUTCOMES

After careful study of the material in Chapter 7, you should be able to:

1. Organize and participate in structured brainstorming and building group consensus, including the nominal group technique (NGT).
2. Demonstrate a basic understanding of the TILMAG process and the use of word associations to stimulate new ideas.
3. Perform a moment-of-truth analysis.
4. Fully explain the PDCA cycle as a planning tool to guide improvement efforts.
5. Construct and explain the use of a radar chart.
6. Use the task cycle process to plan and construct an effective meeting.
7. Demonstrate the use of a check sheet to record raw data.
8. Conduct and properly interpret a Pareto analysis.
9. Construct and use a cause-and-effect diagram.
10. Construct and interpret a histogram from raw data.
11. Understand the benefit of stratifying data.
12. Plot and interpret a scatter diagram and a run chart, being able to identify nonrandom patterns in the data.
13. Define special and common cause sources of variability and describe the use of control charts to differentiate between the two.
14. Define the three basic reasons for collecting control chart data and how each of the three purposes impacts the application of statistical process control tools.
15. Construct and use Gantt charts for monitoring and managing project schedules.
16. Describe the steps required to construct an affinity diagram and explain its use.
17. Describe the steps required to construct a relations diagram and explain its use.
18. Describe the process required to develop a systematic diagram and how it is used.

(continued)

19. Use a matrix format to organize raw data into a structure that facilitates analysis.
20. Identify the principal uses for a process decision program chart and describe the process for constructing one.
21. Construct an arrow diagram to describe a project plan and use it to conduct a PERT analysis.
22. Construct and use flowcharts to analyze process behaviors.
23. Construct a knowledge process map.
24. Describe the ideas behind concept knowledge maps and competency knowledge maps and how they are used.
25. Demonstrate an understanding of the concept behind poka-yoke and how this idea is used to mistake-proof processes and product designs.
26. Describe how a variance-tracking matrix can be used to help identify root causes of variation in a process.
27. Describe how the six honest servants and five whys can be useful in helping stimulate improvement ideas.

The creation of new ideas is facilitated when people work together contributing different viewpoints in a mutually respectful brainstorming environment.

7.1 TOOLS FOR CREATIVITY

7.1.1 Brainstorming and Brainwriting

Structured and unstructured **brainstorming** is used by groups to verbally generate a large number of new ideas within a short period of time, normally with the use of a group facilitator to keep things moving along in an orderly fashion. When ideas are written down, rather than shared verbally, the technique is called **brainwriting**. However, the term *brainstorming* is generically used to refer to either technique.

Brainstorming is used by groups to verbally generate a large number of new ideas within a short period of time.

The difference between structured and unstructured brainstorming/brainwriting is that the former requires each person in turn to contribute an idea when called upon; in the latter case ideas are spontaneous and no one is forced to participate. To be effective the following brainstorming rules of conduct should be followed.

- Free expression, called freewheeling, is encouraged by all.
- All ideas are welcome.
- Criticisms or critiques are strictly forbidden.

There are two types of brainwriting: *nominal* in which ideas are not shared while being generated, and *interacting* where each idea is passed along to the person on the right who has the option of modifying the idea or building on it with a totally new idea. Experience has shown that interactive brainwriting *always* increases the quantity of ideas over nominal brainwriting and structured or unstructured brainstorming.

7.1.2 Nominal Group Technique

The **nominal group technique** (NGT) is a brainwriting method that helps a group reach consensus on the prioritization of a list of ideas or actions. NGT is particularly well suited for fact finding, idea generation, or problem solving. NGT sessions should never be rushed—it is better to have one long meeting than to break up the time into several shorter meetings. One member of the group is normally assigned the role of facilitator and another assumes the duties of group recorder. The facilitator kicks off the session by reading the issue at hand to the group and explaining the following eight steps of the NGT process.

NLT is used to build group consensus.

Step 1 Silent idea generation. During a period of approximately 20 minutes members of the group independently write down as many ideas as they can think of relating to the issue at hand.

Step 2 Round robin. Going around the group each participant in turn is asked to present, but not discuss, one idea. The group recorder captures the data on a flip chart. The process is repeated until all ideas have been exhausted and recorded.

Step 3 Discussion. The facilitator reads each idea to the group and asks if there is a need for clarification, interpretation, or explanation. Brief discussions are permitted and similar ideas are combined whenever possible. When this step is finished, the recorder assigns a unique identifying number to each item on the final list.

Step 4 Preliminary vote. The facilitator asks each participant to select five to seven ideas from the list and write these on 3-by-5-inch index cards—one idea per card. Each participant is then asked to rank the cards in order of importance and to record his or her rankings on each card. Cards are collected and the votes recorded on a flip chart.

Step 5 Break. The facilitator calls for a 10-minute break.

Step 6 Discussion. The facilitator leads the group through a 30-minute discussion of the preliminary vote to identify any inconsistent voting patterns. Focused discussion can be reopened on ideas that appear to have received too few or too many votes.

Step 7 Final vote. The facilitator calls for a final vote on priorities. At this point, step 4 can be repeated, asking participants to again select and rate the top five to seven ideas. As an alternative, participants can be asked to assign a score on a scale of 0 (not important) to 10 (most important). There is no limit to how many ideas a person can give the same score (for example, all ideas on the list could receive a score of 10 if the person thought they all merited top priority).

Step 8 Listing items in priority order. The facilitator collects all scorecards and calculates the aggregate score for each of the items, which are then recorded next to their corresponding item numbers on the flip chart.

7.1.3 Associations and TILMAG

Word or picture associations and biotechniques use words, pictures, and examples from nature to stimulate fresh ideas. These examples are typically unrelated to the issue at hand, and the process is initiated with a list of words, group of pictures, or in the case of biotechniques, some biological system (either animal or vegetable). The idea is to break stereotypes and force the group to look at some issue in totally new ways.

TILMAG is an acronym that is made up of the first letters of the German words that when translated into English mean "transformation of ideal solution elements with a common associations matrix."[4] **TILMAG** uses brainstorming together with word associations to help groups identify solutions by analyzing paired combinations of ideal solution elements.[5] Table 7-1 provides an illustration of how TILMAG was used to create new ideas for increasing market share. The first step in this process is to brainstorm ideas and then, as a group, to identify those attributes, qualities, or properties that must be present in any solution or recommendation.

In the example shown, the group decided that acceptable solutions would have to produce superior quality, yield low-cost and short cycle times, and result in happy customers. These attributes are called the problem's ideal solution elements (ISEs). The group jointly constructed a TILMAG matrix in which the ISEs were placed across the top row and down the left column and all pairwise associations were investigated.

For each pairwise association the group employs picture and word associations to find a person, animal, object, place, event, or activity that represents each association. The final step is to use the TILMAG matrix to stimulate ideas for an analogy chart that links back to the original issue. In this example, the team concluded that a solution based on QIM, better training, and a supportive culture with high-performing work practices would help increase market share.

> **TILMAG is a technique designed to break stereotypes and force a group to look at an issue in a totally new way.**

7.1.4 Moment-of-Truth Analysis

A **moment-of-truth analysis** is a procedure for quickly analyzing the type and quality of responses that result from instantaneous customer contacts. When used by a quality team, this tool can highlight procedures that need improving. There are three steps to the analysis.

Step 1 The team creates a list of all contacts that customers have with the organization. It is sometimes helpful to construct a flowchart to help identify moment-of-truth opportunities. If feasible, involving customers in this step can sometimes provide perspectives that might otherwise be overlooked. Figure 7-1 illustrates a line of questioning that can be useful to stimulate ideas to identify the moments of truth that customers encounter from the time they enter to the time they exit a system.

> **A moment-of-truth analysis tracks how the organization responds to customer contacts.**

Step 2 From the moment-of-truth contacts list, a Truth Cases Map, such as the one shown in Table 7-2, can be created. For each customer contact, the map provides the current and desired situations.

Step 3 As the final step, the team compares the current and desired situations and formulates an action plan to close any observed gaps.

7.2 TOOLS FOR PLANNING

7.2.1 PDCA Cycle

We discussed the **Plan-Do-Check-Act (PDCA)** cycle in Chapter 5, explaining that this is a structured method engineers can use to guide improvement efforts. We briefly revisit this tool here because the four-step process it represents is fundamental to QIM. First appearing as Figure 5-3 the graphical depiction of the PDCA cycle is reintroduced here as Figure 7-2 to reemphasize the significance of each of the four steps.

- **Plan—What do we want to know?** To emphasize its relative importance, in Figure 7-2 we have deliberately allocated more area to the *Plan* step than

> **The Plan step of PDCA sets the agenda and raises relevant questions.**

---

[4] *Transformation idealer Lösungselemente mit Assoziationen und Gemeinsamkeiten,* created by Helmut Schlicksupp.

[5] Ritter, Diane, and Michael Brassard. *Memory Jogger: A Pocket Guide to Creative Thinking.* Salem, NH: GOAL/QPC, 1998.

TABLE 7-1 An Example TILMAG Process

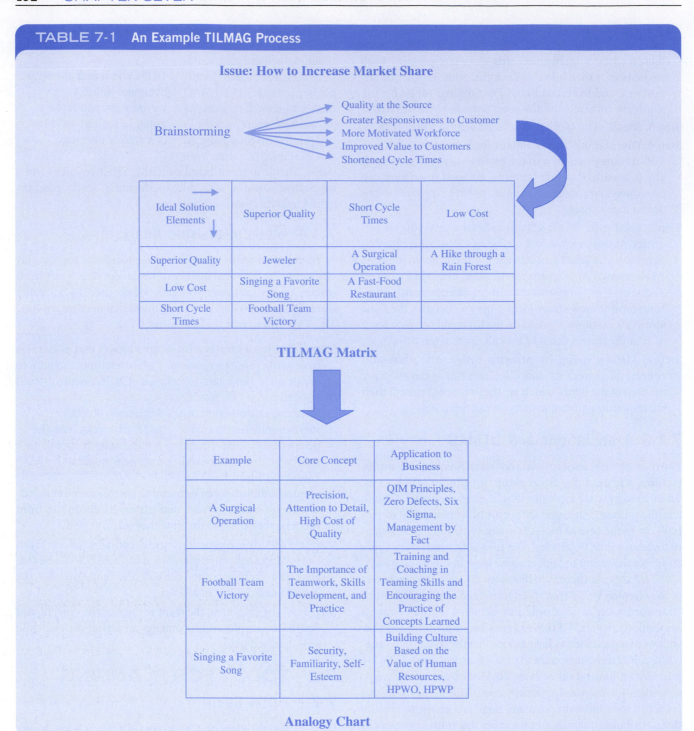

Issue: How to Increase Market Share

Brainstorming →
- Quality at the Source
- Greater Responsiveness to Customer
- More Motivated Workforce
- Improved Value to Customers
- Shortened Cycle Times

Ideal Solution Elements	Superior Quality	Short Cycle Times	Low Cost
Superior Quality	Jeweler	A Surgical Operation	A Hike through a Rain Forest
Low Cost	Singing a Favorite Song	A Fast-Food Restaurant	
Short Cycle Times	Football Team Victory		

TILMAG Matrix

Example	Core Concept	Application to Business
A Surgical Operation	Precision, Attention to Detail, High Cost of Quality	QIM Principles, Zero Defects, Six Sigma, Management by Fact
Football Team Victory	The Importance of Teamwork, Skills Development, and Practice	Training and Coaching in Teaming Skills and Encouraging the Practice of Concepts Learned
Singing a Favorite Song	Security, Familiarity, Self-Esteem	Building Culture Based on the Value of Human Resources, HPWO, HPWP

Analogy Chart

to the others. It is essential that decision makers invest enough time and energy in planning, which runs counter to the quality engineer's penchant for getting on with the "doing." During the planning step, the problem statement should be clearly articulated and as many questions generated as possible. The group then develops a plan on how it intends to get their questions answered. This will typically include a strategy for collecting data, and will specify who will do what and how it will be done.

- **Do—What can we learn and from whom?** The *Do* step is essentially concerned with implementing the plan developed during the Plan step. This can entail a broad range of activities from simple data collection to performing a designed experiment.

The Do step of PDCA executes the plan and answers the relevant questions.

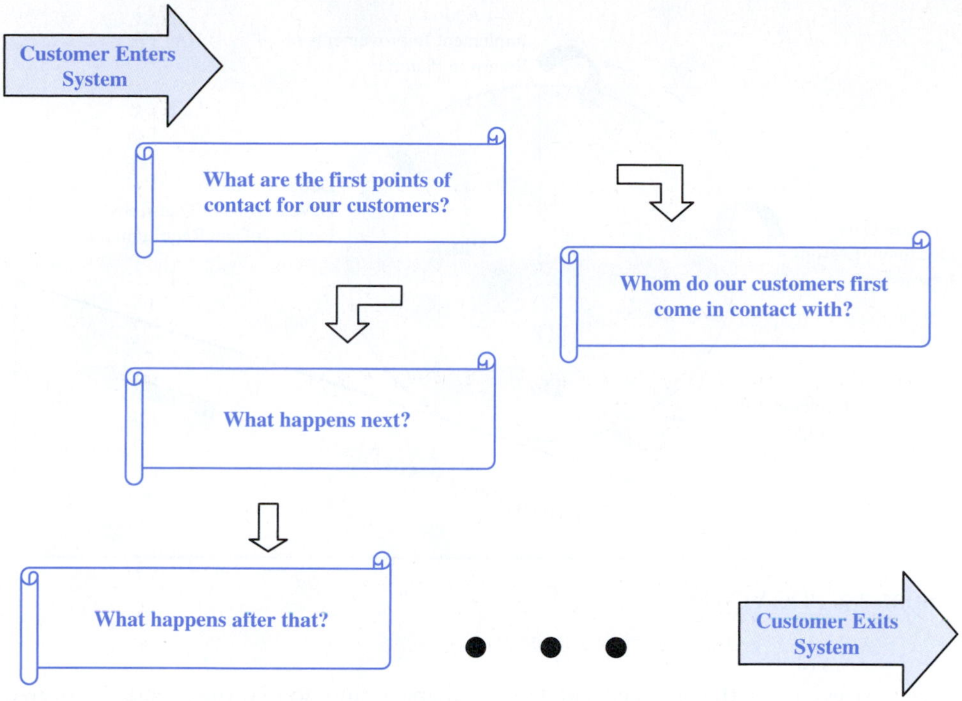

FIGURE 7-1. Moment-of-Truth Analysis

TABLE 7-2 Truth Cases Map for Hotel Service

	Truth Cases Map	
Customer Contacts	**Current Situation**	**Desired Situation**
Speaks with agent on phone to make reservation	Busy at times, occasionally rude, customer placed on hold	Friendly, helpful, welcoming, responsive, immediate attention
First sight of hotel exterior	Some peeling paint, untidy gardens	Immaculate gardens, flowers, manicured lawns, fresh paint
Greeted by porter	Efficient but brusque, personal appearance could improve	Welcoming, friendly, helpful, courteous, well-groomed, neat
First encounter with check-in staff	Long wait for service, personal appearance of staff could improve, reservations sometimes lost or special requests not fulfilled	Immediate service, tidiness and uniformity of appearance, attention to detail, go the extra mile to satisfy request no matter how unusual
First sight of room	Occasional shortages, obvious shortcuts on cleaning, instances of shortage of towels, soap, etc.	Meets expectations as advertised, clean, well stocked, well maintained, bathrooms spick and span
Experience with housekeeping staff	Guests staying over have to wait until late in day for room to be made	Room made up early in day, restocked, squeeky clean
Room service	Delays in service, errors in orders, cold food	Timely, courteous, accurate, food served hot and tasty
Dines in hotel restaurant	Limited menu, pricey, wait for tables, excellent food quality	Variety, competitive prices, timely service, good food
Encounter with check-out staff	Wait for check-out, mistakes in billing, discourteous at times	Express check-out, accurate account, courteous, friendly

- **Check or study—What did we learn?** In the *Check* step, the data are analyzed against the issues and questions raised during the Plan step. The learning that takes place will either provide the basis for action or generate a new set of issues and questions.

The Check step of PDCA compiles and evaluates lessons learned.

- **Act—Act on what was learned.** In the *Act* step, actions can include piloting a new method, implementing new controls, solving the stated problem, or simply recognizing the team was trying to answer the wrong set of questions or the set was incomplete. In the

The Act step of PDCA takes appropriate action based on what was learned.

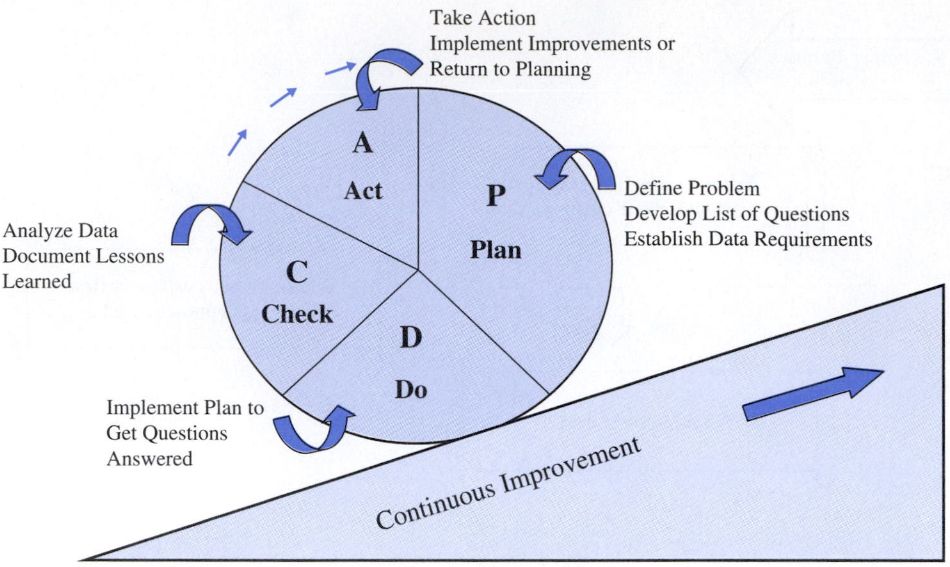

FIGURE 7-2. PDCA Cycle

latter case, the process cycles back to the Plan step and the process repeats.

By iterating through the PDCA cycle, a team learns together and builds a collective problem-solving capability. The quality and depth of the questions under Plan improve, as do the analytical skills used in Do and Check. Hence, the PDCA cycle not only provides a structured process that facilitates problem solving, but it also serves as a framework for enabling the development of teaming and learning skills.

7.2.2 Radar or Spider Chart

A **radar chart,** sometimes referred to as a **spider chart,** is a tool that graphically displays the gaps and their relative magnitudes among 5 to 10 performance areas. This chart helps a team visualize the areas of obvious strengths and weaknesses, and can be useful in strategic planning and quality improvement. An example radar chart is shown in Figure 7-3.

To construct a radar chart, a large circle is drawn and a spoke is constructed from the origin to the perimeter corresponding to each performance category. Each spoke is then labeled with the name of the category and subdivided into the number of increments the team has established for its rating scale. A numerical scale is normally used with numbers that range between 0 (serious gap) to 5, 6, or 7 (exceptional strength).

A radar chart helps decision makers visualize strengths and weaknesses.

7.2.3 Task Cycle

How to run an efficient meeting can be a challenge. Usually the time available for a meeting is limited and, if unproductive, considerable resources can be wasted. It is essential to manage time so that the agenda is covered, everyone has an opportunity to be heard, and, the meeting ends with expected outcomes. Having the group agree ahead of time to a set of norms, similar to the ones shown in Figure 7-4, can help if these are strictly enforced by the group's members. The **task cycle,** illustrated in Figure 7-5, is a preplanning tool that can help keep a meeting on purpose and achieve desired outcomes within the allotted time. The four components of the cycle are purpose-products-process-functioning capability. The *purpose* refers to why the meeting has been convened (i.e., what does the group expect to accomplish by calling this particular meeting at this designated time?). The *products* are included on a concise list of desirable outcomes from the meeting and can serve as the basis for judging whether the meeting was successful. The *process* defines how the meeting will be conducted including the meeting agenda, who will preside, what rules will be in place (e.g., norms), and, if appropriate, assignments for specific agenda items. Finally, the *functioning capability* includes all those resources necessary for the meeting to be a success, including any handouts, specialized equipment, suitable venue, and beverage or food items.

A task cycle helps in planning and conducting an efficient meeting.

7.3 THE BASIC TOOLKIT FOR PROBLEM SOLVING: SEVEN OLD TOOLS

7.3.1 Introduction

Seven basic tools have been universally applied to the control of quality. Because the use of these tools has been prevalent through all (first, second, and third) generations of the

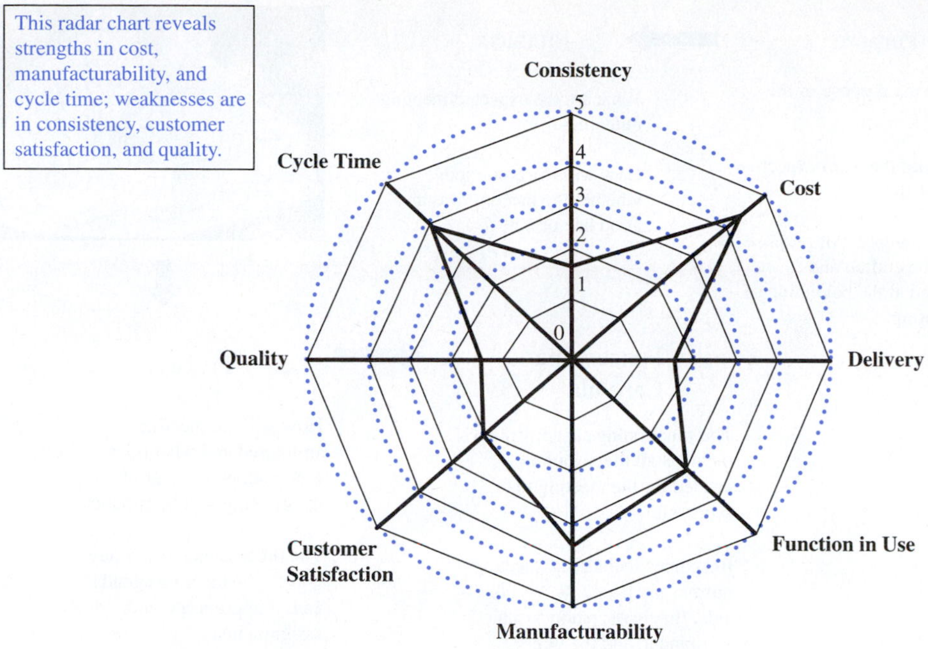

FIGURE 7-3. Radar Chart

1. Candor
We value the principle of expressing what is on our minds in an open environment which is conducive to furthering discussion or generating new ideas.

2. Confidentiality
We value the principle of protecting each individual's right to privacy in order to remove any fears associated with free expression.

3. Equality
We value the principle that bestows peer equity to all members of the group.

4. Truth
We value the principle of seeking and expressing the truth; always clearly articulating any assumptions.

5. Data
We value the principle of seeking to back up all decisions, assumptions, plans, and positions with relevant and reliable data.

6. Learning
We value the principle of learning together; there is no such thing as a stupid question or comment.

7. Respect
We value the principle of holding each of our fellow team members in high esteem; all opinions, including counteropinions are welcome.

8. Listening
We value the principle of listening to both the words and meanings of what others have to say, without being deafened by our eagerness for self-expression.

9. Constructive Participation
We value the principle of constructive criticism, which helps solve a problem or strengthen a weakness; we are intolerant to destructive criticism, which is aimed at blocking or diverting a group momentum or discrediting or belittling individual participants.

10. Teamwork
We value the principle of working together and forming community ideals and objectives; we value the absence of individual competition and the need to impress or succeed at an individual level.

11. Positive Attitude
We value the principle of looking beyond the barriers and constraints, focusing instead on possibilities and opportunities.

FIGURE 7-4. Representative Group Norms

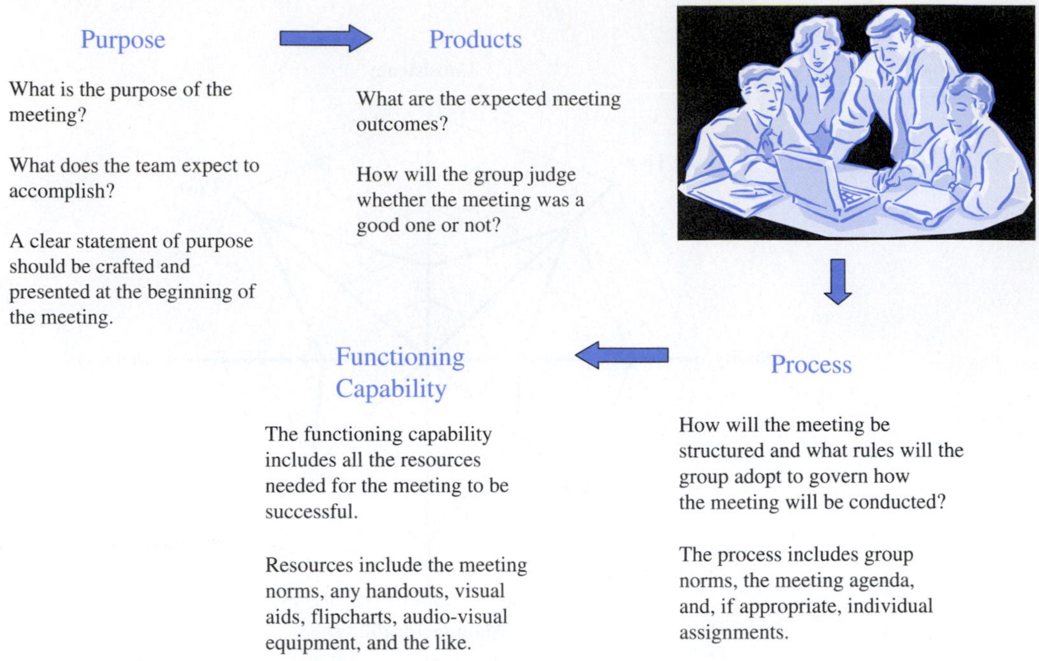

FIGURE 7-5. Task Cycle for Planning an Efficient Meeting

quality movement they have been tagged the "seven old tools"[6] and include

- Check sheet
- Pareto chart
- Cause-and-effect diagram
- Histogram
- Stratification
- Scatter diagram
- Graph or chart, with specific emphasis on the control chart

7.3.2 Check Sheet

Before a problem can be analyzed it needs to be understood. A check sheet, normally in handwritten form on paper, is used to record raw data in a format that can be easily understood by everyone. As an example, Figure 7-6 is a check sheet that was used for categorizing defect types in printed materials. The inspector in this case examined 5,560 printed pages, randomly selected from production skids of a magazine produced over a six-week period.

A check sheet is used to manually record data.

Figure 7-7 is an example of a check sheet that was used to tally up the incidences of line stoppages by machine and shift. An inspector simply made a tally notation on the form each time a stoppage occurred over a 24-hour period.

[6]Ishikawa, K. *What Is Total Quality Control? The Japanese Way*. Upper Saddle River, NJ: Prentice Hall, 1985.

7.3.3 Pareto Chart

The Pareto principle posits that only a few causes (the vital 20%) are responsible for the majority (80%) of problems. Improvement benefits can be leveraged by focusing attention on the key issues (that is, the 20%), and while looking at critical factors it is not uncommon to discover and resolve many of the other lesser important problems by default. When check sheet data are plotted on a **Pareto chart,** the most important problems are revealed. It is customary to plot a pair of graphs—a bar graph that displays item percentages sorted in descending order, and a line graph that plots the cumulative percentage of items on the sorted list. These two graphs are then plotted on the same chart.

Figure 7-8 shows a pair of Pareto charts for the printing operation data presented in Figure 7-6. The *significant (vital) problems* can be separated from the trivial ones by extending a horizontal line from the 80% point on the *y*-axis over to the line graph, and then dropping a vertical line perpendicular to the *x*-axis. In this case it is clear that improvement efforts should target problems A, B, and C (i.e., low density and misregister problems).

A Pareto chart helps separate the significant few from the trivial many.

All types of problems do not necessarily have an equal impact on quality. A variant of the Pareto procedure assigns weights to items according to their relative importance. Figure 7-9 is a weighted Pareto chart showing the incidences of different types of defects in automobile tire manufacturing.

Check Sheet for Quality Problems	
Date: *06/03/2004*	Collected by: *J. Doe*
Description of Operation: *Printing*	
Problem	Frequency
Density Too High—Black Ink	*13,10,9,3,2 = 37*
Density Too Low—Black Ink	*17,6,70,7,7 = 107*
Density Too High—Colors	*7,9,18,35,8 = 77*
Density Too Low—Colors	*242,168,70,21,105 = 606*
Misregister	*60,47,105,140,140 = 492*
Nip Marks	*14,11,14,14,8 = 61*
Scratches	*8,8,6,3,5 = 30*
Incorrect Fold	*2,3,4,1,1 = 11*
Wrinkles	*3,15,8,2,1 = 29*
Scumming	*13,13,9,11,4 = 50*
Piling	*3,4,3,2 = 12*
Scuffing, Belt Marks	*14,7,3,4,7 = 35*
Severe Set-Off	*14,4,7,2,1 = 28*

FIGURE 7-6. Example Check Sheet

Tally of Downtime by Shift and Machine				
Machine	First Shift	Second Shift	Third Shift	Total
Red	/	//	卌 卌 /	14
Green		/	卌 ///	9
Blue	/		卌 卌	11
Gold			卌 卌 //	12
Silver	//		卌 卌 卌	17
Bronze			卌 卌 /	11
Total	4	3	67	74

FIGURE 7-7. Check Sheet for Source of Downtime Problems

Tire defects have been categorized into four severity levels as follows.

- Level 4 severity (the worst) problems are those that result in having to scrap the product. Level 4 problems have been assigned a severity index (weight) of 40.
- Level 3 defects are "blems." A tire with a blem is functional and safe, nevertheless imperfect, and must be sold at a substantial discount. Level 3 defects have been assigned a severity index of 10 or 15.
- Level 2 defects are repairable but costly. Defects in this category have been assigned an index of 3 or 5.
- Level 1 defects (the lowest level) represent a class of minor repairables and have been assigned weightings of 1 or 2.

To understand the impact of the weighted Pareto chart, consider the level 4 defect "wrinkled fabric." This defect ranks

Problem	Total	Code	Percentage	Cumulative Percentage
Density Too Low—Colors	606	A	39.0	39.0
Misregister	492	B	31.7	70.7
Density Too Low—Black Ink	107	C	6.9	77.6
Density Too High—Colors	77	D	5.0	82.6
Nip Marks	61	E	3.9	86.5
Scumming	50	F	3.2	89.8
Density Too High—Black Ink	37	G	2.4	92.1
Scuffing Marks	35	H	2.3	94.4
Scratches	30	I	1.9	96.3
Wrinkles	29	J	1.9	98.2
Set-Off	28	K	1.8	100.0
Totals	1,552		100	

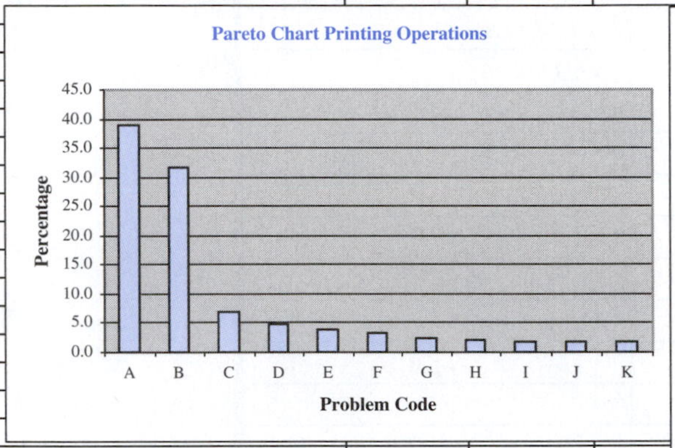

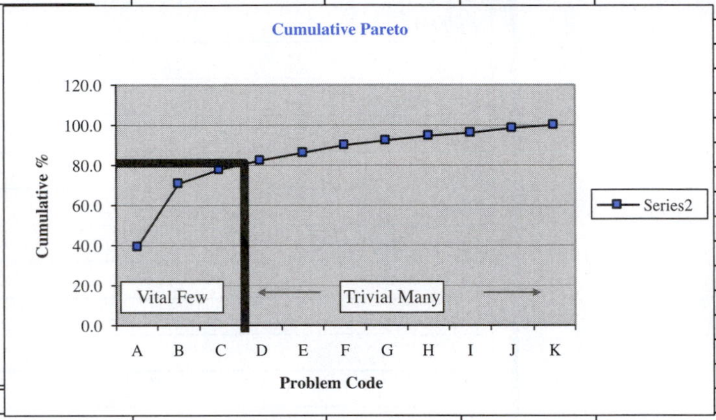

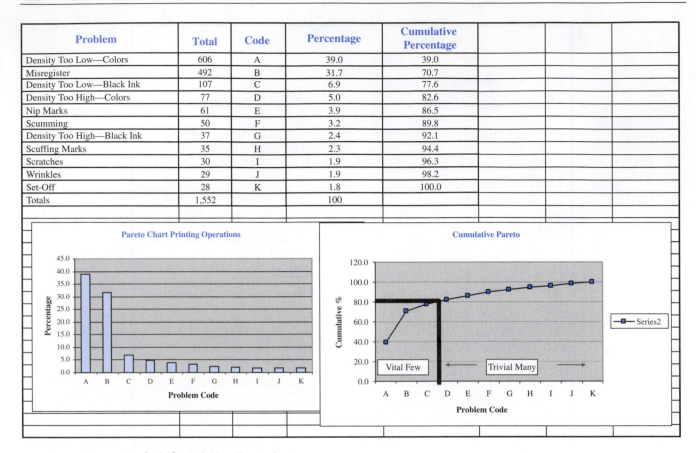

FIGURE 7-8. Pareto Analysis for Printing Operation

Defect Type	Total	Severity Index	Weighted Count		%	Cumulative Percentage
Snaked Component	134	10	1340	L	12.9	12.9
Wrinkled Fabric	32	40	1280	A	12.4	25.3
Exposed First Ply	121	10	1210	K	11.7	37.0
Wrong I.D.	76	15	1140	I	11.0	48.0
Wide Liner Splice	203	5	1015	P	9.8	57.8
Open S/W Splice	22	40	880	E	8.5	66.3
Open Liner Splice	72	10	720	J	6.9	73.2
Off-Centered Bead	102	5	510	M	4.9	78.1
Wrong Component	22	15	330	H	3.2	81.3
Exposed Ply Splice	6	40	240	D	2.3	83.6
Misplaced Splice/Spotting	45	5	225	N	2.2	85.8
Humped W/W Splice	71	3	213	R	2.1	87.8
Flipback Turn-up	5	40	200	B	1.9	89.8
Foreign Material	11	15	165	G	1.6	91.4
Open Ply Splice	4	40	160	C	1.5	92.9
Wrinkled Component	68	2	136	W	1.3	94.2
Wide Ply Splice	24	5	120	O	1.2	95.4
Open W/W Splice	34	3	102	S	1.0	96.4
Split Ply	2	40	80	F	0.8	97.1
Torn or Jagged S/W Edges	25	3	75	T	0.7	97.9
No Builder Number	50	1	50	Z	0.5	98.3
Loose Chafer	44	1	44	Y	0.4	98.8
Air between Plies	19	2	38	V	0.4	99.1
Humped S/W Splice	12	3	36	Q	0.3	99.5
Blister Under S/W	13	2	26	U	0.3	99.7
Miscellaneous	23	1	23	AA	0.2	100.0
Wrinkled Chafer	2	2	4	X	0.0	100.0
	1,242		10,362		100.0	

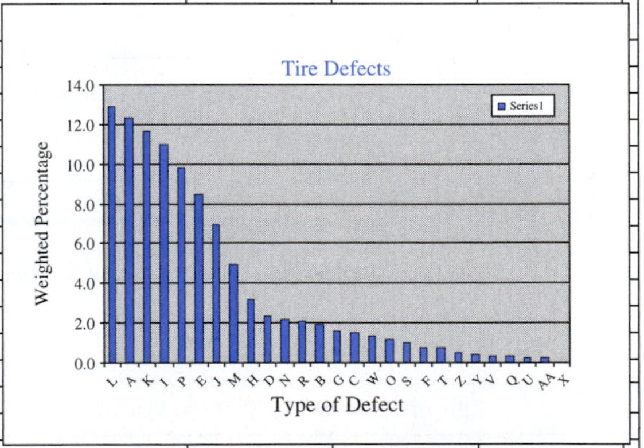

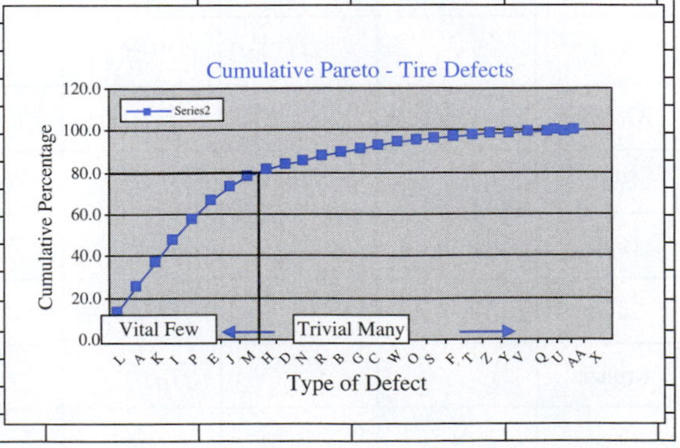

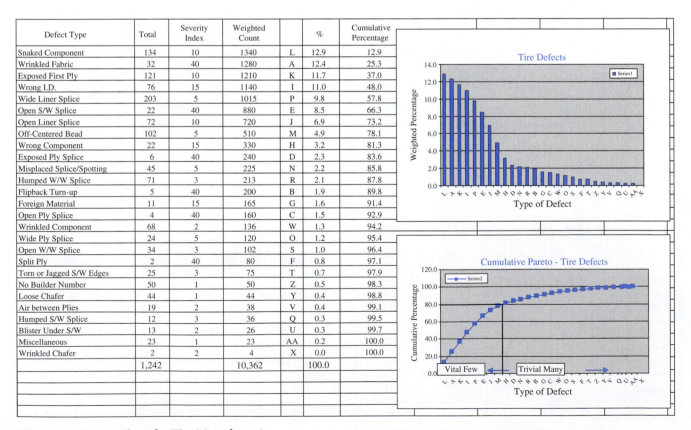

FIGURE 7-9. Pareto Chart for Tire Manufacturing

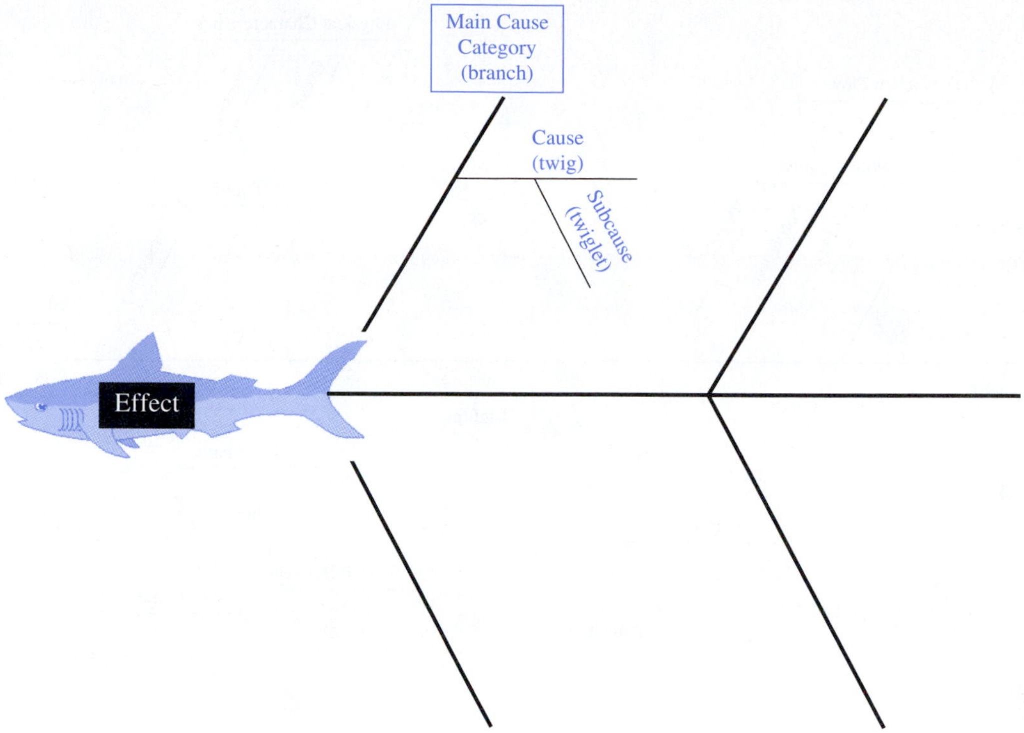

FIGURE 7-10. Fishbone Diagram Structure

19th out of 29 in frequency of occurrence, but 2nd in weighted importance. The Pareto analysis reveals that by focusing attention on only 10 of the 29 types, all major sources of poor quality will be addressed.

7.3.4 Cause-and-Effect Diagram

The **cause-and-effect (CE) diagram** is also called a **fishbone diagram** (due to its similarity to the skeletal structure of a fish), and an **Ishikawa diagram** (in honor of its founder).[7] Once a team decides which problem it wants to solve, possibly from a Pareto analysis, the CE diagram can help it identify candidate causes.

The CE structure is illustrated in Figure 7-10. To construct a CE diagram, the problem, or *effect*, is placed in a box (analogous to the head of the fish). A horizontal line (the backbone) is drawn from the box, and from the backbone angled fishbones are inserted corresponding to each main problem category. This forms the skeleton of the fishbone diagram.

A cause-and-effect diagram helps a team focus on root causes.

Main categories can be anything relevant to the problem, but the typical ones include *materials, methods, personnel*, and *machines*. From each main stem smaller bones (or twigs) are constructed for each candidate cause, and from these, smaller bones (or twiglets) representing subcauses are drawn. The diagram can include as many sublevels as required to get to the root causes.

Two teams in a large metropolitan daily newspaper prepared the fishbone diagrams shown in Figures 7-11 and 7-12. One team was trying to find ways to reduce the number of ads that had to be reworked after the initial layout. The other team was attempting to learn why there were discrepancies between their measurements of paper basis weight and that of their suppliers.

The CE diagram adds structure to brainstormed ideas. It is important that the team understands that all "causes" shown on the diagram are merely *candidates* and have not been proven to be true causes of the stated effect. Consequently, it is essential that all causes suggested by team members should be listed, no matter how bizarre, zany, or unlikely.

A cause-and-effect diagram adds structure to brainstorming.

When the team is satisfied that most of the likely causes have been captured, the next step is to narrow the list to just a few for detailed study.

Four guidelines can be helpful in achieving this end.

- **Guideline 1.** Separate the causes into those that the team can potentially control and those beyond its control. A team in one organization stratified its causes into two fishbones that resembled a pair of kissing fish, as illustrated in Figure 7-13. The fishbone on the right contained those causes that were beyond the group's control,

[7] Ishikawa, K. *Guide to Quality Control.* Tokyo: Asian Productivity Organization, 1982.

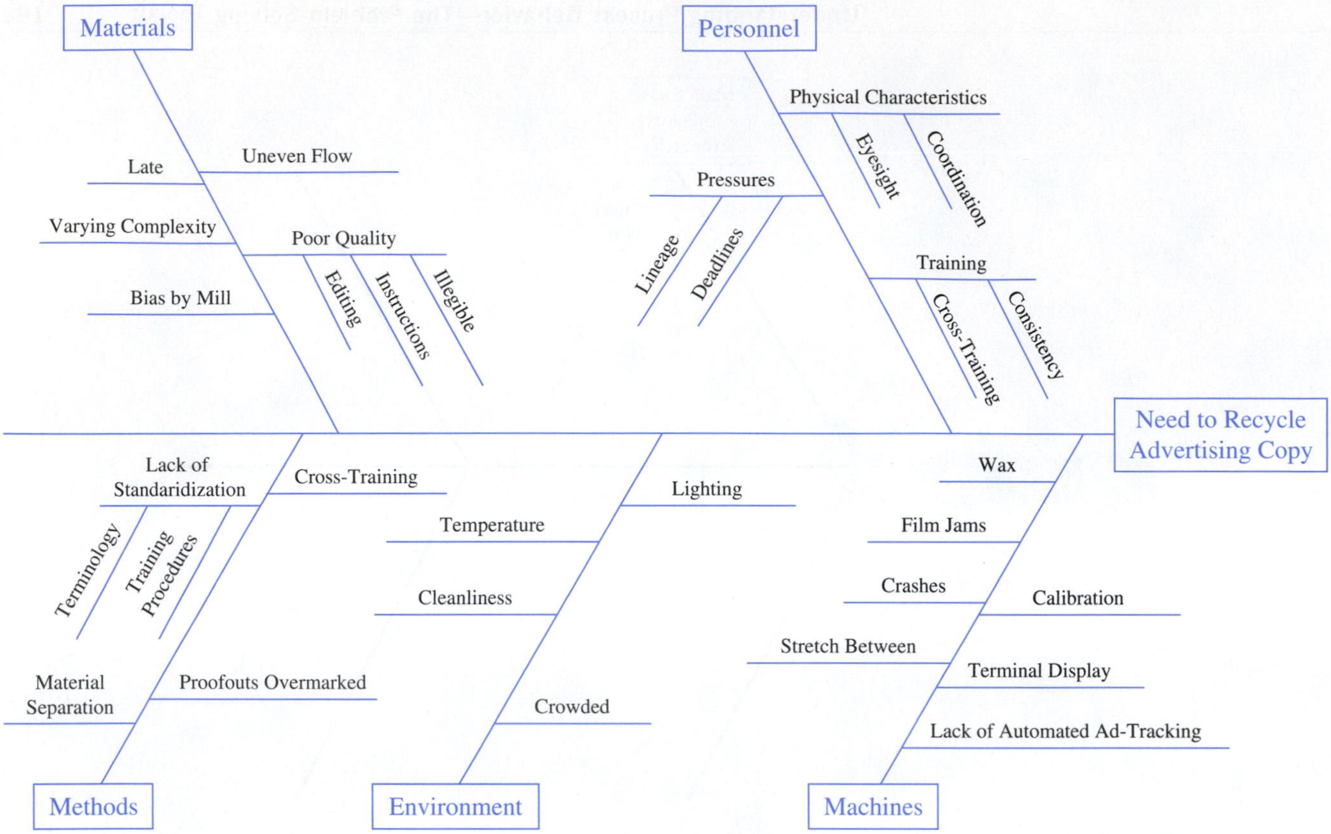

FIGURE 7-11. Fishbone Diagram for Advertising Copy Recycling Problem

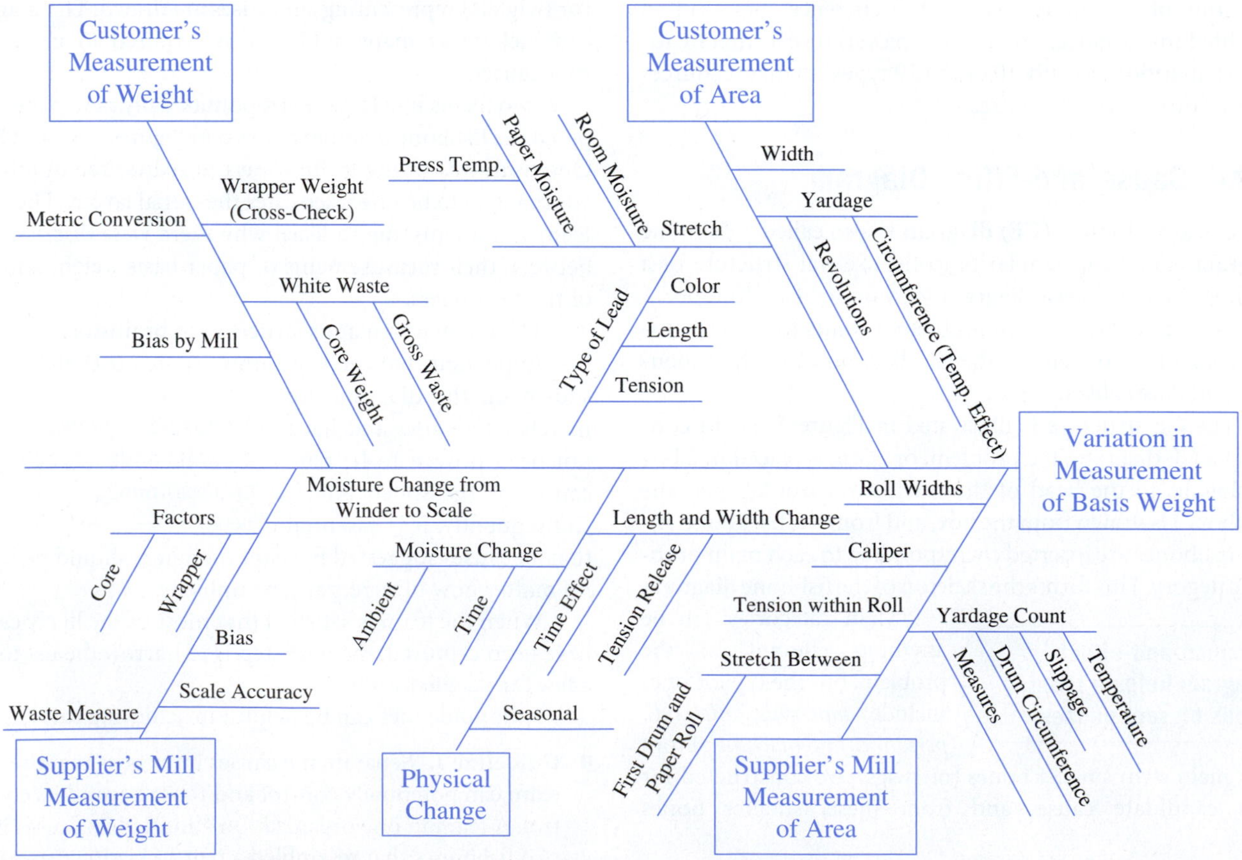

FIGURE 7-12. Fishbone Diagram for Cross-Calibration of Paper Basis Weight

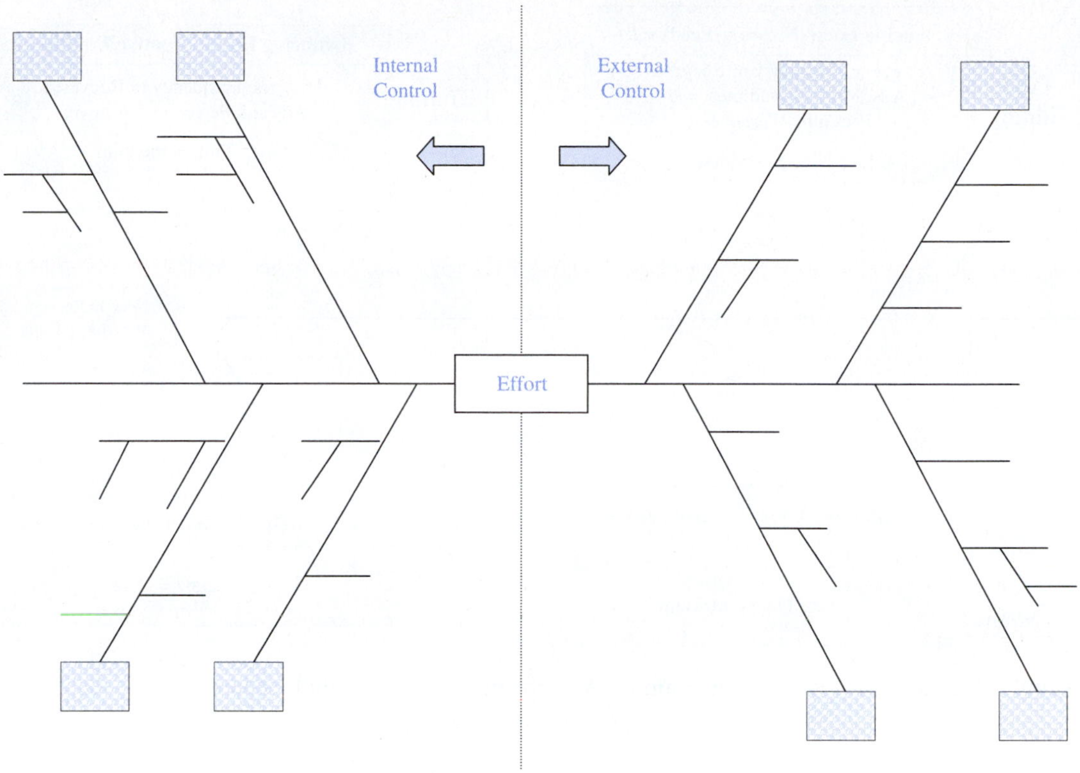

FIGURE 7-13. Dual Fishbone Diagrams

while those on the left were considered to be internal matters that the team could potentially influence.

- **Guideline 2.** Have each team member identify the top three causes—that is, the ones that each individual *believes* are *likely* to be major drivers. This can be done most effectively outside a formal meeting. Next, have the team meet, combine everyone's selections, and try to reach consensus (possibly using the nominal group technique) on a short list of three to five causes to investigate further.
- **Guideline 3.** Determine relationships. Focusing on the short list of selected causes, have the team try to understand if an association exists between the effect and each of the suspected causes, and also between each pair of causes. If data are available, scatter plots (discussed below) and statistical correlation analysis (refer to Section 13.2) can be used to make these determinations.
- **Guideline 4.** Combine systems thinking with the cause-and-effect relationships. The team should question what feedback mechanisms operate between the effect and the candidate cause—that is, is the feedback loop negative (balancing) or positive (reinforcing)? As we discussed in Chapter 2, a negative feedback loop has a self-regulating trigger that causes a modification in behavior. When activated, the cause will induce some desired behavior or result, with respect to the effect. It will in turn have a negative influence on the cause and in essence the roles will reverse. In a balancing cycle, the effect will become a cause—just the opposite of how the relationship was originally perceived. In the case of a positive or reinforcing feedback loop, the effect simply signals the cause to keep doing what it is doing. Whether positive or negative, there is usually a time lapse between when the feedback is received and when the system responds.

Figure 7-14 is an example of a feedback fishbone diagram for the advertising recycle problem that was originally diagrammed in Figure 7-11. Feedback loops have been identified together with their triggers and lag times. A feedback CE diagram is a condensed version of the original, where the main causation branches represent the small group of causes that were identified for further investigation.

To determine the type of feedback loops, the following question is directed at each causation branch: *How is the behavior in the causation branch altered as a result of variability (i.e., increases and decreases) in the measured effect?*

In the example, whenever the defect rate increases, the frequency of ad recycles diminishes the overall productivity of the ad preparation process. When this happens, deadlines, so vital in the newspaper industry, are jeopardized and pressure reverberates back up the value stream. Then, through the trigger mechanisms, the roles reverse. The effect now becomes the driver, activating improvement efforts in training and editing.

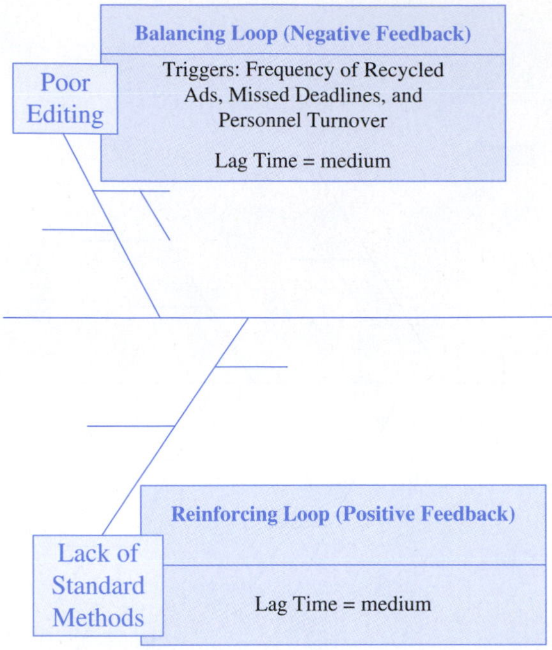

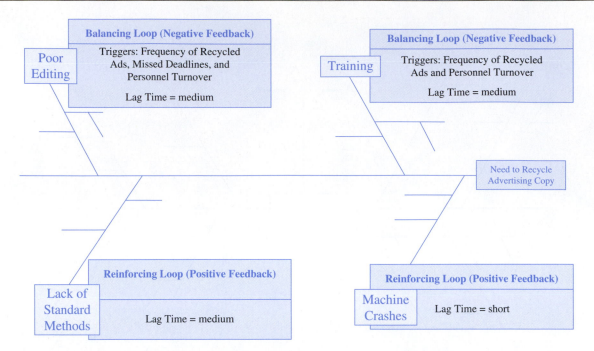

FIGURE 7-14. Feedback Fishbone Diagram for Advertising Copy Recycling Problem

Triggers occur whenever there is a positive trend (or unacceptable level) of ad recycles, or experienced personnel leave and have to be replaced by inexperienced ones. As improvement efforts in training and editing begin to work, the roles reverse again and the trend in ad recycles changes sign and becomes a negative trend. When it reaches an acceptable level the triggers cause a slackening of the improvement efforts, representing yet another reversal of roles.

Machine crashes and lack of standard methods are examples of positive feedback loops. As the number of recycles increases, rework is comingled with new work, and the total workload grows. With increased work and tightening deadlines, workers must increase their pace just to keep up. With mounting pressures, panic sets in and workers are more likely to cut corners (looking for shortcuts) and make mistakes. Inconsistent methods across the work group become even more nonstandard and machine crashes become more prevalent, driving the defect rate even higher.

7.3.5 Histogram

A **histogram** provides a graphical picture of process output. Collecting raw measurements is meaningless unless the data can be organized in a way that aids discovery and analysis. Figure 7-15a displays the measurements of 100 50-pound bags, selected at random and weighed to the nearest 100th of a pound. A plot of the raw data is shown in Figure 7-15b and is difficult to interpret. Figure 7-15c shows the same data organized as a histogram.

To construct this histogram, each of the 100 data points was assigned to 1 of 11 class intervals. A bar chart was then constructed by plotting the frequencies calculated in each class interval. The histogram provides an approximate picture of the process distribution. In this case, it appears that the distribution of bag weights is normal[8] with an average that is close to the 50-pound target.

The specifications for this process require a tolerance of 50 pounds ± 1 pound and from the histogram we can get an immediate insight as to how well the process is performing. It would appear that approximately 5% of the bags are underweight and 4% are overweight. Furthermore, the natural spread of the process is wider (approximately 3 pounds) than the 2-pound tolerance permitted. Therefore *it would appear that the process is not capable* of meeting the specs unless the natural variability can be reduced.

A histogram converts raw data into a picture of process output.

Through pattern recognition, histograms can provide valuable clues leading to improvement opportunities. Figure 7-16 illustrates eight different histogram shapes.

Pattern A represents a symmetrical "bell-shaped" curve that is usually referred to as a *normal* distribution. This is the shape of the output from many industrial processes, if they

[8]The normal distribution, also called the Gaussian distribution, is a common probability distribution encountered in quality management. The frequency distribution is continuous and when plotted is symmetrical around the mean and is bell shaped.

Bag No.	Bag Wt.	Bag No.	Bag Wt.	Bag No.	Bag Wt.	Bag No.	Bag Wt.
1	49.21	26	50.57	51	50.20	76	49.07
2	50.40	27	49.35	52	49.44	77	49.57
3	50.07	28	49.29	53	49.63	78	49.18
4	50.47	29	48.92	54	49.72	79	49.03
5	50.16	30	50.15	55	50.03	80	48.64
6	50.40	31	50.14	56	49.88	81	49.23
7	50.04	32	50.82	57	50.13	82	49.45
8	50.06	33	49.27	58	49.21	83	50.49
9	49.35	34	49.41	59	49.64	84	48.99
10	49.58	35	49.89	60	50.03	85	51.45
11	50.86	36	49.78	61	49.70	86	49.28
12	49.73	37	49.40	62	49.39	87	48.72
13	49.50	38	50.47	63	49.80	88	49.14
14	49.51	39	50.24	64	51.07	89	49.75
15	50.23	40	50.12	65	50.18	90	49.40
16	50.61	41	50.17	66	50.46	91	50.54
17	50.41	42	49.94	67	50.22	92	50.12
18	50.63	43	50.01	68	50.04	93	49.43
19	49.99	44	49.24	69	50.01	94	49.23
20	50.19	45	49.39	70	49.51	95	49.77
21	49.97	46	50.45	71	50.34	96	49.66
22	51.16	47	49.87	72	49.20	97	49.25
23	49.91	48	49.69	73	48.64	98	50.35
24	48.32	49	50.31	74	50.46	99	50.43
25	50.54	50	50.02	75	49.49	100	49.43

(a)

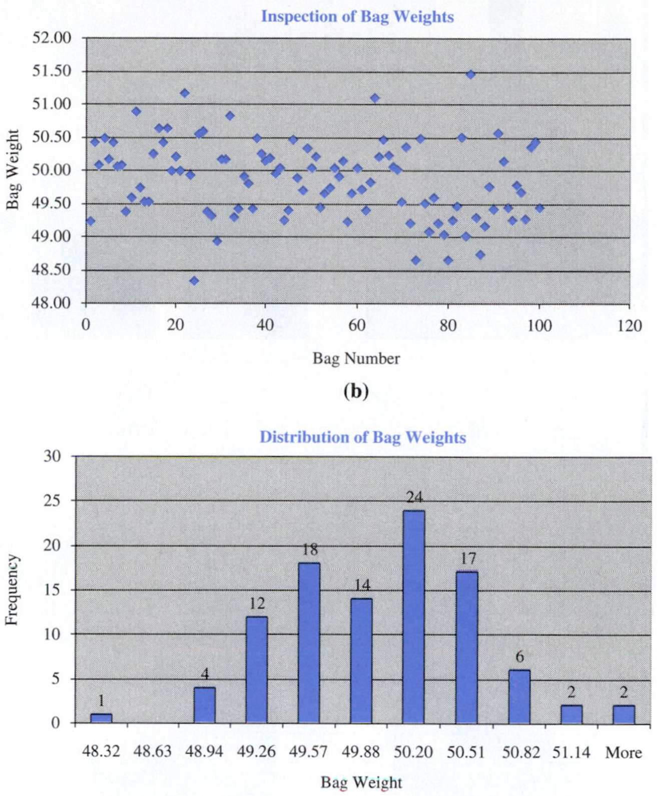

(b)

(c)

FIGURE 7-15. Bag Weight Measurements

are stable and the only sources of variation present are the random fluctuations that are inherent in the system. In a normal distribution, the mean and median are equal—50% of the output lies to the left of the average (or mean) and 50% to the right.

In pattern B, the distribution looks as if it may have originated as a normal distribution, but the middle of the distribution (the highest-quality product) has been carved out. This *gutted normal* is typical of cases where a supplier can sell product that is of the highest quality—produced near the target with small variation—at a higher price than product of a lower grade. Even though all output may fall within the allowed specification range and the *average* is on target, none of the individual units are equal to the target and the variance can be substantially greater than if the original distribution were intact.

The *augmented normal,* shown as pattern C, could result from an inspector reclassifying output that is just below or above the specification limits. Hence, there is a "lumping" effect at the first and last class intervals where the specification cutoffs occur. Pattern D is a *truncated normal.* A pattern such as this could occur if a process is centered on target, its spread is wider than the tolerance limits, and the unacceptable output has not been reported in the data. A distribution can also truncate if a natural barrier, such as zero (when no negative values are possible), is encountered.

Patterns E and F represent distributions that have a skew—that is, they lack the symmetry of the normal distribution. In a *negatively skewed* distribution more than 50% of the distribution and the median lie to the right of the mean. In a *positively skewed* distribution more than 50% of the distribution and the median are to the left of the mean.

A histogram can expose patterns that can lead to improvement opportunities.

Pattern G could have two possible interpretations. One scenario is that defective product was sorted from the good, because the process was incorrectly targeted. Alternatively, if negative numbers are not possible (e.g., weights), *truncation* at a measurement equal to zero can produce this pattern.

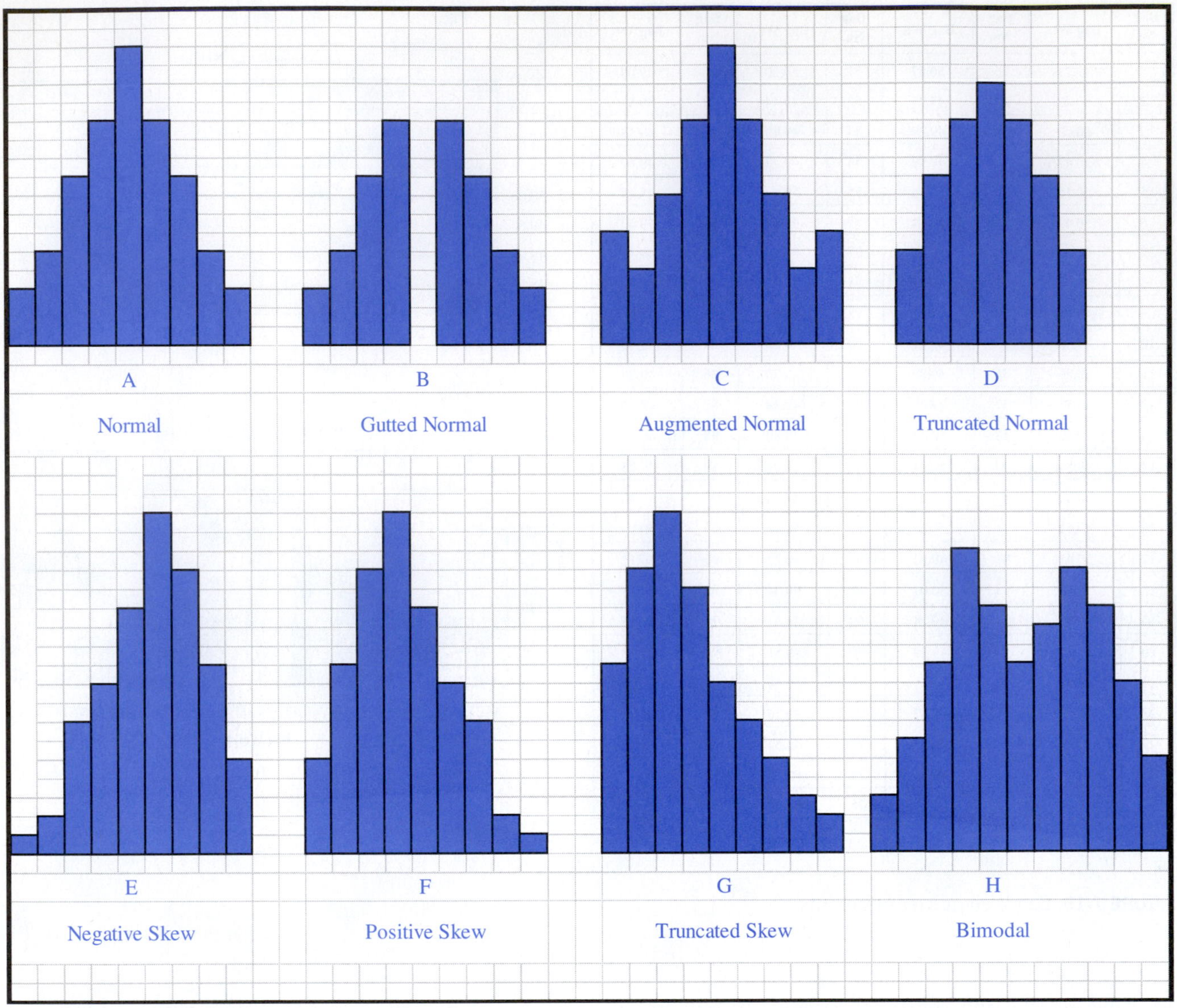

FIGURE 7-16. Histogram Shapes

Pattern H could be the resulting picture when several normal processes are overlaid, called *bimodal*; for example, if samples represent the output from several machines, each having a different process average.

7.3.6 Stratification

Stratification is the process of sorting data into meaningful classifications so that clues as to the who, what, where, when, why, and how relative to an issue at hand can easily be found. For example, a random selection of product at a packing operation might reveal that a certain percentage is outside the specification range, and more disturbing, the data suggest that the process is not capable of meeting specs. Stratifying the data by machine, operator, or shift could lead directly to the source of the unacceptable variability and its subsequent reduction.

Stratification sorts data into meaningful classifications.

The ability to think about possible sources of variation, and to design stratified data collection schemes, is fundamental to continuous improvement. As an example, consider Figure 7-17. A Pareto chart for the printing operations problem is reintroduced. In this case the defects have been classified not only by type but also by shift. After studying this chart we might conclude that Shift 3 is a good prospect for improvement.

7.3.7 Scatter Diagram

Like histograms, **scatter diagrams** aid in pattern recognition. A scatter diagram, also called a **scatterplot**, can be used to gain insights into the relationship between two factors

Problem	Total	Code	Shift 1	Shift 2	Shift 3	Shift 1 %	Shift 2 %	Shift 3 %	%	Cum %
Density Too Low—Colors	606	A	100	50	456	6.4	3.2	29.4	39.0	39.0
Misregister	492	B	129	112	251	8.3	7.2	16.2	31.7	70.7
Density Too Low—Black Ink	107	C	23	34	50	1.5	2.2	3.2	6.9	77.6
Density Too High—Colors	77	D	34	12	31	2.2	0.8	2.0	5.0	82.6
Nip Marks	61	E	12	10	39	0.8	0.6	2.5	3.9	86.5
Scumming	50	F	21	12	17	1.4	0.8	1.1	3.2	89.8
Density Too High—Black Ink	37	G	7	13	17	0.5	0.8	1.1	2.4	92.1
Scuffing Marks	35	H	8	10	17	0.5	0.6	1.1	2.3	94.4
Scratches	30	I	9	8	13	0.6	0.5	0.8	1.9	96.3
Wrinkles	29	J	10	7	12	0.6	0.5	0.8	1.9	98.2
Set-Off	28	K	6	7	15	0.4	0.5	1.0	1.8	100.0
Totals	1,552								100	

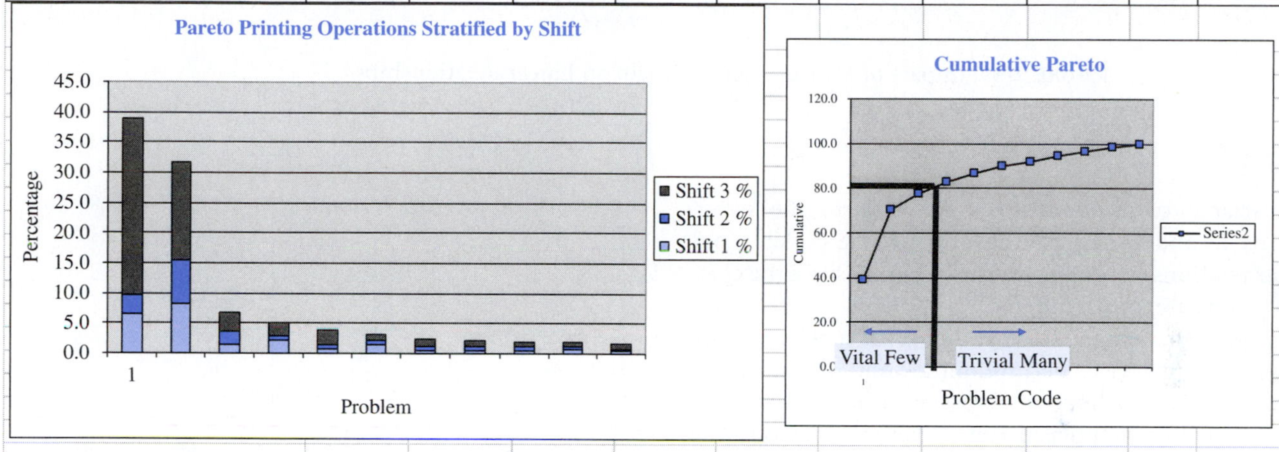

FIGURE 7-17. Stratified Pareto Chart for Printing Operations

(or variables). If a relationship is found, it cannot necessarily be inferred that one variable is the *cause* of the other; however, the scatter diagram can provide graphical evidence that the relationship is real and will provide some knowledge regarding the *strength* of the relationship.

A scatter diagram is constructed by plotting the values of one variable on the horizontal (x) axis, and the corresponding value of the other variable on the vertical (y) axis. A relationship (association) between the two variables is evident if the resulting plot produces some nonrandom pattern of points. The strength of the relationship is determined by the variability of the cluster of points relative to a mathematical expression describing the association. A relationship can be linear or nonlinear.

A scatter diagram can reveal nonrandom relationships.

Figure 7-18 illustrates some scatter diagrams that are linear. Pattern A shows a strong positive association—x and y will move in the same direction. When x increases, y increases and vice versa. The association is strong because the cluster of points is tight (low variability) around the mathematical straight line (called a least-squares line) that best fits the data. Pattern B shows the pattern in reverse—the variables move in opposite directions. An increase in

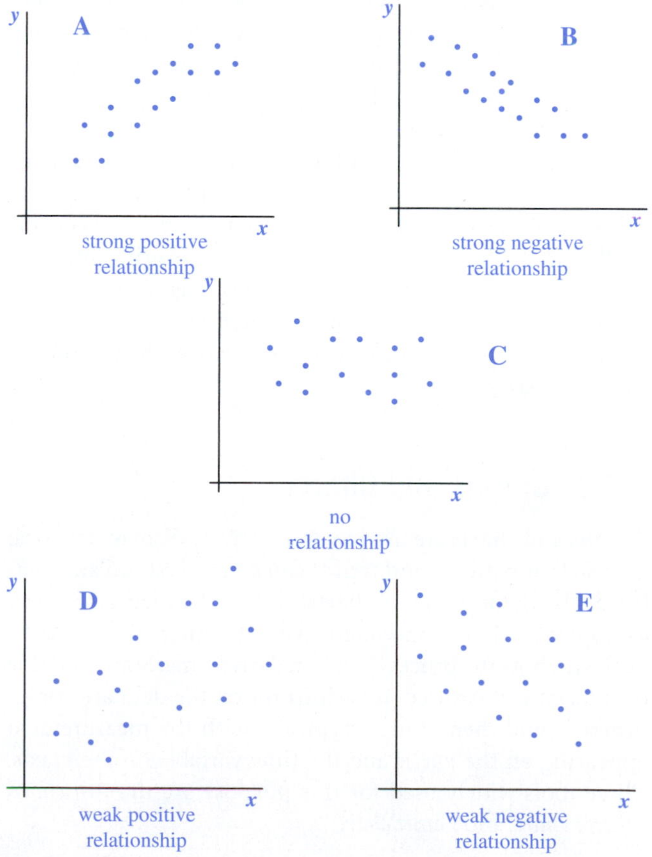

FIGURE 7-18. Linear Relationships

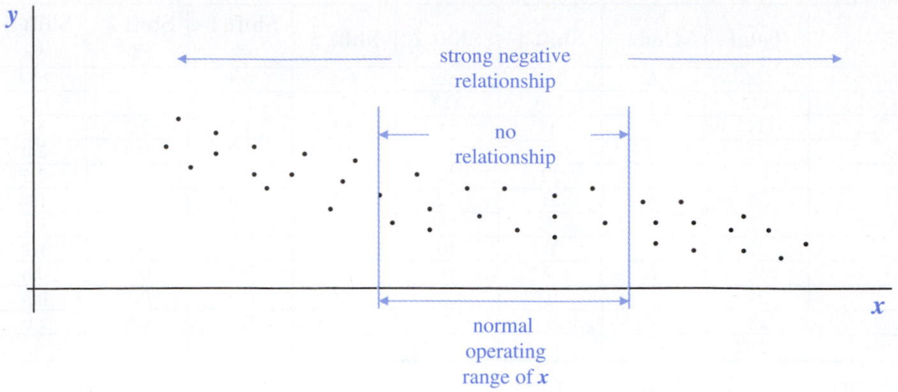

FIGURE 7-19. Impact of Measurement Domain on Linear Relationships

one variable corresponds to a decrease in the other. Patterns D and E suggest the presence of possible but weaker relationships than patterns A and B, since the cluster of points are more scattered.

In pattern C there is no evidence of a relationship. The scatter appears to be random around an imaginary straight line with a slope of zero (i.e., a horizontal line). However, a little skepticism can sometimes pay off. The points on scatter diagrams represent data that typically come from process measurements taken when operating parameters are at their normal set points. This is because scatter diagrams almost always reflect empirical data. Even though there is no evidence to suggest a relationship between x and y in the normal range, by deliberately moving out of the operating domain, one can sometimes expose a relationship as illustrated in Figure 7-19.

Not all associations are linear. Figure 7-20 shows some typical nonlinear patterns in F, G, I, and J. Understanding the nature of a nonlinear association can provide valuable clues in discovering improvement options. Pattern H has an underlying relationship that is linear; however, the variance of y depends on the value of x—as x increases the variance of y also increases.

Charts and graphs reveal patterns in plotted points.

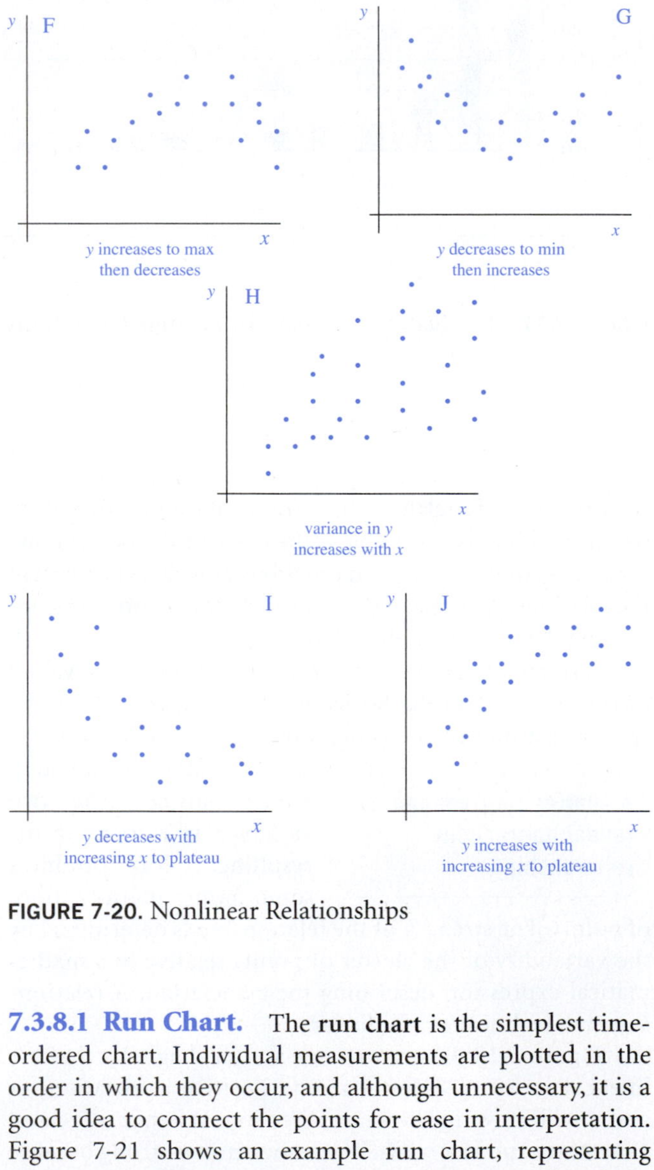

FIGURE 7-20. Nonlinear Relationships

7.3.8 Graphs and Charts

Graphs and **charts** are used to descriptively display statistical data so that patterns and trends can be understood and analyzed. Histograms, Pareto charts, and scatter diagrams are all examples used in quality management. Other charting tools deal with how the process is affected over time. Some variable of interest is measured at various times, the data are "time-ordered" and then plotted, typically with the measurement appearing on the y-axis and the time variable on the x-axis. Three tools widely used for this purpose are the *run chart, control chart,* and *Gantt chart.*

7.3.8.1 Run Chart. The **run chart** is the simplest time-ordered chart. Individual measurements are plotted in the order in which they occur, and although unnecessary, it is a good idea to connect the points for ease in interpretation. Figure 7-21 shows an example run chart, representing 28 measurements of a mineral product that was produced in a continuous line flow process. The average has been computed

Sample	Measurement
1	25.03
2	25.63
3	26.87
4	30.23
5	23.80
6	17.73
7	11.70
8	12.27
9	14.33
10	16.00
11	17.43
12	18.13
13	17.27
14	17.63
15	17.27
16	18.50
17	19.90
18	21.20
19	24.67
20	26.60
21	26.03
22	26.90
23	22.50
24	20.87
25	19.30
26	21.23
27	20.37
28	17.80
	577.19
	16.51

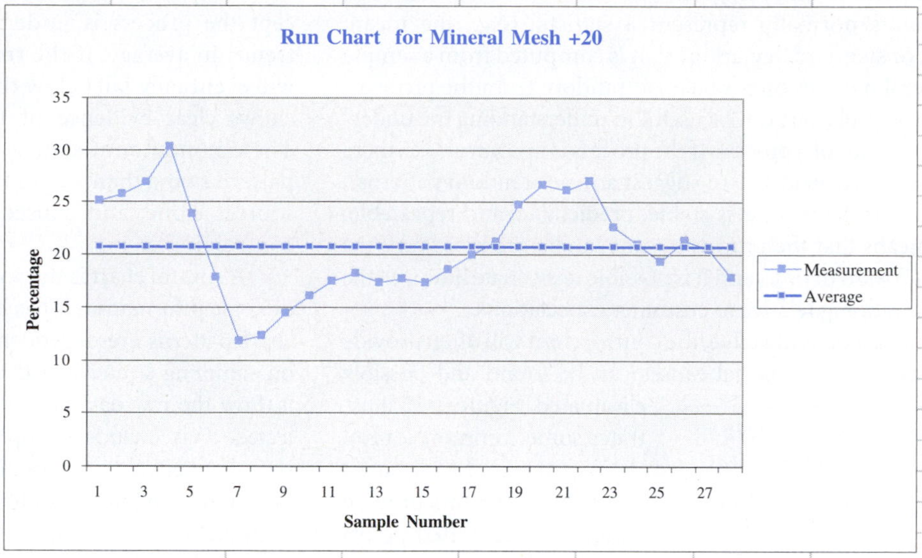

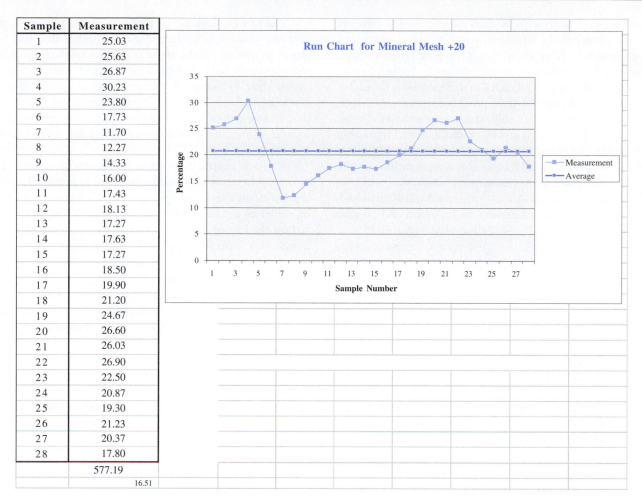

FIGURE 7-21. Run Chart

and drawn on the chart. The run chart is best used to identify sustained shifts or trends in the process average. If the process is not changing over time approximately the same number of points should appear on either side of the average, and be *randomly distributed.* In the mineral example, there is an equal split but the distribution does not look random. It appears that, at certain times the average is higher, and at other times lower, than the overall average of all 28 measurements.

One way to check for a change in average is to count the number of "runs." A run is defined as a buildup of consecutive points on the same side of the average (or centerline). A run length of seven or more points provides statistical evidence that the average has changed. An alternative to the number of runs is a test for trends. Six consecutive points either increasing or decreasing with no reversal in direction provides evidence that the average is trending.

A run chart is a time-ordered plot of individual measurements.

7.3.8.2 Control Chart.

A **control chart** is a run chart with **control limits.** These limits represent the maximum and minimum allowable values for any individual plotted point. Any point that exceeds these limits provides statistical evidence that the *process average has changed* and that the chart's centerline is no longer a reliable approximation. The maximum value is called the upper control limit and the minimum value the lower control limit. The range of permissible values that lie between the two limits represents expected variability that is due to random causes.

Random variability is present in all samples and is aptly called **common cause variability.** If certain sources of variability are present in some samples, but not others, one would expect the variability to exceed the bounds imposed by the control limits—such sporadic fluctuations are called **special cause variability.** The purpose of a control chart is to expose the presence of special cause variation. This can be done by detecting nonrandom patterns or observing any plotted points outside the control limits.

The purpose of a control chart is to test the statistical stability of a process.

There is a fundamental difference between a run chart and a control chart. The points plotted on run charts typically represent individual measurements (e.g., machine downtime, percentage yields, or scrap). Control charts can also be constructed to analyze individual measurements; however, the points normally represent a statistic (e.g., the mean, range, or standard deviation) that is computed from a sample of several measurements taken at random from the process.

A control chart can be useful in understanding the underlying behavior of a process. If the process is in control (i.e., there is no statistical evidence to suggest any nonrandom patterns), we say that the process is **stable**, **predictable**, and **repeatable**. This means that the process output can be predicted forward and backward in time, and it is possible to evaluate how well the process is doing relative to customer expectations.

If a process is unstable, the control chart will often provide hints as to where special causes can be found and possibly eliminated. Figure 7-22 illustrates some common control chart patterns. In all cases, the x-axis represents time. In pattern B the plotted points appear to be randomly distributed about the centerline (average) and all lie within the control limits. Process B, therefore, is in a state of statistical control. The process shown in pattern A appears to have started at an average higher than the centerline and then, approximately halfway through the sampling process, experienced a sustained decrease in average. Even though none of the points exceeded either of the control limits, the pattern is nonrandom and the process is not in control.

A process is in a state of statistical control if it is stable, predictable, and repeatable.

The process depicted by pattern C is like pattern A, except the process is undergoing a continual decrease (or trend) in average. If the trend continues unabated a point will eventually fall below the lower control limit. Pattern D shows clear evidence of the presence of special causes. These nonrandom sources of variability result in a wider pattern swing than would be expected from common cause sources alone, and consequently points fall outside both control limits.

A control chart is the **voice of the process** but will provide only the information it has been designed to provide. Control chart patterns are dependent on sampling strategies—that is, how the raw data are collected. This includes sample size, frequency, and specifically how individual sampling units are selected. An important consideration in determining the sampling plan is the intended purpose for the data. There are three basic reasons for collecting control chart data.

The patterns that appear on control charts depend on how the samples were taken.

- ***To characterize the process output.*** A process in control can be used to estimate the quality characteristics of process output, including the percentage that exceeds or falls short of requirements. Such information can be invaluable in communicating with customers or justifying improvement efforts.

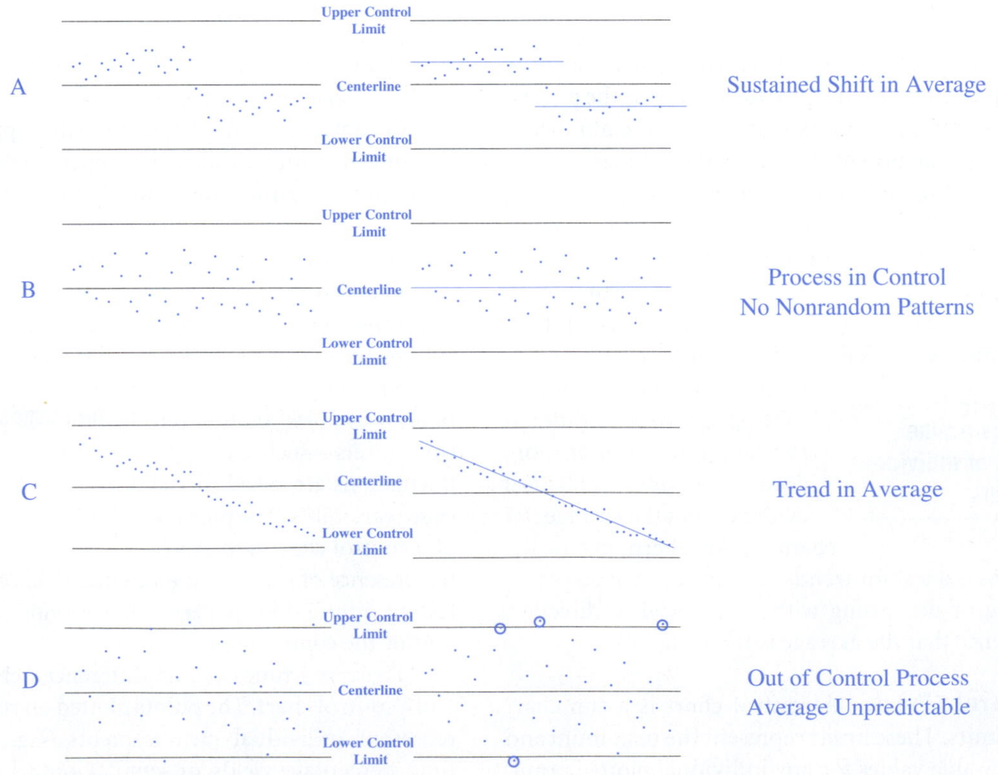

FIGURE 7-22. Control Chart Patterns

- ***To monitor process performance and intervene when necessary.*** A control chart can act as a tracking device and navigational aid. As a tracking device it can validate improvements and provide early warning to an operator if the process starts losing productivity gains. As a navigation tool operators can use the control chart to indicate when they should intervene and "steer the process" and, just as importantly, when to leave the process alone.
- ***To improve a process.*** With a properly designed sample a control chart can expose sources of variability—often converting common causes into special causes—to focus improvement efforts. More information can often be gained from charts not in control than from data that exhibit statistical stability.

Data can be collected in order to characterize process output, to monitor process performance, or to identify improvement opportunities.

Control charts are covered in further detail in Chapters 8 through 11. An important assumption underlying the use of run charts or control charts is statistical independence—that is, there should be no time-related correlation, or **autocorrelation,** in the plotted points. The measured value at one point in time should be unaffected by the value obtained for previous measurements and should not affect the value of the subsequent or future measurements. We discuss how to deal with autocorrelated data in Chapter 12.

7.3.8.3 Gantt Chart.

Henry L. Gantt, an American engineer and social scientist, developed the **Gantt chart** in 1917. Originally designed for production scheduling, the Gantt chart is now frequently used to plan and track progress on projects.

A Gantt chart is constructed with the horizontal (x) axis representing time increments, and the vertical (y) axis representing the individual tasks that make up the project. Horizontal bars (of a length corresponding to the time consumed by each task) are placed on the chart in the time slot scheduled. Where two tasks can occur simultaneously bars can overlap, but not when one task must be completed before another can commence. As a project is executed Gantt charts can be easily adjusted to reflect actual process status. From a Gantt chart the estimated project completion time is always readily available.

A Gantt chart is used to plan and track progress.

Figure 7-23 shows a pair of Gantt charts that were constructed for the manufacture of a product involving 14 tasks. The precedence diagram shows the sequence of operations

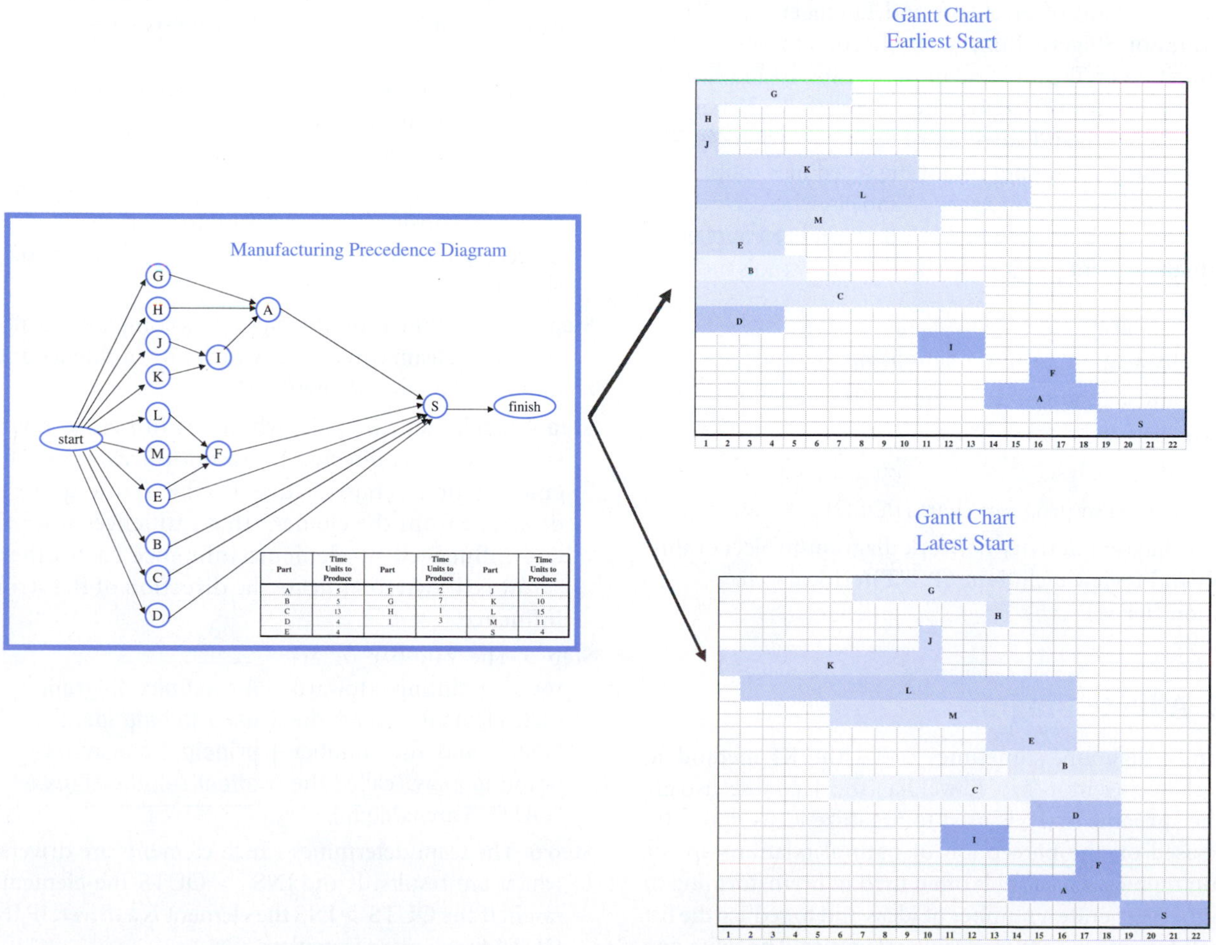

Part	Time Units to Produce	Part	Time Units to Produce	Part	Time Units to Produce
A	5	F	2	J	1
B	5	G	7	K	10
C	13	H	1	L	15
D	4	I	3	M	11
E	4			S	4

FIGURE 7-23. Gantt Charts for Scheduling an Assembly Operation

that must be followed and the task times are shown in the table. One of the Gantt charts shows the earliest activity start times for each task, and the other depicts the latest. By comparing the two charts it is easy to determine how much scheduling flexibility exists for each task. It is also clear that the critical tasks are K, I, A, and S. Any delay at all in any of those activities will postpone the estimated overall completion time of 22 time units.

7.4 SEVEN NEW TOOLS FOR PROBLEM SOLVING

7.4.1 Introduction

In the early 1970s it had become clear to the Japanese that the effective implementation of TQM required that managers know more than simply how to apply basic statistical tools. Although the tools of numerical analysis are essential, the Japanese recognized that much of the information processed by middle to senior management is nonnumeric and often takes the form of verbal data and communications. Consequently in 1972 a committee chaired by Yoshinobu Nayatani was established to develop some new quality control tools for management. The committee's research took more than five years and resulted in a set of techniques called the "seven new tools." In 1978 the Japanese Union of Scientists and Engineers (JUSE) invited Professor Shigeru Mizuno, Professor Emeritus of the Tokyo Institute of Technology, and Professor Yoshio Kondo, head of the faculty of engineering at Kyoto University, to guide the education and advocacy of the new set of tools. The Japanese committee did not invent the seven new tools, most of which were developed in America and elsewhere. However, by the late 1970s most of these tools had not been applied to quality management programs. The set of new tools includes

The seven new tools can be used to process nonnumeric data.

- Affinity diagram
- Relations diagram or interrelationship digraph
- Systematic diagram or dendrogram (tree diagram)
- Matrix diagram
- Matrix data analysis
- Process decision program chart (PDPC)
- Arrow diagram/activity network diagram/project evaluation and review technique (PERT)/critical path method (CPM)

7.4.2 Affinity Diagram

An **affinity diagram,** sometimes called the **KJ method** in honor of its inventor Jiro Kawakita (the Japanese usually write the surname first), is used to organize ideas into categories based on the perception of natural relationships or common themes. This tool is often used in brainstorming to help a group generate a number of ideas and to reduce the list to a smaller number of manageable categories. The following steps are involved in creating an affinity diagram, as depicted in Figure 7-24.

Step 1 Using a brainstorming/brainwriting process participants record their ideas on sticky notes, one idea per note, and then place their notes on a white board or wall.

Step 2 Participants individually try to sort the items into related groups. No discussion between participants is permitted. If a person disagrees with someone's placement of an item they are permitted to relocate it to another category. The sorting and resorting continues until it appears the team has reached some consensus. It is important that this entire step be completed in silence.

An affinity diagram can be used to organize a number of ideas into a small number of categories.

Step 3 The group holds an open discussion trying to agree on headings for the categories. Header cards are created and placed on the chart above the sorted data.

7.4.3 Relations Diagram

A **relations diagram,** also called an **interrelationship digraph,** aids in the discovery of cause-and-effect relationships and in segregating major drivers from minor ones. Figure 7-25 depicts the general format for constructing an interrelationship digraph and employs the following six steps.

Step 1 The team reaches consensus on some issue or problem that requires further study. The issue is stated in clear terms that can be understood by everyone.

Step 2 The team then determines and lists all elements (or factors) it considers to be relevant to the issue or problem. A rectangular symbol is placed on the digraph for each element.

Step 3 Each element on the digraph is compared to all others and the team determines which ones influence and are influenced by that element.

Step 4 Each element pair, where an influence relationship exists, is connected by an *influence arrow*. The element pair is connected with the arrow going in a direction from the element that influences toward the one influenced. If elements influence each other the arrow is drawn to indicate the direction of the stronger influence.

Step 5 The number of arrows pointing toward each element (called the "INS") and the number pointing away (called the "OUTS") are tallied.

A relations diagram is used to help identify principal cause-and-effect relationships.

Step 6 The team determines which elements are drivers and which are results. If the INS > OUTS the element is a *result*. If the OUTS > INS the element is a *driver*. IF INS = OUTS the element is neither.

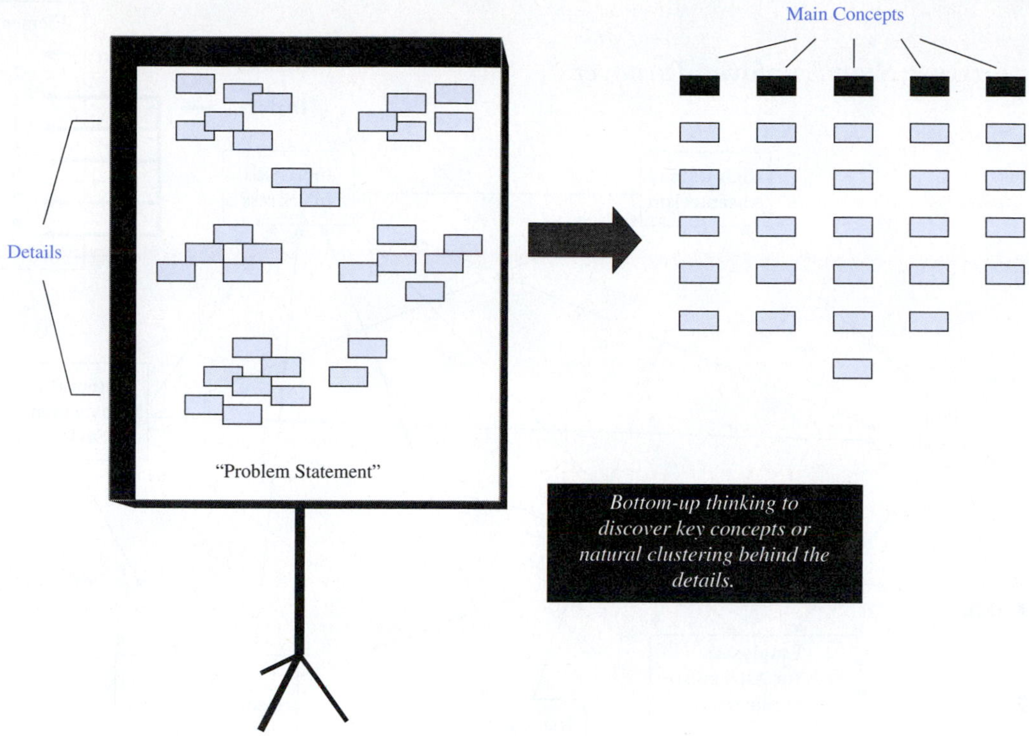

FIGURE 7-24. Affinity Diagram

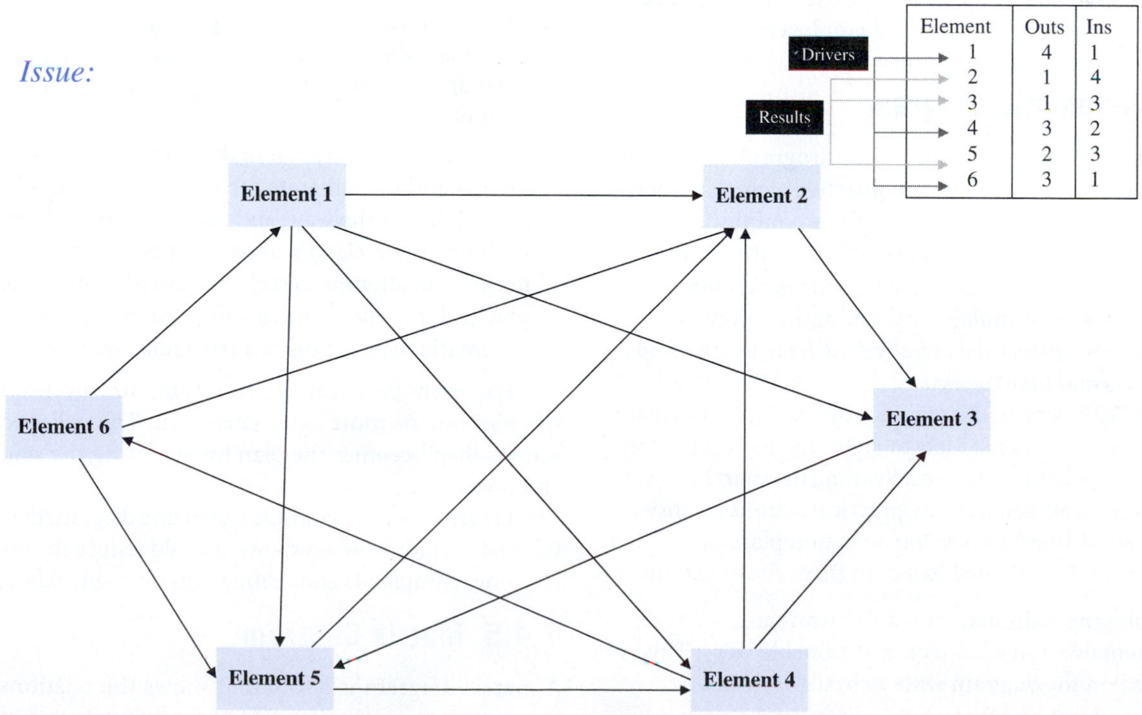

FIGURE 7-25. Interrelationship Digraph

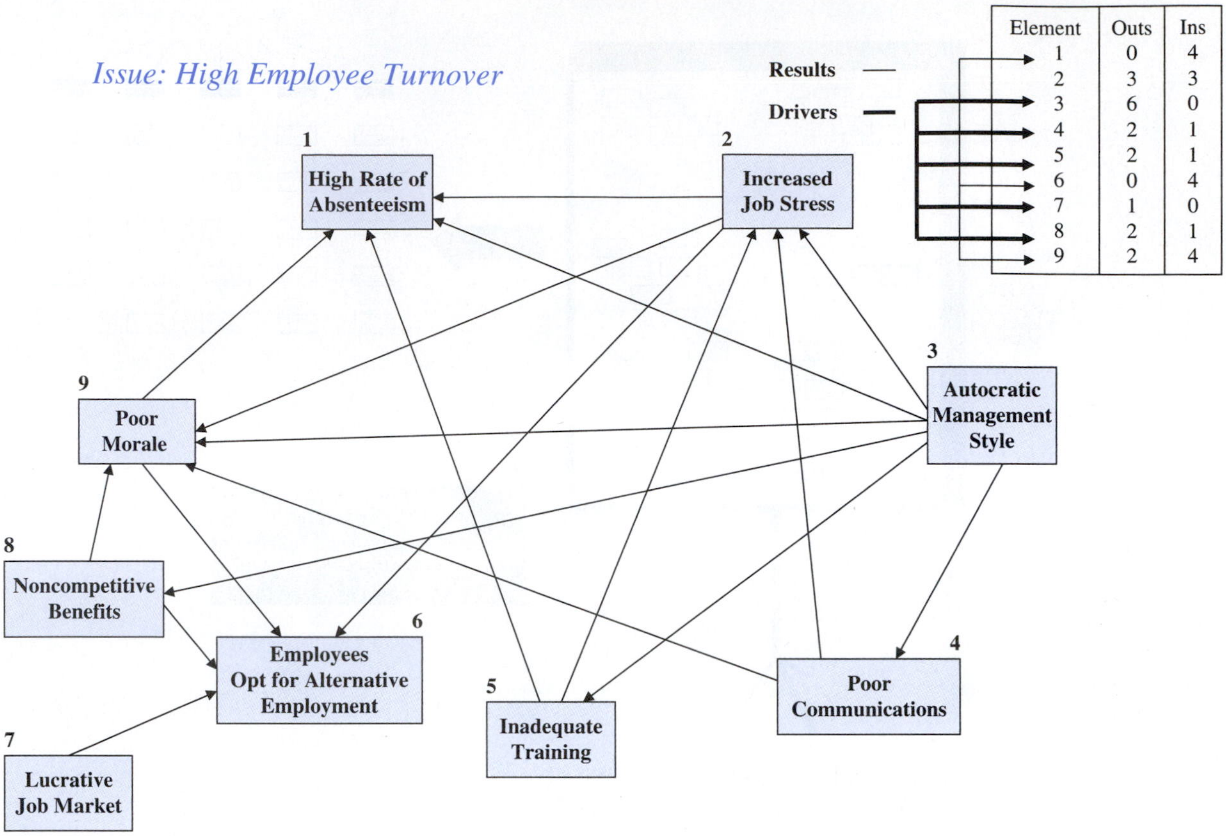

Element	Outs	Ins
1	0	4
2	3	3
3	6	0
4	2	1
5	2	1
6	0	4
7	1	0
8	2	1
9	2	4

FIGURE 7-26. Interrelationship Digraph

Figure 7-26 is an example interrelationship digraph that a team applied to the problem of excessive employee turnover. It was discovered that the principal driver could be traced back to an autocratic management style that failed to recognize the fundamental needs of employees.

7.4.4 Systematic Diagram

A **systematic diagram,** also called a **dendrogram,** is a tree diagram that displays a hierarchy of objectives, which branches out to a hierarchy of ways and means for accomplishing those objectives. This tool is more focused than an affinity or relations diagram. At each level of the tree the generation of the next level down is stimulated by asking questions such as *How will we accomplish the objective?* or *How will we implement the ways and find the means?*

Figure 7-27 depicts the generic format for a systematic diagram. As the tree is read left to right, the logical progression goes from general to specific. Reading the chart backward (specific to general) helps clarify practical actions by answering the question *Why?* To develop an action plan, each of the ways and means is evaluated based on three classifications.

- Practical—the indicated ways and means are implementable and lead to clear actionable steps. This is indicated on the diagram with a circular symbol.
- Impractical—the indicated ways and means are not implementable, at least not at this time. This could be because they are inconsistent with overall corporate strategy, there are insufficient resources or time, or implementation could lead to some negative consequences elsewhere in the system. An impractical classification is indicated on the diagram with the symbol **X**.

A systematic diagram can be used to link ways and means to objectives.

- Uncertain—the decision makers need additional information and to deliberate further before the implementability of the ways and means can be determined. Each uncertain classification must be carefully scrutinized in an attempt to reclassify as either practical or impractical. An uncertain classification is depicted on the systematic diagram using a triangular symbol.

For each practical classification, the decision group specifies one or more actionable item. The collection of all actions then becomes the plan for achieving the stated goals and objectives.

Figure 7-28 is an example systematic diagram that was developed to study how a restaurant could reduce the number of customer complaints concerning eggs on its breakfast menu.

7.4.5 Matrix Diagram

A **matrix diagram** is a tool that shows the relationship between two or more sets of factors. A matrix is a structure that provides rows and columns that represent the factors

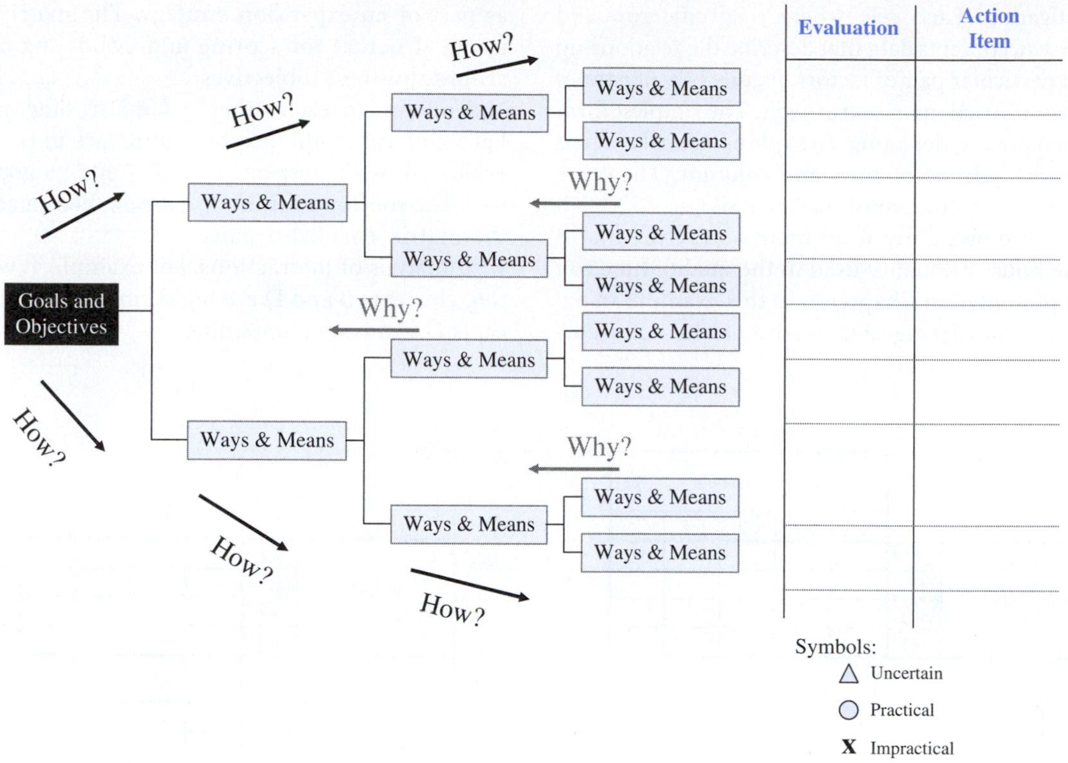

FIGURE 7-27. Format for Systematic Diagram

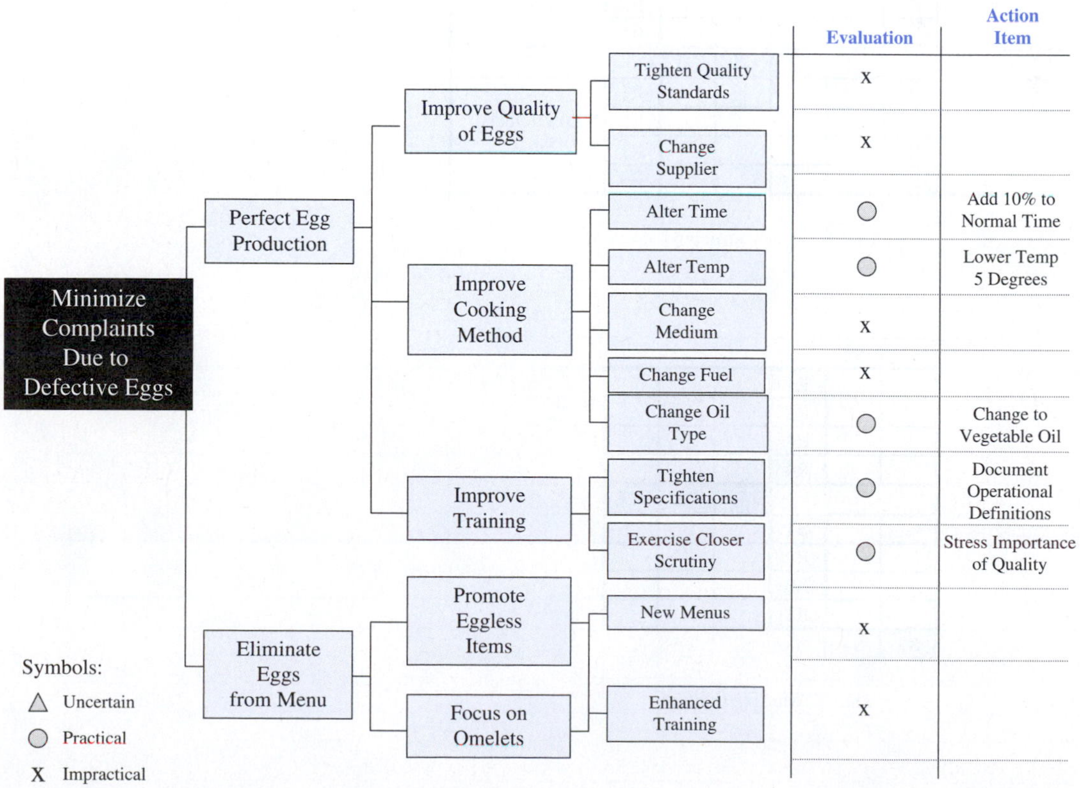

FIGURE 7-28. Systematic Diagram for the Production of Eggs

under investigation. Each cell, where a specific row and column intersect, contains data that describe the relationship between that particular pair of factors. Figure 7-29 illustrates several commonly used matrix structures. The simplest form is the L-shaped matrix, depicting a two-dimensional analysis that contains a single set of rows and columns. The other variations enable a multidimensional analysis.

Figure 7-30 shows a "rooftop" matrix structure that is typical of the house of quality used in the quality function deployment presented in Chapter 5. In this example an organization was considering several new product concepts as part of an expansion strategy. The matrix provided a useful structure for scoring and evaluating each concept against business objectives and relative to each other. Each concept could also be evaluated with respect to risk. The rooftop portion of the matrix enabled a pair-wise analysis of interactions. For example, it would appear that concepts B and D are highly incompatible, while concepts D and E are compatible.

A matrix diagram adds structure to two or more sets of data and their respective relationships.

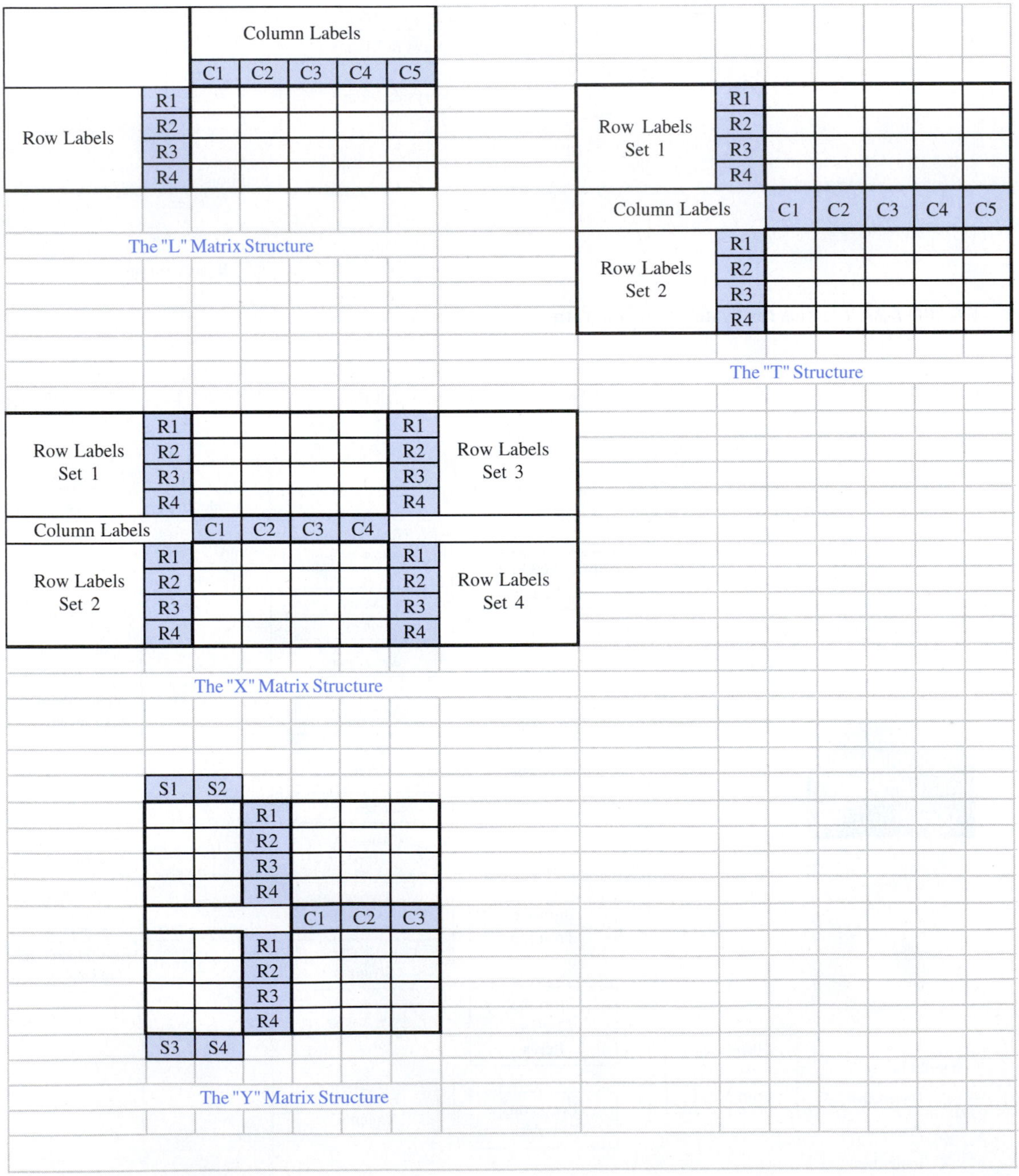

FIGURE 7-29. Matrix Structures

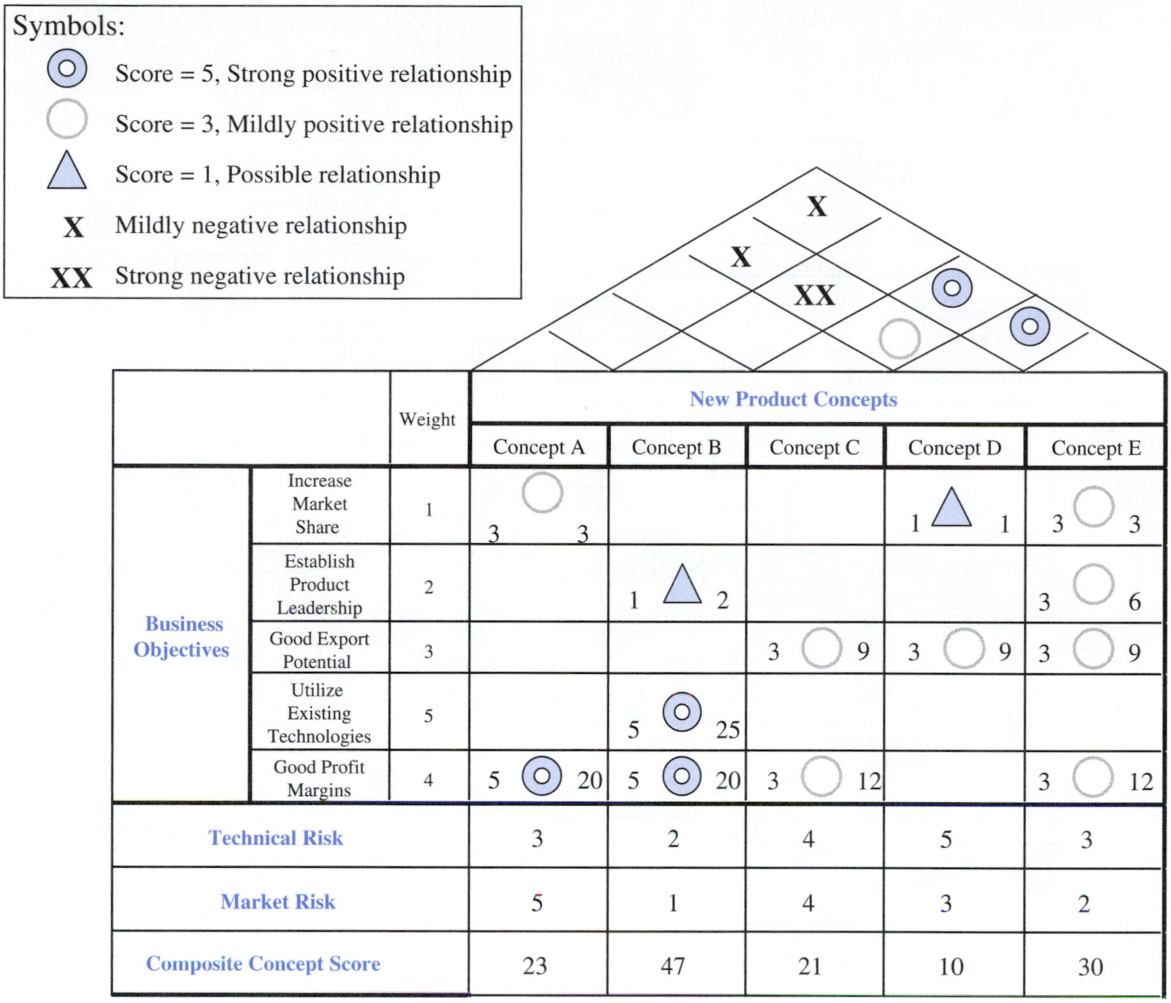

Symbols:
Score = 5, Strong positive relationship
Score = 3, Mildly positive relationship
Score = 1, Possible relationship
X Mildly negative relationship
XX Strong negative relationship
X
X
XX
Weight
New Product Concepts
Concept A
Concept B
Concept C
Concept D
Concept E
Business Objectives
Increase Market Share 1 | 3 3 | | | 1 1 | 3 3
Establish Product Leadership 2 | | 1 2 | | | 3 6
Good Export Potential 3 | | | 3 9 | 3 9 | 3 9
Utilize Existing Technologies 5 | | 5 25 | | |
Good Profit Margins 4 | 5 20 | 5 20 | 3 12 | | 3 12
Technical Risk 3 2 4 5 3
Market Risk 5 1 4 3 2
Composite Concept Score 23 47 21 10 30

FIGURE 7-30. House of Quality Application of Matrix Diagram

7.4.6 Matrix Data Analysis

The most sophisticated of the new tools, as proposed by Mizuno,[9] is called **matrix data analysis,** and is an application of principal components analysis. The idea is to take a matrix of relationships between factors and to convert it into a set of preference vectors (called *eigenvectors*) that consist of sums of fractions of those factors that significantly contribute to process variance. The vectors of composite preferences can then be used to determine which characteristics are important. Matrix data analysis can be useful in determining the product or service features that result in the most customer satisfaction, and those that can be eliminated with little or no customer dissatisfaction.

7.4.7 Process Decision Program Chart

A **process decision program chart** (PDPC) is used to display the many alternative paths that can occur during the execution of a plan, including contingencies, so that strategies for dealing with them can be developed in advance. There is no standard format for a PDPC. Some are shown as tree diagrams while others are displayed as flowcharts. The idea is to graphically display the steps of a process or project and the possible paths that are taken to complete all steps. Where things can go wrong the path may branch at any particular step. Countermeasures (contingency plans), devised to help the process return to purpose, are shown on the chart. There are three principal uses for a process decision program chart.

- To identify all possible contingencies that can occur in the execution of any new plan where risks are involved.
- To plan the implementation of a complex project in which the cost of failure is high.
- To manage a project with an extremely tight time constraint that permits insufficient time to deal with contingencies as they occur.

A PDPC should be constructed by a team that has the experience and competence to define all process steps and possible contingencies. The group places the steps on a chart in a tree diagram, or flowchart, in time and prerequisite sequence. This can be done in either a horizontal or vertical

[9] Mizuno, S. (ed.). *Management for Quality Improvement: The 7 New QC Tools.* Cambridge, MA: Productivity Press, 1988.

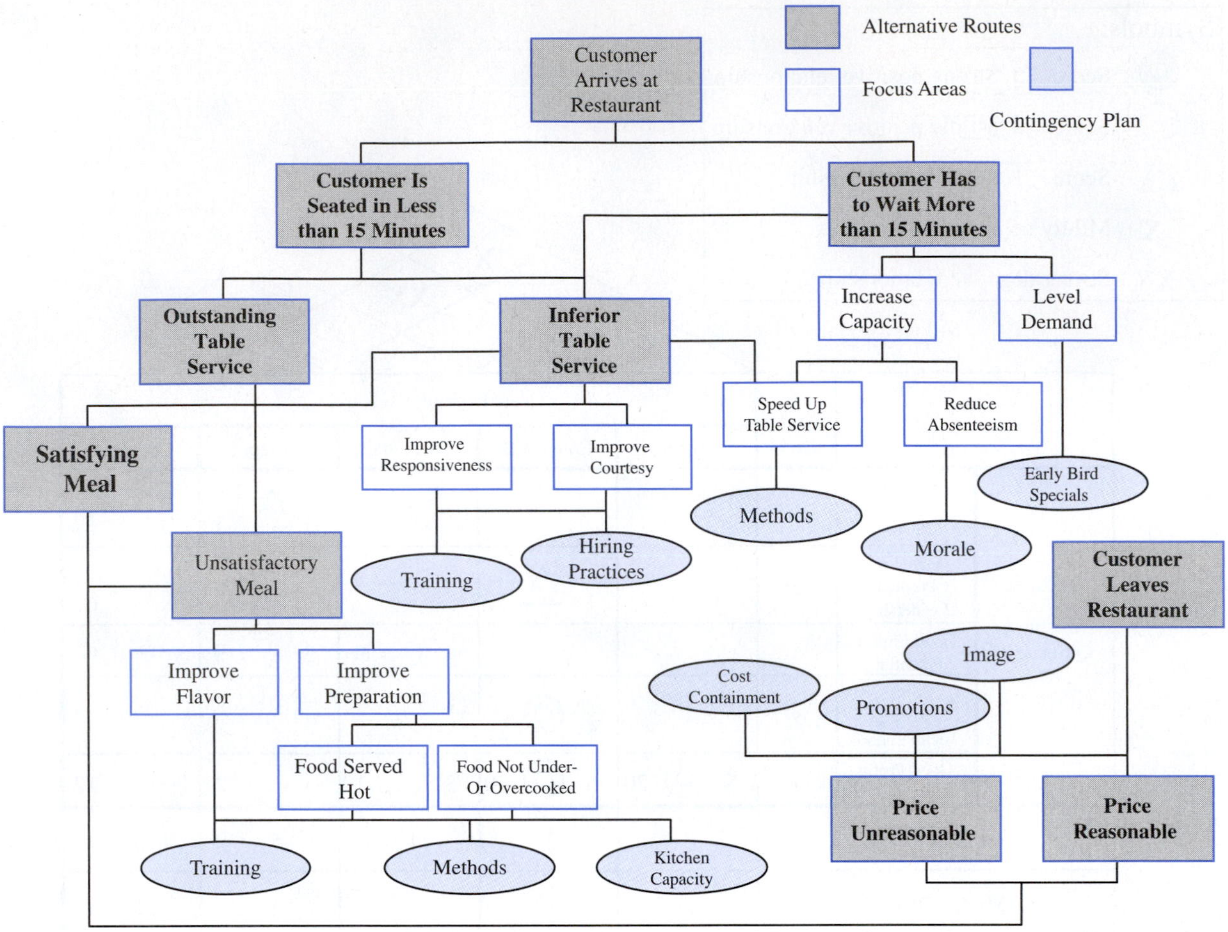

FIGURE 7-31. Process Decision Program Chart

format. The group then scrutinizes each step asking, "What can go wrong here?"

The answers to these questions are contingencies, and are recorded on the chart as possible branches or diversions from the main process flow. Finally the group determines plausible explanations or feasible countermeasures to each contingency and documents those on the chart. Figure 7-31 is an example PDPC for customer service in a restaurant.

A process decision program chart identifies contingencies that can occur during the execution of a plan.

7.4.8 Arrow Diagram

Also known as an activity network diagram, program evaluation and review technique (PERT), or a critical path method (CPM), an **arrow diagram** is a tool borrowed from project management and used to schedule a sequence of complex tasks. There are two conventions used for constructing an arrow diagram—activities on arrows (AOA) and activities on nodes (AON). These are illustrated in Figures 7-32 and 7-33, respectively. The first step in constructing an arrow diagram is to identify all tasks, their prerequisites, and the time required for completion. A variant procedure uses probability distributions for completion times to compute expected times and variances for each task. Another variant estimates the resources consumed by each task.

In the AOA method, nodes are milestone events that represent the completion of all activities preceding an event, and signify the earliest start time for any succeeding activity. In the AON convention the nodes signify accomplishment of tasks and the arrows depict the technological linkages.

An arrow diagram is used to manage a complex project.

Arrow diagrams are useful for planning the schedules of each task, determining the deployment of critical resources, identifying activities that can occur simultaneously, calculating by how much individual tasks can be delayed without affecting the overall project duration, and determining the shortest total time required to complete all tasks (the critical path). Once the arrow diagram is constructed, the critical path can be determined by locating the shortest path through the network from start to finish.

Applications of arrow diagrams include reengineering a production process, planning a complex business venture such as a new product development, establishing a new business unit/organizational structure, or managing a large mul-

Operation	Prerequisites	Time Required	Operation	Prerequisites	Time Required	Operation	Prerequisites	Time Required
A	G, H, I	5	F	E, L, M	2	K	—	10
B	—	5	G	—	7	L	—	15
C	—	13	H	—	1	M	—	11
D	—	4	I	J, K	3	S	A, B, C, D, E, F	4
E	—	4	J	—	1			

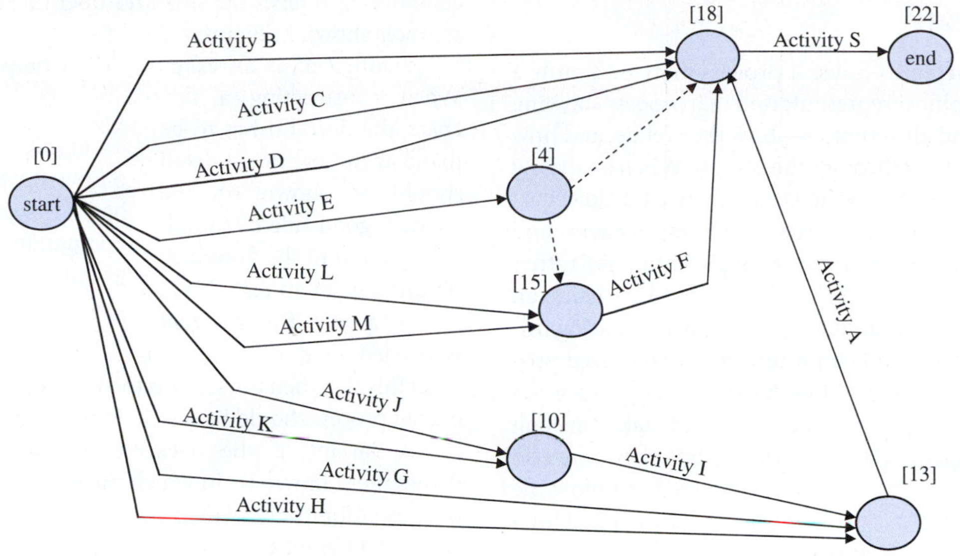

FIGURE 7-32. Activities-on-Arrows Diagram

Operation	Prerequisites	Time Required	Operation	Prerequisites	Time Required	Operation	Prerequisites	Time Required
A	G, H, I	5	F	E, L, M	2	K	—	10
B	—	5	G	—	7	L	—	15
C	—	13	H	—	1	M	—	11
D	—	4	I	J, K	3	S	A, B, C, D, E, F	4
E	—	4	J	—	1			

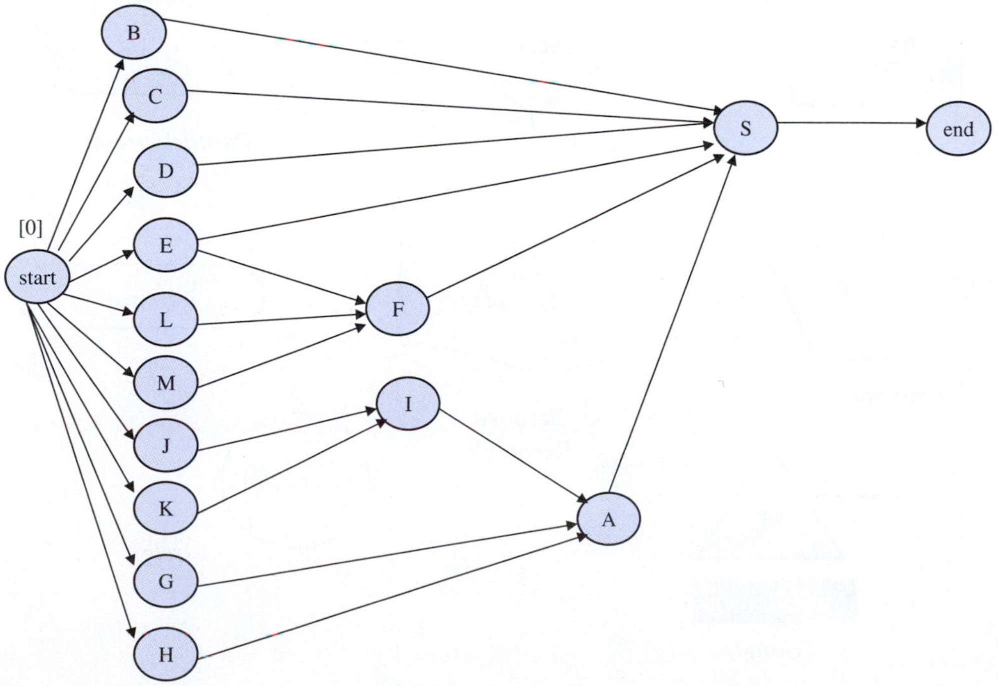

FIGURE 7-33. Activities-on-Nodes Diagram

tidimensional project. Once an arrow diagram has been constructed it is often useful to generate a companion Gantt chart to plan and track progress.

7.5 OTHER USEFUL TOOLS

7.5.1 Flowchart

A **process flowchart,** also called a **process map,** or simply a **flowchart,** is a graphical representation of a process showing inputs, outputs, and all activities—how they relate, and how material or people flow through the system. When analyzing a process, it is always a good idea to construct a flowchart first. It is much easier to conceptualize process behavior and spot improvement opportunities using a visual aid rather than a written description. A flowchart is used to define an existing process (*as is*), redesign a process incorporating improvements (*could be*), and communicate standardized procedures (*operator training*). Flowcharts normally trace the path materials take through a system, but they can also be constructed to follow the movement of people. Flowcharts can be constructed at two levels of detail.

A flowchart is useful in defining a process.

1. The process as a *network of subsystems* (with inputs, processing steps, and outputs).
2. The process as a sequence of steps necessary for the *transformation* of inputs into outputs.

7.5.1.1 Process Flowchart as a Network of Subsystems. Using a flowchart to map a network of subsystems can produce insights into how the components interrelate to produce a finished product or service. Sometimes called an **output process chart,** a system flowchart is a simple graphical tool that utilizes the small number of easily recognizable symbols shown in Figure 7-34.

Figure 7-35 is an example of an output process chart. When constructing a flowchart, the decision has to be made as to how much detail should be shown on the chart. A good rule of thumb is to try and fit the flowchart on one side of an 8.5" × 11" sheet of paper. If more detail is needed than will fit on a sheet this size, then the scope is too broad. A smaller part of the overall process should be selected for study.

An output process chart is a flowchart that shows how all system components interrelate to produce a finished product or service.

A variant of the network systems flowchart is the **deployment flowchart** in which the process is segmented out by responsibility area. For example, if three people are responsible for different steps of the process the chart is divided into three segments, with the name of the responsible individual appearing at the top of each segment. As the chart is constructed, the steps are placed in those segments corresponding to the responsible individual. On a deployment flowchart each line that connects steps across responsibility segments represents a *customer–supplier relationship*. Figure 7-36 illustrates a deployment flowchart for making breakfast. In

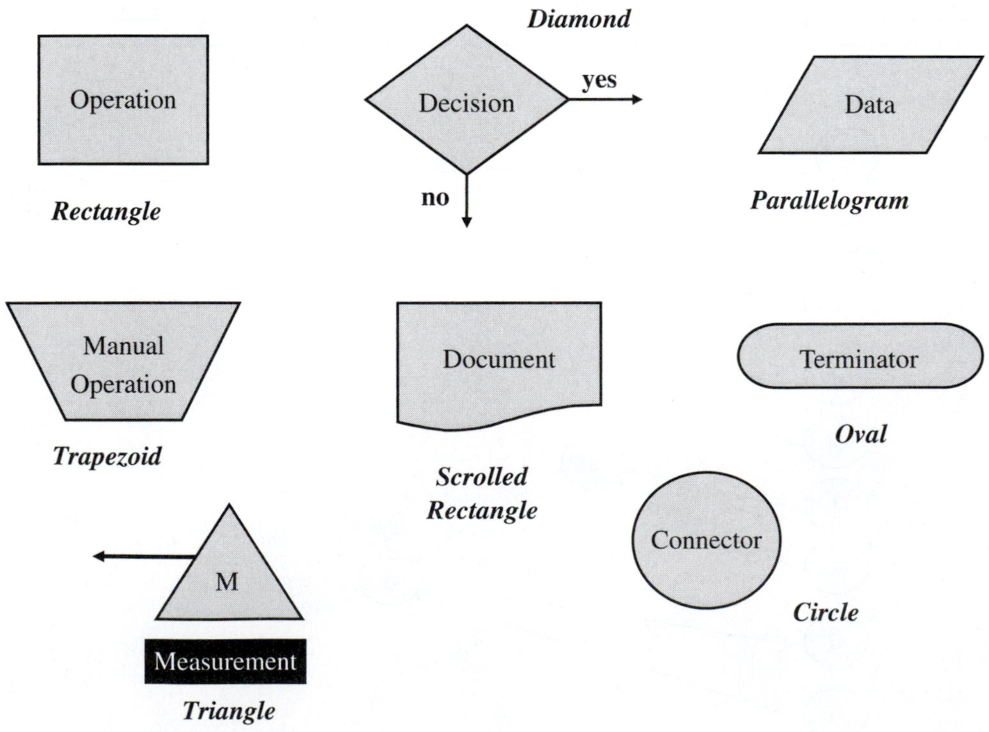

FIGURE 7-34. Symbols Used in Flowcharts

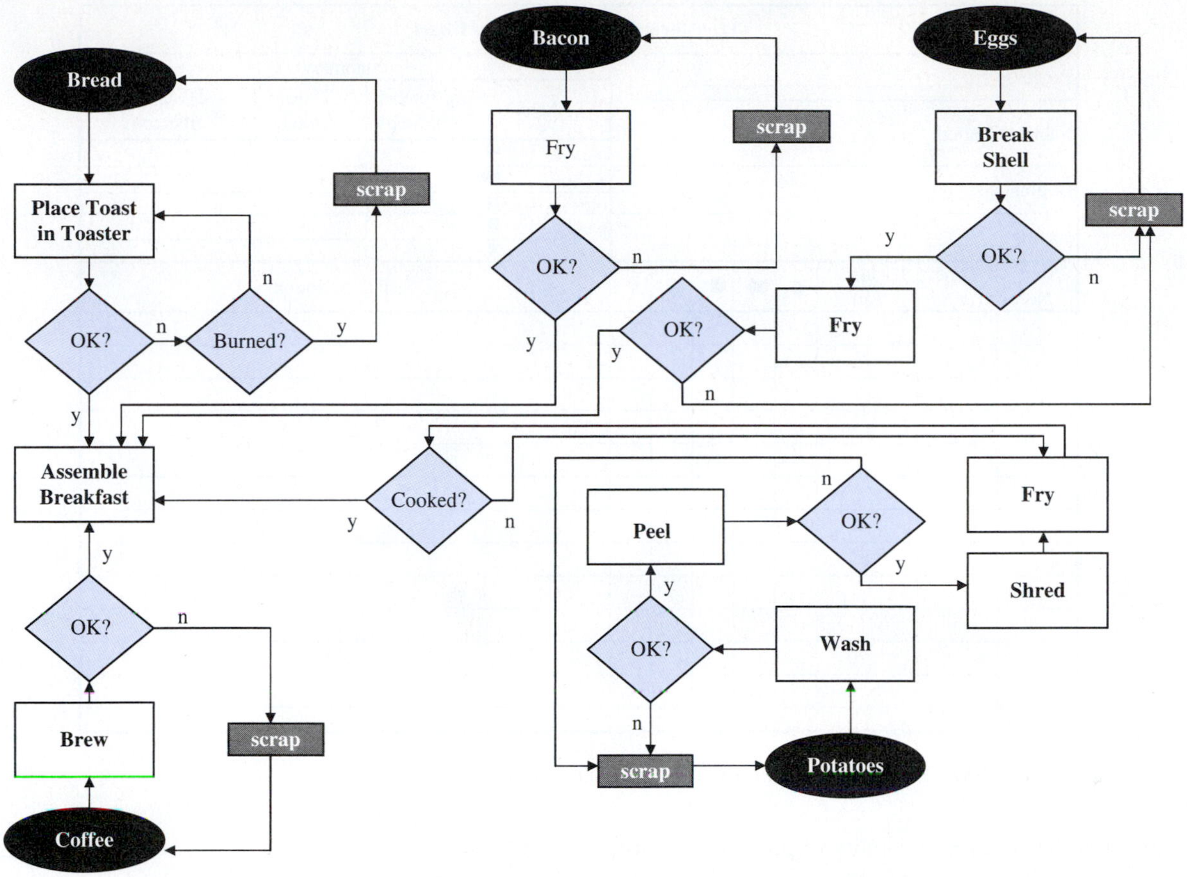

FIGURE 7-35. Output Process Chart for Breakfast Preparation

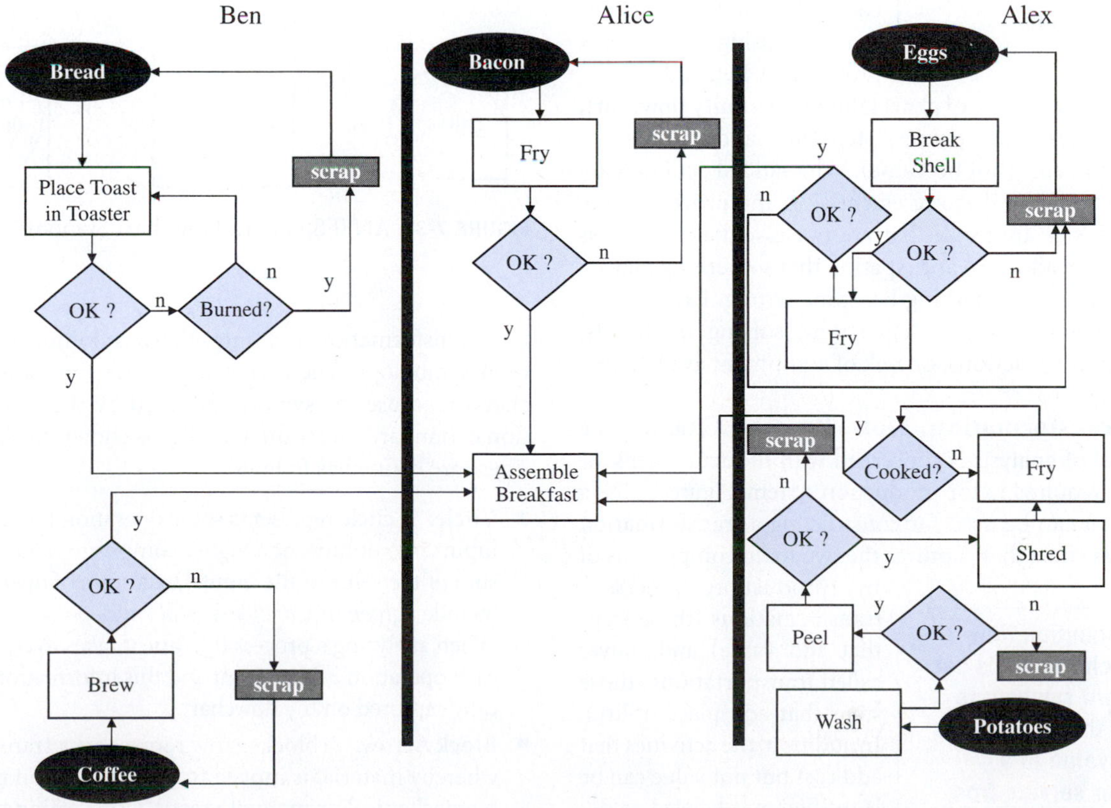

FIGURE 7-36. Deployment Process Chart for Breakfast Preparation

Transformation Flow Process Chart

Process: ______________________

Subject Charted: ______________________

Beginning: ______________________

Ending: ______________________

Summary				
Activity		Number of Steps	Time (min.)	Distance (meters)
Operation	●			
Transport	➡			
Inspect	■			
Delay	D			
Store	▼			

Step No.	Time (min.)	Distance (m)	●	➡	■	D	▼	Step Description

FIGURE 7-37. Transformation Flow Process Chart Form

this example, Ben has the responsibility to prepare the toast and coffee whereas Alex is responsible for the eggs and the potatoes. Alice fries the bacon and assembles the breakfast tray, using the outputs provided by Ben and Alex.

> **A deployment flowchart assigns responsibilities for operations to individuals.**

An alternative type of chart is an **opportunity flowchart,** which separates those process steps that add value, from those that add cost (but no value). Value-added activities are defined to be those that are essential to the process. Cost-adding activities are those that are not essential to process output, but are added in anticipation that something may go wrong or because something has gone wrong. Examples of cost-adding activities are inspections, sorting of defects, awaiting approval actions, or lack of equipment availability.

7.5.1.2 Transformation Flow Process Chart. The second level of analysis is concerned with the actual steps, or operations, required in the production system. Figure 7-37 is a template that can be used for constructing a **transformation flow process chart** that captures the two main components of any production process—transformations (those steps that add value) and moves called transportations (those steps that add place utility). In addition, the activities that add cost but not value can be identified and labeled as delays, inspections, or storages.

> **A transformation flow process chart uses a set of standard symbols to separate those activities that add value to a product or service from those that do not.**

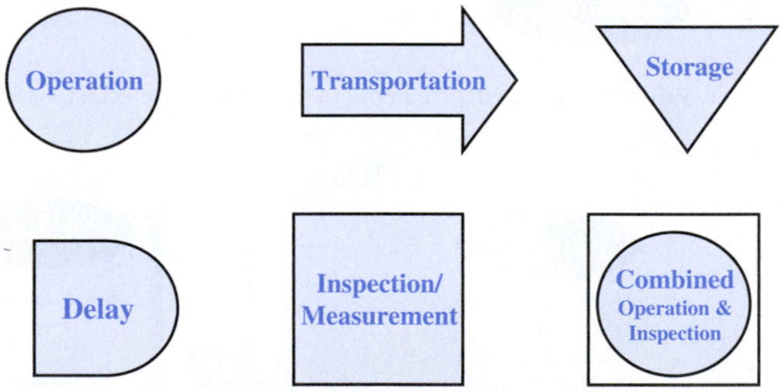

FIGURE 7-38. ANSI Standard Flowchart Symbols

Transformation and output charts do not share a common symbology. The accepted practice for transformation charts is to use the symbols adopted by the American National Standards Institute (ANSI), as shown in Figure 7-38. These are defined as follows:

- **Circle.** A circle represents some operation that transforms inputs into outputs of a higher composite value than the sum of the value of the inputs. Examples of operations include *tighten nut, drill hole, seal envelope,* and *bus table.* When analyzing a process the time it takes to complete each operation is important and this information is typically captured on the flowchart.
- **Block Arrow.** A block arrow represents a transportation whereby material is moved from one location to another by mechanical or manual means. Concerning transportations, one is usually interested in the total number

of moves, the distance traveled, the time required, and the transportation method.

- **Inverted Triangle.** In a flowchart the inverted triangle depicts a storage. This is defined as a *system-designed delay* whereby material is deliberately taken out of the value stream for the purpose of storing it for future use. Storages include inventory buffers where material sits waiting to be sequenced into a subsequent operation. In an ideal process material would flow smoothly from one operation to the next and storage buffers would be unnecessary. Therefore, by identifying where storages exist, improvement opportunities can often be found.
- **Block Uppercase Letter *D*.** This symbol, a large uppercase *D,* identifies delays—those cases where flow is disrupted for some reason other than the explicit intent to create inventory buffers. Examples of delays are *waiting for approvals, waiting for an elevator or scarce equipment,* or *waiting for a material handler*. Once identified, delays should be carefully scrutinized to see if there are ways to reduce their duration and frequency, or to eliminate them altogether.
- **Square.** A square symbol represents an inspection or measurement. Examples of inspections are *monitoring instrumentation* or *control charts, reviewing blueprints,* or *inspecting the quality of outputs.*
- **Combined.** Two or more symbols can be combined if an activity fits more than one of the previous definitions. For example, if operators inspect the quality of their work as an integral part of performing an operation, the process step would be coded as both an operation (circle) and an inspection (square).

Figure 7-39 is a transformation flowchart for a process that manufactures table legs in a large furniture plant.

7.5.2 Knowledge Mapping

Knowledge mapping is a method that is used to break down a broad issue into increasing levels of detail to gain a better understanding of the current knowledge that exists concerning the issue. There are three types of knowledge maps: *process, concept,* and *competency.*

Transformation Flow Process Chart for Furniture Leg Production

Process: __Production of Furniture Legs___

Subject Charted: __Material____

Beginning: ___Operation 10_____

Ending: ___Interim Storage____

Summary				
Activity		Number of Steps	Time (min.)	Distance (meters)
Operation	●	5	4.125	
Transport	➡	6	3.11	153.2
Inspect	■	1		
Delay	D	1	37.6	
Store	▼	1		

Step No.	Time (min.)	Distance (m)	●	➡	■	D	▼	Step Description
10	0.756							Cut lumber to 5.5" width
	0.17	4.6						Move to Operation 20 by roller conveyor
20	0.797							Cut part to 11.5" length
	0.17	4.6						Move to Staging area by roller conveyor
	37.6							Waiting to be transported to Building D
	2.13	127						Transport to Building D and Operation 30 by truck
30	0.839							Drill 1.5" diameter hole using drill press
	0.12	3.1						Move to Operation 40 by roller conveyor
40	0.857							Drill 8.25" diameter by 5/8" holes using drill press
	0.32	7.7						Move to Operation 50 by belt conveyor
50	0.876							Joint edges using jointer
	0.2	6.2						Move to storage area by truck
								Inspect for defects and store

FIGURE 7-39. Transformation Flow Process Chart for Furniture Part

Phase I: *Process Characterization*
Construct Process Flow Chart

Phase II: *Knowledge Mapping*
Identify Knowledge Requirements and Sources

Phase III: *Analysis*
Ask:
- **What knowledge is most critical?**
- **What knowledge is missing?**
- **What knowledge contributes to the value stream and how?**

Phase IV: *Application*
Apply the analysis to the purpose:
- **Planning**
- **Process improvement**
- **Training**

FIGURE 7-40. Constructing a Knowledge Process Map

7.5.2.1 Process Knowledge Map. A **process knowledge map** is a graphical representation of knowledge and its sources, mapped to a business process. Knowledge that either drives the process or results from its operation is portrayed, and includes tacit knowledge (expertise, experience, and intuition), explicit knowledge (codified and documented in procedural manuals), and customer knowledge. Creating a process knowledge map requires an intense team effort, typically during a two- to three-day workshop. A suggested format for the workshop employs four distinct phases as shown in Figure 7-40.

> **A process knowledge map is a transformation flowchart that includes knowledge requirements and sources.**

Knowledge process maps can be applied to planning, process improvement, training programs, the determination of intellectual capital, or tracking of progress on learning objectives. Figure 7-41 is a process knowledge map of the table leg process in the furniture plant introduced earlier.

7.5.2.2 Concept Knowledge Map. A **concept knowledge map** (often referred to simply as a concept map) is a useful tool for organizing and classifying content and visualizing relationships. On the map concepts are connected in a downward-branching hierarchical structure with linking action–verb phrasing that articulate the nature of each relationship. Creating a concept map requires a deep, insightful thought process, and it is sometimes helpful to have group members write basic concepts on index cards or sticky notes so that they can easily be moved around on a table or white board. Creating a concept knowledge map requires the three phases of activity shown in Figure 7-42.

It may be useful to start with a simple, uncluttered map to highlight the key concepts before moving forward with additional detail. Consider a problem that an engineering department in a large university encountered. The faculty were investigating how well its curriculum aligned with the expectations of those companies that recruited its graduates. A concept map was used to identify the core skills required by a practicing engineer and how they were related. Figure 7-43 was the first pass from which one can immediately see that an industrial engineer is concerned with data analysis (statistics), the creation of models, and use of communication and people skills—all for the purpose of solving problems that affect the operation of business systems with the ultimate goal of better satisfying customers. Figure 7-44 was the next map in which additional detail has been added by breaking down some of the key areas. Our understanding has now been enhanced and we can see that the industrial engineer is expected to provide leadership, deal with complexity to come up with solutions that are workable, be able to function as a team player, and have an appreciation for lifelong learning. The third attempt produced an even more detailed version which has been included in Appendix D. Figure 7-45 is an example of a manufacturing application of concept knowledge maps. This book manufacturing plant has mapped the concepts required to produce a quality textbook.

7.5.2.3 Competency Knowledge Map. A **competency knowledge map** can be created for an individual or an organization. For an individual, the map is a competency profile, documenting training, skills, positions, and career path. At the organizational level, the competency map provides a broad documentation of skills, tasks, and core

Process Knowledge Map for Furniture Leg Production

Process: ___Production of Furniture Legs___	**Summary**					**Knowledge Types and Sources**
Subject Charted: __Material____	Activity		No. of Steps	Time (min.)	Distance (meters)	Tacit (expertise, experience, intuition)
	Operation	●	5	4.125		Explicit (codified, formal, standards, procedures)
Beginning: ___Operation 10_____	Transport	➡	6	3.11	153.2	
	Inspect	■	1			
Ending: ___Interim Storage____	Delay	D	1	37.6		Customer (specifications, expectations, instructions)
	Store	▼	1			

Step No.	Time (min.)	Distance (m)	●	➡	■	D	▼	Step Description	Knowledge Required	Knowledge Source
10	0.756		•					Cut lumber to 5.5" width	Skill with table saw, measurement, expertise	Procedure Manual 62-507, Operator Experience
	0.17	4.6		•				Move to Operation 20 by roller conveyor		
20	0.797		•					Cut part to 11.5" length		
	0.17	4.6		•				Move to Staging area by roller conveyor		
	37.6					•		Waiting to be transported to Building D		
	2.13	127		•				Transport to Building D and Op 30 by truck	Skill with forklift truck, expertise	Procedure Manual 65-203, Op Experience/Certification
30	0.839		•					Drill 1.5" diameter hole using drill press	Skill with drill press, expertise	Procedure Manual 62-514, Operator Experience
	0.12	3.1		•				Move to Operation 40 by roller conveyor		
40	0.857		•					Drill 8.25" diam by 5/8" holes using drill press	Skill with drill press, expertise	Procedure Manual 62-514, Operator Experience
	0.32	7.7		•				Move to Operation 50 by belt conveyor		
50	0.876		•					Joint edges using jointer	Skill with jointer, expertise	Procedure Manual 62-482, Operator Experience
	0.2	6.2		•				Move to storage area by truck		
					•		•	Inspect for defects and store	Specifications, perceptions, judgment	Customer, blueprints, inspector training

FIGURE 7-41. Process Knowledge Map

Phase I: *Create List of Main Concepts*

Ask:

- **What are the main ideas relative to the goal or issue at hand?**

Phase II: *Arrange the Concepts from Broadest to Most Specific*

- **Place broadest, most inclusive concept at top of the map.**
- **Add other concepts that appear to be directly linked to it.**
- **Work down the map adding concepts that are linked to second tier.**
- **Continue with a third tier, and so on, until all concepts have been added.**
- **Limit the number of concepts linked to any one concept to three.**

Phase III: *Link Concepts*

- **Link concepts together using straight lines.**
- **On each line write a few linking words that describe the relationship.**

FIGURE 7-42. Constructing a Concept Knowledge Map

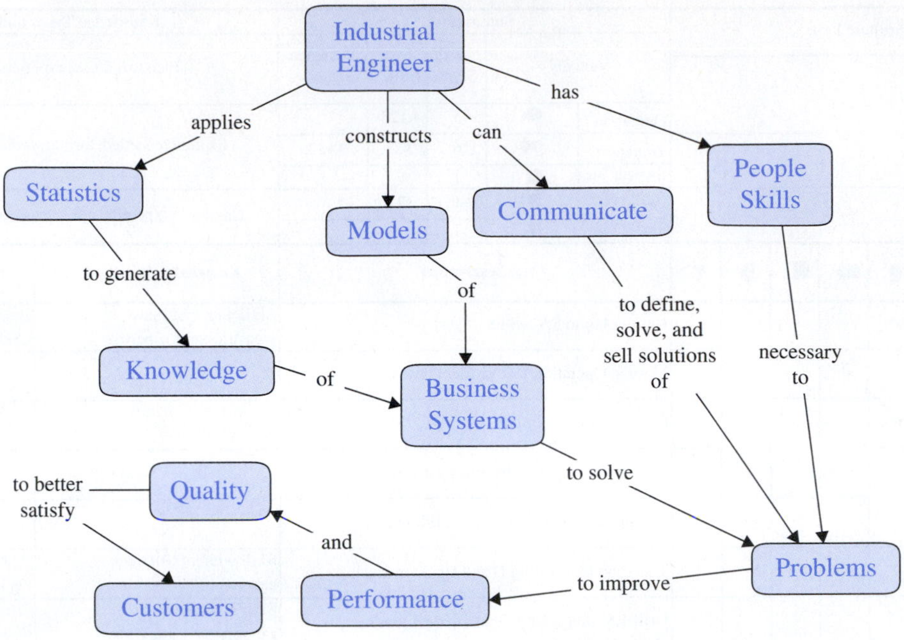

FIGURE 7-43. Brief Concept Map of Industrial Engineer

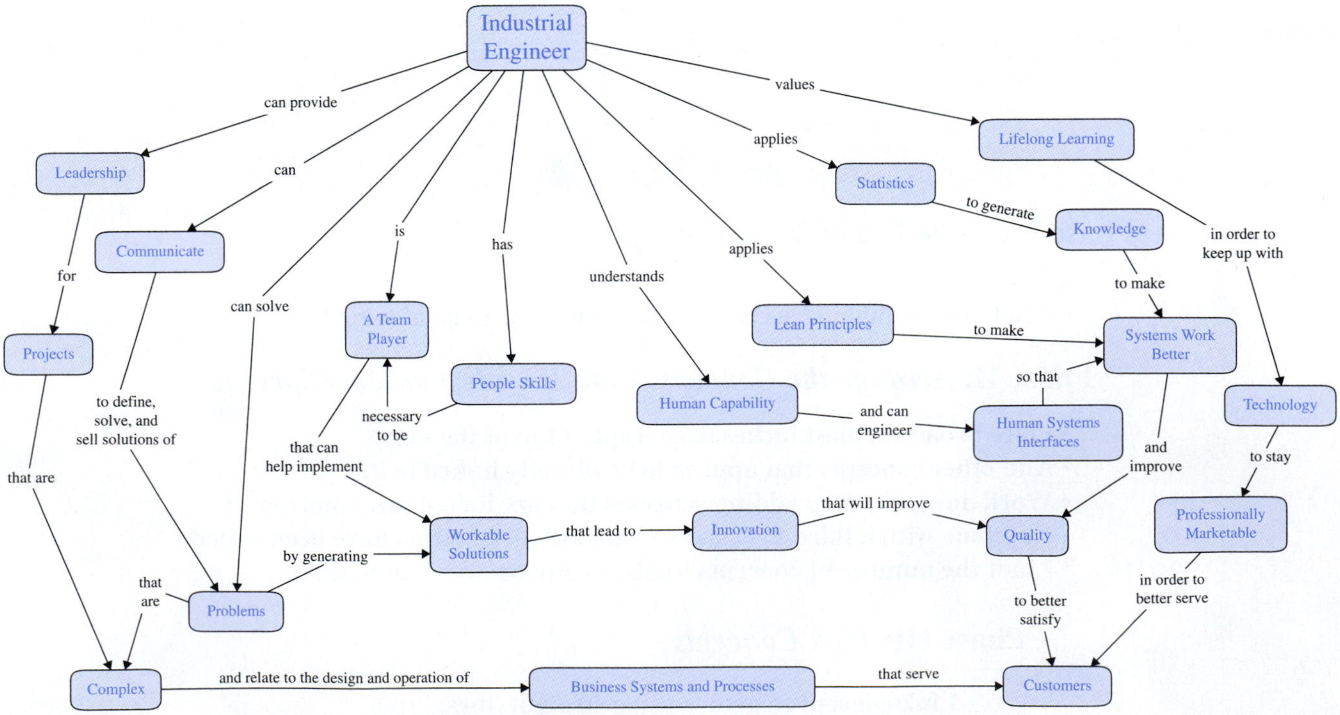

FIGURE 7-44. Simplified Concept Map of Industrial Engineer

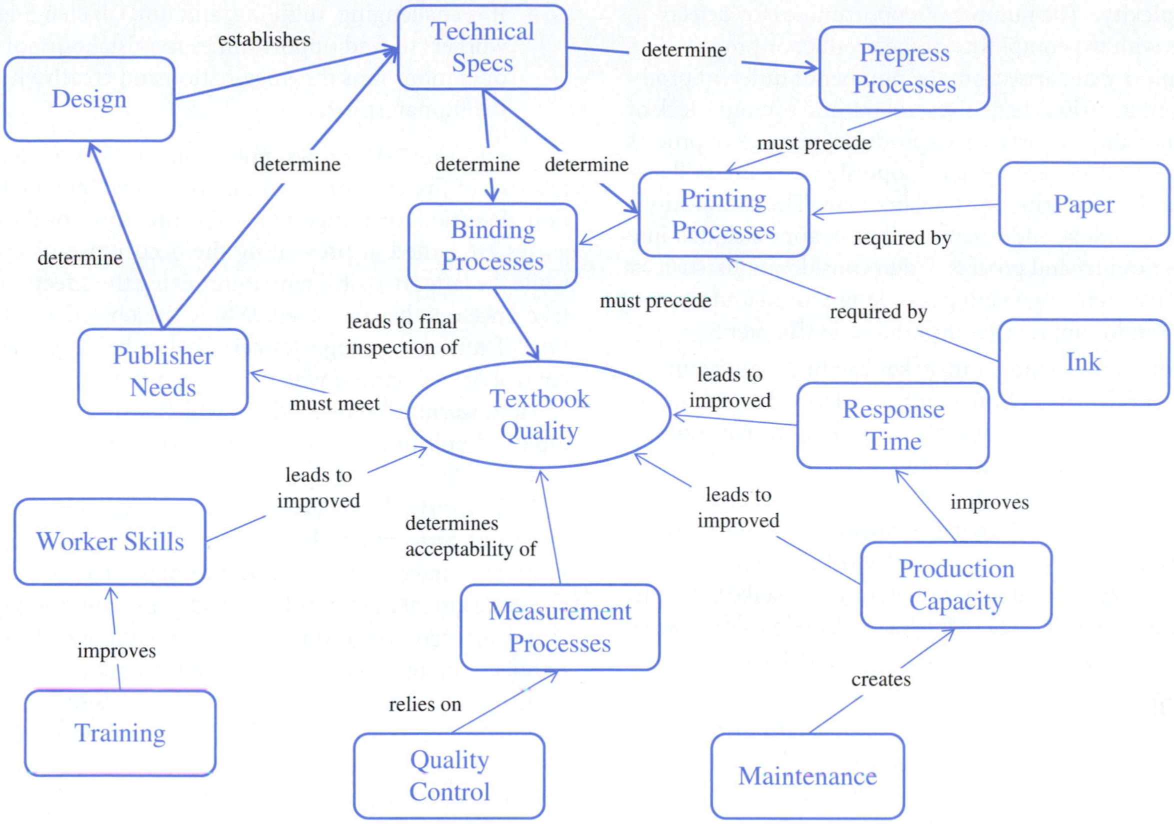

FIGURE 7-45. Concept Map for Textbook Production System

expertise, cross-linked to individuals who possess them. Such a map can provide a ready reference to enable anyone in the organization to seek out expertise in a specific competency area.

7.5.3 Poka-yoke

Poka-yoke (pronounced POH-kay YOH-kay) is a Japanese word that means to avoid (*yokeru*) inadvertent errors (*poke*). As a component in quality design, the use of poka-yoke incorporates mistake-proofing devices intended to either prevent defects from occurring, or to inexpensively inspect process output to determine, *at the source*, whether an error has occurred. This concept is called zero quality control and was pioneered by Shigeo Shingo, a Toyota industrial engineer.[10]

Poka-yoke incorporates mistake-proofing devices in a process or product design.

7.5.3.1 Defect Causes. Defects are created by one of four major sources: cultural factors, random variation, product or process complexity, or human error.

[10]Shingo, Shigeo, translated by A. P. Dillion. *Zero Quality Control: Source Inspection and the Poka-yoke System*. Portland, OR: Productivity Press, 1986.

- **Cultural Factors.** In Chapter 3 we discussed human behavior and the relationship between the workplace culture and quality. Such cultural factors as personal values, attitudes, motivation, commitment, rewards, and growth needs strength (GNS) can impact job performance and quality. The formation of teams, high-performing work practices (HPWPs), good communications, a committed management, and human resource practices that consistently promote the ideals of quality all help reduce the number of defects attributable to culture.
- **Variation.** Process variation can be a principal cause of defects, particularly if the process is inherently incapable of producing output that complies with the design tolerance. The frequency of defect occurrences can be reduced by identifying and removing special cause factors such as differences between machines, operators, shifts, suppliers, and the like. Special cause variation can be managed and decreased using statistical process control techniques. Reducing common cause variation is more difficult. To do so necessitates a fundamental redesign of the process. This requires sophisticated tools such as Taguchi methods and design of experiments.

Defects are caused by cultural factors, random variation, product or process complexity, or human error.

- **Complexity.** The number of opportunities for defects grows with the complexity of the product or process. Complexity can arise from the number of different product configurations (e.g., sizes, colors, and options), lack of commonality of parts across products, number of processing steps, or the technological sophistication and skill levels required for critical processing steps. The complexity issue can best be addressed through an approach that integrates product and process design considerations, such as the concurrent engineering and design for manufacturability (DFM) approaches introduced in Chapter 5.
- **Human Error.** Human mistakes can be a significant source of defects. A human error is defined as an intention that is incorrect, or a correct intention that results in unintended consequences. Human error typically will be part of the common cause system of variation, and generally will not be detected by control charts or other SPC techniques. The idea behind poka-yoke is to prevent mistakes from occurring or to detect a mistake before it becomes a defect. Poka-yoke is also known as mistake-proofing, fool-proofing, or fail-safing. Poka-yoke is a very powerful concept for four reasons.
  1. It is easy to understand. The principal innovators and creators of poka-yoke ideas are often the operators who are closest to their processes.
  2. It makes sense. It is usually easy to get buy-in because poka-yoke is grounded in common sense.
  3. It is inexpensive. Most poka-yoke ideas are easy and inexpensive to implement. No sophisticated technology or system reengineering is normally required.
  4. It is challenging, fulfilling, and fun. Challenging workers to find opportunities to mistake-proof their operations taps the imagination and creative juices of the human spirit.

As Figure 7-46 shows, poka-yoke devices generally fall into one of five categories: elimination, replacement, facilitation, detection, or mitigation.[11] The first three of these categories are aimed at preventing the occurrence of mistakes, while the latter two deal with minimizing the effects of a mistake once one has occurred. Where the objective is prevention of mistakes, categories can further be subdivided into control devices (those which make a mistake an impossibility) and warning devices (those which provide visual or audible symbols or reminders of the correct procedure).

7.5.3.2 Control Devices. Control devices force the worker to perform a task correctly by making it virtually impossible to make a mistake. Examples of poka-yoke control devices are evident in many of the products used every day. As an example, when new technology required unleaded fuel for automobile engines, gasoline stations had to begin stocking both leaded and unleaded products. To prevent customers from accidentally putting leaded gasoline into an unleaded vehicle, the automobile companies collaborated with the oil companies and developed a poka-yoke solution. The aperture in the gas tank on new vehicles was made small enough so that the

[11] Nakajo, T., and H. Kume. "The Principles of Foolproofing and Their Application in Manufacturing." *Reports of Statistical Application Research*, Union of Japanese Scientists and Engineers 32, no. 2 (1985): 10–29.

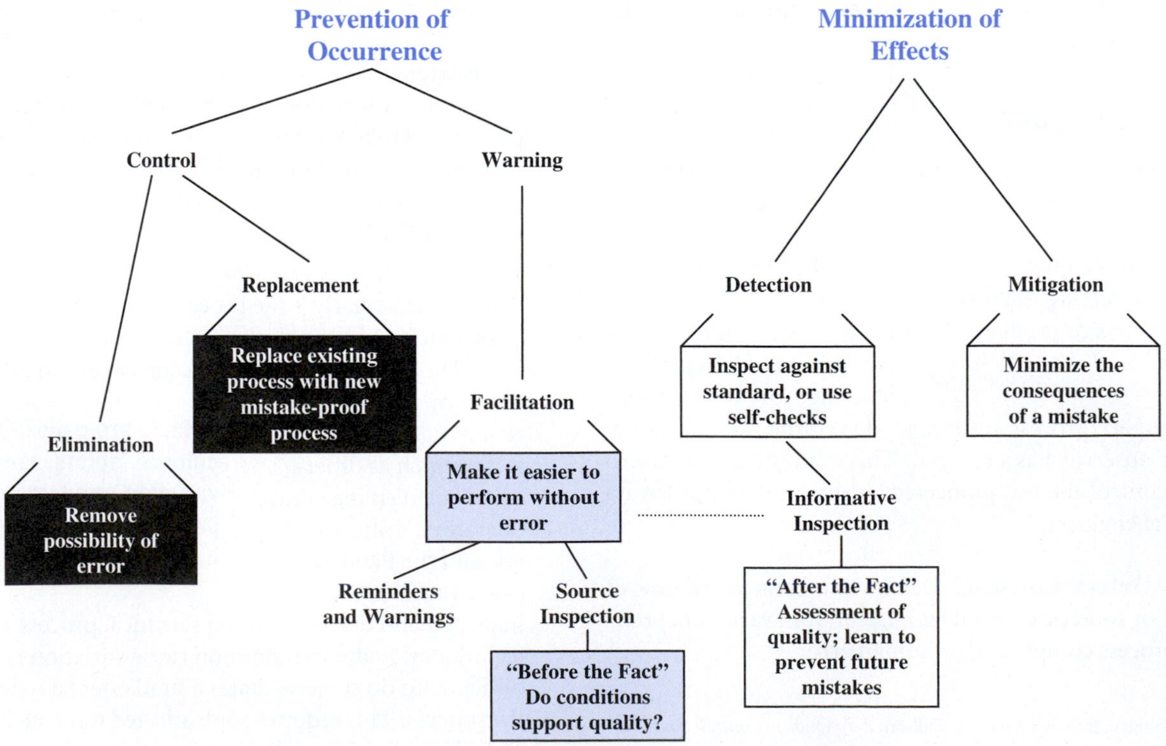

FIGURE 7-46. Categorization of Poka-Yoke Devices

nozzle from the leaded pump would not fit. Only the smaller unleaded pump nozzle can be inserted into the gas tank opening.

Here are two other examples of control poka-yokes. Have you ever tried to put your car in gear when the key was not in the ignition, or insert a floppy disk upside down into a computer drive? You would have had little success. The gearshift lever for an automatic transmission cannot be moved out of Park until the key is in the ignition and turned on. Floppy disks are shaped with a beveled edge that makes it impossible to insert them incorrectly into a computer.

Poka-yoke can be introduced as either a control or warning device.

7.5.3.3 Warning Devices. Warning devices do not force correct behavior but provide visible or audible signals as a reminder. Examples are warning lights in automobiles, color-coded parts in assemblies, clearance bars in parking garages, and park-assist warning systems in automobiles.

7.5.3.4 Application of Poka-yoke. Poka-yoke is likely to work well in the following situations:

- Where labor-intensive operations require manual tasks and worker vigilance
- In tasks where positioning by the worker is critical
- In tasks where adjustments by the operator are required
- Where teams are required to generate common-sense suggestions to reduce defects caused by human errors
- Where SPC is difficult to apply, is apparently ineffective, or appears to have reached its maximum improvement potential
- Where quality is measured on a go/no go attributes basis and not on a variables measurement basis
- Where training costs and employee turnover are high
- Where mixed model scheduling is used
- Where customers can make mistakes in the use of the product or service and place blame on the service provider
- Where special causes, though eliminated, can possibly reoccur
- Where external failure costs are much greater than internal failure costs

On the other hand, poka-yoke is not particularly suited for these situations:[12]

- Where quality is determined using destructive testing
- Where production cycle time is extremely low
- Where process changes occur so frequently that the dynamics exceed the system response capabilities
- Where control charts are so effective that successive inspection and self-checks may be unnecessary

7.5.4 Variance-Tracking Matrix

Figure 7-47 illustrates an L-shaped matrix tool called a **variance-tracking matrix** that can help an organization focus on its core processes and trace root cause analysis. This tool helps a group analyze all activities in the value stream, their

[12]Grout, J. R. "Mistake-Proofing Production." *Production and Inventory Management Journal* 38 no. 3 (1997): 33–37.

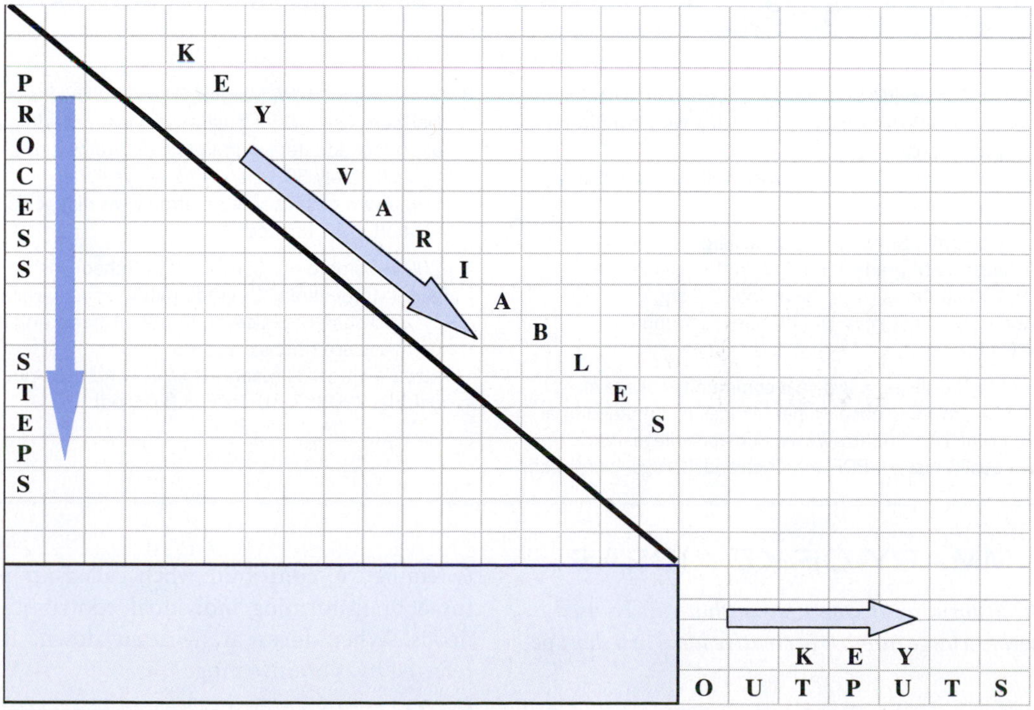

FIGURE 7-47. Format for Variance-Tracking Matrix

associated sources of variability, and how these sources affect downstream processes.

The vertical axis represents the activities on the value stream—that is, the process flow through the production facility. Each major source of variation is assigned a unique number and is listed on the hypotenuse next to the process step where the source occurs. The downstream impact of each source is tracked down the matrix and the variation number is placed next to each downstream process step where there is a likely connection.

The bottom section of the matrix represents key performance characteristics—those things that are important to the customer. The numbers in each performance row correspond to sources of variation that may affect the quality characteristic represented by that row. By tracing the number back up the column, the source of that variation is easily identified along with the process step where the variation is created.

The variance-tracking matrix is usually the product of iterative brainstorming. Once consensus has been reached, the team returns to the brainstorming mode to try to get agreement on a small subset of *key variances*, those that have the most significant impact on key outcomes. Once identified each variance is targeted for reduction or elimination.

The variance-tracking matrix helps pinpoint root causes.

The variance-tracking matrix *facilitates learning*. It is a dynamic tool that evolves as the knowledge base grows, the team becomes more effective, and the analytical process matures. Figure 7-48 represents a variance-tracking matrix that was developed by a large paper plant.

7.5.5 Six Honest Servants and Five Whys

In 1902 Rudyard Kipling[13] unwittingly foreshadowed a learning process when he wrote,

> *I keep six honest serving-men.*
> *They taught me all I knew.*
> *Their names are What and Why and When,*
> *and How and Where and Who.*

These six honest serving-men would later become the building blocks for discovery and improvement. Ideas can often be found by repeatedly applying this tool to each step in a flowchart, with a line of questioning such as the following:

- What is done here? Why is that done there?
- When is that done? Why is it done then?
- How is it done? Why is it done that way?
- Where is it done? Why is it done there?
- Who does it? Why is it done by those persons?

Since each question is followed by "*Why?*" this technique is sometimes referred to as the **five whys** and can be a very effective methodology for challenging convention and stimulating new thinking.

The six honest servants can stimulate improvement ideas.

[13]Kipling, Rudyard. "The Elephant's Child." One of the *Just So Stories for Little Children*, originally published in 1902. Now in the public domain since publication occurred before 1923. Complete text can be accessed at http://www.boop.org/jan/justso/elephant.htm.

SHOWCASE of EXCELLENCE **Winners of the Malcolm Baldrige National Quality Award Education Category**

- 2008: Iredell-Statesville Schools, a K–12 public school system located in southwestern North Carolina within a diverse community and economy. http://iss.schoolwires.com, http:// www.nist.gov/public_affairs/releases/iss_profile.html.
- 2005: Jenks Public Schools, the 11th largest prekindergarten through 12th grade school district in Oklahoma, serving 9,400 students. http://www.jenksps.org, http://www.quality.nist.gov/PDF_files/Jenks_Public_Schools_Profile.pdf.
- 2005: Richland College, a two-year community college in Dallas, Texas, providing educational services to a multicultural student body of 14,500. http://www.richlandcollege.edu, http://www.quality.nist.gov/PDF_files/Richland_College_Profile.pdf.
- 2004: Kenneth W. Monfort College of Business, University of Northern Colorado, a business school providing a wide range of undergraduate and graduate programs and producing approximately 300 graduates per year. http://www.mcb.unco.edu, http://www.nist.gov/public_affairs/Monfort_PDF_final.pdf.
- 2003: Community Consolidated School District 15, a kindergarten through eighth grade public school system serving 12,390 students in seven multicultural municipalities in a northwestern Chicago suburb. http://www.ccsd15.net, http://www.nist.gov/public_affairs/baldrige2003/CCSD_REV_3.3.04.pdf.

7.6 SUMMARY OF KEY IDEAS

Key Idea: *Brainstorming is a technique that can be used by groups to generate a large number of creative ideas in a short period of time.*

Brainstorming can be structured or unstructured and verbal or written. Under structured brainstorming everyone is required to contribute when called upon; with unstructured brainstorming individual contributions are spontaneous. When ideas are written down, the technique is referred to as brainwriting.

Key Idea: *The nominal group technique (NGT) is an effective tool for building group consensus.*

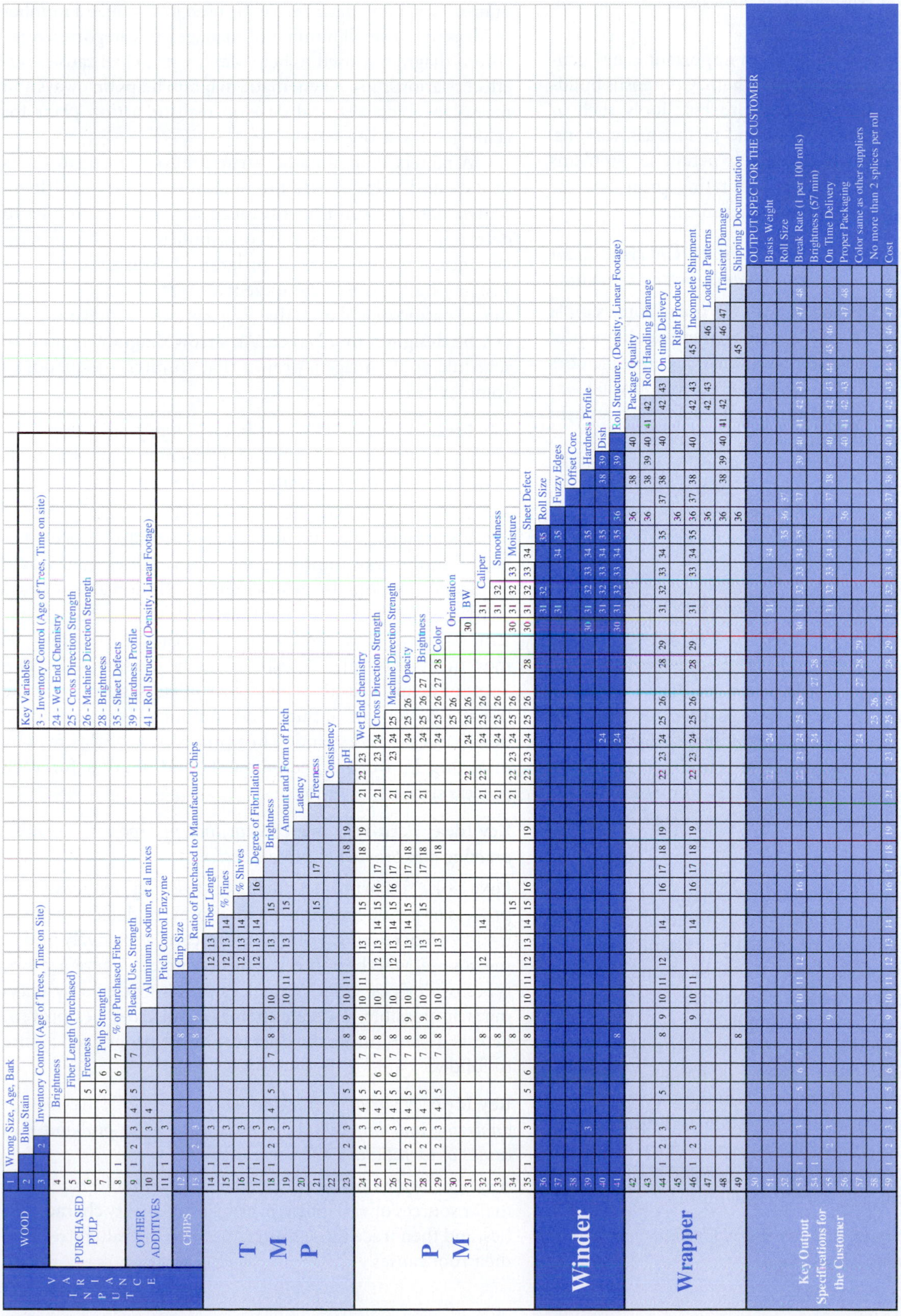

FIGURE 7-48. Example of Variance-Tracking Matrix

The nominal group technique (NGT) is an effective method to help a group reach consensus on fact finding, idea generation, problem solving, and prioritizing of a list.

Key Idea: *Creative tools, such as brainstorming, word associations, and moment-of-truth analysis, help stimulate new ideas and help groups build consensus.*

Creative tools help groups to focus on important issues, generate fresh ideas, and then prioritize them. Consensus building is an important part of this process. Tools such as brainstorming, brainwriting, word and picture associations, and moment-of-truth analyses can help people see problems from different perspectives while facilitating group interaction and making the process more interesting.

Key Idea: *Planning tools, including the PDCA cycle, radar charts, and task cycles, can facilitate organization and structure.*

Good planning can be facilitated with the use of some simple tools. A PDCA cycle is a quality management application of the scientific method (rational decision making) that includes planning, doing (executing the plan), checking on what was learned, and then acting appropriately. A radar chart can help decision makers visualize strengths and weaknesses.

Planning activities can be stalled or prolonged due to the inefficient use of time when groups get together. The productivity of meetings can be improved through more thoughtful planning prior to the event. A task cycle format can help.

Key Idea: *The seven old tools have been time-tested in their usefulness in process control and quality improvement.*

The seven old tools include check sheets, Pareto charts, cause-and-effect diagrams, histograms, stratification, scatter diagrams, and charts. A check sheet is used for manually recording data in a form that can readily be understood by everyone. A Pareto chart is a tool that is useful in separating the significant few from the trivial many so that limited resources can be deployed effectively. A cause-and-effect diagram gives structure to a list of brainstormed causes to help a team focus in a root cause analysis.

A histogram converts raw data into a picture of the underlying distribution of process output. Stratification sorts data into meaningful classifications so that important clues concerning the data will be revealed. A scatter diagram is useful to reveal any meaningful nonrandom relationship that exists between two variables. Charts (and graphs) can reveal any nonrandom patterns of plotted points with respect to time. A run chart is a time-ordered plot of individual measurements that can be used to identify shifts or trends in the process average. The purpose of a control chart is to test the statistical stability of a process. Run and control charts require independence of plotted points. A Gantt chart is used to plan and track progress on projects where precedence requirements must be enforced.

Key Idea: *The seven new tools are useful for managers to analyze systems using nonnumeric data.*

The seven new tools were designed to satisfy a managerial need for tools that could be used to process nonnumeric data. These tools include affinity diagrams, relations diagrams, systematic diagrams, matrix diagrams, matrix data analysis, process decision program charts, and arrow diagrams. An affinity diagram can be used to organize a number of ideas into a small number of manageable categories. A relations diagram is used to help identify principal cause-and-effect relationships. A systematic diagram helps link ways and means to objectives, and develop realistic action plans for achieving those objectives.

A matrix diagram adds structure to two or more sets of data and helps clarify the relationships between them. Matrix data analysis uses advanced statistical techniques to determine those factors that contribute the most to process variance. A process decision program chart is a tool that investigates all possible contingencies that can occur during the execution of a plan so that appropriate countermeasures can be designed in advance. An arrow diagram is used to schedule and manage a project that requires a sequence of complex tasks.

Key Idea: *A process flowchart is a tool that facilitates process analysis by providing a graphical representation of process inputs, outputs, activities, relationships, and flows.*

A flowchart provides the analyst with a picture of the process, which makes it easier to conceptualize process behavior and to spot improvement opportunities than would be possible working from a written description alone. Flowcharts normally trace the path materials take through a system, although they can also be constructed to follow the movement of people. Flowcharts can be constructed at two levels of detail—either as a network of subsystems or as the transformation of inputs into outputs. Transformation flowcharts can be expanded in a process knowledge map by including knowledge requirements and sources for each process step.

Key Idea: *Poka-yoke is a concept that incorporates mistake-proofing into product or process designs.*

Poka-yoke incorporates mistake-proofing devices in a process or product design to either prevent defects (or misuse) from occurring or to determine, inexpensively and at the source, whether an error has occurred. There are two types of devices. Control devices prevent defects by making it impossible to create them; warning devices prevent defects by reminding the operator of the proper procedure to follow.

Key Idea: *A variance-tracking matrix is an effective tool for analyzing and improving processes, as it helps pinpoint root causes of variation.*

A variance-tracking matrix can help problem solvers identify major sources of variability, in critical-to-quality characteristics, and then trace those sources back up the value stream to their root causes.

7.7 MIND EXPANDERS TO TEST YOUR UNDERSTANDING

7-1. How can a moment-of-truth analysis, flowcharting, and the six honest servants be combined and used as part of the PDCA cycle to improve service delivery?

7-2. Construct a process flowchart to describe each of the following:

a. Planning a vacation

b. Hiring a new staff member

c. Preparing a meal

d. Refinishing a piece of old furniture

7-3. The Human Resources Department of AMP Vacuum, Inc. has designed a series of employee e-training courses to support its quality inspired management program. All courses can be taken online; however, some of the courses cannot be taken until other courses are first completed as prerequisites. If all prerequisites are satisfied an employee may enroll in multiple courses simultaneously. The following table shows the required prerequisites and duration for each training course.

	Training	Prerequisite Course	Duration (days)
A	Teaming and People Skills	None	5
B	Tools for Creativity	A	10
C	Tools for Planning	B	5
D	Flowcharting	A	3
E	Knowledge Mapping	D	5
F	Morphological Analysis	A	3
G	MOT Analysis	B, D	1
H	Force Field Analysis	B	1
I	Six Honest Servants and Five Whys	B, D	2
J	SPC: The Basic Toolkit	C, D, I	10
K	Seven New Tools	B, J	10
L	Poka-Yoke	B, D	7
M	Variance-Tracking	J	4

a. Construct an AOA and an AON network that describe the paths available to an individual employee who goes through the entire training program.

b. Calculate the minimum number of days required for an employee to complete all courses.

c. Construct a Gantt chart that can be used to track the earliest start times for scheduling the entire training program.

7-4. The following measurements were collected hourly from two production processes.

Process A		Process B	
6.5	9.0	9.2	4.7
4.1	9.6	6.8	8.6
5.4	9.8	10.6	9.0
5.1	8.1	13.2	10.3
5.3	8.9	13.0	9.1
3.9	9.0	14.3	9.2
5.5	7.2	4.5	9.1
6.1	7.3	9.4	13.4
3.1	8.8	12.7	9.8
5.2	8.1	7.3	9.5
7.3	9.0	8.3	8.7
4.0	5.7	5.8	14.9
4.2	8.4	5.4	12.2
6.3	6.8	7.6	15.9
4.7	7.2	8.1	8.4

a. Construct a run chart for each of the processes.

b. What can you conclude from the pattern of points on these two charts?

7-5. Ace Jet Airlines is losing business because of late flight departures. A quality improvement team has met to discuss the issue. They have decided to construct a fishbone diagram to try and identify some possible reasons why flights fail to depart on time. They have identified five major categories they want to use to structure causes on the diagram. They are *personnel, equipment, procedures, materials*, and *other*. Using these main divisions, construct a fishbone diagram to help this team with their work.

7-6. How can the use of a knowledge map support a commitment to the knowledge worker and facilitate knowledge transfer?

7-7. The following check sheet was constructed on 45 defective motors identified during a final inspection process in a small electronics plant.

Sparks Electric Corporation Defective Motors

Defective Unit	High Turn-on Speed	High Ripple Current	High Leakage	Low Output at Low Speed	Low Output at High Speed	Dead Unit	Bad Regulator	Bad Voltage Setpoint
1	✓				✓			
2		✓					✓	
3	✓	✓						
4	✓	✓					✓	
5			✓					
6		✓		✓			✓	
7	✓	✓	✓				✓	
8		✓						
9	✓	✓		✓			✓	
10	✓	✓						✓
11	✓						✓	
12		✓		✓				
13	✓	✓	✓			✓		
14	✓	✓						
15		✓					✓	
16		✓					✓	
17		✓	✓		✓			
18		✓					✓	
19		✓	✓					
20		✓		✓			✓	
21	✓			✓			✓	
22	✓	✓						
23	✓	✓	✓	✓	✓		✓	✓
24		✓					✓	
25			✓				✓	
26		✓		✓				
27		✓		✓			✓	
28		✓			✓			
29	✓	✓	✓			✓		
30		✓		✓			✓	✓
31		✓					✓	✓
32		✓		✓	✓		✓	
33	✓		✓				✓	
34		✓						
35		✓			✓			
36	✓			✓				
37		✓			✓	✓		
38		✓	✓	✓				✓
39		✓						
40		✓	✓				✓	
41	✓	✓		✓				
42		✓						
43	✓	✓	✓	✓			✓	
44		✓				✓		✓
45	✓	✓		✓			✓	

a. Construct a Pareto chart for these data.

b. Which problems appear to deserve the most attention? Why?

c. Assume that high leakage and low output at low and high speeds is 5 times as severe as high turn-on speed, and that a dead unit or bad voltage setpoint is 10 times as critical as high turn-on speed. Also assume that high turn-on speeds, high ripple current, and a bad regulator are all of equal importance. Construct a weighted Pareto chart and describe how your recommendations would change from your conclusions in part b.

7-8. Suggest a poka-yoke device for each of the following and indicate whether the device is of the warning or control type.

a. Filling an automobile crankcase with the correct amount of oil

b. Ensuring that the polarity is not crossed on an electrical connection

c. Doing a preflight check on an aircraft

d. Detecting a weak battery in a smoke detector

e. Keeping a coffeepot from overheating

f. Ensuring that DVDs are properly inserted into DVD recorders.

CHAPTER EIGHT

STATISTICAL PROCESS CONTROL BY VARIABLES

QUALITY MAXIM #8: *A process in statistical control is stable, repeatable, and predictable, and provides assurance that past behavior is representative of future performance.*

Better Shape Up, Processes! SPC Can Help You Do It

A popular song[1] from the 1978 film version of *Grease* features the lead actors John Travolta (who plays Danny) and Olivia Newton-John (as Sandy). Through its lyrics, Danny admits his attraction to Sandy. She replies that she too has felt the power, but if they are to be together there are conditions. Danny will have to shape up because she has standards[2] that he will have to learn to respect. While this song is about two young people in love, its message is intrinsic to the principles of quality and process management. Quality-at-the-source, which is subsumed under the Japanese practice of jidoka[3], requires an understanding of how a process is performing and the ability to intervene with an appropriate strategy when performance falls short of requirements. In other words, armed with the knowledge of the how and why of processes and a desired standard (goal), operators can apply quality tools to get their processes to "shape up" to meet expectations. Workers on the front-line, due to the nature of their job assignments, are intimately involved (at least technically) with their machines and processes. Using their expertise and instincts they can *sense* when a process is changing. Statistical process control (SPC) tools leverage on this phenomenon by employing statistics to predict and confirm workers' intuition. These tools are the *voice of the process* and provide data that can form patterns and provide early warnings that something is going awry. Seasoned operators equipped with the powerful tools of SPC are the lifeblood of quality—this is where processes can be shaped to protect stakeholder interests.

[1]The song titled "You're the One That I Want" was written by John Farrar for the 1978 film version of the musical *Grease*, and performed by John Travolta and Olivia Newton-John. The film was directed by Randal Kleiser and based on Jim Jacobs and Warren Casey's stage musical of the same name.

[2]The music and lyrics to "You're the One That I Want" can be accessed at http://www.lyricsmania.com/lyrics/grease_lyrics_25338/other_lyrics_55893/youre_the_one_that_i_want_lyrics_571359.html.

[3]Jidoka was described in Chapter 6.

LEARNING OBJECTIVES AND KEY OUTCOMES

After careful study of the material in Chapter 8, you should be able to:

1. Differentiate between nominal, ordinal, interval, and ratio scales of measurement.
2. Discuss the nature of variables measurements and provide examples.
3. Define the statistical terms population, sampling frame, distribution, and sampling statistic.
4. Define statistical control and differentiate between common cause and special cause variation.
5. Apply statistical tests to determine statistical control for control chart startup, including runs tests.
6. Apply statistical tests to determine ongoing operational statistical control, including tests for runs and other nonrandom patterns.
7. Construct and interpret $\bar{X}$ and R charts.
8. Construct and interpret $\bar{X}$ and s charts.
9. Construct and interpret X and MR charts.
10. Apply the Box-Cox transformation to a nonnormal data set.

8.1 MEASUREMENT SCALES

The control of quality is all about measurements. It has been said that *if you can't measure it you can't improve it*; and also *if it is not being measured it is not being managed.* The secret to success is being able to select measures that reflect those features in a product or service that customers care about, and then to develop a measurement process that can consistently produce measurements that are accurate and reliable on

To improve a process one must be able to measure it.

the critical parameters that relate to those features. A measurement process assigns a value to some attribute, and the measured value provides a convenient means for comparing objects with respect to that attribute. The measurement process can select from a hierarchy of four scales that increase in complexity and information. The scales are nominal, ordinal, interval, and ratio.

8.1.1 Nominal Scale

On a **nominal scale,** objects that are assigned equal values are considered to be the same with respect to the measured attribute. The values assigned to an individual object have no numeric meaning. Examples of nominal measures are *part numbers*, *go/no go*, *defective/conforming*, and *good/bad*. Counting is the only arithmetic operation that can be performed on nominal data. For quality purposes the term **attributes measurement** is used when the data collected is nominal. Such data typically reflect a count of the occurrence of some event such as the number of instances or objects observed. For example, in a sample of 100 items from a certain production lot, an inspector might count and record 8 defectives. The measurement for any of these was a nonnumeric label—"defective," and is useful only as a designator to sort product into broad categories.

A nominal scale assigns to objects names or values that have no numeric meaning.

8.1.2 Ordinal Scale

On an **ordinal scale,** objects with a higher value have more of some attribute than objects with a lower value. With ordinal measures, one item can be judged to be greater than, equal to, or less than another. This comparative analysis is the only legitimate mathematical operation that can be performed. The size of the interval between adjacent scale values is indeterminate; therefore the degree of difference between objects cannot be ascertained. For example, item B and item C can both be greater than item A, and item B can be less than item C. One can deduce that item C might be the most preferred and item A the least preferred, but no information is available as to how much more benefit is derived by selecting C over B or either of these options over A. Examples of ordinal measures are items arranged on a prioritization matrix, types of defects plotted on a Pareto chart, and any rank ordering.

An ordinal scale is used to measure relative preferences.

8.1.3 Interval Scale

With an **interval scale**, the degree of difference between objects can be determined because the spacing between adjacent scale values is equal. For example, the difference between 7 and 6 is the same as a difference between 87 and 86. However, ratios on an interval scale are not equivalent since there is no natural "zero" point. Temperature readings in degrees Fahrenheit or Celsius are measurements on an interval scale. One cannot say that if 90 degrees is twice as hot as 45 degrees, then 200 degrees is twice as hot as 100 degrees. In addition to the arithmetic operations that can be performed on nominal and ordinal data, addition and subtraction operations can be applied to interval data. When data for quality purposes are collected using interval measures, the data are said to be *variables* data.

An interval scale can measure degrees of differences.

8.1.4 Ratio Scale

In quality, a **variables measurement** typically involves data collected using a **ratio scale**—the highest level of measurement. Ratio measures are made relative to some rational zero point. On the ratio scale, interval differences are the same and since ratios are equivalent, they can be compared. The ratio of 8:4 is the same as the ratio of 50:25. All mathematical operations including multiplication and division can be applied to ratio data. Examples of ratio measures are lengths, distances, weights, ages, revenues, and pressures.

With a ratio scale, differences can be compared rationally using the full range of mathematical operations.

8.2 POPULATIONS, DISTRIBUTIONS, AND SAMPLE STATISTICS

8.2.1 Statistical Populations

The first step in performing a statistical analysis of variables data is to identify the **population** of interest. In statistics the term *population* is used to define the set of all possible measurements that can be taken from the **universe**, which is the collection of all objects possessing some characteristic of interest. For example, a cement plant might be concerned with the cure rate of a particular concrete product that it produces and packages in 80-pound bags. The universe in this case consists of all 80-pound bags produced, and the population is the measured cure rates (either actual or potential) of all bags in the universe.

The universe is all objects that have some characteristic and the population is the set of all possible values for that characteristic.

A **census** is taken when all members of a population are observed. A census is possible only when the population is finite; even so, it is often impracticable to take a census due to the cost, time, and effort required. In some cases—for example, a Brinell hardness test—the destructive nature of the test rules out the possibility of a census.

In most industrial applications finite populations do not exist because the observable units represent the output from some process that is continuous and dynamic. In the absence of a population a **frame** is defined from which samples are taken. The frame can be thought of as the production that occurs during some designated period of time. To illustrate the difference between a population and a sampling frame consider a process that produces laundry detergent in 2.95-liter plastic bottles. If a question arises concerning the quality of the laundry detergent that was produced on second shift last Thursday, the population of all detergent produced during that period is finite and well defined.[4] However, if the interest is in the quality of detergent produced over time (and how that quality can be improved), a finite population does not exist. Much of the process output was produced in the past and has been consumed or is yet to be produced in the future. In either event a significant number of the process output units cannot be retrieved, measured, and studied. When it is not possible to identify and/or retrieve all members of a population (actual or theoretical) then a sampling frame must be used to proceed with a statistical analysis.[5]

A frame is a statistical representation of a larger population from which samples are taken.

The sampling frame must be representative of the population membership. This requires an element of judgment and depends on the objectives of the study. For example, a sampling frame could represent all of the production from one shift, the output across multiple machines and/or operators, material from several vendors, and so on. All units that are within a frame have a chance of being selected in any sample, while units not included in the frame have no chance of selection. Any statistical inferences that result from analyses of samples can be extrapolated to the frame but not beyond.

[4]This would constitute an enumerative study, as described in Chapter 1.

[5]When processes are studied over time and sampling frames have to be used, the investigation is usually an analytic study as described in Chapter 1.

8.2.2 Distributions

A **distribution**—or more accurately known as a *probability distribution*—describes the values that a random variable can assume and the respective probabilities that these values will occur. The values must cover all possible outcomes and the total of all probabilities must sum exactly to one, or 100%. A **discrete distribution** is one that has a finite number of outcomes each with a positive probability. The top two distributions shown in Figure 8-1 are discrete. The distribution on the left are the probabilities and outcomes associated with a coin toss exercise. A single flip of the coin can result in one of two possible outcomes: heads or tails. Each outcome has a probability of occurrence equal to 0.5. The second distribution at the top of Figure 8-1 shows the probabilities and outcomes associated with throwing a pair of dice. On each throw there are exactly 11 possible outcomes: 2, 3, 4, 5, 6, 7, 8, 9, 10, 11, or 12. The probabilities of these outcomes occurring are 1/36, 1/18, 1/12, 1/9, 5/36,

A parameter is a characteristic that conveys important information. A statistic is used to estimate population parameters.

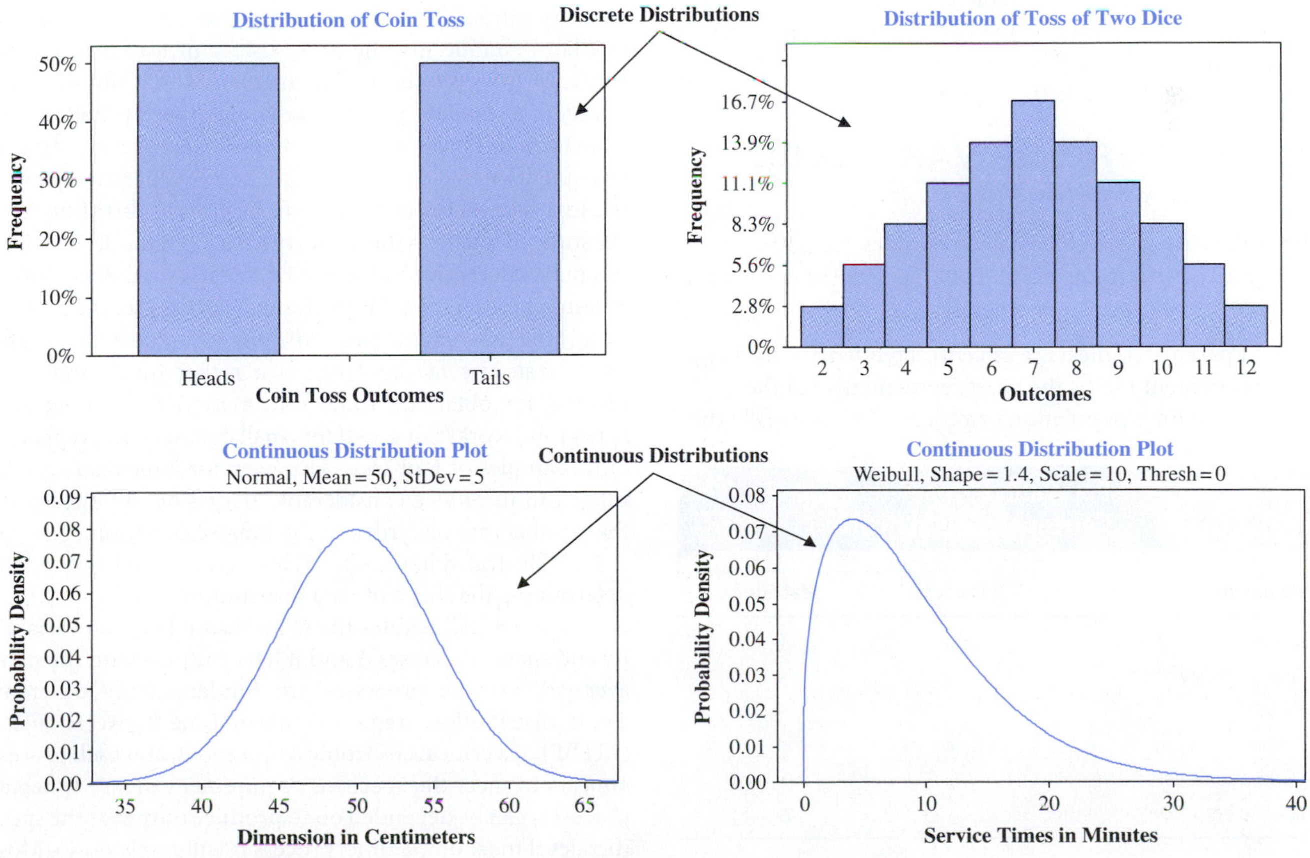

FIGURE 8-1. Distribution Plots

1/6, 5/36, 1/9, 1/12, 1/18, and 1/36, respectively. Notice that these probabilities sum to one.

By contrast a **continuous distribution** describes events over a continuous range of possible outcomes, with the probability of any specific outcome equal to zero. The two distributions shown at the bottom of Figure 8-1 illustrate the probability distributions associated with two continuous variables. The distribution on the left represents the measurements of part dimensions from a machining process and the distribution on the right represents the customer wait times on a customer service 1-800 hotline. In the case of a continuous distribution the total area under the curve (i.e., integrating the probability density function across the domain of measured values) is equal to one.

8.2.3 Sample Statistics

As indicated in Chapter 5, most of the concerns in quality management are addressed through analytic studies, and the occasions appropriate to enumerative studies are rare. Because finite populations do not exist in the case of analytic studies, the relevant underlying distributions must be estimated using sampling techniques. A **statistic** is defined as a numerical quantity calculated from a sample that is used to estimate the parameters of a distribution. A **parameter** is some characteristic of a population or sampling frame that conveys important information. The standard statistical symbology uses Greek letters to designate true process parameters. Corresponding Arabic letters are used to represent the parameter estimates derived from sample statistics. The most common parameters and their corresponding statistics are shown in Table 8-1.

A distribution is a plot showing all the possible values for a random variable and the probabilities of each value.

8.3 NATURE OF VARIABLES MEASUREMENTS

When data are collected using a variables scale (either interval or ratio) two parameters and one characteristic of the underlying distribution are of interest.

- First parameter, **measure of central tendency**—the single measurement that is the most representative of the entire set for a population or process. This is usually the arithmetic mean (average). As indicated in Table 8-1 this is the parameter μ and is usually estimated from the sample statistic $\overline{X}$ (X-bar). Other options are the median or mode.
- Second parameter, **variability**—how dispersed the individual data points are around the measure of central tendency (average). The most common measure of dispersion is the variance or standard deviation. As indicated in Table 8-1 these are the parameters σ^2 and σ, usually estimated from the sample statistics s and s^2, respectively. Other options are the mean absolute deviation (MAD) and the average range $\overline{R}$.
- Distribution characteristic, **shape**—the profile of the entire set of measurements for a population or process (e.g., unimodal, symmetrical, continuous, etc.). The characteristic shape can provide useful information concerning process behavior and possible improvement opportunities.

TABLE 8-1 Population Parameters and Corresponding Statistics

Parameter	Process	Statistic
Mean	μ	$\overline{X}$
Standard Deviation	σ	s
Variance	σ^2	s^2
Proportion	π	p
Correlation Coefficient	ρ	r
Coefficient of Determination	ρ^2	r^2

In most quality applications the arithmetic **mean** is used to measure central tendency and either the **standard deviation** or **range** is used for process dispersion. The mean is defined as the expected value of the variable of interest and, for a particular data set, is computed by adding up all the observations and dividing the sum by the total number of observations. The standard deviation is a measure of the average spread of the data about its mean. The larger the standard deviation the more spread out and dispersed the data; the smaller the standard deviation the more compact (consistent) the data pattern. At times, when the data are excessively skewed, the **median** or **mode** may be more useful measures of central tendency than the mean. The median is that number in a population, or data set, that separates the top half (50%) from the lower half (50%). The mode is the value in a data set or population that occurs most frequently. Due to its symmetry, in a normal distribution – an important distribution in the study of quality – the mean, median, and mode are equal. A simple alternative to the use of the standard deviation to measure process spread is the **range.** In a data set the range is simply the difference between the highest value in the set and the lowest. Use of the range is a rather quick and dirty method for obtaining some information concerning data spread and works fairly well for small data sets (i.e., representative samples of 6 or less). However, for large data sets the range can introduce considerable error since all but two of the numbers are discarded in the range calculation.

As illustrated in Chapter 7, histograms can be useful for determining the shape of data distributions.

Figure 8-2 illustrates the relationship between variability and shape. Processes A and B have outputs with the same averages, yet the processes are fundamentally different. These distributions represent mean time between failure (MTBF) of a critical electronic component, and each process appears to meet the specified requirement of 5.5. Whereas process A can be depended on to produce output at the specified level most of the time, process B will rarely do so. Most

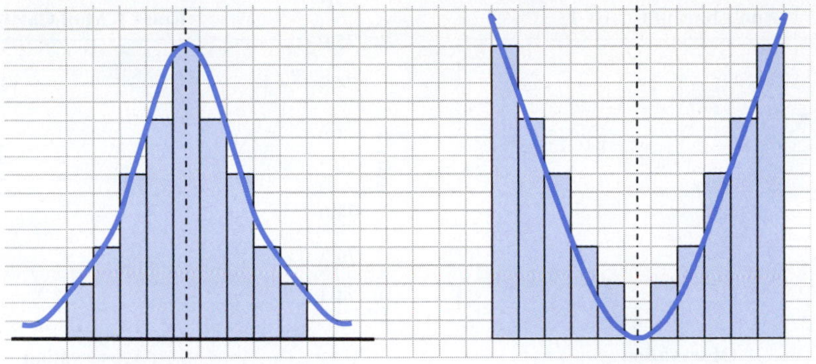

FIGURE 8-2. Two Processes with Equal Means

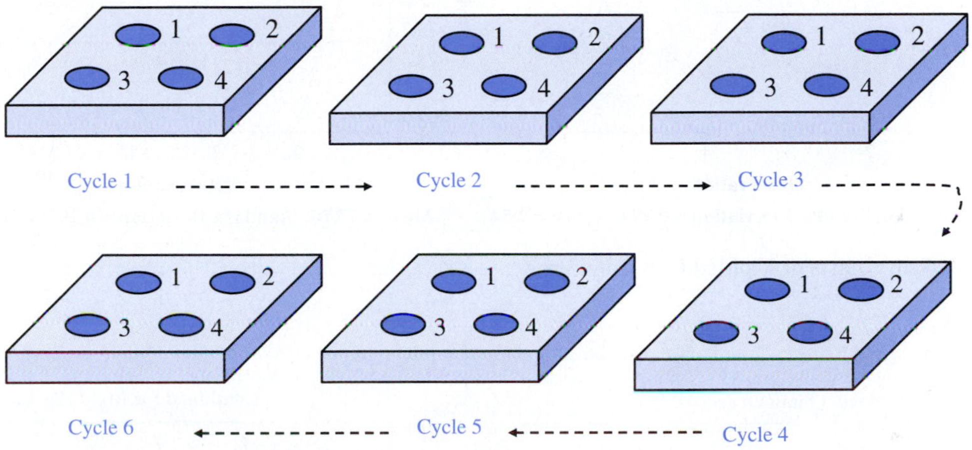

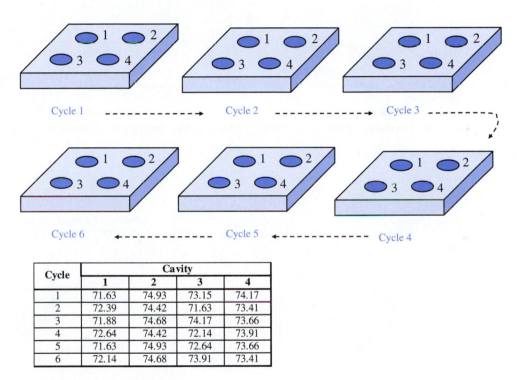

Cycle	Cavity			
	1	2	3	4
1	71.63	74.93	73.15	74.17
2	72.39	74.42	71.63	73.41
3	71.88	74.68	74.17	73.66
4	72.64	74.42	72.14	73.91
5	71.63	74.93	72.64	73.66
6	72.14	74.68	73.91	73.41

FIGURE 8-3. Molded Caps

of the output from process B will either fall short of or exceed the target. The measurements from process A range from a low of 1.5 to a high of 9.5; for process B the measurement range is 0.5 to 10.5. However, the majority of the observations from process B are at the extremes and no measurements at all were recorded at the target value. This results in a standard deviation for process B that is twice that for process A.

To illustrate the nature of variables data, consider the plastic cap molding process illustrated in Figure 8-3. Each cycle of the mold produces four caps. For six consecutive cycles the cap diameters from a representative mold were measured, resulting in a total of 24 measurements as shown. Run charts for each cavity are shown in Figure 8-4 and for the combined data in Figure 8-5. Data fluctuations are obvious and come from numerous sources. One source of variation is *within cavity*, caused by the fact that as the process goes from one cycle to the next each cavity is incapable of perfectly duplicating the product that was produced during the previous cycle. Another important source is variation that occurs *between cavities*, that exists because no two cavities can perfectly replicate each other's outputs. Numerous time-related sources can also exist and have similar impacts on all cavities: factors such as materials, maintenance, and personnel. In addition, samples capture any measurement variability that may be present—that is, the reality that the measurement process is

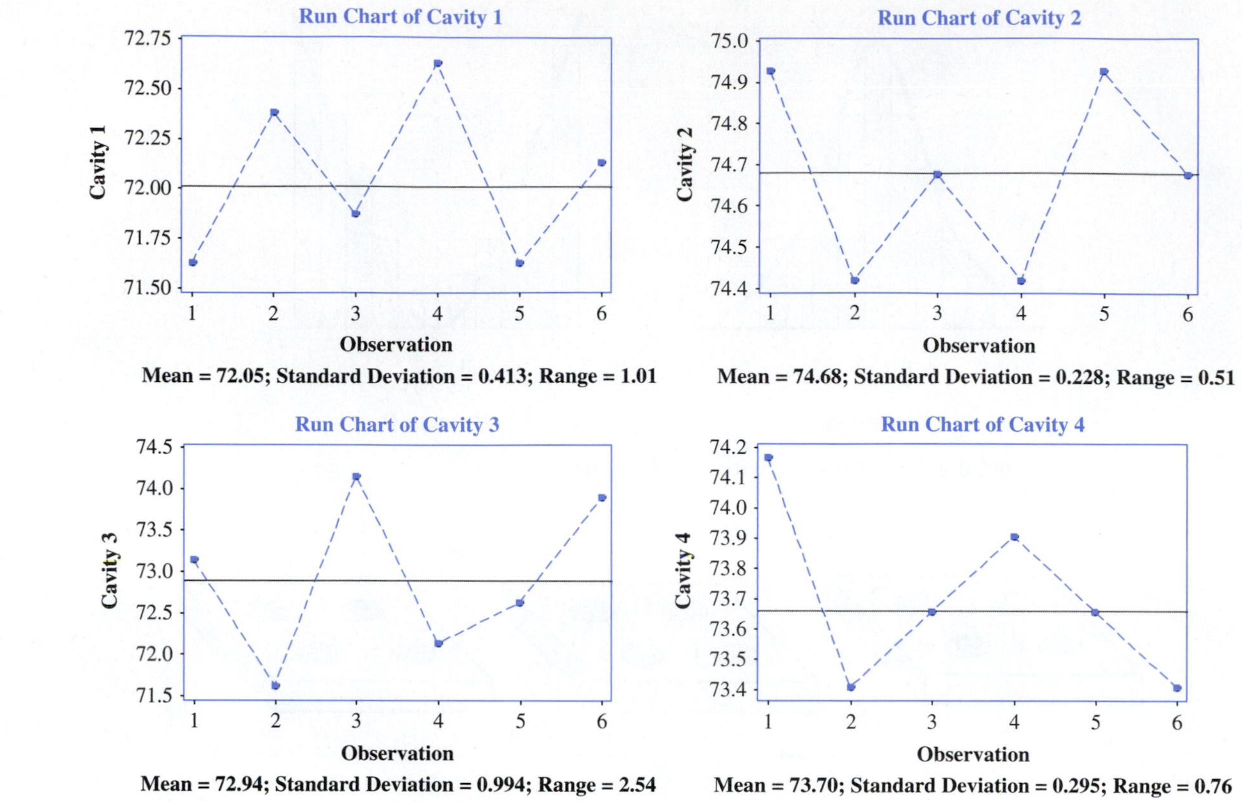

FIGURE 8-4. Run Charts of Molded Cap Data Sets

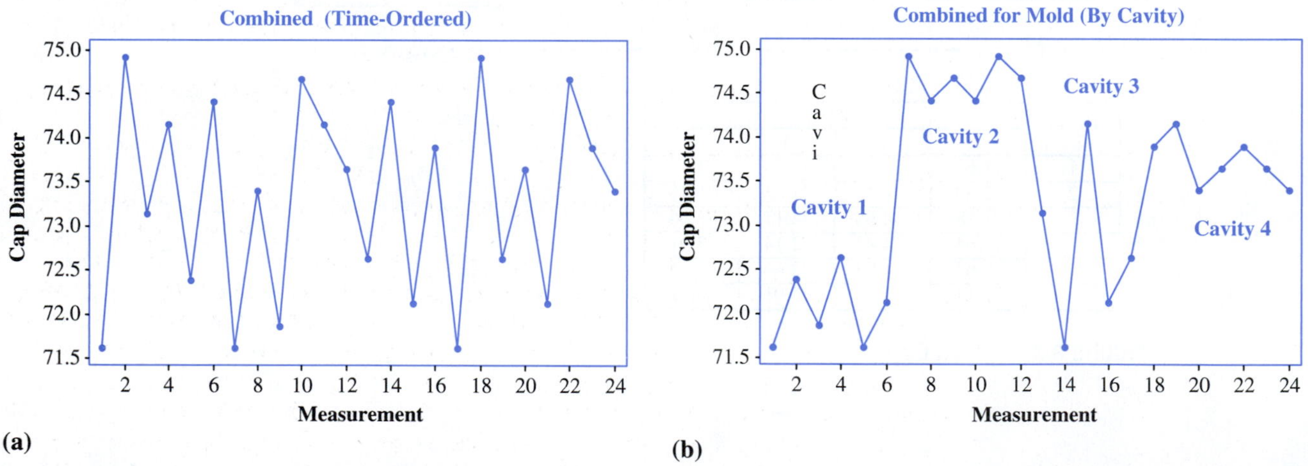

FIGURE 8-5. Run Charts for the Combined Data

incapable of precisely and accurately revealing the value of the true cap diameter.

Figure 8-4 reveals that the four mold positions are different. Cavity 1 appears to have the lowest average dimension and cavity 2 the highest. Cavity 3 has the greatest dispersion around its mean and cavity 2 the least. If the data are plotted in strict time-ordered sequence (that is, in the order that the measurements were taken), as shown in Figure 8-5a, the cavity-to-cavity differences could go unnoticed. Figure 8-5b reveals the apparent differences when the data are plotted by cavity. There are insufficient data to make any judgments regarding the shape of the distributions. Assuming that this process follows a normal distribution, Figure 8-6 provides a cavity-to-cavity comparison of the various distribution shapes.

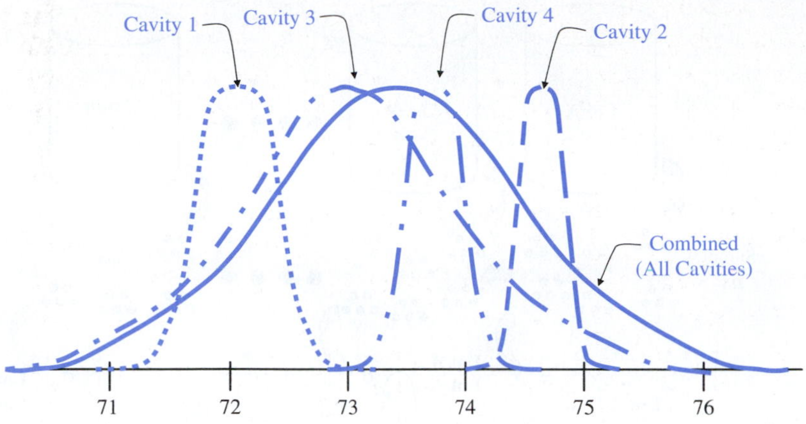

FIGURE 8-6. Molded Cap Distribution Shapes by Cavity

8.4 STATISTICAL CONTROL

8.4.1 The Nature of Statistical Control

In Chapter 7 we briefly discussed the two broad types of variability that are captured in any sampling plan: *common cause* and *special cause*. Whenever samples are collected from a process certain sources of variability (those due to common causes) will be active. Common cause variability is the reason why individual measurements *within a sample* will have different observed values and can be viewed as systemically induced and short term. That is, since a sample is intended to provide a representative snapshot of how a process is behaving at any point in time, the variability captured within a sample is an estimate of the process variability that existed while the sample was being obtained. The time frame is usually relatively short and any observed fluctuations can be attributed to process design considerations. Reducing common cause variation therefore requires improving or reengineering the system. Examples of common causes are measurement error, operator inconsistencies, and within-machine capabilities (i.e., the inability of a machine to exactly replicate performance).

Common cause sources of variability a present in all samples.

Those variability sources that are active in some samples and not in others are due to special, or *assignable*, causes. As samples are collected over time, the fact that these samples can be observed to be different (e.g., have different individual measurements, different means, different standard deviations, etc.) is due to a combination of special cause and common cause variability. Examples of special causes are shift-to-shift differentials, machine-to-machine differences, operator-to-operator discrepancies, and any time-related factors such as materials, maintenance, and production mix.

A process is said to be in a state of **statistical control** when the only sources of variability active are *due to common causes*. This means that all sources of variability have been captured within a sample and each sample is statistically like all other samples. The absence of special causes means that there are no additional active sources that add a between-sample component. A process in control is *stable*, meaning that the element of time is not a factor—a sample taken at any point in time is like a sample taken at any other time. Once a process has reached this level of stability it is said to be *repeatable* and *predictable*, and important process parameters such as mean, dispersion, and shape can be estimated with confidence.

Special causes are present in some samples and not in others.

To better understand statistical control, refer to Figure 8-7. Lot A represents product that has been produced over a finite short time period. This could be a production batch, production output during a time interval (e.g., from nine o'clock to ten o'clock, during a particular shift, on a certain day of the week, output using material from a particular vendor, and so on). As the illustration shows, samples of size 5 have been randomly selected and used to estimate the distribution of the entire lot. At some later time the production process produces lot B. The sampling procedure is repeated and used to estimate the distribution of all items in that lot. If the samples from lot B (and any subsequent lot) predict equivalent overall process distributions, the process is in statistical control. Stated another way—if samples drawn from a dynamic process (one that can potentially change over time) produce results that are similar to what would be expected from a static population (one that remains constant over time), then all the variation observed within a sample is random variation due to common causes, and the process is in a state of statistical control.

A process that is in statistical control is repeatable, predictable, and stable.

Being in statistical control is not the same as producing quality output. Statistical control is simply the first step in process improvement. When a process is in control its parameters can be estimated with calculable risk, and its past, present, and future output can be predicted with a high level of confidence. In the

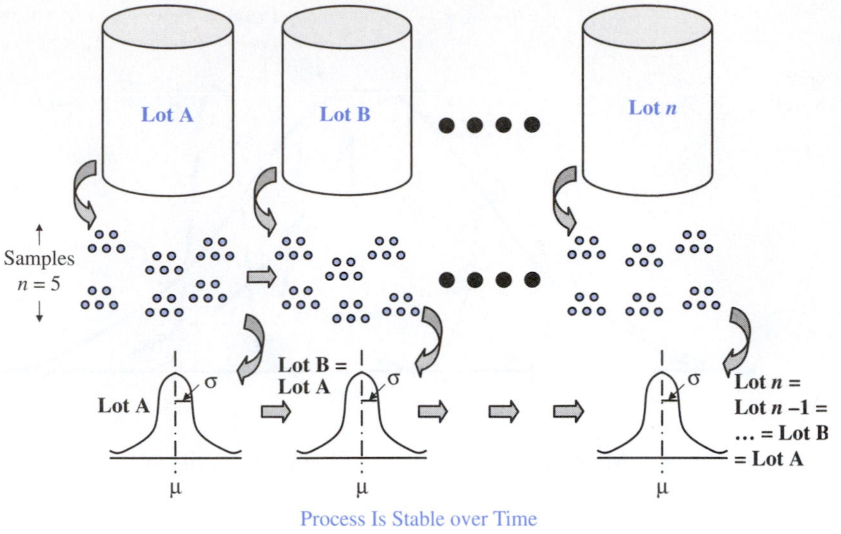

FIGURE 8-7. The Concept behind Statistical Control

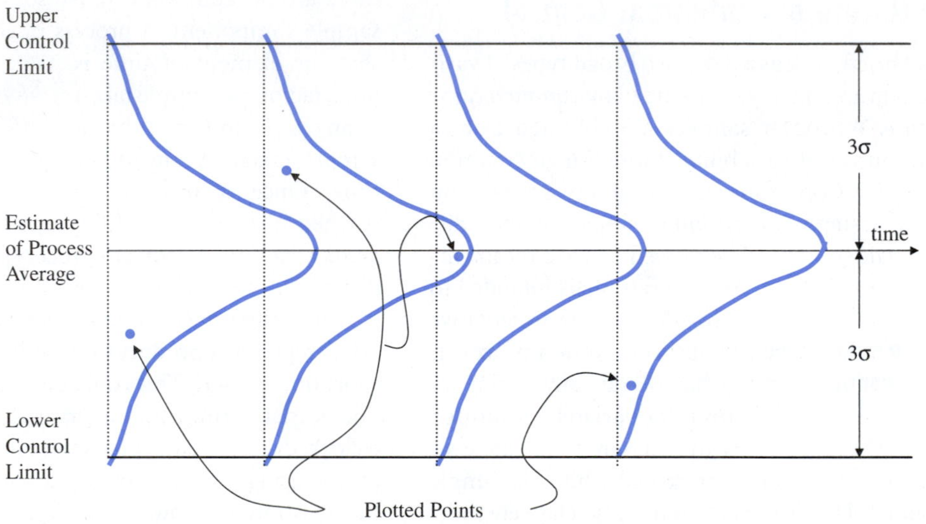

FIGURE 8-8. Control Chart Structure

absence of control, process behavior is sporadic and unpredictable. Before one can determine what steps need to be taken to improve a process it is essential to know how the process is behaving and be able to compare actual with desired performance.

8.4.2 Control Charts

In Chapter 7 we described the control chart as a tool that can reveal the "voice of the process," assist in understanding how processes behave, and help identify improvement opportunities. The greatest benefit comes from data that are plotted in a time-ordered sequence. Therefore it is recommended that control chart construction begin with a run chart and that a minimum of 25 to 30 points be plotted initially. These will be used to estimate the mean and standard deviation of the process and to compute control limits. If the chart shows a state of statistical control, the mean and standard deviation can then be used to estimate process capability and quality of output.

Figure 8-8 shows the generic control chart structure. Sample data points are randomly selected in time sequence from a process that is assumed to be normally distributed. As we learned in Chapter 7 it is also assumed that the points are independent and not autocorrelated. The chart's centerline is equal to the average of all the data plotted (a point estimate of the distribution mean) and the spread of the distribution is estimated using all 25 or 30 samples. Upper and lower control limits are constructed equidistant from the centerline three standard deviations away. These limits represent the largest and smallest measurements that one would expect, assuming that the hypothesized distribution (i.e., normal with mean = centerline and standard deviation = one-third the distance from centerline to control limits) is correct and does not change over time.

Once control is established, the limits and centerline on the control chart are fixed. Process sampling is continued and each subsequent sample is plotted on the chart. As long as the chart shows no evidence of any nonrandom patterns, one can

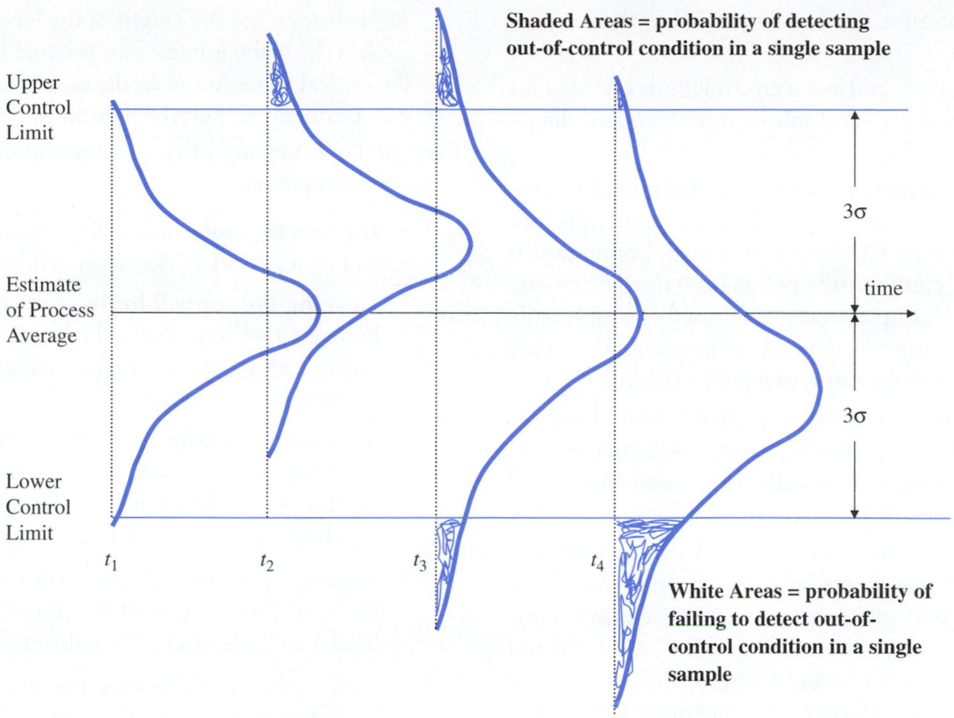

FIGURE 8-9. Out-of-Control Conditions

assume that the process mean and dispersion remain where they were when control was first established. Figure 8-9 illustrates some out-of-control scenarios that can result in nonrandom patterns appearing on a control chart. At time t_1, the process is in control with a mean equal to the centerline, and standard deviation equal to one-third the distance between the centerline and each of the control limits. As long as the process remains in control any new sample has an equal chance of being above or below the centerline (that is, higher or lower than the assumed process average). There is a very small, almost negligible, chance that a point will fall outside either of the control limits (about 1 in 1,000).

By time t_2 the process average has increased. It is now higher than the value represented by the centerline of the chart. At time t_3, the process has returned to its original mean; however, the standard deviation has increased. When the sample is taken at time t_4, the standard deviation is still at the higher level and the mean has decreased. Each of these out-of-control conditions increases the probability of a sampling point occurring outside the control limits. Nevertheless, as Figure 8-9 shows, there is a significant probability that even when the process changes, a sampling point will occur within the control limits and the change will go undetected. To provide some safeguards, there is a battery of statistical tests that can be employed to determine nonrandomness.

8.4.3 Tests for Statistical Control

Statistical control must be achieved before a control chart can be used to effectively monitor a process. A chart in control can provide the basis for knowing *when to intervene* in a process (make adjustments) and as importantly, *when to leave a process alone.* There are several tests for nonrandomness that should be applied, depending on whether the chart is a new startup, or is ongoing.

8.4.3.1 Steps for Initiating a Variables Control Chart. The first step in constructing a control chart is to determine the sample size and frequency of sampling. There is always a trade-off between the size of an individual sample and how often the samples are taken. A good rule of thumb for variables measurements is to *think five.* A sample size of 5 is ideal in most circumstances. The ideal size of 5 might be reduced to 4 or 3 if measurements are costly or difficult to obtain. If, on the other hand, a sample of 5 is insufficient to capture all sources of process variability (e.g., six spindles, seven layers, etc.), the ideal sample size should be increased. As for frequency, samples should be taken at convenient intervals, giving due consideration to process stability and whether changes occur gradually or abruptly. Once these decisions are made the following 10 steps should be followed.

> **Sampling plans should be sufficient to capture all sources of process variability.**

Step 1 Obtain between 25 and 30 random samples from the process.

Step 2 Compute the average and either the range or standard deviation of each sample. Use these data as the basis for calculating estimates of the process mean and standard deviation.

Step 3 Plot the sample points on a chart in time sequence and connect the points using straight lines.

Step 4 Draw a centerline on the chart equal to the mean of the raw data.

Step 5 Construct upper and lower control limits on the chart equal to the mean plus and minus three standard deviations, respectively.

Step 6 If all plotted points are inside the control limits, proceed to step 7. Otherwise, investigate any points that are outside. If (and only if) the out-of-control points can be easily explained, eliminate the points from the data set and return to step 2. Examples of explanations that could justify removing points from the data set are transcription errors, an unusual machine problem, or a power failure.

Step 7 Count the total number of plotted points above the centerline and the total number below the centerline. Let the symbol s represent the smaller of the two counts and the symbol r represent the larger of the two counts.

Step 8 Count the total number of runs. A run consists of a count of consecutive points lying on the same side of the centerline. The start of a new run occurs each time a line connecting the points crosses the centerline. Apply the test for too few runs shown in Table 8-2.[6] If the number of runs is less than or equal to the critical value found in the table, stop. The process is not in control. Otherwise, continue.

Step 9 Count and record the length of each run (that is, the number of consecutive points making up each run). Apply the test for the length of the longest run shown in Table 8-3.[7] If the longest run is equal to or greater than the critical value found in the table, stop. The process is not in control. Otherwise, continue.

Step 10 Look for any obvious nonrandom patterns in the data. Examples are

- **Hugging the centerline**—all (or most) of the points are within one standard deviation of the chart's centerline.
- **Hugging the control limits**—most of the points lie between two and three standard deviations of the mean, but no points (or relatively few) are actually outside the control limits.
- **Stratification**—when each data point is identified by machine, shift, operator, or some other criterion, these identifiers will cluster into distinct groupings, each of which appear to come from a different distribution.
- **Trends**—although there are no points outside the control limits, and neither of the runs tests have been violated, the data appear to be following a trend over time.

If any nonrandom patterns are observed, *stop.* Otherwise, the control chart can be operationalized. To operationalize the control chart, the control limits are fixed at the levels where control was observed and the chart can then be used to plot samples taken from future

[6]The table for critical values for too few runs is also included as Table VI in Appendix B.

[7]The table for critical values for longest run is also included as Table VII in Appendix B.

TABLE 8-2 Critical Values for Too Few Runs

	s															
r	**5**	**6**	**7**	**8**	**9**	**10**	**11**	**12**	**13**	**14**	**15**	**16**	**17**	**18**	**19**	**20**
5	3															
6	3	3														
7	3	4	4													
8	3	4	4	5												
9	4	4	5	5	6											
10	4	5	5	6	6	6										
11	4	5	5	6	6	7	7									
12	4	5	6	6	7	7	8	8								
13	4	5	6	6	7	8	8	9	9							
14	5	5	6	7	7	8	8	9	9	10						
15	5	6	6	7	8	8	9	9	10	10	11					
16	5	6	6	7	8	8	9	10	10	11	11	11				
17	5	6	7	7	8	9	9	10	10	11	11	12	12			
18	5	6	7	8	8	9	10	10	11	11	12	12	13	13		
19	5	6	7	8	8	9	10	10	11	12	12	13	13	14	14	
20	5	6	7	8	9	9	10	11	11	12	12	13	13	14	14	15

Source: Adapted from Swed, F. S. and C. Eisenhart. "Tables for Testing Randomness of Grouping in a Sequence of Alternatives," *Annals of Mathematical Statistics* 14, (1943): 66–87.

The probability of an equal or smaller number of runs than the critical value is less than or equal to .05 when the process is in control.

TABLE 8-3 Critical Values for Longest Run

	s															
r	**5**	**6**	**7**	**8**	**9**	**10**	**11**	**12**	**13**	**14**	**15**	**16**	**17**	**18**	**19**	**20**
5	5															
6	6	6														
7	6	6	6													
8	7	7	7	7												
9	8	7	7	7	7											
10	8	8	7	7	7	7										
11	9	8	8	8	7	7	7									
12	10	9	9	8	8	8	8	8								
13	10	10	9	9	8	8	8	8	8							
14	11	10	10	9	9	8	8	8	8	8						
15	11	11	10	10	9	9	9	8	8	8	8					
16	12	11	11	10	10	9	9	9	9	8	8	8				
17	13	12	11	11	10	10	9	9	9	9	8	8	8			
18	13	12	12	11	11	10	10	9	9	9	9	9	9	9		
19	14	13	12	12	11	11	10	10	9	9	9	9	9	9	9	
20	15	14	13	12	11	11	10	10	10	9	9	9	9	9	9	9

Source: Adapted from Takashima, M. "Tables for Testing Randomness by Means of Length of Runs." *Bulletin of Mathematical Statistics* 6, (1955): 17–23.

production. Control limits or the centerline are not altered unless there is statistical evidence that the process has changed from the initial conditions.

Note that when a rational subgroup cannot be formed (e.g., in the case of a continuous and highly homogeneous product) a sample size of 1 must be selected and an *individuals chart*, discussed later in this chapter, must be used. In this case the 10-step process does not apply.

8.4.3.2 Tests for Operational Control. Sampling points are plotted continually on the control chart as items are randomly selected from the process and measured over time. Numerous statistical tests can be applied to an operational chart to detect out-of-control conditions. Eight rules are commonly used and are easy to employ using a statistical software package such as Minitab. The first four rules, shown in Figure 8-10, were designed by a committee at Western Electric[8] in the mid-1950s and have become accepted practice for detecting nonrandom patterns on control charts. For convenience, the rules have been numbered consistent with the convention used by Minitab. The purpose of these tests is to determine whether statistical evidence exists that would suggest a process change. In applying these tests the control chart area between the control limits is divided into zones.

Zone A includes the area that lies more than two standard deviations from the centerline. Zone B is the area between one and two standard deviations from the mean, and zone C is the area between the centerline and one standard deviation. All of the tests are designed so that there will be less than a 5% probability of being wrong and acting on a false signal. The Western Electric rules are as follows.

- **Rule 1: Out-of-control point.** A single point outside the control limits provides statistical evidence that the process has changed or is unstable.
- **Rule 2: Eight-point-run rule.** A run of eight consecutive points on the same side of the centerline in zone C or beyond provides evidence of a process change. (Note that Minitab uses a Nine-point-run rule).
- **Rule 5: Two-of-three rule.** Two out of three consecutive points on the same side of the centerline in zone A or beyond is evidence that the process has changed.
- **Rule 6: Four-of-five rule.** Four out of five consecutive points on the same side of the centerline in zone B or beyond is evidence of a process change.

In the mid-1980s Nelson proposed that some additional rules be added for detecting nonrandom patterns such as "hugging the centerline" (too little observed variability in the plotted points), "hugging the control limits," trends, over adjustment, and so forth.[9] These rules are illustrated in Figure 8-11 and described as follows:

- **Rule 3: Six-point-trend rule.** Six consecutive points that change value in the same direction, either decreasing or increasing, provide statistical evidence that the data are nonrandom and that the process is trending.

[8]Western Electric Company. *Statistical Quality Control Handbook.* IN: Western Electric, 1958.

[9]Nelson, Lloyd S. "The Shewhart Control Chart—Tests for Special Causes." *Journal of Quality Technology* 16, no. 4 (1984): 237–39.

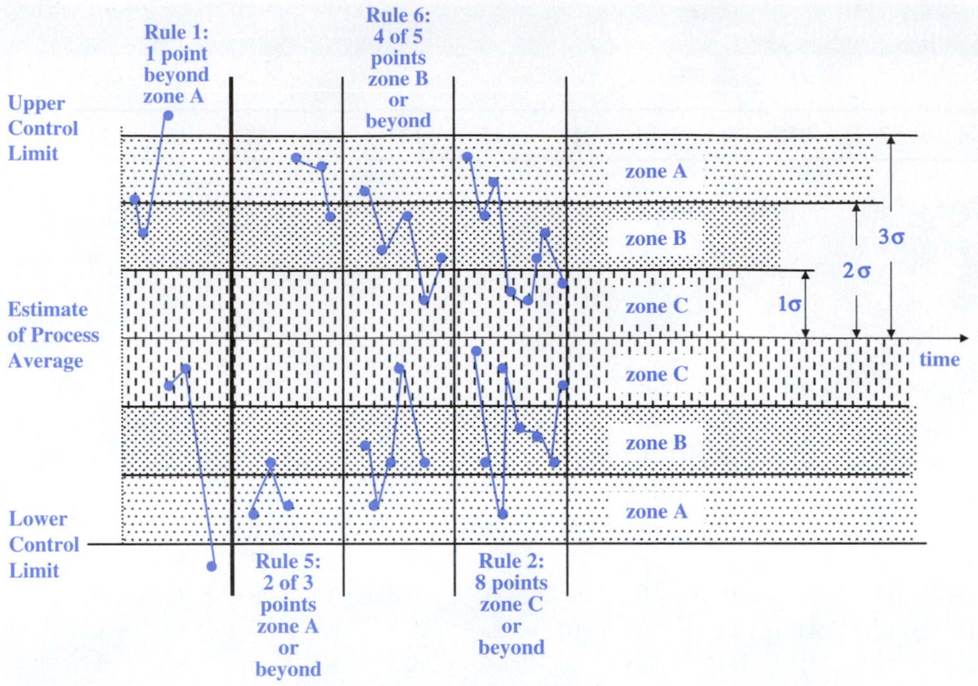

FIGURE 8-10. Zone Rules for Out-of-Control Conditions

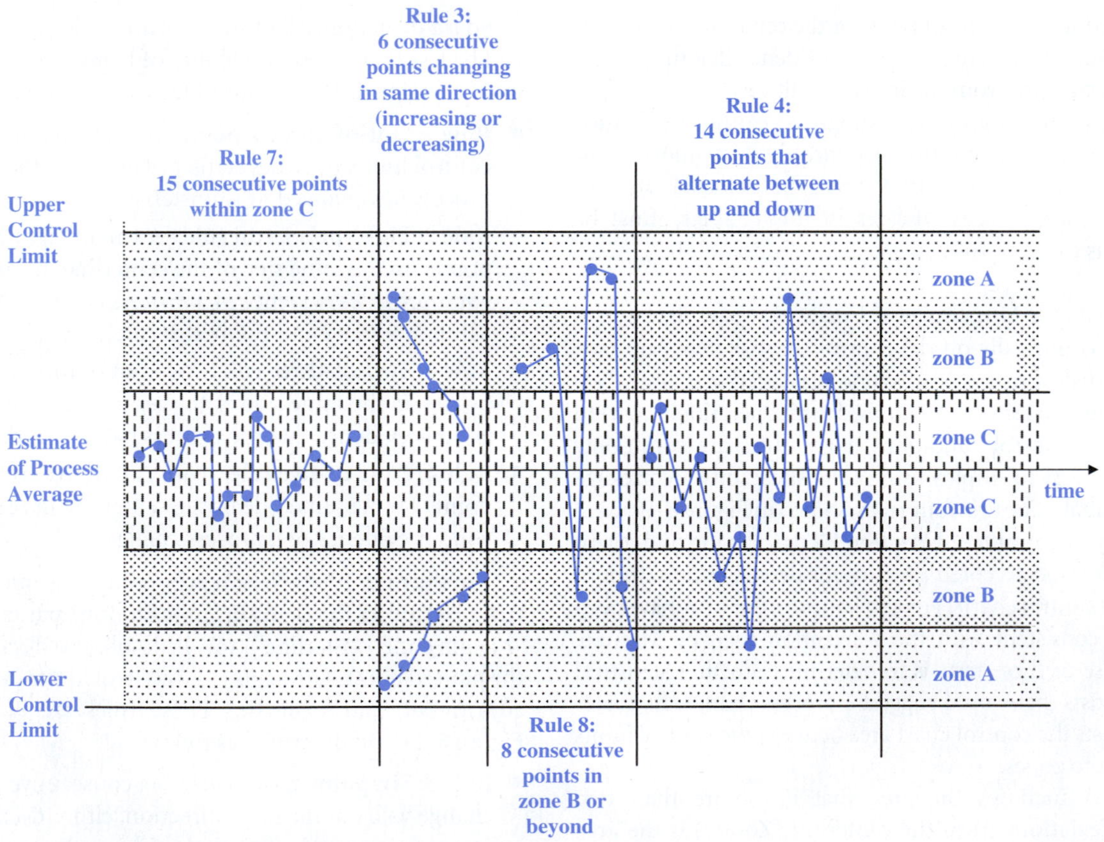

FIGURE 8-11. Additional Rules for Out-of-Control Conditions

- **Rule 4: Zigzag rule.** Fourteen consecutive points that alternatively change direction (up goes down, down goes up) provide statistical evidence of nonrandomness. A random pattern is one that is not predictable; however, a pattern that always reverses its previous direction is predictable to some extent. An example of how this pattern could occur is a phenomenon known as *overcontrolling*. An operator who is trying to steer a process to achieve some target can unwittingly induce variability. When a sample is above the target the operator makes an adjustment to try and reduce the mean. The next sample therefore is below the mean reflecting the reduced average. In response, the operator increases the average. As this process is repeated, a zigzag pattern occurs on the control chart. The process is never stable because of the operator's continual intervention.
- **Rule 7: Hugging the centerline rule.** Fifteen consecutive points within zone A, on either side of the centerline, provide statistical evidence that something about the process has changed. There is less variability in sample results than what is expected—it could be that the overall common cause variability has been reduced or the sampling scheme could be flawed.
- **Rule 8: Hugging control limits rule.** A run of eight consecutive points lying in zone B or beyond, on either side of the centerline, defies the underlying assumption of normality (under which most of a distribution will be concentrated around its mean). This pattern therefore provides evidence of nonrandomness and could be due to the presence of multiple distributions, sampling errors, reclassification of sorted product, or other sources that have the effect of creating a greater-than-expected variability from the mean. If all eight points are on the same side of the centerline, this may be an indication that the process mean has changed. In this case eight points in a row have roughly the same probability of occurrence as a single point beyond zone A (the control limits).

8.5 $\overline{X}$ CHARTS

8.5.1 $\overline{X}$ Charts, Estimating Process Dispersion Using Sample Ranges

When control charts are used to monitor variables data, a *pair* of charts is needed and must be read together to properly interpret data patterns. Since either the process *dispersion* or *mean* (or both) can be out of control, charts are needed to independently evaluate the stability of each parameter. The first of the two charts to be plotted (and interpreted) deals with process dispersion and is called an ***R* chart.** An *R* chart is a time-ordered plot of sample ranges that represent those sources of variability that have been captured *within* a sample, and is best applied in the case of small sample sizes (seven or less). The second chart in the pair is called an $\overline{X}$ (pronounced X-bar) **chart** and is a plot of sample averages. As we shall see, while the *R* chart is primarily concerned with those sources of variability captured *within* a sample the $\overline{X}$ chart is primarily concerned with those sources of variability that are active *between* samples.

Consider the sampling data in Table 8-4. Five measurements ($n = 5$) have been taken from one of the production processes of Acme Fastener Company to form each of 30 samples that have been collected over three shifts for five consecutive days. For each of these, the average ($\overline{X}$) and the range (R) have been computed using Equations 8-1 and 8-2 defined below.

The points on an *R* chart represent sources of variation that have been captured within a sample.

$$\overline{X}_i = \frac{\sum_{j=1}^{n} x_{ij}}{n} \qquad \textbf{8-1}$$

$$R_i = x_i^{high} - x_i^{low} \qquad \textbf{8-2}$$

where,

x_{ij} = the jth measurement of the ith sample
n = the number of measurements within a sample (i.e., sample size)

In developing the formula for control limits for $\overline{X}$ and R charts, consider the following scenario. Let us assume that one takes a very large number of samples of size n from a population with known standard deviation σ. Each time a sample is taken the average and range is computed. In addition, a statistic called the relative range ω, defined by $\frac{R}{\sigma}$, is also computed. Over time and after a very large number of samples are taken, two distributions can be constructed: $f(\omega)$ and $f(R)$. The mean of $f(\omega)$, which we will call d_2, is given by

$$d_2 = E[\omega] = E\left[\frac{R}{\sigma}\right] = \frac{1}{\sigma}E[R] = \frac{\overline{R}}{\sigma} \qquad \textbf{8-3}$$

As the number of samples gets very large (e.g., approaches infinity) for any sample size, d_2 converges to a constant. We can also compute another set of constants, d_3, defined as follows.

$$d_3 = \frac{\sigma_R}{\sigma} \qquad \textbf{8-4}$$

A professor illustrates the use of control charts while his students look on. Control charts are a simple yet powerful tool for analyzing process variability.

TABLE 8-4 Samples from Acme Fastener Process

Sample No.	Measurement					Range	x-bar
1	15.41	13.57	15.85	13.38	14.12	2.47	14.47
2	13.03	18.07	13.55	13.77	15.50	5.04	14.79
3	13.91	15.88	12.88	15.10	14.06	3.00	14.37
4	13.98	13.78	11.63	16.57	12.79	4.94	13.75
5	16.22	15.24	15.91	13.80	16.53	2.73	15.54
6	17.33	12.45	14.59	15.52	14.25	4.89	14.83
7	15.62	19.01	16.31	13.72	14.39	5.29	15.81
8	15.87	14.94	11.86	12.30	14.97	4.02	13.99
9	12.70	16.20	13.60	15.08	15.74	3.50	14.66
10	18.37	14.76	12.05	15.62	13.90	6.32	14.94
11	15.28	14.94	15.87	16.81	14.03	2.78	15.38
12	15.25	15.66	14.58	14.62	14.34	1.32	14.89
13	11.73	15.54	16.05	12.56	13.53	4.32	13.88
14	14.19	14.23	15.09	17.46	13.16	4.30	14.83
15	18.03	15.45	16.17	16.99	12.80	5.24	15.89
16	13.07	11.37	14.97	14.42	18.66	7.29	14.50
17	14.11	16.21	13.71	12.33	13.05	3.88	13.88
18	13.84	14.13	14.13	14.34	14.81	0.97	14.25
19	13.54	13.20	14.99	15.01	12.33	2.68	13.81
20	12.41	13.59	12.46	16.74	13.06	4.33	13.65
21	14.90	15.49	16.95	19.18	16.91	4.28	16.69
22	10.77	12.74	12.09	13.33	15.69	4.92	12.93
23	14.59	12.75	15.01	15.07	17.87	5.12	15.06
24	15.57	13.40	15.03	14.69	13.62	2.17	14.46
25	14.02	15.21	15.31	15.94	13.07	2.87	14.71
26	18.42	18.06	10.79	13.18	16.96	7.63	15.48
27	16.20	13.48	12.92	14.62	12.31	3.89	13.91
28	15.36	11.83	16.61	12.94	15.79	4.78	14.51
29	14.69	14.74	15.43	15.98	16.45	1.76	15.46
30	12.68	15.62	12.68	15.43	14.45	2.94	14.17
Totals						119.64	439.48

where,

σ_R is the standard deviation of the distribution $f(R)$.

Note that d_2 and d_3 are constants for any given value of n. If a process is in control Equation 8-5 can be used to estimate process standard deviation from the average sample range, as follows.

$$\sigma \cong \frac{\overline{R}}{d_2} \qquad \text{8-5}$$

Control limits on $\overline{X}$ and R charts are typically set at a distance that is three standard deviations from the centerline of the chart. Using Equations 8-4 and 8-5 the limits for the R chart can be developed as follows.

$$\textit{Upper Control Limit} = UCL_R = \overline{R} + 3\sigma_R = \overline{R} + 3\sigma d_3$$
$$= \overline{R} + 3\frac{\overline{R}d_3}{d_2}$$

$$\textit{Lower Control Limit} = LCL_R = \overline{R} - 3\sigma_R = \overline{R} - 3\sigma d_3$$
$$= \overline{R} - 3\frac{\overline{R}d_3}{d_2}$$

Formulae for *R* Chart

$$\textit{Centerline} = \overline{R} = \frac{\sum_{i=1}^{k} R_i}{k} \qquad \text{8-6}$$

$$UCL_R = \overline{R}\left[1 + 3\frac{d_3}{d_2}\right] = D_4\overline{R} \quad \textbf{8-7}$$

$$LCL_R = \overline{R}\left[1 - 3\frac{d_3}{d_2}\right] = D_3\overline{R} \quad \textbf{8-8}$$

where

k = total number of samples taken

The d_2, d_3, D_3 and D_4 constants come from Table 8-5 and depend on sample size.

Table 8-5[10] contains control chart constants for sample sizes ranging from 2 to 10. For small sample sizes ($n \leq 7$) $D_3 = 0$. In most cases therefore the $LCL_R = 0$. In reality, for small sample sizes there is no lower control limit for the range.

When an $\overline{X}$ chart is constructed, the central limit theorem, as defined below, provides statistical protection against overreacting to a false signal.

Central Limit Theorem

If samples of size n are taken from a process with mean $= \mu_x$ and variance $= \sigma_x^2$, and the average of all n measurements, $\overline{X}$, is computed for each sample, the distribution of the $\overline{X}$ s will be normal with mean $\mu_{\overline{x}} = \mu_x$ and variance

$$\sigma_{\overline{x}}^2 = \frac{\sigma_x^2}{n} \text{ – provided the sample is large enough.}$$

Note: Experience has shown that the central limit theorem applies as long as $n \geq 3$.

TABLE 8-5 Constants for $\overline{X}$-R Control Charts

Sample Size	$\overline{X}$-R Control Charts				
n	A_2	D_3	D_4	d_2	d_3
2	1.880	0	3.267	1.128	0.853
3	1.023	0	2.574	1.693	0.888
4	0.729	0	2.282	2.059	0.880
5	0.577	0	2.114	2.326	0.864
6	0.483	0	2.004	2.534	0.848
7	0.419	0.076	1.924	2.704	0.833
8	0.373	0.136	1.864	2.847	0.820
9	0.337	0.184	1.816	2.970	0.808
10	0.308	0.223	1.777	3.078	0.797

[10] This table of control chart constants can also be found as Table I in Appendix B.

Three sigma control limits on an $\overline{X}$ chart are developed as follows when the range chart has been used to estimate within subgroup variability.

$$UCL_{\overline{X}} = \overline{\overline{X}} + 3\sigma_{\overline{X}} = \overline{\overline{X}} + 3\frac{\sigma}{\sqrt{n}} = \overline{\overline{X}} + 3\left[\frac{\overline{R}}{d_2\sqrt{n}}\right]$$

$$LCL_{\overline{X}} = \overline{\overline{X}} - 3\sigma_{\overline{X}} = \overline{\overline{X}} - 3\frac{\sigma}{\sqrt{n}} = \overline{\overline{X}} - 3\left[\frac{\overline{R}}{d_2\sqrt{n}}\right]$$

Let $A_2 = \dfrac{3}{d_2\sqrt{n}}$, then

Formulae for $\overline{X}$ Charts based on Average Range

$$Centerline = \overline{\overline{X}} = \frac{\sum_{i=1}^{k}\overline{X}_i}{k} \quad \textbf{8-9}$$

$$UCL_{\overline{X}} = \overline{\overline{X}} + A_2\overline{R} \quad \textbf{8-10}$$

$$LCL_{\overline{X}} = \overline{\overline{X}} - A_2\overline{R} \quad \textbf{8-11}$$

The A_2 constants come from Table 8-5 and depend on sample size.

Applying Equations 8-6 through 8-11 we can construct a pair of control charts for the Acme Fastener Corporation data.

R Chart:

$$Centerline = \overline{R} = \frac{\sum_{i=1}^{k} R}{k} = \frac{119.64}{30} = 3.99$$

$$UCL_R = D_4\overline{R} = (2.114)(3.99) = 8.43$$
$$LCL_R = D_3\overline{R} = (0)(3.99) = 0\ (none)$$

$\overline{X}$ Chart:

$$Centerline = \overline{\overline{X}} = \frac{\sum_{i=1}^{k}\overline{X}_i}{k} = \frac{439.48}{30} = 14.65$$

$$UCL_{\overline{X}} = \overline{\overline{X}} + A_2\overline{R} = 14.65 + (0.577)(3.99) = 16.95$$
$$LCL_{\overline{X}} = \overline{\overline{X}} - A_2\overline{R} = 14.65 - (0.577)(3.99) = 12.35$$

When analyzing a process using $\overline{X}$ and R charts, the R chart should always be constructed and analyzed first. This is because the control limits on the $\overline{X}$ chart depend on the value of $\overline{R}$ as shown by Equations 8-10 and 8-11.

Figure 8-12 is the output from Minitab for the Acme Fastener data. The centerline on the range chart ($\overline{R} = 3.99$) is obtained by computing the average of the ranges of all thirty samples. The chart passes the first test for control as no points are outside the control limits. There are 16 points above the centerline and 14 below. Therefore the runs test will be applied using s = 14 and r = 16. Investigating the

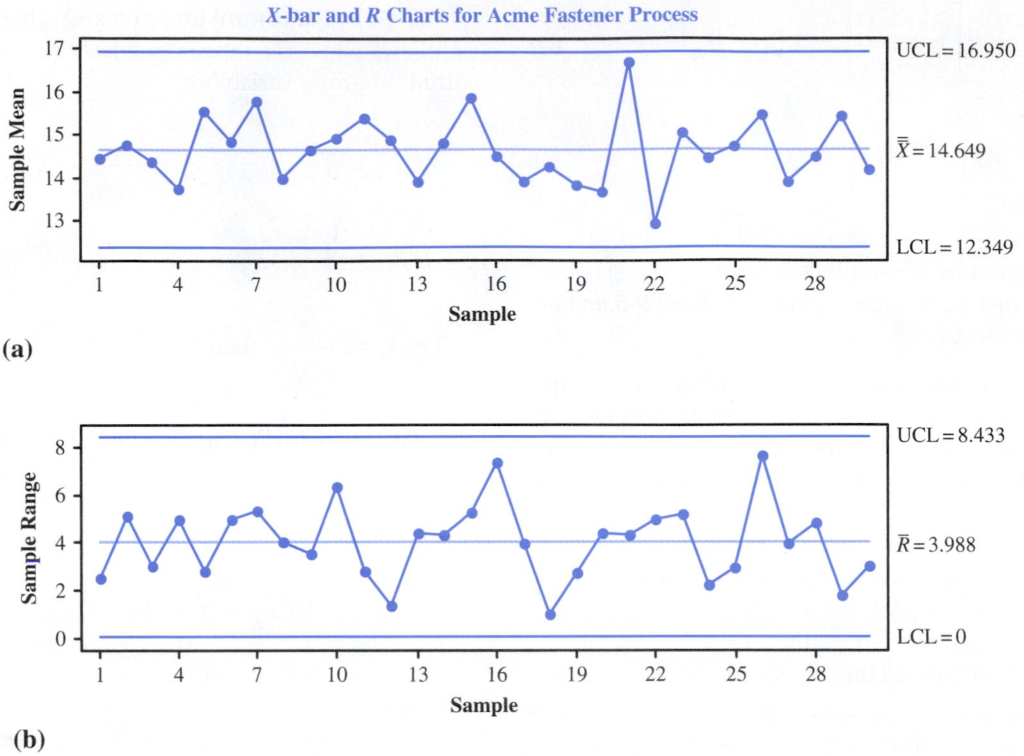

FIGURE 8-12. Minitab Output for Acme Fastener Process

points on the range chart reveals that there are 16 runs. Table 8-2 provides the critical value of 11 for too few runs. Since the total number of runs exceeds the critical value, the range chart passes the test for too few runs.

Table 8-3 shows that the critical value for the length-of-the-longest-run is 8. Since 4 is the longest run length (obtained by points 13–16 and 21–24), the range passes this test as well. Hence one can assume that for this process the range chart is in control, which means that all of the variability captured within a sample is stable and repeatable. Since the range is in control, the within sample variance can be estimated using the following relationship.

$$\sigma_w \cong \frac{\bar{R}}{d_2} \qquad \textbf{8-12}$$

where

σ_w = the within sample variability component

Once the *R* chart is in control the within-sample variability component can be estimated.

In the Acme Fastener case, since the range is in control we can proceed to estimate the variability (standard deviation) of all the sources of variance that are active within a sample.

$$\sigma_w \cong \frac{\bar{R}}{d_2} = \frac{3.99}{2.326} = 1.715$$

Once the range chart is in control attention can focus on the $\bar{X}$ chart. Applying the statistical control tests to the $\bar{X}$ chart results in the following:

- All plotted points are inside the control limits.
- There are an equal number of points above and below the centerline and a total of 17 runs. The critical value for too few runs with $r = 15$ and $s = 15$ is 11. Since 17 is greater than 11, the chart passes the test for too few runs.
- The length of the longest run is 5 (points 17—21). The critical value for longest run is 8. Since 5 is less than 8 the chart passes the test for the length of the longest run.

Because the $\bar{X}$ and R charts each pass all tests it can be concluded that the process is in statistical control. With the $\bar{X}$ chart in control, one can conclude that the process mean is stable and repeatable at 14.65, and that there is no between-sample variability (or that it is negligible). That is, we can assume that all the variability is of the common cause variety and was captured within the subgroups. Therefore it is legitimate to use σ_w as an estimate of the process standard deviation, which is given the designation σ_x (or simply σ).

8.5.2 $\bar{X}$ Charts, Estimating Process Dispersion Using the Sample Standard Deviations

An alternative to the range chart is the *s* **chart.** An *s* chart serves the same purpose as the *R* chart; the difference is that in lieu of the sample ranges, sample standard deviations are computed and plotted on the s chart. The R chart and s chart therefore provide the same information. *It is never necessary to construct both.* However, if sample sizes are large (greater than or equal to 8) the s chart is preferred. In developing $\overline{X}$ and s charts we shall return to our previous scenario in which we take a large number of samples of size n from a population with known standard deviation σ. Now each time a sample is taken, the average $(\overline{X})$ and standard deviation (s) is computed. In addition, this time we will compute a statistic called the relative standard deviation γ, defined by $\frac{s}{\sigma}$. After a very large number of samples are taken, we can construct the distributions: $f(\gamma)$ and $f(s)$. The mean of $f(\gamma)$, represented by the symbol c_4, is given by

$$c_4 = E[\gamma] = E\left[\frac{s}{\sigma}\right] = \frac{1}{\sigma}E[s] = \frac{\bar{s}}{\sigma} \qquad \textbf{8-13}$$

As the number of samples gets very large (e.g., approaches infinity) for any sample size, c_4 converges to a constant. We can also compute another set of constants, c_5, defined as follows.

$$c_5 = \frac{\sigma_s}{\sigma} \qquad \textbf{8-14}$$

Where

σ_s is the standard deviation of the distribution $f(s)$.

Note that c_4 and c_5 are constants for any given value of n. When a process is in control, Equation 8-15 below can be used to estimate process standard deviation from the average sample standard deviation.

$$\sigma \cong \frac{\bar{s}}{c_4} \qquad \textbf{8-15}$$

Equations 8-13 and 8-14 can be used to derive the s chart limits as follows:

$$\textit{Upper Control Limit} = UCL_s = \bar{s} + 3\sigma_s = \bar{s} + 3\sigma c_5$$

$$= \bar{s} + 3\frac{\bar{s}c_5}{c_4}$$

$$\textit{Lower Control Limit} = LCL_s = \bar{s} - 3\sigma_s = \bar{s} - 3\sigma c_5$$

$$= \bar{s} + 3\frac{\bar{s}c_5}{c_4}$$

Formulae for s Chart

$$\textit{Centerline} = \bar{s} = \frac{\sum_{i=1}^{k} s_i}{k} \qquad \textbf{8-16}$$

$$UCL_s = \bar{s}\left[1 + 3\frac{c_5}{c_4}\right] = B_4\bar{s} \qquad \textbf{8-17}$$

$$LCL_s = \bar{s}\left[1 - 3\frac{c_5}{c_4}\right] = B_3\bar{s} \qquad \textbf{8-18}$$

The c_4, c_5, B_3, and B_4 constants come from Table 8-6 and depend on sample size.

TABLE 8-6 Constants for $\overline{X}$-s Control Charts

Sample Size	$\overline{X}$-s Control Charts				
n	A_3	B_3	B_4	c_4	c_5
2	2.659	0	3.267	0.7979	0.6029
3	1.954	0	2.568	0.8862	0.4632
4	1.628	0	2.266	0.9213	0.3888
5	1.427	0	2.089	0.9400	0.3412
6	1.287	0.030	1.970	0.9515	0.3077
7	1.182	0.118	1.882	0.9594	0.2821
8	1.099	0.185	1.815	0.9650	0.2622
9	1.032	0.239	1.761	0.9693	0.2459
10	0.975	0.284	1.716	0.9727	0.2322

Table 8-6[11] contains control chart constants for sample sizes ranging from two to ten. For small sample sizes ($n \leq 5$) $B_3 = 0$. Therefore, as was the case with the range chart, normally the $LCL_s = 0$. In reality, for small sample sizes there is no lower control limit for the range.

In Table 8-7 the Acme Fastener data have been duplicated, and the Range column has been replaced by a column in which the standard deviation for each sample has been computed using Equation 8-19 shown below.

$$s_i = \sqrt{\frac{\sum_{j=1}^{n}(x_{ij} - \overline{X}_i)^2}{n-1}} \qquad \textbf{8-19}$$

Three sigma control limits on an $\overline{X}$ chart are developed as follows when the standard deviation chart has been used to estimate within subgroup variability.

$$UCL_s = \overline{\overline{X}} + 3\sigma_s = \overline{\overline{X}} + 3\frac{\sigma}{\sqrt{n}} = \overline{\overline{X}} + 3\left[\frac{\bar{s}}{c_4\sqrt{n}}\right]$$

$$LCL_s = \overline{\overline{X}} - 3\sigma_s = \overline{\overline{X}} - 3\frac{\sigma}{\sqrt{n}} = \overline{\overline{X}} + 3\left[\frac{\bar{s}}{c_4\sqrt{n}}\right]$$

$$\text{Let } A_3 = \frac{3}{c_4\sqrt{n}}$$

Formulae for $\overline{X}$ Charts based on Average Standard Deviation

$$\textit{Centerline} = \overline{\overline{X}} = \frac{\sum_{i=1}^{k}\overline{X}_i}{k} \qquad \textbf{8-20}$$

$$UCL_s = \overline{\overline{X}} + A_3\bar{s} \qquad \textbf{8-21}$$

$$LCL_s = \overline{\overline{X}} - A_3\bar{s} \qquad \textbf{8-22}$$

The A_3 constants come from Table 8-6 and depend on sample size.

[11]This table of control chart constants can also be found as Table I in Appendix B.

TABLE 8-7 Samples from Acme Fastener Process

Sample No.	Measurement					x-bar	s
1	15.41	13.57	15.85	13.38	14.12	14.47	1.108
2	13.03	18.07	13.55	13.77	15.50	14.79	2.056
3	13.91	15.88	12.88	15.10	14.06	14.37	1.156
4	13.98	13.78	11.63	16.57	12.79	13.75	1.832
5	16.22	15.24	15.91	13.80	16.53	15.54	1.084
6	17.33	12.45	14.59	15.52	14.25	14.83	1.791
7	15.62	19.01	16.31	13.72	14.39	15.81	2.055
8	15.87	14.94	11.86	12.30	14.97	13.99	1.789
9	12.70	16.20	13.60	15.08	15.74	14.66	1.472
10	18.37	14.76	12.05	15.62	13.90	14.94	2.329
11	15.28	14.94	15.87	16.81	14.03	15.38	1.040
12	15.25	15.66	14.58	14.62	14.34	14.89	0.546
13	11.73	15.54	16.05	12.56	13.53	13.88	1.867
14	14.19	14.23	15.09	17.46	13.16	14.83	1.623
15	18.03	15.45	16.17	16.99	12.80	15.89	1.977
16	13.07	11.37	14.97	14.42	18.66	14.50	2.710
17	14.11	16.21	13.71	12.33	13.05	13.88	1.468
18	13.84	14.13	14.13	14.34	14.81	14.25	0.360
19	13.54	13.20	14.99	15.01	12.33	13.81	1.168
20	12.41	13.59	12.46	16.74	13.06	13.65	1.792
21	14.90	15.49	16.95	19.18	16.91	16.69	1.654
22	10.77	12.74	12.09	13.33	15.69	12.93	1.814
23	14.59	12.75	15.01	15.07	17.87	15.06	1.833
24	15.57	13.40	15.03	14.69	13.62	14.46	0.926
25	14.02	15.21	15.31	15.94	13.07	14.71	1.149
26	18.42	18.06	10.79	13.18	16.96	15.48	3.347
27	16.20	13.48	12.92	14.62	12.31	13.91	1.540
28	15.36	11.83	16.61	12.94	15.79	14.51	2.026
29	14.69	14.74	15.43	15.98	16.45	15.46	0.767
30	12.68	15.62	12.68	15.43	14.45	14.17	1.431
Totals						439.48	47.71

Applying Equations 8-16 through 8-22 we can construct a pair of control charts for the Acme Fastener Corporation data.

R Chart:

$$Centerline = \bar{s} = \frac{\sum_{i=1}^{k} s_i}{k} = \frac{47.71}{30} = 1.59$$

$$UCL_s = B_4\bar{s} = (2.089)(1.59) = 3.32$$
$$LCL_s = B_3\bar{s} = (0)(1.59) = 0 \text{ (none)}$$

$\overline{X}$ Chart:

$$Centerline = \overline{\overline{X}} = \frac{\sum_{i=1}^{k} \overline{X}_i}{k} = \frac{439.48}{30} = 14.65$$

$$UCL_s = \overline{\overline{X}} + A_3\bar{s} = 14.65 + (1.427)(1.59) = 16.92$$

$$LCL_s = \overline{\overline{X}} - A_3\bar{s} = 14.65 - (1.427)(1.59) = 12.38$$

Since the standard deviation is in control we can proceed to estimate the variability (standard deviation) of all the sources of variance that are active within a sample, using Equation 8-14.

$$\sigma_w \cong \frac{\bar{s}}{c_4} = \frac{1.59}{0.940} = 1.691$$

Figure 8-13 provides a comparison of the *s* chart with the *R* chart. The pattern of points is roughly the same. The *s* chart is a little more sensitive than the *R* chart to changes that occur to variability within samples. This is evidenced by point number 26 that is out of control on the *s* chart but inside the control

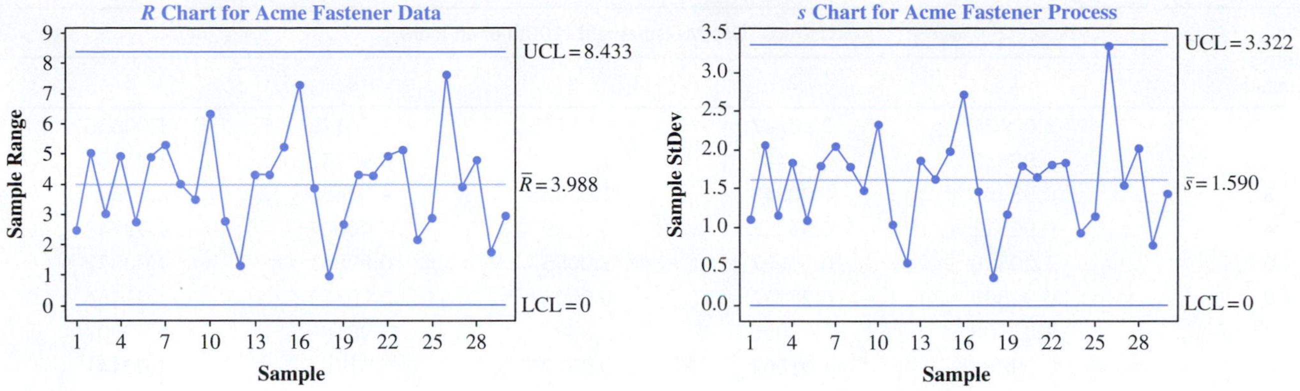

FIGURE 8-13. Comparison of R and s Charts

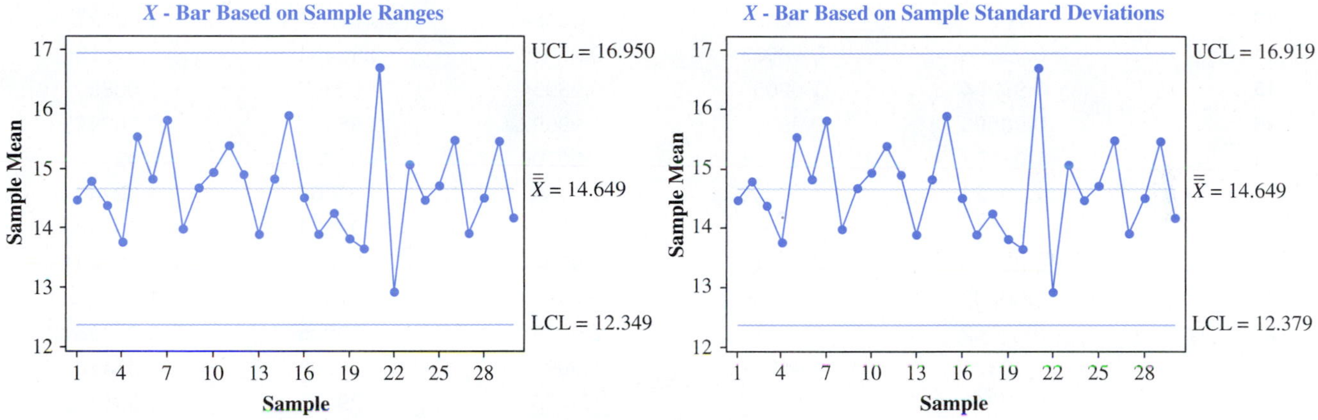

FIGURE 8-14. Comparison of $\overline{X}$ Charts Using R and s Charts

limits on the R chart. Nevertheless, for the small sample sizes that are typical in most applications, the convenience of the R chart (ranges are easier to compute on the factory floor than standard deviations) usually overrides any additional advantage derived from the s chart. Figure 8-14 shows that the control limits on the $\overline{X}$ chart are roughly the same whether they are based on the sample ranges or sample standard deviations.

8.5.3 Using Minitab Software for $\overline{X}$ Charts

Much of the computational drudgery (and vulnerability to human error) associated with plotting control charts can be eliminated with the use of a statistical software package such as Minitab. Consider the data shown in Table 8-8. Over a period of five days, 30 samples of five transmission components were randomly selected from a production line and measures of a critical clearance dimension were taken. These data were then recorded on the spreadsheet shown. Note that the measurements are expressed in one-hundredths of an inch.

The data in Table 8-8 can be pasted directly into a Minitab worksheet as shown in Figure 8-15. Then, as shown in Figure 8-16, click on these successive dropdown menu commands from the main toolbar: *Stat* > *Control Charts* > *Variables Charts for Subgroups* > *Xbar-R* or *Xbar-s*. In our case, since our subgroup size is equal to 5, the range chart option was chosen.

The dialog box shown in Figure 8-17 will appear. Select the option that permits the data to be presented in a row of several columns and designate the columns C1 through C5 as shown. Then click on *Xbar-R Options*. In the dialog box that appears, click on the *Estimate* tab. The dialog box shown on the right side of Figure 8-17 will appear. Place a tick on the *Rbar* option. This will cause Minitab to use the average range as the basis for estimating the within sample standard deviation and the control limits for the $\overline{X}$ chart. After this is done click on the *Tests* tab and the dialog box shown in Figure 8-18 will appear. Place a check next to all of the rules that are to be applied to test for statistical control. Finally, click *OK* in all open dialog boxes.

Minitab output displays the $\overline{X}$ and R charts shown in Figure 8-19. If any of the designated statistical tests had been violated, the point where the violation occurred would have been highlighted in red with the number of the applicable rule. In addition, Minitab produces a report that lists the out-of-control signals, the points where they occurred, and a statement of the rule violated.

TABLE 8-8 Samples of Critical Gap Measurements

Sample Number	Gap Measurement (100th of an inch)				
	1	2	3	4	5
1	1.00426	0.98287	0.98245	1.00178	1.00856
2	0.99652	0.99485	1.00184	0.99447	1.01308
3	0.98456	1.00843	1.00425	1.00685	0.9967
4	0.99511	0.99356	0.9929	1.00607	0.99748
5	1.00326	0.99607	1.0055	0.99841	1.00572
6	0.99333	0.98737	0.99022	0.99533	1.01464
7	1.01808	0.98026	0.98802	0.9818	0.99007
8	0.9988	1.01064	0.99975	1.01767	1.00441
9	1.02482	1.00661	0.99331	0.98143	1.00013
10	1.01181	0.99578	1.00717	1.02245	0.98286
11	1.01064	1.004	1.00442	1.00295	0.99897
12	1.0015	0.99436	1.0249	1.00153	1.01078
13	1.0015	1.00233	1.00384	0.99994	1.00427
14	0.99668	0.98937	0.97539	0.98406	0.99618
15	0.99284	1.0205	0.98954	1.00307	1.0085
16	0.98596	0.99506	0.99997	0.98652	0.98297
17	1.01152	1.01193	1.00706	1.02288	0.991
18	1.00726	1.00705	1.0139	0.98866	1.00745
19	1.00141	1.01061	1.01163	0.99524	1.00198
20	1.00278	0.99615	0.9932	1.0007	1.02024
21	1.00202	0.99234	0.99917	0.97674	1.00229
22	0.99462	1.00211	1.01297	0.99825	0.99798
23	1.00325	1.01072	1.00938	0.98305	1.01472
24	0.99946	0.98568	1.00285	0.99755	0.98645
25	1.01447	1.00491	0.9887	0.98923	1.00245
26	1.00297	0.98026	1.00913	0.98295	1.00317
27	0.99588	1.00091	1.00914	0.9991	1.01667
28	1.00935	0.99639	0.98442	0.97989	1.02036
29	1.01372	1.01284	0.99924	0.99531	1.00218
30	1.00027	1.00328	0.99046	0.99435	0.99595

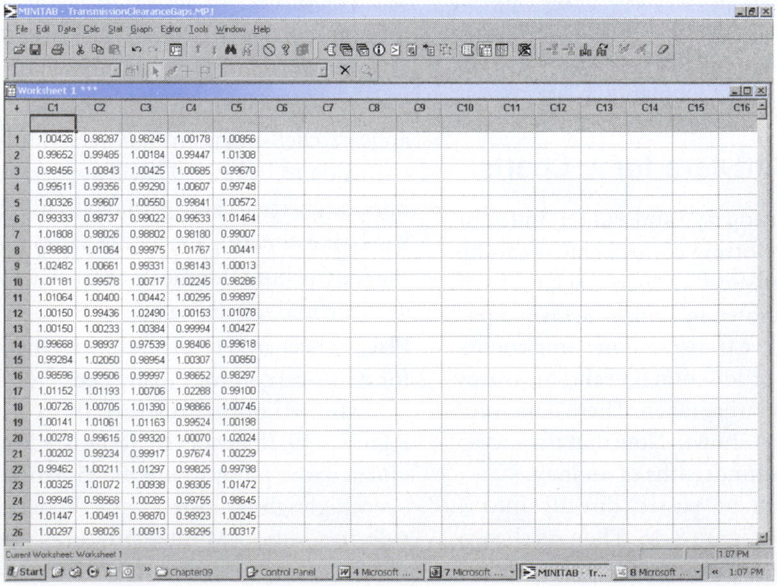

FIGURE 8-15. Transportation Gap Measures on Minitab Worksheet

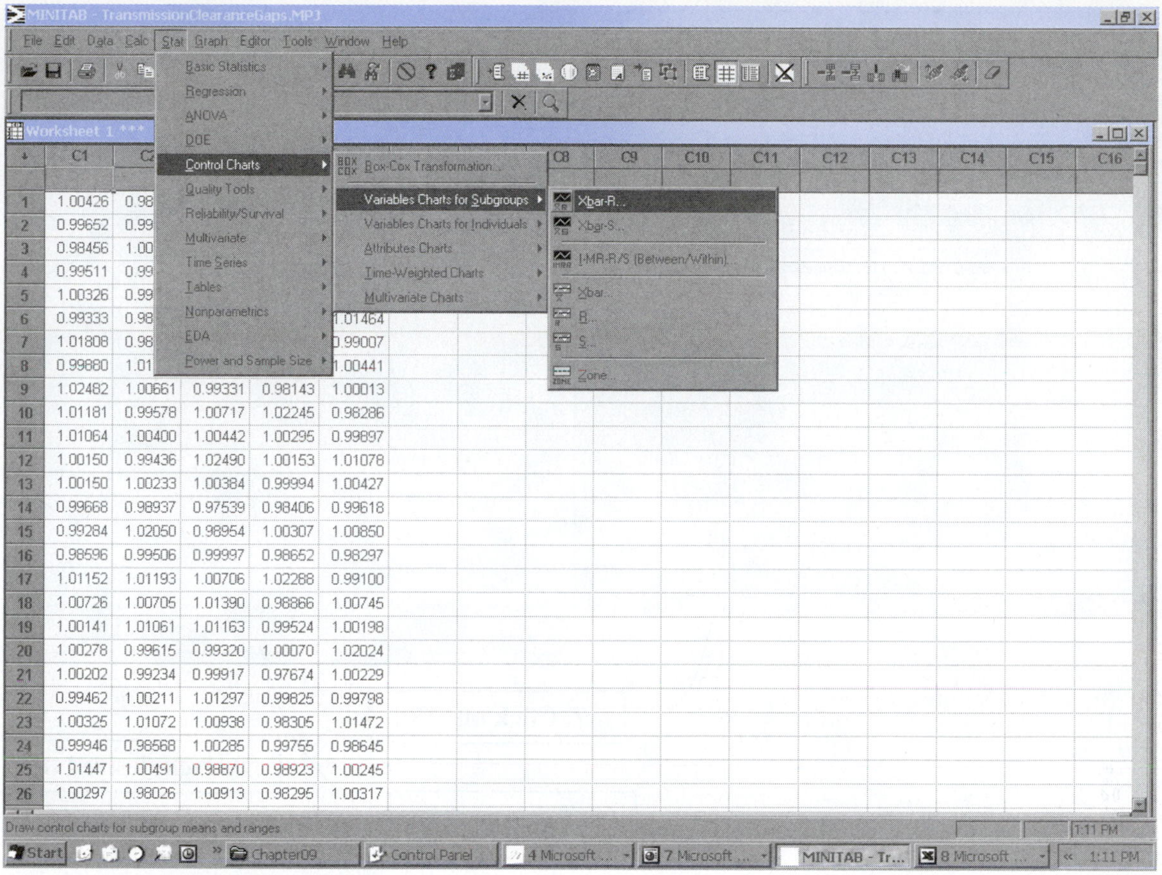

FIGURE 8-16. Minitab Pull-Down Menus for $\overline{X}$ Charts

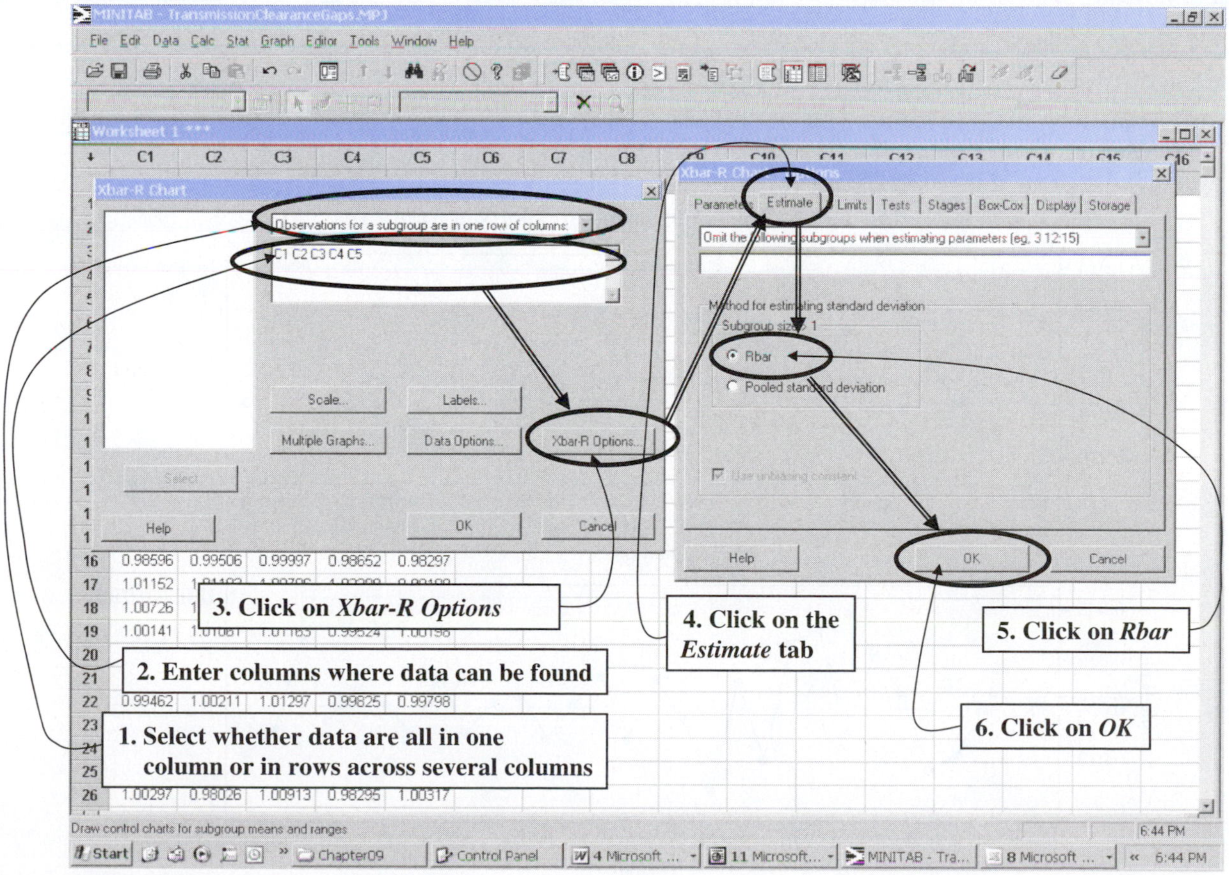

FIGURE 8-17. Minitab $\overline{X}$ and R Parameters

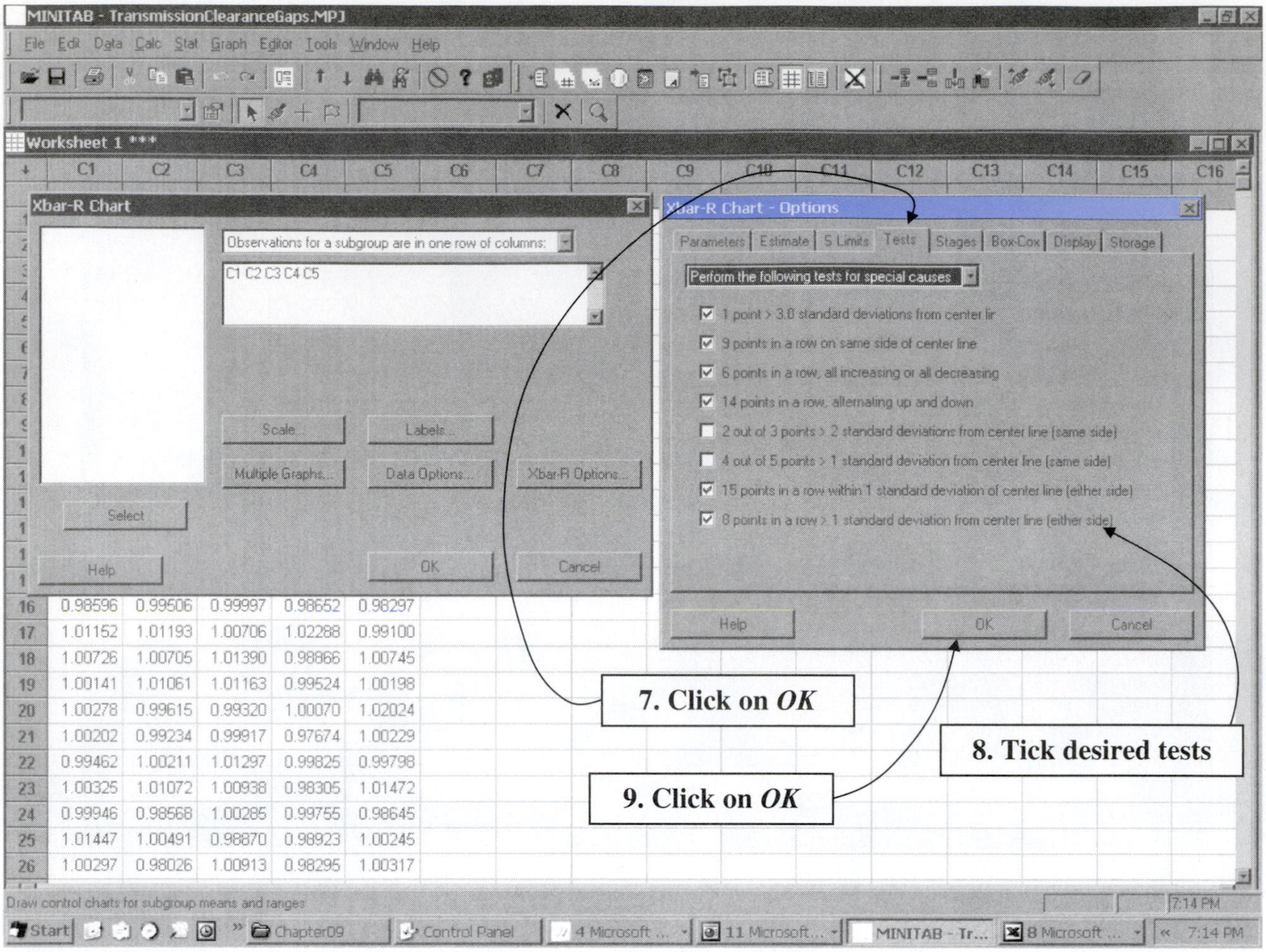

FIGURE 8-18. Minitab $\overline{X}$ and *R* Statistical Tests

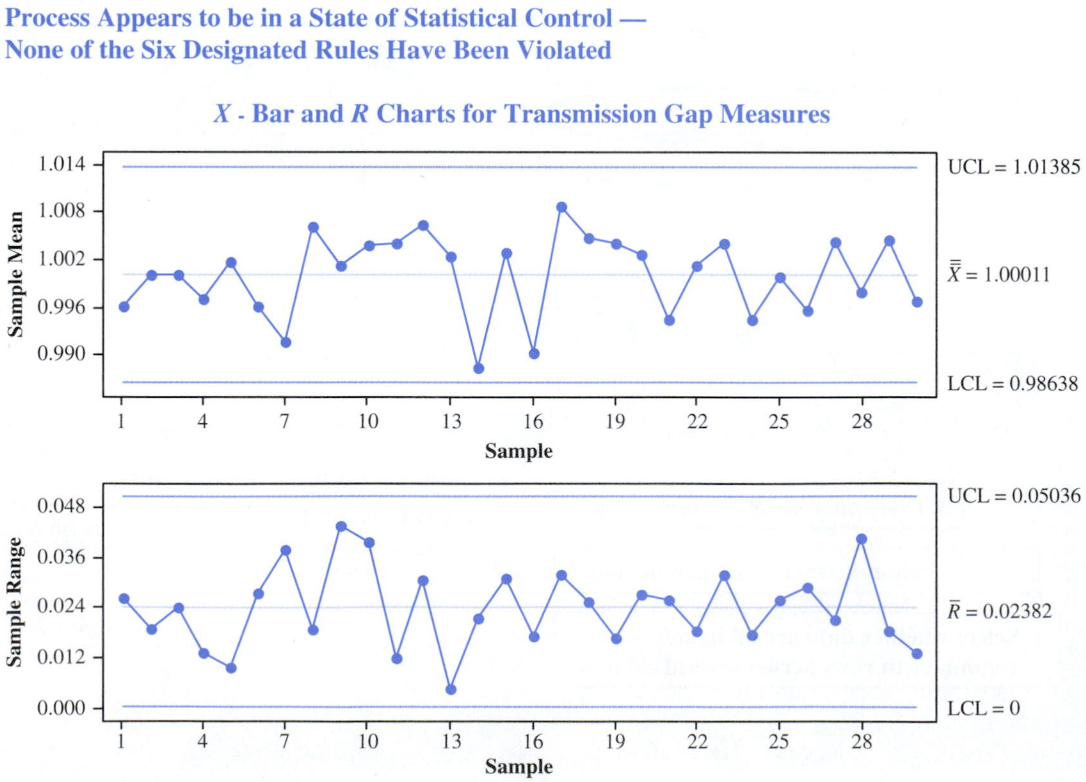

FIGURE 8-19. Minitab Control Charts for Transmission Data

8.6 INDIVIDUALS (X) CHARTS

8.6.1 *X* Charts, Estimating Process Dispersion Using a Moving Range

Certain types of data do not lend themselves to rational subgrouping—for example, a sample that would effectively distinguish between short- and long-term sources of variability. Trying to put together a sample of multiple units would be senseless. Examples of this type of data are money, time, and samples that are drawn from a continuous process that is homogeneous throughout (e.g., flour, paint, roofing shingles, and chemicals).

An individuals chart is necessary when a rational subgroup is not possible.

Consider a steel mill, where iron content is important to the quality of any batch of steel. Table 8-9 shows measurements that were taken at half-hour intervals over a 15-hour period. This type of process does not lend itself to rational subgrouping. If a sample of five were collected consecutively, or even at short intervals, the dominant (if not only) source of variability captured within the sample would be due to measurement error and not the process. In the case of steel-making, process variability is primarily due to time-related factors. As these are *variables data* a pair of charts are needed to properly analyze the process mean and variance. With a sample of size $n = 1$ this is problematic, since it is not possible to calculate a range. To overcome this difficulty a *moving range* is used, which is the calculated difference between the high and low of a group of measurements taken in close proximity. The most common procedure is to use a two-point moving range—that is, each range is computed as the absolute value of the difference between the two most recent consecutive measurements. Obviously, the first moving range value cannot be computed until the second measurement has been taken. The relevant formulae for the *MR* and *X* charts are:

TABLE 8-9 Individuals Data for Steel Mill

Measurement	Iron Content	Moving Range
1	2.446	
2	2.211	0.235
3	2.211	0.000
4	2.539	0.328
5	1.998	0.541
6	2.131	0.132
7	2.290	0.159
8	2.165	0.124
9	1.797	0.368
10	2.272	0.475
11	2.050	0.221
12	2.439	0.388
13	2.029	0.410
14	2.053	0.024
15	2.378	0.325
16	2.248	0.130
17	2.106	0.142
18	1.958	0.148
19	1.860	0.098
20	2.135	0.275
21	2.139	0.004
22	2.037	0.101
23	2.157	0.120
24	2.076	0.082
25	2.217	0.141
26	2.247	0.030
27	2.142	0.105
28	2.098	0.044
29	2.477	0.379
30	2.336	0.141
Totals	65.243	5.672

Formulae for *MR* Chart

$$Centerline = \overline{MR} = \frac{\sum_{i=1}^{k} MR_i}{k-1} \qquad \textbf{8-23}$$

$$UCL_{MR} = \overline{MR} + 3\sigma_{MR} = D_4\overline{MR} \qquad \textbf{8-24}$$

$$LCL_{MR} = \overline{MR} - 3\sigma_{MR} = D_3\overline{MR} \qquad \textbf{8-25}$$

The D_3 and D_4 constants come from Table 8-5 and depend on sample size.

Note: The sample size is normally = 2, in which case $D_3 = 0$ and $D_4 = 3.287$.

Formulae for *X* Charts Based on Moving Range

$$Centerline = \overline{X} = \frac{\sum_{i=1}^{k} X_i}{k} \qquad \textbf{8-26}$$

$$UCL_X = \overline{X} + 3\sigma_X = \overline{X} + 3\frac{\overline{MR}}{d_2} \qquad \textbf{8-27}$$

$$LCL_X = \overline{X} - 3\sigma_X = \overline{X} - 3\frac{\overline{MR}}{d_2} \qquad \textbf{8-28}$$

The d_2 constants come from Table 8-5 and depend on sample size.

Note: The sample size is normally = 2, in which case $d_2 = 1.128$.

Applying these formulae to the steel mill data,

MR chart:

$$Centerline = \overline{MR} = \frac{\sum_{i=2}^{k} MR_i}{k-1} = \frac{5.672}{29} = 0.1956$$

$$UCL_{MR} = D_4\overline{MR} = (3.267)(0.1956) = 0.639$$

$$LCL_{MR} = D_3\overline{MR} = (0)(0.1956) = 0$$

X chart:

$$Centerline = \overline{X} = \frac{\sum_{i=1}^{k} X_i}{k} = \frac{65.243}{30} = 2.1748$$

$$UCL_X = \overline{X} + 3\frac{\overline{MR}}{d_2} = 2.1748 + 3\left[\frac{0.1956}{1.128}\right] = 2.695$$

$$LCL_X = \overline{X} - 3\frac{\overline{MR}}{d_2} = 2.1748 + 3\left[\frac{0.1956}{1.128}\right] = 1.655$$

Figure 8-20 shows the pair of individuals control charts for the iron content measurements. The process appears to be in statistical control (that is, there are no out-of-control points and no evidence of trends or stratification) with a mean approximately equal to 2.1748 and standard deviation approximately equal to 0.173. Since the moving range chart is in statistical control a modified version of Equation 8-12 can be used to estimate the short-term variation for this process.

$$\sigma_{short} = \frac{\overline{MR}}{c_4} \qquad \textbf{8-29}$$

$$\sigma_{short} = \frac{0.1956}{0.7979} = 0.2451$$

where

σ_{short} is the short-term standard deviation and $c_4 = 0.7979$ is the value for $n = 2$ found in Table 8-5.

Now, since the X chart is also in control we can conclude that $\sigma = \sigma_{short}$; that is, the short-term variability reflects the longer-term variability over time and can therefore be used as an estimate of the process standard deviation. However, it is important to note that this conclusion may be flawed.

Evaluating statistical control in the case of X charts is not as straightforward as with $\overline{X}$ and R charts. In the case of the moving range chart, each point is not independent of the others since each measurement (with the exception of the first) is used in the computation of two consecutive moving range values. Therefore *the standard runs tests do not apply.* Another consideration is the placement of the control limits on the X chart. Placing the limits equidistant from the centerline (three standard deviations in each direction) is based on the assumption of normality. This is statistically sound when the points plotted are averages ($\overline{X}$s) of samples, because the central limit theorem ensures that the distribution of the $\overline{X}$s will be approximately normal. However, this may not be true for the distribution of the Xs. In practice it is not uncommon for distributions of individuals data (e.g., time and money) to have a noticeable skew. Given these two considerations the standard runs tests do not apply to X charts.

Standard runs tests do not apply to individuals control charts.

8.6.2 Transformations for Nonnormal Data

The limits on control charts are symmetrically placed at a distance that is three standard deviations from the predicted process mean. This placement is based on the assumption that the process, when in statistical control, will follow a distribution pattern that is approximately normal and that the probability of a point falling beyond the three-sigma control limits for a stable process can be estimated using normal tables (i.e., approximately 1 in 1,000). While the assumption of normality holds for many industrial processes it does not apply to all. Certain attributes for some processes will inherently be **nonnormal**. Classic examples of this involve the measurement of time and dollars, both of which can be expected to produce skewed patterns.

When rational subgroups can be put together, averaged, and plotted on $\overline{X}$ and R charts, the central limit theorem alleviates any concerns of nonnormality since the $\overline{X}$s will be normally distributed. However, when rational subgrouping

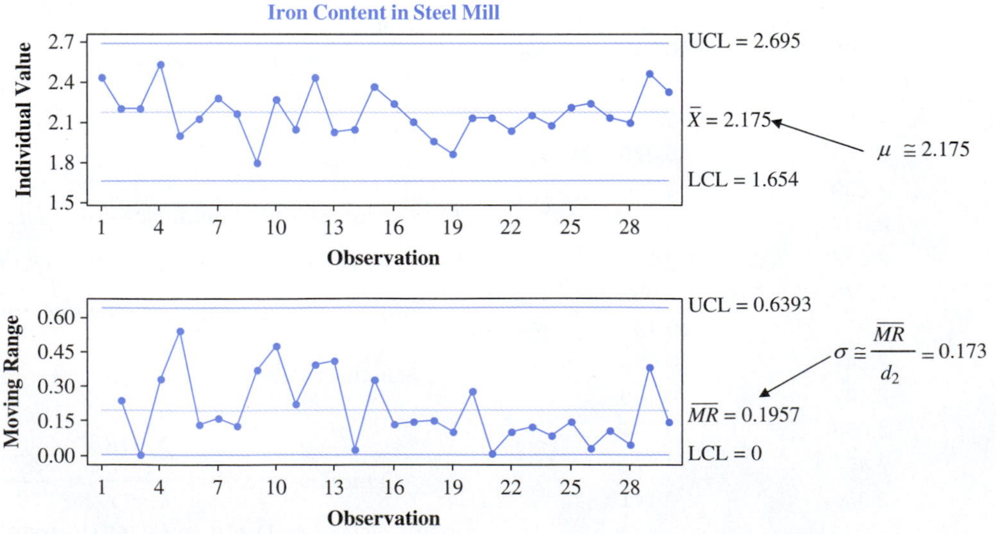

FIGURE 8-20. Individuals Control Charts for Steel Mill Data

is not possible and it is necessary to collect and plot individual measurements there can be no assurance that the normality assumption will hold. Table 8-10 is a record of 100 stopwatch measurements that were taken of teller transactions at a large city bank. The histogram for these data has a pronounced positive skew. The X and MR charts are plotted in Figure 8-21 with control limits symmetrically placed three sigma from the estimated mean. The standard deviation has been estimated using a two-point moving range.

At first glance this process appears to be badly out of control. Twenty-three of the 100 points result in rule failures (specifically rules 1, 2, and 7), and there appears to be excessive white space below the centerline (indicating a nonrandom pattern). However, a study of the histogram pattern

TABLE 8-10 Teller Transaction Times

United Metropolitan Bank

Teller Transaction Times (in min.)

Transaction No.	Time	Transaction No.	Time	Transaction No.	Time	Transaction No.	Time	Transaction No.	Time
1	1.934	21	1.409	41	2.128	61	2.323	81	1.396
2	2.716	22	8.159	42	3.040	62	0.655	82	8.920
3	1.473	23	1.112	43	2.015	63	0.864	83	1.340
4	1.971	24	1.931	44	1.629	64	3.334	84	1.649
5	1.581	25	1.060	45	2.535	65	1.990	85	3.290
6	0.743	26	1.952	46	0.906	66	1.429	86	2.409
7	0.857	27	0.879	47	0.764	67	3.810	87	1.639
8	1.480	28	1.693	48	0.771	68	1.069	88	1.872
9	1.463	29	1.142	49	1.112	69	1.178	89	1.344
10	1.331	30	1.229	50	1.273	70	0.919	90	0.751
11	1.819	31	1.374	51	5.846	71	1.263	91	0.724
12	0.782	32	0.768	52	3.805	72	1.374	92	0.856
13	2.479	33	4.599	53	1.753	73	0.971	93	0.636
14	1.131	34	2.386	54	1.921	74	1.355	94	3.626
15	0.721	35	0.853	55	1.380	75	2.726	95	1.033
16	1.077	36	1.684	56	0.940	76	0.623	96	0.906
17	8.109	37	4.519	57	1.075	77	0.968	97	0.644
18	6.352	38	2.380	58	2.006	78	2.594	98	6.559
19	1.002	39	0.623	59	0.942	79	2.737	99	1.026
20	1.002	40	0.976	60	1.216	80	4.052	100	1.911

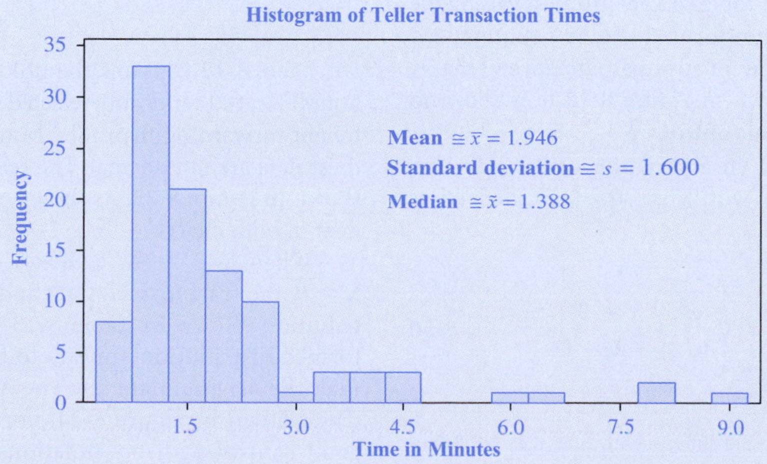

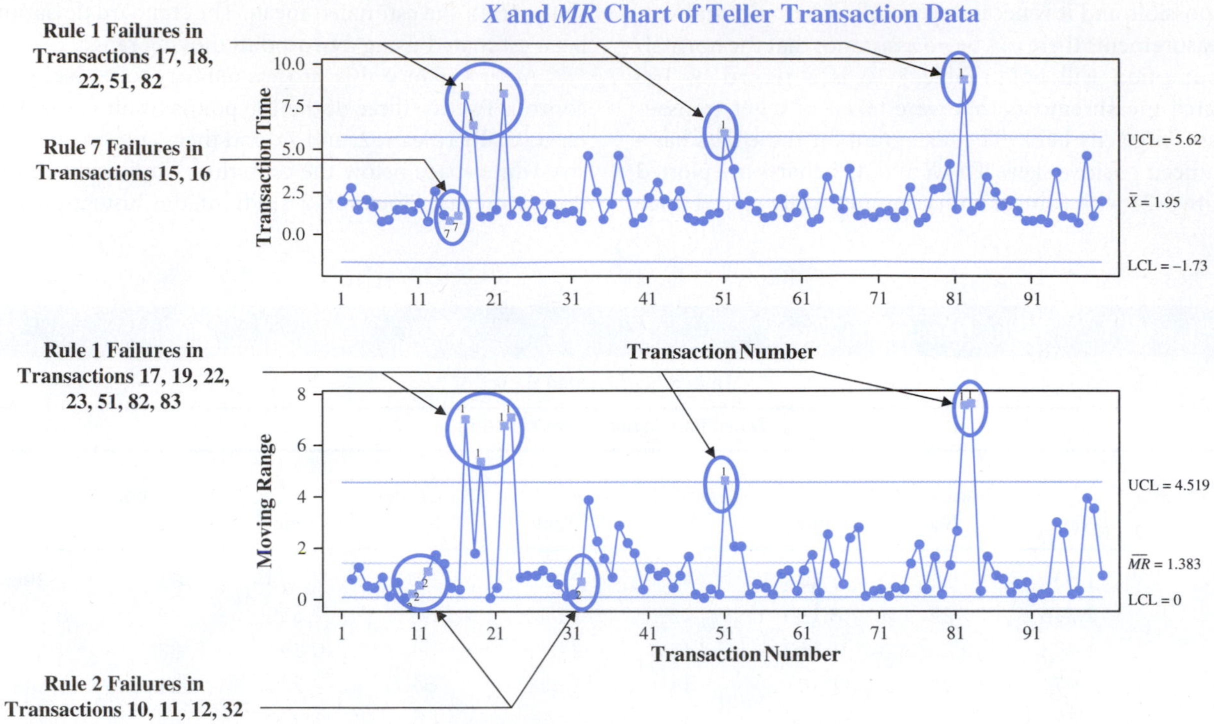

FIGURE 8-21. Control Charts for United Metropolitan Bank

brings into question the legitimacy of basing the control limits on an assumption of normality. First, because of the nature of the data, negative values are not possible. Yet, the lower control limit on the X chart has been established at -1.73. Because in reality that value could never be reached, we might conclude that the clustering pattern observed within zone C is reasonable. In addition, the histogram (and common sense) tells us that the system will occasionally deliver unusually high numbers which will show up as spikes on a runs chart and not be due to the presence of special causes.

The problem of nonnormality can be addressed by using a transformation to convert nonnormal raw data into a surrogate set that does follow an approximate normal distribution. To illustrate, take the United Metropolitan Bank data presented in Table 8-10 and for each x, compute a y using the formula $y = 1/\sqrt{x}$. The transformed data are summarized in Table 8-11 and now follow an approximate normal shape. The X and MR charts shown in Figure 8-22 now show no evidence of lack of statistical control.

This example is based on a variation of the Box–Cox family of transformations (with λ set at -0.5), defined as follows.[12]

$$y = T(x, \lambda) = \begin{cases} x^{\lambda} & \lambda \neq 0 \\ \ln(x) & \lambda = 0 \end{cases} \qquad \textbf{8-30}$$

[12]This is the form used by Minitab and differs slightly from the usual definition that is $T(x, \lambda) = \dfrac{(x^{\lambda} - 1)}{\lambda}$ for $\lambda \neq 0$.

Equation 8-30 can be used by trial and error to find a λ that will transform a data set into the "best" normal distribution, and then to select this (or a more convenient) λ to operationalize. Some convenient choices for λ are the following.

$$\lambda = 2 \Rightarrow y = x^2$$
$$\lambda = 1.5 \Rightarrow y = \sqrt{x^3}$$
$$\lambda = 0.5 \Rightarrow y = \sqrt{x}$$
$$\lambda = 0 \Rightarrow y = \ln x$$
$$\lambda = -0.5 \Rightarrow y = \frac{1}{\sqrt{x}}$$
$$\lambda = -1 \Rightarrow y = \frac{1}{x}$$
$$\lambda = -1.5 \Rightarrow y = \frac{1}{\sqrt{x^3}}$$

Table 8-12 contains the unloading times (expressed in hours) for 100 randomly selected trucks arriving at a certain freight forwarding firm. The histogram shape suggests that these data are not normal. The relevant X and MR charts are shown in Figure 8-23. As shown, statistical instability is evident in both charts.

When we apply a Box–Cox transformation, with $\lambda = 0$, the transformed data and their corresponding distribution shown in Table 8-13 are obtained. The transformed distribution appears to be much closer to normal than the original data set. The X and MR charts for these data, shown in Figure 8-24, portray a process that shows good control with no violations in any of the stipulated operational rules.

TABLE 8-11 Transformed Teller Transaction Times

United Metropolitan Bank

Transformed Teller Transaction Times (min.)

Trans No.	Time	$\lambda = -0.5$	Trans No.	Time	$\lambda = -0.5$	Trans No.	Time	$\lambda = -0.5$	Trans No.	Time	$\lambda = -0.5$
1	1.934	0.719	26	1.952	0.716	51	5.846	0.414	76	0.623	1.267
2	2.716	0.607	27	0.879	1.067	52	3.805	0.513	77	0.968	1.017
3	1.473	0.824	28	1.693	0.768	53	1.753	0.755	78	2.594	0.621
4	1.971	0.712	29	1.142	0.936	54	1.921	0.721	79	2.737	0.604
5	1.581	0.795	30	1.229	0.902	55	1.380	0.851	80	4.052	0.497
6	0.743	1.160	31	1.374	0.853	56	0.940	1.031	81	1.396	0.846
7	0.857	1.080	32	0.768	1.141	57	1.075	0.965	82	8.920	0.335
8	1.480	0.822	33	4.599	0.466	58	2.006	0.706	83	1.340	0.864
9	1.463	0.827	34	2.386	0.647	59	0.942	1.031	84	1.649	0.779
10	1.331	0.867	35	0.853	1.083	60	1.216	0.907	85	3.290	0.551
11	1.819	0.741	36	1.684	0.771	61	2.323	0.656	86	2.409	0.644
12	0.782	1.131	37	4.519	0.470	62	0.655	1.236	87	1.639	0.781
13	2.479	0.635	38	2.380	0.648	63	0.864	1.076	88	1.872	0.731
14	1.131	0.941	39	0.623	1.267	64	3.334	0.548	89	1.344	0.863
15	0.721	1.178	40	0.976	1.012	65	1.990	0.709	90	0.751	1.154
16	1.077	0.964	41	2.128	0.686	66	1.429	0.836	91	0.724	1.175
17	8.109	0.351	42	3.040	0.574	67	3.810	0.512	92	0.856	1.081
18	6.352	0.397	43	2.015	0.705	68	1.069	0.967	93	0.636	1.254
19	1.002	0.999	44	1.629	0.784	69	1.178	0.921	94	3.626	0.525
20	1.002	0.999	45	2.535	0.628	70	0.919	1.043	95	1.033	0.984
21	1.409	0.843	46	0.906	1.051	71	1.263	0.890	96	0.906	1.051
22	8.159	0.350	47	0.764	1.144	72	1.374	0.853	97	0.644	1.246
23	1.112	0.948	48	0.771	1.139	73	0.971	1.015	98	4.559	0.468
24	1.931	0.720	49	1.112	0.948	74	1.355	0.859	99	1.026	0.987
25	1.060	0.971	50	1.273	0.886	75	2.726	0.606	100	1.911	0.723

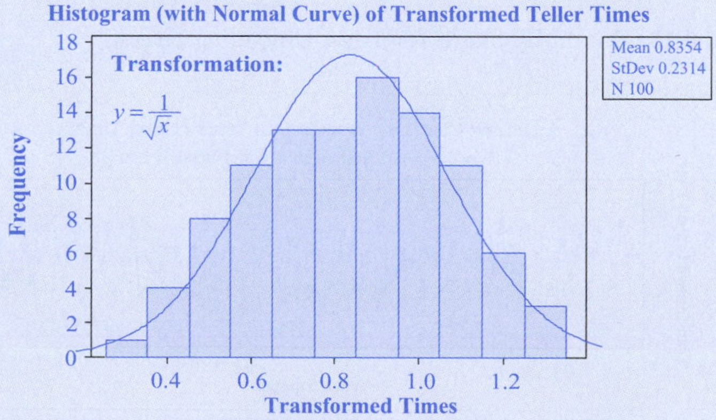

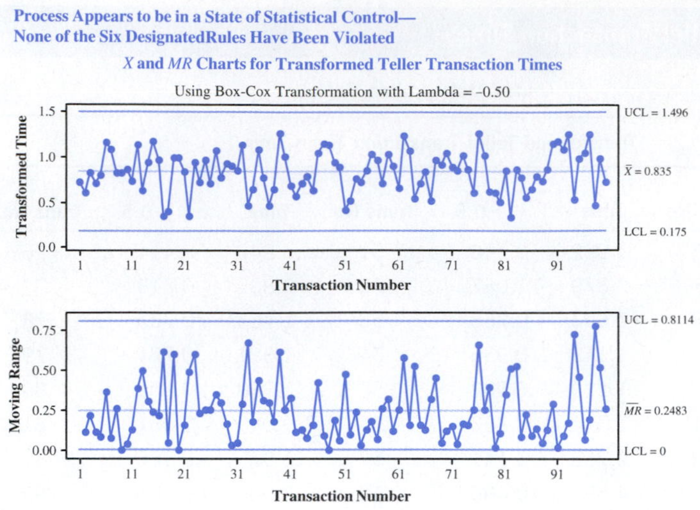

FIGURE 8-22. Charts for Inverse Square Root Transformation for Bank Teller Data

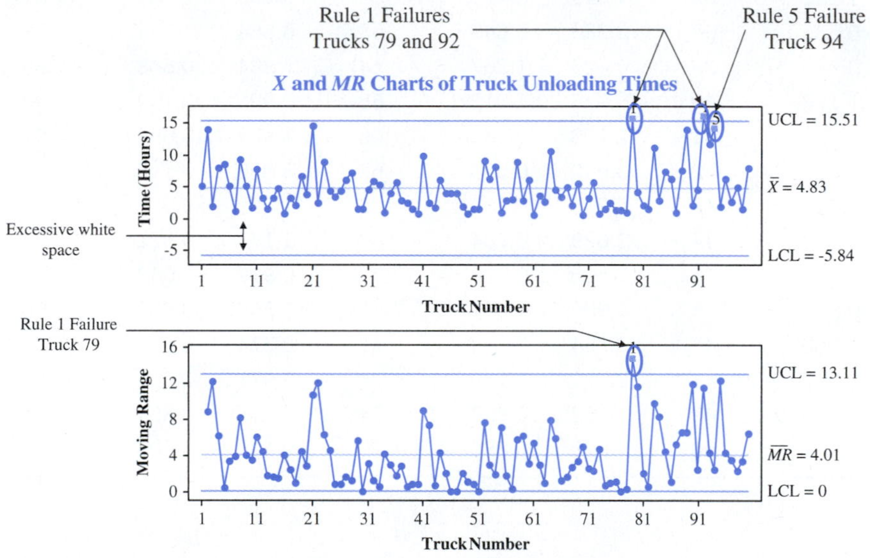

FIGURE 8-23. X and MR Charts for Truck Unloading Times

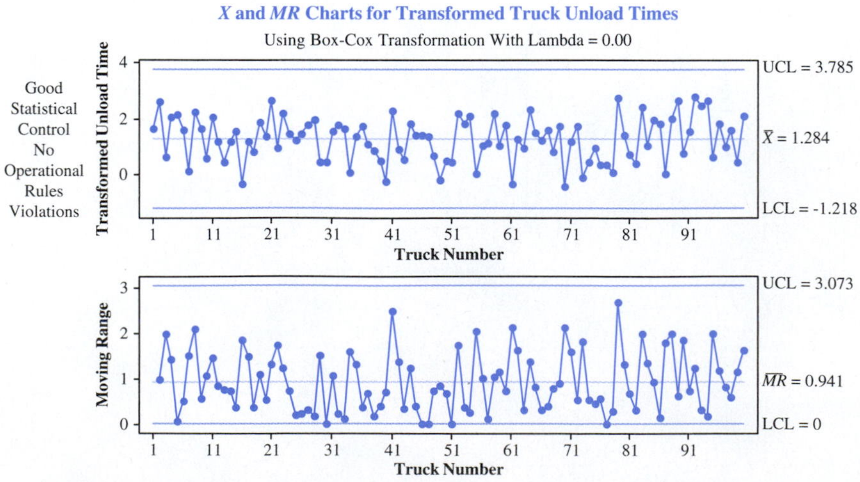

FIGURE 8-24. X and MR Charts for Transformed Truck Unloading Times

TABLE 8-12 Data For Truck Unloading Times

Allied Trucking Line							
Truck Unloading Time (hours)							
Truck No.	Hours	Truck No.	Hours	Truck No.	Hours	Truck No.	Hours
1	5.201	26	4.446	51	1.618	76	1.470
2	14.121	27	6.138	52	9.279	77	1.461
3	1.902	28	7.345	53	6.287	78	1.087
4	8.089	29	1.597	54	8.226	79	15.945
5	8.575	30	1.592	55	1.044	80	4.202
6	5.132	31	4.697	56	2.860	81	2.129
7	1.134	32	5.973	57	3.198	82	1.536
8	9.385	33	5.329	58	9.066	83	11.330
9	5.297	34	1.081	59	2.881	84	2.929
10	1.811	35	4.021	60	6.073	85	7.410
11	7.838	36	5.846	61	0.717	86	6.302
12	3.333	37	2.979	62	3.663	87	1.038
13	1.558	38	2.457	63	2.668	88	7.677
14	3.246	39	1.647	64	10.657	89	14.250
15	4.754	40	0.799	65	4.640	90	2.233
16	0.740	41	9.886	66	3.410	91	4.686
17	3.260	42	2.490	67	5.038	92	16.256
18	2.242	43	1.775	68	2.260	93	11.882
19	6.734	44	6.188	69	5.675	94	14.300
20	3.896	45	4.133	70	0.660	95	1.919
21	14.704	46	4.146	71	3.317	96	6.324
22	2.593	47	4.079	72	5.691	97	2.774
23	9.002	48	1.954	73	0.925	98	5.059
24	4.358	49	0.832	74	1.614	99	1.599
25	3.535	50	1.625	75	2.582	100	8.155

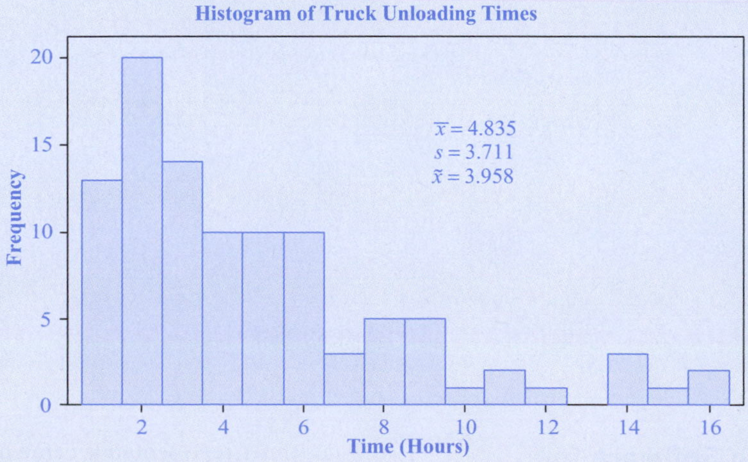

TABLE 8-13 Transformed Data For Truck Unloading Times

Allied Trucking Line											
Truck Unloading Time (hours)											
Trk No.	Hours	ln (Hours)	Trk No.	Hours	ln (Hours)	Trk No.	Hours	ln (Hours)	Trk No.	Hours	ln (Hours)
1	5.201	1.649	26	4.446	1.492	51	1.618	0.481	76	1.470	0.385
2	14.121	2.648	27	6.138	1.814	52	9.279	2.228	77	1.461	0.379
3	1.902	0.643	28	7.345	1.994	53	6.287	1.838	78	1.087	0.084
4	8.089	2.090	29	1.597	0.468	54	8.226	2.107	79	15.945	2.769
5	8.575	2.149	30	1.592	0.465	55	1.044	0.043	80	4.202	1.435
6	5.132	1.635	31	4.697	1.547	56	2.860	1.051	81	2.129	0.756
7	1.134	0.126	32	5.973	1.787	57	3.198	1.163	82	1.536	0.429
8	9.385	2.239	33	5.329	1.673	58	9.066	2.205	83	11.330	2.427
9	5.297	1.667	34	1.081	0.078	59	2.881	1.058	84	2.929	1.075
10	1.811	0.594	35	4.021	1.391	60	6.073	1.804	85	7.410	2.003
11	7.838	2.059	36	5.846	1.766	61	0.717	−0.333	86	6.302	1.841
12	3.333	1.204	37	2.979	1.092	62	3.663	1.298	87	1.038	0.037
13	1.558	0.443	38	2.457	0.899	63	2.668	0.981	88	7.677	2.038
14	3.246	1.177	39	1.647	0.499	64	10.657	2.366	89	14.250	2.657
15	4.754	1.559	40	0.799	−0.224	65	4.640	1.535	90	2.233	0.803
16	0.740	−0.301	41	9.886	2.291	66	3.410	1.227	91	4.686	1.545
17	3.260	1.182	42	2.490	0.912	67	5.038	1.617	92	16.256	2.788
18	2.242	0.808	43	1.775	0.574	68	2.260	0.815	93	11.882	2.475
19	6.734	1.907	44	6.188	1.823	69	5.675	1.736	94	14.300	2.660
20	3.896	1.360	45	4.133	1.419	70	0.660	−0.416	95	1.919	0.652
21	14.704	2.688	46	4.146	1.422	71	3.317	1.199	96	6.324	1.844
22	2.593	0.953	47	4.079	1.406	72	5.691	1.739	97	2.774	1.020
23	9.002	2.197	48	1.954	0.670	73	0.925	−0.078	98	5.059	1.621
24	4.358	1.472	49	0.832	−0.184	74	1.614	0.479	99	1.599	0.469
25	3.535	1.263	50	1.625	0.486	75	2.582	0.949	100	8.155	2.099

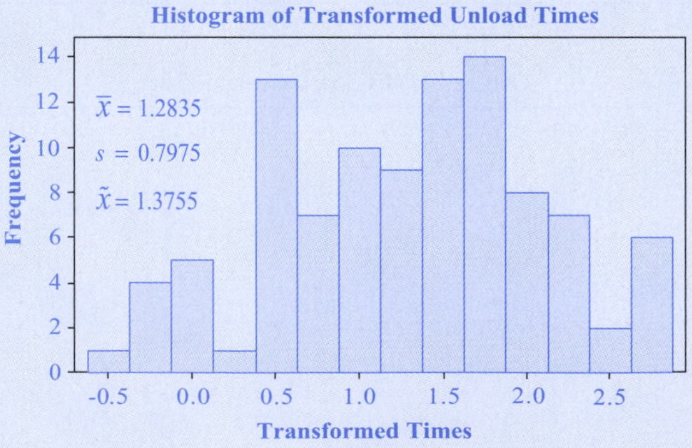

8.6.3 Using Minitab Software for Individuals Charting

The data in Table 8-14 are a sample of 100 randomly selected calls to a toll-free (1-800) customer service hotline operated by a large consumer products company. The times recorded represent the length of time customers had to wait before a customer representative came on line to offer help. We will use Minitab to set up X and MR charts to analyze these data.

The first step is to paste the spreadsheet data shown in Table 8-14 into a Minitab worksheet, placing all the measured values in a single column and then click on the drop-down menus in the following sequence shown in Figure 8-25: *Stat > Basic Statistics > Display Descriptive Statistics.* In the

TABLE 8-14 Data for Telephone Response Times

M. C. Merchandising

Telephone Response Time for Customer Service Hotline

Call No.	Hours	Call No.	Hours	Call No.	Hours	Call No.	Hours
1	0.605	26	0.484	51	0.324	76	0.499
2	0.292	27	0.421	52	0.030	77	0.140
3	0.026	28	0.517	53	0.181	78	0.120
4	0.212	29	0.434	54	0.028	79	0.066
5	0.541	30	0.064	55	0.059	80	0.504
6	0.041	31	0.246	56	0.055	81	0.266
7	0.097	32	0.578	57	0.491	82	0.177
8	0.264	33	0.500	58	0.009	83	0.080
9	0.481	34	0.040	59	0.142	84	0.324
10	0.037	35	0.006	60	0.671	85	0.843
11	0.193	36	0.626	61	0.490	86	0.177
12	0.258	37	0.396	62	0.372	87	0.202
13	0.006	38	0.118	63	0.100	88	0.762
14	0.149	39	0.168	64	0.055	89	0.005
15	0.171	40	0.288	65	0.035	90	0.289
16	0.009	41	0.569	66	0.515	91	0.263
17	0.300	42	0.156	67	0.051	92	0.305
18	0.403	43	0.216	68	0.022	93	0.209
19	0.007	44	0.002	69	0.067	94	0.419
20	0.030	45	0.444	70	0.023	95	0.069
21	0.099	46	0.144	71	0.429	96	0.150
22	0.361	47	0.029	72	0.492	97	0.856
23	0.027	48	0.130	73	0.820	98	0.110
24	0.257	49	0.358	74	0.002	99	0.008
25	0.410	50	0.803	75	0.038	100	0.022

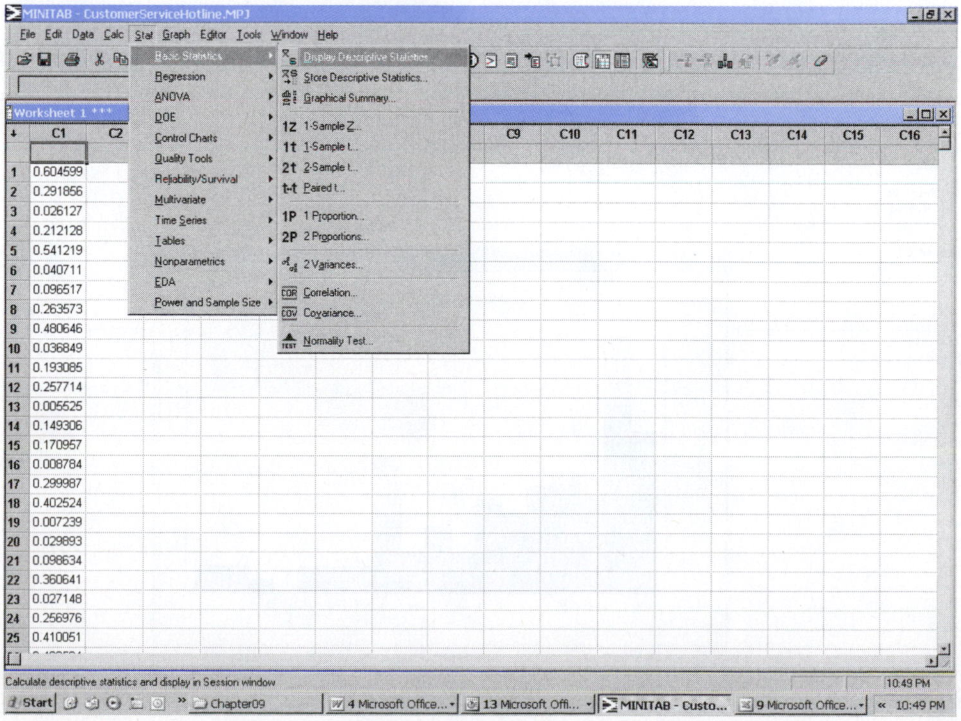

FIGURE 8-25. Minitab Dropdown Menu for Descriptive Statistics

dialog box that appears enter the column number where the data are located as shown. Next, click on *Graphs* and in the dialog box that appears, click on the *Histogram of data* option. At this point the dialog box should look like Figure 8-26. Now click on *OK*.

Click on *OK* again and the Minitab will display the histogram shown in Figure 8-27. Notice that a report containing relevant distribution parameters (e.g., mean, standard deviation, median, etc.) are also displayed. The title and labels can be changed with a double-click of the mouse, and the

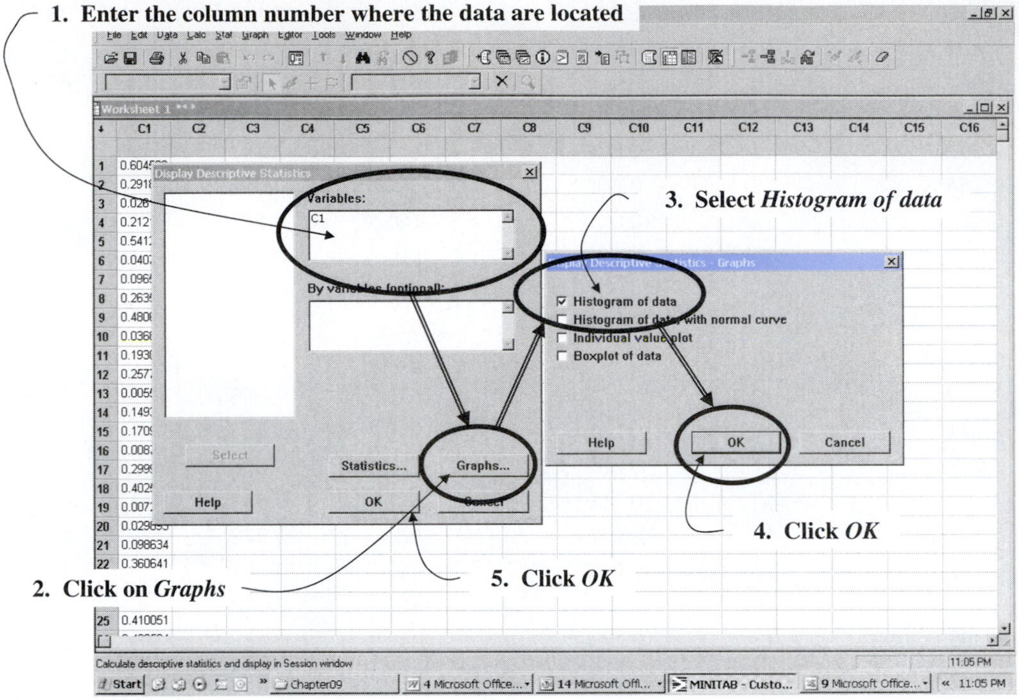

FIGURE 8-26. Minitab Dropdown Menu for Histogram Selection

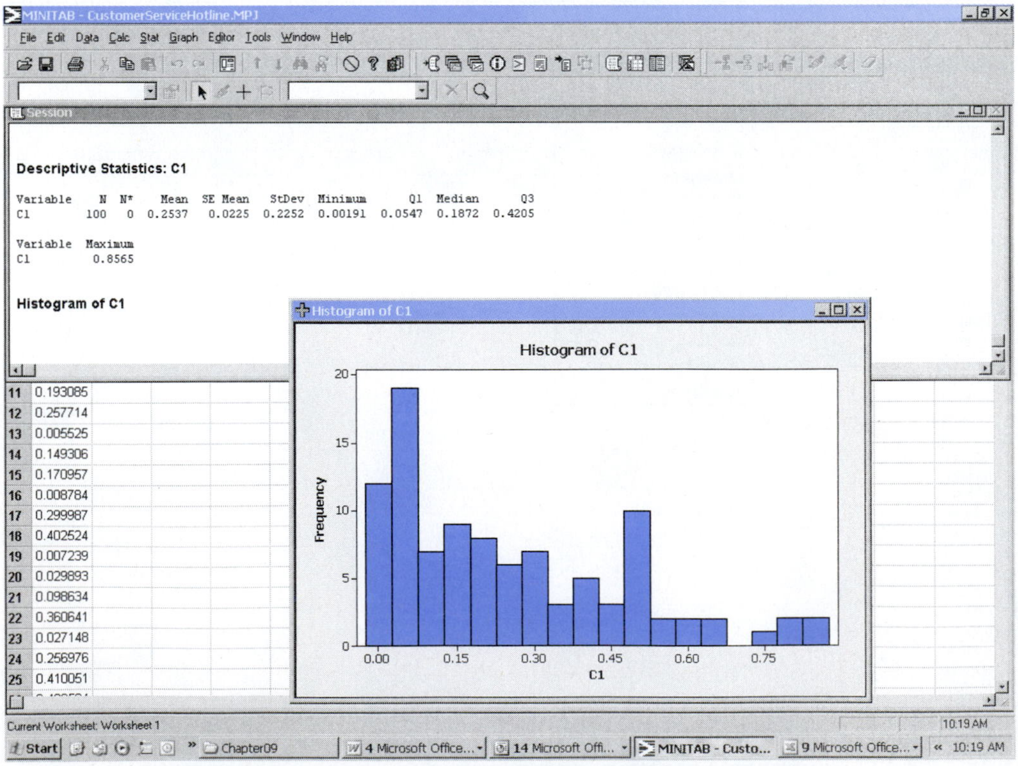

FIGURE 8-27. Minitab Display of Histogram

TABLE 8-15 Data For Telephone Response Times

M. C. Merchandising

Telephone Response Time for Customer Service Hotline

Call No.	Hours	Call No.	Hours	Call No.	Hours	Call No.	Hours	Call No.	Hours
1	0.605	21	0.099	41	0.569	61	0.490	81	0.266
2	0.292	22	0.361	42	0.156	62	0.372	82	0.177
3	0.026	23	0.027	43	0.216	63	0.100	83	0.080
4	0.212	24	0.257	44	0.002	64	0.055	84	0.324
5	0.541	25	0.410	45	0.444	65	0.035	85	0.843
6	0.041	26	0.484	46	0.144	66	0.515	86	0.177
7	0.097	27	0.421	47	0.029	67	0.051	87	0.202
8	0.264	28	0.517	48	0.130	68	0.022	88	0.762
9	0.481	29	0.434	49	0.358	69	0.067	89	0.005
10	0.037	30	0.064	50	0.803	70	0.023	90	0.289
11	0.193	31	0.246	51	0.324	71	0.429	91	0.263
12	0.258	32	0.578	52	0.030	72	0.492	92	0.305
13	0.006	33	0.500	53	0.181	73	0.820	93	0.209
14	0.149	34	0.040	54	0.028	74	0.002	94	0.419
15	0.171	35	0.006	55	0.059	75	0.038	95	0.069
16	0.009	36	0.626	56	0.055	76	0.499	96	0.150
17	0.300	37	0.396	57	0.491	77	0.140	97	0.856
18	0.403	38	0.118	58	0.009	78	0.120	98	0.110
19	0.007	39	0.168	59	0.142	79	0.066	99	0.008
20	0.030	40	0.288	60	0.671	80	0.504	100	0.022

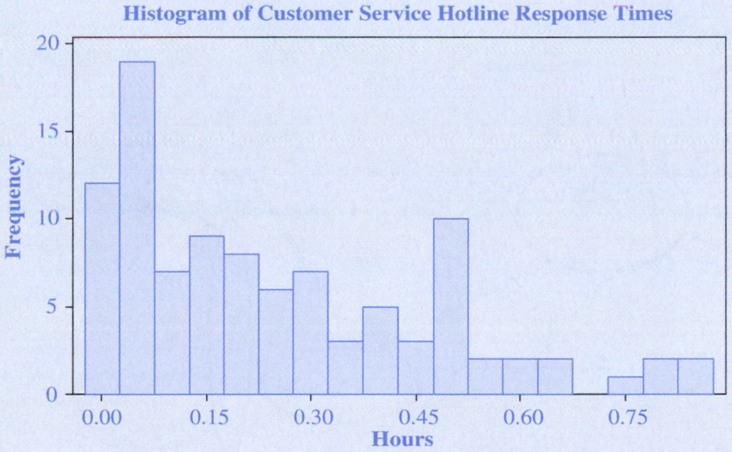

histogram can be copied and pasted into other Microsoft documents—for example, Table 8-15 illustrates how the histogram can be pasted into the spreadsheet containing the original data.

As the histogram suggests the underlying hotline service time distribution is nonnormal; therefore we should consider applying a transformation function. In Minitab we can construct X and MR charts with and without the transformation using the same dropdown menus: *Stats > Control Charts > Variables Charts for Individuals > I-MR,* as illustrated in Figure 8-28. When the dialog box appears follow the sequence of steps shown in Figure 8-29. Click *OK* again and the X and MR charts for the raw data, shown in Figure 8-30, are displayed.

To apply an appropriate Box–Cox transformation to these data we return to the first dialog box: *I-MR Options > Box-Cox* tab as illustrated in Figure 8-31. Follow the steps shown and select the *Optimal lambda* option. Clicking on *OK* twice to close the two dialog boxes will result in the control charts shown in Figure 8-32.

Figure 8-32 indicates that the optimal value of lambda is 0.32, therefore the Box–Cox transformation function is $y = x^{0.32}$.

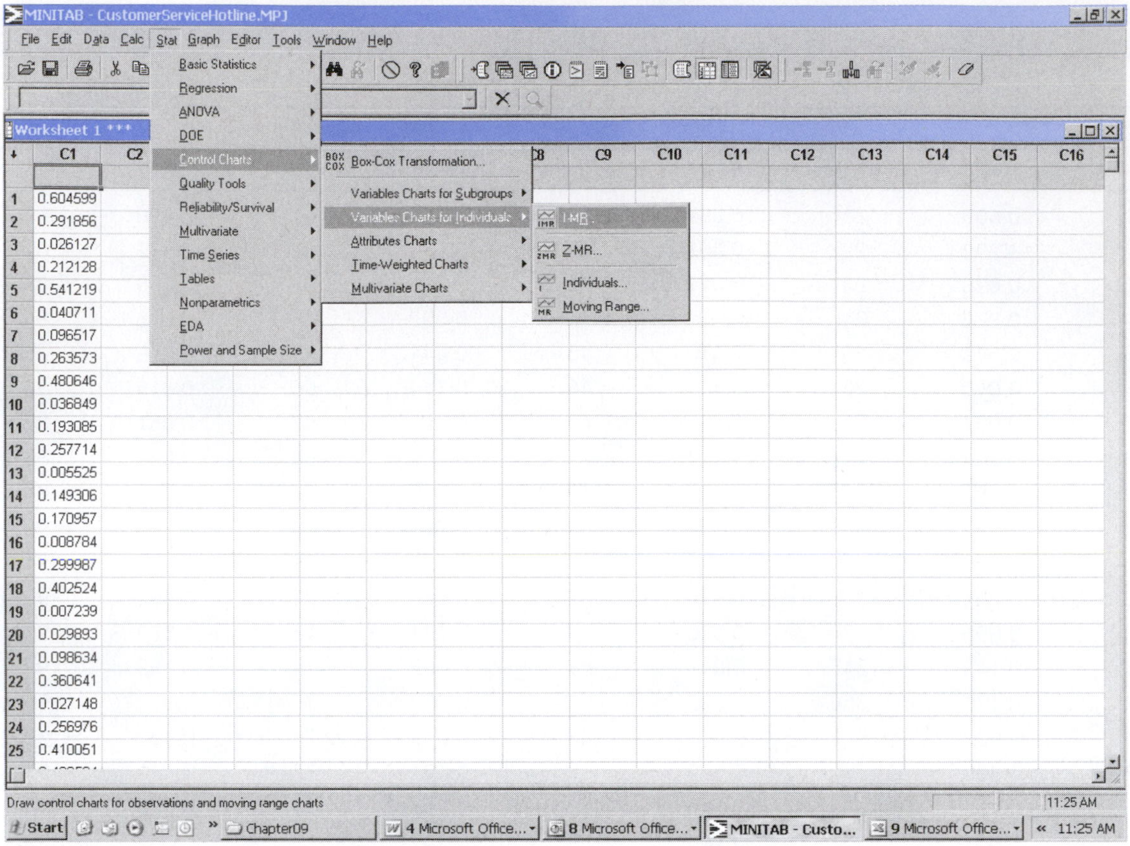

FIGURE 8-28. Minitab Dropdown Menus for *X* and *MR* Charts

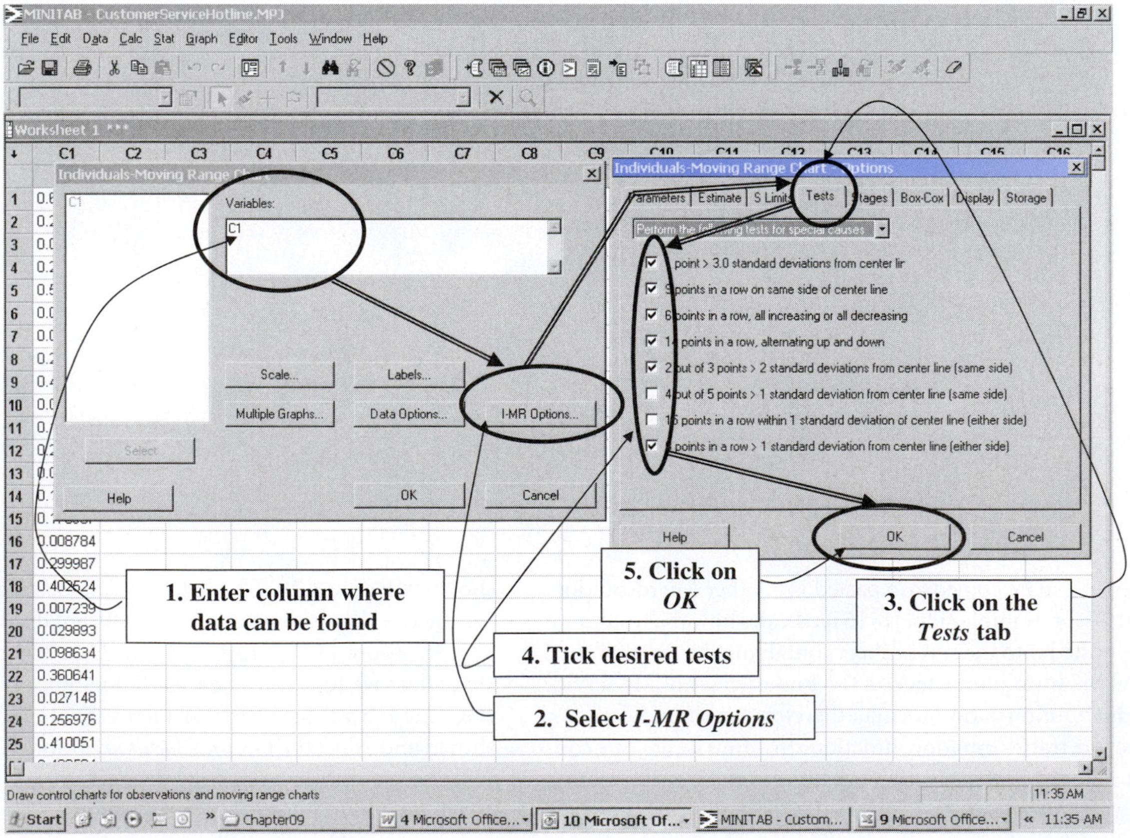

FIGURE 8-29. Minitab *X* and *MR* Statistical Tests

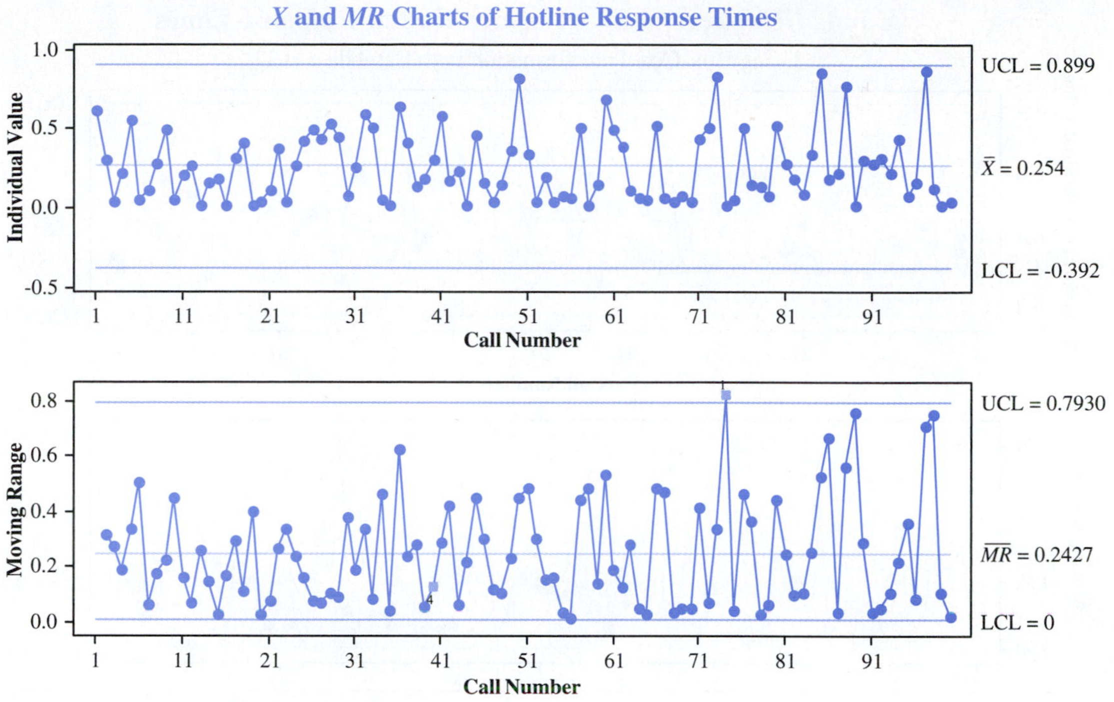

FIGURE 8-30. *X* and *MR* Charts for Hotline Raw Data

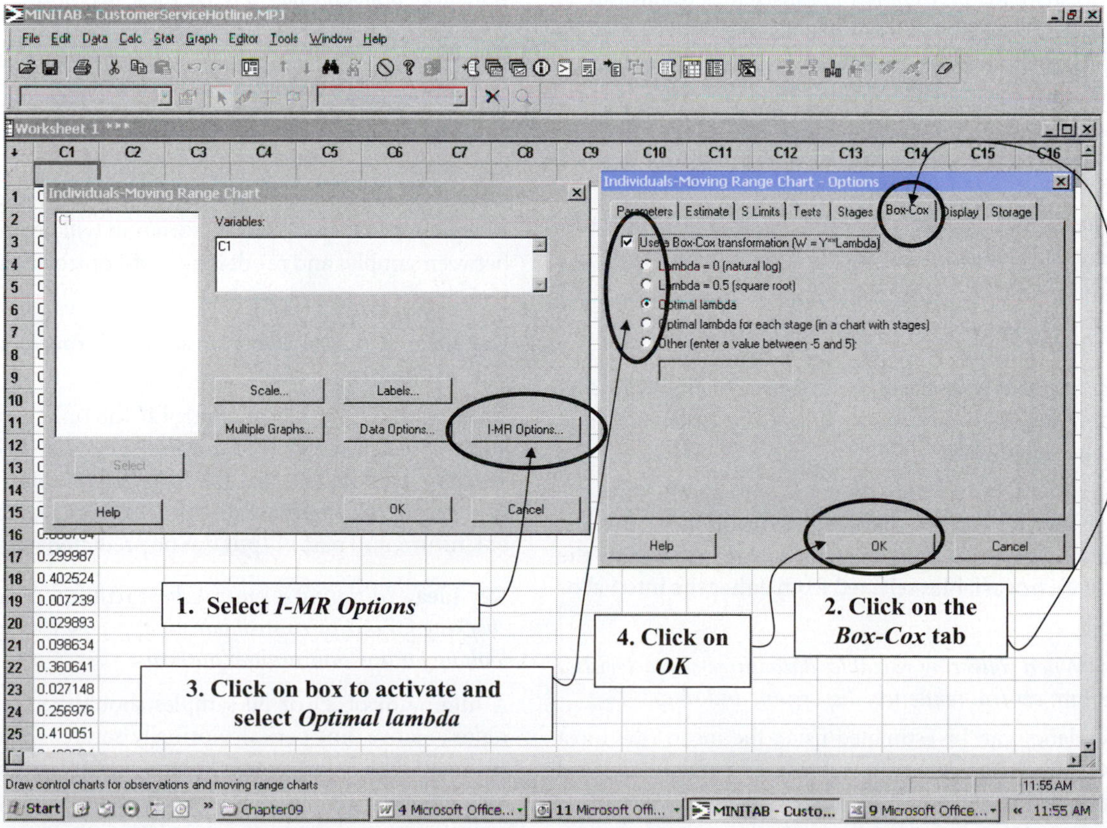

FIGURE 8-31. Minitab Box-Cox Transformation Selection

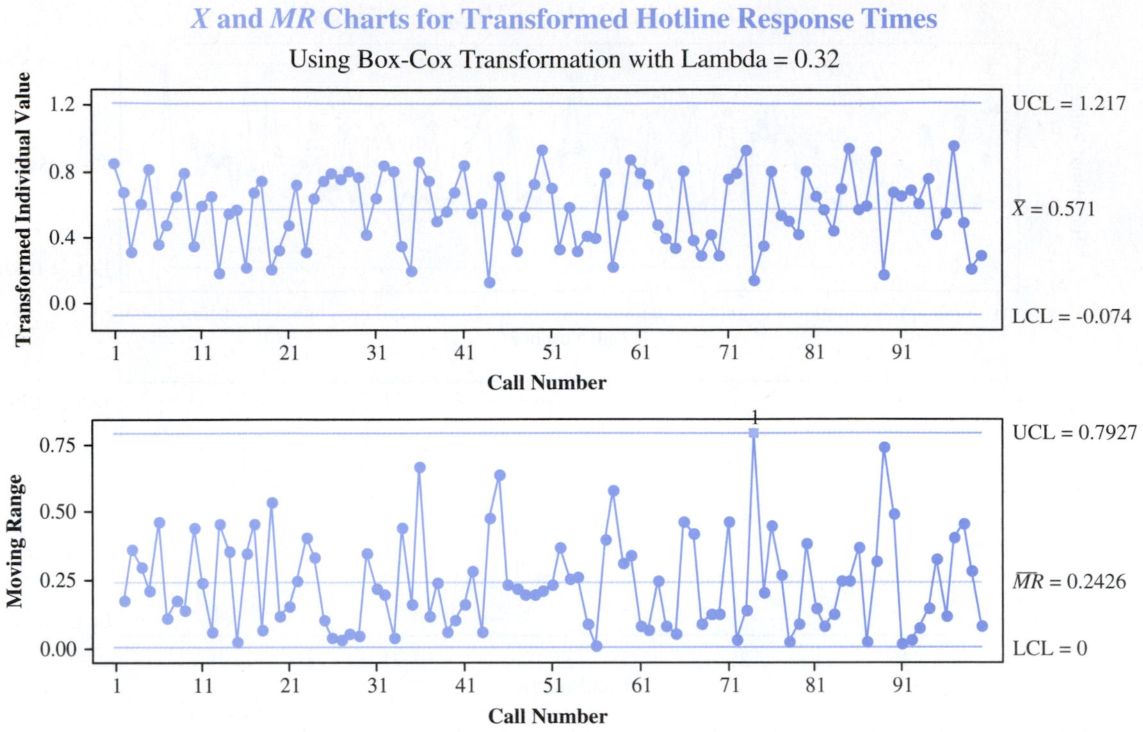

FIGURE 8-32. *X* and *MR* Charts for Transformed Data Using Optimal Lambda

This is an awkward function to operationalize; therefore the optimal lambda is used as a guide to obtain a practical one. It is not necessary to strive for perfection and satisfactory results can be obtained by simply selecting a convenient lambda that is close to the optimal one. In this case, we select $\lambda = 0.5$. This is easily implemented since the transformation simply requires that the square root of each measured value be calculated. To select this option, follow the steps shown in Figure 8-31, only this time select the Box-Cox $\lambda = 0.5$ as shown in Figure 8-33. This sequence will produce the pair of charts shown in Figure 8-34. The transformed data and histogram are shown in Table 8-16.

8.7 SUMMARY OF KEY IDEAS

Key Idea: *Improvement requires measurement and a measuring process that can capture data on one of four scales: nominal, ordinal, interval, or ratio.*

Measures should be selected that are consistent with the product or service features that are important to customers. In general, measures are either attributes selected from the nominal scale, or variables selected from either the interval or ratio scales.

Key Idea: *When collecting variables data, three characteristics are important: central tendency, dispersion, and shape.*

Central tendency can be estimated using the mean (the most widely used parameter), median, or mode. Dispersion or spread can be estimated using either the range or standard deviation. The range is useful only for small sample sizes. The shape of a data distribution can be evaluated by constructing a histogram from raw data. The histogram shape is determined by how many intervals are selected and the interval breakpoints. However, this is an issue that rarely has to be addressed because computer software have built-in rules to make these decisions.

Key Idea: *A variables control chart is a tool for separating special cause from common cause variability.*

Common cause variability is present in all samples whereas special cause variability is active in some samples and not in others. Common cause variability is systemic and short term. The presence of special cause variation will normally show up between samples and result in out-of-control signals on a control chart.

Key Idea: *A process that is in statistical control is stable, predictable, and repeatable.*

A process is in statistical control if the only sources of variability present are due to common causes. The absence of special causes means that the process is stable and that time is not a factor. Therefore, the process is repeatable and present, future, and past output can be predicted with confidence.

Key Idea: *Initiating a control chart requires 25 to 30 samples, checking for out-of-control points, applying two runs tests, and looking for any nonrandom patterns.*

A minimum of 25 or 30 samples should be collected over a representative time period. Normally, samples of size 4 or 5 are sufficient. The averages and either ranges or standard deviations are computed for each sample. These parameters are used to compute control limits and are plotted on control charts. The charts are then checked for any out-of-control points. Next, two runs tests—too-few-runs and the length-of-the-longest-run

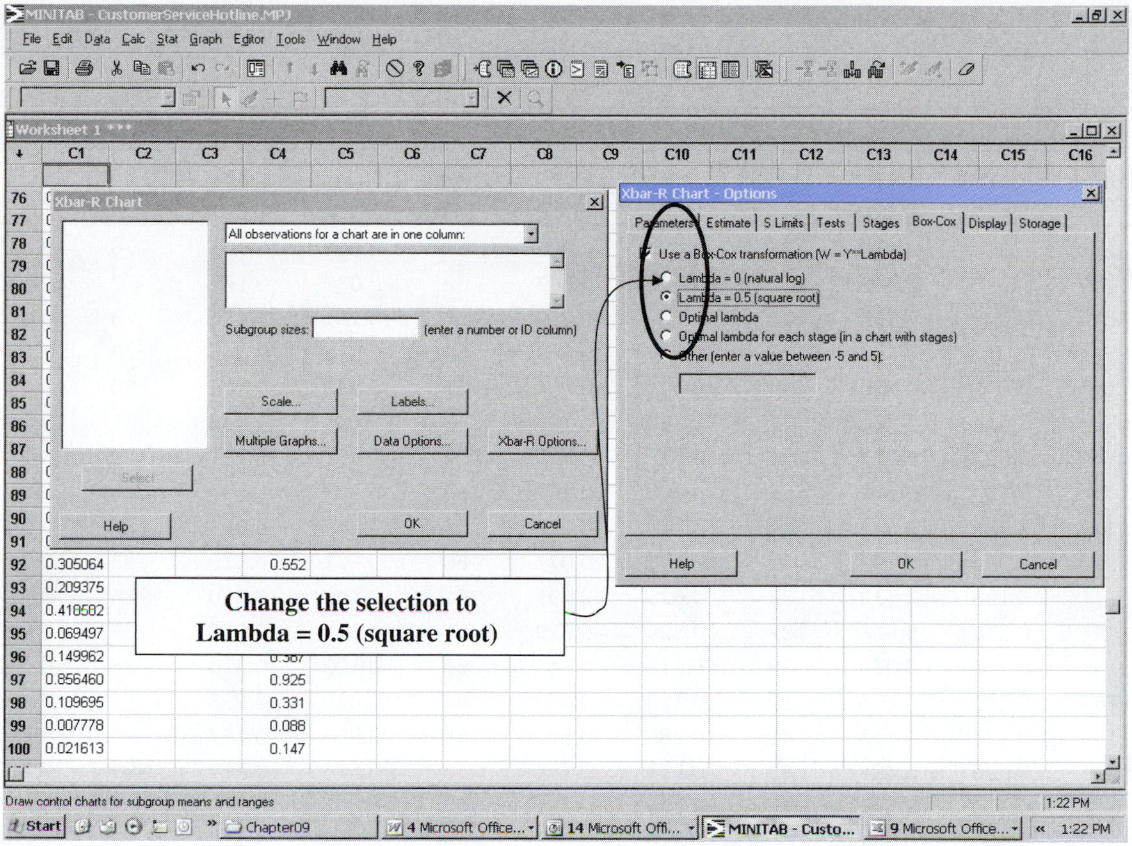

FIGURE 8-33. Selection of Practical Lambda

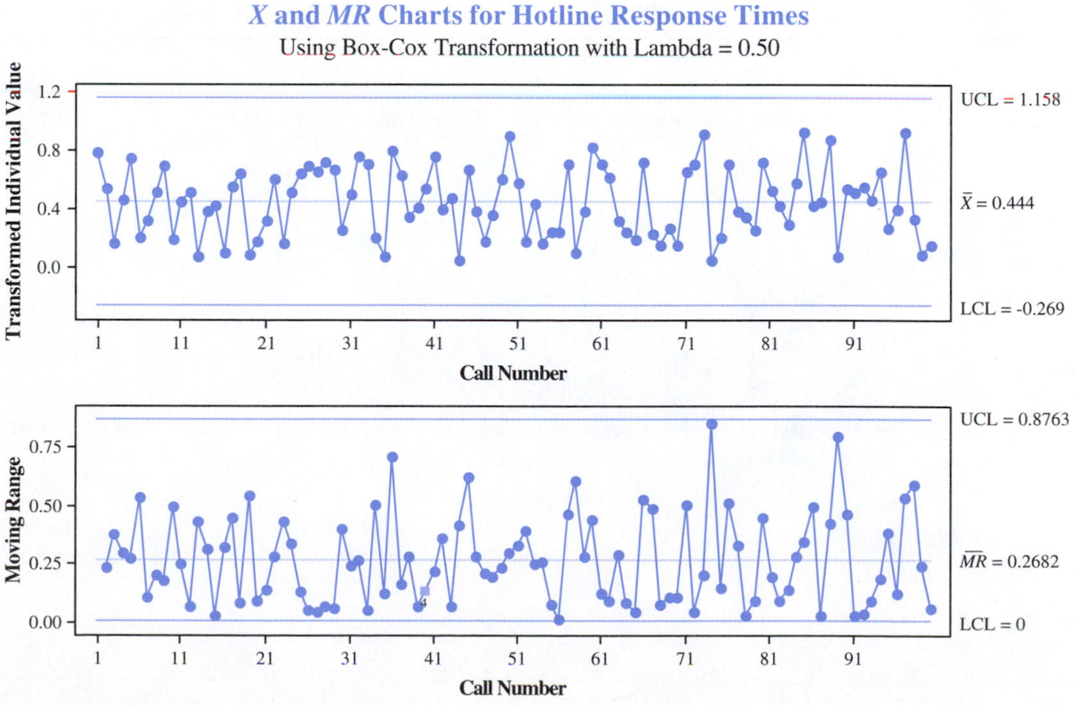

FIGURE 8-34. *X* and *MR* Charts for Final Choice of Lambda

TABLE 8-16 Transformed Data For Telephone Response Times

M. C. Merchandising											
Telephone Response Time for Customer Service Hotline											
Call No.	Hours	SqRt (Hours)	Call No.	Hours	SqRt (Hours)	Call No.	Hours	SqRt (Hours)	Call No.	Hours	SqRt (Hours)
1	0.605	0.778	26	0.484	0.695	51	0.324	0.569	76	0.499	0.706
2	0.292	0.540	27	0.421	0.649	52	0.030	0.173	77	0.140	0.374
3	0.026	0.162	28	0.517	0.719	53	0.181	0.426	78	0.120	0.347
4	0.212	0.461	29	0.434	0.659	54	0.028	0.166	79	0.066	0.257
5	0.541	0.736	30	0.064	0.254	55	0.059	0.244	80	0.504	0.710
6	0.041	0.202	31	0.246	0.496	56	0.055	0.235	81	0.266	0.516
7	0.097	0.311	32	0.578	0.760	57	0.491	0.701	82	0.177	0.421
8	0.264	0.513	33	0.500	0.707	58	0.009	0.093	83	0.080	0.283
9	0.481	0.693	34	0.040	0.199	59	0.142	0.376	84	0.324	0.569
10	0.037	0.192	35	0.006	0.077	60	0.671	0.819	85	0.843	0.918
11	0.193	0.439	36	0.626	0.791	61	0.490	0.700	86	0.177	0.421
12	0.258	0.508	37	0.396	0.629	62	0.372	0.610	87	0.202	0.450
13	0.006	0.074	38	0.118	0.344	63	0.100	0.316	88	0.762	0.873
14	0.149	0.386	39	0.168	0.409	64	0.055	0.233	89	0.005	0.068
15	0.171	0.413	40	0.288	0.537	65	0.035	0.187	90	0.289	0.538
16	0.009	0.094	41	0.569	0.754	66	0.515	0.717	91	0.263	0.513
17	0.300	0.548	42	0.156	0.395	67	0.051	0.226	92	0.305	0.552
18	0.403	0.634	43	0.216	0.464	68	0.022	0.148	93	0.209	0.458
19	0.007	0.085	44	0.002	0.044	69	0.067	0.259	94	0.419	0.647
20	0.030	0.173	45	0.444	0.666	70	0.023	0.151	95	0.069	0.264
21	0.099	0.314	46	0.144	0.380	71	0.429	0.655	96	0.150	0.387
22	0.361	0.601	47	0.029	0.169	72	0.492	0.701	97	0.856	0.925
23	0.027	0.165	48	0.130	0.360	73	0.820	0.906	98	0.110	0.331
24	0.257	0.507	49	0.358	0.598	74	0.002	0.049	99	0.008	0.088
25	0.410	0.640	50	0.803	0.896	75	0.038	0.194	100	0.022	0.147

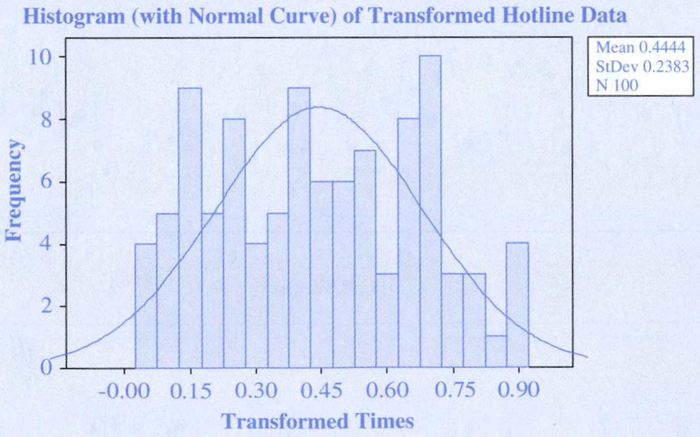

SHOWCASE of EXCELLENCE Winners of the Malcolm Baldrige National Quality Award Service Category

- 2005: DynMcDermott Petroleum Operations Company, a government contractor that operates and maintains the U.S. Strategic Petroleum Reserve representing a cache of up to 700 million barrels of crude oil. http://www.dynmcdermott.com, http://www.quality.nist.gov/PDF_files/DynMcDermott_Profile.pdf.
- 2003: Boeing Aerospace Support, a part of the Boeing Company that provides products and services including aircraft maintenance, modification and repair, and training for aircrews and maintenance staff. http://www.boeing.com, http://www.nist.gov/public_affairs/baldrige2003/Boeing_3.3.04.pdf.
- 1999: BI, a management consulting company that helps other companies achieve their own goals by enhancing the performance of their people. http://www.biperformance.com, http://www.nist.gov/public_affairs/bald99/BI.htm.
- 1997: Xerox Business Services, a provider of document outsourcing and consulting services to businesses worldwide. http://www.consulting.xerox.com, http://www.quality.nist.gov/Xerox_97.htm.
- 1994: AT&T Consumer Communications Services, a provider of long-distance communications services primarily to more than 80 million residential customers. http://www.att.com, http://www.quality.nist.gov/ATT_CCS.htm.

tests—are applied. Even if there are no out-of-control points and the data pass both runs tests, the charts still need to be evaluated for nonrandom patterns such as hugging the centerline or control limits, stratification, and trends.

Key Idea: *The control status of an operational control chart can be continually checked by applying eight standard tests.*

The control chart is a tool that dynamically monitors the stability of a process over time. Eight rules can be applied to detect statistical evidence of nonrandomness. These are a single out-of-control point, 8 (or 9) consecutive points on the same side of the centerline, 2 out of 3 consecutive points in zone A or beyond on the same side of the centerline, 4 out of 5 consecutive points in zone B or beyond on the same side of the centerline, 6 consecutive points that trend in the same direction, 14 consecutive points that alternatively change direction, 15 consecutive points within zone A, or 8 consecutive points in zone B and beyond.

Key Idea: *The central limit theorem is used to justify the placement of control limits at three standard deviations from the centerline.*

The central limit theorem ensures that sampling distributions are normally distributed regardless of the distributions of the raw data of the processes from which the samples were drawn. If the assumption of normality holds there is less than 1 chance in 1,000 that a point will lie further than three standard deviations from the mean. Therefore, if a rule states that a point outside the control limits signals the presence of special causes, there is a very small probability that the application of this rule will lead to a false conclusion.

Key Idea: *When data are not normal, transformations can sometimes be used so that control charts can be constructed with symmetrically spaced control limits.*

Skewed distributions can be misleading when plotted on traditional control charts where the underlying assumption is normality. However, often a simple transformation can be used to convert a nonnormal distribution to a transformed data set that is normal. The Box-Cox family, used by Minitab, provides a useful array of choices.

8.8 MIND EXPANDERS TO TEST YOUR UNDERSTANDING

8-1. Classify each of the following number streams with respect to its corresponding measurement scale: nominal, ordinal, interval or ratio.

a. List of 10 part numbers
b. Hourly temperature readings taken on the Celsius scale
c. Heights of all the members of a basketball team, expressed in feet and inches
d. Number of defective electronic circuits tested
e. Grading of eggs into small, medium, large, and extra-large classifications
f. Amount of time it takes to load each of 12 dump trucks
g. Number designator for a university quality control course
h. Your middle name
i. Where a runner places in the Boston marathon
j. System used to assign grades to students in a university course (e.g., A, B, C, D, F)
k. Cost of a vacation in Europe

8-2. Assemble five of your classmates together as a team with a pair of dice to perform the following procedure: Each team member is to generate five samples of five by rolling the pair of dice 25 times and recording the results. The first 5 rolls will constitute the first sample, the next 5 rolls the second sample, and so on. When all five team members have completed their rolls you should have 25 samples of 5 each. Construct $\overline{X}$ and R charts, either manually or using software such as Minitab. Then answer the following questions.

a. Is the process in a state of statistical control? Look for out-of-control points and compliance with the two runs tests.
b. Is there any evidence to suggest that differences exist between team members? With respect to variance? With respect to average? Explain.
c. What is the estimated process average? How does this estimate compare to the expected value computed using statistical probabilities?
d. What is the estimated process standard deviation? How does this estimate compare to the expected standard deviation computed using statistical probabilities?
e. Construct $\overline{X}$ and s charts for your data and repeat parts a–d. Comment on how the results compare with those obtained using the $\overline{X}$ and R pair of charts.
f. Construct a histogram of your team's results and comment on the shape.

8-3. As an extension to problem 8-2, have each team member generate an additional 3 samples of five, only this time by rolling a single die rather than a pair of dice. This procedure will result in an additional 15 samples for the team. Plot these 15 samples on the original $\overline{X}$ and R charts. Be certain that the control limits and centerlines do not change. Then answer the following questions.
a. Is there evidence to suggest that the last 15 samples came from a process different from the first 25? Explain.
b. Is the process of the single die different with respect to variance, mean, or both? How can you tell?

8-4. Construct $\overline{X}$ and R charts for the 15 samples generated in problem 8-3.
a. Is the single-die process in a state of statistical control? Look for out-of-control points and apply the two runs tests.
b. Is there any evidence to suggest that differences in the single-die data exist between team members? With respect to variance? With respect to average? Explain.
c. What is the estimated process average? How does this estimate compare to the expected value computed using statistical probabilities?
d. What is the estimated process standard deviation? How does this estimate compare to the expected standard deviation computed using statistical probabilities?
e. Construct a histogram of your team's results and comment on the shape.

8-5. Record the length of time it takes you to drive to work (or school) each day for 20 days. Plot the data on an individuals and two-point moving range chart. What is your interpretation of the data? Can you explain any out-of-control points?

8-6. A product is manufactured to meet an upper specification limit of 150.0 and a lower specification limit of 100.0. The product is inspected in subgroups of four and $\overline{X}$ and R charts are maintained to monitor product quality. The process standard deviation has been determined to be equal to 5.00.
a. To ensure that no more than 0.135% of the product fails to meet each specification limit, over what range (high and low) can the process average be permitted to drift?
b. Calculate the upper control limit for the highest process average computed in part a and the lower control limit for the lowest process average.
c. Comment on the pros and cons of using a modified control chart using the limits computed in part b to control this process.
d. What would you expect the values of the centerline and control limits to be for an R chart to control this process?

8-7. A chemical company manufactures an organic solid that is used in certain dye makeups. The following data were collected from this process.

Sample Number	Measurement 1	2	3	4
1	46.25	50.26	53.61	49.00
2	52.54	48.16	51.30	48.95
3	44.14	49.30	43.53	52.14
4	45.61	40.59	48.53	49.56
5	46.23	46.19	47.76	46.76
6	43.46	47.35	46.63	51.02
7	46.21	50.37	48.38	47.67
8	51.29	54.88	47.52	50.06
9	47.73	49.75	45.19	50.70
10	52.35	47.60	48.94	47.80
11	47.88	51.11	44.38	49.67
12	46.98	46.33	53.51	44.39
13	47.56	44.48	52.40	45.87
14	49.62	44.58	41.87	50.08
15	45.54	45.14	51.28	50.91
16	47.23	46.28	43.79	45.01
17	48.86	43.71	48.08	48.48
18	44.84	51.55	44.24	49.39
19	44.87	49.43	51.23	46.83
20	47.77	51.71	46.07	50.30
21	41.92	45.68	51.41	44.64
22	48.66	48.07	43.32	43.51
23	47.97	49.23	43.08	43.31
24	46.15	50.82	42.12	43.46
25	51.15	53.11	49.94	45.86

a. Construct $\overline{X}$ and R charts for these data and comment on whether this process is in statistical control.

b. If appropriate, estimate the value of the short-term standard deviation (that is, the standard deviation captured within a sample). If appropriate, estimate the value of the long-term (process) standard deviation (that is, the within-sample plus between-sample standard deviation).

Assuming that the process variation has been captured within the samples, answer parts c–e.

c. If the mean strength of the organic solid increases by 3.0 units, what is the probability of detecting this shift on the $\overline{X}$ chart on the first sample taken after the shift?

d. If the average value of the sample ranges increases by 25% without a change in process average, what is the probability of detecting this change on the $\overline{X}$ chart on the first sample taken after the shift? How will this show up?

e. If the changes described in parts c and d occur simultaneously, what is the probability of detecting these changes on the $\overline{X}$ chart on the first sample after the shift?

8-8. Androx Chemicals manufactures a powdered compound that is used to coat gunite swimming pools. An important measure is the Dry Brightness (DB) Number, and customers will not accept product that has a DB number less than 90.3. Because it is difficult to obtain a precise reading, a technician takes a grab sample of material, divides it into three parts, and then measures each part. The three measurements are then averaged and plotted on control charts. The following data represent 20 samples that were obtained from this process.

Sample No.	Dry Brightness Reading		
1	91.50	91.30	91.20
2	91.70	91.50	91.30
3	91.50	91.70	91.50
4	91.60	91.50	91.70
5	91.70	91.60	91.50
6	91.90	91.70	91.60
7	91.90	91.90	91.70
8	91.90	91.90	91.90
9	91.90	91.90	91.90
10	91.40	91.80	91.10
11	91.10	91.40	91.80
12	91.80	91.10	91.40
13	91.70	91.80	91.10
14	91.10	91.70	91.80
15	91.60	90.80	90.50
16	92.30	91.10	90.80
17	92.00	91.60	91.10
18	91.60	92.30	91.60
19	91.20	92.00	92.30
20	91.30	91.60	92.00

a. Use control charts to analyze this process. Is the process in control?

b. Does it appear that the process is meeting the minimum specification limit?

c. What would you recommend as the next step?

8-9. The following times were recorded for a bagger at a local grocery store.

Grocery Bagging Times (min.)

Observation	Time	Observation	Time
1	0.76	16	0.61
2	1.02	17	3.10
3	0.87	18	4.28
4	1.04	19	1.76
5	3.22	20	0.34
6	7.10	21	6.33
7	3.21	22	2.18
8	3.28	23	9.60
9	2.24	24	0.36
10	1.13	25	3.20
11	11.74	26	1.21
12	6.37	27	1.02
13	8.02	28	0.30
14	2.77	29	5.86
15	4.12	30	2.35

a. Construct a histogram for these data and comment on the shape.

b. What is your estimate of the mean and standard deviation for the data?

c. Plot an X and MR pair of control charts and comment on the state of control.

d. Use a $\lambda = 0$ Box-Cox transformation to convert the data to a distribution that is approximately normal. Plot the histogram for the transformed data set.

e. Plot an X and MR pair of control charts for the transformed data and comment on the state of control.

8-10. A certain mineral is inspected for particle size. Since particles tend to separate out, a sample is taken from four different locations in the storage silo. The four samples are then averaged. The results of 30 samples are shown in the table.

Sample Number	Measurement 1	2	3	4	$\bar{X}$	R
1	1.92	1.83	1.88	1.98	1.90	0.15
2	1.62	1.94	1.86	1.81	1.81	0.32
3	1.85	1.98	1.80	1.96	1.90	0.17
4	1.92	1.87	1.77	1.72	1.82	0.20
5	1.85	1.83	1.73	1.76	1.80	0.12
6	1.60	1.86	1.87	1.71	1.76	0.27
7	1.82	1.83	1.63	1.92	1.80	0.29
8	2.13	1.73	1.99	1.91	1.94	0.40
9	1.76	1.89	1.80	1.57	1.76	0.32
10	1.74	1.73	2.16	1.76	1.85	0.43
11	1.72	1.75	1.82	1.92	1.80	0.20
12	2.07	1.76	1.92	1.65	1.85	0.42
13	1.83	1.93	1.95	1.76	1.87	0.19
14	1.97	1.89	1.87	1.88	1.90	0.10
15	1.74	1.75	1.53	2.25	1.82	0.72
16	1.85	1.74	1.93	1.83	1.84	0.18
17	1.57	1.73	1.70	1.69	1.67	0.16
18	1.90	2.11	1.79	1.99	1.95	0.32
19	1.95	1.85	1.85	2.00	1.91	0.15
20	2.03	1.79	1.70	1.68	1.80	0.35
21	1.95	2.20	2.21	1.78	2.04	0.43
22	1.78	1.81	1.79	2.01	1.85	0.24
23	1.87	1.96	1.71	1.55	1.77	0.41
24	2.00	1.94	2.03	1.60	1.89	0.43
25	2.04	1.79	1.99	1.96	1.95	0.25
26	1.79	1.57	1.49	1.68	1.63	0.30
27	1.66	2.00	1.81	1.62	1.77	0.38
28	2.08	1.80	1.92	1.86	1.92	0.28
29	1.82	1.78	1.83	1.84	1.82	0.06
30	1.78	1.73	1.56	1.74	1.70	0.22

a. Construct an $\bar{X}$ and R chart for these data and comment on whether this process is in statistical control.

b. In this particular process what does the average range represent?

c. After the data were collected it was discovered that some problems with the test equipment had been experienced when samples 15, 21, and 26 were collected. Eliminate these samples from the data and construct new $\bar{X}$ and R charts. Comment on the state of control.

d. If appropriate, estimate the average particle size and the process variability in particle size. Does any particle size variability exist between samples?

8-11. An important dimension on a tire tread is called the shoulder gauge. For a particular model the design specifications for shoulder gauge is 0.48 in. ± 0.03 in. The following data were collected over a period of one week in a certain tire plant.

Sample Number	Measurement 1	2	3	4	5	$\bar{X}$	R
1	0.493	0.484	0.479	0.487	0.463	0.4816	0.030
2	0.500	0.517	0.482	0.489	0.500	0.4978	0.035
3	0.487	0.481	0.499	0.494	0.500	0.4923	0.019
4	0.507	0.486	0.495	0.488	0.498	0.4946	0.021
5	0.489	0.480	0.497	0.493	0.479	0.4877	0.018
6	0.492	0.492	0.481	0.493	0.469	0.4854	0.024
7	0.496	0.489	0.495	0.483	0.495	0.4915	0.013
8	0.494	0.502	0.490	0.486	0.498	0.4943	0.016
9	0.485	0.483	0.504	0.504	0.501	0.4954	0.021
10	0.499	0.491	0.493	0.475	0.487	0.4890	0.024
11	0.483	0.497	0.490	0.489	0.485	0.4888	0.015
12	0.482	0.502	0.509	0.491	0.475	0.4919	0.033
13	0.487	0.500	0.504	0.491	0.488	0.4940	0.017
14	0.502	0.505	0.479	0.504	0.486	0.4952	0.027
15	0.485	0.503	0.493	0.492	0.481	0.4909	0.022
16	0.495	0.494	0.503	0.478	0.475	0.4890	0.027
17	0.493	0.501	0.489	0.496	0.499	0.4957	0.012
18	0.493	0.476	0.483	0.489	0.486	0.4854	0.017
19	0.491	0.491	0.493	0.484	0.479	0.4874	0.014
20	0.479	0.488	0.496	0.475	0.492	0.4859	0.021
21	0.485	0.490	0.491	0.477	0.491	0.4866	0.014
22	0.493	0.493	0.494	0.478	0.494	0.4902	0.016
23	0.495	0.495	0.486	0.513	0.485	0.4949	0.028
24	0.491	0.498	0.483	0.478	0.498	0.4898	0.020
25	0.493	0.482	0.488	0.499	0.493	0.4911	0.018
26	0.493	0.501	0.504	0.503	0.498	0.4997	0.011
27	0.494	0.499	0.494	0.493	0.501	0.4963	0.008
28	0.487	0.494	0.505	0.487	0.482	0.4911	0.023
29	0.487	0.492	0.482	0.493	0.503	0.4916	0.021
30	0.500	0.493	0.493	0.501	0.472	0.4919	0.029

a. Construct control charts for these data and determine if the process is in a state of statistical control.

b. If appropriate, estimate the process standard deviation.

c. Using the results from parts a and b, estimate the percentage of total process output that will lie outside each specification limit.

d. What would be your recommendations to management concerning this process?

8-12. A bakery produces a certain type of bread in an oven that stacks trays nine high. The specification for the moisture content of finished loaves is 11% ± 2%. Samples are taken by selecting a loaf at random from each tray. The results of 30 such samples are shown in the table.

Sample	Oven Level										
Number	1	2	3	4	5	6	7	8	9	$\overline{X}$	R
1	10.87	9.40	11.75	9.49	9.55	11.86	11.10	11.18	12.49	10.854	1.135
2	11.01	9.01	10.19	10.07	11.10	10.23	13.02	10.52	12.64	10.866	1.272
3	9.91	11.11	9.93	11.70	10.01	9.76	11.84	11.68	13.03	10.998	1.154
4	9.14	11.25	10.30	10.88	11.04	11.51	10.87	12.28	11.65	10.993	0.893
5	9.10	11.21	10.28	10.20	9.56	10.94	10.43	12.25	13.24	10.800	1.296
6	8.98	9.87	10.55	8.83	11.03	10.79	11.42	11.05	13.40	10.657	1.378
7	11.33	11.16	8.78	9.50	10.58	8.78	10.54	12.15	11.64	10.495	1.227
8	11.35	10.09	10.46	9.64	10.02	9.09	11.74	11.47	12.57	10.715	1.131
9	9.66	10.15	10.48	10.77	12.06	9.89	12.85	11.30	12.09	11.027	1.113
10	8.93	9.49	9.76	10.86	12.12	10.52	10.28	9.85	10.68	10.276	0.924
11	10.04	10.09	10.00	10.11	10.12	11.85	12.94	11.18	12.39	10.968	1.161
12	9.47	10.59	11.53	11.68	12.93	13.11	10.57	11.54	12.34	11.527	1.183
13	8.68	10.57	9.96	10.19	11.49	13.12	11.28	11.50	10.16	10.772	1.255
14	9.54	10.20	10.14	11.52	10.43	11.79	11.14	11.07	12.86	10.966	1.014
15	10.49	7.97	9.32	13.43	11.44	12.19	13.62	11.89	13.94	11.587	2.029
16	9.78	10.83	10.51	10.92	9.60	10.69	10.66	11.21	11.90	10.677	0.696
17	9.07	10.06	10.62	9.80	12.00	11.22	12.94	8.99	11.46	10.686	1.339
18	9.13	10.41	8.70	11.93	12.69	13.05	10.52	8.45	11.44	10.702	1.703
19	8.09	11.16	10.62	11.01	10.71	10.81	13.56	11.67	11.43	11.006	1.410
20	10.64	10.41	12.21	10.69	12.56	11.14	11.80	12.02	10.14	11.289	0.877
21	10.51	12.51	10.70	10.95	10.96	11.44	11.45	12.96	12.10	11.508	0.848
22	9.71	10.83	9.78	10.98	11.00	9.85	12.19	11.74	11.56	10.849	0.908
23	10.47	9.71	8.55	9.79	9.38	11.77	11.75	12.27	12.13	10.647	1.367
24	10.74	11.08	9.50	9.02	10.64	12.05	11.59	12.25	12.50	11.042	1.206
25	9.91	10.01	9.82	9.71	10.68	10.00	11.54	13.56	13.21	10.939	1.502
26	7.07	8.89	11.53	10.88	9.39	10.95	11.30	11.78	12.01	10.423	1.640
27	9.26	9.79	12.54	10.34	10.91	10.11	10.93	11.70	12.59	10.908	1.175
28	9.73	9.73	10.87	10.29	11.67	13.11	10.58	10.59	12.36	10.993	1.163
29	9.16	8.65	11.94	9.84	11.09	10.78	10.84	11.39	12.16	10.649	1.206
30	10.67	8.79	10.20	11.96	10.29	10.59	13.01	10.42	13.09	11.002	1.414

a. Construct an $\overline{X}$ and s chart for these data and comment on the control status of the baking process.

b. What sources of variability are captured in the s chart? What sources are captured in the $\overline{X}$ chart?

c. Compute an estimate of the percentage of baked loaves that will be too moist. Compute an estimate of the percentage of loaves that will be too dry.

d. After analyzing the $\overline{X}$ and s charts, the quality director decided to organize the data in a different way—stratifying them by tray level. The following table reflects this restructuring.

   i. Construct $\overline{X}$ and R charts for these data and interpret the results. What should the quality director do with this information?

   ii. Is there any evidence that the within-tray variability is affected by oven level? Explain.

iii. Is there any evidence that the moisture content is affected by oven level? Explain.

iv. Assuming that it is possible to center the process average at the target of 11% for each tray level, what is your estimate of the percentage of loaves that will be too dry and too moist? How does this answer compare to the one you obtained in part c?

Oven Level	Sample Number	Observation 1	2	3	4	5	$\overline{X}$	R
1	1	10.87	11.01	9.91	9.14	9.10	10.006	1.90
	2	8.98	11.33	11.35	9.66	8.93	10.051	2.42
	3	10.04	9.47	8.68	9.54	10.49	9.643	1.81
	4	9.78	9.07	9.13	8.09	10.64	9.341	2.55
	5	10.51	9.71	10.47	10.74	9.91	10.265	1.03
	6	7.07	9.26	9.73	9.16	10.67	9.179	3.60
2	7	9.40	9.01	11.11	11.25	11.21	10.398	2.24
	8	9.87	11.16	10.09	10.15	9.49	10.151	1.67
	9	10.09	10.59	10.57	10.20	7.97	9.884	2.61
	10	10.83	10.06	10.41	11.16	10.41	10.575	1.10
	11	12.51	10.83	9.71	11.08	10.01	10.830	2.80
	12	8.89	9.79	9.73	8.65	8.79	9.170	1.15
3	13	11.75	10.19	9.93	10.30	10.28	10.489	1.81
	14	10.55	8.78	10.46	10.48	9.76	10.006	1.77
	15	10.00	11.53	9.96	10.14	9.32	10.189	2.21
	16	10.51	10.62	8.70	10.62	12.21	10.531	3.51
	17	10.70	9.78	8.55	9.50	9.82	9.670	2.15
	18	11.53	12.54	10.87	11.94	10.20	11.417	2.34
4	19	9.49	10.07	11.70	10.88	10.20	10.468	2.22
	20	8.83	9.50	9.64	10.77	10.86	9.917	2.03
	21	10.11	11.68	10.19	11.52	13.43	11.385	3.33
	22	10.92	9.80	11.93	11.01	10.69	10.870	2.13
	23	10.95	10.98	9.79	9.02	9.71	10.089	1.96
	24	10.88	10.34	10.29	9.84	11.96	10.662	2.12
5	25	9.55	11.10	10.01	11.04	9.56	10.253	1.55
	26	11.03	10.58	10.02	12.06	12.12	11.159	2.10
	27	10.12	12.93	11.49	10.43	11.44	11.280	2.81
	28	9.60	12.00	12.69	10.71	12.56	11.512	3.10
	29	10.96	11.00	9.38	10.64	10.68	10.531	1.61
	30	9.39	10.91	11.67	11.09	10.29	10.670	2.28
6	31	11.86	10.23	9.76	11.51	10.94	10.860	2.10
	32	10.79	8.78	9.09	9.89	10.52	9.814	2.00
	33	11.85	13.11	13.12	11.79	12.19	12.411	1.33
	34	10.69	11.22	13.05	10.81	11.14	11.381	2.36
	35	11.44	9.85	11.77	12.05	10.00	11.022	2.20
	36	10.95	10.11	13.11	10.78	10.59	11.107	3.00

Oven Level	Sample Number	Observation 1	2	3	4	5	$\overline{X}$	R
	37	11.10	13.02	11.84	10.87	10.43	11.453	2.59
	38	11.42	10.54	11.74	12.85	10.28	11.365	2.56
	39	12.94	10.57	11.28	11.14	13.62	11.910	3.05
7	40	10.66	12.94	10.52	13.56	11.80	11.895	3.04
	41	11.45	12.19	11.75	11.59	11.54	11.706	0.74
	42	11.30	10.93	10.58	10.84	13.01	11.333	2.43
	43	11.18	10.52	11.68	12.28	12.25	11.584	1.76
	44	11.05	12.15	11.47	11.30	9.85	11.163	2.31
	45	11.18	11.54	11.50	11.07	11.89	11.435	0.82
8	46	11.21	8.99	8.45	11.67	12.02	10.469	3.57
	47	12.96	11.74	12.27	12.25	13.56	12.558	1.82
	48	11.78	11.70	10.59	11.39	10.42	11.176	1.36
	49	12.49	12.64	13.03	11.65	13.24	12.610	1.58
	50	13.40	11.64	12.57	12.09	10.68	12.078	2.72
	51	12.39	12.34	10.16	12.86	13.94	12.338	3.78
9	52	11.90	11.46	11.44	11.43	10.14	11.273	1.76
	53	12.10	11.56	12.13	12.50	13.21	12.299	1.66
	54	12.01	12.59	12.36	12.16	13.09	12.442	1.08

8-13. Random samples are taken of the number of parts arriving at an assembly point during five-minute time intervals. The table below summarizes observations that were made during 30 intervals randomly selected over an eight-hour shift.

Observation	No. Parts Arriving	Observation	No. Parts Arriving
1	4	16	4
2	5	17	6
3	3	18	8
4	5	19	3
5	2	20	8
6	2	21	7
7	6	22	7
8	2	23	2
9	4	24	6
10	10	25	6
11	9	26	3
12	3	27	3
13	5	28	2
14	6	29	6
15	4	30	8

a. Construct an X and 2-point MR chart for these data and interpret the results.

b. How many parts would you estimate will arrive in the next five-minute interval?

c. What is your estimate of the maximum and minimum number of arrivals that will occur during any five-minute interval?

8-14. A certain liquid product is placed in cans on an automatic filling line. The fill specifications are 10 oz ± 0.5 oz. Thirty samples of size 4 have been collected at random over several hours of production. These results are summarized in the following table.

a. Construct $\overline{X}$ and R charts and assess the state of control of this process.

b. What is your estimate of the process mean and standard deviation?

c. What percentage of containers will be outside the specification limits?

d. How much average overfill is required if it is desired that fewer than 1 container in 1,000 will have a fill volume less than 9.5 oz?

Sample	Fill Volumes (oz) 1	2	3	4	$\overline{X}$	R
1	10.14	10.27	10.59	9.91	10.23	0.68
2	10.27	10.15	10.07	10.03	10.13	0.24
3	10.46	9.94	10.19	9.87	10.12	0.59
4	10.08	10.00	9.97	10.23	10.07	0.26
5	9.80	10.24	10.06	10.33	10.11	0.53
6	10.38	10.06	10.47	10.03	10.23	0.44
7	10.26	10.16	10.25	10.31	10.25	0.15
8	10.16	9.92	10.18	10.15	10.10	0.25
9	10.30	10.08	9.92	9.92	10.05	0.38
10	10.41	9.97	9.78	10.27	10.11	0.63
11	10.13	10.51	10.23	10.52	10.35	0.39
12	10.25	9.99	10.45	10.24	10.23	0.46
13	10.07	10.27	10.34	10.00	10.17	0.33
14	10.21	10.00	10.07	10.12	10.10	0.21
15	10.25	10.37	9.60	10.28	10.13	0.77
16	10.24	10.19	10.36	10.39	10.29	0.20
17	10.14	10.42	10.30	10.45	10.33	0.31
18	10.41	10.12	10.11	10.13	10.19	0.31
19	10.25	10.28	10.35	9.94	10.21	0.41
20	10.38	10.20	10.38	9.97	10.23	0.41
21	10.26	10.29	10.12	10.06	10.18	0.23
22	10.12	10.13	10.62	10.08	10.24	0.55
23	10.25	10.07	9.95	10.02	10.07	0.30
24	10.51	10.58	10.44	9.90	10.36	0.69
25	10.14	10.20	9.98	10.15	10.12	0.22
26	10.26	10.24	9.92	9.90	10.08	0.37
27	10.21	10.43	10.35	9.93	10.23	0.50
28	10.21	10.11	10.56	10.48	10.34	0.44
29	10.38	10.18	10.25	10.23	10.26	0.21
30	10.32	10.01	10.22	10.35	10.22	0.33

8-15. Times were recorded at random intervals for the times required to process customer service calls. The times were for Bob, one of the customer service representatives, and are recorded in this table.

Customer Service Transaction Times (min.) for Bob

Customer	Transaction Times	Customer	Transaction Times
1	10.69	16	23.54
2	9.58	17	7.53
3	2.90	18	3.28
4	3.77	19	8.29
5	1.26	20	9.09
6	0.28	21	2.53
7	0.65	22	11.14
8	5.55	23	2.45
9	8.18	24	0.26
10	13.22	25	3.29
11	8.96	26	9.27
12	4.03	27	7.31
13	1.70	28	4.43
14	7.02	29	17.43
15	1.24	30	1.45

a. Construct an X chart and 2-point MR chart for Bob's data and comment on the control status.

b. Transform Bob's data using a Box-Cox transformation with $\lambda = 0$. Construct an X chart and 2-point MR chart using the transformed data. Is the process in control? Explain any inconsistencies between these conclusions and those in part a.

c. What is your estimate of Bob's average transaction time and Bob's process standard deviation?

d. Information was collected for Judy, a second customer service representative, as shown in the following table.

Customer Service Transaction Times (min.) for Judy

Customer	Transaction Times	Customer	Transaction Times
1	0.62	16	24.91
2	52.31	17	7.25
3	28.02	18	5.34
4	32.77	19	43.50
5	36.13	20	14.86
6	3.40	21	6.16
7	13.10	22	4.88
8	11.98	23	12.16
9	10.49	24	22.98
10	6.38	25	23.01
11	4.77	26	9.67
12	31.23	27	5.44
13	5.73	28	13.01
14	30.45	29	2.12
15	5.67	30	6.10

Convert Judy's data using a Box-Cox transformation with $\lambda = 0$, and then plot Judy's data on Bob's control chart that you constructed in part b. Is there any evidence to suggest that Judy's performance differs from Bob's? Explain.

e. Plot Judy's transformed data on a separate control chart and assess its control status.

f. What is your estimate of Judy's average transaction time and Bob's process standard deviation?

8-16. Harris Hemp Company tests the tensile strength of a certain manufactured industrial grade rope. The specifications for this product stipulate a minimum tensile strength of 440 kilograms (kg). The following data represent samples taken randomly from the production line during the past week.

	Tensile Strength (kg)			
Sample	**1**	**2**	**3**	**4**
1	471	474	495	488
2	476	460	467	526
3	486	495	484	478
4	500	452	488	464
5	493	503	471	482
6	451	494	465	476
7	479	461	484	466
8	490	501	461	497
9	497	464	484	492
10	522	483	472	492
11	499	473	490	476
12	478	488	498	473
13	483	492	495	462
14	465	457	449	491
15	466	485	480	470
16	478	483	495	491
17	488	487	489	493
18	488	489	477	474
19	484	485	485	499
20	463	492	467	474
21	502	487	481	479
22	500	464	474	506
23	490	492	495	467
24	494	475	498	465
25	489	468	491	470

a. Construct control charts to determine if the process is in control.

b. Use the data collected to estimate the process average tensile strength and the standard deviation.

c. What percentage of the process output would you estimate is below the minimum required tensile strength?

d. Assuming that the process standard deviation continues at its current level, where should the process be centered so that less than one rope in one million will fail to meet the minimum specification?

e. Assuming that the process average continues at its current level, by what percentage would the process variance have to be reduced so that less than one rope in one million will fail to meet the minimum specification?

8-17. The Taylor Tool company manufactures hardened tool steel. The specifications for one of its products, A768C, requires a hardness of 550 ± 50. The following data have been collected on steel produced over the past three shifts.

	Brinell Hardness (kg per mm²)									
Sample	**1**	**2**	**3**	**4**	**5**	**6**	**7**	**8**	**9**	**10**
1	533	554	587	531	579	559	617	575	516	569
2	540	549	561	550	528	545	550	582	544	593
3	561	613	542	536	606	550	565	560	540	544
4	529	547	524	513	535	563	560	556	512	563
5	555	572	528	526	566	556	556	559	520	548
6	533	558	602	529	545	574	548	514	536	545
7	539	583	512	574	558	524	562	506	531	526
8	552	559	542	565	532	553	584	585	545	563
9	555	533	536	555	551	539	558	557	562	547
10	537	587	553	546	528	528	545	540	542	490
11	528	589	523	536	543	565	530	597	558	568
12	540	592	555	603	524	499	520	550	522	496
13	575	562	552	575	596	539	529	555	578	555
14	509	541	542	533	548	571	578	616	560	570
15	529	556	542	530	514	531	549	553	551	596
16	536	559	530	528	569	587	581	542	557	584
17	516	544	552	561	580	557	563	566	557	533
18	564	584	545	580	539	573	586	548	537	617
19	584	555	499	566	541	576	545	564	593	578
20	529	569	527	521	565	606	578	559	538	543
21	545	517	532	568	507	532	536	554	524	550
22	569	505	525	586	517	498	576	528	534	566
23	531	522	535	552	552	526	545	560	592	532
24	504	528	597	540	546	536	588	528	544	573
25	571	574	561	553	566	520	485	547	611	534

a. Construct $\overline{X}$ and s charts to determine the state of statistical control.

b. Use the charts constructed in part a to estimate the mean hardness and standard deviation of this process.

c. What percentage of the process output fails to comply with specifications?

d. Construct $\overline{X}$ and R charts for the data and repeat parts a through c.

e. Compare your results using an s chart and an R chart to estimate process variation. Which would you prefer and why?

8-18. A design of experiment was conducted on the process described in problem 8-17. Some process improvements were made and the following data were collected.

a. Validate the improvements by plotting these data on the $\overline{X}$ and s charts you constructed in part a of problem 8-17.

b. Verify process stability by constructing new $\overline{X}$ and s charts for the post-improvement data.

c. What is your estimate of the process average and standard deviation following process improvements?

	Brinell Hardness (kg per mm²) after Improvements									
Sample	1	2	3	4	5	6	7	8	9	10
1	537	550	535	562	551	575	544	524	556	530
2	557	550	535	517	531	540	551	539	523	573
3	555	575	554	547	557	547	534	558	575	586
4	531	552	552	508	581	536	518	539	554	581
5	518	540	519	543	540	575	552	530	579	575
6	574	549	568	577	526	577	537	566	568	533
7	550	581	538	554	559	548	549	566	563	552
8	550	551	541	543	549	565	543	514	549	539
9	530	557	546	543	545	603	542	524	559	538
10	563	530	547	566	544	568	559	546	531	532
11	528	549	553	564	559	561	523	527	560	576
12	542	543	557	553	539	556	549	546	555	530
13	536	523	561	551	523	550	549	551	563	533
14	565	529	536	530	549	526	535	527	528	551
15	529	563	530	559	558	519	568	533	567	563
16	584	546	562	579	516	545	535	524	559	535
17	539	563	535	536	540	570	574	571	576	546
18	539	539	541	546	572	540	552	554	549	546
19	593	537	537	541	544	527	553	527	553	536
20	546	583	542	529	518	557	516	538	533	563
21	550	563	550	516	555	558	514	584	538	550
22	554	559	560	551	557	541	578	539	502	529
23	577	531	559	545	566	562	570	549	523	545
24	508	557	575	551	566	546	561	565	540	511
25	552	538	566	541	538	541	554	566	534	541

8-19. The Road Grip Tire and Rubber Company conducts tests on how well its tires perform under conditions similar to those encountered by its customers. One test focuses on tread wear under a variety of simulated road and driving conditions. The following data were collected on 24 samples of three radial XR20 tires each, collected at random from the manufacturing process over three consecutive shifts of operation.

a. Construct $\overline{X}$ and R charts to determine if the process is in a state of statistical control.

b. What is your estimate of the average treadwear and the standard deviation of treadwear for this product?

c. For any tire selected at random what would you predict to be the minimum and maximum expected treadwear?

	Tread Wear (1/100th in.)				Tread Wear (1/100th in.)		
Sample	1	2	3	Sample	1	2	3
1	41	40	38	13	24	28	15
2	25	18	22	14	20	42	14
3	35	24	19	15	31	20	32
4	20	22	25	16	34	23	38
5	41	31	32	17	34	34	25
6	29	39	36	18	26	20	23
7	29	27	32	19	26	36	45
8	22	35	38	20	36	22	23
9	28	27	32	21	26	17	34
10	34	27	36	22	27	17	32
11	25	47	41	23	26	32	23
12	23	30	22	24	26	28	27

8-20. Thirty successive heats of a steel alloy are tested at the Midwest Steel Works. The resulting data have been recorded in the following table.

Heat Number	Hardness (Coded Data)	Heat Number	Hardness (Coded Data)
1	55	16	56
2	51	17	54
3	57	18	55
4	56	19	58
5	53	20	57
6	59	21	57
7	52	22	57
8	56	23	53
9	57	24	53
10	54	25	56
11	56	26	55
12	54	27	53
13	52	28	58
14	57	29	55
15	54	30	56

a. Use appropriate control charts to determine if the alloy hardness is in a state of statistical control.

b. If hardness must at least equal a coded value of 54, what percentage of the heats fail to comply?

c. By what percentage would the process variance need to be reduced to ensure that no more than 0.1% of the heats fail to comply with the minimum hardness requirement?

8-21. The following times were recorded at random intervals as total drive-thru service time at a local neighborhood Wally Burgers fast-food restaurant. Wally's strives to service its customers accurately and efficiently with a maximum time goal of 4 minutes.

a. Construct a histogram for these data and comment on the shape.

b. Plot an X and MR pair of control charts and comment on whether the drive-thru service times appear to be in a state of control.

c. What is your estimate of the mean and standard deviation for this process? If appropriate, use control chart data as the basis for these estimates.

d. What percentage of the time would you estimate that Wally's service time goal is not being met (i.e., above the upper specification limit)?

Wally Burgers' Drive-Thru Service Time (in min.)

Observation	Time	Observation	Time
1	1.06	16	0.91
2	1.72	17	1.4
3	1.17	18	1.98
4	1.44	19	2.06
5	2.52	20	1.64
6	2.8	21	2.03
7	1.51	22	0.88
8	2.98	23	1.3
9	2.54	24	2.66
10	1.83	25	2.9
11	3.04	26	1.51
12	3.07	27	1.72
13	2.32	28	3.6
14	2.47	29	2.56
15	3.42	30	1.65

e. If the variance of service time remains the same, what would the average time have to be lowered to so that the number exceeding the UTL will not exceed more than 3.4 customers per million?

f. Assuming that the current average service time remains unchanged, by what percentage would the process variance need to be reduced to ensure that no more than 3.4 customers per million fail to meet the maximum time goal?

8-22. The production line at the N&B Rug Company produces a series of 8-by-10-foot rugs. Management performs audits on the lengths of 4 different rugs each hour. The lengths of these rugs must meet a specification range of 120 in. ± 10 in. The data in the following table reflect the results of audits conducted over a 30-hour period.

a. Construct $\overline{X}$ and R charts for these data. Is the process in a state of statistical control?

b. If appropriate, estimate the process parameters.

c. After evaluating the data for the first 30 samples, the production engineer reported that a machine malfunction had occurred creating some excessive product variation around the time that sample number 16 was collected. With this information you decide to discard sample number 16 from the data set. Construct new $\overline{X}$ and R charts for the remaining 29 points. Is the process in a state of statistical control?

d. If appropriate, estimate the process parameters.

e. Estimate the percentage of parts that will be outside the specification limits.

Sample Number	Rug 1	Rug 2	Rug 3	Rug 4	Sample Number	Rug 1	Rug 2	Rug 3	Rug 4
1	117.140	122.139	116.621	119.189	16	128.0142	126.2161	126.8748	121.3915
2	124.990	122.620	118.962	116.154	17	120.5417	116.7137	121.9032	124.4956
3	116.550	116.787	112.710	115.048	18	117.2221	124.8453	122.1636	116.2655
4	117.283	125.639	123.467	113.308	19	123.4000	121.8460	118.3899	120.8717
5	122.768	116.260	117.296	119.238	20	119.1793	127.2177	121.6338	125.0126
6	117.118	121.126	121.018	121.076	21	115.4728	118.7268	122.1513	120.2861
7	120.329	119.072	118.834	127.567	22	109.4799	109.9745	123.7395	114.7177
8	114.324	115.476	119.065	111.459	23	123.3551	116.3492	122.0301	117.2296
9	112.271	115.367	128.864	118.786	24	115.2267	116.1826	111.4129	113.1616
10	113.841	123.198	122.500	116.440	25	118.2741	120.3131	116.6585	118.6717
11	117.893	118.687	120.631	112.980	26	117.8978	124.6000	117.0006	114.1942
12	121.295	126.213	115.757	119.314	27	119.1105	115.0319	120.6676	115.9658
13	116.913	118.350	121.511	113.923	28	120.0245	123.5194	119.0712	113.5043
14	117.316	123.525	121.432	117.208	29	120.9018	115.1035	117.2599	116.7153
15	117.301	110.880	117.954	117.864	30	116.9717	121.1568	118.3422	120.9515

f. After evaluating the previous data, management decided to conduct an additional 20 audits. The results are shown in the following table.

Sample Number	Rug 1	Rug 2	Rug 3	Rug 4
1	122.281	128.456	119.986	122.497
2	117.909	123.261	127.273	123.806
3	126.714	119.136	114.733	124.572
4	124.965	120.523	126.304	131.592
5	121.925	120.049	123.029	127.730
6	125.657	120.881	121.897	116.588
7	118.671	122.770	126.261	128.115
8	129.232	117.041	125.634	124.685
9	124.020	128.908	124.275	117.187
10	122.795	122.012	122.665	123.156
11	125.735	126.577	124.925	119.614
12	126.406	126.790	131.127	116.783
13	116.665	119.339	121.264	119.277
14	118.370	122.055	120.884	122.963
15	115.560	125.139	120.284	129.608
16	131.505	125.773	120.415	122.183
17	123.820	118.773	119.672	125.318
18	121.229	123.447	125.727	122.355
19	120.691	121.188	125.384	123.802
20	124.453	126.568	125.280	122.570

g. Plot these new points on the original $\overline{X}$ and R charts and comment on the state of control. Has the process changed?

h. Construct $\overline{X}$ and R charts of just the new 20 sample points. Are these data in a state of statistical control?

i. Based on the charts you constructed in part h, estimate the percentage of rugs that will be too short. Estimate the percentage that will be too long.

8-23. Tennessee Accessories, Inc. is working to implement new time standards for a particular sewing process in its T-shirt manufacturing facility. This is a very difficult task, however, because the skill of the individual employees and the age of the equipment

Sewing Procedure Time Study (min.)			
Day 1		Day 2	
Time of Day	Sample	Time of Day	Sample
8:30 AM	17.43	8:30 AM	3.26
10:00 AM	4.72	10:00 AM	13.44
10:30 AM	2.9	10:30 AM	2.03
11:00 AM	2.43	11:00 AM	1.26
12:00 PM	1.24	12:00 PM	1.83
12:30 PM	15.69	12:30 PM	4.79
1:30 PM	4.11	1:30 PM	3.93
2:00 PM	2.45	2:00 PM	1.73
3:00 PM	2.42	3:00 PM	4.78
4:00 PM	15.26	4:00 PM	1.43

cause a great deal of variability in the amount of time the process requires. To account for this variability, random time studies were performed at various times during the day with different employees and different machines. An experimental standard time of 6.5 minutes was chosen as an initial hypothesis for the process. The results of the time study are shown in the table.

a. Generate a histogram of the sample data and comment on the normality of the process.
b. Construct a pair of *X* and *MR* charts, and comment on the control.
c. Now, use a Box-Cox transformation with optimal λ to generate a pair of transformed control charts, and comment on the state of control.
d. Estimate the mean and standard deviation of the process, along with the percentage of samples that are greater than the experimental standard time.

8-24. A certain product is manufactured to meet specifications of 36 cm $\pm$ 0.05 cm and is inspected in subgroups of four. The sampling scheme involves taking one measurement each from four different operators. The following data are the results of 20 consecutive samples.

a. Construct $\overline{X}$ and *R* charts from the data, and comment on the state of control. Estimate the parameters (mean and standard deviation) of this process.
b. Estimate the percentage of points that the process will produce outside the specification limits.
c. The following data represent an additional 20 samples from this process. Plot the points on the control chart constructed in part a and comment on whether the control status has changed.
d. Now plot these new 20 samples on separate $\overline{X}$ and *R* charts and comment on the state of control of the process while these samples were collected. Using these data estimate the process parameters (mean and standard deviation) for the last 20 points. Compare these estimates with the results you obtained in part a.
e. What action should be taken concerning this process?

	Operator			
Sample	**1**	**2**	**3**	**4**
1	35.959	36.047	36.001	36.000
2	36.008	35.998	35.968	35.990
3	36.031	35.958	35.968	35.993
4	36.002	36.001	36.037	35.990
5	35.996	36.028	36.044	36.014
6	35.996	36.033	35.993	36.037
7	36.042	36.009	35.989	36.012
8	36.037	36.074	35.962	35.980
9	35.978	36.008	36.024	35.953
10	36.023	35.982	36.029	35.990
11	36.020	35.994	36.001	35.987
12	35.992	36.027	35.992	36.019
13	35.984	35.930	35.976	35.991
14	36.054	35.988	35.992	36.001
15	35.996	35.993	36.005	35.974
16	36.024	35.948	35.957	35.973
17	36.017	35.968	35.972	36.014
18	35.988	36.025	35.989	36.026
19	35.998	36.017	36.038	35.987
20	36.044	35.969	35.976	35.981

	Operator			
Sample	**1**	**2**	**3**	**4**
21	36.030	36.035	36.004	36.046
22	36.027	36.044	36.051	36.065
23	36.077	35.995	36.045	35.987
24	36.018	36.017	36.020	35.987
25	36.016	36.046	36.038	36.027
26	35.988	36.021	36.060	36.032
27	36.035	36.010	36.010	36.031
28	36.031	36.033	36.025	36.011
29	35.989	36.013	35.997	36.048
30	36.033	36.022	36.056	36.051
31	36.048	36.025	36.031	36.064
32	36.032	36.046	36.042	36.031
33	36.015	35.985	36.013	36.026
34	36.019	35.999	36.065	36.031
35	36.013	36.034	36.064	36.009
36	35.990	36.007	36.033	36.057
37	36.027	36.012	36.019	36.056
38	36.050	36.054	35.995	36.046
39	36.025	36.024	35.992	35.960
40	36.054	36.014	36.008	36.028

8-25. In awarding the construction contracts for the new Freedom Tower in New York City, the city contracted with your company to provide a specialized welding service for the steel girders for this project. Since the project must proceed according to a tight schedule, and thousands of welds must be made, welding times are critical. Data have been collected on 50 welding times that were observed during crew training sessions and are reflected in the table below. To receive certification, the requirements for all welders is to be able to produce high-quality welds in no more than 10 minutes.

Welding Time (min.)			
Observation	Time	Observation	Time
1	4.555	26	10.2097
2	3.711	27	19.3935
3	10.950	28	6.6392
4	8.639	29	6.2914
5	15.234	30	16.0297
6	6.383	31	1.4671
7	6.356	32	1.3363
8	6.880	33	8.4815
9	4.020	34	7.1757
10	1.414	35	6.3445
11	7.906	36	16.4467
12	5.438	37	4.243
13	6.002	38	2.0252
14	20.764	39	5.3992
15	13.196	40	9.8326
16	1.373	41	7.4587
17	10.641	42	3.9621
18	6.553	43	3.3363
19	7.196	44	23.2068
20	1.925	45	7.8671
21	3.958	46	1.3904
22	5.290	47	10.6189
23	21.812	48	6.2741
24	9.512	49	5.172
25	1.872	50	1.454

a. Construct a histogram of the raw data. Comment on its normality and whether the process appears to be meeting the maximum welding time requirement.

b. Construct X and MR charts and comment on the state of statistical control.

c. Next, use a Box-Cox transformation with $\lambda = 0$ to generate a transformed pair of control charts and comment on control.

d. Using the raw data, estimate the mean and standard deviation of the process, and the percentage of welds that are likely to exceed 10 minutes. How confident are you in these estimates given your conclusions from part c?

e. Use Minitab to perform a capability analysis, first assuming a normal distribution and then using a Johnson transformation of the data. Compare and discuss how these two approaches and the answer you derived in part d differ.

8-26. Sandy's Hamburger Haven has recently implemented a new policy that is intended to guarantee fast and dependable service. The new policy states that if a customer has to wait more than 30 seconds after placing an order in the drive through, the customer

Observation	Measured	Observation	Measured
1	22.7098	31	24.1986
2	28.5818	32	22.3498
3	22.3013	33	19.9131
4	20.2175	34	21.3818
5	17.6546	35	20.9749
6	16.3910	36	31.3345
7	21.3819	37	23.4646
8	25.6197	38	20.9570
9	21.2647	39	18.0233
10	23.6092	40	24.3035
11	20.0449	41	20.6712
12	27.7050	42	20.2990
13	21.0183	43	24.0512
14	22.6822	44	22.1972
15	24.1459	45	20.1181
16	30.5566	46	21.8725
17	25.1578	47	23.5723
18	17.3187	48	22.2233
19	16.0317	49	23.3868
20	19.3129	50	23.2332
21	16.5825	51	23.5161
22	24.1377	52	26.1952
23	22.8705	53	30.6933
24	22.0960	54	24.8193
25	21.4009	55	19.0917
26	23.5266	56	23.0411
27	23.8764	57	23.8348
28	26.3259	58	28.5909
29	26.7252	59	22.3928
30	23.8997	60	17.3403

will receive the meal for free. Sandy's has contracted a group of industrial engineers to come in and evaluate the times of service at particular locations. The data provided are service times during the peak lunch hour from 11:00 a.m. to 1:00 p.m. All times shown are in seconds.

a. Construct a histogram of the data and comment on its normality.

b. Construct *X* and *MR* charts and comment on the state of statistical control.

c. Using the data and control charts from part b, estimate the mean and standard deviation of the service times.

d. Estimate the percentage of services that are likely to exceed the stipulated time limit.

e. If the process variation cannot be reduced, where would the process have to be centered so that there will be no more that 3.4 failures per million opportunities?

f. Assuming that the average service time remains at the level computed in part c, by what percentage would the process variance have to be reduced so that no more than 3.4 failures per million opportunities are produced?

8-27. A company, Steelten, produces frictionless linear sleeves to slide over guide rods. Listed below are 50 sample measurements from the inside dimensions of the sleeve. These samples were taken by selecting a sleeve at random each 15 minutes from the production process. Drawings state that the inside dimensions of the sleeve must be greater than 5.000007 mm with a +.005 mm tolerance. Management wants to determine whether this process is capable of meeting the stipulated specification.

a. Construct a histogram of these data and comment on their normality.

b. Construct *X* and *MR* charts for the process and comment on the state of statistical control.

c. Use Minitab to construct *X* and *MR* charts using a Box-Cox transformation with optimal λ to generate a transformed pair of control charts. Comment on the state of statistical control for the transformed data. What value of λ was used in the transformation?

d. Using the first 48 data points form 12 subgroups of size 4. Construct $\overline{X}$ and *R* charts and comment on the state of statistical control.

e. Use the raw data in part a and then the results derived from the two pairs of charts (constructed in parts b, c, and d respectively) to estimate the percentage of output that you estimate will exceed the single specification limit. Compare the results of the four approaches. Which one would you recommend and why?

Sample	Observation	Sample	Observation
1	5.001694923	26	5.011476667
2	5.005671712	27	5.018806934
3	5.007475376	28	5.001875063
4	5.023444816	29	5.030795054
5	5.000006991	30	5.001027603
6	5.010427947	31	5.000444375
7	5.026445903	32	5.002040977
8	5.01766181	33	5.000172802
9	5.000966222	34	5.008660108
10	5.004208677	35	5.002222287
11	5.011208867	36	5.015723974
12	5.002558834	37	5.000905723
13	5.002515267	38	5.003497899
14	5.004492611	39	5.000148058
15	5.007460553	40	5.006262018
16	5.00503618	41	5.01177234
17	5.009858766	42	5.010279185
18	5.00078874	43	5.012454576
19	5.004531009	44	5.010487071
20	5.001814537	45	5.011158451
21	5.00173855	46	5.015782673
22	5.000208214	47	5.022627968
23	5.004980822	48	5.002943183
24	5.008198258	49	5.059122706
25	5.012005428	50	5.00459357

8-28. A process at a manufacturing plant produces a two-layer film. The thickness of each layer is measured using a properly calibrated measuring device. The first layer has a target mean specification of 0.055 in. and the second layer has a target mean specification of 0.030 in. The combined finished product has specifications of 0.085 in. ± 0.005 in. The following data were obtained from measurements taken respectively from the processes where each of the layers is manufactured. Four samples were collected each hour from each process over a period of 30 hours. These results are shown below.

a. Construct $\overline{X}$ and *R* charts respectively for each layer. Use these data to determine the standard deviation and variance for each layer. What is the estimated variance of the resultant product?

b. Estimate the percentage of layered films that will be outside the specification limits.

c. What should be done to retarget the components so that the composite product (layer film) production is on target?

Sample	Layer 1			
1	0.0552	0.0565	0.0573	0.0524
2	0.0537	0.0550	0.0568	0.0557
3	0.0561	0.0513	0.0519	0.0560
4	0.0543	0.0570	0.0564	0.0576
5	0.0534	0.0580	0.0543	0.0551
6	0.0559	0.0550	0.0555	0.0555
7	0.0528	0.0576	0.0566	0.0579
8	0.0561	0.0559	0.0541	0.0537
9	0.0537	0.0564	0.0549	0.0515
10	0.0553	0.0555	0.0537	0.0546
11	0.0574	0.0564	0.0531	0.0538
12	0.0539	0.0543	0.0558	0.0539
13	0.0556	0.0559	0.0544	0.0550
14	0.0559	0.0557	0.0531	0.0540
15	0.0540	0.0543	0.0550	0.0544
16	0.0534	0.0560	0.0556	0.0505
17	0.0564	0.0553	0.0565	0.0553
18	0.0533	0.0531	0.0516	0.0583
19	0.0581	0.0556	0.0579	0.0567
20	0.0559	0.0559	0.0553	0.0544
21	0.0543	0.0542	0.0549	0.0571
22	0.0557	0.0530	0.0568	0.0574
23	0.0567	0.0553	0.0527	0.0549
24	0.0512	0.0535	0.0552	0.0560
25	0.0552	0.0563	0.0569	0.0568
26	0.0552	0.0565	0.0559	0.0562
27	0.0537	0.0529	0.0523	0.0532
28	0.0552	0.0552	0.0546	0.0532
29	0.0569	0.0555	0.0560	0.0556
30	0.0538	0.0528	0.0541	0.0565

Sample	Layer 2			
1	0.0259	0.0263	0.0263	0.0281
2	0.0272	0.0271	0.0275	0.0310
3	0.0266	0.0260	0.0257	0.0285
4	0.0289	0.0283	0.0270	0.0268
5	0.0247	0.0267	0.0275	0.0272
6	0.0270	0.0257	0.0267	0.0286
7	0.0294	0.0273	0.0278	0.0253
8	0.0265	0.0279	0.0264	0.0272
9	0.0261	0.0274	0.0275	0.0272
10	0.0296	0.0287	0.0277	0.0278
11	0.0274	0.0285	0.0259	0.0246
12	0.0287	0.0274	0.0257	0.0268
13	0.0284	0.0269	0.0278	0.0275
14	0.0273	0.0281	0.0275	0.0275
15	0.0261	0.0249	0.0290	0.0266
16	0.0280	0.0292	0.0290	0.0269
17	0.0273	0.0273	0.0276	0.0256
18	0.0251	0.0253	0.0259	0.0260
19	0.0259	0.0285	0.0275	0.0278
20	0.0259	0.0260	0.0273	0.0270
21	0.0262	0.0258	0.0261	0.0255
22	0.0276	0.0274	0.0283	0.0268
23	0.0267	0.0247	0.0243	0.0286
24	0.0263	0.0273	0.0258	0.0271
25	0.0260	0.0274	0.0266	0.0265
26	0.0245	0.0260	0.0268	0.0292
27	0.0263	0.0269	0.0259	0.0276
28	0.0265	0.0281	0.0260	0.0275
29	0.0262	0.0221	0.0270	0.0260
30	0.0294	0.0283	0.0274	0.0280

d. If the retargeting suggested in part c can be achieved, estimate the total percentage of output that would be outside the specification limits.

e. If the retargeting suggested in part c can be achieved, how much would the variance of layer 1 have to be reduced in order to meet the composite specifications in the $\pm 3\sigma$ sense?

CHAPTER NINE

ISSUES RELEVANT TO STATISTICAL PROCESS CONTROL BY VARIABLES

QUALITY MAXIM #9: *That which cannot be measured cannot be improved, and the measurement processes used must be accurate, precise, and have good capabilities to discriminate between units of output.*

Beware of Data! Things Are Seldom What They Seem

In Gilbert and Sullivan's *H.M.S. Pinafore*[1] little Buttercup, a commoner who is in love with Pinafore's commanding officer Captain Corcoran, cannot act on her feelings due to the differences in their social standings. When the captain confesses that if it were not for these differences he would return her affections she replies in song that things are not always as they seem, often appearing to be something else. She points out that all that glitters is not gold and one can gild a farthing[2] but it is still a farthing.[3] As the story unfolds it is revealed that, years before, Buttercup had served as the nursemaid for two babies—one a commoner, the other of aristocratic status—and that she had deliberately switched the two. The captain had been the infant who was of common blood and this revelation now paved the way to allow for their ultimate union, thus shattering the class barrier that had separated them. Now, as we fast-forward through time 130 years into the twenty-first century, we find the wisdom of little Buttercup can aptly be applied to process measurement in the modern era. In the business world, data drive behavior: There is something about the availability of numbers that holds mystical powers and the human psyche is quick to accept on face value the authenticity of the number streams generated. However, a measurement does not necessarily reflect the truth. At best a measurement, or series of measurements, can provide only an estimate of the true value being measured and is only as good as the measurement process employed. A large measurement error can lead to two negative consequences. First, the measurement process may not be capable of discerning the differences between two items that in reality are not the same. Second, if the measurement variation is large it can dominate process variation, masking out any underlying trends or nonrandom patterns. This can lead to a vicious cycle of overadjustment in which operators can mistake measurement error for process variation and try to compensate by erroneously making adjustments. The result: Additional variation is unwittingly introduced into the process and an otherwise stable process can be made unstable.

[1] *H.M.S. Pinafore,* with music by Arthur Sullivan and libretto by W. S. Gilbert, is a comic opera in two acts. The fourth operatic collaboration for Gilbert and Sullivan, *H.M.S. Pinafore* was their first big hit, which opened at the Opera Comique in London on *May 25, 1878, and continued for a run of 571 performances.*

[2] A farthing is a British coin under an old currency system—not worth much, just a quarter of a penny.

[3] The lyrics for "Things Are Seldom What They Seem," a duet between Captain Corcoran and Little Buttercup, can be accessed at http://math.boisestate.edu/gas/pinafore/web_opera/pin14.html.

LEARNING OBJECTIVES AND KEY OUTCOMES

After careful study of the material in Chapter 9, you should be able to:

1. **Discuss the importance of measurement processes to the effective control of quality.**
2. **Differentiate between the concepts of accuracy and precision, and describe how measurement processes can be evaluated with respect to each.**
3. **Test for linearity in a measurement process and explain why linearity is important.**
4. **Conduct a measurement study to compute the components of measurement error, including bias, repeatability, and reproducibility.**
5. **Conduct a measurement study using Minitab software.**

(continued)

6. **Compute the signal-to-noise ratio (SNR) and the ratio ρ and apply to the evaluation of measurement variation.**
7. **Differentiate between resolution and discrimination and compute a discrimination ratio to evaluate the suitability of any measurement process.**
8. **Discuss the three fundamental reasons for collecting data from industrial processes and how each reason impacts quality improvement programs.**
9. **Describe how different sampling schemes can affect a process analysis and conclusions concerning process behavior.**
10. **Determine process capability for one-sided and two-sided specifications for normally distributed data and interpret the results.**
11. **Determine process capability for one-sided and two-sided specifications for data that are not normally distributed and interpret the results.**
12. **Use Minitab software to conduct capability analyses.**
13. **Use control charts to isolate and separate sources of process variation.**

9.1 MEASUREMENT PROCESSES

9.1.1 Measurement Is a Process

For each production process with output attributes that are measured for quality control purposes, there is a corresponding **measurement process.** Measurements generated by this process are *not the true values* of the production process attribute being measured. The measurement is simply a *representation* of the truth. If the representation is close to the truth, it can be said that the measurement process is good. If there is much error and the representation is far from the truth, it can be said that the measurement process is poor. Several issues arise concerning the quality of any measurement process.

A measurement is only a representation of the truth.

A quality professional reviews a production report for errors and accuracy. To be credible data collected should be accurate, precise, and unbiased.

1. Does the process, on average, provide the true value of the measured attribute? That is, is the measurement process **accurate**? Figure 9-1 illustrates the concept of accuracy. If a single item is measured repeatedly using an accurate process, the average of all measurements will come close to the true value being measured. If the average is different from the true value, the process is said to contain bias. Accuracy is a calibration issue. When the presence of bias is discovered, the problem can often be corrected by recalibrating the measuring tool.

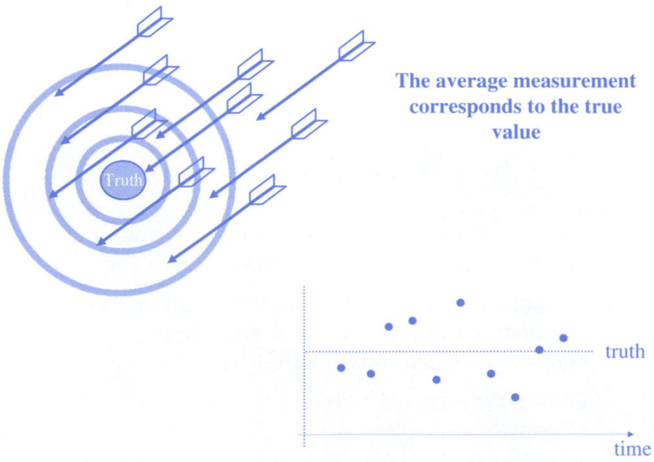

FIGURE 9-1. An Accurate Measurement Process

2. Is the measurement process **precise**? How much variation exists among consecutive measurements of the same item? Figure 9-2 illustrates how a precise process has little variability around its mean. A process can be precise, but inaccurate; accurate, but imprecise; imprecise and inaccurate; or accurate and precise. The last condition is obviously the most desirable.

Precision is the ability of a measurement process to repeat its results.

3. Is the measurement process consistent in its ability to represent the truth—that is, without **bias**? In other words, is the measurement process in statistical control with respect to accuracy and precision?
4. Is the measurement process capable of properly discriminating between different production units, batches, runs, and so on?

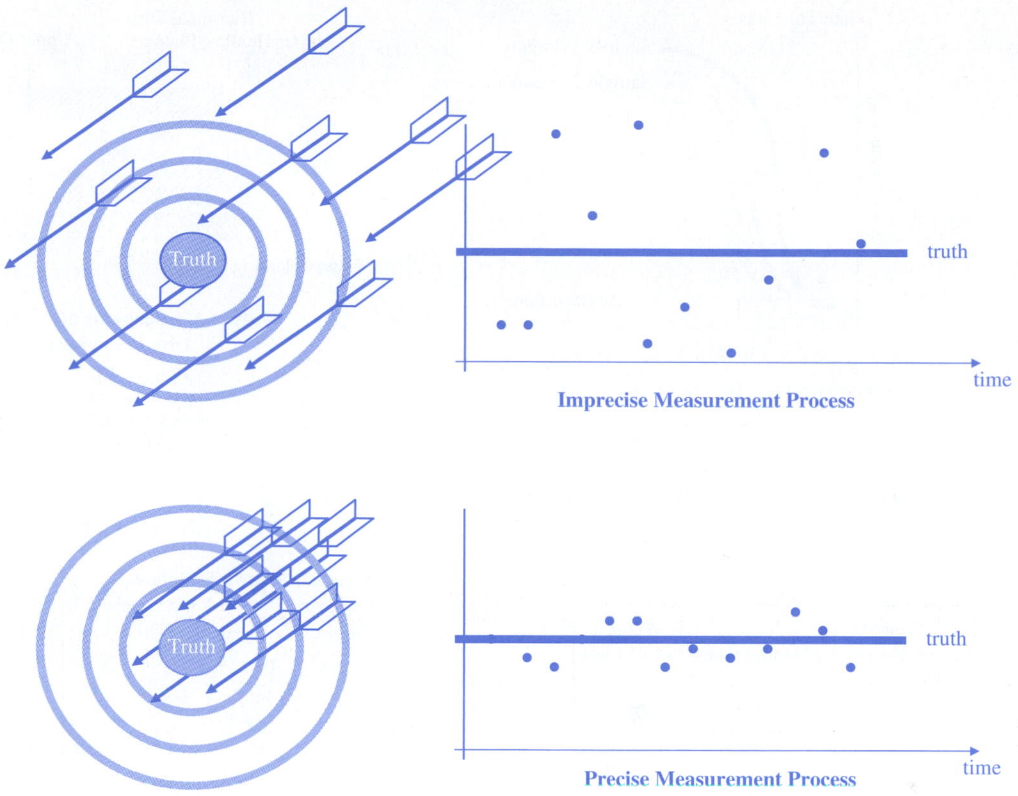

FIGURE 9-2. Measurement Precision

5. How much of the observed process variability is due to measurement error? How much of the design tolerance is consumed by measurement error?

Measurement is a process and can therefore be analyzed like any production process using similar tools of analysis. Before one can properly study a production process it is important to know the capabilities of the process that will be used to evaluate quality. The following true story illustrates why.

During production, automobile tires pass through numerous quality control checks. One inspection, performed early in the process, involves the collection of a rubber sample from each batch coming from the mixing room. The sample is sent to the quality control lab where it goes through a number of tests. One of these is on an apparatus called an oscillating disc rheometer. A certain plant has eight rheometer stations that are used to test each rubber batch for elastic modulus, viscous modulus, tangent delta, and cure rate—all important to the ultimate safety, durability, and performance of automobile tires.

The test shown in Figure 9-3 requires a QC technician to cut a rubber test strip and place the sample on an oscillating disc located between an upper and lower die in the rheometer machine. The machine closes and the disc begins to oscillate, placing stress on the material. The output is a graph that plots torque as a function of time over a three-minute period. If the batch is acceptable the plotted graph will produce an ogive (S) curve that will plot through predetermined gates in three critical zones (lower part of S, steep upward slope, final leveling out). If the graph is not an ogive or fails to plot through the gates, the batch is deemed unacceptable and is rejected.

Due to the pressures of time and workload, the standard lab practice in this plant was for a technician to sheer off a sample piece of material from the master test strip, and the sizes of these specimens could vary significantly. Concerned with whether this variability would impact the test results, the QC director performed the following experiment. From the most recent batch he had a technician cut three test specimens—a small, medium, and large size. The results shown in Figure 9-3 were surprising. *The size of the sample does matter!* The production batch tested could have been either accepted (in the case of the medium size sample) or rejected (in the case of the small or large samples) based on the size of the material specimen tested! When the equipment vendor's operation manual was consulted it was discovered that a properly prepared specimen would consist of two pieces of material, 2" × 2" × 1/4", and layered. Most rheometric equipment manufacturers strongly recommend the use of sample die cutters to keep samples consistent. In this case, by the next day the plant manager had equipped the QC lab with a set of these cutters.

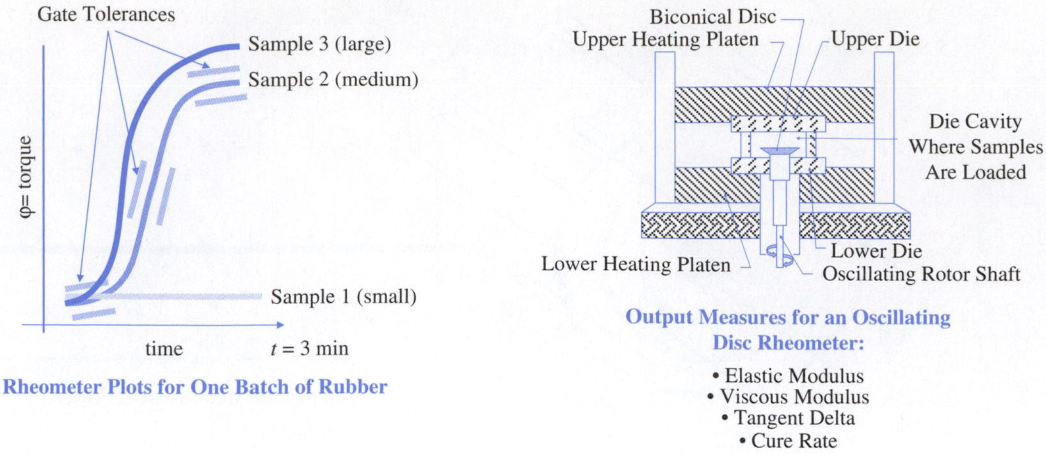

FIGURE 9-3. A Rheometer Test for Tire Rubber

Measurement	Weights of 454 gm Standard	Measurement	Weights of 454 gm Standard
1	456.71	16	453.85
2	453.29	17	454.74
3	455.24	18	455.67
4	453.84	19	453.18
5	455.45	20	454.82
6	455.25	21	457.69
7	455.66	22	455.41
8	455.36	23	454.35
9	455.40	24	454.28
10	453.74	25	455.63
11	456.42	26	456.31
12	454.91	27	455.94
13	455.21	28	454.75
14	457.86	29	454.62
15	456.59	30	453.81

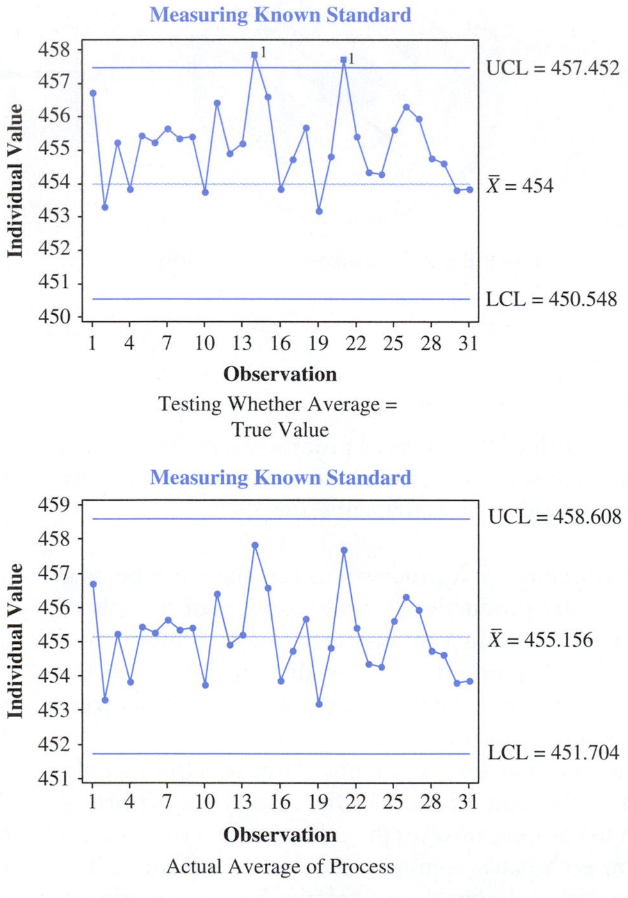

FIGURE 9-4. A Test for Accuracy

9.1.2 Accuracy—A Calibration Issue

Accuracy can be determined only by repeatedly measuring a standard that has a *known true value.* Figure 9-4 shows data that have been collected to determine if a certain weighing device is accurate. The test involves placing a 454-gram standard weight (one that has been certified by the National Institute of Standards and Testing (NIST)) repeatedly on the set of scales and recording the results. A total of 30 measurements were taken and two individuals (X) control charts constructed. The control limits on the charts are based on a two-point moving range chart (not shown), which was in statistical control.

Each of the X charts answers a different question. The first chart is used to determine whether the process is biased. The centerline and control limits on this X chart are those that theoretically would apply under the assumption that the process average is equal to the true value of 454 grams. As this chart is out of control there is statistical evidence to suggest that the *assumed* process average is incorrect. Two points are out of control and there is a violation of the seven-point-run rule. The scales clearly need to be recalibrated, but by how much? This is the question that is answered by the second X chart.

Accuracy can be determined by repeatedly measuring a known standard.

The second chart is constructed to determine whether the process is stable and if its operating parameters can be estimated. Figure 9-4 shows this chart to be in control with a centerline equal to 455.2. One can conclude that the scales being tested are positively biased, producing measurements that on the average are more than one gram heavier than the true value.

Calibration of a measurement process should be performed using standards that are *close to the values that are to be measured.* That is, a weighing process like the one illustrated can be accurate when weighing 1-pound (454-gram) units, but biased when weighing anything considerably lighter or heavier. If a measurement process is to be applied across a wide interval of values it is important to test for **linearity,** as demonstrated in Figure 9-5. A linear process can hold its calibration across a variety of magnitudes. If a measurement process is not linear, it is advisable to restrict its use to those measures where the calibration holds.

A pseudostandard can be used for calibration purposes if a certified standard is unavailable.

A standard is needed to determine accuracy. However, a certified standard does not always exist. In some cases it is

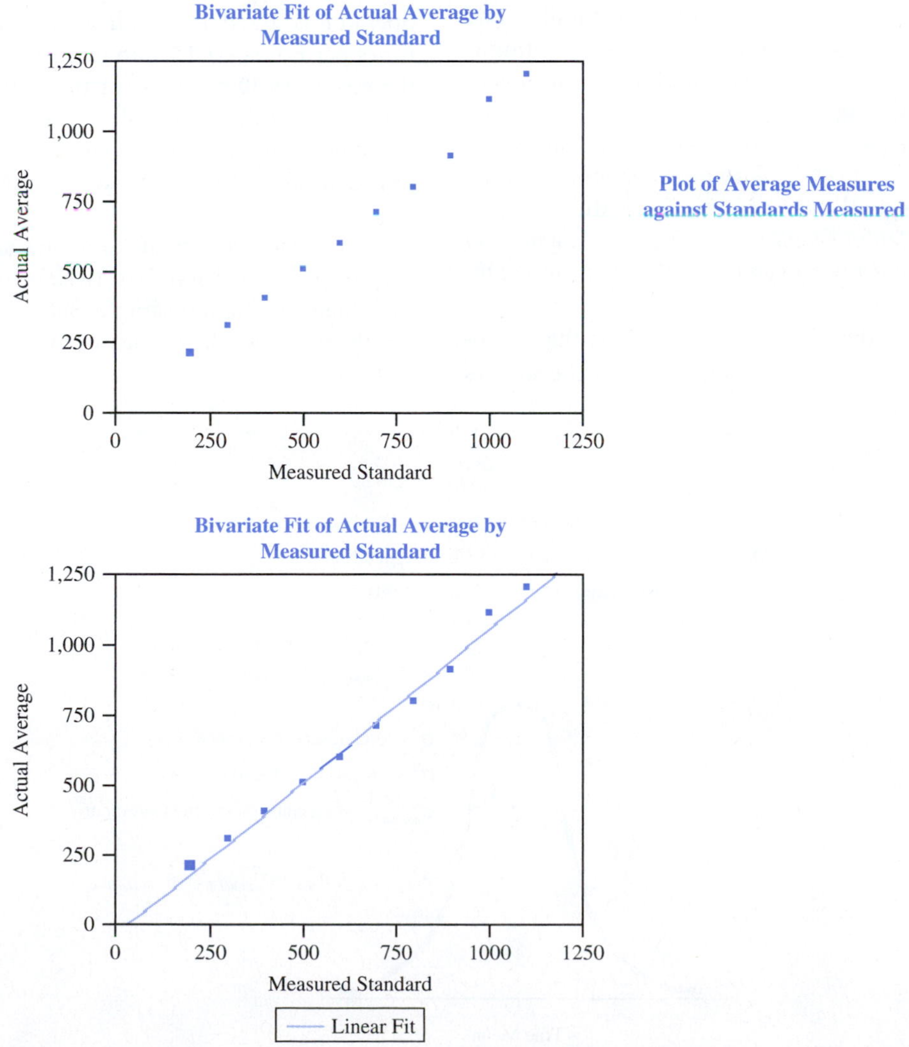

FIGURE 9-5. Test for Linearity

possible to create an in-house *pseudostandard*. This is production material that has been set aside for the purpose of calibrating measurement processes. Use of a pseudostandard does not eliminate bias. However, the pseudostandard will ensure that any bias that is present will be consistent over time.

9.1.3 Precision—How to Capture Measurement Error

Figure 9-6 depicts the components of any measurement. The number that is recorded is made up of the *true mean* of the process plus a *process variability* component plus *measurement error*. Measurement error can be further decomposed into

- The variability that is due to bias
- A component due to using different operators, devices, or locations—called the reproducibility effect
- A component due to replication (or random) effects—called the repeatability effect

All the sources of variability combine (the variances are additive) to form the total observed variance. The concepts of repeatability and reproducibility are simple. **Repeatability** error means that *one person who takes numerous measurements with the same measuring device will not always get the same answer.* **Reproducibility** error means that *different people measuring the same item, or the same person using different gauges, will not always get the same answer.*

> **Measurement error consists of bias, repeatability error, and reproducibility error.**

The total observed variance is the variability that is captured on control charts, and is made up of all the sources cited previously. For example, once an $\overline{X}$ chart shows a state of control it can be assumed that the R chart has captured most (if not all) of the variability. Hence,

$$\sigma_x^2 = \sigma_{observed}^2 \cong \left[\frac{\overline{R}}{d_2}\right]^2 = \sigma_{process}^2 + \sigma_{measurement}^2 \qquad \textbf{9-1}$$

$$\sigma_{measurement}^2 = \sigma_{bias}^2 + \sigma_{repeatability}^2 + \sigma_{reproducibility}^2 \qquad \textbf{9-2}$$

When measurements are taken from a process, one is normally trying to estimate $\sigma_{process}^2$. If a measurement process is properly calibrated $\sigma_{bias}^2 = 0$, in which case $\sigma_{measurement}^2 = \sigma_{repeatability}^2 + \sigma_{reproducibility}^2$. If $\sigma_{measurement}^2$ can be reliably estimated, Equation 9-1 can be used to compute the value of $\sigma_{process}^2$ from $\sigma_{observed}^2$.

To illustrate how this may be done, we shall return to the previous example of the 454-gram standard weight. Assume that in the production environment two operators, Jack and Jane, are required to perform these measurements. Jack works first shift and Jane covers the second shift. It is decided that these two operators will conduct a measurement study as part of their routine inspection duties. Each operator is asked to weigh a production specimen four times per hour—at 15-minute intervals. They each record the results of 30 samples (of the four measures per hour) collected over the course of a week. The results of the combined 60 samples are shown in Table 9-1. These have been mixed up so there is no way to separate Jack's samples from Jane's.

For the measurement study, Jack and Jane were asked to weigh the certified standard twice—immediately after they took their production sample. Special forms were provided for them to use. This measurement data is summarized in Table 9-2.

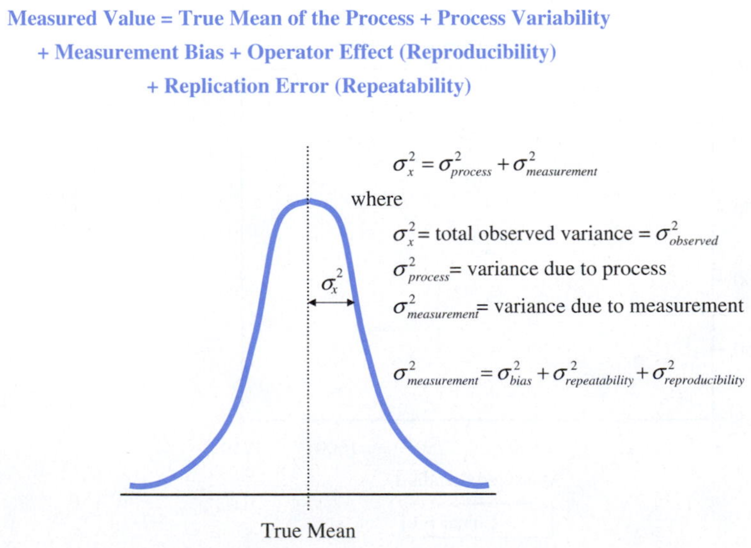

Measured Value = True Mean of the Process + Process Variability
+ Measurement Bias + Operator Effect (Reproducibility)
+ Replication Error (Repeatability)
$\sigma_x^2 = \sigma_{process}^2 + \sigma_{measurement}^2$
where
σ_x^2 = total observed variance = $\sigma_{observed}^2$
$\sigma_{process}^2$ = variance due to process
$\sigma_{measurement}^2$ = variance due to measurement
$\sigma_{measurement}^2 = \sigma_{bias}^2 + \sigma_{repeatability}^2 + \sigma_{reproducibility}^2$
σ_x^2
True Mean

FIGURE 9-6. Anatomy of a Measurement

TABLE 9-1 Production Data—Jack and Jane Combined

Sample No.	Measurement 1	2	3	4	$\bar{X}$	R
1	497.610	455.827	459.757	460.376	468.393	41.783
2	455.705	475.266	478.829	454.069	465.967	24.759
3	474.449	458.563	455.328	455.165	460.876	19.284
4	484.994	445.357	450.138	438.507	454.749	46.487
5	454.806	428.440	454.882	438.055	444.046	26.442
6	441.500	478.927	453.303	477.538	462.817	37.427
7	446.045	453.527	481.531	455.356	459.115	35.485
8	452.707	456.303	469.669	453.342	458.005	16.962
9	457.221	450.113	477.894	486.065	467.823	35.952
10	455.872	486.288	476.131	475.586	473.469	30.416
11	449.157	463.927	460.752	467.379	460.303	18.222
12	472.389	459.808	472.163	429.179	458.385	43.210
13	480.279	463.102	476.918	473.932	473.558	17.177
14	433.015	443.454	468.989	450.994	449.113	35.975
15	451.794	464.348	457.096	450.810	456.012	13.537
16	449.952	481.455	433.902	453.294	454.651	47.553
17	466.558	471.396	458.426	468.884	466.316	12.970
18	451.169	450.902	432.913	453.160	447.036	20.248
19	478.714	474.475	486.642	452.430	473.065	34.212
20	440.295	492.641	471.991	458.072	465.749	52.347
21	436.940	442.886	447.452	454.690	445.492	17.750
22	462.735	438.406	440.491	460.209	450.460	24.329
23	465.222	462.077	472.696	470.068	467.516	10.618
24	466.531	464.942	474.213	468.572	468.565	9.271
25	444.390	470.081	453.813	449.912	454.549	25.691
26	448.860	457.242	479.736	449.042	458.720	30.876
27	472.333	488.490	469.779	440.431	467.758	48.058
28	433.528	468.287	459.093	464.802	456.427	34.759
29	444.773	468.928	446.723	462.145	455.642	24.155
30	467.983	504.264	438.163	484.815	473.806	66.101
31	443.362	465.733	485.368	453.855	462.079	42.006
32	483.099	457.892	457.229	443.436	460.414	39.663
33	459.433	475.792	455.802	460.975	463.000	19.990
34	462.886	447.364	457.707	472.172	460.032	24.809
35	476.539	460.203	475.027	445.365	464.284	31.174
36	461.028	419.037	472.750	476.799	457.404	57.762
37	474.437	454.565	462.815	465.407	464.306	19.872
38	451.111	447.402	451.839	481.573	457.981	34.171
39	462.418	462.871	478.924	432.123	459.084	46.801
40	444.301	453.474	459.522	457.942	453.810	15.221
41	477.872	456.578	434.042	454.074	455.642	43.830
42	440.248	460.140	446.924	460.589	451.975	20.340
43	436.979	438.710	455.648	481.633	453.242	44.655
44	446.406	478.957	470.511	436.820	458.174	42.136
45	452.069	442.732	459.428	459.625	453.464	16.893
46	455.323	476.752	460.619	456.398	462.273	21.429
47	472.699	458.935	438.724	466.962	459.330	33.975
48	452.467	457.670	444.782	466.326	455.311	21.544

(*Continued*)

TABLE 9-1 Production Data—Jack and Jane Combined(*Continued*)

Sample No.	Measurement 1	2	3	4	$\bar{X}$	R
49	471.597	482.368	455.089	450.317	464.843	32.051
50	472.955	468.099	474.697	454.794	467.637	19.903
51	475.506	479.576	481.844	462.335	474.815	19.510
52	470.679	450.697	460.144	459.288	460.202	19.982
53	439.480	450.068	446.515	444.546	445.152	10.588
54	471.841	448.038	474.098	442.865	459.210	31.232
55	466.798	456.749	470.803	479.206	468.389	22.457
56	455.233	459.778	457.596	441.975	453.645	17.803
57	453.925	452.947	458.207	461.935	456.753	8.987
58	476.856	476.352	465.198	478.573	474.244	13.375
59	461.507	477.111	481.693	461.089	470.350	20.604
60	437.523	448.188	436.588	470.265	448.141	33.677

TABLE 9-2 Jack and Jane Combined Measurement Data

Trial	Jack	Jane	Trial	Jack	Jane
1	454.23	456.92	16	454.50	456.61
	455.23	452.38		453.22	454.64
2	455.82	456.35	17	453.79	454.23
	452.95	454.49		452.87	455.33
3	453.72	453.14	18	454.99	453.60
	454.01	454.25		453.08	454.54
4	454.39	452.21	19	454.00	451.94
	453.43	453.12		454.90	452.52
5	452.77	454.23	20	454.92	456.10
	454.92	454.97		453.32	452.49
6	452.61	453.81	21	455.80	452.27
	454.05	454.15		454.32	452.98
7	453.60	452.81	22	455.01	452.97
	454.89	453.55		453.59	453.43
8	453.86	455.85	23	452.93	452.86
	453.99	453.15		453.28	453.84
9	451.05	453.66	24	455.15	453.25
	454.78	453.91		453.95	456.33
10	453.66	455.31	25	452.79	453.22
	454.40	456.12		454.50	455.93
11	453.33	453.88	26	453.39	456.34
	454.51	453.40		452.66	455.84
12	453.70	452.72	27	454.53	450.89
	455.26	452.66		452.96	451.44
13	453.36	456.37	28	451.42	453.32
	455.35	451.68		452.38	453.27
14	455.33	454.64	29	452.97	455.52
	454.65	456.16		454.66	454.53
15	453.89	454.78	30	454.71	453.30
	454.05	455.79		453.01	454.74

To begin the measurement analysis Jack's and Jane's data were combined in 30 samples of size 4 and plotted on $\overline{X}$ and R charts. As Figure 9-7 shows, the Range and the $\overline{X}$ charts are both in control. Since the $\overline{X}$ chart is centered at 454 grams, one can conclude that there is no bias and that the measurement process is accurate. The samples have been designed to include measurements from both operators so the ranges capture both the repeatability and reproducibility components of measurement error. Using Equation 9-2 and the fact that both charts are in control, measurement error can be estimated by

$$\sigma^2_{measurement} = \left[\frac{\overline{R}}{d_2}\right]^2 = \left[\frac{2.553}{2.059}\right]^2 = 1.537$$

The QC director was interested in learning how much of the measurement variation was due to the operator effect and how much is due to the replication effect. The director decided to stratify the data collected, separating Jack's samples from Jane's. The resulting control charts are shown in Figure 9-8. From these charts it appears that, while neither Jane nor Jack is introducing any bias into the measurements, they are different with respect to their individual precision. In this case, it appears that Jane is less precise than Jack.

Observing this, the QC director decided to plot Jack's and Jane's respective data on different control charts. These charts show that Jane is not only less precise than Jack, but her chart also shows evidence that her measurement process is not stable. Refer to Figure 9-9. Jack is a very experienced operator so his results came as no surprise. The QC director therefore decided to use Jack's performance to estimate the gauge repeatability error. From this, the reproducibility error can be easily calculated using Equation 9-2.

$$\sigma^2_{reproducibility} = \sigma^2_{measurement} - \sigma^2_{repeatability}$$

$$\sigma^2_{reproducibility} = 1.537 - (1.08)^2 = 0.371$$

$$\frac{\sigma^2_{reproducibility}}{\sigma^2_{measurement}} = \frac{0.371}{1.537} = 0.241 \Rightarrow 24.1\%$$

This analysis shows that approximately 24% of the measurement error is due to differences between the operators. The next issue is to determine whether the measurement error ($\sigma^2_{measurement} = 1.537$) is large enough to be a concern. Figure 9-10 depicts the $\overline{X}$ and R charts for production data shown in Table 9-1. These are the combined data for Jack and Jane and there is no information as to who collected which sample. As both control charts are in control, $\sigma_{observed} \cong 14.128$. It is important to know whether the measurement error is a significant component of this observed variation, in which case an operator could mistake measurement variation for process variation and unintentionally overcontrol the process. As a result operators could make unneeded process adjustments and unwittingly add variability to the resulting output.

The **signal-to-noise ratio** (*SNR*), computed using Equation 9-3, is a convenient measure of the relative size of a measurement error. Another convenient measure is ρ, defined by Equation 9-4, which represents the total percentage

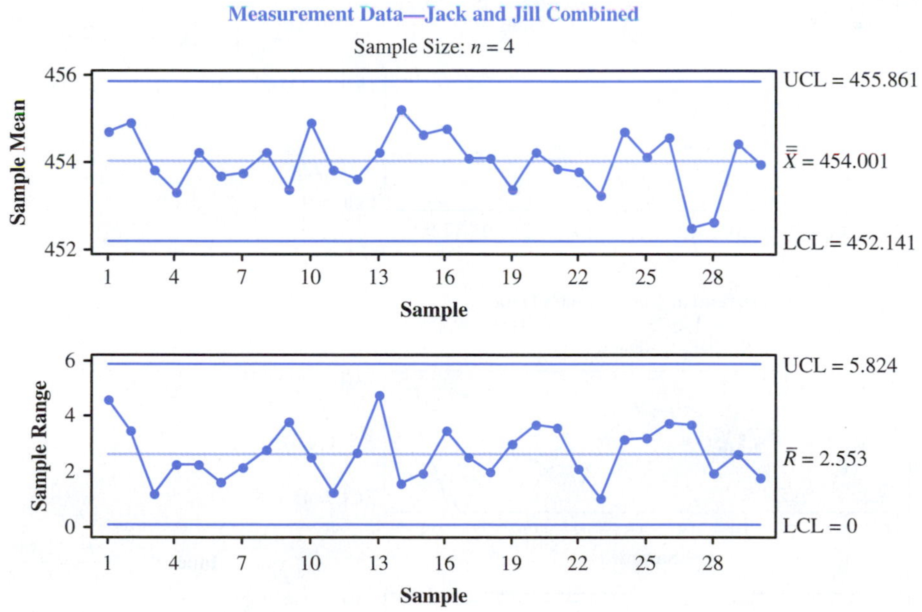

$$\sigma^2_{bias} \cong 0$$

$$\sigma^2_{measurement} \cong \left(\frac{\overline{R}}{d_2}\right)^2 = \left(\frac{2.553}{2.059}\right)^2 = (1.2399)^2 = 1.537$$

$$\sigma_{measurement} \cong \sqrt{1.537} = 1.240$$

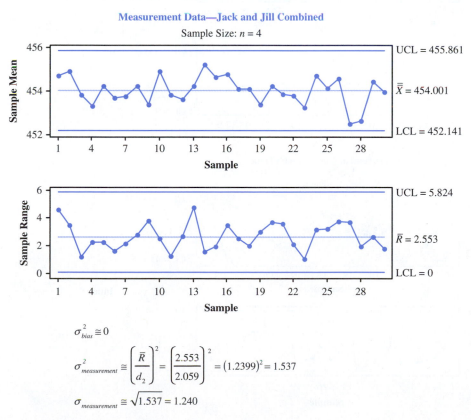

FIGURE 9-7. Measurement Study

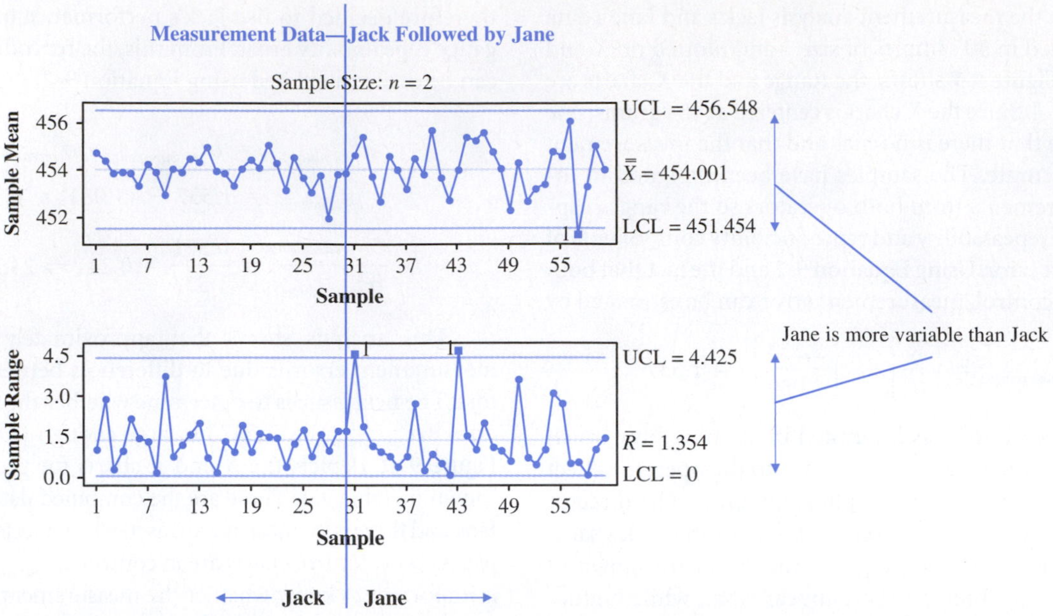

FIGURE 9-8. Combining Two Operators on One Chart

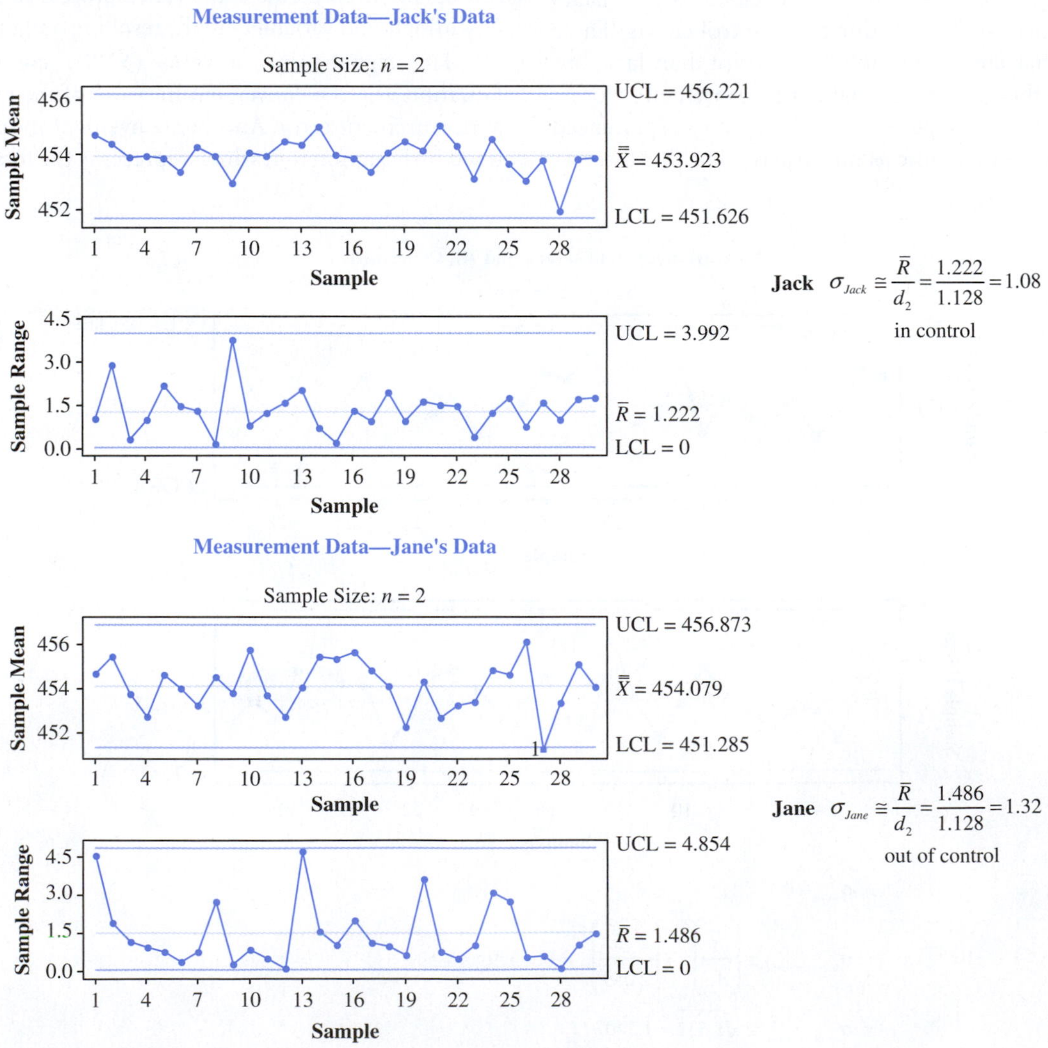

FIGURE 9-9. Independent Charts for Measurement Data

$$SNR = \frac{\sigma_{observed}}{\sigma_{measurement}} = \frac{14.128}{1.240} = 11.39$$

$$\left(\frac{1}{SNR}\right)^2 = \frac{\sigma^2_{measurement}}{\sigma^2_{observed}} = \frac{1.537}{(14.128)^2} = 0.0077 \Rightarrow$$

0.77% of observed variation is due to measurement error

Production Data—Jack and Jane Combined

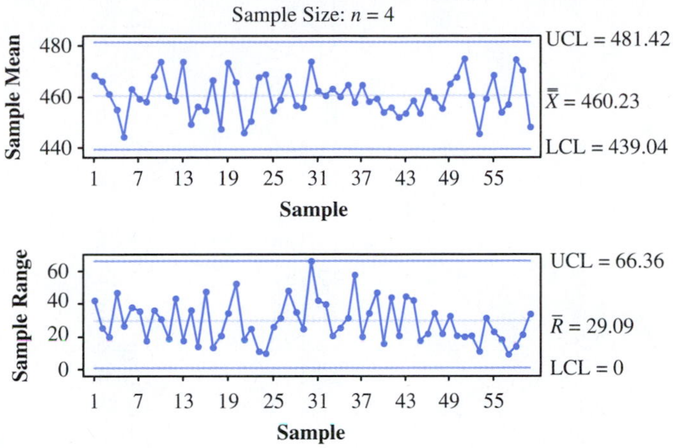

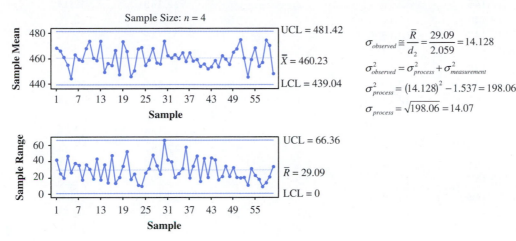

$$\sigma_{observed} \cong \frac{\bar{R}}{d_2} = \frac{29.09}{2.059} = 14.128$$

$$\sigma^2_{observed} = \sigma^2_{process} + \sigma^2_{measurement}$$

$$\sigma^2_{process} = (14.128)^2 - 1.537 = 198.06$$

$$\sigma_{process} = \sqrt{198.06} = 14.07$$

FIGURE 9-10. Production Data—Jack and Jane Combined

of the observed variation that is due to the process. The square of the reciprocal of the *SNR*, $1 - \rho$ defined by Equation 9-5, represents the percentage of the observed variation that is due to measurement error. All of these statistics can provide insights as to whether the size of the measurement error is large enough to cause concern.

A signal-to-noise ratio greater than 10 is ideal.

$$SNR = \frac{\sigma_{observed}}{\sigma_{measurement}} \qquad \textbf{9-3}$$

$$\rho = \frac{\sigma^2_{process}}{\sigma^2_{observed}} \qquad \textbf{9-4}$$

$$1 - \rho = \left[\frac{1}{SNR}\right]^2 = \frac{\sigma^2_{measurement}}{\sigma^2_{observed}} \qquad \textbf{9-5}$$

Rules of Thumb

Rule 1: If $SNR \geq 10$ or $(1 - \rho) \leq 0.01$ the measurement error can be considered negligible with respect to the size of the process variation. The process variation can be estimated from the observed variation provided that the process is stable with respect to average and variability.

Rule 2: If $4.5 \leq SNR < 10$ or $0.01 < (1 - \rho) \leq 0.05$ the measurement error may affect the ability to differentiate between dissimilar units of output. It may be advisable to take at least two measurements of each sampling unit and average them rather than use individual measures. The measurement error should be subtracted from the observed variation to estimate the process variation.

Rule 3: If $SNR < 4.5$ or $0.05 < (1 - \rho)$ the measurement process has an inadequate ability to differentiate between dissimilar units of output.

If the $SNR \geqslant 10$, less than 1% of the total observed variation is due to measurement error. As Figure 9-10 shows, Jack and Jane's combined *SNR* is approximately equal to 11.39, and the measurement error represents only about three-quarters of 1% of the total observed variation. When Equation 9-1 is used to compute the process variability, it can be seen that the measurement error is negligible ($\sigma_{process} = 14.07 \cong \sigma_{observed} = 14.128$). Once it is determined that the $SNR \geqslant 10$ and the measurement error is negligible, it is not necessary to subtract $\sigma_{measurement}$ from $\sigma_{observed}$ in order to estimate $\sigma_{process}$, as it is appropriate to assume that $\sigma_{observed}$ provides a good approximation to $\sigma_{process}$ provided both the relevant $\bar{X}$ and *R* charts are both in a state of statistical control.

Other considerations in determining the quality of a measurement process are *discrimination* and *resolution*. **Discrimination** is the ability to differentiate between units of output. **Resolution** refers to the smallest possible unit of measure. If the resolution is low, discrimination is likely to be poor. Refer to the data in Table 9-3. These data were collected on the diameter of 12-inch wafers at a semiconductor plant. A total of 25 samples were taken of four measurements each. The first column in Table 9-3 represents the measurements taken and recorded to the nearest 1,000th of an inch. The second column are the same measurements, but with the reduced resolution to the nearest 100th of an inch. The final column records the measurements to the nearest 10th of an inch.

TABLE 9-3 Measurement Devices with Different Levels of Resolution

	Measuring Device				Measuring Device		
Sample	Nearest 1,000th of an inch	Nearest 100th of an inch	Nearest 10th of an inch	Sample	Nearest 1,000th of an inch	Nearest 100th of an inch	Nearest 10th of an inch
1	12.140	12.14	12.1	14	12.132	12.13	12.1
	12.146	12.15	12.2		12.144	12.14	12.1
	12.135	12.14	12.1		12.147	12.15	12.2
	12.146	12.15	12.2		12.143	12.14	12.1
2	12.146	12.15	12.2	15	12.134	12.13	12.1
	12.145	12.15	12.2		12.143	12.14	12.1
	12.137	12.14	12.1		12.138	12.14	12.1
	12.144	12.14	12.1		12.145	12.15	12.2
3	12.140	12.14	12.1	16	12.142	12.14	12.1
	12.131	12.13	12.1		12.139	12.14	12.1
	12.136	12.14	12.1		12.142	12.14	12.1
	12.137	12.14	12.1		12.146	12.15	12.2
4	12.149	12.15	12.2	17	12.137	12.14	12.1
	12.141	12.14	12.1		12.132	12.13	12.1
	12.147	12.15	12.2		12.132	12.13	12.1
	12.141	12.14	12.1		12.143	12.14	12.1
5	12.130	12.13	12.1	18	12.133	12.13	12.1
	12.149	12.15	12.2		12.135	12.14	12.1
	12.143	12.14	12.1		12.136	12.14	12.1
	12.148	12.15	12.2		12.148	12.15	12.2
6	12.133	12.13	12.1	19	12.139	12.14	12.1
	12.142	12.14	12.1		12.142	12.14	12.1
	12.140	12.14	12.1		12.131	12.13	12.1
	12.141	12.14	12.1		12.144	12.14	12.1
7	12.130	12.13	12.1	20	12.130	12.13	12.1
	12.135	12.14	12.1		12.134	12.13	12.1
	12.130	12.13	12.1		12.130	12.13	12.1
	12.141	12.14	12.1		12.132	12.13	12.1
8	12.133	12.13	12.1	21	12.135	12.14	12.1
	12.149	12.15	12.2		12.138	12.14	12.1
	12.149	12.15	12.2		12.139	12.14	12.1
	12.140	12.14	12.1		12.134	12.13	12.1
9	12.136	12.14	12.1	22	12.142	12.14	12.1
	12.131	12.13	12.1		12.148	12.15	12.2
	12.147	12.15	12.2		12.138	12.14	12.1
	12.138	12.14	12.1		12.139	12.14	12.1
10	12.135	12.14	12.1	23	12.147	12.15	12.2
	12.147	12.15	12.2		12.131	12.13	12.1
	12.137	12.14	12.1		12.140	12.14	12.1
	12.144	12.14	12.1		12.146	12.15	12.2
11	12.145	12.15	12.2	24	12.138	12.14	12.1
	12.136	12.14	12.1		12.134	12.13	12.1
	12.133	12.13	12.1		12.149	12.15	12.2
	12.131	12.13	12.1		12.144	12.14	12.1
12	12.134	12.13	12.1	25	12.145	12.15	12.2
	12.148	12.15	12.2		12.143	12.14	12.1
	12.132	12.13	12.1		12.137	12.14	12.1
	12.136	12.14	12.1		12.148	12.15	12.2
13	12.145	12.15	12.2				
	12.141	12.14	12.1				
	12.139	12.14	12.1				
	12.133	12.13	12.1				

Figure 9-11 provides a comparison of the charts for the three different measuring devices. Control charts (A), with resolution to the nearest 1,000th of an inch, provide good discrimination of parts. As both (A) charts show good control, the process standard deviation can be estimated by

Discrimination is the ability to differentiate between units; resolution is the smallest possible unit of measure.

$$\sigma_{process} \cong \frac{\overline{R}}{d_2} = \frac{0.0118}{2.059} = 0.00573 \text{ inch}$$

However, charts (B) and (C) are inadequate, with resolutions of 100th of an inch and 10th of an inch, respectively. Notable in these charting pairs are the low number of discreet values that show up on the range charts—in the case of (C), numerous ranges of zero appear. If the measuring process that resulted in the (B) charts were to be used, the estimate for the process variability would be

$$\sigma_{process} \cong \frac{\overline{R}}{d_2} = \frac{0.0140}{2.059} = 0.00680 \text{ inches}$$

$\Rightarrow$ 18.7% higher than chart (A) results.

Rules of Thumb

There is inadequate discrimination if the measurement process produces,

Rule 1: *Three or fewer discreet values on the R chart.*

Rule 2: *Four discreet values on the range chart and 25% or more of the ranges are equal to zero.*

If the (C) process charts were employed, the results would be worse:

$$\sigma_{process} \cong \frac{\overline{R}}{d_2} = \frac{0.0720}{2.059} = 0.03497 \text{ inch}$$

$\Rightarrow$ 510.3% higher than chart (A) results.

The discrimination ratio (*DR*), proposed by Wheeler and Lyday,[4] can be used to evaluate and compare measurement processes, and is calculated using Equations 9-6, 9-7, or 9-8 as follows.

$$DR = \sqrt{(SNR)^2(1+\rho)} \qquad \textbf{9-6}$$

$$DR = \sqrt{\frac{(1+\rho)}{(1-\rho)}} \qquad \textbf{9-7}$$

$$DR = \sqrt{\frac{2\sigma^2_{process}}{\sigma^2_{measurement}} + 1} \qquad \textbf{9-8}$$

Rules of Thumb

Rule 1: *If $DR \geq 14$ the measurement process has very good discrimination properties.*

Rule 2: *If $6 \leq DR < 14$ the measurement process has satisfactory discrimination properties.*

Rule 3: *If $DR < 6$ the discrimination ability of the measurement process is inadequate.*

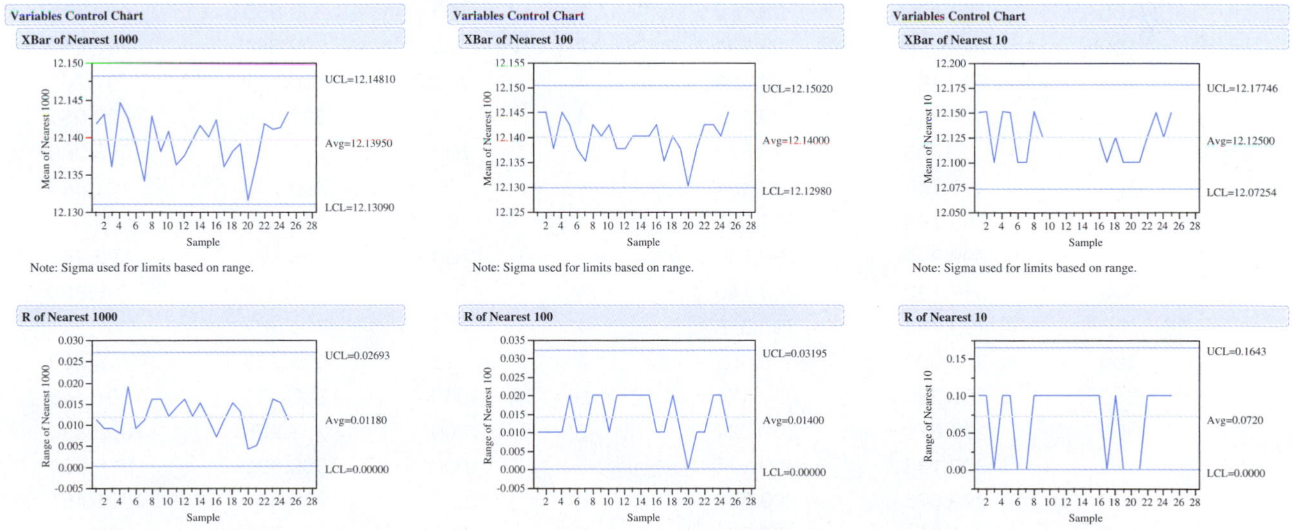

A Resolution to the Nearest 1,000th of an Inch 13 Different Range Values

B Resolution to the Nearest 100th of an Inch 3 Different Range Values

C Resolution to the Nearest 10th of an Inch 2 Different Range Values—28% Zeros

FIGURE 9-11. Charts Showing Different Levels of Discrimination

[4]Wheeler, Donald J. and Richard Lyday, *Evaluating the Measurement Process*, 2ed. Knoxville, TN, SPC Press, 1990.

Returning to the Jack and Jane example,

$$SNR = \frac{14.60}{1.240} = 11.77$$

$$1 - \rho = \left[\frac{1}{11.77}\right]^2 = 0.00721$$

$$\rho = 1 - 0.00721 = 0.99279$$

$$DR = \sqrt{\frac{1 + 0.99279}{1 - 0.99279}} = \sqrt{276.393} = 16.625$$

Since $SNR > 10$, $(1 - \rho) < 0.01$, and $DR > 14$, Jane and Jack's measurement process would be considered very adequate for its current application.

9.1.4 Measurement Studies Using Minitab

Using a statistical software package such as Minitab can facilitate the analysis of measurement processes. Minitab can be used to check linearity and bias, and to estimate the repeatability and reproducibility components of measurement error.

9.1.4.1 Linearity and Bias. Consider a scenario where an inspector wishes to check the linearity and bias of two measurement processes, which we will refer to as A and B. To do this the inspector uses NIST-certified standards with values of 100, 300, 500, 700, and 1,000, measures each of these standards 10 times on each measurement process, and records the results in Table 9-4.

A Minitab worksheet is set up as shown in Figure 9-12. The first column lists the part number, the second column includes the corresponding reference (known) value, and the third column lists the recorded measurement. Next, click on the *Stat* dropdown menu and follow the sequence shown in Figure 9-13. In the dialog box that appears I have entered the respective column names as shown in Figure 9-14. After clicking on *OK* the analysis summary shown in Figure 9-15 will appear.

TABLE 9-4 Data to Check Linearity and Bias

Part No.	Reference	Measurement Process A	Measurement Process B	Part No.	Reference	Measurement Process A	Measurement Process B
1	100	100.927	98.483	3	500	498.881	499.919
1	100	99.608	100.445	3	500	498.371	499.717
1	100	99.339	98.360	3	500	501.131	498.946
1	100	99.965	99.243	3	500	500.931	500.106
1	100	99.365	98.486	3	500	500.880	498.475
1	100	99.958	98.533	4	700	699.554	702.497
1	100	100.483	98.686	4	700	700.131	701.269
1	100	99.341	99.116	4	700	699.946	700.056
1	100	100.303	98.813	4	700	701.189	698.988
1	100	99.788	99.494	4	700	700.040	701.380
2	300	300.195	301.051	4	700	700.349	699.057
2	300	299.904	298.739	4	700	698.891	699.241
2	300	299.732	299.740	4	700	700.816	700.820
2	300	299.784	299.058	4	700	701.387	701.221
2	300	301.002	300.062	4	700	699.763	700.319
2	300	299.233	299.474	5	1000	1000.431	1001.142
2	300	298.873	300.024	5	1000	999.925	1002.830
2	300	300.138	300.177	5	1000	1000.892	999.864
2	300	299.625	299.539	5	1000	998.580	1001.481
2	300	300.518	299.592	5	1000	1001.309	999.718
3	500	499.218	499.575	5	1000	1000.137	1001.997
3	500	499.365	502.243	5	1000	999.291	1002.035
3	500	498.802	501.119	5	1000	1000.207	1001.437
3	500	500.046	500.150	5	1000	1000.830	1000.442
3	500	498.944	499.956	5	1000	999.811	1002.701

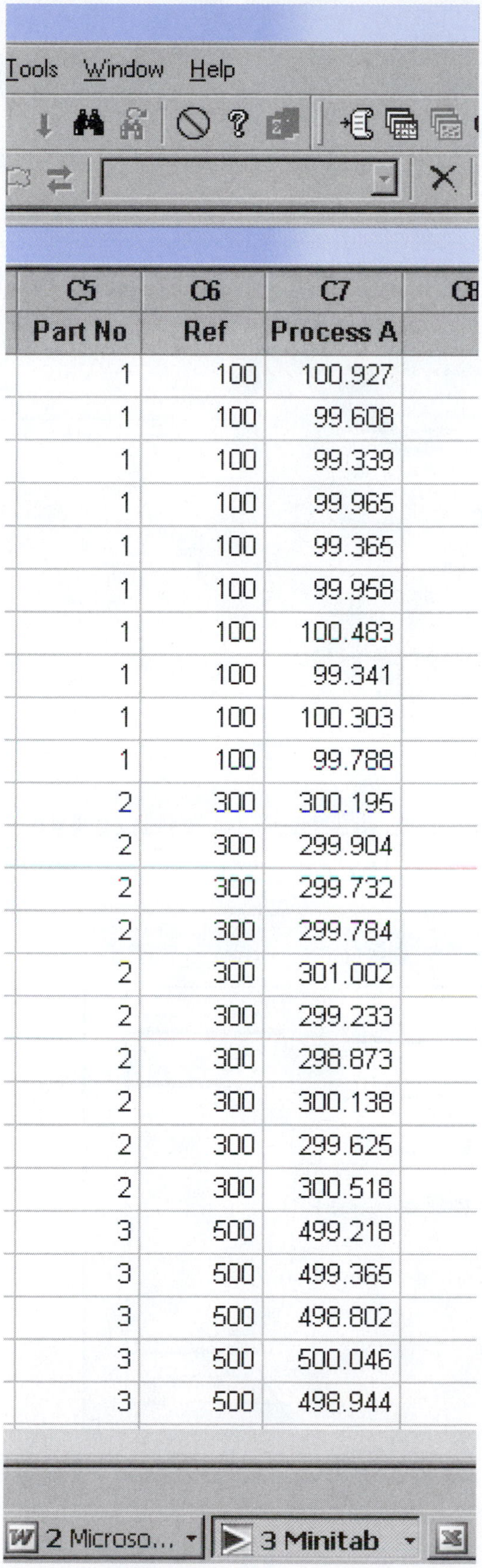

C5	C6	C7
Part No	Ref	Process A
1	100	100.927
1	100	99.608
1	100	99.339
1	100	99.965
1	100	99.365
1	100	99.958
1	100	100.483
1	100	99.341
1	100	100.303
1	100	99.788
2	300	300.195
2	300	299.904
2	300	299.732
2	300	299.784
2	300	301.002
2	300	299.233
2	300	298.873
2	300	300.138
2	300	299.625
2	300	300.518
3	500	499.218
3	500	499.365
3	500	498.802
3	500	500.046
3	500	498.944

FIGURE 9-12. Minitab Worksheet for Bias and Linearity

There are two fundamental questions that this analysis will answer.

- Does the measurement process perform consistently across the range of reference values tested (the linearity part)?
- Is there significant measurement bias at any of the reference values tested (the bias part)?

A clue to the answers to both of these questions lie in an observation of the *p* values shown in Figure 9-15. With respect to the linearity issue the Minitab analysis fits a regression curve to the average measurements calculated across the range of reference values. The slope of this curve is tested for statistical significance. A large *p* value indicates a lack of statistical significance, which means that there is no statistical evidence that the slope of the regression curve differs from zero. A slope of zero means that the measurement process performs consistently for all values tested.

The bias question is tested for each of the certified reference values. A high *p* value indicates the lack of statistical evidence that bias exists. In the case of Figure 9-15 there is no evidence of bias for any of the five values tested.

By comparison, consider the linear and bias analysis for measurement process B shown in Figure 9-16. The small *p* values provide statistical evidence that this process does not perform consistently over the full range of values tested. Use of this process should be restricted to a small range of values around 500.

9.1.4.2 Gauge R&R. Minitab can also be used to check repeatability and reproducibility error components. This is often referred to in practice as a *gage R&R analysis.* To illustrate how Minitab can help with this type of analysis we will return to the production environment of Jack and Jane. Let us assume that 15 items have been selected at random from the production line. Jack and Jane are each asked to measure these 15 items twice and record their results. These data have been summarized in Table 9-5.

A Minitab worksheet can be set up with the format shown in Figure 9-17. Three columns are required: the first contains the item number (there needs to be more than one), the second contains the operator name (there needs to be more than one), and the third column the individual measures.

Once the worksheet is established the sequence of drop-down menus illustrated in Figure 9-18 is accessed: *Stat > Quality Tools > Gage Study > Gage R&R Study [Crossed].* In the dialog box that appears, enter the names of the columns containing the part numbers, operators, and measurement data; and then click on whether the analysis is to be an analysis of variance (ANOVA) or using $\overline{X}$ and *R* charts, as illustrated in Figure 9-19.

Next, click on *OK* in the dialog box and the report shown in Figure 9-20 is produced. The box in the upper right-hand

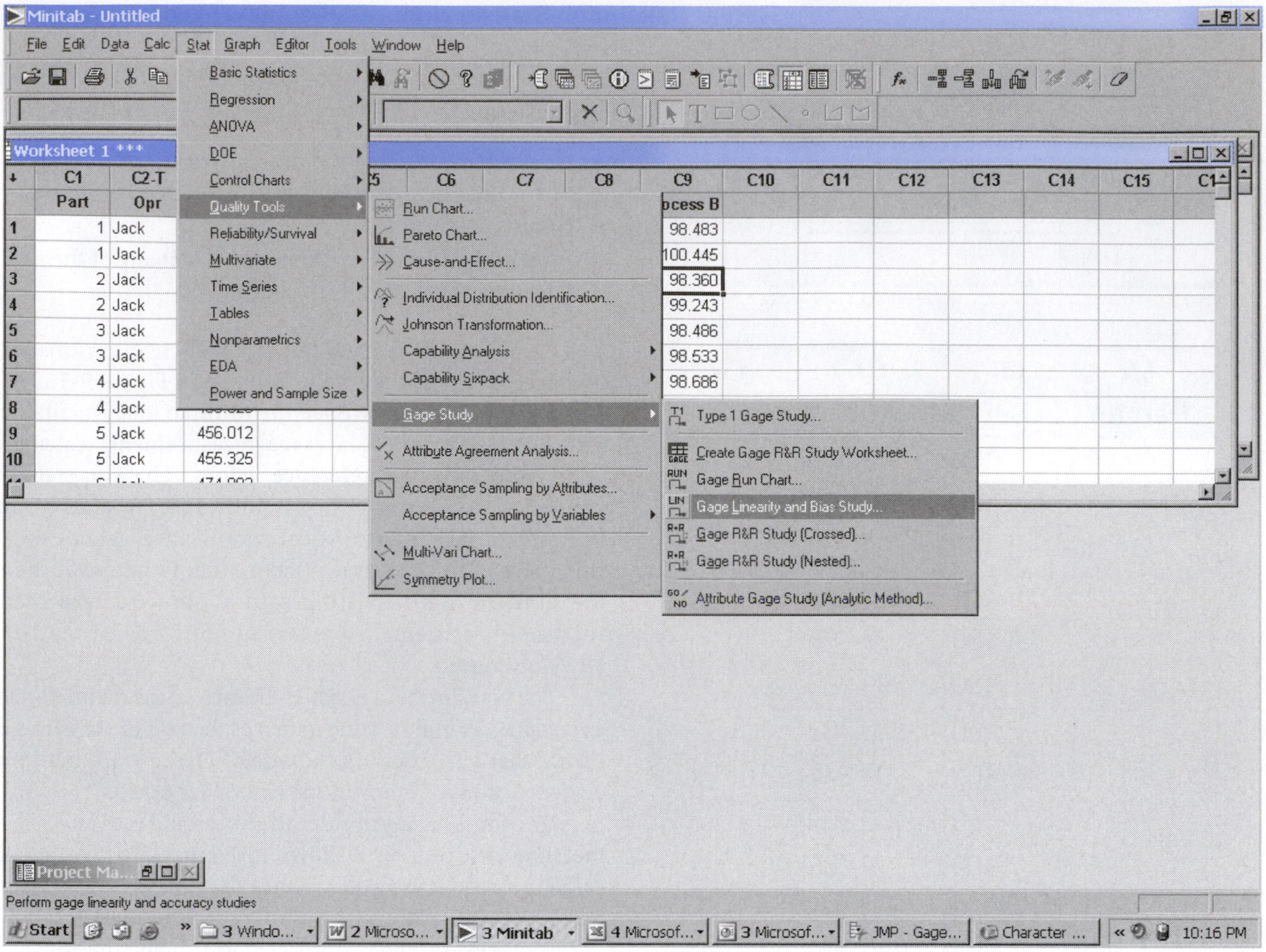

FIGURE 9-13. Dropdown Menu Sequence for Linearity and Bias Test

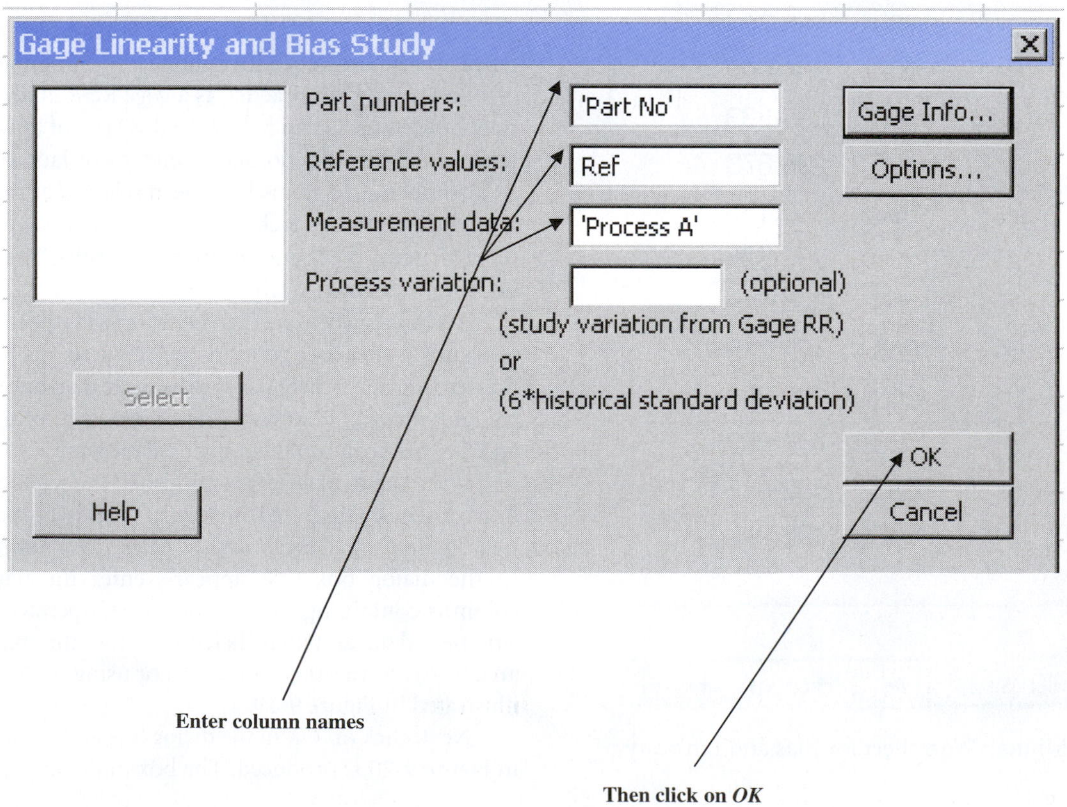

FIGURE 9-14. Linearity and Bias Dialog Box

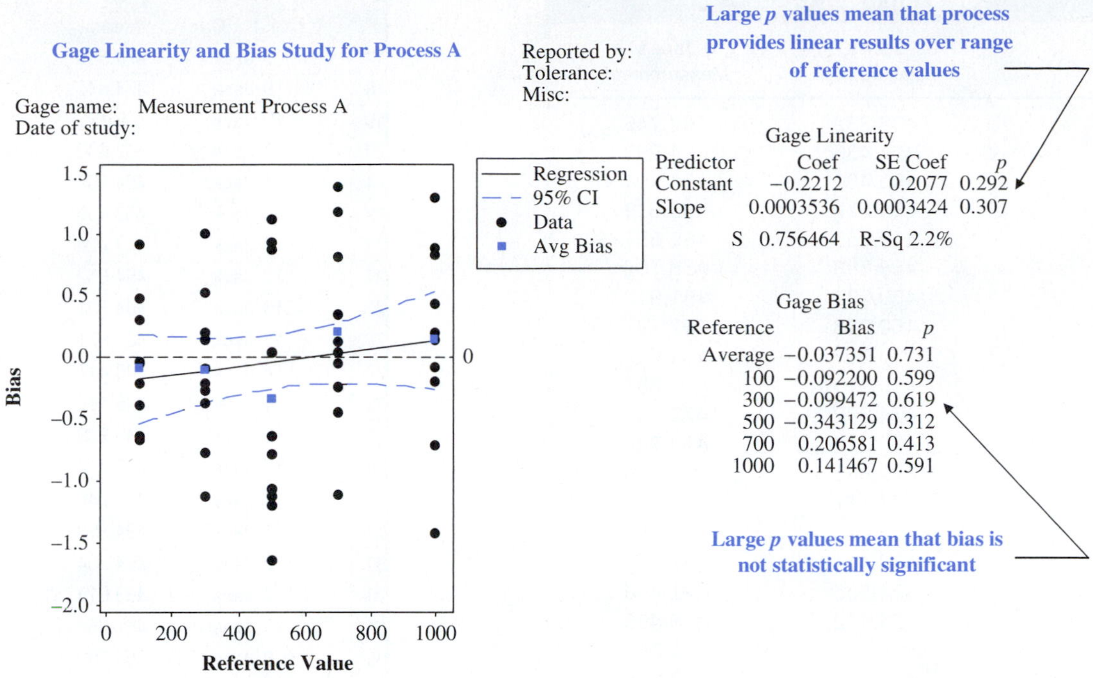

FIGURE 9-15. Linearity and Bias Analysis for Measurement Process A

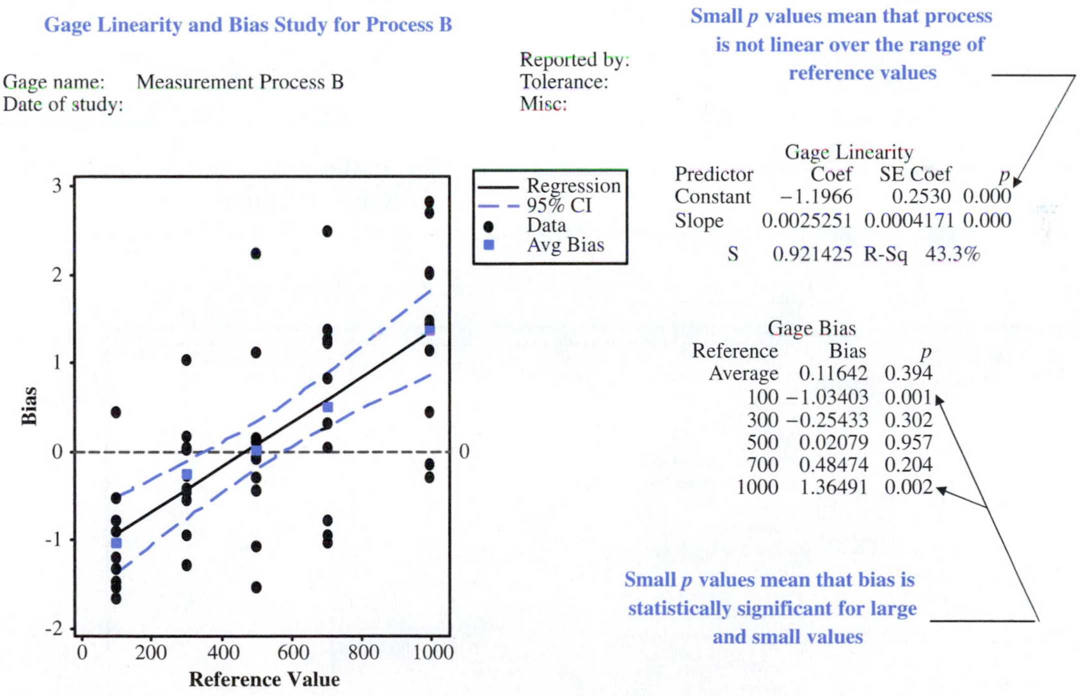

FIGURE 9-16. Linearity and Bias Analysis for Measurement Process B

corner shows the ANOVA data and in the upper left-hand corner the $\overline{X}$ and R analysis results. Reviewing the report and the charts, two conclusions can be drawn. First, measurement error is negligible compared to the magnitude of process variation—measurement variation accounts for only 0.89% of the total variation observed. Second, the reproducibility component of measurement error is not great—accounting for only 0.03% of the total observed variation. From these conclusions it is clear that the measurement process in question is adequate for this production process, and the both inspectors, Jack and Jane, can continue to perform the measurements on their respective shifts with minimal negative consequences.

TABLE 9-5 Jack and Jane Production Measurement Data

Item No.	Jack's Measurement	Jane's Measurement
1	498.332	497.746
	497.436	494.993
2	455.460	454.704
	454.627	453.619
3	458.155	461.851
	458.898	458.702
4	460.237	461.912
	460.028	462.797
5	456.012	455.647
	455.325	457.857
6	474.992	473.033
	475.627	473.788
7	477.631	479.355
	477.177	475.999
8	455.036	454.559
	453.444	457.831
9	474.502	471.469
	474.442	476.455
10	456.435	460.053
	457.670	458.214
11	454.907	454.270
	453.159	457.350
12	454.426	454.778
	454.552	454.048
13	484.501	483.469
	484.871	484.532
14	445.417	445.301
	445.164	444.660
15	449.939	453.108
	450.744	451.979

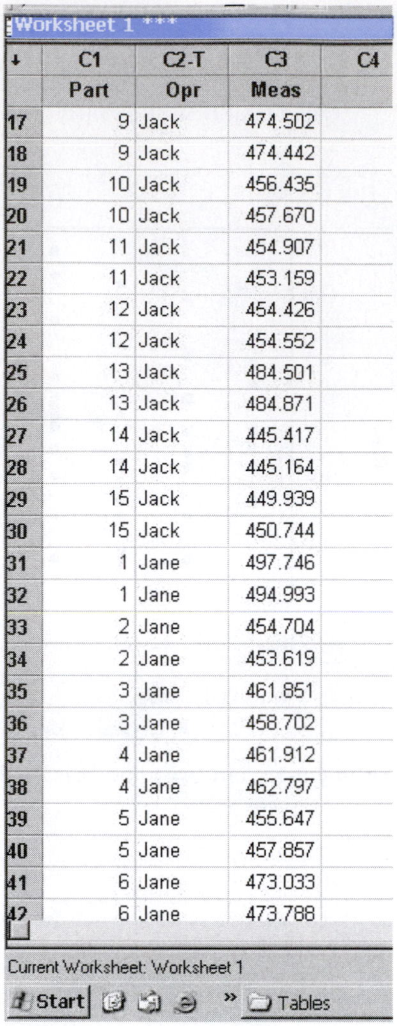

	C1 Part	C2-T Opr	C3 Meas	C4
17	9	Jack	474.502	
18	9	Jack	474.442	
19	10	Jack	456.435	
20	10	Jack	457.670	
21	11	Jack	454.907	
22	11	Jack	453.159	
23	12	Jack	454.426	
24	12	Jack	454.552	
25	13	Jack	484.501	
26	13	Jack	484.871	
27	14	Jack	445.417	
28	14	Jack	445.164	
29	15	Jack	449.939	
30	15	Jack	450.744	
31	1	Jane	497.746	
32	1	Jane	494.993	
33	2	Jane	454.704	
34	2	Jane	453.619	
35	3	Jane	461.851	
36	3	Jane	458.702	
37	4	Jane	461.912	
38	4	Jane	462.797	
39	5	Jane	455.647	
40	5	Jane	457.857	
41	6	Jane	473.033	
42	6	Jane	473.788	

FIGURE 9-17. Minitab Worksheet for Gage R&R Study

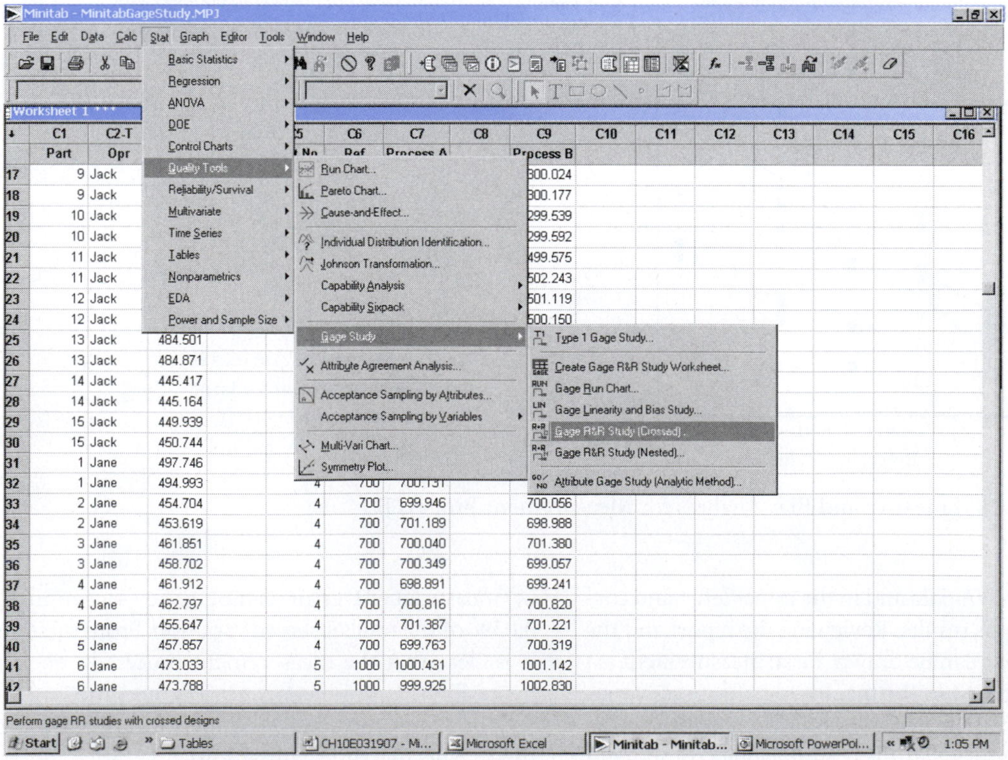

FIGURE 9-18. Dropdown Menu Sequence for Gage R&R Study

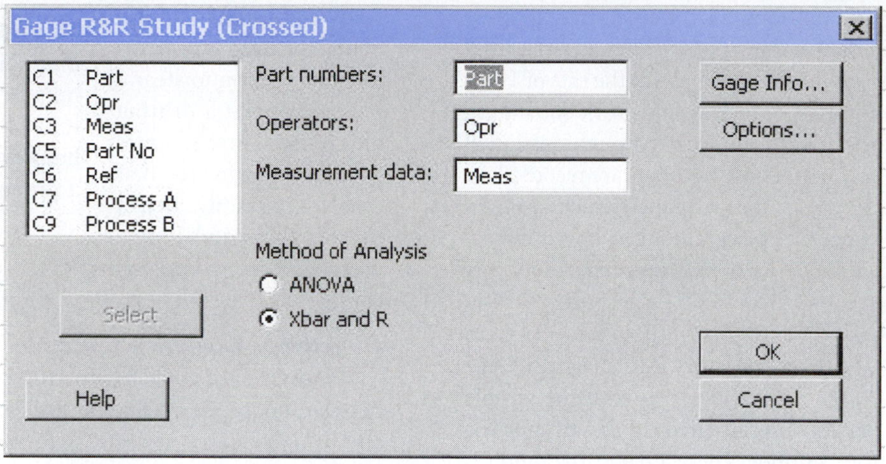

FIGURE 9-19. Gage R&R Dialog Box

Source	StdDev (SD)	(6 * SD)	(%SV)
Total Gage R&R	1.2587	7.5520	8.56
Repeatability	1.2288	7.3727	8.36
Reproducibility	0.2726	1.6358	1.85
Part-To-Part	14.6455	87.8730	99.63
Total Variation	14.6995	88.1969	100.00

Number of Distinct Categories = 16

Total Gage R&R	1.854	0.89
Repeatability	1.790	0.86
Reproducibility	0.064	0.03
Opr	0.064	0.03
Part-To-Part	205.482	99.11
Total Variation	207.336	100.00

Source	StdDev (SD)	Study Var (6 * SD)	%Study Var (%SV)
Total Gage R&R	1.3616	8.1698	9.46
Repeatability	1.3378	8.0269	9.29
Reproducibility	0.2535	1.5209	1.76
Opr	0.2535	1.5209	1.76
Part-To-Part	14.3346	86.0079	99.55
Total Variation	14.3992	86.3950	100.00

Gage R&R (Xbar/R) for Meas

Gage name: Jack and Jane Measurement Process
Date of study:

Reported by:
Tolerance:
Misc:

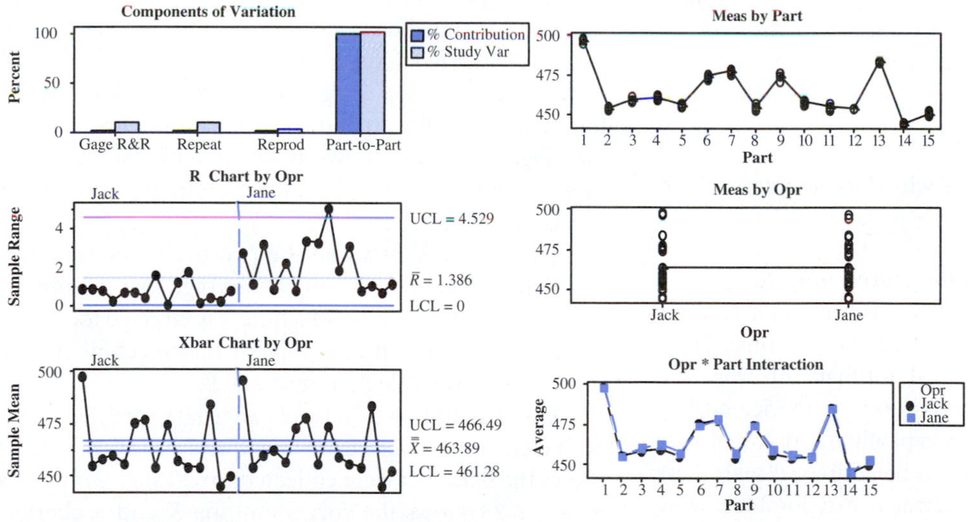

FIGURE 9-20. Gage R&R Minitab Report Form

9.2 SAMPLING STRATEGIES

9.2.1 Random Sampling

Statistical process control by variables requires strategies for measuring a fraction (usually a relatively small proportion) of the output of a process in order to make inferences concerning some important characteristics of the entire distribution of output over time. It is important that samples be selected that are representative of those characteristics to be estimated. In statistical terms, sampling should be taken at random—called **random sampling**—which if taken literally means that a group of objects should be selected for inspection entirely by chance so that each member of the population has the same probability of being chosen.

In practice, such pure randomness is not feasible. To illustrate, let us assume that one wishes to select 5 items per hour at random from an assembly line that produces at the rate of 1,000 per hour. A purely random approach would work something like this. All the items produced during a given hour are assigned a number between 1 and 1,000. Those numbers are recorded on slips of paper which are then placed in a bowl, well mixed, and five numbers randomly drawn. The inspector locates the corresponding items to measure. An alternative approach is to base the sample on random time. The production rate cited equates to approximately one unit every four seconds. There are 900 4-second intervals in each hour. Five random numbers can be generated between 1 and 900 to tell the inspector when to select an item from the line. Neither of these schemes are practicable. In reality, sampling strategies are always designed with convenience in mind. Samples may be selected as consecutive units or spread out. In some processes, sampling is limited to the extraction of product through designated sampling ports or accessible points along the value stream.

Sampling decisions are based on sampling objectives, process volatility, and process knowledge.

In developing a sampling strategy a significant element of judgment is always required. A decision must be made as to whether it is better to take large samples less frequently or small samples more often. An additional concern is how much production time should be built into a sample—is it best to select consecutively produced items or spread them out? The answers to these questions depend on the objectives of the sampling scheme, the volatility of the process, and how much is known about process behavior.

9.2.2 Sampling Objectives

There are three principal reasons why data samples are collected from industrial processes.

Objective 1: To characterize the process output.

Samples are taken from a process in order to draw conclusions concerning process behavior over time. These conclusions provide a statistical estimate of what the customer receives in the way of goods or service quality. Questions concerning process capability and compliance with customer requirements can be addressed with these samples, which are typically taken at an exit point following the last value-adding operation. In designing a sampling scheme to characterize process output it is important to capture all major sources of variability within each sample, especially the presence of any time influences (e.g., multiple machines, operators, shifts, vendors, methods, etc.). It is also essential that these samples represent a homogeneous mixture of process output over time and be truly representative of the product mix delivered to customers.

Objective 2: To monitor and influence process behavior.

Samples are taken from a process at critical operations, with the objectives of monitoring process parameters relative to targets and signaling the need for operator intervention to steer a process when it is drifting off-course. These samples should be designed so that they can readily detect process shifts and provide useful guidance to operators, so that the tendency to overcontrol (needless process adjustments) is avoided.

Samples are taken to characterize process output, monitor and influence process behavior, or to improve a process.

Objective 3: To improve processes.

Samples are collected for the purpose of better understanding process behavior and the root causes of variability and quality defects. Samples for the purpose of improvement may be taken on an ad hoc basis and should be designed to provide information that answers a predetermined set of questions. While it is important that processes be in statistical control for objectives 1 and 2, control charts showing out-of-control signals can often provide the information required for objective 3.

9.2.3 Impact of Various Sampling Schemes

For any process there is no unique sampling plan, and the conclusions and actions can differ dramatically depending on how data are collected. To illustrate, consider the two-machine production process illustrated in Figure 9-21. The product is produced on machines A and B at roughly the same rate, then fed onto a common conveyor belt where the output of both machines is comingled onto an accumulation conveyor where it is packaged and shipped. The specification for a critical quality attribute for this product is 15 mm ± 5 mm. Figure 9-21 also shows three different schemes that have been proposed to sample output from this two-machine process.

Figure 9-22 shows that the two machines are different and that neither is producing an average that is close to the required target of 15 mm. Machine A is targeted too high and machine B too low. It also appears that machine A may be slightly more variable than machine B.

Using sampling scheme 1, alternate samples of size 4 are taken respectively from machines A and B. Table 9-6 shows the data that were collected for 30 such samples, and Figure 9-23 shows the corresponding $\overline{X}$ and R charts. On these charts the odd-numbered samples pertain to machine A and the even-numbered samples are from machine B. The variability captured within a sample, represented on the range chart, is applicable to sources that are solely within each machine. The range chart is in control as no statistical rules have been violated and there is no apparent pattern that would stratify the points according to machine. We therefore conclude that there is no statistical evidence that the machine variances differ, and a reasonable estimate for the standard deviation of each machine is $\sigma \cong 1.84$. The $\overline{X}$ chart tells a different story. There are numerous rule 1

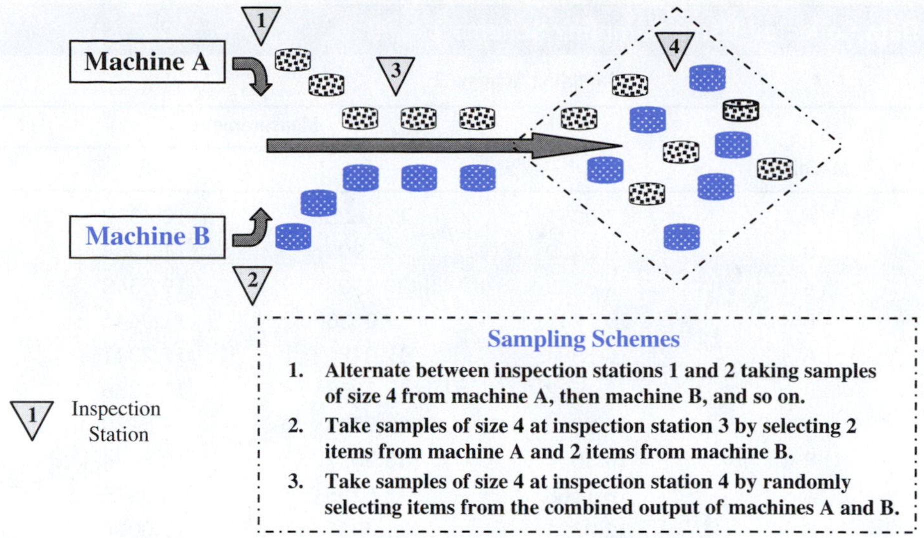

FIGURE 9-21. Various Sampling Schemes from a Two-Machine Production Process

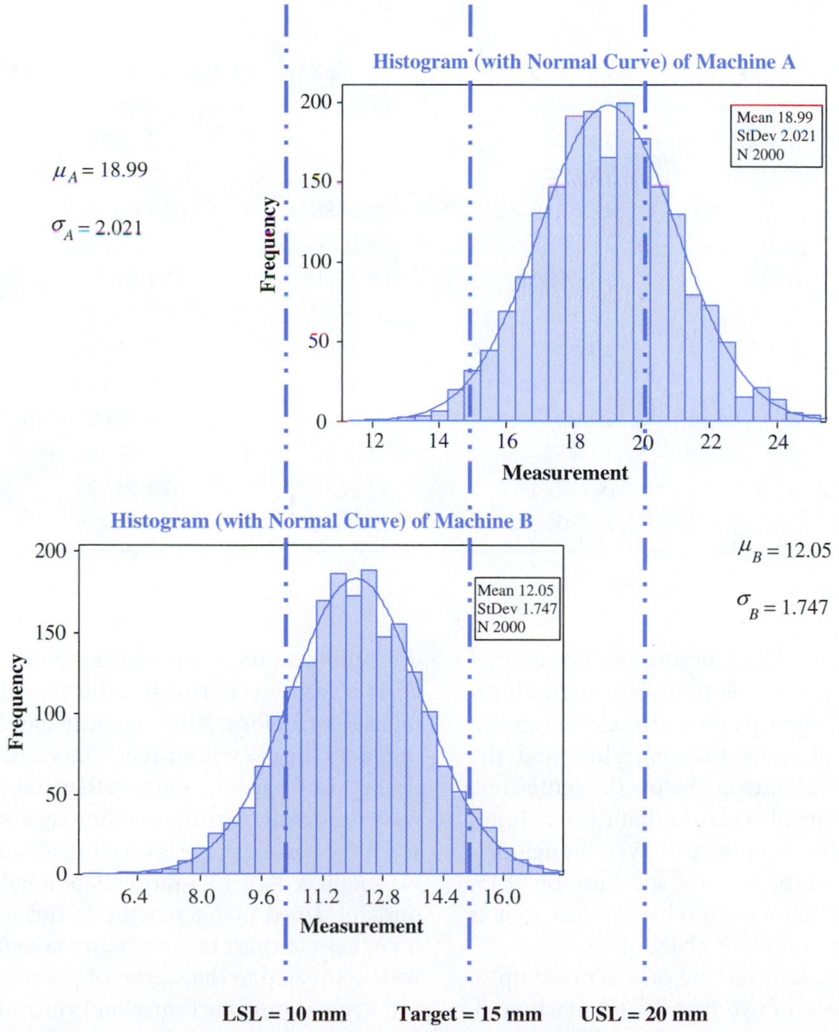

FIGURE 9-22. Machine Distributions

TABLE 9-6 Data Collected Using Sampling Scheme 1

Sampling Scheme 1

Sample Number	Machine	Measurement 1	Measurement 2	Measurement 3	Measurement 4
1	A	17.0961	17.1628	19.4382	17.3498
2	B	10.2420	9.3990	11.1984	13.9864
3	A	17.3588	20.4702	19.5348	21.0678
4	B	15.4745	7.9526	11.9641	15.4262
5	A	16.6884	17.0780	17.2341	18.1421
6	B	12.2632	12.3958	12.1336	13.0321
7	A	18.6395	22.5283	19.5229	24.1495
8	B	10.3534	13.3627	10.8824	11.5475
9	A	18.1168	15.5295	16.9976	16.6177
10	B	10.0394	11.3396	11.0004	9.8772
11	A	17.4942	18.0635	19.4876	18.1324
12	B	11.4107	13.3164	10.7319	9.3422
13	A	18.5411	20.0928	21.1386	19.5953
14	B	12.4250	15.5270	14.2338	11.7708
15	A	22.8389	18.1818	17.8763	21.2016
16	B	10.8349	13.4051	12.0266	11.9992
17	A	18.2176	21.1687	16.6290	15.4234
18	B	13.0847	12.3516	9.9867	11.1321
19	A	19.1641	18.7096	20.0987	16.6983
20	B	12.6407	11.5331	13.1131	9.1861
21	A	17.0919	22.5698	19.0592	19.3357
22	B	11.1861	15.1014	8.2620	10.8434
23	A	19.4995	18.6620	21.2775	22.8685
24	B	12.1362	10.9695	13.4121	11.2375
25	A	19.8709	19.9452	16.4305	17.0456
26	B	10.2136	10.8475	13.4110	12.1045
27	A	24.2021	16.9924	18.2280	20.7764
28	B	11.9727	12.2691	14.4816	12.1892
29	A	21.8276	20.0671	22.0253	18.8602
30	B	15.2019	10.0378	8.9501	10.7741

failures (out-of-control points) signifying the presence of between-sample variability. The points stratify by machine, resulting in the pronounced zigzag pattern. The odd-numbered points (machine A) are all above the centerline and the even-numbered (machine B) ones are below the centerline. This pattern provides statistical evidence that the machines differ in mean. It is interesting to note that even though neither machine is centered near the nominal specification of 15, on the average the two machines are producing close to it as evidenced by the centerline on the $\overline{X}$ chart of 15.46.

In sampling scheme 2, each sample of 4 is made up of two items from machine A and two items from machine B. Table 9-7 summarizes the data for 30 samples collected using scheme 2, and the corresponding $\overline{X}$ and R charts are shown in Figure 9-24. Note that this pair of charts is remarkably different from those constructed under scheme 1. The range chart is again in control, but this time the centerline is much higher than before. This is because the scheme 2 samples capture not only the within-machine variability sources but also the between-machine sources. Remember that this scheme involves selecting two items from each machine. In this case when the average range is used to estimate the total variability, we obtain $\sigma \cong 4.13$—more than double what we had previously obtained under scheme 1. Turning to the $\overline{X}$ chart, we note that the chart looks *almost too controlled.* The limits are wide compared to the scatter of points, which has resulted in some rule 7 (hugging centerline) violations. This has occurred because in scheme 2 the samples are perfectly stratified—that is, every sample contains an equal number of randomly selected objects from each of the two machine distributions.

Rules violations:

Rule 1—23 points
Rule 4—16 points
Rule 8—23 points

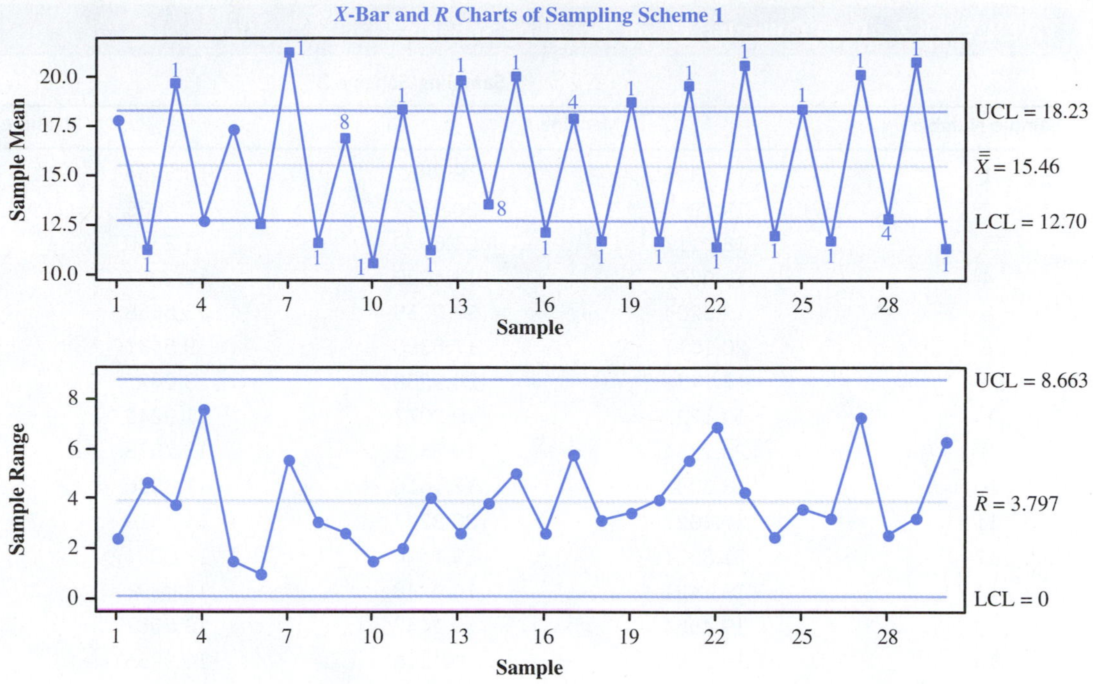

$$\sigma \cong \frac{\bar{R}}{d_2} = \frac{3.797}{2.059} = 1.84$$

FIGURE 9-23. Control Charts for Sampling Scheme 1

Rules violations:

Rule 7—5 points

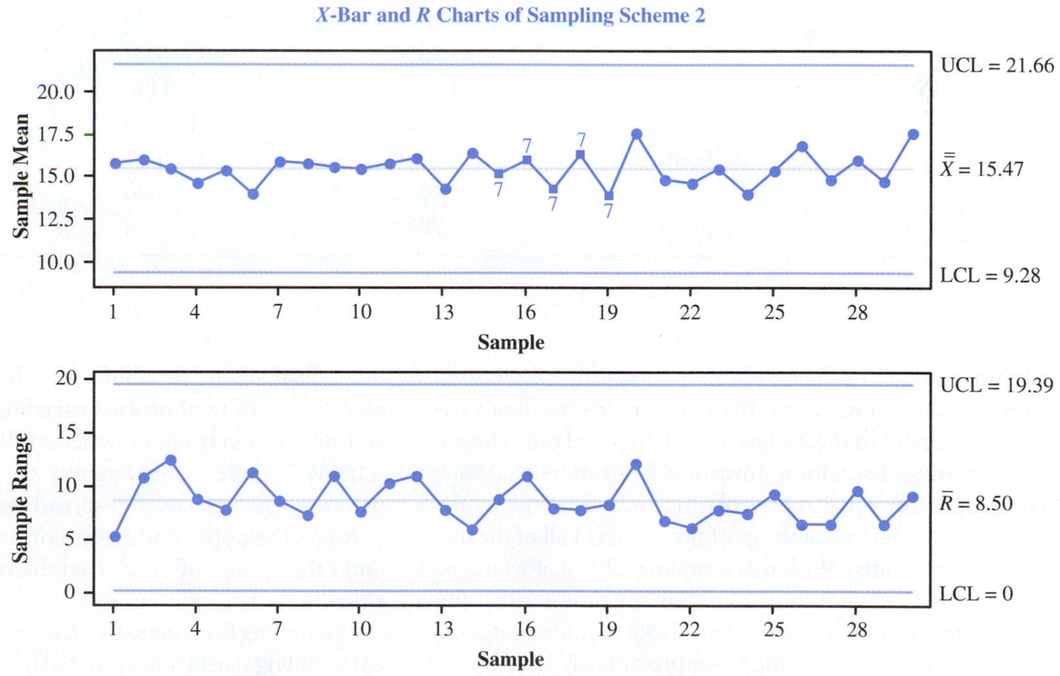

$$\sigma \cong \frac{\bar{R}}{d_2} = \frac{8.5}{2.059} = 4.13$$

FIGURE 9-24. Control Charts for Sampling Scheme 2

TABLE 9-7 Data Collected Using Sampling Scheme 2

	Sampling Scheme 2			
Sample Number	Machine A		Machine B	
1	17.0316	18.6852	13.5454	13.9781
2	19.1903	20.9327	13.6635	10.2652
3	18.9117	21.6370	9.2360	11.9400
4	19.9609	14.4090	11.2947	12.8028
5	19.3204	17.0055	13.5668	11.5206
6	20.1611	17.5350	9.5431	8.8398
7	17.1772	20.9208	13.0677	12.4037
8	18.1222	19.2077	11.9246	13.9404
9	20.1254	19.2413	13.7678	9.2895
10	18.6993	17.1616	11.0836	14.9113
11	20.1027	20.8297	11.4319	10.6266
12	21.3901	19.4516	13.0004	10.4665
13	19.1421	14.3748	12.0696	11.4982
14	19.4554	19.1447	13.5557	13.6142
15	16.5160	19.7216	10.9956	13.1968
16	16.9792	21.6981	10.8351	14.3874
17	19.8614	13.0824	11.9025	12.4756
18	18.8307	20.1693	12.5185	13.7571
19	18.5034	15.1018	11.5032	10.3188
20	23.2737	22.2302	11.2374	13.7759
21	17.4971	17.3652	10.7477	13.8177
22	17.0972	17.1885	11.1915	13.0340
23	19.4013	16.9415	11.6978	13.7451
24	16.3822	17.2883	9.9307	12.1493
25	17.5648	19.8550	13.3227	10.7039
26	20.1509	19.4720	13.8396	14.1814
27	18.4873	16.4001	12.1538	12.2685
28	21.6086	18.1559	12.2675	11.9643
29	17.4448	17.9056	11.5586	11.8872
30	20.2229	21.7077	12.6628	15.9330

While this scheme has been effective in capturing the total variability in the range chart, the machine-to-machine variability component in the $\overline{X}$s has been dampened out. The average of samples containing measures from more than one distribution, with equal representation, will converge to the grand mean—that is, the average of the means of all of the distributions represented. With this sampling scheme it would be extremely difficult, if not impossible, for any time-related special causes to show up in the $\overline{X}$ chart. Note that the centerline on this $\overline{X}$ is the same as scheme 1—approximately 15.5.

The third sampling scheme prescribes taking samples of size 4 from the combined output of machines A and B. Data for 30 samples using scheme 3 are summarized in Table 9-8 and the control charts are depicted in Figure 9-25. Unlike the previous two schemes, both charts show good statistical control. The range chart captures both within and between-machine variability and the average range produces an estimate of total process variability (i.e., $\sigma \cong 4.01$). This is similar to the result obtained under scheme 2. Since under scheme 3 there is no attempt to stratify the sample, the pattern of $\overline{X}$s has a greater spread, consuming all of the space between the upper and lower control limits. Like schemes 1 and 2 the average of the $\overline{X}$s for scheme 3 is also approximately equal to 15.4.

Sampling from the same machine distributions with different sampling schemes has resulted in three very different control chart patterns. Which is the best? This depends on the desired objective. If the objective is to characterize the output (what the customer sees)—previously described as objective 1—scheme 3 is best. One would be able to predict with some degree of confidence that over time the process is delivering to the customer a distribution with a mean equal to approximately

TABLE 9-8 Data Collected Using Sampling Scheme 3

	Sampling Scheme 3			
	Measurements from Combined Stream of Output			
Sample Number	1	2	3	4
1	13.1384	21.7202	18.8461	18.8286
2	19.4498	18.2396	21.9488	9.0142
3	13.4464	14.2554	12.8270	8.6394
4	17.0304	15.8769	12.0451	18.0858
5	12.8094	14.0066	18.5907	12.5016
6	12.8028	18.4550	12.5467	11.2580
7	18.3428	8.0726	20.1027	23.4266
8	10.0744	10.2652	10.4762	15.0285
9	17.0456	20.3530	12.7150	19.0825
10	18.0690	19.9628	20.4411	18.3705
11	21.8540	9.3434	11.8317	12.3784
12	9.5960	19.4013	23.7069	9.0679
13	18.6987	13.4189	12.2675	21.0865
14	13.6096	11.4291	14.9686	14.2854
15	22.6096	22.3384	11.0243	12.1180
16	14.4206	7.9261	10.6566	13.8909
17	11.8807	21.6130	15.5232	13.6959
18	17.1773	17.2618	16.4283	12.5411
19	10.1683	9.3984	17.8881	12.1538
20	11.2203	17.3928	11.1789	17.0519
21	13.3470	16.2867	14.9716	9.7902
22	18.6836	18.3383	9.3670	11.5668
23	17.7775	13.3164	13.8177	20.7884
24	16.3985	17.5295	11.9025	18.3545
25	11.4755	15.8119	17.2152	18.7977
26	22.6852	19.6276	14.9113	12.9247
27	18.1706	9.8849	21.2775	17.4805
28	11.9129	18.5804	19.0058	12.6628
29	13.2352	20.6293	12.9935	15.2491
30	13.3113	21.2016	9.9829	20.0207

15.4 and a standard deviation approximately equal to 4. However, a quick calculation will show that much of the product delivered will be out-of-specifications, since $15.4 \pm 3\sigma = 15.4 \pm 12 = (3.4, 27.4)$. This suggests that customers could receive product as low as 3.4 mm (when the minimum allowable is 10) and as high as 27.4 mm (when the maximum allowable is 20). This concern is confirmed by again referring to Figure 9-22.

If the objective is improvement—previously described as objective 3—scheme 1 highlights a significant improvement opportunity. By recentering each machine to the target dimension and eliminating the machine-to-machine differences, process variance can potentially be reduced to $\sigma \cong 1.84$ and process spread to $15.4 \pm 3\sigma = 15.4 \pm 3(1.84) = (10.0, 20.9)$. Comparing the range charts of scheme 1 and scheme 3, this would represent a 78.9% improvement opportunity:

$$\frac{(4.01)^2 - (1.84)^2}{(4.01)^2} = 0.789$$

This is not enough to eliminate all defects; nevertheless, this action would represent a dramatic improvement in service. Also, improvement potential can be estimated by comparing the range charts of schemes 1 and 2. The range chart for scheme 2 provides a very good estimate of the variability within and between machines, whereas the range chart for scheme 1 includes the within-machine variability only.

If the objective is process monitoring (i.e., objective 2), scheme 1 could be effective once the machine-to-machine differences are eliminated.

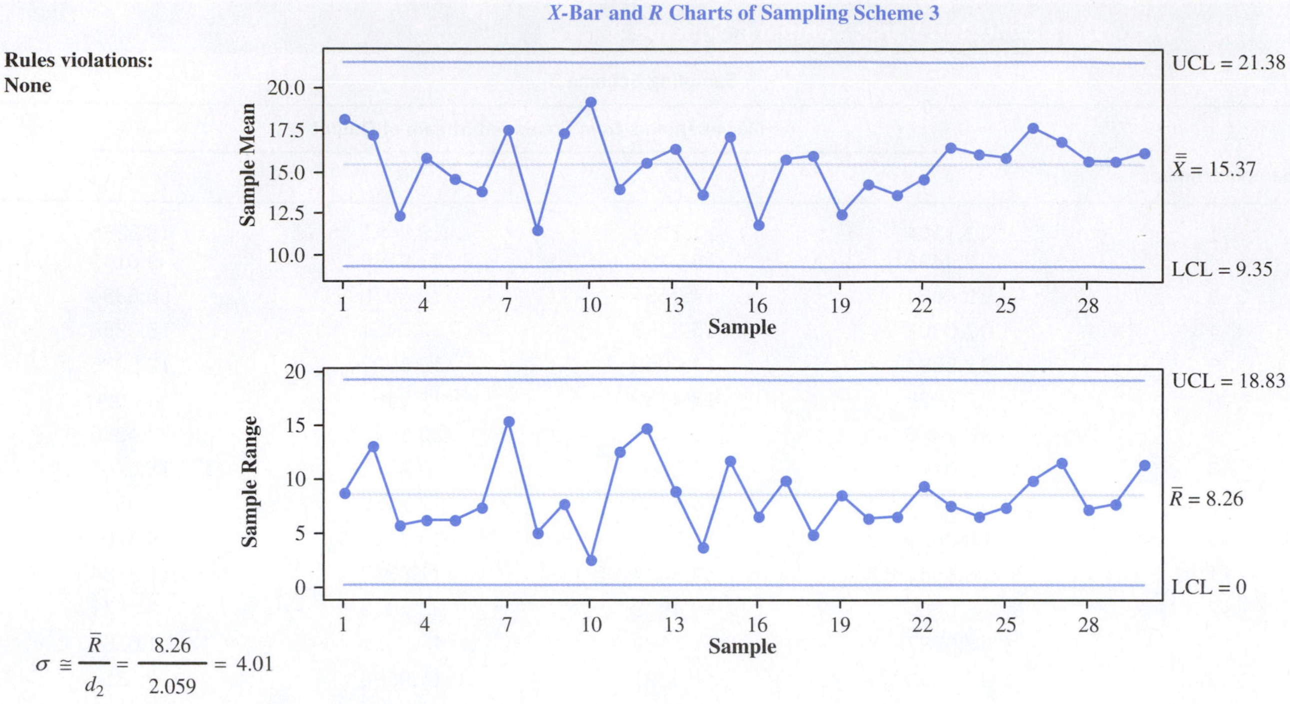

FIGURE 9-25. Control Charts for Sampling Scheme 3

9.3 PROCESS CAPABILITY

9.3.1 Engineering Tolerance, Natural Tolerance, and Process Capability

If a process is in statistical control and producing some unacceptable output, it may be because the process average (target) is too high or too low, or it may be that the process is incapable of consistently producing output that meets requirements. Some basic definitions are necessary to properly understand the concept of **process capability.** The first of these is **engineering tolerance** (*ET*). The *ET* is the difference between the highest acceptable measurement (the upper specification, or tolerance, limit—*USL*) and the lowest acceptable measurement (the lower specification, or tolerance, limit—*LSL*). That is,

> **Engineering tolerance is the difference between the upper and lower specification limits.**

$$ET = USL - LSL \qquad \textbf{9-9}$$

Note that the *ET* is established by design, and has nothing to do with where the process is performing at any point in time. The second important definition is the **natural tolerance** (*NT*). The *NT* is the spread of a process that is in statistical control and is considered to be equal to approximately six process standard deviations. Then,

> **Natural tolerance is six process standard deviations, which approximates the natural process spread.**

$$NT = 6\sigma_{\text{process}} \qquad \textbf{9-10}$$

where,

$\sigma_{process}$ is the long-term sustainable process standard deviation

A process is capable if $NT \leq ET$. If $NT > ET$ some defective product will be produced unless some of the common cause variability sources can be identified and removed. This will usually require fundamental system improvements. Centering the process cannot eliminate defective output. However, a normally distributed process that is not capable will produce the minimum out-of-tolerance product if it is centered exactly between the specification limits. Process capability is illustrated in Figure 9-26.

9.3.2 Capability Indices for Two-Sided Specification Limits

9.3.2.1 The Two-Sided Indices. There are five basic **capability indices** that can be used whenever processes are in control and the process mean and variance can be reliably estimated from control chart data. These are called the *C* indices and are given the following subscript designations: C_p, C_r, C_{pk}, C_{pm}, and C_{pmk}.

1. $C_p = \dfrac{ET}{NT}$ **9-11**

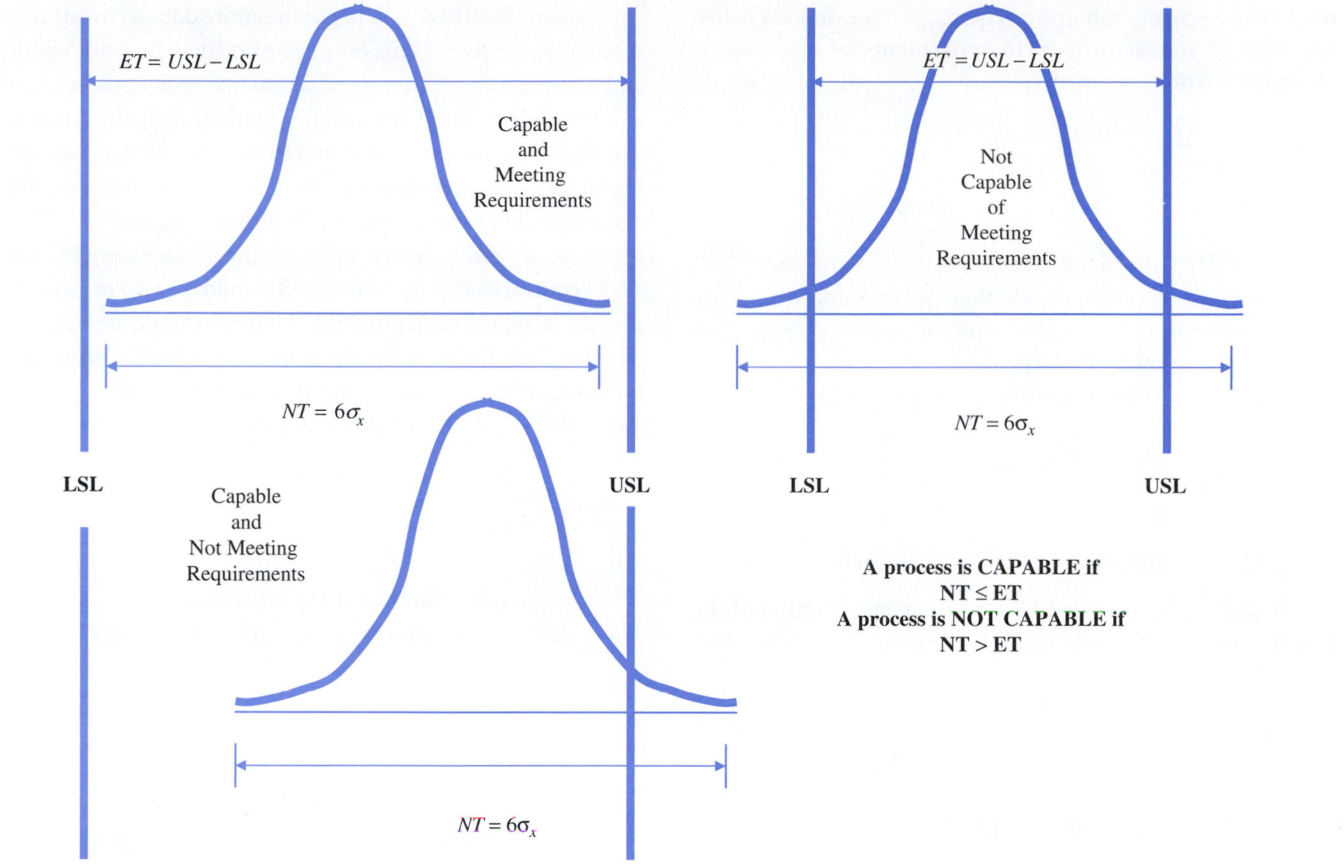

FIGURE 9-26. Process Capability

C_p is a measure of whether the tolerance limits are wider than the process natural tolerance.

If $C_p \geq 1$ the process is said to be capable.

2. $C_r = \dfrac{NT}{ET}$ **9-12**

C_r represents what proportion of the tolerance interval is "consumed" by the process spread.

If $C_r \leq 1$ the process is said to be capable.

3. $C_{pk} = \text{minimum}\ \{C_{pl}, C_{pu}\}$ **9-13**

where,

$$C_{pu} = \frac{|USL - \mu|}{3\sigma_{process}} \text{ and } C_{pl} = \frac{|LSL - \mu|}{3\sigma_{process}}$$

C_{pk} measures not only whether the ET ≥ NT; C_{pk} also measures process centering and whether requirements are actually being met.

C_{pk} is an index that expresses the distance from the process center (mean) to the closest specification limit, expressed in multiples of three standard deviations.

If $C_{pk} \geq 1$ the process is said to be capable.

Some Insights Regarding the C_{pk} Index

$C_{pk} = 1 \Rightarrow$ approximately 2,700 defective parts per million.

$C_{pk} \geq 1 \Rightarrow$ fewer than 2,700 defective parts per million.

$C_{pk} \geq 1.33 \Rightarrow$ fewer than 63 defective parts per million.

$C_{pk} \geq 1.67 \Rightarrow$ fewer than 1 defective parts per 2 million.

$C_{pk} \geq 2 \Rightarrow$ fewer than 1 defective parts per 500 million.

Processes with capability indices greater than one are regarded as capable.

While processes with a C_p or $C_{pk} = 1$ are generally regarded as capable, competitiveness may demand much higher values. The object of the Six Sigma programs, discussed in detail in Chapter 14, is to achieve C_{pk} values of 2 or better for all critical processes.

Two additional capability indices are sometimes usefully employed.

Equation 9-15 computes the standard deviation of the data with respect to the process target. The C_{pm} index can be used to test process centering. The C_{pmk} index defined below provides greater sensitivity to movements of the process mean away from the target value.

4. $$C_{pm} = \frac{ET}{6\sigma_{Target}} \qquad \textbf{9-14}$$

where,

$$\sigma_{Target} = \sqrt{\sigma^2_{process} + (\mu - \text{Target})^2} \qquad \textbf{9-15}$$

C_{pm} is a measure of whether the tolerance limits are wide enough to compensate for a process center that is not equal to the target.

If $C_{pm} \geq 1$ the process is said to be capable.

5. $$C_{pmk} = \min\left\{\frac{|USL - \mu|}{3\sigma_{Target}}, \frac{|LSL - \mu|}{3\sigma_{Target}}\right\} \qquad \textbf{9-16}$$

If $C_{pmk} \geq 1$ the process is said to be capable.

A sixth C index, called $C(u, v)$, is a generalization of the first five, and has been defined as a function of two nonnegative parameters u and v.[5]

6. $$C(u,v) = \frac{d - u|\mu - Midpoint|}{3\sqrt{\sigma^2_{process} + v(\mu - \text{Target})^2}} \qquad \textbf{9-17}$$

where,

$$d = ET/2 = (USL - LSL)/2$$

$$Midpoint = (USL + LSL)/2$$

If $C(u, v) \geq 1$ the process is said to be capable.

When both u and v are set to zero, Equation 9-17 yields the C_p index (Equation 9-11). The C_{pk} index (Equation 9-13) is obtained by setting u equal to one and v equal to zero; C_{pm} (Equation 9-14) results when u equals zero and v is set to one; and finally C_{pmk} (Equation 9-16) is obtained by setting both parameters to one.

9.3.2.2 The Two-Sided *P* Indices. When it is difficult to achieve statistical control, reliable estimates of sustainable process variation cannot be achieved. This might be the case with new products or processes, short runs, or where control has yet to be established. In these circumstances the process standard deviation cannot be estimated from control chart data. The raw sample data must be used as follows.

$$\sigma_{observed} \cong s_{observed} = \sqrt{\frac{\sum_{i=1}^{kn}\left[x_i - \overline{\overline{X}}\right]}{kn - 1}} \qquad \textbf{9-18}$$

where,

n = sample size
k = number of samples taken

$$\mu_{observed} \cong \overline{\overline{X}} = \frac{\sum_{i=1}^{kn} x_i}{kn} = \text{average of entire data set.} \qquad \textbf{9-19}$$

[5]Vannman, K. "A Unified Approach to Capability Indices." *Statistica Sinica* 5, no. 2 (1995): 805–20.

Equations 9-18 and 9-19 use the entire data set to estimate σ and μ respectively. This estimate of σ includes both within and between-sample sources of variability. When processes are in a state of statistical control the standard deviation used to determine capability is estimated from control chart data using either $\overline{R}/d_2$ or $\overline{s}/c_4$. In this case the C indices should be used to determine long-term sustainable process capability. When processes are not stable or there are insufficient data to construct control charts, Equation 9-18 should be used to estimate process standard deviation and P indices should be used for determining capability. Both P and C charts use the same data for estimating the process mean; however, in the case of P charts the long-term stability of the mean is uncertain.

7. $$P_p = \frac{ET}{6\sigma_{observed}} \qquad \textbf{9-20}$$

where,

$\sigma_{observed}$ is computed using Equation 9-18.

If $P_p \geq 1$ the process is thought to be capable.

8. $$P_r = \frac{6\sigma_{observed}}{ET} \qquad \textbf{9-21}$$

where,

$\sigma_{observed}$ is computed using Equation 9-18.

If $P_r \leq 1$ the process is thought to be capable.

9. $$P_{pk} = \text{minimum}\left\{P_{pl}, P_{pu}\right\} \qquad \textbf{9-22}$$

where,

$$P_{pu} = \frac{|USL - \mu_{observed}|}{3\sigma_{observed}} \text{ and } P_{pl} = \frac{|LSL - \mu_{observed}|}{3\sigma_{observed}}$$

$\sigma_{observed}$ is computed using Equation 9-18 and μ_{observed} is computed using Equation 9-19.

If $P_{pk} \geq 1$ process is considered to be capable.

10. $$P_{pm} = \frac{ET}{6\sigma_{Target}} \qquad \textbf{9-24}$$

where,

$$\sigma_{\text{Target}} = \sqrt{\frac{\sum_{i=1}^{n}(x_i - \text{Target})^2}{n - 1}} \qquad \textbf{9-24}$$

If $P_{pm} \geq 1$ the process is said to be capable.

11. $$P_{pmk} = \min\left\{\frac{|USL - \mu_{observed}|}{3\sigma_{Target}}, \frac{|LSL - \mu_{observed}|}{3\sigma_{Target}}\right\} \qquad \textbf{9-25}$$

where,

σ_{Target} is computed using Equation 9-24

If $P_{pmk} \geq 1$ the process is said to be capable.

12. $$P(u, v) = \frac{d - u|\mu_{observed} - Midpoint|}{3\sqrt{\sigma^2_{observed} + v(\mu_{observed} - \text{Target})^2}} \quad \textbf{9-26}$$

where,

$$d = ET/2E = (USL - LSL)/2$$

$$Midpoint = (USL + LSL)/2$$

$\sigma_{observed}$ is computed using Equation 9-18 and μ_{observed} is computed using Equation 9-19.
If $C(u, v) \geq 1$ the process is said to be capable.

9.3.3 Capability Example

Wizard Inc. manufactures a product with a specified design tolerance of 13 ± 3. $\overline{X}$ and R charts have been in place to monitor the process and both charts show good statistical control. The $\overline{X}$ chart has a centerline equal to 13.4 and the R chart centerline is equal to 2.45. The sample size is 4. The standard deviation for the last 10 samples (40 measurements) is equal to 1.29, and the standard deviation for those measurements with respect to the target is equal to 1.44. The average of the last 40 measurements is equal to 13.65.

$$\sigma_{process} = \frac{\overline{R}}{d_2} = \frac{2.45}{2.059} = 1.19$$

$$ET = USL - LSL = 16 - 10 = 6$$

$$NT = 6\sigma_{process} = 6(1.19) = 7.14$$

$$C_p = \frac{ET}{NT} = \frac{6}{7.14} = 0.84 \Rightarrow \text{not capable}$$

$$C_r = \frac{1}{C_p} = \frac{1}{0.84} = 1.19 \Rightarrow \text{not capable}$$

$$C_{pk} = \text{minimum}\left\{C_{pu}, C_{pl}\right\}$$

where,

$$C_{pu} = \frac{|USL - \mu|}{3\sigma_{process}} = \frac{|16 - 13.4|}{3(1.19)} = 0.73$$

$$C_{pl} = \frac{|LSL - \mu|}{3\sigma_{process}} = \frac{|10 - 13.4|}{3(1.19)} = 0.95$$

$$C_{pk} = \text{minimum}\{0.73, 0.95\} = 0.73$$

Process is not capable and not meeting requirements.

$$C_{pm} = \frac{ET}{6\sigma_{Target}} = \frac{6}{6\sqrt{(1.19)^2 + (13.4 - 13)^2}}$$
$$= 0.796 \Rightarrow \text{not capable}$$

$$C_{pmk} = \min\left\{\frac{|USL - \mu|}{3\sigma_{Target}}, \frac{|LSL - \mu|}{3\sigma_{Target}}\right\}$$

$$= \min\left\{\frac{|16 - 13.4|}{3\sqrt{(1.19)^2 + (13.4 - 13)^2}}, \frac{|10 - 13.4|}{3\sqrt{(1.19)^2 + (13.4 - 13)^2}}\right\}$$

$$= \min\left\{\frac{2.6}{3.77}, \frac{3.4}{3.77}\right\} = \min\{0.69, 0.90\}$$

$$= 0.69 \Rightarrow \text{not capable}$$

$$P_p = \frac{ET}{NT} = \frac{6}{6(1.29)} = 0.775 \Rightarrow \text{not capable}$$

$$P_r = \frac{1}{P_p} = \frac{1}{0.775} = 1.29 \Rightarrow \text{not capable}$$

$$P_{pk} = \text{minimum}\left\{P_{pu}, P_{pl}\right\}$$

where,

$$P_{pu} = \frac{|USL - \mu_{observed}|}{3\sigma_{observed}} = \frac{|16 - 13.65|}{3(1.29)} = 0.607$$

$$P_{pl} = \frac{|LSL - \mu_{observed}|}{3\sigma_{observed}} = \frac{|10 - 13.65|}{3(1.29)} = 0.943$$

$$P_{pk} = \text{minimum}\{0.607, 0.943\} = 0.607 \Rightarrow \text{not capable and not meeting requirements}$$

$$P_{pm} = \frac{ET}{6\sigma_{Target}} = \frac{6}{6(1.44)} = 0.694 \Rightarrow \text{not capable}$$

$$P_{pmk} = \text{minimum}\left\{\frac{|USL - \mu_{observed}|}{3\sigma_{Target}}, \frac{|LSL - \mu_{observed}|}{3\sigma_{Target}}\right\}$$

$$= \text{minimum}\left\{\frac{|16 - 13.65|}{3(1.44)}, \frac{|10 - 13.65|}{3(1.44)}\right\}$$

$$= \text{minimum}\{0.54, 0.84\} = 0.54 \Rightarrow \text{not capable}$$

9.3.4 Capability Analysis for One-Sided Specification Limits

When the specification limits are **one-sided**, the equations in Section 9.3.2 do not apply since a well-defined engineering tolerance does not exist. The most widely used capability indices for one-sided specifications are based on using either the upper or lower factor from Equations 9-13, 9-16, 9-22, or 9-25 as appropriate. This leads to the following C and P indices for one-sided specifications.

9.3.4.1 The One-Sided *C* Indices. When statistical control has been established, and the process mean and variance can be reliably estimated using control chart data, the following one-sided indices can be used to evaluate process capability.

1. $$C_{pu} = \frac{|USL - \mu|}{3\sigma_{process}} \text{ or } C_{pl} = \frac{|LSL - \mu|}{3\sigma_{process}} \qquad \textbf{9-27}$$

2. $$C_{pmu} = \frac{|USL - \text{Target}|}{3\sqrt{\sigma^2_{process} + (\mu - \text{Target})^2}}$$

or

$$C_{pml} = \frac{|LSL - \text{Target}|}{3\sqrt{\sigma^2_{process} + (\mu - \text{Target})^2}} \qquad \textbf{9-28}$$

3. $$CPU = \frac{|USL - \text{Target}| - |\text{Target} - \mu|}{3\sigma_{process}}$$

or

$$CPL = \frac{|LSL - \text{Target}| - |\text{Target} - \mu|}{3\sigma_{process}} \qquad \textbf{9-29}$$

The following two families of indices have been proposed as the one-sided counterparts to Equation 9-17.[6] These are generalizations of Equations 9-27 through 9-29 and are functions of the nonnegative parameters u and v.

4. $$C_{pau}(u, v) = \frac{(|USL - \mu|) - u(|\mu - \text{Target}|)}{3\sqrt{\sigma^2_{process} + v(\mu - \text{Target})^2}}$$

or

$$C_{pal}(u, v) = \frac{(|LSL - \mu|) - u(|\mu - \text{Target}|)}{3\sqrt{\sigma^2_{process} + v(\mu - \text{Target})^2}} \qquad \textbf{9-30}$$

where, $u \geq 0, v \geq 0$

5. $$C_{pvu}(u, v) = \frac{(|USL - \text{Target}|) - u(|\mu - \text{Target}|)}{3\sqrt{\sigma^2_{process} + v(\mu - \text{Target})^2}}$$

or

$$C_{pvl}(u, v) = \frac{(|LSL - \text{Target}|) - u(|\mu - \text{Target}|)}{3\sqrt{\sigma^2_{process} + v(\mu - \text{Target})^2}} \qquad \textbf{9-31}$$

where $u \geq 0, v \geq 0$ and $u, v \neq (0, 0)$

Equation 9-27 is obtained by setting u and v equal to zero in Equation 9-30. Setting $u = 1$ and $v = 0$ in Equation 9-31 yields Equation 9-29. Equation 9-28 is derived by setting $u = 0$ and $v = 1$ in Equation 9-30.

9.3.4.2 The One-Sided *P* Indices. When process control cannot be established the standard deviation and mean must be estimated using the sampling data at hand. In this situation the *P*-index counterparts to Equations 9-27 through 9-31 must be applied.

6. $$P_{pu} = \frac{|USL - \mu_{observed}|}{3\sigma_{observed}} \text{ or } P_{pl} = \frac{|LSL - \mu_{observed}|}{3\sigma_{observed}} \qquad \textbf{9-32}$$

7. $$PPU = \frac{|USL - \text{Target}| - |\text{Target} - \mu_{observed}|}{3\sigma_{observed}}$$

or

$$PPL = \frac{|LSL - \text{Target}| - |\text{Target} - \mu_{observed}|}{3\sigma_{observed}} \qquad \textbf{9-33}$$

8. $$P_{pmu} = \frac{|USL - \text{Target}|}{3\sqrt{\dfrac{\sum_{i=1}^{n}(x_i - \text{Target})^2}{n - 1}}}$$

or

$$P_{pml} = \frac{|LSL - \text{Target}|}{3\sqrt{\dfrac{\sum_{i=1}^{n}(x_i - \text{Target})^2}{n - 1}}} \qquad \textbf{9-34}$$

9. $$P_{pau}(u, v) = \frac{|USL - \mu_{observed}| - u|\mu_{observed} - \text{Target}|}{3\sqrt{\sigma^2_{observed} + v(\mu_{observed} - \text{Target})^2}}$$

or

$$P_{pal}(u, v) = \frac{|LSL - \mu_{observed}| - u|\mu_{observed} - \text{Target}|}{3\sqrt{\sigma^2_{observed} + v(\mu_{observed} - \text{Target})^2}} \qquad \textbf{9-35}$$

where, $u \geq 0, v \geq 0$

10. $$P_{pvu}(u, v) = \frac{|USL - \text{Target}| - u|\mu_{observed} - \text{Target}|}{3\sqrt{\sigma^2_{observed} + v(\mu_{observed} - \text{Target})^2}}$$

or

$$P_{pvl}(u, v) = \frac{|LSL - \text{Target}| - u|\mu_{observed} - \text{Target}|}{3\sqrt{\sigma^2_{observed} + v(\mu_{observed} - \text{Target})^2}} \qquad \textbf{9-36}$$

where $u \geq 0, v \geq 0$ and $u, v \neq (0, 0)$

9.3.5 Estimating Percentage or Number of Defective Parts

When a process is stable, its mean and standard deviation can be estimated with statistical confidence, and Equations 9-11 through 9-17 or Equations 9-27 through 9-31 can be employed to determine process capability. When process control has not been established and limited sample data must be used to estimate the process parameters, a somewhat less reliable evaluation of process capability can be achieved using Equations 9-20 through 9-26 or Equations 9-32 through 9-36. If any of these formulae reveal that the process is incapable of producing product within the specification range 100% of the time, the ability to estimate the degree of nonconformance is important. The standard

[6]Vannman, K. "Families of Capability Indices for One-Sided Specification Limits." *Statistics: A Journal of Theoretical and Applied Statistics* 31, no. 1 (1998): 43–66.

practice is to consult standard normal tables to estimate the percentage or number of defective units of output. Such calculations are based on the assumption that most stable processes follow normal distributions and, in order to access the tables, the upper and lower z scores are computed as follows.

$$\text{Upper } z \text{ score} = z_u = \frac{|USL - \mu|}{\sigma_{process}} \qquad \textbf{9-37}$$

$$\text{Lower } z \text{ score} = z_l = \frac{|LSL - \mu|}{\sigma_{process}} \qquad \textbf{9-38}$$

Returning to the Wizard problem that was introduced in Section 9.3.3,

$$z_u = \frac{|16 - 13.4|}{1.19} = 2.18$$

$$z_l = \frac{|10 - 13.4|}{1.19} = 2.86$$

Table II in Appendix B contains the areas under a standard normal curve and can be used for estimating the percentage of output that is above the upper specification limit or below the lower specification limit. Finding $z_u = 2.18$ and $z_l = 2.86$ in Table II, we discover that the $p_u = 0.014629$ and $p_l = 0.002118$. To convert these proportions to percentages we multiply by 100 and conclude that roughly 1.5% of the output of this process is out of spec on the high side and approximately 0.2% is out on the low side. If we subtract the total percentage defective from 100, we have the process yield—in this case, yield = 100—(1.46 + 0.2) = 98.34%.

At first glance this process yield may not seem too bad. However, in a Six Sigma environment quality is measured in terms of the number of **defective parts per million opportunities (DPMO)**. For the Wizard process we can estimate the DPMO by using our computed z_u and z_l to access Table III in Appendix B, where we learn that the process is producing approximately 14,629 parts per million outside the upper specification limit and 2,118 parts per million below the lower specification limit. These numbers are far from the Six Sigma quality objective of one part per million.

> **Process yield is computed by subtracting the total percentage of nonconforming output from 100.**

Sustainability is also an important issue. The previous calculations produce an estimation of short-term capability and beg the question of whether this performance can be sustained over time. In Six Sigma, long-term performance is differentiated from short-term capability through a mean-shift allowance. The assumption is that even for a process in control, a 1.5σ drift in the mean is likely over time. In this example the Six Sigma score is derived by subtracting 1.5 from the smallest z (i.e., 2.18 − 1.5) yielding 0.68. Looking up this revised z score in Table III we discover that in the long-term we could expect the process to produce 248,252 DPMO—nearly one defective in four! Six Sigma is covered in detail in Chapter 14.

> **The Six Sigma score is derived by subtracting 1.5 from the computed *z* score, and is used to estimate the long-term defects per million opportunities.**

In the case of one-sided specifications, either Equation 9-37 or 9-38 as appropriate would be used to estimate process yield.

9.3.6 Capability Analysis on Nonnormal Data

In Chapter 8 we applied **Box-Cox transformations** to the analysis of nonnormal process data. When the process cannot reasonably be approximated using the standard normal distribution, the techniques described earlier should be applied to a transformed data set in order to determine capability. To illustrate, consider the process data shown in Table 9-9. Five readings (in parts per million) of trace contaminants that are found in a certain liquid product were taken at random intervals over a 20-day period. The process target for these contaminants is 75 and the maximum allowable is 150. The histogram in Figure 9-27 shows that the data is not normally distributed and that some of the measurements obtained in the sample exceed the one-sided upper specification limit.

The Box-Cox transformation, using $\lambda = 0$, generates the histogram shown in Figure 9-28. Equation 9-37 can now be applied to the transformed data to estimate the percentage of process output that lies above the upper specification limit.

$$\text{Upper z score} = z_u = \frac{|5.01 - 3.46|}{1.434} = 1.08$$

From Table II in Appendix B we find that $z = 1.08$, which corresponds to approximately 14% of the data above the upper specification limit. Note that had we used the raw data and incorrectly assumed normality we would have computed a z score equal to 1.25 and estimated the percentage out of specifications to be 10.56%—a 25% difference!

> **Nonnormal data should be transformed before performing a capability analysis.**

Figure 9-29 shows the Minitab printout for a capability analysis using the Box-Cox transformation described above. The analysis shown is based on subgrouping the data by day into samples of size 5. Minitab computes capability indices based on the transformed data—the C_{pk} computation uses the transformed within subgroup standard deviation and the P_{pk} and P_{pm} (called the C_{pm} by Minitab) are computed using the transformed overall standard deviation (which includes both within- and between-subgroup variation).

The transformed data is then used to estimate the percentage of total output that is outside the specification limits. Three estimates are displayed: the percentage of the sample data that were actually observed out of spec (in this case 8 out of the 100 points, or 8%), the estimated percentage based on within-subgroup variation only (i.e., 13.65%), and the estimated percentage based on the overall variation (i.e., 13.98%).

TABLE 9-9 Trace Contaminants Found in Liquid Product

Day	Trace Contaminants in Parts Per Million					$\bar{X}$	R
1	63.84	48.69	15.27	35.89	65.82	45.90	50.54
2	26.14	99.47	79.72	1.32	14.26	44.18	98.14
3	269.35	379.95	59.63	147.45	55.98	182.47	323.98
4	14.14	55.23	2.01	11.49	48.24	26.22	53.22
5	93.47	44.01	10.05	2.20	71.15	44.18	91.27
6	285.39	85.90	71.90	53.83	36.88	106.78	248.51
7	75.43	3.22	62.11	20.61	50.33	42.34	72.21
8	37.50	13.97	66.48	0.16	27.72	29.17	66.32
9	51.12	18.25	34.78	96.25	19.38	43.96	78.01
10	9.42	79.16	118.35	86.72	29.40	64.61	108.93
11	11.37	8.32	28.82	50.62	4.52	20.73	46.10
12	32.32	309.94	46.28	77.07	107.95	114.71	277.62
13	2.07	35.42	14.58	119.67	5.37	35.42	117.60
14	40.26	216.67	66.04	5.10	45.58	74.73	211.57
15	9.80	72.91	61.08	2.60	36.38	36.55	70.31
16	89.06	127.42	16.09	224.99	0.14	91.54	224.85
17	9.57	18.20	17.58	69.76	4.79	23.98	64.97
18	23.33	30.45	36.06	32.03	94.79	43.33	71.45
19	177.04	4.85	196.91	100.00	20.09	99.78	192.06
20	74.88	141.60	24.48	141.19	24.72	81.37	117.12

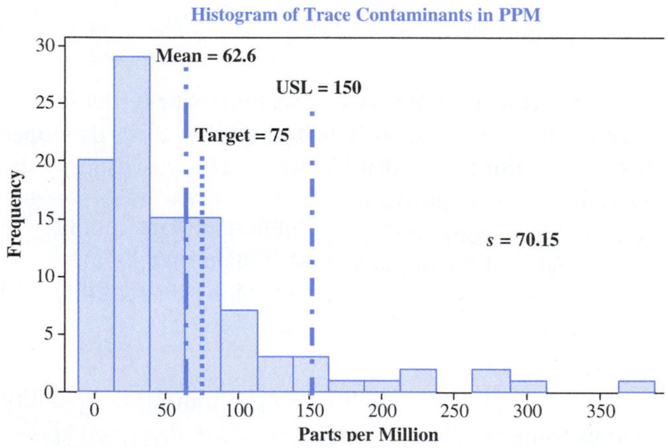

FIGURE 9-27. Histogram of Trace Contaminants

Two lab technicians review data collected during the production run of a new drug in a pharmaceutical plant. In the health care industry process capability is critical to patient recovery, costs, product liability, and corporate image.

9.3.7 Using Minitab for Capability Analyses

To illustrate the use of Minitab for performing capability analyses we return to the Acme Fastener data that we introduced in Chapter 8. The data from 30 samples of size 5 have been reproduced in Table 9-10. The specifications for this particular quality attribute are 15.5 ± 4.5. The nominal dimension of 15.5 has been selected as the target. The first step in using Minitab is to paste this spreadsheet into a Minitab worksheet.

Next, follow this dropdown sequence: *Stat* > *Quality Tools* > *Capability Analysis* > *Normal,* as shown in Figure 9-30. Then follow the steps shown in Figure 9-31. The Box-Cox dialog box is activated only in the case of nonnormal data. Since the Acme data follows a normal distribution we can bypass the Box-Cox step. Figure 9-32 is the Minitab

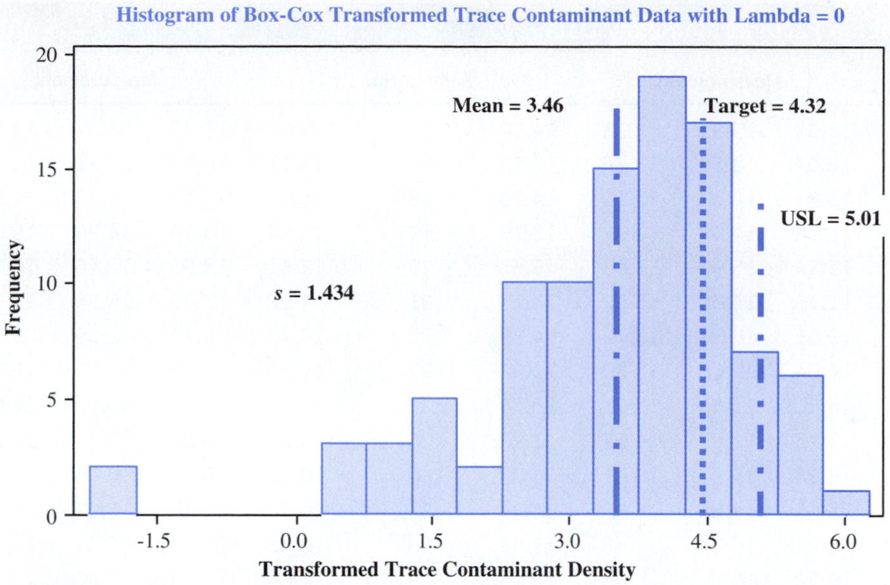

FIGURE 9-28. Transformed Data for Trace Contaminants

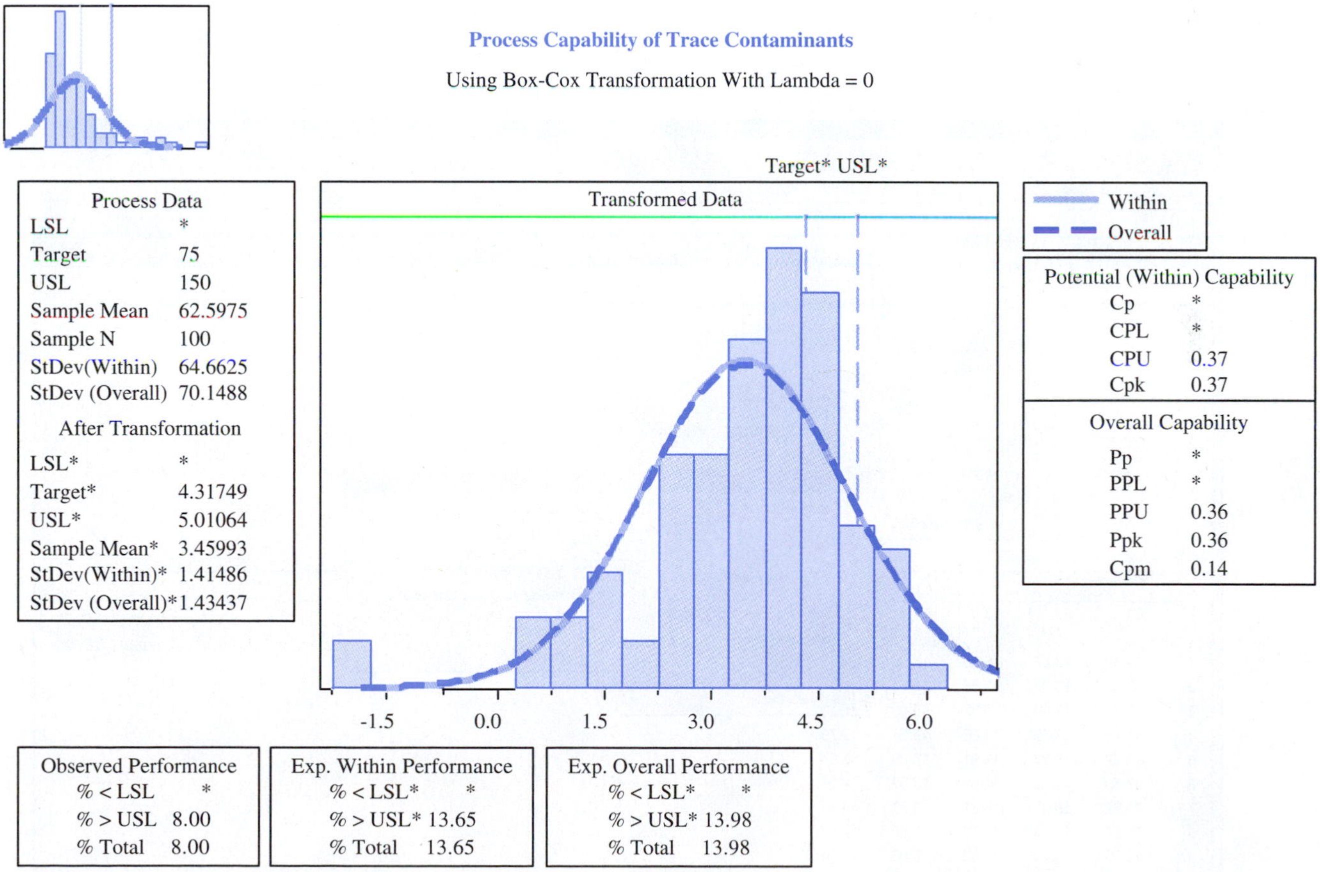

FIGURE 9-29. Minitab Capability Analysis Using Box-Cox Transformation

display of the results. The histogram of the raw sampling data (in this case 150 measurements) is shown together with the target, specification limits, and a normal curve with the mean and standard deviation of the data set. The process data, including the computed mean and standard deviation, is contained in the box to the left of the histogram. Minitab separately computes the within-subgroup standard deviation and the overall standard deviation. Note in this case the two

TABLE 9-10 Acme Fastener Company Data

Sample No.	Measurement					Sample No.	Measurement				
1	15.41	13.57	15.85	13.38	14.12	16	13.07	11.37	14.97	14.42	18.66
2	13.03	18.07	13.55	13.77	15.50	17	14.11	16.21	13.71	12.33	13.05
3	13.91	15.88	12.88	15.10	14.06	18	13.84	14.13	14.13	14.34	14.81
4	13.98	13.78	11.63	16.57	12.79	19	13.54	13.20	14.99	15.01	12.33
5	16.22	15.24	15.91	13.80	16.53	20	12.41	13.59	12.46	16.74	13.06
6	17.33	12.45	14.59	15.52	14.25	21	14.90	15.49	16.95	19.18	16.91
7	15.62	19.01	16.31	13.72	14.39	22	10.77	12.74	12.09	13.33	15.69
8	15.87	14.94	11.86	12.30	14.97	23	14.59	12.75	15.01	15.07	17.87
9	12.70	16.20	13.60	15.08	15.74	24	15.57	13.40	15.03	14.69	13.62
10	18.37	14.76	12.05	15.62	13.90	25	14.02	15.21	15.31	15.94	13.07
11	15.28	14.94	15.87	16.81	14.03	26	18.42	18.06	10.79	13.18	16.96
12	15.25	15.66	14.58	14.62	14.34	27	16.20	13.48	12.92	14.62	12.31
13	11.73	15.54	16.05	12.56	13.53	28	15.36	11.83	16.61	12.94	15.79
14	14.19	14.23	15.09	17.46	13.16	29	14.69	14.74	15.43	15.98	16.45
15	18.03	15.45	16.17	16.99	12.80	30	12.68	15.62	12.68	15.43	14.45

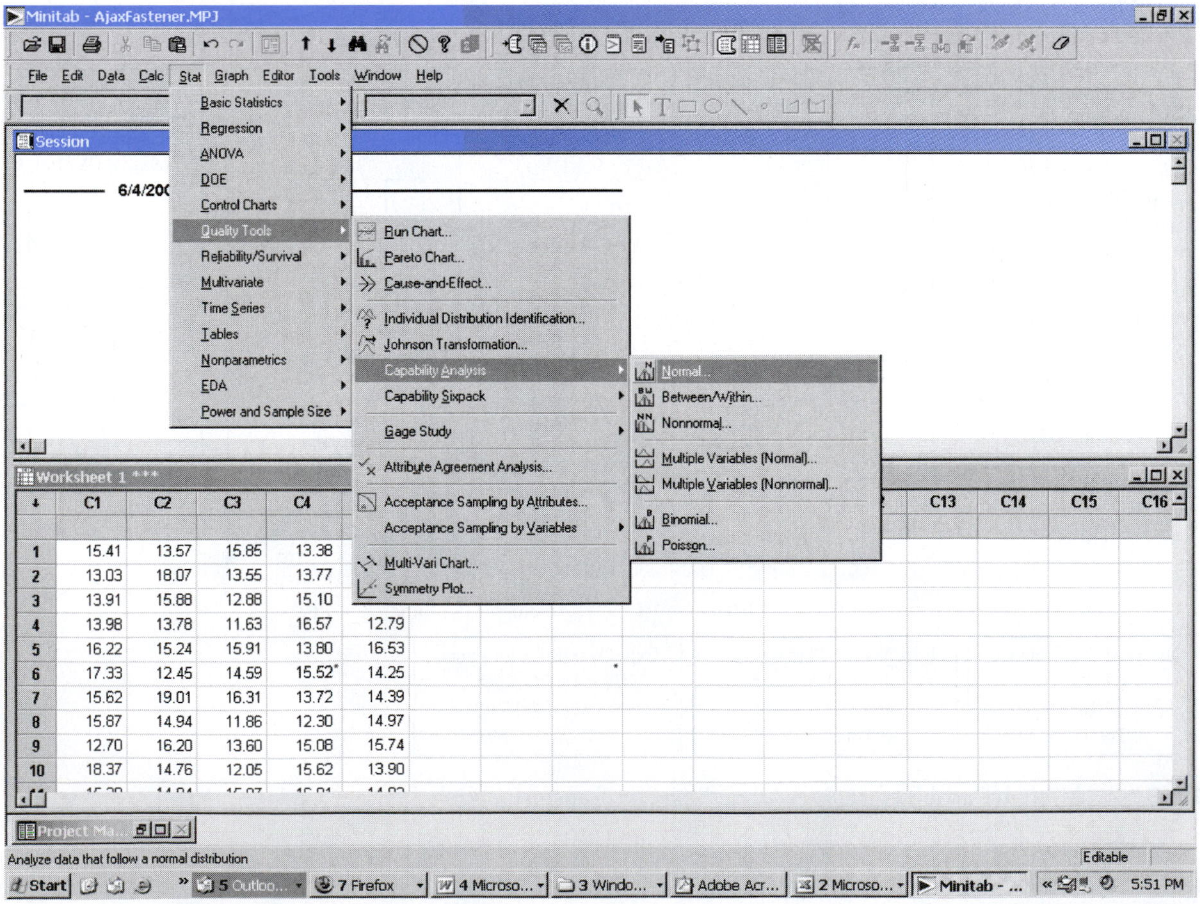

FIGURE 9-30. Minitab Dropdown Sequence for Capability Analysis

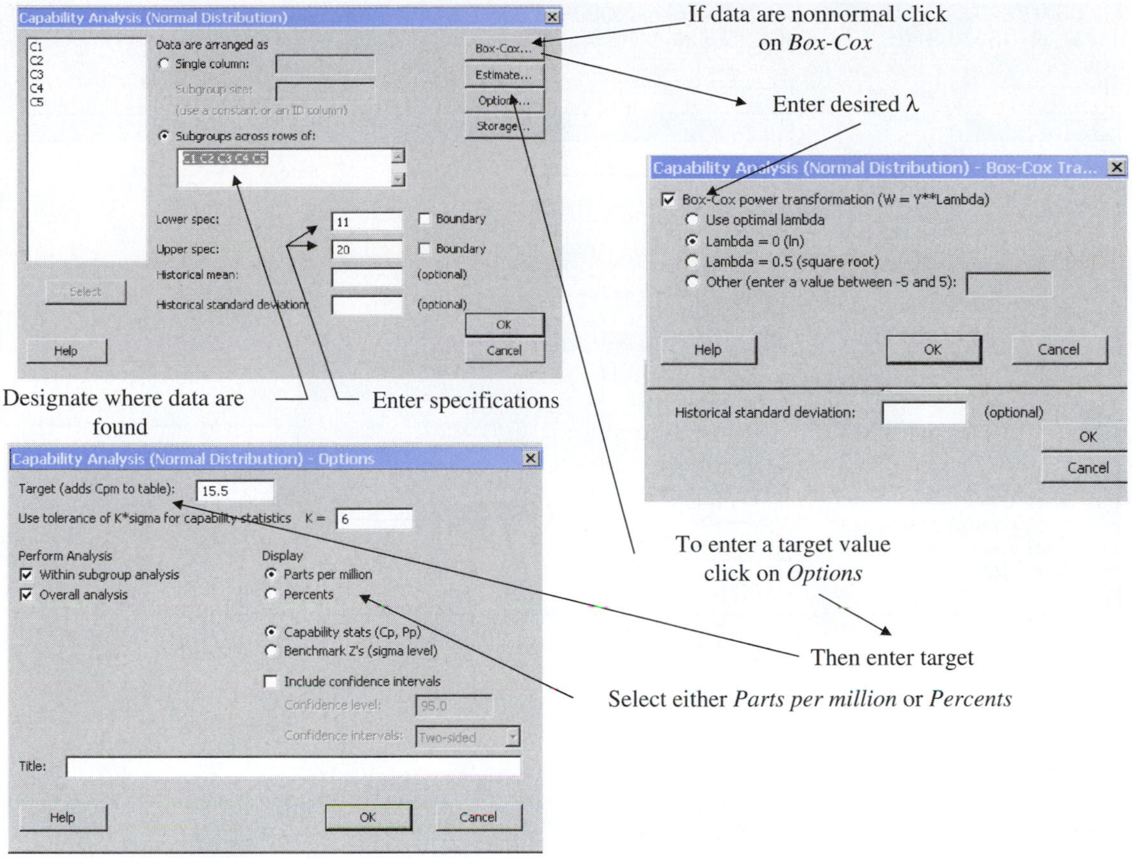

FIGURE 9-31. Minitab Capability Dialog Boxes

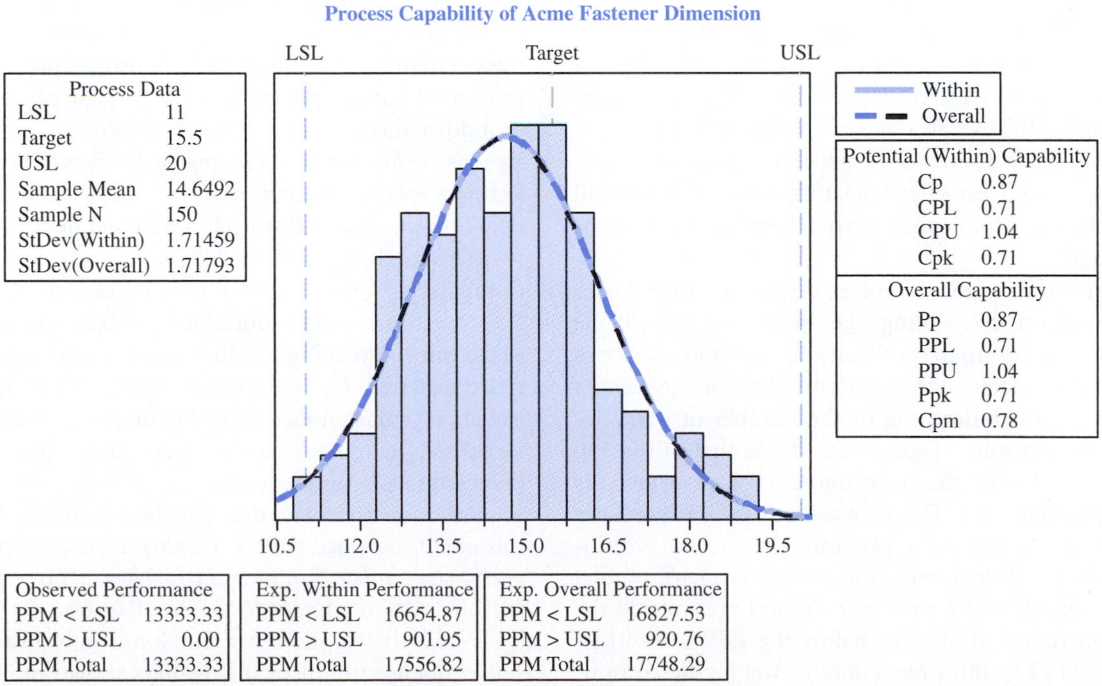

FIGURE 9-32. Minitab Display of Capability Results for Acme Fasteners

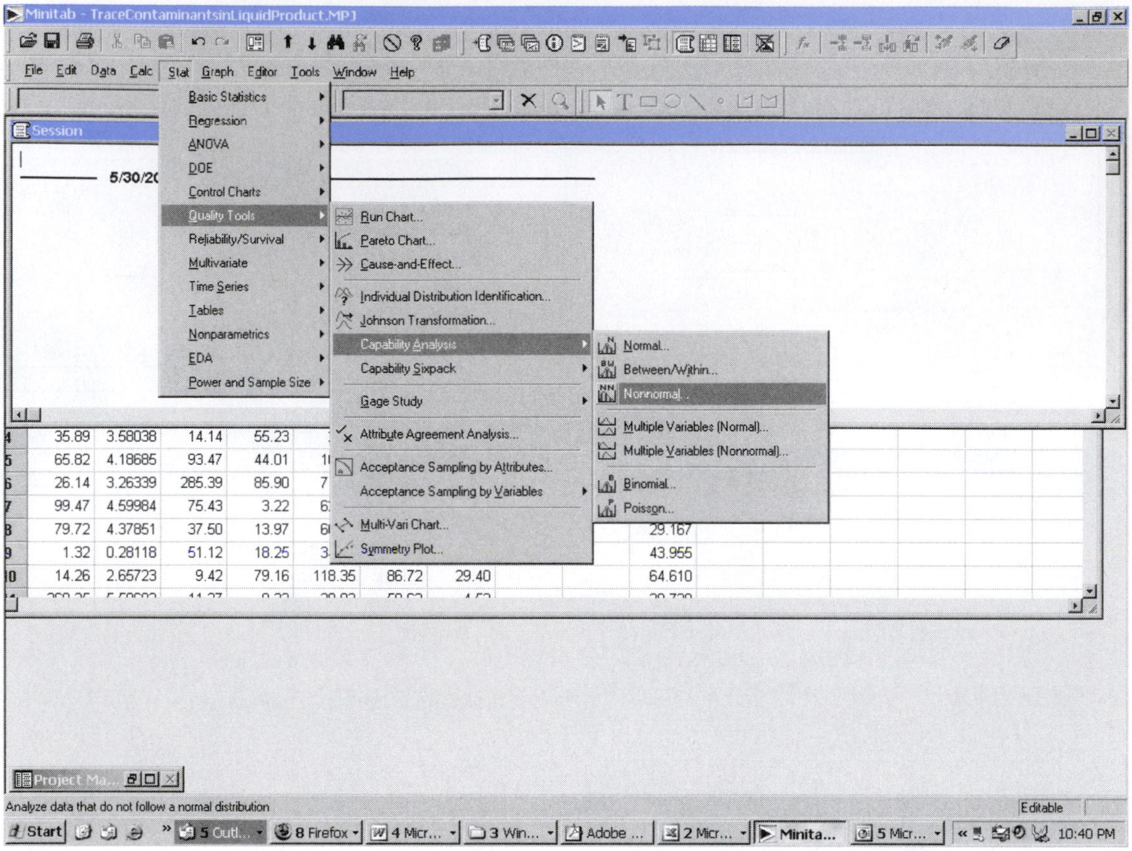

FIGURE 9-33. Dropdown Menu Sequence for Nonnormal Distributions

are nearly the same because, as shown in Chapter 8, this process is in a state of statistical control.

The right-hand side of the display contains the computed capability indices. C_p and C_{pk} are computed using the within-subgroup standard deviation and P_p, P_{pk}, and C_{pm} are computed using the overall standard deviation. With a $C_p = 0.87$ we can infer that, even if the process is centered on target, the process is not capable of meeting the specifications all the time. The $C_{pk} = 0.71$ confirms that the process is not centered on target.

At the bottom of the capability display are three boxes with information concerning the degree to which the process fails to conform to specifications. Minitab offers two options for these data—either in terms of the percentage of output that is nonconforming or the number of parts per million nonconforming. Figure 9-32 shows that of the 150 data points 1.33% of the measurements were below the lower specification limit. This equates to 13,333.33 parts per million. Using the normal approximation and the within-subgroup standard deviation (the second box) we predict that a total of 17,556.82 parts per million are outside the specification range—16,654.87 below the *LSL* and 901.95 above the *USL*. The third box contains the predicted nonconformities using the overall standard deviation—17,748.29 total with 16,827.53 below the *LSL* and 920.76 above the *USL*.

In the previous section the Box-Cox transformation was applied to the trace contaminant data to obtain the Minitab capability display shown in Figure 9-29. We now return to the trace contaminant problem to demonstrate an alternative approach for handling nonnormal data. Figure 9-33 shows the dropdown menu sequence: *Stat* > *Quality Tools* > *Capability Analyis* > *Nonnormal*. This approach offers the option of selecting a Johnson transformation[7] which is useful when the Box-Cox approach fails to achieve a good normal fit.

The resulting Minitab display is shown in Figure 9-34. Comparing Figures 9-29 and 9-34 we observe that the Box-Cox transformation produced a $P_{pk} = 0.36$ and 13.98% projected nonconforming product as compared to the Johnson transformation $P_{pk} = 0.48$ and 9.67%. Note that the projected percentage nonconforming under the Johnson transformation is closer to the percentage nonconforming observed in the sample results.

Additional information can be obtained by following this menu sequence: *Stat* > *Quality Tools* > *Capability Sixpack* > *Normal* or *Nonnormal* (the *Normal* option is selected for information relevant to the Box-Cox transformation and the *Nonnormal* option for the Johnson transformation). This sequence produces the displays shown in Figure 9-35.

[7]Johnson, Norman L. "Systems of Frequency Curves Generated by Methods of Translation." *Biometrika* 36, no. 1/2 (1949): 149–76.

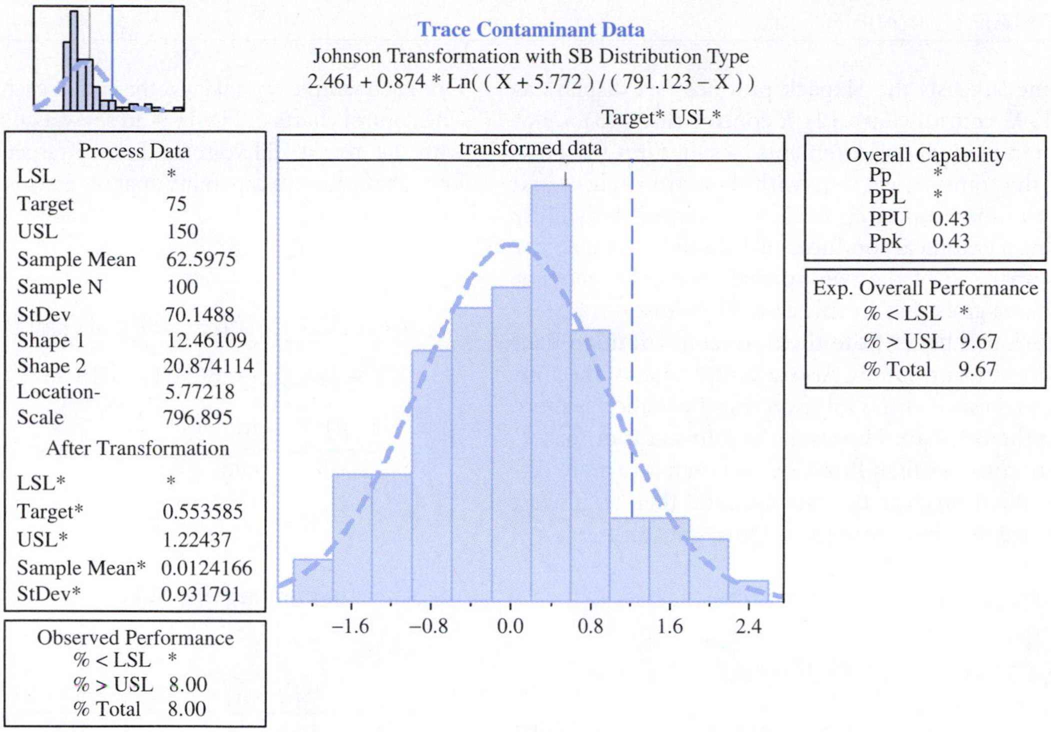

FIGURE 9-34. Capability Display Using Johnson Transformation

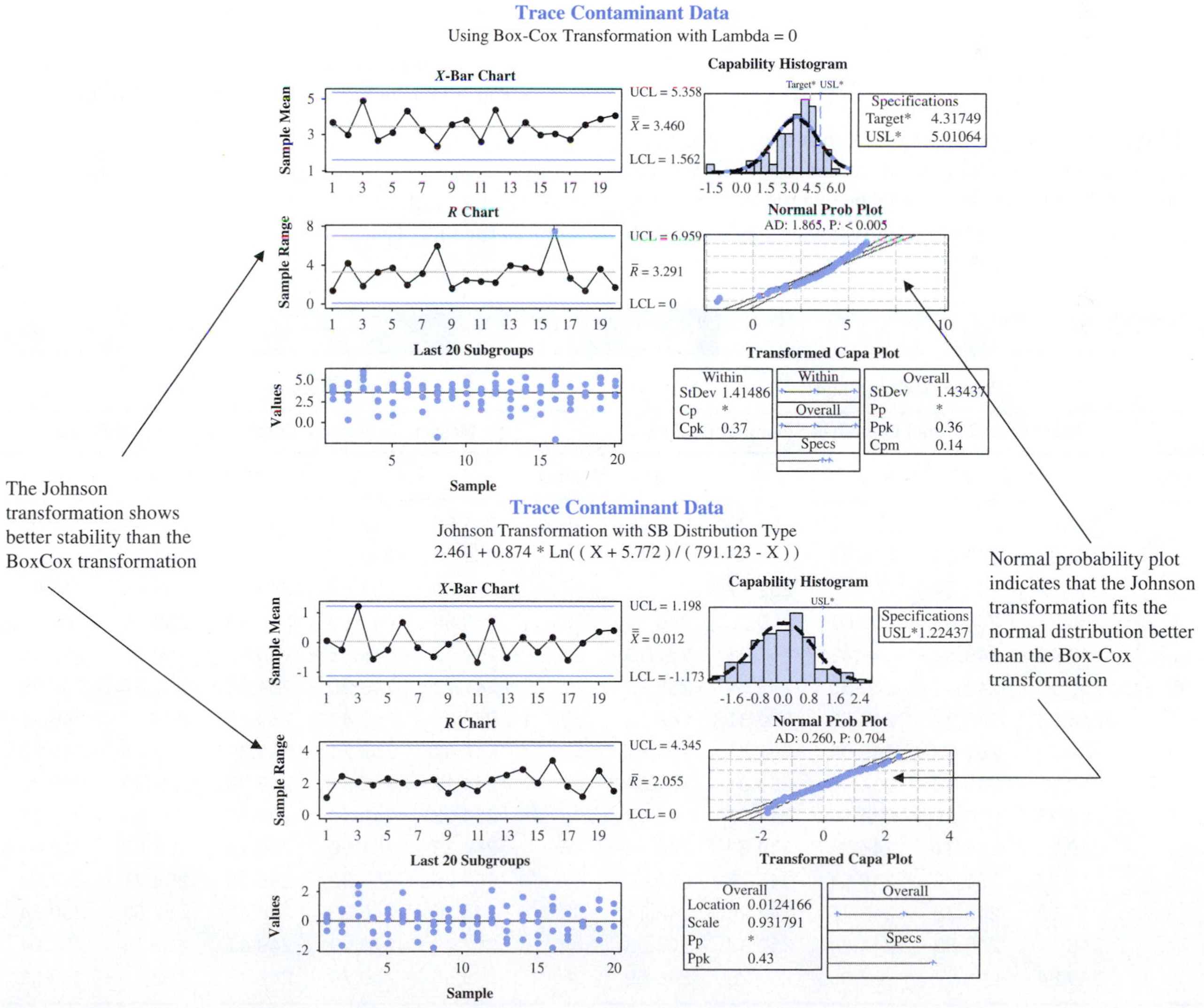

FIGURE 9-35. Comparison of Box-Cox and Johnson Transformations for Trace Contaminant Data

As the name suggests the Sixpack produces six displays as follows: (1) $\overline{X}$ control chart, (2) R control chart, (3) a plot showing all individual measurements by subgroup, (4) a histogram of the transformed data with the normal curve and specification limits superimposed, (5) a normal probability plot showing a graphical goodness-of-fit, and (6) a capability plot. A comparison of the two normal probability plots in Figure 9-35 suggests that in this case the Johnson transformation provides a better fit to the normal distribution than the Box-Cox transformation. Also, a comparison of the control charts suggests that the Johnson transformation is more stable than the Box-Cox. However, the Johnson transformation is more complex than Box-Cox and therefore more difficult to readily transpose the raw data and then to convert the transposed data back into the original data units.

9.4 USE OF CONTROL CHARTS TO ISOLATE SOURCES OF VARIATION

In Section 9.2.3 we illustrated how different sampling schemes, applied to the same data set, can result in dramatically different control chart patterns. Control charts, used in conjunction with creative sampling, can be an effective first line of defense in identifying sources of variation and improvement opportunities.

Consider the data shown in Table 9-11. They were collected from a bottling plant that uses two filling lines to place liquid product in bottles. The specifications for the fill volume are 250 ml ± 10 ml. Samples were taken by randomly selecting and measuring the contents of five bottles at the packaging line. One such sample was taken at the end of each production shift. The control charts in Figure 9-36 show a good state of control with the average fill volume close to target (perhaps slightly low). Performing a capability analysis reveals the following.

$$\sigma_{process} = \frac{\overline{R}}{d_2} = \frac{7.69}{2.326} = 3.31$$

$$ET = USL - LSL = 260 - 240 = 20$$

$$NT = 6\sigma_{process} = 6(3.31) = 19.86$$

$$C_p = \frac{ET}{NT} = \frac{20}{19.86} = 1.01 \Rightarrow \text{barely capable}$$

$$C_r = \frac{1}{C_p} = \frac{1}{1.01} = 0.990$$

$$C_{pk} = \text{minimum}\,\{C_{pu}, C_{pl}\}$$

where,

$$C_{pu} = \frac{|USL - \overline{\overline{X}}|}{3\sigma_{process}} = \frac{|260 - 249.825|}{3(3.31)} = 1.025$$

$$C_{pu} = \frac{|LSL - \overline{\overline{X}}|}{3\sigma_{process}} = \frac{|240 - 249.825|}{3(3.31)} = 0.989$$

$C_{pk} = \text{minimum}\,\{1.025, 0.989\} = 0.989 \Rightarrow$ Process is borderline capable and is not likely to be meeting requirements all the time.

The capability analysis reveals that the process is borderline capable and is likely producing out-of-spec product some of the time. Calculating $z_u = 3.07$ and $z_l = 2.97$, we can

TABLE 9-11 Data Collected At Packaging of Combined Output

	PopFiz Bottling Plant						PopFiz Bottling Plant				
	Measurements from Combined Stream of Product						Measurements from Combined Stream of Product				
Sample Number	1	2	3	4	5	Sample Number	1	2	3	4	5
1	251.82	251.59	250.75	255.28	249.09	16	244.86	248.93	252.71	247.44	251.97
2	243.80	245.04	246.81	255.43	257.41	17	250.31	249.32	246.27	247.70	250.33
3	250.32	245.34	251.87	252.41	253.67	18	249.60	254.55	249.11	249.46	246.54
4	245.96	245.87	251.01	250.05	246.84	19	253.64	241.88	250.79	252.30	243.99
5	248.40	245.52	249.42	253.07	251.10	20	252.17	246.95	256.87	252.01	252.35
6	250.15	248.47	249.35	253.54	255.22	21	247.71	246.30	254.36	250.77	250.88
7	246.23	248.27	248.60	250.31	248.21	22	248.87	250.47	253.38	248.10	246.21
8	252.72	251.03	245.90	251.83	254.81	23	244.61	253.25	249.30	256.26	249.73
9	249.56	245.10	251.98	253.81	251.87	24	244.17	252.37	249.78	252.40	249.39
10	250.93	250.37	244.87	245.37	249.97	25	254.02	248.13	252.72	246.29	251.33
11	248.75	254.71	245.95	249.67	244.48	26	254.33	251.60	249.13	252.62	248.96
12	246.72	250.29	248.04	246.78	254.66	27	254.14	251.32	246.90	253.71	247.98
13	255.62	246.78	243.76	247.26	248.18	28	246.76	246.32	248.45	252.68	250.34
14	248.14	248.88	244.84	250.93	253.06	29	252.29	248.02	249.41	252.14	252.24
15	249.84	246.87	252.95	250.28	246.59	30	249.55	250.74	249.80	252.09	249.02

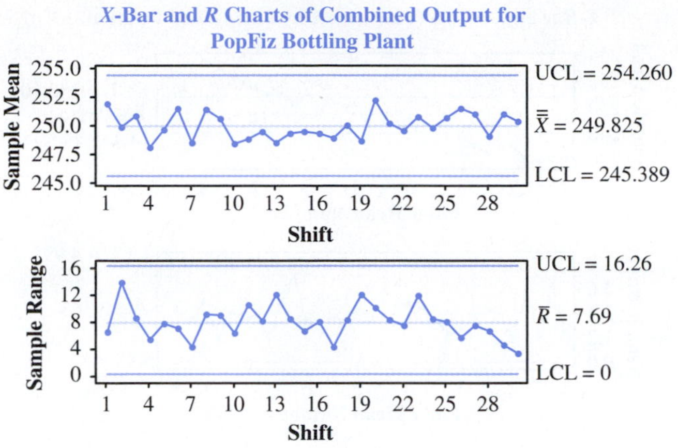

FIGURE 9-36. Control Charts for Fill Volume—Combined Stream

TABLE 9-12 Data Collected from Line 1 Output

PopFiz Bottling Plant

Sample Number	Measurements from Line 1				
	1	2	3	4	5
1	245.75	252.17	247.84	249.08	249.31
2	244.58	253.22	248.04	248.36	250.44
3	246.53	246.52	248.39	249.02	249.96
4	246.96	253.70	254.22	247.33	250.88
5	245.71	246.08	247.41	247.82	253.02
6	244.88	246.05	248.03	248.83	249.38
7	250.60	249.36	247.91	254.06	250.33
8	248.23	247.48	252.73	251.52	252.16
9	245.83	246.78	247.57	249.25	253.44
10	245.89	252.08	250.36	252.57	249.57
11	252.08	250.36	252.57	248.84	250.55
12	246.41	247.91	246.30	251.65	252.44
13	245.02	247.71	249.32	250.37	248.97
14	245.14	250.63	250.65	252.38	250.90
15	246.24	247.03	249.86	249.76	250.64

consult Tables II and III in Appendix B and estimate that 0.107% (1070 DPMO) of the production is overfilled and 0.1489% (1,489 DPMO) is underfilled.

Improvement opportunities can be identified by sampling in different ways so that targeted sources of variation are systematically captured in, or excluded from, the range chart. Tables 9-12 and 9-13 represent data that were collected at the ends of bottling lines 1 and 2, respectively. The control charts for these data are shown in Figure 9-37.

Comparing the two pairs of charts it appears that both lines are centered close to the target of 250 ml, perhaps a little on the low side. It also appears that line 1 may be less variable than line 2, as evidenced by the differences in the centerlines on the respective range charts. However, this observation

TABLE 9-13 Data Collected from Line 2 Output

PopFiz Bottling Plant

Sample Number	Measurements from Line 2				
	1	2	3	4	5
1	251.46	246.47	253.20	249.46	249.69
2	251.37	249.20	241.81	250.25	247.29
3	247.32	252.36	254.33	250.73	244.52
4	256.84	251.79	250.85	246.33	247.86
5	246.44	247.56	249.58	250.13	249.87
6	257.43	245.72	249.77	250.77	249.82
7	253.31	240.90	251.78	250.32	243.66
8	246.98	252.18	251.36	248.29	249.46
9	247.97	257.04	251.17	254.25	248.63
10	248.71	249.67	244.80	243.99	250.38
11	248.83	246.81	254.32	257.37	257.90
12	250.23	246.22	248.03	246.56	251.18
13	248.83	245.30	258.13	257.01	243.71
14	250.96	250.73	245.48	249.12	250.83
15	253.42	245.82	248.38	253.98	250.13

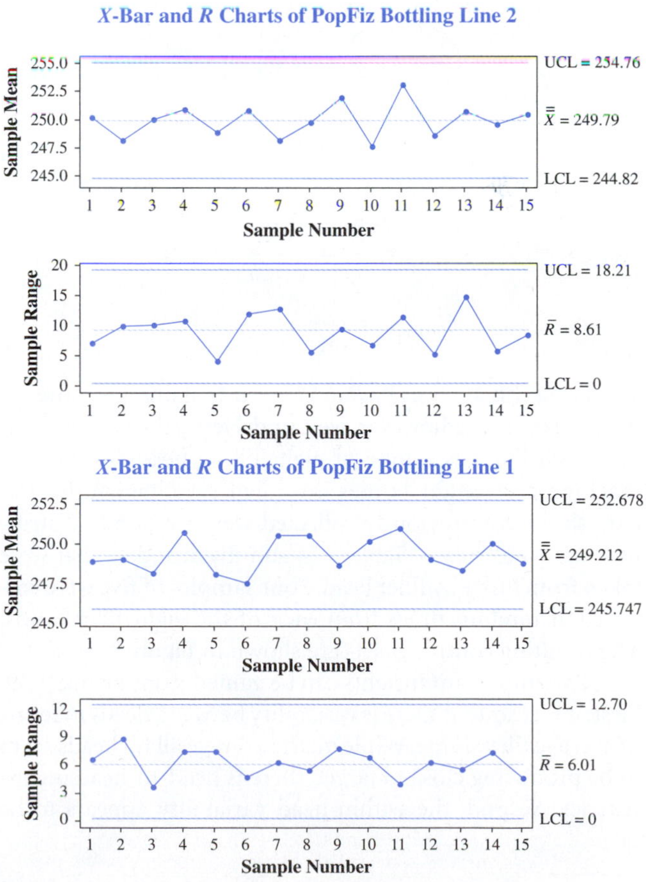

FIGURE 9-37. Comparison of Control Charts by Line

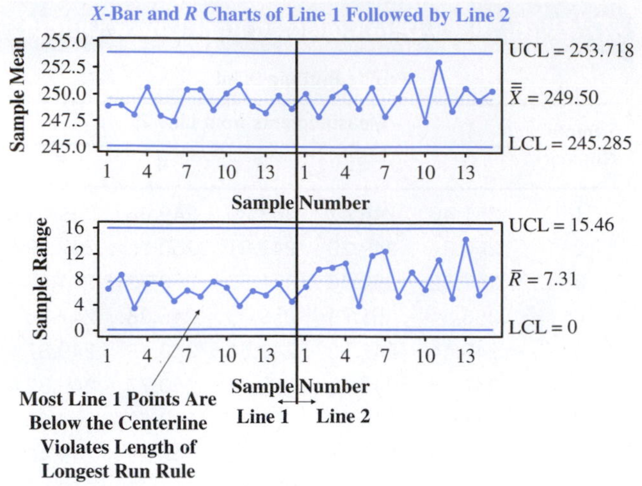

FIGURE 9-38. Line 1 Data Followed By Line 2 Data

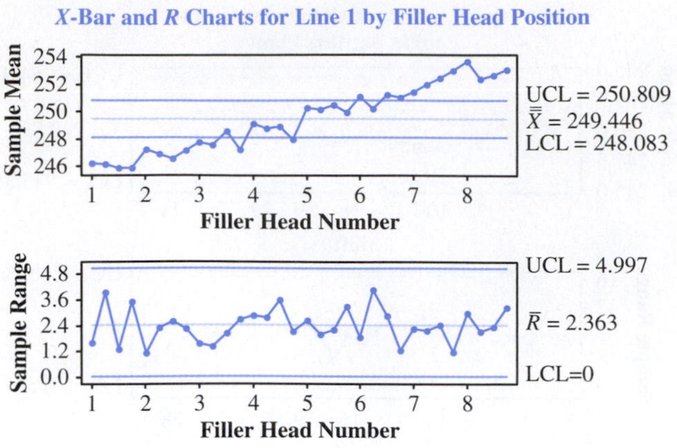

FIGURE 9-39. Line 1 Data Collected by Filler Head

is not sufficient to know if the *variances* are statistically different. We can address the issue of statistical difference by combining the two lines on a single chart as in Figure 9-38. The points are plotted in a way that will highlight any stratification. All of line 1 points have been plotted consecutively, followed by the line 2 data. We can observe that most of the line 1 points are below the range chart centerline, and the length-of-the-longest-run rule is violated.[8] Acting on this observation we can perform a quick estimate of the relevant standard deviations.

$$\sigma_{Line\ 1} = \frac{\bar{R}}{d_2} = \frac{6.01}{2.326} = 2.58$$

$$C_{p(Line\ 1)} = \frac{20}{6(2.58)} = 1.29$$

and

$$\sigma_{Line\ 2} = \frac{\bar{R}}{d_2} = \frac{8.61}{2.326} = 3.70$$

$$C_{p(Line\ 2)} = \frac{20}{6(3.70)} = 0.90$$

From this analysis it would appear that line 1 is more capable than line 2; however, we can delve further into this issue. Each line has a total of eight filling heads so on each machine cycle eight bottles are filled simultaneously. The data that were previously collected were randomly distributed across all heads. Table 9-14 summarizes data that were taken from line 1 by filler head. Four samples of five were collected at random times from each of the eight filler heads. The resulting control charts are shown in Figure 9-39.

Two important insights can be gained from Figure 9-39. First, it is clear that there is variability between heads in terms of average fill volume. While on the average all fill heads seem to be producing close to target, there is head-to-head inconsistency. Second, the within-head variability appears to be consistent as evidenced by an in-control range chart. The within-head variability for all heads can then be estimated as follows:

$$\sigma_{within-head-Line\ 1} = \frac{\bar{R}}{d_2} = \frac{2.363}{2.326} = 1.016$$

$$C_{p(Line\ 1)} = \frac{20}{6(1.016)} = 3.28$$

Hence, the opportunity exists to improve the process from one that is borderline capable to one that is Six Sigma capable by correcting the average fill volumes of each filler head and making them all the same. Further data collected by head are summarized in Figures 9-40a and 9-40b.

Centering the filler heads on the nominal dimension could potentially achieve the following quality level,

$$C_{pk} = \frac{|260 - 250|}{3(1.016)} = 3.28$$

9.5 SUMMARY OF KEY IDEAS

Key Idea: *Measurement is a process that parallels the production process from which measurements are taken.*

The effective control of quality is dependent on the measurements that are available for analysis. Measurements are not the truth but rather a representation of the truth. How far a measure deviates from reality is determined by the measurement error that is inherent in the measurement process, and consists of components that relate to bias and precision. Before analyzing a production process it is imperative that a good understanding of measurement error and the consistency of the measurement process be obtained. This can be achieved using many of the same tools that are applied to production operations.

Key Idea: *A good measurement system is one that can discriminate adequately between items of output and has a relatively low measurement error; measurement error consists of bias, repeatability, and reproducibility errors.*

[8] With $r = 14$ and $s = 11$, Table 9-12 provides a critical value for longest run of 6. The longest run in Figure 9-38 is 7.

TABLE 9-14 Data Collected From Line 1 By Filler Head

PopFiz Bottling Plant

Filler Head	Sample Number	Measurements from Line 1				
		1	2	3	4	5
1	1	245.75	245.69	247.25	245.95	246.20
	2	244.58	245.57	246.45	248.52	245.34
	3	246.53	245.52	245.28	245.29	246.48
	4	246.96	243.95	245.57	247.42	245.31
2	1	247.59	247.47	246.59	247.70	246.77
	2	247.87	245.60	247.52	247.40	245.74
	3	246.52	246.14	245.56	246.22	248.14
	4	246.31	247.07	247.26	248.52	246.36
3	1	246.90	247.23	248.31	248.44	247.84
	2	247.70	246.66	248.04	247.16	248.04
	3	249.24	247.25	248.72	249.18	248.39
	4	247.23	247.41	246.01	246.71	248.65
4	1	249.08	247.93	250.75	248.17	249.46
	2	248.36	248.41	248.24	248.04	250.79
	3	249.02	249.86	246.38	249.21	249.97
	4	247.33	248.83	247.88	246.76	248.80
5	1	249.31	250.01	248.99	251.36	251.59
	2	250.44	249.81	249.17	250.18	251.14
	3	249.96	249.60	250.32	251.79	250.77
	4	250.88	249.21	251.06	250.62	247.81
6	1	251.04	251.86	250.02	251.44	250.93
	2	248.28	249.53	249.25	251.59	252.36
	3	249.39	252.24	251.13	251.91	251.43
	4	251.22	251.53	250.71	250.32	251.48
7	1	252.17	251.74	249.97	252.06	251.37
	2	250.85	250.91	252.18	253.00	252.93
	3	253.08	252.93	251.31	251.27	253.66
	4	253.70	252.56	252.62	253.08	252.98
8	1	252.79	252.95	255.76	253.23	253.76
	2	253.22	252.81	251.14	252.66	252.04
	3	251.53	252.11	253.83	252.77	253.05
	4	254.22	251.26	252.03	253.44	254.50

Discrimination is the ability to differentiate between dissimilar units of output and is affected by measurement variation and also the resolution of the particular measuring device being used. Resolution refers to the smallest quantity unit that can be measured. Low resolution usually results in poor discrimination. Measurement variation consists of the summation of variances due to bias, repeatability, and reproducibility. A process is biased if it consistently delivers an average result that differs from the true result. A repeatability error is the amount of variation in results that one person experiences while taking multiple measurements with the same device. A reproducibility error is the variation that occurs due to different people measuring the same item, or the same person measuring an item using different measuring devices.

Key Idea: *For any production process, there is no unique sampling scheme; a sampling strategy should be designed to address specific issues or answer a predetermined set of questions.*

Process data are collected for one of three fundamental purposes: to characterize an output distribution, to monitor and control process behavior, or for process improvement. The sampling will differ depending on the purpose. When taking random samples there will always be a natural separation between short-term sources of variability (those

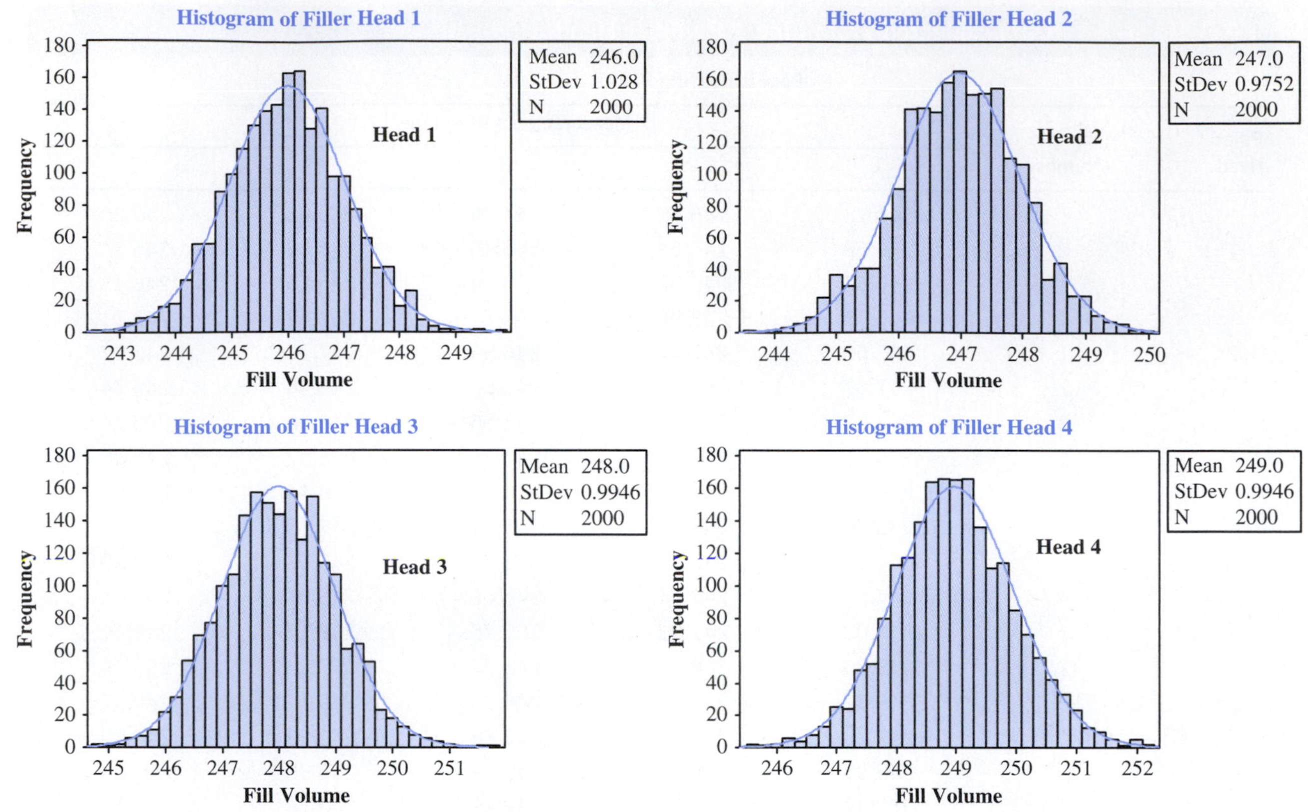

FIGURE 9-40a. Histograms of Fill Volumes—Heads 1 to 4

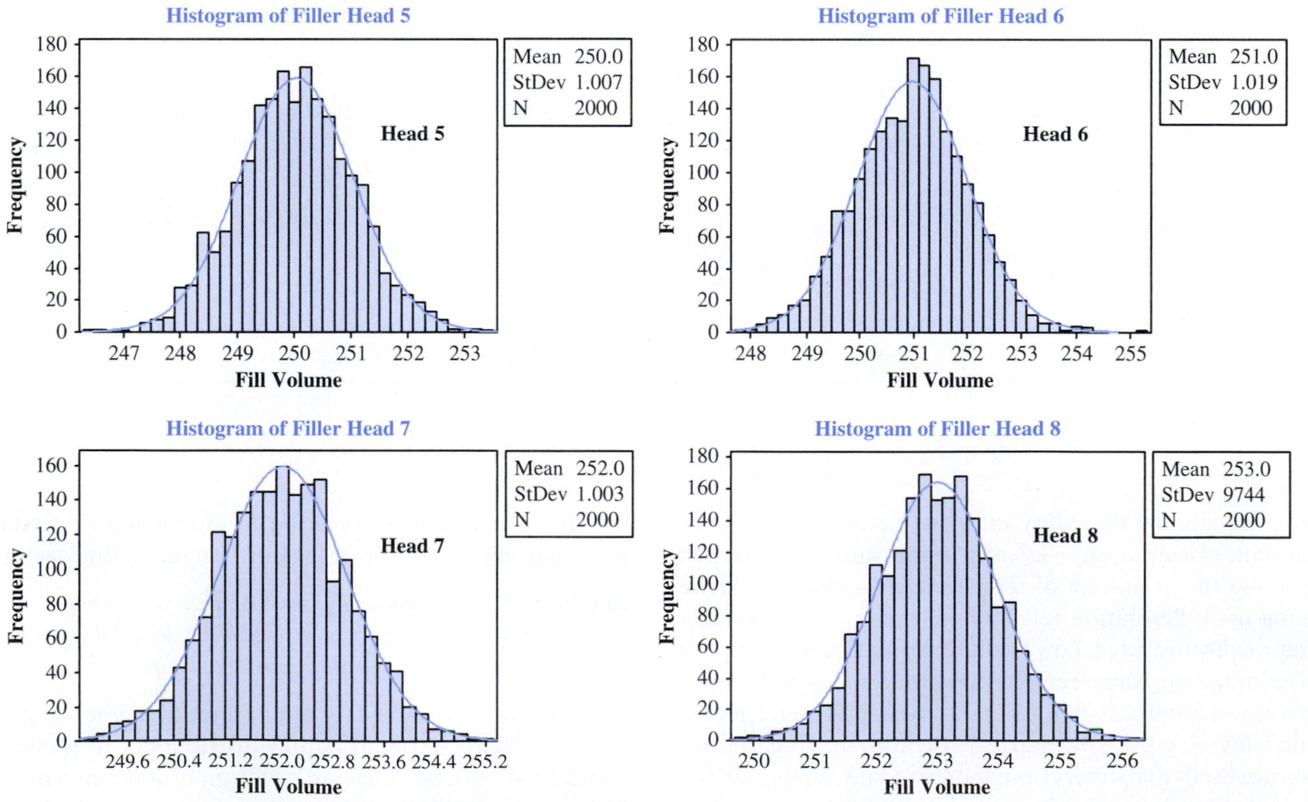

FIGURE 9-40b. Histograms of Fill Volumes—Heads 5 to 8

SHOWCASE OF EXCELLENCE **Winners of the Malcolm Baldrige National Quality Award Mixed Categories**

- 2007: City of Coral Springs (Not-For-Profit), a city in Broward County of southern Florida with a council-manager form of government.
  http://www.coralsprings.org, http://www.quality.nist.gov/PDF_files/Coral_Springs_Profile.pdf.
- 2007: U.S. Army Armament Research, Development, and Engineering Center (Not-For-Profit), one of the specialized research, development, and engineering centers within the U.S. Army Material Command.
  http://www.pica.army.mil, http://www.quality.nist.gov/PDF_files/ARDEC_Profile.pdf.
- 2001: Chugach School District (Education), a school district in south central Alaska providing programs ranging from preschool to postsecondary education.
  http://www.chugachschools.com, http://www.nist.gov/public_affairs/chugach.htm.
- 2001: Pearl River School District (Education), provider of K–12 education for 2,500 students 20 miles north of New York City.
  http://www.pearlriver.k12.ny.us, http://www.nist.gov/public_affairs/peralriver.htm.
- 2001: University of Wisconsin–Stout (Education), one of 13 publicly supported universities in the University of Wisconsin System with approximately 8,000 students.
  http://www.uwstout.edu/mba,http://www.nist.gov/public_affairs/uwstout.htm.

captured within a sample) and long-term sources (those that show up between samples). The design of a sampling plan will generally define the common cause system of variation to be those causes which are captured within each sample. In reducing variation, the actions required to eliminate common causes can be quite different from those intended to remove special causes.

Key Idea: *When two-sided specifications are involved, a capable process is one that can be centered on target, is in statistical control, and has a natural tolerance that is less than the engineering tolerance.*

Process capability is important in assessing whether design specifications can and are being met. A process is inherently capable if the natural spread, defined by $6\sigma_{process}$, is less than the engineering design tolerance. However, a process can be inherently capable and still produce defective output if the process is improperly centered or if the process is unstable. Numerous capability indices are available to provide information concerning capability.

Key Idea: *When one-sided specifications are involved a capable process is one that can be targeted and sustained in a way that will not exceed the specification limit most of the time.*

In the case of a single specification limit the concept of an engineering tolerance does not apply. In this case capability is generally defined as the ability to target, or center, a process at a distance in excess of three standard deviations from the specification limit of interest, and to be able to sustain this performance over time. This requires statistical control of the process mean and standard deviation.

Key Idea: *Control charts can be effective tools to provide useful information concerning sources of variation and improvement opportunities.*

Control charts, used in conjunction with creative sampling, can be an effective first line of defense in identifying sources of variation and improvement opportunities. By sampling in ways that deliberately stratify data according to suspected sources of variation—such as by machine, operator, shift, and so on—out-of-control signals will often expose improvement opportunities.

9.6 MIND EXPANDERS TO TEST YOUR UNDERSTANDING

9-1. The specifications on a certain bagged product are set at 50.0 lb ± 0.5 lb. The control charts below have been constructed for this process. The centerlines on the range and averages charts are 0.3175 and 50.2798.

a. What percentage of the bags produced by this process will be over weight? What percentage will be under weight?

b. Calculate the C_p and C_{pk} indices for this process. Is the process capable of meeting these specification limits? Explain.

c. Assume that production management has set 49.8 lb as the target weight. Calculate the C_{pm} and C_{pmk} indices. Is the process capable of achieving this target and still meet the specification limits? Explain.

d. If the average increases to 50.3 lb, what is the probability of getting an out-of-control signal on the next sample after the change?

e. At the increased average, what percentage of the bags produced will be overweight? What percentage will be underweight?

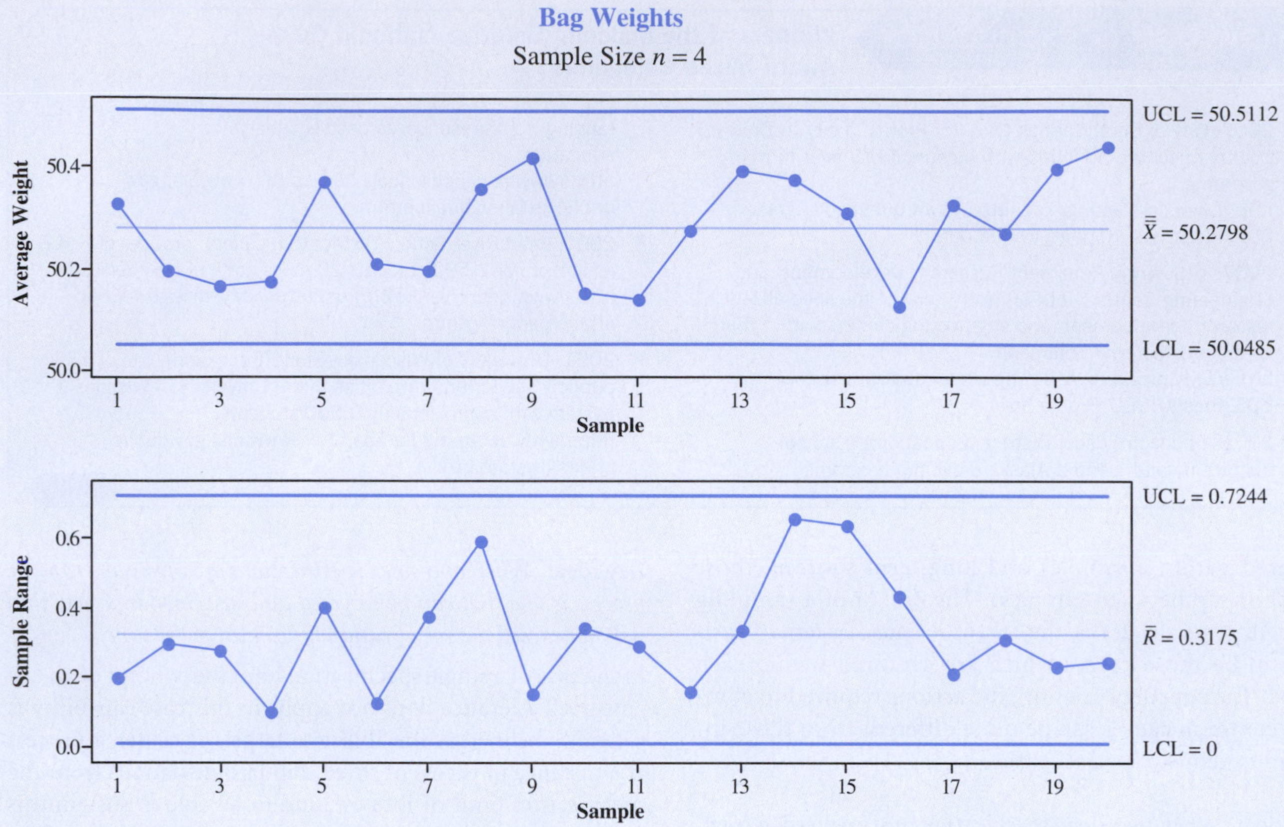

9-2. A beverage plant has a filling line that places product in 20 oz bottles. Consumers expect each bottle to contain exactly 20 oz of product as stated on the label; however, the plant is within the legal requirements if the output is within the range 19–21 oz. Fill volumes are monitored using random samples of five bottles, selected once each hour, and plotted on $\bar{X}$ and R charts. Both charts show good control. The following capability indices have been computed: $C_p = 1.23$ and $C_{pk} = 0.69$.

a. What is the average overfill (in excess of 20 oz) for this process? Estimate the percentage of bottles that are above the label volume.

b. What are the centerline and control limits on the R chart?

c. What are the centerline and control limits on the $\bar{X}$ chart?

d. If the process is centered on a target of 20 oz, by what percentage does the fill variance have to be reduced in order to achieve a $C_{pk} = 1.5$? How would you recommend that the plant manager move forward to accomplish this goal?

9-3. A dial plug gauge is used to measure the diameter of drilled holes that have a design tolerance of 10 mm ± 0.1 mm. To check calibration the quality control technician uses it to measure a certified 10 mm standard at the beginning of each shift. The table below summarizes the technician's measurements of this standard over 30 days. On the production line the practice has been to maximize the opportunity for rework by targeting to the low side of the specification range—9.98 mm.

Day	Measurement	Day	Measurement
1	9.9992	16	9.9996
2	10.0036	17	10.0008
3	10.0043	18	10.0035
4	10.0020	19	10.0026
5	10.0039	20	10.0016
6	9.9987	21	10.0053
7	10.0026	22	10.0052
8	10.0040	23	10.0039
9	10.0047	24	10.0053
10	10.0025	25	10.0028
11	10.0058	26	10.0041
12	10.0055	27	10.0066
13	10.0021	28	10.0034
14	10.0009	29	10.0034
15	10.0030	30	10.0056

a. Use X and two-point MR charts to determine the control status of the use of this gauge and if the gauge needs to be calibrated.

b. Assuming that a single technician took all the measurements, estimate the repeatability error.

c. Assume the gauge was calibrated and used to produce the production control charts shown below. Compute the signal-to-noise ratio and comment on the adequacy of this gauge for measuring these 10 mm holes.

***X*-Bar and *R* Charts for 10-mm-Diameter Holes**

Sample Size $n = 4$

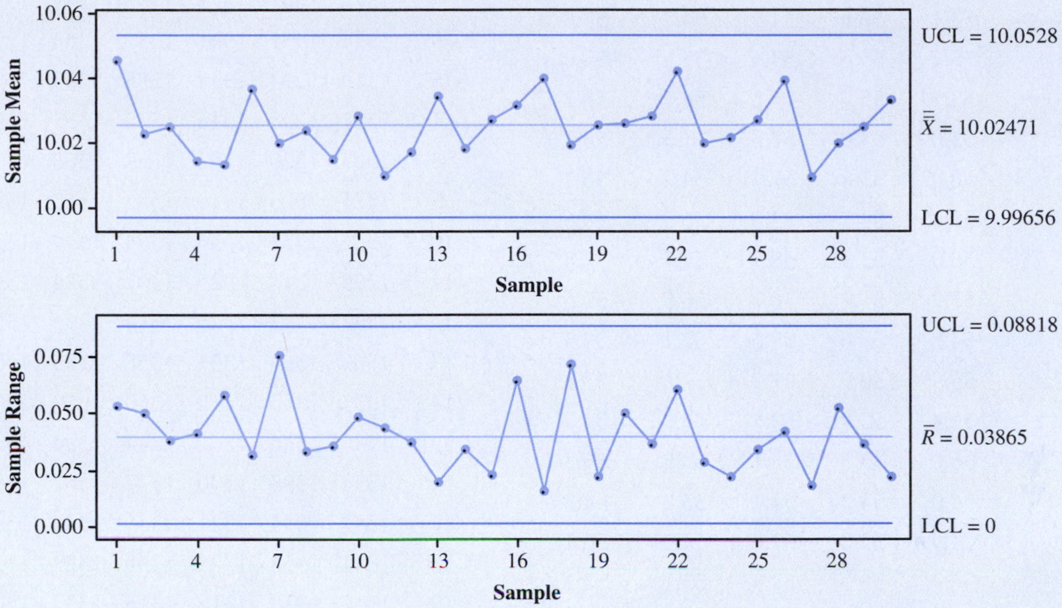

d. Compute the C_p and C_{pk} for this process and comment on its capability to meet the design specifications.

e. What percentage of the holes, if any, would you expect to be outside the specification limits?

f. What is your estimate of the number of defects per million opportunities for this drilling operation?

g. The control charts in part c were constructed from the sampling data shown in the following table. Using these data, compute the C_{pm} and P_{pk} indices. Comment on any similarities or differences you observe.

Sample	Measurement				Sample	Measurement			
No	1	2	3	4	No	1	2	3	4
1	10.0141	10.0508	10.0666	10.0492	16	10.0124	10.0445	10.0018	10.0655
2	10.0315	10.0369	9.9877	10.0325	17	10.0463	10.0436	10.0318	10.0355
3	10.0169	10.0452	10.0274	10.0080	18	10.0124	9.9841	10.0551	10.0238
4	10.0334	10.0187	9.9931	10.0095	19	10.0310	10.0131	10.0218	10.0344
5	10.0085	10.0054	9.9903	10.0473	20	10.0033	10.0333	10.0123	10.0526
6	10.0176	10.0482	10.0443	10.0343	21	10.0157	10.0451	10.0409	10.0095
7	10.0516	10.0336	10.0161	9.9766	22	10.0060	10.0582	10.0356	10.0660
8	10.0087	10.0414	10.0097	10.0325	23	10.0024	10.0187	10.0255	10.0300
9	10.0209	9.9946	10.0291	10.0137	24	10.0346	10.0206	10.0132	10.0166
10	10.0300	10.0068	10.0197	10.0544	25	10.0390	10.0309	10.0306	10.0056
11	9.9897	9.9924	10.0325	10.0227	26	10.0509	10.0396	10.0115	10.0524
12	10.0076	10.0207	10.0004	10.0365	27	10.0140	10.0007	10.0015	10.0178
13	10.0356	10.0321	10.0242	10.0428	28	10.0339	10.0419	9.9904	10.0104
14	10.0276	10.0010	10.0076	10.0344	29	10.0157	10.0265	10.0098	10.0451
15	10.0187	10.0409	10.0248	10.0217	30	10.0442	10.0334	10.0298	10.0227

9-4. A consumer products manufacturer has collected the following 25 samples of a particular composite measure. The sample size was 5. The specifications for this product are 53 ± 5.5.

	Composite Reading				
Sample	1	2	3	4	5
1	53.1	56.6	54.9	54.7	57.8
2	56.0	53.6	53.6	55.2	55.0
3	57.5	55.6	54.8	54.7	58.2
4	53.0	53.0	54.3	51.3	56.7
5	56.0	55.8	57.7	54.2	56.2
6	55.4	54.7	52.8	56.5	54.6
7	56.0	52.9	58.9	54.7	52.8
8	53.2	51.2	56.7	56.5	57.6
9	54.8	55.0	58.0	57.1	52.3
10	55.9	50.9	57.7	57.6	57.1
11	56.4	52.9	53.7	56.0	55.7
12	56.0	55.5	55.1	53.8	55.3
13	54.0	53.2	54.1	53.3	54.8
14	54.9	52.8	57.6	54.1	56.1
15	56.7	55.1	53.8	53.1	57.1
16	53.6	53.0	55.9	57.0	52.5
17	57.3	55.1	54.0	52.8	53.6
18	58.0	53.8	56.5	55.9	56.5
19	55.8	54.7	55.0	55.3	54.3
20	53.2	55.1	57.4	53.4	58.6
21	56.1	55.4	58.7	54.1	52.4
22	55.3	55.2	53.1	56.4	55.2
23	56.0	55.3	57.5	53.5	50.2
24	55.0	54.9	55.0	53.3	54.8
25	55.4	55.3	56.2	56.2	57.2

a. Construct a pair of control charts to determine if this process is in control. Is it? Explain.

b. Is the process on target? Explain.

c. Is the process capable? Justify your answer by computing both the C_p and C_{pk} indices.

d. What immediate action would you recommend? What long-term action would you recommend?

9-5. Viscosity is one of the critical-to-quality characteristics for a slurry product at a plant that produces minerals for an industrial market. The target viscosity is 1,300, and the customer has agreed to accept viscosities in the range 1,100–1,500. To determine how well the process is performing an operator collects 6 samples per shift and sends the material to a quality control lab where viscosity is determined. The data from 20 shifts are shown in the table below.

Sample	Measured Specimen							
NO.	1	2	3	4	5	6	$\overline{X}$	R
1	1340	1316	1329	1343	1324	1297	1324.8	46
2	1325	1305	1322	1335	1323	1308	1319.7	30
3	1346	1326	1298	1295	1329	1307	1316.8	51
4	1376	1361	1373	1338	1349	1380	1362.8	42
5	1324	1351	1341	1337	1284	1298	1372.5	67
6	1315	1293	1319	1388	1393	1382	1348.3	100
7	1321	1298	1316	1332	1353	1322	1323.7	55
8	1305	1320	1302	1337	1288	1315	1311.2	49
9	1321	1326	1310	1333	1313	1329	1322.0	23
10	1357	1357	1356	1354	1339	1312	1345.8	45
11	1295	1318	1325	1302	1324	1340	1317.3	45
12	1357	1312	1345	1315	1295	1350	1329.0	62
13	1355	1363	1351	1358	1335	1309	1345.2	54
14	1335	1370	1372	1360	1359	1307	1350.5	65
15	1344	1346	1361	1344	1394	1372	1360.2	50
16	1354	1395	1390	1377	1362	1368	1374.3	41
17	1347	1344	1321	1330	1335	1366	1340.5	45
18	1341	1358	1328	1335	1329	1326	1336.2	32
19	1317	1310	1313	1316	1333	1306	1315.8	27
20	1297	1312	1316	1354	1358	1313	1325.0	61

a. What sources of variability are captured in the R and $\overline{X}$ charts, respectively, using this method of sampling?

b. Is this process in control? What would you recommend to management?

c. The plant manager made improvement of this process a priority. A team was assembled with representatives from process engineering, maintenance, and operators from each of the three shifts. A cause-and-effect diagram was constructed and

Sample No.	$\overline{X}$	R	Sample No.	$\overline{X}$	R
21	1310.0	45	31	1309.8	17
22	1312.0	33	32	1301.0	25
23	1293.2	40	33	1288.5	32
24	1288.8	14	34	1305.7	35
25	1302.3	12	35	1308.8	20
26	1305.8	20	36	1296.0	15
27	1308.5	35	37	1302.0	35
28	1295.0	47	38	1294.2	40
29	1300.2	30	39	1306.8	25
30	1301.2	22	40	1294.0	15

several possible causes were tagged for investigation. A problem with a valve in the volumetric feeder for the CMC additive was discovered. At times, intermittently and unpredictably, the valve would stick open and the quantity of cellulose (a viscosity builder) released into the mixing tank was excessive. The feeder was repaired by replacing a part, and then the new data shown in the table at the bottom of page 336 were collected. The same sampling plan was used as before.

Did the replacement valve fix the problem? Is the process capable? Is the process on target? Is the process producing to specifications? Use control charts and capability indices to justify your answers.

9-6. There are two lab technicians that the plant in problem 9-5 employs to perform the viscosity tests. Sue has a total of 20 years' experience and works the day shift. Bob, who has been performing the tests for only 2 years, covers the night shift. For a period of 30 days, each of these technicians has taken a sample from a pseudostandard batch, split it, and measured the viscosity for each half. Their results are shown in the table.

	Sue			Bob	
Day	**Meas. 1**	**Meas. 2**	**Day**	**Meas. 1**	**Meas. 2**
1	1299	1301	1	1302	1298
2	1301	1299	2	1300	1299
3	1300	1298	3	1299	1300
4	1301	1301	4	1301	1296
5	1297	1300	5	1301	1300
6	1299	1298	6	1300	1299
7	1300	1299	7	1298	1301
8	1300	1301	8	1300	1301
9	1300	1297	9	1300	1297
10	1301	1301	10	1302	1300
11	1302	1300	11	1300	1300
12	1300	1299	12	1300	1302
13	1300	1301	13	1299	1299
14	1301	1301	14	1302	1300
15	1301	1300	15	1295	1300
16	1299	1298	16	1302	1299
17	1298	1302	17	1300	1299
18	1301	1302	18	1301	1298
19	1300	1301	19	1298	1299
20	1300	1300	20	1299	1301
21	1299	1298	21	1300	1302
22	1299	1300	22	1300	1301
23	1299	1300	23	1298	1301
24	1300	1300	24	1299	1300
25	1300	1300	25	1301	1299
26	1301	1300	26	1300	1301
27	1301	1302	27	1295	1295
28	1300	1298	28	1300	1300
29	1299	1300	29	1302	1297
30	1300	1299	30	1300	1299

a. Compute the measurement error. What proportion is due to reproducibility? What proportion is due to repeatability?

b. Compute the signal-to-noise ratio.

c. Compute the discrimination ratio.

d. Is measurement error a concern for this process? Explain.

9-7. An alternative measurement study was conducted on the two technicians described in problem 9-6. Sample material was collected from 15 production batches and split into four portions. Each was coded and divided up between Bob and Sue so that each inspector received exactly two portions from each batch (although they had no way of identifying each sample with a batch number). They were then asked to measure the viscosity of the 30 samples they had been given and to record the results. Using the code numbers the sample results were then reorganized as shown in the table.

a. Perform a gauge R&R analysis to estimate the total variation (process variation plus measurement error).

b. What proportion of the total variation is due to repeatability error?

c. What proportion of the total variability is due to reproducibility error?

d. Is this measurement process adequate? Explain.

9-8. For this classroom measurement study, the class is organized into small groups with four to six students assigned to each group. Then, each group identifies some object that each member of the group will measure and the measuring device and resolution to be used. For example, each member of group A could contribute a ring and the group task would be to measure the inside ring diameters using a set of inside vernier calipers measured to the nearest 1,000th of an inch; group B could be asked to each donate a wristwatch

Production Batch	Bob		Sue	
	1	2	1	2
1	1310	1311	1306	1309
2	1308	1309	1305	1308
3	1321	1320	1315	1318
4	1321	1321	1318	1322
5	1301	1303	1299	1303
6	1290	1292	1292	1294
7	1301	1301	1300	1303
8	1288	1289	1287	1290
9	1304	1305	1300	1304
10	1318	1319	1315	1320
11	1300	1301	1298	1301
12	1304	1303	1305	1300
13	1286	1286	1288	1286
14	1286	1284	1287	1286
15	1307	1307	1310	1305

and tasked to measure the thickness of the watch face using a micrometer to the nearest 10,000th of an inch; group C could be asked to each donate a pen (ballpoint or fountain) and tasked to measure the length using a vernier height gauge to the nearest 1,000th of an inch; and so on. Once the groups and their respective tasks have been defined the following data collection procedure will be followed.

Each member of the group measures the first in their collection (e.g., the firstring), carefully recording the item number, the person who measured it, and the measurement. This process should be repeated until one measurement has been made by each group member of the stipulated dimension of each item. Then the entire process is replicated twice, resulting in three measurements by each group member of each to the items in the group's collection. Following this procedure avoids bias that could arise if any one person were permitted to take three consecutive measurements of any of the items.

When the groups have finished collecting their data each group should have $3m$ total raw data points, where $m =$ the number of members of the group. Each group will then be instructed to organize the data on a spreadsheet, use Minitab to conduct a gauge R&R analysis, and answer the following questions.

a. What is the total observed variation of the process that the group was tasked to measure?

b. What percentage of the total observed variation is due to measurement?

c. What percentage of the total observed variation is due to replication (repeatability) error?

d. What percentage of the total observed variation is due to a reproducibility error?

e. What is the signal-to-noise ratio for the group's particular process?

f. Based on the signal-to-noise ratio, is the measurement process that was used adequate? Explain.

g. What is the discrimination ratio for the group's particular process?

h. Based on the discrimination ratio, is the measurement process that was used adequate? Explain.

9-9. Krypton Ltd. manufactures silicon wafers for the semiconductor industry. The business is highly competitive with pressure from customers to design wafers with more elements per surface area. There are two ways that this can be done: larger diameter wafers and thinner lines on the wafer surface. The industry standard has been 200 mm diameters, but there is a migration by many in the industry to move toward a larger 300 mm. Such a move requires major capital and retooling.

The other alternative, thinner lines, is attractive to many customers because the chips (made from the wafers) can be smaller and the circuits operate at faster speeds, consume less energy, and generate less heat. These are attractive attributes for computers and many electronic products. The attribute, line width, is expressed in microns (μM). The goal is to produce a wafer with a line width of 0.11 μM.

Wafer production is a continuous process, and the production steps create many sources of variation. On the wafer itself the line width measure depends on where the measurement is taken. Wafers are produced in cassettes (25 in a batch) and the location within the cassette can influence the measure. In addition, cassettes may be different. The following data represent line width measurements that were taken from 30 wafers. Three wafers were selected at random from 10 cassettes (also selected at random). Each wafer was measured at five locations: top, left, center, right, and bottom. For ease in analysis the results have been recorded by multiplying the micron measure by 10.

Use control charts to determine the following:

a. The average of the overall process and its variability.

b. The proportion of the total variability that is due to location differences within the wafer.

c. The proportion of the total variability that is due to location differences with a cassette.

d. The proportion of the total variability that is due to different cassettes.

e. Assuming normality, what is the widest and narrowest line width that can be expected from this process?

Cassette No.	Wafer No.	Wafer Line Width, Measured in microns (μM) × 10				
		Top	Left	Center	Right	Bottom
1	1	3.20	2.25	2.07	2.41	2.39
	2	2.65	2.00	1.86	2.14	1.98
	3	2.89	2.06	1.63	2.30	2.33
2	4	3.16	2.52	2.07	2.29	2.12
	5	2.06	2.22	1.47	1.68	1.90
	6	2.35	2.17	1.54	1.97	2.25
3	7	2.20	1.73	1.36	1.67	1.43
	8	2.23	1.56	1.52	2.07	1.78
	9	2.24	1.75	1.37	1.62	1.54
4	10	2.93	2.04	1.79	1.98	2.16
	11	2.86	2.10	1.92	2.02	2.23
	12	3.23	2.90	2.17	3.04	3.19
5	13	3.05	2.51	1.95	2.47	2.58
	14	3.86	3.35	2.53	3.19	3.36
	15	3.69	3.40	2.96	2.95	3.47
6	16	2.94	2.53	1.94	2.77	2.38
	17	3.22	2.30	2.26	2.65	2.42
	18	3.18	2.85	1.60	2.81	2.90
7	19	2.17	2.03	1.67	1.66	2.31
	20	2.91	2.32	1.85	2.39	2.19
	21	2.20	3.32	2.70	1.96	2.51
8	22	1.96	1.36	0.97	1.95	1.64
	23	2.36	1.76	1.17	2.23	1.31
	24	2.42	1.99	1.40	2.00	2.14
9	25	2.19	2.29	1.70	1.93	2.06
	26	2.36	1.80	1.24	1.68	1.85
	27	2.01	1.52	0.79	2.00	1.35
10	28	2.83	2.50	1.94	2.35	2.31
	29	3.07	2.06	1.79	1.86	1.96
	30	3.07	2.29	1.87	2.48	2.02

9-10. A certain machining process is being checked to determine its capability to achieve the required tolerance on a part dimension that has specifications of 22 mm ± 0.05 mm. No control chart information is available. A total of 50 parts were run off, measured, and recorded as shown in the following table.

Measurements (mm)									
Part	Meas.	Part	Meas.	Part	Meas.	Part	Meas.	Part	Meas.
1	21.9531	11	21.9978	21	21.9668	31	21.9739	41	21.9895
2	21.9767	12	21.9258	22	21.9604	32	22.0224	42	21.9896
3	21.9987	13	21.9841	23	21.9759	33	21.9956	43	21.9899
4	21.9425	14	21.9653	24	22.0024	34	21.9908	44	21.9849
5	21.9607	15	21.9901	25	21.9913	35	21.9838	45	21.9903
6	21.9864	16	21.9725	26	21.9608	36	21.9867	46	21.9688
7	21.9718	17	21.9718	27	21.9980	37	22.0026	47	21.9898
8	21.9812	18	21.9556	28	22.0286	38	21.9572	48	21.9478
9	22.0073	19	22.0017	29	21.9791	39	21.9594	49	21.9612
10	21.9664	20	21.9756	30	21.9849	40	21.9687	50	21.9543

a. Compute the P_{pk} index for these data and comment on whether the process is capable of meeting the specifications.

b. Using the data in part a, estimate the natural tolerance of the process and comment on whether the natural process spread exceeds the engineering tolerance.

The machine in question was used to complete a production run of 5,000 parts. During the run 30 samples of size 4 were collected and documented in the table below.

c. Construct $\overline{X}$ and R charts for this process. Is the process in control? Explain.

d. Using the data in part c, estimate the process mean and standard deviation if appropriate, and use them to estimate the natural tolerance of the process. How do these results compare to your answer in part b? How do you explain any differences?

e. Using your calculations in part d, compute the C_{pk} index. How does this compare with the P_{pk} index you computed in part a?

f. If this process is to be used for this part in the future, what would you suggest management focus on to improve it?

	Measurement					
Sample	1	2	3	4	$\overline{X}$	R
1	21.9960	22.0241	21.9750	21.9686	21.9909	0.0555
2	21.9886	21.9480	21.9429	21.9963	21.9689	0.0534
3	21.9590	21.9901	21.9959	21.9876	21.9832	0.0369
4	21.9813	21.9982	21.9766	22.0098	21.9915	0.0332
5	21.9794	21.9741	22.0120	21.9545	21.9800	0.0575
6	21.9883	21.9600	22.0042	21.9982	21.9877	0.0442
7	21.9655	21.9983	22.0199	21.9797	21.9908	0.0544
8	21.9464	21.9701	21.9635	21.9728	21.9632	0.0264
9	21.9950	22.0040	21.9815	21.9995	21.9950	0.0225
10	22.0048	21.9860	22.0003	21.9943	21.9964	0.0188
11	21.9941	21.9740	21.9638	21.9746	21.9766	0.0303
12	21.9788	21.9663	21.9594	21.9529	21.9643	0.0259
13	22.0263	21.9607	21.9896	22.0071	21.9959	0.0656
14	21.9556	21.9998	22.0000	21.9741	21.9824	0.0443
15	21.9817	21.9765	21.9869	21.9994	21.9861	0.0229
16	22.0117	21.9742	21.9569	21.9701	21.9782	0.0548
17	21.9808	21.9793	21.9814	21.9718	21.9783	0.0097
18	21.9889	21.9895	21.9453	21.9829	21.9766	0.0442
19	21.9703	21.9583	21.9852	21.9502	21.9660	0.0350
20	22.0026	21.9643	21.9804	22.0223	21.9924	0.0581
21	21.9542	21.9977	21.9950	21.9751	21.9805	0.0435
22	21.9540	21.9972	21.9963	21.9763	21.9810	0.0432
23	22.0074	21.9515	21.9969	21.9784	21.9835	0.0559
24	21.9833	21.9926	21.9772	22.0006	21.9884	0.0234
25	21.9905	22.0008	21.9930	21.9958	21.9950	0.0103
26	21.9571	21.9453	21.9795	21.9438	21.9564	0.0357
27	21.9943	21.9697	21.9687	21.9382	21.9677	0.0560
28	21.9764	21.9583	21.9974	22.0080	21.9850	0.0497
29	22.0091	21.9958	21.9641	21.9513	21.9801	0.0578
30	21.9730	21.9766	21.9703	21.9918	21.9779	0.0216

Sample	Flatness (μ in.)				
1	12.79	17.83	8.18	11.30	18.07
2	12.34	24.97	7.87	9.83	13.55
3	20.55	20.01	15.40	19.06	16.47
4	16.18	20.16	16.47	13.25	15.59
5	22.53	25.23	14.22	13.56	14.90
6	19.98	16.91	20.89	15.10	23.42
7	14.39	14.57	17.27	15.72	15.15
8	18.24	4.16	17.43	11.65	12.98
9	9.18	11.17	14.53	10.43	19.30
10	15.91	18.50	20.11	18.56	15.73
11	10.69	14.06	13.34	9.56	23.94
12	19.60	18.12	17.10	18.61	6.54
13	11.24	7.95	16.24	8.97	20.73
14	9.80	14.31	11.44	8.62	18.97
15	13.91	16.39	10.95	13.34	11.75
16	15.10	19.05	14.44	5.46	18.73
17	12.00	20.36	16.44	17.84	16.38
18	11.11	17.24	16.34	14.93	8.74
19	20.23	15.36	21.11	15.86	19.65
20	18.12	10.06	13.51	13.15	17.92
21	24.67	18.33	14.27	18.77	13.30
22	13.43	14.38	18.79	19.47	8.13
23	15.68	15.85	19.36	12.02	13.75
24	14.26	16.75	12.16	13.02	13.38
25	17.94	13.56	23.88	12.04	17.73
26	11.53	12.17	16.86	17.74	12.56
27	15.33	14.45	9.17	18.22	13.45
28	13.36	7.90	15.45	15.90	16.31
29	18.52	15.99	14.52	13.87	13.21
30	14.42	13.67	18.31	12.48	20.45

9-11. In a certain grinding operation a metal plate has a stipulated specification on flatness that is not to exceed 30 microinches (μ in.)[9]. The grinding process uses 10 μ in. as a target. Five plates were selected at random each hour over a 30-hour period and the measured flatness is recorded in the above table.

a. Is this process in a state of statistical control? Explain.

b. Use Minitab to verify that the raw data distribution is approximately normal.

c. Compute the C_{pk}, P_{pk}, and C_{pm} indices. Is this process capable of meeting the specifications? Explain.

d. Estimate the percentage of total plates produced that will exceed the stipulated specification. How many parts per million (PPM) is this?

e. What would you recommend management do to decrease the PPM that exceed the maximum allowable flatness?

9-12. The iron (Fe) content of steel is measured each hour in the Slagsburg Steel Works, Mill Number 4. The results of 30 hours of sampling are shown in the following table. The steel-making process has a one-sided specification limit for Metallic Fe that should not exceed 40%.

Sample No.	Percentage Metallic Fe Content	Sample No.	Percentage Metallic Fe Content
1	21.02	16	12.96
2	29.05	17	50.94
3	2.03	18	49.08
4	2.38	19	36.57
5	21.25	20	23.79
6	12.46	21	22.94
7	30.22	22	28.09
8	34.42	23	4.84
9	7.8	24	37.88
10	5.27	25	3.03
11	5.37	26	3.19
12	14.19	27	5.63
13	4.3	28	34.96
14	42.8	29	58.48
15	15.91	30	36.07

a. Perform a capability analysis on this process assuming normality. What is the C_{pk} index? What is your estimate of the percentage of output that will exceed the maximum of 40%? Using the Minitab capability sixpack facility and observing the histogram and normal probability plot, do the data appear to be normal? Explain.

b. Perform a capability analysis using a Box-Cox transformation and optimal λ. What is the C_{pk} index? What is your estimate of the percentage of output that will exceed the maximum of 40%? Using the Minitab capability sixpack facility and observing the histogram and normal probability plot, do the Box-Cox transformed data appear to be normal? Explain.

c. Perform a capability analysis using a Johnson transformation. What is the C_{pk} index? What is your estimate of the percentage of output that will exceed the maximum of 40%? Using the Minitab capability sixpack facility and observing the histogram and normal probability plot, do the Johnson transformed data appear to be normal? Explain.

d. Compare your solutions in parts a through c. Use your best judgment to answer the following questions.

i. Is the steel-making process at Mill 4 in statistical control with respect to metallic iron content?

ii. What percentage of the total steel-making production would you estimate is out of specification?

iii. How many parts per million are out of specification?

iv. What would you recommend management do to improve the performance of this process?

9-13. Mastic Pty. Ltd. is a company that produces mastics, caulks, and adhesives for the construction industry. Caulking products are blended in large mixers, pumped into individual tubes, and then capped. The production supervisor for the caulking brands is concerned about the filling process for a particular product—sealer 7098. This sealer has a stipulated label weight of 8 oz, and the production fill specifications are 8 oz ± 0.5 oz. The following data have been collected at random from the filling lines during the past week.

a. Is this process in a state of statistical control? Explain.

b. Use Minitab to verify that the raw data distribution is approximately normal.

c. Compute the C_{pk}, P_{pk}, and C_{pm} indices. Is this process capable of meeting the specifications? Explain.

[9]A microinch is one-millionth of an inch.

Sample	Caulk Tube Fill (oz) 1	2	3	4	$\overline{X}$	R
1	8.11	8.27	8.27	7.77	8.10	0.51
2	8.06	8.56	8.39	8.17	8.30	0.50
3	7.91	8.25	8.20	8.34	8.18	0.43
4	7.96	8.36	8.07	8.13	8.13	0.40
5	8.23	8.22	8.22	8.41	8.27	0.20
6	7.83	8.34	8.30	8.30	8.19	0.51
7	8.00	8.01	8.24	8.36	8.15	0.36
8	8.05	8.16	7.90	8.24	8.09	0.33
9	8.11	7.77	8.10	8.35	8.08	0.57
10	8.21	8.36	8.16	8.10	8.21	0.27
11	8.42	8.18	8.16	8.22	8.24	0.25
12	8.05	7.88	8.10	8.12	8.04	0.24
13	8.27	8.21	8.25	8.20	8.23	0.07
14	8.25	8.34	8.09	8.32	8.25	0.25
15	8.13	8.37	8.04	8.00	8.14	0.38
16	8.43	8.15	8.13	8.40	8.28	0.30
17	7.82	8.06	8.34	8.13	8.09	0.52
18	8.11	8.12	7.84	7.88	7.99	0.28
19	8.26	7.73	8.31	8.14	8.11	0.58
20	8.09	8.11	8.16	8.48	8.21	0.39
21	8.27	8.22	8.26	8.35	8.27	0.12
22	8.44	8.17	8.31	8.02	8.23	0.42
23	7.83	8.00	8.27	8.25	8.09	0.43
24	8.22	8.37	8.36	8.31	8.31	0.15
25	7.69	8.25	8.50	8.33	8.19	0.82

Sample	Caulk Tube Fill (oz) after Improvements 1	2	3	4	$\overline{X}$	R
26	8.00	8.00	8.07	7.99	8.02	0.08
27	8.07	8.02	8.11	7.93	8.03	0.18
28	7.99	8.03	8.07	8.01	8.03	0.09
29	7.99	7.99	8.02	7.98	8.00	0.03
30	8.15	8.04	8.05	8.10	8.08	0.12
31	7.94	7.99	8.01	7.89	7.96	0.12
32	7.83	8.13	8.18	8.00	8.04	0.35
33	8.12	8.03	8.08	7.86	8.02	0.26
34	8.01	8.19	7.96	8.02	8.04	0.22
35	8.03	8.00	7.92	8.01	7.99	0.10
36	7.98	8.11	8.23	7.98	8.08	0.25
37	7.96	8.02	7.96	7.98	7.98	0.06
38	7.95	8.01	8.11	7.98	8.01	0.15
39	7.91	7.79	7.82	8.08	7.90	0.29
40	7.97	7.90	8.04	8.20	8.03	0.31
41	7.99	8.07	7.95	8.19	8.05	0.24
42	7.86	8.08	7.96	7.93	7.96	0.23
43	8.08	8.10	8.00	7.98	8.04	0.12
44	7.99	8.04	7.98	8.21	8.06	0.23
45	8.00	7.96	7.97	7.86	7.95	0.14
46	7.81	8.06	7.92	7.91	7.93	0.25
47	8.10	8.08	7.95	7.94	8.02	0.16
48	7.91	8.15	7.89	7.96	7.98	0.26
49	8.13	8.00	7.92	8.00	8.01	0.20
50	7.92	7.90	7.97	7.80	7.89	0.16

d. Estimate the percentage of tubes filled that will exceed the stipulated specifications. How many parts per million is this?

e. If the material cost for this product is $0.20 per ounce and 10,000 tubes per day are produced, estimate the total cost of overfills.

f. Concerned by the results in part e, the production supervisor formed a kaizen blitz team and ordered a study to improve the filling process. Some improvements resulted and 25 more samples were taken from the process. These data are shown in the following table.

   i. By plotting these new data on the original control charts from part a can you conclude that the improvements have worked? Explain.

   ii. Repeat parts a through e using the postimprovement data. Estimate the total cost savings that have resulted from the improvement efforts.

9-14. Anchors Away, Inc. manufactures a variety of nautical anchors. One of its products is the Neptune-200, a 200-lb anchor that is popular among boating enthusiasts who own medium-sized speedboats or yachts. Due to shipping regulations the company must be certain the anchors produced are 200 lb and it is essential that the measurement process for controlling weights be adequate. The actual weight specifications for the Neptune-200 is that anchors must be within the tolerance range of 200 lbs ± 10 lbs. To determine if the scale used to test the anchors is correctly calibrated, Anchors Away had a quality inspector measure a known 200-lb standard weight twice 30 times throughout the day. The results are shown in the following table.

a. Construct $\overline{X}$ and R charts for the measurement study and answer the following questions.

   i. Is the measurement process in a state of statistical control?

	Measurement			Measurement	
Trial	A	B	Trial	A	B
1	199.109	199.213	16	197.732	197.257
2	200.799	199.565	17	200.884	200.924
3	197.089	199.955	18	199.775	200.205
4	198.817	197.734	19	199.575	199.682
5	202.126	199.219	20	197.582	201.684
6	201.969	200.073	21	199.556	200.309
7	199.773	200.316	22	200.792	200.900
8	201.238	197.854	23	199.156	201.108
9	199.516	199.964	24	199.817	198.486
10	199.889	201.420	25	201.129	202.628
11	201.736	197.994	26	199.938	198.800
12	199.778	200.864	27	201.599	196.927
13	202.363	201.586	28	202.282	200.419
14	200.404	200.273	29	200.135	199.840
15	199.297	201.407	30	198.419	200.604

ii. Do the charts give you sufficient information to determine whether the measurement process needs to be recalibrated? Explain.

iii. How precise is the measurement device?

b. Anchors Away has also collected random samples of anchor weights as they came out of the casting mold in batches of four, 30 times throughout the day. The data, shown in the table below, were obtained using the measurement process evaluated in part a. Assume the same operator used the same scale. Construct $\overline{X}$ and R charts for these data and answer the following questions.

i. Is the production process in a state of statistical control?

ii. What is the total observed variance?

iii. What proportion of the total variance is due to reproducibility? What proportion is due to bias? What proportion is due to repeatability? What proportion is due to measurement?

iv. What proportion of the total variance is due to process?

v. Calculate the signal-to-noise ratio. Based on this ratio, would you say that the measurement process is adequate?

vi. Calculate the discrimination ratio. Based on this ratio, would you say that the measurement process is adequate?

vii. Is the process meeting the target regulation weight of 200 lb?

viii. What proportion of the total anchors produced would you estimate are too heavy? What proportion of anchors are too light?

ix. What can be concluded about the process based on the signal-to-noise and discrimination ratios?

	Measurement			
Trial	A	B	C	D
1	205.902	206.711	207.410	208.820
2	207.799	209.638	208.695	209.399
3	207.909	206.565	207.412	207.797
4	205.277	209.482	208.415	207.869
5	207.885	206.926	206.932	205.777
6	205.837	206.335	208.628	204.455
7	207.803	205.730	206.882	206.205
8	205.539	206.337	207.975	210.548
9	206.594	207.276	205.903	207.166
10	209.386	208.780	205.495	206.921
11	207.139	204.069	209.329	206.149
12	208.394	207.931	204.666	206.271
13	204.793	206.794	206.880	206.596
14	206.979	206.643	205.613	207.082
15	205.036	207.076	209.440	208.065
16	205.872	206.228	208.446	208.200
17	210.264	204.709	206.586	208.721
18	209.717	207.873	207.185	203.168
19	208.530	206.833	207.508	208.727
20	205.336	208.085	208.088	208.281
21	207.421	208.279	206.820	205.593
22	205.713	206.340	206.995	207.046
23	206.065	205.627	208.664	207.265
24	207.007	210.730	207.151	205.243
25	208.092	204.339	204.457	208.125
26	208.223	208.165	209.509	206.818
27	208.575	203.547	207.323	204.569
28	205.616	208.638	208.981	207.227
29	207.836	208.793	207.265	207.009
30	206.349	205.423	208.586	208.316

c. What recommendations would you make to the management of Anchors Away, Inc.?

9-15. A certain production operation is performed by two machines that are used to fill bottles with a liquid product. An important quality characteristic of this product is the viscosity, which is measured in units called pascal-seconds (Pa s). The output from both machines is merged onto an accumulation conveyor where it travels to a packaging area to be combined, shrink wrapped, packaged, and then packed into cartons. As a new quality engineer you have been asked to design a sampling scheme to control viscosity for this two-machine operation. Your approach is to test several options.

a. Your first idea was to take a sample that consists of viscosity measurements from four bottles taken at the exit point of machine 1 for the first sample, four from the exit point of machine 2 for the second sample, and then to repeat this process alternating between machines until 30 samples have been collected. The results of this scheme are shown in the following table.

   i. Using this sampling scheme what sources of variability have been captured in the R chart? What sources have been captured in the $\overline{X}$ chart?

   ii. Construct $\overline{X}$ and R charts and interpret the results.

b. Your second idea was to randomly select two bottles from machine 1 and two bottles from machine 2 for each sample. The results of 30 samples using this scheme are shown in the table.

   i. Using this sampling scheme, what sources of variability have been captured in the R chart? What sources have been captured in the $\overline{X}$ chart?

   ii. Construct $\overline{X}$ and R charts and interpret the results.

		Viscosity Measurements (Pa s × 10^{-4})			
Sample	Machine	1	2	3	4
1	1	4.8913	4.1118	4.7754	4.5574
2	2	7.3746	3.7497	7.8673	5.0037
3	1	4.2920	5.2452	3.7951	2.4663
4	2	6.0261	4.9049	5.7173	4.8776
5	1	2.7932	5.8414	3.6671	3.8276
6	2	6.3780	6.2039	4.6254	4.7056
7	1	3.8569	3.3405	3.3465	3.3688
8	2	4.9944	4.6507	5.2948	7.0960
9	1	4.5798	5.6698	4.6101	6.1596
10	2	4.8186	7.7765	6.7731	6.8076
11	1	3.9606	5.5382	3.4224	2.9359
12	2	6.0074	5.4975	6.8410	5.5035
13	1	2.4570	4.0544	4.8498	3.6845
14	2	4.2957	6.9404	6.7271	3.8254
15	1	5.2700	3.3726	5.5947	5.7015
16	2	5.5634	6.1873	6.0841	6.4279
17	1	4.2988	3.0089	2.7400	3.8638
18	2	7.3280	6.3587	5.6123	5.4296
19	1	2.8562	5.6038	4.0388	3.9358
20	2	6.3457	6.7773	5.8940	5.2966
21	1	4.6763	5.3924	3.8046	4.1385
22	2	6.3001	6.6313	5.7550	7.3034
23	1	4.7152	3.7158	3.4253	3.6787
24	2	5.9624	5.6566	5.5403	5.0097
25	1	2.8218	3.1325	3.2963	4.9997
26	2	3.4240	2.5027	4.7506	5.9661
27	1	3.5550	3.6236	5.1860	2.3477
28	2	5.9451	5.1573	5.1135	5.5836
29	1	3.2639	3.3576	2.8275	3.0814
30	2	5.8067	6.8623	5.2209	5.0364

	Viscosity Measurements (Pa s × 10^{-4})			
Sample	1	2	1	2
1	6.1576	4.7682	4.3503	5.9068
2	3.4620	5.5978	4.6538	5.8934
3	1.9656	5.8575	4.7441	4.0444
4	4.6347	3.9422	2.5251	6.5728
5	3.1902	5.3946	4.1692	5.2204
6	4.0454	6.2608	4.6862	6.0559
7	4.2710	4.9436	3.0092	7.1869
8	3.9626	5.1236	4.5369	6.2242
9	3.7778	7.3627	3.3750	6.9625
10	4.0031	6.5562	6.5079	6.3370
11	2.7934	5.6210	4.5969	4.9311
12	3.2642	6.5737	3.3497	5.4387
13	3.6012	5.5629	5.0629	7.0194
14	3.6271	5.5965	3.8379	5.3882
15	4.0524	6.5102	4.7418	6.2417
16	3.7086	5.5645	5.6421	5.2413
17	5.2163	7.5013	4.6693	7.6324
18	4.7564	5.0264	3.6388	8.1647
19	2.9337	6.0328	5.5993	5.9766
20	4.7212	7.0292	3.9013	6.9989
21	2.6239	7.5338	4.6963	5.0481
22	4.3936	6.0692	3.6294	5.2017
23	5.2406	5.3692	4.2626	6.5541
24	5.0168	5.6927	4.4437	4.1506
25	4.2718	5.7160	4.7276	3.6260
26	6.6889	5.6974	2.7037	7.6951
27	2.0384	5.7716	4.6916	5.8555
28	2.8104	7.8905	4.6343	5.3867
29	3.9453	5.2634	3.4079	5.6179
30	5.0103	7.9171	2.9929	6.9557

c. A third idea was to sample at the packaging area by randomly selecting four units from the accumulation conveyor. Using this scheme there would be no way of knowing which machine produced the units that were selected for any particular sample. The data collected under this plan are shown in the following table.

i. Using this sampling scheme, what sources of variability have been captured in the R chart? What sources have been captured in the $\overline{X}$ chart?

ii. Construct $\overline{X}$ and R charts and interpret the results.

d. Of the three sampling plans, which plan would likely do the best job in estimating the distribution of viscosity in the eyes of the customer? What is your estimate of the mean and standard deviation of this distribution?

e. Is there any evidence to suggest that the two machines have different production variances (i.e., is the within-machine variability different for each machine)? Explain. Which sampling scheme is likely to provide the most help in answering this question?

f. Is there any evidence to suggest that the two machines have different production means (i.e., does each machine appear to be targeted at a different level)? Explain. Which sampling scheme is likely to provide the most help in answering this question?

	Viscosity Measurements (Pa s $\times 10^{-4}$)			
Sample	**1**	**2**	**3**	**4**
1	4.6862	7.3627	1.9656	3.7778
2	2.7932	6.0559	5.3946	3.6271
3	6.0692	5.4387	5.8414	5.8575
4	5.2413	4.0454	4.6507	2.4663
5	5.5035	3.8638	5.5634	4.6254
6	5.5645	4.9436	7.3746	5.7160
7	2.8104	5.9766	2.5251	2.6239
8	6.5079	4.7056	3.2642	4.5969
9	4.7276	3.8046	5.7716	4.8186
10	3.0814	5.2966	7.3280	6.3457
11	7.0960	5.3692	6.7731	6.0841
12	5.9451	6.9625	5.5403	2.0384
13	6.5728	4.8498	5.5382	7.3034
14	4.6916	5.3882	2.9337	3.6294
15	5.2948	5.6421	6.5102	5.5993
16	3.8569	5.2406	6.8076	5.2163
17	4.2957	3.8254	4.9997	6.3001
18	5.7173	5.0264	5.3924	7.5338
19	5.0481	5.5629	6.5541	3.7158
20	5.7015	2.9359	3.3750	3.1325
21	6.3587	5.0097	6.9557	5.1135
22	4.9944	3.8276	6.2039	3.4079
23	2.7934	4.0388	3.8379	2.3477
24	3.1902	6.8623	6.1873	6.0328
25	4.7506	3.6388	5.0103	3.3405
26	5.6038	2.8218	6.9404	2.7037
27	5.0168	5.6179	6.0074	3.4253
28	4.6101	2.5027	4.1506	5.4296
29	5.1573	5.8555	2.8275	5.6210
30	4.8776	3.4240	3.0089	5.2634

9-16. Westview Candleworks, Inc. is preparing to offer a new product in its Premier line—the Tuscan Taper, a buff-colored product designed to grace the most exquisite of candelabra. The Tuscan Taper has a design specification on height of 75 cm ± 5 cm. To try and determine process capability a pilot run is conducted during which the quality engineer pulled at random 25 samples of four taken over the duration of the run. These results are recorded in the table shown.

a. Use these data to construct $\overline{X}$ and R charts and comment on the state of control.

Sample Number	Measurement			
1	77.6	72.3	71.2	76.0
2	73.8	77.1	75.0	76.7
3	65.8	69.7	70.0	70.1
4	74.0	75.3	72.9	77.3
5	76.5	74.8	75.5	78.0
6	75.0	76.1	73.2	75.8
7	77.4	74.1	72.8	73.2
8	78.7	72.5	75.1	76.1
9	71.1	73.6	73.0	81.4
10	65.8	69.7	70.0	70.1
11	75.0	74.3	73.8	73.1
12	76.2	78.6	76.6	76.0
13	74.7	72.4	73.6	77.5
14	75.2	75.2	73.0	73.0
15	74.7	70.2	75.4	76.0
16	76.5	74.8	75.2	75.4
17	74.7	77.2	74.9	73.2
18	76.5	77.8	75.0	74.8
19	67.0	67.0	67.0	67.0
20	77.6	78.2	72.9	76.4
21	76.4	77.7	75.5	79.4
22	74.5	72.1	73.1	75.1
23	75.2	76.8	74.8	74.8
24	72.1	74.5	76.5	77.6
25	77.4	78.1	73.2	73.1

b. Calculate P_p and P_{pk} indices for the pilot run data. Does it appear that the process is capable of producing the Tuscan Taper within the height specification range? Explain.

c. During the pilot run, process engineers discovered some technical problems that were causing excessive variation in individual candle heights across the duration of the run. Process design changes were made and the product was ramped up to full-scale production. For the purpose of quality control samples of size 4 were collected each 15 minutes. The results of 25 samples are recorded in the following table.

Sample Number	Measurement			
1	75.988	76.414	77.558	74.057
2	77.046	77.045	75.782	77.502
3	76.446	75.373	75.267	75.284
4	74.261	76.134	74.664	74.513
5	75.376	73.614	76.120	75.989
6	79.500	73.375	76.267	74.079
7	76.944	76.280	75.871	74.720
8	79.385	73.276	75.093	75.267
9	74.936	72.339	74.692	71.581
10	80.215	73.046	76.420	76.728
11	76.394	73.616	75.823	74.729
12	73.927	75.852	71.897	76.888
13	75.889	72.075	73.760	76.039
14	75.870	73.419	76.247	72.382
15	75.709	74.217	75.542	72.047
16	73.549	73.805	77.875	75.848
17	78.179	75.436	73.252	75.030
18	70.845	77.051	73.602	76.281
19	74.428	75.630	75.881	73.755
20	78.114	76.578	74.350	75.430
21	78.195	74.764	74.406	74.727
22	72.004	73.560	72.272	77.368
23	78.235	74.150	75.483	76.718
24	75.024	73.968	75.387	72.978
25	70.378	74.361	73.518	74.016

i. Construct $\overline{X}$ and R charts for the production data and determine whether the process is in a state of control.

ii. Now, compute C_p and C_{pk} indices. Is the process capable of meeting the height specifications? Explain.

iii. By how much would the process variance need to be reduced in order to ensure that no more than 3.4 candles per million produced would be either too short or too tall (i.e., $C_{pk} = 2$). Assume the average stays where it is on the charts plotted in part c(i).

9-17. The A&P Wireless Network has begun production of its new line of extended life cell phone batteries. Thirty samples with a sample size of 4 have been collected at random throughout the first few days of manufacturing. The target talk time for each battery is 240 minutes and A&P widely touts this in its marketing materials. For manufacturing purposes, the specifications for talk time are 240 min. ± 55 min.

	Talk Time (min.)			
Samples	Battery 1	Battery 2	Battery 3	Battery 4
1	228.9760	235.4048	214.8863	233.7728
2	277.5794	245.4841	251.8144	215.3348
3	237.4299	239.6758	261.1716	213.2420
4	227.6591	249.4535	247.9534	212.4509
5	238.2617	226.5729	276.2791	243.5897
6	267.9728	242.9797	254.2453	235.9931
7	224.2749	209.4283	220.6360	229.1729
8	240.8200	232.2464	243.3356	240.9652
9	235.0082	234.4837	234.3501	226.9345
10	224.3337	258.4311	233.7101	224.7768
11	238.8137	242.7098	227.6584	245.1574
12	247.1537	230.9114	223.5222	263.5699
13	254.9279	266.1170	236.3007	239.4287
14	226.6227	235.3221	255.4939	250.3539
15	241.5015	251.3779	238.9225	216.2759
16	189.2993	236.7145	225.2183	226.7815
17	217.9847	262.3997	267.1382	248.6659
18	221.8791	246.3003	257.7557	226.7788
19	227.0974	253.0430	258.2243	243.6264
20	241.1327	234.1984	219.2993	232.1199
21	208.2145	209.9358	242.8495	215.6454
22	224.0399	266.3195	236.6367	204.2011
23	264.2231	234.0916	264.5267	257.0702
24	223.5178	241.6940	221.1516	240.4169
25	244.7766	224.8368	251.0133	238.5098
26	262.7761	218.8325	250.5417	219.1620
27	252.4674	251.9162	238.5688	225.3205
28	222.0424	242.9383	237.2528	237.3021
29	273.8904	229.9225	224.7283	294.0730
30	242.1451	253.1485	208.9359	245.2188

a. Construct $\overline{X}$ and R charts for these data and use these charts to determine if the process producing cell batteries is in a state of control with respect to battery life.

b. Determine if this process is capable by computing C_p, C_{pk}, P_p, P_{pk}, and C_{pm} indices. Discuss the implications of each of these.

c. What percentage of the output of the cell battery manufacturing process will fail to meet the specification requirements? What percentage will have expected talk time lives that will be less than the target?

d. How far can the mean value of the talk time deviate from the target value without creating a $C_{pm} < 1$?

9-18. Joe and Kate operate a wire thinning machine in a tire manufacturing company. The machine thins the wire to four different diameters. During their shift, Joe and Kate take turns measuring the diameter of the finished product before it moves on to the next station. The measurements are made independently in order to ensure that the measurements are not biased. First, Joe measures the diameters of four different wires, and then Kate measures the same wires. They repeat this entire process two more times, for a total of four samples with three measurements each per person. Use the data in the table below and Minitab to conduct a gauge R&R analysis. Then answer the questions that follow.

	Wire Number	Measurement 1	2	3
Joe	1	0.902	0.891	0.819
	2	0.805	0.777	0.856
	3	0.860	0.842	0.819
	4	0.756	0.872	0.769
Kate	1	0.813	0.834	0.833
	2	0.748	0.748	0.748
	3	0.854	0.853	0.847
	4	0.729	0.728	0.725

a. What is the total observed variation for this process?

b. What percentage of the total observed variation is due to measurement?

c. What percentage of the total observed variation is due to replication (repeatability) error?

d. What percentage of the total observed variation is due to a reproducibility error?

e. What is the signal-to-noise ratio for this process?

f. Based on the signal-to-noise ratio is the measurement process adequate? Explain.

g. What is the discrimination ratio for Joe and Kate's measurement process?

h. Based on the discrimination ratio is the measurement process adequate? Explain.

i. What action does management need to take to improve this measurement process?

9-19. Glazed Products Conglomerates operates three manufacturing facilities that are used to mass produce a number of dinnerware products. One of the critical quality measures is plate diameter. Since plates are usually not completely round, the procedure used is to select plates at random and measure the diameter at six different locations around the plate's circumference. Below are the results of measurements taken on thirty plates that were selected during the production run of one particular pattern. These plates were selected at 30-minute intervals over two production shifts.

Plate Number	Measurement Locations 1	2	3	4	5	6
1	21.1419	24.9408	22.2230	20.5149	16.7347	22.6937
2	16.1244	19.0037	17.4116	18.6971	20.4205	16.9109
3	15.2677	17.4596	21.1376	14.6630	16.0754	17.5136
4	14.1900	11.6606	13.8349	9.1828	13.1401	14.7871
5	17.4971	16.0398	16.1080	9.9519	14.1656	17.1329
6	16.0459	16.4860	15.6038	18.9632	17.4677	17.3676
7	18.0726	13.9655	16.3473	14.3596	17.7405	17.8751
8	17.7782	19.1714	14.5812	16.2434	15.8489	13.3621
9	14.2527	12.7476	14.9235	12.1210	7.1476	12.4490
10	16.5752	18.6753	17.2756	17.1445	13.9300	15.2794
11	11.2404	10.1817	14.0355	16.8696	18.0254	15.4528
12	22.8960	22.8616	20.8760	21.4651	15.2326	21.4049
13	15.0166	16.6438	15.9978	17.5205	15.7221	17.8091
14	15.9112	17.2718	16.1350	16.4213	14.5627	21.0249
15	9.7860	13.0439	11.7448	13.3574	11.6880	16.7314
16	9.6004	12.6547	16.2479	15.4493	10.8989	11.6172
17	13.5771	18.1639	13.8015	14.7064	17.3088	20.7239
18	18.4406	15.8016	17.9552	20.7093	19.8879	23.6181
19	16.0853	15.2035	15.7418	16.6020	17.9664	13.8573
20	11.2585	15.9238	10.8371	13.2465	15.1454	12.7784
21	13.4521	10.7076	13.3846	10.8929	14.7848	11.7505
22	20.1508	19.2589	25.5451	19.4803	19.3232	21.4040
23	18.4606	18.0627	21.3420	21.3289	19.4657	21.9017
24	13.0967	18.1804	18.0482	19.2777	16.4347	18.1713
25	20.0066	19.4286	22.3388	20.3960	20.1351	20.2425
26	24.6170	20.2148	24.6260	20.7424	21.2133	23.2789
27	19.3771	19.7494	20.4645	21.6304	18.7319	20.6282
28	22.4569	14.5227	20.8012	24.3502	19.0964	18.8143
29	21.5790	21.4821	19.8548	19.3992	19.5242	19.5181
30	11.3535	16.5953	17.3014	14.0590	10.9272	15.4918

a. Use these data to construct $\overline{X}$ and R charts to determine if the process is in a state of control.

b. What sources of variation would you expect to capture in the R chart using this method of sampling? What sources of variation would you expect to capture in the $\overline{X}$ chart?

c. If appropriate, use the charts you constructed in part a to estimate the variance from all sources captured within each subgroup. What does this variance represent?

Data were also collected over 15-minute intervals during a production run. The procedure used was to select four plates at random each quarter of an hour and measure each plate diameter once, at a random location. The results obtained for 30 samples are shown in the table.

Sample	Plate Number			
	1	2	3	4
1	14.6992	19.3843	15.2100	21.2453
2	17.4727	12.8546	18.6930	21.9663
3	14.0459	20.5769	15.3547	18.5188
4	15.2639	20.4771	15.4502	16.7740
5	16.8944	13.2142	14.7760	21.3877
6	13.1265	19.9588	15.0267	18.7286
7	17.3723	16.9630	14.4804	18.8662
8	17.3674	14.9940	16.0525	19.3359
9	14.5378	13.1829	21.8571	11.7743
10	18.7675	17.2014	15.5563	17.6225
11	14.1333	19.6364	15.8337	19.4837
12	11.0347	15.1709	13.0703	22.1533
13	17.0333	14.4424	15.0511	13.4392
14	19.1504	11.9798	21.3365	17.4999
15	15.4911	19.4526	19.4305	18.2408
16	17.3325	16.2247	15.8738	15.0876
17	18.3775	18.8970	16.4896	15.0391
18	10.9726	16.3048	15.1047	15.5979
19	15.8760	18.5875	18.3642	15.8201
20	17.7419	19.1104	15.8108	17.1652
21	19.1949	17.0940	22.8740	18.6536
22	21.3146	17.0008	16.9420	17.1430
23	17.5985	19.5372	16.8218	19.1684
24	18.9185	24.1223	20.8515	15.7093
25	16.7728	13.0469	14.5414	12.5970
26	15.6443	13.5464	17.8254	16.0646
27	15.0624	18.0352	14.6134	16.4916
28	14.4776	12.7980	17.7582	15.3802
29	14.3864	14.8899	17.3458	12.8869
30	14.3454	23.8164	18.8246	15.8100

d. Construct $\overline{X}$ and R charts for this sampling scheme and determine if the process is in a state of control.

e. If appropriate estimate the process mean and standard deviation.

f. Use your answers in parts c and e to determine the percentage of total observed variation that is within each plate and the percentage that is between plates. Assume that measurement variation is negligible.

A third sampling scheme used by Glazed Products involves selecting one plate at random from each of four cartons each day prior to the cartons being sealed for shipment to customers. This procedure was repeated over 30 days to obtain the data shown in the following table.

Sample	Plate Number			
	1	2	3	4
1	17.9950	14.8964	19.8972	17.1493
2	17.7548	17.1063	16.7789	17.0554
3	21.1908	20.1025	20.0091	17.3396
4	15.7095	7.3827	16.6293	12.6708
5	11.6807	16.9150	16.5036	19.5377
6	17.9608	14.7022	22.0541	11.2181
7	20.7332	19.6314	15.1173	21.2203
8	19.1607	21.6452	17.1388	19.2369
9	20.0360	17.6865	11.0447	25.7693
10	20.4552	20.2615	20.2836	19.1623
11	9.8574	16.6203	19.9141	15.5116
12	12.1873	16.1566	19.7859	18.7261
13	14.5052	17.1242	13.3214	22.8965
14	20.8879	18.4732	14.8509	21.4674
15	14.9480	21.9756	19.2823	23.3419
16	18.9553	15.2159	17.5313	19.0604
17	20.5262	17.9771	18.2237	18.6781
18	13.9525	16.5901	19.3156	25.1039
19	13.2055	20.0510	16.5831	17.2328
20	17.9823	18.0033	18.5737	14.6032
21	15.3293	21.3885	17.2559	23.8627
22	13.0137	12.6306	20.6194	9.9778
23	23.5770	17.3225	18.8049	21.6233
24	16.7901	15.5052	17.9178	11.4736
25	15.5123	18.6737	16.4252	17.4951
26	15.4022	16.2063	24.2727	17.6247
27	16.1506	12.2037	24.8557	21.5532
28	18.4034	16.9604	15.9884	17.6537
29	18.5026	20.8530	22.0071	18.2573
30	17.1785	20.2879	16.0800	17.0837

g. Construct $\overline{X}$ and R charts for this sampling scheme and determine if the process is in a state of control.

h. If appropriate estimate the process mean and standard deviation.

i. How do your answers in parts d and e compare to your answers in parts g and h? How can you explain any differences?

j. With this new information and assuming a negligible measurement error, estimate the percentage of the total process variability that is due to within-plate locational (out of roundness) variation, between-plate short-term variation, between-plate long-term (within a batch) variation, and long-term between-batch variation.

9-20. Anaco Ltd. builds car components for major European, Asian, and American automobile manufacturers. The company is contemplating the purchase of a new welding robot that will be used to produce bumpers for certain models for a Japanese customer. Anaco plants located in North Carolina and Arizona each produce this particular bumper; however, each of these plants utilizes a different type of industrial robot. The North Carolina plant uses a Robotech 2000 while the Arizona plant uses a MachineCo IS29. An industrial engineer was sent to each of these plants to perform time studies on the cycle time required to produce bumpers on the welding operation. The engineer's first data collection strategy involved taking measurements from each machine as it welded four consecutive bumpers. These data are shown below:

a. Given these data, what sources of variability have been captured in the R chart? What sources have been captured in the $\overline{X}$ chart? Construct the R and $\overline{X}$ charts for this process and determine the state of statistical control. If appropriate, estimate the process mean and standard deviation.

b. Next, the industrial engineer devised a sampling scheme whereby the times required by each robot for two consecutive bumpers respectively were combined into subgroups of four as shown in the following table.

Given these data, what sources of variability have been captured in the R chart? What sources have been captured in the $\overline{X}$ chart? Construct the R and $\overline{X}$ charts for this process and determine the state of statistical control. If appropriate, estimate the process mean and standard deviation.

		Sample						Sample			
Sample	Robot Model	1	2	3	4	Sample	Robot Model	1	2	3	4
1	Robotech 2000	29.349	35.372	33.214	29.831	20	MachineCo IS29	41.316	42.349	40.826	33.170
2	MachineCo IS29	36.955	37.281	39.875	40.785	21	Robotech 2000	29.298	36.797	28.832	30.054
3	Robotech 2000	32.488	37.208	31.001	29.494	22	MachineCo IS29	37.298	41.344	37.953	33.666
4	MachineCo IS29	35.294	39.873	38.077	37.113	23	Robotech 2000	35.293	28.804	39.886	35.816
5	Robotech 2000	35.435	39.899	33.076	28.969	24	MachineCo IS29	46.490	37.414	37.371	47.657
6	MachineCo IS29	37.852	34.288	36.768	41.110	25	Robotech 2000	36.631	35.728	36.866	35.617
7	Robotech 2000	33.971	33.253	30.136	37.324	26	MachineCo IS29	40.887	36.669	45.745	38.750
8	MachineCo IS29	41.154	38.591	40.675	42.477	27	Robotech 2000	34.868	31.524	30.431	34.513
9	Robotech 2000	33.218	37.337	35.746	32.791	28	MachineCo IS29	41.258	38.314	42.436	37.999
10	MachineCo IS29	41.191	40.219	40.500	43.025	29	Robotech 2000	28.206	30.865	30.975	29.952
11	Robotech 2000	34.794	39.859	32.776	39.767	30	MachineCo IS29	46.098	37.231	39.778	38.686
12	MachineCo IS29	40.107	40.509	43.882	46.131	31	Robotech 2000	24.416	26.190	36.343	32.358
13	Robotech 2000	32.801	34.151	32.452	35.538	32	MachineCo IS29	37.720	40.413	39.802	41.773
14	MachineCo IS29	45.629	39.764	39.681	44.024	33	Robotech 2000	28.655	35.269	31.534	29.154
15	Robotech 2000	27.527	26.512	31.051	35.255	34	MachineCo IS29	41.868	35.108	37.873	38.807
16	MachineCo IS29	38.137	42.062	41.143	41.912	35	Robotech 2000	32.267	35.935	28.761	27.927
17	Robotech 2000	28.829	35.931	27.740	37.222	36	MachineCo IS29	38.681	39.442	40.294	38.851
18	MachineCo IS29	38.080	39.748	40.711	37.800	37	Robotech 2000	31.448	30.212	31.263	32.722
19	Robotech 2000	33.876	30.238	32.355	34.648	38	MachineCo IS29	43.741	40.472	45.417	36.475

Sample	Robotech 2000	Robotech 2000	MachineCo IS29	MachineCo IS29
1	32.872	36.550	47.221	44.385
2	37.017	36.706	39.796	41.663
3	28.714	34.471	38.916	39.204
4	31.195	30.919	38.721	42.172
5	33.133	31.612	37.023	41.785
6	34.237	32.770	47.945	43.650
7	36.694	32.273	40.266	40.623
8	30.111	34.543	43.696	36.193
9	30.847	26.909	43.351	41.802
10	39.651	34.323	39.579	39.445
11	33.029	32.817	40.748	40.199
12	30.147	35.902	38.276	42.235
13	30.224	38.562	42.697	41.558
14	31.412	31.536	39.996	36.058
15	33.687	34.202	42.974	45.159
16	31.495	35.155	34.679	37.466
17	32.607	37.389	39.341	40.968
18	32.054	32.164	42.369	38.135
19	30.684	34.591	42.145	39.297
20	31.270	34.241	42.560	45.560
21	36.813	37.374	40.501	42.253
22	32.233	36.447	38.811	41.272
23	30.932	32.457	38.876	35.805
24	31.322	29.460	37.043	45.348
25	30.366	29.918	45.651	48.269

Sample	Machine Data Randomly Comingled from Both Welders			
1	39.681	36.866	37.371	31.001
2	40.623	37.324	39.297	33.971
3	36.694	34.202	36.631	45.159
4	42.062	28.969	35.255	36.550
5	32.414	28.832	39.778	29.298
6	37.999	30.975	39.875	37.023
7	40.219	34.868	35.372	41.785
8	45.417	40.675	32.872	38.135
9	42.436	29.952	39.899	32.770
10	26.909	41.802	33.253	45.651
11	46.490	42.235	36.343	31.263
12	29.154	43.696	42.697	40.266
13	39.579	36.797	33.029	45.560
14	32.817	35.108	30.366	30.136
15	35.728	39.796	33.076	34.679
16	42.349	26.512	37.389	31.524
17	43.882	37.043	30.212	38.686
18	31.536	33.810	33.218	35.294
19	32.054	40.199	33.876	30.224
20	34.151	37.050	35.902	37.281
21	41.868	30.147	37.337	32.355
22	39.859	32.791	30.054	44.727
23	39.802	36.447	34.794	39.886
24	36.955	34.323	29.349	41.191
25	38.591	32.358	38.077	29.494

c. As a third and final sampling scheme the engineer put together subgroups of four by randomly picking the robots to be timed. This scheme resulted in some samples containing more measurements from the Robotech 2000 machine and others more measurements from the MachineCo IS29 machine. Some samples may have an equal number of observations from each robot model. The resulting data are shown in the table below.

Given these data, what sources of variability have been captured in the R chart? What sources have been captured in the $\overline{X}$ chart? Construct the R and $\overline{X}$ charts for this process and determine the state of statistical control. If appropriate, estimate the process mean and standard deviation.

d. Which of the three sampling schemes is best? Explain.

9-21. At the Callaway Golf Company, a new long-distance golf ball is manufactured to a standard diameter of 45.7 mm in order to remain above the minimum standard diameter of 42.7 mm approved by the USGA. The quality control department has been given the responsibility to identify the sources of variation due to process variability and measurement error. The director of QC has expressed a concern for measuring the three components of measurement error, and requests that the operator performance be computed and interpreted accordingly. Two operators, Carol and Chris, conduct a measurement study to determine the sources of variation. Carol is assigned the opening production shift and Chris is assigned the closing production shift. They both are instructed to use an outside micrometer to measure the outside diameter of each golf ball. Carol measures a ball from production every 10 minutes for five hours. After the first measurement she measures each ball a second time, writes a number on the ball and places it in a box. On the second shift Chris is instructed to take the box of balls and, in numerical order, measure each ball in the box twice, carefully recording the two measurements. As a first step in the measurement study, Carol and Chris' data are combined as shown below.

Golf Ball #	Carol	Chris	Golf Ball #	Carol	Chris
1	47.335	47.604	16	44.021	43.962
	47.441	47.513		44.295	43.289
2	45.570	44.745	17	46.661	45.965
	45.770	44.916		46.967	47.402
3	45.159	45.291	18	45.884	45.706
	45.176	45.554		45.576	46.280
4	45.723	44.941	19	43.960	43.165
	45.782	44.514		44.423	44.514
5	45.702	46.250	20	45.718	45.423
	45.556	45.643		45.604	46.187
6	44.655	45.169	21	45.284	44.936
	45.070	44.681		45.504	45.948
7	47.315	47.005	22	45.434	44.763
	47.259	47.836		45.060	44.286
8	46.956	46.601	23	44.489	44.688
	46.188	46.353		44.587	44.460
9	45.487	45.976	24	45.766	46.102
	45.404	45.181		45.519	46.390
10	46.127	45.637	25	44.654	44.674
	45.858	46.158		44.573	44.818
11	44.037	44.170	26	44.650	44.750
	44.252	44.788		44.785	44.809
12	46.245	47.135	27	48.280	48.908
	46.868	46.305		47.962	48.469
13	45.730	45.689	28	45.915	46.521
	45.731	45.488		45.372	46.632
14	48.097	47.357	29	44.028	43.977
	48.233	47.718		43.983	44.912
15	45.310	45.039	30	45.189	46.439
	45.342	45.029		45.404	44.915

a. Use $\overline{X}$ and R charts to estimate the total measurement error in this process. Assume that measurement bias is negligible.

b. Next, control chart Carol's and Chris's measurement data independently, and use these charts to make a judgment whether there are differences between the two.

c. Based on your conclusions from part b compute the repeatability and reproducibility errors.

d. Use the data and any appropriate control charts to estimate the percentage of the total observed variation that is due to measurement. What percentage is due to repeatability error? What percentage is due to reproducibility error?

e. Use Minitab to conduct a gauge R&R study. How do these results compare with parts a through d?

f. Compute the signal-to-noise ratio and the discrimination ratio. Is the measurement process adequate for this process? Explain.

9-22. SunTech Co. is developing a new window film that will prevent windows from shattering upon impact. The specifications on the film thickness are 370 μm ± 10 μm. The film is made up of three layers: a dyed polyester film, metallic lining, and a scratch-resistant coating. Data collected from the production of each layer component are presented in the table below. The specifications for each component are as follows (all measurements shown are in μm):

Polyester film	100 ± 5
Metallic lining	120 ± 2
Scratch-resistant coating	150 ± 3

Polyester Layer

Sample	Observation 1	2	3	4	5
1	107.1397	99.2086	100.9042	100.0536	102.4504
2	100.8674	102.3427	100.6385	96.3919	100.1238
3	100.3446	100.3614	102.1023	101.2086	101.2238
4	103.2022	98.9172	104.1252	100.0115	98.3616
5	104.7148	100.7343	101.6746	100.3884	104.4914
6	100.8449	103.0214	102.8825	103.3160	102.7463
7	99.1435	100.3274	100.6272	98.9970	101.8641
8	98.8761	99.3938	99.2531	101.7511	99.3864
9	102.7593	100.6944	97.4749	98.8227	104.2506
10	96.3279	100.3979	101.6386	102.2754	100.7074

a. Construct $\overline{X}$ and R charts based on the production data and comment on the state of control of each component.

b. If appropriate, estimate the mean, variance and standard deviation of the final assembled film thicknesses based on the data.

Metallic Lining

Sample	Observation 1	2	3	4	5
1	118.9965	119.4768	120.1195	121.3607	119.5365
2	119.9275	118.5908	119.0238	120.1219	119.7238
3	121.0552	120.7206	120.5632	120.0192	118.4176
4	120.4201	121.724	118.9609	120.8402	119.2126
5	120.0643	119.5225	120.0839	119.8735	119.9827
6	120.4571	119.2172	119.1524	119.5827	120.6158
7	118.7643	118.5731	119.9608	118.6721	118.0054
8	119.5879	120.8373	121.3822	119.2283	119.0834
9	120.6605	121.2018	118.9951	120.5087	118.8764
10	119.3968	120.8473	120.6957	120.9016	118.3809

Scratch-Resistant Coating

Sample	Observation 1	2	3	4	5
1	147.4471	150.0942	150.9554	150.8616	151.3109
2	153.342	154.1938	149.4951	150.4375	150.8502
3	149.8125	151.6689	149.6252	151.7198	149.5837
4	152.2016	150.9156	150.873	152.1419	149.4748
5	149.2191	149.002	151.3166	148.825	150.9926
6	150.5601	150.2813	150.0309	151.1049	150.8316
7	150.5922	151.0496	148.2893	151.0391	148.8795
8	151.8168	152.6681	148.701	149.4111	151.992
9	150.4225	151.3513	147.2515	148.5842	148.5399
10	151.0835	149.8955	153.0184	150.0018	147.9945

c. For each component, calculate the estimated percentage of the relevant production output that will be outside the stipulated specification limits.

d. Estimate the percentage of the assembled films that will be outside the stipulated specification limits. How do you reconcile this answer with your answer in part c?

e. Assuming that the production distributions for each of the three components can be centered on their respective target specifications, recommend a set of revised assembly specifications that will ensure that no more than 1% of the films will be too thick and no more than 1% of the films will be too thin.

f. Assuming that a $C_{pk} = 1$ could be achieved for each of the three components, how would your answer to part e change?

9-23. Quality managers at Shea Enterprises are concerned about the calibration and effectiveness of their measurement processes. They had their top operator measure a standard twice that was NIST-certified to a true value of 250. The results of 30 such samples are shown in the table below.

Sample Number	Measurement #1	Measurement #2
1	250.0128	250.2800
2	267.4249	228.5290
3	246.5536	245.9324
4	245.1880	244.6992
5	240.4488	257.5699
6	251.0193	230.7222
7	274.0861	247.7872
8	235.5541	241.7373
9	247.0407	255.2091
10	275.2505	229.1230
11	246.1878	246.0441
12	263.5506	221.8329
13	229.1763	242.5462
14	270.6591	257.1511
15	272.5148	244.9429
16	275.3717	277.1027
17	249.0805	252.0093
18	249.2743	243.5419
19	232.0104	252.3836
20	243.9453	265.9753
21	239.7296	275.5325
22	249.8736	255.1597
23	236.3297	245.8322
24	241.7323	266.5706
25	254.6401	246.4936
26	265.5903	244.0194
27	260.7349	256.7252
28	254.0772	249.3391
29	250.4862	279.2046
30	243.1398	259.4464

a. Construct $\overline{X}$ and R charts for these data and estimate the measurement error. Comment on whether the charts are in control or not.

b. Use the results from part a to comment on the accuracy and precision of the measurement process for the operator who took the measurements.

c. Management used the information developed in part b to calibrate the process and institute some training and new maintenance procedures for the measurement process. Once these actions were completed an additional 30 samples of two measurements on the standard were taken by the same operator as before. Those data are presented in the following table. Construct $\overline{X}$ and R charts for the new data and comment on whether you believe management's actions have had an impact on measurement variation for this process.

Sample Number	Measurement #1	Measurement #2
1	259.167	244.950
2	252.298	251.341
3	251.422	256.481
4	243.461	266.862
5	255.192	253.873
6	262.084	258.422
7	247.454	252.634
8	244.180	224.641
9	246.219	252.888
10	257.755	236.550
11	253.914	252.081
12	234.828	258.883
13	265.347	235.833
14	248.369	251.594
15	257.379	250.904
16	256.626	239.737
17	240.522	231.814
18	244.111	259.653
19	239.287	252.255
20	272.375	243.612
21	257.339	258.982
22	250.381	250.147
23	242.297	251.855
24	257.845	244.262
25	233.442	257.306
26	244.035	235.027
27	265.461	265.228
28	250.154	259.777
29	268.620	247.691
30	243.140	259.446

d. The operator, using the measurement process described above, has taken 30 samples of size 4 from the production line over time, as shown in the table below.

	Measurement			
Sample	1	2	3	4
1	224.2049	273.7448	233.3339	257.9898
2	307.9549	233.7363	205.7783	283.0985
3	278.5704	264.8903	309.0668	280.0854
4	197.3284	306.4442	278.3346	265.5885
5	217.4347	260.6557	276.1048	239.7879
6	236.2441	231.2572	249.9189	256.4436
7	228.3399	208.4859	278.5430	232.2953
8	260.4318	225.2673	238.7179	246.6358
9	258.3697	261.2308	269.3814	275.7765
10	300.6458	238.0774	283.0957	247.6345
11	245.7903	284.0077	251.0791	268.9268
12	247.5736	228.0621	294.0911	287.4512
13	280.7654	258.6742	296.0933	253.0263
14	253.7945	237.5738	260.0130	262.2726
15	196.5277	280.8686	252.5527	268.9720
16	277.0418	235.0226	262.7454	278.0818
17	249.8199	286.3814	258.3653	255.8920
18	260.2927	209.0062	226.3027	215.8781
19	248.4438	263.1994	245.2672	237.9386
20	256.9607	272.7323	270.9780	185.2605
21	257.8158	255.8612	217.8889	245.2885
22	275.0417	256.0066	245.3079	251.3061
23	280.8023	261.5766	257.1571	250.5958
24	319.7759	265.9681	234.9655	258.3335
25	236.7504	213.3363	249.5419	227.9901
26	268.2115	298.5090	255.6334	277.2691
27	243.3733	253.0867	261.4494	236.3459
28	248.6302	239.3650	283.1586	241.6452
29	240.2615	240.0727	278.5546	258.9025
30	255.3799	198.3590	259.6345	279.9303

Construct $\overline{X}$ and R charts for these production data to determine whether the process is in a state of control. Use the information on these charts to compute the signal-to-noise ratio and the discrimination ratio. Is the measurement process adequate for this production process? Explain.

9-24. Jim and Jack were both out one night, getting suitably smashed on Ole Rocky Top stubbies, when they started to debate the process in which the cans (stubbies) were formed and how their diameters differed. As the night progressed and the heat of the debate heightened, they brought out a set of digital micrometers to perform a *scientific* analysis, not only on the

Ole Rocky Top brand, but also on three other favorite products (the contents of which they regretfully had to empty before proceeding with their study). They obtained the results shown in the following table (all measurements are in inches).

	Operators	
Product	Jim	Jack
Ole Rocky Top	4.2096	4.1699
	4.1719	4.1625
	4.1812	4.1893
	4.0517	4.0322
	4.1095	4.1355
Great Southern Ale	4.0417	4.0785
	4.0071	3.9857
	4.0667	4.0579
	3.9848	3.9866
	3.9919	3.9995
Blue Pacific	3.9320	3.9183
	4.0292	4.0415
	3.8263	3.7997
	3.9758	3.9939
	3.8278	3.8158
Bulldog Suds	3.8439	3.8436
	3.8075	3.7765
	3.7222	3.7597
	3.7658	3.7462
	4.0942	4.0933

a. Use Minitab to perform a gauge R&R study for Jim's and Jack's data. What percentage of the total observed variance is due to the process? What percentage is due to the measurement process? How is the measurement error distributed between repeatability and reproducibility errors?

b. Use the information developed in part a to compute the signal-to-noise and discrimination ratios. Do you believe the gauge used for these measurements is adequate to differentiate between different products? Explain.

c. After performing the analysis in parts a and b, Jack told Jim he thought they should check their gauge for bias and linearity. They were able to locate some certified standards that were certified to be 1", 2", 3", 4", and 6". They then used their digital micrometers to measure each of the standards 10 times, producing the data shown in the table above. Use Minitab to perform a linearity and bias analysis. Is there any evidence that there is measurement bias across the range of measurements? Explain.

Actual Value	Measurement	Part No.	Actual Value	Measurement	Part No.
1	1.0045	1	3	3.0002	3
1	0.9924	1	3	3.0018	3
1	0.9849	1	3	2.9954	3
1	0.9853	1	3	2.9975	3
1	0.9869	1	3	2.9959	3
1	0.9912	1	4	3.9966	4
1	0.9881	1	4	4.0179	4
1	0.9964	1	4	4.0090	4
1	0.9998	1	4	4.0012	4
1	0.9949	1	4	3.9996	4
2	2.0185	2	4	3.9994	4
2	1.9922	2	4	3.9977	4
2	1.9868	2	4	3.9916	4
2	1.9993	2	4	4.0009	4
2	1.9873	2	4	3.9878	4
2	1.9916	2	6	6.0109	5
2	2.0097	2	6	6.0005	5
2	1.9868	2	6	5.9913	5
2	2.0061	2	6	6.0118	5
2	1.9958	2	6	5.9919	5
3	3.0105	3	6	5.9935	5
3	2.9974	3	6	6.0070	5
3	2.9906	3	6	6.0105	5
3	3.0006	3	6	6.0027	5
3	2.9947	3	6	6.0002	5

9-25. A factory produces jars of a food item, popular with Korean nationals, called Kim-Chi. The net weight specification of each filled jar of Kim-Chi is 500 g ± 5 g. To check how well the process is performing, a total of 25 samples of size 4 have been collected at random over several hours of packing. The resulting data are shown in the table above.

a. Construct $\overline{X}$ and R charts for these data and comment on whether this process is in a state of statistical control.

Sample	Weight (g)			
	1	2	3	4
1	497	502	500	499
2	504	499	506	504
3	502	499	503	501
4	496	501	497	502
5	498	501	497	503
6	505	506	498	497
7	500	500	495	498
8	503	508	505	507
9	502	500	499	496
10	499	500	503	495
11	498	497	499	500
12	505	499	497	507
13	498	498	496	498
14	504	498	495	505
15	505	502	504	499
16	501	495	498	500
17	504	496	500	503
18	502	505	504	500
19	504	505	498	496
20	497	494	502	505
21	504	502	495	497
22	505	499	496	508
23	498	504	497	499
24	501	496	500	499
25	500	502	505	500

Sample	Weight (g)			
	1	2	3	4
1	497	502	500	499
2	504	499	506	504
3	502	499	503	501
4	496	501	497	502
5	498	501	497	503
6	505	506	498	497
7	500	500	495	498
8	503	498	505	495
9	502	500	499	496
10	499	500	503	495
11	498	497	499	500
12	505	499	497	507
13	498	498	496	498
14	504	498	495	505
15	505	502	504	499
16	501	495	498	500
17	504	496	500	503
18	502	505	504	500
19	504	505	498	496
20	497	494	502	505
21	504	502	495	497
22	505	499	496	508
23	498	504	497	499
24	501	496	500	499
25	500	502	505	500

b. Calculate P_p and P_{pk} indices for this process and comment on capability.

c. When studying the process it was discovered that one of the filling heads was occasionally sticking resulting in an inconsistent metering of material. The head was repaired and another 25 samples were taken, resulting in the data shown in the above table. Construct $\overline{X}$ and R charts for these new data and comment on whether this process is in a state of statistical control.

d. Use the data generated in part c to perform a capability analysis again, only this time compute the C_p and C_{pk} indices.

e. To achieve a $C_{pk} = 2$, by what percentage would the process variance have to be reduced?

9-26. You are an employee at Dineco Plate Company, and you want to make sure that a new line of plates is being manufactured according to specifications before it is sold. The tolerance for the diameter of the plates is 254 mm ± 5 mm. To be sure the plates are in spec, you have captured data on the plate diameters in three ways, as described below.

Thirty plates were collected one day at random from the production line. The diameter of each plate was measured six times, at different points on each plate. In the sampling scheme #1 table shown below is the recorded data for those measurements .

a. Construct $\overline{X}$ and R charts for these data and comment on the state of statistical control for the process. Can you explain any out-of-control condition?

b. Using this sampling scheme, what sources of variability have been captured in the R chart? In the $\overline{X}$ chart? Use these data to estimate the average plate

Diameter Measurement (mm): Sampling Scheme #1						
Plate	Location 1	Location 2	Location 3	Location 4	Location 5	Location 6
1	254	251	253	252	252	252
2	250	249	252	248	249	251
3	254	255	257	254	253	255
4	253	253	254	251	252	253
5	253	254	254	251	254	254
6	255	255	255	255	255	256
7	255	255	256	255	253	258
8	253	255	255	256	254	255
9	252	252	255	253	255	254
10	253	254	255	253	254	255
11	252	254	251	252	252	252
12	252	253	254	250	253	254
13	254	251	251	252	252	253
14	255	257	254	254	256	254
15	254	253	253	255	252	255
16	257	253	255	256	255	254
17	254	253	253	254	253	254
18	256	256	258	254	254	256
19	253	252	253	252	254	251
20	254	256	256	253	253	253
21	254	253	252	252	253	253
22	254	256	254	252	255	254
23	254	253	252	252	252	255
24	252	254	253	254	252	252
25	254	255	254	254	254	254
26	255	253	257	254	252	257
27	255	252	253	257	254	253
28	257	256	257	255	254	254
29	254	254	253	255	255	256
30	253	251	254	254	251	251

diameter and standard deviation of diameters. What do these numbers represent and how confident are you in the estimates?

The next sampling plan involved selecting and measuring four plates every 15 minutes from the production line. The plates were selected at random and each plate in the sample was manufactured at approximately the same time. The sampling scheme #2 table below contains the data collected over one shift.

c. Using this sampling scheme, what sources of variability have been captured in the R chart? In the $\overline{X}$ chart? Use these data to estimate the average plate diameter and standard deviation of diameters. What do these numbers represent and how confident are you in the estimates?

The third sampling scheme involved going into the warehouse and randomly selecting four plates per day over a period of 30 days. These data are shown in the sampling scheme #3 table shown below.

Time	Diameter Measurement (mm): Sampling Scheme #2 Plate 1	Plate 2	Plate 3	Plate 4
8:00 a.m.	254	253	256	256
8:15 a.m.	253	253	254	252
8:30 a.m.	252	254	255	255
8:45 a.m.	252	255	251	253
9:00 a.m.	252	256	255	253
9:15 a.m.	252	253	254	255
9:30 a.m.	252	253	255	254
9:45 a.m.	253	255	254	254
10:00 a.m.	257	253	254	254
10:15 a.m.	252	258	256	252
10:30 a.m.	255	254	251	255
10:45 a.m.	256	255	254	256
11:00 a.m.	255	252	254	256
11:15 a.m.	254	256	254	252
11:30 a.m.	254	254	253	254
11:45 a.m.	253	253	255	251
12:00 p.m.	252	257	255	251
12:15 p.m.	255	251	254	253
12:30 p.m.	253	256	256	253
12:45 p.m.	254	256	250	257
1:00 p.m.	256	254	254	252
1:15 p.m.	249	254	253	252
1:30 p.m.	255	253	256	253
1:45 p.m.	256	254	255	251
2:00 p.m.	251	251	254	257
2:15 p.m.	257	254	252	256
2:30 p.m.	254	253	254	256
2:45 p.m.	255	255	253	255
3:00 p.m.	254	253	253	251
3:15 p.m.	256	255	255	253

Day	Diameter Measurement (mm): Sampling Scheme #3 Plate 1	Plate 2	Plate 3	Plate 4
1	253	255	252	254
2	257	254	252	252
3	252	255	254	255
4	254	254	252	256
5	253	254	256	253
6	254	255	256	252
7	252	255	254	252
8	253	254	255	254
9	253	255	254	254
10	254	250	255	253
11	252	256	254	252
12	254	255	252	253
13	253	255	255	253
14	255	254	256	253
15	253	256	254	254
16	254	256	254	256
17	253	252	254	253
18	254	256	254	253
19	255	252	253	255
20	255	255	252	255
21	254	255	254	256
22	253	254	257	253
23	252	258	254	253
24	254	255	253	252
25	256	257	255	251
26	256	253	255	256
27	252	252	256	256
28	254	254	254	258
29	253	255	256	250
30	251	252	255	253

d. Using this sampling scheme, what sources of variability have been captured in the R chart? In the $\overline{X}$ chart? Use these data to estimate the average plate diameter and standard deviation of diameters. What do these numbers represent and how confident are you in the estimates?

e. Given your results for a through d, estimate the percentage of the total observed variation that is due to within-plate differences, between-plate differences within a batch, and between-batch differences. Assume that measurement variability is negligible.

9-27. The management of Greenfield Tire and Rubber has asked you to review the gum ply and wire reinforce thickness at one of its tire manufacturing facilities. The rework shop is receiving more than normal jobs that were not meeting the overall CTA (critical to assembly) specification of 0.25 in. ± 0.1 in. Meeting these specifications on each assembly is important so that parts do not exceed the limitations of the curing mold. Each assembly consists of two layers of material: a gum ply layer and a wire reinforce layer. You have been provided with production data for the gum ply and wire reinforce processes respectively. These data are shown in the following two tables.

Sample Number	Gum Ply Layer Thickness (in.) 1	2	3	4
1	0.1700	0.1087	0.0961	0.0856
2	0.1888	0.1362	0.2005	0.1371
3	0.1559	0.1320	0.1465	0.1610
4	0.1149	0.1566	0.1381	0.1219
5	0.1289	0.0913	0.1586	0.0903
6	0.1768	0.1150	0.1615	0.1278
7	0.1678	0.1357	0.1942	0.0928
8	0.1927	0.0879	0.1197	0.1379
9	0.1130	0.1590	0.1527	0.1363
10	0.1221	0.1814	0.1434	0.0931
11	0.1621	0.1392	0.0877	0.1392
12	0.1248	0.1322	0.1294	0.1434
13	0.1586	0.1777	0.1601	0.1533
14	0.1617	0.1146	0.1659	0.1385
15	0.1668	0.1311	0.1612	0.1847
16	0.1857	0.1360	0.1416	0.1249
17	0.1216	0.1292	0.1060	0.1207
18	0.1171	0.1406	0.1481	0.1827
19	0.1164	0.1064	0.1283	0.1832
20	0.1366	0.1189	0.1427	0.1566
21	0.1980	0.1784	0.1656	0.1574
22	0.1090	0.1083	0.1340	0.0707
23	0.1291	0.1076	0.1424	0.1833
24	0.1248	0.1285	0.1329	0.1407
25	0.0932	0.1474	0.1659	0.1623
26	0.1485	0.1483	0.1159	0.1278
27	0.1261	0.1458	0.0956	0.1718
28	0.1801	0.1698	0.1403	0.1184
29	0.1628	0.1056	0.1511	0.1421
30	0.1543	0.0906	0.1405	0.1685

Sample Number	Wire Reinforce Layer Thickness (in.) 1	2	3	4
1	0.1118	0.1045	0.0889	0.0700
2	0.0981	0.0586	0.0949	0.0920
3	0.1114	0.0800	0.0990	0.1619
4	0.0834	0.1154	0.0911	0.0780
5	0.0858	0.1000	0.0930	0.0977
6	0.1100	0.1092	0.1197	0.0637
7	0.1085	0.1139	0.0927	0.1145
8	0.1237	0.1173	0.0687	0.0580
9	0.0875	0.1089	0.0482	0.1220
10	0.0795	0.0886	0.1373	0.1029
11	0.0877	0.0765	0.1174	0.0706
12	0.1346	0.1176	0.1059	0.1225
13	0.1095	0.1078	0.1172	0.0938
14	0.1058	0.0819	0.1343	0.1169
15	0.0919	0.1144	0.1256	0.1128
16	0.0878	0.0682	0.0950	0.0898
17	0.1004	0.0923	0.1074	0.0944
18	0.0903	0.1339	0.1195	0.0886
19	0.0969	0.1177	0.0743	0.0864
20	0.0736	0.0914	0.1014	0.0658
21	0.1245	0.1025	0.0774	0.1293
22	0.1181	0.0704	0.1012	0.0984
23	0.1216	0.1003	0.0776	0.1094
24	0.1086	0.0727	0.0844	0.0783
25	0.0855	0.0849	0.0536	0.0916
26	0.0820	0.1075	0.1387	0.1110
27	0.0924	0.1594	0.0867	0.0811
28	0.1142	0.0695	0.1081	0.0759
29	0.0944	0.1082	0.0974	0.0936
30	0.1282	0.1164	0.1130	0.1306

a. Construct $\overline{X}$ and R charts for the production data for the two components of the gum ply wire reinforce assembly. Are these processes in a state of statistical control?

b. Estimate the parameters (μ and σ) of the distribution of the final thicknesses of individual gum ply wire reinforce assemblies.

c. Estimate the percentage of assemblies that will be outside the specification limits. How many parts per million is this?

d. What would you recommend that the management of Greenfield Tire and Rubber do in order to get the gum ply wire reinforce thickness closer to the nominal specification?

e. If management is successful in achieving the target specification, what percentage can be expected to be outside specifications assuming that the process variances for each component remain unchanged? How many parts per million is this?

f. Assuming that both components are on target, how much would the variance of the both components have to be reduced to in order to achieve no more than 3.4 parts per million outside either specification limit? Assume that the variances of each component will be in the same proportion of the total after the change as before.

9-28. A certain machining process is being checked to determine its capability to achieve the required tolerance on a part dimension that has specifications of 100 mm ± 28 mm. The following data have been collected.

Sample No.	Measurements (mm) A	B	C	D
1	90.741	102.300	98.642	106.069
2	102.711	100.882	103.314	98.569
3	104.066	105.620	96.165	100.412
4	106.602	95.978	96.265	95.869
5	100.904	108.558	94.882	98.573
6	104.922	100.243	97.053	111.042
7	102.738	108.145	98.679	103.788
8	102.388	104.159	100.204	99.328
9	97.825	95.209	96.273	93.430

a. Construct an $\overline{X}$ and R chart for this process. Is the process in control? Explain.

b. Compute the C_{pm} index for these data and comment on whether the process is capable of meeting the specification.

c. Estimate the percentage of output that is likely to exceed the specification limit.

d. How much can the mean of the process be permitted to deviate from the target and still produce a $C_{pm} \geq 1$.

e. Do you have any concern about your conclusions since only 9 samples were included in the analysis? Discuss the merits of increasing the number of samples in your study.

f. Would you reach the same conclusions if you computed the C_{pk} instead of the C_{pm}? Explain.

9-29. Bart and Homer are two engineers who work in the quality control lab at Schneider's Pretzel Co. in Handover, Germany. Their job is to make sure all of the pretzels are made to a consistent size. Oversized pretzels means they're giving away free dough, but undersized pretzels can make for some angry customers. The company certainly doesn't want to ruin its reputation for quality, especially during Oktoberfest! Every hour, Bart and Homer each measure a "standardized" pretzel using one of the same measuring devices used by the production crew. *Note:* In lieu of a certified standard, companies sometimes adopt a "pseudostandard" that can be used for measurement studies. Bart's and Homer's measurements over a 30-hour period are shown below.

Measurements of Standard Pretzels (in.)

Part	Bart		Homer	
1	1.146	1.142	1.142	1.141
2	1.145	1.143	1.141	1.142
3	1.145	1.146	1.141	1.144
4	1.145	1.143	1.142	1.143
5	1.146	1.145	1.141	1.142
6	1.143	1.145	1.141	1.143
7	1.145	1.143	1.144	1.141
8	1.146	1.144	1.146	1.142
9	1.143	1.144	1.143	1.143
10	1.144	1.145	1.143	1.143
11	1.145	1.144	1.143	1.144
12	1.143	1.144	1.145	1.138
13	1.144	1.143	1.145	1.144
14	1.144	1.143	1.144	1.144
15	1.144	1.143	1.142	1.143
16	1.142	1.142	1.142	1.143
17	1.142	1.144	1.142	1.143
18	1.142	1.143	1.144	1.142
19	1.144	1.142	1.145	1.148
20	1.143	1.143	1.143	1.142
21	1.143	1.141	1.144	1.143
22	1.143	1.142	1.144	1.146
23	1.145	1.142	1.145	1.143
24	1.142	1.143	1.146	1.145
25	1.141	1.143	1.145	1.145
26	1.141	1.142	1.145	1.141
27	1.144	1.143	1.145	1.143
28	1.142	1.143	1.146	1.149
29	1.141	1.142	1.143	1.140
30	1.141	1.143	1.145	1.137

a. Using Bart's and Homer's data, develop $\overline{X}$ and R charts to determine estimates of the repeatability (within operator) error and reproducibility (between operator) errors.

b. Every fifteen minutes Bart measures two pretzels selected at random from production line A and Homer measures two pretzels selected at random from a parallel production line B. The four measures are then combined into a single sample. The table below contains the data collected after 30 production samples have been taken. Construct $\overline{X}$ and R charts for the production data.

c. Compute the signal-to-noise and discrimination ratios for the measurement process. Comment on whether you believe the measuring process is adequate for this production process and why.

Measurements of Production Pretzels (in.)

Part	Bart		Homer	
1	1.079	1.142	1.127	1.181
2	1.146	1.174	1.206	1.181
3	1.090	1.123	1.130	1.154
4	1.093	1.156	1.116	1.120
5	1.146	1.171	1.116	1.135
6	1.087	1.162	1.108	1.136
7	1.082	1.174	1.128	1.103
8	1.048	1.107	1.146	1.139
9	1.165	1.188	1.060	1.186
10	1.135	1.169	1.189	1.114
11	1.112	1.230	1.096	1.134
12	1.123	1.165	1.124	1.138
13	1.101	1.132	1.175	1.173
14	1.123	1.183	1.085	1.140
15	1.110	1.175	1.159	1.099
16	1.139	1.174	1.131	1.095
17	1.218	1.170	1.106	1.179
18	1.178	1.145	1.158	1.145
19	1.150	1.149	1.094	1.128
20	1.165	1.116	1.151	1.204
21	1.142	1.125	1.132	1.198
22	1.139	1.098	1.171	1.136
23	1.144	1.123	1.144	1.112
24	1.210	1.154	1.136	1.135
25	1.118	1.148	1.176	1.168
26	1.146	1.124	1.153	1.171
27	1.166	1.102	1.135	1.163
28	1.107	1.148	1.126	1.149
29	1.087	1.112	1.150	1.140
30	1.153	1.108	1.094	1.137

9-30. American Sugarplums, Inc. has a factory that produces a variety of confectionary products for the holiday season. One of its high-volume items is a candycane that has a specified tolerance on length of 10 cm ± 1 cm. This product is produced continuously over several weeks on a dedicated line in large batches. To ensure quality an inspector selects four canes at random from the line every 15 minutes with a total of 30 samples per shift. After several shifts into the run the and R charts show good control with $\overline{\overline{X}} = 10.01$ and $\overline{R} = 0.7225$.

a. What is the process standard deviation (σ) for the distribution of candycane lengths?

b. Is the manufacturing process capable of producing candycanes within the specified tolerance range? Compute C_p and C_{pk} indices to justify your answer.

c. What are the upper and lower control limits on the $\overline{X}$ and R charts?

d. What percentage of candycanes would you estimate are too long? What percentage are too short?

e. Toward the middle of a run a technical malfunction leads to a one-third standard deviation increase in the average length (i.e., $\overline{\overline{X}}_{new} = \overline{\overline{X}}_{old} + {}^{\sigma}\!/_{3}$). The process variance was unaffected.

i. If the change in average goes uncorrected, how will the percentage of candycanes that are too long or too short be affected?

ii. What is the probability that this shift in average will be detected on the first sample taken after the change has occurred?

9-31. Universal Commodities, Inc. builds a product with an upper tolerance limit of 102.5 units and a lower tolerance limit of 97.5 units. The process mean is 100.4 units which is equal to the nominal design specification. Both the $\overline{X}$ and R charts show control and the sample size is 4.

a. If $\overline{R}$ is equal to 1.915, and the $\overline{X}$s show control, estimate the process standard deviation.

b. Compute the C_p for this process. Is the process capable?

c. Estimate the percentage of process output that will exceed the tolerance limits.

d. What is the probability of detecting a 0.5σ shift in mean on the $\overline{X}$ chart on the first sample following the shift?

CHAPTER TEN

STATISTICAL PROCESS CONTROL BY ATTRIBUTES

QUALITY MAXIM #10: *If the process is right the results will take care of themselves.*[1]

The Hammer of Quality—Knowledge-Driven Decisions

A 1960s Peter, Paul, and Mary hit song "If I Had a Hammer"[2] became an anthem symbolizing the American Civil Rights Movement. The song lyrics use a hammer, bell, and song to respectively symbolize justice, freedom, and brotherly love—those ingredients necessary to right the social injustices of the time.[3] Is it possible to metaphorically identify the hammer, bell, and song of the quality movement? Quality is not a random outcome. Quality, defined as satisfying stakeholder needs, is the consequence of carefully planned and executed strategies based on a profound understanding of what is required, an organization's capabilities to provide it, and a philosophy of continuous improvement. Much of the knowledge comes from the application of statistical tools to effectively collect, analyze, and interpret data, and the assimilation of these results into the corporate knowledge base to facilitate learning. This is the quality hammer. The bell is wisdom—the political will to use the knowledge most effectively to serve the best interests of all stakeholders. And finally, the corporate song is the culture that bonds all participants to a common purpose. It is the culture where relationships are built and where the issues of trust, respect, integrity, loyalty, and equity are judged. In this chapter we cover quality by attributes measurement, a process that predicts and attempts to control the frequency of errors, defects, or defective units of output. If perfection is the goal, one could argue that, at best, attributes measurement is an interim step since ultimately there should be no defects to track. As quality improves, the cost of implementing an attributes plan will increase because to control the level of defects, defects must be picked up within the samples. As the occurrence of defects becomes rarer, the sample size must increase so that some defects will be captured. Eventually it may cease to be economically viable to continue with the attributes method of control, and one could argue that this may be the ultimate aim.

[1]Takashi Osada.

[2]"If I Had a Hammer" was written by Pete Seeger and Lee Hays in 1949 and first recorded by The Weavers, a folk music quartet that consisted of Seeger, Hays, Ronnie Gilbert, and Fred Hellerman. Peter, Paul, and Mary, a folk-singing trio consisting of Peter Yarrow, Noel "Paul" Stookey, and Mary Travers, made the song a hit when they rerecorded it in 1962.

[3]The music and lyrics for "If I Had a Hammer" can be accessed at http://www.lyricsfreak.com/p/peter%2C+paul+%26+mary/if+i+had+a+hammer_20107670.html.

LEARNING OBJECTIVES AND KEY OUTCOMES

After careful study of the material in Chapter 10, you should be able to:

1. Discuss the lessons learned from the bead box experiment, how the bead box simulates process sampling, and how the bead box is dissimilar to industrial sampling procedures.
2. Develop an operational definition for a quality standard and discuss why operational definitions are important.
3. Recognize applications for *p* and *np* charts and design appropriate attributes data collection strategies.
4. Construct and interpret *p* and *np* charts for constant sample sizes.
5. Construct and interpret *p* charts for variable sample sizes, using four approaches: variable control limits, control limits based on average sample size, max/min control limits, and standardized control limits.
6. Calculate the minimum sample size required to satisfy the ninety percent, the min five, the min two, and the fifty percent detection rules.
7. Construct and interpret *c* and *u* charts.
8. Properly differentiate between applications that are appropriate to *p* or *np* charts from those that are appropriate for *c* or *u* charts.
9. Demonstrate an understanding of the role of acceptance sampling by attributes in the achievement of quality goals.

(continued)

10. Construct and interpret operating characteristic (OC) curves for single and double sampling plans.
11. Construct and interpret average outgoing quality (AOQ) and average total inspection (ATI) curves and compute the average outgoing quality limit (AOQL) for single and double sampling plans.
12. Construct and interpret an average sample number (ASN) curve for double sampling plans.
13. Use Grubbs's approach to design a single sampling plan.
14. Use Grubbs's approach to design a double sampling plan.

10.1 NATURE OF ATTRIBUTES MEASUREMENTS

Attributes data are collected using the nominal scale of measurement and a physical count is taken of some feature of interest. In summarizing the data only a single characteristic—*the central tendency* (average count)—is of interest. A typical application involving the collection of attributes data is to determine what percentage of a particular production lot, or process, is defective on the average. To answer this question random samples from the lot or process are inspected and the percentage of defectives found in each are then used to estimate the percentage defective in the entire lot or output distribution.

In attributes measurement there are only two possible outcomes—good or bad.

A warehouse worker retrieves an item from inventory. Effective quality control by the relevant supplier can ensure the product is ready to use and eliminate the need for any further inspection at the receiving site.

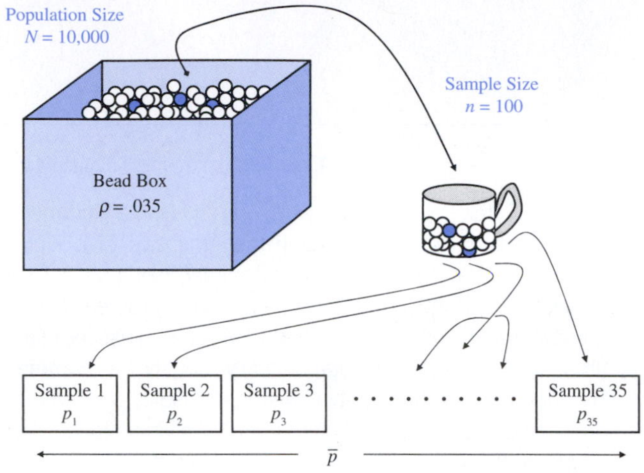

FIGURE 10-1. Bead Box Experiment

A simple experiment illustrates some important concepts concerning the nature of attributes measurement. Figure 10-1 presents a box that contains 10,000 small wooden beads. The exact number is not important provided it is large and does not change over the course of the experiment. Of the 10,000 beads, we shall assume that 350 are red and 9,650 are white. The beads are representative of the output of a production process: red beads are defective product and white beads represent product that is good enough to ship to customers. This bead box therefore simulates a process that is 3.5% defective. There are only two possible states for any individual unit of output—it is either white (good) or red (defective). We will try to estimate the percentage of defective units produced by repeatedly taking small samples from the box (process) and recording the number or percentage of inspected units that are defective. The underlying statistical distribution for data of this type is the **binomial**,[4] which has two parameters—the size of each sample (n), and the probability of a defective appearing in any given sample (π). In the bead box experiment just described $\pi = 0.035$.[5] One would seldom if ever know the true value of π. On occasion it might be possible to perform a 100% inspection and learn the *truth*; however, even this approach is not foolproof as sampling error can taint the results.

Since one cannot know for sure the population mean, it can be estimated using random samples. As shown in Figure 10-1 we have drawn 35 samples of 100 from the bead box. It is

[4]The characteristics of the binomial distribution can be found in Appendix E.
[5]The distribution describing the occurrence of good product is also binomial with $\pi = 1 - 0.035 = 0.965$.

TABLE 10-1 Attributes Data Generated from Beadbox

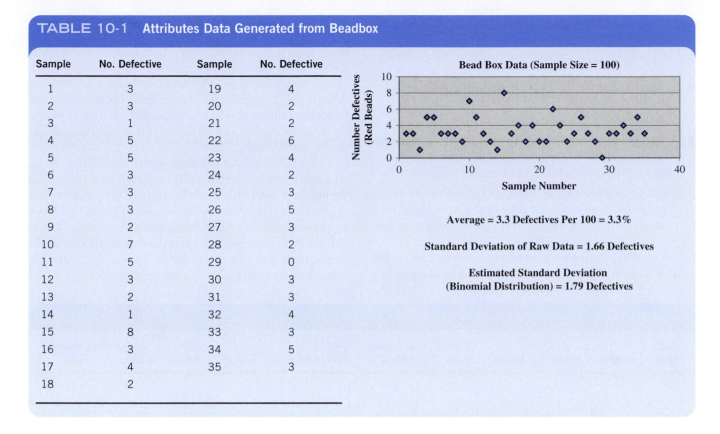

Sample	No. Defective	Sample	No. Defective
1	3	19	4
2	3	20	2
3	1	21	2
4	5	22	6
5	5	23	4
6	3	24	2
7	3	25	3
8	3	26	5
9	2	27	3
10	7	28	2
11	5	29	0
12	3	30	3
13	2	31	3
14	1	32	4
15	8	33	3
16	3	34	5
17	4	35	3
18	2		

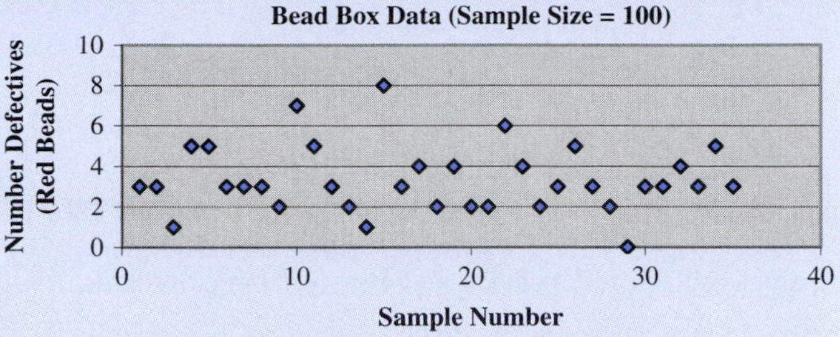

Average = 3.3 Defectives Per 100 = 3.3%

Standard Deviation of Raw Data = 1.66 Defectives

Estimated Standard Deviation (Binomial Distribution) = 1.79 Defectives

important that we mix the beads well between each draw. To be on safe statistical grounds, we should return each sampling unit to the box before selecting the next unit. However, *with large* populations, or if we are sampling from an industrial process, the results will not significantly differ if one samples without replacement. Normal practice is just to set aside the samples as they are collected and tally the results as shown in Table 10-1.

Population parameters can be estimated using random samples.

Even though the population is 3.5% defective, in this experiment individual samples ranged from a low of 0% defective (sample number 29) to a high of 8% (sample number 15). Therefore our first observation is: *Any single sample can be a very poor estimate of the true population parameter.* Nevertheless, the average of all 35 samples at 3.3% is much closer to the true value of 3.5%. The variable pattern across all samples is evidenced in the run chart shown. Our second observation is: *The average of a large number of samples provides a good approximation of the true mean.*

Since the bead box population is static, a third observation can be made. *All variation observed in sampling results is due to sampling error.* This is not process variation. When collecting attributes data the variation captured in samples is sampling error only unless the percentage of defectives is changing over time. If the percentage is changing over time the process is *unstable*, out-of-control signals will expected on applicable control charts, and it is not possible to estimate its mean with any confidence.

10.2 OPERATIONAL DEFINITIONS AND INSPECTOR CONSISTENCY

A significant source of inspection error can be the lack of a common understanding by all inspectors of how the data should be collected and the requirements. This is particularly critical in attributes measurement where inspectors are often required to exercise judgments on whether a unit "passes" or "fails." Therefore, before any measurement takes place it is essential that *operational definitions* be developed and be integrated into the training programs for all personnel involved.

Operator inconsistencies can be a source of reproducibility error.

An operational definition is a clear, concise, detailed statement of what it takes for a particular measure to conform to requirements. There should be no room for ambiguities or different interpretations. Adjectives such as *smooth, reliable, good, safe, red*, and *clean* are often used in the definition of quality attributes. Such words have no communicable meaning unless they are operationally defined in terms of sampling and test results. For example, what does one mean by the word *clean?* What is clean enough for a table in a restaurant is unsatisfactory for an operating room or the manufacture of semiconductor wafers. And how should cleanliness be measured? Will it be determined by merely taking samples from an operating surface (tabletop) or do air quality samples need to be taken as well? Over what time period? One hour, one day, one week? Is cleanliness to be defined as an average condition or an instantaneous one (i.e., anytime a sample is taken)?

Operational Definition

- A clear, concise, communicable written statement, including tangible examples if appropriate, of what it takes for a particular attribute to conform to requirements.
- A set of decision criteria that can be used to determine whether an attribute conforms to requirements that include specified sampling plans, test methods, and measurement procedures.
- The ability to clearly differentiate between acceptable and unacceptable results, so that a clear *yes/no* decision can always be made on the basis of the requirements and the decision criteria.

A certain printing plant has inspectors stationed at the end of a casing-in line to check finished books prior to packing. These inspectors look for problems with endpapers, covers, or trimming. One inspection criterion is to check for wrinkles in endpapers. To management's surprise it was found that not all inspectors could agree on what constituted a wrinkle—all having differing and personal definitions. A wrinkle had never been operationally defined. Such differences in understanding can lead to operator inconsistencies and drive up the magnitude of the measurement error.

How can **inspector consistency** be checked? The QC director in a certain assembly plant selected 20 parts—which she numbered and then carefully inspected herself. According to her criteria, she rejected parts numbered 1, 4, 8, 11, 14, and 16 and accepted the rest. After recording her personal results she had three qualified inspectors independently inspect the same sample. If any part was rejected, the inspector was asked to indicate by code the reason for the rejection. Table 10-2

TABLE 10-2 Reconciling Inspector Consistency

Test Unit	Inspector	Accept	Reject	Reason Code	QC Reason Code	Test Unit	Inspector	Accept	Reject	Reason Code	QC Reason Code
1	1		x	32	32	**11**	1		x	40	41
	2		x	36			2		x	40	
	3	x					3		x	40	
2	1	x				12	1	x			
	2	x					2	x			
	3		x	36			3		x	45	
3	1	x				13	1	x			
	2	x					2	x			
	3	x					3	x			
4	1	x			32	**14**	1	x			46
	2	x					2	x			
	3	x					3	x			
5	1	x				15	1	x			
	2	x					2	x			
	3	x					3	x			
6	1	x				**16**	1		x	32	32
	2		x	23			2		x	32	
	3		x	23			3		x	31	
7	1	x				17	1	x			
	2	x					2	x			
	3	x					3		x	25	
8	1		x	41	41	18	1	x			
	2	x					2	x			
	3	x					3	x			
9	1	x				19	1	x			
	2	x					2	x			
	3	x					3	x			
10	1	x				20	1	x			
	2	x					2		x	23	
	3	x					3	x			

summarizes the data that had been compiled at the end of the experiment. These results were enlightening, and the QC director was able to use this knowledge as the basis for improved training by focusing on specific areas of inconsistency.

As an example, two of the three inspectors rejected part number 1; however, their reasons for rejection differed. Only inspector number 1 agreed with the QC director on the reason for the rejection. Inspector number 3 passed part number 1. Part numbers 4 and 14 were rejected by the director and accepted by all three inspectors. On the other hand, part number 6 was acceptable to the director but rejected by two of the inspectors. Only one of the three inspectors agreed with the director on the unacceptability of part number 8. Inspector number 3 thought that parts 12 and 17 should be defective, and inspector number 2 erroneously rejected part number 20. All inspectors agreed on the rejection of part number 11, but for different reasons.

Operational definitions are essential to ensure inspector consistency.

Much can be learned from studying data such as these. Misunderstandings are highlighted and training needs can be readily identified. Even if an organization thinks that sufficient definitions are in place it is a good idea to test them to ensure that they are understood the same way by those whom the organization trusts for its quality control.

10.3 *p* CHARTS AND *np* CHARTS

Unlike variables charting, when control charts are used to analyze attributes data, only one chart is needed and that chart is used to track the single parameter of interest (the mean). As with variables charts it is essential that points on an attributes chart be independent.

To illustrate an attributes chart application consider a plant that produces cooking oil. Bottles filled with oil proceed through a labeling operation prior to packaging and packing. An inspector randomly selects 100 bottles from a shift's production and inspects for defective labels (such as missing, wrinkled, torn, or crooked) or supplier problems (e.g., printing defects). All defective bottles, including the reason code, are recorded. A Pareto analysis by reason code can help reduce defectives in the future.

Table 10-3 is a summary of inspection data that were collected over 30 consecutive shifts. For each sample of 100, the number of defective labels has been recorded. An extra column has been added to the spreadsheet that expresses the number of defectives as a proportion of sample size—for example, 3 defectives, when expressed as a proportion, is 0.03 of the sample. For control chart purposes the standard notation is to assign the symbol p to the proportion of defectives, and the symbol np to the number of defectives.

Since the sample size is constant (every sample contains exactly 100 bottles), there are two ways to chart these data.

TABLE 10-3 Inspection Summary for Defective Labels

Sample Number	Sample Size	Number of Defectives, *np*	Percentage of Defectives, *p*
1	100	2	0.02
2	100	2	0.02
3	100	4	0.04
4	100	3	0.03
5	100	5	0.05
6	100	4	0.04
7	100	7	0.07
8	100	5	0.05
9	100	4	0.04
10	100	7	0.07
11	100	2	0.02
12	100	4	0.04
13	100	2	0.02
14	100	2	0.02
15	100	3	0.03
16	100	3	0.03
17	100	4	0.04
18	100	2	0.02
19	100	4	0.04
20	100	2	0.02
21	100	3	0.03
22	100	6	0.06
23	100	1	0.01
24	100	2	0.02
25	100	4	0.04
26	100	5	0.05
27	100	4	0.04
28	100	3	0.03
29	100	2	0.02
30	100	3	0.03
Totals	3000	104	

Figure 10-2 is a ***p* chart** and Figure 10-3 is an ***np* chart.** All points on both charts are inside the respective control limits and one can easily verify that these charts pass the statistical tests for too-few and length-of-longest run.[6] It can therefore be concluded with a high degree of confidence that this process is in statistical control and is producing approximately 3.5% defective labels on the average. The formulae that are used to compute the control limits on p and np charts are as follows:

[6]With $r = 16$ and $s = 14$, Tables 8-2 and 8-3 provide critical values for too-few runs and length of the longest run equal to 11 and 8, respectively. The number of runs in Figures 10-2 and10-3 is 15 and the longest run is 6.

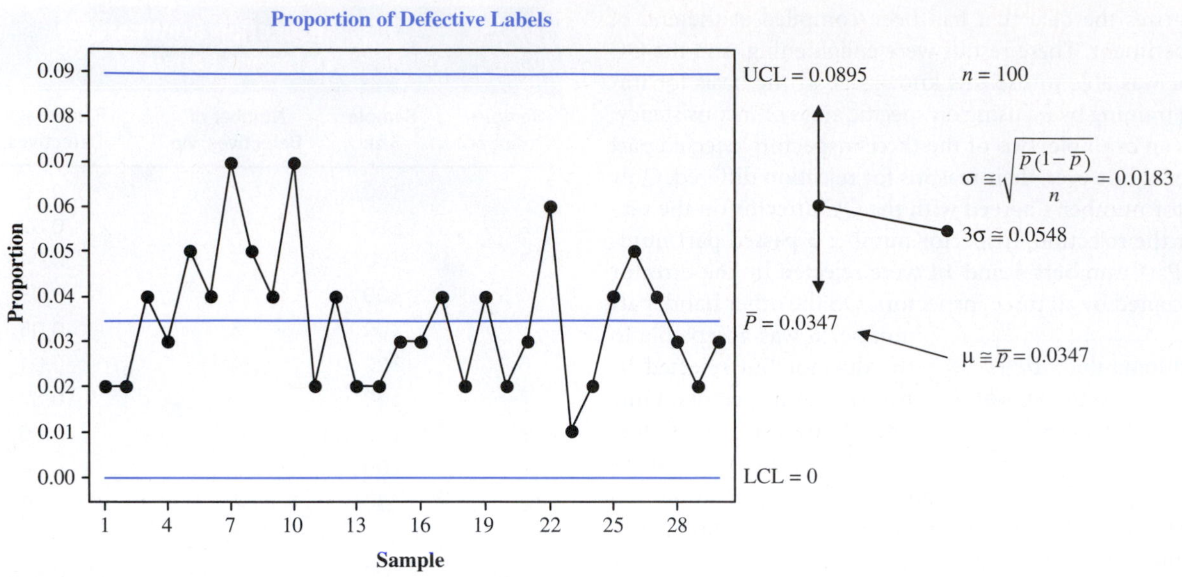

FIGURE 10-2. *p* Chart for Defective Labels

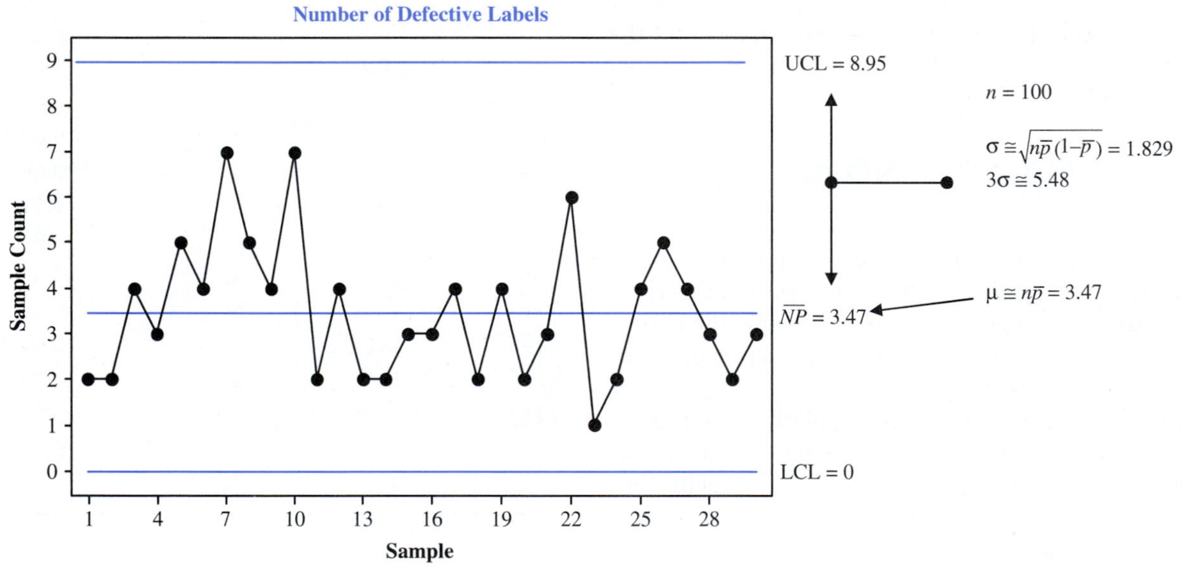

FIGURE 10-3. *np* Chart for Defective Labels

Formulae for *p* Charts

$$Centerline = \bar{p} = \frac{Total\ number\ of\ defectives}{Total\ number\ of\ items\ inspected} \qquad \textbf{10-1}$$

$$UCL = \bar{p} + 3\sigma_p = \bar{p} + 3\sqrt{\frac{\bar{p}(1-\bar{p})}{n}} \qquad \textbf{10-2}$$

$$LCL = \max\left\{\bar{p} - 3\sqrt{\frac{\bar{p}(1-\bar{p})}{n}}, 0\right\} \qquad \textbf{10-3}$$

where,

n = sample size

Formulae for *np* Charts

$$Centerline = n\bar{p} = n\left[\frac{Total\ number\ of\ defectives}{Total\ number\ of\ items\ inspected}\right] \qquad \textbf{10-4}$$

$$UCL = n\bar{p} + 3\sqrt{n\bar{p}(1-\bar{p})} \qquad \textbf{10-5}$$

$$LCL = \max\left\{n\bar{p} + 3\sqrt{n\bar{p}(1-\bar{p})}, 0\right\} \qquad \textbf{10-6}$$

where,

n = sample size

When n is constant either a p chart or an np chart may be used.

From the previous equations, it is obvious that the control limits depend on the size of the sample. *As the sample size increases, the control limits tighten; as it decreases, the control limits widen.* This makes sense when one considers that, as the sample size increases, the sample average is more likely to be a better estimate of the process average. For a constant sample size the formulae yield constant control limits.

10.3.1 Variable Sample Sizes

What happens if the sample size is not constant? A **variable sample size** results when random samples selected from a process are of differing sizes, and is rare in the case of variables measurement but is a common occurrence with attributes measurement, particularly where 100% inspection is involved. Table 10-4 contains data collected from a plant that produces printed circuit boards. An inspection station performs a 100% inspection of all boards produced. The inspection involves a test in which a board is plugged into a device that determines whether current can pass through the circuit. If current flows the board is passed; otherwise it is rejected as defective. For control chart purposes the inspection data are compiled daily, and since daily production rates vary, the sample size is not constant. When n is variable, an np chart is not an option and a p chart must be used. There are four approaches to calculating control limits for a p chart when the sample size varies: variable limits, limits based on the average sample size, a max/min approach, and standardized limits.

TABLE 10-4 Data for Testing an Electric Circuit

Sample Number	No Defectives np	Sample Size n	Proportion Defective p
1	21	1002	0.0210
2	27	1109	0.0243
3	20	998	0.0200
4	27	1234	0.0219
5	15	1087	0.0138
6	19	1056	0.0180
7	15	1187	0.0126
8	16	1108	0.0144
9	11	998	0.0110
10	26	1007	0.0258
11	25	1254	0.0199
12	18	1343	0.0134
13	21	1045	0.0201
14	19	1253	0.0152
15	16	1023	0.0156
16	22	987	0.0223
17	19	1157	0.0164
18	17	1178	0.0144
19	21	1039	0.0202
20	12	1039	0.0115
21	26	978	0.0266
22	21	1167	0.0180
23	18	1243	0.0145
24	20	1198	0.0167
25	14	1203	0.0116
26	15	1180	0.0127
27	20	1095	0.0183
28	19	1009	0.0188
29	18	1087	0.0166
30	8	1165	0.0069
Totals	566	33429	

10.3.1.1 Variable Limits. Figure 10-4a illustrates a p chart with variable limits. Using the circuit board data and Equations 10-1 through 10-3, the chart centerline and control limits relative for each sample have been computed. This approach has some disadvantages. One problem is that revised control limits must be computed each time a new sample is collected. A second concern is the ability to communicate chart interpretations. It is sometimes difficult to explain to someone who is unfamiliar with control charts why a point that was in control yesterday would be out of control today (or vice versa). It could appear to the uninitiated that the limits are being arbitrarily set.

No points are out of control in Figure 10-4a and the chart passes all runs tests ($s = 13, r = 17$). The critical value for too-few runs is 10 (the actual number of runs is 16) and for the longest run is 9 (the actual length of the longest run is 4). As the process exhibits control we are confident that the process is producing, and will continue to produce, approximately 16.93% defective output in the future. This will continue to be the estimate of process performance until the control chart shows evidence that the process average has changed. When improvement actions are taken the control chart can be used to validate that such actions produced the intended results. For example, rules violations might be expected, signifying that the improvements have worked and the process average has decreased.

10.3.1.2 Limits Based on Average Sample Size. Figure 10-4b is an alternative approach that works well provided the sample sizes do not vary greatly. Instead of calculating a set of control limits for each sample, constant limits have been constructed based on the average sample size $\overline{n}$. Constant limits can be used based on $\overline{n}$ provided the sample size range (maximum–minimum) is no more than $\pm$ 25% of $\overline{n}$. Then in the future, as additional samples are collected, actual limits need only be computed for points that lie close to the $\overline{n}$ limits. If a point falls just inside the limits, the control limit (either upper or lower) needs to be computed based on the actual n only if $n > \overline{n}$ (that is, the actual limits are tighter than the $\overline{n}$ limits). If a point is just outside the limits, the upper or lower control limit needs to be recomputed based on the actual n if and only if $n < \overline{n}$ (that is, the actual limits are wider than those based on the $\overline{n}$ limits).

10.3.1.3 Max/Min Limits. A third approach, illustrated in Figure 10-5, involves plotting a pair of control

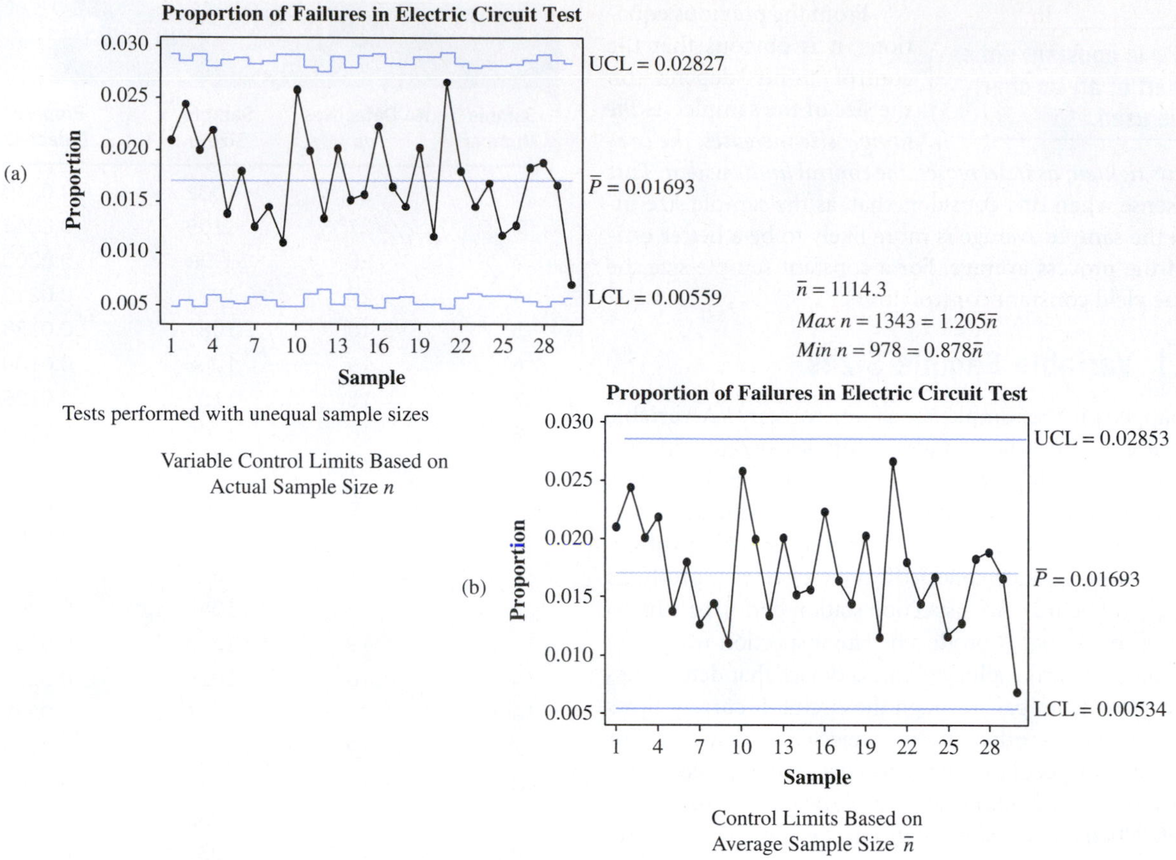

FIGURE 10-4. *p* Charts for Variable Sample Sizes

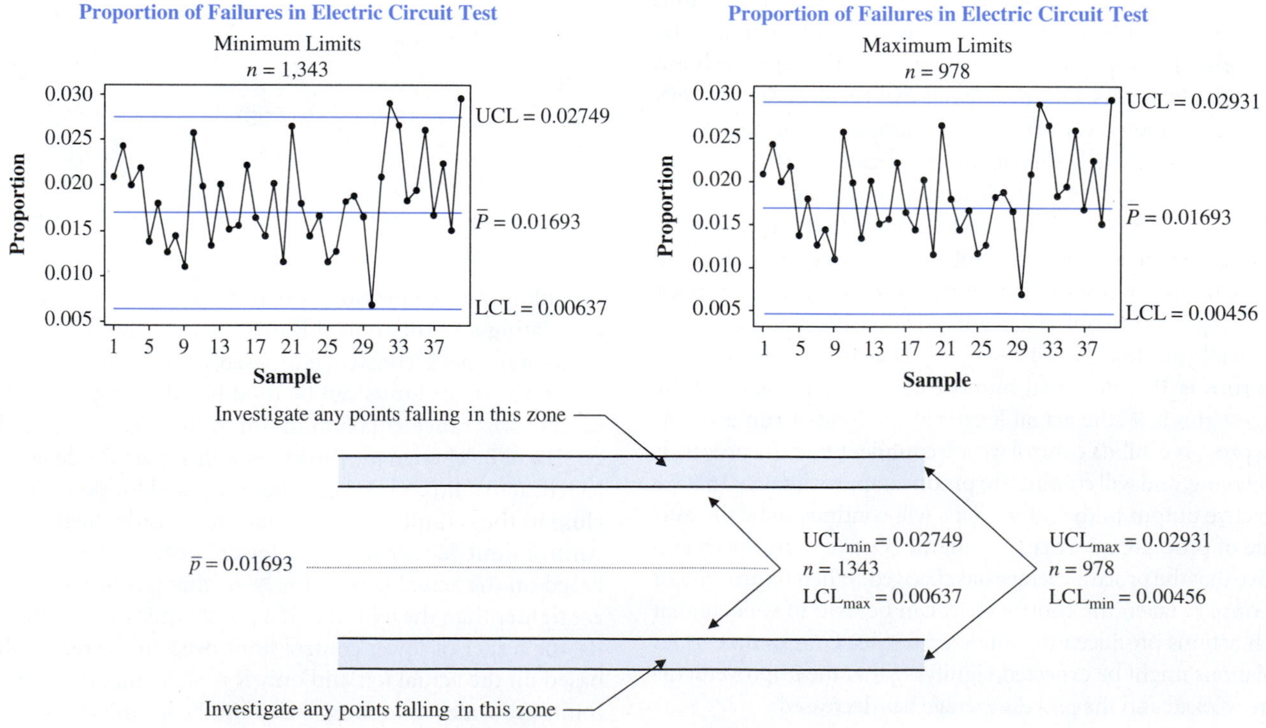

FIGURE 10-5. *p* Chart Control Limits Based on Maximum and Minimum Sample Sizes

limits. One set of limits is based on the largest sample size likely to be encountered. This set of limits is the tightest that might ever be expected, and any point that falls within them will be in control no matter what the actual sample size. The other set of limits is based on the smallest expected sample size, and are the widest limits that are likely to be encountered. Any point that lies outside the widest limits is out of control regardless of the actual sample size. The zone that lies between the two sets of limits is the region where those true limits (based on actual sample sizes) would theoretically be constructed. Therefore, *it is necessary to investigate and possibly construct actual limits only for those points that fall within this zone.* Typically this procedure will not involve many points, and only one control limit (either the upper or lower) will ever have to be actually computed for any particular sample.

Table 10-5 shows the measurements for the circuit tester collected over the next 10 days. These points have been added to the max/min control charts in Figure 10-5 and the expanded chart is displayed as Figure 10-6. The process average appears to have increased. The average of the most recent 10 samples is 22.5% as compared to the average of the first 30 samples of 16.9%. Point 32 falls within the investigation zone and the exact upper control limit needs to be computed. Equation 10-2 produces an upper control limit of 0.0286; therefore the actual measurement of 0.0291 is out of control. At 0.0296 point 40 lies beyond the widest control limit of 0.0293 so there is no necessity to compute the exact limit. Point 40 is clearly out of control. At 0.00687 point 30 lies within the tightest lower control limit and is clearly in control.

TABLE 10-5 Data for Testing an Electric Circuit: The Next 10 Days

Sample Number	No. Defectives np	Sample Size n	Proportion Defective p
31	24	1145	0.0210
32	32	1098	0.0291
33	28	1050	0.0267
34	21	1145	0.0183
35	19	976	0.0195
36	34	1303	0.0261
37	20	1190	0.0168
38	27	1200	0.0225
39	15	995	0.0151
40	33	1115	0.0296
Totals	253	11217	

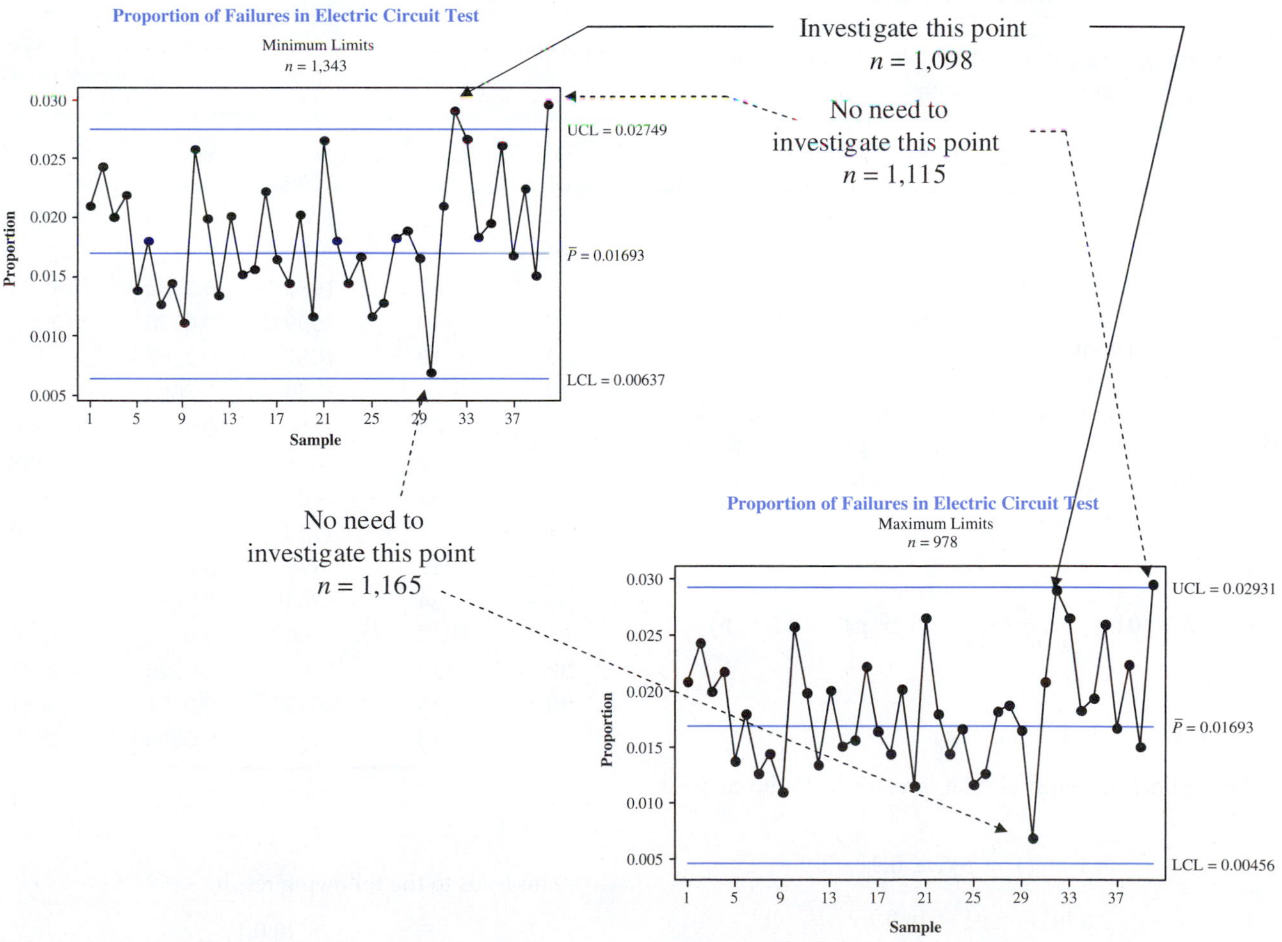

FIGURE 10-6. Operational p Chart with Max/Min Limits for Circuit Breaker Data

In addition to the two out-of-control points, had point 37 not occurred slightly below the centerline, the rule of seven would have been violated in measurements 31 through 37. The next steps would include:

- Revising the limits to reflect the current status of the process, or
- Taking action in the short term to try and reduce the average, thereby restoring statistical control.

10.3.1.4 Standardized Control Chart. A fourth option is to plot a standardized control chart in which each data point is converted to the number of standard deviation units from the average and the control limits are set at ± 3. Equation 10-7 is used to convert the data as follows.

$$z_i = \frac{p_i - \bar{p}}{\sigma_p} = \frac{p_i - \bar{p}}{\sqrt{\dfrac{\bar{p}(1-\bar{p})}{n}}} \qquad \textbf{10-7}$$

In Table 10-6 we have added a column in which the standardized data have been computed. The resulting control chart is shown in Figure 10-7.

TABLE 10-6 Standardized Data for Testing an Electric Circuit

Sample Number	No. Defectives *np*	Sample Size *n*	Proportion Defective *p*	*z*
1	21	1002	0.0210	0.8326
2	27	1109	0.0243	1.7363
3	20	998	0.0200	0.6099
4	27	1234	0.0219	1.1710
5	15	1087	0.0138	−0.9323
6	19	1056	0.0180	0.1200
7	15	1187	0.0126	−1.2796
8	16	1108	0.0144	−0.7786
9	11	998	0.0110	−1.5623
10	26	1007	0.0258	2.0109
11	25	1254	0.0199	0.6556
12	18	1343	0.0134	−1.1471
13	21	1045	0.0201	0.6378
14	19	1253	0.0152	−0.6328
15	16	1023	0.0156	−0.4555
16	22	987	0.0223	1.1454
17	19	1157	0.0164	−0.2817
18	17	1178	0.0144	−0.8052
19	21	1039	0.0202	0.6645
20	12	1039	0.0115	−1.4644
21	26	978	0.0266	2.1643
22	21	1167	0.0180	0.1268
23	18	1243	0.0145	−0.8137
24	20	1198	0.0167	−0.2147
25	14	1203	0.0116	−1.5525
26	15	1180	0.0127	−1.2562
27	20	1095	0.0183	0.1909
28	19	1009	0.0188	0.3203
29	18	1087	0.0166	−0.2385
30	8	1165	0.0069	−2.7693
31	24	1145	0.0210	0.8908
32	32	1098	0.0291	2.9398
33	28	1050	0.0267	2.2627
34	21	1145	0.0183	0.2148
35	19	976	0.0195	0.4666
36	34	1303	0.0261	2.3629
37	20	1190	0.0168	−0.1845
38	27	1200	0.0225	1.3185
39	15	995	0.0151	−0.5851
40	33	1115	0.0296	3.0776

10.3.2 Sample Size Determination

While the determination of sample size is generally not critical in the case of variables measures due to the central limit theorem, determining an adequate sample size is very important with attributes data. Samples must be large enough so that defective units are captured in most if not all of the samples. Some general guidelines can be helpful in determining minimum sample sizes.

Most attributes samples should capture at least one defective.

10.3.2.1 Ninety Percent Rule.[7] This guideline requires that samples be large enough so there is a 90% chance that each sample will contain at least one defective unit. An operational guideline can be derived as follows.

The binomial probability that a sample contains at least one defective is equal to one minus the probability that the sample contains zero defectives. Mathematically this can be derived as follows.

$$prob(X \geq 1) = 1 - prob(x = 0)$$

$$prob(X = 0) = \frac{n!}{0!(n-0)!}p^0(1-p)^n = (1-p)^n$$

Therefore,

$$prob(X \geq 1) = 1 - (1-p)^n$$

The guideline requires that $prob(X \geq 1)$ be at least equal to 0.9.

$$1 - (1-p)^n \geq 0.9$$

$$n \ln(1-p) \geq \ln 0.1$$

This leads to the following result.

$$n \geq \frac{\ln 0.1}{\ln(1-p)} \qquad \textbf{10-8}$$

[7] Rice, W. B. *Control Charts in Factory Management*. New York: Wiley, 1947.

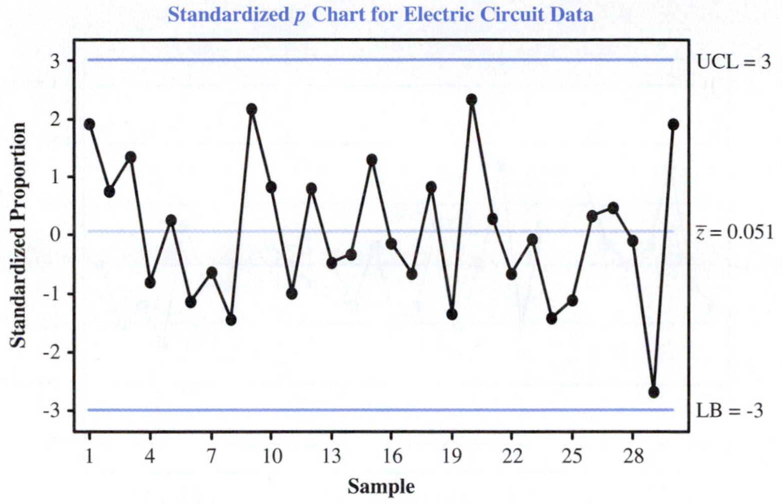

FIGURE 10-7. Standardized Control Chart for Electric Circuits Data

To illustrate the use of Equation 10-8, assume that a certain process is producing an average of 1% defective. The minimum sample size required is

$$n \geq \frac{\ln 0.1}{\ln(1 - 0.01)}$$

$$n \geq \frac{\ln 0.1}{\ln 0.99} = \frac{-2.3026}{-0.0101} = 229.1$$

Using the ninety percent rule, one should select a sample size equal to 230 or higher.

10.3.2.2 Min Five Rule.[8] This guideline requires that each sample be large enough to capture a minimum of five defective units. In a sample of size n, taken from a population producing an average rate of defectives equal to p, the expected number of defectives in each sample will be np. From this we can derive the following operational rule.

$$np \geq 5$$

$$n \geq \frac{5}{p} \qquad \textbf{10-9}$$

Applying Equation 10-9 to the previous example yields the following result.

$$n \geq \frac{5}{0.01}$$

$$n \geq 500$$

Applying the min five rule would require a sample size of at least 500 in this case.

10.3.2.3 Min Two Rule.[9] This guideline requires that samples be large enough so that at least two defectives, on the average, are captured in a prescribed *group of samples*. This rule can be operationalized as follows.

$$\frac{n\bar{p}_1 + n\bar{p}_2 + \cdots + n\bar{p}_k}{k} \geq 2$$

$$\sum_{j=1}^{k} n\bar{p}_j \geq 2k$$

$$n\sum_{j=1}^{k} \bar{p}_j \geq 2k$$

$$\text{If } \bar{p}_1 = \bar{p}_2 = \cdots = \bar{p}_k = \bar{p}$$

$$nk\bar{p} \geq 2k$$

$$n \geq \frac{2}{\bar{p}} \qquad \textbf{10-10}$$

For our example, in which $\bar{p} = 0.01$, Equation 10-10 for the min two guideline would result in a minimum sample size of 200.

10.3.2.4 Fifty Percent Detection Rule.[10] Equations 10-2 and 10-3 and Equations 10-5 and 10-6 define the relationship that exists between sample size and control limits for p and np charts respectively. As the sample size increases control limits tighten, making the control chart more sensitive to changes in the process average. The fifty percent detection rule requires the sample size to be large enough so that the probability of detecting a shift in the process average, by some stipulated amount, over some specified number of consecutive samples k, is 50%. The application of this guideline is dependent on how large of a shift one wants to detect with 50% probability.

Figure 10-8a is a p chart of plug gauge failures for a critical dimension in an automotive plant. Limits have been drawn at the 3σ, 2σ, and 1σ distances from the centerline.

---

[8]Ryan, T. P. *Statistical Methods for Quality Improvement*. New York: Wiley, 1989.

[9]Levine, David M., Patricia P. Ramsey, and Mark L. Berenson. *Business Statistics for Quality and Productivity*. Upper Saddle River, NJ: Prentice Hall, 1995.

[10]Duncan, A. J. *Quality Control and Industrial Statistics*. 5th ed. Homewood, IL: Irwin, 1986.

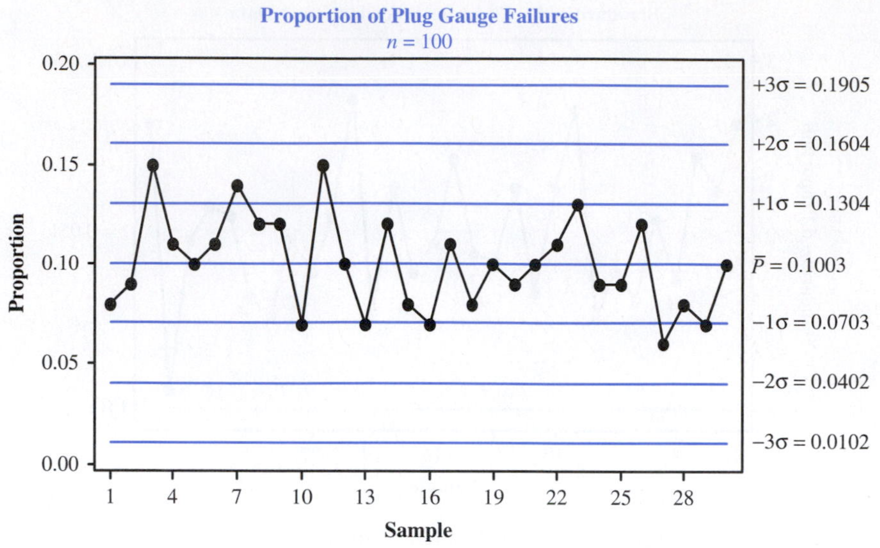

FIGURE 10-8a. *p* Chart for Plug Gauge Failures, $n = 100$

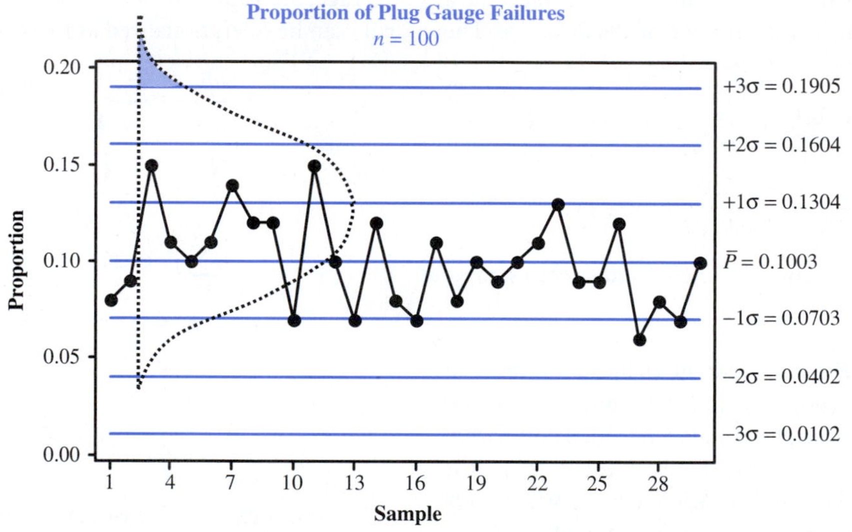

FIGURE 10-8b. *p* Chart for Plug Gauge Failures Showing a 1σ Shift

The sample size for these data was 100. The probability of detecting a shift in average within the next *k* samples is equivalent to one minus the product of the probabilities of failing to detect the shift over *k* consecutive samples. If we wish this probability to be 50%, then we set

$$(\textit{probability of nondetection})^k = 0.5$$

In the example shown in Figure 10-8a, the upper control limit is 0.1905. A 1-sigma shift would result in a new process average of 0.1304. The shaded area on Figure 10-8b represents the probability of detecting a point out of control on the next sample taken after the shift has occurred. To calculate this area we compute a *z* score as follows.

$$z = \frac{0.1905 - 0.1304}{\sqrt{\frac{0.1304(1 - 0.1304)}{100}}} = 1.785$$

From Table II in Appendix B we find,
Probability of detection $= \Phi(1.79) = 0.0367$
Therefore, the probability of nondetection $=$
$1 - 0.0367 = 0.9633$.
Probability of nondetection in *k* consecutive samples $= 0.9633^k = 0.5$.
$k \ln 0.9633 = \ln 0.5$

$$k = \frac{\ln 0.5}{\ln 0.9633} = \frac{-0.6931}{-0.0374} = 18.5 \cong 19$$

Therefore, with a sample size of 100 it will take a run of 19 consecutive points (samples) before the cumulative probability of detecting a 1-sigma shift will reach 0.5.

Now we consider Figure 10-9 showing data that were collected from the same plug gauge process—with the exception that the sample size for these data is 250. If a shift of the

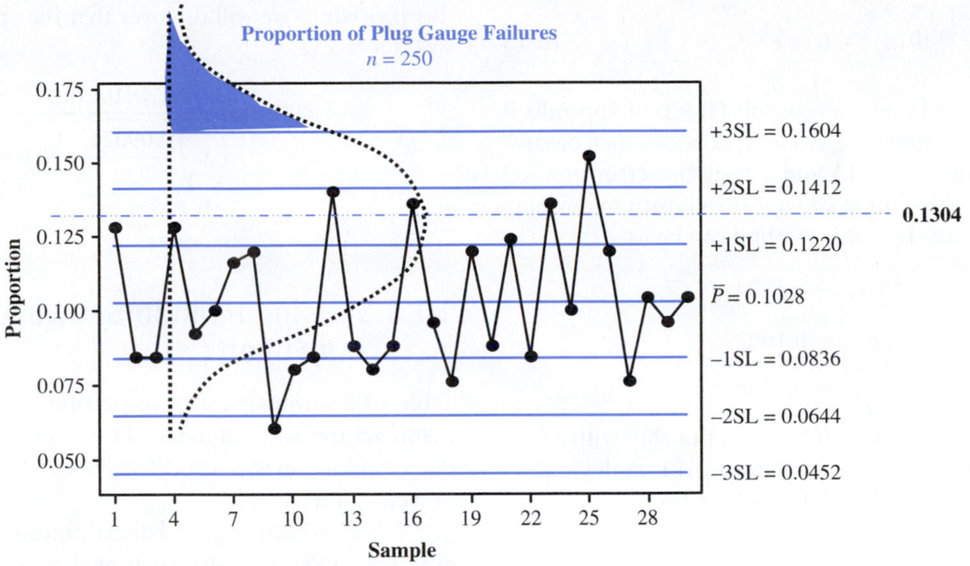

FIGURE 10-9. p Chart for Plug Gauge Failures, $n = 250$

same magnitude—that is, 0.1304—were to occur, the question is how much more sensitive is this chart in detecting the shift than in the previous chart where $n = 100$? The following analysis illustrates the impact on chart sensitivity.

$$z = \frac{0.1604 - 0.1304}{\sqrt{\dfrac{0.1304(1 - 0.1304)}{250}}} = 1.409$$

From Table II in Appendix B we find,
Probability of detection $= \Phi(1.41) = 0.793$
Therefore, the probability of nondetection $=$ $1 - 0.0793 = 0.9207$.
Probability of nondetection in k consecutive samples $= 0.9207^k = 0.5$.
$k \ln 0.9207 = \ln 0.5$

$$k = \frac{\ln 0.5}{\ln 0.9207} = \frac{-0.6931}{-0.0826} = 8.4 \cong 9$$

Thus by increasing the sample size from 100 to 250 the number of consecutive samples required to secure a 50% probability of detecting a shift from 0.1 to 0.13 (a 30% increase) will drop from 19 to 9.

For a specified value of k we can derive an expression for calculating the minimum sample size n.

Let $\bar{p}_{new}$ = the new proportion defective
$\bar{p}_{old}$ = the old proportion defective

The probability of detecting the shift from $\bar{p}_{old}$ to $\bar{p}_{new}$ with any single sample can be estimated by calculating the z score and then consulting normal tables.

$$z = \frac{UCL_p - \bar{p}_{new}}{\sqrt{\dfrac{\bar{p}_{new}(1 - \bar{p}_{new})}{n}}}$$

From Equation 10-2,

$$UCL_p = \bar{p}_{old} + 3\sqrt{\frac{\bar{p}_{old}(1 - \bar{p}_{old})}{n}}$$

Therefore,

$$z = \frac{\bar{p}_{old} + 3\sqrt{\dfrac{\bar{p}_{old}(1 - \bar{p}_{old})}{n}} - \bar{p}_{new}}{\sqrt{\dfrac{\bar{p}_{new}(1 - \bar{p}_{new})}{n}}} \qquad \textbf{10-11}$$

From Equation 10-11 we can obtain

$$z^2(\bar{p}_{new}(1 - \bar{p}_{new})) = n\left(\bar{p}_{old} + 3\sqrt{\frac{\bar{p}_{old}(1 - \bar{p}_{old})}{n}} - \bar{p}_{new}\right)^2$$

We then solve this expression for the required sample size as a function of z and $\bar{p}_{new}$.

$$n = \frac{z^2(\bar{p}_{new}(1 - \bar{p}_{new}))}{\left(\bar{p}_{old} + 3\sqrt{\dfrac{\bar{p}_{old}(1 - \bar{p}_{old})}{n}} - \bar{p}_{new}\right)^2} \qquad \textbf{10-12}$$

The obvious difficulty with solving Equation 10-12 is that the variable n appears on both sides of the equation. However, for a stipulated value of k we can solve for the normal deviate z and then use trial and error to converge on the appropriate value for n. First we note from the fifty percent detection rule that

$$[1 - \Phi(z)]^k = 0.5$$

which can be used to solve for $\Phi(z)$ as a function of k.

$$k \ln[1 - \Phi(z)] = \ln(0.5)$$

$$\ln[1 - \Phi(z)] = \frac{\ln(0.5)}{k} = \frac{-0.693}{k}$$

$$1 - \Phi(z) = e^{\frac{-0.693}{k}}$$

$$\Phi(z) = 1 - e^{\frac{-0.693}{k}} \qquad \textbf{10-13}$$

For any specified k one can consult Table II in Appendix B and locate the corresponding z. The z score can then be substituted into Equation 10-13 and a trial-and-error method used to compute the sample size n. Let us return to our plug gauge example to see how this method can be used.

$$UCL_p = 0.1905$$

$$\bar{p}_{old} = 0.1003$$

$$\bar{p}_{new} = 0.1304$$

Let us assume that we wish to detect this shift with a 50% probability within the first seven consecutive samples following the shift—that is, let $k = 7$. From Equation 10-13,

$$\Phi(z) = 1 - e^{\frac{-0.693}{7}} = 0.0943$$

From Appendix B, Table II, $z \cong 1.315$

We now apply Equation 10-12 iteratively. We will start with $n = 100$.

$$100 = \frac{(1.315)^2(0.1304(0.8696))}{\left(0.1003 + 3\sqrt{\frac{0.1003(0.8997)}{100}} - 0.1304\right)^2}$$

$$100 \neq \frac{0.196}{0.0036} = 54.43$$

Since $100 > 54.43$ we need to try again. Because the numerator on the right side of the equation is independent of n we can change only the denominator. Since the left hand side is less than the right hand side we need to decrease the denominator and this can be done only by increasing the sample size. Therefore, on the second attempt we will set $n = 200$.

$$200 = \frac{0.196}{\left(3\sqrt{\frac{0.09024}{200}} - 0.0301\right)^2}$$

$$200 \neq \frac{0.196}{0.00113} = 173.36$$

A sample size of 200 is still not enough, so we try $n = 240$.

$$240 = \frac{0.196}{\left(3\sqrt{\frac{0.09024}{240}} - 0.0301\right)^2}$$

$$240 \neq \frac{0.196}{0.00079} = 248.71$$

A sample size of 240 is too high, so we now know that the correct sample size to satisfy the fifty percent detection rule is between 200 and 240. We can surmise from our previous results that the correct sample size is closer to 240 than 200 and can start a trial-and-error search back from 240. Continuing this procedure we will discover that the appropriate sample size is $n = 232$.

$$232 = \frac{0.196}{\left(3\sqrt{\frac{0.09024}{232}} - 0.0301\right)^2}$$

$$232 \cong \frac{0.196}{0.000845} = 231.99$$

10.3.3 Using Minitab Software for *p* and *np* Charts

Table 10-7 summarizes the inspection data of a factory that manufactures surgical masks. Over a period of 30 days each day's production was inspected and defective masks were removed and recorded.

Table 10-7 can be pasted directly into a Minitab worksheet as shown in Figure 10-10. Then, as shown in Figure 10-11, we click on three successive dropdown menus: *Stat > Control Charts > Attributes Charts > either p or np.* In our case, since our sample sizes are unequal we must use the p chart option.

The dialog box shown in Figure 10-12 will appear. Enter the columns where the data have been stored for the defective counts and samples sizes as shown. In this case we have placed the count of defectives in column C1 and the corresponding sample sizes in column C2. Note that, had the sample size been constant we would have entered the sample size in the dialog box in lieu of the column C2. By clicking on *Label* we can place a title on our p chart as shown in Figure 10-13. When we click on *OK* the control chart shown in Figure 10-14 appears.

TABLE 10-7 Inspection Data for Surgical Masks

Universal Medical Supply

Surgical Masks Inspection

No. Defectives	Sample Size	No. Defectives	Sample Size
24	1051	24	801
20	1003	24	1081
18	855	17	1236
18	841	22	1049
12	915	17	1114
25	777	18	1090
24	762	23	1125
26	831	22	1183
29	1176	25	867
13	913	18	839
27	859	17	1026
17	830	28	1158
12	803	19	1015
24	935	10	776
23	902	19	997

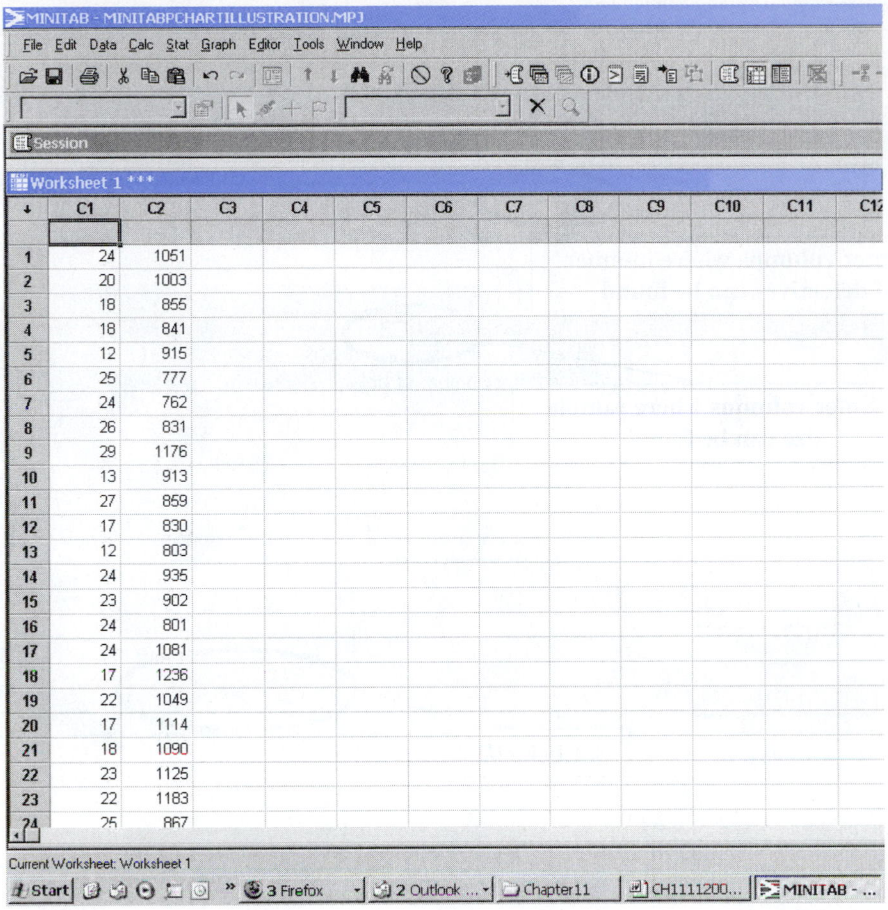

FIGURE 10-10. Minitab Worksheet for Surgical Mask Data

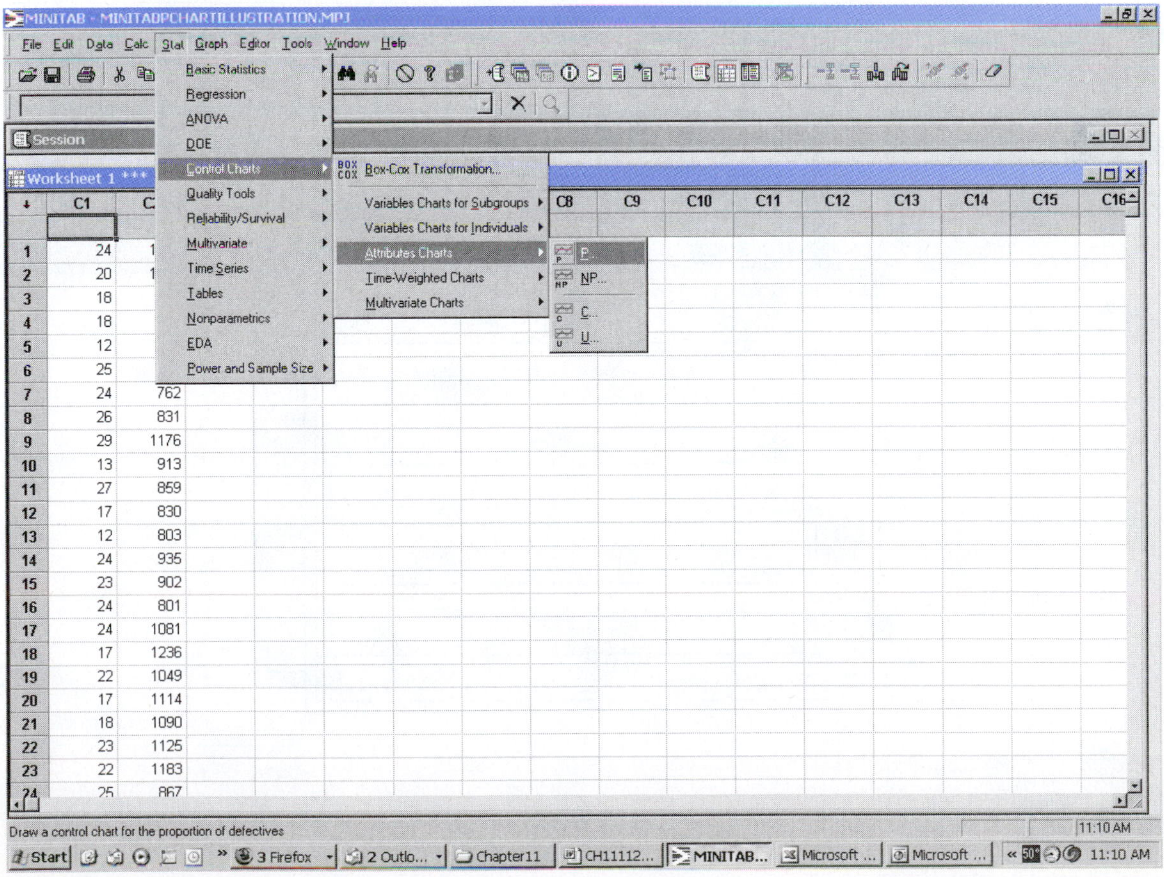

FIGURE 10-11. Minitab Dropdown Menus for Attributes Charts

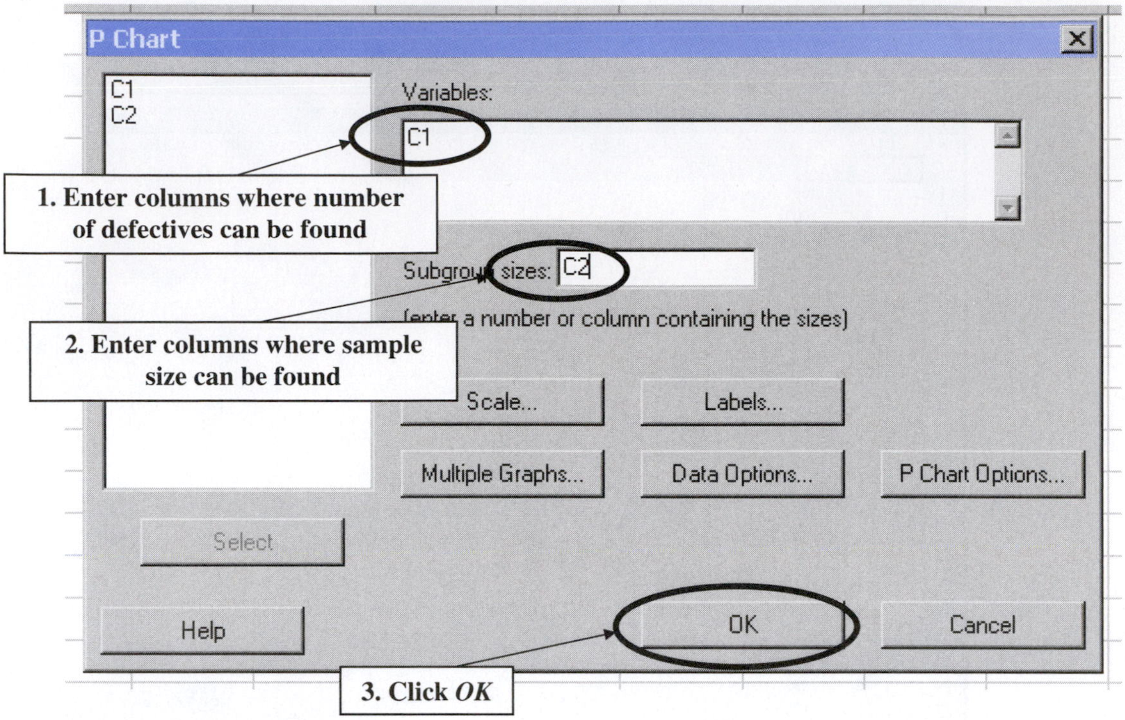

FIGURE 10-12. *p* Chart Dialog Box

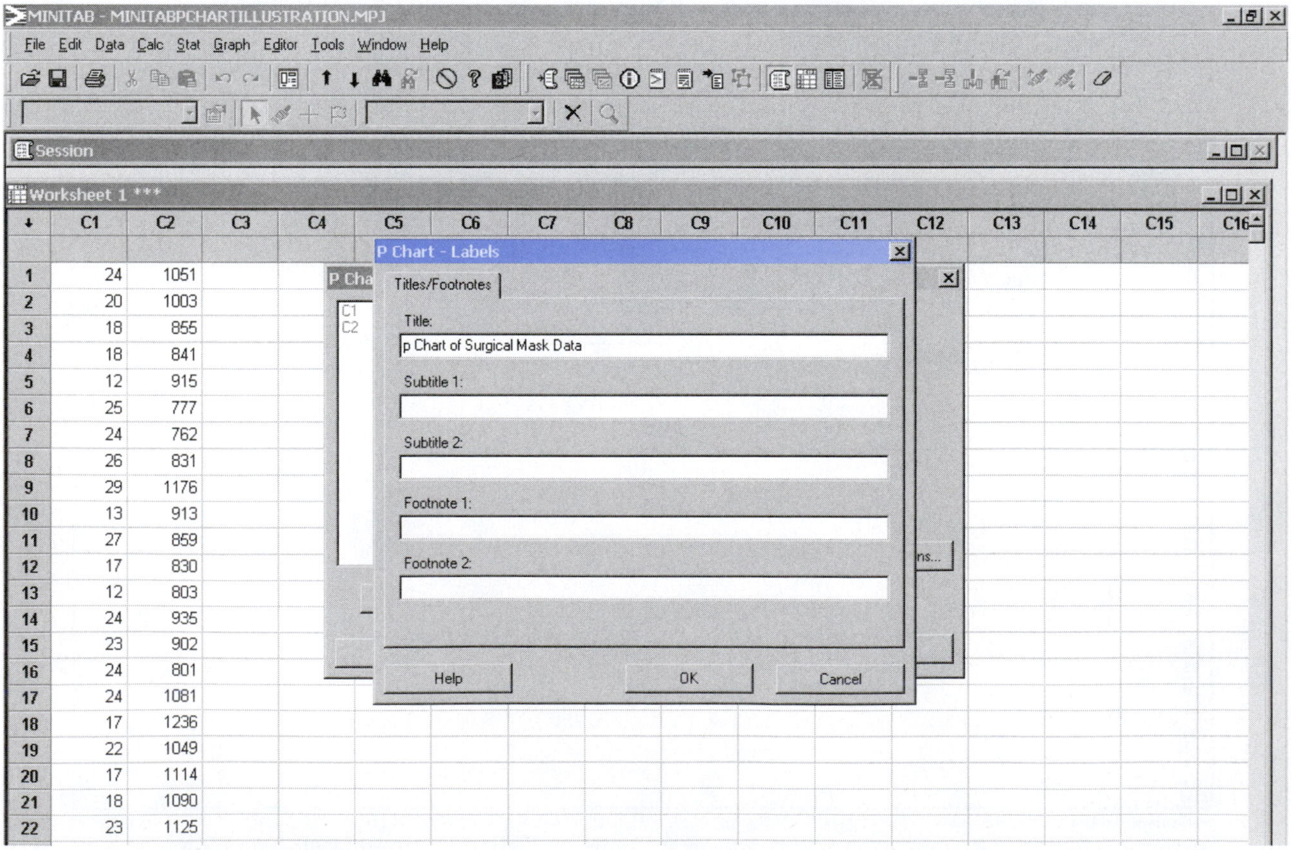

	C1	C2
1	24	1051
2	20	1003
3	18	855
4	18	841
5	12	915
6	25	777
7	24	762
8	26	831
9	29	1176
10	13	913
11	27	859
12	17	830
13	12	803
14	24	935
15	23	902
16	24	801
17	24	1081
18	17	1236
19	22	1049
20	17	1114
21	18	1090
22	23	1125

FIGURE 10-13. Title Box for *p* Chart

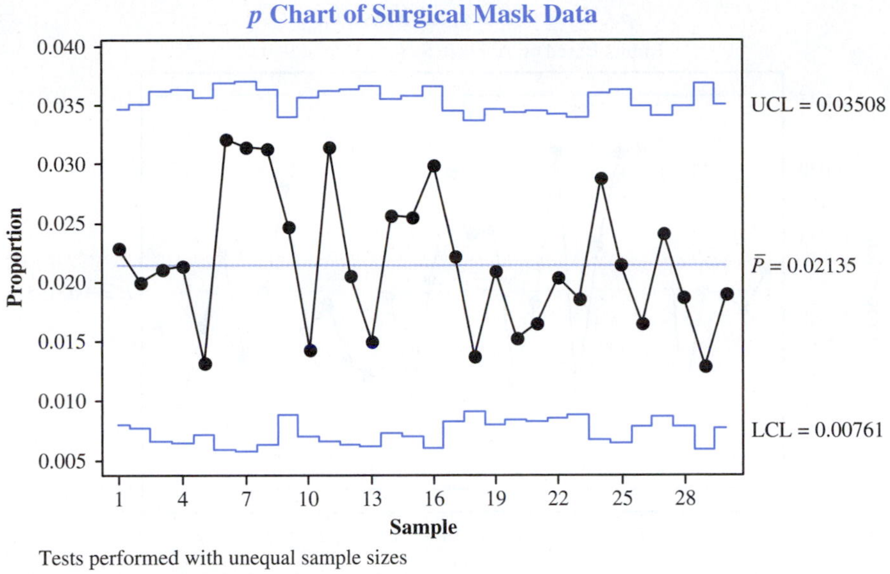

FIGURE 10-14. *p* Chart for Surgical Mask Data with Variable Limits

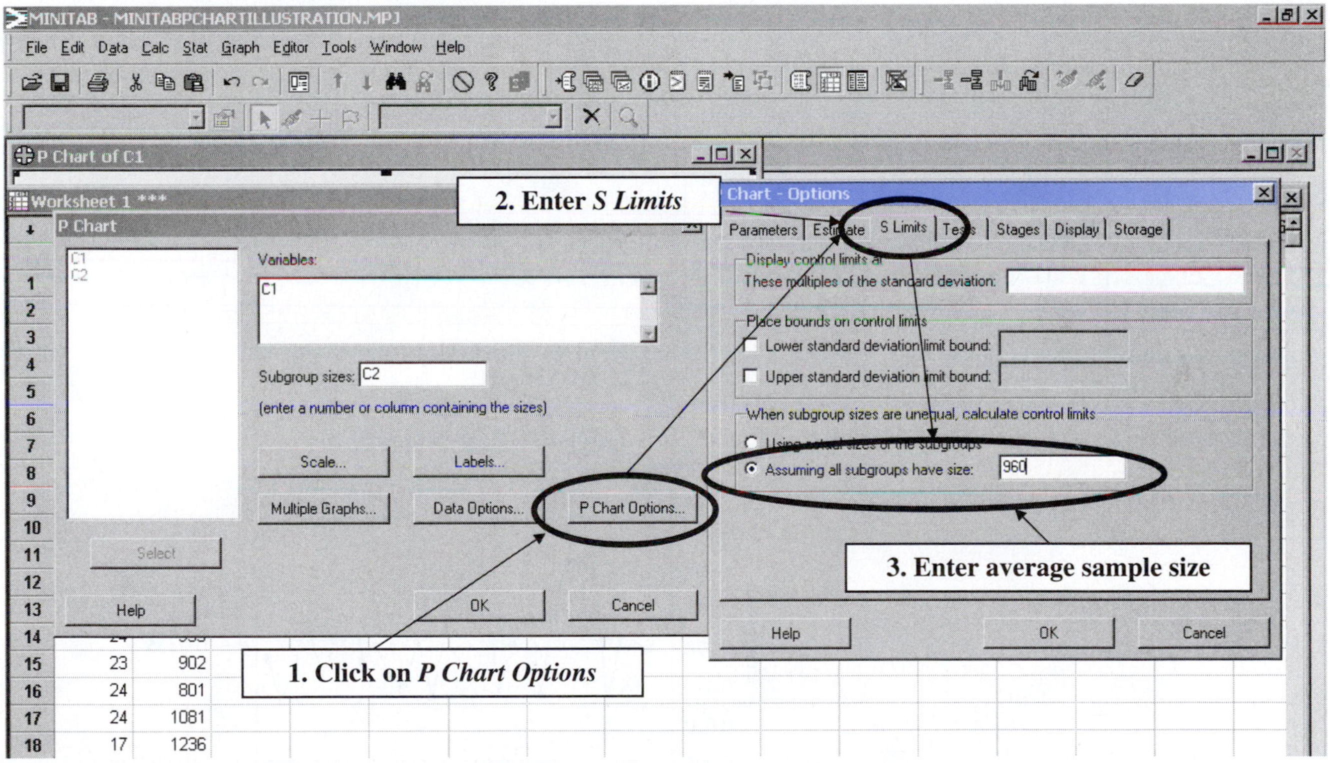

FIGURE 10-15. Entering Average Sample Size into Minitab

We can establish a control chart based on the average samples size ($\bar{n} = 960$) by clicking on *P Chart Options* > *S Limits*, and then entering the average sample size as shown in Figure 10-15. Clicking on *OK* in both open dialog boxes produces the control chart shown in Figure 10-16.

The procedure used for control limits based on average sample size can be repeated for the max/min limits procedure by entering the maximum and minimum sample sizes in lieu of the average sample size. In the data shown in Table 10-7 the smallest sample size recorded was 762 and the largest was 1,236. If we select as the maximum 1,250 and the minimum 750, the pair of charts shown in Figure 10-17 is produced.

To use Minitab to plot a standardized *p* chart we need to use a different technique. Table 10-8 contains the standardized data for the surgical masks process. We first paste the

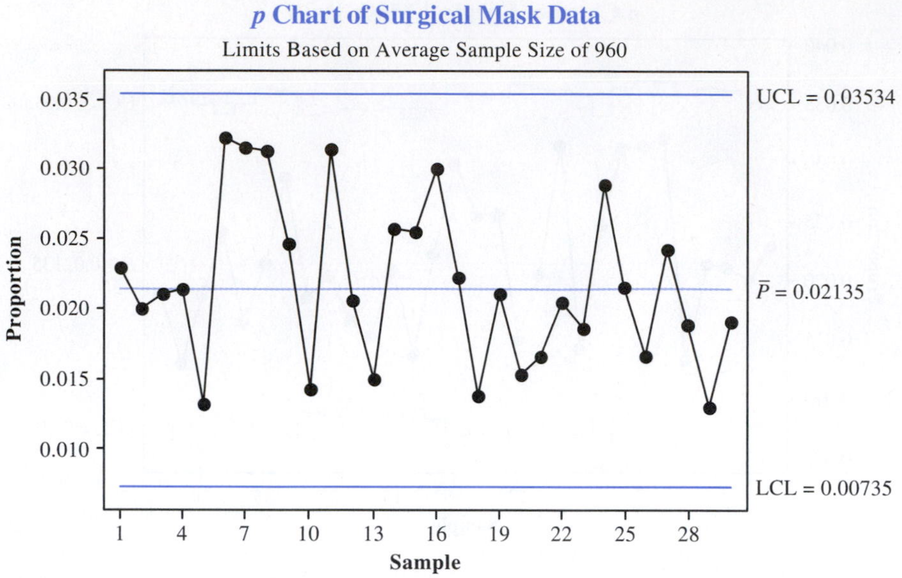

FIGURE 10-16. *p* Chart for Surgical Mask Data Based on Average Sample Size

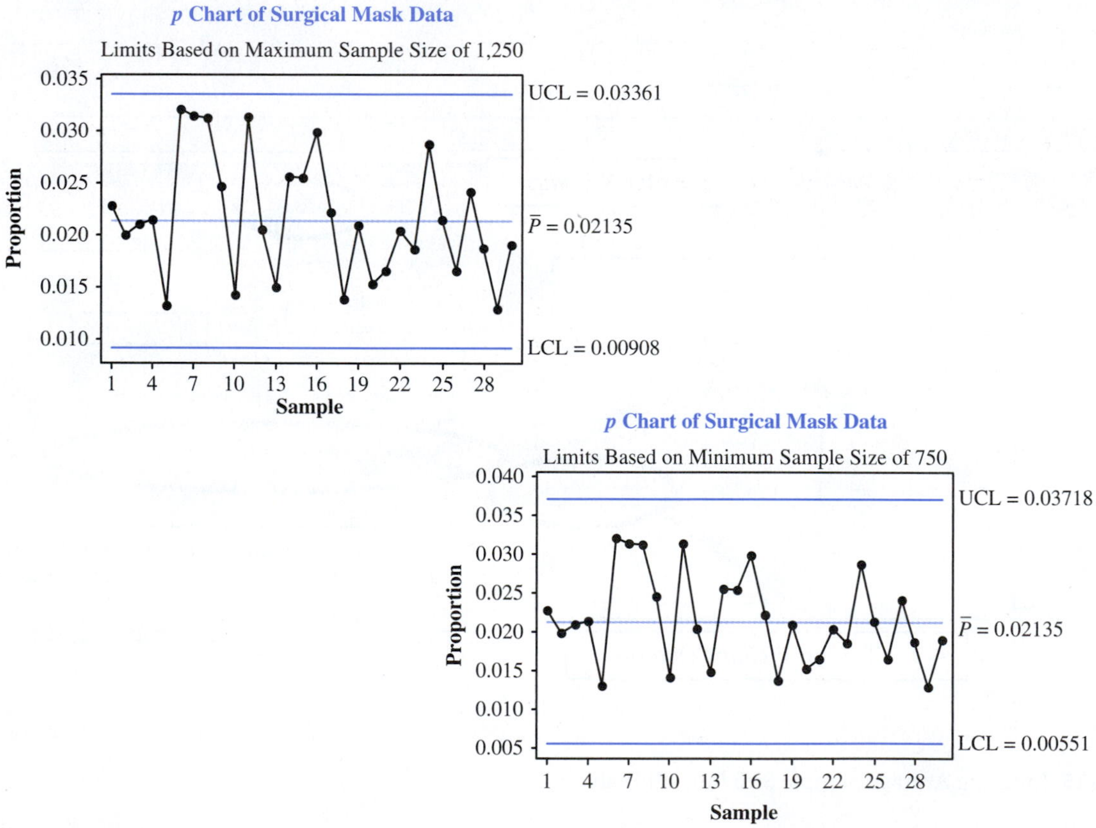

FIGURE 10-17. *p* Charts with Control Limits Based on Max/Min Sample Sizes

standardized data into the Minitab worksheet. Then we follow the dropdown menu sequence shown in Figure 10-18. Note that we treat the standardized chart as an individuals chart, and we choose the option that permits us to plot the individuals chart without the accompanying moving range chart.

By clicking on *I Chart Options* > *S Limits* the fixed upper and lower control limits of +3 and −3 respectively can be established as shown in Figure 10-19. Clicking *OK* in the two open dialog boxes produces the standardized charts shown in Figure 10-20.

TABLE 10-8 Standardized Inspection Data for Surgical Masks

Universal Medical Supply						
Surgical Masks Inspection						
No. Defectives	**Sample Size**	**z**	**No. Defectives**	**Sample Size**	**z**	
24	1051	0.3339	24	801	1.6871	
20	1003	−0.3082	24	1081	0.1945	
18	855	−0.0595	17	1236	−1.8468	
18	841	0.0113	22	1049	−0.0839	
12	915	−1.7228	17	1114	−1.4055	
25	777	2.0883	18	1090	−1.1039	
24	762	1.9384	23	1125	−0.2094	
26	831	1.9826	22	1183	−0.6544	
29	1176	0.7861	25	867	1.5255	
13	913	−1.4859	18	839	0.0215	
27	859	2.0450	17	1026	−1.0588	
17	830	−0.1724	28	1158	0.6670	
12	803	−1.2553	19	1015	−0.5792	
24	935	0.9143	10	776	−1.6305	
23	902	0.8628	19	997	−0.5002	
312	13453		303	15357	0.02135	0.14454

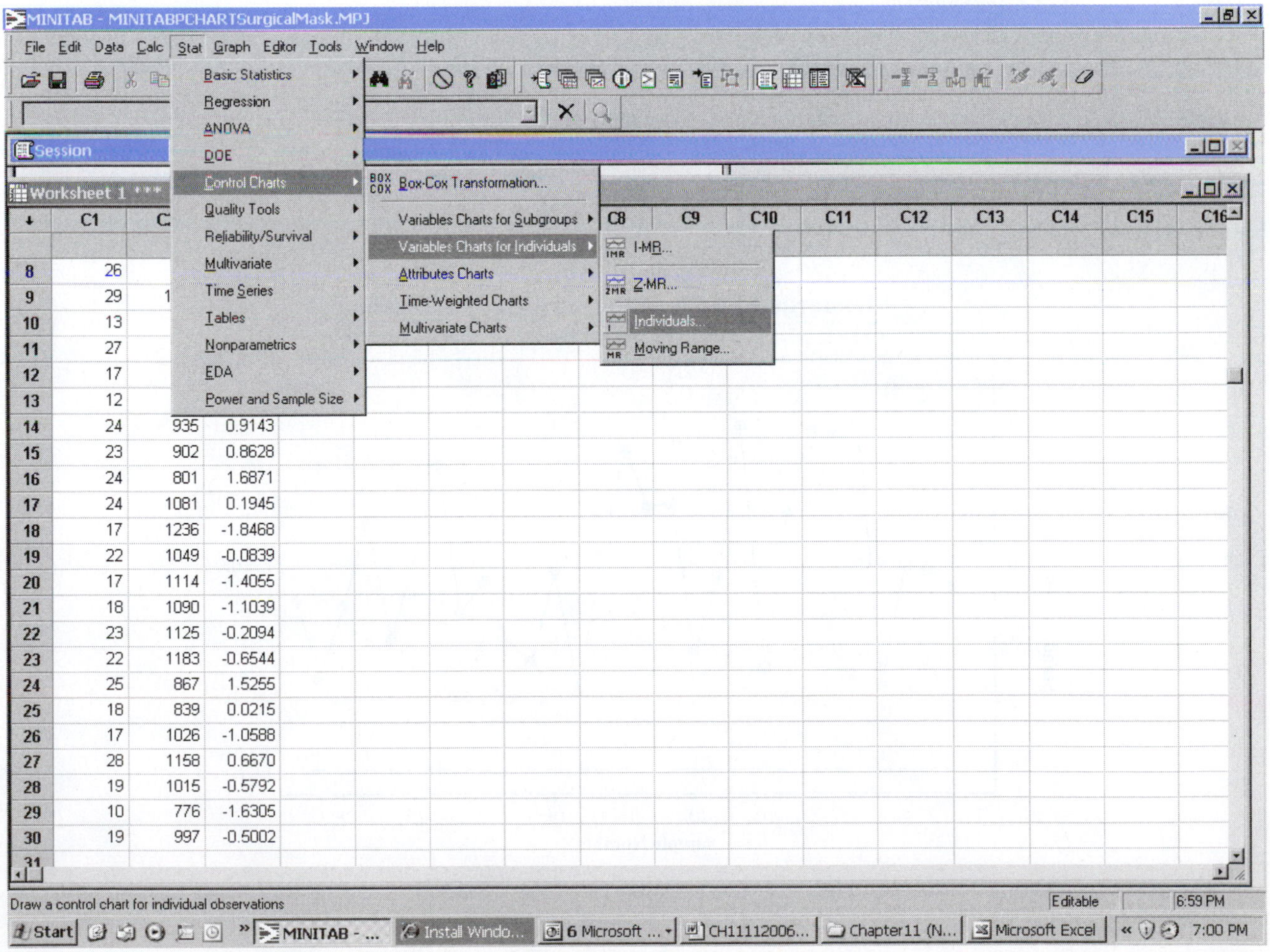

FIGURE 10-18. Dropdown Menu Sequence for Standardized *p* Chart

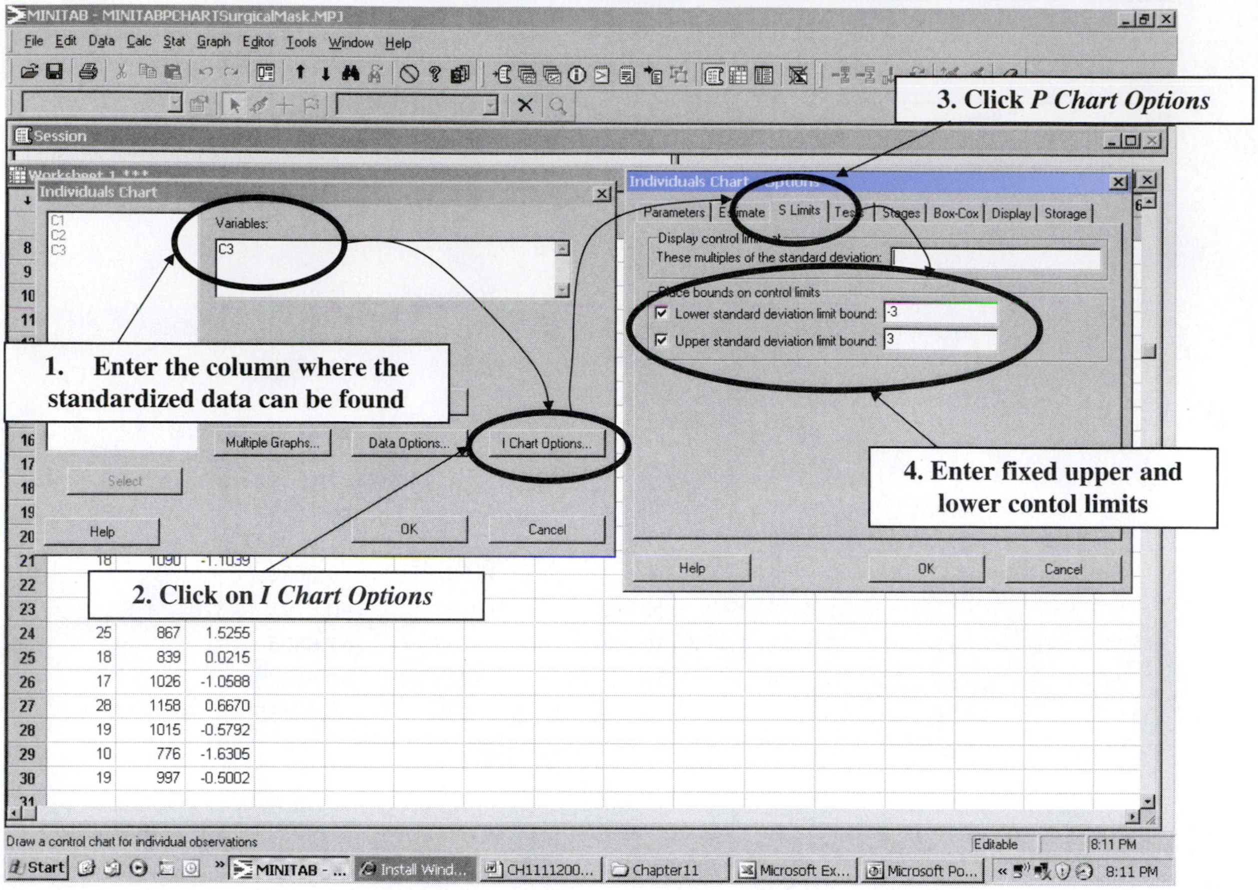

FIGURE 10-19. Placing Fixed Limits on Standardized Control Chart

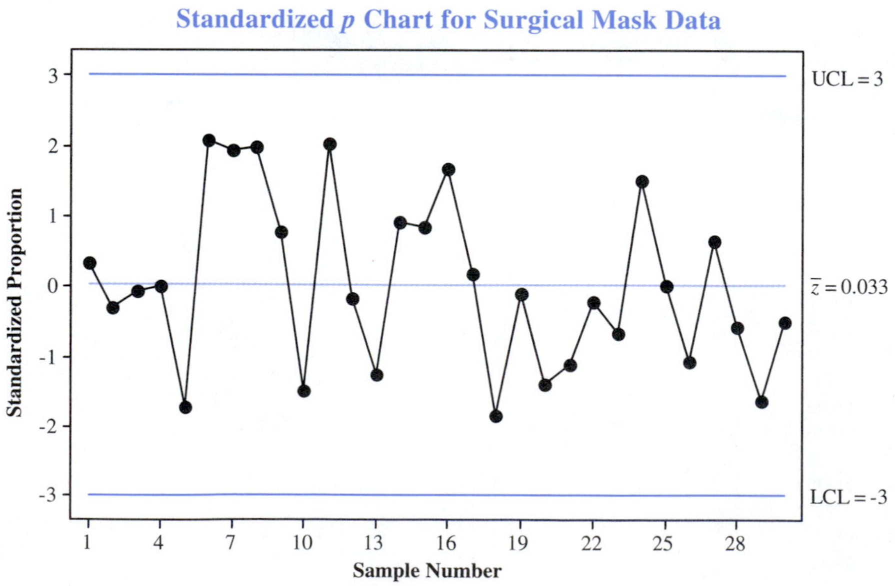

FIGURE 10-20. Standardized p Chart for Surgical Mask Data

10.4 *c* CHARTS AND *u* CHARTS

10.4.1 The *c* Chart for Fraction Defectives

A *p* chart tracks defectives. With *p* chart data the number of items failing to pass inspection is counted, and the total count for each sample is plotted on the *p* chart as a proportion of the sample size. An item in the sample can fail to pass inspection for any number of reasons and each cause of failure is called a *defect.* A defect is a quality attribute that when measured fails to meet minimum requirements. It is an error—some flaw that creates a less-than-perfect condition. Some defects (imperfections) are so minor that by themselves they would not cause a production unit to be labeled as defective.

A defective unit of output may contain more than one defect.

When the interest is in tracking defects rather than defectives, ***c* charts** or ***u* charts** are required. Table 10-9 contains data that have been collected in a factory that manufactures fine-quality pool tables. Prior to shipment, each table produced is closely inspected for any defects such as blemishes in the finish, missing parts, or felt bed imperfections. The inspectors are meticulously trained in how to spot a defect and, for each table inspected, record the number and type of defect. The data in Table 10-9 represent the total count for all types of defects for 30 consecutive tables inspected.

The sample size is defined differently for *c* charts as compared to *p* charts. When *defects* are counted, the concept of a sample size relates to the *opportunity space* within which a defect can occur, rather than to a discreet number of sample units. The fixed opportunity for occurrence is defined as the *inspection unit.* An inspection unit can be an item (e.g., radio, table, dozen eggs), a unit of area (e.g., 100 square meters, 3 lineal feet), a unit of volume (e.g., 1,000 cubic centimeters, 50-gallon drum), some span of time (hour, shift, week), and so on. The number of defects per inspection unit follows a Poisson distribution[11] if the probability of a defect is constant and the occurrence of defects is an independent process. A *c* chart is used when the number of inspection units is constant from sample to sample. The equations for *c* charts are as follows.

TABLE 10-9 Data for the Inspection of Pool Tables

Table No.	No. of Defects c	Table No.	No. of Defects c
1	1	16	2
2	2	17	2
3	0	18	2
4	4	19	1
5	2	20	1
6	3	21	2
7	2	22	5
8	3	23	4
9	2	24	5
10	1	25	2
11	0	26	1
12	4	27	6
13	1	28	4
14	3	29	2
15	2	30	2

Formulae for *c* Charts

$$Centerline = \bar{c} = \frac{\text{Total number of defects}}{\text{Total number of inspection units}} = \frac{\text{Total number of defects}}{k} \qquad \textbf{10-14}$$

$$UCL = \bar{c} + 3\sqrt{\bar{c}} \qquad \textbf{10-15}$$

$$LCL = \max\{\bar{c} - 3\sqrt{\bar{c}}, 0\} \qquad \textbf{10-16}$$

where,

k = number of inspection units

Applying Equations 10-14 through 10-16 to the pool table inspection process,

$$Centerline = \bar{c} = \frac{71}{30} = 2.37$$

$$UCL = 2.37 + 3\sqrt{2.37} = 6.98$$

$$LCL = \max\{2.37 - 3\sqrt{2.37},0\} = \max\{-2.25,0\} = 0$$

Figure 10-21 is a plot of the *c* chart for the pool table data. Since the chart shows a state of statistical control, one can be confident in predicting that the next pool tables produced will, on the average, contain approximately 2.37 defects per table.

10.4.2 The *u* Chart for the Number of Defects per Inspection Unit

When the number of inspection units is not the same from sample to sample, a *u* chart must be used. Table 10-10 is a summary of the data collected on errors in purchase orders over a 30-day period. The number of purchase orders processed is not the same day-to-day, so the opportunity for errors to occur is not constant. An inspection unit was defined as 10 purchase orders. For the purpose of the *u* chart the sample size *n* is the total number of purchase orders processed each day divided by 10. The following equations are used to construct a *u* chart.

Formulae for *u* Charts

$$Centerline = \bar{u} = \frac{\text{Total number of defects}}{\text{Total number of inspection units}} = \frac{\text{Total number of defects}}{k} \qquad \textbf{10-17}$$

$$UCL = \bar{u} + 3\sqrt{\frac{\bar{u}}{n}} \qquad \textbf{10-18}$$

$$LCL = \max\left\{\bar{u} - 3\sqrt{\frac{\bar{u}}{n}}, 0\right\} \qquad \textbf{10-19}$$

where,

k = total number of inspection units

n = number of inspection units in a particular sample

[11]The Poisson distribution and its characteristics are covered in Appendix E.

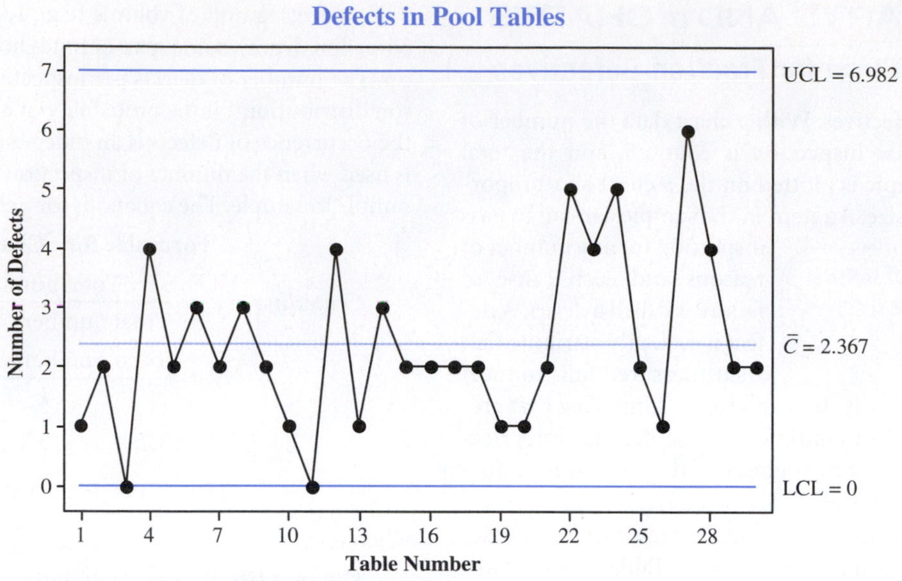

FIGURE 10-21. *c* Chart for Pool Table Defects

TABLE 10-10 Errors in Purchase Orders over a 30-Day Period

Day	No. of Purchase Order Errors	No. of Purchase Orders	No. of Inspection Units
1	12	76	7.6
2	13	78	7.8
3	8	79	7.9
4	7	77	7.7
5	15	93	9.3
6	8	71	7.1
7	12	83	8.3
8	4	89	8.9
9	11	82	8.2
10	7	93	9.3
11	17	81	8.1
12	17	80	8
13	10	93	9.3
14	8	86	8.6
15	13	88	8.8
16	11	85	8.5
17	11	90	9
18	13	89	8.9
19	8	68	6.8
20	6	71	7.1
21	6	91	9.1
22	9	77	7.7
23	7	83	8.3
24	10	79	7.9
25	4	77	7.7
26	8	70	7
27	12	99	9.9
28	4	59	5.9
29	9	88	8.8
30	7	84	8.4

Applying Equation 10-17 to the purchase order problem,

$$\textit{Centerline} = \bar{u} = \frac{287}{245.9} = 1.167$$

As Equations 10-18 and 10-19 show, the limits on a *u* chart differ for each sample. For example on the first day, with 76 purchase orders (7.6 inspection units), we can apply Equations 10-18 and 10-19 to construct the following limits.

$$UCL = 1.167 + 3\sqrt{\frac{1.167}{7.6}} = 2.343$$

$$LCL = \max\left\{1.167 - 3\sqrt{\frac{1.167}{7.6}},0\right\}$$

$$= \max\{-0.0085,0\} = 0$$

On day 27 with 9.9 inspection units,

$$UCL = 1.167 + 3\sqrt{\frac{1.167}{9.9}} = 2.205$$

$$LCL = \max\left\{1.167 - 3\sqrt{\frac{1.167}{9.9}},0\right\}$$

$$= \max\{0.146,0\} = 0.146$$

Figure 10-22 is the *u* chart for the entire 30-day period. Since the *u* chart shows a state of control it can be concluded that the process is stable and is producing on the average approximately 1.167 errors per 10 purchase orders.

10.4.3 Using Minitab Software for *c* and *u* Charts

Table 10-11 provides data that were taken for the inspection of randomly selected rolls of product in a plant that manufactures wallpaper. Rolls were examined for printing defects

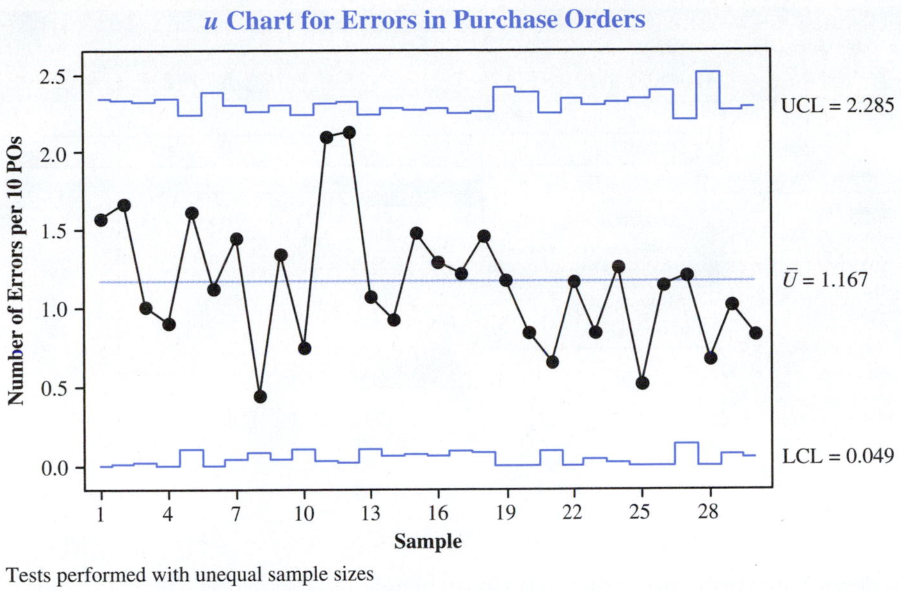

FIGURE 10-22. *u* Chart for Purchase Order Errors

TABLE 10-11 Inspection Data for Wallpaper Manufacturer

Roll No.	No. of Defects	No. of Sq. Ft.	Roll No.	No. of Defects	No. of Sq. Ft.
1	14	126	16	8	112
2	11	126	17	8	112
3	4	126	18	8	126
4	10	100	19	8	112
5	8	112	20	3	112
6	9	126	21	6	126
7	10	112	22	4	126
8	13	112	23	10	126
9	15	140	24	12	112
10	6	112	25	23	126
11	5	100	26	8	112
12	15	140	27	5	100
13	5	100	28	6	126
14	14	126	29	2	112
15	8	112	30	8	140

such as discoloration, spots, lines, creases, and poor color matching. After pasting the spreadsheet data into a Minitab worksheet, we click on the dropdown menu sequence shown in Figure 10-23. When we click on *OK* the dialog box shown in Figure 10-24 appears. Following the sequence shown will produce the *u* chart shown in Figure 10-25.

Figure 10-25 shows an out-of-control process, with an excessive number of defects showing up in roll 25. If possible, production output and conditions around the time that roll 25 was produced should be investigated.

10.5 DIFFERENTIATING BETWEEN *p* CHART AND *c* CHART DATA

Whether data should be plotted on a *p* (*np*) or *c* (*u*) chart can be perplexing. Table 10-12 provides a comparison of the differentiating characteristics of the two chart types. The easiest test is to consider the possible range of values that can occur in any sample. In the case of a *p* chart, the minimum sample value is zero (i.e., no defectives). This is also the minimum value for a *c* chart (i.e., no defects). However, in the case of the *p* chart there is a maximum (*n*)—meaning all are defective, but there is no maximum for the *c* chart (e.g., what is the maximum number of defects that can occur on a sheet of glass?).

A second test is based on the inspection objective. If the objective is to make a final disposition decision on the units inspected—for example, to determine if they can be shipped to customers or forwarded to the next processing station—then the data can be tracked using a *p* (or *np*) chart. If, on the other hand, the objective has nothing to do with whether or not a production unit can be shipped—instead focusing on those qualities that are less-than-perfect—the data are best tracked using a *c* (or *u*) chart.

10.6 ACCEPTANCE SAMPLING BY ATTRIBUTES

10.6.1 Overview and Notation

Acceptance sampling is a statistical tool that can be used to estimate the quality of any fixed supply of material by measuring a small representative percentage of it. This method is often used in certification programs to compare suppliers. There are two types of acceptance sampling plans—those based on attributes measures and those based on variables

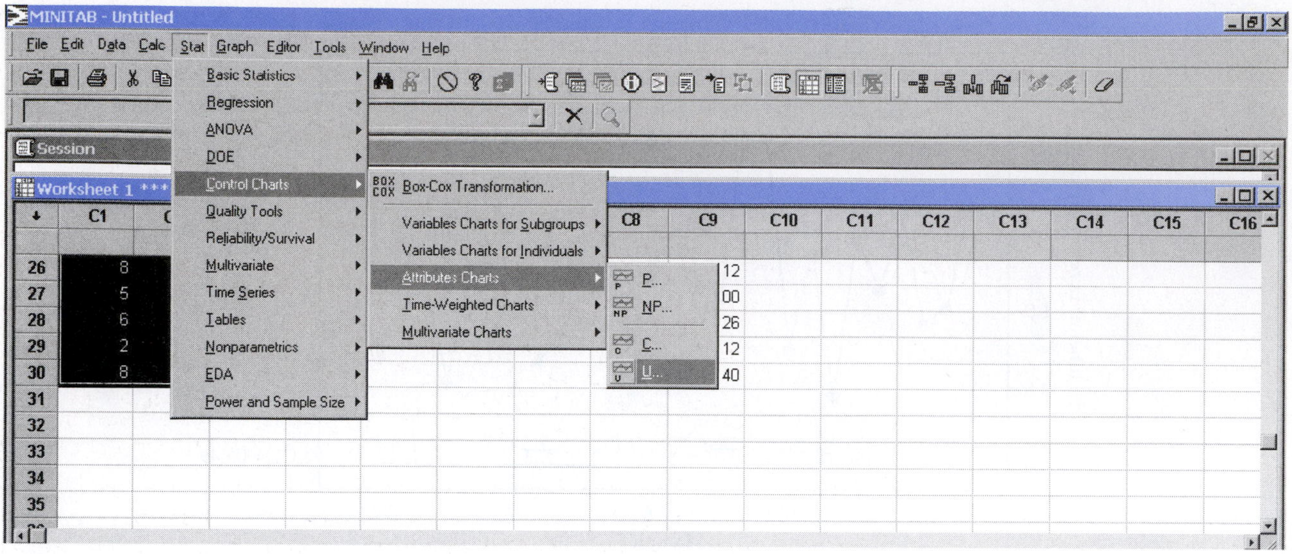

FIGURE 10-23. Dropdown Menu Sequence for *u* Chart Construction

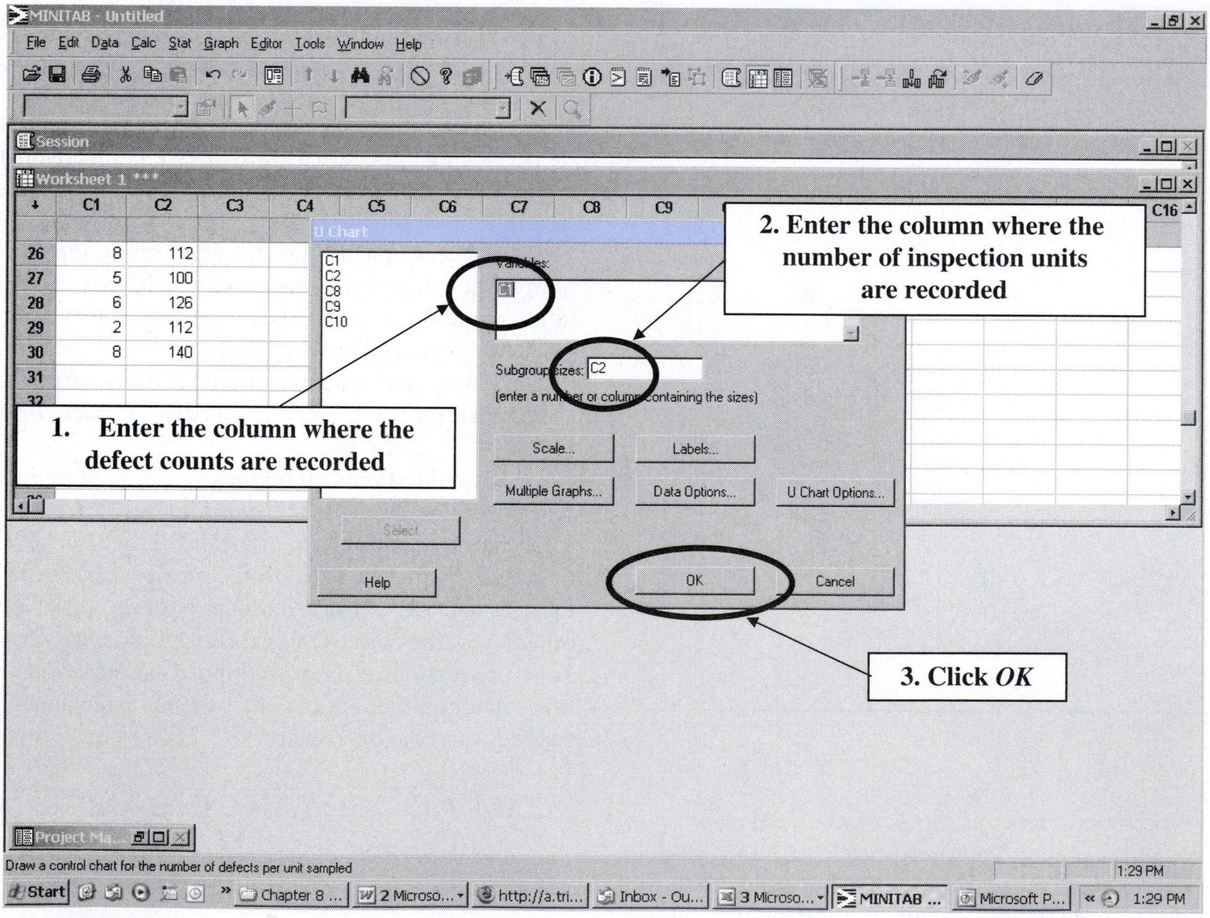

FIGURE 10-24. Constructing a *u* Chart for Wallpaper Data

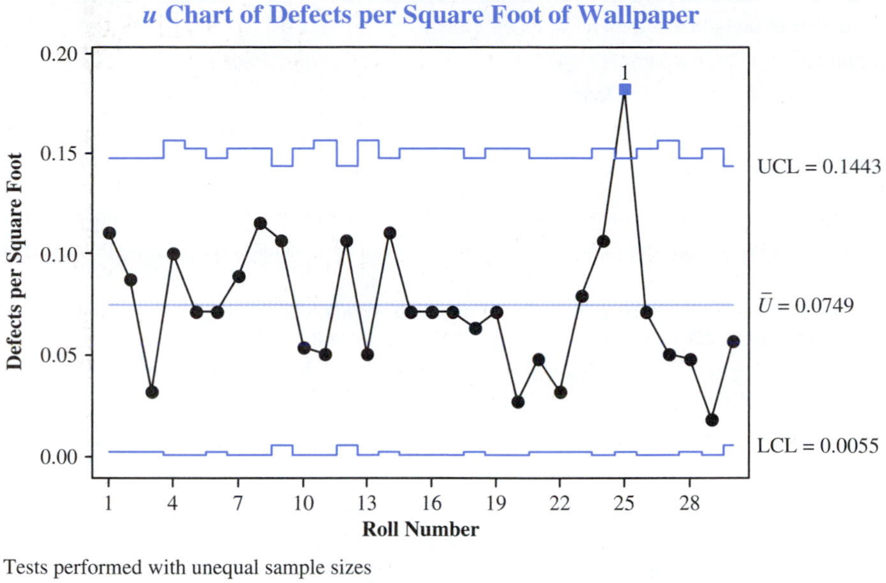

FIGURE 10-25. *u* Chart for Defects in Wallpaper Manufacturing

TABLE 10-12 Guidelines on Differentiating between *p* and *c* Charts

Differentiating Characteristic	*p* Chart	*c* Chart
Minimum count in sample	0	0
Maximum count in sample	*n*	No Limit
Sample size	Discreet production units	Inspection unit (defect opportunity)
Inspection decision	Is item acceptable and can it be shipped?	Are there any imperfections in the inspection unit?

measures. The former can be applied to qualitative or quantitative characteristics, while the latter is restricted to characteristics that are measured quantitatively. This textbook will limit the discussion to attributes plans which are the most widely applied form of acceptance sampling.

In its simplest form, acceptance sampling occurs at the interface between a supplier and customer. When the customer receives a shipment from the supplier a random sample is drawn and, on the basis of the number of nonconforming items found, the entire shipment is either accepted or rejected.

Acceptance sampling results in the acceptance or rejection of a shipment based on a random sample.

Some critical issues arise concerning the sampling and inspection transactions that occur between supplier and customer: *How many units should be included in the sample? What number of nonconforming items in the sample will result in the shipment being accepted? What number of nonconforming items in the sample will result in the shipment being rejected? What is considered to be an acceptable fraction nonconforming? Why not inspect the entire shipment?* The answers to these questions can differ dramatically depending on whether they are posed from the producer's or supplier's point of view.

An acceptance sampling plan answers the first three questions by specifying the sample size and the maximum **acceptance number**—the number of nonconforming items that can be present in a sample if the lot is to be accepted. Developing a plan that is satisfactory to the consumer (who would like to accept only shipments with 100% conformance) and also the producer (who would like to have 100% of the shipments accepted) often requires compromise. The producer must be prepared to accept some risk that a *good* shipment will be rejected, and the consumer must accept some risk that a *bad* shipment will be accepted. The notation shown in Figure 10-26 will help clarify what differentiates a *good* shipment from a *bad* one.

The producer risks that good shipments will be rejected; the consumer risks that bad shipments will be accepted.

Using this notation, we will define the risk assumed by the producer and the consumer. The **producer's risk,** α, is the probability that a shipment of good quality, defined to be the AQL, will be rejected. Similarly, the **consumer's risk,** β, is the probability that a shipment of poor quality, defined to be the LTPD, will be accepted.[12]

[12]Note that this is consistent with the definitions of α and β for statistical hypothesis testing, where the null hypothesis is expressed as: H_0: *the shipment is of good quality.* In this context, α is the Type I error—the probability of rejecting H_0 when it is true. β is the Type II error—the probability of failing to reject H_0 when it is false.

N = total number of units in a shipment (or lot).

n = total number of units in a sample.

d = number of nonconforming units found in a sample of size *n*.

c = acceptance number: the maximum number of nonconforming units that will enable a sample of size *n* to be accepted.

p = proportion of nonconforming units: for small single shipments this will be the proportion nonconforming in the shipment; for large shipments, this will be the process average proportion nonconforming.

AQL = acceptable quality level: the poorest quality that a consumer considers acceptable; this is expressed as a process average percent nonconforming.

LTPD = lot tolerance percent defective: the poorest quality a consumer will accept in a single lot; this is also referred to as the ***LQL*** (lower quality limit) or ***RQL*** (rejectable quality level).

AOQ = average outgoing quality: the expected average percent nonconforming outgoing quality for a given percent nonconforming incoming quality.

AOQL = average outgoing quality limit: the worst possible average percent nonconforming in the outgoing product stream.

FIGURE 10-26. Acceptance Sampling Notation for Attributes Inspection

10.6.2 The Role of Acceptance Sampling

Acceptance sampling generally takes place at the interface between the customer and the supplier and well after production has been completed. For this reason it is an audit tool used to ensure the quality of product by providing guidelines for determining acceptance or rejection based on clearly defined sampling rules. Acceptance sampling is not intended to estimate product quality. What the procedure does well is define the probability of accepting lots at defined quality levels, either the AQL or LTPD, when applied to a series of shipments.

Acceptance sampling is not intended to estimate quality and is not a substitute for SPC.

Acceptance sampling cannot substitute for statistical process control. It does not measure process stability and cannot lead to improvements. The objective is simply to determine the degree of conformance of a fixed supply of material, and acceptance sampling should not be used as a misguided attempt to inspect quality into the product. By the time a lot is presented to an acceptance sampling station it is too late to influence the outcome. It is far better to prevent the occurrence of defects through good quality control policies at the point of production than to try and detect defectives after the fact.

It is generally unwise to rely on acceptance sampling for quality assurance. In lean production environments there is no time for corrective actions should a shipment be rejected. If high-quality materials are not available when and where they are needed, production runs may have to be aborted, schedules altered, or commitments may be missed. Supplier certification procedures normally require SPC documentation (on an ongoing basis) to ensure that suppliers are controlling quality at the source. If a supplier's process is capable, on-target, and in statistical control, there is little need for incoming inspection.

The role of acceptance sampling is essentially to provide some realistic compromise between 100% inspection and no inspection at all when there is insufficient process history to effectively use SPC—for example, in the case of new products or purchased materials from new suppliers. In such cases, acceptance sampling may be implemented as a temporary measure until adequate process history can be accumulated. Acceptance sampling can also be useful to monitor the output from processes that are unstable, not capable of meeting requirements, or both. Although such processes are being improved, acceptance sampling might offer an attractive alternative to 100% inspection so that the release of defective material to customers can be controlled.

Compared to 100% inspection, acceptance sampling has some advantages. It is less time-consuming and costly to inspect a small sample instead of an entire shipment. If the sampling procedure involves destructive testing, such as fracturing a structural beam, 100% sampling is not feasible. Inspector fatigue is also a factor that favors sampling inspection. The higher the workload for inspectors the more likely the inspectors will grow weary and make mistakes. Acceptance sampling can also provide improvement incentives. If a lot undergoes a 100% inspection, usually it is only the defective units that are returned to the supplier. However, with acceptance sampling, if the quality is bad enough the entire shipment is rejected and returned to the supplier. Faced with this prospect a supplier may take greater care to ensure the quality is at least good enough to pass the acceptance criteria.

Acceptance sampling can play an important role where there is inadequate process history.

On the negative side, acceptance sampling carries with it the producer's and consumer's risks that were mentioned previously. Under 100% inspection, there are no α and β risks, and no uncertainty because (barring any inspection error) the true fraction nonconforming is known at the conclusion of the inspection procedure. Acceptance sampling is also somewhat harder to administer than a 100% inspection procedure. Sampling schemes must be carefully planned and thoroughly documented. If entire lots are inspected there is no need for detailed planning and the only documentation required is a lot-by-lot count of the numbers and types of defects encountered.

10.6.3 Randomness in Sampling

Acceptance sampling requires that inspection be performed on a **random sample.** This begs the question: *What constitutes a random sample*? Suppose a shipment consists of 2,000 identical components and that the sampling plan specifies a sample size of 100 units. If the sample is truly random each of the 2,000 components has an equal probability of inclusion in the sample thereby ensuring that the sample is representative of the lot.

Acceptance sampling depends on the ability to take a random sample.

However, a sample may be far from representative. Humans draw samples and, if improperly trained, will seek the easiest and most efficient means to achieve results. Suppose that a shipment of 2,000 components arrives on 20 pallets. The receiving agent removes the first pallet from the truck and proceeds to use the 100 units on that pallet to form the required sample of 100. If the lot is rejected the clerk need only place the pallet back on the truck and send the entire shipment back to the supplier. While minimizing the time and effort required at receiving, this practice may not produce random samples. The 100 components on the first pallet have a 100% probability of inclusion in the sample, but the remaining 1,900 components have 0% probability of inclusion.

If appropriate practices are not followed, acceptance sampling can produce risk factors that are dramatically different from those stipulated in the sampling design. In the previous example, if the components were known to have been thoroughly mixed prior to packaging, palletizing, and loading, a random sample could be obtained by selecting 100 units from anywhere in the lot—including taking the entire sample from the first pallet. However, in general components are packaged and palletized in the order of their production. It is impractical to try and deliberately homogenize the product prior to shipment. Therefore, drawing all the sampling units from a single point in the shipment could result in a biased sample.

While it appears that the receiving agent could be blamed for taking a poor sample, the fault often can be traced to the lack of proper training. Inspection personnel must be thoroughly indoctrinated on why random sampling is necessary and how representative samples can be obtained without wasting unnecessary time and effort. It is essential that operational definitions be established to clearly communicate to all inspectors how samples should be selected, what criteria are to be used, and how conformance/nonconformance decisions are to be made.

For small lots a true random sample can be constructed using the following procedure. Continuing with our example of the shipment of 2,000, one would assign a unique number (from 1 to 2,000), to each of the components in the shipment. The sample is formed using a random number generator to select 100 of the 2,000 numbers. The components corresponding to these numbers are the ones that will be pulled for inspection.

For large lots it may be impractical to assign a unique number to each unit. In this case, the lot can be *stratified* into small groups with equal numbers of units in the sample taken from each group. In our example, there are 100 units on each of 20 pallets. A stratified sample of $n = 100$ would include five components from each pallet. Suppose that each pallet consists of five layers each of 20 units. The lot can then be further subdivided into layers of product, with one unit from each layer selected for the sample. While this stratification is not random in a pure sense, it does increase a sample's overall representation by ensuring that the sample is taken from different points throughout the lot.

10.6.4 Types of Attributes Acceptance Sampling Plans

There are four types of attributes acceptance sampling plans: **single sampling, double sampling, multiple sampling,** and **sequential sampling.** A single sampling plan requires that a sample of size n be selected from a shipment. If the number of nonconforming units in the sample, d, is less than or equal to the acceptance number, c, the lot is accepted; if $d > c$, the lot is rejected.

Double sampling plans are more complex and may require that a second sample be taken before the disposition of a lot can be determined. Under this type of plan, a shipment may be accepted, rejected, or subjected to a second sample on the basis of the results of the first sample with size n_1. If a second sample of size n_2 is required, the results of both samples are combined to determine whether the lot is accepted or rejected. Depending on the required sample sizes, a double sampling plan may require inspection of fewer units than single sampling provided a decision can be reached on the first sample. Generally, when a second sample is required double sampling requires more inspection than a single sampling plan.

Under a double sampling plan the decision to accept or reject may be deferred until a second sample is taken.

Multiple sampling extends the concept of double sampling by requiring, under certain circumstances, that more than two samples be inspected before a decision can be reached. These plans are similar to double sampling in that a decision to accept the lot, reject the lot, or take another sample is made as each subsequent sample is collected and inspected. The sampling continues until a clear decision to accept or reject can be reached, or until a prespecified number of units have been inspected. Whereas multiple sampling plans are more complicated, they may result in the inspection of fewer units on the average than

Under multiple sampling a disposition decision may be deferred until a prespecified number of units or samples have been inspected.

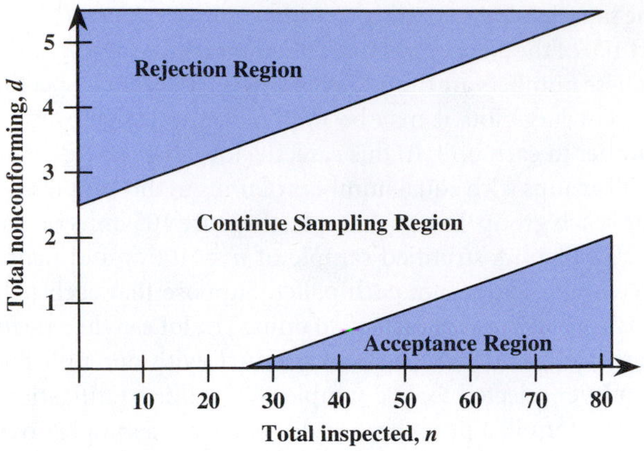

FIGURE 10-27. Sampling Regions in Sequential Sampling

would be required under comparable single or double sampling plans.

Sequential sampling is another extension of double sampling. Under this type of sampling plan, individual units are inspected and, after each unit, a decision to accept, reject, or continue sampling is made. As the total number of units sampled increases, the criteria for accepting and rejecting are changed such that, theoretically, the sampling could continue until all units in the shipment have been inspected. An example of the decision regions for sequential sampling is shown in Figure 10-27.

Under sequential sampling a disposition decision may be possible on a unit-by-unit inspection basis.

To illustrate how attributes acceptance sampling works we have restricted our discussion to single and double sampling plans. For additional information concerning multiple and sequential sampling plans, or to understand how acceptance sampling plans can be designed for variables measurement, students should consult supplemental resource materials.[13]

10.6.5 Single Sampling Plans

Single sampling is the simplest form of lot-by-lot acceptance based on attributes. A single sampling plan requires three parameters: the shipment (or lot) size (N), the sample size (n), and the acceptance number (c). If the number of nonconforming units found in the sample, d, is less than or equal to c, the lot is accepted. If $d > c$, the lot is rejected. For example, suppose the shipment size is $N = 2{,}000$, the sample size is $n = 100$, and the acceptance number is $c = 2$. Under this plan, a random sample of 100 units is selected from the shipment. If the number of nonconforming units in the sample is less than or equal to 2, the shipment is accepted; otherwise, the shipment is rejected.

[13]For example, Grant, Eugene, and Richard Leavenworth. *Statistical Quality Control*. 7th ed. New York: McGraw-Hill, 1996.

10.6.5.1 Producer's Risk, Consumer's Risk, and OC Curves.

An **operating characteristic** (OC) **curve** is a plot of the probability of accepting a shipment, P_A, versus the fraction nonconforming, p. Assuming the shipment is large or that it is a part of a continuing series of shipments, the binomial distribution applies and the binomial probability function[14] can be used to calculate P_A over a range of values for fraction nonconforming.

An OC curve is a plot of the probability of acceptance against the quality of the lot.

Example OC Curve for a Single Sampling Plan

Consider a single sampling plan for shipment size $N = 2{,}000$, sample size $n = 100$, and acceptance number $c = 2$. For any fraction nonconforming p, the binomial probability of acceptance is

$$P_A = \sum_{d=0}^{2}\left(\frac{100!}{d!(100-d)!}p^d(1-p)^{100-d}\right)$$

For a fraction nonconforming $p = 0.025$, the probability of acceptance is

$$P_A = \sum_{d=0}^{2}\left(\frac{100!}{d!(100-d)!}(0.025)^d(1-0.025)^{100-d}\right)$$
$$= 0.5422$$

This procedure is repeated, computing P_A for a number of different p values ranging from 0 to 1. The OC curve, constructed by plotting P_A against p, is shown in Figure 10-28.

The concepts of *producer's risk*, α, and *consumer's risk*, β, were previously defined respectively as the probabilities of rejecting a good shipment and of accepting a bad shipment. Single sampling plans are usually designed so that they simultaneously satisfy a *producer's criterion*, $P_A \geq 1 - \alpha$ at $p_1 = \text{AQL}$, and a *consumer's criterion*, $P_A \leq \beta$ at $p_2 = \text{LTPD}$. This requires a plan with an OC curve that passes through these two points. To achieve this, the two nonlinear equations (in two unknowns) shown as Equations 10-20 and 10-21 must be solved.

$$\alpha = 1 - \sum_{d=0}^{c}\frac{n!}{d!(n-d)!}p_1^d(1-p_1)^{n-d} \quad \textbf{10-20}$$

$$\beta = \sum_{d=0}^{c}\frac{n!}{d!(n-d)!}p_2^d(1-p_2)^{n-d} \quad \textbf{10-21}$$

The solution of this system of equations requires a complex iterative approach, where a value of c is assumed and values for n that satisfy the producer's risk and values that satisfy the consumer's risk are determined. If no sample size n can be found that simultaneously satisfies both producer and consumer risks, then it is necessary to select a larger acceptance number and repeat the process. Since the cost of sampling increases with sample size, the objective is to determine the plan with the smallest sample size that satisfies both the stipulated producer and consumer risks. The plan selected will typically meet one of the risk criteria and exceed the other.

[14]Details on the binomial distribution can be found in Appendix E.

p	P_A
0.00	1.0000
0.01	0.9206
0.02	0.6767
0.03	0.4198
0.04	0.2321
0.05	0.1183
0.06	0.0566
0.07	0.0258
0.08	0.0113
0.09	0.0048
0.10	0.0019

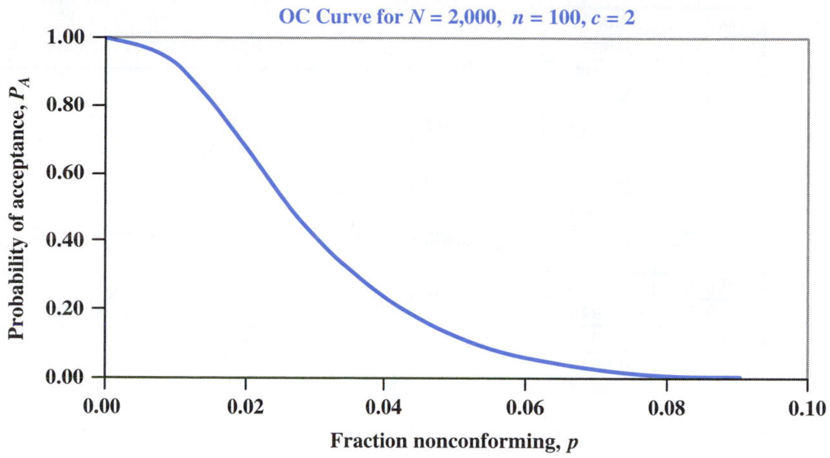

FIGURE 10-28. An OC Curve

A number of approaches have been devised to facilitate the development of such plans, including Larson's binomial nomograph,[15] Grubbs's approach,[16] MIL-STD-105D,[17] and the Dodge-Romig tables.[18]

10.6.5.2 Measures of Plan Performance—AOQL and ATI. Under single sampling, an entire shipment is either accepted or rejected on the basis of one sample of n units. In an environment of JIT, lean production, or Six Sigma, a customer does not normally have the luxury of rejecting incoming materials. In the absence of on-hand inventories, in sufficient quantities to buffer against quality problems, returning a shipment to the vendor would likely result in lost production. Some sampling plans provide for *rectifying inspection* to allow the customer to go ahead and use the conforming units within a rejected shipment.

Under rectifying inspection, if a lot is not accepted, it is subjected to a 100% screening inspection operation to separate the conforming materials from the nonconforming. Conforming material can then proceed to production while the nonconforming material is returned to the supplier to be replaced with good product. Even when a shipment is accepted, any nonconforming units found in the sample are also replaced by the supplier with conforming product. This has the effect of increasing the average quality level (the AOQ) accepted over several shipments because all rejected lots become 100% conforming following screening inspection and replacement. For shipments of size N, samples of size n, and incoming fraction nonconforming p, the AOQ for a single sampling plan can be calculated using Equations 10-22 and 10-23 as follows.

Under rectifying inspection rejected lots undergo 100% inspection to screen out defective units.

$$AOQ = \frac{P_A p(N - n)}{N} \qquad \textbf{10-22}$$

$$AOQ = \frac{\sum_{d=0}^{c} \frac{n!}{d!(n-d)!} p^{d+1}(1-p)^{n-d}(N-n)}{N} \qquad \textbf{10-23}$$

An AOQ curve is generated by solving Equation 10-23 for several values of p, then plotting AOQ against p. The average outgoing quality limit, AOQL, is the maximum value of the curve and represents the *worst average level of quality that the customer would receive under rectifying inspection.* AOQL is one of the important metrics for evaluating alternative sampling plans. It may be determined visually from the AOQ curve or approximated by taking the maximum AOQ value from the table used to plot the curve.

Example AOQ Curve and AOQL for Single Sampling Plan

Consider a single sampling plan for shipment size $N = 2{,}000$, sample size $n = 100$, and acceptance number $c = 2$. Applying Equation 10-23, the AOQ for fraction nonconforming $p = 0.04$ is

$$\text{AOQ} = \frac{\sum_{d=0}^{2} \frac{100!}{d!(100-d)!} 0.04^{d+1}(0.96)^{100-d}(1900)}{2000}$$

$$= 0.0088$$

By repeating this procedure, and calculating the AOQ for a number of p values (ranging from 0 to 1), an AOQ curve can be constructed (a plot of AOQ versus p). The AOQ curve for the example cited previously is shown in Figure 10-29, and indicates that the AOQL $\cong$ 0.0130.

---

[15]Larson, H. R. "A Nomograph of the Cumulative Binomial Distribution." *Western Electric Engineer,* April 1965.

[16]Grubbs, F. E. "On Designing Single Sampling Plans." *Annals of Mathematical Statistics,* 1949.

[17]American Society for Quality Control, American National Standard-Sampling Procedures and Tables for Inspection by Attributes, ANSI/ASQC Z1.4-1981, 1981.

[18]Dodge, H. F., and H. G. Romig. *Sampling Inspection Tables—Single and Double Sampling.* 2nd ed. New York: Wiley, 1959.

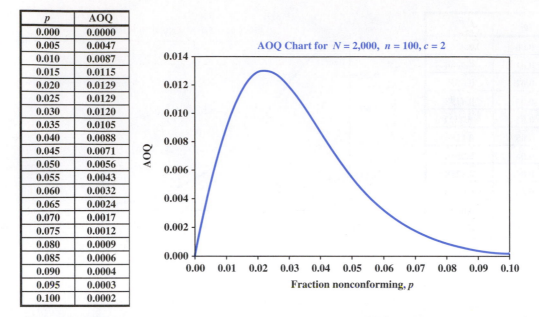

p	AOQ
0.000	0.0000
0.005	0.0047
0.010	0.0087
0.015	0.0115
0.020	0.0129
0.025	0.0129
0.030	0.0120
0.035	0.0105
0.040	0.0088
0.045	0.0071
0.050	0.0056
0.055	0.0043
0.060	0.0032
0.065	0.0024
0.070	0.0017
0.075	0.0012
0.080	0.0009
0.085	0.0006
0.090	0.0004
0.095	0.0003
0.100	0.0002

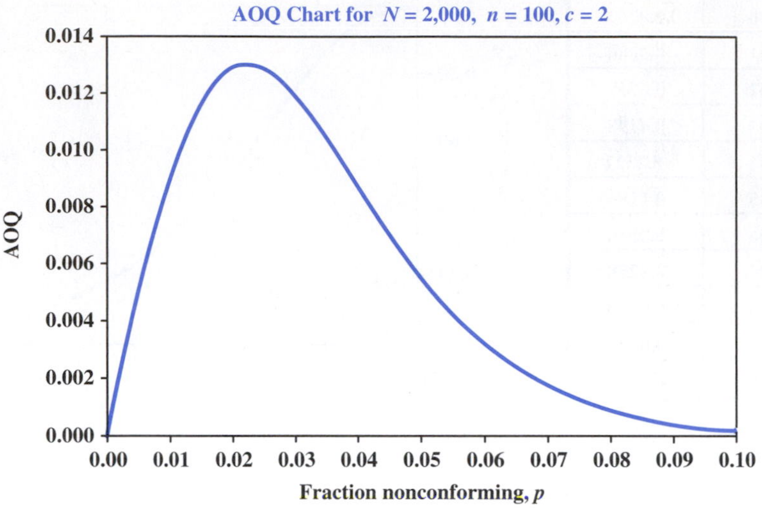

FIGURE 10-29. An AOC Curve

The **average total inspection (ATI)** is another important metric for evaluating a sampling plan. Under rectifying inspection, the ATI represents the average number of units inspected per shipment over a series of shipments. It is calculated from the lot size, the sample size, and the probability of acceptance, and is dependent on sample size, fraction nonconforming, and acceptance number. Equations 10-24 and 10-25, shown below, can be used to calculate the ATI for any given plan and an assumed quality level p.

$$ATI = n + (1 - P_A)(N - n) \qquad \textbf{10-24}$$

$$ATI = n + \left[1 - \left(\sum_{d=0}^{c}\left(\frac{n!}{d!(n-d)!}\right)p^d(1-p)^{n-d}\right)\right](N-n) \qquad \textbf{10-25}$$

Example ATI Curve for Single Sampling Plan

Consider a single sampling plan for shipment size $N =$ 2,000, sample size $n = 100$, and acceptance number $c = 2$. Applying Equation 10-25, the ATI for fraction nonconforming $p = 0.025$ is approximately 970 units:

$$ATI = 100 + \left[1 - \left(\sum_{d=0}^{2}\left(\frac{100!}{d!(100-d)!}\right)0.025^d (0.975)^{n-d}\right)\right](1900) = 969.8$$

The ATI curve is developed by plotting ATI against fraction nonconforming over a range of values for p. The curve for this example is shown in Figure 10-30.

The AOQL and ATI each provide different information about the performance of a sampling plan. The previous example showed that AOQ is low for high quality lots (i.e., those with low p). Assuming random sampling, the AOQ can be no greater than the incoming fraction nonconforming, p. As quality decreases (i.e., p increases), the AOQ will approach p—until the probability of acceptance decreases to such a level that a significant proportion of the shipments are subjected to rectifying inspection. Recall that for those lots subjected to 100% inspection the fraction nonconforming after inspection will be zero. As p continues to increase, the number of lots subjected to rectifying inspection will increase, resulting in a decrease in AOQ. For large values of fraction nonconforming, AOQ approaches zero. The AOQL provides an indication to the customer of the worst average fraction nonconforming that would be received under a particular sampling plan after rectifying inspection has been performed. Customers prefer low values of AOQL.

The ATI curve indicates the relationship between incoming fraction nonconforming and the average number of units that would be inspected for that fraction nonconforming under a particular sampling plan. The cost of sampling is important because sampling is an *appraisal cost*, one of the four cost-of-quality (COQ) categories.[19] An advantage that acceptance sampling offers over 100% inspection is the reduction in sampling time and cost—the larger the number of units sampled, the greater the sampling cost. Thus a sampling plan that minimizes the number of units inspected for a particular level of incoming quality is preferred. The minimum ATI is the size of the sample, n. Shipments of 100% conformance have 100% probability of acceptance. As fraction nonconforming increases a greater proportion of lots are subjected to rectifying inspection, where all N units in the shipment must be inspected. Thus for $p = 0$, ATI $= n$, but as p increases, it will reach a point where the ATI will approach N.

[19] The four cost-of-quality categories are appraisal costs, prevention costs, internal failure costs, and external failure costs.

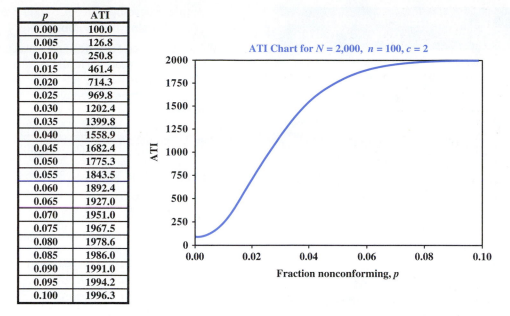

p	ATI
0.000	100.0
0.005	126.8
0.010	250.8
0.015	461.4
0.020	714.3
0.025	969.8
0.030	1202.4
0.035	1399.8
0.040	1558.9
0.045	1682.4
0.050	1775.3
0.055	1843.5
0.060	1892.4
0.065	1927.0
0.070	1951.0
0.075	1967.5
0.080	1978.6
0.085	1986.0
0.090	1991.0
0.095	1994.2
0.100	1996.3

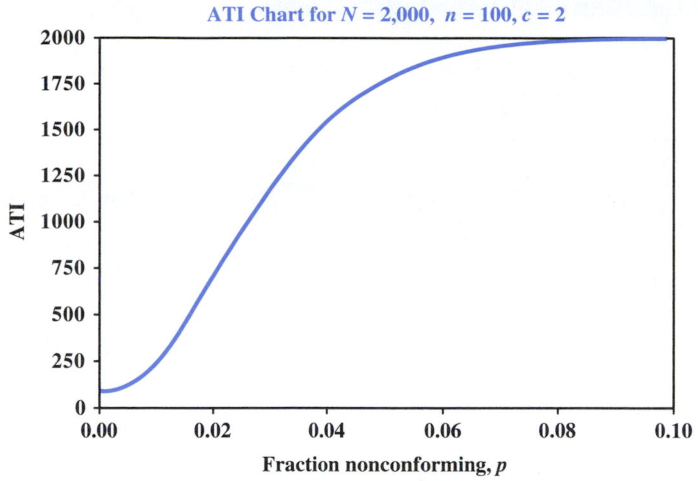

FIGURE 10-30. An ATI Curve

10.6.5.3 Curtailment Inspection for Rejected Shipments.

If the sole purpose of acceptance sampling is to make decisions regarding acceptance or rejection one would cease inspection of a sample and reject the lot as soon as the acceptance number is exceeded in the sample. This practice is known as **curtailment** and would result in the inspection of fewer units for rejected lots. For a sample of size $n = 100$ with acceptance number $c = 2$, if the third nonconforming item is found on the 22nd item inspected, only 22 units would be inspected for that shipment, and the shipment would be rejected and returned to the supplier.

> Under curtailment inspection the inspection process is stopped as soon as a decision can be made.

In recent times, curtailment has fallen out of favor and is not widely practiced. There are two reasons for this. First, curtailment and rectifying inspection are mutually exclusive—if inspection is curtailed after the 22nd unit and the shipment is rejected, the remaining $(N - 22)$ units are not inspected. Under curtailment, the shipment would be returned to the producer and none of the components would be available until the shipment is replaced. Rectifying inspection allows the customer to use all of the conforming components, which are subsequently identified through 100% screening thus avoiding a potential stock-out and the disruption of production.

The other major shortcoming is that curtailment does not provide the customer with a reliable estimate of supplier quality. For $n = 100$ and $c = 2$, suppose that the 3rd nonconforming item is found on the 6th unit inspected. Under curtailment, 50% of the units inspected are nonconforming. Using these sampling results to infer that the entire lot is 50% defective could be quite misleading.

10.6.5.4 Grubbs's Approach—Single Sampling Plans.

We previously demonstrated that, in order to simultaneously satisfy the producer's and consumer's risk criteria, an inspection plan's OC curve must pass through two points: $[P_A = (1-\alpha), p = \text{AQL}]$ and $[P_A = \beta, p = \text{LTPD}]$. Rather than trying to *exactly* satisfy both criteria, a plan can be designed that meets a stipulated producer's risk and comes close to meeting the consumer's risk. Alternatively, the plan can be designed to precisely meet the stipulated consumer's risk and come close to meeting the producer's risk. Table 10-13, developed by F. E. Grubbs,[20] can be used to design single sampling plans for producer's risk of $\alpha = 0.05$ and consumer's risk of $\beta = 0.10$.

A Grubbs table is used by first computing the ratio, R_0, of the stipulated LTPD and AQL. Tabulated values of R are shown in Table 10-13 as $R_{Tab} = p_2/p_1$. Larger ratios of R yield smaller acceptance numbers c, and smaller sample sizes n. A large R results from a large relative difference between *AQL* and *LTPD*. For a large R, small samples and acceptance numbers are sufficient to determine the disposition of a shipment. Smaller ratios, on the other hand, require larger sample sizes and acceptance numbers.

The sampling plan is developed by locating the two tabulated $R = p_2/p_1$ values in the right-most column that bracket the actual calculated p_2/p_1. For each of these, the corresponding acceptance number can be found in the left column of the table. If the calculated ratio is near one of the tabulated R values, only one acceptance number need be considered. The two columns, labeled np_1 and np_2 respectively, are used to calculate a sample size corresponding to each of the candidate acceptance numbers. The np_1 column is used for calculating

[20] Grubbs, F. E. "On Designing Single Sampling Inspection Plans." *Annals of Mathematical Statistics* 20 (1949): 256.

TABLE 10-13 Grubbs Table for Single Sampling Plans, $\alpha = 0.05$, $\beta = 0.10$

Acceptance Number, c	1 − Producer's Risk $1-\alpha = 0.95$ np_1	Consumers Risk $\beta = 0.10$ np_2	R_{Tab} = LTPD/AQL $= p_2/p_1$
0	0.051	2.303	44.84
1	0.355	3.890	10.96
2	0.818	5.322	6.51
3	1.366	6.681	4.89
4	1.970	7.944	4.06
5	2.613	9.274	3.55
6	3.286	10.532	3.21
7	3.981	11.771	2.96
8	4.695	12.995	2.77
9	5.426	14.206	2.62
10	6.169	15.407	2.50
11	6.924	16.598	2.40
12	7.690	17.782	2.31
13	8.464	18.958	2.24
14	9.246	20.128	2.18
15	10.035	21.292	2.12

Source: Grubbs, F. E., "On Designing Single Sampling Inspection Plans." *Annals of Mathematical Statistics* 20 (1949): 256. Reprinted with permission of the Institute of Mathematical Statistics.

a sample size that meets the stipulated producer's risk and the np_2 column is used to calculate a sample size that meets the stipulated consumer's risk.

Recall that the producer's risk is defined as the probability of *failing to accept* (or rejecting) a shipment if the process average nonconformance p, is equal to the *AQL* (p_1), and the consumer's risk is defined as the probability of *accepting* a shipment if the process average nonconformance p, is equal to the *LTPD* (p_2). For a given acceptance number, the producer favors a smaller sample size as this reduces the probability that shipments of good quality will be rejected. Similarly, the consumer favors a larger sample size because this decreases the probability that lots of inferior quality will be accepted. For calculated $R = p_2/p_1$ values falling in between the tabulated R values, the row indexed by the larger tabulated R will yield a sampling plan that meets the stipulated producer's criterion, but will have consumer's risk in excess of $\beta = 0.10$ (when the np_1 column is used to calculate n). If the np_2 column is used the plan will meet the stipulated consumer's criterion but will have producer's risk larger than $\alpha = 0.05$. If the smaller tabulated R is used to index the row, then both producer's risk and consumer's risk criteria can be met, but the calculated sample size and the acceptance number will be larger. The following example illustrates the use of a Grubbs table for developing a single sampling plan.

Example Development of Single Sampling Plan using the Grubbs Table

We wish to develop a single sampling plan for shipment size $N = 1{,}000$ that meets the following criteria: AQL = 0.5%, LTPD = 3.0%, $\alpha = 0.05$, $\beta = 0.10$. We will use Table 10-13 to determine the sample size, n, and acceptance number, c.

Step 1. Compute $R_0 = \dfrac{LTPD}{AQL} = \dfrac{0.03}{0.005} = 6.00$

Step 2. Identify in Table 10-13 the rows that bracket R_0. $R_{TAB} = 6.51$, $c = 2$ and $R_{TAB} = 4.89$, $c = 3$ bracket R_0.

Step 3. Design sampling plans for each R_{TAB}.

For $c = 2$, $np_1 = 0.818$ and

$$n = \frac{np_1}{p_1} = \frac{0.818}{0.005} = 163.6 \cong 164.$$

$np_2 = 5.322$ and

$$n = \frac{np_2}{p_2} = \frac{5.322}{0.03} = 177.4 \cong 177.$$

For $c = 3$, $np_1 = 1.366$ and

$$n = \frac{np_1}{p_1} = \frac{1.366}{0.005} = 273.2 \cong 273.$$

$np_2 = 6.681$ and

$$n = \frac{np_2}{p_2} = \frac{6.681}{0.03} = 222.7 \cong 223.$$

Step 4. Compute Producer's and Consumer's Risks for each plan using Equations 10-20 and 10-21.

$$\alpha = 1 - \sum_{d=0}^{c} \frac{n!}{d!(n-d)!} p_1^d (1-p_1)^{n-d}$$

$$\beta = \sum_{d=0}^{c} \frac{n!}{d!(n-d)!} p_2^d (1-p_2)^{n-d}$$

Plan 1: $c = 2$, $n = 164$, $p_1 = 0.005$, $p_2 = 0.03$

$$\alpha = 1 - \left[\sum_{d=0}^{2}\left(\frac{164!}{d!(164-d)!} 0.005^d (0.995)^{164-d}\right)\right]$$

$$= 1 - .9501 = 0.0499$$

Since $\alpha < 0.05$ Plan 1 meets stipulated Producer's risk.

$$\beta = \left[\sum_{d=0}^{2}\left(\frac{164!}{d!(164-d)!} 0.03^d (0.97)^{164-d}\right)\right]$$

$$= 0.12766$$

Since $\beta > 0.10$ Plan 1 fails to meet stipulated Consumer's risk.

Plan 2: $c = 2$, $n = 177$, $p_1 = 0.005$, $p_2 = 0.03$

$$\alpha = 1 - \left[\sum_{d=0}^{2}\left(\frac{177!}{d!(177-d)!} 0.005^d (0.995)^{177-d}\right)\right]$$

$$= 1 - .94004 = 0.05996$$

Since $\alpha > 0.05$ Plan 2 fails to meet stipulated Producer's risk.

$$\beta = \left[\sum_{d=0}^{2}\left(\frac{177!}{d!(177-d)!}0.03^d(0.97)^{177-d}\right)\right]$$
$$= 0.09738$$

Since $\beta < 0.10$ Plan 2 meets stipulated Consumer's risk.

Plan 3: $c = 3,\ n = 273,\ p_1 = 0.005,\ p_2 = 0.03$

$$\alpha = 1-\left[\sum_{d=0}^{3}\left(\frac{273!}{d!(273-d)!}0.005^d(0.995)^{273-d}\right)\right]$$
$$= 1 - .9509 = 0.0491$$

Since $\alpha < 0.05$ Plan 3 meets stipulated Producer's risk.

$$\beta = \left[\sum_{d=0}^{3}\left(\frac{273!}{d!(273-d)!}0.03^d(0.97)^{273-d}\right)\right]$$
$$= 0.03528$$

Since $\beta < 0.10$ Plan 3 more than meets stipulated Consumer's risk.

Plan 4: $c = 3,\ n = 223,\ p_1 = 0.005,\ p_2 = 0.03$

$$\alpha = 1-\left[\sum_{d=0}^{3}\left(\frac{223!}{d!(223-d)!}0.005^d(0.995)^{223-d}\right)\right]$$
$$= 1 - .97349 = 0.02651$$

Since $\alpha < 0.05$ Plan 4 more than meets stipulated Producer's risk.

$$\beta = \left[\sum_{d=0}^{3}\left(\frac{223!}{d!(223-d)!}0.03^d(0.97)^{223-d}\right)\right]$$
$$= 0.09597$$

Since $\beta < 0.10$ Plan 4 meets stipulated Consumer's risk.

Step 5. Select a sampling plan. A plan with an acceptance number, $c = 2$, that meets both producer's and consumer's risk criteria cannot be achieved. The sampling plans $c = 3$ will meet or exceed producer's and consumer's risks, but the corresponding sample sizes are considerably larger. The plans that should be considered are $c = 2$, $n = 177$ (slightly exceeds producer's risk and meets consumer's risk) and $c = 3$, $n = 223$ (meets both producer's and consumer's risks). To assist in making the final decision, Equation 10-25 can be used to construct the ATI curves for the two candidate plans, as shown in Figure 10-31. While the plan for $c = 3$ requires a larger sample size, the ATI curves for the two plans are very similar. Therefore, if rectifying inspection is to be implemented, there is little difference between the two plans in the average number of units to be inspected per shipment. Either plan could be selected.

10.6.6 Double Sampling Plans

Double sampling plans are an extension of single sampling plans, in which the "accept" or "reject" decision can be deferred pending the results of a second sample. Rather than the binary accept or reject options offered by single plans, double sampling provides a third alternative that permits taking a second sample. The procedure begins by collecting and inspecting the first sample of size n_1. If the number of nonconforming units d_1 in the first sample is less than or equal to the first acceptance number c_1, then the lot is accepted. If d_1 exceeds the second acceptance number c_2, the lot is rejected. If $c_1 < d_1 \leq c_2$, then a second sample of size n_2 is required. The numbers of nonconforming units in both samples are added together, and if $(d_1 + d_2) \leq c_2$, the lot is accepted. Otherwise, the lot is rejected on the basis of the combined samples.

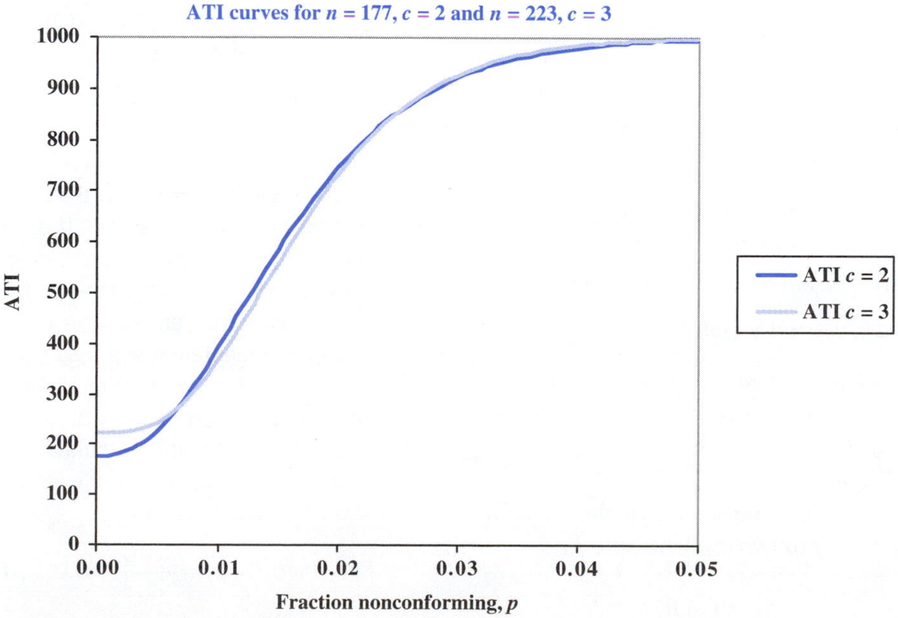

FIGURE 10-31. ATI Curves for Candidate Plans

Double sampling can result in lower total sampling costs than single sampling.

The double sampling plan provides a second chance for a shipment to be accepted. Since, under rectifying inspection, rejection requires 100% screening inspection a double sampling plan often results in a lower ATI (and less sampling costs) than a single sampling plan with identical criteria.

10.6.6.1 Operating Characteristic Curves for Double Sampling Plans. Reduced total inspection under double sampling comes with a price—*increased complexity*. Rather that accepting or rejecting on the basis of a single sample, the inspector will (1) accept on the basis of the first sample, (2) reject on the basis of the first sample, (3) accept after taking a second sample, or (4) reject after taking a second sample. Rectifying inspection may be implemented in lieu of rejecting a sample. Before proceeding, I need to introduce some additional terminology.

$P_{A,1}$: *Probability of accepting on the first sample.*
$P_{R,1}$: *Probability of rejecting on the first sample.*
$P_{A,2}$: *Probability of accepting on the second sample.*
$P_{R,2}$: *Probability of rejecting on the second sample.*
P_A: *Overall probability of acceptance.* $P_A = P_{A,1} + P_{A,2}$
P_R: *Overall probability of rejection.* $P_R = P_{R,1} + P_{R,2}$
P_2: *Total probability of taking a second sample* $= 1 - (P_{A,1} + P_{R,1})$ $P_A + P_R = P_{A,1} + P_{A,2} + P_{R,1} + P_{R,2} = 1.00$

The provision for a second sample complicates the probability calculations that are required to develop OC, ATI, and AOQ curves. As Equation 10-26 shows, the probability of acceptance on the first sample is computed the same for double sampling as for single sampling.

$$P_{A,1} = \sum_{d_1=0}^{c_1}\left(\frac{n_1!}{d!(n_1-d_1)!}p^{d_1}(1-p)^{n_1-d_1}\right) \qquad \textbf{10-26}$$

Determining the probability of acceptance on the second sample requires the calculation of joint probabilities—that is, the probability of acceptance on a second sample is conditioned upon the requirement to take a second sample.

$P_{A,2}$ = *Probability that the sum of defectives on the first and second samples is less than the second acceptance number, given that a second sample is required.*

This is written as a conditional probability as follows:

$$P_{A,2} = \Pr\{d_1 + d_2 \le c_2 | \textit{Second sample is required}\}$$

This conditional probability statement can be rewritten in terms of outcomes of the first and second samples as follows:

$$P_{A,2} = \Pr\{d_2 \le (c_2 - d_1) | c_1 < d_1 \le c_2\} \qquad \textbf{10-27}$$

The multiplication rule from statistical theory defines the conditional probability of two events, *A* and *B*,

$$\Pr\{A|B\} = \frac{\Pr\{A \cap B\}}{\Pr\{B\}} \qquad \textbf{10-28}$$

where,

$\Pr\{A|B\}$ = *Probability that Event A will occur given that Event B has occurred.*
$\Pr\{A \cap B\}$ = *The joint probability that Events A and B will both occur.*
$\Pr\{B\}$ = *The probability that Event B will occur.*

If Events A and B are independent,

$$\Pr\{A \cap B\} = P\{A\}P\{B\} \qquad \textbf{10-29}$$

Let B represent the event where a second sample has to be taken because $c_1 < d_1 \leqslant c_2$, and let A represent the event that the lot is accepted on the second sample (i.e. $d_2 < (c_2 - d_1)$. Then Equation 10-27 becomes Equation 10-28. We can note that

$$\Pr\{(\text{acceptance} | d_1) \cap \text{second sample is required}\}$$
$$= \Pr\{d_2 \le (c_2 - d_1)\}\Pr\{d_1\}$$

where,

$$c_1 < d_1 \leqslant c_2$$

The total probability of accepting a lot on the second sample is

$$P_{A,2} = \sum_{d_1=c_1+1}^{c_2}\sum_{d_2=0}^{c_2-d_1}\Pr\{d_1\}\Pr\{d_2\} \qquad \textbf{10-30}$$

This following example illustrates how these probabilities can be computed.

Example Probability of Acceptance for a Double Sampling Plan

A double sampling plan calls for $n_1 = 40$, $c_1 = 1$, $n_2 = 40$, and $c_2 = 3$. For fraction nonconforming $p = 0.03$, we can apply Equation 10-26 to obtain the probability of acceptance on the first sample:

$$P_{A,1} = \sum_{d_1=0}^{1}\left(\frac{40!}{d_1!(40-d_1)!}0.03^{d_1}(0.97)^{40-d_1}\right)$$
$$= 0.66154$$

There are a finite number of outcomes for the first and second samples that would result in acceptance on the second sample. First we shall consider the outcomes for the first sample. If $d_1 = 0$ or 1, the shipment is accepted without taking a second sample. If $d_1 \geqslant 4$, the lot is rejected. For the intermediate numbers, $d_1 = 2$ or 3, a decision will be deferred pending the results of a second sample.

Next, consider possible outcomes for the second sample. Acceptance will result if $d_1 + d_2 \leqslant c_2$. Therefore, if $d_1 = 2$, $d_2 = 0$ or 1 will result in acceptance; $d_2 \geqslant 2$, rejection. If $d_1 = 3$, only $d_2 = 0$ will lead to lot acceptance. We will calculate the probability of these outcomes for each sample and then apply Equation 10-30 to compute $P_{A,2}$.

$$\Pr\{d_1 = 2\} = \frac{40!}{2!\,38!}0.03^2(0.97)^{38} = 0.2206$$

$$\Pr\{d_2 = 0\} = \frac{40!}{0!\,40!}0.03^0(0.97)^{40} = 0.2957$$

$$\Pr\{d_2 = 1\} = \frac{40!}{1!\,39!} 0.03^1(0.97)^{39} = 0.3658$$

$$\Pr\{d_1 = 3\} = \frac{40!}{3!\,37!} 0.03^3(0.97)^{37} = 0.864$$

$$P_{A,2} = \Pr\{d_1 = 2\}\Pr\{d_2 = 0\} + \Pr\{d_1 = 2\} \times \Pr\{d_2 = 1\} + \Pr\{d_1 = 3\}\Pr\{d_2 = 0\}$$

$$P_{A,2} = (0.2206)(0.2957) + (0.2206)(0.3658) + (0.0864)(0.2957) = 0.1715$$

The overall probability of acceptance $P_A = P_{A,1} + P_{A,2}$:

$$P_A = P_{A,1} + P_{A,2} = 0.6615 + 0.1715 = 0.833$$

As with single sampling plans, the OC curve for a double sampling plan is a plot of P_A versus p. The probabilities for the plan presented earlier are shown in Table 10-14, and the plot of the corresponding OC curve is shown in Figure 10-32.

10.6.6.2 Double Sampling Performance—ASN, AOQL, and ATI. Both the AOQL and ATI are measures of performance for single and double sampling plans. A third metric, the **average sample number** (ASN), is applicable to double and multiple sampling plans. The ASN is similar to the ATI in that both provide a measure of the average number of units inspected over a series of shipments (for a particular level of incoming quality). The metrics differ in that ATI calculations include the 100% screening under rectifying inspection while ASN does not.

Under double sampling, where the lot may be accepted on the first sample, rejected on the first sample, accepted after the second sample, or rejected after the second sample, the minimum ASN is n_1, the size of the first sample. If a decision cannot be reached on the first sample, the number of items inspected is $(n_1 + n_2)$. The majority of high-quality shipments (e.g., those with $p <$ AQL) will be accepted on the first sample, resulting in ASN values near n_1. Similarly, for poor quality (those with $p >$ LQL), most shipments would be rejected on the first sample, resulting in small ASN values. Incoming quality at intermediate levels, between the AQL and LQL, would result in a

ASN measures the average number of units inspected under schemes that do not permit rectifying inspection.

TABLE 10-14 OC Probabililties for a Double Sampling Plan with $n_1 = n_2 = 40$, $c_1 = 1$, $c_2 = 3$

p	$P_{A,1}$	$Pr\{d_1 = 2\}$	$Pr\{d_1 = 3\}$	$Pr\{d_2 = 0\}$	$Pr\{d_2 = 1\}$	$P_{A,2}$	P_A
0.00	1.00000	0.00000	0.00000	1.00000	0.00000	0.00000	1.00000
0.01	0.93926	0.05324	0.00681	0.44752	0.36163	0.04613	0.98539
0.02	0.80954	0.14479	0.03743	0.19865	0.32432	0.08316	0.8927
0.03	0.66154	0.22063	0.08643	0.08745	0.21636	0.07459	0.73613
0.04	0.52098	0.26456	0.13963	0.03817	0.12723	0.04909	0.57006
0.05	0.39906	0.27767	0.18511	0.01652	0.06954	0.02695	0.42602
0.06	0.29904	0.26746	0.21624	0.00708	0.03617	0.0131	0.31214
0.07	0.22006	0.24246	0.23116	0.00301	0.01813	0.00582	0.22588
0.08	0.15945	0.21	0.2313	0.00127	0.00882	0.00241	0.16186
0.09	0.11397	0.17545	0.21979	0.00053	0.00418	0.00094	0.11491
0.1	0.08047	0.14233	0.20032	0.00022	0.00194	0.00035	0.08082
0.11	0.05619	0.11264	0.17635	0.00009	0.00088	0.00013	0.05632
0.12	0.03883	0.08726	0.15072	0.00004	0.00039	0.00004	0.03887

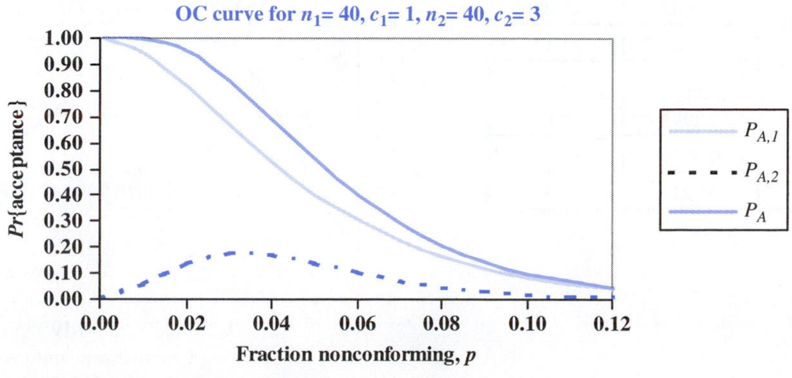

FIGURE 10-32. OC Curve for Double Sampling Plan

larger ASN. Equation 10-31 is the formula used for computing the ASN for a double sampling plan.

$$ASN = n_1(P_{A,1} + P_{R,1}) + (n_1 + n_2)(1 - (P_{A,1} + P_{R,1}))$$

$$ASN = n_1 + n_2(1 - (P_{A,1} + P_{R,1})) \qquad \textbf{10-31}$$

Example Average Sample Number for a Double Sampling Plan

A double sampling plan calls for $n_1 = 40$, $c_1 = 1$, $n_2 = 40$, and $c_2 = 3$. For fraction nonconforming $p = 0.03$, applying Equation 10-26 and recognizing that the probability of rejecting the lot on the first sample is $\Pr\{d_1 > 3\} = 1 - \Pr\{d_1 \le 3\}$, we can calculate P_{A1} and P_{R1} and then compute the ASN from Equation 10-31.

$$P_{A,1} = \sum_{d_1=0}^{1}\left(\frac{40!}{d_1!(40-d_1)!}0.03^{d_1}(0.97)^{40-d_1}\right)$$

$$= 0.66154$$

$$P_{R,1} = 1 - \sum_{d_1=0}^{3}\left(\frac{40!}{d_1!(40-d_1)!}0.03^{d_1}(0.97)^{40-d_1}\right)$$

$$= 0.03140$$

$$ASN = 40 + 40(1 - (0.66154 + 0.03140))$$

$$= 52.2825$$

The ASN curve for this sampling plan is constructed by plotting the ASN against p for various levels of incoming quality. A table of the ASN values and the resulting curve are shown in Figure 10-33.

Recall that the AOQ is the average quality level that a customer will accept, following rectifying inspection and replacement if warranted. As with single sampling plans, if the incoming lot quality is very good (e.g., low p), most shipments will be accepted on the first sample with a quality level equal to p. For the case of poor quality ($p >$ LQL), shipments will likely be subjected to 100% screening inspection and all nonconforming units replaced, resulting in a quality level near $p = 0$. If the incoming quality is between the AQL and LQL, the quality level accepted by the customer will be somewhere between 0 and p. The maximum value on the AOQ curve is the average outgoing quality limit (AOQL) which is the worst-case level of quality accepted by the customer for a sampling plan. Because a shipment may be accepted on either the first or second sample the calculation of AOQ is complicated. Equation 10-32 is the formula used for these calculations.

$$AOQ = \frac{p(P_{A,1}(N - n_1) + P_{A,2}(N - n_1 - n_2))}{N} \qquad \textbf{10-32}$$

Example AOQ Curve and AOQL for Double Sampling Plan

A double sampling plan calls for $n_1 = 40$, $c_1 = 1$, $n_2 = 40$, and $c_2 = 3$ for a series of shipments of size $N = 1{,}000$. Applying Equation 10-32 to a lot with $p = 0.03$,

$$AOQ = \frac{0.03(0.6615(1000 - 40) + 0.1715(1000 - 40 - 40))}{1000}$$

$$= 0.0238 \text{ or } 2.38\%$$

Table 10-15 contains the probabilities necessary to plot the AOQ curve shown in Figure 10-34. Based on the tabulated values and the AOQ curve, the AOQL is approximately 0.0261 for the case of incoming fraction nonconforming $p = 0.04$.

The average total inspection (ATI) for a double sampling plan indicates the number of units on average that will be inspected for a specific quality level. The ATI is important because of its implications on sampling cost. With the potential to accept or reject a shipment on either the first or second sample, calculating ATI is more complex for double than for single sampling plans. Equation 10-33 is the formula for calculating the ATI for a double sampling plan.

$$ATI = n_1P_{A,1} + (n_1 + n_2)P_{A,2} + N(1 - P_A) \qquad \textbf{10-33}$$

p	$P_{A,1}$	$P_{R,1}$	ASN
0.00	1.0000	0.0000	40.00
0.02	0.8095	0.0082	47.29
0.04	0.521	0.0748	56.17
0.06	0.299	0.2173	59.35
0.08	0.1594	0.3993	57.65
0.1	0.0805	0.5769	53.71
0.12	0.0388	0.7232	49.52

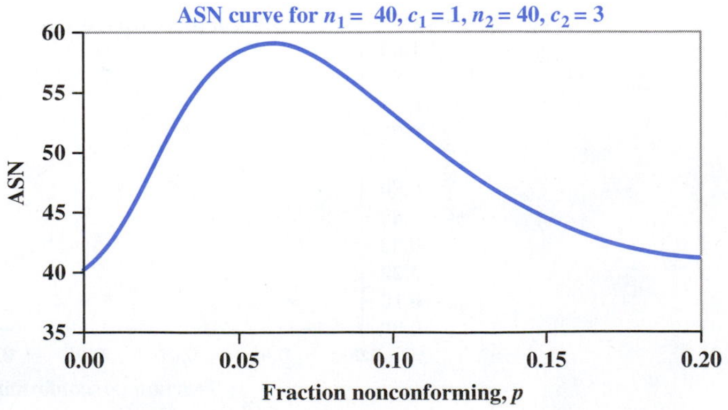

FIGURE 10-33. Example ASN Curve

TABLE 10-15 AOQ Probabilities

p	$P_{A,1}$	$Pr\{d_1 = 2\}$	$Pr\{d_1 = 3\}$	$Pr\{d_2 = 0\}$	$Pr\{d_2 = 1\}$	$P_{A,2}$	P_A	AOQ
0.00	1.0000	0.0000	0.0000	1.0000	0.0000	0.0000	1.0000	0.000
0.02	0.8095	0.1448	0.0374	0.4457	0.3638	0.1339	0.9434	0.018
0.04	0.521	0.2646	0.1396	0.1954	0.3256	0.1651	0.6861	0.0261
0.06	0.299	0.2675	0.2162	0.0842	0.2149	0.0982	0.3972	0.0226
0.08	0.1594	0.21	0.2313	0.0356	0.1238	0.0417	0.2012	0.0153
0.1	0.0805	0.1423	0.2003	0.0148	0.0657	0.0144	0.0949	0.0091
0.12	0.0388	0.0873	0.1507	0.006	0.0328	0.0043	0.0431	0.0049
0.14	0.018	0.0496	0.1022	0.0024	0.0156	0.0011	0.0192	0.0026
0.16	0.0081	0.0265	0.0639	0.0009	0.0071	0.0003	0.0083	0.0013

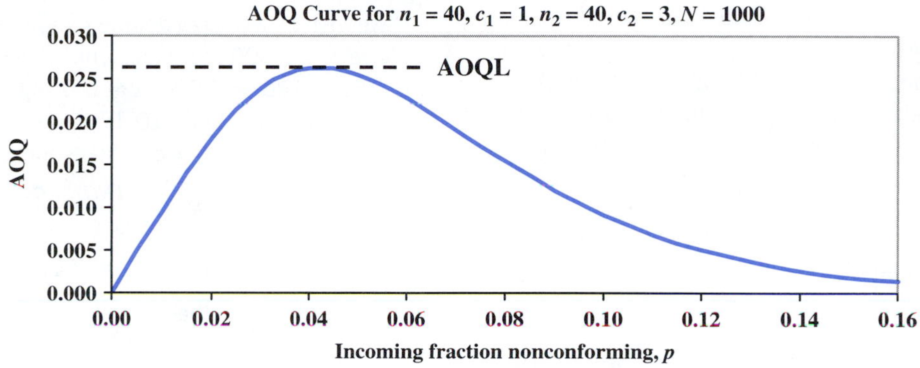

FIGURE 10-34. Example AOQ Curve

TABLE 10-16 ATI Probabilities

p	$P_{A,1}$	$Pr\{d_1 = 2\}$	$Pr\{d_1 = 3\}$	$Pr\{d_2 = 0\}$	$Pr\{d_2 = 1\}$	$P_{A,2}$	P_A	ATI
0.00	1.0000	0.0000	0.0000	1.0000	0.0000	0.0000	1.0000	40.00
0.02	0.8095	0.1448	0.0374	0.4457	0.3638	0.1339	0.9434	99.66
0.04	0.521	0.2646	0.1396	0.1954	0.3256	0.1651	0.6861	347.96
0.06	0.299	0.2675	0.2162	0.0842	0.2149	0.0982	0.3972	622.59
0.08	0.1594	0.21	0.2313	0.0356	0.1238	0.0417	0.2012	808.55
0.1	0.0805	0.1423	0.2003	0.0148	0.0657	0.0144	0.0949	909.48
0.12	0.0388	0.0873	0.1507	0.006	0.0328	0.0043	0.0431	958.77
0.14	0.018	0.0496	0.1022	0.0024	0.0156	0.0011	0.0192	981.66
0.16	0.0081	0.0265	0.0639	0.0009	0.0071	0.0003	0.0083	992.01

Example ATI Curve for Double Sampling Plan

A double sampling plan calls for $n_1 = 40$, $c_1 = 1$, $n_2 = 40$, and $c_2 = 3$ for a series of shipments of size $N = 1000$. For fraction nonconforming $p = 0.03$, the ATI is

$$ATI = 40(0.6615) + (40 + 40)(0.1715)$$
$$+ 1000(1 - 0.8331) = 207.13$$

Table 10-16 contains the probabilities for the ATI curve shown in Figure 10-35.

10.6.6.3 Grubbs's Approach—Double Sampling Plans. Grubbs developed tables to assist with the determination of double sampling plans for producer's risk $\alpha = 0.05$ and consumer's risk $\beta = 0.10$. Grubbs tables cover the two most common conventions for double sampling plans, where either $n_2 = n_1$ or $n_2 = 2n_1$. The procedure for the use of these tables is similar to that for single sampling plans. The ratio $R = p_2/p_1$ is used to determine the acceptance numbers, c_1 and c_2. As with single sampling plans, the calculated R will usually fall somewhere between two tabulated R values resulting in two candidate sets of acceptance numbers. For an approximate R value,

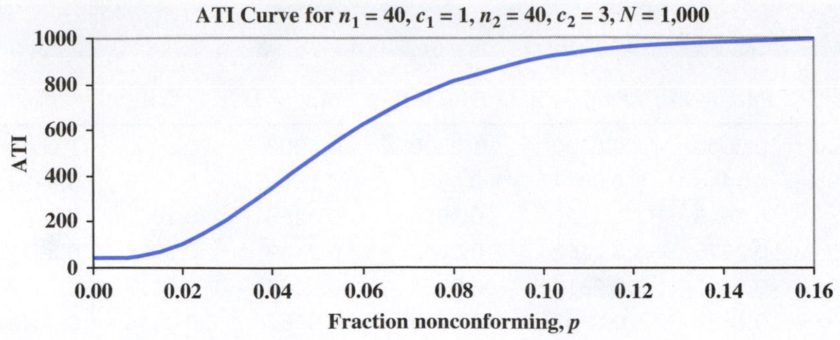

FIGURE 10-35. Example ATI Curve

the column for $P_A = 0.95$ will provide an n_1p that meets the stipulated producer's risk, and the column for $P_A = 0.10$ will produce a plan that meets the specified consumer's risk. As with single sampling plans, two adjacent rows will bracket the computed p_2/p_1. The row having a tabulated $R < p_2/p_1$ will produce plans that either meet the producer's risk while having $\beta > 0.10$, or meet the consumer's risk while having $\alpha > 0.05$. The row with tabulated $R < p_2/p_1$ will produce plans that simultaneously satisfy both producer's and consumer's risk criteria, but require larger sample sizes. Tables 10-17 and 10-18 are Grubbs tables for $n_2 = n_1$ and for $n_2 = 2n_1$, respectively.

Example Development of Double Sampling Plan Using Grubbs Tables

We wish to develop a double sampling plan for shipment size $N = 1{,}000$ that meets the following criteria: AQL = 0.5%, LTPD = 3.0%, $\alpha = 0.05$, $\beta = 0.10$, with $n_2 = n_1$. We will use Table 10-17 to determine sample sizes, n_1 and n_2, and the acceptance numbers, c_1 and c_2.

Step 1. Compute $R_0 = \dfrac{LTPD}{AQL} = \dfrac{p_2}{p_1} = \dfrac{0.03}{0.005} = 6.00$

Step 2. Identify in Table 10-17 the rows that bracket R_0. R_0 falls between Plan 3 and Plan 4

TABLE 10-17 Grubbs Table for Double Sampling Plans $n_2 = n_1$, $\alpha = 0.05$, $\beta = 0.10$

				Approx. n_1p	
Plan #	$R = p_2/p_1$	c_1	c_2	$1 - \alpha = 0.95$	$\beta = 0.10$
1	11.9	0	1	0.21	2.5
2	7.54	1	2	0.52	3.92
3	6.79	0	2	0.43	2.96
4	5.39	1	3	0.76	4.11
5	4.65	2	4	1.16	5.39
6	4.25	1	4	1.04	4.42
7	3.88	2	5	1.43	5.55
8	3.63	3	6	1.87	6.78
9	3.38	2	6	1.72	5.82
10	3.21	3	7	2.15	6.91
11	3.09	4	8	2.62	8.1
12	2.85	4	9	2.9	8.26
13	2.6	5	11	3.68	9.56
14	2.44	5	12	4	9.77
15	2.32	5	13	4.35	10.08
16	2.22	5	14	4.7	10.45
17	2.12	5	16	5.39	11.41

Source: Chemical Corps Engineering Agency, *Manual No. 2: Master Sampling Plans for Single, Duplicate, Double, and Multiple Sampling.* MD: Edgewood Arsenal, Army Chemical Center, 1953.

TABLE 10-18 Grubbs Table for Double Sampling Plans $n_2 = 2n_1$, $\alpha = 0.05$, $\beta = 0.10$

				Approx. n_1p	
Plan #	$R = p_2/p_1$	c_1	c_2	$1 - \alpha = 0.95$	$\beta = 0.10$
1	14.5	0	1	0.16	2.32
2	8.07	0	2	0.3	2.42
3	6.48	1	3	0.6	3.89
4	5.39	0	3	0.49	2.64
5	5.09	1	4	0.77	3.92
6	4.31	0	4	0.68	2.93
7	4.19	1	5	0.96	4.02
8	3.6	1	6	1.16	4.17
9	3.26	2	8	1.68	5.47
10	2.96	3	10	2.27	6.72
11	2.77	3	11	2.46	6.82
12	2.62	4	13	3.07	8.05
13	2.46	4	14	3.29	8.11
14	2.21	3	15	3.41	7.55
15	1.97	4	20	4.75	9.35
16	1.74	6	30	7.45	12.96

Source: Chemical Corps Engineering Agency, *Manual No. 2: Master Sampling Plans for Single, Duplicate, Double, and Multiple Sampling.* MD: Edgewood Arsenal, Army Chemical Center, 1953.

For Plan 3, $R_{TAB} = 6.79$, $c_1 = 0$ and $c_2 = 2$

For Plan 4, $R_{TAB} = 5.39$, $c_1 = 1$ and $c_2 = 3$

Step 3. Design sampling plans for each R_{TAB}.

Plan 3: For $P_A = 1 - \alpha = 0.95$ $np_1 = 0.43$ and

$$n = \frac{np_1}{p_1} = \frac{0.43}{0.005} = 86$$

$$n_1 = n_2 = 86$$

For $P_A = \beta = 0.10$ $np_1 = 2.96$ and

$$n = \frac{np_1}{p_1} = \frac{2.96}{0.03} = 98.67 \cong 99$$

$$n_1 = n_2 = 99$$

Plan 4: For $P_A = 1 - \alpha = 0.95$ $np_1 = 0.76$ and

$$n = \frac{np_1}{p_1} = \frac{0.76}{0.005} = 152$$

$$n_1 = n_2 = 152$$

For $P_A = \beta = 0.10$ $np_1 = 4.11$ and

$$n = \frac{np_1}{p_1} = \frac{4.11}{0.03} = 137$$

$$n_1 = n_2 = 137$$

Step 4. Compute Producer's and Consumer's Risks for each plan using Equations 10-26 and 10-30 and the observation that, $\alpha = 1 - \Pr\{Accept \mid p_1\}$ and $\beta = \Pr\{Accept \mid p_2\}$

Plan 3: $c_1 = 0$, $c_2 = 2$, $n_1 = n_2 = 86$

$$\text{At } p_1 = 0.005,\ P_{A,1} = \frac{86!}{0!\,86!}0.005^0(0.995)^{86} = 0.6498$$

$$P_{A,2} = \left(\frac{86!}{1!\,85!}0.005^1(0.995)^{85}\right)\left(\left(\frac{86!}{0!\,86!}0.005^0(0.995)^{86}\right) + \left(\frac{86!}{1!\,85!}0.005^1(0.995)^{85}\right)\right) + \left(\frac{86!}{2!\,84!}0.005^2(0.995)^{84}\right)\left(\frac{86!}{0!\,86!}0.005^0(0.995)^{86}\right)$$

$$P_{A,2} = (0.2808)(0.6498 + 0.2808) + (0.05997)(0.6498) = 0.3003$$

$$P_A = 0.6498 + 0.3003 + 0.9501$$

$\alpha = 1 - 0.9501 = 0.049900 < 0.05 \Rightarrow$ meets stipulated Producer's Risk

At $p_2 = 0.030$, $P_{A,1} = 0.0728$, $P_{A,2} = 0.0702$, $P_A = 0.0728 + 0.0702 = 0.1430$

$\beta = 0.1430 > 0.10 \Rightarrow$ fails to meet stipulated Consumer's Risk

Plan 3: $c_1 = 0$, $c_2 = 2$, $n_1 = n_2 = 99$

At $p_1 = 0.005$, $P_{A,1} = 0.6088$, $P_{A,2} = 0.3215$, $P_A = 0.6088 + 0.3215 = 0.9303$

$\alpha = 1 - 0.9303 = 0.0697 > 0.05 \Rightarrow$ fails to meet stipulated Producer's Risk

At $p_2 = 0.030$, $P_{A,1} = 0.0490$, $P_{A,2} = 0.0410$, $P_A = 0.0490 + 0.0410 = 0.0900$

$\beta = 0.0900 < 0.10 \Rightarrow$ meets stipulated Consumer's Risk

Plan 4: $c_1 = 1$, $c_2 = 3$, $n_1 = n_2 = 152$

At $p_1 = 0.005$, $P_{A,1} = 0.8233$, $P_{A,2} = 0.1272$, $P_A = 0.8233 + 0.1272 = 0.9505$

$\alpha = 1 - 0.9505 = 0.0495 < 0.05 \Rightarrow$ meets stipulated Producer's Risk

At $p_2 = 0.030$, $P_{A,1} = 0.0556$, $P_{A,2} = 0.0076$, $P_A = 0.0556 + 0.0076 = 0.0632$

$\beta = 0.0632 < 0.10 \Rightarrow$ meets stipulated Consumer's Risk

Plan 4: $c_1 = 1$, $c_2 = 3$, $n_1 = n_2 = 137$

At $p_1 = 0.005$, $P_{A,1} = 0.8497$, $P_{A,2} = 0.1141$, $P_A = 0.8497 + 0.1141 = 0.9638$

$\alpha = 1 - 0.9638 = 0.0362 < 0.05 \Rightarrow$ meets stipulated Producer's Risk

At $p_2 = 0.030$, $P_{A,1} = 0.0807$, $P_{A,2} = 0.0140$, $P_A = 0.0807 + 0.0140 = 0.0947$

$\beta = 0.0947 < 0.10 \Rightarrow$ meets stipulated Consumer's Risk

Step 5. Select a sampling plan. Plan 3, with $R = 6.79 > p_2/p_1$ yields double sampling plans that do not simultaneously meet both producer's and consumer's risk criteria. Plan 4, with $R = 5.39 > p_2/p_1$ yields plans that are capable of meeting both producer's and consumer's risk, but larger sample sizes are required.

Example Development of Double Sampling Plan Using Grubbs Tables

We wish to develop a double sampling plan for shipment size $N = 1{,}000$ that meets the following criteria: AQL = 0.5%, LTPD = 3.0%, $\alpha = 0.05$, $\beta = 0.10$, with $n_2 = 2n_1$. We will use Table 10-18 to determine sample sizes, n_1 and n_2, and the acceptance numbers, c_1 and c_2.

Step 1. Compute $R_0 = \dfrac{LTPD}{AQL} = \dfrac{p_2}{p_1} = \dfrac{0.03}{0.005} = 6.00$

Step 2. Identify in Table 10-18 the rows that bracket R_0.

R_0 falls between Plan 3 and Plan 4

For Plan 3, $R_{TAB} = 6.48$, $c_1 = 1$ and $c_2 = 3$

For Plan 4, $R_{TAB} = 5.39$, $c_1 = 0$ and $c_2 = 3$

Step 3. Design sampling plans for each R_{TAB}.

Plan 3: For $P_A = 1 - \alpha = 0.95$ $np_1 = 0.60$ and

$$n = \frac{np_1}{p_1} = \frac{0.60}{0.005} = 120$$

$$n_1 = 120,\ n_2 = 2n_1 = 240$$

For $P_A = \beta = 0.10$ $np_1 = 3.89$ and

$$n = \frac{np_1}{p_1} = \frac{3.89}{0.03} = 129.67 \cong 130$$

$$n_1 = 130,\ n_2 = 2n_1 = 260$$

Plan 4: For $P_A = 1 - \alpha = 0.95$ $np_1 = 0.49$ and

$$n = \frac{np_1}{p_1} = \frac{0.49}{0.005} = 98$$

$$n_1 = 98,\ n_2 = 2n_1 = 196$$

For $P_A = \beta = 0.10$ $np_1 = 4.11$ and

$$n = \frac{np_1}{p_1} = \frac{2.64}{0.03} = 88$$

$$n_1 = 88,\ n_2 = 2n_1 = 176$$

Step 4. Compute Producer's and Consumer's Risks for each plan using Equations 10-26 and 10-30 and the observation that,

$$\alpha = 1 - \Pr\{Accept \mid p_1\} \text{ and } \beta = \Pr\{Accept \mid p_2\}$$

Plan 3: $c_1 = 1,\ c_2 = 3,\ n_1 = 120$ $n_2 = 240$
At $p_1 = 0.005$, $P_{A,1} = 0.8784$,
$P_{A,2} = 0.0713$,
$P_A = 0.8784 + 0.0713 = 0.9497$

$\alpha = 1 - 0.9497 = 0.0503 \cong 0.05 \Rightarrow$ meets stipulated Producer's Risk

At $p_2 = 0.030$, $P_{A,1} = 0.1218$,
$P_{A,2} = 0.0011$,
$P_A = 0.1218 + 0.0011 = 0.1229$

$\beta = 0.1229 > 0.10 \Rightarrow$ fails to meet stipulated Consumer's Risk

Plan 3: $c_1 = 1,\ c_2 = 3,\ n_1 = 130,\ n_2 = 260$
At $p_1 = 0.005$, $P_{A,1} = 0.8617$,
$P_{A,2} = 0.0756$,
$P_A = 0.8617 + 0.0756 = 0.9373$

$\alpha = 1 - 0.9373 = 0.0627 > 0.05 \Rightarrow$ fails to meet stipulated Producer's Risk

At $p_2 = 0.030$, $P_{A,1} = 0.0957$,
$P_{A,2} = 0.0006$, $P_A = 0.0957 + 0.0006 = 0.0963$

$\beta = 0.0963 < 0.10 \Rightarrow$ meets stipulated Consumer's Risk

Plan 4: $c_1 = 0,\ c_2 = 3,\ n_1 = 98,\ n_2 = 196$
At $p_1 = 0.005$, $P_{A,1} = 0.6119$,
$P_{A,2} = 0.3374$, $P_A = 0.6119 + 0.3374 = 0.9493$

$\alpha = 1 - 0.9493 = 0.0507 \cong 0.05 \Rightarrow$ meets stipulated Producer's Risk

At $p_2 = 0.030$, $P_{A,1} = 0.0505$,
$P_{A,2} = 0.0146$,
$P_A = 0.0505 + 0.0146 = 0.0651$

$\beta = 0.0651 < 0.10 \Rightarrow$ exceeds stipulated Consumer's Risk

Plan 4: $c_1 = 0,\ c_2 = 3,\ n_1 = 88$ $n_2 = 176$
At $p_1 = 0.005$, $P_{A,1} = 0.6433$,
$P_{A,2} = 0.3199$,
$P_A = 0.6433 + 0.3199 = 0.9632$

$\alpha = 1 - 0.9632 = 0.0368 < 0.05 \Rightarrow$ exceeds stipulated Producer's Risk

At $p_2 = 0.030$, $P_{A,1} = 0.0685$,
$P_{A,2} = 0.0272$,
$P_A = 0.0685 + 0.0272 = 0.0957$

$\beta = 0.0957 < 0.10 \Rightarrow$ meets stipulated Consumer's Risk

A plot of the *ATI* curves for the four double sampling plans developed, ($c_1 = 0$, $c_2 = 2$, $n_1 = n_2 = 99$), ($c_1 = 1$, $c_2 = 3$, $n_1 = n_2 = 137$), ($c_1 = 1$, $c_2 = 3$, $n_1 = 130$, $n_2 = 260$), and ($c_1 = 0$, $c_2 = 3$, $n_1 = 88$, $n_2 = 176$), is shown in Figure 10-36. Note that the curves for these plans are similar, but the plan with $c_1 = 0$, $c_2 = 3$, $n_1 = 88$, $n_2 = 176$ has the lowest ATI for product quality near the AQL.

10.7 COMPARISON OF SINGLE AND DOUBLE ACCEPTANCE SAMPLING PLANS

The ATI and AOQL are the usual measures for comparing single and double sampling plans and deciding which is preferred. The single and double plans with the lowest ATI's (at the AQL), in the previous exercise, were selected and compared using the ATI and AOQ charts shown in Figure 10-37.

Note that the double sampling plan requires fewer units to be inspected on average for superior incoming quality. However, the AOQ comparison indicates that the single sampling plan will have a lower AOQL. Ultimately, the decision will be based on a trade-off between cost—a function of the ATI and the complexity of the sampling system—and protection from poor quality—a function of the AOQL. In this case, protection from poor quality favors the single sampling plan. It is not possible without further information (e.g., the actual cost of sampling and documentation costs for the two plans) to determine whether double sampling or the single plan is less costly. The ATI favors double sampling; however record-keeping, tracking, and documentation are all simpler and will normally be less costly for single sampling.

10.8 SUMMARY OF KEY IDEAS

Key Idea: *Attributes data collection is used to estimate the proportion of defectives or defects in a population or process using the results of a random sample.*

There are two types of attributes data: data that are suitable for p (or np) charts and data that are appropriate for c (or u) charts. With p chart data a random sample is used to estimate the proportion of a process or population that fails to conform

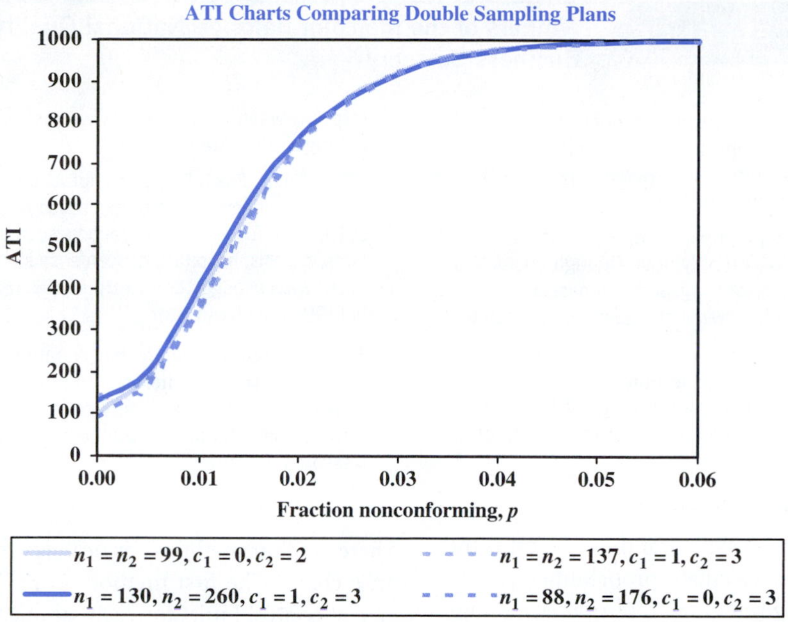

FIGURE 10-36. ATI Curves Comparing Double Sampling Plans

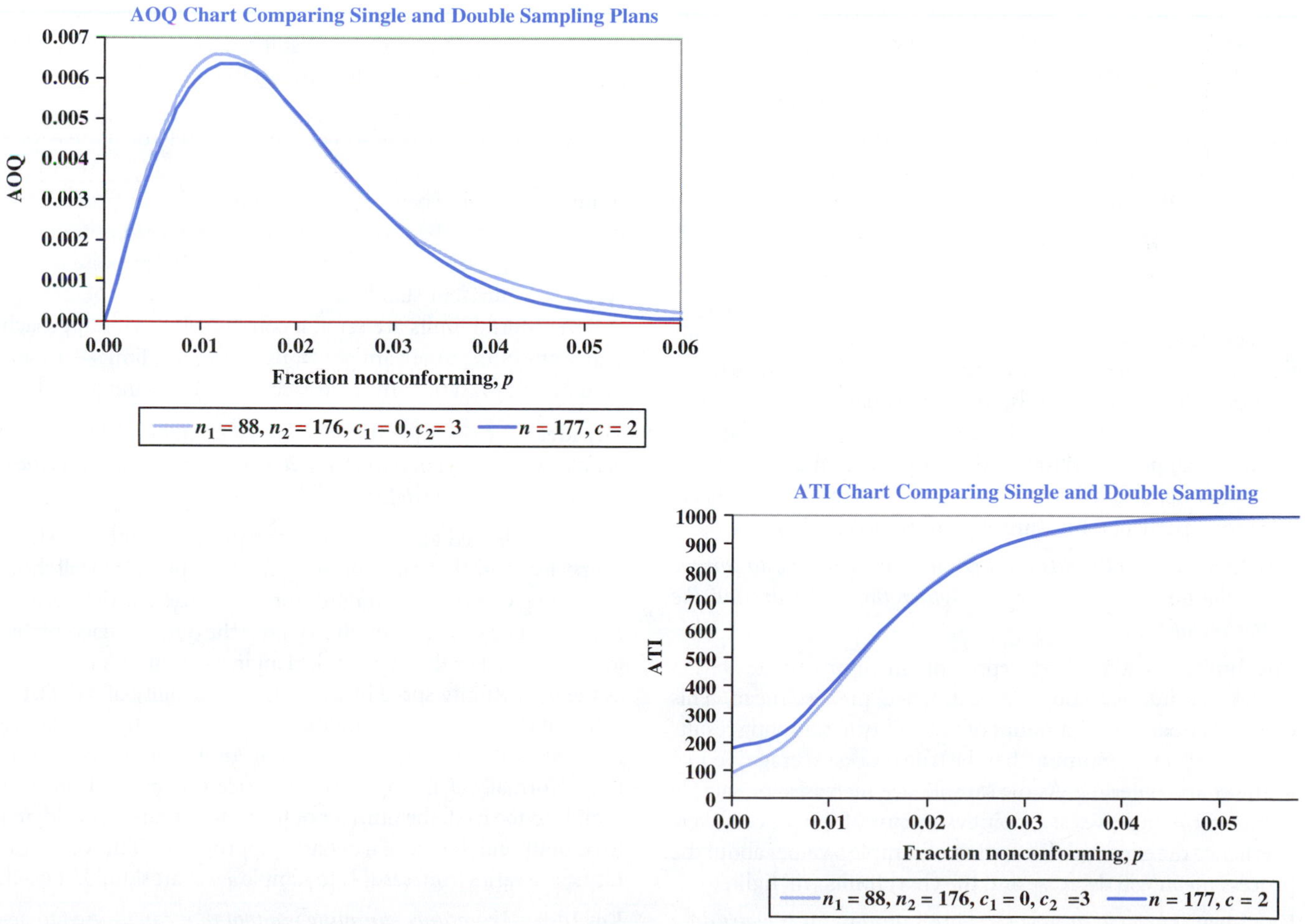

FIGURE 10-37. Comparison of Single and Double Sampling Plans

SHOWCASE of EXCELLENCE **Winners of the Malcolm Baldrige National Quality Award Small Business Category**

- 2002: Branch-Smith Printing Division, provider of a wide range of designing, printing, binding, and mailing services. http://branchsmithprinting.com, http://www.nist.gov/public_affairs/releases/branch.htm.
- 2001: Pal's Sudden Service, a privately owned, quick-service restaurant chain that serves primarily drive-through customers at 17 locations in the vicinity of Kingsport, Tennessee. http://www.PalsWeb.com, http://www.nist.gov/public_affairs/pals.htm.
- 2000: Los Alamos National Bank, the third largest independent bank in New Mexico, providing a full range of financial services to consumer, commercial, and government markets in northern New Mexico. http://www.lanb.com, http://www.nist.gov/publicaffairs/baldrige00/LosAlamos.htm.
- 1999: Sunny Fresh Foods, manufacturer and distributer of more than 160 different types of egg-based food products to more than 1,200 U.S. foodservice operations, such as quick service restaurants, schools, hospitals, convenience stores, and food processors. http://www.usaeggs.com, http://www.nist.gov/public_affairs/bald99/sunnyfresh.htm.
- 1996: Custom Research Inc., a national marketing research firm that assists its clients with new product developments in consumer, medical, and service businesses. http://cresearch.com, http://www.quality.nist.gov/Custom_Research_96.htm.

to requirements. The underlying statistical distribution is the *binomial* that can be used to calculate probabilities and to compute minimum sample sizes. When the sample size is constant, the use of an *np* chart eliminates the need to compute an individual proportion for each sample.

In the case of *c* chart data the objective is to estimate the number of defects, or errors, that occur within some opportunity space. The Poisson probability mass function is used to describe the distribution of defect occurrences. When the opportunity space varies from sample to sample a *u* chart must be used in lieu of the *c* chart.

Either type of data can be organized by defect type or defective cause in a Pareto chart to help focus improvement efforts.

Key Idea: *A significant source of inspection error with attributes data collection is the lack of a common understanding by all inspectors of how the data should be collected and the quality requirements.*

With attributes measurement inspectors are often required to exercise judgment on whether a unit conforms or fails to conform to requirements. This can lead to inconsistencies across inspectors. To minimize this, it is essential that *operational definitions* be developed and integrated into the training programs for all personnel involved. An operational definition is a clear, concise, detailed, and unambiguous definition of what it takes for a particular measure to conform to requirements.

Key Idea: *Control limits on a* p *chart are sensitive to sample size—the larger the sample the tighter the limits around the centerline of the chart.*

The limits on a *p* chart represent an approximate 99.7% confidence interval about the estimated process mean. This confidence band is the amount of variability in sampling results that are expected assuming that the true process average is equal to the chart centerline. As the sample size increases the sample parameter *p* improves as a point estimator of the process average; hence the expected fluctuation of sample *p* values about the process mean will decrease and the chart limits will tighten.

Key Idea: *Variable sample sizes are common for* p *chart data collection schemes and require special techniques of analysis. Four methods are available.*

There are four methods for dealing with variable sample sizes on *p* charts. The first method involves calculating a different pair of control limits for each sample. This method has some obvious disadvantages. Apart from the effort required in continuously computing new control limits, variable limits can be perceived by some to be somewhat arbitrary. A second method, which is practicable as long the maximum and minimum sample sizes are within 25% of the average, is to compute a single set of control limits based on the average sample size. With this technique individual limits will still have to be computed for any identified points that lie close to the control limits, either on the inside or outside.

A third approach is to compute two sets of limits—one based on the maximum sample size and one based on the minimum sample size. Then, only certain points that lie between the two limits need to be further investigated. A fourth option is to plot a standardized control chart in which each data point is converted to equivalent standard deviation units from the average and the control limits are set at a constant $\pm$ 3. This approach shares one of the disadvantages of using variable limits—a standard deviate must be computed each time a new sample is taken.

Key Idea: *A* c *chart (or* u *chart) is used to monitor the number of defects per inspection unit and a defect may or may not cause an output unit to be defective.*

The data collected for *c* and *u* charts represent counts of defects (imperfections) that cause the output from a process to fall short of meeting world class standards for excellence. The defects may or may not be serious enough to render the output unacceptable to customers. For this type of data an inspection unit is defined as the opportunity space in which the probability of a defect is constant (following a Poisson distribution), and the sample size is stated as the number of inspection units considered at one time. Normally, if the opportunity space is constant from one sample to the next, the number of inspection units is considered to be unity and the use of a *c* chart is appropriate. If the opportunity space varies from sample to sample a *u* chart should be used.

Key Idea: *Acceptance sampling is a tool that can be used to determine whether a fixed supply of material meets minimum stipulated requirements.*

Acceptance sampling is not a substitute for statistical process control and cannot determine ongoing process predictability or identify improvement opportunities. Its major role is to provide a temporary quality measure in cases where there is insufficient process history to use SPC techniques or where a process is unstable or incapable of meeting stipulated requirements.

Key Idea: *In any acceptance sampling scheme it is essential that the sample taken be random.*

While it may facilitate data collection to choose items for a sample that can be conveniently accessed, such a sample may not be representative of the entire fixed lot of material. A true random sample assigns an equal probability of selection to each item in the lot. Since most lots are packaged and packed in the order in which they were produced (at least roughly) sampling units should be selected across the entire lot; otherwise time-related sources of variability will not be captured within the sample.

Key Idea: *Acceptance sampling plans can be compared using producer's and consumer's risks, operating characteristic (OC) curves, average outgoing quality limit (AOQL), average total inspection (ATI), and average sample number (ASN).*

For any specific acceptance sampling plan, the operating characteristic (OC) curve is a plot of the probability of acceptance against the proportion nonconforming in a lot. From the OC curve the producer's and consumer's risks can be determined. The producer's risk is the probability of rejecting a lot that is at least as good as some specified level p, called the acceptable quality level (AQL). The consumer's risk is the probability that some lot that is as bad as some specified level p, called the lot tolerance percent defective (LTPD), will be accepted.

Under rectifying inspection the average outgoing quality, AOQ, improves as good items are substituted for defective ones. The average outgoing quality limit, AOQL, represents the worst average level of quality that the customer would expect to receive.

The average total inspection, ATI, is the average number of units per lot that must be inspected, including the rejected lots that undergo 100% screening. A related metric that applies to double or multiple sampling plans is the average sample number, ASN, which is the average number of units inspected prior to making a decision. Unlike the ATI the ASN does not take into account rectification inspections. Sampling plans with a low AOQL, ATI, and/or ASN are preferred.

10.9 MIND EXPANDERS TO TEST YOUR UNDERSTANDING

10-1. A certain manufacturing plant is having problems meeting its delivery commitments. One of the problems has been traced to purchased materials which are frequently late, received in the wrong quantities, or the vendor ships the wrong parts in error. The purchasing director has been tasked to see if errors in purchase orders are contributing to the problem. Ten purchase orders per day are selected at random over a 20-day period. Each purchase order is carefully scrutinized and the total number of errors are recorded. The aggregate total for all 10 purchase orders sampled is then computed. The results are shown below.

a. Is the process of purchase order error generation stable? What is your best estimate of the number of errors that will be on the next purchase order cut? Support your answers with control chart data.

b. The following data were collected for a 20-day period following the implementation of some improvements in the purchasing department. Using the control chart that you constructed for part a, demonstrate whether there is statistical evidence to validate the improvement.

Errors in 10 Randomly Selected Purchase Orders

Day	Errors Recorded	Day	Errors Recorded	Day	Errors Recorded	Day	Errors Recorded
1	8	11	15	21	9	31	2
2	11	12	11	22	7	32	4
3	7	13	10	23	5	33	6
4	19	14	13	24	3	34	3
5	7	15	16	25	3	35	4
6	12	16	13	26	3	36	5
7	11	17	16	27	4	37	5
8	6	18	7	28	7	38	3
9	10	19	15	29	5	39	4
10	6	20	13	30	4	40	2

c. Construct a control chart for the process following the improvements. Is the process in control? What is your best estimate of the number of errors in the next purchase order cut?

d. Management wishes to reduce the sampling effort from 10 to 5 per day. What centerline and control limits should be placed on the control chart?

10-2. Chip Technologies manufactures personal computers for large chain electronic retail outlets. Each day several orders are selected at random and each computer carefully inspected against a check list. The total defects are recorded and plotted on a control chart. Since orders vary in size, the total number of computers inspected differs each day. The results for 20 days of inspection are shown in the table below.

Day	Number of Computers Inspected	Number of Defects	Day	Number of Computers Inspected	Number of Defects
1	15	35	11	18	21
2	27	27	12	16	31
3	18	70	13	32	98
4	36	69	14	16	31
5	25	78	15	17	15
6	16	14	16	37	147
7	37	71	17	19	32
8	23	65	18	37	106
9	19	41	19	32	34
10	26	82	20	36	101

a. Analyze this process using an appropriate control chart. What would be your recommendations to the plant manager?

b. A Six Sigma team was formed to investigate the defect problem. Some changes were implemented, aimed at stabilizing the process. After all workers had been thoroughly trained in the new procedures, data were collected for a 20-day period and are summarized in the table.
Have the changes improved the process? Justify your answer using control chart data.

Day	Number of Computers Inspected	Number of Defects	Day	Number of Computers Inspected	Number of Defects
1	20	12	11	3	2
2	35	45	12	12	5
3	16	13	13	16	10
4	20	11	14	27	34
5	19	30	15	17	9
6	8	3	16	42	45
7	9	5	17	38	35
8	17	20	18	25	28
9	7	4	19	12	7
10	6	3	20	32	19

10-3. The controller of a large chemical refinery is concerned about the late processing of some accounts payable, thereby failing to take advantage of discounts that are available for timely payments. The following data represent performance over the most recent 30-month period.

Month	Number of Accounts Paid Late	Total Number of Accounts Processed
January	94	1032
February	101	1176
March	44	653
April	99	1055
May	97	1110
June	42	640
July	56	768
August	78	746
September	64	516
October	89	1153
November	61	571
December	133	1031
January	181	934
February	90	575
March	160	947
April	142	856
May	145	1243
June	88	766
July	131	1014
August	150	1163
September	151	1099
October	103	939
November	225	1218
December	214	1165
January	220	1104
February	100	784
March	63	836
May	44	555
June	85	1199
July	39	588

a. Construct a control chart to analyze this process, using control limits based on the average sample size. Is it appropriate to use the average sample size in this case? Explain.
b. Identify the points that should be investigated based on the limits you calculated in part a. How many points did you need to investigate? Is the process in control?
c. Construct another control chart using the max/min approach described in Section 10.3.1.3. How many points do you need to investigate using this approach?
d. Over the next few months the controller challenged the employees to find ways to improve the process. Many suggestions were made and acted upon. Data for the next 30 months are shown below. Has the process improved? Where does it appear that the process has stabilized at a new level of performance? Construct a new control chart for this point onward and estimate the average percentage of payments that are now late.

After Improvements

Month	Number of Accounts Paid Late	Total Number of Accounts Processed
August	87	865
September	97	1080
October	48	598
November	71	1010
December	37	731
January	27	669
February	38	751
March	58	1021
April	24	553
May	56	1070
June	33	567
July	37	745
August	43	774
September	57	1060
October	31	642
November	29	607
December	51	886
January	52	985
February	44	989
March	61	1110
April	40	860
May	30	828
June	28	583
July	31	690
August	32	568
September	50	1102
October	48	1046
November	36	606
December	45	921
January	33	804

Sample No.	No. of Defects	Sample No.	No. of Defects	Sample No.	No. of Defects
1	5	11	11	21	23
2	26	12	17	22	5
3	27	13	7	23	6
4	15	14	19	24	12
5	8	15	18	25	18
6	20	16	13	26	26
7	9	17	14	27	9
8	6	18	25	28	10
9	27	19	5	29	26
10	22	20	27	30	24

10-4. A plant conducts final inspection on samples of 10 automobile radios selected at random. A summary of the most recent 30 samples is shown below.

a. Construct an appropriate control chart and comment on whether this process is in statistical control.
b. If the number of radios in a sample is reduced from 10 to 5 what impact does this have on the control chart (i.e., the centerline and control limits)?

10-5. A food processor sells canned vegetables to supermarket chains. The supermarkets consider dented cans to be nonconforming since consumers are reluctant to buy them. The supermarkets would like to develop an acceptance sampling plan with a consumer's risk of 10% and an LTPD of 5%. The processor wants a producer's risk of 5% with an AQL of 0.75%. Shipments are made in lots of five pallets. Each pallet contains five layers of six cartons, and each carton contains 48 cans.

a. Design a single sampling plan using Grubbs table.
b. Construct the OC curve for this plan.
c. Construct the AOC curve for this plan. Use the AOC curve to estimate the AOQL. What does the AOQL represent?
d. Construct the ATI curve for this plan. Use the ATI curve to estimate the average number of cans inspected per shipment. What assumptions do you make in this calculation?

10-6. Consider the food processor described in problem 10-5.

a. Design a double sampling plan using the Grubbs table.
b. Construct the OC curve for this plan.

c. Construct the AOC curve for this plan. Use the AOC curve to estimate the AOQL.

d. Construct the ATI curve for this plan. Use the ATI curve to estimate the average number of cans inspected per shipment.

e. Compute the ASN for this plan. Why is the ASN different from the ATI computed in part d?

f. Compare this double sampling plan with the single sampling plan designed in problem 10-5. Which plan would you choose? Why?

10-7. A common single sample acceptance sampling plan ($n = 100, c = 1$) has been used by a manufacturer and one of its customers to accept and reject production lots.

a. Compute the probability of acceptance for several values of p and sketch an approximate operating characteristic curve.

b. If the producer's and consumer's risks, α and β, are both defined to be 5%, use your OC curve and the binomial probabilities to estimate the values of the corresponding AQL and LTPD.

10-8. A textile manufacturer uses SPC to monitor and control imperfections in bolts of cloth. An inspection unit of 100 square meters has been selected. The following table represents data that have been collected for 30 randomly selected bolts from production during the last week.

Bolt No.	No. Imperfections	Square Meters	Bolt No.	No. Imperfections	Square Meters
1	5	100	16	10	120
2	10	120	17	12	150
3	9	150	18	6	100
4	7	100	19	6	100
5	3	100	20	5	180
6	12	150	21	8	150
7	16	180	22	6	120
8	13	180	23	8	120
9	9	150	24	7	100
10	6	100	25	9	150
11	3	100	26	11	180
12	2	100	27	6	120
13	18	180	28	5	180
14	6	150	29	10	120
15	5	120	30	3	150

a. Construct an appropriate control chart and evaluate the control status of this process.

b. The following data were collected for 10 additional bolts of cloth. Has the control status changed? Explain. What would you recommend?

Bolt Number	No. Imperfections	Square Meters
31	6	100
32	7	100
33	12	150
34	18	180
35	12	180
36	12	120
37	9	120
38	10	100
39	9	100
40	9	100

10-9. The specifications on the bagged product described in problem 9-1 are set at 50.0 lb ± 0.5 lb. To check the consistency of the bagging operation in meeting these specs, an inspector randomly selects seven pallets from each lot number and weighs all the bags on the selected pallets. The pallets each contain 40 bags. Any bag found to be either overweight or underweight is set aside and counted as a defective. The inspection results for 20 lots are recorded below.

Lot Number	Number of Defective Bags	Lot Number	Number of Defective Bags
1	22	11	13
2	19	12	23
3	20	13	17
4	21	14	23
5	25	15	20
6	25	16	23
7	25	17	21
8	17	18	18
9	11	19	30
10	19	20	24

a. Is the bagging process in a state of statistical control? Justify your answer by plotting a control chart appropriate to these data.

b. What is your estimate of the percentage of bag weights that do not meet specifications?

c. The next 10 lots inspected yielded the following results: 30, 38, 24, 33, 33, 21, 22, 29, 31, 26. How do you interpret these new data?

d. We know from part d of problem 9-1 that the process mean increased to 50.3. At that level what sample size would be required in order to detect this change with the same probability as you calculated in problem 9-1, part d?

10-10. Evergreen Electric Company manufactures strings of holiday lights. As part of final quality control testing 500 strings are randomly selected and plugged into a power source. The total number of unlit bulbs are counted and recorded. These are considered to be nonconformities and the inspectors are instructed to replace any unlit bulbs with good ones prior to passing a string on to packaging. Any strings that fail to light or blink properly are tagged as defective, recorded, and placed in a reject bin. The data collected from 30 consecutive batches are shown in the following table.

Batch	No. of Strings with Lighting or Blinking Failures	No. of Unlit Bulbs	Batch	No. of Strings with Lighting or Blinking Failures	No. of Unlit Bulbs
1	12	43	16	17	35
2	13	43	17	9	39
3	6	31	18	9	39
4	3	44	19	9	47
5	6	37	20	6	40
6	8	33	21	14	34
7	11	35	22	10	40
8	11	58	23	9	49
9	10	51	24	10	45
10	16	46	25	8	40
11	12	45	26	14	31
12	8	37	27	10	36
13	9	34	28	13	45
14	9	41	29	12	41
15	12	39	30	12	52

a. Assuming that each of the strings tested contains 100 bulbs, construct a p chart to test whether the number of strings with lighting or blinking failures is in a state of statistical control. Explain why this is a p chart rather than a c chart application? Now, construct an np chart for these data and compare your results with the p chart.

b. Estimate the percentage of light strings in the process output that are defective.

c. Assuming that each of the strings tested contains 100 bulbs, construct an appropriate control chart to test whether the number of unlit bulbs is in a state of statistical control.

d. Estimate the percentage of unlit bulbs that would be expected in each sample of 500 light strings.

e. Modify the centerline and limits on the control chart you constructed in part c to accommodate a new sampling scheme where only 100 strings per batch are to be tested.

10-11. Evergreen Electric manufactures holiday light strings in three lengths: 50, 100, and 150 lights per string. Samples are taken at random but in such a way that the number of strings tested varies batch to batch. The data shown in the table below is the sampling results for 30 recent batches.

Batch	No Strings Tested	No of Lights per String	No of Strings With Lighting or Blinking Failures	No. of Unlit Bulbs
1	500	100	12	43
2	550	150	13	43
3	435	50	6	31
4	600	100	3	44
5	489	50	6	37
6	611	100	8	33
7	526	100	11	35
8	433	150	11	58
9	543	100	10	51
10	567	150	16	46
11	398	100	12	45
12	455	50	8	37
13	542	100	9	34
14	500	50	9	41
15	507	100	12	39
16	457	150	17	35
17	569	150	9	39
18	500	50	9	39
19	456	100	9	47
20	498	50	6	40
21	503	150	14	34
22	587	150	10	40
23	566	100	9	49
24	588	150	10	45
25	499	50	8	40
26	455	150	14	31
27	483	100	10	36
28	534	150	13	45
29	522	100	12	41
30	515	100	12	52

a. Construct a series of p charts to test whether the number of lighting or blinking failures is in a state of statistical control. To make this determination construct the following charts.
   i. A p chart with variable limits.
   ii. A p chart with limits based on average sample size. Comment on the appropriateness of using this chart.
   iii. A p chart with limits based on maximum and minimum sample sizes.
   iv. A standardized p control chart.

b. If appropriate, estimate the percentage of light strings in the process output that are defective.

c. Construct an appropriate control chart to test whether the number of unlit bulbs is in a state of statistical control. If appropriate, estimate the percentage of unlit bulbs that would be expected in each light string.

d. What action would you recommend that the leadership of Evergreen Electric take concerning these processes?

10-12. Problem 9-13 presented data on sealer 7098, a caulking product produced by Mastic Pty. Ltd. This product is also inspected for properly sealed caps. The tubes of caulking compound are packaged in boxes of 144. From each batch a box is selected randomly for inspection. The number of leaking tubes is recorded as shown in the following table.

Sample	No. Defective Tubes	Sample	No. Defective Tubes
1	0	16	1
2	2	17	0
3	0	18	3
4	1	19	3
5	2	20	5
6	1	21	1
7	1	22	3
8	6	23	0
9	2	24	0
10	0	25	2
11	1	26	0
12	1	27	1
13	1	28	2
14	0	29	3
15	1	30	1

a. Construct an appropriate control chart to determine whether the capping process is in a state of statistical control.

b. It was discovered that some problems with pneumatic pressure were encountered during the time that samples 8 and 20 were being manufactured. This was considered to be an assignable cause of variation. Eliminate these two samples from the data, reconstruct the control chart, and determine the state of statistical control.

c. The boxes selected at random are also inspected for the presence of lumps. Lumps in caulk can present difficulties for customers who wish to dispense a smooth bead. The following data were collected for the boxes inspected.

Sample No.	No. of Lumps	Sample No.	No. of Lumps
1	7	16	7
2	12	17	14
3	12	18	12
4	9	19	11
5	7	20	5
6	13	21	14
7	17	22	16
8	12	23	15
9	7	24	12
10	14	25	13
11	15	26	12
12	10	27	9
13	11	28	10
14	17	29	12
15	10	30	11

   i. Construct an appropriate control chart and determine if the process of producing lumps is in a state of statistical control.
   ii. What is the maximum number of lumps that you might expect to observe in any given box of tubes? What is the minimum expected number?
   iii. What is the expected number of lumps per box? What is the expected number of lumps per tube?

10-13. Deglingo Enterprises Ltd. manufactures door handle components for the automobile industry. Components at the end of an assembly operation are subjected to 100% inspection on several critical characteristics. The data shown in the following table have been collected for the most recent 40 shifts.

Data from DEGLINGO Enterprises

Shift	Production Batch Size	No. of Defectives	Shift	Production Batch Size	No. of Defectives
1	296	20	21	258	23
2	328	11	22	233	8
3	335	22	23	278	13
4	304	28	24	259	12
5	280	13	25	315	16
6	316	17	26	330	20
7	279	8	27	304	19
8	284	1	28	344	20
9	296	16	29	347	9
10	342	14	30	302	20
11	260	11	31	330	15
12	295	19	32	370	14
13	314	16	33	282	18
14	363	30	34	234	2
15	320	16	35	274	10
16	282	17	36	345	22
17	340	5	37	325	13
18	282	14	38	331	14
19	234	11	39	251	8
20	313	19	40	352	25

a. Plot the data on a p chart with variable limits and determine the state of control.

b. Plot the data on a p chart with the control limits based on the average sample size.
   i. Identify those points that require further investigation and explain the reason why.
   ii. Is the process in control? Explain.
   iii. Is it appropriate to use the average sample size in this case? Explain.

c. Plot the data on a p chart using the maximum and minimum control limits method. Use $n = 370$ for the maximum production batch size and $n = 230$ for the minimum batch size. Identify those points that need further investigation and explain why. Determine if the process is in a state of control.

d. Plot the data on a standardized control chart. Assess the control status of the process and compare your results with the answers you got in parts a–c.

10-14. The management of American Sugarplums, Inc., a candycane manufacturer that was introduced in problem 9-30, is considering a faster and less expensive method for checking candycane lengths, as an alternative to $\overline{X}$ and R charts. The proposed method would be to use a device that can quickly check lengths on a go/no go basis and reject those that are either too short or too long. (Candycanes that fail to comply with length specifications cause problems in the packaging operation.)

a. If the process is performing at the level described in problem 9-30, what would be your expected centerline for a p chart using the proposed inspection device in lieu of the $\overline{X}$ and R charts currently in use?

b. What minimum sample size would be required so that the probability of detecting a one-third standard deviation shift in average length (on the first sample following the shift) would be at least 1 in 20 (i.e., 0.05)? Compare your results with the result you obtained in problem 9-30, part e (ii).

10-15. Orange Inc. prepares for the holidays by manufacturing its most popular products such as laptops, touch phones, and MP3 players. Its most recent product is the Orange Touch MP3. Since the popularity and the demand for this product is high, Orange Inc. is producing the Orange Touch MP3 in large batches with long production runs. To check for defects, 30 packing orders are selected at random for inspection during the first shift production run. Each packing order consists of 100 MP3 players. Each MP3 player contains 30 gigabytes for data storage. Prior to shipment, each MP3 player in the packing order selected is inspected for defective gigabytes (defective gigabytes result in a slowdown of data

Packing Order	No. of Defects in Each Order	Packing Order	No. of Defects in Each Order
1	1	16	4
2	3	17	1
3	2	18	7
4	8	19	1
5	3	20	3
6	5	21	3
7	4	22	6
8	4	23	3
9	5	24	3
10	4	25	4
11	2	26	3
12	7	27	4
13	2	28	6
14	3	29	7
15	1	30	2

transfer). Those players that contain defects are sent back to production for repair. The packing order number and corresponding number of defects are recorded in the table located at the bottom of page 409 .

a. Is the production of defects in MP3 players a process that is in a state of statistical control? Use an appropriate control chart to support your answer.

b. As a cost-cutting measure the management of Orange Inc. decided to reduce the inspection effort by inspecting only one-half of a packing order. Using the reduced sampling scheme, an additional 15 packing orders were selected, divided in half, and 50 MP3 players were inspected for defects. These results are shown in the following table. Add these new data to the original control chart and determine if the process has changed. (*Hint:* Make certain you account for the fact that the inspection effort has been reduced; this changes the opportunity for defects to occur.)

Packing Order	No. of Defects in Each Order	No. of MP3 Inspected
1	2	50
2	2	50
3	3	50
4	3	50
5	5	50
6	2	50
7	4	50
8	2	50
9	2	50
10	1	50
11	3	50
12	3	50
13	3	50
14	2	50
15	4	50

c. Plot the new data on a separate control chart and comment on the state of control.

d. What would you recommend to the management of Orange Inc.?

10-16. Pacific Pen Co. manufactures a variety of writing instruments. One thousand pens of a particular model were inspected from each of 30 production batches and each pen was inspected for defects. For each pen containing any defects, if the inspector judged the defects to be bad enough, the pen was scrapped and recorded as a defective. The number of defective pens recorded for these samples is shown in the following table.

Sample Number	No. of Defective Pens	Sample Number	No. of Defective Pens
1	31	16	36
2	34	17	32
3	40	18	28
4	35	19	30
5	24	20	23
6	29	21	38
7	29	22	38
8	23	23	29
9	33	24	38
10	25	25	39
11	38	26	25
12	34	27	25
13	30	28	24
14	24	29	37
15	16	30	22

a. Construct an appropriate control chart to determine if the process that is producing defective pens is stable.

b. What is your estimate of the percentage of this model of pens that will be defective in the next production run? Explain.

c. Pens of this particular model are inspected for three types of defect: ill fit of tops to pen bodies, inability to write, and cracked bodies. The table located on page 411 provides details on the types of defects discovered in the 30 samples described previously.

   i. Construct a Pareto chart and use this chart to identify which of the defect types should be the initial focus for improvement. Assume that the customer impacts of all defect types are equal.

   ii. Construct a u chart for the defect type that occurs most frequently depicting the average number of defects of that type per defective pen. Define one defective pen as an inspection unit. Assess whether this process is in control.

   iii. Construct a c chart for the defect type that occurs most frequently, depicting the average number of defects of that type per inspection unit of 1,000 pens. Assess whether this process is in control.

   iv. Compare your answers to parts (c (ii)) and (c(iii)). Discuss the implications.

d. What should management do next in order to improve this process?

10-17. Brian's bakery bakes cookies each day, but the output varies depending on how much dough is created each day. Every cookie is carefully inspected after being baked to ensure it is not burned before being put up for sale. The table below shows the sample size (production batch size) and the corresponding number of defective cookies each day.

a. Construct a p chart using variable limits, and comment on the state of control.

b. Construct a p chart with limits based on the average sample size. Is using the average sample size for a single set of control limits appropriate? Explain. Which points on the chart need to be investigated further? Is the process in control?

c. Construct a p chart with two sets of limits—one based on the largest possible sample size and the other based on the smallest. Which points need to be investigated further? Is the process in a state of control?

d. Use the data provided to construct a stabilized control chart. Is the process in control?

Sample Size	No. of Defective Cookies	Sample Size	No. of Defective Cookies	Sample Size	No. of Defective Cookies	Sample Size	No. of Defective Cookies
408	46	600	38	466	43	624	51
415	44	495	54	398	55	432	51
585	61	618	59	427	52	532	52
387	38	487	57	444	58	381	59
531	44	482	47	609	55	511	45
623	54	376	37	542	57	521	57
486	47	432	41	568	53	571	54
426	48	529	56	490	45	602	50
617	41	556	49	576	43	547	48
460	57	412	54	617	59	597	48
405	50	495	45	570	60	567	37
584	58	381	52	433	43	570	66
441	47	452	49	553	51	428	54
523	48	420	41	563	49	461	47
508	46	446	55	378	54	461	44
494	42	491	37	485	48	430	56
518	50	498	51	621	52	444	38
384	48	469	46	518	40	432	51
505	46	515	40	510	66	562	47
440	57	399	64	555	42	591	57
507	49	623	52	505	52	515	56
376	45	475	49	403	50	535	42
443	56	493	48	559	43	376	50
476	45	596	63	507	52	394	43
586	47	586	51	507	50	624	54

10-18. A company produces a part for which four types of defects are possible. An occurrence of any one of the defect types causes an automatic rejection of the part. Below is a chart of the defectives and types of defects that were found in 30 samples of 1,000.

Sample Number	No. of Defectives	Sample Size	Defects			
			Defect 1	Defect 2	Defect 3	Defect 4
1	47	1000	8	28	4	12
2	50	1000	10	30	0	12
3	50	1000	11	27	4	12
4	45	1000	9	38	4	25
5	50	1000	10	27	0	17
6	53	1000	13	31	3	22
7	47	1000	11	27	2	15
8	51	1000	5	34	1	26
9	56	1000	12	27	5	16
10	60	1000	4	41	5	25
11	51	1000	6	28	1	30
12	58	1000	9	30	8	26
13	41	1000	10	24	3	19
14	52	1000	8	36	0	21
15	47	1000	6	22	1	22
16	54	1000	11	24	4	25
17	54	1000	14	35	2	12
18	56	1000	15	48	5	24
19	62	1000	11	40	1	24
20	40	1000	8	26	1	17
21	53	1000	4	36	2	23
22	57	1000	14	39	3	24
23	52	1000	8	36	7	19
24	52	1000	2	34	0	18
25	54	1000	7	20	3	25
26	50	1000	8	42	0	14
27	52	1000	2	38	7	13
28	49	1000	3	32	8	12
29	47	1000	9	24	2	12
30	44	1000	8	27	4	18

a. Plot an *np* chart to show whether or not the process is in statistical control.

b. If appropriate, estimate the percentage of output that is defective.

c. Talley up the count of all defects, by defect type, that were found in the 30 samples. Construct a Pareto chart, and use that chart to identify which defect type should be focused on first.

d. Construct a *u* chart for the defect type identified in part c and assess whether or not the process is in control with respect to that defect type.

10-19. Assume that you work for a major aluminum company and your boss has asked you to analyze the number of defects that are present on aluminum cans that are sent down an assembly line. Lots of 50 cans selected at random are to be inspected. Thirty samples of 50 cans have been inspected and the number of defects recorded. These data are presented in the table below.

a. Construct a *c* chart for these data and comment on the state of control.

b. Upon investigation it was found that a production problem was creating an unusual number

Sample	No. Defects	Sample	No. Defects
1	5	16	14
2	5	17	3
3	6	18	7
4	8	19	4
5	5	20	5
6	4	21	3
7	4	22	3
8	2	23	4
9	7	24	4
10	7	25	3
11	2	26	11
12	9	27	4
13	7	28	4
14	7	29	7
15	5	30	1

of defects around the time that sample number 16 was taken. Eliminate that sample from your data and construct a revised *c* chart and comment on whether the process is in a state of control excluding sample number 16.

c. Management has decided to reduce the inspection effort by half, inspecting cans in smaller lots of 25. At this level of reduced inspection an additional 15 samples are taken and the new data are presented in the table below. Plot these samples on the original chart and comment on the state of control. (*Note:* The definition of an inspection unit has changed so the data must be converted to a *u* chart before the state of control can be determined.)

Sample	No. Defects
31	4
32	3
33	2
34	2
35	5
36	4
37	2
38	3
39	0
40	0
41	2
42	2
43	1
44	2
45	3

10-20. The management of Universal Commodities, Inc., described in problem 9-31, is considering changing from variables measurement to a simplified system where this part would be inspected using go/no go gauging. What sample size would be necessary so that the *p* chart provides the same probability of detecting a 0.5σ shift in mean as the probability you computed in problem 9-31d?

10-21. A manufacturing plant specializes in making size AA batteries. Its batteries are 100% inspected to make sure they will produce a charge. The batteries are plugged into an apparatus that turns on a light if there is a charge; if the light fails to come on the batteries are marked defective and thrown out. The factory recorded data over one month, as shown below.

Day	Sample Size	Number Defective	Day	Sample Size	Number Defective
1	125	25	19	94	12
2	113	16	20	107	5
3	96	19	21	101	16
4	116	15	22	93	18
5	99	16	23	82	17
6	106	17	24	123	18
7	95	15	25	78	23
8	84	15	26	98	18
9	120	6	27	107	16
10	80	17	28	86	3
11	76	22	29	119	15
12	121	15	30	79	15
13	110	16	31	89	19
14	88	18	32	116	16
15	108	15	33	122	15
16	82	12	34	87	18
17	124	15	35	96	20
18	80	16	36	103	18

a. Using these data construct a *p* chart with variable limits. Comment on the state of statistical control.

b. Construct a *p* chart with limits based on the average sample size and identify which points need to be investigated to determine the state of control. Show that using the average sample size is appropriate.

c. Construct a *p* chart with two sets of limits—one set based on the largest expected sample size and the other set based on the smallest expected sample size. Using this approach, identify which points need to be investigated further to determine the state of control.

d. Construct a standardized control chart, and comment on the state of control.

CHAPTER ELEVEN

PROCESS IMPROVEMENT IN LEAN PRODUCTION ENVIRONMENTS

QUALITY MAXIM #11: *The identification and removal of sources of waste improves work flow, quality, and responsiveness to customer needs.*

Improvement Tools to Lean On!

The 1989 movie *Lean on Me*[1] was based on the true story of a rebel principal, Joe Clark, who used unconventional methods to turn around a downtrodden, crime-ridden New Jersey high school. Bucking tradition and facing a tide of political opposition, this real-life principal found resistance everywhere. Parents fought him, teachers resented him, and his superiors doubted him. Ironically, the students loved him. It was only through his perseverance and unwillingness to sacrifice principle for political expediency that his tactics eventually triumphed. The movie title was inspired by the song of the same name[2] which offers a sage reminder that change can be good, and that it often pays to have faith in those who are genuinely committed to bringing about sustainable reform. Two years after the release of the movie and nearly two decades after the song was first recorded, lean production made its debut. Although the word *lean* in lean production refers to a new-era business philosophy, some important parallels can be drawn with the message in the movie. In both cases lean is about radical change and focuses more on streamlining processes than on shoring up bottom lines. Lean is about eliminating wasteful nonproductive activities that expend resources but add no stakeholder value. The lean philosophy takes a system's view and, as in the movie, change can be met with stiff internal opposition as traditions, once held sacred, are threatened, and accountability is sharpened and shared. The transformation from tradition to lean can be tough, but fortunately numerous tools are available on which *management can lean to become lean*. The key is to transform the culture top-to-bottom so that lean becomes part of the corporate mind-set and a litmus test for all policies, procedures, and decisions. In the movie Joe Clark tells his students that if they do not succeed in life, not to blame where they came from or the establishment. He says the fault for lack of success must be borne by the students themselves as they have the power to shape their own lives. How many corporations are caught up in a similar dilemma? When changes in the economy, financial markets, consumer tastes, or government regulations threaten sustainability, does a company turn to excuses (e.g., blaming others, seeking bailouts, etc.), or, does it shoulder the responsibility for its troubles, turn its plight into an opportunity to reinvent itself, and learn from the experience?

Sustainability in the competitive retail trades depends on finding and eliminating the sources of waste without compromising on stakeholder value.

[1] *Lean on Me*—written by Michael Schiffer, directed by John G. Avildsen, and starring Morgan Freeman—was a 1989 movie that was loosely based on the story of Joe Clark, an inner-city high school principal in Paterson, New Jersey.

[2] "Lean on Me" was a hit song written and performed by Bill Withers in 1972. It was first recorded on the album *Still Bill* and is one of only nine songs to have hit number one with versions recorded by two different artists. The lyrics and music can be accessed at http://www.rhapsody.com/bill-withers/lean-on-me-the-best-of-bill-withers/lean-on-me/lyrics.html.

LEARNING OBJECTIVES AND KEY OUTCOMES

After careful study of the material in Chapter 11, you should be able to:

1. **Describe the lean production philosophy and the house of the Toyota production system, and explain how its two pillars support the primary goal of lean production.**
2. **Define the terms *flow time, throughput time, cycle time,* and *work-in-progress.***

(continued)

3. Apply Little's law.
4. Differentiate between push and pull scheduling systems and explain how the concept of pull scheduling supports the goals of QIM.
5. Describe how one-card and two-card kanban systems work and how these systems support pull scheduling and the objectives of lean production.
6. Define the concept of one-piece flow and describe the key requirements for creating flow.
7. Describe how the Glenday sieve is a helpful tool for creating flow in production systems.
8. Recognize some of the barriers to flow and suggest means for removing them.
9. Discuss the application of lean production to the service sector, drawing comparisons between the service and manufacturing sectors.
10. Discuss the application of lean production to the public sector.
11. Discuss the concept of lean consumption.
12. Construct and interpret DNOM or DTAR control charts for product families that are produced in small batch quantities.
13. Test whether the process variance is constant for all parts within a product family.
14. Construct and interpret stabilized control charts for product families produced in small batches with unequal process variances.
15. Construct and interpret Z control charts for product families produced in small batches with unequal but known process variances.
16. Demonstrate an understanding of the proper use for andons and other visual control devices.
17. Compute takt time and explain the advantages that can be anticipated by using takt time to balance the tasks along a production line.
18. Compute process yields and line efficiencies for a production line that involves sequential operations.
19. Describe the concept of jidoka (autonomation) and how it contributes to continuous improvement.
20. Demonstrate an understanding of the relationship between visual controls and poka-yoke in the prevention of product defects.
21. Apply the principles of heijunka and takt time to design efficient production sequences for mixed model production cells.
22. Differentiate between setup and changeover times and describe steps that can be taken to dramatically decrease changeover times.
23. Describe various methods for inspecting processes to prevent the occurrence of defects.
24. Describe the procedure for organizing and conducting a kaizen blitz project.

11.1 LEAN PRODUCTION

11.1.1 Background

Variations of the lean production philosophy, which is rooted in an approach called the **Toyota production system (TPS)**,[3] have been implemented under numerous names including **just-in-time** (JIT), **world class manufacturing (WCM)**, **stockless production**, **demand flow technology** (DFT), and the generic moniker **lean production** (or manufacturing).[4] The architects of TPS recognized the value of combining attitudes, themes, and specific techniques into an integrated sociotechnical system for manufacturing. There is a significant overlap between the goals of QIM and the prevailing themes underlying the lean production philosophy, which emphasizes the elimination of waste, the reduction of inventories, workforce participation, and the use of systems thinking in planning and operational decisions.

The lean principles are encapsulated in a structure called the house of the Toyota production system (or **house of quality, HOQ**), illustrated in Figure 11-1. The principal goal of lean is to be responsive to customers and this is achieved through three main objectives: high quality, low cost, and short **delivery lead times**.[5] These objectives are supported by two pillars, just-in-time (JIT) and jidoka. Just-in-time means producing only what is required and at the time it is required, and jidoka ensures quality at the source by empowering workers to

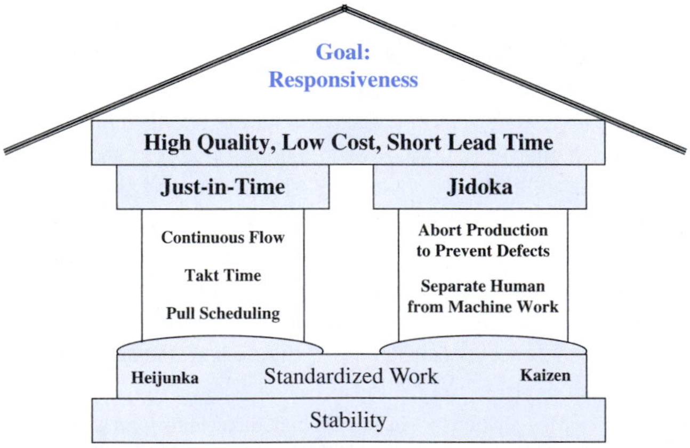

FIGURE 11-1. The House of the Toyota Production System

[3]Ohno, Taiichi. *Toyota Production System: Beyond Large-Scale Production.* Cambridge, MA: Productivity Press, 1995.

[4]Womack, James P., Daniel T. Jones, and Daniel Roos. *The Machine That Changed the World: The Story of Lean Production.* New York: Harper Perennial, Reprint edition, 1991.

[5]The delivery lead time is the total time elapsed between the creation of a customer order and the receipt of the completed order by the customer

intervene whenever they observe or even suspect problems. The operation and improvement of a lean manufacturing environment is an iterative process—driven by problem-solving methodologies such as the PDCA cycle, scientific method, or DMAIC—and strives to standardize and improve work through kaizen events. When DMAIC is the driver, the philosophy is called **Lean Six Sigma (LSS)**, or Lean Sigma which is discussed in more detail in Chapter 14.

> **Kaizen, just-in-time, jidoka, the 5S workplace discipline, and the elimination of the seven deadly wastes are all lean production tools.**

In Chapter 6 we investigated some lean tools that have been successfully applied to the maintenance function (to achieve lean maintenance). Those tools are equally applicable to lean production and include *kaizen, just-in-time, jidoka*, the *5S workplace discipline*, and *elimination of the seven deadly wastes*. It is almost impossible to develop a comprehensive list of all the tools that have been effective in helping organizations become leaner, and new tools are continually under development—all aimed at acquiring a better understanding of processes, the discovery of causal factors, the visualization of system relationships and solutions, and the provision of improved documentation. Most of the tools covered throughout this text satisfy those criteria and can legitimately be considered to be part of the lean toolkit. The discussion here will consequently be restricted to some additional tools that have not been discussed elsewhere or that require further elaboration.

> **The goal of lean production is responsiveness.**

11.1.2 Just-in-Time and Inventory Management

As previously stated the primary goal of lean production is *responsiveness*. Customers want to do business with companies that can deliver high-quality products and services in the time frame required, and provide fast response to unusual demands or changing requirements. The JIT philosophy forces manufacturers to receive and ship materials and products more frequently, in smaller quantities, and to be able to change an order on short notice. These requirements place stresses on the supply chain that do not exist under traditional practices.

Since deliveries are scheduled just before the actual need there can be no margin for error—in the composition, quality, or quantity of an order. Some companies no longer inspect incoming shipments, trusting that their suppliers will get it right. This JIT environment has also necessitated shorter **production batch** sizes. The **economic batch size** formulae of yesteryear have now been replaced by more progressive thinking such as looking for ways to shorten changeover times and designing manufacturing cells to produce a family of products.

Quality and speed are not the same things but are closely related. Producing scrap, or even reworkable defects, have a multiplier effect and can significantly slow down a factory. For example, a 90% yield (10% scrap) can double a factory's cycle time. This is due to reworkable material having to buck the flow and backtrack up the value stream for teardown and repair, additional material having to be scheduled to compensate for scrap losses, operators and inspectors aborting production runs to fix quality problems, and workstation slowdowns when bottleneck operations run out of material.

One of the most obvious sources of waste (muda) is an excessive investment in inventories. Nothing that is produced can benefit a company's financial goals (e.g., profit or revenue) until it is sold and the customer pays the account. At that stage of the transaction the good or service is converted to revenue and contributes to cash flow. There are four categories of inventories: **raw stocks** (also called **raw materials**); **maintenance, repair, and operating (MRO) supplies**; **work-in-process (WIP)**; and **finished goods.** The application of sound supply chain management principles can control the levels of the raw stocks, MRO supplies, and finished goods. WIP, on the other hand, must be carefully managed internally. As the name suggests WIP represents orders that were started but have not yet been completed, and the levels can be significant. It is not uncommon to find WIP stacked around a factory and when tallied up, is found to represent the equivalent of months of production. In addition to the capital that must be tied up in such inventories excessive WIP can contribute to manufacturing inefficiencies. In-process materials must be stacked somewhere and this often means that aisles are blocked, line-of-sight is obstructed, and handlers have a difficult time moving materials in and out of workstations in a timely manner. Workers can suffer claustrophobia and, since workers up and down the value stream are unable to see neighboring workstations, production problems may go unnoticed.

11.1.2.1 Definitions. The relationship that exists between WIP, cycle time, and order processing time helps explain how large volumes of WIP can slow down a factory. First we need some definitions.

> **Muda is defined as anything that does not add value or detracts from it.**

- **Flow time (FT)**, alternatively called throughput time, production cycle time, or **sojourn time**, is the *total time an order is in the value stream*. The time starts when the order reaches the first value-adding workstation, and stops when the order rolls off the last station ready for shipment. Flow time is expressed in time units (e.g., days, weeks).
- **Throughput (TP)** is the *average rate of output* from a production process. This can apply to the entire plant or a single workstation, or even a single machine. Throughput is expressed in the number of units per unit time (e.g., number of units per hour).
- **Cycle time (CT)** is the reciprocal of throughput, and can be interpreted as the time between successive output units produced. Cycle time is expressed in time units per unit output (e.g., number of hours per unit).
- Work-in-process (WIP) is expressed in units and represents the total number of items that have been through at least one value-adding workstation and have not yet completed

the entire production routing. When WIP has completed all the steps in the value chain, its status changes from WIP to finished goods.

11.1.2.2 Little's Law. Little's law defines the relationship that exists among the four variables defined previously.

Little's Law

$$FT = WIP \times CT \quad \text{11-1}$$

$$WIP = FT \times TP \quad \text{11-2}$$

Little's law can be illustrated by the following scenario. Let us assume that a plant manager is interested in estimating how much WIP is scattered around his factory. An order has just begun at the first processing point. The component is marked in a location that will be visible in the finished part. The operators at the final assembly point are asked to be on the lookout for a part containing this particular marked component and, once they have completed it, to bring it immediately to the plant manager's office. On the 21st working day an operator delivered the completed assembly to the plant manager. The flow time was then known to be equal to 21.

Meanwhile, the plant manager had asked one of the plant engineers to stand at the end of the line (at the last station) and study the output rate. After 100 observations or so the engineer reported that it appeared that the line was delivering a unit of output on the average of once every 30 seconds or 2 units per minute. The line in question was scheduled for one 8-hour shift. Using Little's law (Equation 11-2) the plant manager was able to quickly make the following calculation.

$$WIP = FT \times TP$$

$$WIP = (21\ days) \times \left(\frac{2\ units}{minute}\right) \times \left(\frac{60\ minutes}{hour}\right) \times \left(\frac{8\ hours}{day}\right)$$

$$WIP = 20{,}160$$

11.1.3 Pull Scheduling

In a traditional factory, individual work schedules originate with a **master production schedule** (MPS). Advanced materials and capacity plans are generated from the MPS and these plans in turn are used to trigger orders on the shop floor where production is commenced. The flow of work follows a particular production routing (sequence) that defines how the product is to be built. Work begins at the first activity shown on the routing sheet and is then "pushed" from workstation to workstation until all activities are complete. Under **push scheduling** the common procedure is to release jobs to individual departments using time increments called **buckets**. The typical duration for a bucket is one week. As an example of the time impact of push scheduling, consider a particular routing that requires the work of nine departments. The flow time under a push system would be a minimum of nine weeks—*regardless of how much value-adding make time is required.* The consequence of using push as the basis for scheduling is that inventory can (and usually does) stockpile between departments.

In a push system jobs are forward-scheduled using time increments called buckets.

In **pull scheduling** the discipline is reversed. The triggering mechanism for work order releases is a signal from one workstation to the previous workstation that more material is needed. Under a pull system no work is produced until it is needed by the next workstation in the value chain. Pull is an internal requisitioning procedure that starts at the back end of the value stream and triggers orders, step by step, toward the front end. The term "pull scheduling" is actually a misnomer as the pull methodology is not a scheduling system and does not eliminate the need for advanced materials and aggregate planning—functions that are performed well by software programs that contain **manufacturing resources planning** (MRPII) modules.[6] Pull scheduling can be more appropriately characterized as a system for shop-floor controls.

In a pull scheduling system work is triggered from the backend and production is generated only as needed.

There are numerous benefits to pull scheduling. The most dramatic is a decrease in inventories. Inventory reduction does not eliminate the need for WIP, but with pull scheduling WIP levels can be tightly managed, and used only as required to buffer against process variability or to keep bottleneck operations fully utilized.

Under the right conditions pull scheduling can be an effective system for controlling material flows on the factory floor. A card, called a **kanban**, is the usual medium for establishing shop-floor communications. *Kanban* is a term that is the combination of two Japanese words: *kan* meaning visual, and *ban* meaning card or board. So a kanban is a visual signaling device that usually (though not necessarily) involves a physical medium such as a card or board.

11.1.4 Kanban System

In a kanban system, the flow of materials is controlled using a method by which workstations can exchange signals when one is about to run out of material. By *pulling* rather than *pushing* material through the plant, work is produced only as needed. There is no longer the time–bucket constraint and an entire production batch does not need to be completed before material can be transferred to a subsequent workstation. Kanban containers of a size that can be manually transported between workstations permit small quantities of materials to be transferred to the next operation as soon as a container is filled.

Two types of kanban card systems are common—the single-card and the two-card systems. The single-card system, illustrated in Figure 11-2, works well when two sequential workstations, or manufacturing cells, are in close proximity. The sequence works as follows.

1. When the assembly cell empties a container of parts, an operator returns the empty container to the storage area and withdraws a full container.

[6] For example, SAP.

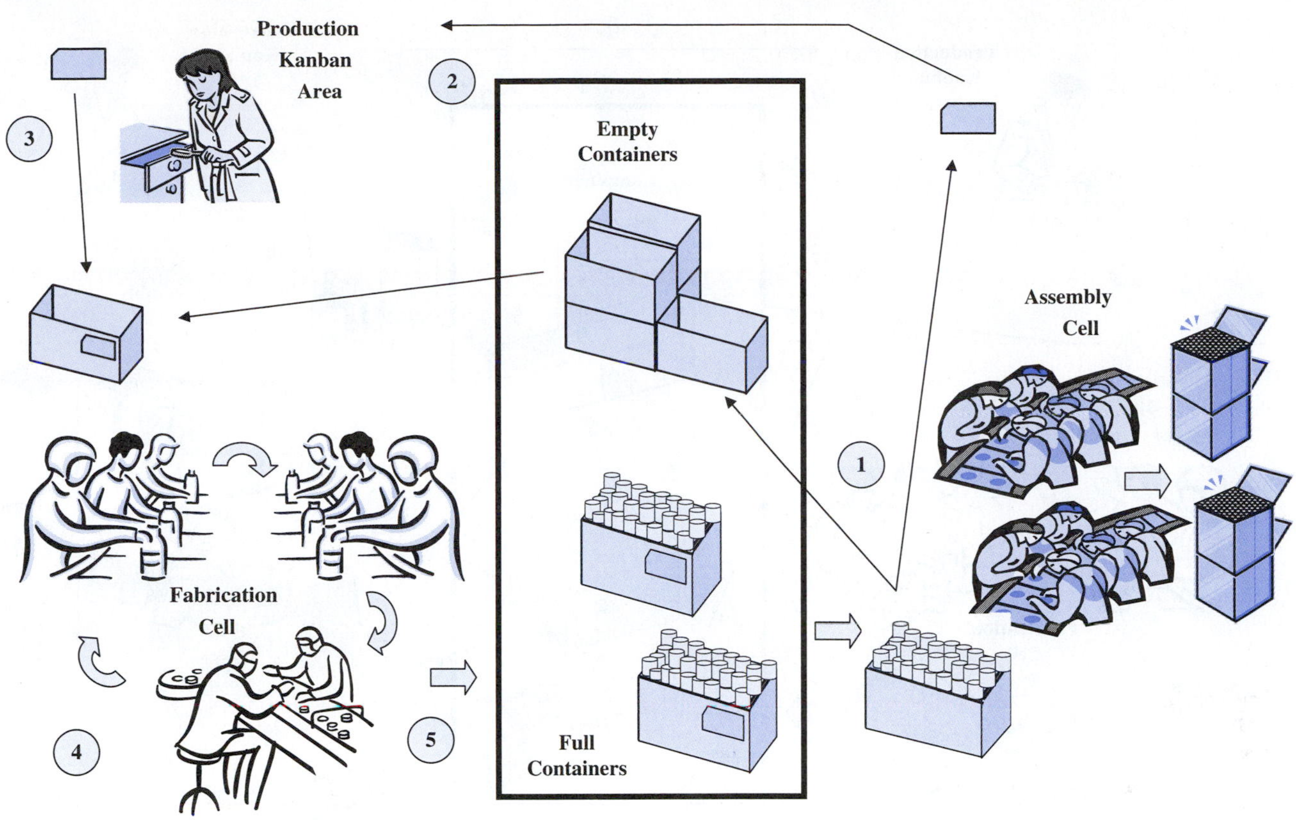

FIGURE 11-2. Single-Card Kanban System

2. The kanban card is removed from the container and returned to the production kanban post as a signal to the upstream fabrication cell to produce more parts.
3. An operator from the fabrication cell will take the card from the production kanban area and withdraw an empty container from the storage area. The card is then attached to the carton.
4. The fabrication cell completes the order.
5. When the container is full it is moved, with the card still attached, to the staging area waiting to be withdrawn by the assembly cell.

A kanban system controls the flow of materials on the shop floor by triggering material releases only when materials are actually needed.

A two-card system is illustrated in Figure 11-3 and is necessary when the two sequential operations are not close. This system uses a *conveyance* card and a *production* card. The two-card sequence works as follows.

1. When the assembly cell starts to work on a container the conveyance card is removed and taken to the conveyance card post, where it is retrieved by a material handler.
2. The material handler then goes to the storage area where a full kanban container is withdrawn and removes the production card from the container. The handler then replaces it with a conveyance card.
3. The full container, with conveyance card attached, is then transported to the material staging area in advance of the assembly cell.
4. The production card is taken to the production post.
5. The production card is withdrawn by a fabrication cell operator as a signal to produce more parts. The operator withdraws an empty container from the storage area and places the production card on the container.
6. The fabrication cell then proceeds to fill the order.
7. When complete, the full container with production card attached is transported to the storage area.

A kanban signal does not have to be a card nor does the signaling medium even have to be written. Any kind of visual signal can suffice. Rolling colored golf balls down a chute, hoisting a flag, or using a semaphore style code can all be effective means for communicating. One simple form of kanban involves painting a square on the floor or taping off a designated area on a workbench between two operators. When material occupies the square the upstream operator does not produce. When the material occupying the area has been depleted, the upstream operator knows that it is okay to start producing again.

Any suitable visual signaling device can be used as a kanban trigger.

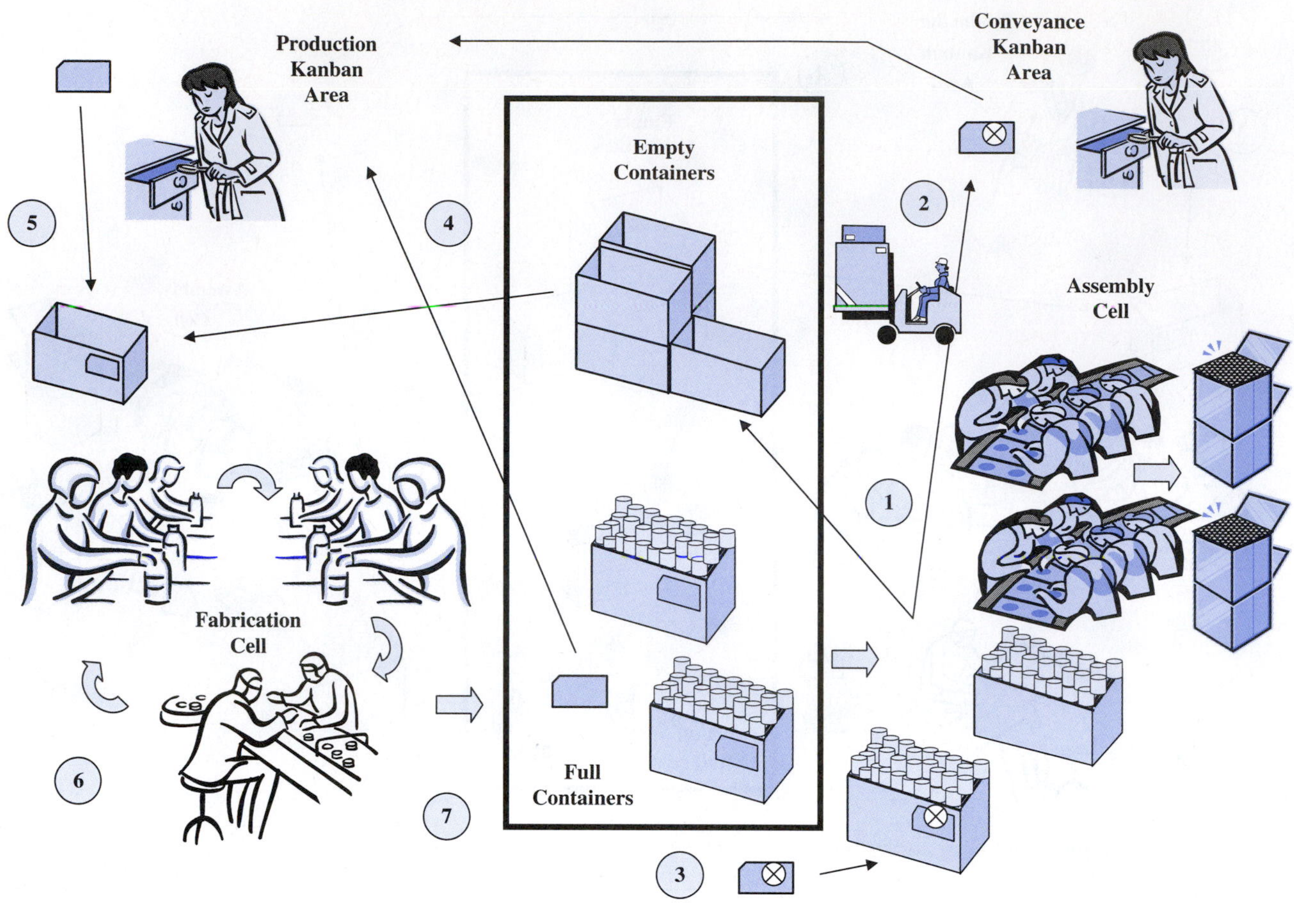

FIGURE 11-3. Two-Card Kanban System

A variation of kanban assigns jobs a priority using a system of distinguishing colors. For example, if red is used to indicate top priority, operators would be instructed to process red cards ahead of other colors.

11.1.5 Creating Flow

By eliminating waste, lean tools are aimed at improving process efficiencies and thereby improving effectiveness. Even though this is essential to the lean philosophy, lean is not only about waste reduction. The foundation principle of lean is to *create flow*. The term **flow** can be defined as the smooth, progressive movement of materials, products, information, and people through a sequence of activities, without interruption and without the creation of waste. Flow and lack of flow in a production process is like the difference between water gushing freely through a hose versus an interrupted flow caused by kinks; or the comparison of traffic speeding along a freeway versus traffic delays caused by accidents or traffic jams. The **flow rate**, also known as the **throughput rate**, is a measure of the average rate at which units of product flow past a specified point in a process. The maximum flow rate is the **process capacity**.

The foundation principle of lean is to create flow.

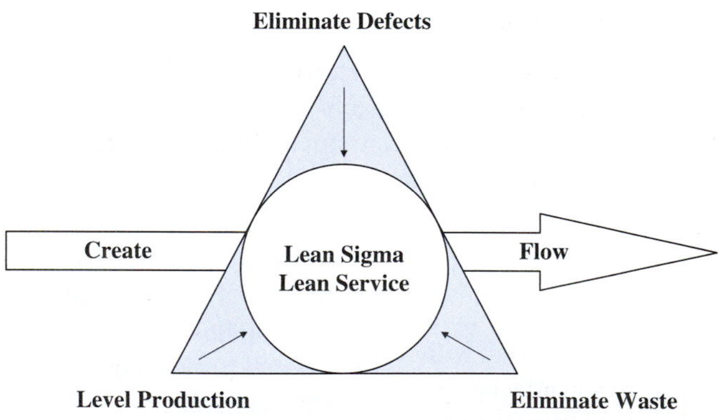

FIGURE 11-4. The Three Requirements for Production Flow

As Figure 11-4 illustrates there are three requirements to flow creation: the elimination of waste is one element; however, the elimination of defects and the leveling of production are also essential. When flow is created the result is "lean . . . (*blank*)." Filling in this blank is application specific, and examples are lean production, lean service, lean sigma, lean health care, lean government, lean maintenance, lean banking, lean construction, and so on. We even hear of lean energy and lean consumption. The list of potential applications is boundless.

The ideal that lean process engineers try to achieve within work cells is called **one-piece flow**, meaning that one work piece at a time is transferred smoothly and continuously between the successive workstations within a cell. Where this is not feasible or economical the process flow will be designed around a small transfer batch instead of a single work piece. Achieving this concept in practice generally requires leveled production so that **economies of repetition**[7] can be realized. Under leveled production, like products follow an identical sequence in the same volume on the same equipment at the same time every production cycle. This implied rigidity of production schedule would seem to defy the desire for flexibility and responsiveness. However, through a sound engineering approach achieving some degree of leveled production is not as difficult as it may seem.

In Chapter 7 the Pareto principle was used as a tool for partitioning a large set of items into two subsets—that is, separating the vital few from the trivial many. The application of this principle to inventory management is well known, where individual stock-keeping units (SKUs) are categorized as A, B, or C depending on whether they are fast, medium, or slow movers. Once classified, tight management controls are placed on the A items (which generally represent a relatively small percentage of the total) and fairly loose controls are placed on the C items (the slow movers).

As it happens, the Pareto principle also applies in many production scenarios where a large product mix is concerned. In these cases it can be observed that a relatively small percentage of the total product range accounts for a substantial percentage of total production volume. The Glenday sieve[8] (named for its founder) is a method similar to the ABC inventory system and categorizes SKUs according to sales volume. The Glenday sieve uses four classifications and provides a system of color coding as shown in Table 11-1.

With a Glenday sieve analysis the initial problem of trying to create flow for 100% of the product mix is dramatically reduced to focusing on a mere 3% to 8% of the total product line. Identifying these **green product lines** enables a detailed analysis aimed at discovering and removing sources of waste. Designing high-volume efficient dedicated green product lines can result in one-piece continuous flows with higher quality and shorter cycle times for these products.

Green product lines can be designed to achieve one-piece flow resulting in higher quality and shorter production cycle times.

The yellow coded SKUs represent opportunities for improvement. Whereas the product mix ratio (calculated by dividing the % volume by the % SKUs) is typically 8 to 10 for green codes, and drops to around unity for the yellow codes. This means that the yellow category mix is much greater relative to the demand. To introduce these products into the fixed cycle engineers concentrate on ways to reduce changeover times and to implement *heijunka* production smoothing methods, explained later in this chapter. Once these products are successfully introduced into the production cells the percentage of the total product range that can be run on a fixed cycle can be as high as 95%.

TABLE 11-1 Glenday Sieve

Cumulative % of Volume	Cumulative % of SKUs	Color Code
45%–55%	3%–8%	Green
80%–95%	40%–60%	Yellow
93%–99%	70%–75%	Blue
100% (Last 1%–7%)	100% (Last 25%–30%)	Red

Source: Adapted from Glenday, Ian. "Moving to flow." *Manufacturing Engineer,* 84, no. 2 (2005): 21, Fig. 1.

Creating flow for blue and red SKUs is more difficult—however, combined these products typically represent 25% to 30% of the total SKUs, but only 5% to 7% of the total volume. Close scrutiny of the items contained in these two categories can often lead to one of two conclusions.

1. With some slight modifications to materials, packaging, or process an SKU can sometimes be introduced into the production cycle creating flow for that particular item. For example, an item may be classified as "blue" with respect to its volume but by placing it in a bottle that is the same size and shape as those for green or yellow items, the specific SKU can be processed on a mixed line with other items having much larger volumes. To differentiate these products different colors or labeling can be used.
2. Some SKUs, particularly those in the "red" category, may not be worthwhile continuing as part of the product mix. If the production costs outweigh the perceived benefits, the prudent decision could be to drop the line. Management should face the reality that supporting those red and blue items that cannot flow requires a production philosophy that is different from those that do flow. It may not be possible to satisfy the interest of stakeholders if in fact there are two businesses sharing the same resources and trying to operate under the same set of parameters—one dedicated to lean practices where material flows smoothly, and the other more characteristic of a customized small-batch job-shop operation where flow patterns are inconsistent and jumbled.

Consideration should be given to discontinuing those red or blue items that cannot flow.

Creating flow in an environment that has traditionally focused on large-scale batch production can be daunting, as lean requires a new way of thinking. A lean implementation can face an uphill battle in the face of resistance by the naysayers who can mount a plethora of excuses as to why the lean concept simply is incompatible with "*our business.*" On its corporate blog a certain consulting firm has listed 10 of

[7]Glenday, Ian. "Moving to Flow." *Manufacturing Engineer* 84, no. 2 (2005): 20–23.
[8]Ibid.

the most common reasons given by its clients for not being able to implement one-piece flow.[9] Included here is a suggested *lean response* to such "excuses."

Excuse #1: Materials are unavailable in the quantities, quality, or time required.

Response: Having the right materials in the right quantities at the right time, and of suitable quality, is a requirement before an organization can make any headway with a lean implementation. Therefore this problem needs to be fixed first. However, there are times when important problems do not get solved simply because they do not receive the attention or resources needed. If this is the case, a one-piece-flow approach can be used to expose these problems in the same manner that lowering the water in a lake might expose navigation hazards lurking beneath the surface. With one-piece flow, when the line stops for lack of materials the problem becomes visible to all and is likely to become a management priority for resolution.

Lean production requires good materials control.

Excuse #2: Equipment is unreliable and breakdowns could cause line stoppages when downstream processes run out of parts or components.

Response: Sustained capacity through lean maintenance practices is also a prerequisite for a lean implementation. Equipment reliability must be high and predictable. If the experience with machine breakdowns is high, and particularly if the number and/or frequency of breakdowns are unstable, this problem must be solved. As with excuse #1, if machine reliability is a problem due to lack of interest by management to solve it, a move to one-piece flow will expose the problem when breakdowns deprive downstream operations of needed parts.

Lean production requires reliable equipment.

Excuse #3: The culture is not ready to accept such wide-sweeping change.

Response: Cultural change comes slowly and is driven by two principal factors—a committed and impassioned leadership and the process of continual learning. Education, political will, empathy, and open communications are the keys to removing resistance. It is important that the message be consistent and clear. Confusion and ambiguity feed fear, and fear heightens resistance.

Lean production requires education, political will, empathy, and open communications.

Excuse #4: The workforce is too specialized, and one-piece flow would require mass cross-training of personnel.

Response: As demonstrated in Chapter 3 there is motivation potential in taking the repetitiveness out of work by enlarging or enriching job content. Increasing the motivating potential score (MPS) of a job and matching the job to employees with high growth needs strengths (GNS) can enhance productivity by providing new meaning to the work and increased job satisfaction. All employees may not respond positively at first, but a learning organization is characterized by leadership that is concerned with the development of its personnel. This includes the proactive encouragement of everyone to continually learn new things and a reward system that appropriately recognizes those that show a willingness to do so.

Lean production requires a commitment to human resource development.

Excuse #5: Changeover times are too long to effectively achieve one-piece flow.

Response: If changeover times are great batch sizes may need to be large, making it uneconomical to achieve one-piece flow. Sometimes jobs can be sequenced in a way that the number and frequency of major (long-duration) changeovers can be minimized and flow can be achieved in between. Such major changeovers should be targeted for reduction using the methodologies described in Section 11.6

One-piece flow on a mixed model line requires small changeover times.

Excuse #6: There is too much distance between processes to move one at a time.

Response: This excuse demonstrates a fundamental lack of understanding of how material under one-piece flow gets physically moved from one workstation to the next. Where the workstations are noncontiguous there is usually a small transfer batch that an operator can move without the need for specialized equipment. When implementing the one-piece-flow concept it may be desirable to make some layout changes to reduce the distances between sequential workstations.

One-piece flow depends on the transfer of small batches of material between workstations.

Excuse #7: The process produces defects that will stop the line too frequently, so we need inventory buffers between workstations.

Response: Refer to the response to Excuse #1.

Excuse #8: Individual process cycle times are unstable or variable, creating imbalance among workers.

[9] http://www.gembapantarei.com/2007/04/ten_reasons_why_one_piece_flow.html.

Response: Attempts should be made to balance production lines using techniques outlined in Chapter 5. In analyzing each process step, tasks should be broken down into the smallest work elements, wasteful effort eliminated, and the remaining elements recombined and packaged according to a target cycle time. Variation will always corrupt the ability to balance a line of interdependent activities. The variation along the line will affect individual workstation capacities and can result in inventory accumulating ahead of some workstations and other workstations having to stop work for lack of material. Pull scheduling can overcome some of these problems by avoiding overproduction. Continuous improvement methodologies such as kaizen and Six Sigma can help reduce process variation.

The presence of variation will always create some line imbalances.

Excuse #9: Machines were not designed for one-piece flow.

Response: There may be some truth to this statement since many machines and processes have been designed with the intent of running minimum batch quantities. As examples, a bakery with an oven capacity of 160 loaves would find scheduling batches of bread smaller than that a waste of energy; and a shirt factory has a numerical controlled pattern cutter that is designed to cut out 350 shirts at a time. These factories would find it difficult to bake a single loaf or cut out the pattern for only one shirt. However, it might be possible to mix model schedule and simultaneously bake loaves of different products (e.g., white, wheat, rye, etc.) in a single bake, or to cut the patterns for numerous fabrics, prints, and colors from a common pattern. If processes and/or machines cannot be designed to accommodate a true one-piece flow, then the designed batch quantity becomes the transfer quantity creating a small "planned" WIP between the constraining process and the next process along the value stream.

Excuse #10: Recurring-but-not-every-cycle work is an interrupting influence on maintaining flow.

Response: In a lean production system, creating jobs that will explicitly take care of those tasks that do not occur during every cycle—such as replenishing materials, removing filled containers, and bringing in or setting up empty cartons—can help maintain a smooth, uninterrupted work flow. Such a position is what the Japanese call a **mizusumashi—water spider** in English. The word *mizusumashi* literally means to "purify the water" and gets its name from a small water beetle that moves around in a clear water environment. It is not known whether the broomlike fibers on the hind legs of this small insect are used to actually create the clean habitat or simply to maintain it in its pristine state. Either way, entomologists can confirm that the water spider resides only in a crystal-clear environment. By analogy, the human water spider maintains the state of a lean environment by pitching in whenever and wherever necessary to keep the flow lines clean and smooth. This may entail anything from materials support to jumping into a cell and helping with the production effort as required. Therefore, the human water spider is much more than a materials handler and must have an intimate and thorough knowledge of the materials, tools, and production methods supported, and is typically also cross-trained in many of the relevant production skills.

In lean production a water spider pitches in whenever and wherever necessary to maintain flow, including assisting with the production tasks if required.

11.1.6 Lean Service

Lean thinking is about getting more with less, and a fundamental question where services are concerned is whether the nature of pure services lends itself to this goal. After all, is not the idea of good service to wow the customer and provide more in order to get more (value)? From a lean perspective, one might query whether it is possible to create flow at all in a services environment.

Lean is getting more with less.

This discussion can begin with the recognition that not all service environments are the same, and mixed results have been achieved in trying to bring lean thinking to a number of service industries. For example, **lean service** has been successfully applied in some instances to insurance,[10] health care,[11] government,[12] and administrative offices.[13] Progress in introducing the lean philosophy to service areas has been spotty at best, and has been slow to catch on in some areas such as government and the hospitality industry.

Perhaps one problem is that some service organizations, notably those in the public sector, are bureaucracies with mechanistic structures. In a bureaucracy the work is typically organized by functional department and passes from one department to the next with minimal inter-departmental interaction. The employees in these organizations are not conditioned to think in terms of value streams, cycle times, response times, lead times, work in process, customer satisfaction, or similar terms that are so familiar to their counterparts who have experienced lean transformations in the private sector. Most government employees have never even heard of the term *takt* time.

Even though there may be no profit motive (e.g. in government agencies) these organizations do have budgets and the resource providers (e.g. taxpayers) have every right to demand

---

[10]Swank, Cynthia Karen. "The Lean Service Machine." *Harvard Business Review* 81, no. 10 (October 2003).

[11]Miller, Diane, ed. "Going Lean in Healthcare." *Innovation Series 2005.* Cambridge, MA: Institute for Healthcare Improvement, 2005.

[12]Gupta, V. "Why Every Government Agency Should Embrace the Lean Process." *American Public Works Association Reporter,* April 2004.

[13]Banks, Sarah, and Karen Juroff. "Office Get Lean." *Universal Advisor,* no. 3 (2004).

better management and accountability concerning the use of their (e.g. public) funds. Sadly it seems to be widely accepted that government at all levels is wasteful and that the private sector can usually provide the same services better, faster, and more efficiently. The typical government response to growing demand by its citizens for increased services has been to expand government. The U.S. Bureau of Labor Statistics has forecasted that during the next decade federal, state, and municipal government employment combined will increase by more than 2.5 million jobs.[14] On top of this was the 2009 American Recovery and Reinvestment Act[15], a $789 billion plan designed to create an additional 3.5 million jobs, a number of which will be in the public sector. Most of the growth is expected at the state and municipal levels. An important question is: *Does simply adding people lead to better government?* We have learned in manufacturing systems that adding capacity only magnifies the costs of operating inefficient systems. Lean thinking is a critical-eye approach where everything that is done can be scrutinized and is fair game for change or elimination. All activities should be justified on the basis of adding value, and work sequences should be balanced in a manner that creates flow. The one-piece-flow concept is often difficult to implement in administrative environments where activities are functionally compartmentalized; managers or administrators experience frequent and unpredictable interruptions such as meetings, visits, and phone calls; and these same managers and administrators often find it convenient to batch-work together and process it at once (such as approval actions).

Another factor that acts as a barrier to flow is the difficulty of maintaining a first-in-first-out (FIFO) discipline for work. Administrators sometimes have the tendency to work from the top of an in-basket thus invoking a last-in-first-out (LIFO) discipline, or hand selecting those jobs first that can be dealt with more quickly or easily. This has the negative impact of increasing the variability on cycle time, and thereby decreasing the overall quality of service.

Lean service begins with the recognition that the service provider system is similar to systems that produce tangible goods.

Applying lean in any service environment begins with the recognition that the service provider system is comprised of components that are similar to those of goods manufacturers and that achieving purpose is accomplished through a primary and secondary value stream network as shown in Figure 11-5. Generically, the primary value stream in any lean service organization consists of five major components: inputs, design, make/serve, support, and consume. Numerous other supporting value streams exist to provide those resources necessary for the efficient and effective completion of the primary value stream tasks—in areas such as engineering and maintenance, finance, human resources, logistics, research and development, marketing, materials control, and the various dimensions of planning (e.g., capacity, strategic, improvement).

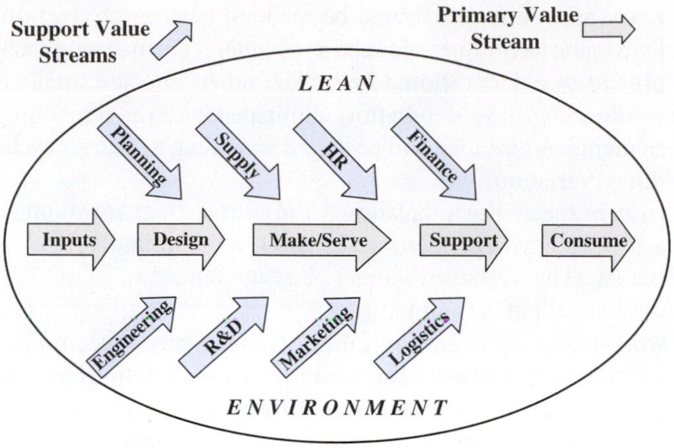

FIGURE 11-5. Lean Service Value Stream

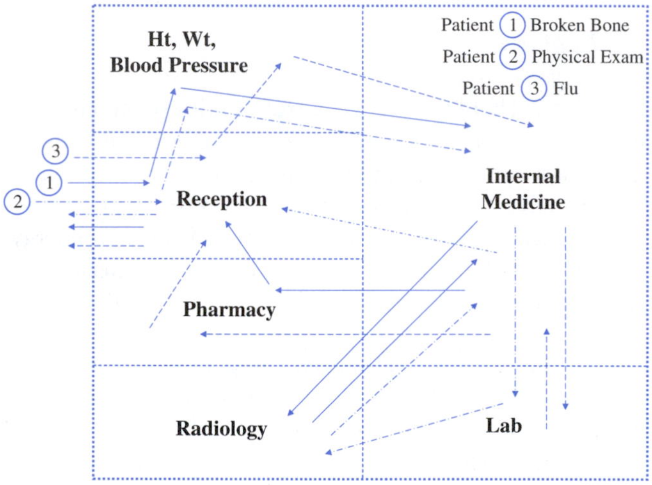

FIGURE 11-6. Healthcare Clinic

The lean philosophy carefully evaluates the work that people do, either in a primary or support role, with respect to the quality objectives of the service or product provided. Where wasteful non-value-adding activities and human energy are identified lean thinking strives to identify the root causes and either eliminate them or nullify their effects. As with manufacturing, the application of the lean philosophy in service industries is not a zero-sum game. Efficiencies breed more efficiencies. Smooth flows without waste lead to better service and happier customers. Customer satisfaction, together with efficient and smooth work processes, in turn lead to a more harmonious, happier, and more productive workplace environment.

In considering how one might create flow in a service environment consider these two scenarios: a health care clinic and a municipal government. A typical health care clinic is organized as shown in Figure 11-6. Patients enter the system at reception where they checked in, and their medical records

[14]U.S. Department of Labor, Bureau of Labor Statistics. "BLS 2002-12 Labor Projections." http://stats.bls.gov/news.release/ecopro.t01.htm.

[15]Bill H.R.1 for the 111th Congress signed into law by President Obama on February 17, 2009. http://frwebgate.access.gpo.gov/cgi-bin/getdoc.cgi?dbname=111_cong_bills&docid=f:h1enr.pdf.

are pulled and updated if necessary. When called, patients are taken to an area where their height, weight, and blood pressure are checked and they are briefly interviewed by a nurse practitioner. Next, the patients are examined by a doctor and, depending on the particular malady, they may be sent to the lab department, radiology department, or both. Then the patients return to the examining area for further consultation with the medical staff. At this point medication may be prescribed and the patients sent to the pharmacy. Patients then depart the system through reception where they pay their bill, arrange for insurance to be filed, make follow-up appointments, and so on. Between each of these steps the patients may be required to wait until those ahead of them in the queue are served. Figure 11-6 illustrates the jumbled flow patterns and variance in paths that must be followed by individual patients who enter the system with different ailments.

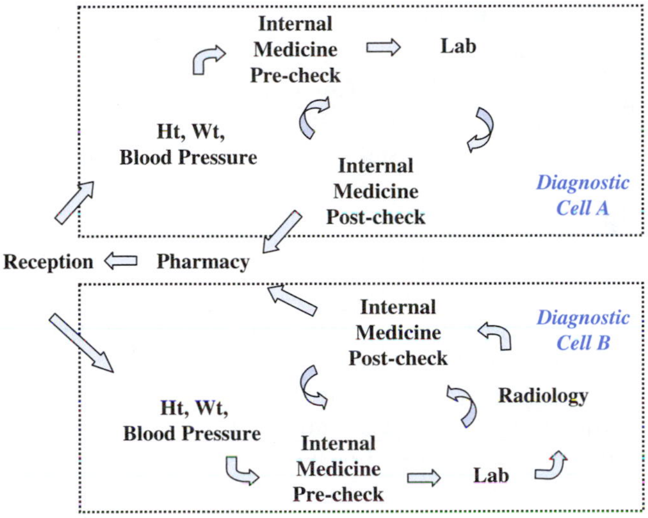

FIGURE 11-7. Creating Flow in HealthCare

Figure 11-7 shows the same health clinic organized for flow. The functions have been arranged in diagnostic cells and patients are routed through a cell depending on their particular requirements. For example, diagnostic cell B contains radiology capabilities, so all patients requiring X-rays must be routed through this cell. Having all functions self-contained within each cell helps reduce patient waiting time and create flow.

Now consider Figure 11-8 which illustrates a typical process for obtaining a building permit from a local municipal government. A total of nine steps are required in this process and each is performed by a different local government department. As an application moves from department to department it joins a queue with other applications waiting to be processed. As shown, this system offers a total of eight opportunities for the application to be delayed as work piles up waiting for some government official to act or for all the necessary paperwork to be in place in order to proceed with the next step.

One-piece flow can be created by grouping several of the functions together in cells. That way an application advances faster through the system with less chance for delays. Figure 11-9 depicts a cellular design for this same process. In this design an application enters the system in the Preliminary Review cell where it is checked for completeness, is assigned a street address, and compliance with all relevant fire and health codes is verified. The application then proceeds to engineering for the scheduling of a site visit where it encounters the first opportunity for delay. Once the site visit has been completed, the application is then forwarded to the Final

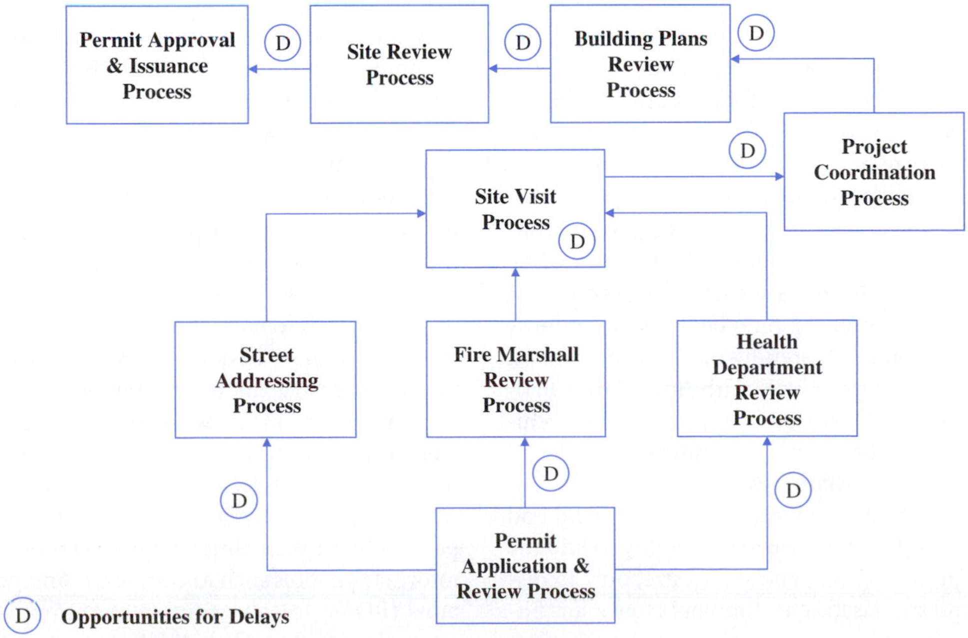

FIGURE 11-8. Typical Municipal Permit Issuing Process

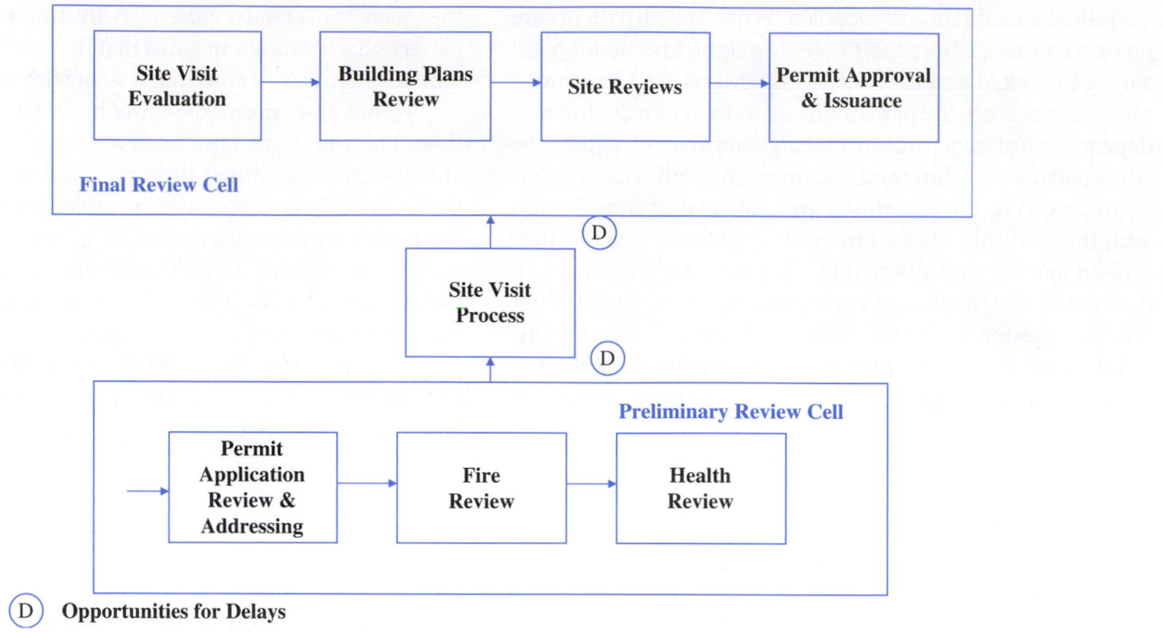

FIGURE 11-9. Lean Permit Issuing Process

Review Cell where it encounters the second opportunity for a delay, which will occur if there is a queue of work waiting to be processed. Within the Final Review Cell all the specialties are represented that are necessary to complete the process. This includes the site report review, determination of the number of site reviews required, building plans review, conducting of the necessary site reviews, and final approval (or disapproval) and issuance of the permit.

11.1.7 Lean Consumption

Shopping is a universal pastime and is something that is shared by and unites cultures the world over. Some claim to derive recreational value from the shopping experience but many find the process daunting, frustrating, and time-consuming. Did you ever stop and think about how much time and money is spent—by manufacturers and retailers—persuading consumers to buy things they do not need, or do not need now, and in quantities that exceed their needs? Advertising and marketing efforts seem to focus more on propaganda and offering economic incentives than on providing useful information. As consumers we have all been conditioned to believe that shopping should be based on impulse rather than on a need that is carefully planned in advance. Retailers that load their warehouses with unsold merchandise or car dealerships that pack their lots with unsold vehicles find that they must offer financial inducements to rid themselves of these excess inventories.

How much time must the average consumer spend and how many shoe stores does she have to visit trying to find the *Lily Ashner* dress pump in pink, and in her size, only to discover that in the end she has to settle for blue? How many billions of hours are wasted by consumers each year in shopper research—thumbing through catalogs, surfing the Internet, making telephone calls, or going from store to store searching for the item they need? Ironically, the stores that fail to have the items desired often have inventories brimming with items their customers do not need.

With major purchases such as automobiles or furniture, consumers are provided incentives for taking the inventory that is physically situated in the dealer's showroom. Never mind if the color is not quite what you wanted, or the style is contemporary and your home is colonial, or you have no need for a 10-track CD player or sophisticated navigation system that has been preinstalled as an option in your automobile. Consumer economics sway heavily in favor of taking what you can get rather than waiting for what you want or need. If you *are* willing to wait and decide to place an order with the factory for a product that precisely meets your needs, you are likely to pay a substantial financial penalty and it may take weeks to get what you ordered.

Lean consumption is concerned with eliminating wasted time, hassle, and lack of responsiveness experienced by consumers.

And, if this were not bad enough, it is becoming more and more common for customers, under the guise of convenience, to actively participate in the transaction—performing a certain amount of "unpaid work" on behalf of retailers. For example, increasingly customers are entering data into Web-based order forms and tracking the progress of their orders online. And in this age of technocrats consumers are spending a greater proportion of their time and energy with technological and information tools such as computers and personal data assistants (PDAs) to solve routine consumption problems—for themselves and on behalf of their service providers. Advanced industrial economies have focused on self-service, and the applications where consumers can actively participate in the service experience are expanding. Seventy-five years ago it would have been unheard of for a customer to handpick food items from a grocer's shelves. Fifty years ago it would have been rare to pump one's own gasoline. Twenty years ago self-service checkout lanes were nonexistent and automated teller machines (ATMs) were just making the scene. Ten years ago it was not common for one to bank and pay bills online without the intervention of a human.

Technological change has dramatically impacted societal values and has transformed lifestyles in ways that create additional pressures on time and energy. In two-wage and single-parent households there is little time to even perform routine chores such as laundry, cooking, and shopping—let alone being able to properly manage consumption. In addition, aging members of the population who may be approaching their retirement years, are confronted with an expanding array of choices and find that they lack the energy and technical expertise to address them. All the time, energy, and money invested on the one hand by providers trying to off-load inventories, and by consumers on the other hand searching for what they need and purchasing what they do not need or in advance of when they need it, are part of consumption cycle waste. Eliminating the sources of this waste is the subject of an emerging area of interest called **lean consumption**,[16] and begs the legitimate question of whether the principles of lean thinking that have worked so well in the manufacturing sector can now be applied to eliminate waste that accrues during the consumption process.

When consumers make purchases that fall short of their requirements or spend excessive time trying to locate what they need, they are helping produce consumption waste.

Just as lean production requires a paradigm shift in manufacturing management, lean consumption requires some dramatic changes in attitudes and behaviors. We begin by challenging some of the premises that have historically been taken for granted. For example, if one believes that shopping is largely driven by impulse, then a recognized need constitutes an instantaneous emotional event, and the timing for meeting that need is *now*. The assumptions underlying this premise are that consumers are incapable of using advanced planning techniques to identify future requirements, that they are unwilling or unable to wait for those requirements to be filled, and that providers are incapable of organizing their supply chains so that production schedules and stock replenishments can effectively be pulled from consumer demand.

As we turn our attention to lean and the elimination of muda these assumptions are worthy of challenge. Do consumers in truth generate needs on impulse and require instant gratification? Some items, particularly large-ticket purchases, are not usually bought this way. There will always be crisis situations such as a tree falling on a car, a flood, a theft, or a total accident loss that can substantially shorten the lead time that the consumer is willing to tolerate. However, in general consumers are in a position where they can potentially plan ahead for the next car, or replacing the dining room furniture, or installing a pool. For such large items, consumers can be *reprogrammed* to place orders months (even years) in advance of when the item is actually needed. The reprogramming process can be facilitated through financial incentives (rewards) offered to the consumer for committing to a purchase in advance. This process has worked in the travel industry for decades. Airlines and cruise lines offer cheaper fares for customers who book in advance and help reduce the risk of flying with empty seats or sailing without a full passenger complement.

For those on the provider side, how great would it be to know with, say 90% confidence, what has actually been sold three months in advance? In this utopian-like scenario, the supply chain would be able to respond by making and distributing exactly to demand, thereby eliminating some of the major sources of waste that exist in the current system. If such a system could be implemented for high-value items, then it is not such a stretch to extend this thinking to other smaller and more routine purchases. What about groceries? I spent two years as a supply officer on a ship in the U.S. Navy. On a naval ship a cycle menu is used to plan all meals for the officers and crew a month in advance, and the provisions are ordered to support the menu. (There are no grocery stores on the high seas.) Arrangements have to be made with the suppliers of those items that are perishable with short shelf lives like bread, milk, and butter for frequent just-in-time deliveries in small quantities. On a ship such items are usually purchased from local vendors in ports of call.

Lean concepts such as just-in-time and advanced materials planning are not solely applicable to manufacturing. Some consumers are already making use of the Internet to place advance and bulk orders for groceries, accessing companies such as Netgrocer.com,[17] GroceriesExpress.com,[18] Peapod,[19] and even Amazon.com.[20] It is not at all inconceivable that, with the proliferation of laptop computers and PDAs, consumers could be offered incentives and education that would result in better consumption management practices. One can only envisage a future where most of the provisioning in our lives will be planned and ordered in advance, from staples like food, clothing, and energy to discretionary spending in areas such as electronics, vacations, and entertainment.

There are six guiding principles of lean consumption:[21]

Guiding Principles of Lean Consumption

1. Satisfy customers' needs completely by ensuring that goods and services delivered fulfill the function intended, and that goods and services work together compatibly.

---

[16]Womack, James P., and Daniel T. Jones. "Lean Consumption." *Harvard Business Review* 83, no. 3 (March 2005).

[17]http://www.netgrocer.com/.

[18]http://www.groceries-express.com/.

[19]http://www.peapod.com/.

[20]http://www.amazon.com/b?ie=UTF8&node=16310211.

[21]Womack and Jones, "Lean Consumption."

2. Do not waste the customers' time.
3. Provide exactly what the customers want.
4. Provide what's wanted exactly where it is wanted.
5. Provide what's wanted exactly when it is wanted.
6. Continually work toward developing aggregate solutions to reduce the customers' time and hassle.

A seventh principle should be added that trumps the other six:

7. Place a high value on developing and maintaining strong customer relationships.

Lean consumption relies on customers having their needs fulfilled with minimal waste on the part of the consumer who is the ultimate recipient of the goods and services concerned and also on the part of all the supply chain participants who help to make it happen. With the first six principles, we may be placing too much responsibility on the consumer to define his or her needs. On occasion the consumer really does not know what his or her needs are precisely or exactly when those needs should be satisfied. The provider network can assist in these matters, but only if a solid relationship built on mutual trust is established first. To deliver at a world standard of excellence it is necessary for a provider to understand its customers well enough to be able to guide them through the difficult and sometimes complex technological labyrinth of choices.

11.2 VISUAL CONTROLS AND MISTAKE-PROOFING

11.2.1 Visual Displays and Controls

Lean production relies heavily on **visual displays and controls** to facilitate communications on the shop floor using signals that are instantly identifiable and understood by all. Visual controls include signs, information displays, layouts, material storage and handling tools, color-coding, and poka-yoke devices.

The main purpose of using visual controls is to make it possible for anyone (workers, supervisors—even those unfamiliar with the technical process) to be able to tell in a matter of minutes whether things are going well without the help of an expert or sophisticated tools. Visual controls can generally be classified in one of two ways: actual or analog.

Visual controls can be classified as either actual or analog.

Actual items include such things as designating locations or positions (such as shadow boards for tool storage), indicating maximum inventory levels (such as a two bin system or physical reorder markings), distinguishing one item from another (such as distinctive color coding to differentiate parts going through a manufacturing cell), or using paper documents to provide specific instructions.

Analog items, sometimes combined with actual items, appeal to the physical senses to trigger desired human responses. These items include such things as colors, shapes and contours, symbols, characters, numbers, graphs, lights, sounds, touch, smells, and taste.

VC systems can be further divided into two types: displays and controls. A visual display is designed to convey information and data that workers should be aware of, but do not necessarily invoke a required response. Examples are charts that show monthly production figures or quality performance (e.g., waste and defect reports). In contrast, a visual control is intended to guide the actions of some targeted group. Examples in society are stop signs, no-smoking signs, and no-parking zones. In the workplace colors, sights, or sounds can trigger desired employee responses, such as stopping a line, reordering material, changing a production routing, and so forth.

Visual control systems are either displays or controls.

When implemented properly, visual controls can be the effective means, devices, or mechanisms to

- Expose problems, abnormal operations, or excessive waste so that immediate corrective interventions can be taken (e.g., to show when something is out of place or missing).
- Display operating status in a highly visible easy-to-read-and-comprehend format.
- Provide instructions (e.g., workplace safety, production throughput, quality standards).
- Convey information.
- Provide timely feedback.

Figure 11-10 highlights some of the benefits that can be expected following a VC system implementation.

Some examples of visual control applications are color-coded pipes and wires; painted areas on the factory floor designating raw materials, finished goods, and scrap; shadow boards for tool storage; indicator lights on machine and gauge controls; display and status boards; and direction of flow indicators.

A VC tool that is widely used in pull systems is called **andon**, the Japanese name for a paper lantern. Andons, used in antiquity

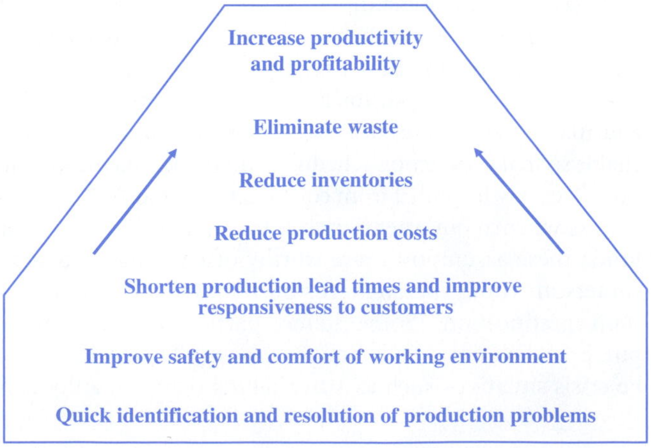

FIGURE 11-10. Benefits of Implementing Visual Controls

to light the way, were like inverted lampshades with candles, and were used in the same way as the flashlight of today. In manufacturing plants andons are used as visual signaling devices. A common use is a lighting system that operates similar to a red, yellow, and green traffic light. Production proceeds smoothly when the light is green. If something goes wrong an operator can summon help by turning on the yellow light, or can stop the line by activating the red light. Andons can also be used for other purposes such as signaling a material handler to replenish the supply of a certain type of material.

An andon is a visual signaling device.

11.2.2 Mistake-Proofing

In Chapter 7 we introduced the concept of poka-yoke as a means for preventing human mistakes. Poka-yoke plays an essential role together with visual controls in preventing human errors and providing the stimulus for human interactions in a lean production environment. For example, a pressure gauge may have a red (danger) zone indicated so that the operator can readily tell when pressure has reached an unacceptable level and take immediate appropriate action. Color markings on scales can provide instant information to prevent under- or overpacking of containers. Alarms and flashing lights can signal that acceptable limits have been exceeded or a machine has malfunctioned. Putting together kits (called kitting) of components and materials can ensure that the correct number of the right part numbers is used in an assembly. Poka-yoke devices, sometimes called fail-safe procedures, are designed to achieve one of two primary objectives:

1. To prevent a worker from making an error that would lead to a product defect
2. To provide quick feedback to operators of any abnormalities in a process in time for corrective action to be taken before defects are produced

11.3 PULL SCHEDULING AND AUTONOMATION

Pull scheduling works best if demand is relatively stable and it is possible to maintain an approximate balance in capacity between workstations. As material will be produced only when required, an unbalanced line leads to inefficiencies. On the other hand, efficiencies accrue when the entire value stream is rhythmic. When work is performed at a common pace there is little waste and resource utilization is high. In Chapter 5 *takt time* was defined as the lean tool used to instill such rhythm. The word *takt* comes from the German word that is used for the baton a conductor uses to direct an orchestra. Just as the conductor uses rhythm to create precise timing, balancing a factory line to takt time establishes a common beat for all activities along the line. Takt time is defined as the cycle time (CT)—that is, the elapsed time between successive units of output—that is required to meet customer demand. Takt time is computed by dividing the total available time (less any breaks, team meetings, and allowances for nonproductive downtime) by the customer demand that must be satisfied during that period. Mathematically, this is expressed as Equation 11-3.

$$\text{Takt time} = \frac{\text{Time available}}{\text{Demand to satisfy}} \qquad \textbf{11-3}$$

As an example, assume a plant must ship 120 units of a particular product per day. Taking into account allowances for nonproductive time we shall assume that there are 450 minutes available per day for production. Using Equation 11-3 takt time can be computed as follows.

Takt time is the per-unit time allowed to meet the customer's time frame.

$$\begin{aligned}\text{Takt time} &= 450 \text{ minutes}/120 \text{ units}\\ &= 3.75 \text{ minutes/unit}\end{aligned}$$

Therefore to meet this demand one unit needs to be completed every 3.75 minutes. Figures 11-11 and 11-12 contrast two production scenarios. The first illustrates the use of a push system to fill the order in question and the second the use of pull scheduling to fill the same order. The order received was for a total of 600 units of a part that has to pass through six departments. In the push system, the expected yield losses across the line require that the material for 641 parts be introduced into the line—an overall process yield of 600/641 = 93.6%. The capacities, expressed as the number of units per hour, also vary and are shown for each department. These differences lead to slight line imbalances which cause a decrease in utilization as follows.

$$\text{Line efficiency} = \frac{\#\text{ productive value-add hours}}{(\#\text{ workstations})(\text{bottleneck time})} \qquad \textbf{11-4}$$

$$\text{Line efficiency} = \frac{(42.7 + 45.4 + 56.5 + 56.3 + 43.6 + 46.5)}{(6)(56.5)}$$

$$\text{Line efficiency} = \frac{291 \text{ value-add hours}}{339 \text{ scheduled hours}} = 85.4\%$$

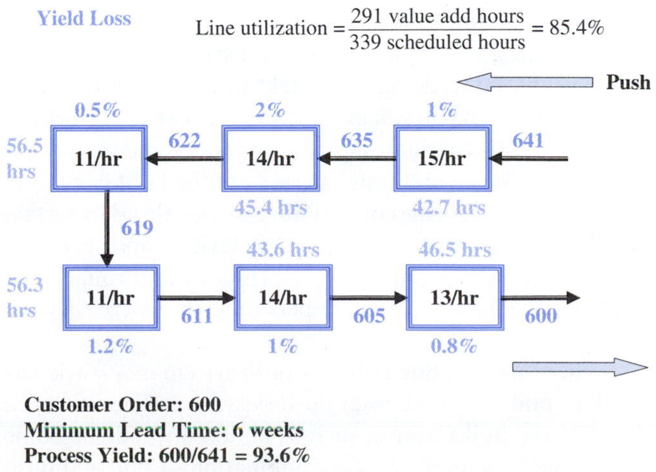

FIGURE 11-11. A Push Production Line

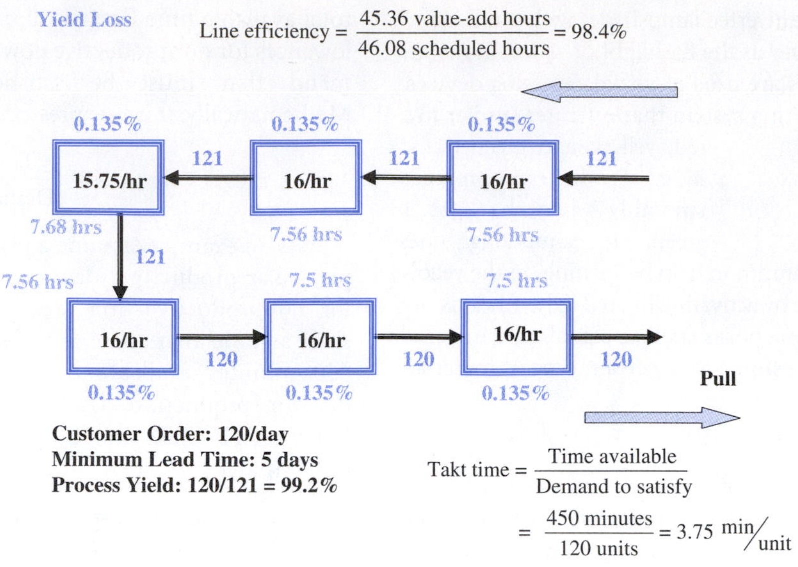

FIGURE 11-12. A Pull Production Line

For this particular order the overall utilization of the resources across the line is only 85.4%. The entire order will be produced as a batch and will show up in successive weeks on the production schedule for each department in turn. Departments are responsible for handing off the completed batch to the next department by the end of the week. Therefore the minimum lead time for pushing the job through all six departments is six weeks.

In the pull system, the large order of 600 has been broken down into smaller daily kanban quantities of 120. The line has been better balanced using the takt time of 3.75 minutes as the target. One of the benefits of the pull system is its emphasis on participation and quality improvement. Teams in each of the six departments have worked together and succeeded in achieving substantial improvements in yields. The pull system has increased line utilization to 98.4%, reduced lead time to 5 days—a significant improvement.

Balancing a line to takt time will not eliminate variabilities due to processing times and capacity imbalances.

Balancing the capacities in a pull system to takt time is the ideal; however, designing to takt time will not eliminate variability, which will still exist within and between workstations due to fluctuations in processing times and in capacity imbalances. Such variations can be smoothed out by placing small kanban quantities of WIP inventories ahead of certain workstations and by flexing capacity. **Flexible capacity** can be achieved by investing in backup machines and through multitasking (being able to move personnel to where they are needed).

There are situations where a push system may work better than pull. However, even push systems can benefit from some of the pull features, such as andon and jidoka. Some companies have found that a combination of pull and push is necessary to accommodate product variety and different types of production processes.

As we previously learned, the house of TPS has two principal pillars—just-in-time (pull scheduling) and jidoka. *Jidoka* is the Japanese word for autonomation and defines a policy under which operators are required to stop a production line anytime they see a problem. Jidoka is human intervention in an automated process. The English name *autonomation* comes from the two root words *autonomous* and *automation*. The idea is to empower people to autonomously detect problems and stop production. However, jidoka is more than simply stopping the line. It also means tracking down the root cause and solving the problem. This means that operators must be trained to spot problems and be empowered to stop a line to fix them when they occur. It is much better to halt production than to continue to produce poor quality. Jidoka can be intimidating to some employees who are fearful of having to accept such an awesome responsibility. Halting an entire production line is a big deal! One solution to this is to use a three-color andon system. If an employee is uncertain whether a problem is serious enough to warrant a shutdown he or she can activate a yellow light to summon a supervisor.

Jidoka requires operators to halt production whenever they see a problem.

The 5S workplace discipline discussed in Chapter 6 is also fundamental to the effective implementation of a pull scheduling system.

11.4 MIXED MODEL SCHEDULING AND PRODUCTION SMOOTHING

In a lean environment, manufacturing cells are typically designed to produce product families in small batch quantities. Within a family, demand for a specific product may be sporadic. To sequence products in a way that leads to smoother production flow a technique called **heijunka** is sometimes helpful. *Heijunka* is a Japanese word that means to make flat and level, and this method has been used successfully to absorb fluctuations in demand without increasing inventory buffers. This is accomplished by breaking down the total volume of orders during some predetermined demand period (e.g., 13 weeks) into small scheduling intervals (e.g., days). A repetitive production sequence is calculated for a scheduling interval, and a mixed model schedule is established for a given production line. A kanban system can then be used to control the correct sequence of materials through the value stream.

Heijunka helps smooth out sporadic demand.

To illustrate how heijunka works consider the following example. A consumer electronics plant produces three products—the Honker, the Wolfer, and the Squeeler. Demands are frozen six weeks out, and the aggregate demand for all three models for the next six-week period is 36,720. Broken down into individual products this translates to 18,360 Honkers, 12,240 Wolfers, and 6,120 Squeelers. The plant operates three shifts, five days per week. The six-week demand will have to be completed over 90 shifts (6 weeks × 5 days × 3 shifts). A scheduling interval at this plant is one shift. Therefore the demand per shift is as follows:

$$\text{Honkers} = \frac{18{,}360}{90} = 204$$
$$\text{Wolfers} = \frac{12{,}240}{90} = 136$$
$$\text{Squeelers} = \frac{6{,}120}{90} = 68$$
$$\text{Total Demand per Shift} = 408$$

Using Equation 11-3 we can compute takt time using 450 minutes of available time as follows.

$$\textit{Takt time} = \frac{\text{450 minutes available}}{\text{408 items produced}} = 1.103 \text{ minutes per item}$$

The ratios of each product to the total are computed as follows:

$$\text{Honker: } \frac{204}{408} = \frac{3}{6}$$
$$\text{Wolfer: } \frac{136}{408} = \frac{2}{6}$$
$$\text{Squeeler: } \frac{68}{408} = \frac{1}{6}$$

Using the least common denominator of 68, we find that a production sequence can be designed with a cycle equal to 6 takt time units and this cycle would repeat every (6)(1.103) = 6.618 minutes for a total of 68 times during the shift. Within each cycle numerous sequences are possible and permissible, so long as the ratios of the product mix computed above are maintained—it does not matter how this is done, since the sequence cycle is so short. Two possible sequences are as follows:

Honker—Honker—Honker—Wolfer—Wolfer—Squeeler

Honker—Wolfer—Honker—Squeeler—Honker—Wolfer

By running small and even batches of the models over these short intervals materials move smoothly between workstations and inventories can be maintained at low levels.

11.5 CHANGEOVER REDUCTION

When mixed products are manufactured on the same production line, adjustments must be made when changing from the production of one product to the production of another. This could involve anything from a die change to adjustments in settings, speeds, feeds, or crewing. Changeovers add cost but no value. While the machine or line is being set up no output is produced. This constitutes part of the fixed cost of a run and has been used as one of the variables in Equation 11-5, the formula commonly used for calculating the economic batch quantity.

$$EBQ = \sqrt{\frac{2(\textit{Setup Costs})(\textit{Demand})}{\textit{Inventory Cost}\left(1 - \dfrac{\textit{Demand}}{\textit{Production Rate}}\right)}} \qquad \textbf{11-5}$$

Equation 11-5 computes the minimum cost batch quantity by trading off the fixed cost of setting up for a production run with the cost of overproducing and holding inventories. Large inventory costs will drive down the **economic batch quantity (size)**; large setup costs will drive it up. In a lean environment the combination of flexibility and low inventories dictates small batch sizes. In many cases the desirable batch size is less than the quantity that would result from applying the formula in Equation 11-5. For example, customers requiring just-in-time deliveries will likely insist on receiving their monthly requirements on a daily basis instead of all at once.

Changeover time is a measure of how long it takes to produce acceptable product from the commencement of a setup.

With a changeover reduction strategy it may be possible to have it both ways—that is, short batch quantities and minimal costs. First, a distinction between a setup and a changeover needs to be made. A **setup** includes all the activities necessary before a production run can be commenced. A changeover is the total time between producing good product on the previous run and producing good product on the next run at the target speed. **Changeover** is the sum of setup time and runup time. **Runup** is the initial production period when the operator is fine-tuning machine adjustments and trying to get the quality right. Changeover time represents a period during which production is potentially lost. **Production loss** experienced during setup and runup times, respectively, is illustrated in Figure 11-13.

The purpose of changeover reduction is to shrink the time devoted to changeovers so more time is available for production. The seminal work in changeover reduction is

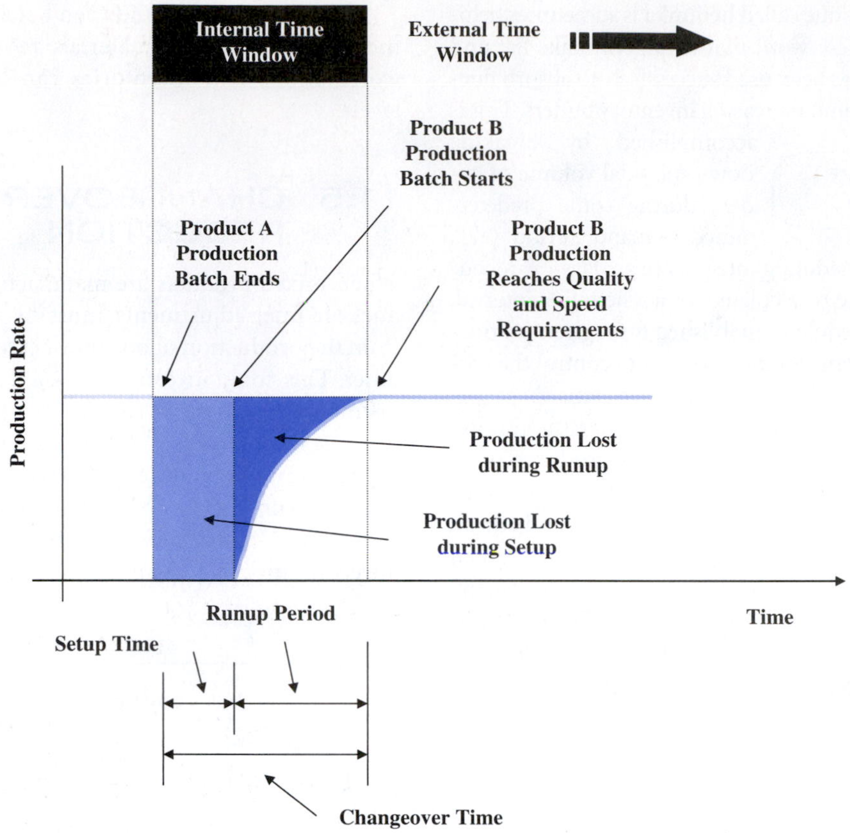

FIGURE 11-13. Changeover and Loss of Production

due to Shigeo Shingo, who in pioneering the single minute exchange of dies (SMED) procedure asserted that changeovers should take no longer than 10 minutes.[22] The basic idea is that this can be achieved by separating all those activities associated with setup time into two categories—internal and external. **External activities** include all the things that an operator can do in order to get ready for the next job while the current job is still running—like fetching and prepositioning the necessary tools.

Internal activities are those that cannot be carried out while the line is up and running, such as changing out dies. Once all activities have been placed into one category or the other the three-prong strategy illustrated in Figure 11-14, and described below, can be employed to try and reduce the total changeover time.

1. Try to convert some internal activities to external ones. Can extra tooling be available so that some time-consuming prep work can be accomplished ahead of time?
2. Try to reduce the time required to perform the internal tasks. Can design changes be used to take advantage of features like quick-release fasteners, alignment keyhole slots, and standardized bolts?
3. Develop a pit-crew mentality. Can methods such as videotaping and critiquing actual changeovers be used to find improvements?

Once changes have been made the operators should practice and compete against themselves to better their times in much the same way that a NASCAR pit crew would strive to improve performance.

Changeover reduction can be facilitated through the use of manufacturing cells designed to produce a family of products. The products that go through a cell are so similar that mixed model scheduling methods can often be employed. In mixed model scheduling each consecutive item can theoretically be different—that is, the hypothetical manufacturing batch size equals one. Group technology, the method discussed in Chapter 5, groups products based on manufacturing similarities and is used to form families with production differences so slight (e.g., a hole drilled in slightly different locations) that rapid changeovers are possible. The cells used to produce these families consist of autonomous work groups, tasked to perform jobs that have been enlarged and enriched. Their ability to multitask permits them to flex capacity to meet changing needs and their personal sense of responsibility for quality has the effect of reducing the error rate.

[22]Shingo, Shigeo. *A Revolution in Manufacturing: The SMED System*. Stamford, CT: Productivity Press, 1985.

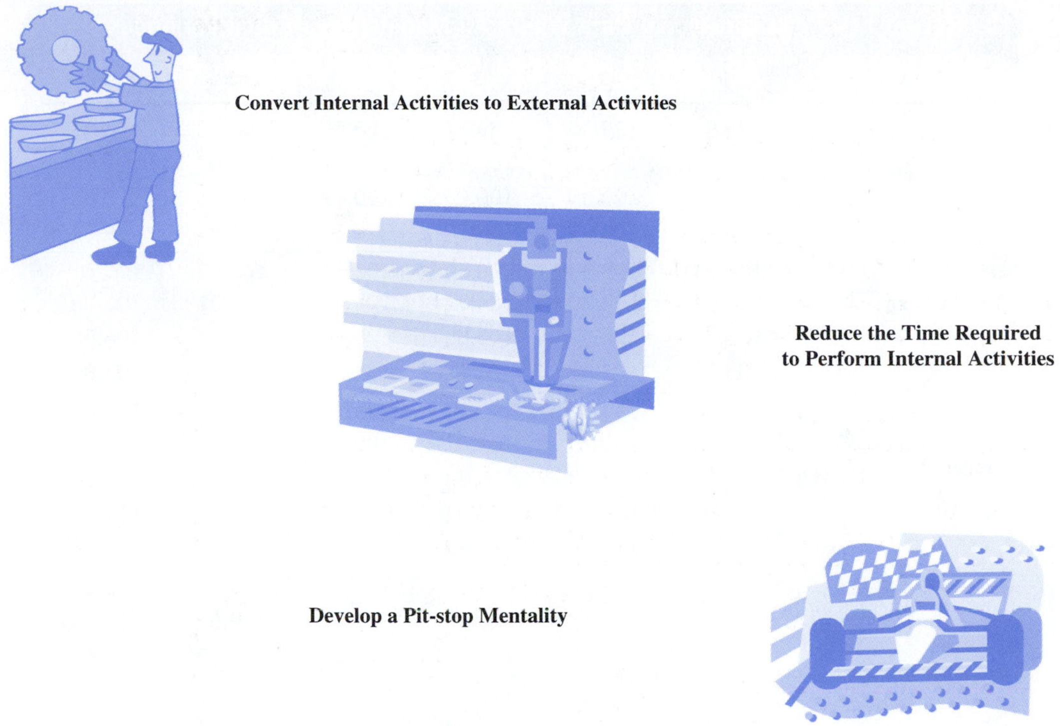

Convert Internal Activities to External Activities
Reduce the Time Required to Perform Internal Activities
Develop a Pit-stop Mentality

FIGURE 11-14. Changeover Reduction Strategies

11.6 SHORT-RUN PROCESS CONTROL

Before steps can be taken to improve a process it is important to know how a process is currently performing. That is not possible (with any confidence) unless the process is in statistical control. Once a process exhibits stability and the major erratic and assignable causes of variation have been removed an improvement team can go to work on the underlying system with the aim of identifying and correcting the most significant sources of variation and thereby enacting sustainable improvements. The use of control charts is a tool that works best in a repetitive production environment with batch runs that are large; and a principal advantage in its use is the need to measure only relatively small samples of output.

A minimum of about 100 observations are necessary to construct a control chart.

To establish control most quality advocates recommend that a minimum of 25 to 30 consecutive samples, with a minimum sample size of 4, pass all eight rules for statistical control.[23] This means that a minimum of 100 or so observations are necessary for control chart startup. However, a control chart is a tool that continues to validate process stability over time and some have argued that many more than 25 subgroups may be necessary to predict process behavior. For example, it has been suggested that for any sample size n, at least $400/(n-1)$ samples be taken before variables control charts can be interpreted with confidence.[24]

When batch quantities are small, which is the case in many cellular designs, the production run can be over before enough data have been collected to establish control. For our purposes, *short-run production* is a term that is applied to any value stream, or a component of one, in which multiple products or services, each with a small individual batch size, are produced in some predefined sequence using common workstations, personnel, and equipment. A typical kanban quantity can easily fall short of the 100 to 120 items of output necessary to generate 25 to 30 points on an $\overline{X}$ and R chart. If attributes measures are used it is possible that the production quantities required for control chart construction could be in the thousands, or even tens of thousands.

11.6.1 Deviation from Nominal (DNOM) or Target (DTAR) Charts

When a **product family** shares a common production routing it is possible to combine the output of the entire family on the same chart. In this instance a family is defined as a collection of products or services that require common tasks. For example, a number of different part numbers may require drilling operations—differing only in the location, number,

[23]For example, Montgomery, D. C. *Introduction to Statistical Quality Control.* 3rd ed. New York: Wiley, 1996.

[24]Quesenberry, C. P. "On Properties of Q Charts for Variables." *Journal of Quality Technology* 27, no. 3 (1995): 184–203.

TABLE 11-2 Mixed Model Data Where All Models Have Equal Variances

	Cycle	1	2	3	4	5	6
Model	MZ Target = 100	99.87	99.30	99.37	98.55	98.13	98.03
		96.75	101.27	102.47	100.17	99.00	102.20
		98.60	100.38	100.93	100.42	99.43	98.82
		99.52	100.09	101.10	98.17	101.14	99.54
	XQ Target = 105	104.70	103.53	106.59	103.67	103.09	105.21
		105.39	105.30	104.77	104.63	106.31	102.08
		104.28	105.90	105.85	108.86	106.13	106.52
		105.22	103.50	104.15	104.32	100.80	103.86
	YT Target = 108	108.81	107.72	110.73	103.78	108.52	109.60
		105.78	106.79	107.82	109.28	107.89	108.10
		110.56	108.48	105.88	108.27	107.40	106.88
		108.42	108.21	105.94	107.02	111.29	111.13
	PZ Target = 112	110.86	110.09	113.36	112.90	111.10	110.28
		108.83	110.99	110.68	111.67	115.33	110.60
		110.49	112.12	113.80	111.67	109.66	114.16
		113.05	111.55	111.03	112.48	109.70	111.47
	AU Target = 115	114.49	113.50	116.47	117.13	115.01	114.82
		114.27	114.62	113.93	112.56	115.57	116.23
		117.34	115.85	117.05	114.20	113.53	115.67
		115.60	118.92	114.91	113.57	115.45	115.30

and size of holes to be drilled. In a mixed model scheduling scenario such as this, all units produced can be combined on a single pair of control charts using a simple data transformation.[25,26,27,28] Mathematically each transformed value is derived by subtracting the respective nominal value from the measurement taken. When it is inappropriate to use a nominal dimension—for example, in the case of a one-sided specification—a target value can be substituted in the transformation formula, which is computed as follows:

$$y_{ij} = x_{ij} - T_i \qquad \textbf{11-6}$$

where,

y_{ij} = transformed data for the jth measurement of the ith part

x_{ij} = the jth raw measurement of the ith part

T_i = the nominal, or target, value for the ith part

Note that the transformation will not affect the ranges. Since for each product type i, T_i is a constant and the range for a sample of raw data will be the same as the range for the sample of transformed data. Therefore, an important assumption underlying the use of the DNOM or DTAR chart is that the process standard deviation is constant across all product types.

To illustrate DNOM (DTAR) control charts, consider the data shown in Table 11-2. Small batch quantities of five different models are produced in sequence as shown. During each batch production samples of size 4 are randomly selected and measured. While each model has a different target specification for the measured attribute, the standard deviation is the same for all models.

Table 11-3 shows the transformed data in DNOM format, and Figure 11-15 is the corresponding pair of DNOM control charts.

A simple method for validating variance equality is the following. After 5 to 6 samples of each product type have been taken, compute the average range, $\overline{R}_i$, for each product i and the average aggregate range $\overline{R}$.

If Equation 11-7, as shown, is satisfied for each product i, then it can be assumed that operationally any differences in product variances are negligible and all parts can be combined on a common pair of DNOM charts.[29]

$$0.67 \le \frac{\overline{R}_i}{\overline{R}} \le 1.33 \qquad \textbf{11-7}$$

[25]Bothe, D. R. *SPC for Short Production Runs*, prepared for U.S. Army Armament, International Quality Institute Inc., Northville, MI, 1988.

[26]Montgomery, *Introduction to Statistical Quality Control.*

[27]Burr, I. W. *Engineering Statistics and Quality Control.* New York: McGraw-Hill, 1953.

[28]Pyzdek, T. "Process Control for Short and Small Runs." *Quality Progress*, April 1993, 51–60.

[29]Bothe, D. R. "A Powerful New Control Chart for Job Shops." *ASQC Quality Congress Transactions.* Toronto, May 1989, 265–67.

TABLE 11-3 DNOM Transformed Data for Mixed Models with Equal Variances

	Cycle	1	2	3	4	5	6
Model	MZ	−0.13	−0.70	−0.63	−1.45	−1.88	−1.97
		−3.26	1.27	2.47	0.17	−1.00	2.20
		−1.40	0.38	0.92	0.42	−0.57	−1.18
		−0.48	0.09	1.10	−1.83	1.14	−0.46
	XQ	−0.30	−1.47	1.59	−1.33	−1.91	0.21
		0.39	0.30	−0.23	−0.37	1.31	−2.92
		−0.72	0.90	0.85	3.86	1.13	1.52
		0.22	−1.50	−0.85	−0.68	−4.20	−1.14
	YT	0.81	−0.28	2.73	−4.22	0.52	1.60
		−2.22	−1.21	−0.18	1.28	−0.11	0.10
		2.56	0.48	−2.12	0.27	−0.60	−1.12
		0.42	0.21	−2.06	−0.98	3.29	3.13
	PZ	−1.14	−1.91	1.36	0.90	−0.90	−1.72
		−3.17	−1.01	−1.32	−0.33	3.33	−1.40
		−1.51	0.12	1.80	−0.33	−2.34	2.16
		1.05	−0.45	−0.97	0.48	−2.30	−0.53
	AU	−0.51	−1.50	1.47	2.13	0.01	−0.18
		−0.73	−0.38	−1.07	−2.44	0.57	1.23
		2.34	0.85	2.05	−0.80	−1.47	0.67
		0.60	3.92	−0.09	−1.43	0.45	0.30

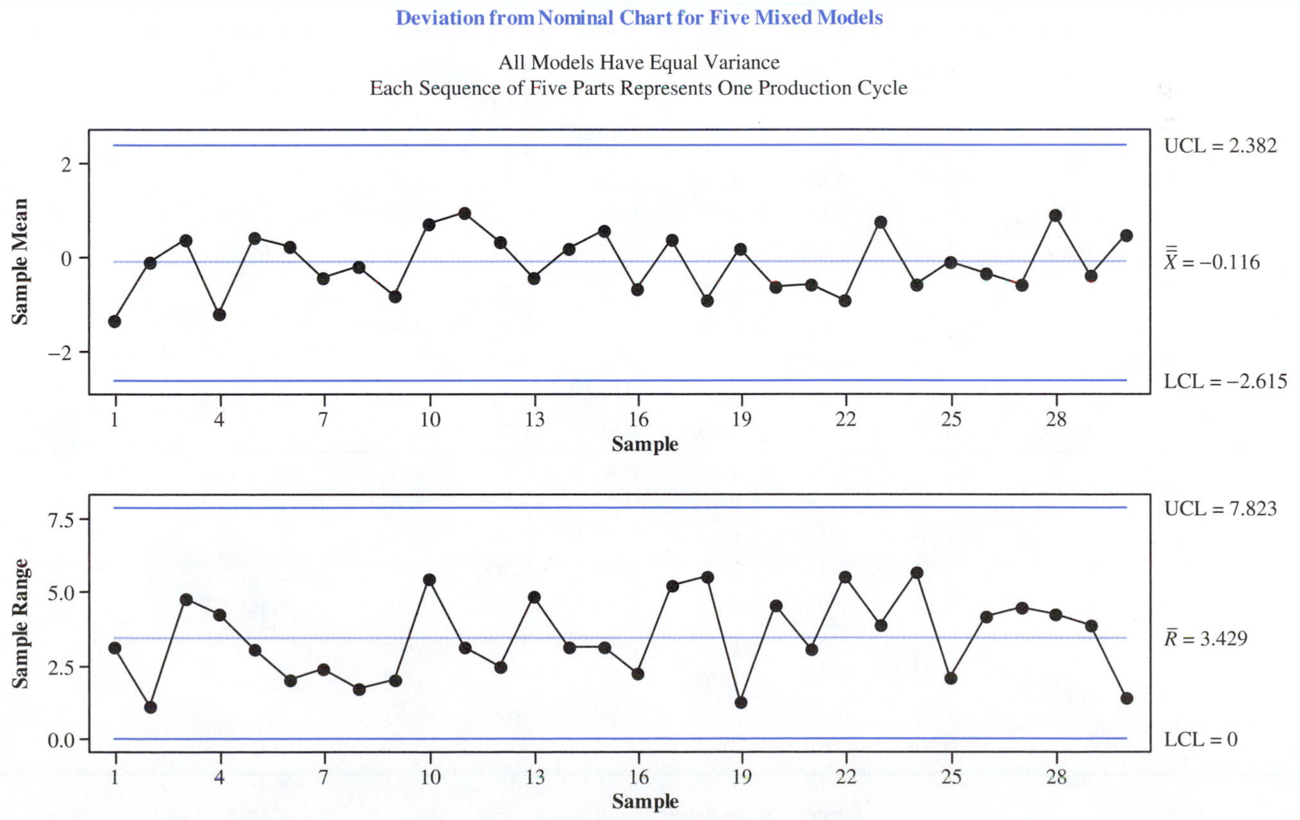

FIGURE 11-15. DNOM Charts for Mixed Model Manufacturing Cell

Otherwise, if for any product i, $\frac{\overline{R}_i}{\overline{\overline{R}}} < 0.67$ or $\frac{\overline{R}_i}{\overline{\overline{R}}} > 1.33$, the ith product is significantly different from the other parts and therefore should be plotted independently.

Table 11-4 shows the individual sample ranges for the previous example. The following ratios suggest that all five products can be appropriately combined on the same pair of charts.

$$\frac{\overline{R}_{MZ}}{\overline{\overline{R}}} = \frac{2.94}{3.45} = 0.852$$

$$\frac{\overline{R}_{XQ}}{\overline{\overline{R}}} = \frac{3.52}{3.45} = 1.020$$

$$\frac{\overline{R}_{YT}}{\overline{\overline{R}}} = \frac{4.16}{3.45} = 1.206$$

$$\frac{\overline{R}_{PZ}}{\overline{\overline{R}}} = \frac{3.36}{3.45} = 0.974$$

$$\frac{\overline{R}_{AU}}{\overline{\overline{R}}} = \frac{3.27}{3.45} = 0.948$$

11.6.2 Stabilized Charts

The use of DNOM or DTAR charts is limited to parts that share a common process standard deviation. When there are significant differences in variability among the product family, short-run $\overline{X}$ and R charts called stabilized charts can be used. In Chapter 8 the following criterion for statistical control on a range chart was developed.

$$D_3\overline{R} \leq R_i \leq D_4\overline{R} \text{ for any sample } i \qquad \textbf{11-8}$$

If we divide Equation 11-8 through by $\overline{R}$ we obtain the following upper and lower limits that are independent of $\overline{R}$.

$$D_3 \leq \frac{R_i}{\overline{R}} \leq D_4 \text{ for any sample } i \qquad \textbf{11-9}$$

When there are multiple products, each having a different variability (resulting in different $\overline{R}$s) Equation 11-9 can be modified as follows.

$$\text{Let } \omega_{ij} = \frac{R_{ij}}{\overline{R}_j} \qquad \textbf{11-10}$$

$$D_3 \leq \omega_{ij} \leq D_4 \qquad \textbf{11-11}$$

TABLE 11-4 DNOM Transformed Data for Mixed Models with Range Calculation

	Cycle	1	2	3	4	5	6	Average Range
Model	MZ	−0.13	−0.70	−0.63	−1.45	−1.88	−1.97	
		−3.26	1.27	2.47	0.17	−1.00	2.20	
		−1.40	0.38	0.92	0.42	−0.57	−1.18	
		−0.48	0.09	1.10	−1.83	1.14	−0.46	
	Range	3.13	1.97	3.10	2.25	3.02	4.17	2.94
	XQ	−0.30	−1.47	1.59	−1.33	−1.91	0.21	
		0.39	0.30	−0.23	−0.37	1.31	−2.92	
		−0.72	0.90	0.85	3.86	1.13	1.52	
		0.22	−1.50	−0.85	−0.68	−4.20	−1.14	
	Range	1.11	2.40	2.44	5.19	5.51	4.44	3.52
	YT	0.81	−0.28	2.73	−4.22	0.52	1.60	
		−2.22	−1.21	−0.18	1.28	−0.11	0.10	
		2.56	0.48	−2.12	0.27	−0.60	−1.12	
		0.42	0.21	−2.06	−0.98	3.29	3.13	
	Range	4.78	1.69	4.85	5.50	3.89	4.25	4.16
	PZ	−1.14	−1.91	1.36	0.90	−0.90	−1.72	
		−3.17	−1.01	−1.32	−0.33	3.33	−1.40	
		−1.51	0.12	1.80	−0.33	−2.34	2.16	
		1.05	−0.45	−0.97	0.48	−2.30	−0.53	
	Range	4.22	2.03	3.12	1.23	5.67	3.88	3.36
	AU	−0.51	−1.50	1.47	2.13	0.01	−0.18	
		−0.73	−0.38	−1.07	−2.44	0.57	1.23	
		2.34	0.85	2.05	−0.80	−1.47	0.67	
		0.60	3.92	−0.09	−1.43	0.45	0.30	
	Range	3.07	5.42	3.12	4.57	2.04	1.41	3.27
				Overall Average Range				3.45

where,

ω_{ij} is the relative range of the ith sample of the jth product

R_{ij} is the range of the ith sample of the jth product

$\overline{R}_j$ is the cumulative average range for all samples taken of the jth product

One criterion for statistical control on an $\overline{X}$ chart is the following.

$$\overline{\overline{X}} - A_2\overline{R} \leq \overline{X}_i \leq \overline{\overline{X}} + A_2\overline{R} \text{ for any sample } i \qquad \textbf{11-12}$$

By manipulating Equation 11-12 upper and lower bounds on the $\overline{X}$ s can be generated that are independent of the ranges.

$$-A_2 \leq \frac{\overline{X}_i - \overline{\overline{X}}}{\overline{R}} \leq +A_2 \text{ for any sample } i \qquad \textbf{11-13}$$

To accommodate multiple products with different average ranges Equation 11-13 can be appropriately modified as follows.

$$\text{Let } \gamma_{ij} = \frac{\overline{X}_{ij} - \overline{\overline{X}}_j}{\overline{R}_j} \qquad \textbf{11-14}$$

$$-A_2 \leq \gamma_{ij} \leq +A_2 \text{ for any sample } i \text{ of the } j^{th} \text{ product} \qquad \textbf{11-15}$$

where,

γ_{ij} is the transformed (stabilized) average of the ith sample of the jth product

$\overline{X}_{ij}$ is the average of the ith sample of the jth product

$\overline{\overline{X}}_j$ is the cumulative aggregate average for all samples of the jth product

$\overline{R}_j$ is the cumulative average range for all samples taken of the jth product

To illustrate the concept of stabilized control charts consider the data in Table 11-5 that have been collected from a manufacturing cell that assembles four different product models in sequence as shown. Table 11-6 contains the data transformed into DNOM format, and Figure 11-16 shows the pair of corresponding DNOM control charts, on which a few statistical control rule failures are noted. However, when the data are grouped by product and replotted, as in Figure 11-17, the differences in product variance are apparent.

In Table 11-7, Equations 11-10 and 11-14 have been used to transform the data into a format suitable for plotting on a pair of stabilized control charts. The resulting charts are displayed in Figure 11-18.

11.6.3 *Q* Statistics and *Z* Charts

When the standard deviation for each part is known (i.e., can be validated over time through stable R or s charts), a transformation that differs slightly from the ones cited previously can be used. This transformation, known as a standard Q statistic,[30] is calculated by subtracting each part's critical dimension from its nominal dimension and subsequently dividing the difference by the process standard deviation. Mathematically a Q statistic transformation is derived as follows.

$$Q_{ijk} = \frac{x_{ijk} - T_i}{\sigma_i} \qquad \textbf{11-16}$$

[30]Quesenberry, "On Properties of Q Charts."

TABLE 11-5 Mixed Model Data Where All Models Have Unequal Variances

	Cycle	1	2	3	4	5	6	7
Model	AQM	71.80	67.82	71.00	69.12	70.35	68.46	72.21
		61.86	71.90	67.08	65.81	70.07	65.22	75.56
		68.48	68.06	70.11	73.52	73.15	67.26	69.42
		68.69	65.46	72.41	71.17	71.92	66.28	72.02
	AQB	51.11	51.02	47.27	51.61	52.23	50.09	50.60
		51.40	51.86	52.77	50.76	50.00	49.00	51.14
		48.78	49.72	47.78	48.50	48.70	47.54	54.02
		50.21	49.75	49.77	49.82	50.56	46.14	46.67
	BCL	41.19	40.08	39.98	38.79	40.18	40.52	39.46
		39.71	39.08	38.82	41.48	39.56	40.89	41.34
		39.42	41.96	40.08	40.74	41.79	39.39	38.64
		39.89	39.49	40.78	39.42	40.76	39.32	39.23
	BCM	35.28	35.07	34.56	34.82	35.98	35.41	34.69
		35.37	35.04	35.53	34.27	34.41	33.61	36.63
		35.11	34.46	34.51	36.00	34.26	34.76	35.64
		35.40	35.22	34.49	36.13	36.41	33.15	34.77

TABLE 11-6 DNOM Transformed Data for Mixed Models with Unequal Variances

Cycle		1		2		3		4		5		6		7	
		Raw Meas	Dev from Nom	Raw Meas	Dev from Nom	Raw Meas	Dev from Nom	Raw Meas	Dev from Nom	Raw Meas	Dev from Nom	Raw Meas	Dev from Nom	Raw Meas	Dev from Nom
Model	AQM	71.80	1.80	67.82	−2.18	71.00	1.00	69.12	−0.88	70.35	0.35	68.46	−1.54	72.21	2.21
		61.86	−8.14	71.90	1.90	67.08	−2.92	65.81	−4.19	70.07	0.07	65.22	−4.78	75.56	5.56
		68.48	−1.52	68.06	−1.94	70.11	0.11	73.52	3.52	73.15	3.15	67.26	−2.74	69.42	−0.58
		68.69	−1.31	65.46	−4.54	72.41	2.41	71.17	1.17	71.92	1.92	66.28	−3.72	72.02	2.02
	AQB	51.11	1.11	51.02	1.02	47.27	−2.73	51.61	1.61	52.23	2.23	50.09	0.09	50.60	0.60
		51.40	1.40	51.86	1.86	52.77	2.77	50.76	0.76	50.00	0.00	49.00	−1.00	51.14	1.14
		48.78	−1.22	49.72	−0.28	47.78	−2.22	48.50	−1.50	48.70	−1.30	47.54	−2.46	54.02	4.02
		50.21	0.21	49.75	−0.25	49.77	−0.23	49.82	−0.18	50.56	0.56	46.14	−3.86	46.67	−3.33
	BCL	41.19	1.19	40.08	0.08	39.98	−0.02	38.79	−1.21	40.18	0.18	40.52	0.52	39.46	−0.54
		39.71	−0.29	39.08	−0.92	38.82	−1.18	41.48	1.48	39.56	−0.44	40.89	0.89	41.34	1.34
		39.42	−0.58	41.96	1.96	40.08	0.08	40.74	0.74	41.79	1.79	39.39	−0.61	38.64	−1.36
		39.89	−0.11	39.49	−0.51	40.78	0.78	39.42	−0.59	40.76	0.76	39.32	−0.68	39.23	−0.77
	BCM	35.28	0.28	35.07	0.07	34.56	−0.44	34.82	−0.18	35.98	0.98	35.41	0.41	34.69	−0.31
		35.37	0.37	35.04	0.04	35.53	0.53	34.27	−0.73	34.41	−0.59	33.61	−1.39	36.63	1.63
		35.11	0.11	34.46	−0.54	34.51	−0.49	36.00	1.00	34.26	−0.74	34.76	−0.24	35.64	0.64
		35.40	0.40	35.22	0.22	34.49	−0.51	36.13	1.13	36.41	1.41	33.15	−1.85	34.77	−0.23

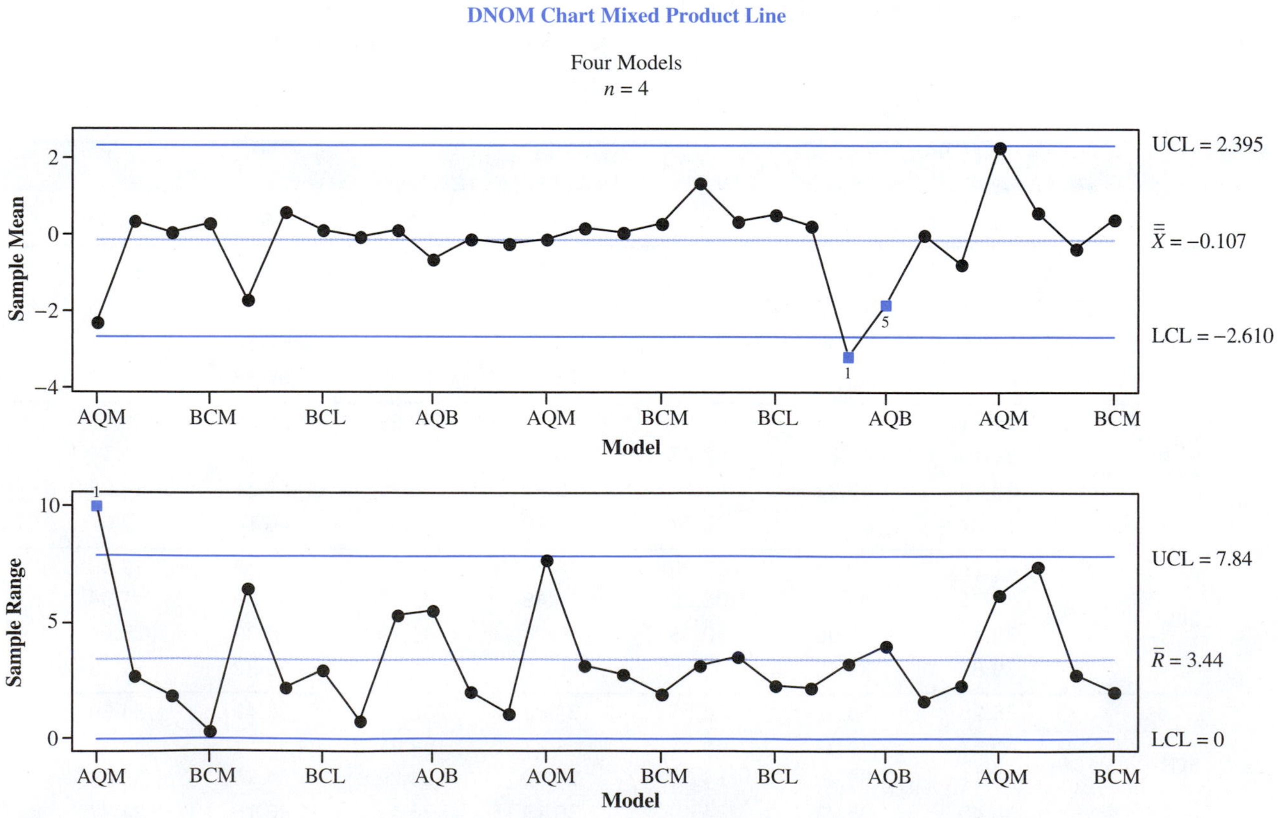

FIGURE 11-16. DNOM Chart for Mixed Product Line with Unequal Variances

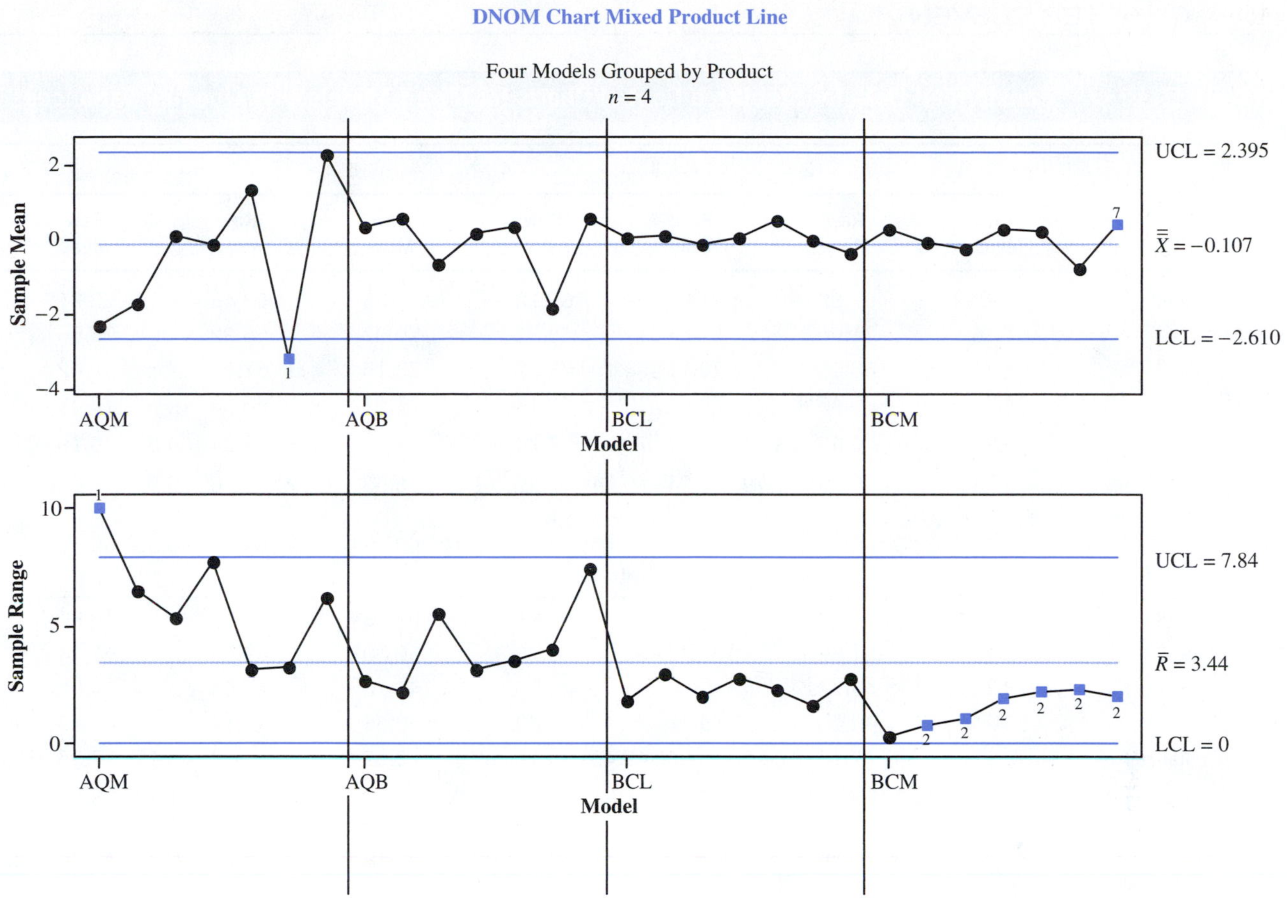

FIGURE 11-17. DNOM Chart for Mixed Product Line with Unequal Variances, Grouped by Product

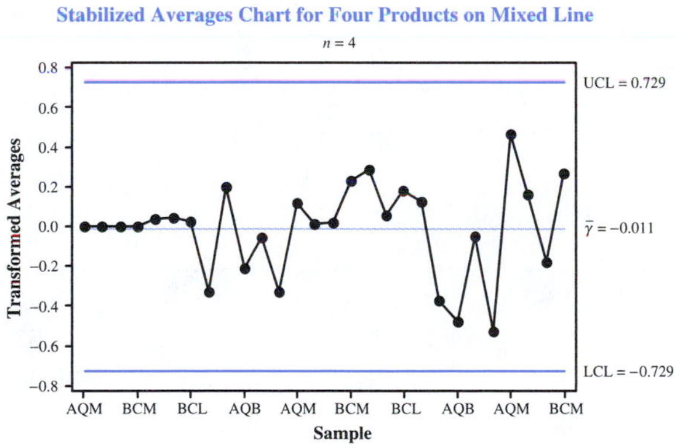

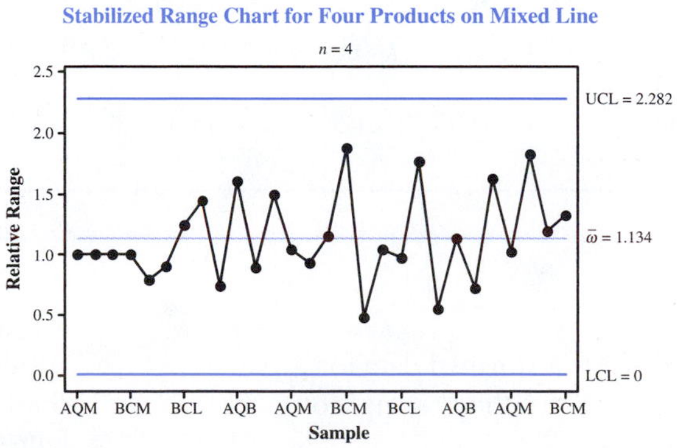

FIGURE 11-18. Stabilized Control Charts for Product Line with Unequal Variances

TABLE 11-7 Transformed Data for Stabilized Control Charts for Mixed Models with Unequal Variances

		1		2		3		4		5		6		7	
	Cycle	**Raw Meas**		**Raw Meas**		**Raw Meas**		**Raw Meas**		**Raw Meas**		**Raw Meas**		**Raw Meas**	
	AQM	71.80		67.82		71.00		69.12		70.35		68.46		72.21	
		61.86		71.90		67.08		65.81		70.07		65.22		75.56	
		68.48		68.06		70.11		73.52		73.15		67.26		69.42	
		68.69		65.46		72.41		71.17		71.92		66.28		72.02	
	Range/Avg	9.94	67.71	6.45	68.31	5.33	70.15	7.70	69.90	3.08	71.37	3.24	66.81	6.15	72.30
	Cum Range/ ω Stat	9.94	1.00	8.19	0.79	7.24	0.74	7.35	1.05	6.50	0.47	5.96	0.54	5.98	1.03
	Cum Average/ γ Stat	67.71	0.00	68.01	0.04	68.72	0.20	69.02	0.12	69.49	0.29	69.04	−0.38	69.51	0.47
	AQB	51.11		51.02		47.27		51.61		52.23		50.09		50.60	
		51.40		51.86		52.77		50.76		50.00		49.00		51.14	
		48.78		49.72		47.78		48.50		48.70		47.54		54.02	
		50.21		49.75		49.77		49.82		50.56		46.14		46.67	
	Range/Avg	2.62	50.38	2.15	50.59	5.50	49.40	3.11	50.17	3.52	50.37	3.94	48.19	7.35	50.61
	Cum Range/ ω Stat	2.62	1.00	2.39	0.90	3.42	1.61	3.35	0.93	3.38	1.04	3.48	1.13	4.03	1.82
	Cum Average/ γ Stat	50.38	0.00	50.48	0.04	50.12	−0.21	50.13	0.01	50.18	0.06	49.85	−0.48	49.96	0.16
Model	**BCL**	41.19		40.08		39.98		38.79		40.18		40.52		39.46	
		39.71		39.08		38.82		41.48		39.56		40.89		41.34	
		39.42		41.96		40.08		40.74		41.79		39.39		38.64	
		39.89		39.49		40.78		39.42		40.76		39.32		39.23	
	Range/Avg	1.76	40.05	2.88	40.15	1.95	39.91	2.68	40.11	2.24	40.57	1.56	40.03	2.70	39.67
	Cum Range/ ω Stat	1.76	1.00	2.32	1.24	2.20	0.89	2.32	1.16	2.30	0.97	2.18	0.72	2.25	1.20
	Cum Average/ γ Stat	40.05	0.00	40.10	0.02	40.04	−0.06	40.06	0.02	40.16	0.18	40.14	−0.05	40.07	−0.18
	BCM	35.28		35.07		34.56		34.82		35.98		35.41		34.69	
		35.37		35.04		35.53		34.27		34.41		33.61		36.63	
		35.11		34.46		34.51		36.00		34.26		34.76		35.64	
		35.40		35.22		34.49		36.13		36.41		33.15		34.77	
	Range/Avg	0.29	35.29	0.76	34.95	1.04	34.77	1.86	35.31	2.15	35.26	2.26	34.23	1.94	35.43
	Cum Range/ ω Stat	0.29	1.00	0.53	1.44	0.70	1.49	0.99	1.88	1.22	1.76	1.39	1.62	1.47	1.32
	Cum Average/ γ Stat	35.29	0.00	35.12	−0.33	35.00	−0.33	35.08	0.23	35.12	0.12	34.97	−0.53	35.04	0.27

where,

Q_{ijk} = standardized transformed data for the kth measurement of the jth sample of the ith part

x_{ijk} = the jth raw measurement of the ith part

T_i = the nominal, or target, value for the ith part

σ_i = the standard deviation of the ith part

As with stabilized control charts, products with different standard deviations can be combined on a single Z control chart. Table 11-8 shows the Q statistic transformations for the four-model mixed production data from Table 11-5. The corresponding Z charts are presented in Figure 11-19. Note that the Z range chart represents the variation of computed Q values within each sample and the $Z\overline{X}$ chart represents the between-sample variation of the computed Q values.

TABLE 11-8 Q Statistics for Z Chart Construction of Mixed Models with Unequal Variances

		1		2		3		4		5		6		7	
	Cycle	Raw Meas	Q Stat	Raw Meas	Q Stat	Raw Meas	Q Stat	Raw Meas	Q Stat	Raw Meas	Q Stat	Raw Meas	Q Stat	Raw Meas	Q Stat
Model	AQM	71.80	0.60	67.82	−0.73	71.00	0.33	69.12	−0.29	70.35	0.12	68.46	−0.51	72.21	0.74
	$\sigma = 3$	61.86	−2.71	71.90	0.63	67.08	−0.97	65.81	−1.40	70.07	0.02	65.22	−1.59	75.56	1.85
	$T_{AQM} = 70$	68.48	−0.51	68.06	−0.65	70.11	0.04	73.52	1.17	73.15	1.05	67.26	−0.91	69.42	−0.19
		68.69	−0.44	65.46	−1.51	72.41	0.80	71.17	0.39	71.92	0.64	66.28	−1.24	72.02	0.67
	AQB	51.11	0.56	51.02	0.51	47.27	−1.37	51.61	0.81	52.23	1.11	50.09	0.04	50.60	0.30
	$\sigma = 2$	51.40	0.70	51.86	0.93	52.77	1.38	50.76	0.38	50.00	0.00	49.00	−0.50	51.14	0.57
	$T_{AQB} = 50$	48.78	−0.61	49.72	−0.14	47.78	−1.11	48.50	−0.75	48.70	−0.65	47.54	−1.23	54.02	2.01
		50.21	0.11	49.75	−0.13	49.77	−0.12	49.82	−0.09	50.56	0.28	46.14	−1.93	53.67	1.83
	BCL	41.19	1.19	40.08	0.08	39.98	−0.02	38.79	−1.21	40.18	0.18	40.52	0.52	39.46	−0.54
	$\sigma = 1$	39.71	−0.29	39.08	−0.92	38.82	−1.18	41.48	1.48	39.56	−0.44	40.89	0.89	41.34	1.34
	$T_{BCL} = 40$	39.42	−0.58	41.96	1.96	40.08	0.08	40.74	0.74	41.79	1.79	39.39	−0.61	38.64	−1.36
		39.89	−0.11	39.49	−0.51	40.78	0.78	39.42	−0.59	40.76	0.76	39.32	−0.68	39.23	−0.77
	BCM	35.28	0.37	35.07	0.09	34.56	−0.59	34.82	−0.24	35.98	1.30	35.41	0.55	34.69	−0.42
	$\sigma = 0.75$	35.37	0.50	35.04	0.05	35.53	0.70	34.27	−0.97	34.41	−0.78	33.61	−1.85	36.63	2.17
	T_{BCM} 35	35.11	0.15	34.46	−0.72	34.51	−0.65	36.00	1.34	34.26	−0.99	34.76	−0.32	35.64	0.86
		35.40	0.54	35.22	0.29	34.49	−0.68	36.13	1.51	36.41	1.88	33.15	−2.47	34.77	−0.30

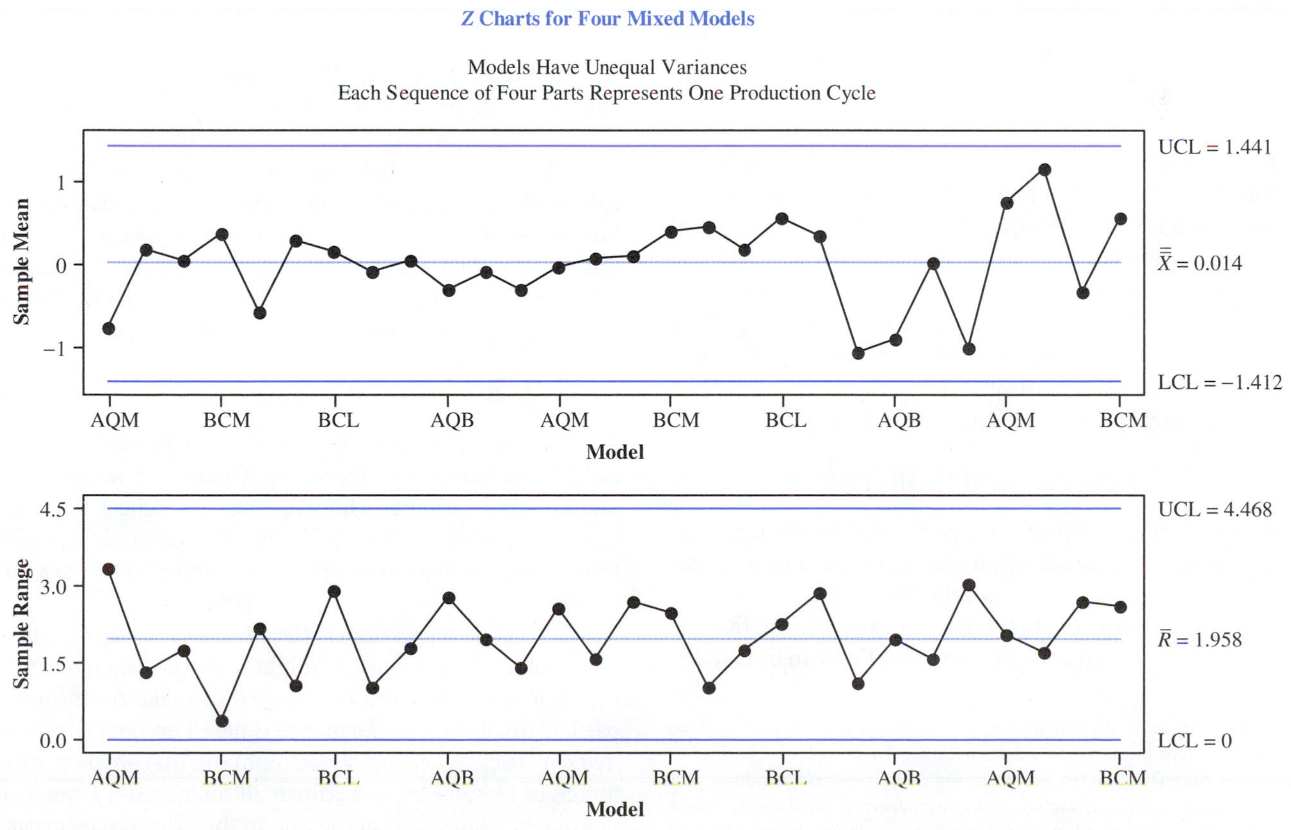

FIGURE 11-19. *Z* Charts for Four-Model Mixed Production Problem

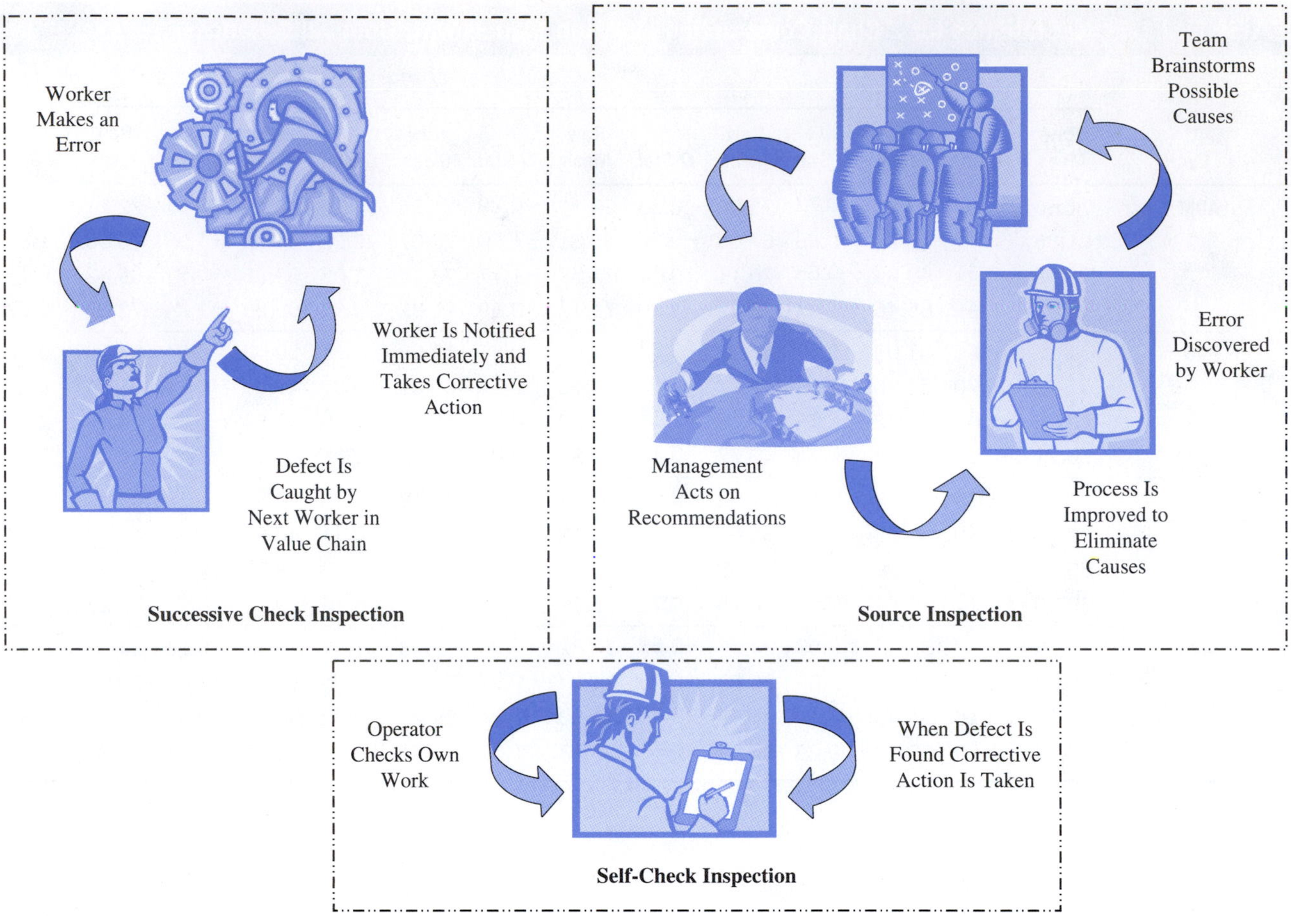

FIGURE 11-20. Inspection Techniques

11.7 INSPECTION TO PREVENT DEFECTS

In a QIM lean environment we can legitimately make a distinction between an error and a defect. Defects occur because people make errors. Even though errors are to some extent inevitable,[31] defects can be prevented if proper feedback can lead to immediate corrective action soon after the error is made. Unlike SPC which relies on samples, such feedback and action requires some form of 100% inspection. Three types of inspection, as illustrated in Figure 11-20, can be employed: successive check, self-check, and source.

11.7.1 Successive Check Inspection

A **successive check inspection** is performed by the next person in the value stream or by an objective evaluator such as a supervisor or group leader. If a defect is detected, feedback is immediately provided to the responsible operator so that the error can be corrected and appropriate action can be taken to prevent recurrence. In implementing a system of successive checks it is important that management is aware of some cultural issues; otherwise interpersonal relationships on the shop floor can potentially be undermined. Unless an atmosphere is created in which everyone fully understands that human errors (which are assumed to be inadvertent) are more easily detected by others than by those committing them, a climate of animosity can result in which workers resent having peers criticize or critique their work.

11.7.2 Self-Check Inspection

A **self-check inspection** requires that an individual worker check his or her own performance. This system works well for many quality attributes. However, successive check inspection is preferred when sensory judgments are required, such as the level of severity of a defect or the matching of colors to a sample swatch.

Self-check inspection suffers from two inherent flaws. One problem is that workers might be too lax when scrutinizing their own work and be tempted to make compromises, particularly if their performance is based on quality productivity metrics. A second shortcoming is that workers might forget, or be too lazy, to perform all the necessary checks on their own. Again, this can be due to the existence of negative incentives resulting from performance-based rewards.

---

[31]An acquiescence to the inevitability of human error is debatable. One would not accept human errors that could cause a commercial jetliner to crash or a patient to die on the operating table; similarly, a culture that adopts as a core value a zero tolerance for human error can, through proper training, attention to detail, and the pursuit of excellence, reduce the incidence error-induced defects.

The installation of poka-yoke devices can help address these problems. An example of a poka-yoke-assisted self-check inspection system involves the checking of a critical diameter of an automobile part. The specifications require that the part be machined to a tolerance of 12 mm ± 0.5 mm. In the production line two gauge/guides were installed, one at a height equivalent to the upper specification limit of 12.5 mm and the other at a height equal to the lower specification limit of 11.5 mm. As product moves under the two gauge/guides those parts that cannot clear the 12.5 mm guide are too large and an out-of-spec alarm sounds as the rejected parts are shunted into a rework barrel. Parts that can not clear the second, 11.5 mm guide, are within the specification range and are routed to the next processing station, while those that can clear the guide (too small) continue ahead to drop into a scrap barrel, again triggering an out-of-spec alarm.

11.7.3 Source Inspection

Like self-check inspection, **source inspection** is also performed by the individual worker performing the task; however, the emphasis goes beyond simply checking for defects. In source inspection the worker is concerned with identifying the errors that created the defects and then discovering and correcting the root causes of those errors. The theory is that errors cannot turn into defects if feedback and corrective action can take place at an early-enough (error source) stage. Six Sigma project teams can be effective in analyzing a process for error proneness and prevention. Management plays a crucial role also in this process as many root causes are likely systemic and their elimination has procedural, process design, and training implications. Root cause analysis tools include Pareto analysis (Chapter 7), cause-and-effect diagrams (Chapter 7), six honest servants and five whys (Chapter 7), and failure mode and effects analysis (FMEA—Chapter 6).

An inexpensive and simple poka-yoke device that has been widely used to help in error prevention is a checklist. This has been especially useful during the process design phase of a product development cycle. The FMEA procedure represents a formalized use of check sheets.

11.8 KAIZEN BLITZ: TEAM-BASED RADICAL IMPROVEMENT

Empowered self-managed work teams (SMWTs) or self-directed work teams (SDWTs), as described in Chapter 3, are the building blocks of lean production. From process improvement to manufacturing cells the lean paradigm relies on broad participation of and shared ownership by the entire workforce. One powerful lean improvement tool that relies on effective teamwork is kaizen blitz.

In Chapter 6 the word *kaizen* was defined to mean "continuous improvement." The literal Japanese meaning is "to take apart and put back together in a better way." The word *blitz* is short for *blitzkrieg*—a derivative formed from two German words: *blitz* meaning "lightning" and *krieg* meaning "war." *Blitzkrieg* is therefore a war (or surprise offensive) conducted with great speed and overpowering force—in short, a sudden overpowering attack. When *blitz* and *kaizen* are combined, **kaizen blitz** can be defined as a sudden overpowering event that aims to tear down some process (or system, product, service, etc.) and put it back together in an improved way. The Japanese term for kaizen blitz is **kaikaku**. Kaizen blitz has been a technique used by some firms in the early stages of a lean production (or Lean Six Sigma) implementation to force a cultural shift and produce some positive results quickly.

A kaikaku event is a focused intensive one-week project in which a team works together to create a sudden radical change in a production process or system. The pace is fast and furious, and the methodology employed combines lean tools such as the 5S workplace discipline, cellular organization, pull scheduling, kanban control, changeover reduction, and line balancing. In deploying such tools teams are empowered to use creative brainstorming and problem-solving methods to rapidly develop and implement viable improvement options. Table 11-9 shows a typical timetable

A kaikaku event focuses on radical breakthrough improvements.

TABLE 11-9 A Typical Kaizen Blitz Project Timetable

Day	Activity
Wednesday	Select and train a small group (3 or 4) cross-functional facilitators in the tools of kaizen blitz and lean production.
Thursday	Identify the problem, process, or system to be studied. Select members for kaizen blitz team.
Friday	Collect and assemble all available data on problem, process, or system as it currently exists.
Monday	Training is provided for the entire team on kaizen blitz and lean production techniques and tools.
Tuesday	Team begins a detailed mapping and analysis of the current situation, including defining the impacts of suppliers and customers.
Wednesday	Team brainstorms new and radical ideas, and begins to work together to develop an "ideal" process map that eliminates bottlenecks, waste, and inefficiencies, and improves materials flow. An action list is created assigning individual responsibilities. Senior management briefings are scheduled.
Thursday	The team begins to pilot some ideas for new methods to determine what might and might not work.
Friday	The team makes final decision on changes and works to fine-tune and debug the new methods so that the changes go into effect the following Monday morning. Thirty-day and 60-day action plans are developed to review and address any issues that arise.

of activities where a week of intensive activity is preceded by three days of preplanning.

A kaizen blitz event begins with a vision of the desired future state (e.g., zero defects, variance reduction of 50%, Six Sigma capability, etc.). The team then works backward to develop an action plan that will bring it close to the desired future goal. The process is data driven, and success depends on the team's ability to collect reliable data concerning the current state of the process under study. Many of the tools described in Chapter 7 for generating ideas, mapping and analyzing processes, and developing improvement options are used to gain an in-depth and consensual understanding of the problem at hand and relevant process parameters. To generate viable ideas it is essential that the team clearly understand the nature and nuances of the current situation. What makes a kaizen blitz event somewhat different from other team-based brainstorming activities is that kaikaku is not just about generating ideas for change. Kaikaku is about effecting change and bringing it about quickly. Within the short span of one week a blitz team must size up the situation, analyze the data, generate ideas, decide on what to change, develop an implementation strategy, and then execute it. This is a tall order, which on the surface may seem unrealistic. However, the advantage is that intensity and urgency overcomes the natural cultural resistance to change and the propensity to politicize proposed alternatives. People have little time to think of reasons to delay or to oppose if there is an executive mandate to deliver dramatic change within such a short time frame. On the positive side, the pressure creates unity, teamwork, enthusiasm, and pride.

Kaizen blitz is a sudden overpowering event that tears down a process and then rebuilds it in an improved manner.

There are some fundamental ingredients that will help make a kaizen blitz event successful.

1. **Fact-based decision making**—Intuition and opinion can have no place in a kaizen blitz project. All decisions must be based on relevant hard data that are collected from reliable sources and which plausibly answer a set of questions previously defined by the team. Analytical tools should be employed that will provide insights and understanding so that new ideas and decisions will be based on real knowledge, not hearsay, belief, politics, or tradition.
2. **Team vision**—The team needs to be able to collectively set its sights on what would constitute an ideal state, and use this as a compass to guide idea generation and action plans.
3. **Inclusion and broad participation**—For a kaizen blitz to be successful everyone who has a vested interest in the outcomes should be involved. This may require a disruption of normal production schedules while personnel are pulled off the line to participate in the kaizen blitz event. Although five days is usually allotted for the event, these days need not be contiguous. In many cases it is practicable to break the time up into smaller one- or two-day segments.
4. **Team preparedness**—Training all team members in how to conduct a kaizen blitz and on the use of appropriate tools is essential to success. Since radical change challenges old habits and tradition and will test established paradigms, the people involved could find the experience to be stressful if not properly prepared. The team members need to know that their organization's senior leadership is solidly behind them and has faith in their ability to succeed. Team members must also believe in themselves and their fellow team members. Everyone must be willing to enthusiastically assume whatever responsibilities the team assigns and maintain a positive upbeat attitude.
5. **Target selection**—Choosing the right problem, process, or system is critical. It is unwise to try and tackle the organization's toughest problem where success is unlikely. A topic should be selected that will have a big and visible impact on the organization and on the people in it. Project scope should be consistent with the one-week time

SHOWCASE of EXCELLENCE

Winners of the Malcolm Baldrige National Quality Award Small Business Category

- 1996: Trident Precision Manufacturing Inc., a manufacturer of precision sheet-metal components, electromechanical assemblies, and custom products, mostly in the office equipment, medical supply, computer, and defense industries. http://www.tridentprecision.com, http://www.quality.nist.gov/Trident_96.htm.
- 1994: Wainwright Industries, Inc., a manufacturer of stamped and machined parts for U.S. and foreign customers in the automotive, aerospace, home security, and information processing industries. http://wainwrightindustries.com, http://www.quality.nist.gov/Wainwright_Industries.htm.
- 1993: Ames Rubber Corporation, a manufacturer of rubber rollers used to feed paper, transfer toner, and fuse toner to paper in office machines such as copiers, printers, and typewriters. http://www.amesrubber.com, http://www.quality.nist.gov/Ames_Rubber_Corporation.htm.
- 1992: Granite Rock Company, a producer of rock, sand, and gravel aggregates; ready-mix concrete; asphalt; road treatments; and recycled road-base materials. http://www.graniterock.com, http://www.quality.nist.gov/Granite_Rock_Co.htm.
- 1991: Marlow Industries Inc., a company that processes raw materials into thermoelectric semiconductors, assembles these devices into thermoelectric coolers, and integrates the coolers into heat exchangers for commercial and defense applications. http://www.marlow.com, http://www.quality.nist.gov/Marlow_91.htm.

constraint. Selecting a project with too broad a scope will lead to failure. If the scope is too narrow, the project risks failing to have a significant enough impact to effect radical change.

Kaikaku is not appropriate for all improvement projects. However, where it can be used it can be a very powerful mechanism for demonstrating how the people within an organization can work together and truly make a difference. By pulling team members off production lines to work on the kaizen blitz project the senior leadership demonstrates its commitment to change in a highly visible and tangible way. As previously discussed, kaikaku is an exceptional way to kick-start change, particularly in the early stages of a lean implementation. It can also be applied to the resolution of a long-standing problem, the quick reengineering of a process, to breathe new energy into a change program that has stalled, and finally to quickly reap the benefits of solving an organization's easy and obvious problems (i.e., harvesting the low-hanging fruit).

11.9 SUMMARY OF KEY IDEAS

Key Idea: *The principle objective of lean production is responsiveness.*

The lean philosophy is illustrated by the house of the Toyota production system that depicts three components of responsiveness: high quality, low cost, and short lead times. Achieving these goals requires the implementation of two core principles (pillars)—just-in-time and jidoka. Just-in-time means delivering exactly what is needed, exactly when needed, and in the quantity needed. To do this requires production flow, leveled production, and matching line capacity to customer demand.

Jidoka is a philosophy that makes quality the responsibility of the front-line worker, thereby ensuring quality at the source. To be successful the two pillars must be implemented in an environment that is stable and consistent and where there is a commitment to continuous improvement.

Key Idea: *Jidoka is a concept that ensures quality at the source in a lean production environment.*

Jidoka, also called autonomation, is one of the principal pillars of the Toyota production system and is the policy that requires operators to stop a production line anytime a problem is observed. This practice ensures that quality problems are corrected before they result in defective product. Modified versions of jidoka involve the use of andon three-light systems. The andon visual controls make it possible to summon help when a problem is suspected but relieves the individual operators of the ultimate responsibility to halt production.

Key Idea: *Excessive inventories represent a significant source of muda.*

Four types of inventories can be found in most factories: raw stocks; maintenance, repair, and operating (MRO) supplies; work-in-process (WIP); and finished goods. Effective supply chain management practices can control the levels of the raw stocks, MRO supplies, and finished goods. Levels of WIP must be carefully managed internally to avoid waste due to the unnecessary investment of working capital or the creation of manufacturing inefficiencies.

Key Idea: *Little's law helps explain why excessive work-in-process inventories can significantly slow down a factory.*

Little's law states that flow time (the time a job remains in the value stream) is equal to the product of the level of work-in-process in the factory and the manufacturing cycle time (the rate at which output units exit the value stream).

Key Idea: *Pull scheduling is a lean production tool that controls materials flow by matching production to demand, resulting in improved responsiveness through production efficiencies, shorter cycle times, reduced inventory, and improved quality.*

Pull scheduling is an internal requisitioning procedure that triggers order releases from the end of the production system back toward the beginning. A communication system, called kanban, is used by downstream workstations to signal the materials needs from internal upstream suppliers. Work is performed only as needed and in small transfer batch quantities that typically can be manually moved between workstations by operators as required. Consequently, there is no need for large stockpiles of inventory between workstations and, since the entire order quantity does not move from workstation to workstation, cycle times can be dramatically reduced. There are numerous benefits to pull scheduling including improvements in cost, quality, and customer response.

Key Idea: *A kanban system is a shop-floor control procedure that supports pull scheduling by using a system of simple signaling devices to trigger work activity up the value stream.*

In a kanban system, the flow of materials is controlled using a method by which workstations can exchange signals when one of them is about to run out of material. By *pulling* rather than *pushing* material through the plant, work is produced only as needed. Kanban containers, of a size that can be manually transported between workstations, permit small quantities of materials to be transferred to the next operation as soon as a container is filled. The most common type of kanban system uses cards as the triggering medium. When sequential workstations or cells are in close proximity, a one-card system can be used. When the workstations are located some distance apart, the use of a two-card system is required.

Key Idea: *When implementing lean thinking the primary objective is to design the system so that a smooth and continuous flow is created.*

Flow is defined as the smooth progressive movement of materials, products, information, and people through a sequence of activities without interruption and without the creation of waste. The ideal is to create *one-piece flow* where one work piece at a time is transferred smoothly and continuously between the successive workstations within a cell. There are three requirements for flow creation: the elimination of waste, the elimination of defects, and the leveling of production. Where one-piece flow is not feasible or economical, process flow should be designed around the smallest practicable transfer batch.

Key Idea: *The Glenday sieve is a tool that can help focus on those products that should be introduced into a lean production cycle.*

The Glenday sieve uses the Pareto principle to sort products by production volume. Those with the highest volume become the focus for lean implementations. The idea is that it is easier to design systems that are capable of creating flow for 3% to 8% of the total product mix than it is to try and create flow for the entire mix.

Key Idea: *In lean systems a water spider (mizusumashi) position can be invaluable in maintaining production flow.*

In a lean production system a water spider (*mizusumashi*) is a person who is responsible for maintaining a smooth, uninterrupted flow by taking care of those tasks that do not occur every cycle, such as replenishing materials, removing filled containers, and bringing in or setting up empty cartons.

Key Idea: *Lean service is the application of lean thinking in service environments and fundamentally employs the same tools that are applied to the production of goods.*

The primary value stream in any lean service organization consists of five major components: inputs, design, make/serve, support, and consume. Numerous other supporting value streams exist to provide those resources necessary to the efficient and effective completion of the primary value stream tasks—in areas such as engineering and maintenance, finance, human resources, logistics, research and development, marketing, materials control, and the various dimensions of planning (e.g., capacity, strategic, improvement). The lean philosophy evaluates the work that people do and eliminates wasteful non-value-adding activities.

Key Idea: *Lean consumption is concerned with eliminating the waste, hassle, and stress from consumers during the process of acquiring and using the products and services they need.*

We live in a consumptive throw-away society that is largely influenced by advertising and marketing propaganda. Consumers are bombarded with information trying to entice them to buy things that they do not need at all, or in advance of when they need them, in quantities in excess of what is needed, or to be willing to settle for something less than what is required. All these issues create waste and some of the underlying assumptions concerning consumer behavior and the consumption cycle are worthy of challenge. *Lean consumption* is a term used to describe the application of lean thinking to the elimination or reduction of much of the wasted effort, time, money, and frustrations suffered by consumers trying to acquire the things they need to support the quality of life they have chosen.

Key Idea: *Visual controls enhance quality at the source by facilitating shop-floor communications.*

Visual controls make it possible for anyone to instantaneously assess whether a process is doing well or needs attention. These controls are classified as actual or analog, and are either displays or controls, depending on whether the device is a physical part of the work design or designed to trigger human responses based on the physical senses; and whether the device is designed simply to convey information or to invoke operator intervention.

Key Idea: *Poka-yoke is a lean production technique that plays an essential role together with visual controls in preventing human errors and providing the stimulus for human interactions.*

Poka-yoke devices are designed to prevent a worker from making an error that would lead to a product defect, or to provide quick feedback to operators of any abnormalities in a process in time for corrective action to be taken before defects are produced.

Key Idea: *Balancing a line to its takt time minimizes waste and maximizes line utilization.*

The takt time is the short cycle time required to meet customer demand. Balancing the line to this time and using a pull scheduling system minimizes the need for inventory buffers between workstations and eliminates inefficiencies created by excess capacity.

Key Idea: *Autonomation (jidoka) is the policy under which operators are required to stop production anytime they spot a problem.*

The idea with autonomation is to empower people to halt production and deal with problems as they occur. This requires tracking down the root cause and solving the problem. To be effective, autonomation requires proper training and cultural change.

Key Idea: *Heijunka is a technique used to absorb fluctuations in demand without having to increase inventory buffers between workstations.*

Heijunka is a production-smoothing tool for a mixed model cell so that materials can move smoothly between workstations with low levels of inventories. The technique generates sequences of small and even batches of each model over short production intervals. A repetitive production sequence is established for each interval that produces small kanban batches of each member of a product family.

Key Idea: *Changeover reduction is a key factor in achieving manufacturing flexibility and being able to produce in small batch quantities.*

Changeover reduction requires a different way of thinking. By grouping the changeover activities into two groupings—internal and external—much of the work for setting up the next job can be done while the preceding job is still running. Dramatic improvements also require a change in attitude. There are three prongs to an effective changeover reduction strategy—converting internal activities into external ones, reducing the time required to perform external activities, and developing a pit-stop mentality with an emphasis on improvement.

Key Idea: *When a product family shares a common production routing and the variances for all products within the family are equal, the combined output for the family can be combined on a DNOM or DTAR chart.*

A product family is defined as a collection of products or services that require common tasks. When it can be assumed that the process variance for all part numbers within the family are equal, the combined output for the product mix can be combined on one pair of charts by transforming the raw measurements into a set of numbers that represent how far the measurements deviate from either a nominal (DNOM) or target (DTAR) dimension. Since the variances across products are equal, sample ranges are unaffected by the transformation, and the range chart can be used to estimate the common process standard deviation. Typically a target value is used in lieu of a nominal dimension in the case of one-sided specifications.

Key Idea: *A stabilized control chart can be used to monitor the behavior of a process family where the process variances are unknown and unequal for all products belonging to the family.*

With a stabilized chart the raw data is transformed to a statistic, called the relative range, which permits the use of constant control limits on the range chart. The relative range is the computed ratio of each sample range to the cumulative average range for all samples within a product family. For the $\overline{X}$ chart the data are transformed into a statistic that measures the ratio of the difference between each sample average and the cumulative average for all samples within the family and the cumulative average range for all samples across the family. The centerline on stabilized $\overline{X}$ and R charts in control will be around one and zero, respectively, with no nonrandom patterns or out-of-control points.

Key Idea: *When the process variances are unequal but known for a product family a Z chart can be used to monitor process behavior.*

When the process variances are known standard Q statistics can be computed by subtracting each part's critical dimension from its nominal dimension and subsequently dividing the difference by the process standard deviation. The Q statistics are combined on a single pair of $\overline{X}$ and R charts, called Z charts, where the range is the variation in Q values within each sample and the $\overline{X}$s represent the between-sample variation of computed Q values.

Key Idea: *Human error is inevitable but defects can be prevented.*

Even though errors are to some extent inevitable, defects can be prevented with proper feedback that can lead to immediate corrective action soon after the error is made. To achieve this some form of 100% inspection is required, employing some combination of successive checks, self-checks, and source inspection.

Key Idea: *A kaizen blitz is an intensive short-term kaikaku event aimed at creating a sudden and radical change in a production process or system.*

The typical duration for a kaizen blitz project is one week and the pace is fast and furious. Projects are conducted by teams and the process is data driven, utilizing many of the planning and analysis tools of SPC and lean production. Due to the intensity and urgency created by the short time frames, the process can successfully overcome internal resistance to change, inertia, and politics. The pressure helps teams coalesce by creating unity, teamwork, enthusiasm, and pride.

11.10 MIND EXPANDERS TO TEST YOUR UNDERSTANDING

11-1. A family of four parts is produced by a manufacturing cell. The parts are scheduled in small batch quantities in sequence. Three parts are selected at random from each batch and measured for a critical characteristic. The sampling results for six production cycles are shown in the following table. The production personnel assume that the process variance is the same for all products within the family.

		Production Cycle					
Product	**Nominal**	**1**	**2**	**3**	**4**	**5**	**6**
A	15	15.024	15.125	15.009	14.728	14.831	15.176
		15.138	14.807	15.224	14.925	15.234	14.836
		14.774	14.964	14.857	15.069	15.160	14.823
B	18	18.374	17.938	17.671	17.977	18.134	17.885
		18.148	17.863	17.756	17.961	18.081	17.857
		17.720	17.807	18.120	17.719	17.956	17.743
C	20	19.943	20.097	19.935	19.634	19.653	20.360
		19.849	19.587	19.618	20.036	19.951	20.150
		20.158	20.118	20.206	19.740	20.291	19.846
D	21	21.273	20.938	20.984	21.147	21.011	20.836
		21.197	21.190	20.950	20.799	20.941	20.640
		20.828	20.560	21.059	20.979	20.720	21.113

a. Construct a DNOM $\overline{X}$ and R chart for this product family and assess the state of statistical control.

b. Apply Equation 11-7 to show the validity of the assumption that the process variance is equal for all products.

11-2. The factory described in problem 11-1 produces a second family of four parts in a manufacturing cell that processes the parts in small batch quantities in sequence. Three parts are selected at random from each batch and measured for a critical characteristic. The sampling results for six production cycles are shown in the following table. It is known that the process variance is not the same for all products within the family.

	Production Cycle					
Product	**1**	**2**	**3**	**4**	**5**	**6**
Q	15.029	15.146	15.232	15.535	14.884	15.129
	14.808	15.000	14.929	14.968	15.240	15.140
	15.138	15.119	15.009	15.184	15.064	15.076
R	20.059	19.940	20.324	20.157	19.621	20.065
	20.257	20.273	20.042	20.654	19.765	20.379
	19.717	20.282	20.089	20.561	20.134	19.693
S	23.892	24.408	24.214	23.688	24.152	25.125
	23.352	23.191	23.484	25.498	23.715	23.285
	23.883	24.120	23.965	23.643	23.629	24.080
T	30.265	28.523	30.007	28.090	30.217	29.037
	29.871	28.866	29.495	29.386	28.398	31.110
	30.008	29.525	28.332	30.248	29.533	31.092

a. Apply Equation 11-7 to these data to verify that the process variances cannot be assumed to be the same for all product numbers.

b. Construct a pair of stabilized control charts and determine the control status of this process.

c. The control charts are continued for an additional six cycles. The data are summarized in the table below. Has the control status for this process changed? Explain.

	Production Cycle					
Product	**7**	**8**	**9**	**10**	**11**	**12**
Q	15.178	14.998	14.717	14.925	15.105	14.913
	15.099	15.355	14.877	15.048	15.310	15.226
	14.985	15.079	15.127	15.049	14.855	15.017
R	19.300	20.048	19.634	20.232	19.746	20.695
	20.533	20.006	19.324	20.141	19.705	20.113
	19.913	20.125	19.285	20.411	19.834	20.345
S	24.351	24.177	23.932	23.764	23.315	24.091
	24.262	23.256	23.127	24.564	23.929	22.941
	23.057	23.473	24.347	24.423	22.498	23.859
T	30.686	31.109	29.700	30.221	30.348	28.993
	31.348	29.119	31.175	30.744	30.537	30.458
	29.648	29.854	30.980	33.055	29.891	29.794

11-3. Consider the data collected in the first six production cycles of the process described in problem 11-2. Assume that the standard deviations for each of the four product models are known and the targets and standard deviations are as follows.

Model	Target	Standard Deviation
Q	15	0.2
R	20	0.4
S	24	0.6
T	30	1.0

a. For each observation compute the Q statistics and construct a pair of Z charts. Interpret the control status of this process.

b. Using the charts you constructed in part a, add the data for the next 6 cycles shown in problem 11-2c. Has the control status changed? Explain.

c. How do your conclusions in parts a and b compare with the conclusions you reached for this process in problem 11-2b and 11-2c? Can you explain any inconsistencies?

11-4. A consumer products manufacturer operates a filling line that places liquid product in containers. Three different products are filled on this line and the fill specifications are as follows.

Product	Minimum Fill (ozs)
Blue Ozone	8
Peaceful Abode	10
Perk-up	12

Data for some recent runs of these products are shown in the following table. From experience the production supervisor believes that the variation in fills is the same for all three products.

a. Using the minimum specified fill quantities as targets, construct a DTAR control chart and determine the control status of this process. From this analysis, what are your preliminary conclusions?

Model	Sample	Observation 1	2	3	4
Blue Ozone	1	8.066	7.864	7.942	8.078
	2	8.219	8.041	8.067	8.057
	3	7.929	7.965	7.943	8.099
	4	7.885	7.762	7.911	7.776
	5	7.877	7.952	8.058	7.987
	6	8.094	8.068	8.080	8.042
	7	8.009	8.065	7.908	7.984
	8	7.999	7.900	7.947	7.912
	9	8.021	8.100	8.006	7.938
	10	7.946	7.998	8.181	8.080
Peaceful Abode	1	9.979	10.035	10.078	10.039
	2	9.990	9.977	10.138	10.052
	3	10.101	10.005	9.789	9.760
	4	9.939	9.939	9.937	10.036
	5	10.147	9.947	9.902	10.318
	6	9.967	10.175	10.009	10.152
	7	9.906	9.958	9.983	9.913
	8	9.909	9.950	9.993	9.981
	9	9.936	10.044	10.031	9.934
	10	9.888	9.906	9.866	9.902
Perk-up	1	12.309	12.121	12.328	12.231
	2	12.145	12.338	12.366	12.065
	3	12.217	12.462	12.276	12.213
	4	12.259	12.206	12.246	12.092
	5	12.210	12.218	12.168	12.110
	6	12.204	12.362	12.180	12.176
	7	12.117	12.031	12.205	12.201
	8	12.181	12.476	12.093	12.152
	9	12.211	12.138	12.141	12.134
	10	12.260	12.144	12.237	12.103

b. Using the charts constructed in part a comment on whether you think that the production supervisor was correct in assuming that process variability is the same for all three products.

c. Assuming a normal distribution of fills, how much average overfill is required to ensure that no more than approximately one-tenth of one percent of all containers are filled at below the stipulated minimum?

d. The production operators make some adjustments on the line to try and avoid underfills while stabilizing the process. After the adjustments have been made the data summarized in the following table are collected. Have the adjustments worked? Is the process in statistical control? Explain.

Model	Sample	Observation after Adjustment 1	2	3	4
Blue Ozone	1	8.154	8.429	8.423	8.343
	2	8.374	8.363	8.260	8.373
	3	8.393	8.348	8.226	8.359
	4	8.142	8.252	8.297	8.416
	5	8.341	8.360	8.442	8.357
	6	8.405	8.349	8.343	8.369
	7	8.247	8.261	8.285	8.308
	8	8.274	8.372	8.248	8.364
	9	8.549	8.284	8.153	8.252
	10	8.204	8.279	8.309	8.274
Peaceful Abode	1	10.207	9.960	10.246	10.312
	2	10.239	10.201	10.427	10.239
	3	10.424	10.525	10.253	10.382
	4	10.256	10.277	10.168	10.168
	5	10.349	10.290	10.313	10.293
	6	10.248	10.456	10.266	10.396
	7	10.305	10.379	10.238	10.115
	8	10.341	10.349	10.429	10.328
	9	10.190	10.217	10.425	10.324
	10	10.308	10.190	10.344	10.442
Perk-up	1	12.391	12.315	12.331	12.268
	2	12.443	12.495	12.464	12.236
	3	12.353	12.304	12.245	12.305
	4	12.246	12.524	12.203	12.174
	5	12.538	12.304	12.141	12.367
	6	12.367	12.150	12.372	12.274
	7	12.132	12.138	12.291	12.426
	8	12.205	12.442	12.422	12.597
	9	12.307	12.216	12.406	12.168
	10	12.101	12.274	12.213	12.262

e. Estimate the average overfill for each product after the adjustments described in part d.

f. Assuming a normal distribution, estimate the percentage of each of the three products that will be underfilled after the adjustments described in part d.

11-5. ABC Ltd. manufactures three parts: X, Y, and Z. Orders are on hand to deliver a total of 45,000 parts to various customers over the next four-week period. The part number breakdown is 4,500 of X; 22,500 of Y; and 18,000 of Z. The plant producing these parts works two eight-hour shifts (with a 10% allowance for downtime), five days per week.

a. Design a heijunka production sequence for these parts, assuming cellular manufacturing and a scheduling unit of one shift.
b. What is the takt time for this process?
c. What is the length of a production cycle for this process?
d. How many times will this cycle be repeated during a shift?

11-6. A certain customer has placed an order for 50,000 gizmos that need to be produced during the next five days. If the factory operates a single eight-hour shift, what is the takt time for this order, allowing 10% nonproductive time for breaks and machine downtime?

11-7. A customer has placed an order for 30,000 first-aid kits that must be delivered within 20 business days. The kits are to be assembled on the four-workstation line shown below and in kanban quantities equal to 1 hour of demand. The workstations are manned a total of 7.5 hours per day.

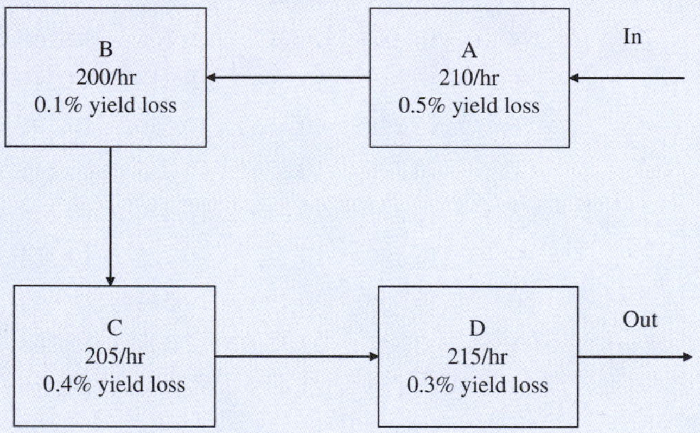

a. What is the takt time for this customer's order?
b. How many kits must be started at workstation A each hour in order to deliver the required quantities from workstation D? What is the overall process yield for the line?
c. What is the bottleneck operation?
d. What is the overall line utilization?

11-8. Is lean production a requirement for achieving third-generation quality? Can lean production be implemented in a push scheduling environment?

11-9. Why is changeover reduction important to the success of a heijunka procedure? What role should changeover time play in the sequencing of different product models on a mixed model line?

11-10. A manufacturer of small residential dehumidifiers has been receiving numerous customer complaints concerning quality problems and backorders. Apparently, the demand for dehumidifiers, which is seasonal, is difficult to accurately forecast, and when customers run low on inventories and reorder, the lead times are so long that the season may be over before replenishment stocks arrive. Quality is also a problem and many units have been returned from consumers for warranty repairs. A Pareto analysis has revealed that the two most frequently occurring problems are circuit board failures and compressor seizures. When these problems occur consumers are unhappy because their units are out of service for repairs during most of the season.

a. What would be the advantages to forming a kaizen blitz team to investigate these problems?
b. Who should be on the team?
c. If you were on the team, how would you suggest the team proceed, given that improvement recommendations are due in one week?
d. Do you think that there is any difference between the characteristics of a kaizen blitz team and any other team? Explain.

11-11. The Magellan Machine Shop produces a certain machined part on its milling machine at the rate of one every 25 seconds. This is a continuous operation that produces parts seven hours per day five days per week. The milling machine is fed parts from an upstream grinding operation. A physical count has revealed that, on the average, there are a total of 1,700 parts ahead of the milling operation waiting to be processed. Use Little's law to determine how long it takes for a part to be milled after it exits the grinding operation.

11-12. A company that produces custom cabinetry is trying to improve customer responsiveness. Currently the company can promise its customers only a five-week delivery as it takes that long on the average for an order to be processed by all of the necessary work stations in the factory. Assuming that the production rate remains unchanged, by what percentage would the company have to reduce its WIP in order to reduce its lead time from five weeks to two weeks?

11-13. Why do you think that lean production stresses the elimination of WIP? Under what conditions is WIP advantageous? How can its levels be controlled? How can Little's formula help?

11-14. Visual controls, such as andon and kanban, are an important part of lean production strategies. How do such controls help achieve the objectives of Lean Six Sigma? How does the reduction in WIP facilitate the use of visual controls?

11-15. The demand for a certain customer is 50,000 gizmos to be produced during the next five days. If the factory operates a single eight-hour shift, what is the takt time for this order, allowing 10% nonproductive time for breaks and machine downtime?

CHAPTER TWELVE

CONTROL OF PROCESS IN SERVICE INDUSTRIES

QUALITY MAXIM #12: *Delivering a quality service requires sincerity, integrity, and empathy—attributes that cannot be directly measured.*

To Serve with Love—Love to Serve!

"To Sir, with Love"[1] was the title song of a film that chronicled the experiences of Mark Thackeray, an unemployed engineer who secured a teaching position in a London school as a temporary measure while he continued to seek employment in his chosen field. He was quick to learn that his pupils were mainly undisciplined rejects from other schools, coming mostly from difficult home environments, and that their unruliness had led to the resignation of their former teacher. Other teachers at the school seemed not to care and their feelings ranged from outright contempt to fear. Mr. Thackeray's unconventional approach was to replace the school's set curriculum with life lessons—teaching refinement, etiquette, and mutual respect. His consistent and genuine concern for his students as people eventually won over their hearts. Ultimately, when Mr. Thackery was offered a lucrative engineering job, he opted to remain in his new profession of teaching. In appreciation for all he had done for them the students proclaim in song that they are willing to do anything to repay him, including trying to give him the moon or writing across the sky in thousand-foot-high letters.[2] The song lyrics express a key quality principle that especially applies to pure services. The word *serve* is a proactive verb—a dare to the provider who is committed to quality to genuinely seek ways to go beyond what is minimally expected to pleasantly surprise customers—"you want the sky—I will try to give it to you!" Evaluating the merits of a service experience is not easy as it is based on a subjective perception that has to be measured on a customer-by-customer basis. It is therefore imperative that service providers master people skills and be trained to project a sincere interest in discovering and satisfying customer needs on an individual basis. This does not imply that customers are always right or are never demanding or difficult, and at times a provider may find it necessary to trade off the needs of one customer for the sake of another. Nevertheless, just as the key to Mr. Thackeray's success was a genuine caring for the welfare of his students, it has been shown that the perception of service quality is higher when customers believe that the service provider deeply cares about their needs. As with tangible goods, service quality is about value and service value can be ensured or diminished at the instantaneous contact a customer has with service provider personnel—a phenomenon called the *moment of truth*. Service quality is also about trust. Customers need to believe that the provider will deliver on its promises, is committed to the principles of excellence, and will do whatever it takes to flexibly adapt the system to address the uniqueness of their requirements. This is the true spirit of service.

---

[1]"To Sir, with Love," written by Don Black and Marc London, was performed by Lulu as the title song of a 1967 British film of the same name, produced and directed by James Clavell and starring Sidney Poitier. The film was based on a novel by E. R. Braithwaite. The screenplay was written by Clavell.
[2]The music and lyrics for "To Sir, with Love" can be accessed at http://fr.truveo.com/To-Sir-With-Love-Lulu-with-Lyrics/id/1189490965.

LEARNING OBJECTIVES AND KEY OUTCOMES

After careful study of the material in Chapter 12, you should be able to:

1. **Demonstrate an understanding of the growing importance of the service sector on the U.S. economy.**
2. **Describe how service quality differs from product quality.**
3. **Describe and demonstrate an understanding of the gap model of service quality.**
4. **Demonstrate an understanding of the importance of internal communications and define some of the tools that are available for improving an internal communications system.**
5. **Discuss the various criteria customers use to determine their individual levels of customer satisfaction.**
6. **Discuss the five behavioral science principles that can be applied to help shape customer attitudes toward service quality.**
7. **Define and discuss various tools and instruments that can be used to measure customer satisfaction.**

(continued)

8. Demonstrate an understanding of the principles of good survey questionnaire design.
9. Discuss some of the possible pitfalls when designing a customer satisfaction survey.
10. Discuss the importance of the design and structure of survey questions and define some common mistakes that, if not avoided, can lead to unintended bias in survey responses.
11. Describe the SERVQUAL methodology and how it differs from other survey approaches for gathering information on customer satisfaction.
12. Construct and interpret control charts for service performance data, including the application of moving ranges to capture systemic time-related variability and the use of transformations to handle nonnormal data.
13. Check raw data for the presence of autocorrelation; if found, fit an appropriate time series model to the data, and use the model to monitor and control the process.

12.1 HOW SERVICE QUALITY DIFFERS FROM PRODUCT QUALITY

12.1.1 The Growing Importance of the Service Sector on the U.S. Economy

Since the 1980s the U.S. economy has experienced a shift from the production of goods to service-providing industries. A recent report by the U.S. Bureau of Labor Statistics predicts that during the decade 2004–2014 service jobs will account for 18.7 million of the 18.9 million new jobs created.[3] Thirty percent of these are projected to be in health care, social assistance, or the private education sector. The next fastest growing quality sector will be professional and business services, then leisure and hospitality, followed by information technologies. By contrast, goods production is expected to decline by 0.4% during the same period, with job losses in manufacturing, agriculture, and mining. The only goods sector projecting an increase is construction, which is expected to grow by 11.4%.

12.1.2 The Nature of Service Quality

As we saw in Chapter 1, quality has its roots in the production of goods, but the changing orientation to service provides the impetus for adapting to service applications those theories and techniques that have worked so well for manufacturing. While service and manufacturing operations are similar in some ways—for example, both employ processes that are designed to satisfy customer needs—there are some important distinctions. These include the nature of outputs, the underlying production processes, and the interaction between the producer (provider) and the customer (consumer). Unlike most products, services are intangible, inseparable, variable, and perishable.

The U.S. economy is experiencing a shift from the production of goods to the provision of services.

As an intangible a service cannot be seen, tasted, felt, heard, or smelled before it is purchased. When the service is completed there is usually no tangible object to serve as evidence that the transaction took place. The customer carries away feelings and perceptions and the major outcome is an experience. If quality is to be measured a subjective assessment must be used to determine whether the experience was a pleasant or an unpleasant one. Feelings on the positive side can be elation, delight, or satisfaction; on the negative side anger, disappointment, or dissatisfaction.

Services tend to be labor intensive and inseparable from the service provider. As the sale takes place the service is typically simultaneously produced and consumed. Unlike most manufactured goods, the ultimate consumer comes face-to-face with the producer who often participates to some degree in both the provision and the consumption of the service. At this *moment-of-truth* contact, the provider is well positioned to obtain immediate firsthand feedback on customer satisfaction.

Window shopping is a favorite pastime of many and can shape customer expectations. Quality is defined as how well customers perceive those expectations have been met.

[3]http://www.bis.gov/oco/oco2003.htm.

The natural variability of all processes has been emphasized throughout this text. However, while the omnipresence of variability acts as a corrupting influence in the quality of manufactured goods, variability presents an even greater threat in the service sector where it is much harder to control. In the case of goods, the source of production is typically centralized with a relatively few (often a single) manufacturing facilities. Within the production system variability can arise as a result of multiple lines, shifts, and operators; however, training is also centralized and production methods are easy to monitor and straightforward to control. Services, on the other hand, are usually decentralized due to the need for service providers to locate close to customers and the point of consumption.

Services are intangible, inseparable, variable, and perishable.

Customers may be directly involved in the provision of a service.

Service quality depends on who provides the service as well as when, where, and how service is delivered. Customers are often directly involved and the perceived level of customer service depends on the rapport that is instantaneously created between provider and consumer. In addition, customer expectations also tend to be more variable in the service sector since customers have a direct one-on-one transactional experience with providers. Hence, a company can have thousands of "moments of truth" in a single day, distributed from Maine to California through hundreds of service outlets.

Each service transaction is a moment of truth.

A service cannot be inventoried, purchased in quantity lot sizes, or backordered. A customer cannot be protected against receiving a defective service through effective sampling inspection. An unsatisfactory service cannot be restored to a satisfactory state through repair or rework. Perhaps the most important distinction is the inexact nature of defining and measuring the quality of a service.

12.1.3 The Gap Model of Service Quality

As with products, in a service environment customers have certain expectations. The quality of service is defined by how each individual customer perceives that the service actually received measures up to what was expected. If the quality does not meet expectations the difference is called a **gap**, and the assessment of quality can be based on measuring the width of this gap.[4]

12.1.4 Service Quality Gaps

Whenever the service that is actually delivered differs from what was required, that difference constitutes a *service quality gap*. Quality gaps can derive from seven sources, as illustrated in the service quality gap model[5] shown in Figure 12-1. Each of these gaps is further discussed below.

Gap 1

The difference between customer expectations and management's perception of customer expectations.

Gap 1 can occur as a result of either the failure to understand customers and markets or poor communications. The former can be caused by inadequate market scanning, market research, or quality function deployment. The latter can be caused by poor vertical communications, too many layers of management, or the lack of a single value stream owner.

It is important to translate customer expectations into service quality specifications that can be widely understood.

Gap 2

The difference between management's perception of customer expectations and the translation of those expectations into service quality specifications.

Gap 2 can occur if management does not have the commitment to do what it takes to place the customer first, if cost considerations become more important than the creation of customer value, if there is no standardized continuous improvement program, or if quality does not play a key role in strategic planning.

Customer expectations and what management thinks customers expect should be aligned.

Gap 3

The difference between service quality specifications and the quality of service delivered.

Gap 3 is a process issue as it relates to the capability of the service delivery system to meet the stipulated requirements. If system outputs fall short of requirements it can be due to inadequacies in training, staffing, technology deployment, management controls, supportive culture, teamwork, or motivation.

The quality of service delivered should meet specifications.

Gap 4

The difference between the service quality delivered and what is communicated externally to customers.

Gap 4 can be caused by poor internal communications between those actually providing services and those communicating and/or making promises to customers.

Communications with customers should be honest and realistic based on capabilities and actual performance.

[4]Parasuraman, A., V. A. Zeithaml, and L. L. Berry. "SERVQUAL—A Multiple-Item Scale for Measuring Consumer Perceptions of Service Quality." *Journal of Retailing* 64, no. 1 (Spring 1988): 12–40.

[5]Shahin, A. "SERVQUAL and Service Quality Gaps: A Framework for Determining and Prioritizing Critical Factors in Delivering Service Quality." http://www.qmconf.com/Docs/0077.pdf.

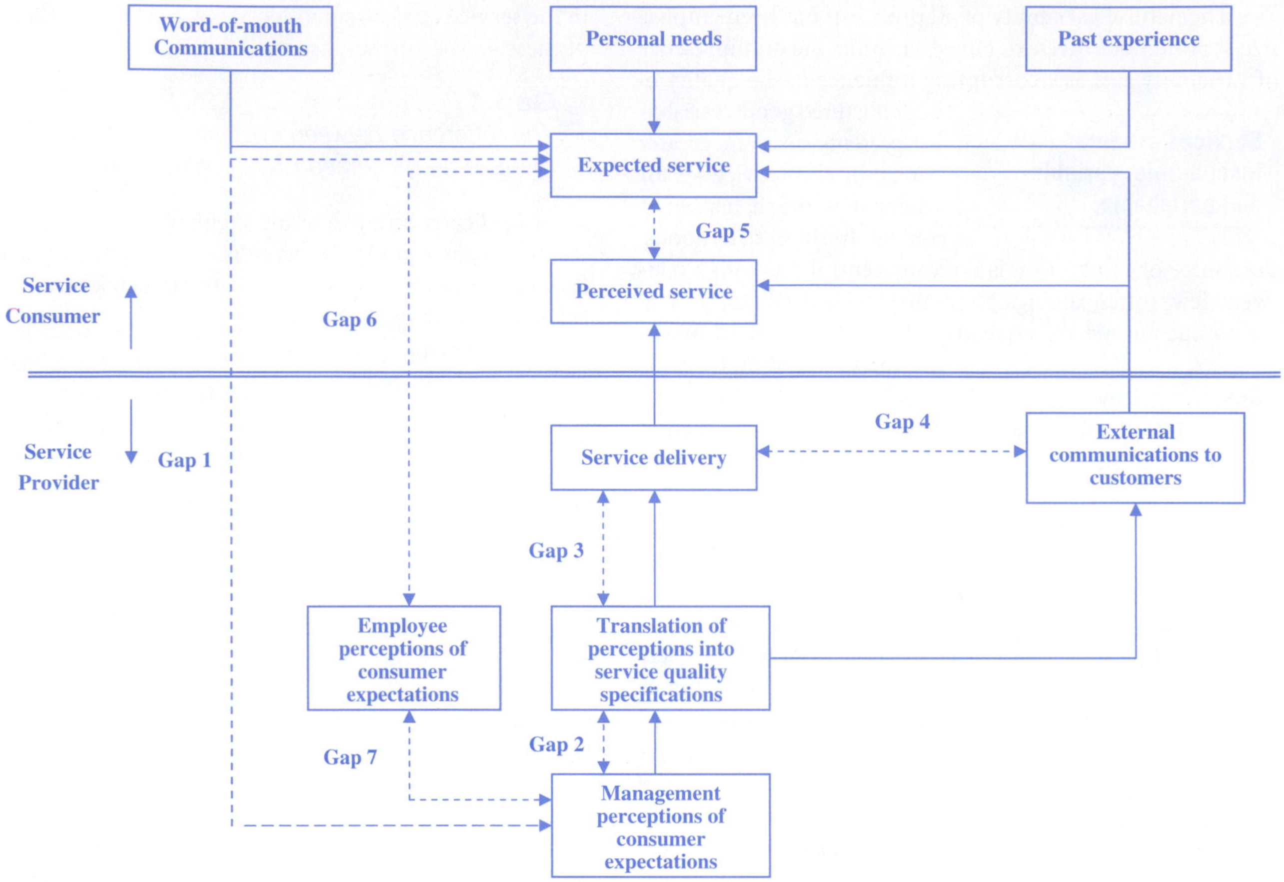

FIGURE 12-1. Model of Service Quality Gaps

Source: Luk, S. T. K., and R. Layton. "Perception Gaps in Customer Expectations: Managers Versus Service Providers and Customers." *Service Industries Journal* 22, no. 2 (2002). Reprinted by permission of Taylor and Francis.

There also may be the propensity to make unrealistic promises to customers in order to secure a sale.

Gap 5

The difference between customer expectations and customer perceptions of the service quality delivered.

Gap 5 is a measure of customer satisfaction—that is, how well customers perceive that the provider has fulfilled its promises and lived up to the customers' expectations. Closing gap 5 is the ultimate challenge of the service provider, as customer expectations are continually changing. As the model in Figure 12-1 shows, customer expectations are influenced by word-of-mouth communications, individual personal needs and attitudes, and past experience. Customer expectations are individual value judgments, and managing gap 5 requires meeting each customer on a personal level.

The quality of service delivered should meet customer expectations.

Gap 6

The difference between customer expectations and the perception of customer expectations by front-line service providers.

Gap 6 is an internal communications and training issue. Employees who interact face-to-face with customers must have thorough knowledge of service specifications and be carefully trained on how to serve all customers. It is important that service specifications be satisfied completely and consistently across the service network by all front-line providers.

Front-line service providers need to fully understand customer expectations.

Gap 7

The difference between employees' perceptions of customer expectations and management's perceptions of customer expectations.

Gap 7 is the failure to align management perceptions with those of front-line service providers, and reflects a breakdown in top-down communications or training.

Front-line providers' and management's understanding of customer expectations should be aligned.

The model depicted in Figure 12-1 shows that only gap 5 resides entirely with the customer or consumer. The other six gaps are either internal within the service provider's system, or at the interface between the provider and the customer. These six gaps (gap 1, 2, 3, 4, 6, and 7) are therefore consequences of the way in which services are delivered. Only gap 5 pertains solely to the customer and can therefore be regarded as the true measure of service quality.

12.1.5 The Importance and Improvement of Internal Communications

Gaps 1, 4, 6, and 7 highlight the important role that **internal communications** play in the delivery of service quality. The open systems model presented in Chapter 2 portrays communications as the fabric that ties all system components together and facilitates synergy and achievement of purpose. Many options are available to the organization that values open disclosure and free and continual dialogue with its stakeholders. These include

- Use of **team meetings** to make announcements, share business news, and solicit anecdotal feedback on experiences with customers
- Use of **storyboard displays** for posting important written communications (e.g., memos, letters, reports, etc.) for highlighting projects (including objectives, methodology, and progress), and for displaying written customer feedback (e.g., letters of appreciation or dissatisfaction, results of customer satisfaction surveys, etc.)
- Use of **company newsletters** to convey information concerning corporate strategy, business decisions, customer profiles and requirements, new technologies, and quality goals and initiatives
- Use of **bulletin boards** to display values, goals, and continuous improvement achievements
- Use of **customer quality award** schemes in goal setting and to clarify customer service quality expectations
- Use of **management staff meetings** to share insights and information and achieve consistency on quality perceptions and goals

12.2 CUSTOMER SATISFACTION IN THE SERVICE SECTOR

Gap 5 in the gap model of service quality reflects the degree with which customers are dissatisfied with the service delivered. A large gap implies a very unhappy customer, a small gap is indicative of a moderate to slightly dissatisfied customer, a gap of zero is consistent with a satisfied customer, and a negative gap means that a customer is delighted with a level of service that has clearly exceeded expectations. Determining the size of gap 5 requires a reliable measure of customer satisfaction which can be difficult to achieve. A starting point is to try and identify the customer satisfaction components.

12.2.1 Criteria for Customer Satisfaction

A 2001 study of 150 individuals in the United Kingdom concluded that customer satisfaction is based on four principal criteria.[6]

1. Delivering what was promised—customers **trust** the service provider to provide quality.
2. Providing a **personal touch**—customers believe the provider cares about them and their needs.
3. Going the extra mile and providing the unexpected—customers believe the provider will do what it takes to earn their **loyalty** and trust.
4. Effectively handling problems and queries—customers believe the provider is committed to **excellence** and will do what it takes to understand them and adapt to their changing needs.

A more recent survey of 115 bank customers investigated how customers perceive service according to five criteria—quality, satisfaction, value, attitude, and performance.[7] The respondents of this study corroborated the 2001 findings and were also consistent with the literature[8] which has argued that the following six principal elements interact to influence a customer's perception of service quality.

1. **Accessibility and flexibility**—Customers favor services that can be easily accessed and can be adjusted and adapted to fit their particular and unique needs.
2. **Reputation, credibility, and the establishment of personal and caring relationships**—Customers desire a service provider who genuinely listens to their needs, whom they can trust, who can create a relaxed environment that treats the customer as a person and focuses squarely on the importance of meeting each customer's needs, and who is consistently courteous.
3. **Professionalism and skills**—Customers desire a service provider who knows the job and has the technical competency to perform it in a professional manner.
4. **Recovery and efficacy of results**—Customers desire a service experience that is speedy, flexible (when required), produces a positive outcome for them, and is delivered by an individual who has the will and authority to take immediate action to correct anything that goes wrong.
5. **Reliability, security, and trustworthiness**—Customers desire services for which they feel safe and secure, provided by someone whom they can trust to make good on all that has been promised.

---

[6]Johnston, Robert. "Towards a Better Understanding of Service Excellence." *Managing Service Quality* 14, no. 2/3 (2004): 129–33.

[7]Dano, Florence, Sylvie Llosa, and Chiara Orsingher. "Words, Words, Mere Words? An Analysis of Services Customers' Perception of Evaluative Concepts." *Quality Management Journal* 13, no. 2 (2006).

[8]Grönroos, C. *Service Management and Marketing*. Lexington, MA: Lexington Books, 1990.

6. **Attitudes, behaviors, and a pleasant environment**—Customers favor services that are provided in a pleasant, comfortable, clean, and friendly setting.

Customers favor providers that are easy to work with and who can resolve problems swiftly.

The study concluded that satisfaction is linked with the degree to which individual expectations are fulfilled and to the competence, friendliness, personalization, and trust of the service provider. When asked about attitude, the respondents generally agreed that this is a judgment that derives from the rapport that an individual customer is able to establish with a contact service provider. Negative attitudes were thought to lead to negative perceptions of quality, and are rooted in three components:[9]

- A cognitive component shapes attitudes based on the accumulation of knowledge acquired through experience. A cognitive process involves conscious intellectual activity such as thinking, reasoning, and remembering.
- An affective component puts a negative or positive spin on an instantaneous event. An affective process is driven by feelings and emotions.
- A behavioral component that constitutes the human response to the affective evaluation or an event. The response may be to take no concrete action.

Not surprisingly, poor service seems to be characterized in terms that are directly opposite to those that customers associate with excellence—that is, failing to deliver on promises; impersonal treatment; an apparent lack of effort or will to please; and the poor handling of problems, queries, or complaints.

In the creation of value, the service sector is similar to manufacturing. A customer's assessment of value is based on perceived benefits traded off against perceived sacrifices. A customer purchasing a low-cost "no-frills" service package in general does not have unrealistic expectations. However, customers tend to make sharp distinctions between a low level of service and no service. A key service attribute is "ease of doing business." Although this is somewhat difficult to measure, customers do favorably respond to organizations that are easy to work with and shy away from those that are difficult.

The "easy to work with" criterion can also be correlated to convenience and **customer empowerment**. A 2002 study of Finnish online banking customers[10] revealed that the creation of value is highly influenced by a customer's freedom to choose the time and place for service delivery. Also cited as important was the ability for customers to have an input into the service-providing process and to be offered an array of services from which to choose.

Other researchers have proposed that the four main criteria from the 2001 UK study be broken down into more detailed subcriteria. For example, Singh and Deshmukh[11] have suggested the following breakdown.

- Under "delivering what was promised" and/or going an extra mile:
  1. Prompt attention from the service attendant
  2. An understanding by the service provider of what the customer wants
  3. Efficiency in the delivery of service
  4. Accuracy in the end result of service
- Under personal touch:
  5. Complete and undivided attention by the service provider
  6. Courteous and polite treatment
  7. Expression of interest in the customer
  8. An explanation of procedures
  9. Expression of proactive helpfulness
  10. Expression of pleasure in serving the customer
- Under handling problems:
  11. Attention to complaints
  12. Resolution of complaints to the customer's satisfaction
  13. Acceptance of responsibility for personal or company errors

12.2.2 Principles for Shaping Customer Perception

An alternative view comes from the disciplines of cognitive psychology, behavioral decision theory, and social psychology.[12] In contrast to the typical approach of retrospectively evaluating customer satisfaction after-the-fact, and asking customers how they felt about a service experience, the behavioral sciences can provide the basis for shaping customer opinions through an in-depth understanding of how people form perceptions and judgments. This view has led to the formulation of the following five principles:

1. *Give bad news first, save good news for last.* When confronted with a good news–bad news situation people will opt to get the bad news out of way first and forestall the good news in anticipation of concluding an experience on a positive note.
2. *Avoid communicating near misses.* Nearly achieving success can create stress and cause customers to anguish over what might have been—for example, the customer just missed out on a substantial sale that ended yesterday.
3. *Relinquish maximum control to customers.* Behavioral literature teaches that customers are less likely to complain about a service where they were able to exercise some control such as self-service situations.

[9]Ibid.

[10]Heinonen, Kristina. "Reconceptualizing Customer Perceived Value: The Value of Time and Place." *Managing Service Quality* 14, no. 2/3 (2004): 205–15.

[11]Singh, Saurabh, and S. G. Deshmukh. "Quality Initiatives in the Service Sector: A Case." *Total Quality Management* 10, no. 1 (January 1999): 5–16.

[12]Chase, Richard B. "It's Time to Get to First Principles in Service Design." *Managing Service Quality* 14, no. 2/3 (2004): 126–28.

4. *Segment the pleasure, combine the pain.* When any experience is broken down into small, clearly identifiable pieces, humans perceive the duration of an experience to have been longer than it actually was. Therefore the pleasant aspects should be broken into small segments that customers perceive to be long and of greater value, and unpleasant aspects should be lumped together. An example of the former is a dinner spread out over numerous courses; an example of the latter is a single waiting room at a health clinic versus requiring the patient to wait in multiple places.
5. *Finish strong.* Behavioral research suggests that how well a service starts is not nearly as important to the customer as how well it finishes. Customers are prone to evaluate their satisfaction level based how they feel as they exit the service facility, and not on the "average" or "highs and lows" experienced during the service.

12.3 CHALLENGES IN DATA COLLECTION AND ANALYSIS

Gaps 1 and 2 of the service quality gap model are addressed during the design phase using an integrated approach such as quality function deployment discussed in Chapter 5. Gaps 4, 6, and 7 can be closed through improved communications. It is with respect to gaps 3 and 5 where quality metrics and associated measures apply.

Gap 3 measures deal with system effectiveness—that is, how well system outcomes correspond with specified requirements. Such measures are typically quantitative and are collected on either an attributes or variables basis. For example, a restaurant might track how long it takes to seat customers, a hair styling salon might collect data on the percentage of repeat customers, or a bank might measure the average length of the queue waiting for teller service. When variables measures are collected in a service environment, it is usually not possible to form a rational subgroup. Therefore analysis tools suitable to individual (X) measurements must be employed. This is discussed further in the next section.

> **Gap 3 measures can be tracked using variables measures; gap 5 measures are typically subjective in nature.**

Gap 5 measures are the ultimate test of how well the service system is working. These also provide the greatest challenge to service providers as they are typically subjective in nature and often require extraordinary effort to collect reliable data.

12.3.1 Data Collection Instruments for Customer Satisfaction

There are numerous tools and instruments that have been used to collect information on customer satisfaction. Among these are the following:

1. **Surveys.** Surveys are the most widely used instrument for gathering data on customer opinions. Surveys can be self-administered, or professional interviewers can be used. A shortcoming of this technique is a typical low response rate that can potentially lead to biased conclusions.
2. **Focus group.** A small group of individuals is selected to meet periodically to discuss issues concerning requirements and service delivery. The typical size is between 3 and 12, and individuals are chosen from a cross section of customer profiles. Focus group meetings usually last no more than two hours.
3. **Targeted interviews.** A small random (or stratified) sample (representative of customers) is selected for face-to-face interviews. Individual interviews usually last no more than one hour and are conducted by an expert with a carefully designed set of structured questions. This technique is time-consuming and costly, but can achieve a high response rate close to 100%.
4. **Comment cards.** Customers are invited to return a card indicating their level of satisfaction listing any specific complaints, and providing any suggestions or comments that can be used to improve service. This tool can provide a useful source of customer feedback.
5. **Dissatisfaction information system.** Some service providers have formal systems in place to capture instances of customer dissatisfaction. Methods are used to document complaints, claims, returns, refunds, recalls, repeat services, litigation, replacements, downgrades, warranties, mistakes, and so on.
6. **Mystery shoppers.** An anonymous shopper evaluates the services provided by the company and by its competitors.

12.3.2 Customer Survey Methodology

The methodology used for customer surveys involves the use of an instrument (questionnaire) that consists of standardized questions that are either closed structure or open-ended. A **closed structure question** is one that elicits a directed response. Examples are multiple choice, multiple response, dichotomies (questions that can be answered yes or no), or rank lists (questions that ask for judgments using a Likert scale). An **open-ended question** is one in which the respondent writes in an answer in his or her own words.

A special skill set is required to properly design and conduct a survey. It is therefore a good idea to recruit the services of an expert who can mentor the organization through the process.

12.3.2.1 Survey Media and Formats. Surveys can be administered using a variety of media, including **face-to-face interviews**, **mail**, **telephone**, and the **Internet**. For a media choice that is self-administered (e.g., federal postal service and Internet) it is important that the survey layout be nonintimidating, user-friendly, uncluttered, and attractive. Some rules of thumb regarding format include:

- Design the questionnaire with ample white space to separate questions and sections, and facilitate readability.
- Make response areas and format consistent and easy to record an intended response. For example, boxes that can be checked off are clearer than blanks or lines.

- Place response areas on the right side of the form since the majority of respondents will be right-handed.
- Use no more than two colors to distinguish between sections and questions.

The following should be avoided because they inhibit readability:

- Multiple font types or sizes
- Small or fancy font types
- Italics
- Distracting color schemes
- Animations or blinking fonts
- Single-spacing or crowding of too much text on a single page

12.3.2.2 Length of Survey. Surveys should be long enough to ask a minimum of two or three questions on each variable (issue) of interest. However, the length should not be so long that completing it exceeds the average respondent's attention span. A well-structured instrument breaks down the list of questions into smaller sections. Completing sections creates a sense of progress and reduces survey fatigue. Typically attention spans range from about 10 minutes for a telephone survey to approximately one hour for a face-to-face interview.

12.3.2.3 Survey Question Design. The design of individual survey questions is also important. Longer and verbose questions increase the likelihood that respondents will misunderstand and provide an incorrect or no response. As a general rule questions should be constrained to 25 words or less and should avoid the use of conjunctions such as *and, or, except, unless, with, but, neither-nor, either-or*, and so on.

Poorly worded or constructed questions can lead to bias. The most common pitfalls to avoid are ambiguity, lack of mutual exclusivity, the creation of a nonexhaustive response set, requiring respondents to rank list too many items, failing to buffer socially favored positions, loaded questions, leading questions, unfamiliar terms and jargon, requiring information for which respondents do not have access, multidimensional questions, compound clauses, requiring memory recall, unnecessary complexity, weak grammatical format, use of hypotheticals, use of inappropriate assumptions, use of gender-specific language, and mismatched response sets.[13]

Survey questions should generally be less than 25 words in length.

- **Ambiguity.** Questions should be clear, concise, and specific. Generalities should be avoided. For example, the question "How would you rate the cleanliness of your room?" can lead to different interpretations of the term *clean*. Does it mean the bed was made, bathroom well stocked, a protector strip placed on the toilet seat, chocolates on the pillow, or all of the above?
- **Mutual exclusivity.** In multiple-response questions, where only one response is permitted, it is important that all available responses be mutually exclusive.
- **Nonexhaustive response sets.** Care must be taken not to omit valid choices that respondents might wish to make. For example, it may be desired that demographic data include information on a respondent's ethnicity. A number of choices are presented including "Caucasian, African American, Asian, Native American, Hispanic, etc." Missing are choices covering respondents of mixed ethnicity, such as "Mixed—Caucasian and Asian" or "Mixed—Asian and African American." Another example of a nonexhaustive response set is forcing the respondent to render an opinion rather than providing a "Neutral," "No opinion," or "Do not know" option.
- **Rank listed questions.** Questions that require respondents to rank list (e.g., provide a score on a scale from 1 to 10) are harder for respondents to answer than other structured formats. A general rule is to limit the number of questions requiring ranking to between four and six to avoid the tendency for respondents to grow weary and start assigning rankings in an arbitrary manner.
- **Socially favored positions.** Questions that deal with beliefs for which there are socially favored positions should be buffered with lead-in phrases. For example, instead of asking "Do you think that courteous and efficient service is important?" the question might be phrased, "Some people think that courtesy is the most important factor in customer service. Others believe that it is more important that the customer service representative fulfill the customer's request efficiently. Which do you believe is the most important factor?"
- **Loaded questions.** Questions that use implicit endorsements or labels to deliberately influence respondent choice should be avoided. A question that reads, "Most discerning customers have given our dining services a five-star rating. How do you rate them?" is an example of a question that carries with it an implicit endorsement from a prestigious class.
- **Leading questions.** Questions that do not legitimize all possible responses should be eliminated. A question such as "Do you believe that the wait staff should be rewarded for good service?" does not legitimize a negative response which may appear cruel and unfair to a staff member who is working hard under trying conditions. Such a question will weight the responses more heavily in the affirmative direction even if service has been less than satisfactory.
- **Unfamiliar terms and jargon.** Avoid terms like *tacit knowledge, stakeholder, supply chain,* and so forth that may not be familiar to all respondents. Try and stick to simple and widely understood terms and jargon.
- **Inaccessible information.** Avoid asking questions that require knowledge that respondents are unlikely to possess. An example would be a question such as, "Was the chef well skilled in how to properly prepare your selected entrée?"
- **Multidimensionality.** Questions that require multidimensional judgments should be avoided. For example, a question might ask the respondent to opine whether a certain service provider gave less than satisfactory service

[13] http://www.statisticssolutions.com/survey_research.htm.

due to a poor attitude. This question requires two parts: first, an opinion on whether the service was poor; and second, whether the employee concerned displayed a poor attitude. It would be better to break down these two factors into two questions.

- **Compound questions.** Avoid compound clauses in questions that introduce unnecessary complexity. The question "Did the attendant behind the desk give you courteous and prompt service that lived up to your expectations and satisfied your needs in a manner that made you feel at ease and have confidence that you and your baggage would reach their destination safely and on time?" would be better to be broken down into numerous short questions asking specifically how the respondent feels about whether the service delivered lived up to expectations, satisfied needs, put the customer at ease, and instilled confidence.
- **Memory recall.** Questions should refrain from asking respondents to recall events that occurred in the past, specifically more than six months ago.
- **Unnecessary complexity.** Complex phrasing, terminology, double negatives, and so on should be avoided because such complexity can overtax the respondent's memory burden.
- **Weak grammatical format.** Questions should be crisp and structured so that the content of the question is presented prior to the response alternatives.
- **Hypotheticals.** Questions should be avoided that require respondents to use their imagination regarding hypothetical situations. An example of this would be questions that begin with phrases such as "What if . . . ?" or "If . . . how would you react?"
- **Inappropriate assumptions.** Questions should refrain from containing language that is based on arguable premises. For example, if asking the question "How much would service quality improve if multiple queues were replaced by a single queue that feeds all servers?" responses such as "Substantially," "A little," "A very little," or "None" assume that there would be no negative effect on quality by redesigning the system as indicated.
- **Gender bias.** Special care should be taken to avoid wording that could be interpreted by some to be gender specific.
- **Mismatched response sets.** Avoid mismatching response sets with questions. For example, if asking "How does our service compare with our competitors?" the responses "Exceeds expectations," "Meets expectations," "Below expectations," "Significantly below expectations" do not match the question, which calls for respondents to opine how service quality compares with those of competitors.

12.3.2.4 Pretesting. It is always a good idea to conduct a pilot survey on a small scale (a technique called **pretesting**) before rolling out the full-scale survey. Pretests can validate survey length and flesh out question ambiguities and other problems such as weak questions that should be eliminated.

Pretesting is a process whereby respondent reactions are tested—for instance, to what extent does the instrument capture and hold interest, what needs to be clarified (instructions or specific questions), are any questions sensitive or politically incorrect, and is enough space available for open-ended questions? Pretesting can also help fine-tune the factors that will be used in a statistical analysis.

12.3.2.5 Survey Pitfalls. Designing and conducting a survey is a process that can produce valuable information. However, an ill-conceived or poorly designed survey can lead to negative outcomes. When contemplating a customer satisfaction survey an organization should be aware and wary of the following possible pitfalls.[14]

- Designing survey form improperly
- Defining survey issues poorly
- Designing sampling techniques/sampling errors poorly
- Failing to properly acknowledge nonresponses
- Treating perceptions as objective measures
- Using incorrect methods of analysis
- Treating a survey as an event rather than a process
- Asking nonspecific questions
- Failing to ask the right questions
- Ignoring the results or using them incorrectly
- Failing to provide feedback when necessary
- Asking too many questions (25–35 after pretesting recommended)
- Using part-time, temporary, or untrained personnel to collect data

12.3.3 SERVQUAL

SERVQUAL is a structured survey methodology for measuring service quality and relies on a customer's ability to qualitatively "score" performance in five areas: *assurance, empathy, reliability, responsiveness*, and *tangibles*.[15] These are briefly defined as follows.

- **Assurance**—the ability of service providers to inspire confidence and trust. Assurance includes courtesy, knowledge, competence, credibility, and security.
- **Empathy**—the projection of a spirit of caring and the delivery of individualized attention. Empathy includes ease of access, effectiveness of communications, and how well the customer is understood by the service provider.
- **Reliability**—the ability to deliver what was promised in a dependable manner and without errors.
- **Responsiveness**—the willingness to help customers, be flexible to their unique requirements, and deliver prompt service.
- **Tangibles**—the physical facilities, equipment, and overall appearance of service providers.

Examples of SERVQUAL applications have included the determination of how patrons assess desired, adequate, and perceived levels of service at four types of restaurants (Chinese,

[14]*Certified Six Sigma Green Belt (CSSGB) Primer*. West Terre Haute, IN: Quality Council of Indiana, 2006.

[15]Parasuraman, A., Leonard L. Berry, and Valarie A. Ziethaml, "Refinement and Reassessment of the SERVQUAL Scale," *Journal of Retailing*, 67 (4) Winter, 1991, 420-450.

TABLE 12-1A SERVQUAL Survey Instrument

	Perceptions (P)								Expectations (E)								
	Directions:								**Directions:**								
	The following set of statements relate to your feelings about XYZ Company's service. For each statement, please show the extent to which you believe XYZ has the feature described by the statement. Score the statement "1" if you strongly disagree that XYZ has that feature, and a "7" if you strongly agree. Any numbers in the middle may be used to indicate how strong your feelings are. There are no right or wrong answers—we are only interested in the number that best shows your perceptions about XYZ's service.								Based on your experience as a customer of companies in this business, please think about the kind of company that would deliver excellent quality of service. Think about the kind of company with which you would be pleased to do business. Please show the extent to which you believe such a company would possess the feature described by each statement listed below. If you believe a statement is not at all essential for excellent companies score the statement "1." If you feel the feature is absolutely essential for an excellent company score the feature "7." If your feelings are less strong, provide a score somewhere in the middle. There are no right or wrong answers—we are only interested in the number that truly reflects your feelings regarding companies that would deliver excellent quality of service.								**Q = P−E**
	Tangibles	1	2	3	4	5	6	7	**Tangibles**	1	2	3	4	5	6	7	
1	XYZ Company has modern-looking equipment.								Excellent companies in this business will have modern-looking equipment.								
2	The facilities at XYZ Company are visually appealing.								The physical facilities at excellent companies in this business will be visually appealing.								
3	The employees of XYZ Company are neat-appearing.								Employees of excellent companies in this business will be neat-appearing.								
4	Materials associated with the service (such as pamphlets or statements) are visually appealing in XYZ Company.								Materials associated with the service (such as pamphlets or statements) will be visually appealing in an excellent company in this business.								
	Reliability	1	2	3	4	5	6	7	**Reliability**	1	2	3	4	5	6	7	
5	When the XYZ Company promises to do something by a certain time, it does so.								When excellent companies in this business promise to do something by a certain time, they will do so.								
6	When you have a problem, XYZ Company shows a sincere interest in solving it.								When customers have a problem, excellent companies in this business will show a sincere interest in solving it.								
7	The XYZ Company performs the service right the first time.								Excellent companies in this business will perform the service right the first time.								
8	The XYZ Company provides its services at the time it promises to do so.								Excellent companies in this business will provide their services at the time they promise to do so.								
9	The XYZ Company insists on error-free records.																

Source: Parasuraman, A., V. A. Zeithaml, and L. L. Berry. "SERVQUAL—A Multiple-Item Scale for Measuring Consumer Perceptions of Service Quality." *Journal of Retailing* 64, no. 1 (Spring 1988): 12–40.

casual dining, full service, and quick service) at the Hong Kong International Airport,[16] how to improve the quality of service in the pest control industry,[17] as a catalyst to bring about cultural and structural change in the Australian Public Sector,[18] to determine service levels in banking,[19] how to generate quality initiatives for the consultancy and technology transfer wing of a major educational institution in India,[20] and to determine the quality of catering services.[21]

The SERVQUAL measuring tool is a survey that uses a seven-point Likert scale to capture customer expectations,

[16]Heung, Vincent C. S., M. Y. Wong, and Hailin Qu. "Airport-Restaurant Service Quality in Hong Kong: An Application of SERVQUAL." *Cornell Hotel and Restaurant Administration Quarterly* 41, no. 3 (June 2000): 86–96.

[17]Ninichuck, Bryan. "Service Quality Is the Key to Good Business." *Pest Control* 69, no. 9 (September 2001): 12–13.

[18]McDonnell, John, and Terry Gatfield. "SERVQUAL as a Cultural Change Agent in the Australian Public Sector." *Case Study in Marketing and Society*. Brisbane, Australia: Queensland University of Technology and Griffith University, 1998.

[19]Hussey, Michael K. "Using the Concept of Loss: An Alternative SERVQUAL Measure." *The Service Industries Journal* 19, no. 4 (October 1999): 89–101.

[20]Singh and Deshmukh, "Quality Initiatives."

[21]Brysland, A., and A. Curry. "Service Improvements in Public Services Using SERVQUAL." *Managing Service Quality* 11, no. 6 (2001): 389–401.

TABLE 12-1B SERVQUAL Survey Instrument

	Perceptions (P)								Expectations (E)								
	Directions: The following set of statements relate to your feelings about XYZ Company's service. For each statement, please show the extent to which you believe XYZ has the feature described by the statement. Score the statement "1" if you strongly disagree that XYZ has that feature, and a "7" if you strongly agree. Any numbers in the middle may be used to indicate how strong your feelings are. There are no right or wrong answers—we are only interested in the number that best shows your perceptions about XYZ's service.								**Directions:** Based on your experience as a customer of companies in this business, please think about the kind of company that would deliver excellent quality of service. Think about the kind of company with which you would be pleased to do business. Please show the extent to which you believe such a company would possess the feature described by each statement listed below. If you believe statement is not at all essential for excellent companies score the statement "1." If you feel the feature is absolutely essential for an excellent company score the feature "7." If your feelings are less strong, provide a score somewhere in the middle. There are no right or wrong answers—we are only interested in the number that truly reflects your feelings regarding companies that would deliver excellent quality of service.								**Q = P−E**
	Responsiveness	1	2	3	4	5	6	7	**Responsiveness**	1	2	3	4	5	6	7	
10	Employees of XYZ Company tell you exactly when services will be performed.								Employees of excellent companies in this business will tell customers exactly when services will be performed.								
11	Employees of XYZ Company give you prompt service.								Employees of excellent companies in this business will give prompt service to customers.								
12	Employees of XYZ Company are always willing to help you.								Employees of excellent companies in this business will always be willing to help customers.								
13	Employees of XYZ Company are never too busy to respond to your requests.								Employees of excellent companies in this business will never be too busy to respond to customer requests.								
	Assurance	1	2	3	4	5	6	7	**Assurance**	1	2	3	4	5	6	7	
14	The behavior of employees of XYZ Company instills confidence in customers.								The behavior of employees of excellent companies in this business will instill confidence in customers.								
15	You feel safe in your transactions with XYZ Company.								Customers of excellent companies in this business will feel safe in transactions.								
16	Employees of XYZ Company are consistently courteous toward you.								Employees of excellent companies in this business will be consistently courteous towards customers.								
17	Employees of XYZ Company have the knowledge to answer your questions.								Employees of excellent companies in this business will have the knowledge to answer customers' questions.								

Source: Parasuraman, A., V. A. Zeithaml, and L. L. Berry. "SERVQUAL—A Multiple-Item Scale for Measuring Consumer Perceptions of Service Quality." *Journal of Retailing* 64, no. 1 (Spring 1988): 12–40.

and what the customer perceives was actually delivered. Tables 12-1a, 12-1b, and 12-1c illustrate a typical SERVQUAL survey instrument.

12.4 USE OF STATISTICAL PROCESS CONTROL TOOLS

12.4.1 Use of Control Charts in a Services Environment

Where a gap 3 metric is involved and it is possible to use quantitative criteria, control charts can be used in a services environment in the same way they are used in a manufacturing setting. For example, if a measure of customer service is *on-time deliveries* (products or services delivered when promised), data could be collected and analysed as shown in Figure 12-2. These data were collected from a random sample of five shipments per day over a 30-day period. In this case early deliveries were considered to be just as bad as late deliveries. Therefore the deviations from the target (equal to zero) were computed for each delivery selected for the samples. An analysis of the control charts shown in Figure 12-2 reveals the target of zero is being missed on the average by 0.43 day with a standard deviation of approximately 1.73 days ($\bar{R}/d_2 = 4.03/2.326$). In this case the company made a commitment to its customers that it will deliver on time within 1 day on either side of the promised delivery date. This 2-day time

TABLE 12-1C SERVQUAL Survey Instrument

	Perceptions (P)								Expectations (E)								
	Directions: The following set of statements relate to your feelings about XYZ Company's service. For each statement, please show the extent to which you believe XYZ has the feature described by the statement. Score the statement "1" if you strongly disagree that XYZ has that feature, and a "7" if you strongly agree. Any numbers in the middle may be used to indicate how strong your feelings are. There are no right or wrong answers—we are only interested in the number that best shows your perceptions about XYZ's service.								**Directions:** Based on your experience as a customer of companies in this business, please think about the kind of company that would deliver excellent quality of service. Think about the kind of company with which you would be pleased to do business. Please show the extent to which you believe such a company would possess the feature described by each statement listed below. If you believe a statement is not at all essential for excellent companies score the statement "1." If you feel the feature is absolutely essential for an excellent company score the feature "7." If your feelings are less strong, provide a score somewhere in the middle. There are no right or wrong answers—we are only interested in the number that truly reflects your feelings regarding companies that would deliver excellent quality of service.								**Q =** P−E
	Empathy	1	2	3	4	5	6	7	**Empathy**	1	2	3	4	5	6	7	
18	The XYZ Company gives you individual attention.								Excellent companies in this business will give customers individual attention.								
19	The XYZ Company has operating hours convenient to all its customers.								Excellent companies in this business will have operating hours convenient to all their customers.								
20	The XYZ Company has employees who give you personal attention.								Excellent companies in this business will have employees who give customers personal attention.								
21	The XYZ Company has your best interests at heart.								Excellent companies in this business will have the customer's best interests at heart.								
22	Employees of XYZ Company understand your specific needs.								The behavior of employees of excellent companies in this business will instill confidence in customers.								

Point-Allocation Questions

	Directions: Listed below are five features pertaining to companies and the services they offer. We would like to know how important each of these features is to you when you evaluate a company's quality of service. Please allocate a total of 100 points among the five features according to how important each feature is to you—the more important a feature is to you, the more points you should allocate to it. Please ensure that the points you allocate to the five features add up to 100.	**Points**
1	The appearance of the company's physical facilities, equipment, personnel, and communications materials.	
2	The ability of the company to perform the promised service dependably and accurately.	
3	The willingness of the company to help customers and provide prompt service.	
4	The knowledge and courtesy of the company's employees and their ability to convey trust and confidence.	
5	The caring, individualized attention the company provides its customers.	
	TOTAL POINTS ALLOCATED	**100**

Source: Parasuraman, A., V. A. Zeithaml, and L. L. Berry. "SERVQUAL—A Multiple-Item Scale for Measuring Consumer Perceptions of Service Quality." *Journal of Retailing* 64, no. 1 (Spring 1988): 12–40.

window commitment can be thought of as the engineering tolerance for delivery time. The natural tolerance for lead time (i.e., a measure of the variability in lead time) is a concept called *span*. For this company, span = NT = 6σ = (6) (1.73) = 10.38 days. With ET = 2 and NT = 10.38, the application of Equations 9-11 and 9-13 results in a C_p = 0.193 and a C_{pk} = 0.109, respectively. This company has much work to do if it is to live up to its commitments.

The type of data collected in service industries may not lend itself to rational subgrouping and may not be normally distributed. This is often true for data that are expressed in units of time or money. The data in Figure 12-3 were collected on transaction times for tellers at a local bank. As the histogram and scatter plot show, the data are not normal. Note how the points are more dispersed above the mean than below.

Span is a measure of variability in lead time.

If we subgroup the data into samples of five each and construct $\overline{X}$ and R charts we obtain Figure 12-4a. The R

Sample	Number of Days			Sample	Number of Days		
	Promised	Actual	Deviation		Promised	Actual	Deviation
1	20	21	1	6	20	24	4
	1	3	2		9	10	1
	20	20	0		13	13	0
	3	4	1		27	28	1
	22	20	-2		25	24	-1
2	4	2	-2	7	10	12	2
	18	20	2		1	4	3
	4	5	1		16	20	4
	5	5	0		16	16	0
	7	6	-1		13	13	0
3	14	15	1	8	14	15	1
	1	3	2		29	28	-1
	8	8	0		7	7	0
	6	6	0		20	21	1
	18	19	1		9	9	0
4	3	3	0	9	24	25	1
	8	6	-2		21	22	1
	16	16	0		22	24	2
	7	7	0		11	12	1
	3	6	3		4	5	1
5	18	14	-4	10	12	13	1
	13	13	0		3	3	0
	30	32	2		23	21	-2
	7	6	-1		2	5	3
	1	5	4		11	12	1

Data for first 10 Days

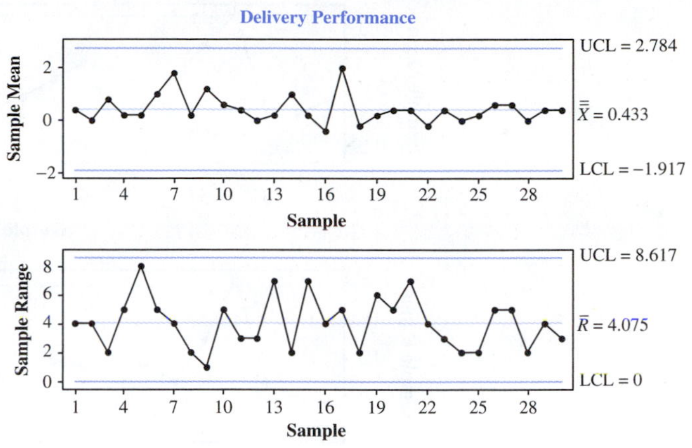

Control Charts for All 30 Days

FIGURE 12-2. Delivery Performance

Teller Transaction Times (min)					Range	$\bar{X}$
6.23	4.04	5.07	2.63	1.72	4.51	3.94
1.22	2.89	2.81	5.75	0.12	5.63	2.56
2.83	0.18	1.26	8.20	0.24	8.02	2.54
10.40	3.34	3.80	27.77	10.70	24.43	11.20
13.00	7.10	17.18	3.43	2.38	14.80	8.62
0.86	8.02	3.87	9.01	7.50	8.14	5.85
11.30	2.15	2.31	0.16	0.62	11.14	3.31
0.55	0.49	7.45	1.62	0.81	6.96	2.18
2.92	0.73	11.12	10.32	10.39	10.39	7.10
10.90	2.07	0.85	2.52	12.27	11.43	5.72
1.13	0.38	0.45	1.98	16.83	16.45	4.15
3.40	0.97	1.82	8.18	2.05	7.21	3.28
13.35	4.09	8.98	4.86	4.09	9.26	7.07
1.01	1.58	9.97	1.13	8.71	8.96	4.48
4.59	5.30	1.72	12.16	4.38	10.44	5.63
0.24	1.31	7.63	1.07	0.81	7.39	2.21
3.03	0.52	0.18	0.18	6.83	6.65	2.14
17.32	1.43	2.87	4.54	17.54	16.10	8.74
0.83	1.83	1.03	1.46	3.26	2.42	1.68
5.31	0.46	2.28	2.60	3.92	4.85	2.92

$\bar{X} = 4.767$

$s = 4.986$

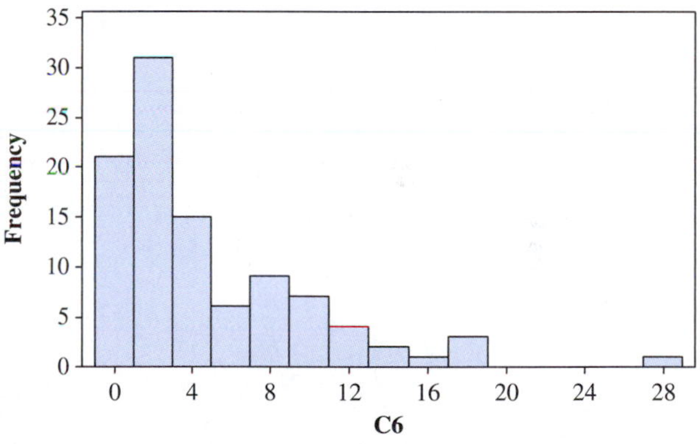

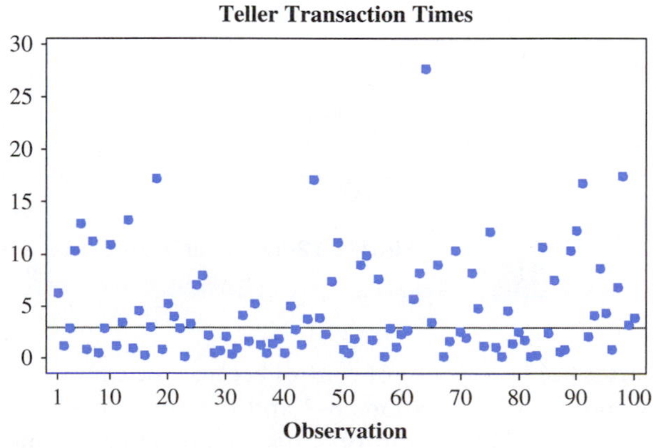

FIGURE 12-3. Teller Transaction Times at First National Bank

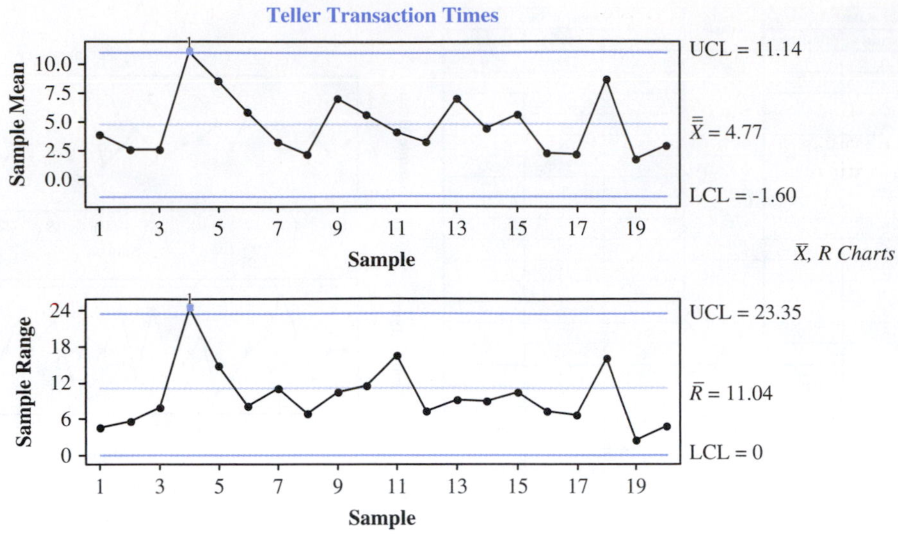

FIGURE 12-4a. Teller Transaction Times

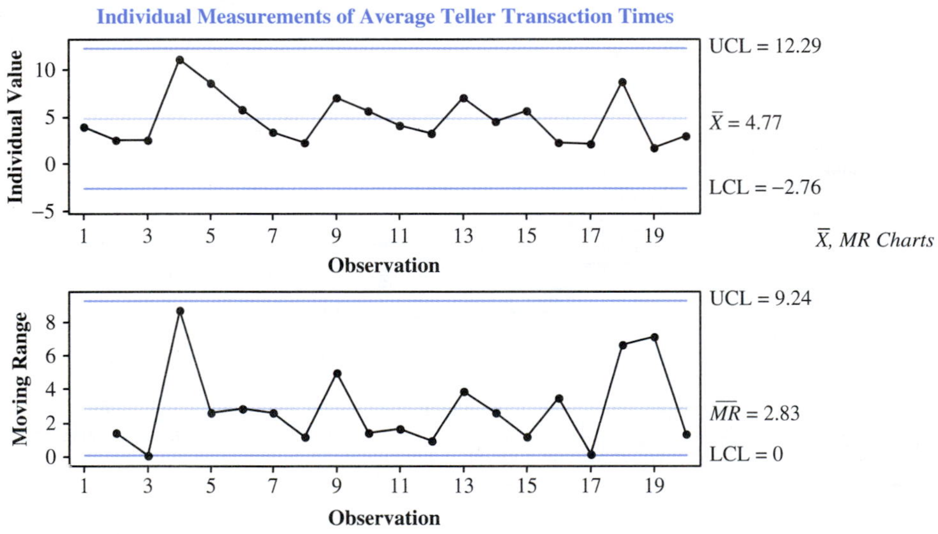

$$\sigma_{\bar{x}} = \frac{\overline{MR}}{d_2} = \frac{2.828}{1.128} = 2.507 \qquad \sigma_{\bar{x}}^2 = \frac{\sigma_x^2}{n}$$

$$(2.507)^2 = \frac{\sigma_x^2}{5} \qquad \sigma_x^2 = 31.4276$$

$$\sigma_x = 5.606$$

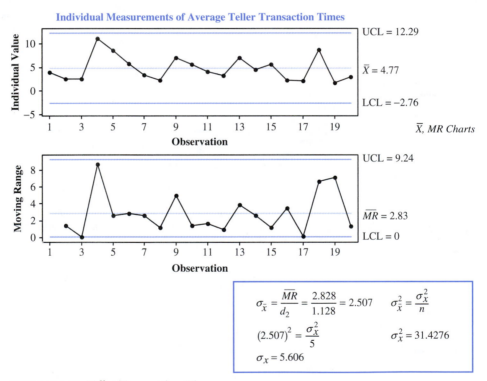

FIGURE 12-4b. Teller Transaction Times

chart is predictably out of control because the control limits are too narrow. Equations 8-7 and 8-8 (the equations for the upper and lower control limits for a range chart) underestimate the maximum and minimum ranges that would be expected for distributions that are not normal. The average range (the centerline on the *R* chart) is also an underestimate of the total process variability. Since the data set is not uniformly and randomly distributed about the process mean, an occasional sample will have a range that exceeds that which

Equations 8-7 and 8-8 would predict. This source of variability will show up between samples rather than within samples. In this case the between-sample variability is not due to special causes but is systemic.

The approach shown in Figure 12-4b can be used to capture the common cause variability. In this approach the $\overline{X}$s are treated as individuals and a 2-point moving range chart is used to capture the variance. A control chart, constructed using the average moving range and Equations 8-26 and 8-27 to establish the limits on the $\overline{X}$ chart, exhibits statistical control. The following relationships can then be used to estimate the process standard deviation.

$$\sigma_{\overline{X}} = \frac{\overline{MR}}{d_2} = \frac{\overline{MR}}{1.128} \quad \textbf{12-1}$$

$$\sigma_X^2 = n\sigma_{\overline{X}}^2 \quad \textbf{12-2}$$

$$\sigma_x = \sqrt{\sigma_X^2}$$

This approach can be applied whenever there is reason to believe that variation between samples is a natural expectation of the process—that is, the process variance is systemic and not due to special causes.

An alternative approach is to plot the individual measurements rather than the averages of subgroups. Figure 12-5a depicts the X and MR charts for the 100 teller transaction measurements. Clearly these charts are out of control and, due to the nonnormal shape of the histogram, we can assume that the data are not randomly distributed about the mean. Using these data to estimate the process standard deviation yields the following results.

$$\sigma_X \cong \frac{\overline{MR}}{d_2} = \frac{4.68}{1.128} = 4.149$$

Since the charts are not in a state of statistical control, this is likely an underestimate of the actual process standard deviation and is less than the estimate shown in Figure 12-4b. Figure 12-5b shows X and MR charts for data using the Box-Cox transformation $y = \ln x$ as described in Chapter 8.

We can see that the transformed data appear to be in good statistical control. The approach used in Figure 12-5b is an effective tool in monitoring the process to detect the presence of special causes. We also demonstrated in Chapter 9 the usefulness of transformations in performing capability analyses and in determining the proportion of nonconforming process output. The difficulty arises in trying to convert the transformed data on the control charts back into the original units. In the previous example,

$$y = \ln x \text{ and } x = e^y$$

However,

$$\overline{X} \neq e^{\overline{Y}} \text{ and } \bar{s}_x \neq e^{\bar{s}_y}$$

In general,

If $y = g(x)$ *and* $x = g^{-1}(y)$

Then $\bar{x} \neq g^{-1}(\bar{y})$ *and* $\bar{s}_x \neq g^{-1}(\bar{s}_y)$

When a normal transformation suggests process stability, the best way to estimate the process parameters is to use the raw data, as shown in Figure 12-3. That is,

$$\mu \cong \overline{X} = 4.767$$

$$\sigma \cong s = 4.986$$

12.4.2 Autocorrelated Data

Variables data, collected for the purpose of process monitoring, should be time ordered if possible to expose any time-related sources of variation. As was previously established in Chapters 7, 8, and 10, basic Shewhart control charts are appropriate provided each data point is independent. However, situations arise when the assumption of independence is not valid and the data is time correlated—for example, if one measurement is higher than the previous measurement, then the next measurement is likely to be higher than the present one.

Consider the data shown in Table 12-2, which were collected randomly on the time customers had to stand in queues at a supermarket checkout stand. The data were collected at 15-minute intervals. Because customer traffic is not uniformly distributed across the day the data are expected to be autocorrelated. This can easily be tested using the Minitab software package. Figure 12-6 shows the dropdown menu sequence used. The data in Table 12-2, have been pasted into a Minitab worksheet. Then the dropdown sequence *Stat > Time Series >Autocorrelation* is selected. This will bring up the dialog box shown in Figure 12-7. By selecting the *Default number of lags* option, Minitab will evaluate all correlations from a time interval lag of one to a lag equal to $n/4$—in this case 9.

Clicking on *OK* in the dialog box shown in Figure 12-7 produces the autocorrelation function graph illustrated in Figure 12-8. This analysis shows a significantly strong one-lag correlation and a weaker but still significant two-interval correlation. Due to the correlated nature of these data conventional Shewhart X and MR charts will not work as illustrated in Figure 12-9.

When data are autocorrelated, a two-step approach must be taken to assess and maintain process stability. The first step is to employ **time series** analysis to find a suitable forecasting model that describes how the data are serially correlated and that can be used to forecast future measurements. Time series is a term used to describe a sequence of plotted data points, measured typically at successive times and commonly spaced at uniform time intervals. Time-ordered control chart data fit this definition. Once a forecasting model has been established that can reliably be used to predict process behavior, the second step is to use the model to generate a data series of residuals. Each time a measurement is taken from the process a residual variable, ρ, is computed by taking the difference between the actual observed value and the value predicted by the model. Specifically,

$$\rho_t = y_t - \psi_t \quad \textbf{12-3}$$

where,

ρ_t = residual at time t

y_t = observation at time t

ψ_t = forecast at time t

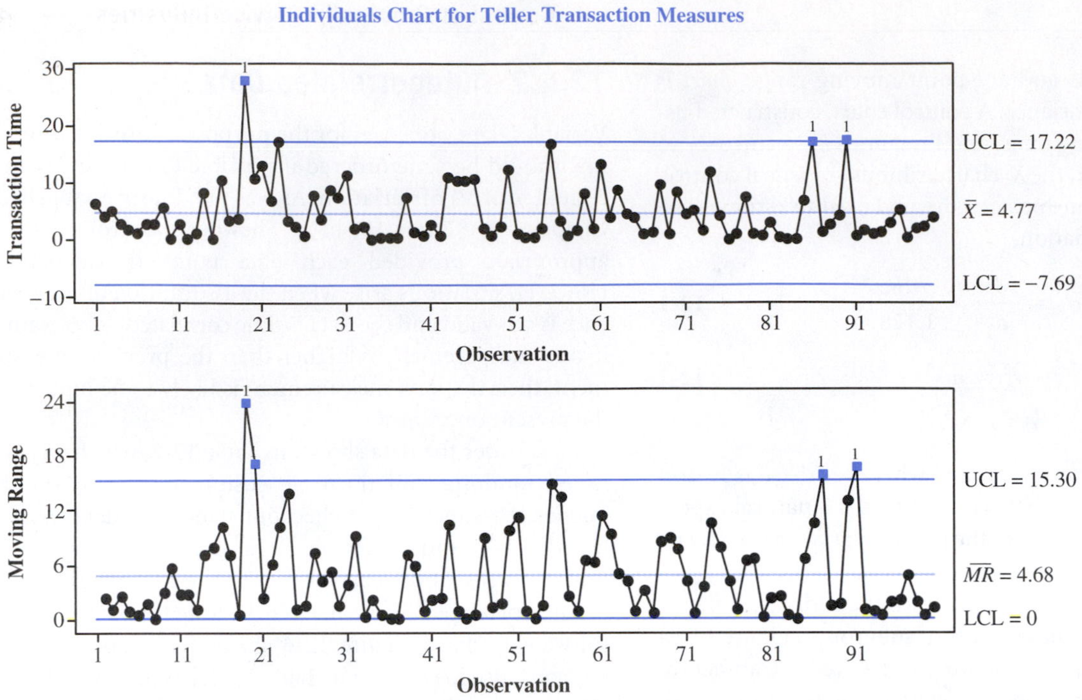

$$\sigma_x \cong \frac{\overline{MR}}{d_2} = \frac{4.68}{1.128} = 4.149$$

$\sigma_x > 4.149$ because charts are out of control.

FIGURE 12-5a. Individuals Chart for Teller Transaction Times

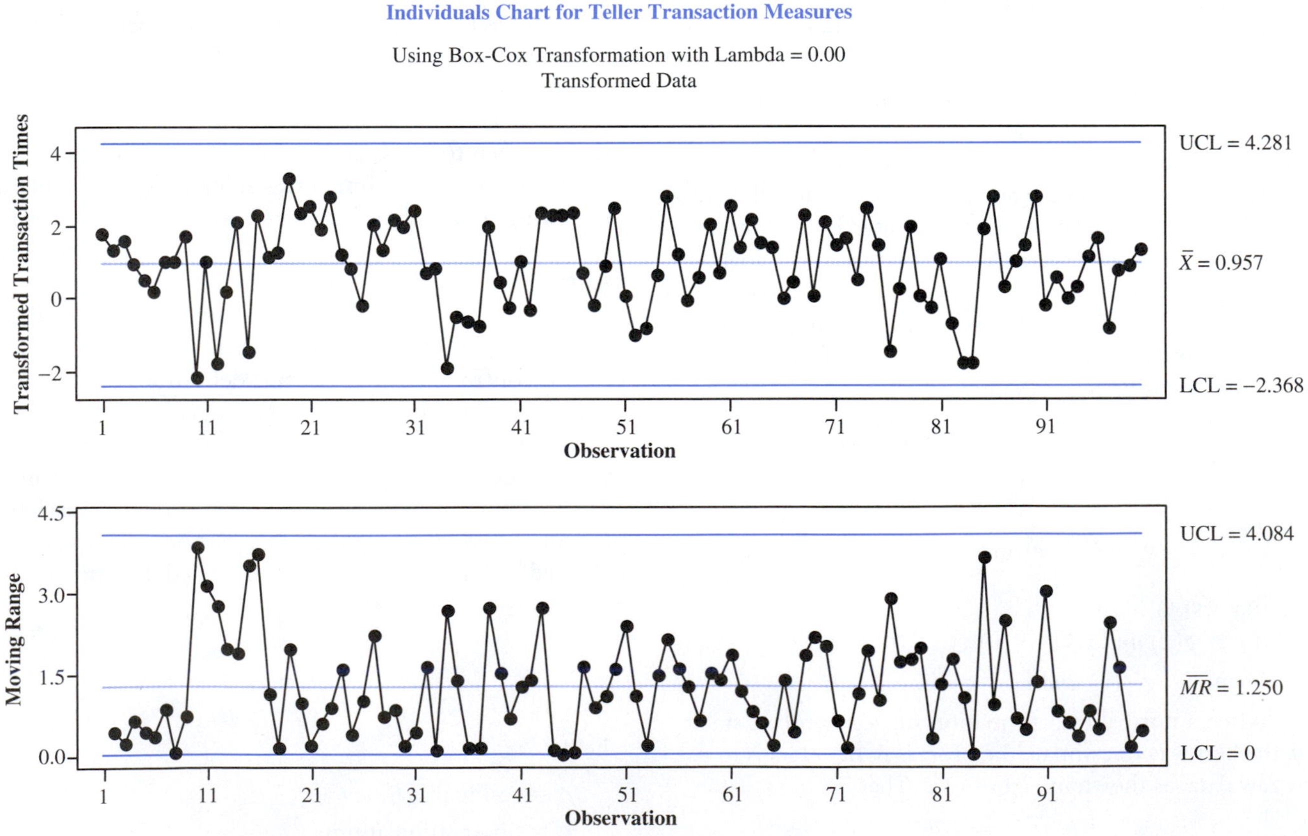

FIGURE 12-5b. Individuals Chart for Transformed Teller Times

TABLE 12-2 Waiting Times in Checkout Queues

Time	Waiting Time in Queue (min)	Time	Waiting Time in Queue (min)
10:00	1.1	2:15	4.68
10:15	2.8	2:30	4.25
10:30	3.78	2:45	3.87
10:45	4.21	3:00	3.3
11:00	4.33	3:15	2.9
11:15	4.45	3:30	2.2
11:30	5.15	3:45	1.5
11:45	4.92	4:00	0.9
12:00	4.8	4:15	1.2
12:15	4.5	4:30	1.6
12:30	4.2	4:45	1.98
12:45	3.9	5:00	2.3
1:00	4.1	5:15	2.1
1:15	4.32	5:30	1.97
1:30	4.33	5:45	1.86
1:45	4.67	6:00	1.54
2:00	4.95	6:15	1.22

If the data series is stationary—that is, the mean and variance are constant over time—the residuals will be normally distributed with a mean of zero. Also, conventional control charts of the residuals will exhibit good statistical control. If an out-of-control signal is detected one can infer that the series is no longer stationary—that is, either the mean or variance has changed (or both)—and the model no longer provides reliable forecasts.

A detailed discussion of time series modeling is beyond the scope of this text, and for the purpose of illustration the autoregressive integrated moving average (ARIMA) model used by Minitab will be used. To access ARIMA use the commands *Stat > Time Series > Single Exp Smoothing,* as shown in Figure 12-10. ARIMA fits a Box-Jenkins[22] model to the data series that can incorporate patterns that may not be visible in plotted data. This methodology is an iterative process that is capable of filtering out and explaining all but random noise variation.

[22]Box, G. E. P., and G. M. Jenkins. *Time Series Analysis, Forecasting, and Control.* 2nd ed. San Francisco: Holden Day, 1976.

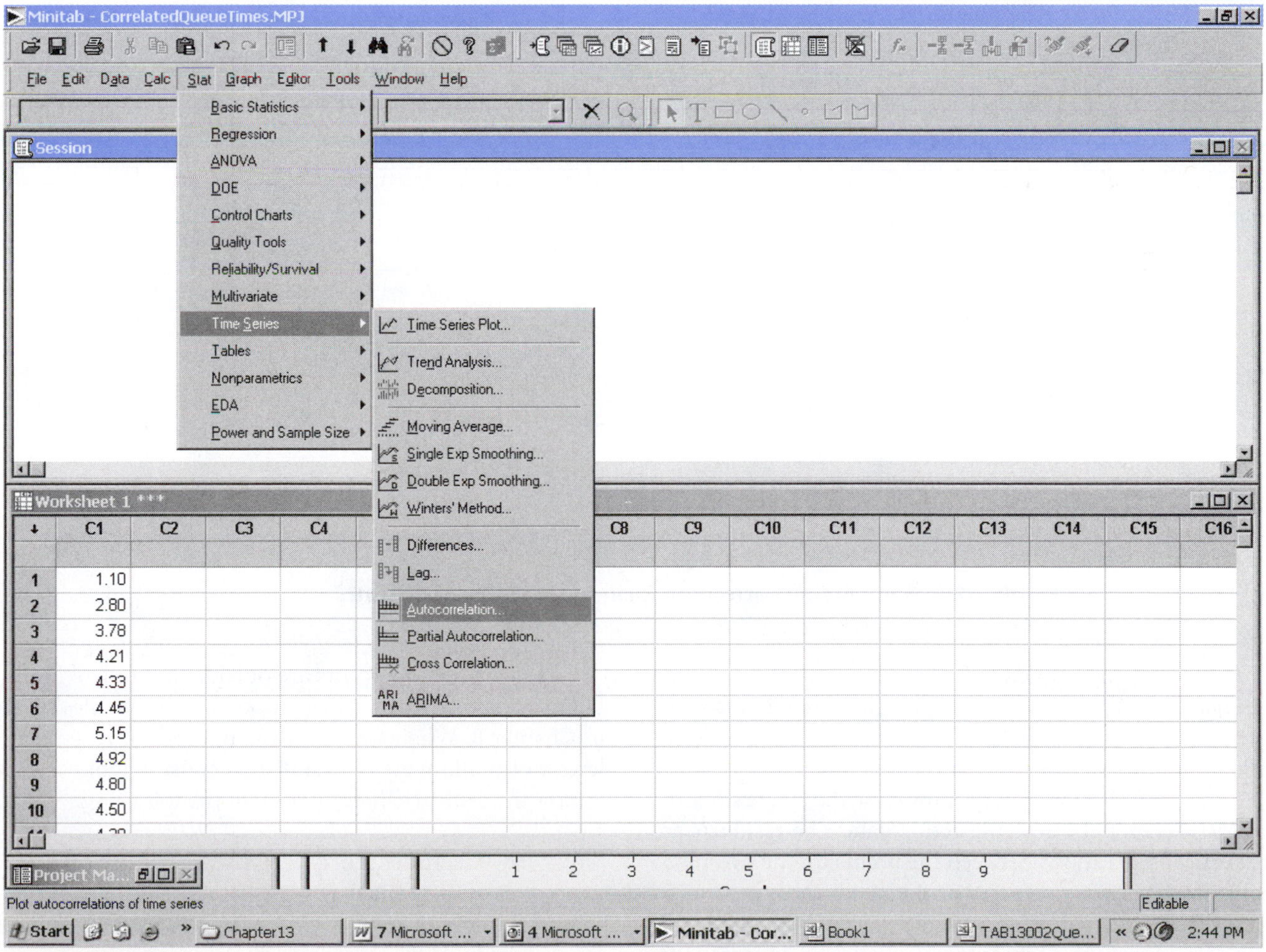

FIGURE 12-6. Dropdown Minitab Menus for Testing Autocorrelation

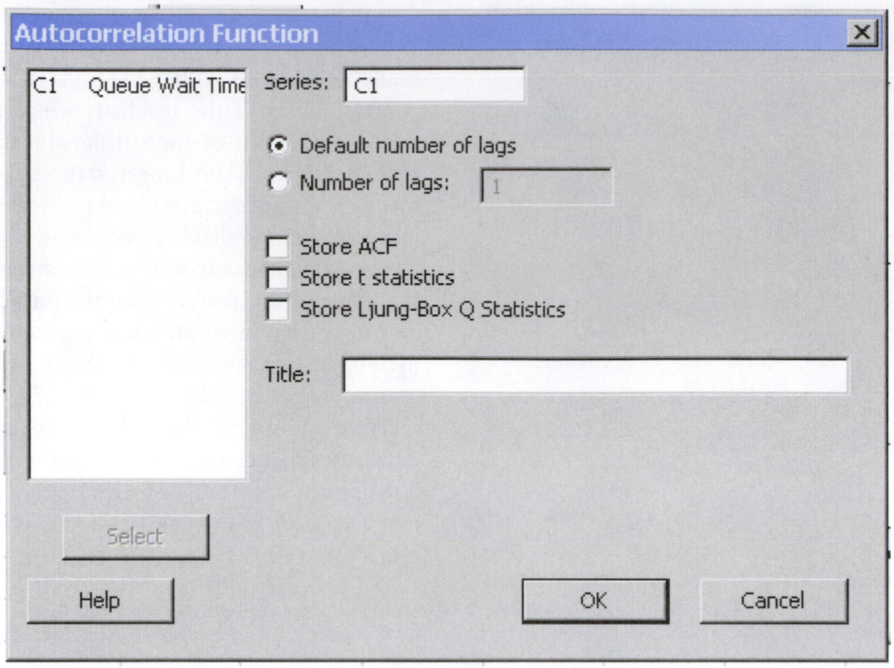

FIGURE 12-7. Minitab Autocorrelation Dialog Box

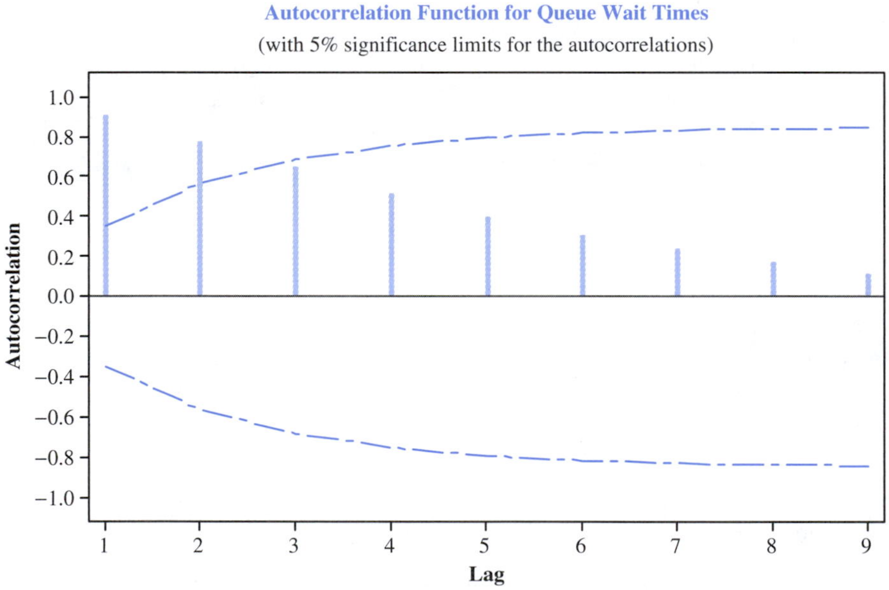

FIGURE 12-8. Minitab Autocorrelation Analysis for Queue Wait Times

Following the sequence in Figure 12-10 and selecting *Optimal ARIMA* for the smoothing weight, and the *Graphs* option, we can produce the menu sequence shown in Figure 12-11 and the graphs displayed in Figures 12-12 and 12-13.

In Figure 12-12 we can see how close the single smoothing ARIMA model tracks the actual data. The residuals are the vertical separations between the points on the two plots. In Figure 12-13 we can see that the residuals appear to be normally distributed. With Minitab we can stipulate a column in the worksheet where the residuals will be stored. We can then construct control charts on the data in this column as shown in Figure 12-14.

Even though no points are outside the control limits in Figure 12-14 we observe a violation in the runs rules covered in Chapter 8. When we count the number of runs and the length of the longest run, and then consult Tables 8-2 and 8-3 for the critical values, we obtain the following.

For the *MR* chart:
$r = 19, s = 15$
Number of runs = 16; critical value (from Table 8-2) $= 12 \Rightarrow$ OK
Length of longest run = 4; critical value (from Table 8-3) $= 9 \Rightarrow$ OK

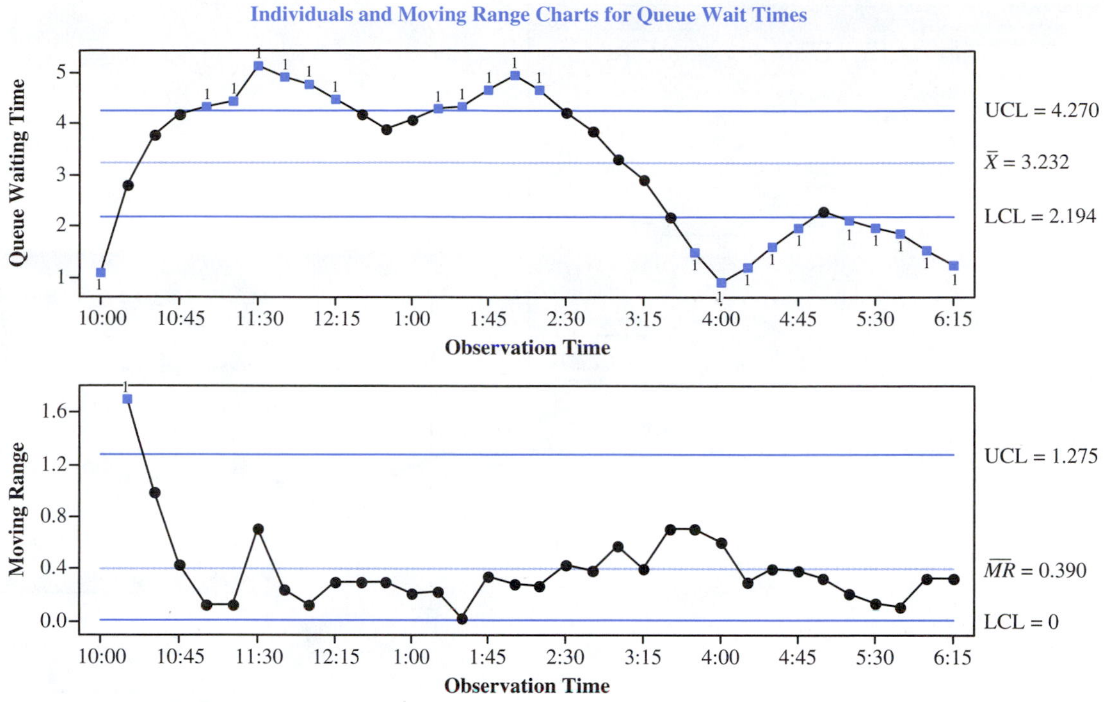

FIGURE 12-9. Individuals and Moving Range Chart for Autocorrelated Queue Wait Times

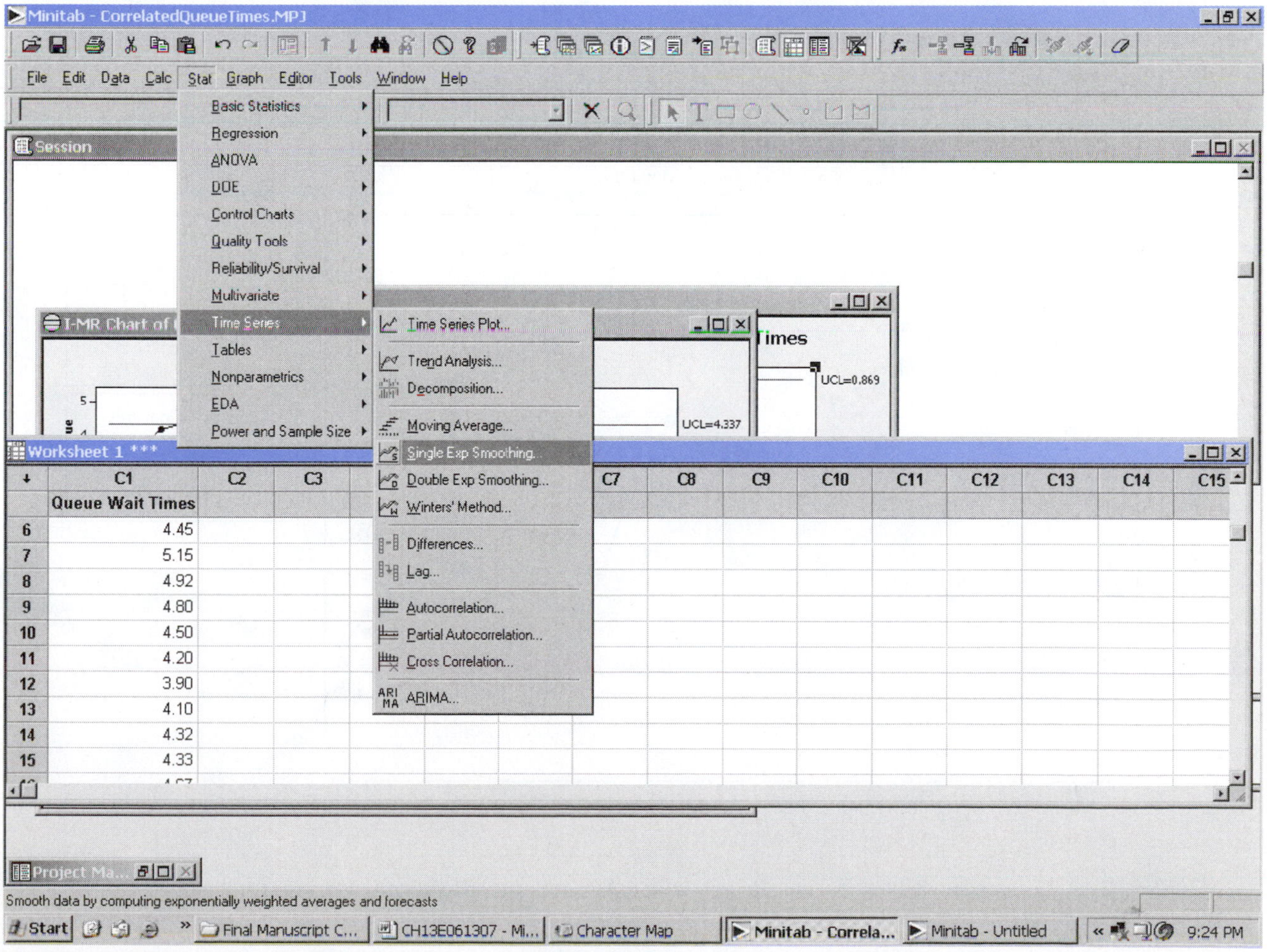

FIGURE 12-10. Dropdown Menu Sequence for ARIMA

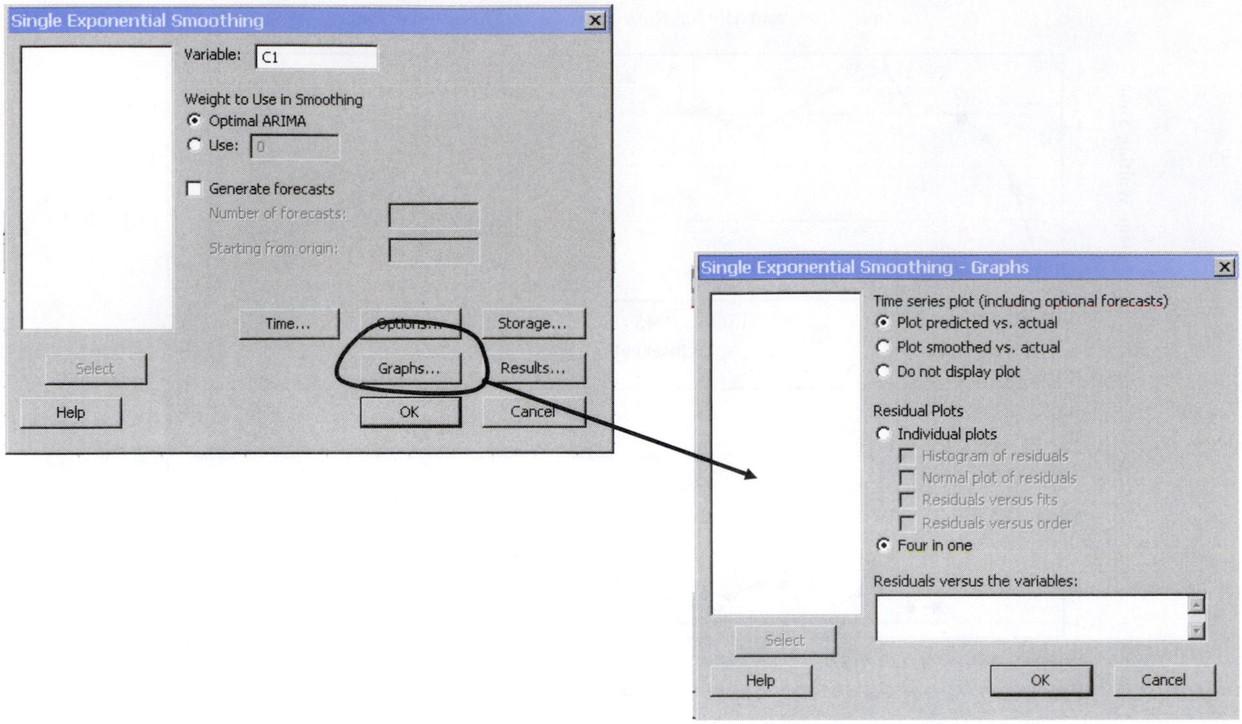

FIGURE 12-11. Minitab ARIMA Menus

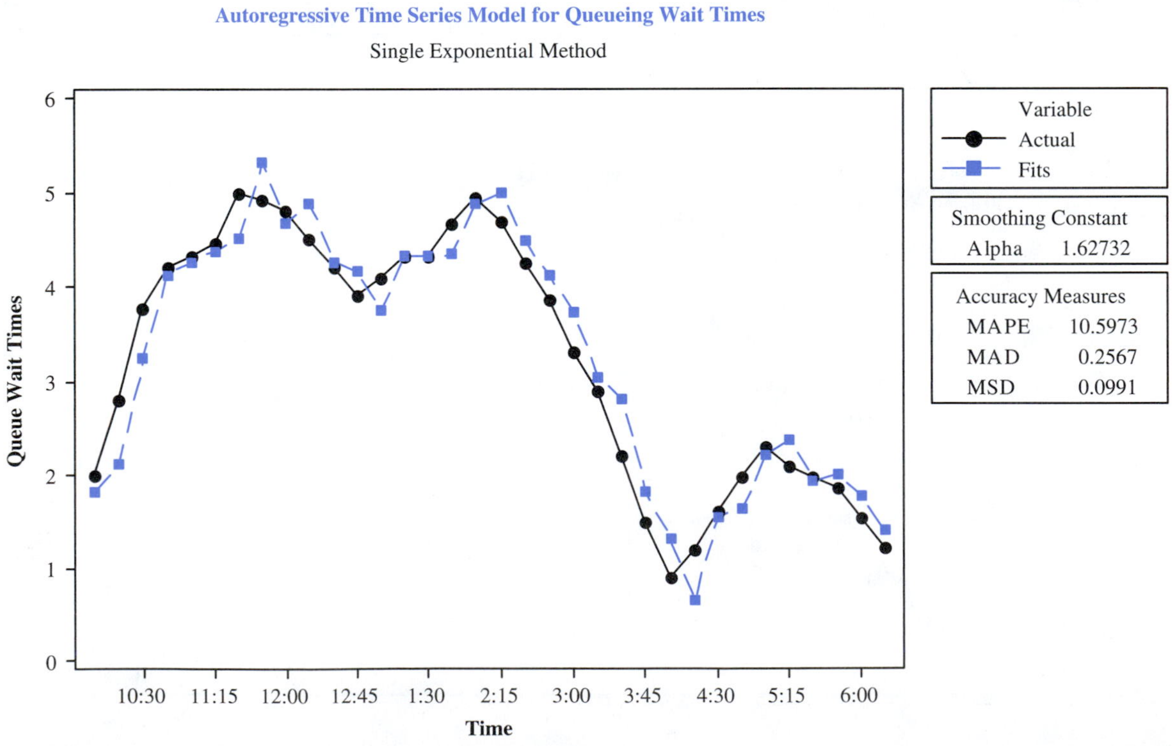

FIGURE 12-12. Time Series Plot of Actual versus Predicted Points for ARIMA Single Smoothing Model

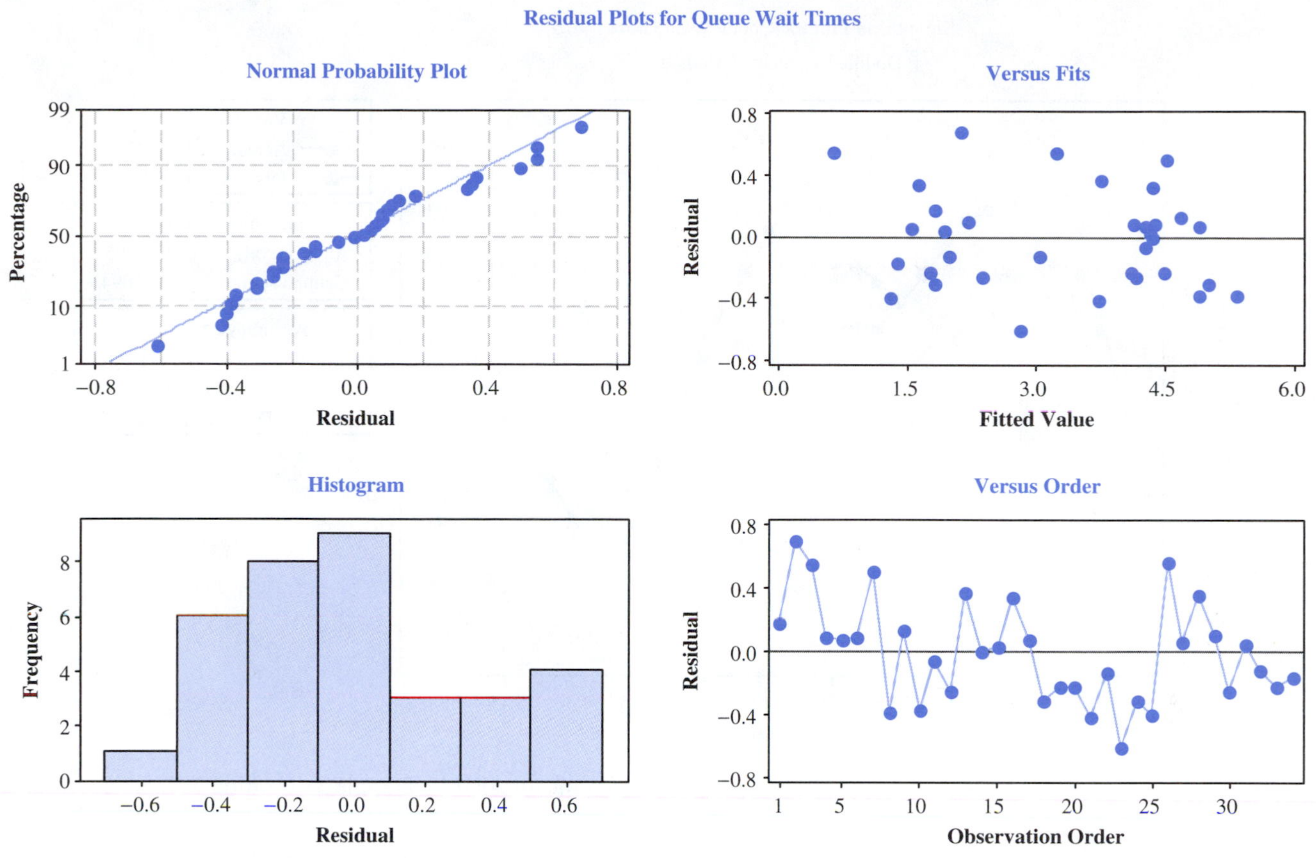

FIGURE 12-13. Residual Plots for ARIMA Single Smoothing Model

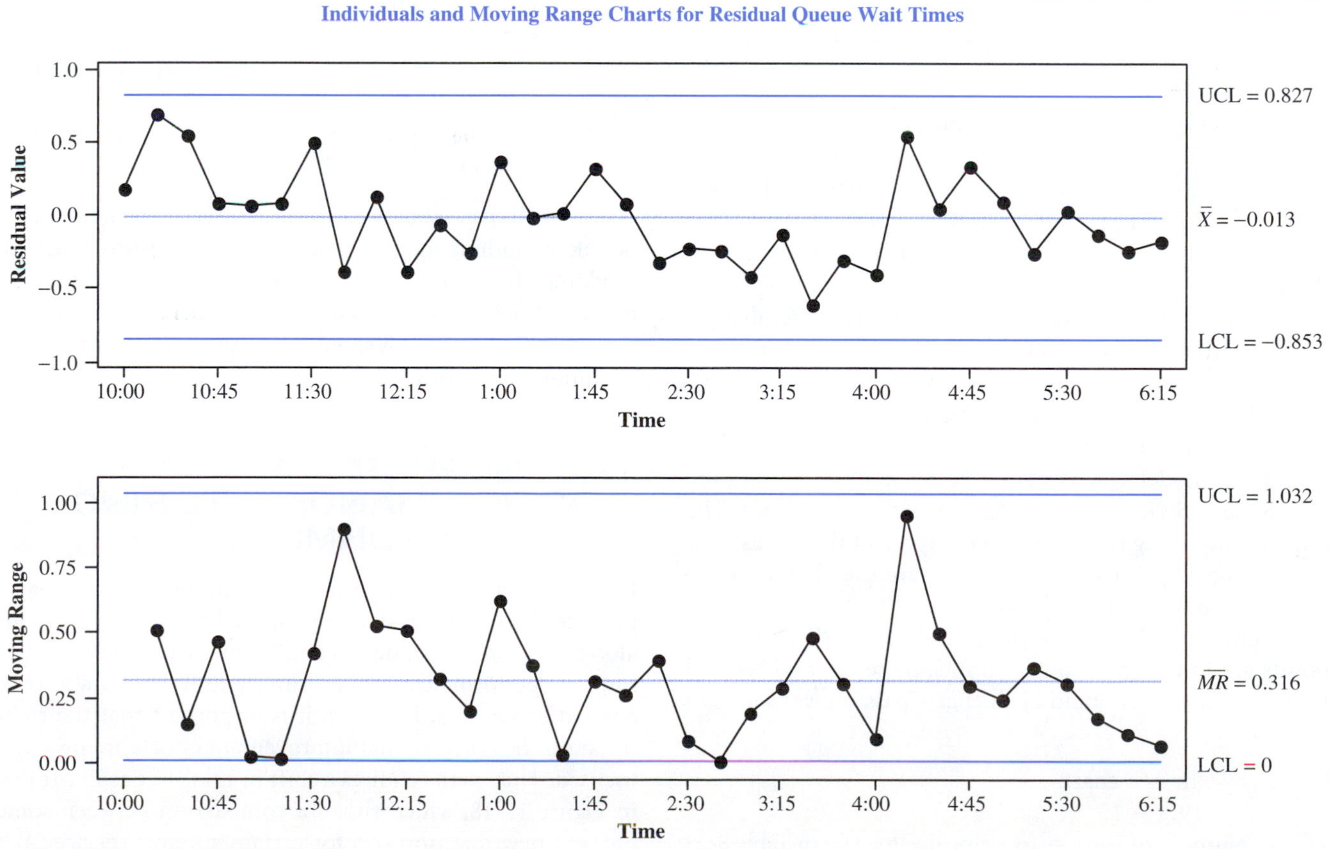

FIGURE 12-14. Individuals and Moving Range Chart for Residuals—ARIMA Single Smoothing Model

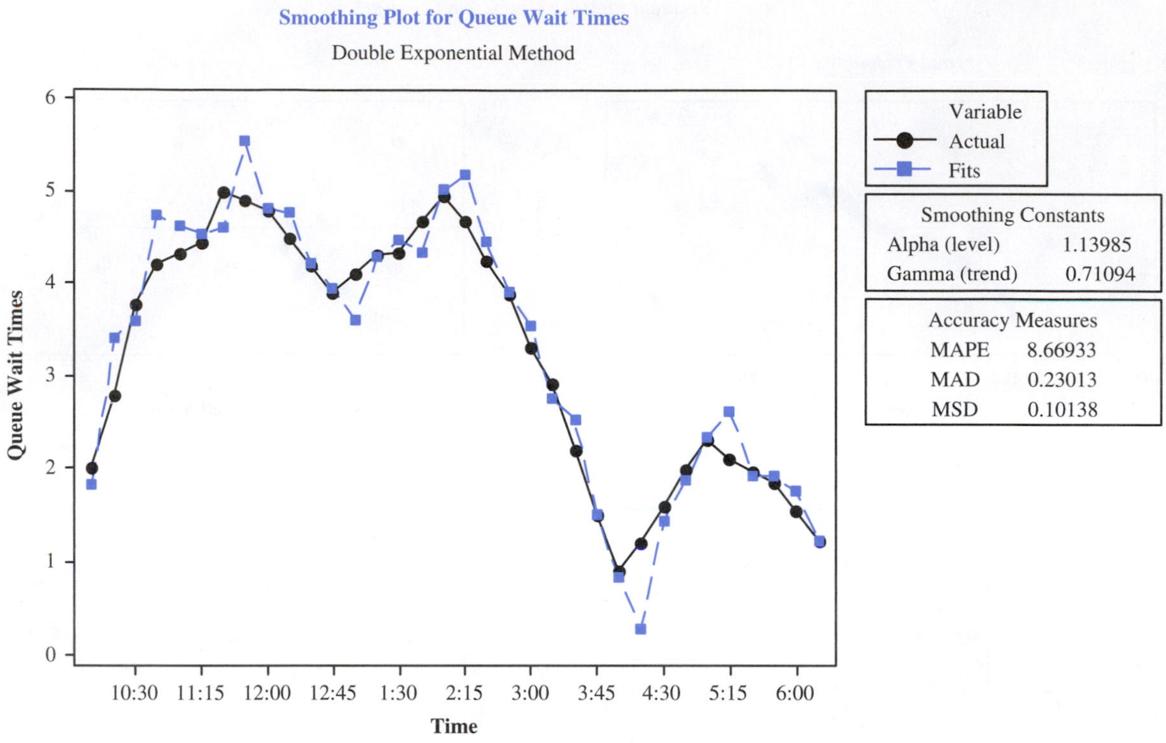

FIGURE 12-15. Time Series Plot of Actual versus Predicted Points for ARIMA Double Smoothing Model

For the *X* chart:
$r = 17, s = 17$
Number of runs = 12; critical value (from Table 8-2) = 12 ⇒ Violation
Length of longest run = 8; critical value (from Table 8-3) = 7 ⇒ Violation

Therefore we have a violation on the *X* chart in both the too-few-runs test and the length-of-the-longest-run rest. This is because, if we refer again to Figure 12-12, we observe that during the early morning period (from 10:00 a.m. to 11:30 a.m.) the model consistently underpredicts the actual queue time; and in the afternoon period (from 2:15 p.m. to 4:00 p.m.) the model consistently overpredicts the actual time. We can infer that the residual errors using this model are not randomly distributed about a mean of zero. However, the *MR* chart shows good control, indicating stability in the forecasting error over time.

To improve our model we can use an ARIMA double smoothing model which can be accessed in Minitab using the *Stat > Time Series > Double Exp Smoothing* sequence. The double smoothing model adds a component that adjusts for trends. Figures 12-15 through 12-17 are the charts for the ARIMA double smoothing model.

Figure 12-17 shows good control on both charts. No points are outside the control limits, there are no observed nonrandom patterns, and both charts pass the runs tests as follows.

For the *MR* chart:
$r = 19, s = 15$
Number of runs = 16; critical value (from Table 8-2) = 12 ⇒ OK
Length of longest run = 4; critical value (from Table 8-3) = 9 ⇒ OK

For the *X* chart:
$r = 20, s = 14$
Number of runs = 21; critical value (from Table 8-2) = 12 ⇒ OK
Length of longest run = 6; critical value (from Table 8-3) = 7 ⇒ OK

As both charts show good control we can conclude that the double smoothing model should provide a consistent fit, with randomly distributed errors, as long as the series remains stationary. Should the process change the model will no longer be a good predictor and this should show up as an out-of-control condition on one or both of the residuals control charts.

12.5 SUPPORT SERVICES IN A MANUFACTURING ENVIRONMENT

In a manufacturing environment, **support value streams** provide those services that are required by the workstations along the **primary value stream**. Many of the tools used to improve performance in the primary functions can also be applied to services. However, it is important that the right questions be asked so that improvement efforts are properly focused. This often requires a shift in mind-set as illustrated in Figure 12-18, which shows a comparison between some old and new question sets for certain support services at a large oil refining company.

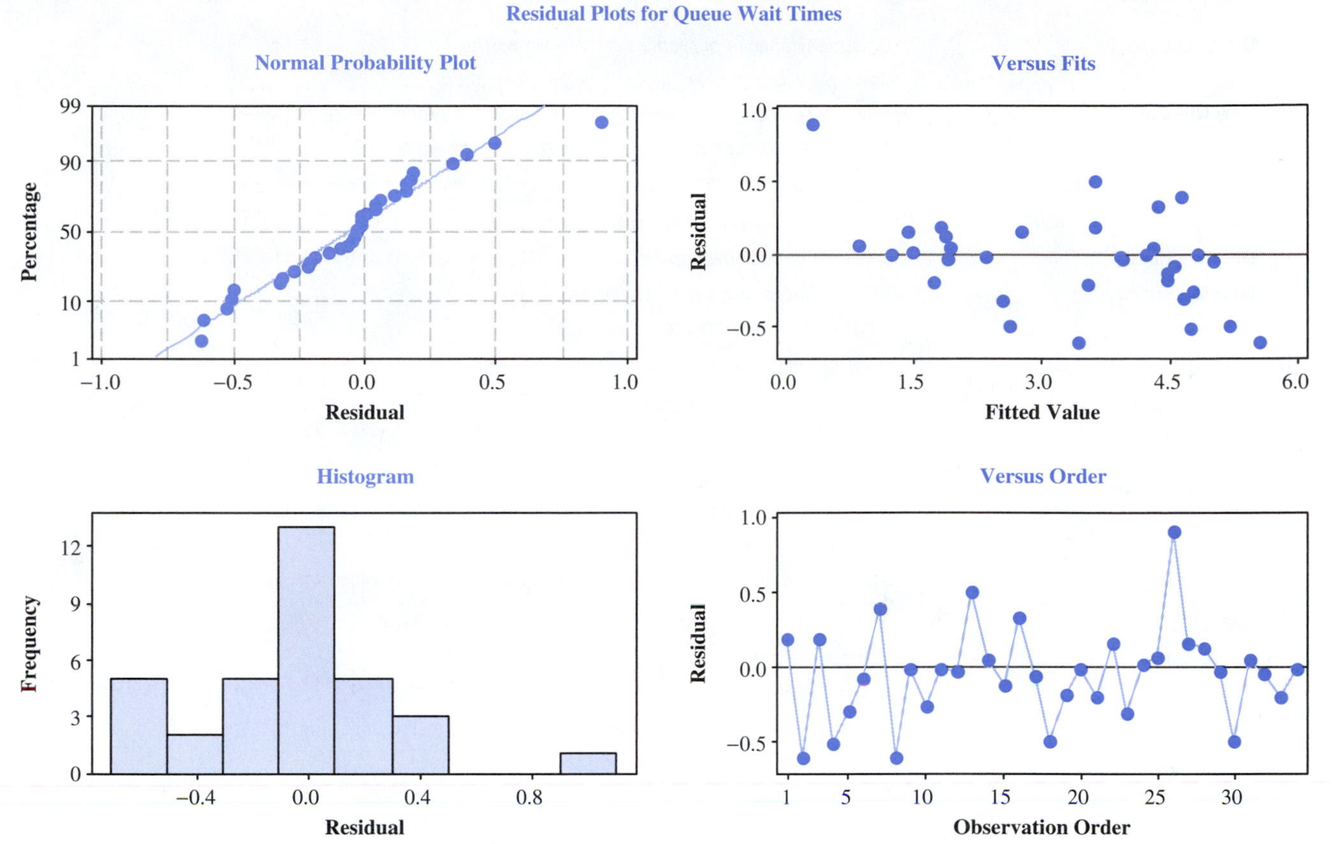

FIGURE 12-16. Residual Plots for ARIMA Double Smoothing Model

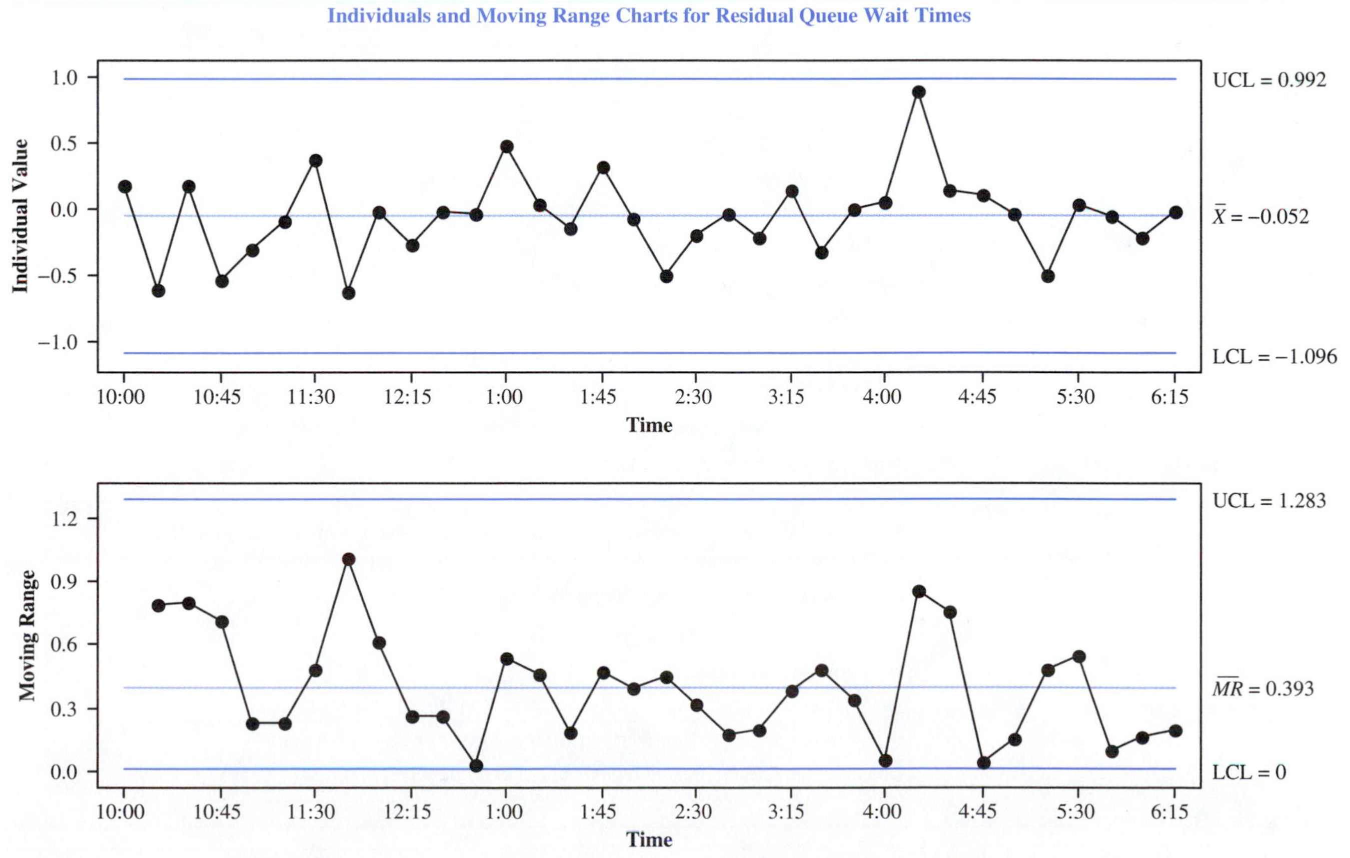

FIGURE 12-17. Individuals and Moving Range Chart for Residuals—ARIMA Double Smoothing Model

The Importance of Asking the Right Questions	
Accounts Payable	
Old Questions	*How many checks per labor-hour are we writing?*
	Can't we use temporary labor to write these checks?
New Questions	*How do we know that everyone understands the procedures?*
	What is the average error rate and is it statistically stable?
	What new procedures are we going to try to lower the average error rate?
Inventory Management	
Old Questions	*Can't we cut our inventories?*
New Questions	*What is the probability of use of the item?*
	What is the cost of being without it?
	What is the optimum number to hold in inventory?
Purchasing	
Old Questions	*Who has the lowest price?*
	Who has the best terms?
New Questions	*Is this supplier using statistical methods to ensure that we get a reliable component?*
	What is the supplier's capability index?
	Does its senior management use statistical thinking to make decisions?
Maintenance	
Old Questions	*Can't you replace that later?*
	How many people can you cut back?
New Questions	*How does the cost of an on-line failure compare to the cost of an off-line one?*
	Do we have reliable data on component life?
	Is there an optimal replacement strategy?
Scheduling	
Old Questions	*How long does that take and can't you get it to us more quickly?*
	What is the average delay?
New Questions	*How much variability is there in that time and what causes it?*
	Is the time in statistical control?
Customer Relations	
Old Questions	*What is the average number the customers will buy?*
	What color do most customers buy?
	What percentage of the customers prefer our product?
New Questions	*How much training do the interviewers get?*
	Will a postcard mail-out give us an adequate response rate?
	How complete is our sampling frame?
	What receipt control system is used?
	Are nonrespondents contacted to assess nonresponse bias?
	Do the data entry staff use a code book?
Personnel	
Old Questions	*Why are those benefit checks late?*
	How will we rank all of these people?
New Questions	*How were the procedures for paying benefit checks developed and how do we know that everyone is following them?*
	Do we have a method of identifying truly outstanding staff?
Sales	
Old Questions	*Why are sales falling off?*
New Questions	*Have we compared the key entry error rates from those new terminals?*
	Do all of our sales staff understand what is meant by statistical control?
	Does our sales force receive regular updates on our SPC progress?

FIGURE 12-18. Asking Relevant Questions of Support Value Chains

SHOWCASE of EXCELLENCE **Winners of the Malcolm Baldrige National Quality Award Service Category**

- 2006: Premier Inc., a health care alliance that provides group purchasing and supply chain management, insurance and risk management, and informatics and performance improvement services. http://www.PremierInc.com, http://www.quality.nist.gov/PDF_files/Premier_Profile.pdf.
- 2003: Caterpillar Financial Services, the financial services business unit within Caterpillar Inc., provides financing for the complete line of Caterpillar's products and financing to users and dealers of Caterpillar equipment and the Caterpillar Business Units. http://www.CatFinancial.com/Baldrige, http://www.nist.gov/public_affairs/baldrige2003/Catrpllar_3.3.04.pdf.
- 2000: Operations Management International Inc., an employee-owned company that operates small to midsized publicly owned wastewater and water treatment plants. http:// www.wateronline.com. http://www.nist.gov/public_affairs/baldrige00/OMI.htm.
- 1999 and 1992: Ritz-Carlton Hotel Company, developer and operator of luxury hotels in North America, Europe, Asia, Australia, the Middle East, Africa, and the Caribbean. http://www.ritzcarlton.com, http://www.nist.gov/public_affairs/bald99/ritz.htm. http://www.quality.nist.gov/Ritz_Carlton_Hotel_Co.htm.
- 1997: Merrill Lynch Credit Corporation, provider of real estate and securities-based consumer credit products, including home financing, personal credit, investment financing, and commercial real estate financing primarily to affluent individuals. http://mlcc.ml.com, http://www.quality.nist.gov/Merrill_Lynch_97.htm.

12.6 SUMMARY OF KEY IDEAS

Key Idea: *There are some important distinctions between quality as it pertains to a manufactured product and the quality of service delivery.*

Unlike most products, services are intangible, inseparable, variable, and perishable. When the service is completed there is usually no tangible object to serve as evidence that the transaction took place. The major outcome is an experience, and quality is based on individual perceptions of whether the experience was pleasant or unpleasant.

Services are also inseparable from the service provider, and unlike most manufactured goods, the ultimate consumer comes face-to-face with the producer who may participate in both the provision and the consumption of the service. Service quality depends on who provides the service as well as when, where, and how service is delivered.

A service cannot be inventoried, repaired, or reworked. The provider normally has only a single chance at earning customer satisfaction, trust, and loyalty.

Key Idea: *The service gap model defines seven gaps that can lead to poor quality; however, only the gap between customers' expectations and customers' perceptions of the quality actually delivered is a reliable measure of service quality.*

The only true measure of service quality is how well customers perceive that providers have fulfilled their promises and lived up to customers' expectations. Customer expectations are influenced by numerous factors, including word-of-mouth communications, individual personal needs and attitudes, and past experience. Since customers make value judgments on an individual basis it is important that providers strive to interact with each customer on a personal level.

Key Idea: *If management is committed to full disclosure and open internal communications, it has numerous options available for improvement.*

Open communications with all stakeholders is fundamental to the open systems approach and to quality inspired management. If management has the will to better communicate it can make full use of such tools as team meetings, storyboard displays, newsletters, electronic media, bulletin boards, quality awards, and staff meetings.

Key Idea: *Customer satisfaction is complex and individual. Studies have revealed that satisfaction is more likely when customers believe that service providers have delivered on their promises, care about them as individuals with personal needs, and are willing to do whatever it takes to satisfy those needs.*

Customer satisfaction is dependent on how customers perceive the competence, friendliness, personalization, and trust of the service provider. Customers have also indicated that they favor providers who are easy to work with and who provide services that are easily accessible and flexible.

Key Idea: *The application of some principles from the behavioral sciences can help service providers positively shape customers' attitudes toward satisfaction.*

Five principles for influencing customers' attitudes are rooted in the disciplines of behavioral science and cognitive and social psychology. The first principle states that customers prefer to receive bad news first and good news last so that a service experience can end on a positive note. The second principle is to avoid communicating "near" successes. The anguish created over what might have been can override any positive aspects of a service transaction. The third principle is to provide customers the maximum practicable control over outcomes, and the fourth principle is to break down pleasurable aspects into small pieces so that customers perceive them to last longer. Unpleasant aspects should be combined. The fifth and final principle is to conclude a service experience on a strong positive note, as customers' attitudes are influenced more by how they feel as they exit the service than how they felt on the average or the highs and lows for the duration.

Key Idea: *Customer satisfaction has been measured using surveys, focus groups, targeted interviews, comment cards, dissatisfaction information systems, and mystery shoppers.*

Surveys are the most widely used instrument as they can be administered directly to a representative cross section of customers. Focus groups may be used when a small representative group of individuals can be identified who are willing to meet periodically to discuss service-related issues. One of the problems with surveys is that response rates are typically low. This can be overcome through targeted interviews in which a small sample of customers is selected for face-to-face interviews. With comment cards customers are invited to make comments on what they liked or disliked concerning their service experience. A dissatisfaction information system is a structured database that stores various types of information concerning customer satisfaction, such as formal complaints, claims, returns, refunds, recalls, repeat services, litigation, replacements, downgrades, warranties, and mistakes. Mystery shoppers are personnel employed by the service provider to shop and report on their services as well as those of their competitors.

Key Idea: *Customer satisfaction surveys can be administered using a variety of media, including face-to-face interviews, federal postal service, telephone, and the Internet. It is important that the survey layout be nonintimidating, user-friendly, uncluttered, and attractive.*

The primary instrument used to collect survey data is the questionnaire. Questionnaires should be designed to enhance readability including ample white space to separate questions and sections, response areas and format consistent and designed in a way that makes it easy for intended responses to be recorded on the right-hand-side of the form, and the use of no more than two colors to distinguish sections and questions. Multiple or fancy font types or sizes, italics, distracting color schemes, animations, single-spacing, or crowding of too much material on one page should be avoided.

Key Idea: *Poorly worded or constructed questions can lead to bias.*

The most common pitfalls to avoid are ambiguity, lack of mutual exclusivity, the creation of a nonexhaustive response set, requiring of respondents to rank list too many items, failure to buffer socially favored positions, loaded questions, leading questions, unfamiliar terms and jargon, requiring of information for which respondents do not have access, multidimensional questions, compound clauses, requiring of memory recall, unnecessary complexity, weak grammatical format, use of hypotheticals, use of inappropriate assumptions, use of gender-specific language, and mismatched response sets.

Key Idea: *SERVQUAL is a structured survey methodology for measuring service quality in five areas: assurance, empathy, reliability, responsiveness, and tangibles.*

SERVQUAL is an approach that requires respondents to score how well a service provider performs in the five areas and, as a comparison, respondents are asked to provide the scores that they perceive the top providers in the same industry would receive in each of these same areas. Assurance measures the ability to inspire confidence and trust; empathy measures the degree to which the provider can project a sense of caring for, understanding of, and responding to individual needs; reliability measures the ability to deliver what was promised; responsiveness measures willingness and flexibility; and tangibles measure the quality of physical facilities, equipment, and overall appearance.

Key Idea: *In a service environment quantitative measurements may not be normally distributed.*

When a data set shows a marked deviation from normality (e.g., severe skew, uniform flatness, multimodality, etc.) transformations should be investigated to determine if the nonnormal data can be converted into a normally distributed data. If so, the transformed data are plotted on conventional control charts. The difficulty arises when one wishes to convert control chart information back into the original unit, which is not easily accomplished.

Key Idea: *Control charts require independence of plotted points and cannot be directly applied to raw data that are correlated in time.*

When autocorrelation is present a time series model should be fitted to the data. The most widely used models for this purpose are called autoregressive integrated moving average (ARIMA) models and are based on Box-Jenkins methodologies. The time series model can be used to forecast past and future measurements in the series. For autocorrelated data a process is in a state of statistical control provided the time series is stationary—that is, the mean and variance are constant. Once a suitable model is found statistical control can be monitored by plotting the residuals (i.e., forecast errors) on conventional control charts. If the process is in control the residuals will be normally and randomly distributed about a mean of zero. Any out-of-control conditions with regard to the residual plots means that some fundamental change in either the mean or variance (or both) has occurred in the process and the model used no longer provides reliable forecasts.

12.7 MIND EXPANDERS TO TEST YOUR UNDERSTANDING

12-1. Differentiate between the following.

a. Quality of an automobile and quality of an automobile repair facility

b. Quality of a textbook and quality of an education

c. Quality of a blouse and quality of service at a clothing store

d. Quality of a fishing rod and quality of a fishing experience on the lake

e. Quality of a steak and quality of a dining out experience at a steak restaurant

12-2. Which of the seven gaps described by the service gap model might occur as a result of each of the following?

a. A movie producer chooses new projects based solely on the input of critics, the media, and management perceptions regarding what movies the public will support.

b. To secure a contract, a construction company promises a client that a project will be completed within six weeks, while the critical path for the project is nine weeks.

c. A health care provider, concerned about flagging profits, has implemented across-the-board cuts in the operating budgets for all departments.

d. A customer service representative fails to implement the company policy that dictates *service with a smile, courtesy above all, and when in doubt call a supervisor*. As a result an unpleasant experience with a customer became much worse.

e. Guests check into an exclusive hotel and perceive that the room is unclean, discover that the TV reception is poor, and are surprised to find that in-the-room complimentary coffee service is not provided.

12-3. Imagine that you have recently been hired by a company that is committed to the principles of QIM and has implemented a number of initiatives to empower employees and be more customer focused. An employee survey has revealed that a significant number of personnel do not trust that management is open and honest with them. Even middle managers have complained that the senior leaders never tell them anything. As a new employee you have been tasked to evaluate the internal communications system and make some recommendations. What process would you follow and what are some of the important factors you would want to investigate?

12-4. Consider the following service providers. For each one, what do you believe are the most important components of customer satisfaction and why?

a. Post office window employees

b. Bank loan officer

c. Enrollment clerk at a university

d. Investment counselor

e. University professor

f. Airline ticket agent

g. Building contractor

12-5. Which of the five behavioral science principles do you think are applicable in shaping positive attitudes in the following service scenarios?

a. Doctor's consultation following a diagnostic procedure

b. Amusement ride at a theme park

c. Self-check lane at supermarket

d. Five-course meal at a fancy restaurant

12-6. The president of First National Bank is evaluating the service at the drive-through teller window at one of the branch banks. Service times were recorded each 15 minutes over a 25-hour period and then subgrouped by hour. The results are shown in the following table.

Hour	Observed Time (secs)				$\bar{X}$	R
1	123.38	105.14	78.63	92.68	99.96	44.76
2	86.29	95.31	116.85	86.32	96.19	30.55
3	105.37	99.35	100.87	97.24	100.71	8.13
4	62.26	100.12	106.86	103.13	93.09	44.60
5	88.57	93.68	88.68	94.90	91.46	6.33
6	100.37	80.87	78.02	91.73	87.75	22.35
7	112.76	101.75	98.73	90.41	100.91	22.36
8	101.65	106.24	87.87	95.76	97.88	18.36
9	84.08	92.86	128.88	69.72	91.39	69.16
10	82.83	117.01	104.50	84.95	97.32	34.17
11	79.15	111.27	104.51	97.19	98.03	32.12
12	93.36	78.96	85.27	72.52	82.53	20.83
13	104.45	99.17	98.88	78.60	95.27	25.85
14	92.31	83.42	126.16	85.10	96.75	42.75
15	60.14	110.25	74.55	95.60	85.14	50.11
16	95.64	93.58	94.35	96.99	95.14	3.41
17	77.06	78.50	96.33	88.38	85.07	19.26
18	75.61	86.20	97.90	98.34	89.51	22.73
19	132.57	85.88	98.27	96.34	103.26	46.69
20	85.79	97.39	109.40	104.41	99.25	23.61
21	99.08	82.92	88.23	75.51	86.44	23.57
22	108.06	102.03	74.85	72.73	89.42	35.33
23	86.69	119.63	94.77	82.68	95.94	36.95
24	109.16	67.22	96.37	82.48	88.81	41.94
25	78.79	105.29	87.00	80.67	87.94	26.50

a. Is the process in control? Use control charts to justify your answer.

b. Is there any evidence that the time required to service a customer at the drive-through window depends on the hour in which the customer arrives? Explain.

c. What percentage of the customers would you estimate take in excess of two minutes to serve? What percentage would you estimate will be served in less than one minute?

12-7. A courier service tracks the delivery times on its packages. Forty packages were selected at random and carefully monitored from the time of receipt from the customer to the time of delivery. The results are reported in the following table.

Shipment	Delivery Time (days)	Shipment	Delivery Time (days)
1	2.17	21	1.27
2	1.19	22	5.04
3	1.93	23	1.21
4	2.94	24	2.24
5	2.65	25	1.85
6	1.75	26	1.30
7	7.29	27	2.58
8	3.26	28	2.81
9	2.71	29	1.39
10	2.22	30	2.83
11	2.81	31	5.15
12	3.89	32	2.73
13	3.61	33	1.91
14	2.17	34	1.47
15	3.05	35	3.35
16	1.03	36	2.41
17	1.36	37	1.54
18	1.25	38	6.56
19	2.55	39	1.58
20	2.46	40	1.55

a. Show that these data are not normally distributed.

b. Use a Box-Cox transformation with $\lambda = 0$ to determine if the process is in a state of statistical control.

c. The courier advertises that shipments will be delivered within three days for this class of service. What percentage of packages would you estimate will take longer than three days to reach their destinations?

d. What is your best estimate of the mean and standard deviation for this process?

e. The logistics manager was not too pleased with the results you obtained in part c and formed a Six Sigma team to improve performance. After some changes were implemented, 40 more samples were collected and are shown in the following table.

Shipment	Delivery Time (days)	Shipment	Delivery Time (days)
41	1.75	61	2.31
42	1.89	62	2.13
43	1.62	63	1.58
44	2.90	64	2.03
45	4.59	65	1.98
46	1.25	66	1.67
47	1.78	67	3.05
48	3.56	68	1.83
49	2.02	69	2.26
50	3.43	70	1.53
51	3.45	71	1.75
52	2.13	72	1.45
53	1.11	73	1.31
54	2.13	74	1.06
55	1.99	75	1.48
56	3.40	76	1.33
57	1.27	77	1.99
58	4.55	78	2.69
59	2.68	79	1.53
60	3.19	80	2.03

Did the changes work? What percentage of the packages shipped now will take longer than three days to reach their destination?

12-8. A milling operation is used to reduce the thickness on the end of a metal plate. As the tool wears the average thickness increases; therefore, between tool replacements the thickness measurements are autocorrelated. The following data represent 18 samples of size 4 that were selected at random during a production run using the same cutting tool.

a. Plot conventional $\overline{X}$ and R charts for these data. Explain why these charts are meaningless for controlling this process.

b. Use Minitab to fit an appropriate ARIMA model to the data. Store the residuals in a column on the Minitab worksheet.

c. Plot a conventional $\overline{X}$ and R chart for the residual data collected in part b. Is the process in statistical control? Explain.

Sample No.	Thickness (mm)	Sample No.	Thickness (mm)	Sample No.	Thickness (mm)
1	0.9200	7	0.8980	13	0.9295
	0.6267		0.8259		0.9697
	0.6720		0.8457		0.9291
	0.6971		0.8349		0.9188
2	0.6778	8	0.8344	14	0.9590
	0.7309		0.8746		0.9387
	0.7323		0.8334		0.9385
	0.8320		0.8635		0.9888
3	0.7306	9	0.8935	15	0.9584
	0.7789		0.8728		0.9784
	0.8067		0.9232		0.9783
	0.8139		0.8823		0.9681
4	0.7800	10	0.8819	16	0.9780
	0.8296		0.8816		0.9476
	0.8069		0.8610		0.9677
	0.8465		0.9114		1.0381
5	0.7933	11	0.8807	17	0.9977
	0.8229		0.9210		1.0177
	0.8319		0.9106		0.9974
	0.8000		0.9407		0.9973
6	0.8912	12	0.9000	18	0.9972
	0.8186		0.8998		1.0173
	0.8484		0.9299		0.9970
	0.8272		0.8994		1.0473

d. The residuals computed for the next 9 samples are recorded in the following table.

Sample No.	Residual Thickness (mm)	Sample No.	Residual Thickness (mm)	Sample No.	Residual Thickness (mm)
19	0.0339	22	0.0496	25	0.0731
	0.0699		−0.0329		0.0339
	0.1249		0.0532		0.0507
	0.1583		−0.0477		0.0948
20	0.0225	23	0.0608	26	0.0460
	0.1363		0.0393		0.1505
	−0.0053		0.0577		0.0120
	−0.0598		0.0378		0.0570
21	−0.0523	24	−0.0886	27	0.0672
	0.0662		0.0565		0.0382
	0.0658		0.1077		0.1009
	−0.0832		0.0321		0.1012

Plot these points on the control charts you constructed in part c. Is there evidence that the process has changed? Explain.

12-9. At 15-minute intervals customer service times at a favorite coffee house were recorded over a 10-hour period on a certain day. The analyst began timing a customer when he or she approached the counter and finished timing when the customer turned to leave, having been served. The observed times are shown in the following table.

Time	Wait + Svc Time (min)	Time	Wait + Svc Time (min)	Time	Wait + Svc Time (min)
8:00	1.8833	11:30	2.7333	3:00	1.7833
8:15	1.7000	11:45	1.8167	3:15	3.7167
8:30	2.1833	12:00	1.6500	3:30	8.5833
8:45	4.5833	12:15	1.4833	3:45	4.2667
9:00	3.6667	12:30	1.9667	4:00	4.8333
9:15	0.5833	12:45	2.5000	4:15	1.6333
9:30	4.1833	1:00	1.6500	4:30	2.3500
9:45	2.9500	1:15	3.0333	4:45	5.0667
10:00	6.2167	1:30	2.9167	5:00	2.7167
10:15	5.9667	1:45	1.3833	5:15	4.9833
10:30	3.8667	2:00	1.8833	5:30	2.8000
10:45	3.1333	2:15	4.3833	5:45	1.4667
11:00	1.6167	2:30	5.5333	6:00	1.0833
11:15	3.1667	2:45	2.0500	6:15	1.3333

a. Construct *X* and *MR* charts and comment on the state of control for the coffee house service process.

b. Use Minitab to construct a histogram for the data. Do the data appear to be normal? Would you expect them to be? Does the histogram cast some doubt as to the validity of the conclusions you reached in part a? Explain.

c. Use Minitab to test the data for the presence of autocorrelation. Is there evidence of autocorrelation? Is it statistically significant? Does this test cast some doubt as to the validity of the conclusions you reached in part a? Explain.

d. With Minitab, use a Box-Cox transformation with an optimal λ to determine the state of control of the service process. Have your conclusions changed from those reached in part a? Explain.

e. On closer investigation the time recorded for the 9:15 time slot was challenged. It was suspected that the observer made an error in either writing down the time or in recording it. Consequently the 9:15 time was deleted from the data set. With the modified data set repeat parts a–d. Based on this analysis, what is your assessment of the state of control of the process?

12-10. When the data in problem 12-9 were subjected to close scrutiny, several of the entries that had originally been made were questionable. Those that were in doubt were replaced with times taken during the applicable time intervals on a similar day. The resulting data are shown in the following table.

Time	Wait + Svc Time (min)	Time	Wait + Svc Time (min)	Time	Wait + Svc Time (min)
8:00	1.8833	11:30	2.7333	3:00	1.7833
8:15	1.7000	11:45	1.8167	3:15	2.7167
8:30	2.1833	12:00	1.6500	3:30	3.5833
8:45	4.5833	12:15	1.4833	3:45	3.2667
9:00	3.6667	12:30	1.9667	4:00	4.1333
9:15	3.5833	12:45	2.5000	4:15	3.6333
9:30	4.1833	1:00	1.6500	4:30	2.3500
9:45	2.9500	1:15	3.0333	4:45	3.0667
10:00	4.2167	1:30	2.9167	5:00	2.7167
10:15	5.9667	1:45	1.3833	5:15	3.9833
10:30	4.8667	2:00	1.8833	5:30	2.8000
10:45	3.1333	2:15	2.3833	5:45	2.4667
11:00	2.6167	2:30	2.5333	6:00	2.0833
11:15	3.1667	2:45	2.0500	6:15	2.3333

a. Construct *X* and *MR* charts for the revised data and comment on the state of control for the coffeehouse service process.

b. Use Minitab to test the data for the presence of autocorrelation. Is there evidence of autocorrelation? Is it statistically significant? Does this test cast some doubt as to the validity of the conclusions you reached in part a? Explain.

c. Use Minitab to fit an appropriate ARIMA model to the data. Store the residuals in a column on the Minitab worksheet.

d. Plot conventional *X* and *MR* charts for the residual data collected in part c. Is the process in statistical control? Explain.

e. Are the residuals normally distributed? Do the residuals appear to be randomly distributed about a mean equal to zero?

PART
FOUR

QUALITY BELOW THE SURFACE

CHAPTER THIRTEEN

USING ANOVA AND DOE TO DETERMINE WHAT MAKES A PROCESS TICK

QUALITY MAXIM #13: *Opinions, arguments, and hypotheses mean nothing unless the conclusions can be verified by experimental science. Verifiable conclusions lead to discovery and discovery leads to sustainable improvement.*

The Process Pretender—Changing Pretense to Understanding

"The Great Pretender"[1] is a song about someone who has gone through life masquerading as someone else—trying to be something he is not, and the lyricist warns that things are not always as they seem. While the song laments the plight of a broken spirit, the idea of seeming to be something quite different from reality is a fundamental truth underlying many an industrial process. Even with years of experience, seasoned veterans on the front lines of production may fail to understand the intricate mysteries of the processes they command. A quality consultant with a metropolitan newspaper had just completed a tour of the pressroom when he received an urgent call from the press supervisor, who expressed concern that the consultant might not appreciate that printing is *more art than science*. This supervisor was merely perpetuating the myth that excellence is influenced more by finesse than technical knowledge, and can explain why operators on different shifts will react to production problems in different ways. Such a lack of profound understanding as to what *should* be done leads to operator inconsistencies and contributes to process variability. Observing *how* processes behave does not explain *why* they behave the way they do, *what* intervention strategies are necessary to influence behavior, *when* interventions are necessary, and *when* they are not. Operators are not to be faulted for their lack of understanding which can only come from the application of some advanced statistical methods such as design of experiments (DOE), analysis of variance (ANOVA), and correlation analysis. Using sound statistical methodologies to delve into uncharted territories can expose the true process drivers, how they interact, and the extent to which they individually or collectively impact quality output. There are several obvious downsides to conducting such in-depth studies. First, there is the cost factor. A properly designed experiment is expensive, time-consuming, and disruptive, and can take capacity away from where it is needed to satisfy customer commitments. Second, planning and conducting industrial experiments require specialized skills which may not be available in-house; and the use of costly external consultants who may be unfamiliar with the processes, can potentially produce suboptimal results. Nevertheless, the use of SPC tools alone can go only so far in improving quality at the source. DOE and ANOVA provide the capabilities to unlock process secrets and help an organization learn which process settings are best, the relative importance of critical process and input variables, and gain an understanding of which variables should be monitored. Quality depends on ending the pretense that operations have mysterious aspects that are unknown and unknowable; striving for improved and shared knowledge; and using process knowledge to optimize performance.

[1]The music and lyrics for "The Great Pretender" were written by Buck Ram, the manager and producer for The Platters, a top vocal group in the early days of rock and roll, who recorded the song as a single on November 3, 1955. These can be accessed at http://www.links2love.com/love_lyrics_134.htm.

LEARNING OBJECTIVES AND KEY OUTCOMES

After careful study of the material in Chapter 13, you should be able to:

1. **Compute and interpret a linear correlation coefficient and coefficient of determination for paired data sets.**
2. **Use a *t* test to determine the statistical significance of a linear correlation coefficient.**
3. **Use computer software to construct and properly interpret an ANOVA table.**
4. **Use an analysis of variance approach to develop and test the fit of a regression model for predicting a process outcome from one or more process variables.**
5. **Use an ANOVA table to determine the statistical significance of a regression model.**

(continued)

6. Use a one-way ANOVA to identify improvement opportunities by determining differences in process means.
7. Formulate a fixed effects two-level design of experiment for the purpose of process improvement.
8. Use computer software to run screening designs to eliminate inferior variables from a designed experiment.
9. Use computer software to run and analyze complete and fractional two-level fixed effects design of experiments.

13.1 PROCESS IMPROVEMENT THROUGH CORRELATION ANALYSIS AND ANOVA

13.1.1 Linear Correlation

Correlation is the study of associations between pairs of variables. A correlation analysis will show the nature (e.g., linear, quadratic), strength, and direction of any observed association. Correlation does not imply causality; however, knowing that two or more variables are strongly related can be the source of useful process knowledge. Conversely, there are times when, based on technical knowledge, there is the expectation that two or more variables are correlated. If the data fail to support this expectation, an investigation as to the reason for this can also lead to insights and discovery.

Correlation does not imply causality.

Consider the data in Table 13-1 which were taken from an extrusion line in a tire plant. The data presented represent two critical measurements taken from tire tread profiles: one called the center gauge measurement and the other the shoulder gauge measurement.

Figure 13-1 depicts the $\overline{X}$ charts for the two variables measured using samples of size 4. The two charts are almost identical. When the center gauge measures low, the shoulder gauge measures low; when the center gauge is high, the shoulder gauge is also high. The conclusion can be drawn that these two variables are not independent and are positively correlated.

The observed association between the two variables and the relative strength of the association can be statistically tested. The first step is to designate one variable as y and the other as x and then plot the xy pairs on a common graph. The resulting graph is called a scatterplot, or scatter diagram. The scatterplot for the tire tread data is shown in Figure 13-2. The first level of analysis is to examine the scatterplot for any noticeable nonrandom patterns. In Figure 13-2 there appears to be a strong positive linear relationship between the x and y variables. The association is positive because we observe that when the x variable increases, the y variable also increases; conversely, y decreases with a decrease in x.

A worker sorts lumber at a pulp mill. Through the use of advanced statistical tools including ANOVA and DOE, these mill workers can better understand how variability in size, species, age, and grade can affect the production processes.

A linear **correlation coefficient**, $-1 \le r \le 1$, can be computed as a measure of the relative strength of association. The sign of r shows the direction of the association (positive or negative) and the closer to plus or minus one the stronger the relationship. The linear correlation coefficient can be computed using the following formula.

$$r = \frac{\sum_{i=1}^{n} x_i y_i - n\bar{x}\bar{y}}{(n-1)s_x s_y} \qquad \textbf{13-1}$$

where,

r = linear correlation coefficient
$\bar{x}$ = average x measurement
$\bar{y}$ = average y measurement
s_x = standard deviation of x measurements
s_y = standard deviation of y measurements
n = total number of measurements

In Table 13-2 the tire data spreadsheet has been expanded to include a column for the cross-product term xy to facilitate the calculation of Equation 13-1.

These calculations are shown as follows:

$$r = \frac{13{,}895{,}673 - (100)(454.38)(304.74)}{(99)(26.403)(18.803)}$$
$$= \frac{48{,}896.88}{49{,}149.11} = 0.9949$$

An alternative measure of the strength of an association is called the **coefficient of determination**, $0 \le r^2 \le 1$. Values of r^2 close to one indicate strong correlation, whereas those close to zero indicate a weak or no association. For the tire tread example,

$$r^2 = (0.9949)^2 = 0.99$$

The coefficient of determination also has an interesting interpretation. The total observed variance of y is s_y^2, and the coefficient of

r^2 is the proportion of the variance of y that can be explained by its association with x.

TABLE 13-1 Two Dimensions on Selected Tire Treads

Tread Number	Center Gauge	Shoulder Gauge	Tread Number	Center Gauge	Shoulder Gauge	Tread Number	Center Gauge	Shoulder Gauge
1	335	498	34	334	496	67	303	451
2	272	414	35	334	496	68	314	466
3	278	419	36	269	411	69	298	443
4	283	424	37	328	492	70	286	425
5	311	463	38	288	433	71	310	460
6	311	464	39	302	449	72	285	424
7	312	464	40	305	455	73	274	416
8	282	422	41	301	445	74	346	513
9	313	465	42	301	448	75	306	456
10	304	451	43	291	434	76	304	454
11	328	492	44	344	505	77	305	455
12	290	434	45	298	441	78	295	439
13	340	505	46	321	477	79	319	474
14	297	440	47	287	429	80	288	433
15	323	482	48	332	494	81	320	474
16	302	448	49	332	495	82	320	475
17	334	495	50	277	417	83	304	454
18	280	421	51	283	423	84	317	470
19	314	466	52	301	448	85	294	439
20	300	443	53	325	490	86	315	466
21	312	464	54	327	491	87	298	441
22	313	465	55	308	458	88	303	451
23	262	388	56	309	458	89	315	467
24	310	461	57	288	432	90	316	470
25	300	445	58	291	438	91	297	441
26	280	420	59	275	416	92	317	471
27	329	494	60	319	473	93	318	473
28	293	438	61	300	444	94	305	455
29	322	479	62	296	440	95	287	428
30	322	481	63	257	388	96	307	457
31	273	415	64	310	462	97	308	458
32	325	484	65	311	463	98	309	459
33	303	450	66	265	408	99	306	456
						100	323	481

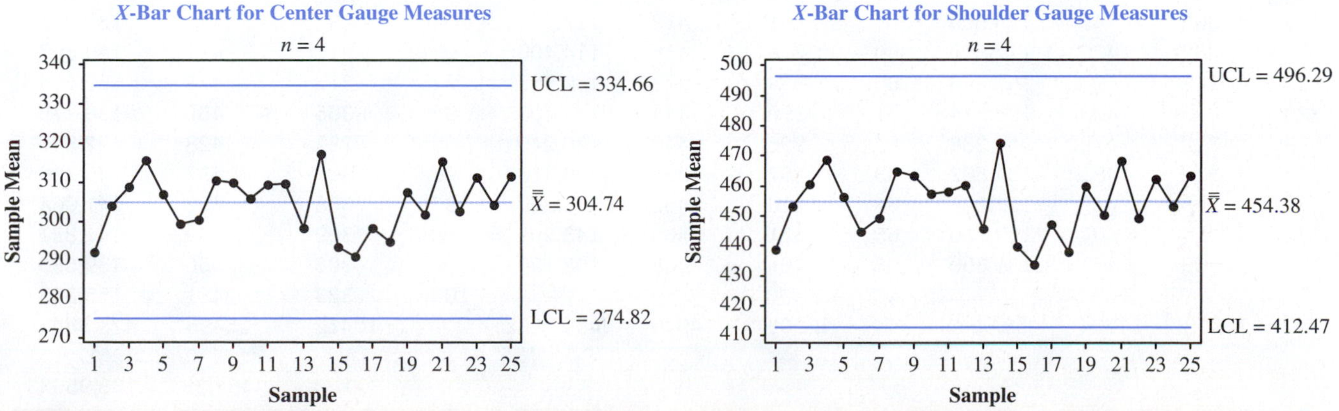

FIGURE 13-1. Strong Positive Correlation between Control Charts for Two Measures

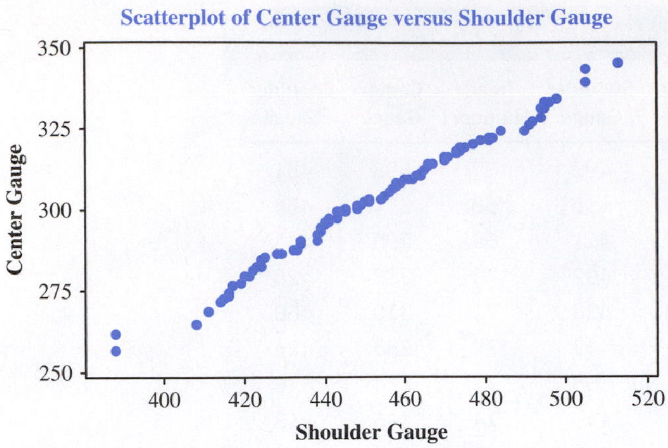

FIGURE 13-2. Scatterplot of Two Tire Tread Dimensions

determination, r^2, represents the total proportion of s_y^2 that can be explained by its linear association with the variable x.

Many statistical software packages will provide an r^2 calculation as part of the output, and will also fit a line to the data. Figure 13-3 shows a line fit and the r^2 calculation using Minitab. The equation shown for the function fitting the data is called a **regression line**. However, a word of caution is in order. Normally a regression line is used to predict a dependent variable (response or y) from an independent variable (x) where it is assumed that the independent variable is a major driver affecting the value of the dependent variable. In many quality control applications a correlation analysis is performed on two dependent variables, as is the case with the tire tread measures. Neither of the dimensions plotted affects the other—both are affected by the same set of process variables

TABLE 13-2 Two Dimensions on Selected Tire Treads—Correlation Coefficient Calculation

Trd No.	Ctr Gge y	Shldr Gge x	xy	Trd No.	Ctr Gge y	Shldr Gge x	xy	Trd No.	Ctr Gge y	Shldr Gge x	xy
1	335	498	166,830	34	334	496	165,664	67	303	451	136,653
2	272	414	112,608	35	334	496	165,664	68	314	466	146,324
3	278	419	116,482	36	269	411	110,559	69	298	443	132,014
4	283	424	119,992	37	328	492	161,376	70	286	425	121,550
5	311	463	143,993	38	288	433	124,704	71	310	460	142,600
6	311	464	144,304	39	302	449	135,598	72	285	424	120,840
7	312	464	144,768	40	305	455	138,775	73	274	416	113,984
8	282	422	119,004	41	301	445	133,945	74	346	513	177,498
9	313	465	145,545	42	301	448	134,848	75	306	456	139,536
10	304	451	137,104	43	291	434	126,294	76	304	454	138,016
11	328	492	161,376	44	344	505	173,720	77	305	455	138,775
12	290	434	125,860	45	298	441	131,418	78	295	439	129,505
13	340	505	171,700	46	321	477	153,117	79	319	474	151,206
14	297	440	130,680	47	287	429	123,123	80	288	433	124,704
15	323	482	155,686	48	332	494	164,008	81	320	474	151,680
16	302	448	135,296	49	332	495	164,340	82	320	475	152,000
17	334	495	165,330	50	277	417	115,509	83	304	454	138,016
18	280	421	117,880	51	283	423	119,709	84	317	470	148,990
19	314	466	146,324	52	301	448	134,848	85	294	439	129,066
20	300	443	132,900	53	325	490	159,250	86	315	466	146,790
21	312	464	144,768	54	327	491	160,557	87	298	441	131,418
22	313	465	145,545	55	308	458	141,064	88	303	451	136,653
23	262	388	101,656	56	309	458	141,522	89	315	467	147,105
24	310	461	142,910	57	288	432	124,416	90	316	470	148,520
25	300	445	133,500	58	291	438	127,458	91	297	441	130,977
26	280	420	117,600	59	275	416	114,400	92	317	471	149,307
27	329	494	162,526	60	319	473	150,887	93	318	473	150,414
28	293	438	128,334	61	300	444	133,200	94	305	455	138,775
29	322	479	154,238	62	296	440	130,240	95	287	428	122,836
30	322	481	154,882	63	257	388	99,716	96	307	457	140,299
31	273	415	113,295	64	310	462	143,220	97	308	458	141,064
32	325	484	157,300	65	311	463	143,993	98	309	459	141,831
33	303	450	136,350	66	265	408	108,120	99	306	456	139,536
								100	323	481	155,363
	10053	14994	458,6566		10009	14949	455,5262		10412	15495	475,3845
									$\bar{y}$	$\bar{x}$	Σxy
									304.74	454.38	138,95673

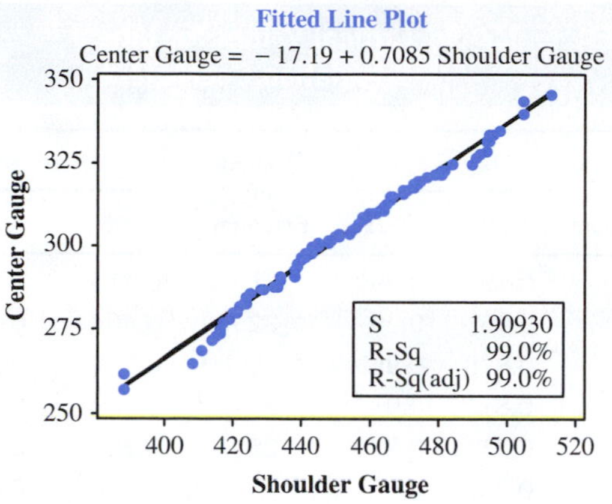

FIGURE 13-3. Linear Fit of Scatterplot for Tread Data

(the true xs). Our purpose of fitting a line to these data is not to predict one dimension from the other—that would be senseless. Our purpose is simply to obtain a visual image of the strength of any association that exists. In this case the association is so strong it seems a waste of resources to continue to measure and plot both dimensions. This insight offers the prospect of replacing the need to measure and chart the numerous features that are specified on drawings, with only a few critical ones that are tightly controlled and monitored.

When the correlation coefficient r is computed, two issues are relevant:

A correlation analysis performed on a set of dependent variables cannot be used to predict one variable from the others.

1. Is r statistically significant?
2. Is r large enough to be operationally important?

The first issue (i.e., statistical significance) is a question of whether enough data have been collected to reasonably conclude that any observed association is real and cannot be attributed to chance. The second issue is whether the observed strength of an association warrants action.

To answer the first question, we can compute a test statistic defined as follows:

$$t = \frac{r}{\sqrt{\dfrac{1-r^2}{n-2}}} \qquad \textbf{13-2}$$

The statistical distribution of t is called the "student's t" with $n - 2$ degrees of freedom. We can consult Table IV in Appendix B to find values of t, designated t_α, such that the area under the distribution beyond t (that is, between t and infinity) is equal to a stated parameter α. This is illustrated in Figure 13-4 where it can be seen that the probability of $t \geq t_\alpha$ is equal to α.

Statistical significance means that sufficient data have been collected to make reliable inferences.

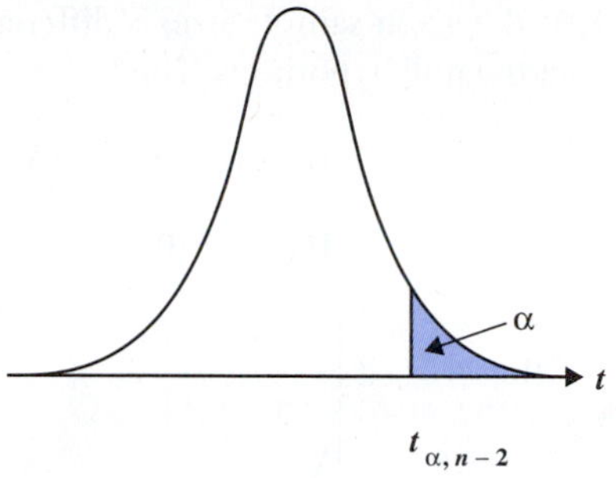

FIGURE 13-4. Student's *t* Distribution

Table IV in Appendix B provides values of t_α for $\alpha = 0.0005, 0.001, 0.005, 0.01, 0.025, 0.05$, or 0.1. Note that as the number of degrees of freedom (and sample size) approaches infinity, the t distribution converges to the standard normal distribution. (That is, for any stipulated α, the $t_{\alpha,n-2}$ values converge to the normal z scores.) The parameter α is the risk that one is willing to take in falsely inferring an association when none exists. This is what the statisticians call a Type I error, or alternatively an alpha error. Referring again to the tire tread data in Table 13-1, if we set $\alpha = 0.05$, from Table IV we find that $1.658 < t_{\alpha,98} < 1.671$. When we apply Equation 13-2 we obtain

$$t = \frac{r}{\sqrt{\dfrac{1-r^2}{n-2}}} = \frac{0.9949}{\sqrt{\dfrac{(1-0.99)}{98}}} = 97.64$$

Since $t \gg t_{0.05,98}$[2] we know that the chance of erroneously concluding that the observed association between the two variables is real, and not due to chance, is very much less than 0.05. In this case the association is so strong that the actual risk of obtaining the r value due to chance alone is only 2.2×10^{-99}!

For any stipulated confidence level $(1-\alpha)$ we can convert Table IV in Appendix B into a table of the minimum r values that would ensure that (at the accepted risk α) our observed results are real and not due to chance causes. To illustrate how this works, suppose we have a sample size $n = 23$ and we wish to be 95% confident that an observed linear association is real. To answer this question we can perform a statistical hypothesis test that is set up as two hypotheses. The first hypothesis, which is the one we will test, is called the **null hypothesis**, designated by the symbol H_0. In hypothesis testing we attempt to build a statistical case that will provide evidence strong enough to reject the null hypothesis; therefore we will state the null in terms that are the opposite of the question we are trying to answer. In this case the H_0 is *there is no association*, which will be tested against the alternative hypothesis, designated H_A *an association exists*. If no association exists, the true value of r, which we designate using the Greek letter equivalent ρ, is equal to zero. So what we will test is whether the calculated

[2]The symbol >> means "very much greater than."

value of r, at the given sample size, is different enough from zero to reject the null hypothesis. The test is set up as follows:

$$H_0 : \rho = 0$$

$$H_A : \rho \neq 0$$

$$\text{Reject } H_0 \text{ if } \left| \frac{r}{\sqrt{\frac{1-r^2}{n-2}}} \right| \geq t_{\frac{\alpha}{2},n-2} \qquad \textbf{13-3}$$

where,

H_0 = the null hypothesis to be disproven
H_A = the alternative hypothesis
ρ = the underlying true correlation coefficient
α = the Type I error : the acceptable risk of falsely rejecting a true H_0
$1 - \alpha$ = the confidence that conclusions reached are correct
n = sample size
$n - 2$ = degrees of freedom for t distribution

Equation 13-3 is a two-tail test which enables us to test the null hypothesis against either positive or negative values of r. This means that the probability of being wrong (that is, the Type I error of falsely rejecting a true null hypothesis) is equally divided between the left and right tails of the t distribution. Hence, Table IV in Appendix B was entered to obtain a critical value of t corresponding to $\alpha/2$ and degrees of freedom equal to $23 - 2 = 21$. The test is carried out as follows:

$$t_{\frac{\alpha}{2},n-2} = t_{0.025,21} = 2.080$$

$$\left| \frac{r}{\sqrt{\frac{1-r^2}{n-2}}} \right| \geq 2.080 \Rightarrow \frac{r^2}{\frac{1-r^2}{21}} \geq 4.3264$$

$$\frac{r^2}{1-r^2} \geq 0.2060 \Rightarrow r^2 \geq \frac{0.2060}{1.2060} = 0.1708$$

$$|r| \geq 0.413$$

These results tell us that, for a sample size of 23, if the absolute value of the computed r is at least 0.413 one can conclude with 95% confidence that the results are statistically significant (real). Table 13-3 contains the critical r scores for α values of 0.05 and 0.01 for degrees of freedom ranging from 1 ($n = 3$) to 100 ($n = 102$). It is obvious that small r values can be statistically significant for large sample sizes; conversely, when n is small, a relatively large r value is required to conclude that an observed association is not due to random chance causes.

As the sample size decreases, larger *r* values are necessary to establish statistical significance.

TABLE 13-3 Critical Values for Linear Correlation Coefficient

Degrees of Freedom	α 0.05	α 0.01	Degrees of Freedom	α 0.05	α 0.01
1	0.997	0.999	21	0.413	0.526
2	0.95	0.99	22	0.404	0.515
3	0.878	0.959	23	0.396	0.505
4	0.811	0.917	24	0.388	0.496
5	0.754	0.874	25	0.381	0.487
6	0.707	0.834	26	0.374	0.479
7	0.666	0.798	27	0.367	0.471
8	0.632	0.765	28	0.361	0.463
9	0.602	0.735	29	0.355	0.456
10	0.576	0.708	30	0.349	0.449
11	0.553	0.684	35	0.325	0.418
12	0.532	0.661	40	0.304	0.393
13	0.514	0.641	45	0.288	0.372
14	0.497	0.623	50	0.273	0.354
15	0.482	0.606	60	0.25	0.325
16	0.468	0.59	70	0.232	0.302
17	0.456	0.575	80	0.217	0.283
18	0.444	0.561	90	0.205	0.267
19	0.433	0.549	100	0.195	0.254
20	0.423	0.537			

13.1.2 Nonlinear Correlation and Analysis of Variance

Table 13-4 contains data that were collected to test the linearity of a certain gauge in an automobile assembly plant. The scatterplot for these data, shown in Figure 13-5, suggests that the underlying relationship between the measurement increment and the average measurement taken at that increment may be nonlinear. A polynomial regression model can be used to estimate the underlying relationship between the two variables as follows:

$$\hat{y}_j = \beta_0 + \beta_1 x_j + \beta_2 x_j^2 + \cdots + \beta_k x_j^k \qquad \textbf{13-4}$$

where,

$\hat{y}_j$ = estimate of y based on x_j
$\beta_0, \beta_1, \beta_2, \cdots, \beta_k$ = regression model coefficients
k = order of the polynomial ($k = 1 \Rightarrow$ linear; $k = 2 \Rightarrow$ quadratic)

To test different models for goodness of fit we will use a statistical technique called **analysis of variance (ANOVA)**, which uses a six-column construct called an ANOVA table.

TABLE 13-4 Measurements to Test Gauge Linearity

Measurement Increment	Average Measurement $n = 5$	Measurement Increment	Average Measurement $n = 5$	Measurement Increment	Average Measurement $n = 5$
1	0.30	21	1.18	41	1.65
2	0.56	22	1.30	42	1.77
3	0.61	23	1.28	43	1.80
4	0.70	24	1.32	44	1.81
5	0.66	25	1.40	45	1.85
6	0.62	26	1.38	46	1.78
7	0.80	27	1.46	47	1.85
8	0.77	28	1.50	48	1.87
9	0.75	29	1.42	49	1.90
10	0.82	30	1.55	50	2.00
11	0.89	31	1.58	51	2.10
12	0.76	32	1.60	52	1.98
13	0.95	33	1.57	53	1.97
14	0.99	34	1.56	54	1.99
15	0.86	35	1.67	55	2.20
16	1.10	36	1.63	56	2.00
17	1.05	37	1.65	57	2.15
18	1.15	38	1.70	58	2.10
19	1.20	39	1.72	59	1.98
20	1.25	40	1.73	60	2.05

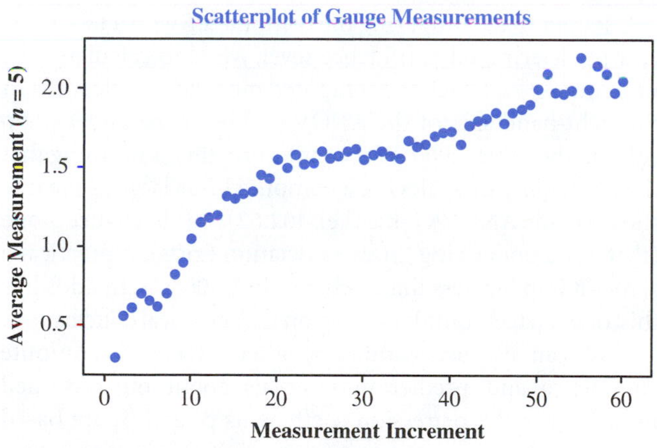

FIGURE 13-5. Scatterplot of Gauge Linearity Data

The six columns are sources of variation, degrees of freedom, sums of squares, mean squares, F score, and p value.

The tabular format for a regression ANOVA table is shown in Table 13-5. Some of the terms used in this table are defined as follows:

TABLE 13-5 ANOVA Table for Regression Models

ANOVA Table

Sources of Variation	Degrees of Freedom	Sum of Squares	Mean Square	F Score	p Value
Model	$c - 1$	SSR	MSR = SSR/$(c - 1)$	MSR/MSE	
Error	$n - c$	SSE	MSE = SSE/$(n - c)$		
Total	$n - 1$	SST	MST = SST/$(n - 1)$		

Terms in ANOVA Table

$$SSR = \text{sum of squares regression} = \sum_{i=1}^{n} (\hat{y}_i - \bar{y})^2 \quad \textbf{13-5}$$

$$SSE = \text{sum of squares error} = \sum_{i=1}^{n} (y_i - \hat{y}_i)^2 \quad \textbf{13-6}$$

$$SST = \text{sum of squares total} = \sum_{i=1}^{n} (y_i - \bar{y})^2 \quad \textbf{13-7}$$

c = the number of coeffecients (β_j) in the regression model

n = the total number of data points

Figure 13-6 is an output from the Minitab software package, Release 15, providing a linear model to fit the gauge data. This output provides some important insights. First we can see a visual plot of the predictive model and observe how well the model fits the plotted points on the scatter diagram. In this case the fit looks reasonable; however, we might conclude that a nonlinear model would provide a better fit.

Next, our attention focuses on the actual predictive equation, which is shown using the names that were input for the x and y variables, respectively. We can see that

$$\beta_0 = 0.7844, \beta_1 = 0.02402$$

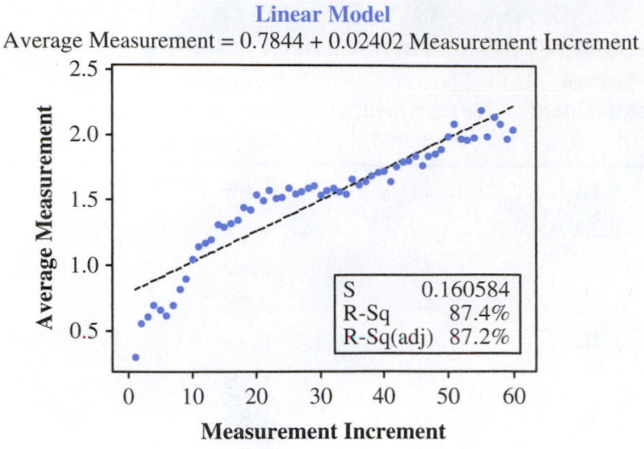

S = 0.160584 R-Sq = 87.4% R-Sq(adj) = 87.2%

Analysis of Variance

Source	DF	SS	MS	F	P
Regression	1	10.3824	10.3824	402.62	0.000
Error	58	1.4957	0.0258		
Total	59	11.8781			

FIGURE 13-6. Linear Model for Gauge Data

yielding,

$$y_i = 0.7844 + 0.02402x_i$$

The corresponding ANOVA table shows that

$$SSR = 10.3824$$

$$SSE = 1.4957$$

$$SST = 11.8781$$

As we previously stated the coefficient of determination, r^2, represents that proportion of the total observed variation in y (the measurement) that can be explained due to its association with x (the increment). This can be expressed as a ratio of the relevant sums of squares, as follows:

$$r^2 = \frac{SSR}{SST} \tag{13-8}$$

Thus, in the case of the gauge study,

$$r^2 = \frac{10.3824}{11.8781} = 0.874$$

$$r = \sqrt{r^2} = 0.935$$

As shown in Figure 13-6, the r^2 coefficient is part of the Minitab output. In absolute terms the r^2 appears to be large; however, the issue of whether the computed r^2 is statistically significant must be resolved. Consulting Table 13-3, we find the critical value for r for $\alpha = 0.01$ and 59 degrees of freedom is approximately equal to 0.325. Since $r = 0.935$ is considerably greater than the critical value one can conclude that r is significant at $\alpha << 0.01$[3].

An alternative method that can apply to nonlinear models as well as linear ones is illustrated by the Minitab output, and involves a test based on the F distribution. As Table 13-5 shows, for each source of variability listed in the ANOVA table, the mean square is calculated by dividing the sum of squares by the relevant number of degrees of freedom. The F statistic is then computed by taking the ratio of the mean square regression (MSR) to the mean square error (MSE). At a designated level of confidence (α), F tables, such as those found in Table V of Appendix B, can be consulted for a critical value of F. The F table is entered using the two applicable degrees of freedom—the columns represent the numerator (MSR) degrees of freedom and the rows represent the denominator (MSE) degrees of freedom. A different F table must be consulted for each predetermined confidence level, α. If the computed F exceeds the critical (tabular) value the results are statistically significant. In the case at hand,

$$F_{computed} = \frac{MSR}{MSE} = \frac{\frac{SSR}{c-1}}{\frac{SSE}{n-c}} = \frac{\frac{10.382407}{2-1}}{\frac{1.495693}{60-2}}$$

$$F_{computed} = \frac{10.382407}{0.025788} = 402.6$$

$$F_{0.01,58} \cong 7.077\,(\textit{Appendix B, Table V})$$

Since $F_{computed} > F_{0.01,58}$ the model is statistically significant at $\alpha << 0.01$.

Rather than comparing the computed F with a tabular F, at a predetermined confidence level, Minitab calculates the minimum α for which the computed model is significant.[4] On the right-hand side of the ANOVA table we see under the p column the value 0.000. We can interpret this as meaning that the probability of achieving a computed F as large as the one shown in the ANOVA table (i.e., 402.62), due to chance alone (that is, no underlying linear association exists), is practically zero—it is in fact less than 1 chance in 10,000. With odds like this, one can feel confident to act on the statistical conclusions.

We can also use Minitab or other software to compute confidence and prediction intervals about our assumed model. Since the regression coefficients β_0 and β_1 are based on sample data uncertainty is involved. If one were to collect and use an alternative set of 60 data points to estimate a linear association, the chances are good that the resultant regression coefficients will differ from those shown on Figure 13-6. In Figure 13-7 the 95% **confidence interval** (CI) is the tight band around the linear prediction line and represents the region within which 95% of all regression lines computed from the underlying process will lie. The wider (outside) bands form the **prediction interval** and represents the region within which 95% of all individual data points collected from the process will fall.

[3]The symbol "<<" means "very much less than."

[4]Minitab and many other software packages such as JMP and Excel refer to this as the p value.

A second level of analysis involves a consideration of how well the model can predict future data points. For a given set of independent variables a **residual** is defined as the difference between what y the model predicts for a given value of x and the value of y that actually does occur. That is,

$$residual_i = y_i - \hat{y}_i \qquad \textbf{13-9}$$

Residuals represent the vertical distances between each point on the scatterplot and the regression line. If the actual point lies above the line the corresponding residual will be positive; if the point lies below the line the residual will be negative. If the model is consistent in its ability to predict across the range of x values in the data set, the residuals will be normally distributed with a mean of zero. In addition, if the model is consistent the points on the scatterplots (residuals plotted against the x's or the residuals plotted against the predicted y's) will be randomly distributed about the horizontal line corresponding to a residual of zero. Any observed nonrandom patterns indicate model inconsistencies and can provide insights into where a particular model has the propensity to under or over estimate. Figure 13-8 displays four residual plots of the gauge model in Figure 13-6. The histogram and normal probability plot suggest the residuals follow an approximate normal distribution.[5] The residual plots suggest that the model is a consistent predictor for measurement increments above 30 and predicted values above 1.5. For measurement increments between 10 and 30 (and predicted values between 1.0 and 1.5) the model appears to underestimate and for measurement increments less than 10 (and predicted values less than 1.0) the model appears to overestimate.

A residual is the difference between the y value that the model predicts and the y value that actually occurs.

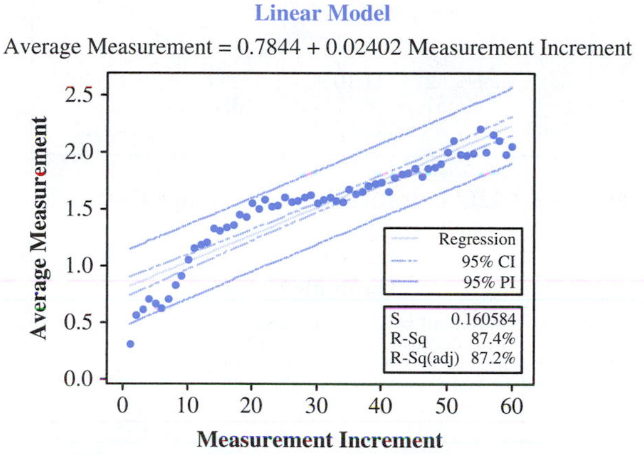

FIGURE 13-7. Confidence and Prediction Intervals for Linear Model

The coefficient of determination r^2 has an inherent weakness. In Equation 13-8, the denominator is fixed regardless of the model or the number of independent variables. As more variables are added to a model only the numerator can change and only in an increasing direction. As an additional term is added to the model it cannot decrease the numerator and will in all likelihood increase it, albeit slightly, resulting in a larger r^2. This will happen even if the mean square error (MSE) increases, making the revised model less efficient.[6]

---

[5] A normal distribution will plot along a straight line on a normal probability plot.

[6] Adding a term to the model reduces the SSE degrees of freedom by one. If the corresponding reduction in SSE is small, the ratio MSE = SSE/d.f. may actually increase.

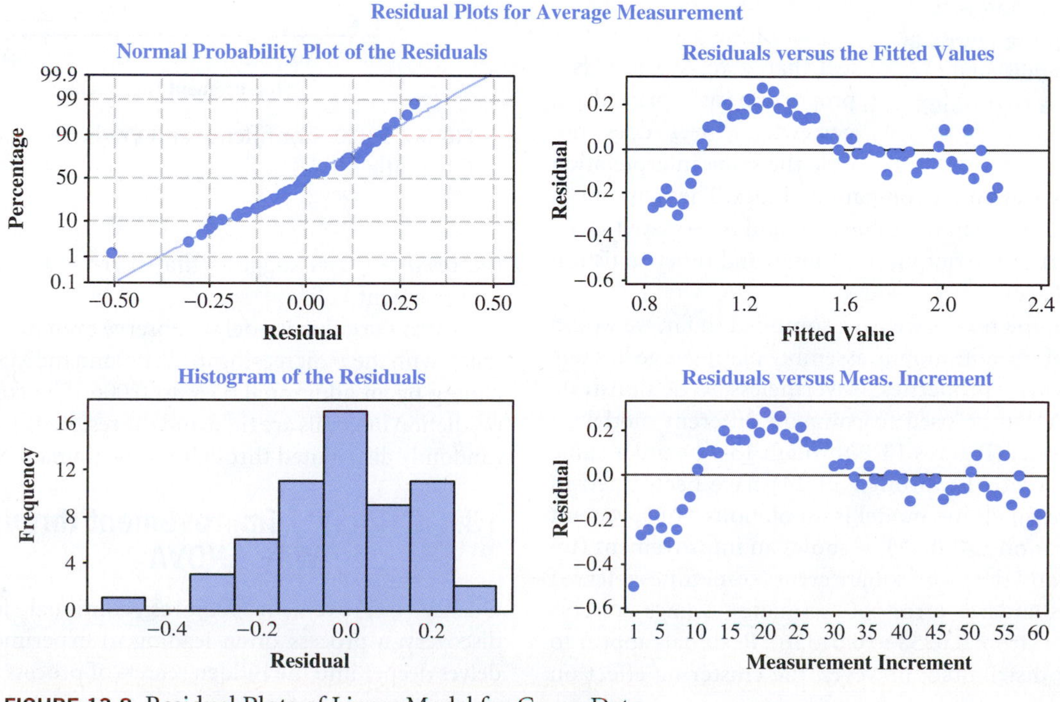

FIGURE 13-8. Residual Plots of Linear Model for Gauge Data

Minitab and other statistical software produce an alternative index called the adjusted r^2, designated r^2_{adj}, and it is an attempt to address the shortcomings of r^2 by adjusting both the numerator and denominator by their respective degrees of freedom. The adjusted r^2 is computed as follows:

$$r^2_{adj} = 1 - (1 - r^2)\left(\frac{n-1}{n-c}\right) \qquad \textbf{13-10}$$

Thus for the gauge model,

$$r^2_{adj} = 1 - (1 - 0.874)\left(\frac{60-1}{60-2}\right) = 0.872$$

An alternative and sometimes easier method for calculating the adjusted r^2 is as follows:

$$r^2_{adj} = 1 - \left(\frac{MSE}{MST}\right) = 1 - \left(\frac{\dfrac{SSE}{n-c}}{\dfrac{SST}{n-1}}\right) \qquad \textbf{13-11}$$

Thus,

$$r^2_{adj} = 1 - \left(\frac{MSE}{MST}\right) = 1 - \left(\frac{\dfrac{1.495693}{58}}{\dfrac{11.8781}{59}}\right)$$

$$= 1 - 0.128 = 0.872$$

Unlike r^2 the adjusted index r^2_{adj} can decline in value when an additional variable is added to the model. This can occur if the additional contribution by the new variable to the explained variation is not enough to compensate for the potential increase in the MSE caused by the loss of a degree of freedom. Also, unlike r^2, r^2_{adj} can have a value that is less than zero. While r^2 is a proportion (or expressed as a percentage), r^2_{adj} does not have the same interpretation and should be viewed as a comparative index. The value of r^2_{adj} will always be less than the value of r^2 and is displayed along with r^2 as part of the output of Minitab and other statistical software packages.

For small sample sizes r^2_{adj} is better than r^2 for determining the merits of adding an additional variable to a regression model.

Based on the results we have compiled so far, we would conclude that the automobile assembly plant gauge has satisfactory linearity properties. Nevertheless, to demonstrate how ANOVA can be used to compare different models, a quadratic model (Figures 13-9 through 13-11) and a cubic model (Figures 13-12 through 13-14) have been fitted to the data. The quadratic model is an obvious improvement over the linear one. At 0.935, r^2 shows an improvement (up from 0.874) and the mean square error (sometimes referred to as the standard error of estimate) shows a 47% improvement (from 0.0258 to 0.01356). Residuals appear to be normally distributed; however, the clustering effect on the residual plots suggests that there is still potential for improvement.

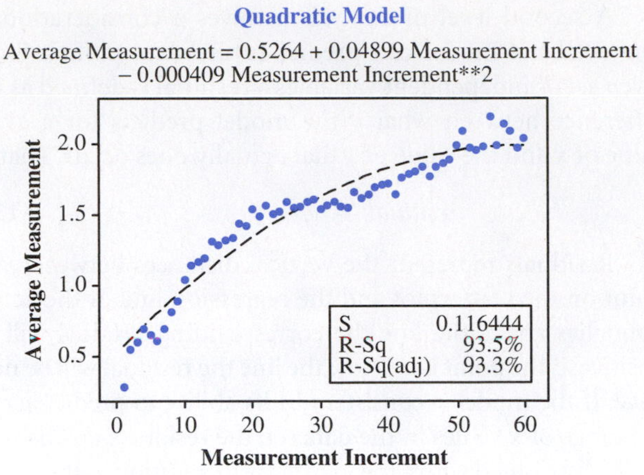

S = 0.116444 R-Sq = 93.5% R-Sq(adj) = 93.3%

Analysis of Variance

Source	DF	SS	MS	F	P
Regression	2	11.1052	5.55259	409.51	0.000
Error	57	0.7729	0.01356		
Total	59	11.8781			

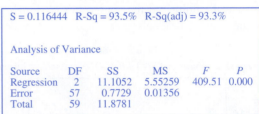

FIGURE 13-9. Quadratic Model for Gauge Data

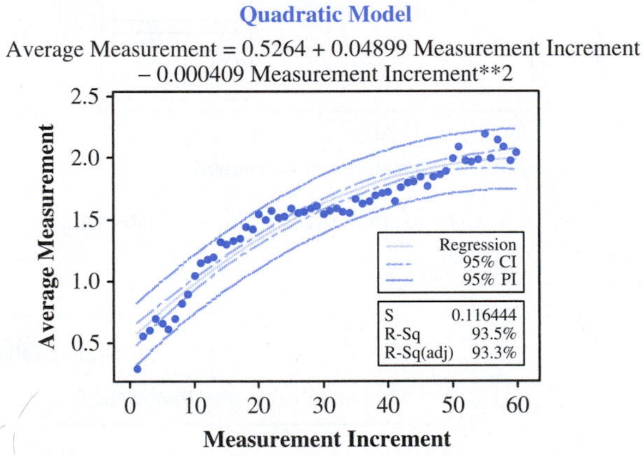

FIGURE 13-10. Confidence and Prediction Intervals for Quadratic Model

With the cubic model we observe even further improvement, with the r^2 increasing to 97.1% and the MSE further declining by an additional 55% to 0.006. The confidence and prediction intervals are tight and the residuals are normal and randomly distributed throughout the domain of measure.

13.1.3 Quality Improvement through One-Way ANOVA

Quality improvement requires continual learning and discovery, a process often leading to experimentation that delves deeper into the hidden secrets of process behavior and

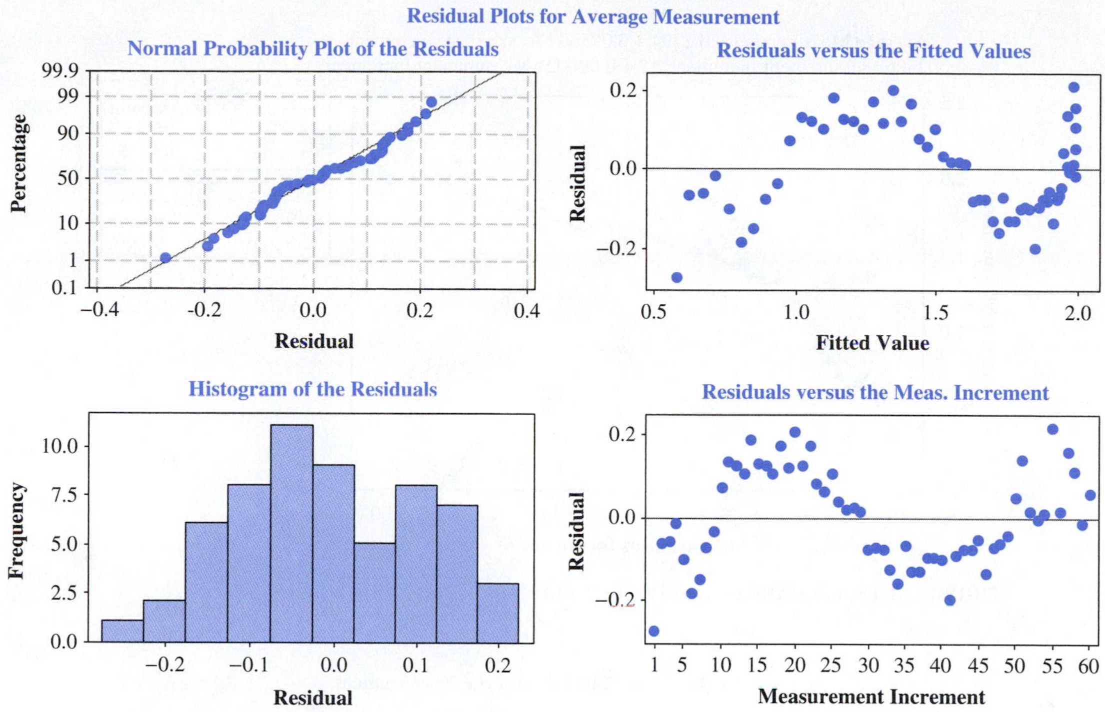

FIGURE 13-11. Residual Plots for Quadratic Model

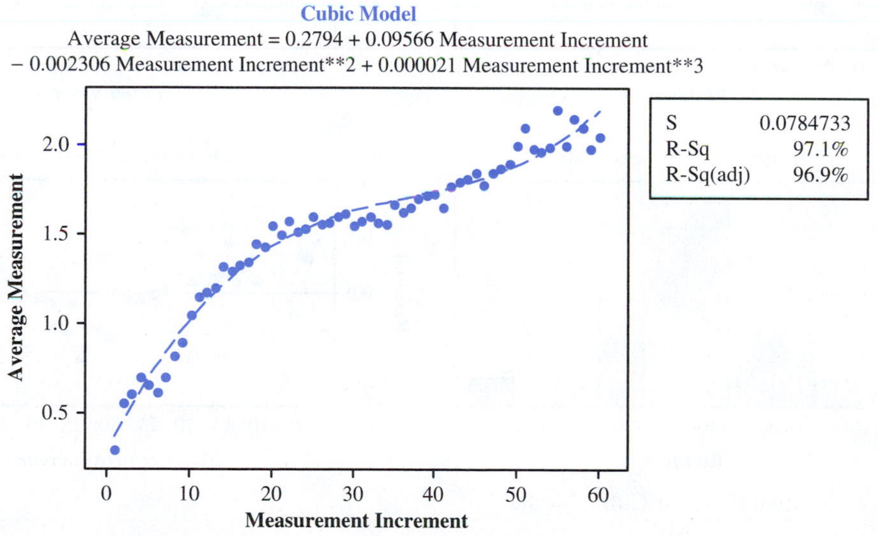

S = 0.0784733 R-Sq = 97.1% R-Sq(adj) = 96.9%

Analysis of Variance

Source	DF	SS	MS	*F*	*P*
Regression	3	11.5332	3.84440	624.29	0.000
Error	56	0.3449	0.00616		
Total	59	11.8781			

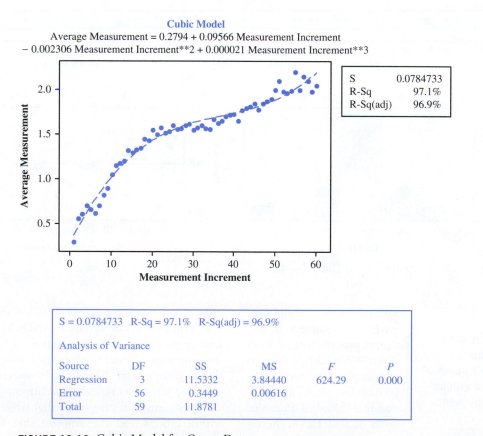

FIGURE 13-12. Cubic Model for Gauge Data

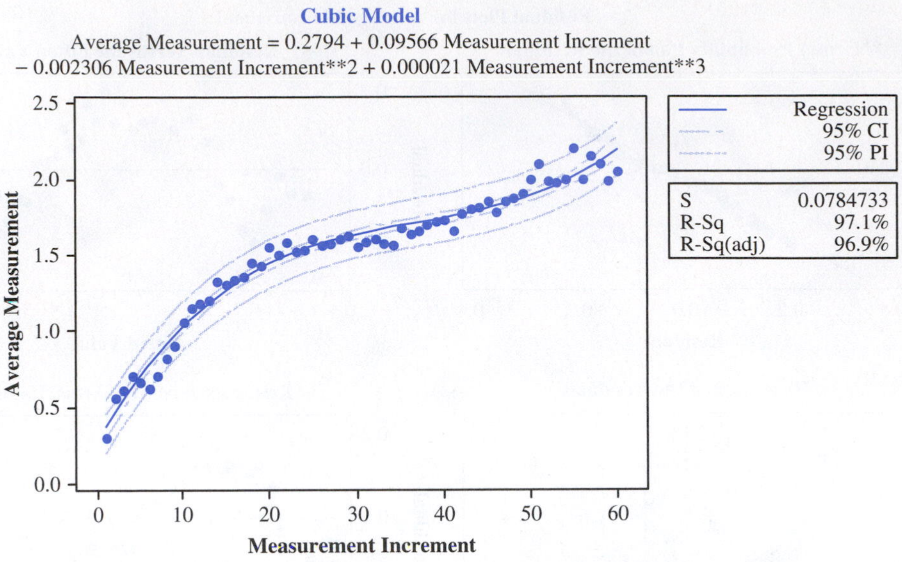

FIGURE 13-13. Confidence and Prediction Intervals for Cubic Model

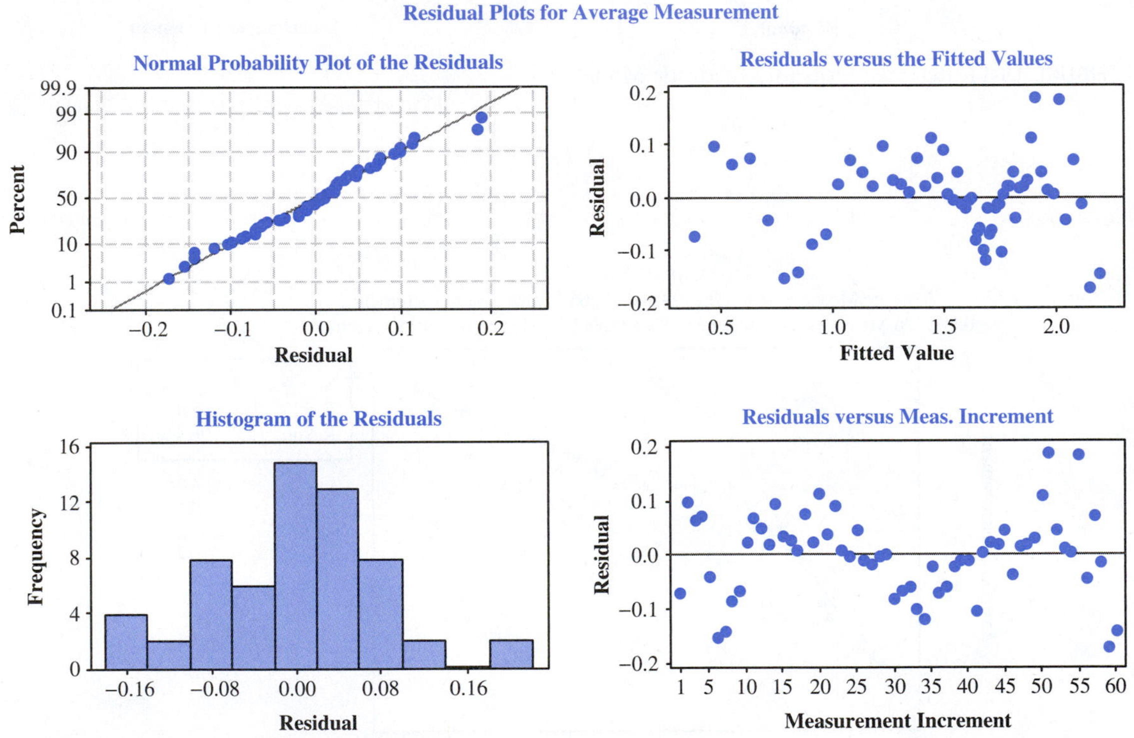

FIGURE 13-14. Residual Plots for Cubic Model

ANOVA analyzes group mean differences by comparing the amount of variation within groups with variation between groups.

relationships than would otherwise be possible with statistical process control tools alone. When a single factor is of interest and subjects or experimental units are randomly assigned to groups or levels of that factor, the resulting experimental design is called a completely randomized, or one-way ANOVA, model. The term **analysis of variance (ANOVA)** is somewhat misleading since its objective is to statistically evaluate differences in group *means*. However, the methodology employed investigates group mean differences by systematically analyzing observed variation in the data partitioning the observed total variability into two categories: that which can be explained within groups and the amount that occurs among the groups. In Chapter 9 the limited use of control charts to address this issue was demonstrated; one-way ANOVA provides results that are more conclusive and statistically defendable.

Let us assume that we have c groups (or levels of a factor under study) that represent subprocesses, the output measures of which are randomly and independently taken. We will also assume that the respective process distributions are normal and have equal variances. The ANOVA test seeks evidence to suggest the null hypothesis that all group means are equal is false and the alternative that at least one mean differs from the others is true. Mathematically, this is stated as follows:

ANOVA Hypothesis Test

$$\mathbf{H}_0: \mu_1 = \mu_2 = \cdots = \mu_c \qquad \textbf{13-12}$$

$$\mathbf{H}_A: \text{there exists a } k \text{ such that } \mu_k \neq \mu_j;\ j \neq k;\ j = 1, 2, \ldots, c$$

In performing an ANOVA test to determine if there is any evidence that the c groups come from processes with different means (i.e., the null hypothesis is false) the total variation in all measurements are subdivided into two parts: that which can be attributed (explained) among the groups and the inherent variation within the groups (error). The tabular format for a one-way ANOVA table is shown in Table 13-6. Some of the terms in the table are defined as follows:

Terms in One-Way ANOVA Table

$$SSA = \text{sum of squares among groups} = \sum_{j=1}^{c} n_j(\overline{X}_j - \overline{\overline{X}})^2 \qquad \textbf{13-13}$$

$$SSW = \text{sum of squares within groups} = \sum_{j=1}^{c}\sum_{i=1}^{n_j} (x_{ij} - \overline{X}_j)^2 \qquad \textbf{13-14}$$

$$SST = \text{sum of squares total} = \sum_{j=1}^{c}\sum_{i=1}^{n_j} (x_{ij} - \overline{\overline{X}})^2 \qquad \textbf{13-15}$$

where,

c = the number of groups or factor levels

n_j = the number of observation in groups or level j

n = the total number of data points across all groups or levels $= \sum_{j=1}^{c} n_j$

x_{ij} = the ith observation in group or level j

$\overline{X}_j$ = the mean of the n_j units in the group j sample

$\overline{\overline{X}}$ = the overall (grand) mean of all n observations $= \dfrac{\sum_{j=1}^{c}\sum_{i=1}^{n_j} x_{ij}}{n}$

TABLE 13-6 ANOVA Table for One-Way Completely Randomized Analysis of Variance

ANOVA Table

Sources of Variation	Degrees of Freedom	Sum of Squares	Mean Square	F Score	p Value
Among Groups	$c - 1$	SSA	MSR = SSA/(c − 1)	MSA/MSW	
Within Groups	$n - c$	SSW	MSW = SSW/(n − c)		
Total	$n - 1$	SST	MST = SST/(n − 1)		

The PopFiz Bottling Plant, which we first introduced in Chapter 9, will be used to illustrate these concepts. Table 13-7[7] shows the data that were collected from bottling line one and have been stratified by filler head. Setting up this problem as a one-way ANOVA, the fills are the measured responses and the head number is the factor. What we will be trying to determine is what proportion of the total observed variation in fills (across all eight heads) can be explained due to the variation in means across heads. That is, we will be looking for statistical evidence that the mean fill rates are not the same for all heads.

Observe that we do not need to use Equation 13-15 and go through the tedious computations to derive SST. As an alternative we use Excel, or similar software, to compute $s_y = 2.5716$ and then use this result and the observed relationship $SST = (n - 1)s_y^2$.

Thus,

$$SST = 159s_y^2 = (159)(2.5716)^2 = 1051.52$$

$$\begin{aligned} SSA = \sum_{j=1}^{c} (\overline{X}_j - \overline{\overline{X}})^2 &= 20(245.979 - 249.446)^2 \\ &+ 20(246.917 - 249.446)^2 + 20(247.755 \\ &- 249.446)^2 + 20(248.663 - 249.446)^2 \\ &+ 20(250.199 - 249.446)^2 + 20(250.884 \\ &- 249.446)^2 + 20(252.218 - 249.446)^2 \\ &+ 20(252.955 - 249.446)^2 \\ &= 240.402 + 127.917 + 57.190 + 12.262 \\ &+ 11.340 + 41.357 + 153.680 + 246.262 \\ &= 890.41 \end{aligned}$$

$$SSW = SST - SSA = 1051.52 - 890.41 = 161.11$$

The completed ANOVA table is shown in Table 13-8.

Our conclusion is that the results are statistically significant with very negligible margin of error. When we consult Table V in Appendix B with $\alpha = 0.01$, $\nu_1 = 7$, and $\nu_2 = 159$

[7]These data were first introduced as Table 9-14.

TABLE 13-7 Data Collected from Line 1 by Filler Head

PopFiz Bottling Plant

Measurements from Line 1

Filler Head	Sample Number	1	2	3	4	5	Head Avgs
	1	245.75	245.69	247.25	245.95	246.20	
	2	244.58	245.57	246.45	248.52	245.34	
1	3	246.53	245.52	245.28	245.29	246.48	245.979
	4	246.96	243.95	245.57	247.42	245.31	
	1	247.59	247.47	246.59	247.70	246.77	
	2	247.87	245.60	247.52	247.40	245.74	
2	3	246.52	246.14	245.56	246.22	248.14	246.917
	4	246.31	247.07	247.26	248.52	246.36	
	1	246.90	247.23	248.31	248.44	247.84	
	2	247.70	246.66	248.04	247.16	248.04	
3	3	249.24	247.25	248.72	249.18	248.39	247.755
	4	247.23	247.41	246.01	246.71	248.65	
	1	249.08	247.93	250.75	248.17	249.46	
	2	248.36	248.41	248.24	248.04	250.79	
4	3	249.02	249.86	246.38	249.21	249.97	248.663
	4	247.33	248.83	247.88	246.76	248.80	
	1	249.31	250.01	248.99	251.36	251.59	
	2	250.44	249.81	249.17	250.18	251.14	
5	3	249.96	249.60	250.32	251.79	250.77	250.199
	4	250.88	249.21	251.06	250.62	247.81	
	1	251.04	251.86	250.02	251.44	250.93	
	2	248.28	249.53	249.25	251.59	252.36	
6	3	249.39	252.24	251.13	251.91	251.43	250.884
	4	251.22	251.53	250.71	250.32	251.48	
	1	252.17	251.74	249.97	252.06	251.37	
	2	250.85	250.91	252.18	253.00	252.93	
7	3	253.08	252.93	251.31	251.27	253.66	252.218
	4	253.70	252.56	252.62	253.08	252.98	
	1	252.79	252.95	255.76	253.23	253.76	
	2	253.22	252.81	251.14	252.66	252.04	
8	3	251.53	252.11	253.83	252.77	253.05	252.955
	4	254.22	251.26	252.03	253.44	254.50	
	Column Avgs	241.82	241.63	241.77	242.28	242.40	
						$\bar{\bar{X}}$	249.446
						s	2.5716
						s^2	6.613347

we obtain $2.639 < F_{0.01,7,159} < 2.792$. With a computed $F = 120$, the actual p value is much less than 0.01 [in fact < 0.0001]. Figure 13-15 is the ANOVA table produced by Minitab. Figures 13-16 and 13-17 are the Minitab plots of the regression model and residuals, respectively.

What does this analysis mean? The ANOVA procedure reveals that around 85% ($r^2 = 84.68$) of the total variability observed in fill volumes can be explained by the differences in mean fill volumes among the eight heads. This is a valuable insight. If plant management wants to lower the variability in overall fill volumes improvement efforts should focus on why there is so much head-to-head variability and try to find ways to make all the heads the same.

TABLE 13-8 ANOVA Table for One-Way Analysis of PopFiz Co.

ANOVA Table

Sources of Variation	Degrees of Freedom	Sum of Squares	Mean Square	*F* Score	*p* Value
Among Groups	7	890.41	127.2014	120.0088	< 0.0001
Within Groups	152	161.11	1.0599		
Total	159	1051.52	6.6133		

13.2 PROCESS IMPROVEMENT THROUGH DOE

13.2.1 Two-Level Two-Factor Fixed Effects Designs

Superior Surgical Supply (SSS) Inc. is a supplier of medical devices and surgical kits to hospitals and health care providers. To avoid contamination during sterilization and handling the integrity of each packaging seal is critical. The factory sealing process involves the use of a heated platen tool that is used to thermally produce an airtight join between the two open edges of a medical package. Seal integrity is determined by a measure called peel strength. SSS management

```
One-Way ANOVA: Fill versus Head

Source    DF       SS       MS       F       P
Head       7   890.45   127.21  120.04   0.000
Error    152   161.07     1.06
Total    159  1051.52

S = 1.029   R-Sq = 84.68%   R-Sq(adj) = 83.98%
                        Individual 95% CIs for Mean Based on Pooled StDev
Level   N    Mean    StDev   —+————————+————————+————————+————
1      20  245.98    1.04   (-*-)
2      20  246.92    0.86       (—*-)
3      20  247.75    0.88           (-*-)
4      20  248.66    1.17               (-*—)
5      20  250.20    1.00                    (-*-)
6      20  250.88    1.09                        (-*—)
7      20  252.22    1.02                             (-*-)
8      20  252.96    1.13                                 (-*-)
                         —+————————+————————+————————+————
                         246.0    248.0    250.0    252.0

Pooled StDev = 1.03
```

FIGURE 13-15. Minitab ANOVA Printout

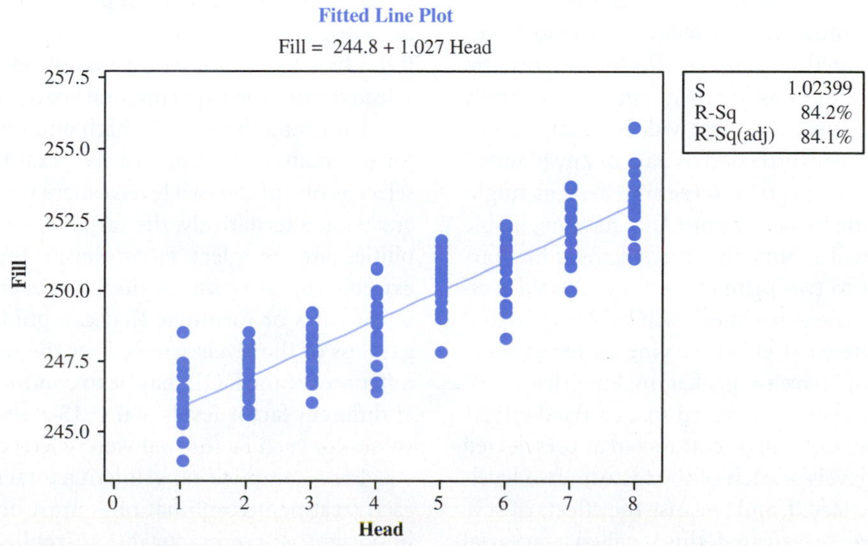

FIGURE 13-16. Plot of Response (Fill) by Factor (Head Number)

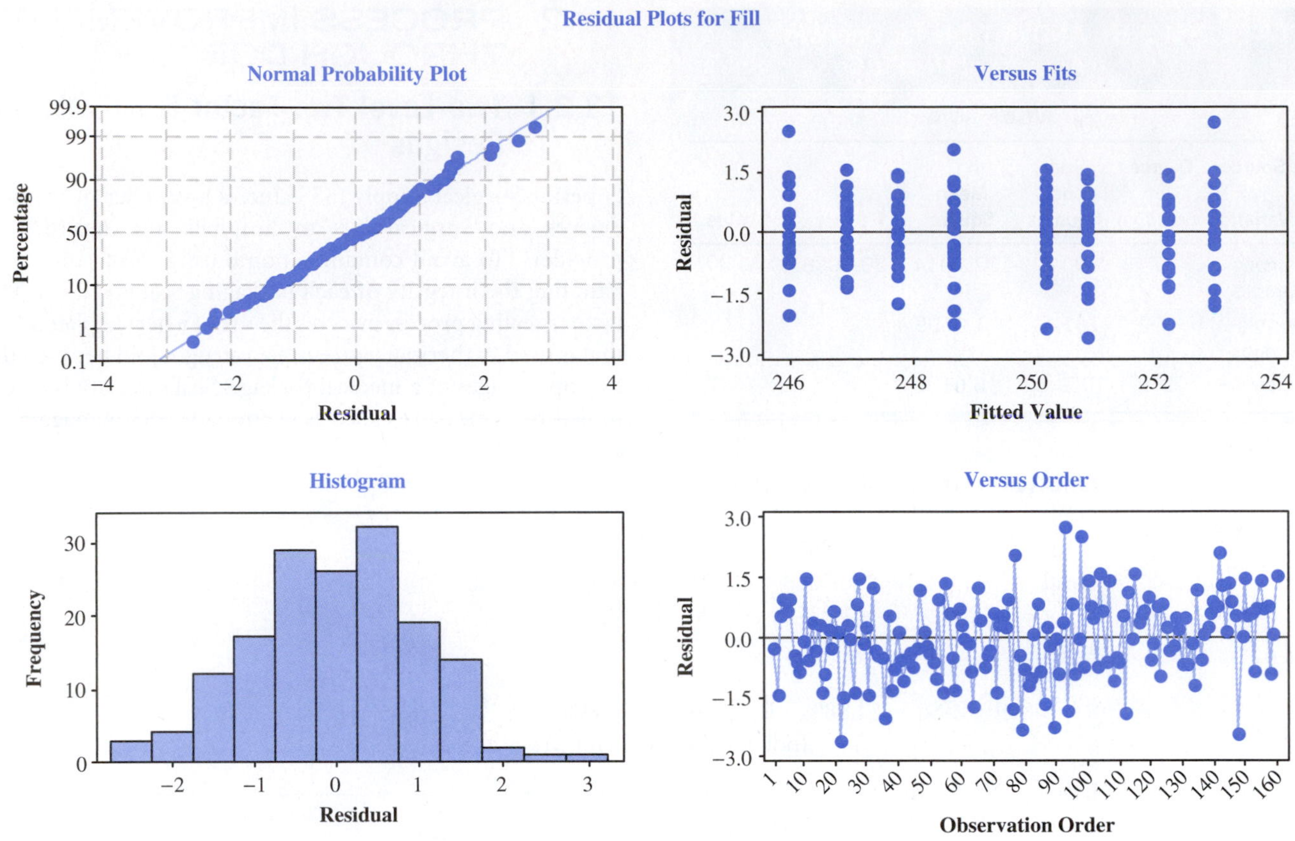

FIGURE 13-17. Residual Plots for PopFiz Bottling Line

would like to know how key process variables affect peel strength in final packaging. To address this issue a **design of experiments** (DOE) approach will be used.

Any designed experiment has some essential elements. The first of these is one or more identified **measured responses**—these are output variables that are used as quality determinants. In this case there is a single measured response—peel strength—and for this particular response higher numbers are generally regarded as more desirable than lower numbers. A second element in any DOE is the identification of factors thought to be related to and to be key influencers of the measured response(s). **Factors** are process drivers and can be thought of as independent, and generally controllable, variables. Typically, there will be many candidate factors that could be investigated as part of any planned experiment. If the number is great, **screening designs** might be employed to eliminate those variables that have negligible effects. In this case we will assume that management has narrowed its investigation to two primary factors—*platen temperature* and *seal time*. The experiment will be designed and conducted to determine what effect varying either of these two factors (or both) will have on packaging integrity.

This type of experiment is referred to as a **fixed effects model**, since the experiment will be conducted at preselected (rather than random) levels of each of the factors. Two levels for each factor will be selected, and because the effects of several factors at a time are investigated, this is called a **factorial design**. In general, a two-level design with k factors is referred to as a 2^k factorial experiment, where 2^k represents the total number of different combinations of factors and levels (called **treatment combinations**) that are possible. In the case of SSS, Inc. $k = 2$ and there are a total of $2^2 = 4$ different treatment combinations possible: low temperature, low time; low temperature, high time; high temperature, low time; and high temperature, high time. Two levels—a high and low—must be selected for each of the two variables from an **experimental space** that can potentially provide a large number of options as shown in Figure 13-18. Figure 13-19 illustrates how two arbitrarily different sets of four points can be selected from the experimental space.

Choosing the specific high and low values for each factor is a matter of judgment. For example, the analyst could select as one of the two levels where the process normally operates or, alternatively, the target specification. Other possibilities are to select factor levels that coincide with the extreme corner points of the experimental space. There is no set of rules or formulae that can guide such decisions. Regardless of the levels selected for the initial study, one of the outcomes of the DOE may be to conduct further experiments at different factor levels. Table 13-9 shows the high and low values for each factor that were selected by SSS.

For a **complete experiment** a total of four runs—one for each treatment combination—must be performed. It is also important where practicable to **replicate** the runs for each

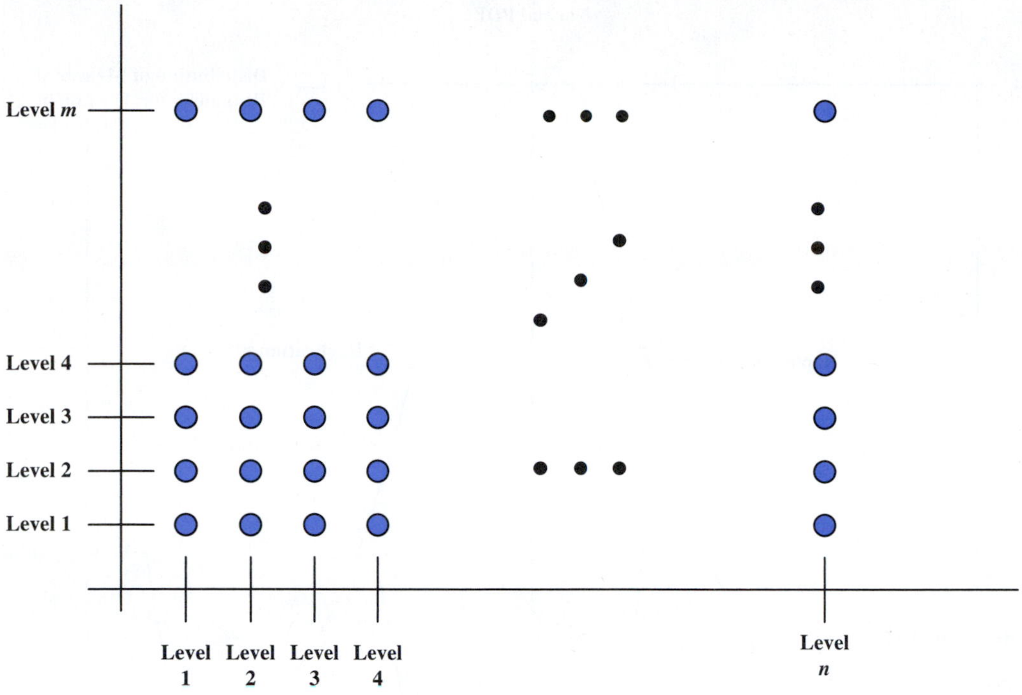

FIGURE 13-18. Experimental Space

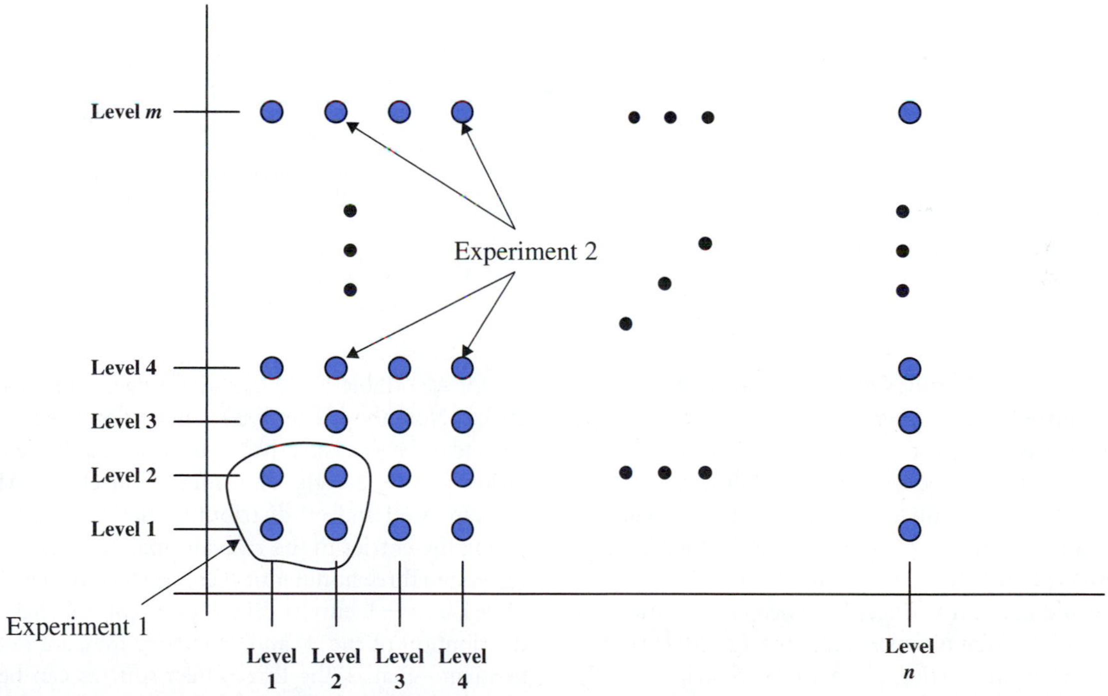

FIGURE 13-19. Examples of Fixed Effects Designs

TABLE 13-9 Low and High Values for Two Process Variables

Factor	Low Level	High Level
Platen Temperature °F	220	300
Seal Time (secs)	0.5	2.0

treatment. Placing too much confidence in a single run could bias the legitimacy of the entire experiment. As Figure 13-20 shows, each replication represents a sample from the *population* distribution of responses for a particular treatment. The average of a number of replications will converge toward the actual expected response for the treatment. With five replications each of the four treatments is studied five times for a total of 20 experimental runs. Ideally, to minimize

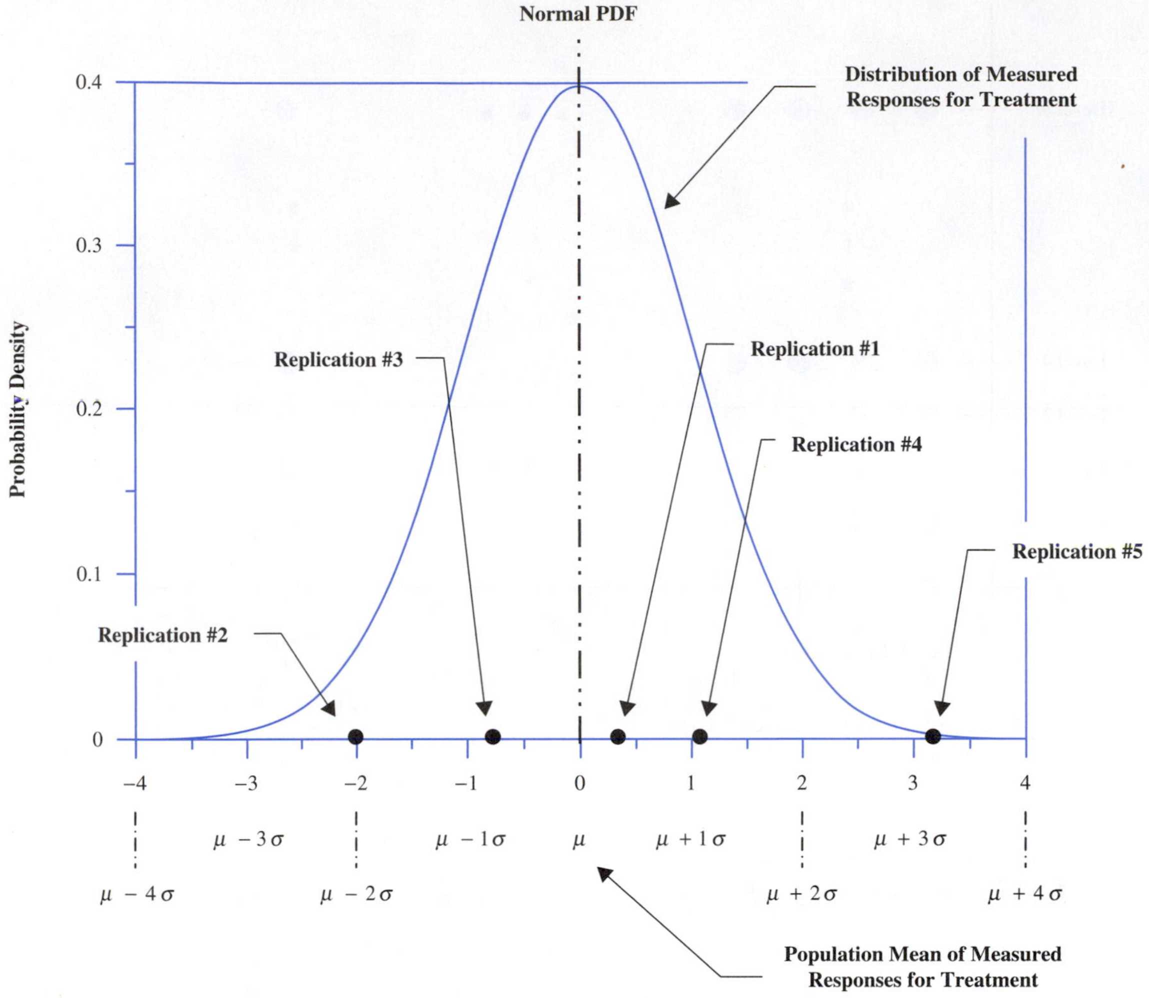

FIGURE 13-20. Advantage of Replicating Runs

experimental error, the 20 runs should be performed in a completely **randomized order**—there will be 20 different setups with one measured response per run. Table 13-10 lists one possible order that could be used for scheduling these runs. Note the convenient notation that uses a "−1" to represent the low level for a factor and a "+1" to represent the high level.

The results of the SSS, Inc. study are shown in Table 13-11, and the row and column averages have been summarized in Table 13-12. At this point two questions are relevant: Do either of the factors, or the interaction of the two factors, significantly affect the variability of measured responses? Do either of the factors, or the interaction of the two factors, significantly affect the mean response? A quick and effective way to answer the first question is to plot an *s* chart of the treatment standard deviations as shown in Figure 13-21. This chart shows no statistical evidence that would suggest that the within-run standard deviations (i.e., package-to-package) are affected by moving from one treatment combination to another.

There are three effects that should be investigated: the **main effect** of time (factor A), the main effect of temperature (factor B), and the **interaction effect** of time and temperature (factor AB). Table 13-13 contains a column for each of these effects. Note the +1 and −1 convention used for high and low factor levels respectively; this is called the **geometric notation** for the design. The interaction effect (AB) column contains a +1 and −1 distribution that is obtained by multiplying the entries in the relevant main effects columns. The results are three nonidentical columns with an equal number of +1 and −1 entries. Since each column has a different distribution of the + and − entries they are linearly independent—that is, the three effect sources can be separated statistically. A convenient **shorthand notation** using a series of lowercase letters is often used to label the various treatment combinations. If a lowercase letter is present the corresponding factor is set at its high level for that particular treatment combination; if absent the corresponding factor is run at its low level. For example, the label "a" represents a treatment combination in which factor A is set at its high level and factor B at its low level. If both factors are set at their respective, low levels the treatment combination is given the label "(l)." Each treatment combination in Table 13-13 has been labeled using this convention.

TABLE 13-10 Randomized Order of Experimental Runs

	Factor		
Experiment Number	Time	Temperature	Run Order
1	−1	−1	15
2	−1	+1	7
3	+1	−1	1
4	+1	+1	2
5	−1	−1	8
6	−1	+1	13
7	+1	−1	18
8	+1	+1	20
9	−1	−1	9
10	−1	+1	3
11	+1	−1	14
12	+1	+1	4
13	−1	−1	10
14	−1	+1	16
15	+1	−1	5
16	+1	+1	11
17	−1	−1	19
18	−1	+1	17
19	+1	−1	12
20	+1	+1	6

TABLE 13-11 Design of Experiment Results

	DOE Results				
Run No.	Time −1 Temp −1	Time +1 Temp −1	Time −1 Temp +1	Time +1 Temp +1	
1	5.998	7.056	11.120	16.088	
2	5.375	6.994	10.257	16.011	
3	5.698	7.733	10.398	15.650	
4	5.059	7.199	9.567	15.974	
5	4.229	8.385	10.475	16.594	
Totals	26.359	37.367	51.817	80.317	
Treatment Averages	5.272	7.473	10.363	16.063	s_{Total}
Treatment Standard Deviations	0.680511	0.586888	0.554903	0.3405067	4.181733

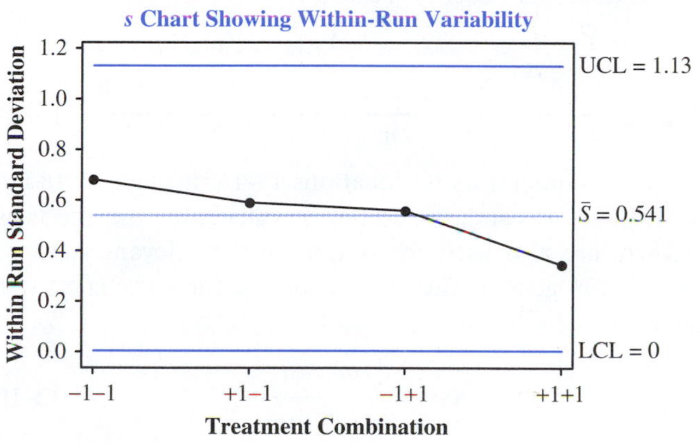

FIGURE 13-21. *s* Chart to Test Equality of Within-Run Variability across Treatments

TABLE 13-12 Treatment Averages

	Time		
Temperature	Low (−1)	High (+1)	Row Averages
Low (−1)	5.272	7.056	6.164
High (+1)	10.363	16.063	13.213
Column Averages	7.8175	11.5595	9.6885

TABLE 13-13 Effects Table

	Main Effects		Interactive Effects		
	A	B	AB		
Treatment Label	Time	Temperature	Time X Temperature	Total Response	Average Response
(1)	−1	−1	+1	26.359	5.272
b	−1	+1	−1	51.817	10.363
a	+1	−1	−1	37.367	7.473
ab	+1	+1	+1	80.317	16.063
Avg @ +1	11.768	13.213	10.668		
Avg @ −1	7.818	6.373	8.918	Overall Average	
Difference	3.951	6.841	1.750	9.793	

In Table 13-13 each effect has been computed by taking the difference between the average measured response at the high (+1) level and the average measured response at the low (−1) level. For example, to compute the main effect for the time factor, we compute the average response obtained when time was set at the high level ($\frac{(7.056)+(16.063)}{2} = 11.560$) minus the average response obtained when time was set at the low level ($\frac{(5.272)+(10.363)}{2} = 7.818$). This difference results in a calculated effect equal to +3.742. All three effects are shown in Table 13-13. Effects can be either positive or negative. The following equations provide expressions for the three effects of interest in terms of the lower case labeling convention.

$$\text{Main Effect for Factor } A = \bar{\bar{Y}}_{Factor\,A@+1} - \bar{\bar{Y}}_{Factor\,A@-1}$$

$$= \frac{a+ab}{2n} - \frac{b+(1)}{2n} = \frac{a+ab-b-(1)}{2n} \qquad \textbf{13-16}$$

$$\text{Main Effect for Factor } B = \bar{\bar{Y}}_{Factor\,B@+1} - \bar{\bar{Y}}_{Factor\,B@-1}$$

$$= \frac{b+ab}{2n} - \frac{a+(1)}{2n} = \frac{b+ab-a-(1)}{2n} \qquad \textbf{13-17}$$

Interaction Effect for Factors AB

$$= \bar{\bar{Y}}_{Factors\,A\,\&\,B@\,same\,levels} - \bar{\bar{Y}}_{Factors\,A\,\&\,B@\,apposite\,levels}$$

$$= \frac{(1)+ab}{2n} - \frac{a+b}{2n} = \frac{(1)+ab-a-b}{2n} \qquad \textbf{13-18}$$

The numerators in Equations 13-16 through 13-18 are called *contrasts* and, in addition to calculating the effects as shown, are also used in computing the relevant sums of squares. In general, the sum of squares for some effect ξ is given by the following[8]:

$$SS\xi = \frac{[contrast_{\xi}]^2}{(2^{k-q})n} \qquad \textbf{13-19}$$

where,

k = the number of factors

q = an integer indicating the fraction of treatments run

$q = 0 \Rightarrow$ complete experiment

$q = 1 \Rightarrow$ half experiment

$q = 2 \Rightarrow$ quarter experiment

Applying Equation 13-19 we can calculate the sums of squares for each of the effects as follows:

$$SSA = \frac{[a+ab-b-(1)]^2}{4n} \qquad \textbf{13-20}$$

$$SSB = \frac{[b+ab-a-(1)]^2}{4n} \qquad \textbf{13-21}$$

$$SSAB = \frac{[(1)+ab-a-b]^2}{4n} \qquad \textbf{13-22}$$

Table 13-14 is the ANOVA table format relative to the 2^2 factorial design, and Table 13-15 is the completed ANOVA for Superior Surgical Supply, Inc.

There is strong evidence, as judged by the p values < 0.00001, to suggest that both time and temperature (main effects) influence the response (peel strength). In addition, there is also evidence to suggest the presence of a significant interaction effect between time and temperature. We can evaluate the magnitude of the three effects by comparing each effect with its estimated standard error $\hat{\sigma}_{effect}$[9]. In computing $\hat{\sigma}_{effect}$ we observe that each effect is calculated by taking the difference of two averages, each of which is computed from exactly one-half of the total measurements. Applying the central limit theorem we can derive an expression for the mean square error of each effect.

$$\hat{\sigma}^2_{effect} = \hat{\sigma}^2_{\bar{Y}(measurements\,@\,high)} + \hat{\sigma}^2_{\bar{Y}(measurements\,@\,low)}$$

$$\hat{\sigma}^2_{effect} = \frac{\hat{\sigma}^2_{y(within\ treatments)}}{\frac{n2^k}{2}} + \frac{\hat{\sigma}^2_{y(within\ treatments)}}{\frac{n2^k}{2}}$$

$$= \frac{\hat{\sigma}^2_{y(within\ treatments)}}{n2^{k-2}} \qquad \textbf{13-23}$$

To obtain $\hat{\sigma}^2_{effect}$ we need to obtain a reliable estimate of $\sigma^2_{y(within\ treatment)}$. The *s* chart that was constructed in Figure 13-21 contains four different point estimates of $\sigma_{y(within\ treatment)}$—one corresponding to each treatment combination. The in-control status of that chart suggests that $\sigma_{y(within\ treatment)}$ is the same for all treatment combinations; therefore, it is logical to

[8]The concept of a fractional experiment, represented in the formula by $q > 0$, is covered in the next section.

[9]The carat symbol (^) over a parameter indicates an estimate of the true value of that parameter.

TABLE 13-14 ANOVA Table for 2^2 Factorial Design

ANOVA Table

Sources of Variation	Degrees of Freedom	Sum of Squares	Mean Square	*F* Score	*p* Value
Factor A (Time)	1	SSA	MSA = SSA	MSA/MSE	
Factor B (Temp)	1	SSB	MSB = SSB	MSB/MSE	
Interaction (AB)	1	SSAB	MSAB = SSAB	MSAB/MSE	
Error	$4(n-1)$	SSE	MSE = SSE/$4(n-1)$		
Total	$4(n-1)$	SST	MST = SST/$4(n-1)$		

TABLE 13-15 ANOVA Table for 2^2 Factorial Design

ANOVA Table

Sources of Variation	Degrees of Freedom	Sum of Squares	Mean Square	*F* Score	*p* Value
Factor A (Time)	1	$[37.367+80.317-26.359-51.817]^2/(4\times5) =$ 1560.8821/20 = 78.0441	78.0441	78.0441/0.3079 = 253.4722	0.00000
Factor B (Temp)	1	$[51.817+80.317-26.359-37.367]^2/(4\times5) =$ 4679.6545/20 = 233.9827	233.9827	233.9827/0.3079 = 759.9308	0.00000
Interaction (AB)	1	$[26.359+80.317-51.817-37.367]^2/(4\times5) =$ 305.9701/20 = 15.2985	15.2985	15.2985/0.3079 = 49.6866	0.00000
Error	16	4.9256	4.9256/16 = 0.3079		
Total	19	(19)(4.181733)2 = 332.2509	332.2509/19 = 17.4869		

average (pool) the four individual estimates to obtain an overall estimate that we believe will be a closer approximation to the true underlying variance. In general, the 2^k variance estimates can be averaged into a pooled estimate of $\sigma^2_{y(within\ treatment)}$ (which we will estimate using s_y^2). If we define s_t^2 to be the sample variance of measured responses for n replications of treatment combination t,

$$\hat{\sigma}^2_{y(within\ treatment)} = s_y^2 = \sum_{t=1}^{2^k} \frac{s_t^2}{2^k} = MSE \qquad \textbf{13-24}$$

To illustrate the application of Equation 13-24 for the SSS, Inc. case,

$$\hat{\sigma}^2_{y(within\ treatment)} = \frac{((0.680511)^2 + (0.586888)^2 + (0.554903)^2 + (0.340507)^2)}{4}$$

$$\hat{\sigma}^2_{y(within\ treatment)} = \frac{1.231395}{4} = 0.307849$$

Note that, as we would expect, our calculation for $\hat{\sigma}^2_{y(within\ treatments)}$ is the same as the mean square error (MSE) that we computed in the ANOVA table shown in Table 13-15. We can now apply Equation 13-23 to obtain $\hat{\sigma}^2_{effect}$.

$$\hat{\sigma}^2_{effect} = \frac{\hat{\sigma}^2_{y(within\ treatments)}}{n2^{k-2}} = \frac{0.307849}{5(2^{2-2})} = 0.0615698$$

From $\hat{\sigma}^2_{effect}$ we can now estimate the effect standard error, $\hat{\sigma}_{effect}$, which we have assumed to be the same for all effects.

$$\hat{\sigma}_{effect} = \sqrt{\frac{\hat{\sigma}^2_{y(within\ treatments)}}{n2^{k-2}}} = \sqrt{0.0615698} = 0.248133$$

We can now test the significance of each effect by setting up the following hypothesis test:

$$H_0 : \varepsilon = 0$$
$$H_A : \varepsilon \neq 0$$

$$\text{Let } t = \frac{\hat{\varepsilon}}{\hat{\sigma}_{effect}} \qquad \textbf{13-25}$$

$$\text{Reject } H_0 \text{ if } t \geq t_{\frac{\alpha}{2},2^k(n-1)}$$

where,

$\hat{\varepsilon}$ = estimated effect for an experimental factor
H_0 = the null hypothesis to be disproven is that there is no effect
H_A = the alternative hypothesis is that the effects differs from zero
$\hat{\sigma}_{effect}$ = estimate of the standard error for the effect
$2^k(n-1)$ = degrees of freedom for t distribution

Equation 13-25 shows that a *t* statistic is computed by dividing the effect estimate by the estimate of the effect standard error. This computed t is then compared to a critical tabular t value with $2^k(n-1)$ degrees of freedom, obtained by consulting Table IV in Appendix B with a stipulated confidence level $(1 - \alpha)$. If the computed t exceeds the critical t one can conclude that the measured effect is real (i.e., the effect differs from zero) and not due to random causes.

Table 13-16 summarizes the SSS, Inc. results and confirms our earlier conclusions. Since all p values < 0.0001 there is compelling evidence that all effects differ from zero. An alternative perspective on this issue is to compute a confidence interval about each effect. This is accomplished by placing the effect at the center of the interval and adding and subtracting m standard deviations. The selection of m depends on the level of confidence. For example, $m = 2$ defines a 95% confidence interval. That is, by defining an interval that is $\varepsilon \pm 2\sigma_{effect}$ one can be 95% confident that this interval captures the true value of ε. An alternative way to look at this is to realize that there is only a 5% chance (i.e., 1 chance in 20) that the true value of ε lies outside the interval. If these odds are unacceptable a wider interval can be constructed

TABLE 13-16 *t* **Tests on Significance of Effects**

Effect	Estimate of Effect $\hat{\varepsilon}$	Standard Error $\hat{\sigma}_{effect}$	*t* Statistic $\frac{\hat{\varepsilon}}{\hat{\sigma}_{effect}}$	Degrees of Freedom $2^k(n-1)$	*p* Value	95% Confidence Interval
A	3.951	0.248	15.931	16	< 0.0001	[3.455, 4.447]
B	6.841	0.248	27.585	16	< 0.0001	[6.345, 7.337]
AB	1.750	0.248	7.056	16	< 0.0001	[1.254, 2.246]

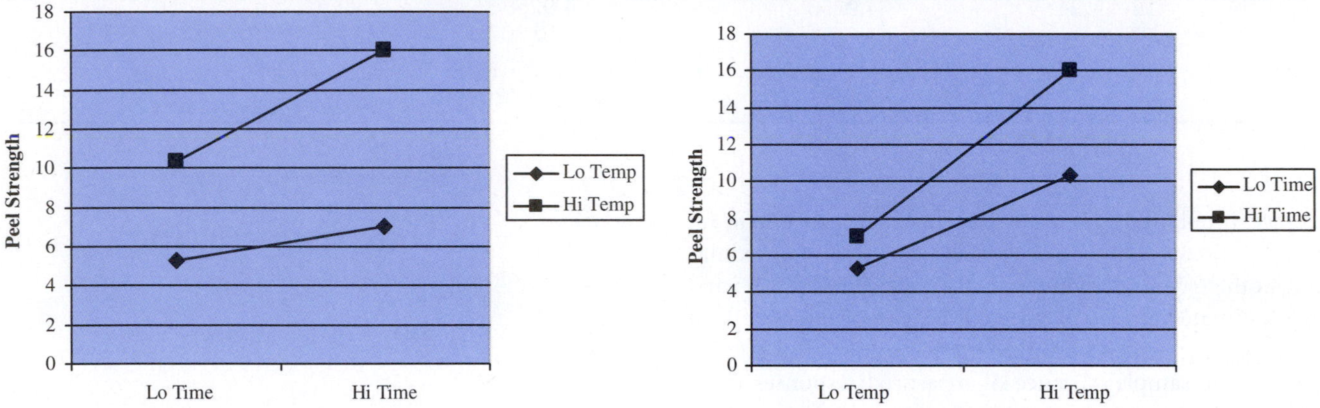

FIGURE 13-22. Effect Charts

with a larger value of m. If the confidence interval so constructed fails to capture zero (that is a zero effect is not a possibility for ε) one can conclude (with 95% confidence) that the observed effect is real. These confidence intervals have been computed and presented in the right hand column of Table 13-16. As zero is not contained in any of the intervals this corroborates our previous conclusions.

Figure 13-22 illustrates another tool that is often useful in analyzing 2^2 designs. The pairs of effect graphs can provide quick visual (though crude) evidence of whether an effect is present and significant. Performing a *t* test on the main effects is equivalent to judging whether the slopes on these graphs statistically differ from zero and whether the vertical point spread is significant. Testing for an interaction effect is equivalent to judging whether or not the graphs are parallel.

In the left side graph of Figure 13-22 the pair of line plots provide visual representations of how moving the time variable from its low setting to its high setting affects peel strength. The higher of the two plots shows the effect when temperature is set at its high level and the lower plot shows the effect on peel strength when temperature is set to its low level. The observed positive slope on both plots would suggest that moving time from low to high has an effect on peel strength no matter what temperature. The vertical separation of the graphs at the low and high levels of time would suggest that moving temperature from low to high affects peel strength no matter what time. The observed nonparallelism of the pair of graphs suggests an interaction between the two factors.

Similar observations can be made with respect to the right-side graph where the pair of plots illustrate the effects of changing time from low to high and when temperature is maintained at its low and high temperature respectively.

13.2.2 Fractional Two-Level Factorial Designs

When designing an experiment it is not uncommon to identify a number of factors that need to be investigated. As we have seen, a two-level k factor design generates a total of 2^k treatment combinations. For small values of k (say no more than 4) it may be feasible to run an experiment in which data are collected for each treatment; however, as we have seen there is a trade-off between the number of treatment combinations run, the number of replications run per treatment combination, and the ability to navigate through the experimental space to investigate responses at different high and low levels of the dominant factors. Consider the following industrial example.[10] In a certain semiconductor plant a spin coater is used to apply a uniform

[10]Montgomery, Douglas C. *Design and Analysis of Experiments.* 3rd ed. New York: Wiley, 1991, 383–84.

TABLE 13-17 Factors in Silicon Wafer Production

Response: Photoresist Thickness

	Factor	Low Level	High Level
A	Final Spin Speed	6650 rpm	7300 rpm
B	Acceleration Rate	5	20
C	Volume of Resist Applied	3 cc	5 cc
D	Time of Spin	14 sec	6 sec
E	Resist Batch Variation	Batch 1	Batch 2
F	Exhaust Pressure	Cover Off	Cover On

photoresist layer on bare silicon wafers. Coating thickness and the variability of thickness are critical factors to downstream operations. The six factors shown in Table 13-17 with their respective high and low values have been selected for study.

To run a full factorial experiment would require the investigation of $2^6 = 64$ different treatment combinations. The complete design is shown in Tables 13-18a through 13-18h. Note that each column is unique and represents a different effect. We can compute the number of effects as follows:

Number of Effects Possible

$$\text{Main Effects} = \text{k (one for each factor)} \qquad \textbf{13-26}$$

$$\text{Interaction Effects: } \frac{k!}{x!(k-x)!} \qquad \textbf{13-27}$$

where,

k = number of factors

x = number of factors interacting

We can use Equations 13-26 and 13-27 to compute the total number of effects for the semiconductor data.

Main Effects: 6

$$\textit{Two-way Interactions: } \frac{6!}{2!(6-2)!} = \frac{720}{(2)(24)} = 15$$

$$\textit{Three-way Interactions: } \frac{6!}{3!(6-3)!} = \frac{720}{(6)(6)} = 20$$

$$\textit{Four-way Interactions: } \frac{6!}{4!(6-4)!} = \frac{720}{(24)(2)} = 15$$

$$\textit{Five-way Interactions: } \frac{6!}{5!(6-5)!} = \frac{720}{(120)(1)} = 6$$

$$\textit{Six-way Interactions: } \frac{6!}{6!(6-6)!} = \frac{720}{(720)(1)} = 1$$

Summing these up we discover that there are a total of 63 different effects (i.e., 63 columns in Tables 13-18a through 13-18h). Therefore we have a design comprising 63 columns and 64 rows (a row for each treatment combination). Note that each column has an equal number of pluses and minuses—32 of each.[11] What might not be so obvious is that the pattern of plus and minus signs is unique for each column. This means that each column is independent of the others. Column independence is important to interpreting the results. Since each column is independent of the others we are able to separate the 63 effects and determine how each one influences the measured response. However, the down side is that in order to do this we would need to conduct a complete experiment (full factorial design) that would require running all 64 treatment combinations.

In practice, with so many factors it is usually impractical or uneconomical to conduct a full experiment. With k factors, if we do not wish to include 2^k runs in the experiment, we can achieve our objectives with only a fraction of the total possible runs, using a 2^{k-q} **fractional design**, where q is a positive integer $< k$ that specifies the proportion of the total number of possible treatment combinations that will be run. For example, if $q = 1$, half of the treatment combinations will be run, for $q = 2$ one-quarter of the treatments will be investigated, and so on. Our procedure is to design a full factorial experiment for $k-q$ factors, and *alias* the remaining q factors with high-order interactions that are assumed to be insignificant. When two or more effects are aliased their individual influences on the response variable are **confounded**—that is combined—so that these effects cannot be separated. It is generally assumed that high-order interactions have a negligible effect on the response; therefore main effects and two-way interactions can be aliased with three-way interactions and above without introducing significant error in the effect estimates that matter.

When two or more effects are combined they are said to be confounded. An alias structure defines those effects that are confounded together.

The first step in a fractional design is to determine how many treatment combinations will be explicitly investigated (i.e., the value of q) and then develop some design generators. To illustrate, by setting $q = 2$ a design will be generated containing one-fourth of the total possible treatments—for the semiconductor case this will be a $2^{6-2} = 2^4 = 16$ design. The **design generator** will determine the alias structure—that is, which effects are confounded and cannot be separated in the analysis.

The following design generators: E = ABC and F = BCD will be used.[12] This leads to the following observation. Since each column in our design matrix contains only -1s and $+1$s, if we multiply any column by itself a column containing

[11]For convenience the 1s have been eliminated from the design matrix, but remain by implication. Therefore, a "$-$" is interpreted as "-1" and a "$+$" is interpreted as "$+1$."

[12]We have chosen these design generators to deliberately confound these two main effects with three-way interactions, thought to be negligible relative to the two main effects. It is generally true that the higher the order of interaction (i.e., the more factors interacting) the more negligible the effect.

TABLE 13-18a Main Effects and Two-Way Interactions, First Half of 2^6 Experiment

Treatment Combination		Main Effects						Two-Way Interactions														
		A	B	C	D	E	F	AB	AC	AD	AE	AF	BC	BD	BE	BF	CD	CE	CF	DE	DF	EF
1	(1)	−	−	−	−	−	−	+	+	+	+	+	+	+	+	+	+	+	+	+	+	+
2	a	+	−	−	−	−	−	−	−	−	−	−	+	+	+	+	+	+	+	+	+	+
3	b	−	+	−	−	−	−	−	+	+	+	+	−	−	−	−	+	+	+	+	+	+
4	ab	+	+	−	−	−	−	+	−	−	−	−	−	−	−	−	+	+	+	+	+	+
5	c	−	−	+	−	−	−	+	−	+	+	+	−	+	+	+	−	−	−	+	+	+
6	ac	+	−	+	−	−	−	−	+	−	−	−	−	+	+	+	−	−	−	+	+	+
7	bc	−	+	+	−	−	−	−	−	+	+	+	+	−	−	−	−	−	−	+	+	+
8	abc	+	+	+	−	−	−	+	+	−	−	−	+	−	−	−	−	−	−	+	+	+
9	d	−	−	−	+	−	−	+	+	−	+	+	+	−	+	+	−	+	+	−	−	+
10	ad	+	−	−	+	−	−	−	−	+	−	−	+	−	+	+	−	+	+	−	−	+
11	bd	−	+	−	+	−	−	−	+	−	+	+	−	+	−	−	−	+	+	−	−	+
12	abd	+	+	−	+	−	−	+	−	+	−	−	−	+	−	−	−	+	+	−	−	+
13	cd	−	−	+	+	−	−	+	−	−	+	+	−	−	+	+	+	−	−	−	−	+
14	acd	+	−	+	+	−	−	−	+	+	−	−	−	−	+	+	+	−	−	−	−	+
15	bcd	−	+	+	+	−	−	−	−	−	+	+	+	+	−	−	+	−	−	−	−	+
16	abcd	+	+	+	+	−	−	+	+	+	−	−	+	+	−	−	+	−	−	−	−	+
17	e	−	−	−	−	+	−	+	+	+	−	+	+	+	−	+	+	−	+	−	+	−
18	ae	+	−	−	−	+	−	−	−	−	+	−	+	+	−	+	+	−	+	−	+	−
19	be	−	+	−	−	+	−	−	+	+	−	+	−	−	+	−	+	−	+	−	+	−
20	abe	+	+	−	−	+	−	+	−	−	+	−	−	−	+	−	+	−	+	−	+	−
21	ce	−	−	+	−	+	−	+	−	+	−	+	−	+	−	+	−	+	−	−	+	−
22	ace	+	−	+	−	+	−	−	+	−	+	−	−	+	−	+	−	+	−	−	+	−
23	bce	−	+	+	−	+	−	−	−	+	−	+	+	−	+	−	−	+	−	−	+	−
24	abce	+	+	+	−	+	−	+	+	−	+	−	+	−	+	−	−	+	−	−	+	−
25	de	−	−	−	+	+	−	+	+	−	−	+	+	−	−	+	−	−	+	+	−	−
26	ade	+	−	−	+	+	−	−	−	+	+	−	+	−	−	+	−	−	+	+	−	−
27	bde	−	+	−	+	+	−	−	+	−	−	+	−	+	+	−	−	−	+	+	−	−
28	abde	+	+	−	+	+	−	+	−	+	+	−	−	+	+	−	−	−	+	+	−	−
29	cde	−	−	+	+	+	−	+	−	−	−	+	−	−	−	+	+	+	−	+	−	−
30	acde	+	−	+	+	+	−	−	+	+	+	−	−	−	−	+	+	+	−	+	−	−
31	bcde	−	+	+	+	+	−	−	−	−	−	+	+	+	+	−	+	+	−	+	−	−
32	abcde	+	+	+	+	+	−	+	+	+	+	−	+	+	+	−	+	+	−	+	−	−

all +1s results. Such a column will all +1s is called an identity vector, is given the symbol I, and will be of the order 2^{k-q}.

From the design generators stated previously, we determine a **defining relation** as follows:

- Multiply each design generator through by its relevant main effect.
- Multiply all possible products of the resulting identity relations.

The design generator determines the aliasing structure.

For our silicon wafer example, the defining relation is computed in the following sequence.

First we multiply each design generator through by its relevant main effect.

$$E = ABC$$

$$(E)(E) = ABCE = I$$

$$F = BCD$$

$$(F)(F) = BCDF = I$$

Next, we multiply all products of the identity relations.

$$I^2 = (ABCE)(BCDF) = AB^2C^2DEF$$

Since $I^2 = I$, $B^2 = I$, and $C^2 = I$ we have,

$$I = ADEF$$

TABLE 13-18b Main Effects and Two-Way Interactions, Second Half of 2^6 Experiment

Treatment Combination		Main Effects						Two-Way Interactions														
		A	B	C	D	E	F	AB	AC	AD	AE	AF	BC	BD	BE	BF	CD	CE	CF	DE	DF	EF
33	f	−	−	−	−	−	+	+	+	+	+	−	+	+	+	−	+	+	−	+	−	−
34	af	+	−	−	−	−	+	−	−	−	−	+	+	+	+	−	+	+	−	+	−	−
35	bf	−	+	−	−	−	+	−	+	+	+	−	−	−	−	+	+	+	−	+	−	−
36	abf	+	+	−	−	−	+	+	−	−	−	+	−	−	−	+	+	+	−	+	−	−
37	cf	−	−	+	−	−	+	+	−	+	+	−	−	+	+	−	−	−	+	+	−	−
38	acf	+	−	+	−	−	+	−	+	−	−	+	−	+	+	−	−	−	+	+	−	−
39	bcf	−	+	+	−	−	+		−	+	+	−	+	−	−	+	−	−	+	+	−	−
40	abcf	+	+	+	−	−	+	+	+	−	−	+	+	−	−	+	−	−	+	+	−	−
41	df	−	−	−	+	−	+	+	+	−	+	−	+	−	+	−	−	+	−	−	+	−
42	adf	+	−	−	+	−	+	−	−	+	−	+	+	−	+	−	−	+	−	−	+	−
43	bdf	−	+	−	+	−	+	−	+	−	+	−	−	+	−	+	−	+	−	−	+	−
44	abdf	+	+	−	+	−	+	+	−	+	−	+	−	+	−	+	−	+	−	−	+	−
45	cdf	−	−	+	+	−	+	+	−	−	+	−	−	−	+	−	+	−	+	−	+	−
46	acdf	+	−	+	+	−	+	−	+	+	−	+	−	−	+	−	+	−	+	−	+	−
47	bcdf	−	+	+	+	−	+	−	−	−	+	−	+	+	−	+	+	−	+	−	+	−
48	abcdf	+	+	+	+	−	+	+	+	+	−	+	+	+	−	+	+	−	+	−	+	−
49	ef	−	−	−	−	+	+	+	+	+	−	−	+	+	−	−	+	−	−	−	−	+
50	aef	+	−	−	−	+	+	−	−	−	+	+	+	+	−	−	+	−	−	−	−	+
51	bef	−	+	−	−	+	+	−	+	+	−	−	−	−	+	+	+	−	−	−	−	+
52	abef	+	+	−	−	+	+	+	−	−	+	+	−	−	+	+	−	−	−	−	−	+
53	cef	−	−	+	−	+	+	+	−	+	−	−	−	+	−	−	+	+	+	−	−	+
54	acef	+	−	+	−	+	+	−	+	−	+	+	−	+	−	−	−	+	+	−	−	+
55	bcef	−	+	+	−	+	+	−	−	+	−	−	+	−	+	+	−	+	+	−	−	+
56	abcef	+	+	+	−	+	+	+	+	−	+	+	+	−	+	+	−	+	+	−	−	+
57	def	−	−	−	+	+	+	+	+	−	−	−	+	−	−	−	−	−	−	+	+	+
58	adef	+	−	−	+	+	+	−	−	+	+	+	+	−	−	−	−	−	−	+	+	+
59	bdef	−	+	−	+	+	+	−	+	−	−	−	−	+	+	+	−	−	−	+	+	+
60	abdef	+	+	−	+	+	+	+	−	+	+	+	−	+	+	+	−	−	−	+	+	+
61	cdef	−	−	+	+	+	+	+	−	−	−	−	−	−	−	−	+	+	+	+	+	+
62	acdef	+	−	+	+	+	+	−	+	+	+	+	−	−	−	−	+	+	+	+	+	+
63	bcdef	−	+	+	+	+	+	−	−	−	−	−	+	+	+	+	+	+	+	+	+	+
64	abcdef	+	+	+	+	+	+	+	+	+	+	+	+	+	+	+	+	+	+	+	+	+

Therefore the defining relation is

$$I = ABCE = BCDF = ADEF \qquad \textbf{13-28}$$

From the defining relation we can determine the entire aliasing structure. To determine the confounding of any effect we simply multiply that effect through the defining relation. For example, if we wish to learn what other effects the two-way interaction effect AB is confounded with we simply multiply Equation 13-28 through by AB to achieve the following.

$$(AB)(I) = (AB)(ABCE) = (AB)(BCDF) = (AB)(ADEF)$$

$$AB = A^2B^2CE = AB^2CDF = A^2BDEF$$

$$AB = CE = ACDF = BDEF$$

We can see that the two-way interaction effect AB is confounded with another two-way interaction effect (CE) and two four-way interaction effects (ACDF and BDEF). Each group of letters in the defining relation is called a *word* and the number of letters the *word length*. The length of the shortest word in the defining relation is the **resolution** of the design and describes the degree to which the estimates of main effects are aliased (confounded) with two-level, three-level, and higher interactions.

A resolution II design, designated R_{II}, aliases main effects with other main effects and is therefore an undesirable approach. A resolution III design (R_{III}) does not alias main effects with each other, but does alias mains with two-way interactions. R_{III} designs are typically used when a large

TABLE 13-18c Three-Way Interactions, First Half of 2^6 Experiment

Treatment		Three-Way Interactions																			
Combination		ABC	ABD	ABE	ABF	ACD	ACE	ACF	ADE	ADF	AEF	BCD	BCE	BCF	BDE	BDF	BEF	CDE	CDF	CEF	DEF
1	(1)	−	−	−	−	−	−	−	−	−	−	−	−	−	−	−	−	−	−	−	−
2	a	+	+	+	+	+	+	+	+	+	+	−	−	−	−	−	−	−	−	−	−
3	b	+	+	+	+	−	−	−	−	−	−	+	+	+	+	+	+	−	−	−	−
4	ab	−	−	−	−	+	+	+	+	+	+	+	+	+	+	+	+	−	−	−	−
5	c	+	−	−	−	+	+	+	−	−	−	+	+	+	−	−	−	+	+	+	−
6	ac	−	+	+	+	−	−	−	+	+	+	+	+	+	−	−	−	+	+	+	−
7	bc	−	+	+	+	+	+	+	−	−	−	−	−	−	+	+	+	+	+	+	−
8	abc	+	−	−	−	−	−	−	+	+	+	−	−	−	+	+	+	+	+	+	−
9	d	−	+	−	−	+	−	−	+	+	−	+	−	−	+	+	−	+	+	−	+
10	ad	+	−	+	+	−	+	+	−	−	+	+	−	−	+	+	−	+	+	−	+
11	bd	+	−	+	+	+	−	−	+	+	−	−	+	+	−	−	+	+	+	−	+
12	abd	−	+	−	−	−	+	+	−	−	+	−	+	+	−	−	+	+	−	−	+
13	cd	+	+	−	−	−	+	+	+	+	−	−	+	+	+	+	−	−	−	+	+
14	acd	−	−	+	+	+	−	−	−	−	+	−	+	+	+	+	−	−	−	+	+
15	bcd	−	−	+	+	−	+	+	+	+	+	+	−	−	−	−	+	−	−	+	+
16	abcd	+	+	−	−	+	−	−	−	−	−	+	−	−	−	−	+	−	−	+	+
17	e	−	−	+	−	−	+	−	+	−	+	−	+	−	+	−	+	+	−	+	+
18	ae	+	+	−	+	+	−	+	−	+	−	−	+	−	+	−	+	+	−	+	+
19	be	+	+	−	+	−	+	−	+	−	+	+	−	+	−	+	−	+	−	+	+
20	abe	−	−	+	−	+	−	+	−	+	−	+	−	+	−	+	−	+	+	+	+
21	ce	+	−	+	−	+	−	+	+	−	+	+	−	+	+	−	+	−	+	−	+
22	ace	−	+	−	+	−	+	−	−	+	−	+	−	+	+	−	+	−	+	−	+
23	bce	−	+	−	+	+	−	+	+	−	+	−	+	−	−	+	−	−	+	−	+
24	abce	+	−	+	−	−	+	−	−	+	−	−	+	−	−	+	−	−	−	−	+
25	de	−	+	+	−	+	+	−	−	+	+	+	+	−	−	+	+	−	+	+	−
26	ade	+	−	−	+	−	−	+	+	−	−	+	+	−	−	+	+	−	+	+	−
27	bde	+	−	−	+	+	+	−	−	+	+	−	−	+	+	−	−	−	+	+	−
28	abde	−	+	+	−	−	−	+	+	−	−	−	−	+	+	−	−	−	+	+	−
29	cde	+	+	+	−	−	+	+	−	+	+	−	−	+	−	+	+	+	−	−	−
30	acde	−	−	−	+	+	+	−	+	−	−	−	−	+	−	+	+	+	−	−	−
31	bcde	−	−	−	+	−	−	+	−	+	+	+	+	−	+	−	−	+	−	−	−
32	abcde	+	+	+	−	+	+	−	+	−	−	+	+	−	+	−	−	+	−	−	−

number of factors need to be screened and reduced in preparation for future experimentation.

Resolution IV (R_{IV}) designs do not alias mains with mains, or mains with two-way interactions, but do alias two-way interactions with other two-way interactions. Our design generators described earlier produce a 2^{6-2} design that is a resolution IV, which will be seen from the aliasing structure developed below. A resolution V (R_V) design does not alias mains with mains, mains with two-way interactions, or two-way interactions with other two-way interactions. An R_V design is excellent when factor interactions are a concern.

A full factorial design does not have a design resolution since it does not confound any of the effects. The resolution is stated as "full factorial" or resolution infinity (R_∞).

Equation 13-28 can used to determine the complete aliasing structure for the 63 effects for the silicon wafer problem.

Main effects:

$$A = BCE = DEF = ABCDF$$

$$B = ACE = CDF = ACDEF$$

$$C = ABE = BDF = ACDEF$$

$$D = BCF = AEF = ABCDE$$

$$E = ABC = ADF = BCDEF$$

$$F = BCD = ADE = ABCEF$$

TABLE 13-18d Three-Way Interactions, Second Half of 2^6 Experiment

Treatment Combination		Three-Way Interactions																			
		ABC	ABD	ABE	ABF	ACD	ACE	ACF	ADE	ADF	AEF	BCD	BCE	BCF	BDE	BDF	BEF	CDE	CDF	CEF	DEF
33	f	−	−	−	+	−	−	+	−	+	+	−	−	+	−	+	+	−	+	+	+
34	af	+	+	+	−	+	+	−	+	−	−	−	−	+	−	+	+	−	+	+	+
35	bf	+	+	+	−	−	−	+	−	+	+	+	+	−	+	−	+	−	+	+	+
36	abf	−	−	−	+	+	+	−	+	−	−	+	+	−	+	−	−	−	+	+	+
37	cf	+	−	−	+	+	+	−	−	+	+	+	+	−	−	+	+	+	−	−	+
38	acf	−	+	+	−	−	−	+	+	−	−	+	+	−	−	+	+	+	+	−	+
39	bcf	−	+	+	−	+	+	−	−	+	+	−	−	+	+	−	+	+	−	−	+
40	abcf	+	−	−	+	−	−	+	+	−	−	−	−	+	+	−	−	+	−	−	+
41	df	−	+	−	+	+	−	+	+	−	+	+	−	+	+	−	+	+	−	+	−
42	adf	+	−	+	−	−	+	−	−	+	−	+	−	+	+	−	+	+	−	+	−
43	bdf	+	−	+	−	+	−	+	+	−	+	−	+	−	−	+	−	+	−	+	−
44	abdf	−	+	−	+	−	+	−	−	+	−	−	+	−	−	+	−	+	−	+	−
45	cdf	+	+	−	+	−	−	−	+	−	+	−	+	−	+	−	+	−	+	−	−
46	acdf	−	−	+	−	+	−	+	−	+	−	−	+	−	+	−	+	−	+	−	−
47	bcdf	−	−	+	−	−	+	−	+	−	+	+	−	+	−	+	−	−	+	−	−
48	abcdf	+	+	−	+	+	−	+	−	+	−	+	−	+	−	+	−	−	+	−	−
49	ef	−	−	+	+	−	+	+	+	+	−	−	+	+	+	+	−	+	+	−	−
50	aef	+	+	−	−	+	−	−	−	−	+	−	+	+	+	+	−	+	+	−	−
51	bef	+	+	−	−	−	+	+	+	+	−	+	−	−	−	−	+	+	+	−	−
52	abef	−	−	+	+	+	−	−	−	−	+	+	−	−	−	−	+	+	+	−	−
53	cef	+	−	+	+	+	−	−	+	+	−	+	−	−	+	+	−	−	−	+	−
54	acef	−	+	−	−	−	+	+	−	−	+	+	−	−	+	+	−	−	−	+	−
55	bcef	−	+	−	−	+	−	−	+	+	−	−	+	+	−	−	+	−	−	+	−
56	abcef	+	−	+	+	−	+	+	−	−	+	−	+	+	−	−	+	−	−	+	−
57	def	−	+	+	+	+	+	+	−	−	−	+	+	+	−	−	−	−	−	−	+
58	adef	+	−	−	−	−	−	−	+	+	+	+	+	+	−	−	−	−	−	−	+
59	bdef	+	−	−	−	+	+	+	−	−	−	−	−	−	+	+	+	−	−	−	+
60	abdef	−	+	+	+	−	−	−	+	+	+	−	−	−	+	+	+	−	−	−	+
61	cdef	+	+	+	+	−	−	−	−	−	−	−	−	−	−	−	−	+	+	+	+
62	acdef	−	−	−	−	+	+	+	+	+	+	−	−	−	−	−	−	+	+	+	+
63	bcdef	−	−	−	−	−	−	−	−	−	−	+	+	+	+	+	+	+	+	+	+
64	abcdef	+	+	+	+	+	+	+	+	+	+	+	+	+	+	+	+	+	+	+	+

Two-way interactions:

AB = CE = ACDF = BDEF

AC = BE = ABDF = CDEF

AD = EF = BCDE = ABCF

AE = BC = DF = ABCDEF

AF = DE = BCEF = ABCD

BD = CF = ACDE = ABEF

BF = CD = ACEF = ABDE

Three-way interactions:

ABC = E = ADF = BCDEF

ABD = CDE = ACF = BEF

ABE = C = BDF = ACDEF

ABF = CEF = ACD = BDE

ACE = B = CDF = ABDFE

ADE = F = BCD = ABCFE

AEF = D = BCF = ABCDE

BCE = A = DEF = ABCDF

TABLE 13-18e Four-Way Interactions, First Half of 2^6 Experiment

Treatment Combination		Four-Way Interactions														
		ABCD	ABCE	ABCF	ABDE	ABDF	ABEF	ACDE	ACDF	ACEF	ADEF	BCDE	BCDF	BCEF	BDEF	CDEF
1	(1)	+	+	+	+	+	+	+	+	+	+	+	+	+	+	+
2	a	−	−	−	−	−	−	−	−	−	−	+	+	+	+	+
3	b	−	−	−	−	−	−	+	+	+	+	−	−	−	−	+
4	ab	+	+	+	+	+	+	−	−	−	−	−	−	−	−	+
5	c	−	−	−	+	+	+	−	−	−	+	−	−	−	+	−
6	ac	+	+	+	−	−	−	+	+	+	−	−	−	−	+	−
7	bc	+	+	+	−	−	−	−	−	−	+	+	+	+	−	−
8	abc	−	−	−	+	+	+	+	+	+	−	+	+	+	−	−
9	d	−	+	+	−	−	+	−	−	+	−	−	−	+	−	−
10	ad	+	−	−	+	+	−	+	+	−	+	−	−	+	−	−
11	bd	+	−	−	+	+	−	−	−	+	−	+	+	−	+	−
12	abd	−	+	+	−	−	+	+	+	−	+	+	+	−	+	−
13	cd	+	−	−	−	−	+	+	+	−	−	+	+	−	−	+
14	acd	−	+	+	+	+	−	−	−	+	+	+	+	−	−	+
15	bcd	−	+	+	+	+	−	+	+	−	−	−	−	+	+	+
16	abcd	+	−	−	−	−	+	−	−	+	+	−	−	+	+	+
17	e	+	−	+	−	+	−	−	+	−	−	−	+	−	−	−
18	ae	−	+	−	+	−	+	+	−	+	+	−	+	−	−	−
19	be	−	+	−	+	−	+	−	+	−	−	+	−	+	+	−
20	abe	+	−	+	−	+	−	+	−	+	+	+	−	+	+	−
21	ce	−	+	−	−	+	−	+	−	+	−	+	−	+	−	+
22	ace	+	−	+	+	−	+	−	+	−	+	+	−	+	−	+
23	bce	+	−	+	+	−	+	+	−	+	−	−	+	−	+	+
24	abce	−	+	−	−	+	−	−	+	−	+	−	+	−	+	+
25	de	−	−	+	+	−	−	+	−	−	+	+	−	−	+	+
26	ade	+	+	−	−	+	+	−	+	+	−	+	−	−	+	+
27	bde	+	+	−	−	+	+	+	−	−	+	−	+	+	−	+
28	abde	−	−	+	+	−	−	−	+	+	−	−	+	+	−	+
29	cde	+	+	−	+	−	−	−	+	+	+	−	+	+	+	−
30	acde	−	−	+	−	+	+	+	−	−	−	−	+	+	+	−
31	bcde	−	−	+	−	+	+	−	+	+	+	+	−	−	−	−
32	abcde	+	+	−	+	−	−	+	−	−	−	+	−	−	−	−

Four-way interactions:

ABCD = ED = AF = BCEF

ABCE = ADEF = BCDF

ABCF = EF = AD = BCDE

ABDE = CD = ACEF = BF

ABDF = CDEF = AC = BE

ABEF = CF = ACDE = BD

ACDF = BDEF = AB = CEF

Five-way interactions:

ABCDE = D = AEF = BCF

ABCDF = DEF = A = BCE

ABCEF = F = AED = BCD

ABDEF = CDF = ACE = B

ACDEF = BDF = ABE = C

BCDEF = ADF = E = ABC

Six-way interaction:

ABCDEF = DF = AE = BC.

TABLE 13-18f Four-Way Interactions, Second Half of 2^6 Experiment

Treatment Combination		Four-Way Interactions														
		ABCD	ABCE	ABCF	ABDE	ABDF	ABEF	ACDE	ACDF	ACEF	ADEF	BCDE	BCDF	BCEF	BDEF	CDEF
33	f	+	+	−	+	−	−	+	−	−	−	+	−	−	−	−
34	af	−	−	+	−	+	+	−	+	+	+	+	−	−	−	−
35	bf	−	−	+	−	+	+	+	−	−	−	−	+	+	+	−
36	abf	+	+	−	+	−	−	−	+	+	+	−	+	+	+	−
37	cf	−	−	+	+	−	−	−	+	+	−	−	+	+	−	+
38	acf	+	+	−	−	+	+	+	−	−	+	−	+	+	−	+
39	bcf	+	+	−	−	+	+	−	+	+	−	+	−	−	+	+
40	abcf	−	−	+	+	−	−	+	−	−	+	+	−	−	+	+
41	df	−	+	−	−	+	−	−	+	−	+	−	+	−	+	+
42	adf	+	−	+	+	−	+	+	−	+	−	−	+	−	+	+
43	bdf	+	−	+	+	−	+	−	+	−	+	+	−	+	−	+
44	abdf	−	+	−	−	+	−	+	−	+	−	+	−	+	−	+
45	cdf	+	−	+	−	+	−	+	−	+	+	+	−	+	+	−
46	acdf	−	+	−	+	−	+	−	+	−	−	+	−	+	+	−
47	bcdf	−	+	−	+	−	+	+	−	+	+	−	+	−	−	−
48	abcdf	+	−	+	−	+	−	−	+	−	−	−	+	−	−	−
49	ef	+	−	−	−	−	+	−	−	+	+	−	−	+	+	+
50	aef	−	+	+	+	+	−	+	+	−	−	−	−	+	+	+
51	bef	−	+	+	+	+	−	−	−	+	+	+	+	−	−	+
52	abef	+	−	−	−	−	+	+	+	−	−	+	+	−	−	+
53	cef	−	+	+	−	−	+	+	+	−	+	+	+	−	+	−
54	acef	+	−	−	+	+	−	−	−	+	−	+	+	−	+	−
55	bcef	+	−	−	+	+	−	+	+	−	+	−	−	+	−	−
56	abcef	−	+	+	−	−	+	−	−	+	−	−	−	+	−	−
57	def	−	−	−	+	+	+	+	+	+	−	+	+	+	−	−
58	adef	+	+	+	−	−	−	−	−	−	+	+	+	+	−	−
59	bdef	+	+	+	−	−	−	+	+	+	−	−	−	−	+	−
60	abdef	−	−	−	+	+	+	−	−	−	+	−	−	−	+	−
61	cdef	+	+	+	+	+	+	−	−	−	−	−	−	−	−	+
62	acdef	−	−	−	−	−	−	+	+	+	+	−	−	−	−	+
63	bcdef	−	−	−	−	−	−	−	−	−	−	+	+	+	+	+
64	abcdef	+	+	+	+	+	+	+	+	+	+	+	+	+	+	+

Tables 13-19a through 13-19d include the 16 runs that make up the fractional subset for this design. Observe that in these tables all columns are no longer independent. The columns corresponding to effects that are aliased are now identical. For example the column for the main effect A (Table 13-19a) contains a minus sign (−) in the first eight rows followed by a plus sign (+) in the last eight rows. This pattern is repeated in the DEF and BCE columns (Table 13-19b) and the ABCDF column (Table 13-19d), as these four affects are all confounded. Therefore in our analysis when we estimate the magnitude of the main effort of factor A, our calculations will also include the interactive influences of factors D, E, and F interacting, factors B, C, and E interacting, and factors A, B, C, D, and F interacting. Our assumption will be that the three-way and five-way interaction effects will be negligible compared to the main effect and that the confounding will not affect the validity of our conclusions.

Table 13-20 contains the actual data that were collected on the photoresist thickness for the 16 treatment runs. Three replications for each treatment were obtained by measuring left, right, and center locations on the wafers selected. The main effects for each of the six factors have been estimated by computing the difference between the photoresist thickness when the factor was at its high level and the thickness when it was at its low level. The last two columns in Table 13-20 contain the standard deviations and respective variances for each of the 16 runs of three replicates. The standard deviations have been plotted on the *s* chart in Figure 13-23. Since the chart shows a good state of control there is no reason to conclude that the within-treatment variances are not all equal.

TABLE 13-18g Five- and Six-Way Interactions, First Half of 2^6 Experiment

	Treatment Combination	Five-Way Interactions						Six-Way Interaction
		ABCDE	ABCDF	ABCEF	ABDEF	ACDEF	BCDEF	ABCDEF
1	(1)	−	−	−	−	−	−	+
2	a	+	+	+	+	+	−	−
3	b	+	+	+	+	−	+	−
4	ab	−	−	−	−	+	+	+
5	c	+	+	+	−	+	+	−
6	ac	−	−	−	+	−	+	+
7	bc	−	−	−	+	+	−	+
8	abc	+	+	+	−	−	−	−
9	d	+	+	−	+	+	+	−
10	ad	−	−	+	−	−	+	+
11	bd	−	−	+	−	+	−	+
12	abd	+	+	−	+	−	−	−
13	cd	−	−	+	+	−	−	+
14	acd	+	+	−	−	+	−	−
15	bcd	+	+	−	−	−	+	−
16	abcd	−	−	+	+	+	+	+
17	e	+	−	+	+	+	+	−
18	ae	−	+	−	−	−	+	+
19	be	−	+	−	−	+	−	+
20	abe	+	−	+	+	−	−	−
21	ce	−	+	−	+	−	−	+
22	ace	+	−	+	−	+	−	−
23	bce	+	−	+	−	−	+	−
24	abce	−	+	−	+	+	+	+
25	de	−	+	+	−	−	−	+
26	ade	+	−	−	+	+	−	−
27	bde	+	−	−	+	−	+	−
28	abde	−	+	+	−	−	+	+
29	cde	+	−	−	−	+	+	−
30	acde	−	+	+	+	+	+	+
31	bcde	−	+	+	+	+	−	+
32	abcde	+	−	−	−	−	−	−

Applying Equation 13-19 to our 2^{6-2} fractional design we can compute all of the relevant sums of squares for the main effects as follows:

$$SSA = \frac{[ae + adef + acf + acd + abf + abd + abce + abcdef - (1) - df - cef - cde - bef - bde - bc - bcdf]^2}{(3)(16)}$$

$$SSA = \frac{[-1{,}808]^2}{48} = 68{,}101.33$$

$$SSB = \frac{[bef + bde + bc + bcdf + abf + abd + abce + abcdef - (1) - df - cef - cde - ae - adef - acf - acd]^2}{(3)(16)}$$

$$SSB = \frac{[-4{,}302]^2}{48} = 385{,}566.75$$

$$SSC = \frac{[cef + cde + bc + bcdf + acf + acd + abce + abcdef - (1) - df - ae - adef - bef - bde - abf - abd]^2}{(3)(16)}$$

$$SSC = \frac{[-1{,}192]^2}{48} = 29{,}601.33$$

TABLE 13-18h Five- and Six-Way Interactions, Second Half of 2^6 Experiment

Treatment Combination		Five-Way Interactions ABCDE	ABCDF	ABCEF	ABDEF	ACDEF	BCDEF	Six-Way Interaction ABCDEF
33	f	−	+	+	+	+	+	−
34	af	+	−	−	−	−	+	+
35	bf	+	−	−	−	+	−	+
36	abf	−	+	+	+	−	−	−
37	cf	+	−	−	+	−	−	+
38	acf	−	+	+	−	+	−	−
39	bcf	−	+	+	−	−	+	−
40	abcf	+	−	−	+	−	+	+
41	df	+	−	+	−	−	−	+
42	adf	−	+	−	+	+	−	−
43	bdf	−	+	−	+	−	+	−
44	abdf	+	−	+	−	+	+	+
45	cdf	−	+	−	−	+	+	−
46	acdf	+	−	+	+	+	+	+
47	bcdf	+	−	+	+	+	−	+
48	abcdf	−	+	−	−	−	−	−
49	ef	+	+	−	−	−	−	+
50	aef	−	−	+	+	+	−	−
51	bef	−	−	+	+	−	+	−
52	abef	+	+	−	−	+	+	+
53	cef	−	−	+	−	+	+	−
54	acef	+	+	−	+	−	+	+
55	bcef	+	+	−	+	+	−	+
56	abcef	−	−	+	−	−	−	−
57	def	−	−	−	+	+	+	−
58	adef	+	+	+	−	−	+	+
59	bdef	+	+	+	−	+	−	+
60	abdef	−	−	−	+	−	−	−
61	cdef	+	+	+	+	−	−	+
62	acdef	−	−	−	−	+	−	−
63	bcdef	−	−	−	−	−	+	−
64	abcdef	+	+	+	+	+	+	+

$$SSD = \frac{[df + cde + bde + bcdf + adef + acd + abd + abcdef - (1) - cef - bef - bc - ae - acf - abf - abce]^2}{(3)(16)}$$

$$SSD = \frac{[110]^2}{48} = 252.08$$

$$SSE = \frac{[cef + cde + bef + bde + ae + adef + abce + abcdef - (1) - df - bc - bcdf - acf - acd - abf - abd]^2}{(3)(16)}$$

$$SSE = \frac{[-186]^2}{48} = 720.75$$

$$SSF = \frac{[df + cef + bef + bcdf + adef + acf + abf + abcdef - (1) - cde - bde - bc - ae - acd - abd - abce]^2}{(3)(16)}$$

$$SSF = \frac{[-836]^2}{48} = 14{,}560.33$$

TABLE 13-19a Main Effects and Two-Way Interactions, 2^{6-2} Fractional Design

			Main Effects						Two-Way Interactions															
Run	Treatment Combination		A	B	C	D	E	F	AB	AC	AD	AE	AF	BC	BD	BE	BF	CD	CE	CF	DE	DF	EF
1	1	(1)	−	−	−	−	−	−	+	+	+	+	+	+	+	+	+	+	+	+	+	+	+
2	41	df	−	−	−	+	−	+	+	+	−	+	−	+	−	+	−	−	+	−	−	+	−
3	53	cef	−	−	+	−	+	+	+	−	+	−	−	−	+	−	−	+	+	+	−	−	+
4	29	cde	−	−	+	+	+	−	+	−	−	−	+	−	−	−	+	+	+	−	+	−	−
5	51	bef	−	+	−	−	+	+	−	+	+	−	−	−	−	+	+	+	−	−	−	−	+
6	27	bde	−	+	−	+	+	−	−	+	−	−	+	−	+	+	−	−	−	+	+	−	−
7	7	bc	−	+	+	−	−	−	−	−	+	+	+	+	−	−	−	−	−	−	+	+	+
8	47	bcdf	−	+	+	+	−	+	−	−	−	+	−	+	+	−	+	+	−	+	−	+	−
9	18	ae	+	−	−	−	+	−	−	−	−	+	−	+	+	−	+	+	−	+	−	+	−
10	58	adef	+	−	−	+	+	+	−	−	+	+	+	+	−	−	−	−	−	−	+	+	+
11	38	acf	+	−	+	−	−	+	−	+	−	−	+	−	+	+	−	−	−	+	+	−	−
12	14	acd	+	−	+	+	−	−	−	+	+	−	−	−	−	+	+	+	−	−	−	−	+
13	36	abf	+	+	−	−	−	+	+	−	−	−	+	−	−	−	+	+	+	−	+	−	−
14	12	abd	+	+	−	+	−	−	+	−	+	−	−	−	+	−	−	−	+	+	−	−	+
15	24	abce	+	+	+	−	+	−	+	+	−	+	−	+	−	+	−	−	+	−	−	+	−
16	64	abcdef	+	+	+	+	+	+	+	+	+	+	+	+	+	+	+	+	+	+	+	+	+

TABLE 13-19b Three-Way Interactions, 2^{6-2} Fractional Experiment

			Three-Way Interactions																			
Run	Treatment Combination		ABC	ABD	ABE	ABF	ACD	ACE	ACF	ADE	ADF	AEF	BCD	BCE	BCF	BDE	BDF	BEF	CDE	CDF	CEF	DEF
1	1	(1)	−	−	−	−	−	−	−	−	−	−	−	−	−	−	−	−	−	−	−	−
2	41	df	−	+	−	+	+	−	+	+	−	+	+	−	+	+	−	+	+	−	+	−
3	53	cef	+	−	+	+	+	−	−	+	+	−	+	−	−	+	+	−	−	−	+	−
4	29	cde	+	+	+	−	−	+	+	−	+	+	−	−	+	−	+	+	+	−	−	−
5	51	bef	+	+	−	−	−	+	+	+	+	−	+	−	−	−	−	+	+	+	−	−
6	27	bde	+	−	−	+	+	+	−	−	+	+	−	−	+	+	−	−	−	+	+	−
7	7	bc	−	+	+	+	+	+	+	−	−	−	−	−	−	+	+	+	+	+	+	−
8	47	bcdf	−	−	+	−	−	+	−	+	−	+	+	−	+	−	+	−	−	+	−	−
9	18	ae	+	+	−	+	+	−	+	−	+	−	−	+	−	+	−	+	+	−	+	+
10	58	adef	+	−	−	−	−	−	−	+	+	+	+	+	+	−	−	−	−	−	−	+
11	38	acf	−	+	+	−	−	−	+	+	−	−	+	+	−	−	+	+	+	+	−	+
12	14	acd	−	−	+	+	+	−	−	−	−	+	−	+	+	+	+	−	−	−	+	+
13	36	abf	−	−	−	+	+	+	−	+	−	−	+	+	−	+	−	−	−	+	+	+
14	12	abd	−	+	−	−	−	+	+	−	−	+	−	+	+	−	−	+	+	−	−	+
15	24	abce	+	−	+	−	−	+	−	−	+	−	−	+	−	−	+	−	−	−	−	+
16	64	abcdef	+	+	+	+	+	+	+	+	+	+	+	+	+	+	+	+	+	+	+	+

TABLE 13-19c Four-Way Interactions, 2^{6-2} Fractional Design

Run	Treatment Combination		Four-Way Interactions														
			ABCD	ABCE	ABCF	ABDE	ABDF	ABEF	ACDE	ACDF	ACEF	ADEF	BCDE	BCDF	BCEF	BDEF	CDEF
1	1	(1)	+	+	+	+	+	+	+	+	+	+	+	+	+	+	+
2	41	df	−	+	−	−	+	−	−	+	−	+	−	+	−	+	+
3	53	cef	−	+	+	−	−	+	+	+	−	+	+	+	−	+	−
4	29	cde	+	+	−	+	−	−	−	+	+	+	−	+	+	+	−
5	51	bef	−	+	+	+	+	−	−	−	+	+	+	+	−	−	+
6	27	bde	+	+	−	−	+	+	+	−	−	+	−	+	+	−	+
7	7	bc	+	+	+	−	−	−	−	−	−	+	+	+	+	−	−
8	47	bcdf	−	+	−	+	−	+	+	−	+	+	−	+	−	−	−
9	18	ae	−	+	−	+	−	+	+	−	+	+	−	+	−	−	−
10	58	adef	+	+	+	−	−	−	−	−	−	+	+	+	+	−	−
11	38	acf	+	+	−	−	+	+	+	−	−	+	−	+	+	−	+
12	14	acd	−	+	+	+	+	−	−	−	+	+	+	+	−	−	+
13	36	abf	+	+	−	+	−	−	−	+	+	+	−	+	+	+	−
14	12	abd	−	+	+	−	−	+	+	+	−	+	+	+	−	+	−
15	24	abce	−	+	−	−	+	−	−	+	−	+	−	+	−	+	+
16	64	abcdef	+	+	+	+	+	+	+	+	+	+	+	+	+	+	+

TABLE 13-19d Five- and Six-Way Interactions, 2^{6-2} Fractional Experiment

Run	Treatment Combination		Five-Way Interactions						Six-Way Interaction
			ABCDE	ABCDF	ABCEF	ABDEF	ACDEF	BCDEF	ABCDEF
1	1	(1)	−	−	−	−	−	−	+
2	41	df	+	−	+	−	−	−	+
3	53	cef	−	−	+	−	+	+	−
4	29	cde	+	−	−	−	+	+	−
5	51	bef	−	−	+	+	−	+	−
6	27	bde	+	−	−	+	−	+	−
7	7	bc	−	−	−	+	+	−	+
8	47	bcdf	+	−	+	+	+	−	+
9	18	ae	−	+	−	−	−	+	+
10	58	adef	+	+	+	−	−	+	+
11	38	acf	−	+	+	−	+	−	−
12	14	acd	+	+	−	−	+	−	−
13	36	abf	−	+	+	+	−	−	−
14	12	abd	+	+	−	+	−	−	−
15	24	abce	−	+	−	+	+	+	+
16	64	abcdef	+	+	+	+	+	+	+

TABLE 13-20 2^{6-2} Fractional Design Experimental Data

Run	Treatment Combination		Main Effects A	B	C	D	E	F	Resist Thickness Left	Center	Right	Average	Stand. Dev.	Variance
1	1	(1)	−	−	−	−	−	−	4647	4687	4660	4664.67	20.4042	416.33
2	41	df	−	−	−	+	−	+	4650	4670	4645	4655.00	13.2288	175.00
3	53	cef	−	−	+	−	+	+	4653	4635	4612	4633.33	20.5508	422.33
4	29	cde	−	−	+	+	+	−	4666	4695	4672	4677.67	15.3080	234.33
5	51	bef	−	+	−	−	+	+	4565	4496	4563	4541.33	39.2726	1542.33
6	27	bde	−	+	−	+	+	−	4542	4547	4538	4542.33	4.5092	20.33
7	7	bc	−	+	+	−	−	−	4470	4492	4495	4485.67	13.6504	186.33
8	47	bcdf	−	+	+	+	−	+	4494	4430	4472	4465.33	32.5167	1057.33
9	18	ae	+	−	−	−	+	−	4666	4695	4682	4681.00	14.5258	211.00
10	58	adef	+	−	−	+	+	+	4514	4591	4580	4561.67	41.6453	1734.33
11	38	acf	+	−	+	−	−	+	4581	4591	4576	4582.67	7.6376	58.33
12	14	acd	+	−	+	+	−	−	4611	4634	4630	4625.00	12.2882	151.00
13	36	abf	+	+	−	−	−	+	4420	4433	4417	4423.33	8.5049	72.33
14	12	abd	+	+	−	+	−	−	4496	4502	4482	4493.33	10.2632	105.33
15	24	abce	+	+	+	−	+	−	4393	4306	4302	4333.67	51.4231	2644.33
16	64	abcdef	+	+	+	+	+	+	4355	4388	4343	4362.00	23.3024	543.00
Avg @ +1			4507.83	4455.88	4520.67	4547.79	4541.63	4528.08				Pooled Variance	S_p^2	598.38
Avg @ −1			4583.17	4635.13	4570.33	4543.21	4549.38	4562.92						49.86
Difference			−75.333	−179.25	−49.667	4.58333	−7.75	−34.833						

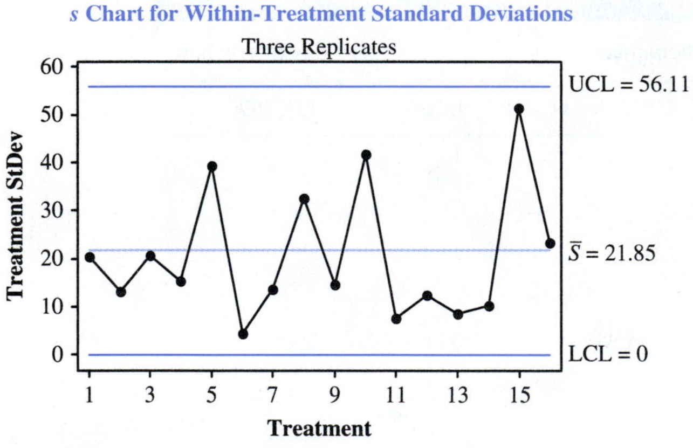

FIGURE 13-23. *s* Chart Showing Equality of Within-Treatment Variances

From the complete ANOVA table shown in Table 13-21 it appears that factors A (final spin speed), B (acceleration rate), C (volume of resist applied), and F (exhaust pressure) have a significant effect while factors D (time of spin) and E (resist batch variation) do not. These conclusions are based on the application of the *F* test, and the resultant *p* values.

Since the *s* chart in Figure 13-23 shows good control Equation 13-24 can be applied to estimate the common pooled variance for each treatment as shown in Table 13-20.

Next, an appropriately modified version of Equation 13-23 can be applied to obtain an estimate of σ^2_{effect}.

$$\hat{\sigma}^2_{effect} = \frac{\hat{\sigma}^2_{y(within\ treatment)}}{n2^{k-q-2}} = \frac{598.38}{(3)(2^{6-2-2})} = \frac{598.38}{12} = 49.86$$

$$\hat{\sigma}_{effect} = \sqrt{49.86} = 7.061$$

The *t* tests shown in Table 13-22 confirm our previous conclusions. In the case of factors A, B, C, and F the 95% confidence intervals (that capture the true effects) fail to capture zero. In fact the *p* values suggest that a confidence level in excess of 99.99% would be required for the interval estimates to be wide enough to include zero as a possibility.

13.2.3 Screening Designs

When a large number of factors are being investigated efficient screening designs can be used to pare out those factors that have negligible effects on the measured response. Such designs are typically of resolution R_{III} and screen on the main effects only, which are heavily confounded with two-way interactions and above. **Plackett-Burman designs**, named after their cofounders,[13] are commonly used in screening applications. This family consists of economical designs that specify the number

[13]Plackett, R. L., and J. P. Burman. "The Design of Optimum Multifactorial Experiments." *Biometrika* 33, no. 4 (June 1946): 305–25.

TABLE 13-21 ANOVA Table for 2^{6-2} Factorial Design

ANOVA Table

Sources of Variation	Degrees of Freedom	Sum of Squares	Mean Square	*F* Score	*p* Value
A (including aliased effects)	1	68,101.33	68,101.33	38.72	< 0.0001
B (including aliased effects)	1	385,566.75	385,566.75	219.21	< 0.0001
C (including aliased effects)	1	29,601.33	29,601.33	16.83	0.0002
D (including aliased effects)	1	252.08	252.08	0.143	0.7073
E (including aliased effects)	1	720.75	720.75	0.410	0.5255
F (including aliased effects)	1	14,560.33	14,560.33	8.28	0.0063
Error	41	72,113.43	1,758.86		
Total	47	570,916.00	12,147.15		

TABLE 13-22 *t* Tests on Significance of Effects for 2^{6-2} Fractional Experiment

	Estimate of Effect	Standard Error	*t* Statistic	Degrees of Freedom		
Effect	$\hat{\varepsilon}$	$\hat{\sigma}_{effect}$	$\frac{\hat{\varepsilon}}{\hat{\sigma}_{effect}}$	$2^{k-q}(n-1)$	*p* Value	95% Confidence Interval
A	−75.33	7.061	−10.669	32	< 0.0001	[−89.46, −61.21]
B	−179.25	7.061	−25.386	32	< 0.0001	[−193.37, −165.13]
C	−49.67	7.061	−7.034	32	< 0.0001	[−63.79, −35.55]
D	4.58	7.061	0.649	32	0.521	[−9.54, 18.70]
E	−7.75	7.061	−1.098	32	0.280	[−21.87, 6.37]
F	−34.83	7.061	−4.933	32	< 0.0001	[−48.95, −20.71]

of experimental runs in multiples of four (as compared to powers of two in 2^k designs). *The number of Plackett-Burman runs is one more than the number of factors in the design.*

Designs are produced starting with a **generating vector** (column) for an n-run design. The generating vector is used to generate all remaining columns to complete the design.

The following are the Plackett-Burman generating vectors for 8 to 24 runs.

$n = 8$ (+ + + − + − −)

$n = 12$ (+ + − + + + − − − + −)

$n = 16$ (+ + + + − + − + + − − + − − −)

$n = 20$ (+ + − − + + + + − + − + − − − − + + −)

$n = 24$ (+ + + + + − + + − − + + − − + − + − − − −)

The generating vector is used for the first $(n - 1)$ rows of column A. Then column B is created by taking the last value in column A, making it the first value in column B, then sliding the remaining values in A into the second through the $(n - 1)$ row positions of column B. Column C is created by placing the last value in column B in the first position of column C, and then sliding the remainder of the values in column B into the second to $(n - 1)$ row positions of column C.

TABLE 13-23 Plackett-Burman Design for $n = 8$ Runs and 7 Factors

	Factor						
Run	A	B	C	D	E	F	G
1	+	−	−	+	−	+	+
2	+	+	−	−	+	−	+
3	+	+	+	−	−	+	−
4	−	+	+	+	−	−	+
5	+	−	+	+	+	−	−
6	−	+	−	+	+	+	−
7	−	−	+	−	+	+	+
8	−	−	−	−	−	−	−

This procedure is continued until all $(n - 1)$ columns are complete. At this point there will also be a total of $(n - 1)$ rows. Add an nth row consisting of all −1s. Table 13-23 illustrates the Plackett-Burman design for seven factors and eight runs. The complete set of designs for the generating vectors shown previously can be found in Tables I through V in Appendix F.

To illustrate the Plackett-Burman procedure consider the following industrial scenario. A manufacturer of automobile components is experiencing difficulties with part shrinkage of a certain plastic molded part. The following seven production factors have been selected for study.

A. Cycle time
B. Mold temperature
C. Holding pressure
D. Gate size
E. Cavity thickness
F. Holding time
G. Screw speed

An eight-run Plackett-Burman designed experiment was conducted and four parts were produced at each of the specified treatment runs. The results of this study are summarized in Table 13-24. The engineers conducting this DOE had two questions: (1) What effect does each of the factors have on average part shrinkage? (2) What effect does each of the factors have on the variation in part shrinkage?

The first question can be addressed by performing an analysis of variance on the main effects. From Table 13-25 (the ANOVA table) it appears that cycle time (factor A) is the only factor of the seven studied that has a significant effect on average part shrinkage. The computed p value is 0.019 which is the probability that our conclusion could be wrong (that is, cycle time does not have any effect on part shrinkage).

TABLE 13-24 Example of Plackett-Burman Design for $n = 8$ Runs and 7 Factors

		Factor							Replicate							
Run	**Treatment**	**A**	**B**	**C**	**D**	**E**	**F**	**G**	**1**	**2**	**3**	**4**	**Total Response**	**Average Response (Shrinkage)**	**Within-Treatment Std Dev**	**Within-Treatment Variance**
1	adfg	+	−	−	+	−	+	+	2.6	4.7	3.6	1.5	12.4	3.10	1.369	1.873
2	abeg	+	+	−	−	+	−	+	3.5	3.6	3.5	3.5	14.1	3.53	0.050	0.003
3	abcf	+	+	+	−	−	+	−	4.5	2.4	2.7	5.1	14.7	3.68	1.328	1.763
4	bcdg	−	+	+	+	−	−	+	3.6	1.0	3.3	0.9	8.8	2.20	1.449	2.100
5	acde	+	−	+	+	+	−	−	2.5	2.4	2.5	2.3	9.7	2.43	0.096	0.009
6	bdef	−	+	−	+	+	+	−	2.4	2.5	2.3	2.4	9.6	2.40	0.082	0.007
7	cefg	−	−	+	−	+	+	+	0.8	3.0	0.8	3.2	7.8	1.95	1.330	1.770
8	(1)	−	−	−	−	−	−	−	2.6	2.7	2.8	2.8	10.9	2.73	0.096	0.009
Avg y @ +1		3.18	2.95	2.56	2.53	2.58	2.78	2.69								
Avg y @ −1		2.32	2.6	2.94	2.97	2.93	2.72	2.81					s_t	1.041		
Effect on y		0.86	0.40	−0.38	−0.44	−0.35	0.06	−0.11					SST	33.6		
Avg s^2 @ +1		3.65	3.87	5.64	3.99	1.79	5.41	5.75								
Avg s^2 @ −1		3.89	3.66	1.89	3.54	5.74	2.12	1.79								
$F = \frac{S^2_{\text{larger}}}{S^2_{\text{smaller}}}$		1.07	1.06	2.98	1.13	3.21	2.55	3.21								
p value		0.45	0.46	0.04	0.42	0.03	0.06	0.03								

TABLE 13-25 ANOVA Table for Plackett-Burman Screening 8-Run Design

ANOVA Table

Sources of Variation	Degrees of Freedom	Sum of Squares	Mean Square	F Score	p Value
A (including aliased effects)	1	5.951	5.951	6.32	0.01910
B (including aliased effects)	1	1.28	1.28	1.36	0.25500
C (including aliased effects)	1	1.125	1.125	1.19	0.2862
D (including aliased effects)	1	1.531	1.531	1.626	0.2145
E (including aliased effects)	1	0.98	0.98	1.041	0.3178
F (including aliased effects)	1	0.031	0.031	0.03	0.8639
G (including aliased effects)	1	0.101	0.101		
Error	24	22.601	0.94		
Total	31	33.60	1.08		

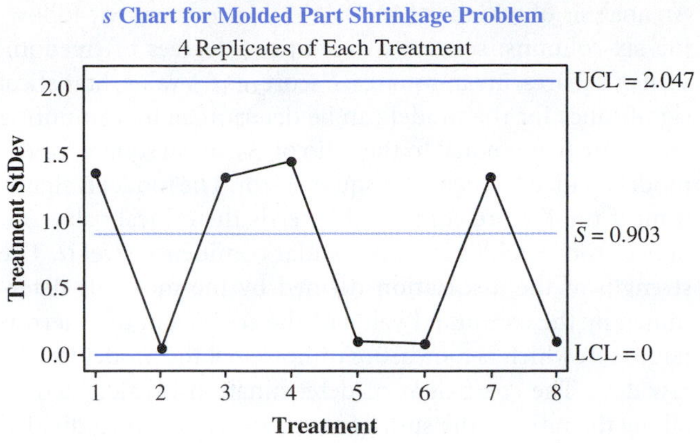

FIGURE 13-24. *s* Chart for Plackett-Burman Runs of Molded Plastic Parts

The second question can be addressed by either plotting an *s* chart or performing a statistical *F* test. The *s* chart shown in Figure 13-24 suggests that part shrinkage variation is unaffected by treatment combination (i.e., factor levels). However, the *s* chart contains only eight points so the *F* test provides a better result (in the sense that it is more believable). For each main effect the average within-treatment sample variance is computed for the high level and the low level. An *F* statistic is then computed by taking the ratio of the larger average variance to the smaller. Then, consulting the *F* distribution tables (Table V in Appendix B) a critical value of *F* can be retrieved.

In accessing the *F* tables, the numerator and denominator degrees of freedom are calculated using the following formula.

$$\text{degrees of freedom} = \frac{(n - 1)(\textit{number of runs})}{2} \qquad \textbf{13-29}$$

Applying Equation 13-29 to the part shrinkage problem we obtain

$$\textit{degrees of freedom} = \frac{(3)(8)}{2} = 12$$

If the computed *F* statistic exceeds the tabular value of the relevant factor we can conclude that moving the factor from high to low, or vice versa, has a significant effect on part shrinkage variance (at the stipulated confidence level). Alternatively, at the calculated value of *F*, we can search the *F* tables and derive the approximate value of *p* for which the results are significant. This latter approach has been taken in Table 13-24, from which we can conclude that factors C (holding pressure), E (cavity thickness), F (holding time), and G (screw speed) appear to influence variability. From this analysis, we can clearly eliminate from further consideration the two factors that seem to have no effect on either the mean response or the variance of response: factors B (mold temperature) and D (gate size). The engineers would be wise to perform further studies on the remaining five factors.

13.3 SUMMARY OF KEY IDEAS

Key Idea: *A correlation analysis determines whether an association between two variables exists. What it does not do is to establish the presence of a causal relationship.*

If two variables are linearly associated a correlation analysis will determine the strength and direction of the association. The measure used to summarize the association is a correlation coefficient, *r*. Weak associations will result in values of *r* close to zero. Strong associations will produce values of *r* close to +1 or −1. The sign on *r* depicts the direction of the association. In performing a correlation analysis it is important that the sample size be large enough to produce a statistically significant *r*. However, the fact that *r* is statistically significant does not mean that the association is operationally important and therefore the basis for management action.

SHOWCASE of EXCELLENCE **Winners of the Malcolm Baldrige National Quality Award Manufacturing Category**

- 1992: AT&T Network Systems Group, Transmission Systems Business Unit, a company that develops, manufactures, markets, and services systems for transporting data, voice, and images over public and private telecommunications networks. http://www.quality.nist.gov/AT&T_NSG.htm.
- 1992: Texas Instruments Incorporated, Defense Systems & Electronics Group, a company that designs and manufactures precision-guided weapons, airborne radar systems, infrared vision equipment and other electro-optic systems, and electronic warfare systems. http://www.quality.nist.gov/Texas_Instruments_Inc_DSEG.htm.
- 1991: Zytec Corporation, the manufacturer of power supplies for original equipment manufacturers of computers and electronic office, medical, and testing equipment. http://www.quality.nist.gov/Zytec_91.htm.
- 1990: Cadillac Motor Car Company, the flagship division of General Motors and manufacturer of nine car models in four manufacturing plants. http://www.cadillac.com. http://www.quality.nist.gov/Cadillac_90.htm.
- 1990: IBM Rochester–AS/400 Division, a manufacturer of intermediate computer systems and hard disk drives to serve world markets. http://www.ibm.com, http://www.quality.nist.gov/IBM_90.htm.

Two variables can be nonlinearly associated, in which case the computation of a correlation coefficient, r, is meaningless. In these cases the coefficient of determination, r^2, can be computed using analysis of variance techniques. The r^2 represents the proportion of the total variation in one variable that can be explained by its association with the other. Therefore strong associations produce r^2 values close to one.

In quality management it is possible and proper to conduct correlation analyses on two dependent (output) variables. However, it is not appropriate to use the model to try and predict the value of one from the value of the other.

Key Idea: *The strength and direction of a linear association between two variables can be assessed by computing a linear correlation coefficient.*

The linear correlation coefficient is a number between -1 and $+1$. Coefficients close to $+1$ or -1 indicate strong linear associations and those close to zero imply weak associations. The sign of the coefficient provides information concerning the direction of the association. A minus sign means that an increase in one variable is accompanied by a corresponding decrease in the other. A plus sign indicates that the two variables move in the same direction.

A simple t test can be used to determine if a computed correlation coefficient is statistically significant.

Key Idea: *An association that is statistically significant is not necessarily operationally important.*

Statistical significance means that the sample size has been large enough so that any observed association is real and not due to random variation. However, an observed association may be too weak to warrant any management action.

Key Idea: *The coefficient of determination can be used to measure the strength of linear and nonlinear associations.*

The coefficient of determination, r^2, has a value that ranges from zero to $+1$. Larger coefficients are indicative of stronger relationships. The coefficient of determination is a measure of the proportion of the total observed variation in the dependent variable y that can be explained due to the association between y and one or more independent variables x.

Key Idea: *Correlation analysis can be performed on any set of variables; however, regression makes sense only when a relationship between a dependent variable and a set of independent variables is being investigated.*

In quality management applications it is sometimes useful to determine the correlation between pairs of dependent variables. For example, both variables could be quality attributes that are process outcomes. In this instance, developing a regression model that could be used to predict one variable from the other (or a set of variables) would be meaningless.

Key Idea: *An analysis of variance (ANOVA) table is a tool that can help determine the nature, strength and statistical significance for linear and nonlinear associations between a dependent variable (the response) and one or more independent variables (factors).*

An analysis of variance (ANOVA) table contains the following six columns: sources of variation, degrees of freedom, sum of squares, mean squares, F score, and p value. Statistical significance for the model can be determined by computing an F score that is equal to the ratio of the mean square for the model divided by the mean square error. The model is significant if the F score computed exceeds the critical value obtained from F tables for a particular confidence level α. The strength of the association defined by the model is determined by the computed value of the coefficient of determination, r^2, which is a measure of how well the model fits the raw data. The coefficient of determination is calculated by taking the ratio of the sum of squares for the model divided by the total sum of squares.

Key Idea: *Residuals will be normally distributed with a mean of zero for a model that fits the data well across the entire measurement domain.*

For a given set of independent variables a residual is defined as the difference between the dependent variable that the model predicts will occur and the actual value that does occur. Residuals represent the vertical distances between each point on the scatterplot and the regression line. If the actual point lies above the line the corresponding residual will be positive; if the point lies below the line the residual will be negative. If the model is consistent in its ability to predict across the entire domain of x the residuals will be normally distributed and randomly distributed about zero when plotted against the xs or the predicted ys. Any nonrandom patterns can provide some insights into where a particular model may have the propensity to under or overestimate.

Key Idea: *Analysis of variance (ANOVA) is a useful tool that can be applied to determine if statistical differences exist between data sets.*

The ANOVA methodology investigates group mean differences by systematically analyzing observed variation in the data. This is accomplished by partitioning the observed total variability into two categories: that which can be explained within groups and that which occurs among the groups. A statistical difference is concluded when the proportion of total variability explained between groups is significant. This methodology can help identify improvement opportunities or validate process improvement initiatives.

Key Idea: *A design of experiments (DOE) is used to determine from a number of candidate factors which ones are the key influencers of some measured response.*

A designed experiment is a structured methodology that determines the main effects and interaction effects that a set of preselected controllable factors have on some measured response. Factor levels are deliberately manipulated while changes in the response variable are observed. These changes are then analyzed with respect to magnitude and statistical significance. The DOE approach can be used to determine those process settings that will achieve optimal process outcomes.

Key Idea: *When the number of factors are so large that a complete experiment is not practicable or is uneconomical, a fractional design offers an efficient methodology for evaluating the significance and magnitude of how the factors influence the response variable.*

Even a modest number of factors can result in the need to investigate a large number of treatment combinations. By confounding main and two-way effects with higher-order effects (that are assumed to be negligible) the number of treatment combinations that must be studied can be substantially reduced. The confounded effects are fully defined by the alias structure which is determined from the particular experimental design.

Key Idea: *It is important that replicates be conducted for each treatment combination studied and that runs be conducted in random order.*

All data contain random variation and replicates are necessary to estimate the within-effect variance and to dampen this source of variability in estimating the magnitude of each effect. This is accomplished by averaging the observations from all replicates of each treatment combination. If possible it is important to conduct a separate setup for each replicate and to randomize the order of all runs to avoid any time-induced bias.

Key Idea: *When a large number of factors are to be investigated, screening designs can be used to reduce the number by identifying those factors that clearly have negligible effects on the response.*

Screening designs, such as the Plackett-Burman family, are efficient resolution III designs that do not confound main effects, but may confound main effects with any two-way or higher-order effects. Such designs make it possible to efficiently estimate main effects and eliminate any negligible factors.

13.4 MIND EXPANDERS TO TEST YOUR UNDERSTANDING

13-1. The following data were collected in an automobile tire plant to determine if any correlation exists between a process outcome called dynamic balance (DB) and four critical process attributes: linear splice width (LSW), ply splice width (PSW), ply splice placement (PSP), and tread splice placement (TSP).

Obs	Dynamic Balance	Linear Splice Width	Ply Splice Width	Ply Splice Placement	Tread Splice Placement	Obs	Dynamic Balance	Linear Splice Width	Ply Splice Width	Ply Splice Placement	Tread Splice Placement
1	2.275	0.2	0.2	0	0.00	21	2.800	0.2	0.6	0	0.30
2	2.775	0.2	0.2	0	0.15	22	3.075	0.2	0.6	90	0.00
3	2.200	0.2	0.2	0	0.30	23	2.350	0.2	0.6	90	0.15
4	2.825	0.2	0.2	90	0.00	24	2.950	0.2	0.6	90	0.30
5	3.225	0.2	0.2	90	0.15	25	3.278	0.2	0.6	180	0.00
6	3.944	0.2	0.2	90	0.30	26	2.175	0.2	0.6	180	0.15
7	1.925	0.2	0.2	180	0.00	27	3.600	0.2	0.6	180	0.30
8	2.825	0.2	0.2	180	0.15	28	2.700	0.3	0.2	0	0.00
9	1.900	0.2	0.2	180	0.30	29	1.611	0.3	0.2	0	0.15
10	2.100	0.2	0.3	0	0.00	30	2.719	0.3	0.2	0	0.30
11	2.125	0.2	0.3	0	0.15	31	2.625	0.3	0.2	90	0.00
12	1.925	0.2	0.3	0	0.30	32	3.275	0.3	0.2	90	0.15
13	2.025	0.2	0.3	90	0.00	33	3.222	0.3	0.2	90	0.30
14	2.525	0.2	0.3	90	0.15	34	2.150	0.3	0.2	180	0.00
15	3.000	0.2	0.3	90	0.30	35	2.500	0.3	0.2	180	0.15
16	2.755	0.2	0.3	180	0.00	36	3.375	0.3	0.2	180	0.30
17	2.525	0.2	0.3	180	0.15	37	2.500	0.3	0.3	0	0.00
18	2.200	0.2	0.3	180	0.30	38	2.500	0.3	0.3	0	0.15
19	2.800	0.2	0.6	0	0.00	39	2.600	0.3	0.3	0	0.30
20	2.300	0.2	0.6	0	0.15	40	2.725	0.3	0.3	90	0.00

Obs	Dynamic Balance	Linear Splice Width	Ply Splice Width	Ply Splice Placement	Tread Splice Placement	Obs	Dynamic Balance	Linear Splice Width	Ply Splice Width	Ply Splice Placement	Tread Splice Placement
41	2.750	0.3	0.3	90	0.15	61	1.700	0.6	0.2	180	0.00
42	2.550	0.3	0.3	90	0.30	62	2.350	0.6	0.2	180	0.15
43	2.725	0.3	0.3	180	0.00	63	3.444	0.6	0.2	180	0.30
44	3.200	0.3	0.3	180	0.15	64	1.825	0.6	0.3	0	0.00
45	3.025	0.3	0.3	180	0.30	65	2.725	0.6	0.3	0	0.15
46	3.000	0.3	0.6	0	0.00	66	2.825	0.6	0.3	0	0.30
47	2.400	0.3	0.6	0	0.15	67	2.550	0.6	0.3	90	0.00
48	2.275	0.3	0.6	0	0.30	68	2.475	0.6	0.3	90	0.15
49	2.000	0.3	0.6	90	0.00	69	3.275	0.6	0.3	90	0.30
50	2.975	0.3	0.6	90	0.15	70	2.700	0.6	0.3	180	0.00
51	2.475	0.3	0.6	90	0.30	71	3.225	0.6	0.3	180	0.15
52	2.375	0.3	0.6	180	0.00	72	2.525	0.6	0.3	180	0.30
53	2.325	0.3	0.6	180	0.15	73	2.575	0.6	0.6	0	0.00
54	4.375	0.3	0.6	180	0.30	74	2.725	0.6	0.6	0	0.15
55	2.975	0.6	0.2	0	0.00	75	3.225	0.6	0.6	0	0.30
56	3.175	0.6	0.2	0	0.15	76	2.550	0.6	0.6	90	0.00
57	3.275	0.6	0.2	0	0.30	77	2.450	0.6	0.6	90	0.15
58	1.900	0.6	0.2	90	0.00	78	3.175	0.6	0.6	90	0.30
59	2.125	0.6	0.2	90	0.15	79	2.500	0.6	0.6	180	0.00
60	2.950	0.6	0.2	90	0.30	80	3.700	0.6	0.6	180	0.15

a. Fill in the blank cells in the following correlation matrix by computing all pairwise correlation coefficients.

	DB	LSW	PSW	PSP	TSP
DB	1.00				
LSW		1.00			
PSW			1.00		
PSP				1.00	
TSP					1.00

Correlation Matrix

b. In the correlation matrix you developed in part a, place an asterisk in each cell where the correlation coefficient is statistically significant at $\alpha = 0.05$.

c. Assume that dynamic balance is a legitimate response variable. Develop a regression model that can be used to predict dynamic balance (DB) using the other four variables (LSW, PSW, PSP, and TSP) as process variables.

i. Is the regression model significant? Support your answer with an ANOVA table. You may either perform this analysis manually or use a software program such as Minitab.

ii. What percentage of the total variability in dynamic balance can be explained by its relationship to the other four variables?

iii. Are the residuals normally distributed? Why is this important?

iv. Do you think that this model is useful to management? Why or why not?

13-2. An airline company wishes to improve its service to customers who call in on its toll-free reservations phone line. The airline executives fear that, if customers calling in are placed on hold for too long, they may give up and book flights with competitors. The table below shows data that were collected over a six-week period. For each 24-hour period the total number of incoming calls was recorded. For each call, the total time (in minutes) from the time each customer connected to the automated switchboard until the time the customer was able to talk to a customer service representative was computed. Then the average hold time for each day was calculated.

Day	Total No. Calls	Avg. Mins on Hold	Day	Total No. Calls	Avg. Mins on Hold
1	12533	17.44	23	12134	17.90
2	11112	16.45	24	9123	16.30
3	8033	11.10	25	8288	11.85
4	7294	13.09	26	10345	13.56
5	10145	16.14	27	7374	11.58
6	5915	12.71	28	6697	11.88
7	9876	14.33	29	9841	15.17
8	6033	10.27	30	6208	11.85
9	10009	14.96	31	6133	12.36
10	7218	12.50	32	7567	11.35
11	7162	12.39	33	8790	13.08
12	9432	13.97	34	6369	11.45
13	8967	13.01	35	7689	13.45
14	5262	12.15	36	6181	12.87
15	7562	11.29	37	10035	14.20
16	9654	15.02	38	5789	12.60
17	6458	11.85	39	6945	10.35
18	7658	11.96	40	8194	10.35
19	11947	16.05	41	9120	11.89
20	6063	11.45	42	6505	14.06
21	7713	13.09	43	8888	14.52
22	9832	15.90	44	6818	11.31

a. Construct the scatter diagram for these data.

b. Compute the linear correlation coefficient for the two variables and determine if the correlation is statistically significant.

c. Use a computer software program to determine the regression equation that could be used to predict the average telephone hold time from the number of incoming calls. Determine if the regression is statistically significant.

d. Use the regression equation developed in part c to predict the holding time for a day in which there are 9,500 incoming calls.

e. Compute the coefficient of determination r^2 and interpret its meaning in the context of this problem.

f. Compute and plot the residuals for this problem. How do you interpret the residual plot?

g. Compute the adjusted coefficient of determination, r^2_{adj}, and compare your results with those you obtained in part e.

13-3. A steel company produces cylindrical bars of a certain alloy. Five bars are selected at random, each from a different production batch. Each bar is cut in two at ten predetermined locations and the cross-sectional diameter is measured at each cut. The results are summarized in the following table.

Cross-Section Segment	Rod Diameter				
	1	2	3	4	5
1	67.59	66.66	69.49	71.07	66.19
2	70.42	66.42	72.45	72.82	64.15
3	73.10	68.98	69.23	70.81	65.61
4	70.89	70.43	65.45	65.66	67.39
5	70.35	69.96	66.86	72.32	65.46
6	64.53	70.95	67.82	68.37	68.83
7	70.82	69.16	64.76	70.33	65.80
8	66.56	70.49	67.82	66.09	68.56
9	70.09	68.27	68.31	71.19	68.73
10	67.35	69.71	66.82	74.82	70.45
11	73.75	68.04	69.80	71.09	67.89
12	66.75	72.81	67.41	71.12	69.90
13	71.20	68.01	64.35	69.18	66.54
14	66.43	67.18	70.76	70.36	68.48
15	69.49	70.37	69.72	67.85	69.49

a. Use one-way analysis of variance to determine if the five batches are different. Prepare an ANOVA table to support your conclusion. What is the relevant p value? How do you interpret it?

b. What proportion of the total variation in rod diameter that customers see is due to variation between rods? What proportion is due to variation within rods?

13-4. An article in *Quality Engineering* reported an experimental design study to improve the subscription response rate for the direct-mail marketing campaign of a bimonthly magazine.[14] The following seven factors and levels were selected for the study.

Factor	Level	
	Low	High
A	No act now insert	Act now insert
B	No credit card	Credit card
C	Hard offer	Harder offer
D	No bumper sticker	Bumper sticker
E	Guarantee	Stronger guarantee
F	No testimonials	Testimonials
G	Daring and uninhibited	Tough and courageous

[14]Ledolter, Johannas, and Arthur J. Swersey. "Using a Fractional Factorial Design to Increase Direct Mail Response at *Mother Jones* Magazine." *Quality Engineering* 18, no. 4 (2006): 469–75.

The following 2^{7-3} fractional design was selected with design generators E = ABC, F = BCD, and G = ACD. The average responses are also shown in the table below.

Run	A	B	C	D	E	F	G	Response (%)
1	–	–	–	–	–	–	–	2.08
2	+	–	–	–	+	–	+	2.76
3	–	+	–	–	+	+	–	2.36
4	+	+	–	–	–	+	+	3.04
5	–	–	+	–	+	+	+	2.36
6	+	–	+	–	–	+	–	2.52
7	–	+	+	–	–	–	+	2.64
8	+	+	+	–	+	–	–	2.64
9	–	–	–	+	–	+	+	2.40
10	+	–	–	+	+	+	–	2.52
11	–	+	–	+	+	–	+	3.24
12	+	+	–	+	–	–	–	2.12
13	–	–	+	+	+	–	–	2.12
14	+	–	+	+	–	–	+	3.12
15	–	+	+	+	–	+	–	1.96
16	+	+	+	+	+	+	+	3.20

a. What is the defining relation and resolution for this design?

b. Generate the complete aliasing structure for the design.

c. Do these results provide evidence that any of the seven factors significantly affect the direct-mail response rates? Which ones?

d. For all of the significant factors identified in part c, plot interaction graphs for all possible pairs and comment on any observed possible two-way interactions.

13-5. Set up the following experimental design at home. Purchase some bulk popcorn kernels from a local supermarket (at least 4 cups). You will set up an experimental design to determine the effect that various identified factors have on a measured response—the number of kernels that fail to pop. You are to use a Plackett-Burman screening design to determine the main effects of each of the factors listed in the preceding table.

Popcorn Experimental Design Study

Factors	Levels: Low	Levels: High
Size of saucepan	small	large
Material of saucepan	aluminum	glass
Quantity of popcorn	1/3 cup	1/2 cup
Type of oil	canola	olive
Quantity of oil	1/4 cup	1/3 cup
Heat setting	medium	high
Amount of salt	none	1/4 tsp

For each treatment combination the pan is to be heated up from a cold start until the first kernels begin to pop. At that point the pan is to be shaken continuously until the popping stops. You will judge that the popping has ceased when an interval of five seconds passes without any additional popping. The contents of popped corn will then be emptied into a container where the popped kernels will be separated from the unpopped ones. The unpopped kernel count is the measured response.

a. Is there evidence to suggest that the within-treatment variances are the same for all treatments? Explain.

b. List estimates of the main effects for each of the seven factors and their respective p values. What are your conclusions?

c. Can you use the information from this study to develop an operational procedure that should lead to optimum results? What would you recommend?

13-6. A study is conducted on the drilling operation in a white goods plant. Two controllable process variables have been identified—drill bit diameter (x_1) in millimeters and the rate at which the drill is permitted to penetrate the material (x_2) in millimeters per revolution. At each predetermined value of the process variables two response variables are measured—the axial force on the drill shaft (y_1) and the torque force on the bit (y_2). Fifteen data points are shown in the following table.

Drill Bit Diameter (mm)	Feed Rate (mm/revolution)	Axial Force on Drill (kg)	Torque (m-kg)
6	0.1524	104.42	0.1384
8	0.1524	170.25	0.2906
8	0.3302	258.78	0.5258
6	0.3302	170.25	0.2906
3	0.2286	127.12	0.1384
6	0.127	102.15	0.1522
10	0.2286	263.32	0.5258
6	0.4318	256.51	0.4705
6	0.2286	181.6	0.3044
6	0.2286	181.6	0.2906
6	0.2286	172.52	0.2906
6	0.2286	172.52	0.2629
10	0.3302	221.67	0.4824
3	0.4318	195.65	0.2222

a. Fit a linear regression model to these data that can be used to predict axial force from drill bit diameter. Compute the r^2 coefficient for this model and construct a residual plot.

b. Fit a linear regression model to this data that can be used to predict axial force from feed rate. Compute the r^2 coefficient for this model and construct a residual plot.

c. Fit a linear regression model to this data that can be used to predict axial force from both drill bit diameter and feed rate. Compute the r^2 coefficient for this model and construct a residual plot.

d. Compare the models in parts a through c in their complexity and ability to predict.

e. Use the model developed in part c to predict the axial force that would result from using a 10 mm drill bit and a feed rate of 0.1524 mm/revolution.

f. Repeat parts a through d, substituting torque for axial force as the dependent variable.

13-7. A winery has experienced problems with labels falling off bottles of wine. A design of experiment was conducted to improve the adhesive properties of the label application process. Four factors were used in the study: (A) the amount of glue applied, (B) the temperature of the glue pot, (C) postapplication cool-down rate, and (D) the pressure applied. A complete experiment was conducted with three replications for each treatment combination. The following results were obtained.

Treatment Combination	$\bar{y}$	s	Treatment Combination	$\bar{y}$	s
(1)	4.64	0.196	d	4.02	0.703
a	5.3	0.345	ad	3.45	0.346
b	4.33	0.224	bd	5.61	0.578
ab	5.61	0.198	abd	5.72	0.421
c	4.83	0.357	cd	3.34	0.305
ac	5.9	0.447	acd	5.27	0.396
bc	5.94	0.632	bcd	6.31	0.239
abc	6.45	0.208	abcd	7.41	0.522

a. Is it reasonable to assume that the within-treatment variance is equal for all treatment combinations? Support your answer.

b. How many main effects are there in this study? How many two-way interactions? How many three-way interactions? How many four-way interactions?

c. Which factors (main and interaction) are significant?

d. Which factors appear to have a large enough effect on adhesive force to be of operational importance?

e. Based on this study what would be your recommendations to the management of this winery?

13-8. The manager of Heritage Merchandise Mart wishes to learn whether the type of background music piped through the store has any effect on sales volume. Over a period of six months, six different types of music were played and the total sales recorded. The types of music played were hard rock, country, jazz, rhythm and blues, classical, and soft rock. To eliminate any bias created by time and order, the order was randomized throughout each six-day cycle. The data collected are summarized in the following table.

	Daily Sales					
Day	Hard Rock	Country	Jazz	Rhythm & Blues	Classical	Soft Rock
1	2325	2982	3041	2507	3785	2472
2	2635	2768	3112	2462	3249	2734
3	2513	2757	3095	2449	3680	2715
4	2623	3004	3215	2463	3360	2464
5	2347	2779	3428	2373	3575	2995
6	2364	2944	3449	2441	3507	3113
7	2635	2846	3092	2653	3677	3034
8	2592	3044	2907	2294	3638	2590
9	2447	2804	3255	2450	3366	2696
10	2537	2674	3365	2112	3438	2848
11	2692	2882	3479	2178	3643	2706
12	2579	3025	3026	2478	3658	2693
13	2498	2950	2609	2416	3341	2850
14	2709	2858	3012	2370	3513	2642
15	2362	3086	3082	2551	3645	2769
16	2458	3095	3156	2401	3462	2744
17	2310	2790	2978	2799	3438	2654
18	2396	3486	3201	2542	3330	2814
19	2456	2850	3478	2542	3683	2815
20	2582	2975	3106	2530	3100	2614
21	2553	2913	3469	2497	3540	2912
22	2311	2798	3386	2364	3570	2925
23	2259	3022	2939	2561	3422	2923
24	2456	2806	3285	2443	3366	2709
25	2459	3018	3256	2322	3379	2772
26	2499	2976	3117	2489	3592	2628
27	2551	3099	2985	2275	3429	2953
28	2761	3141	2994	2338	3233	2747
29	2557	2975	3249	2356	3944	2727
30	2358	2941	3096	2333	3639	2757

a. Use one-way analysis of variance to determine if there is sufficient evidence to suggest that the type of background music played affects sales. Prepare an ANOVA table to support your conclusion. What is the relevant *p* value? How do you interpret it?

b. What proportion of the total variation in sales volume is due to variation between music types? What proportion is due to variation within music types?

c. Is it reasonable to assume that the standard deviation of sales is the same for each background music type? Support your answer by constructing an *s* chart.

13-9. Set up the following experimental design at home to investigate which factors have a significant effect on the time it takes to boil four cups of water. Use the following five factors.

Water Boiling Experimental Design Study

Factors		Levels: Low	High
A	Size of saucepan	3 qt	4 qt
B	Stove setting	Next to highest	Highest
C	Salt	None	2 tbs
D	Water source	Chilled	Room temperature
E	Cover	Without	With

The following experimental procedure should be followed. The quantity of water (i.e., 4 cups) should be consistent for each run. It is not necessary to use a graduated measuring cup—any container of similar size will do. However, the same container should be used to measure the quantity for each run.

Start each run with a dry saucepan and a fresh measure of water. Then center the saucepan on the burner by observation. Use whatever stove you have—gas, electric, glass-top burners, and so on. Make certain that the same burner is used for each run to eliminate any between-burner variation. Saucepans should be cooled down completely to room temperature between runs. For treatment combinations requiring the addition of salt, a measuring spoon should be used for level measures (use a knife to screed of the top of the spoon). Add the salt to the liquid and stir for 20 seconds prior to placing the saucepan on the burner.

The measured experimental response is the time it takes from the time the stove is turned on until the onset of boiling is noted. A consistent and reliable timing device, such as an electronic timer, should be used and started when the stove is first turned to the "on" position. The onset of boiling is defined as the time when bubbles first start escaping into the atmosphere (not when bubbles begin to cluster on the sides of the container).

a. Construct a 2^{5-2} fractional experimental design, using design generators E = ABC and D = BC. Which factors appear to have the greatest effect on the time to boil water? What are the relevant *p* values for these effects?

b. What is the defining relation and the complete aliasing structure for this design?

c. What is the resolution for this design?

d. What higher-order effects are confounded with the effects you identified as in part a? What are the management implications of this?

13-10. A study was conducted on 35 nursing homes in eastern Tennessee[15] to determine the relationship between nurse staffing hours and four independent variables: the number of nursing home beds, the number of nursing home residents, whether or not the nursing home accepts Medicaid recipients, and whether or not the nursing home is a profit institution. The variables were further defined as follows. Nurse staffing hours were defined as the number of hours worked by the nursing staff per resident per day. The number of beds was the total number of beds that are certified in each nursing home studied, and the number of residents was the total number of people occupying certified beds in each nursing home. A dummy 0–1 variable was used to represent Medicaid participation, where a 1 was used for homes that allow Medicaid to pay for long-term care for low-income people and a 0 was used for homes that do not admit Medicaid recipients. A dummy 0–1 variable was also used to represent type of ownership (profit or not-for-profit). A value of 1 was assigned to profit-seeking enterprises and a 0 to nursing homes that are not. All homes that are not-for-profit were either religious or governmental entities. The data compiled are shown in the following table.

a. Fit a linear regression model to these data that can be used to predict nurse staffing hours from the four independent variables specified.

[15] Deen, Isatta E. "An Investigation on Key Drivers of Nursing Home Costs." Term Project, *Statistical Methods in Industrial Engineering.* Knoxville: University of Tennessee, 2003.

Nursing Home Facility	Nurse Staffing Hours	No. Beds	No. Residents	Medicaid	Ownership Type
ASBURY PLACE AT MARYVILLE	2.78	181	171	1	0
BLOUNT MEMORIAL TRANS CARE CTR	4.88	60	58	1	0
BRAKEBILL NURSING HOME INC.	2.9	222	197	1	1
BRIARCLIFF HEALTH CARE CENTER	3	120	112	1	1
COLONIAL HILLS NURSING CENTER	2.57	203	194	1	1
FAIRPARK HEALTHCARE CENTER	3.11	75	71	1	1
FORT SANDERS SEVIER NURSING HOME	3	54	51	1	0
FORT SANDERS TCU	6.18	24	24	0	0
HARRIMAN CARE & REHAB CENTER	2.84	180	159	1	1
HILLCREST NORTH	3.26	340	307	1	0
HILLCREST SOUTH	4.17	95	85	1	0
HILLCREST WEST	5.11	145	131	1	0
HOLSTON HEALTH CARE CENTER	3.44	110	88	1	0
JEFFERSON CITY HEALTH AND REHAB CENTER	2.67	186	152	1	1
JEFFERSON COUNTY NURSING HOME	3.23	135	132	1	0
LAKE CITY HEALTH CARE CENTER	3.26	146	109	1	1
LCC OF JEFFERSON CITY	3.17	121	111	1	1
LOUDON HEALTH CARE CENTER	2.73	192	163	1	1
MARINER HEALTH CARE OF NORRIS	3.21	103	103	1	1
MARSHALL C. VOSS HEALTH CENTER	3.24	130	101	1	1
MARYVILLE HEALTHCARE AND REHAB	2.85	189	161	1	1
NHC HEALTHCARE, FARRAGUT	4.14	60	55	1	1
NHC HEALTHCARE, FT SANDERS	3.03	172	162	1	1
NHC HEALTHCARE, KNOXVILLE	3.62	134	128	1	1
NHC HEALTHCARE, OAK RIDGE	3	128	122	1	1
NORTHHAVEN HEALTH CARE CENTER	2.77	96	90	1	1
PIGEON FORGE CARE & REHAB CTR	3.25	120	104	1	1
RIDGEVIEW TERRACE OF LIFE CARE	3.07	132	118	1	1
ROCKWOOD CARE & REHAB CE	3.22	157	132	1	1
SERENE MANOR MEDICAL CTR.	2.67	79	76	1	0
SEVIER CO HEALTH CARE CTR	3.45	149	143	1	1
SHANNONDALE HEALTH CARE CENTER	3.87	200	192	1	0
ST MARY'S TRANSITIONAL CARE UNIT	4.94	25	24	0	0
SUNBRIDGE CARE MAYNARDVILLE	2.44	77	73	1	1
WINDWOOD HEALTH AND REHAB CENT	2.55	120	116	1	1

Compute the r^2 coefficient for this model and construct a residual plot. Is the model statistically significant? What is the p value?

b. Use a software package like Minitab to determine which of the four independent variables are statistically significant. What are the relevant p values?

c. Use your software package to complete the following correlation matrix. Place an asterisk in each cell that is statistically significant.

	Staff Hrs	No. Beds	No. Residents	Medicaid	Own. Type
Staff Hrs	1.00				
No. Beds		1.00			
No. Residents			1.00		
Medicaid				1.00	
Own. Type					1.00

Correlation Matrix

d. With increased Medicaid participation would you expect nurse staffing hours to increase or decline? Explain.

e. Use the model you developed in part a to predict the number of nurse staffing hours required for an institution that is not-for-profit, accepts Medicaid patients, has 200 certified beds, and 150 residents.

CHAPTER FOURTEEN

CONTINUOUS IMPROVEMENT: PURSUIT OF SIX SIGMA QUALITY

QUALITY MAXIM #14: *Continuous improvement is the relentless pursuit of a better tomorrow. Unity of purpose is the beginning, staying on purpose is progress, working together to achieve purpose is success.*

Imagine—The Possibilities Are Endless!

John Lennon[1] had a talent for conveying his feelings and thoughts through music and he often chose topics with contemporary social messages. In his hit song "Imagine"[2] Lennon was able to skillfully paint a musical picture of what he thought a more perfect world might be like—one just needed to stop and imagine the possibilities. Lennon wanted to see change in the world and could imagine how things might be if only those constraints and barriers hostile to the goals of unity could be removed. Quality improvement is like that. The relentless pursuit of excellence is fraught with hazards and roadblocks that can inhibit smooth learning curves and improved performance. The soul of innovation is imagination. Sometimes it really is impossible to see the forest for all the trees. Nadler and Chandon[3] have suggested an ideals approach in which an organization that is facing the necessity for change—rather than going through a problem-solving methodology to improve the current system—should adopt instead a forward-looking visionary approach. They argue that it is far more productive to imagine an ideal solution (future ideal solution target [FIST]), expand the creative thinking space appropriately, build a team with inclusive representation, and then work toward developing a path that will accomplish the FIST. To be globally competitive in today's world incremental improvements are not enough. Periodic breakthroughs are needed and achieving these can test the limits of human imagination.

[1]John Lennon was one of four members of the Beatles, a very successful rock and pop band from Liverpool, England, that was formed in 1960 and disbanded in 1970. After the Beatles, Lennon enjoyed a very successful solo music career but was controversial in his work as a peace activist and artist.

[2]"Imagine" was written and performed by John Lennon and was first released on his 1971 album by the same name. The music and lyrics can be accessed at http://www.lyricsmode.com/lyrics/j/john_lennon/imagine.html.

[3]Nadler, Gerald, and William J. Chandon. "Making Changes: The FIST Approach." *Journal of Management Inquiry* 13, no. 3 (2004): 239–46. Available at http://jmi.sagepub.com/cgi/content/abstract/13/3/239.

LEARNING OBJECTIVES AND KEY OUTCOMES

After careful study of the material in Chapter 14, you will be able to:

1. Define the term *continuous improvement.*
2. Define the theory of constraints cycle and discuss how the TOC process can help an organization improve its ability to deliver sustained value to its customers.
3. Describe the drum-buffer-rope scheduling system of the TOC methodology, discuss the role that inventories play in protecting constraint activities, and demonstrate an understanding of what scenarios are best suited for the DBR approach.
4. Define Six Sigma and describe the features that have made this approach successful in many companies.
5. Describe the concept of a core business process and where these processes fit within the open systems framework and in Six Sigma.

14.1 WHAT IT MEANS TO CONTINUOUSLY IMPROVE

Continuous improvement has been defined as a quality philosophy that assumes it is always possible to do better, and that processes should be continuously reevaluated and improvements implemented;[4] the ongoing evaluation and change of processes, products, programs, and services to make them work better;[5] and a management approach that involves continuously searching for ways to improve processes and the goods and services they produce.[6] A thesaurus provides the following synonyms for the word *continuous:* ceaseless, continual, incessant, unbroken, unceasing, uninterrupted, and unremitting. Given these definitions, does this mean that a commitment to a continuous improvement program implies the relentless pursuit of improvement of every process, every task, and each and every outcome forever and ever?

The reality is that organizations have limited resources and the best results can be expected by directing those resources to the most critical problems and those that, if solved, will have the greatest impact on organizational performance. Many of the early efforts with TQM failed because personnel became burned out or frustrated after persistent efforts failed to produce any significant results or management recognition. By contrast, the theory of constraints[7] (TOC) takes a systems view by focusing on the identification and removal of constraints that are impeding quality, and Six Sigma[8] uses a structured approach to select projects and develop a business case that justifies the expenditure of resources on expected results.

An organization should focus continuous improvement efforts where they will produce the greatest benefits for its stakeholders.

Continuous improvement is an intelligent stakeholder-focused approach that uses a systems view to allocate resources to improvement opportunities that are likely to yield the greatest benefits.

In a QIM organization the idea of continuous improvement is inculcated as a core value and is viewed from the system's perspective. As we discussed in Chapter 2, at the enterprise level performance should be measured using metrics that are classified using a balanced scorecard framework. To be competitive, it is essential that an organization commit to continuously improve in every area that either directly or indirectly serves its various stakeholders. As we also learned in Chapter 2 the organization is a large, complex open system of numerous interdependent subsystems that interact in complex ways. It is within these systems and at the system interfaces that continual improvement must take place. The old convention of making everyone in the plant responsible for improvement was not only frustrating, confusing, and wasteful; in many cases this policy was counterproductive. In the absence of a systems view, changes could be made with little or no appreciation for system consequences. Hence, localized improvements often caused inefficiencies, waste, and additional costs in upstream, downstream, or support operations.

A systems view should be employed when allocating resources to continuous improvement efforts.

Management owns all systems—therefore it is only management who can effect system changes. When a culture embraces continuous improvement as a core value, a high premium is placed on business excellence. That means that each member of the organization is expected to strive for perfection continuously, carefully following established procedures, and giving proper attention to details. Due diligence means sounding an alert when something does not appear right and always acting in the stakeholders' interests and behaving in a way that reflects credit on the company and coworkers.

From the perspective of the workforce, continuous improvement means creating a work environment characterized by proper training, appropriate and well-maintained tools and equipment, positive attitudes, enthusiasm, energy, willingness and opportunities to learn, mutual respect, and a genuine concern for customers. From the perspective of management, continuous improvement means increasing competitiveness by getting better at producing the mix of goods, services, and results that each stakeholder group trusts the organization to provide.

14.2 THEORY OF CONSTRAINTS

14.2.1 The Theory of Constraints Improvement Cycle

The **theory of constraints** (TOC), also called **constraint management,** is the quintessence of the work of Avraham Y. Goldratt,[9] whose research

A constraint is anything that limits the optimal achievement of performance metrics.

---

[4]http://accuracybook.com/glossary.htm#C.

[5]http://www3.gov.ab.ca/env/air/Info/definitions.html.

[6]http://www.fiu.edu/~pie/sec8appglossary.htm.

[7]Goldratt, Eliyahu M., and Jeff Cox. *The Goal: A Process of Ongoing Improvement.* Great Barrington, MA: North River Press, 1992.

[8]Pande, Peter S., Robert P. Neuman, and Roland R. Cavanagh. *The Six Sigma Way: How GE, Motorola, and Other Top Companies Are Honing Their Performance.* New York: McGraw-Hill, 2000.

[9]Goldratt and Cox, *The Goal.*

revealed that approximately 40% of untapped potential is locked within the constraints of the business system. The four-step TOC process is illustrated in Figure 14-1.

A *constraint* is defined as anything that limits high performance relative to purpose. Constraints can be found anywhere within the open business system, including the market, supply chain, internal value streams, organizational structure, capabilities, or culture. When a constraint is removed or broken it is said to have been *elevated*. If not elevated a constraint continues to produce a less-than-desirable future. Two important assumptions are implicit in how we have defined a constraint.

1. The organization has defined its system and purpose.
2. The organization knows how to measure its performance to determine if it is achieving its stated purpose.

Breaking a constraint can often occur with little or no investment. During step one of the TOC cycle (**Identify** the constraints) an organization will strive to identify factors that are inhibiting success. Once the constraints are identified, steps two through four of the TOC cycle (**Exploit**, **Subordinate**, and **Elevate**) focus attention on exploring intervention strategies that shape a future more ideal than the one predicted under the current system. Options can include either removing (elevating) a constraint, or effectively managing around it (exploiting and subordinating). This methodology can help the organization identify those constraints that are roadblocks to a chosen strategy. As an example, TOC can help management identify any new competencies that should be developed or those existing competencies that need improvement. For each competency so identified, the next step is to determine the inhibitors—those factors that stand in the way of excellence.

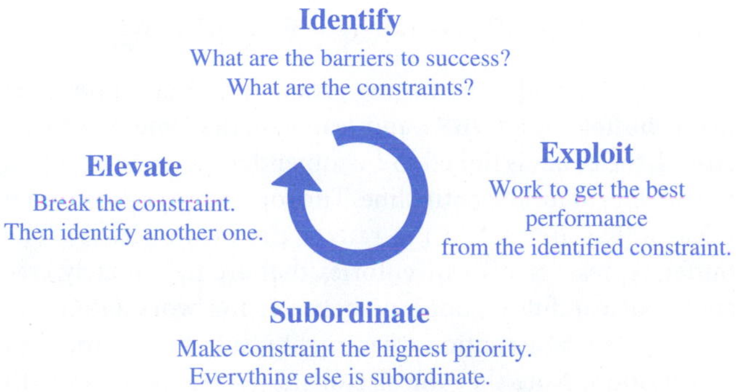

FIGURE 14-1. Theory of Constraints Cycle

To illustrate how this might work at the strategic level, suppose an organization considers the following constraints to be key barriers to success:

1. Lack of ability to acquire and maintain cutting-edge technologies
2. Lack of agility due to an internal resistance to change
3. A passive-reactive response to an escalating public interest in protecting natural resources

The line of questioning shown in Figure 14-2 defines a process that could possibly lead to some ideas that could help break one or more of the constraints.

Within the system, where the focus is on process performance, the idea of creating value is developed around three principal parameters: throughput, inventory, and operating expense. These terms are defined unconventionally as follows:

- **Throughput** (T): the rate at which money is generated through sales. Throughput is considered to be the value-added contribution of a production system.
- **Inventory** (I): what is spent on the materials and other items the organization intends to convert to throughput.
- **Operating Expense** (OE): what is spent in order to convert inventory to throughput.

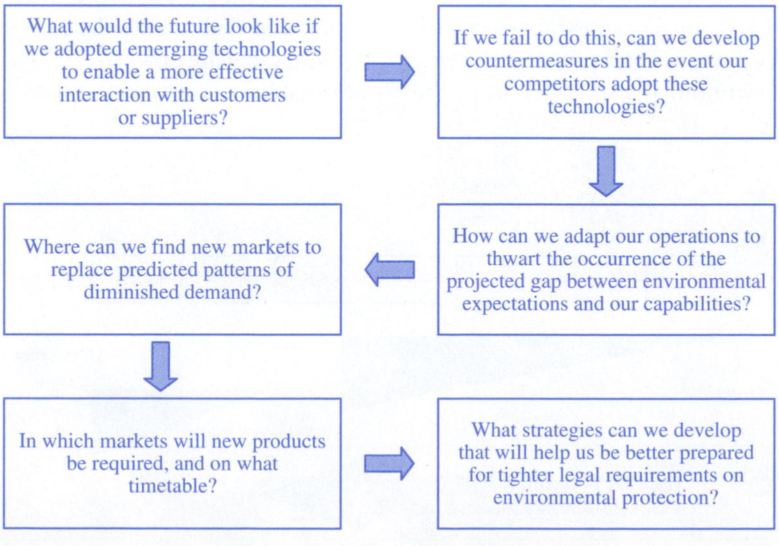

FIGURE 14-2. Thought Process Used to Elevate Strategic Constraints

In terms of these definitions throughput is money flowing into the business, operating expense is money flowing out, and inventory represents money that resides within the business. Profit then, is simply T − OE (incoming minus outgoing); and the investors' return on investment (ROI) is (T − OE)/I (the ratio of profits to assets). Of the three key parameters conventional management wisdom usually places the highest priority on OE, followed by I, with the lowest priority placed on T. That is, the traditional philosophy is to focus more energy on reducing costs than on the process that converts raw materials into products that sell and produce revenue streams into the organization. With the TOC approach these priorities are reversed—the highest priority is place on T and the lowest on OE.

Under TOC the highest management priority is to maximize throughput (T).

When throughput is a top priority, management perceives that the key to improvement is to find ways to increase it. The improvement of throughput is consistent with four guiding principles:

1. All process steps will not have the same capacity due to line imbalances, disruptions, and process variation.
2. Classical approaches to scheduling usually amplify problems caused by disruptions, variation, and imbalances and should be avoided.
3. A hybrid scheduling approach that combines pull and push technologies is required.
4. Shop-floor management attention should focus on bottleneck activities.

The four steps of the TOC cycle shown in Figure 14-1 are concerned with the identification and elevation of constraints. The theory is that a system will be improved provided its weakest link (the constraint) is identified and fixed, then the next weakest link is identified and fixed, and so on. However, even though improvement efforts strive to eventually remove a constraint, the second two steps of the TOC cycle should not be overlooked. Once a constraint has been identified everything else becomes second priority. The emphasis is on getting the maximum performance from the constraining activity. On a production line the constraint is referred to as a bottleneck operation. The TOC philosophy teaches that *an hour lost in a bottleneck is an hour lost to the entire system*, and conversely *an hour saved in a nonbottleneck operation is a mirage*—that is, it is of no value.

An hour lost in a bottleneck is an hour lost to the system.

Therefore, following the third step of the cycle and subordinating all else to the bottleneck, the bottleneck operation sets the pace (cadence) for the entire line. The constraint activity is aptly called the **drum**. The drum determines the line requirements and schedule from customer demand data. The drum activity then communicates these requirements to the entry activity in the line called the **gate**. The schedules for the drum and gate activities are therefore based on a pull philosophy since they are driven by demand. All other activities are based on a push philosophy. The gate will push material to the next activity and this will continue down the line toward the drum. The drum will push material toward the end of the line and shipping. This hybrid process of pull–push scheduling is called **synchronous manufacturing** and is illustrated in Figure 14-3.

14.2.2 Drum-Buffer-Rope Scheduling

The TOC method of managing production is also known as **drum-buffer-rope (DBR)** and **buffer management**. As discussed, the drum is linked to customer demand and provides the drumbeat for the entire line. The rope represents material releases that are tied to the rate of the drum activity. The buffer represents WIP inventories that are deliberately created (but carefully managed) between the workstations to protect against disruptions that might result due to problems or variation. Note that the size of the buffer increases with proximity to the bottleneck. Since the bottleneck is the constraining factor, it should never be starved of material. The inventory buffers help guard against this ever happening. Nevertheless, the levels of inventory need to be carefully monitored and managed. Since variation can also occur in activities that are downstream of the drum, inventory buffers are established between the drum and shipping to ensure that

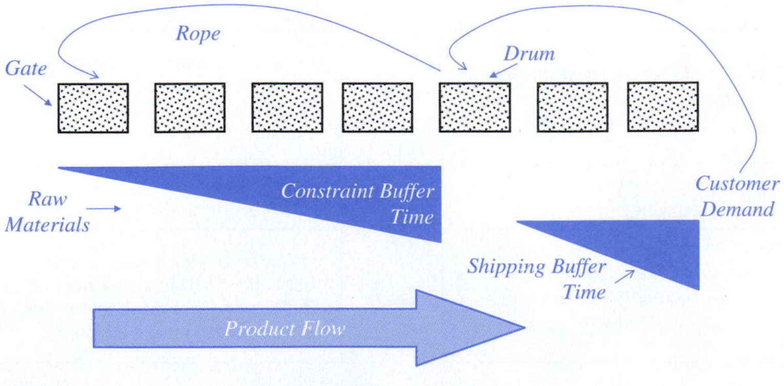

FIGURE 14-3. Synchronized DBR Flow of TOC

delivery commitments will not be missed. The idea is to keep the bottleneck operation running—throughput cannot be created until inventory is converted. For all activities the culture must take on a **roadrunner mentality**[10]—when there is work to process operators go full speed ahead working as fast as is physically possible. When there is no work, operators can relax and recover until the next batch arrives.

Under TOC inefficiencies are tolerated if they do not impact a bottleneck.

A DBR production system is highly immune to disruptions, avoids excessive WIP, and produces small batch sizes to minimize delivery lead times. Even so, disruptions do occur that require special management attention. In such cases, buffer management is used to mitigate and often prevent delays or shutdowns. Not all production scenarios lend themselves to a successful implementation of the DBR scheduling approach. These systems work best in lean low-inventory environments where the manufacturing operations are capable of consistently meeting delivery dates 95% of the time or better, there is a potential for cutting lead times and inventory levels in half, and there is minimal need for expediting and rescheduling.

14.3 SIX SIGMA QUALITY

14.3.1 Six Sigma Defined

Six Sigma is "a comprehensive and flexible system for achieving, sustaining and maximizing business success," and "uniquely driven by close understanding of customer needs, disciplined use of facts, data, and statistical analysis, and diligent attention to managing, improving, and reinventing business processes."[11] This could just as easily be a definition of QIM, as quality inspired managers acquire systems knowledge concerning personnel, customers, stakeholders (other than customers or personnel), competitors, and processes, and then apply what they have learned to better adapt their business systems to meet the needs of those who depend on them. What the Six Sigma methodology provides is a vehicle of deployment for the QIM paradigm. Six Sigma has succeeded where other implementation strategies have failed due to the structure and discipline of Six Sigma and the following three features.

1. In Six Sigma, individual project goals are tied directly to *bottom-line results*. Expressing a project in terms of financial outcomes is well understood by senior managers and will garner support from many stakeholder groups, especially stockholders.
2. In Six Sigma, projects are prioritized on the basis of *expected benefits*. This differs from other approaches that attest that everyone is working continuously on quality improvement efforts. While in a QIM culture everyone is personally accountable for quality, making all personnel responsible for continuous improvement can be counterproductive and fail to direct attention and leverage resources to those areas offering the greatest opportunities. Some processes have more improvement potential than others. Expecting continuous improvements in all workstations across the board can place unfair expectations on those processes that are already lean and provide little if any low-hanging fruit. An overzealous use of teams, without guidelines and a method for prioritization, can eventually falter because teams can become frustrated and demoralized. After dedicated hard work with little result, people can soon grow weary and lose interest.
3. A successful Six Sigma program requires management participation. A champion and a sponsor, both from the management ranks, are assigned oversight responsibility for each project and can direct the necessary resources where needed to achieve project objectives. Senior management selects the projects and decides who will be on the project team. Through tollgate reviews, the champion and sponsor are tied to the project from the beginning and can ensure that the team stays on purpose.

Six Sigma is successful because it is benefit driven and requires top management participation.

The basic idea behind Six Sigma stems from the statistical principle that a six standard deviation (*sigma*) spread will capture almost all of any process distribution. Therefore, if a process can be controlled so that its mean is on target and its natural variability (Six Sigma) is less than the tolerance allowed (the difference between the upper and lower specification limits), customer requirements will be met or exceeded.[12] Experience however has shown that there is a difference between short-term performance and the behavior that a process can sustain long term. In other words, a process that is capable in the short term may be unable to sustain its capability over the long haul. Figure 14-4 illustrates the Six Sigma concept and shows the process drift that can be expected to occur over time. In Chapter 9 we demonstrated how this shift can be up to 1.5 standard deviations which can make a considerable difference in performance. Long-term Six Sigma capability takes this drift into account when estimating the defective parts per million opportunities (DPMO).[13]

Six Sigma strives to improve efficiency (doing the thing right) and effectiveness (doing the right thing) at the same time to help an organization work smarter by setting and

[10]This term refers to a cartoon character created for Warner Brothers in 1948 by Chuck Jones. The roadrunner is a large brown bird of the cuckoo family who, in the cartoon series, is continuously outwitting and outrunning his adversary Wile E. Coyote. When cornered he had the knack of immediately going from a velocity of zero to breakneck speed to get him out of harms way—this is the attribute referred to as roadrunner mentality in TOC.

[11]Pande, Neuman and Cavanagh. *The Six Sigma Way.*

[12]This was the definition given to capability in Chapter 9.

[13]DPMO can be estimated by consulting Table III in Appendix B.

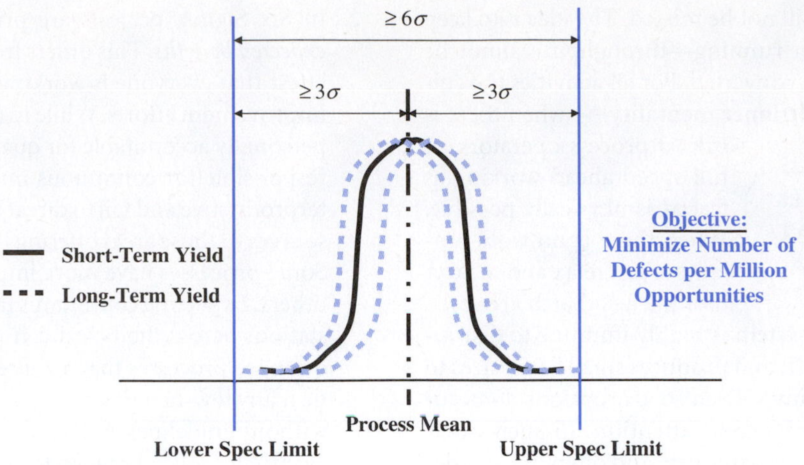

FIGURE 14-4. Conceptual Aim of Six Sigma

aggressively pursuing goals that matter to stakeholders—and to do so without error or waste. Six Sigma is about fact (statistical analysis), perfection, and culture. It provides discipline, focus, accountability, and an implementation framework where there are no spectators, only participants. Improvement teams set specific objectives and are held accountable for reaching them. The culture is transformed as the organization's structure aligns itself to the achievement of the Six Sigma goals. Individual responsibilities and expectations are clearly delineated, and there is a clear and widely understood training regimen to provide the skills set needed.

Six Sigma strives to improve efficiency and effectiveness at the same time.

Perhaps the most distinctive difference between Six Sigma and its forebears is that Six Sigma defines a strategy for getting work done. Management owns all primary and support value streams, as these are components of the larger organizational system. In 1985 Porter introduced the concept of a value chain as a sequence of related work activities,[14] which was shown in Chapter 2 to be the transformation engine governing how, where, when, and by whom quality will be built. The value chain, or stream, was defined as comprising three key elements that must work together in harmony. These are *individual attributes, group processes,* and *work design*. A value stream can be viewed as either a fragmented chain of workstations or an integrated stream of related activities. The former view defines the value stream by function, observing that work flows from one department to the next, each contributing to the overall value of system output. This view is consistent with traditional vertical organizations segmented by function, and also functional layouts, such as the flexible flow process designs discussed in Chapter 5.

Six Sigma defines a value stream as a core business process.

[14]Porter, Michael E. "Competitive Advantage." New York: *Free Press*, 1985.

The alternative view, and the one advocated by Six Sigma, is to define a value stream as a core business process. Core business processes are those primary value streams that directly contribute to providing products, services, or outcomes to stakeholders. Any activity that is not primary is part of a support value stream and does not contribute to a core business process.

Under Six Sigma support functions are of secondary importance. When the idea of value stream management was first introduced in Chapter 2, we stressed the importance of appointing a single person as the process owner. Such a structure defies the traditional chain-of-command organizational structure since the primary value stream will almost always cut across functional boundaries.

In Six Sigma organizations, **process owners** are usually selected from management but not always. In some cases it might be more appropriate to appoint someone from the nonmanagement ranks. What is most important is that the process owner be technically knowledgeable about the process, be respected by the people who operate the process, have a personal stake in the success of the process, and finally, be someone who is a critical thinker and has an improvement mindset.

Core business processes are broad in scope, and will typically be a list of no more than five to eight value streams that are critical to mission. These will include processes such as

- Order fulfillment
- New product/service development
- Customer service
- Customer cultivation
- Environmental protection

Support processes are those value streams that serve as enablers to the core processes, and include such areas as human resources development and acquisition, finance and budgeting, facilities, evaluation and rewards, and information technologies.

14.3.2 Incremental Improvement through the DMAIC Methodology

In Chapter 5 we introduced three fundamental models for guiding improvement efforts. These were the scientific method, the PDCA/PDSA cycle, and the DMAIC tactics. The oldest of these is the scientific method and its documented use can be traced back to the ancient Greeks and Egyptians who used logic and deductive reasoning to explain the mysteries of science. The scientific method is taught in colleges and universities to engineering and science majors as a systematic methodology for scientific investigation and discovery.

During the last half century SPC, TQM, and other quality initiatives have helped popularize the PDCA (Plan-Do-Check-Act) cycle, also referred to as the Deming wheel. The Six Sigma model uses DMAIC, an acronym that stands for the interconnected steps of Define-Measure-Analyze-Improve-Control. The three models are all grounded in the same philosophy—*use data to create knowledge relative to a problem and then act on the knowledge to solve the problem.*

Improvement models use data to create knowledge to solve the problem at hand.

The diagrams originally introduced as Figures 5-4 and 5-5 are reintroduced here as Figures 14-5 and 14-6.

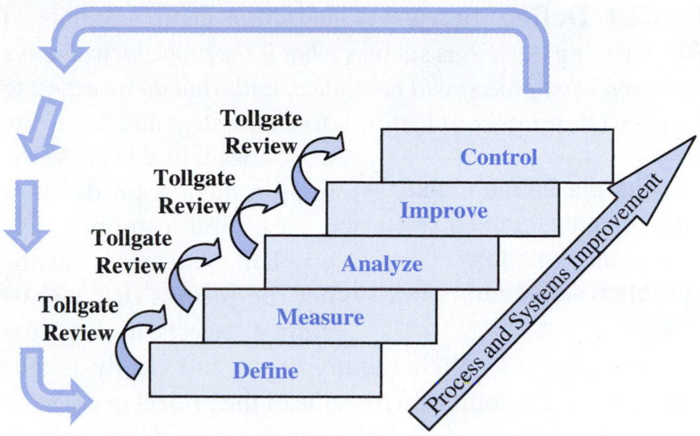

FIGURE 14-5. Six Sigma Improvement Model

Figure 14-5 illustrates the steps of the DMAIC process and Figure 14-6 shows the relationship between DMAIC and the other two basic improvement models.

A feature that distinguishes DMAIC from other improvement models is the provision for "tollgates" between each of the phases. The tollgates represent milestone events during which the improvement team meets with senior management and other project participants to review progress and make certain the project stays on target and is aligned with corporate goals.

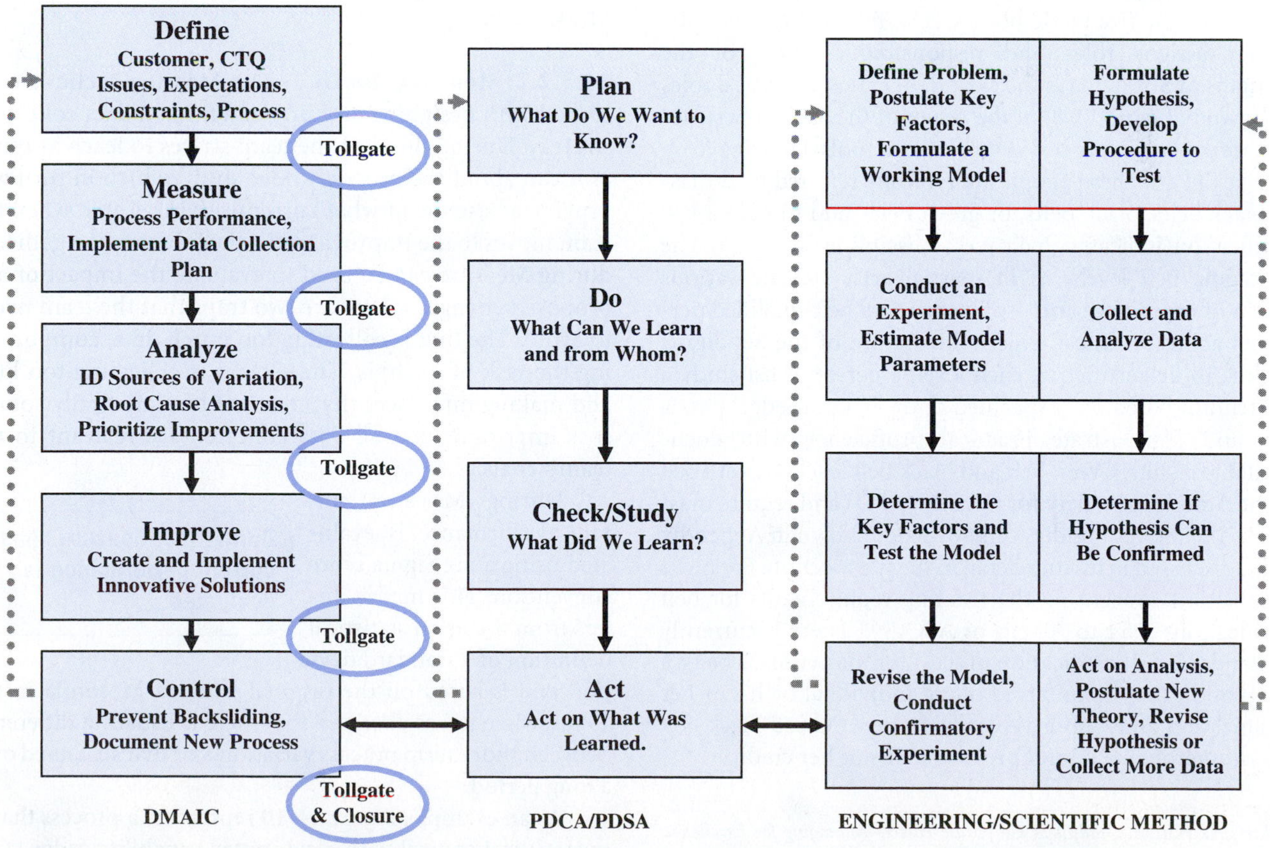

FIGURE 14-6. Improvement Models

14.3.2.1 Define Tactic. The **Define** tactic attempts to ask clarifying questions such as *what is the problem, who is the customer, what process will be studied, and what do we expect to achieve*? The purpose of Define is to set the stage and justify the work ahead. In this phase the improvement team develops a clear and communicable problem statement, identifies the process that will be studied, selects its key customers and defines their requirements, and clearly assigns roles and responsibilities. The value of the project needs to be confirmed and a set of project goals and boundaries drafted.

During the Define phase the team develops a clear and communicable problem statement.

The Define tactic requires tools that can be used to document and justify. The first and most important of these is the team charter which makes the business case—a few sentences to explain why the project has a high priority and that *links expected outcomes to the firm's strategic plan and financial objectives*. The charter also contains a short *concise problem statement*, written in nonthreatening terms, taking care not to attach blame, and advocating no specific cause or solution. Next, the charter contains some *project-specific objectives*, such as a "reduction in the number of defects by 50%," or "shrinkage of cycle time by 15%." These objectives can be revised later if necessary. Figure 14-7 is an illustration of a typical Six Sigma team charter.

As Figure 14-7 shows the **team charter** contains a timeline for project completion. This too can be revised later if necessary, but it is important that the team work to deadlines for each of the five tactic phases. As a final component, the charter assigns roles and responsibilities. One of the strengths of Six Sigma is the clear delineation of critical roles as shown in Figure 14-8. At the center of the Six Sigma structure are experts in process improvement tools. Drawing from the martial arts these people are typically referred to as master black belts, black belts, or green belts, and to hold a *belt* certification indicates dedication, discipline, and skill. The Six Sigma belt levels, as in martial arts, indicate various depths of experience and proficiency, and belt-qualified personnel act as technical experts in the use of the Six Sigma toolkit. To be certified at each level, a person must study a curriculum based on a specified body of knowledge, pass a test, and demonstrate practical proficiency with documentable results. Green belt and black belt levels are certified by the American Society for Quality (ASQ) and require mastery of the relevant bodies of knowledge shown in Appendix C and discussed in further detail in the next section. For more information concerning the training requirements for belt certifications refer to Aikens or ASQ.[15,16] There is currently no standardized certification process for master black belts, a designation that is conferred on an individual by his or her organization and who is typically an experienced black belt with numerous successful projects to his or her credit.

During the Define phase the improvement team performs a customer needs analysis. Each process has key customers. The particular process that will be targeted for study will either have been explicitly identified in the charter, or will be implicit in its problem statement. The first step is to obtain as much information as possible on customers and to state their expectations in unambiguous terms, so that each member of the team will understand process requirements. The compilation of this information is called the voice of the customer (VOC), a term borrowed from the house of quality of the QFD methodology we discussed in Chapter 5.

The third and final step of the Define stage is to develop a high-level process chart, such as a SIPOC map. SIPOC is an acronym that stands for Suppliers-Inputs-Processes-Outputs-Customers. The SIPOC map expands the scope of the flowcharting tool we introduced in Chapter 7 to include suppliers and customers. This provides a bird's-eye view of the value stream and its linkages to the environment. Figure 14-9 is an example of a SIPOC map.

A final step in the Define tactic is a tollgate review—a meeting that is held for the purpose of reviewing progress. The meeting is led by the team leader (usually a black belt), and is attended by team members, the project sponsor, the project champion, a master black belt, and the process owner. The review focuses on the project goals (and whether they have changed), the project justification (with particular reference to how the project is expected to benefit the bottom line), the projected resource requirements, and the team's plans for proceeding to the next tactic—which is Measure.

14.3.2.2 Measure Tactic. The **Measure** tactic is concerned with designing and implementing a data collection strategy. During this step the team strives to learn as much as it can about the process under study, which in turn will provide a baseline of what currently exists. Later, when the team moves to the Improve tactic, the information gathered during Measure can be used to evaluate the impact of any proposed changes. There are two traps that the team needs to avoid. The first is collecting too much data, compounding the task of analysis. The second is collecting too little and making inferences that may not be statistically sound. It is important that all data collected are relevant to the team's goals.

During Measure, current performance is evaluated using a **Six Sigma score** convention. This metric differs from the usual statistical definition of a standard deviation and is based on the original work at Motorola in the 1980s, when it was observed that there is usually a difference between short-term process variation and that sustained over a long period.

During the Measure phase current performance is evaluated.

As an example, Figure 14-10 represents a process that is in statistical control with a computed capability index (C_{pk}) equal to 1. Under a traditional capability analysis, one would

[15]Aikens, C. Harold. *Quality: A Corporate Force—Managing for Excellence.* Upper Saddle River, NJ: Prentice Hall, 2006.

[16]http://www.ASQ.org.

Grinder Gear Works

Six Sigma Team Charter

PROBLEM STATEMENT

Thirty-three percent of all MPD-5000 master cogs shipped to customers fall short of meeting requirements. Twenty percent are rejected at customers' receiving stations and are returned. These customers claim that according to their measurements the parts fail to meet specifications. Ten percent of parts shipped are technically acceptable but are delivered late. Many of our customers are scheduling production based on just-in-time deliveries and depend on orders arriving when promised. The remaining 3% are not caught and eventually end up in products and cause failures. Grinder Gear is currently facing litigation in two major product liability cases. These quality problems are not only hurting our image in the marketplace but the expedited delivery (when we know a shipment would otherwise be late), scrap and rework, processing returns, and having to ship replacement by expedited means are costing the company $550,000 per month. This does not take into account the loss of goodwill or the cost if Grinder should lose either of the pending lawsuits (one suit is seeking $10 million and the other $4.5 million).

GOAL STATEMENT

Cut late deliveries in half (down to 5%) by the end of October. Reduce the number of out-of-spec parts by 75% by the end of November so that results can be reported to stockholders at their annual meeting in December.

CONSTRAINTS

Team members will be expected to devote 10 to 20 hours per week to the project. The project Champion will be requested to provide backup support to team members to ensure that their regular production duties are covered and that their project responsibilities are not neglected.

ASSUMPTIONS

The focus of the project will be on improving the existing processes, not on designing a new one. However, the team will not be predisposed to any specific cause or solution on the front end. All ideas will be fair game.

TEAM GUIDELINES

The team will meet on a regular schedule between 9 a.m. and 10 a.m. on Fridays. Special team meetings can be convened at other times as needed. Decisions will be by consensus, using group techniques such as brainstorming and nominal group techniques to broaden participation and buy-in. If consensus cannot be reached the team leader will make the final decision.

Grinder Gear Works

Six Sigma Team Charter

Page 2

TEAM MEMBERS

Allan Dutwidle—Grinder Shop
Jim Walker—Black Belt and Team Leader
Kelly Peterson—Order Fulfillment
Maria Hoskins—Shipping
Robert Napoli—Engineering
Sam Nelson—Technical Sales
Ernest Childs—Machining

Other key players:

John Crenshaw—Champion and VP of Customer Service
Wayne Masterson—Master Black Belt and Technical Support Group
Judy Henshaw—IT Group

PRELIMINARY PROJECT PLAN

The team has set a project completion date of December 31. To reach that date the team will have to work aggressively and transition smoothly through each phase of DMAIC. The team is setting the following milestone dates for tollgate meetings.

DEFINE:	July 15
MEASURE:	August 15
ANALYZE:	September 15
IMPROVE:	October 15
CONTROL:	November 15

FIGURE 14-7. Example Six Sigma Team Charter

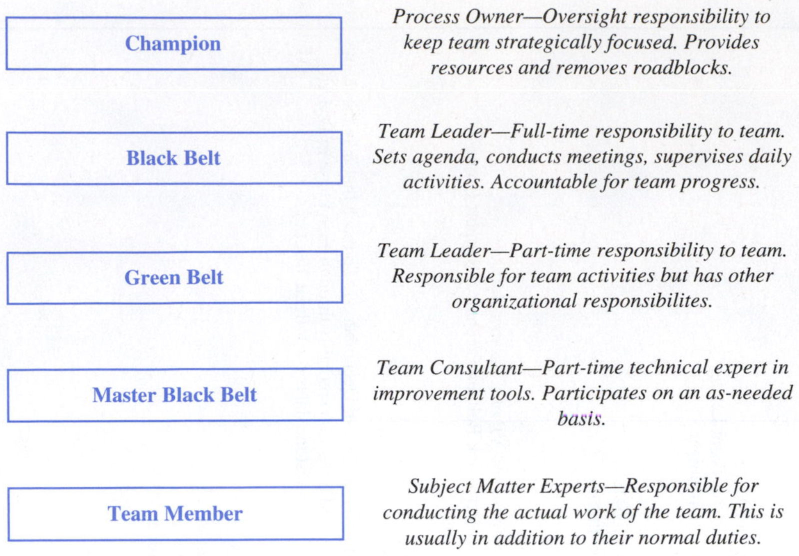

Champion
Process Owner—Oversight responsibility to keep team strategically focused. Provides resources and removes roadblocks.
Black Belt
Team Leader—Full-time responsibility to team. Sets agenda, conducts meetings, supervises daily activities. Accountable for team progress.
Green Belt
Team Leader—Part-time responsibility to team. Responsible for team activities but has other organizational responsibilities.
Master Black Belt
Team Consultant—Part-time technical expert in improvement tools. Participates on an as-needed basis.
Team Member
Subject Matter Experts—Responsible for conducting the actual work of the team. This is usually in addition to their normal duties.

FIGURE 14-8. Roles Critical to the Success of Six Sigma

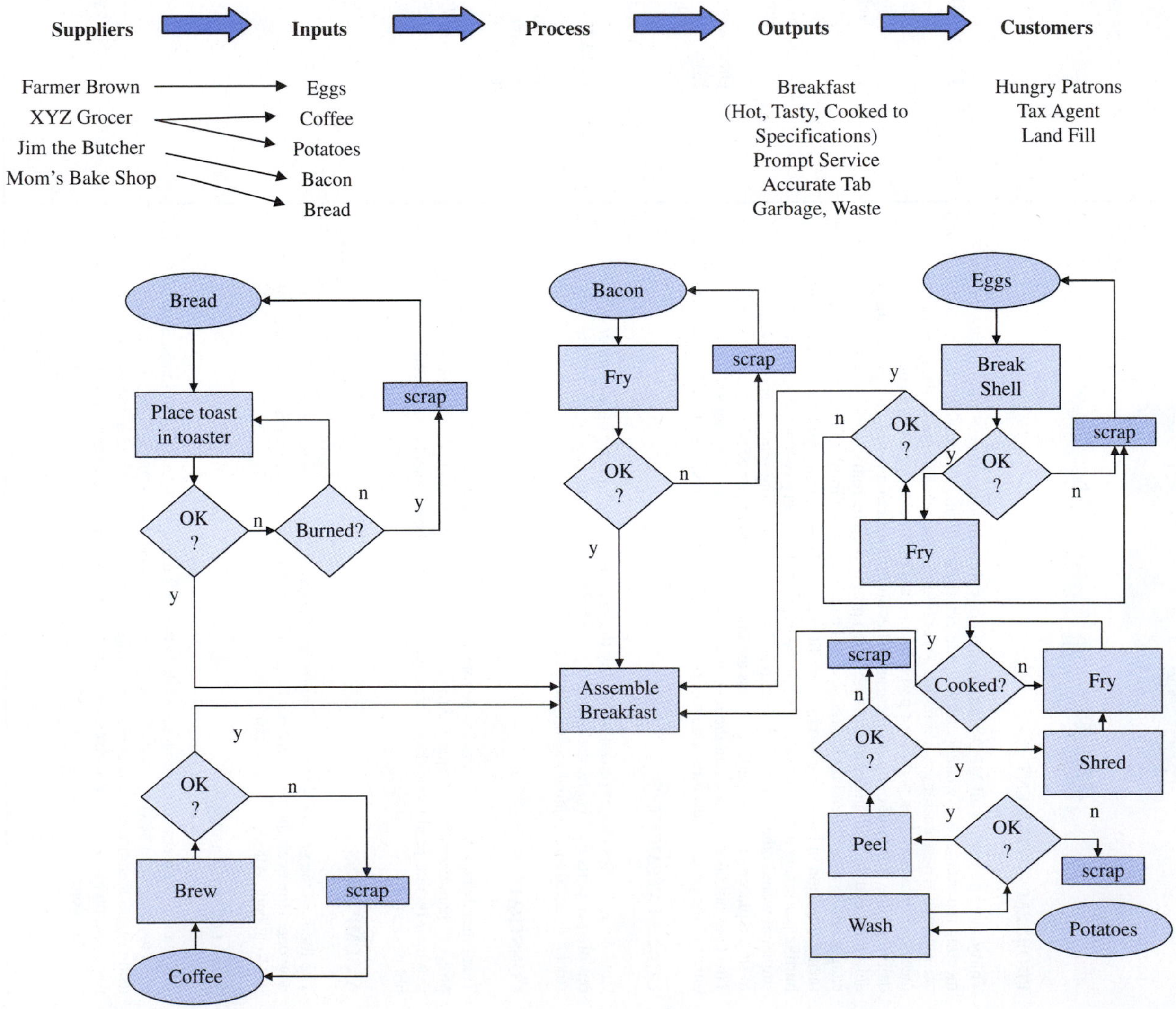

FIGURE 14-9. SIPOC Map

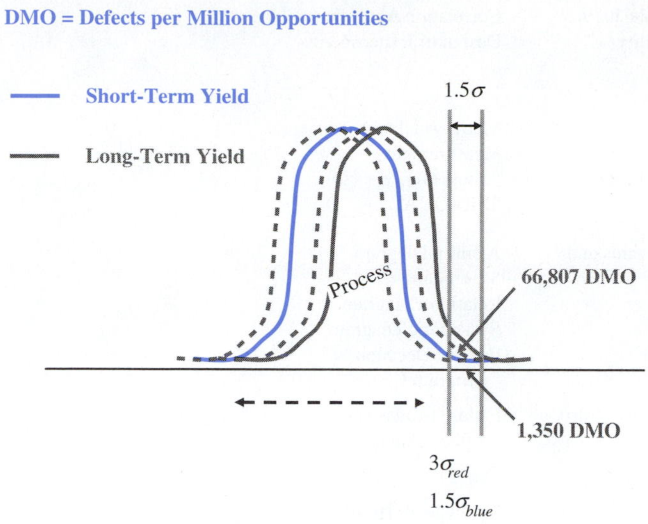

FIGURE 14-10. Six Sigma Scoring

TABLE 14-1 Six Sigma Conversion Chart

Capability Index (C_{pk})	Short-Term Process Sigma	Long-Term Process Sigma	Process Yield	Defectives per Million
2	6	4.5	99.99966	3.4
1.83	5.5	4	99.9968	32
1.67	5	3.5	99.98	233
1.5	4.5	3	99.87	1,350
1.33	4	2.5	99.4	6,210
1.17	3.5	2	97.7	22,750
1	3	1.5	93.3	66,807
0.83	2.5	1	84.1	158,655
0.67	2	0.5	69.1	308,538
0.5	1.5	0	50.0	500,000
0.33	1	−0.5	30.9	691,462
0.17	0.5	−1	15.9	841,345
0	0	−1.5	6.7	933,193

correctly estimate that the process is meeting the specification limit shown 99.865% of the time. Out of 1 million parts produced, 1,350 will be defective. This estimate is obtained by observing that the specification limit is 3 standard deviations from the process mean. The Six Sigma "score"[17] computes a 1.5 standard deviation shift, adjusting for the likelihood that over time even a stable system moves around. Therefore, in this case, the Six Sigma score is $3\sigma - 1.5\sigma = 1.5\sigma$. At the adjusted score of 1.5σ the process long-term yield is reduced to 93.3193% or 66,807 defective parts per million produced. These numbers are based on cumulative standard normal tables and are obtained by consulting Table III in Appendix B.

Table 14-1 converts computed C_{pk} indices into corresponding Six Sigma scores. To achieve quality in the defects-per-million-opportunities (DPMO) range, it is clear from the chart that a minimum $C_{pk} = 2$ is required. At that level a stable process is capable of achieving a long-term yield of 99.99966%, or 3.4 DPMO. This is a near-perfect state and statistically equates to a process average that is 4.5σ from the specification limit that was used to compute the C_{pk}.

The Measure tactic utilizes many of the tools that were described in Chapters 7 through 13 to build knowledge progressively. The team first uses some simple tools and, as information is gathered, introduces more sophisticated tools as appropriate. During Measure the team has five objectives:

1. Describe the process.
2. Focus on significant factors.
3. Generate and organize new ideas.
4. Collect data and ensure accuracy.
5. Understand and reduce variation.

Figure 14-11 illustrates these steps and the appropriate collage of tools that might be applicable to each.

[17]The Six Sigma score was previously defined in Chapter 9.

At the end of the Measure phase, the tollgate review is concerned with the justification of the sequence and set of tools selected to guide the search for information. The team must explain its logic in deciding what information to collect, how it went about collecting it, how it was analyzed, and what conclusions were drawn from the results. The review—called the **Measure tollgate**—can also challenge any part of the data or the accuracy of the measurement system. After all questions concerning measurements and information are satisfactorily answered, the team will outline its plans for proceeding with the Analyze tactic.

14.3.2.3 Analyze Tactic. Once measurements have been made the data collected are used to focus the team on the factors that are most likely causing the problem. Simply narrowing the scope does not prove causality. That requires additional data analysis, which is the objective of the **Analyze** tactic. The tools used for causal analysis can be categorized two ways: those that cause process variation and those that cause lead time variation.

One of the most important Analyze tools is the fishbone diagram. During the Analyze step the team brainstorms until it is satisfied that most of the candidate causes have been identified. Then the team might decide to investigate relationships between the effect and any number of the candidate causes, or alternatively, the interest might be whether any of the candidate causes are correlated with each other. There also may be some concerns that the stated causes are actually effects.

> **During the Analyze phase measurements are applied to gain a better understanding of the problem.**

To conduct a **causal analysis,** the team can use control charts of selected process variables to identify and separate

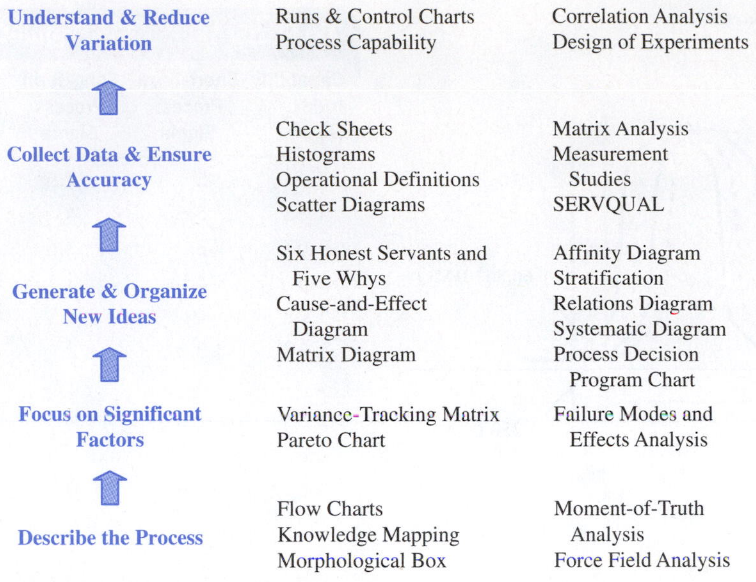

Understand & Reduce Variation
Runs & Control Charts
Process Capability
Correlation Analysis
Design of Experiments
Collect Data & Ensure Accuracy
Check Sheets
Histograms
Operational Definitions
Scatter Diagrams
Matrix Analysis
Measurement Studies
SERVQUAL
Generate & Organize New Ideas
Six Honest Servants and Five Whys
Cause-and-Effect Diagram
Matrix Diagram
Affinity Diagram
Stratification
Relations Diagram
Systematic Diagram
Process Decision Program Chart
Focus on Significant Factors
Variance-Tracking Matrix
Pareto Chart
Failure Modes and Effects Analysis
Describe the Process
Flow Charts
Knowledge Mapping
Morphological Box
Moment-of-Truth Analysis
Force Field Analysis

FIGURE 14-11. Measure Tactic Tools

general sources of variability, such as machine-induced, shift effect, method, and so on. Honing in on specific causes usually requires the use of some of the advanced statistical tools covered in Chapter 13. Analysis of variance (ANOVA), correlation analysis, regression analysis, and experimental design can be conducted with the help of the team's black belt and master black belt who should be proficient in the use of such tools. The object of the Analyze tactic is to select a few causes that will be tagged for elimination in the Improve phase.

The **Analyze tollgate** provides the opportunity for the team to defend the causes selected for improvement. During the review the team presents its methodology and retraces the logic followed. It will also be called on to demonstrate how eliminating each selected cause will add value to the team's objective, and to reveal how the team plans to eliminate the cause during the Improve phase.

14.3.2.4 Improve Tactic.

During the Analyze tactic, cause-and-effect relationships were investigated and confirmed. The purpose of the **Improve** tactic is to generate workable solutions that will eliminate the confirmed causes or nullify their effects. During the phases leading up to Improve, the team conditions itself to think in the broadest possible terms—and no idea is rejected out of hand. The emphasis is on creativity and thinking beyond accepted assumptions. *In Improve the approach changes dramatically.* Since the process has narrowed to a few (perhaps only one) causes, the team now must become focused and come up with solutions that can be implemented. The group must therefore be able to adjust from an imaginative mindset to a practical one.

The Improve phase is concerned with generating solution ideas, testing each idea, and finally selecting one for implementation. Many of the tools that were useful in the phases leading up to Improve can be applied again to help with the stimulation of new ideas and to evaluate proposed solutions. For example, creative tools such as brainstorming and the nominal group technique can help. The use of flow charts can aid in pinpointing where in the process the cause is occurring, and the six honest servants and five whys can sometimes result in good ideas. In some cases it is desirable to use more advanced tools such as design of experiments or simulation to evaluate the impact of alternatives being considered. There are a number of tools that are part of the lean production toolkit that can also be considered at this point as a means for eliminating waste.

During the Improve phase workable solution options are generated.

The **Improve tollgate** continues to focus on justification, logic, and linkages between causes. During the review, the team will present to the panel what improvements were considered, which ones were selected, and how the improvements address specific confirmed causes to help solve the problem. The team is also required to answer questions such as how the chosen improvements were tested and how the team can be certain that the benefits claimed will be achieved. At this point if management approves, the team can move to the final step, Control.

14.3.2.5 Control Tactic.

The **Control** tactic is about sustainability. Through the Analyze, Measure, and Improve phases the team will have reached the point where specific defendable recommendations can be made for process improvements. The job is not complete until those improvements can be implemented and the baton passes from the improvement team to the process owner. New procedures must be carefully documented in unambiguous language and all workers, their supervisors, and the process owner need to receive consistent training in the new methods.

The tools that are appropriate to Control are those that can help the team document the improved method, determine the appropriate measures to monitor ongoing performance, and collect and analyze process data during implementation to validate training efforts and perform a postaudit evaluation of actual benefits. Many of the same tools that were used previously in the cycle will be used again in Control, only the emphasis will shift from problem solving to process surveillance. For example, flowcharting may be used to illustrate a revised sequence of steps to include in the new documentation. As another example, control charts may be useful to determine stability, identify any inconsistencies in training or understanding, and validate new levels of process behavior.

During the Control phase improvements are implemented and standardized.

The **Control tollgate** is the final review and formally closes the project. It is therefore essential that this review ensures sustainability of improvements and leverage of any learnings that took place. This final tollgate also provides senior management the opportunity to formally and publicly recognize the contributions of all team players and to express appreciation. During the Control review there are several key issues that management will want to know. For example, it is important that the team explains what measures have been identified to track ongoing progress, their rationale, and the measurement process to be used. Individual roles and responsibilities also need to be clarified. *Who will be responsible for collecting what data, how will the data be analyzed, and who will take what action when the data provide what (specific) information?* As for sustainability, the team describes the training program that has been carried out, including who received what training, and the plan for providing this training to new employees.

Last, disseminating the knowledge accumulated during DMAIC is important. As a final step in the review process, the team will reflect on the lessons learned and whether any best practices were borrowed from others. It will then outline a plan for sharing this information outside the small circle of personnel who formed the project team. With this last step, Six Sigma becomes a vehicle to facilitate the learning process company-wide so that future projects do not necessarily have to start from scratch.

14.3.3 ASQ Six Sigma Certifications

As stated, the American Society for Quality (ASQ) certifies individual competency in the knowledge and application of Six Sigma tools at two levels—green belt and black belt.

ASQ defines a certified green belt as an individual who works in support of or under the supervision of a Six Sigma black belt, analyzes and solves quality problems, and is involved in quality improvement projects.[18] To become certified a person must have three years of work experience in an area related to the green belt body of knowledge and pass a four-hour written examination consisting of 100 multiple-choice questions.

The Six Sigma black belt is typically someone who is leading Six Sigma project teams on a full-time basis. To become certified as a black belt an individual must complete two Six Sigma projects documented by signed affidavits from a company official, or alternatively complete one project with a signed affidavit and three years of work experience within the black belt body of knowledge. In addition, black belts must pass a four-hour written examination that consists of 150 multiple-choice questions. Exams are given in most major cities across the United States several times a year.

14.3.3.1 Certified Six Sigma Green Belt Body of Knowledge.[19] Companies are beginning to show a greater interest in recruiting graduates from engineering and business schools that can make immediate contributions and do not have to be trained. Fresh graduates who already have a green belt certification are particularly attractive for those organizations that have Six Sigma or Lean Six Sigma programs in place. All that is needed to achieve this is for undergraduate programs to cover the green belt BOK and encourage students who have the requisite work experience to sit for the examination prior to graduation.[20] The green belt BOK is organized into five main categories that closely follow the DMAIC process as shown in Appendix C. There we have listed the relevant topics under each of the main areas, and for each topic we have shown the cognitive level at which examination questions are written and a linkage to the Chapters in this text where each topic is covered. The cognitive levels shown follow Bloom's taxonomy as shown in Figure 14-12.[21]

14.3.3.2 Certified Six Sigma Black Belt Body of Knowledge.[22] It is somewhat more difficult to incorporate the black belt BOK into an undergraduate business or engineering curriculum due to the requirements that candidates for this certification must first earn a green belt and must also have relevant work experience and complete documented projects. However, this is not impossible and some institutions are beginning to consider how they can rise to the challenge and find creative ways of reforming their respective curricula. Certifiable projects can come from industrial partnerships, or can be performed within the academic institution. A possibility is to incorporate black belt opportunities in graduate curricula. The black belt BOK is organized into nine principle categories—five that relate to the DMAIC process and

[18]http://www.asq.org.

[19]The Certified Six Sigma Green Belt Body of Knowledge, with reference links to relevant sections of this textbook, is included in Appendix C. Each topic within the BOK is shown with the corresponding intended level of difficulty of the questions asked on the examination.

[20]For those who do not meet the work experience requirement, the Institute of Industrial Engineers (IIE) offers a green belt certificate that only requires passing a written examination.

[21]Anderson, L. W., & D. R. Krathwohl, eds. *A Taxonomy for Learning, Teaching, and Assessing: A Revision of Bloom's Taxonomy of Educational Objectives.* New York: Longman, 2001.

[22]The Certified Six Sigma Black Belt Body of Knowledge, with reference links to relevant sections of this textbook, is included in Appendix C. Each topic within the BOK is shown with the corresponding intended level of difficulty of the questions asked on the examination.

Bloom's Taxonomy	Green/Black Belt BOK
Evaluation appraise, argue, assess, attach, choose, compare, defend, estimate, judge, predict, rate, core, select, support, value, evaluate.	**Evaluate** Make judgments about the value of proposed ideas or solutions by comparing proposals to specific criteria or standards.
Synthesis arrange, assemble, collect, compose, construct, create, design, develop, formulate, manage, organize, plan, prepare, propose, set up, write.	**Synthesis** Put parts or elements together in such a way as to reveal a pattern or structure not clearly there before; identify which information from a complex set is appropriate to examinefurther.
Analysis analyze, appraise, calculate, categorize, compare, contrast, criticize, differentiate, discriminate, distinguish, examine, experiment, question, test.	**Analyze** Break down information into its constituent parts and recognize their relationships and how they are organized; identify sublevel factors or salient data from a complex scenario.
Application apply, choose, demonstrate, dramatize, employ, illustrate, interpret, operate, practice, schedule, sketch, solve, use, write.	**Apply** Know when and how to use ideas, procedures, methods, formulae, principles, theories.
Understanding classify, describe, discuss, explain, express, identify, indicate, locate, recognize, report, restate, review, select, translate.	**Understand** Read and understand descriptions, communications, reports, tables, diagrams, directions, regulations.
Knowledge arrange, define, duplicate, label, list, memorize, name, order, recognize, relate, recall, repeat, reproduce, state.	**Remember** Recall or recognize terms, definitions, facts, ideas, materials, patterns, sequences, methods, principles.

FIGURE 14-12. Bloom's Taxonomy

four that cover enterprise-wide issues, process management, team management, and DFSS, as shown in Appendix C. As with the green belt BOK we have listed the relevant topics under each of the main areas along with the Bloom's cognitive level of testing.

14.4 SIX SIGMA METHODOLOGIES FOR NEW PRODUCT DESIGNS AND RADICAL IMPROVEMENT

The adoption of Six Sigma usually means that a company's management is committed to DMAIC to achieve incremental improvements in value stream performance. This approach has proven to be an effective tool in facilitating the implementation of lean production to identify and eliminate sources of waste.

A much smaller number of companies have moved the Six Sigma concept to the next level, applying its principles to the design of new products and to radical breakthrough process improvements. The acronym used to describe these applications is DFSS (design for Six Sigma) and, unlike the Six Sigma DMAIC methodology, there are no universally recognized set of steps or phases—companies applying DFSS will typically customize the process to suit its individual needs or adopt a version that is advocated by the particular consulting company that is assisting with the deployment. As a result DFSS should be considered to be an approach rather than a well-defined methodology.

In Chapter 5 some integrated approaches to new product development were discussed, including integrated product and process design (IPPD), design for manufacturability (DFM), concurrent engineering (CE) and design for supportability (DFS). DFSS can be considered to be a structured approach that supports any of these three concepts, where stakeholder considerations take precedence over technical factors, individual personal biases, or the freedom of creative genius.

Figure 14-13 shows some of the various DFSS frameworks found in practice. The first methodology shown (labeled I) is popular because it contains the same number of phases as DMAIC and can be pronounced in similar fashion—that is, DMADV (duh-mad-vee). The five phases for this methodology are Define-Measure-Analyze-Design-Verify, defined as follows:

1. **Define:** Clearly define the project goals and customer requirements. Consider the needs of both internal and external customers.
2. **Measure:** Apply measurements where applicable to customer needs and translate in terms that are meaningful to production processes. Identify industry benchmarks and use them for comparative analyses with competitors and against industry-wide standards.
3. **Analyze:** Generate process alternatives that can be deployed to meet customer requirements.
4. **Design:** Generate detailed product/process designs that can efficiently meet the defined customer requirements.
5. **Verify:** Confirm that the design product/processes perform as expected and meet customer requirements.

The DMADOV methodology (labeled II on Figure 14-13) is a slight modification of methodology I, with an included

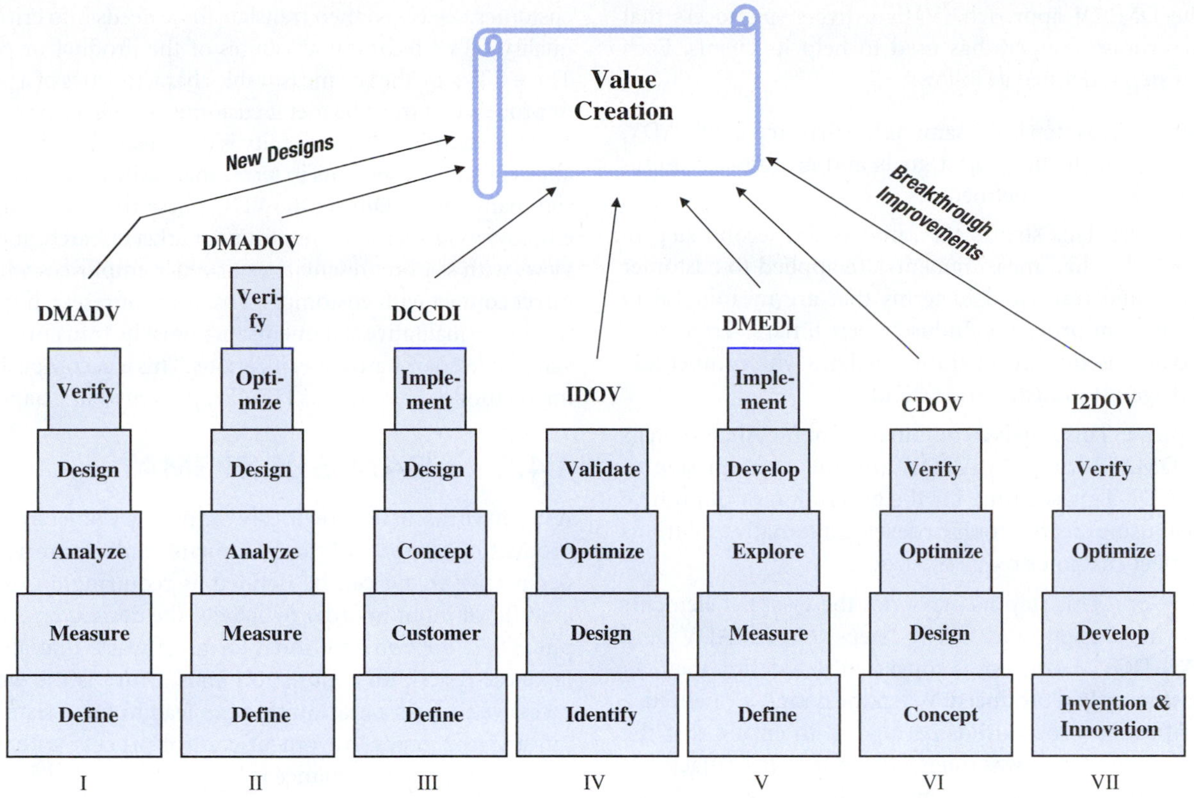

FIGURE 14-13. Design for Six Sigma Improvement Models

Optimization step that is performed between the Design and Verification phases. Methodologies III–VII represent other approaches that have been popularized in recent years. The DCCDI[23] process involves five steps, described as follows:

1. **Define:** The Define phase under DCCDI is similar to the Define step under DMADV, during which the project goals are clearly articulated and understood by all.
2. **Customer:** During the Customer phase a thorough understanding of the customer and the competitive marketplace is acquired. This includes a detailed analysis of external customer requirements, intelligence on major competitors, and a compilation of internal customer strengths, weaknesses, and needs.
3. **Concept:** The Concept phase is an integrated activity that brings together all interests (e.g., manufacturing, marketing, financial, customer, etc.) to develop workable concepts that are subsequently reviewed, debated, and a final one selected.
4. **Design:** The Design phase makes a reality out of the concept, developing a product design that can be efficiently reproduced in quantity and that meets customer and business requirements. This phase often includes prototyping and testing.
5. **Implement:** During this final phase the design is fully commercialized. This step includes the design of the manufacturing process that will be needed to produce the product and all tooling and fixtures. Once the process design has been finalized the product is formally launched through marketing and sales.

Methodology IV in Figure 14-13 (IDOV) has been widely applied in manufacturing. As the acronym suggests, this approach employs only four steps.

1. **Identify:** The intended market and its customers are clearly identified. Then an analysis is made of customer needs and defined in terms and specifications that clearly identify those attributes that are critical to quality (CTQ).
2. **Design:** During the Design phase the CTQ attributes are translated into functional requirements, and design concepts that will meet those requirements are generated. Finally, a selection process narrows the list of alternative concepts to a few which are closely scrutinized leading to a final "best" design alternative.
3. **Optimize:** Once the final design has been selected, advanced statistical and operations research tools are employed to make design modifications based on predicting and optimizing product performance.
4. **Validate:** The validation step ensures that the final optimized design does in fact meet customer requirements with respect to all stipulated CTQ attributes.

[23]Tennant, Geoff. *Design for Six Sigma: Launching New Products and Services Without Failure*, Reprint ed. Burlington, VT: Gower Publishing, 2002.

The DMEDI approach (V) is a five-step process that PricewaterhouseCoopers has used to help its clients. Each DMEDI step is defined as follows:

1. **Define:** This step is the same as the first step in DMADV, during which the project goals and customer requirements are clearly defined.
2. **Measure:** This step is the same as the second step in DMADV when measurements are applied to customer needs and translated in terms that are meaningful to production processes. Industry benchmarks are identified and used for comparative analyses with competitors and against industry-wide standards.
3. **Explore:** This step is a combination of the Analyze steps of DMADV and DMADOV, and the Concept step of DCCDI. This is a time for the generation of innovative ideas using creative tools to develop alternative solutions to meet customer requirements.
4. **Develop:** This step incorporates the essential elements of the Design and Verify steps of DMADV and DMADOV. A concept is converted to a design that can be efficiently reproduced. When the design is finalized, a verification procedure is performed to ensure that the design does what was intended—that is, it satisfactorily meets the stipulated customer requirements.
5. **Implement:** This is the same as the final step in the DCCDI approach, during which the design is commercialized and all factors of production are designed and configured to reproduce the design in volume.

The last two models in Figure 14-13, the CDOV (VI) and the I2DOV (VII), are similar. These steps are defined as follows:

1. **Concept/Invention & Innovation:** During this step the problem (need) is defined and creative ideas are generated as possible solutions.
2. **Design/Develop:** As with other approaches this step translates the concept into a detailed design that can be manufactured.
3. **Optimize:** Advanced statistical and operations research tools are used to modify the design in order to improve its ability to meet customer requirements at minimal cost.
4. **Verify:** As with other approaches there is a need to confirm that the final modified design performs as expected, meets customer requirements and cost predictions.

DFSS is an integrated approach that can be used to reengineer processes to achieve quantum improvements in performance. Many of the quality tools that have previously been introduced throughout this text can be applied in the DFSS methodology. For example, design for supportability (DFS) concepts are often employed during DFSS to integrate total life-cycle effects into the design process, and the house of quality (HOQ) framework (quality function deployment [QFD]) can be applied during the Define and Measure steps.

All of the variations of DFSS commence by utilizing some version of a voice of the customer (VOC) analysis to determine customer needs and then translate these needs into critical-to-quality CTx[24] technical attributes of the product or process. These CTxs are the key measurable characteristics of a product or process that must be met if customers (internal or external) are to be satisfied. The difficulty is that customers do not usually express requirements in terms that easily translate into design parameters. During the VOC phase the Six Sigma team employs tools such as focus groups, market research, and interviews with sales representatives or other employees who have direct contact with customers. This form of research typically produces qualitative statements that must be transformed into actionable quantitative specifications. This is accomplished using methodologies such as DFMEA presented in Chapter 6.

14.5 LEAN SIX SIGMA

As we have discussed previously the primary objective of lean production is to identify and eliminate *muda*. From a systems perspective *muda* can be defined as requiring an excessive amount of input in order to achieve the desired level of output. There are two additional forms of waste that also have Japanese descriptors: *mura* and *muri*. **Mura** is the word for excessive process variation that can lead to inconsistent output and outcomes. An out-of-control process with unpredictable future performance is an example of mura.

While lean production stresses the elimination of muda, Six Sigma focuses on the elimination of mura.

Since Six Sigma incremental improvement projects typically target variability issues, it is fair to say that a principal goal of Six Sigma is the elimination of mura.

There is really no inherent conflict between the principles of *lean* and those of Six Sigma. The former attacks muda and its sources; the latter goes after mura and its sources. The domains of interest obviously overlap. Lean views variability as a form of muda and Six Sigma will often go after waste that is not necessarily related to variability (mura) issues. Additionally, a consequence of *mura* can be the creation of more muda (e.g., inventory, wasted effort, idle time, etc.). Lean and Six Sigma are not only kindred doctrines but in many ways inseparable. Therefore, the term *Lean Six Sigma* does not in reality contribute any new perspective or philosophy to a QIM-aspiring company. It is difficult to imagine a scenario where a Six Sigma enterprise is not also committed to and using the lean tool kit. It is more likely that a lean organization has been successful in applying lean tools to eliminate sources of waste without the benefits offered by the Six Sigma framework. In this latter case the organization could potentially benefit by evolving from lean to Lean Six Sigma.

The third category of waste is called **muri**, a condition under which a system is overburdened or placing undue

[24]In Chapter 2 the CTx characteristics were defined as those performance factors that are important to the customer. The most common CTx factors are CTQ—critical to quality, CTD—critical to delivery, CTC—critical to cost, CTP—critical to process, and CTS—critical to safety.

stress on people and/or machines to deliver unreasonable outputs or in unrealistic timeframes. From a systems viewpoint muri is created when too little input is provided for the expected level of output. An example of muri would be expecting a line with a design capacity of 1,000 units per hour to produce 1,500 units. Scheduling 1,500 for production on this line is irrational and will result in the accumulation of inventory, which is a form of muda. Another example of muri is a poor ergonomic work design that requires operators to risk injury lifting heavy loads or subjecting their bodies to undue stresses and strains. Muri can occur wherever commitments are made that exceed a system's capabilities, stressing the performance of all components beyond their design limits. This can in turn generate additional waste in the form of muda and mura.

Efficient and effective process designs are free from muri.

14.6 COST OF QUALITY

In Chapter 5 we studied how cost of quality (COQ) considerations impact the design of products or services, and we introduced Taguchi's loss function, which propounds that there is always a cost associated when quality deviates from the stipulated target requirements. Four principles were presented as guidelines for creating robust designs that will minimize the variation thereby minimizing the Taguchi loss. Our attention will now turn to a somewhat different view of COQ—the one used by Philip Crosby who suggested that COQ is a term that should be ascribed to all those costs accrued in either ensuring that quality goals are met or the consequences of failing to meet those goals. In this context COQ has two components—the cost of conformance and the cost of nonconformance[25]—and can be thought of as the financial sacrifice an organization is compelled to make if its product or service is not produced or delivered correctly the first time every time. From scrap to rework to litigation to lost sales, companies continually lose money due to poor quality, and it has been estimated that the COQ can be as large as 30% of an organization's total operating costs.[26]

Cost of quality (COQ) is the cost of conformance plus the cost of nonconformance.

The investment that an organization makes in quality assurance and process improvement represents the cost of conformance component of COQ. This includes all the costs that can be attributed to operating the quality program and includes such activities as sampling, testing, inspections, training, reviews, audits, documentation, and testing. The cost of releasing personnel from production tasks to spend time on Six Sigma, kaizen blitz, or problem-solving teams is also included as part of the cost of conformance.

The cost of nonconformance includes the total costs associated with the failure to achieve a satisfactory level of quality. This includes scrap and rework and postdelivery undesirable outcomes such as warranty claims, loss of business, product recalls, and lawsuits. In Lean Six Sigma environments exceptional efforts are often taken to minimize the cost of nonconformance, particularly those in-process costs that would result in reduced process yields due to scrap and rework. Two metrics commonly used to track such cost-drivers are first time yield (FTY) and rolled throughput yield (RTY).

14.6.1 First Time Yield

First time yield (FTY), sometimes referred to as first pass yield, is defined as the number of acceptable units of output from a process divided by the number of units introduced to the process as input. FTY is an aggregate measure of the amount of rework and scrap that is produced by a value stream.

First Time Yield (FTY)

$$FTY = \frac{\text{\# of acceptable units produced including units reworked}}{\text{Total number of units produced}} \qquad \textbf{14-1}$$

Figure 14-14 illustrates a stream that consists of a four-process production routing for a particular product. Material for 500 parts is introduced as input to process A. Only 450 parts emerge from process A as acceptable on the first pass through. While producing the batch of 500 it was necessary to scrap 50 of the parts. Twenty of the parts required some reworking but could be salvaged as acceptable parts. The FTY for process A is 450 good parts divided by 500 processed parts = 0.9. Of the 450 good parts entering process B only 425 satisfactory parts were produced on the first pass—25 were scrapped outright and 15 parts had to be reworked. Therefore the FTY for process B is 425 divided by 450 or 0.944. Of the 425 good units entering process C, 9 parts were scrapped and only 416 good parts are produced first time through. None had to be reworked. Finally, of the 416 parts presented to process D, 6 were scrapped, and 4 were reworked, resulting in 410 good parts produced on the first pass. The FTYs for processes C and D are therefore 0.979 and 0.986, respectively.

The total yield for the process sequence (value stream) is the product of all the individual FTYs. That is, the total yield = (0.9)(0.944)(0.979)(0.986) = 0.82. The total yield can also be computed by dividing the output from the last process, process D, of 410 by the input to the first process, process A, of 500. That is, 410 divided by 500 = 0.82.

14.6.2 Rolled Throughput Yield

While FTY considers yield losses based on the ratio of inputs to outputs by process, the **rolled throughput yield** (RTY) metric represents the probability that a part can pass through a series of processes defect free, and takes rework into consideration. Figure 14-15 shows the yields that each of the four processes achieve when rework is factored out. For example, the actual yield of any process is equal to the satisfactory units produced (input − scrap − rework) divided by input. Process A was able to net only 430 parts without rework, so

[25] Crosby, Philip B. *Quality Is Free: The Art of Making Quality Certain.* New York: McGraw-Hill, 1979.

[26] http://www.isixsigma.com/dictionary/Cost_Of_Quality-497.htm.

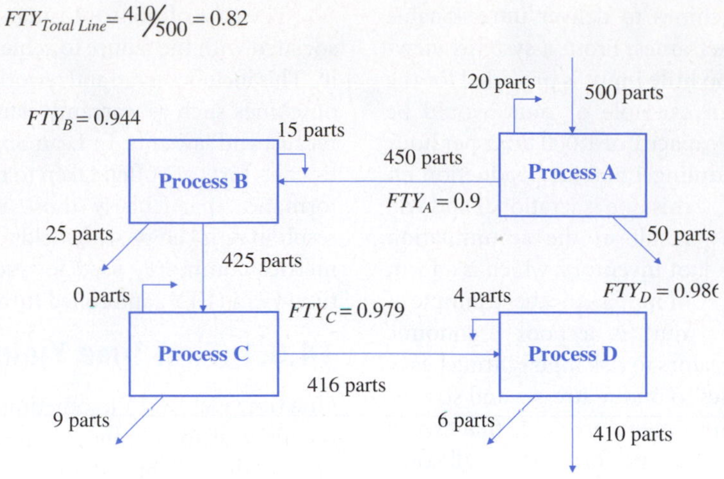

FIGURE 14-14. First Time Yields

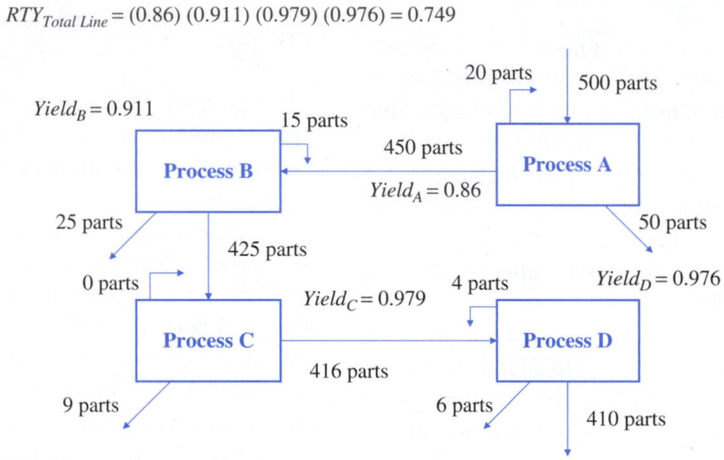

FIGURE 14-15. Rolled Throughput Yield

SHOWCASE of EXCELLENCE

Winners of the Malcolm Baldrige National Quality Award Mixed Categories

- 2004: Robert Wood Johnson University Hospital Hamilton (Healthcare), a health care facility that offers an innovative approach to patient care that includes medical, surgical, obstetric, cardiac, orthopedic, emergency, intensive care, and outpatient cancer care services, as well as wellness services including community health education, fitness, and physical therapy. http://www.rwjhamilton.org, http://www.nist.gov/public_affairs/RWJUHH_PDF_final.pdf.
- 2003: Baptist Hospital Inc. (Healthcare), the largest, most comprehensive health care system in the Florida Panhandle operating two hospitals and an ambulatory care complex that offers a wide array of outpatient and diagnostic services. http://www.ebaptisthealthcare.org, http://www.nist.gov/public_affairs/baldrige2003/Baptist_3.3.04.pdf.
- 2003: Saint Luke's Hospital of Kansas City (Healthcare), a comprehensive teaching and referral health care organization affiliated with the Protestant Episcopal Church providing 24-hour coverage in all health care disciplines. http://www.saintlukeshealthsystem.org, http://www.nist.gov/public_affairs/baldrige2003/St.Lukes_3.3.04.pdf.
- 2002: SSM Health Care (Healthcare), the provider of a wide range of medical services including emergency care, surgery, oncology, mental health, obstetrics, cardiology, orthopedic, pediatric, and rehabilitative care. http://www.ssmhc.com, http://www.nist.gov/public_affairs/releases/ssmhealth.htm.
- 2000: Federal Express Corporation (Service), an air-express shipping company that has 90,000 employees at more than 1,650 sites that handle a daily volume of 1.5 million packages that are tracked in a central information system and transported with the world's largest air cargo fleet. http://fedex.com, http://www.quality.nist.gov/FederalExpress_90.htm.

the actual yield for process A is 430 divided by 500, or 86%. For process B the actual yield is 410 (parts produced without rework) divided by 450, or 0.911. Similarly processes C and D had yields equal to 0.979 and 0.976, respectively.

To obtain the RTY one multiplies all of the true yields along the line together—in this case, (0.86)(0.911)(0.979)(0.976) = 0.749. The RTY of 74.9% is a better indication of true line capacity than the line FTY = 82% that included the inefficiencies associated with rework.

Rolled Throughput Yield (RTY)

$$RTY = \prod_{i=1}^{m} \left(\frac{\overline{\omega}_i - \omega_i - \varphi_i}{\overline{\omega}_i} \right) \qquad \textbf{14-2}$$

where,

$\overline{\omega}_i$ = total number of units produced by operation i

ω_i = total number of units scrapped by operation i

φ_i = total number of units reworked by operation i

14.7 SUMMARY OF KEY IDEAS

Key Idea: *A systems view is important when allocating resources to continuous improvement efforts.*

It is inefficient and can actually be counterproductive to create a quality culture that makes continuous improvement a part of everyone's job. This can lead to frustrations, burnout, and suboptimization. Competitive QIM organizations are committed to continuous improvement, but from a systems perspective, directing its valuable and limited resources to those activities that provide the greatest leverage to creating stakeholder value. Change simply for the sake of change, in the absence of a complete understanding of system consequences, can produce unintended outcomes such as inefficiencies, waste, and additional costs having deleterious effects on customer value.

Key Idea: *The theory of constraints (TOC) is a methodology that focuses management attention and resources on maximizing the conversion of inventory into throughput.*

Under TOC the attention is on identifying constraints that impede a company's ability to perform. In a production system constraints are bottlenecks that define line capacity, and throughput can be increased by identifying, exploiting, and elevating bottleneck activities. Since a line's capacity is defined by its bottlenecks, extraordinary measures should be taken to maximize bottleneck performance. This can be done through inventory buffering to ensure that bottlenecks are never starved for material or through the use of flexible crewing and proactive maintenance strategies to minimize the incidences of bottleneck break-downs.

Key Idea: *Under theory of constraints all nonconstraining activities are subordinated to the constraining activities.*

Once a constraint is identified, all management attention is directed to exploiting the performance of the constraint (trying to get the most out of it) and toward elevating it (effecting improvements that will eliminate the constraint). At that point the TOC cycle repeats and another constraint—an activity that may have been nonconstraining before—is identified.

Key Idea: *The drum-buffer-rope scheduling system is a hybrid pull–push system designed to ensure that the constraint activity is paced to meet demand and never has to shut down due to lack of material.*

The DBR system is also called synchronous manufacturing and uses a combination of pull and push scheduling relative to the bottleneck, or constraint, activity within the value stream. The constraint activity is called the drum because this activity, which determines the overall line capacity, sets the pace for all other workstations along the line. The drum pulls demand via the rope from the customer and the gate activity (the first activity in the line) pulls demand from the drum. The gate then pushes material toward the drum and the drum pushes material to activities further down the line. All activities are protected from the effects of variability through carefully managed inventory buffers. The principal objective is to keep the bottleneck operation continuously running since throughput cannot be created until inventory is converted. All activities in the line work to a "roadrunner mentality" working as fast as is physically possible when orders are present. Otherwise, operators can be idle or focus efforts on activities such as routine maintenance or team meetings until the next order arrives.

Key Idea: *Six Sigma is a comprehensive system that ties improvement efforts directly to business success.*

The success of Six Sigma is largely attributable to three key elements: individual project goals are tied directly to bottom-line results, improvement projects are carefully selected and prioritized on the basis of how projected benefits will impact key performance metrics, and the requirement that top management play an active role.

Key Idea: *Six Sigma is concerned with improving the efficiency and effectiveness of an organization's core business processes.*

An organization's core business processes are those primary value streams that directly contribute to providing products or services to customers. The number of core processes will not be great, usually less than 8 or 10, and a member of the top management team is assigned the responsibility of process owner. This person usually serves as the champion for any project relevant to his or her process.

Key Idea: *The Six Sigma improvement model is called DMAIC and is based on the creation of knowledge to solve a problem. Step-transition activities called tollgates ensure the active participation by top management and help keep teams on track.*

DMAIC is an acronym that stands for the interconnected steps of Define-Measure-Analyze-Improve-Control and is not fundamentally different from the PDCA and scientific methods. One feature that distinguishes DMAIC from other approaches is the provision for "tollgates" between each of the phases. Tollgates represent milestone events during which the improvement team meets with senior management and other project participants to review progress and make certain the project stays on target and continues to be aligned with the goals stated in its charter.

Key Idea: *Six Sigma distinguishes between short-term variation and long-term variation.*

During the Measure phase of the DMAIC process current performance is evaluated using a Six Sigma score. The score represents the predicted long-term sustainable variation for a process that is in statistical control. The percentage of defective product is estimated by subtracting 1.5 standard deviations from *z* scores prior to consulting normal tables such as Table II in Appendix B.

Key Idea: *When Six Sigma is applied to the design of new products or radical breakthroughs in process performance, the methodology is called design for Six Sigma (DFSS).*

Design for Six Sigma (DFSS) is the application of the Six Sigma philosophy to any of the integrated product design and development approaches, including integrated product and process design (IPPD), design for manufacturability (DFM), concurrent engineering (CE), and design for supportability (DFS). Unlike the Six Sigma DMAIC methodology, there are no universally recognized set of steps or phases—many variations of DFSS exist as organizations will adopt an approach that best fits individual needs or is advocated by an external consultant that has been contracted to assist with implementation.

Key Idea: *Lean Six Sigma is the formal recognition of the natural compatibility of lean production and Six Sigma.*

Lean production is a philosophy that identifies and eliminates sources of waste (muda), while the main emphasis of Six Sigma is the identification and elimination of sources of variability (mura). A third source of waste is system overburden (muri), which is the result of poor system design, poor capacity planning, or poor scheduling. All three sources of waste are interrelated and each source can create waste in one of the others. Therefore, an integrated quality program must be one that incorporates the elements of both lean and Six Sigma thinking.

Key Idea: *Cost of quality (COQ) is cost of conformance plus the cost of nonconformance.*

Cost of quality (COQ) can be thought of as the financial sacrifice an organization is forced to make if its product or service is not produced or delivered correctly the first time every time. The cost of nonconformance includes such areas as scrap loss, rework, warranties, and litigation. The cost of conformance includes all the costs associated with the operation of a formal quality program to ensure quality at the source. This includes all the inspection costs, quality control software, improvement teams, supplier certification programs, training, and the like.

Key Idea: *First time yield (FTY) and rolled throughput yield (RTY) are measures used as a basis for estimating the internal component of the cost of nonconformance.*

The first time yield (FTY) measures the number of acceptable units of output that a process or work center produces as a percentage of the number of units worked on. FTY only accounts for scrap as yield loss. Rolled throughput yield (RTY) represents the probability that a part can pass through a series of processes defect free, and both scrap and rework are factored into yield losses.

14.8 MIND EXPANDERS FOR YOUR UNDERSTANDING

14-1. You have been hired as a consultant to evaluate the quality program at the Town Crier, the major newspaper in a large midwestern city. The editor in chief takes pride in the paper's history of progressiveness and customer responsiveness. TQM has been practiced for 15 years and control charts are visible in many areas from composition to printing to distribution. All production employees are organized into continuous improvement (CI) teams that meet for 45 minutes each Friday morning to share continuous improvement ideas. Nevertheless, for the last two years there has been a noticeable downward trend in profits, as productivity has declined, advertising revenues have dropped, and production costs have increased. The editor in chief is perplexed and does not understand why the CI program is not working as intended. As his consultant, what are some of the issues that you would want to investigate first?

14-2. How does the theory of constraints (TOC) philosophy maximize the efficiency of a production process? What role does inventory play? What role does process improvement play? Is TOC scheduling a pull or a push system? Does TOC apply in a pure services environment? Explain.

14-3. Why is a progressive maintenance strategy important to the success of Six Sigma? How does the 5S workplace discipline support TPM?

14-4. In a minerals plant, viscosity is an important quality characteristic of a particular product. The target viscosity is 1,300 and the customer has agreed to accept viscosities in the range 1,100 to 1,500. Six samples are taken per shift and sent to the quality control lab for testing. The averages and ranges of all six samples are plotted on $\overline{X}$ and R charts, respectively. The $\overline{X}$ chart shows good control with a centerline $\overline{\overline{X}} = 1435$ and the range chart is in control with a centerline $\overline{R} = 46.88$.

a. What is the Six Sigma score for this process?

b. What is the long-term process yield?

c. Compute the DFMO.

14-5. Lean production and Six Sigma have been politicized as two different manufacturing doctrines. Some have proposed Lean Six Sigma as a third approach that integrates the concepts of lean and Six Sigma.

a. Do you think it is possible to successfully implement the concepts of Six Sigma without also implementing the lean production philosophy? Explain.

b. Do you think it is possible to successfully implement the concepts of lean production without also implementing Six Sigma? Explain.

14-6. Do you believe it is necessary for a Six Sigma organization to also be an HPWO? Explain.

14-7. In a typical TQM organization the focus historically was on mass training, where all employees received instruction in a variety of quality tools, such as those presented in Chapter 7. By comparison, Six Sigma emphasizes a certification program that explicitly defines the body of knowledge required to achieve recognized proficiency at established levels. Discuss what you believe to be the relative merits of each approach.

14-8. Sertex Sealants, Inc. has been in business for 33 years producing sealing strips for automobiles, and its rubber seals can be found around door moldings, trunk lids, and windshields of many popular cars and trucks. The company operates two plants—an extrusion plant where production starts, and a finishing plant, where holes are drilled, fabric added, and fasteners attached depending on specifications. Manufacturing cells consisting of between six and eight workers perform various finishing operations. A job will typically go through several cells before it is ready for shipment.

During the extrusion operation a color stripe is painted on each part that is distinctive of that particular part number. When the cell receives a batch of work they can immediately tell if this batch is of *reds*, *greens*, *blues*, and so on. The cell team has become so proficient with the activities for each color that team members can change from one color (part number) to another with minimal changeover time. Each batch comes to the cell in a kanban unit—a moveable cart on wheels that contains 150 strips. It takes the team about one hour on the average to work off a kanban batch.

All of Sertex's customers require daily deliveries. Trucks start departing the loading dock at 10 a.m. each day and continue until 4 p.m. when the last truck leaves. At 4:30 p.m. Hank Dunlop, a material expeditor, receives a report on orders that for whatever reason missed the last truck. Hank then gets on the phone and arranges special shipments by air freight. In the last 12 months, Sertex paid a total of $800,000 for premium shipments by air.

Jerry Crenshaw, general manager, cannot understand why his company does so poorly in meeting customer commitments on time. After all, advanced orders are taken right off the electronic data interchange (EDI) system that provides his company with a six-week frozen production schedule for each product. Three years ago the Sertex process improvement team—an internal consulting group consisting of three engineers—designed a kanban system to operate between the two plants. In addition, the plant manager of the finishing plant, Dick Wainwright, holds a production meeting each day at 9 a.m. to discuss any high-priority jobs in the pipeline and to finalize the day's production schedule. Each department head also conducts a weekly meeting with the team leader of each production cell to review control charts and discuss any quality problems. Nevertheless, recently the number of customer complaints has been increasing. Jerry has decided to bring in some experts from outside the company for advice.

A team from CPR Consulting, headed by Joyce Brady, arrived bright and early Monday morning. By noon, they were interviewing personnel and pouring over documentation. On Wednesday they reported the following findings to the general manager.

I. The kanban system is not working. It was designed to requisition materials from the extrusion plant for the *current* day's orders, not the *following* day's requirements. As a result inventories build for materials that are not needed.

II. The daily production meetings are a farce. Department heads discuss the daily requirements and agree on a schedule. Then following the meeting, some of these supervisors will check on-hand inventory levels in the warehouse and make their own decisions on what should be produced. In many cases inventory in the warehouse has already been earmarked for shipment.

III. Large discrepancies were found between physical counts in the warehouse and computer records.

IV. An extraordinary number of full kanban containers were observed stacked around the plant. Consultants were told that these were orders that were in process, but no one could provide clear documentation as to the exact details of the work. A quick tally revealed that this material represents approximately six months' production.

V. Approximately 70% of the workforce has had some training in quality tools including basic SPC and pull scheduling. However, the application of these tools appears to vary widely across the plant. There appears to be little conscientious effort at ongoing quality improvement.

VI. Orders that are complete and ready for shipment are loaded onto a truck and transported to finished goods inventory in a warehouse two miles from the plant. Trucks can sit at the plant dock for hours before being moved. Completed orders are added to computerized inventory records by Mary White at the warehouse, based on receiving documentation handed to her by warehouse staff after they have verified the material with respect to part number and quantity.

VII. Sertex has six customers; four of them account for 83% of the business. Demand can be forecasted out to 10 weeks in the future and with reasonable certainty 6 weeks ahead. Demand patterns are stable. Last-minute changes sometimes occur, but this is the exception rather than the rule.

VIII. Even with the large annual cost for premium freight, Sertex still only met 88% of its on-time delivery commitments to customers during the last 12 months. During this same period, 5% of its shipments were returned with quality problems. Each returned shipment had to be 100% inspected and the defective material reworked. The company is losing credibility with its customers who depend on it for quality and dependability.

IX. The manufacturing cells seem to be working well. Cell teams can change seamlessly from one part number to another, and all members of the team are multiskilled and can readily adjust whenever a team member is absent. The only problem is related to machine breakdowns resulting in 15% lost productive time. When a breakdown occurs, the cell has to wait for maintenance to send someone to fix the problem.

Drawing from the principles you have learned from this text, get together with some of your classmates and form a small group to brainstorm the following thought-provoking questions.

a. How can the principles of Lean Six Sigma help Sertex? If you were part of the CPR consulting team what would be your recommendations to Jerry Crenshaw?

b. Based on the CPR report, is Sertex's culture ready for an implementation of Lean Six Sigma? How can Jerry know and what can he do to prepare the organization? Is it important to try and implement lean production before Six Sigma, or vice versa? Can they be implemented simultaneously, or does it not matter?

c. Is the Sertex facility appropriate for pull scheduling? What approach should be taken to determine its appropriateness and to implement a workable system?

d. What training requirements should Sertex consider as it prepares for a Lean Six Sigma implementation?

e. What do you suspect is the reason that there is so much WIP inventory on the factory floor? What steps would you recommend Sertex take to investigate this problem and determine how the huge backlog of WIP can be eliminated and then kept to a minimum?

f. What role if any do you think TOC could play in helping Sertex solve its production problems?

g. What role if any do you think progressive maintenance strategies such as TPM could play in helping to resolve some of Sertex's problems?

APPENDIX A

DEMING'S FOURTEEN POINTS

Point 1: Create Constancy of Purpose Creating a constancy of purpose requires a total commitment, starting with top leadership, which is prepared to place quality above quantity and bottom line results. This involves taking a long-range versus short-term view of the organization and the willingness to learn and change.

Point 2: Adopt the New Philosophy Adopting the new philosophy is the organizational realization that the world has changed and continues to change; and that the internal culture needs to adapt to the change including an intolerance to errors and a commitment to fact-based decision making. The new philosophy recognizes that there are no quick fixes to systemic problems and that quality and total costs can be negatively correlated.

Point 3: Cease Dependence on Mass Inspection Inspection adds costs but no value. Quality must be designed and then built into products and process. Quality is only as good as the capability of the process used to produce it and improvement requires that system deficiencies be identified and corrected. Each worker must become responsible for his or her contribution to quality.

Point 4: End the practice of awarding business on the basis of price tag alone The cheapest source can ultimately be the most costly. Reliability, durability, quality, and consistency are important attributes that should be considered as well as cost. It is important that vendors be quality-certified.

Point 5: Continually improve systems The route to continuous improvement involves more than simply identifying and solving problems. Processes can only perform up to their design capabilities, and management alone has the ability to reengineer systems and processes.

Point 6: Institute modern methods of training Each employee must clearly understand his or her job and be thoroughly trained to do it properly. This requires consistent work methods and a formal structured training program.

Point 7: Institute leadership Quality is built by leading, not driving. Good leadership understands the importance of the human component and the need for profound knowledge of systems capability and the human–system interface. Effective leaders also value clear, concise, open, and consistent communications.

Point 8: Drive out fear Fear is pervasive and omnipresent in all organizations. Fear has many sources—for example, fear of supervisors, fear of failure, fear of change and the unknown, job insecurities, internal competition for rewards and recognition, and performance reviews. Quality depends on an environment of trust and full disclosure. Effective communications is a key factor.

Point 9: Break down barriers among departments Quality depends on cross-functional teamwork and cooperation so that systems can be effectively analyzed, operated, and improved. This requires a reward system that supports organizational rather than functional goals.

Point 10: Eliminate numerical goals, targets, and slogans Slogans and campaigns unfairly imply that the responsibility for quality resides at the worker level. Quality cannot simply be "willed" or "mandated" from above, and setting unattainable goals for workers can be counterproductive. If the entire workforce culture embraces the principles of excellence, and workers are given the necessary methods and tools, then slogans and campaigns are irrelevant.

Point 11: Eliminate numerical quotas Management by the numbers fails to consider quality requirements or implications, and production standards typically do not address causal systems or improvement issues. At leadership levels, formal programs such as management by objectives (MBO) can be contests that feed internal competition between functional areas and can therefore act as barriers to quality.

Point 12: Remove barriers that take away pride Management should treat people with respect, regarding them as participants and not opponents, and recognize that the most valuable assets are an organization's human resources—what the people carry around in their heads, and their ability to work together. The three major inhibitors to quality in any firm are the performance appraisal, the daily production report, and the financial management system, as these three tools take away pride by making numbers more important than people.

Point 13: Institute a vigorous program of education and self-improvement Quality requires the willingness by management to invest in people. When human resources are viewed as an asset this makes good business sense. Education and self-improvement programs should emphasize exercising and stimulating the mind and need not be job related. If the quality system is based on broad participation and intellectual input is valued, it is important that programs be initiated that will help develop mental capabilities to reach peak potentials.

Point 14: Put everyone to work on the transformation An implementation strategy requires careful planning and patience. Ultimately everyone in the organization must be involved. Cultural change and new paradigms cannot be expected quickly and must evolve over time.

CROSBY'S FOURTEEN-STEP APPROACH TO QUALITY IMPROVEMENT

Step 1: Management Commitment Management must recognize and accept the need for quality improvement, especially the importance of defect prevention. A management-supported quality policy should state that each individual is expected to perform exactly in conformance to requirements or cause the requirements to officially be changed to what the customer needs. Management must also agree that quality improvement is a practical way to achieve profits. A personal commitment by management raises the level of visibility for quality and ensures everyone's cooperation.

Step 2: Quality Improvement Team The second step is to bring together representatives of each department to form a quality improvement team. Team members should be carefully selected to include those who can speak for their departments and have the authority to commit their departments to specific actions. All team members should clearly understand their respective roles. It works well to appoint one of the members as the team chairperson.

Step 3: Quality Measurement It is necessary to obtain baseline data to determine the status of quality throughout the company. Quality measures must be established where they do not exist and reviewed where they do. Measurements should be well documented to show where improvement is possible and where corrective action is necessary. Do not forget nonproduction areas such as engineering, accounting, marketing, and purchasing. Formalizing an organization's measurement system strengthens the inspection and test functions and ensures proper measurement. Establishing a good documentation system, including highly visible charts, provides the foundation for the entire quality program.

Step 4: Cost of Quality Evaluation Cost of quality (COQ) is not an absolute performance measure. COQ measures where corrective action will be profitable—the higher the cost, the more corrective action that needs to be taken. Integrating the concept of COQ into an organization's formal accounting system establishes quality management performance as part of the overall company performance system.

Step 5: Quality Awareness It is now time to begin to raise the awareness of all employees on the impacts of poor quality. Supervisors should be oriented first and then used to further orient and communicate quality objectives to their employees. This should be done company-wide and everyone should participate. Open communications set the stage for everyone adopting the habit of talking positively about quality and for being receptive to corrective actions and error cause removals.

Step 6: Corrective Action As employees are encouraged to talk about the problems they are experiencing with their processes, opportunities for corrective action are identified—not only those that have been found through sampling inspection, but also less obvious ones that the workers themselves have observed. These opportunities can be brought to the attention of supervisors who can pass them up the chain of command for review and action.

Step 7: Zero Defects Planning Three or four members of the quality improvement team should be selected to investigate the zero defects concept and ways to implement the program. The idea should clearly be communicated that zero defects means doing things right the first time every time.

Step 8: Supervisor Training All levels of management should be thoroughly indoctrinated prior to implementation of the formal quality program. Each manager must understand the program well enough to explain it to his or her subordinates. The idea is that once managers are in tune with the program objectives and realize its value, they will then focus their actions on supporting its implementation.

Step 9: Zero Defects Day A single day should be set aside for the company-wide implementation. That way, everyone will understand the program the same way and recognize that this day marks the beginning of a new mind-set. Marking a day on the calendar as the day to commence zero defects creates a milestone that gives the program an emphasis and memory that will be long-lasting.

Step 10: Goal Setting Meetings are held by supervisors with their employees who are asked to establish goals they would like to achieve within 30, 60, and 90 day time frames. All goals should be specific and measurable.

Step 11: Error Cause Removal All workers are asked to describe on one page any problem that inhibits their performance of error-free work. These problems are forwarded to the appropriate department, such as industrial engineering, for study. It is important that all problems submitted are acknowledged within 24 hours. The objective is to build trust between management and workers, letting them know that their input is important and will be heard, seriously considered, and answered.

Step 12: Recognition Those who meet their goals or perform outstanding acts should be recognized. Prizes or awards should not be financial, but represent genuine praise and recognition so that the personnel involved feel appreciated.

Step 13: Quality Councils Form a quality council that brings quality professionals and team chairpersons together regularly to interact and determine actions necessary to improve the quality implementation. The quality council is the best source of information on the status of programs and ideas that have been vetted for refinements.

Step 14: Do It Over Again The program is never over, but changing situations and personnel turnover require continual training and revitalization. The anniversary of Zero Defects Day can be set aside as a special annual event that could be marked by a special luncheon for all employees. Repetition and re-dipping makes the program perpetual and ingrained as part of the culture.

DEMING'S SEVEN DEADLY DISEASES

Disease 1: Lack of Constancy of Purpose—to plan a marketable product or service to keep the company in business and provide jobs.

Disease 2: Emphasis on Short-Term Profits

Disease 3: Personal Evaluation Appraisal—applied to management personnel the effects can be devastating.

Disease 4: Mobility of Management—rotation of leadership as they hop from job to job.

Disease 5: Use of Visible Figures by Management—giving little consideration to figures that are unknown or unknowable.

Disease 6: Excessive Medical Costs

Disease 7: Excessive Warranty Costs—these can be fueled by lawyers who work on contingency fees.

APPENDIX B

Table I Control Chart Constants

	$\bar{X}$-*R* Control Charts					$\bar{X}$-*s* Control Charts				
Sample Size *n*	A_2	D_3	D_4	d_2	d_3	A_3	B_3	B_4	c_4	c_5
2	1.880	0	3.267	1.128	0.853	2.659	0	3.267	0.7979	0.6029
3	1.023	0	2.574	1.693	0.888	1.954	0	2.568	0.8862	0.4632
4	0.729	0	2.282	2.059	0.880	1.628	0	2.266	0.9213	0.3888
5	0.577	0	2.114	2.326	0.864	1.427	0	2.089	0.9400	0.3412
6	0.483	0	2.004	2.534	0.848	1.287	0.030	1.970	0.9515	0.3077
7	0.419	0.076	1.924	2.704	0.833	1.182	0.118	1.882	0.9594	0.2821
8	0.373	0.136	1.864	2.847	0.820	1.099	0.185	1.815	0.9650	0.2622
9	0.337	0.184	1.816	2.970	0.808	1.032	0.239	1.761	0.9693	0.2459
10	0.308	0.223	1.777	3.078	0.797	0.975	0.284	1.716	0.9727	0.2322

Table II Areas under Standard Normal Curve

$z = \dfrac{\left\lvert \text{Specification Limit} - \bar{X} \right\rvert}{\sigma_x}$	0.00	0.01	0.02	0.03	0.04	0.05	0.06	0.07	0.08	0.09
6.0	0.000000	0.000000	0.000000	0.000000	0.000000	0.000000	0.000000	0.000000	0.000000	0.000000
5.9	0.000000	0.000000	0.000000	0.000000	0.000000	0.000000	0.000000	0.000000	0.000000	0.000000
5.8	0.000000	0.000000	0.000000	0.000000	0.000000	0.000000	0.000000	0.000000	0.000000	0.000000
5.7	0.000000	0.000000	0.000000	0.000000	0.000000	0.000000	0.000000	0.000000	0.000000	0.000000
5.6	0.000000	0.000000	0.000000	0.000000	0.000000	0.000000	0.000000	0.000000	0.000000	0.000000
5.5	0.000000	0.000000	0.000000	0.000000	0.000000	0.000000	0.000000	0.000000	0.000000	0.000000
5.4	0.000000	0.000000	0.000000	0.000000	0.000000	0.000000	0.000000	0.000000	0.000000	0.000000
5.3	0.000000	0.000000	0.000000	0.000000	0.000000	0.000000	0.000000	0.000000	0.000000	0.000000
5.2	0.000000	0.000000	0.000000	0.000000	0.000000	0.000000	0.000000	0.000000	0.000000	0.000000
5.1	0.000000	0.000000	0.000000	0.000000	0.000000	0.000000	0.000000	0.000000	0.000000	0.000000
5.0	0.000000	0.000000	0.000000	0.000000	0.000000	0.000000	0.000000	0.000000	0.000000	0.000000
4.9	0.000000	0.000000	0.000000	0.000000	0.000000	0.000000	0.000000	0.000000	0.000000	0.000000
4.8	0.000001	0.000001	0.000001	0.000001	0.000001	0.000001	0.000001	0.000001	0.000001	0.000001
4.7	0.000001	0.000001	0.000001	0.000001	0.000001	0.000001	0.000001	0.000001	0.000001	0.000001
4.6	0.000002	0.000002	0.000002	0.000002	0.000002	0.000002	0.000002	0.000002	0.000001	0.000001
4.5	0.000003	0.000003	0.000003	0.000003	0.000003	0.000003	0.000003	0.000002	0.000002	0.000002
4.4	0.000005	0.000005	0.000005	0.000005	0.000004	0.000004	0.000004	0.000004	0.000004	0.000004
4.3	0.000009	0.000008	0.000008	0.000007	0.000007	0.000007	0.000007	0.000006	0.000006	0.000006
4.2	0.000013	0.000013	0.000012	0.000012	0.000011	0.000011	0.000010	0.000010	0.000009	0.000009
4.1	0.000021	0.000020	0.000019	0.000018	0.000017	0.000017	0.000016	0.000015	0.000015	0.000014

(Continued)

Table II Areas under Standard Normal Curve *(Continued)*

$z = \dfrac{\lvert \text{Specification Limit} - \bar{\bar{X}} \rvert}{\sigma_x}$	0.00	0.01	0.02	0.03	0.04	0.05	0.06	0.07	0.08	0.09
4.0	0.000032	0.000030	0.000029	0.000028	0.000027	0.000026	0.000025	0.000024	0.000023	0.000022
3.9	0.000048	0.000046	0.000044	0.000042	0.000041	0.000039	0.000037	0.000036	0.000034	0.000033
3.8	0.000072	0.000069	0.000067	0.000064	0.000062	0.000059	0.000057	0.000054	0.000052	0.000050
3.7	0.000108	0.000104	0.000100	0.000096	0.000092	0.000088	0.000085	0.000082	0.000078	0.000075
3.6	0.000159	0.000153	0.000147	0.000142	0.000136	0.000131	0.000126	0.000121	0.000117	0.000112
3.5	0.000233	0.000224	0.000216	0.000208	0.000200	0.000193	0.000185	0.000178	0.000172	0.000165
3.4	0.000337	0.000325	0.000313	0.000302	0.000291	0.000280	0.000270	0.000260	0.000251	0.000242
3.3	0.000483	0.000466	0.000450	0.000434	0.000419	0.000404	0.000390	0.000376	0.000362	0.000349
3.2	0.000687	0.000664	0.000641	0.000619	0.000598	0.000577	0.000557	0.000538	0.000519	0.000501
3.1	0.000968	0.000935	0.000904	0.000874	0.000845	0.000816	0.000789	0.000762	0.000736	0.000711
3.0	0.001350	0.001306	0.001264	0.001223	0.001183	0.001144	0.001107	0.001090	0.001035	0.001001
2.9	0.001866	0.001807	0.001750	0.001695	0.001641	0.001589	0.001538	0.001489	0.001441	0.001395
2.8	0.002555	0.002477	0.002401	0.002327	0.002256	0.002186	0.002118	0.002052	0.001988	0.001926
2.7	0.003467	0.003364	0.003264	0.003167	0.003072	0.002980	0.002890	0.002803	0.002718	0.002635
2.6	0.004661	0.004527	0.004396	0.004269	0.004135	0.004025	0.003907	0.003793	0.003681	0.003573
2.5	0.006210	0.006037	0.005868	0.005703	0.005543	0.005386	0.005234	0.005085	0.004940	0.004799
2.4	0.008198	0.007976	0.007760	0.007549	0.007344	0.007143	0.006947	0.006756	0.006569	0.006387
2.3	0.010724	0.010444	0.010170	0.009903	0.009642	0.009387	0.009138	0.008894	0.008656	0.008424
2.2	0.013903	0.013553	0.013209	0.012874	0.012546	0.012225	0.011911	0.011604	0.113040	0.011011
2.1	0.017864	0.017429	0.017003	0.016586	0.016177	0.015778	0.015386	0.015003	0.014629	0.014262

$z = \dfrac{\lvert \text{Specification Limit} - \bar{\bar{X}} \rvert}{\sigma_x}$	0.00	0.01	0.02	0.03	0.04	0.05	0.06	0.07	0.08	0.09
2.0	0.022750	0.022216	0.021692	0.021173	0.020675	0.020182	0.019699	0.019226	0.018763	0.018309
1.9	0.028717	0.028067	0.027429	0.026803	0.026190	0.025588	0.024998	0.024419	0.023852	0.023296
1.8	0.035930	0.035148	0.034380	0.033625	0.032884	0.032157	0.031443	0.030742	0.030054	0.029379
1.7	0.044566	0.043633	0.042716	0.041815	0.040930	0.040059	0.039204	0.038364	0.037538	0.036727
1.6	0.054799	0.053699	0.052616	0.051551	0.050503	0.049472	0.048457	0.047460	0.046479	0.045514
1.5	0.066807	0.065522	0.064256	0.063008	0.061780	0.060571	0.059380	0.058208	0.057053	0.055917
1.4	0.080757	0.079270	0.077804	0.076359	0.074934	0.073529	0.072145	0.070781	0.069437	0.068112
1.3	0.096801	0.095098	0.093418	0.091759	0.090123	0.088508	0.086915	0.085344	0.083793	0.082264
1.2	0.115070	0.113139	0.111232	0.109349	0.107488	0.105650	0.103835	0.102042	0.100273	0.098525
1.1	0.135666	0.133500	0.131357	0.129238	0.127143	0.125072	0.123024	0.121001	0.119000	0.117023
1.0	0.158655	0.156248	0.153864	0.151505	0.149170	0.146859	0.144572	0.142310	0.140071	0.137857
0.9	0.184060	0.181411	0.178786	0.176186	0.173609	0.171056	0.168528	0.166023	0.163543	0.161087
0.8	0.211855	0.208970	0.206108	0.203269	0.200454	0.197663	0.194895	0.192152	0.189430	0.186733
0.7	0.241964	0.238852	0.235765	0.232695	0.229650	0.226627	0.223627	0.220650	0.217695	0.214764
0.6	0.274253	0.270931	0.267629	0.264347	0.261086	0.257846	0.257627	0.251429	0.248252	0.245097
0.5	0.308538	0.305026	0.301532	0.298056	0.294599	0.291160	0.287740	0.284339	0.280957	0.277595
0.4	0.344578	0.340903	0.337243	0.333598	0.329969	0.326355	0.322758	0.319178	0.315614	0.312067
0.3	0.382089	0.378280	0.374484	0.370700	0.366928	0.363169	0.359424	0.355691	0.351973	0.348268
0.2	0.420740	0.416834	0.412936	0.409046	0.405165	0.401294	0.397432	0.393580	0.389739	0.385908
0.1	0.460172	0.456205	0.452242	0.448283	0.444330	0.440382	0.436441	0.432505	0.428576	0.424655
0.0	0.500000	0.496011	0.492022	0.488034	0.484047	0.480061	0.476078	0.472097	0.468119	0.464144

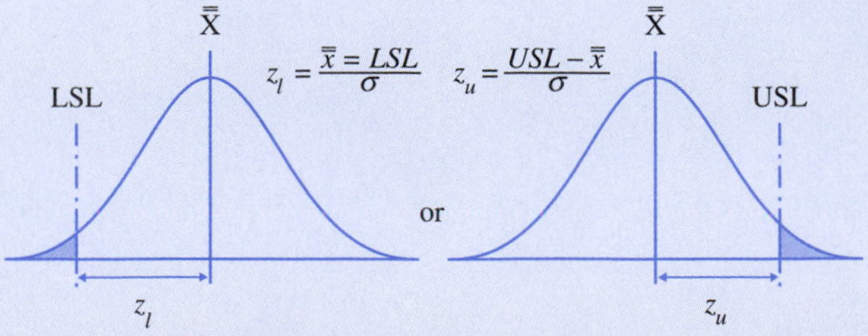

Table III Defects per Million Opportunities

$z = \dfrac{\left\lvert \text{Specification Limit} - \overline{\overline{X}} \right\rvert}{\sigma_x}$	0.00	0.01	0.02	0.03	0.04	0.05	0.06	0.07	0.08	0.09
6.0	0.0008	0.0007	0.0007	0.0006	0.0006	0.0005	0.0005	0.0004	0.0003	0.0002
5.9	0.0017	0.0016	0.0015	0.0015	0.0014	0.0013	0.0012	0.0011	0.0010	0.0009
5.8	0.0032	0.0031	0.0029	0.0027	0.0025	0.0024	0.0022	0.0021	0.0020	0.0018
5.7	0.0060	0.0056	0.0053	0.0050	0.0047	0.0044	0.0041	0.0039	0.0037	0.0034
5.6	0.0107	0.0101	0.0095	0.009	0.0084	0.0080	0.0075	0.0071	0.0067	0.0063
5.5	0.0189	0.0179	0.0169	0.0160	0.0151	0.0142	0.0134	0.0127	0.0120	0.0113
5.4	0.0333	0.0314	0.0297	0.0281	0.0266	0.0251	0.0238	0.0225	0.0212	0.0201
5.3	0.0579	0.0548	0.0518	0.0491	0.0464	0.0439	0.0416	0.0393	0.0372	0.0352
5.2	0.0996	0.0944	0.0894	0.0847	0.0802	0.0760	0.0720	0.0681	0.0645	0.0611
5.1	0.1698	0.1610	0.1527	0.1448	0.1373	0.1302	0.1234	0.1170	0.1109	0.1051
5.0	0.2866	0.2721	0.2583	0.2452	0.2327	0.2209	0.2096	0.1989	0.1887	0.1790
4.9	0.4791	0.4553	0.4327	0.4111	0.3906	0.3710	0.3524	0.3347	0.3179	0.3018
4.8	0.7933	0.7546	0.7177	0.6826	0.6491	0.6172	0.5869	0.5579	0.5304	0.5041
4.7	1.3007	1.2385	1.1792	1.1226	1.0685	1.0170	0.9679	0.9211	0.8764	0.8339
4.6	2.1124	2.0133	1.9187	1.8283	1.7420	1.6596	1.5810	1.5059	1.4343	1.3660
4.5	3.3976	3.2413	3.0919	2.9491	2.8127	2.6822	2.5576	2.4386	2.3248	2.2162
4.4	5.4125	5.1685	4.9350	4.7116	4.4979	4.2935	4.0979	3.9109	3.7321	3.5611
4.3	8.5399	8.1627	7.8014	7.4554	7.1241	6.8068	6.5031	6.2123	5.9339	5.6675
4.2	13.3457	12.7685	12.2151	11.6845	11.1759	10.6885	10.2213	9.7736	9.3446	8.9336
4.1	20.6575	19.7829	18.9436	18.1381	17.3653	16.6237	15.9123	15.2300	14.5754	13.9477

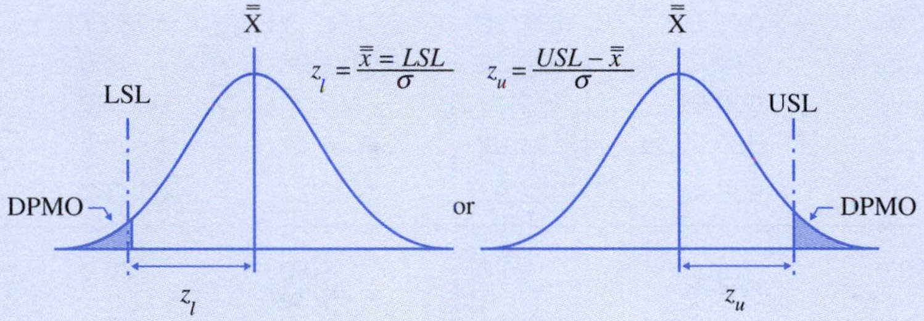

(Continued)

Table III Defects per Million Opportunities *(Continued)*

$z = \dfrac{\lvert \text{Specification Limit} - \bar{\bar{X}} \rvert}{\sigma_x}$	0.00	0.01	0.02	0.03	0.04	0.05	0.06	0.07	0.08	0.09
4.0	31.6712	30.3593	29.0990	27.8884	26.7255	25.6088	24.5363	23.5065	22.5178	21.5686
3.9	48.096	46.148	44.274	42.473	40.741	39.076	37.475	35.936	34.458	33.037
3.8	72.348	69.483	66.726	64.072	61.517	59.059	56.694	54.418	52.228	50.122
3.7	107.800	103.730	99.611	95.740	92.010	88.417	84.957	81.624	78.414	75.324
3.6	159.109	153.099	147.302	141.711	136.319	131.120	126.108	121.275	116.617	112.127
3.5	232.629	224.053	215.773	207.780	200.064	192.616	185.427	178.491	171.797	165.339
3.4	336.929	324.814	313.106	301.791	290.857	280.293	270.088	260.229	250.707	241.51
3.3	483.424	466.480	450.087	434.230	418.892	404.058	389.712	375.841	362.429	349.463
3.2	687.138	663.675	640.953	618.951	597.649	577.025	557.061	537.737	519.035	500.937
3.1	967.603	935.437	904.255	874.032	844.739	816.352	788.846	762.195	736.375	711.364
3.0	1349.898	1306.238	1263.873	1222.769	1182.891	1144.207	1106.685	1070.294	1035.003	1000.782
2.9	1865.81	1807.14	1750.16	1694.81	1641.06	1588.87	1538.20	1489.00	1441.24	1394.89
2.8	2555.13	2477.08	2401.18	2327.40	2255.68	2185.96	2118.21	2052.36	1988.38	1926.21
2.7	3466.98	3364.16	3264.10	3166.72	3071.96	2979.76	2890.07	2802.81	2717.94	2635.40
2.6	4661.19	4527.11	4396.49	4269.24	4145.30	4024.59	3907.03	3792.56	3681.11	3572.60
2.5	6209.67	6036.56	5867.74	5703.13	5542.62	5386.15	5233.61	5084.93	4940.02	4798.80
2.4	8197.5	7976.3	7760.3	7549.4	7343.6	7142.8	6946.9	6755.7	6569.1	6387.2
2.3	10724.1	10444.1	10170.4	9903.1	9641.9	9386.7	9137.5	8894.0	8656.3	8424.2
2.2	13903.4	13552.6	13209.4	12873.7	12545.5	12224.5	11910.6	11603.8	11303.8	11010.7
2.1	17864.4	17429.2	17003.0	16585.8	16177.4	15777.6	15386.3	15003.4	14628.7	14262.1

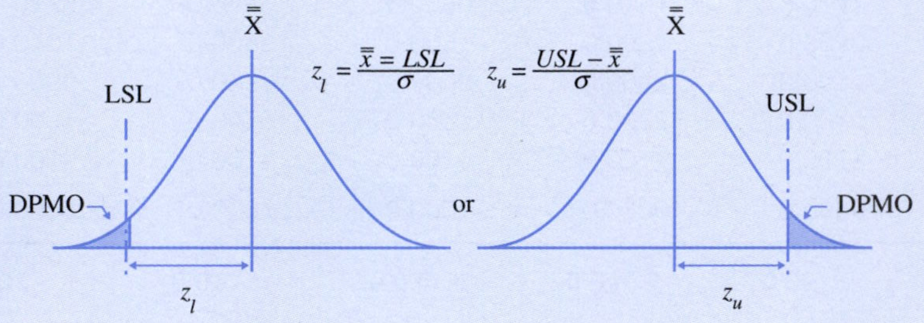

Table III Defects per Million Opportunities

$z = \dfrac{\lvert \text{Specification Limit} - \bar{X} \rvert}{\sigma_x}$	0.00	0.01	0.02	0.03	0.04	0.05	0.06	0.07	0.08	0.09
2.0	22750.1	22215.6	21691.7	21178.3	20675.2	20182.2	19699.3	19226.2	18762.8	18308.9
1.9	28716.6	28066.6	27428.9	26803.4	26189.8	25588.1	24997.9	24419.2	23851.8	23295.5
1.8	35930.3	35147.9	34379.5	33625.0	32884.1	32156.8	31442.8	30741.9	30054.0	29379.0
1.7	44565.5	43632.9	42716.2	41815.1	40929.5	40059.2	39203.9	38363.6	37538.0	36727.0
1.6	54799.3	53698.9	52616.1	51550.7	50502.6	49471.5	48457.2	47459.7	46478.7	45514.0
1.5	66807.2	65521.7	64255.5	63008.4	61780.2	60570.8	59379.9	58207.6	57053.4	55917.4
1.4	80756.7	79269.8	77803.8	76358.5	74933.7	73529.3	72145.0	70780.9	69436.6	68112.1
1.3	96800.5	95097.9	93417.5	91759.1	90122.7	88508.0	86915.0	85343.5	83793.3	82264.4
1.2	115069.7	113139.4	111232.4	109348.6	107487.7	105649.8	103834.7	102042.3	100272.6	98525.3
1.1	135666.1	133499.5	131356.9	129238.1	127143.2	125071.9	123024.4	121000.5	119000.1	117023.2
1.0	158655.2	156247.6	153864.2	151505.0	149170.0	146859.1	144572.3	142309.7	140071.1	137856.6
0.9	184060	181411	65521.7	176186	173609	171056	168528	166023	163543	161087
0.8	211855	208970	206108	203269	200454	197663	194895	192150	189430	186733
0.7	241964	238852	235762	232695	229650	226627	223627	220650	217695	214764
0.6	274253	270931	267629	264347	261086	257846	254627	251429	248252	245097
0.5	308538	305026	301532	298056	294599	291160	287740	284339	280957	277595
0.4	344578	340903	337243	333598	329969	326355	322758	319178	315614	312067
0.3	382089	378280	374484	370700	366928	363169	359424	355691	351973	348268
0.2	420740	416834	412936	409046	405165	401294	397432	393580	389739	385908
0.1	460172	456205	452242	448283	444330	440382	436441	432505	428576	424655
0.0	500000	496011	492022	488034	484047	480061	476078	472097	468119	464144

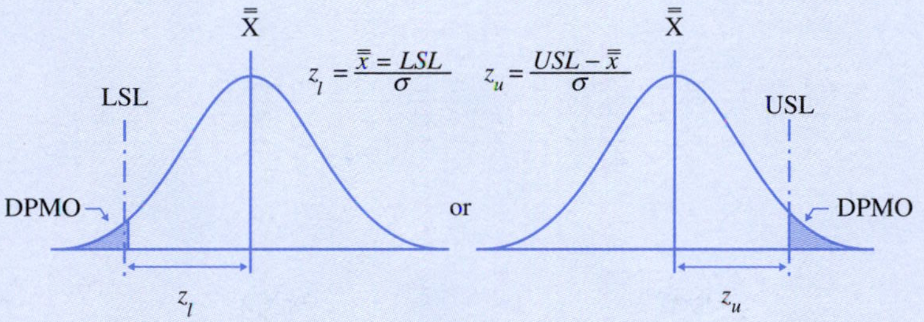

Table IV t Distribution Table

Degrees of Freedom	α						
	0.1	0.05	0.025	0.01	0.005	0.001	0.0005
1	3.078	6.314	12.706	31.821	63.656	318.289	636.578
2	1.886	2.920	4.303	6.965	9.925	22.328	31.600
3	1.638	2.353	3.182	4.541	5.841	10.214	12.924
4	1.533	2.132	2.776	3.747	4.604	7.173	8.610
5	1.476	2.015	2.571	3.365	4.032	5.894	6.869
6	1.440	1.943	2.447	3.143	3.707	5.208	5.959
7	1.415	1.895	2.365	2.998	3.499	4.785	5.408
8	1.397	1.860	2.306	2.896	3.355	4.501	5.041
9	1.383	1.833	2.262	2.821	3.250	4.297	4.781
10	1.372	1.812	2.228	2.764	3.169	4.144	4.587
11	1.363	1.796	2.201	2.718	3.106	4.025	4.437
12	1.356	1.782	2.179	2.681	3.055	3.93	4.318
13	1.350	1.771	2.160	2.650	3.012	3.852	4.221
14	1.345	1.761	2.145	2.624	2.977	3.787	4.140
15	1.341	1.753	2.131	2.602	2.947	3.733	4.073
16	1.337	1.746	2.120	2.583	2.921	3.686	4.015
17	1.333	1.740	2.110	2.567	2.898	3.646	3.965
18	1.330	1.734	2.101	2.552	2.878	3.610	3.922
19	1.328	1.729	2.093	2.539	2.861	3.579	3.883
20	1.325	1.725	2.086	2.528	2.845	3.552	3.850
21	1.323	1.721	2.080	2.518	2.831	3.527	3.819
22	1.321	1.717	2.074	2.508	2.819	3.505	3.792
23	1.319	1.714	2.069	2.500	2.807	3.485	3.768
24	1.318	1.711	2.064	2.492	2.797	3.467	3.745
25	1.316	1.708	2.06	2.485	2.787	3.450	3.725
26	1.315	1.706	2.056	2.479	2.779	3.435	3.707
27	1.314	1.703	2.052	2.473	2.771	3.421	3.689
28	1.313	1.701	2.048	2.467	2.763	3.408	3.674
29	1.311	1.699	2.045	2.462	2.756	3.396	3.66
30	1.310	1.697	2.042	2.457	2.750	3.385	3.646
60	1.296	1.671	2.000	2.390	2.660	3.232	3.460
120	1.289	1.658	1.980	2.358	2.617	3.160	3.373
∞	1.282	1.645	1.960	2.326	2.576	3.091	3.291

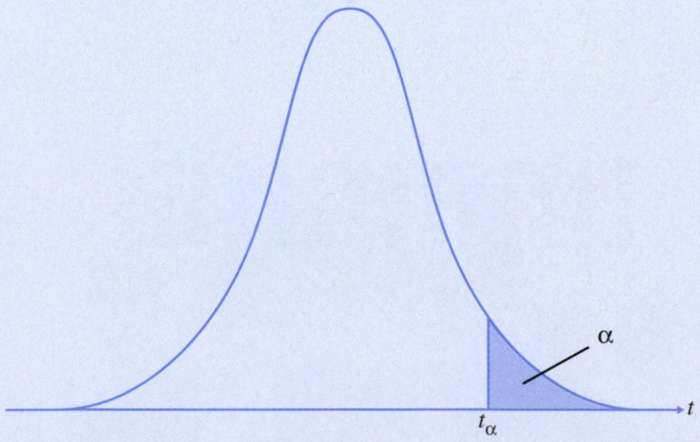

Table V F Distribution Table, $\alpha = 0.05$

Denominator Degrees of Freedom, ν_2	Numerator Degrees of Freedom, ν_1																		
	1	2	3	4	5	6	7	8	9	10	12	15	20	24	30	40	60	120	∞
1	161.45	199.50	215.71	224.58	230.16	233.99	236.77	238.88	240.54	241.88	243.91	245.95	248.01	249.05	250.10	251.14	252.20	253.25	254.31
2	18.513	19.000	19.164	19.247	19.296	19.330	19.353	19.371	19.385	19.396	19.413	19.429	19.446	19.454	19.462	19.471	19.479	19.487	19.496
3	10.128	9.552	9.277	9.117	9.014	8.941	8.887	8.845	8.812	8.786	8.745	8.703	8.660	8.639	8.617	8.594	8.572	8.549	8.526
4	7.709	6.944	6.591	6.388	6.256	6.163	6.094	6.041	5.999	5.964	5.912	5.858	5.803	5.774	5.746	5.717	5.688	5.658	5.628
5	6.608	5.786	5.410	5.192	5.050	4.950	4.876	4.818	4.773	4.735	4.678	4.619	4.558	4.527	4.496	4.464	4.431	4.399	4.365
6	5.987	5.143	4.757	4.534	4.387	4.284	4.207	4.147	4.099	4.060	4.000	3.938	3.874	3.842	3.808	3.774	3.740	3.705	3.669
7	5.591	4.737	4.347	4.120	3.972	3.866	3.787	3.726	3.677	3.637	3.575	3.511	3.445	3.411	3.376	3.340	3.304	3.267	3.230
8	5.318	4.459	4.066	3.838	3.688	3.581	3.501	3.438	3.388	3.347	3.284	3.218	3.150	3.115	3.079	3.043	3.005	2.967	2.928
9	5.117	4.257	3.863	3.633	3.482	3.374	3.293	3.230	3.179	3.137	3.073	3.006	2.937	2.901	2.864	2.826	2.787	2.748	2.707
10	4.965	4.103	3.708	3.478	3.326	3.217	3.136	3.072	3.020	2.978	2.913	2.845	2.774	2.737	2.700	2.661	2.621	2.580	2.538
11	4.844	3.982	3.587	3.357	3.204	3.095	3.012	2.948	2.896	2.854	2.788	2.719	2.646	2.609	2.571	2.531	2.490	2.448	2.405
12	4.747	3.885	3.490	3.259	3.106	2.996	2.913	2.849	2.796	2.753	2.687	2.617	2.544	2.506	2.466	2.426	2.3842	2.3410	2.2962
13	4.667	3.806	3.411	3.179	3.025	2.915	2.832	2.767	2.714	2.671	2.604	2.533	2.459	2.420	2.380	2.339	2.297	2.252	2.206
14	4.600	3.739	3.344	3.112	2.958	2.848	2.764	2.699	2.646	2.602	2.534	2.463	2.388	2.349	2.308	2.266	2.223	2.178	2.131
15	4.543	3.682	3.287	3.056	2.901	2.791	2.707	2.641	2.588	2.544	2.475	2.403	2.328	2.288	2.247	2.204	2.160	2.114	2.066
16	4.494	3.634	3.239	3.007	2.852	2.741	2.657	2.591	2.538	2.494	2.425	2.352	2.276	2.235	2.194	2.151	2.106	2.059	2.010
17	4.451	3.592	3.197	2.965	2.810	2.699	2.614	2.548	2.494	2.450	2.381	2.308	2.230	2.190	2.148	2.104	2.058	2.011	1.960
18	4.414	3.555	3.160	2.928	2.773	2.661	2.577	2.510	2.456	2.412	2.342	2.269	2.191	2.150	2.107	2.063	2.017	1.968	1.917
19	4.381	3.522	3.127	2.895	2.740	2.628	2.544	2.477	2.423	2.378	2.308	2.234	2.156	2.114	2.071	2.026	1.980	1.930	1.878
20	4.351	3.493	3.098	2.866	2.711	2.599	2.514	2.447	2.393	2.348	2.278	2.203	2.124	2.083	2.039	1.994	1.946	1.896	1.8432
21	4.325	3.467	3.073	2.840	2.685	2.573	2.488	2.421	2.366	2.321	2.250	2.176	2.096	2.054	2.010	1.965	1.917	1.866	1.812
22	4.301	3.443	3.049	2.817	2.661	2.549	2.464	2.397	2.342	2.297	2.226	2.151	2.071	2.028	1.984	1.938	1.889	1.838	1.783
23	4.279	3.422	3.028	2.796	2.640	2.528	2.442	2.375	2.320	2.275	2.204	2.128	2.048	2.005	1.961	1.914	1.865	1.813	1.757
24	4.260	3.403	3.009	2.776	2.621	2.508	2.423	2.355	2.300	2.255	2.183	2.108	2.027	1.984	1.939	1.892	1.842	1.790	1.733
25	4.242	3.385	2.991	2.759	2.603	2.490	2.405	2.337	2.282	2.237	2.165	2.089	2.008	1.964	1.919	1.872	1.822	1.768	1.711
26	4.225	3.369	2.975	2.743	2.587	2.474	2.388	2.321	2.266	2.220	2.148	2.072	1.990	1.946	1.901	1.853	1.803	1.749	1.691
27	4.210	3.354	2.960	2.728	2.572	2.459	2.373	2.305	2.250	2.204	2.132	2.056	1.974	1.930	1.884	1.836	1.785	1.731	1.672
28	4.196	3.340	2.947	2.714	2.558	2.445	2.359	2.291	2.236	2.190	2.118	2.041	1.959	1.915	1.869	1.820	1.769	1.714	1.654
29	4.183	3.328	2.934	2.701	2.545	2.432	2.346	2.278	2.223	2.177	2.105	2.028	1.945	1.901	1.854	1.806	1.754	1.698	1.638
30	4.171	3.316	2.922	2.690	2.534	2.421	2.334	2.266	2.211	2.165	2.092	2.015	1.932	1.887	1.841	1.792	1.740	1.684	1.622
40	4.085	3.232	2.839	2.606	2.450	2.336	2.249	2.180	2.124	2.077	2.004	1.925	1.839	1.793	1.744	1.693	1.637	1.577	1.509
60	4.001	3.150	2.758	2.525	2.368	2.254	2.167	2.097	2.040	1.993	1.917	1.836	1.748	1.700	1.649	1.594	1.534	1.467	1.389
120	3.920	3.072	2.680	2.447	2.290	2.175	2.087	2.016	1.959	1.911	1.834	1.751	1.659	1.608	1.554	1.495	1.429	1.352	1.254
∞	3.842	2.996	2.605	2.372	2.214	2.099	2.010	1.938	1.880	1.831	1.752	1.666	1.571	1.517	1.459	1.394	1.318	1.221	1.000

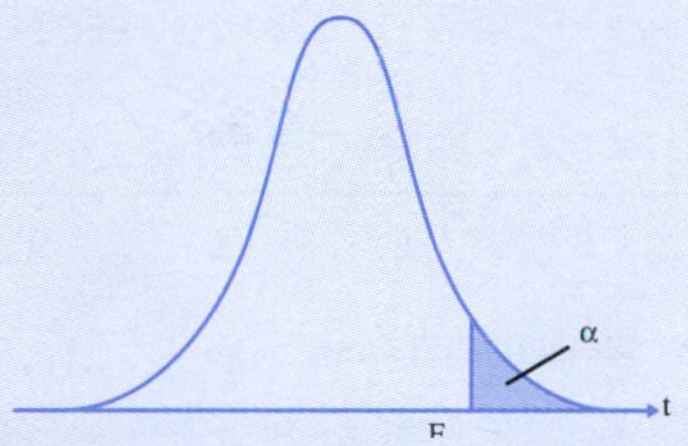

(Continued)

Table V F Distribution Table, $\alpha = 0.25$ *(Continued)*

Denominator Degrees of Freedom, ν_2	Numerator Degrees of Freedom, ν_1: 1	2	3	4	5	6	7	8	9	10	12	15	20	24	30	40	60	120	∞
1	647.79	799.50	864.16	899.58	921.85	937.11	948.22	956.66	963.28	968.63	976.71	984.87	993.10	997.25	1001.41	1005.60	1009.80	1014.02	1018.26
2	38.506	39.000	39.166	39.248	39.298	39.332	39.355	39.373	39.387	39.398	39.415	39.431	39.448	39.456	39.465	39.473	39.481	39.490	39.498
3	17.443	16.044	15.439	15.101	14.885	14.735	14.624	14.540	14.473	14.419	14.337	14.253	14.167	14.124	14.081	14.037	13.992	13.947	13.902
4	12.218	10.649	9.979	9.605	9.365	9.197	9.074	8.980	8.905	8.844	8.751	8.657	8.560	8.511	8.461	8.411	8.360	8.309	8.257
5	10.007	8.434	7.764	7.388	7.146	6.978	6.853	6.757	6.681	6.619	6.525	6.428	6.329	6.278	6.227	6.175	6.123	6.069	6.015
6	8.813	7.260	6.599	6.227	5.988	5.820	5.696	5.600	5.523	5.461	5.366	5.269	5.168	5.117	5.065	5.012	4.959	4.904	4.849
7	8.073	6.542	5.890	5.523	5.285	5.119	4.995	4.899	4.823	4.761	4.666	4.568	4.467	4.415	4.362	4.309	4.254	4.199	4.142
8	7.571	6.060	5.416	5.053	4.817	4.652	4.529	4.433	4.357	4.295	4.200	4.101	4.000	3.947	3.894	3.840	3.784	3.728	3.670
9	7.209	5.715	5.078	4.718	4.484	4.320	4.197	4.102	4.026	3.964	3.868	3.769	3.667	3.614	3.560	3.505	3.449	3.392	3.333
10	6.937	5.456	4.826	4.468	4.236	4.072	3.950	3.855	3.779	3.717	3.621	3.522	3.419	3.365	3.311	3.255	3.198	3.140	3.080
11	6.724	5.256	4.630	4.275	4.044	3.881	3.759	3.664	3.588	3.526	3.430	3.330	3.226	3.173	3.118	3.061	3.004	2.944	2.883
12	6.554	5.096	4.474	4.121	3.891	3.728	3.607	3.512	3.436	3.374	3.277	3.177	3.073	3.019	2.963	2.906	2.848	2.787	2.725
13	6.414	4.965	4.347	3.996	3.767	3.604	3.483	3.388	3.312	3.250	3.153	3.053	2.948	2.893	2.837	2.780	2.720	2.659	2.595
14	6.298	4.857	4.242	3.892	3.663	3.501	3.380	3.285	3.209	3.147	3.050	2.949	2.844	2.789	2.732	2.674	2.614	2.552	2.487
15	6.200	4.765	4.153	3.804	3.576	3.415	3.293	3.199	3.123	3.060	2.963	2.862	2.756	2.701	2.644	2.585	2.524	2.461	2.3950
16	6.115	4.687	4.077	3.729	3.502	3.341	3.219	3.125	3.049	2.986	2.889	2.788	2.681	2.625	2.568	2.509	2.447	2.383	2.316
17	6.042	4.619	4.011	3.665	3.438	3.277	3.156	3.061	2.985	2.922	2.825	2.723	2.616	2.560	2.502	2.442	2.380	2.315	2.247
18	5.978	4.560	3.954	3.608	3.382	3.221	3.100	3.005	2.929	2.866	2.769	2.667	2.559	2.503	2.445	2.384	2.321	2.256	2.187
19	5.922	4.508	3.903	3.559	3.333	3.172	3.051	2.956	2.880	2.817	2.720	2.617	2.509	2.452	2.394	2.333	2.270	2.203	2.133
20	5.872	4.461	3.859	3.515	3.289	3.128	3.007	2.913	2.837	2.774	2.676	2.573	2.465	2.408	2.349	2.287	2.223	2.156	2.085
21	5.827	4.420	3.819	3.475	3.250	3.090	2.969	2.874	2.798	2.735	2.637	2.534	2.425	2.368	2.308	2.246	2.182	2.114	2.042
22	5.786	4.383	3.783	3.440	3.215	3.055	2.934	2.839	2.763	2.700	2.602	2.498	2.389	2.332	2.272	2.210	2.145	2.076	2.003
23	5.750	4.349	3.751	3.408	3.184	3.023	2.902	2.808	2.731	2.668	2.570	2.467	2.357	2.299	2.239	2.176	2.111	2.041	1.968
24	5.717	4.319	3.721	3.379	3.155	2.995	2.874	2.779	2.703	2.640	2.541	2.437	2.327	2.269	2.209	2.146	2.080	2.010	1.935
25	5.686	4.291	3.694	3.353	3.129	2.969	2.848	2.753	2.677	2.614	2.515	2.411	2.301	2.242	2.182	2.118	2.052	1.981	1.9060
26	5.659	4.266	3.670	3.329	3.105	2.945	2.824	2.729	2.653	2.590	2.491	2.387	2.276	2.217	2.157	2.093	2.026	1.954	1.878
27	5.633	4.242	3.647	3.307	3.083	2.923	2.802	2.707	2.631	2.568	2.469	2.364	2.253	2.195	2.133	2.069	2.002	1.930	1.853
28	5.610	4.221	3.626	3.286	3.063	2.903	2.782	2.687	2.611	2.547	2.448	2.344	2.232	2.174	2.112	2.048	1.980	1.907	1.829
29	5.588	4.201	3.607	3.267	3.044	2.884	2.763	2.669	2.592	2.529	2.430	2.325	2.213	2.154	2.092	2.028	1.959	1.886	1.807
30	5.568	4.182	3.589	3.250	3.027	2.867	2.746	2.651	2.575	2.511	2.412	2.307	2.195	2.136	2.074	2.009	1.940	1.866	1.787
40	5.424	4.051	3.463	3.126	2.904	2.744	2.624	2.529	2.452	2.388	2.288	2.182	2.068	2.007	1.943	1.875	1.803	1.724	1.637
60	5.286	3.925	3.343	3.008	2.786	2.627	2.507	2.412	2.334	2.270	2.169	2.061	1.945	1.882	1.815	1.744	1.667	1.581	1.482
120	5.152	3.805	3.227	2.894	2.674	2.515	2.395	2.299	2.222	2.157	2.055	1.945	1.825	1.760	1.690	1.614	1.530	1.433	1.310
∞	5.024	3.689	3.116	2.786	2.567	2.408	2.288	2.192	2.114	2.048	1.945	1.833	1.709	1.640	1.566	1.484	1.388	1.268	1.000

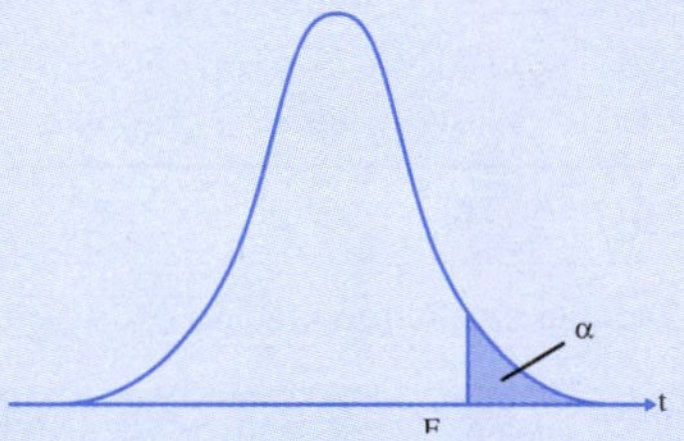

Table V F Distribution Table, $\alpha = 0.01$

Denominator Degrees of Freedom, ν_2	Numerator Degrees of Freedom, ν_1: 1	2	3	4	5	6	7	8	9	10	12	15	20	24	30	40	60	120	∞∝
1	4052.18	4999.50	5403.35	5624.58	5763.65	5858.99	5928.36	5981.07	6022.47	6055.85	6106.32	6157.29	6208.73	6234.63	6260.65	6286.78	6313.03	6339.39	6365.86
2	98.503	99.000	99.166	99.249	99.299	99.333	99.356	99.374	99.388	99.399	99.416	99.433	99.449	99.458	99.466	99.474	99.482	99.491	99.499
3	34.116	30.817	29.457	28.710	28.237	27.911	27.672	27.489	27.345	27.229	27.052	26.872	26.690	26.598	26.505	26.411	26.316	26.221	26.125
4	21.198	18.000	16.694	15.977	15.522	15.207	14.976	14.799	14.659	14.546	14.374	14.198	14.020	13.929	13.838	13.745	13.652	13.558	13.463
5	16.258	13.274	12.060	11.392	10.967	10.672	10.456	10.289	10.158	10.051	9.888	9.722	9.553	9.466	9.379	9.291	9.202	9.112	9.020
6	13.745	10.925	9.780	9.148	8.746	8.466	8.260	8.102	7.976	7.874	7.718	7.559	7.396	7.313	7.229	7.143	7.057	6.969	6.880
7	12.246	9.547	8.451	7.847	7.460	7.191	6.993	6.840	6.719	6.620	6.469	6.314	6.155	6.074	5.992	5.908	5.824	5.737	5.650
8	11.259	8.649	7.591	7.006	6.632	6.371	6.178	6.029	5.911	5.814	5.667	5.515	5.359	5.279	5.198	5.116	5.032	4.946	4.859
9	10.561	8.022	6.992	6.422	6.057	5.802	5.613	5.467	5.351	5.257	5.111	4.962	4.808	4.729	4.649	4.567	4.483	4.398	4.311
10	10.044	7.559	6.552	5.994	5.636	5.386	5.200	5.057	4.942	4.849	4.706	4.558	4.405	4.327	4.247	4.165	4.082	3.996	3.909
11	9.646	7.206	6.217	5.668	5.316	5.069	4.886	4.744	4.632	4.539	4.397	4.251	4.099	4.021	3.941	3.860	3.776	3.690	3.602
12	9.330	6.927	5.953	5.412	5.064	4.821	4.640	4.499	4.388	4.296	4.155	4.010	3.858	3.780	3.701	3.619	3.535	3.449	3.361
13	9.074	6.701	5.739	5.205	4.862	4.620	4.441	4.302	4.191	4.100	3.960	3.815	3.665	3.587	3.507	3.425	3.341	3.255	3.165
14	8.862	6.515	5.564	5.035	4.695	4.456	4.278	4.140	4.030	3.939	3.800	3.656	3.505	3.427	3.348	3.266	3.181	3.094	3.004
15	8.683	6.359	5.417	4.893	4.556	4.318	4.142	4.004	3.895	3.805	3.666	3.522	3.372	3.294	3.214	3.132	3.047	2.959	2.868
16	8.531	6.226	5.292	4.773	4.437	4.202	4.026	3.890	3.780	3.691	3.553	3.409	3.259	3.181	3.101	3.018	2.933	2.845	2.753
17	8.400	6.112	5.185	4.669	4.336	4.102	3.927	3.791	3.682	3.593	3.455	3.312	3.162	3.084	3.003	2.920	2.835	2.746	2.653
18	8.285	6.013	5.092	4.579	4.248	4.015	3.841	3.705	3.597	3.508	3.371	3.227	3.077	2.999	2.919	2.835	2.749	2.660	2.566
19	8.185	5.926	5.010	4.500	4.171	3.939	3.765	3.631	3.523	3.434	3.297	3.153	3.003	2.925	2.844	2.761	2.674	2.584	2.489
20	8.096	5.849	4.938	4.431	4.103	3.871	3.699	3.564	3.457	3.368	3.231	3.088	2.938	2.859	2.778	2.695	2.608	2.517	2.421
21	8.017	5.780	4.874	4.369	4.042	3.812	3.640	3.506	3.398	3.310	3.173	3.030	2.880	2.801	2.720	2.636	2.548	2.457	2.360
22	7.945	5.719	4.817	4.313	3.988	3.758	3.587	3.453	3.346	3.258	3.121	2.978	2.827	2.749	2.667	2.583	2.495	2.403	2.305
23	7.881	5.664	4.765	4.264	3.939	3.710	3.539	3.406	3.299	3.211	3.074	2.931	2.781	2.702	2.620	2.535	2.447	2.354	2.256
24	7.823	5.614	4.718	4.218	3.895	3.667	3.496	3.363	3.256	3.168	3.032	2.889	2.738	2.659	2.577	2.492	2.403	2.310	2.211
25	7.770	5.568	4.675	4.177	3.855	3.627	3.457	3.324	3.217	3.129	2.993	2.850	2.699	2.620	2.538	2.453	2.364	2.270	2.169
26	7.721	5.526	4.637	4.140	3.818	3.591	3.421	3.288	3.182	3.094	2.958	2.815	2.664	2.585	2.503	2.417	2.327	2.233	2.131
27	7.677	5.488	4.601	4.106	3.785	3.558	3.388	3.256	3.149	3.062	2.926	2.783	2.632	2.552	2.470	2.384	2.294	2.198	2.097
28	7.636	5.453	4.568	4.074	3.754	3.528	3.358	3.226	3.120	3.032	2.896	2.753	2.602	2.522	2.440	2.354	2.263	2.167	2.064
29	7.598	5.420	4.538	4.045	3.725	3.499	3.330	3.198	3.092	3.005	2.868	2.726	2.574	2.495	2.412	2.325	2.234	2.138	2.034
30	7.562	5.390	4.510	4.018	3.699	3.473	3.304	3.173	3.067	2.979	2.843	2.700	2.549	2.469	2.386	2.299	2.208	2.111	2.006
40	7.314	5.179	4.313	3.828	3.514	3.291	3.124	2.993	2.888	2.801	2.665	2.522	2.369	2.288	2.203	2.114	2.019	1.917	1.805
60	7.077	4.977	4.126	3.649	3.339	3.119	2.953	2.823	2.718	2.632	2.496	2.352	2.198	2.115	2.028	1.936	1.836	1.726	1.601
120	6.851	4.787	3.949	3.480	3.174	2.956	2.792	2.663	2.559	2.472	2.336	2.192	2.035	1.950	1.860	1.763	1.656	1.533	1.381
∞	6.635	4.605	3.782	3.319	3.017	2.802	2.639	2.511	2.407	2.321	2.185	2.039	1.878	1.791	1.696	1.592	1.473	1.325	1.000

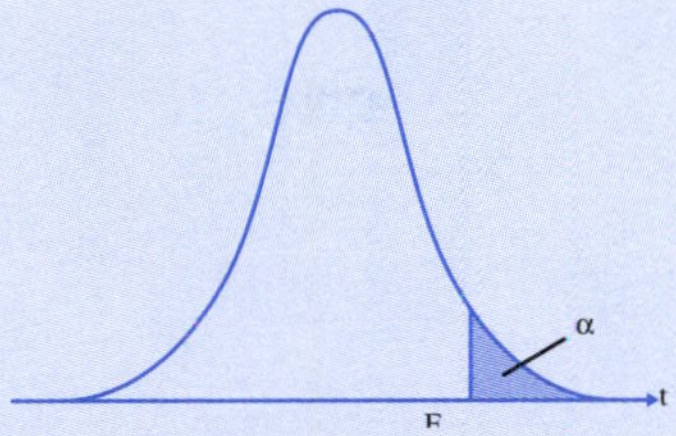

Table VI Critical Values for Too Few Runs

r	s = 5	6	7	8	9	10	11	12	13	14	15	16	17	18	19	20
5	3															
6	3	3														
7	3	4	4													
8	3	4	4	5												
9	4	4	5	5	6											
10	4	5	5	6	6	6										
11	4	5	5	6	6	7	7									
12	4	5	6	6	7	7	8	8								
13	4	5	6	6	7	8	8	9	9							
14	5	5	6	7	7	8	8	9	9	10						
15	5	6	6	7	8	8	9	9	10	10	11					
16	5	6	6	7	8	8	9	10	10	11	11	11				
17	5	6	7	7	8	9	9	10	10	11	11	12	12			
18	5	6	7	8	8	9	10	10	11	11	12	12	13	13		
19	5	6	7	8	8	9	10	10	11	12	12	13	13	14	14	
20	5	6	7	8	9	9	10	11	11	12	12	13	13	14	14	15

Source: Adapted from Swed, F. S., and Eisenhart C. "Tables for Testing Randomness of Grouping in a Sequence of Alternatives." *Annals of Mathematical Statistics* 14, (1943): 66–87.
The probability of an equal or smaller number of runs than the critical value is less than or equal to .05 when the process is in control.

Table VII Critical Values for Longest Run

r	s = 5	6	7	8	9	10	11	12	13	14	15	16	17	18	19	20
5	5															
6	6	6														
7	6	6	6													
8	7	7	7	7												
9	8	7	7	7	7											
10	8	8	7	7	7	7										
11	9	8	8	8	7	7	7									
12	10	9	9	8	8	8	8	8								
13	10	10	9	9	8	8	8	8	8							
14	11	10	10	9	9	8	8	8	8	8						
15	11	11	10	10	9	9	9	8	8	8	8					
16	12	11	11	10	10	9	9	9	9	8	8	8				
17	13	12	11	11	10	10	9	9	9	9	8	8	8			
18	13	12	12	11	11	10	10	9	9	9	9	9	9	9		
19	14	13	12	12	11	11	10	10	9	9	9	9	9	9	9	
20	15	14	13	12	11	11	10	10	10	9	9	9	9	9	9	9

Source: Adapted from Takashima, M. "Tables for Testing Randomness by Means of Length of Runs." *Bulletin of Mathematical Statistics* 6 (1955): 17–23.

APPENDIX C

CERTIFIED SIX SIGMA GREEN BELT BODY OF KNOWLEDGE

Section I: Overview—Six Sigma and the Organization—15 Questions (15% of the test)

Subsection	Cognitive Level	Content	Chapter Reference
Six Sigma and Organization Goals	**Understand**	Value of Six Sigma to organizations, history of the quality movement, how an organization operates as an open system with active feedback loops that provide it with the information it needs for sustainability.	**Chapter 1**
	Understand	Be able to identify an organization's key performance indicators—those that drive profit, market share, customer satisfaction, efficiency, product differentiation—and be able to know how key metrics are developed and how they impact the organization.	**Chapter 2**
	Understand	Be able to describe the process that an organization uses to select Six Sigma projects, determine the appropriateness of the DMAIC process compared to other methodologies, and confirm that a particular project supports organizational goals.	**Chapter 14**
Lean Principles	**Understand**	Be able to define and describe lean concepts (such as value stream, flow, pull scheduling, kanban) and a variety of tools that are widely used to eliminate sources of waste (such as kaizen, 5S, poka-yoke, value stream mapping).	**Chapter 11**
	Understand	Know how to differentiate between value-added and non-value-added activities, including excess inventories, space, test inspections, rework, transportation, storage; and the value of reduced cycle time to improve throughput.	**Chapter 5 and Chapter 6**
	Understand	Be able to describe the theory of constraints.	**Chapter 14**
Design for Six Sigma in the Organization	**Understand**	Be able to describe quality function deployment (QFD).	**Chapter 5**
	Analyze	Be able to perform FMEAs and distinguish between a design FMEA (DFMEA) and a process FMEA (PFMEA).	**Chapter 6**
	Understand	Be able to describe design for Six Sigma (DFSS) methodologies and how they relate to DMAIC.	**Chapter 14**

CERTIFIED SIX SIGMA GREEN BELT BODY OF KNOWLEDGE

Section II: Six Sigma—Define—25 Questions (25% of the test)

Subsection	Cognitive Level	Content	Chapter Reference
Process Management for Projects	**Analyze**	Be able to define and describe process components and boundaries and to identify process improvement efforts	**Chapters 2 and 6**
	Apply	Be able to identify process owners, identify and classify internal and external customers, identify other stakeholders, be able to show how projects will impact all customers, and be able to effectively collect quantitative and qualitative customer data.	**Chapters 2, 7, 9, and 14**
	Analyze	Be able to use graphical, statistical, and quantitative tools to analyze customer feedback.	**Chapters 7–13**
	Apply	Be able to assist in translating customer feedback into project goals, identifying critical to quality (CTQ) characteristics, and using QFD to translate customer requirements into performance measures.	**Chapters 2 and 5**
Project Management Basics	**Apply**	Be able to develop a project charter, including a problem statement; assist in the development of project scope using tools such as Pareto charts and process maps; assist with the development of process metrics such as cycle time, quality, and cost; establish key metrics that relate to the voice of the customer; use planning tools such as Gantt charts, critical path and PERT network techniques; and be able to provide input and select the proper vehicle for presenting project documentation.	**Chapters 7 and 14**
	Understand	Be able to describe the purpose and benefit of a project risk analysis	**Chapter 14**
	Apply	Be able to describe achievements and apply project learnings to identify other improvement opportunities.	**Chapter 14**
Management and Planning Tools	**Apply**	Be able to define, select, and use the following tools: affinity diagrams, interrelationship digraphs, tree diagrams, prioritization matrices, matrix diagrams, process decision program charts (PDPC), and activity network diagrams.	**Chapter 7**
Business Results for Projects	**Analyze**	Be able to calculate performance metrics including defects per unit (DPU), rolled throughput yield (RTY), cost of quality (COQ), defects per million opportunities (DPMO), process variances, and process capability indices; and define and describe the failure mode and effects analysis (FMEA) methodology including the ability to calculate risk priority numbers (RPN).	**Chapters 5–6, 8–12 and 14**
Team Dynamics and Performance	**Understand**	Be able to define and describe the process of team building and the stages of team evolution.	**Chapter 3**
	Apply	Be able to describe and define the roles and responsibilities of participants on Six Sigma teams, including black belts, master black belts, green belts, champions, executives, coaches, facilitators, team members, sponsors, and process owners; define and apply creative tools such as brainstorming, nominal group technique (NGT), and multivoting; and use effective and appropriate communication techniques to overcome any barriers to project success.	**Chapters 7 and 14**

CERTIFIED SIX SIGMA GREEN BELT BODY OF KNOWLEDGE

Section III: Six Sigma—Measure—30 Questions (30% of the test)

Subsection	Cognitive Level	Content	Chapter Reference
Process Analysis and Documentation	**Analyze**	Be able to review process maps, written procedures, work instructions, flowcharts, etc., be able to identify process input and output variables using a SIPOC chart, and to document the relationships of these variables using cause and effect diagrams, relational matrices, etc.	**Chapters 7 and 14**
Probability and Statistics	**Apply**	Be able to distinguish between enumerative (descriptive) and analytical (inferential) studies, and distinguish between a population parameter and a sample statistic; be able to define the central limit theorem and its significance in inferential statistics and the construction of control charts.	**Chapters 1 and 8**
	Apply	Be able to describe and apply basic probability concepts such as independence, mutual exclusivity, multiplication rules, etc.	**Implicit in Chapters 8–14**
Collecting and Summarizing Data	**Analyze**	Be able to identify and classify continuous (variables) and discrete (attributes) data; and to describe and define nominal, ordinal, interval and ratio measurement scales.	**Chapters 8 and 10**
	Apply	Be able to define and apply methods for collecting data such as check sheets, coded data, etc.; define and apply various sampling techniques including random sampling, stratefied sampling, sample homogeneity, etc.	**Chapters 7–10**
	Analyze	Be able to define, compute, and interpret measures of dispersion and central tendency, and construct and interpret frequency distributions and cumulative frequency distributions.	**Chapters 7–10**
	Create	Be able to depict relationships by constructing, applying, and interpreting diagrams and charts such as stem-and-leaf plots, box-and-whisker plots, run charts, scatter diagrams, Pareto charts, etc.; be able to depict distributions by constructing, applying and interpreting diagrams such as histograms, normal probability plots, etc.	**Chapters 8–14**
Probability Distributions	**Apply**	Be able to describe and interpret normal, binomial, Poisson, chi square, Student's t, and F distributions.	**Chapters 8–13**
Measurement System Analysis	**Evaluate**	Be able to calculate, analyze, and interpret measurement system capability using repeatability and reproducibility (gage R&R), measurement correlation, bias, linearity, percent agreement, and precision/tolerance.	**Chapter 9**
Process Capability and Performance	**Evaluate**	Be able to identify, describe, and apply the elements of designing and conducting process capability studies, including identifying characteristics, identifying specifications and tolerances, developing sampling plans, and verifying stability and normality; distinguish between natural process tolerance and engineering tolerance, and calculate process performance metrics such as percent defectives.	**Chapters 4 and 9**
	Evaluate	Be able to define, select, and calculate capability indices to assess process capability. These indices include C_p, C_{pk}, P_p, P_{pk}, and C_{pm}.	**Chapter 9**
	Evaluate	Be able to describe the assumptions and conventions that are appropriate when only short-term data are collected and when only attributes data are available; be able to describe the changes in relationships that occur when long-term data are used, and interpret the relationships between long- and short-term capability as it relates to a 1.5 sigma shift.	**Chapter 14**
	Apply	Be able to describe the relationship between attributes data and P_p and P_{pk} indices.	**Chapters 9 and 10**

CERTIFIED SIX SIGMA GREEN BELT BODY OF KNOWLEDGE

Section IV: Analyze—15 Questions (15% or the test)

Subsection	Cognitive Level	Content	Chapter Reference
Exploratory Data Analysis	**Create**	Be able to create and interpret multivar studies to interpret the difference between positional, cyclical, and temporal variation; to apply sampling plans to investigate the largest sources of variation.	**Chapter 9**
	Evaluate	Be able to interpret the correlation coefficient and determine its statistical significance (*p* value); be able to recognize the difference between correlation and causation; be able to interpret the linear regression equation and determine its statistical significance (*p* value); and be able to use regression models for estimation and prediction.	**Chapter 13**
Hypothesis Testing	**Apply**	Be able to define and distinguish between statistical and practical significance and apply tests for significance level, power, Type I and Type II errors; be able to determine appropriate sample size for various tests.	**Chapter 13 and Implicit in Chapters 7–10**
	Apply	Be able to describe and apply basic probability concepts such as independence, mutual exclusivity, multiplication rules, etc.; be able to define, compare, and contrast statistical and practical significance in statistical tests concerning means, variances, and proportions.	**Implicit in Chapters 8–14**
	Understand	Be able to define and describe paired-comparison parametric hypothesis tests.	**Implicit in Chapters 8–14**
	Apply	Be able to define terms related to one-way ANOVA and interpret their results and data plots.	**Chapter 13**
	Analyze	Be able to define and interpret chi square and use it to determine statististical significance.	**Implicit in Chapter 13**

Section V: Six Sigma—Improve and Control—15 Questions (15% of the test)

Subsection	Cognitive Level	Content	Chapter Reference
Design of Experiments (DOE)	**Understand**	Be able to define and describe basic DOE terms such as independent and dependent variables, factors and levels, response, treatment, error, repetition, and replication.	**Chapter 13**
	Apply	Be able to interpret main effects and interaction plots.	**Chapter 13**
Statistical Process Control	**Analyze**	Be able to describe the objectives and benefits of SPC, including controlling process performance, identifying special and common causes, etc.	**Chapter 7**
	Understand	Be able to define and describe how rational subgrouping is used.	**Chapters 8 and 9**
	Apply	Be able to identify, select, construct, and apply the following types of charts: *x*-bar/*R*, *x*-bar/*s*, individuals and moving range, median, *p*, *np*, *c*, and *u*.	**Chapters 8–10**
	Analyze	Be able to interpret control charts and distinguish between common and special causes using rules for determining statistical control.	**Chapters 7–11**
Implement and Validate Solutions	**Create**	Be able to use various improvement methods such as brainstorming, main effects analysis, multivari studies, FMEA, and measurement system capability analysis to identify, implement, and validate solutions through *F* test, *t* test, etc.	**Chapters 6, 7, 9, and 13**
Control Plan	**Apply**	Be able to assist in developing a control plan to document and hold the gains, and assist in implementing controls and monitoring systems.	**Chapters 8–14**

CERTIFIED SIX SIGMA BLACK BELT BODY OF KNOWLEDGE

Section I: Enterprise-Wide Deployment—9 Questions (9% of the test)

Subsection	Cognitive Level	Content	Chapter Reference
Enterprise-Wide View	**Remember**	Be able to describe the origins of continuous improvement and its impact on other improvement models.	**Chapter 1**
	Understand	Be able to describe the value of Six Sigma, its philosophy, history, and goals; describe the value of lean production, its philosophy, history, and goals; and describe the relationship between Six Sigma and lean production.	**Chapters 6, 11 and 14**
	Understand	Be able to describe the relationship among various business processes (design, production, purchasing, accounting, sales, etc.) and the impact these relationships can have on business systems.	**Chapters 2–6**
	Understand	Be able to describe how Six Sigma and lean production tools are applied to processes in all types of enterprises: manufacturing, service, transactional, product and process design, innovation, etc.	**Chapters 7–14**
Leadership	**Understand**	Be able to describe the responsibilities of executive leaders and how they affect the deployment of Six Sigma in terms of providing resources, managing change, communicating ideas, etc.	**Chapters 2 and 14**
	Apply	Be able to describe the impact an organization's culture and inherent structure can have on the success of Six Sigma, and how deployment failure can result from the lack of resources, management support, etc.; be able to identify and apply various techniques to overcome these barriers; be able to describe and use various techniques for facilitating and managing organizational change.	**Chapters 2 and 3**
	Apply	Be able to describe how projects and kaizen events are selected, when to use Six Sigma instead of other problem-solving approaches, and the importance of aligning Six Sigma objectives with organizational goals.	**Chapters 11–14**
	Understand	Be able to describe the roles and responsibilities of Six Sigma participants: black belt, master black belt, green belt, champion, process owners and project sponsors.	**Chapter 14**

Section II: Organizational Process Management and Measures—9 Questions (9% of the test)

Subsection	Cognitive Level	Content	Chapter Reference
Impact on Stakeholders	**Understand**	Be able to describe the impact Six Sigma projects can have on customers, suppliers and other stakeholders.	**Chapters 1–14**
Critical to *x* (CT*x*) Requirements	**Apply**	Be able to define and describe various CT*x* requirements critical to quality (CTQ), cost (CTC), process (CTP), safety (CTS), delivery (CTD), etc. and the importance of aligning projects with those requirements.	**Chapters 2 and 14**
Benchmarking	**Apply**	Be able to define and distinguish between various types of benchmarking, including best practices, competitive, collaborative, etc.	**Chapter 2**
Business Performance Measures	**Understand**	Be able to define and describe various business performance measures, including balanced scorecard, key performance indicators (KPIs), the financial impact of customer loyalty, etc.	**Chapter 2**
Financial Measures	**Apply**	Be able to define and use financial measures, including revenue growth, market share, margin, cost of quality (COQ), net present value (NPV), return on investment (ROI), cost–benefit analysis, etc.	**Chapters 2 and 14**

CERTIFIED SIX SIGMA BLACK BELT BODY OF KNOWLEDGE

Section III: Team Management—16 Questions (16% of the test)

Subsection	Cognitive Level	Content	Chapter Reference
Team Formation	**Apply**	Be able to define and describe various types of teams (e.g., formal, informal, virtual, crossfunctional, self-directed, etc.), and determine what team model will work best for a given situation; be able to identify constraining factors including geography, technology, schedules, etc.	**Chapter 3**
	Understand	Be able to define and describe various team roles and responsibilities, including leader, facilitator, coach, individual member, etc.	**Chapters 3, 11, and 14**
	Apply	Be able to define and describe various factors that influence the selection of team members, including required skills sets, subject matter expertise, availability, etc.; be able to identify and describe the elements required for launching a team, including having management support, establishing clear goals, ground rules and timelines, and how these elements can affect the team's success.	**Chapters 3 and 14**
Team Facilitation	**Apply**	Be able to describe and apply techniques that motivate team members and support and sustain their participation and commitment; be able to facilitate the team through the classic stages of development: forming, storming, norming, performing and adjourning; be able to identify and use appropriate communication methods (both within the team and from the team to various stakeholders) to report progress, conduct milestone reviews and support the overall success of the project.	**Chapter 3**
Team Dynamics	**Evaluate**	Be able to identify and use various techniques (e.g., coaching, mentoring, intervention, etc.) to overcome various group dynamic challenges, including overbearing/dominant or reluctant participants, feuding and other forms of unproductive disagreement, unquestioned acceptance of opinions as facts, groupthink, floundering, rushing to accomplish or finish, digressions, tangents, etc.	**Chapter 3**
Time Management for Teams	**Apply**	Be able to select and use various time management techniques including publishing agendas with time limits on each entry, adhering to the agenda, requiring prework by attendees, ensuring that the right people and resources are available, etc.	**Chapter 3**
Team Decision-Making Tools	**Apply**	Be able to define, select and use tools such as brainstorming, nominal group technique, multivoting, etc.	**Chapter 7**
Management and Planning Tools	**Apply**	Be able to define, select and apply the following tools: affinity diagrams, tree diagrams, process decision program charts (PDPC), matrix diagrams, interrelationship digraphs, prioritization matrices and activity network diagrams.	**Chapter 7**
Team Performance Evaluation and Reward	**Analyze**	Be able to measure team progress in relation to goals, objectives and other metrics that support team success and reward and recognize the team for its accomplishments.	**Chapters 2 and 3**

(Continued)

CERTIFIED SIX SIGMA BLACK BELT BODY OF KNOWLEDGE

Section IV: Define—15 Questions (15% of the test)

Subsection	Cognitive Level	Content	Chapter Reference
Voice of the Customer	**Apply**	Be able to segment customers for each project and show how the project will impact both internal and external customers.	**Chapter 5**
	Apply	Be able to identify and select the appropriate data collection method (surveys, focus groups, interviews, observation, etc.) to gather customer feedback to better understand customer needs, expectations and requirements; be able to ensure that the instruments used are reviewed for validity and reliability to avoid introducing bias or ambiguity in the responses.	**Chapters 5 and 12**
	Apply	Be able to define, select and use appropriate tools to determine customer requirements, such as CTQ flow-down, quality function deployment (QFD) and the Kano model.	**Chapter 5**
Project Charter	**Create**	Be able to develop and evaluate the problem statement in relation to the project's baseline performance and improvement goals.	**Chapter 14**
	Analyze	Be able to develop and review project boundaries to ensure that the project has value to the customer.	**Chapter 14**
	Apply	Be able to develop the goals and objectives for the project on the basis of the problem statement and scope.	**Chapter 14**
	Analyze	Be able to identify and evaluate performance measurements (e.g., cost, revenue, schedule, etc.) that connect critical elements of the process to key outputs.	**Chapters 2 and 14**
Project Tracking	**Create**	Be able to identify, develop and use project management tools, such as schedules, Gantt charts, tollgate reviews, etc., to track project progress.	**Chapters 7 and 14**

Section V: Measure—26 Questions (26% of the test)

Subsection	Cognitive Level	Content	Chapter Reference
Process Characteristics	**Evaluate**	Be able to identify input and output process variables and evaluate their relationships using SIPOC and other tools.	**Chapters 2, 5, and 14**
	Evaluate	Be able to evaluate process flow and utilization to identify waste and constraints by analyzing work in progress (WIP), work in queue (WIQ), touch time, takt time, cycle time, throughput, etc.	**Chapters 5, 11, and 14**
	Analyze	Be able to analyze processes by developing and using value stream maps, process maps, flowcharts, procedures, work instructions, spaghetti diagrams, circle diagrams, etc.	**Chapter 7**
Data Collection	**Evaluate**	Be able to define, classify and evaluate qualitative and quantitative data, continuous (variables) and discrete (attributes) data and convert attributes data to variables measures when appropriate.	**Chapters 8–12**
	Apply	Be able to define and apply nominal, ordinal, interval and ratio measurement scales.	**Chapter 8**
	Evaluate	Be able to define and apply the concepts related to sampling (e.g., representative selection, homogeneity, bias, etc.); be able to select and use appropriate sampling methods (e.g., random sampling, stratified sampling, systematic sampling, etc.) that ensure the integrity of data.	**Chapters 8–12**
	Apply	Be able to develop data collection plans, including consideration of how the data will be collected (e.g., check sheets, data coding techniques, automated data collection, etc.) and how it will be used.	**Chapters 7 and 8**
Measurement Systems	**Understand**	Be able to define and describe measurement methods for both continuous and discrete data.	**Chapters 4 and 9**
	Evaluate	Be able to use various analytical methods (e.g., repeatability and reproducibility [R&R], correlation, bias, linearity, precision to tolerance, percent agreement, etc.) to analyze and interpret measurement system capability for variables and attributes measurement systems.	**Chapter 9**

CERTIFIED SIX SIGMA BLACK BELT BODY OF KNOWLEDGE

Section V: Measure—26 Questions (26% of the test) *(Continued)*

Subsection	Cognitive Level	Content	Chapter Reference
	Understand	Be able to identify how measurement systems can be applied in marketing, sales, engineering, research and development (R&D), supply chain management, customer satisfaction and other functional areas.	Chapter 4 and 7–14
	Understand	Be able to define and describe elements of metrology, including calibration systems, traceability to reference standards, the control and integrity of standards and measurement devices, etc.	Chapter 4
Basic Statistics	Apply	Be able to define and distinguish between population parameters and sample statistics (e.g., proportion, mean, standard deviation, etc.).	Chapter 8
	Apply	Be able to describe and use the central limit theorem and be able to apply the sampling distribution of the mean to inferential statistics for confidence intervals, control charts, etc.	Chapter 8
	Evaluate	Be able to calculate and interpret measures of dispersion and central tendency and construct and interpret frequency distributions and cumulative frequency distributions; be able to construct and interpret diagrams and charts, including box-and-whisker plots, run charts, scatter diagrams, histograms, normal probability plots, etc.	Chapters 8–12
	Evaluate	Be able to define and distinguish between enumerative (descriptive) and analytic (inferential) statistical studies and evaluate their results to draw valid conclusions.	Chapter 9
Probability	Apply	Be able to describe and apply probability concepts such as independence, mutually exclusive events, multiplication rules, complementary probability, joint occurrence of events, etc.	Chapters 8–12
	Evaluate	Be able to describe, apply and interpret the following distributions: normal, Poisson, binomial, chi square, Student's t and F distributions.	Chapters 8–13
	Apply	Be able to describe when and how to use the following distributions: hypergeometric, bivariate, exponential, lognormal and Weibull.	Chapters 8–13
Process Capability	Evaluate	Be able to define, select and calculate C_p and C_{pk} indices to assess process capability; be able to define, select and calculate P_p, P_{pk} and C_{pm} indices to assess process performance.	Chapter 9
	Evaluate	Be able to describe and use appropriate assumptions and conventions when only short-term data or attributes data are available and when long-term data are available; be able to interpret the relationship between long-term and short-term capability.	Chapters 9–11 and 14
	Apply	Be able to identify non-normal data and determine when it is appropriate to use Box-Cox or other transformation techniques.	Chapter 9
	Apply	Be able to calculate the process capability and process sigma level for attributes data.	Chapters 9 and 10
	Evaluate	Be able to describe and apply elements of designing and conducting process capability studies, including identifying characteristics and specifications, developing sampling plans and verifying stability and normality.	Chapters 4 and 8–12
	Evaluate	Be able to distinguish between natural process limits and specification limits, and calculate process performance metrics such as percent defective, parts per million (PPM), defects per million opportunities (DPMO), defects per unit (DPU), process sigma, rolled throughput yield (RTY), etc.	Chapters 9 and 14

CERTIFIED SIX SIGMA BLACK BELT BODY OF KNOWLEDGE

Section VI: Analyze—24 Questions (24% of the test)

Subsection	Cognitive Level	Content	Chapter Reference
Measuring and Modeling Relationships Between Variables	**Analyze**	Be able to calculate and interpret the correlation coefficient and its confidence interval, and describe the difference between correlation and causation. Serial correlation will not be tested.	**Chapter 13**
	Evaluate	Be able to calculate and interpret regression analysis, and apply and interpret hypothesis tests for regression statistics. Use the regression model for estimation and prediction, analyze the uncertainty in the estimate, and perform a residuals analysis to validate the model. Models that have non-linear parameters will not be tested.	**Chapter 13**
	Analyze	Be able to use and interpret multivariate tools such as principal components, factor analysis, discriminant analysis, multiple analysis of variance (MANOVA), etc., to investigate sources of variation.	**Not covered**
	Analyze	Be able to use and interpret charts of multivari studies and determine the difference between positional, cyclical and temporal variation.	**Chapter 9**
	Analyze	Be able to analyze attributes data using logit, probit, logistic regression, etc., to investigate sources of variation.	**Not covered**
Hypothesis Testing	**Evaluate**	Be able to define and interpret the significance level, power, type I and type II errors of statistical tests; be able to define, compare and interpret statistical and practical significance.	**Chapters 8–13**
	Apply	Be able to calculate sample size for common hypothesis tests (e.g., equality of means, equality of proportions, etc.).	**Chapters 8–12**
	Evaluate	Be able to define and distinguish between confidence and prediction intervals; be able to define and interpret the efficiency and bias of estimators; be able to calculate tolerance and confidence intervals; be able to use and interpret the results of hypothesis tests for means, variances and proportions; be able to select, calculate and interpret the results of ANOVAs.	**Chapters 8–13**
	Evaluate	Be able to define, select and interpret the results of goodness-of-fit (chi-square) tests.	**Chapters 8–13**
	Evaluate	Be able to select, develop and use contingency tables to determine statistical significance.	**Not covered**
	Evaluate	Be able to select, develop and use various non-parametric tests, including Mood's Median, Levene's test, Kruskal-Wallis, Mann-Whitney, etc.	**Not covered**
Failure Mode and Effects Analysis (FMEA)	**Evaluate**	Be able to describe the purpose and elements of FMEA, including risk priority number (RPN), and evaluate FMEA results for processes, products and services; be able to distinguish between design FMEA (DFMEA) and process FMEA (PFMEA), and interpret results from each.	**Chapter 6**
Additional Analysis Methods	**Analyze**	Be able to use various tools and techniques (gap analysis, scenario planning, etc.) to compare the current and future state in terms of predefined metrics.	**Chapters 8–14**
	Evaluate	Be able to define and describe the purpose of root cause analysis, recognize the issues involved in identifying a root cause, and use various tools (e.g., the five whys, Pareto charts, fault tree analysis, cause-and-effect diagrams, etc.) for resolving chronic problems.	**Chapter 7**
	Analyze	Be able to identify and interpret the 7 classic wastes (overproduction, inventory, defects, overprocessing, waiting, motion and transportation) and other forms of waste such as resource underutilization, etc.	**Chapters 6 and 11**

CERTIFIED SIX SIGMA BLACK BELT BODY OF KNOWLEDGE

Section VII: Improve—23 Questions (23% of the test)

Subsection	Cognitive Level	Content	Chapter Reference
Design of Experiments (DOE)	**Understand**	Be able to define basic DOE terms, including independent and dependent variables, factors and levels, response, treatment, error, etc.	**Chapter 13**
	Apply	Be able to define and apply DOE principles, including power and sample size, balance, repetition, replication, order, efficiency, randomization, blocking, interaction, confounding, resolution, etc.	**Chapter 13**
	Evaluate	Be able to plan, organize and evaluate experiments by determining the objective, selecting factors, responses and measurement methods, choosing the appropriate design, etc.	**Chapters 13 and 14**
	Evaluate	Be able to design and conduct completely randomized, randomized block and Latin square designs and evaluate their results; be able to design, analyze and interpret two-level fractional factoral experiments and describe how confounding affects their use; be able to design, conduct and analyze full factorial experiments.	**Chapter 13**
Waste Elimination	**Analyze**	Be able to select and apply tools and techniques for eliminating or preventing waste, including pull systems, kanban, 5S, standard work, poka-yoke, etc.	**Chapters 6, 7, 11, and 14**
Cycle-Time Reduction	**Analyze**	Be able to use various tools and techniques for reducing cycle time, including continuous flow, single-minute exchange of die (SMED), etc.	**Chapter 11**
Kaizen and Kaizen Blitz	**Apply**	Be able to define and distinguish between kaizen and kaizen blitz methods and apply them in various situations.	**Chapters 6 and 11**
Theory of Constraints (TOC)	**Understand**	Be able to define and describe the theory of constraints (TOC) concept and its uses.	**Chapter 14**
Implementation	**Evaluate**	Be able to develop plans for implementing the improved process (i.e., conduct pilot tests, simulations, etc.), and evaluate results to select the optimum solution.	**Chapter 14**
Risk Analysis and Mitigation	**Apply**	Be able to use tools such as feasibility studies, SWOT analysis (strengths, weaknesses, opportunities and threats), PEST analysis (political, environmental, social and technological), and consequential metrics to analyze and mitigate risk.	**Chapters 2 and 14**

CERTIFIED SIX SIGMA BLACK BELT BODY OF KNOWLEDGE

Section VIII: Control—21 Questions (21% of the test)

Subsection	Cognitive Level	Content	Chapter Reference
Statistical Process Control (SPC)	**Understand**	Be able to define and describe the objectives of SPC, including monitoring and controlling process performance, tracking trends, runs, etc., and reducing variation in a process.	**Chapters 7–12**
	Apply	Be able to identify and select critical characteristics for control chart monitoring; be able to define and apply the principle of rational subgrouping; be able to select and use the following control charts in various situations: *x*-bar/*R*, *x*-bar/*s*, *p*, *np*, *c*, *u*, short-run SPC and moving average.	**Chapters 8–12**
	Analyze	Be able to interpret control charts and distinguish between common and special causes using rules for determining statistical control.	**Chapters 8–12**
Other Control Tools	**Understand**	Be able to define the elements of total productive maintenance (TPM) and describe how it can be used to control the improved process.	**Chapter 6**
	Understand	Be able to define the elements of a visual factory and describe how they can help control the improved process.	**Chapter 11**
Maintain Controls	**Evaluate**	Be able to review and evaluate measurement system capability as process capability improves, and ensure that measurement capability is sufficient for its intended use.	**Chapter 9**
	Apply	Be able to develop a control plan for ensuring the ongoing success of the improved process including the transfer of responsibility from the project team to the process owner.	**Chapter 14**
Sustain Improvements	**Apply**	Be able to document the lessons learned from all phases of a project and identify how improvements can be replicated and applied to other processes in the organization.	**Chapters 2 and 14**
	Apply	Be able to develop and implement training plans to ensure continued support of the improved process.	**Chapter 3**
	Apply	Be able to develop or modify documents including standard operating procedures (SOPs), work instructions, etc., to ensure that the improvements are sustained over time.	**Chapter 2**
	Apply	Be able to identify and apply tools for ongoing evaluation of the improved process, including monitoring for new constraints, additional opportunities for improvement, etc.	**Chapters 2 and 14**

Section IX: Design for Six Sigma (DFSS) Frameworks and Methodologies—7 Questions (7% of the test)

Subsection	Cognitive Level	Content	Chapter Reference
Common DFSS Methodologies	**Understand**	Be able to identify and describe the DMADV (define, measure, analyze, design, and validate) and DMADOV (define, measure, analyze, design, optimize, and validate) methodologies.	**Chapter 14**
Design for *X* (DFX)	**Understand**	Be able to describe design constraints, including design for cost, design for manufacturability and producibility, design for test, design for maintainability, etc.	**Chapter 5**
Robust Design and Process	**Apply**	Be able to describe the elements of robust product design, tolerance design, and statistical tolerancing.	**Chapters 4 and 5**
Special Design Tools	**Understand**	Be able to describe how Porter's five forces analysis, portfolio architecting, and other tools can be used in strategic design and planning.	**Chapter 2**
	Apply	Be able to describe and use the theory of inventive problem solving (TRIZ), systematic design, critical parameter management, and Pugh analysis in designing products or processes.	**Chapter 5**

APPENDIX D

DETAILED CONCEPT MAP OF INDUSTRIAL ENGINEER

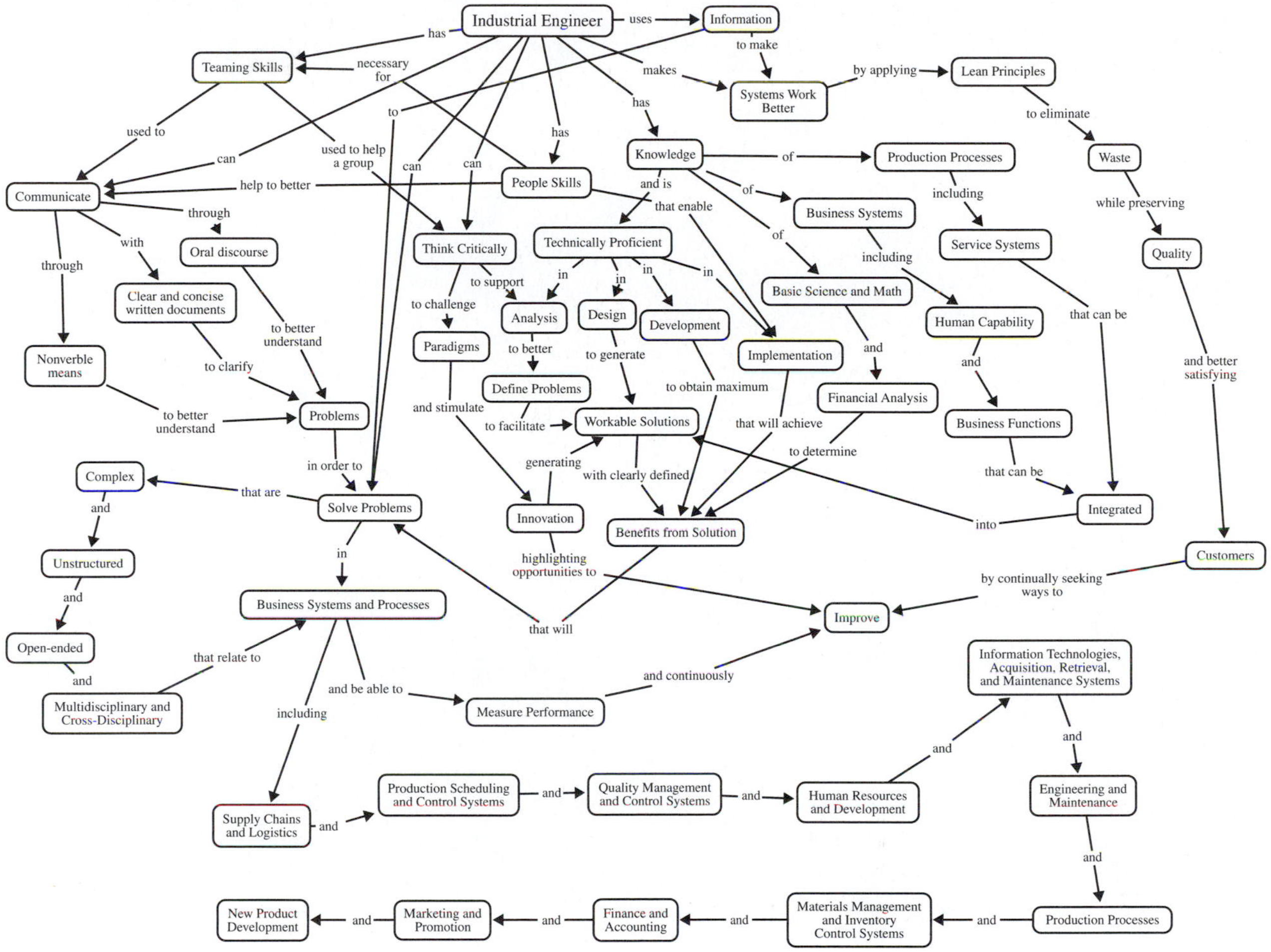

APPENDIX E

DISCRETE DISTRIBUTIONS IMPORTANT TO STATISTICAL PROCESS CONTROL

I. Binomial Distribution

Consider the scenario where only two outcomes are possible during an inspection process (for example, the outcomes could be go/no go, good/defective, pass/fail, etc.), and the following conditions hold:

1. There is a constant chance of a particular outcome (e.g., defective) in each inspected unit.
2. The outcomes are independent—that is, the fact that an item in the sample is defective does not alter the probability that the next item inspected will be defective.

If we let the variable X represent the number of defectives (or no-goes, failures, etc.) in a sample of n items, X has a binomial distribution with parameters n and p, where p is the constant probability of any item selected at random being defective. The binomial distribution is a discrete probability distribution with the following characteristics.

Binomial Distribution

Probability function:

$$prob(X = x) = f(X) = \begin{cases} \dfrac{n!}{x!(n-x)!}p^x(1-p)^{n-x} & \text{for } x = 0, 1, \ldots, n \\ 0 & \text{otherwise} \end{cases}$$

where,

n is a positive integer and $0 < p < 1$

$n! = n(n-1)(n-2)\ldots(3)(2)(1)$ and $0! = 1$

Mean:

$$\mu = E[X] = \sum_{x=0}^{n} x\frac{n!}{x!(n-x)!}p^x(1-p)^{n-x} = np$$

Variance:

$$\sigma^2 = Var[X] = \sum_{x=0}^{n}(x-np)^2 \times \frac{n!}{x!(n-x)!}p^x(1-p)^{n-x} = np(1-p)$$

II. Poisson Distribution

Consider the scenario where a particular outcome (for example, defects, errors, failures, etc.) is being counted over a very large number of trials (i.e., opportunities) and the following conditions hold:

1. The occurrence of the outcome is extremely rare so that the probability of the outcome occurring at any particular opportunity is very small.
2. The trials are independent—that is, the fact that a defect is found does not alter the probability that another defect will be found at the next opportunity.
3. An inspection (sampling) unit has been defined in terms of time or space (for example, an inspection unit could be a lot of n items, n square feet or n linear yards of material, n hours of production, etc.). The inspection process involves selecting inspection units at random and counting the number of occurrences found. The number of occurrences (defects) per inspection unit follows a Poisson distribution which has the following characteristics:

If we let the variable X represent the number of defects (or failures) per inspection unit, X has a Poisson distribution with parameter λ, where λ is the product of the constant probability that an occurrence will occur at any opportunity times the number of opportunities per inspection. The Poisson distribution is a discrete probability distribution with the following characteristics.

Poisson Distribution

Probability function:

$$prob(X = x) = f(X) = \begin{cases} \dfrac{e^{-\lambda}\lambda^x}{x!} & \text{for } x = 0, 1, 2, \ldots \\ 0 & \text{otherwise} \end{cases}$$

where,

$\lambda > 0$

$x! = x(x-1)(x-2)\ldots(3)(2)(1)$ and $0! = 1$

Mean:

$$\mu = E[X] = \sum_{x=0}^{\infty} x\frac{e^{-\lambda}\lambda^x}{x!} = \lambda$$

Variance:

$$\sigma^2 = Var[X] = \sum_{x=0}^{\infty}(x-\lambda)^2\frac{e^{-\lambda}\lambda^x}{x!} = \lambda$$

Note that an interesting feature of the Poisson distribution is that the mean and variance are equal.

APPENDIX F

Table I Plackett-Burman Design for $n = 8$ Runs and 7 Factors

Run	Factor A	B	C	D	E	F	G
1	+	−	−	+	−	+	+
2	+	+	−	−	+	−	+
3	+	+	+	−	−	+	−
4	−	+	+	+	−	−	+
5	+	−	+	+	+	−	−
6	−	+	−	+	+	+	−
7	−	−	+	−	+	+	+
8	−	−	−	−	−	−	−

Table II Plackett-Burman Design for $n = 12$ Runs and 11 Factors

Run	Factor A	B	C	D	E	F	G	H	I	J	K
1	+	−	+	−	−	−	+	+	+	−	+
2	+	+	−	+	−	−	−	+	+	+	−
3	−	+	+	−	+	−	−	−	+	+	+
4	+	−	+	+	−	+	−	−	−	+	+
5	+	+	−	+	+	−	+	−	−	−	+
6	+	+	+	−	+	+	−	+	−	−	−
7	−	+	+	+	−	+	+	−	+	−	−
8	−	−	+	+	+	−	+	+	−	+	−
9	−	−	−	+	+	+	−	+	+	−	+
10	+	−	−	−	+	+	+	−	+	+	−
11	−	+	−	−	−	+	+	+	−	+	+
12	−	−	−	−	−	−	−	−	−	−	−

Table III Plackett-Burman Design for n = 16 Runs and 15 Factors

	Factor														
Run	A	B	C	D	E	F	G	H	I	J	K	L	M	N	O
1	+	−	−	−	+	−	−	+	+	−	+	−	+	+	+
2	+	+	−	−	−	+	−	−	+	+	−	+	−	+	+
3	+	+	+	−	−	−	+	−	−	+	+	−	+	−	+
4	+	+	+	+	−	−	−	+	−	−	+	+	−	+	−
5	−	+	+	+	+	−	−	−	+	−	−	+	+	−	+
6	+	−	+	+	+	+	−	−	−	+	−	−	+	+	−
7	−	+	−	+	+	+	+	−	−	−	+	−	−	+	+
8	+	−	+	−	+	+	+	+	−	−	−	+	−	−	+
9	+	+	−	+	−	+	+	+	+	−	−	−	+	−	−
10	−	+	+	−	+	−	+	+	+	+	−	−	−	+	−
11	−	−	+	+	−	+	−	+	+	+	+	−	−	−	+
12	+	−	−	+	+	−	+	−	+	+	+	+	−	−	−
12	−	+	−	−	+	+	−	+	−	+	+	+	+	−	−
14	−	−	+	−	−	+	+	−	+	−	+	+	+	+	−
15	−	−	−	+	−	−	+	+	−	+	−	+	+	+	+
16	−	−	−	−	−	−	−	−	−	−	−	−	−	−	−

Table IV Plackett-Burman Design for n = 20 Runs and 19 Factors

	Factor																		
Run	A	B	C	D	E	F	G	H	I	J	K	L	M	N	O	P	Q	R	S
1	+	−	+	+	−	−	−	−	+	−	+	−	+	+	+	+	−	−	+
2	+	+	−	+	+	−	−	−	−	+	−	+	−	+	+	+	+	−	−
3	−	+	+	−	+	+	−	−	−	−	+	−	+	−	+	+	+	+	−
4	−	−	+	+	−	+	+	−	−	−	−	+	−	+	−	+	+	+	+
5	+	−	−	+	+	−	+	+	−	−	−	−	+	−	+	−	+	+	+
6	+	+	−	−	+	+	−	+	+	−	−	−	−	+	−	+	−	+	+
7	+	+	+	−	−	+	+	−	+	+	−	−	−	−	+	−	+	−	+
8	+	+	+	+	−	−	+	+	−	+	+	−	−	−	−	+	−	+	−
9	−	+	+	+	+	−	−	+	+	−	+	+	−	−	−	−	+	−	+
10	+	−	+	+	+	+	−	−	+	+	−	+	+	−	−	−	−	+	−
11	−	+	−	+	+	+	+	−	−	+	+	−	+	+	−	−	−	−	+
12	+	−	+	−	+	+	+	+	−	−	+	+	−	+	+	−	−	−	−
12	−	+	−	+	−	+	+	+	+	−	−	+	+	−	+	+	−	−	−
14	−	−	+	−	+	−	+	+	+	+	−	−	+	+	−	+	+	−	−
15	−	−	−	+	−	+	−	+	+	+	+	−	−	+	+	−	+	+	−
16	−	−	−	−	+	−	+	−	+	+	+	+	−	−	+	+	−	+	+
17	+	−	−	−	−	+	−	+	−	+	+	+	+	−	−	+	+	−	+
18	+	+	−	−	−	−	+	−	+	−	+	+	+	+	−	−	+	+	−
19	−	+	+	−	−	−	−	+	−	+	−	+	+	+	+	−	−	+	+
20	−	−	−	−	−	−	−	−	−	−	−	−	−	−	−	−	−	−	−

Table V Plackett-Burman Design for n = 24 Runs and 23 Factors

	Factor																						
Run	A	B	C	D	E	F	G	H	I	J	K	L	M	N	O	P	Q	R	S	T	U	V	W
1	+	−	−	−	−	+	−	+	−	−	+	+	−	−	+	+	−	+	−	+	+	+	+
2	+	+	−	−	−	−	+	−	+	−	−	+	+	−	−	+	+	−	+	−	+	+	+
3	+	+	+	−	−	−	−	+	−	+	−	−	+	+	−	−	+	+	−	+	−	+	+
4	+	+	+	+	−	−	−	−	+	−	+	−	−	+	+	−	−	+	+	−	+	−	+
5	+	+	+	+	+	−	−	−	−	+	−	+	−	−	+	+	−	−	+	+	−	+	−
6	−	+	+	+	+	+	−	−	−	−	+	−	+	−	−	+	+	−	−	+	+	−	+
7	+	−	+	+	+	+	+	−	−	−	−	+	−	+	−	−	+	+	−	−	+	+	−
8	−	+	−	+	+	+	+	+	−	−	−	−	+	−	+	−	−	+	+	−	−	+	+
9	+	−	+	−	+	+	+	+	+	−	−	−	−	+	−	+	−	−	+	+	−	−	+
10	+	+	−	+	−	+	+	+	+	+	−	−	−	−	+	−	+	−	−	+	+	−	−
11	−	+	+	−	+	−	+	+	+	+	+	−	−	−	−	+	−	+	−	−	+	+	−
12	−	−	+	+	−	+	−	+	+	+	+	+	−	−	−	−	+	−	+	−	−	+	+
13	+	−	−	+	+	−	+	−	+	+	+	+	+	−	−	−	−	+	−	+	−	−	+
14	+	+	−	−	+	+	−	+	−	+	+	+	+	+	−	−	−	−	+	−	+	−	−
15	−	+	+	−	−	+	+	−	+	−	+	+	+	+	+	−	−	−	−	+	−	+	−
16	−	−	+	+	−	−	+	+	−	+	−	+	+	+	+	+	−	−	−	−	+	−	+
17	+	−	−	+	+	−	−	+	+	−	+	−	+	+	+	+	+	−	−	−	−	+	−
18	−	+	−	−	+	+	−	−	+	+	−	+	−	+	+	+	+	+	−	−	−	−	+
19	+	−	+	−	−	+	+	−	−	+	+	−	+	−	+	+	+	+	+	−	−	−	−
20	−	+	−	+	−	−	+	+	−	−	+	+	−	+	−	+	+	+	+	+	−	−	−
21	−	−	+	−	+	−	−	+	+	−	−	+	+	−	+	−	+	+	+	+	+	−	−
22	−	−	−	+	−	+	−	−	+	+	−	−	+	+	−	+	−	+	+	+	+	+	−
23	−	−	−	−	+	−	+	−	−	+	+	−	−	+	+	−	+	−	+	+	+	+	+
24	−	−	−	−	−	−	−	−	−	−	−	−	−	−	−	−	−	−	−	−	−	−	−

GLOSSARY

Abstract mental simulation A decision skill that includes the ability to take a set of outcomes and then imagine a plausible process that could have created them.

Acceptable quality level (AQL) The maximum percentage defective in a sampling plan that will be accepted 95% of the time; lots at or better than the AQL will be accepted at least 95% of the time and rejected at most 5% of the time.

Acceptance number The maximum number of defective units in a sample that will permit acceptance of the inspection lot.

Accuracy How close a measurement conforms to the true value of the attribute being measured.

Activity-based costing (ABC) A cost accounting methodology that identifies the cost drivers of processes and allocates the true costs back to the products or services produced. ABC avoids the distortions in product costing that result from the arbitrary allocations of indirect costs used by traditional accounting methods.

Affective domain The collection of personality factors that influence human behavior, and include motivation, attitudes, perceptions, stereotypes, values, and feelings.

Affinity diagram A tool used to organize ideas into categories based on the perception of natural relationships or common themes. Often used in brainstorming to help a group generate a number of ideas and to reduce the list to a smaller number of manageable categories. The affinity diagram, sometimes called the KJ method, is one of the seven new tools of quality management.

Alias structure In a design of experiments, this structure determines which effects (main and interaction) are confounded with each other; confounded effects cannot be separated in the analysis.

Alpha error Sometimes called the Type I error, in statistical analyses this is the probability of rejecting as false a hypothesis that is actually true.

Ambiguity The state in which a word, sentence, or message is capable of being understood in two or more possible senses or ways.

Analogues A decision skill that includes the ability to mentally create situations similar to the decision scenario for the purpose of generating improved decision options.

Analysis of variance (ANOVA) The term used to refer to a collection of statistical models and procedures that partitions the total observed variance in a response variable into components that are due to explanatory sources.

Analytical decision tools Decision tools that come from statistics and operations research that help the decision maker better understand the decision environment, to analyze hypothesized scenarios, perform "what if?" analyses, and aid in the design of optimum policies.

Analytical processes The methodologies used to convert raw measurement data into useful information.

Analytic hierarchy process (AHP) A multiple-criteria decision process that provides a methodology whereby teams of decision makers can build consensus by assigning relative importance scores to criteria and alternatives, using subjective and personal preferences.

Analytic study An investigation that is aimed at trying to understand how a process behaves and to define the principal process drivers in an effort to predict future behavior and be able to improve the process.

Analyze The third tactic in the Six Sigma DMAIC model during which the factors that are most likely causing the defined problem are isolated and the root causes and sources of variation are identified.

Analyze tollgate A meeting that is held at the conclusion of the Analyze step in the DMAIC Six Sigma process; the purpose is to review progress; the meeting is led by the team leader (usually a black belt), and is attended by team members, the project sponsor, the project champion, a master black belt, and the process owner. The main purpose is to defend the causes selected for improvement.

Andon A term used to refer to a system whereby workers are empowered to halt production and notify management, maintenance, or other workers of a quality or process problem; the basis of jidoka, one of the essential tools of lean production.

Angularity The condition of a surface, axis, or center plane that is at a specified angle from a data plane or axis.

Anticipatory failure determination (AFD) A technique used in design for reliability to identify and remove the root causes for failure; reverse logic is used by treating failures as desired outcomes and trying to create designs in which failures will occur with reliable predictability.

Arrow diagram A tool, borrowed from project management, used to schedule a sequence of complex tasks. Two conventions are used: activities on arrows (AOA) and activities on nodes (AON). The arrow diagram is one of the seven new tools of quality management.

Artificial neural network (ANN) A network of interconnecting artificial neurons that are programmed to mimic the properties of biological neurons (i.e., the human brain). This tool is used to gain an understanding of biological neural networks or for solving artificial intelligence problems.

Assessment The act of determining how well system outcomes match system objectives.

Associations Processes that involve forming mental connections or bonds between sensations, ideas, or memories; word or picture associations are often used as a means to stimulate fresh ideas.

Assurance The ability of service providers to inspire confidence and trust. Assurance includes courtesy, knowledge, competence, credibility, and security; one of the five areas surveyed using the SERVQUAL instrument.

Attributes acceptance sampling Also called lot acceptance sampling, a technique that uses random samples from a much larger lot to make a decision as to whether the lot should be accepted or rejected. The method is a middle of the road approach between no inspection and 100% inspection.

Attributes measurement A measurement process whereby data is taken on a nominal scale and a count is maintained of each of the nominal values, usually referred to as events or occurrences. The data are discrete random variables.

Autocorrelation The strength of a relationship between observations as a function of the time separation between them; a measure of the lack of statistical independence in time-ordered process data.

Autonomation A policy that requires operators to stop a production line anytime they see a problem, and track down and eliminate the root causes; also known by its Japanese word, jidoka.

Autonomy The quality or state of self-governing; self-directing freedom and moral independence; also, one of the five factors used in computing the motivating potential score (MPS) of a job that represents the freedom a worker has to determine pace, sequencing, and work methods; the other factors are skill variety, task identity, task significance, and feedback.

Autoregressive integrated moving average (ARIMA) A mathematical model that is fitted to time series data to either better understand the data or to predict future points in the series. Generally referred to as an ARIMA (p, d, q) model where p, d, and q are integers greater than or equal to zero and refer to the order of the autoregressive, integrated, and moving average parts of the model respectively.

Availability A maintenance performance metric that is defined as a percentage measure of the degree to which machines and equipment are in an operable and committable state at the point in time when needed. Availability measures combine the concepts of reliability (the probability that a process will not fail) and maintainability (the probability that a process is successfully restored following a failure).

Average outgoing quality The expected average percentage nonconforming of outgoing product for a given value of percentage nonconforming incoming product quality.

Average outgoing quality limit (AOQL) The worst possible average percentage nonconforming in an outgoing product lot.

Average sample number (ASN) The average number of units inspected over a series of shipments (for a particular level of incoming quality). Unlike the average total inspection (ATI) metric, the ASN calculation does not include the 100% screening under rectifying inspection.

Average total inspection (ATI) The average number of units inspected per shipment that results over a series of shipments under a rectifying inspection scheme. It is calculated from the lot size, the sample size, and the probability of acceptance, and is dependent on sample size, fraction nonconforming, and acceptance number.

Balanced scorecard A strategic planning and management tool that is used extensively in business to align business activities to the vision and strategy of the organization, improve internal and external communications, and monitor organization performance against strategic goals.

Behavior management A management process that focuses on enriching the quality of all processes associated with the human resource management system—that is, transforming the mix of inputs into desirable behaviors.

Benchmarking The term used to describe a continuous and formal process for defining the best practices—usually but not necessarily within the same industry sector.

Beta error Sometimes called the Type II error, in statistical analyses this is the probability of failing to reject (i.e., accepting as true) a hypothesis that is actually false.

Bias The amount by which the average measurement of multiple measures of the same object deviates from the true value of the feature being measured; an accurate process is one that has no bias.

Binomial distribution A discrete probability distribution of the number of successes in a sequence of n independent Bernouili trials (trials with only two outcomes: success or failure) each of which yields a successful outcome with probability p and probability of failure $1-p$.

Bore gauge An instrument used for accurately measuring the dimensions of holes.

Bottleneck The workstation in a sequential production routing with the lowest capacity; the bottleneck determines the overall line capacity.

Box-Cox transformations A family of transformations used to transform nonnormal process data into a normally distributed data distribution.

Brainstorming A group creativity technique designed to generate a large number of ideas within a short period of time.

Brainwriting A brainstorming method in which the ideas generated are written down rather than shared verbally.

Bucket Under push scheduling the minimum time increment used for releasing jobs to processing departments; a bucket is typically one week.

Buffer management A term sometimes applied to theory of constraints that acknowledges the managed use of inventory buffers to absorb capacity variations that occur along a value stream to ensure that the drum (constraint) activity never has to lower its productivity due to material starvation.

Bulletin boards A conventional information source that employs software programs that allow users to connect to a computer system and perform such functions as downloading software or data, listening to news and music, playing games, reading news, and exchanging messages with other users. As a part of internal communications, system can be manual or computerized to display values, goals, and continuous improvement achievements.

Business intelligence Sometimes called executive information systems, business intelligence provides computer information, using graphical displays and user-friendly interfaces that facilitate the internal and external information needs that senior executives require in order to effectively execute an organization's strategy.

Business sustaining costs Costs that cannot be directly allocated under activity-based costing (ABC) methods and are lumped together to be allocated using traditional accounting practices.

Calibration A measurement concept that refers to how far the average measurement resulting from a measurement process deviates from the true measurement as verified by a certified standard.

Caliper A measuring instrument that is used to measure the distance between two points with great precision.

Capability indices The measures used to determine the capability of a process to meet its required specifications for some product feature. A number of options are available: the C-indices (C_p, C_r, C_{pk}, C_{pm}, and C_{pmk}) are used for stable processes when control chart data can be used to estimate the process variance, and the P-indices (P_r, P_{pk}, P_{pm}, and P_{pmk}) are used when the process variance has to be estimated from sample data.

Case-based reasoning An artificial intelligence information source that uses computer learning to solve new problems based on the experiences gained from the solution of similar problems in the past.

Causal analysis Methodologies used to identify and remove the root causes of some problem of interest; leads to permanent solutions rather than treating symptoms.

Cause-and-effect (CE) diagram One of the seven old tools of statistical process control, the cause-and-effect diagram provides a structure for exploring all the potential or real causes (or inputs) that result in a single effect (or output). Potential causes are typically arranged using four major categories—manpower, materials, methods, and machinery are common for manufacturing applications; and equipment, policies, procedures and people are common for service applications. The cause-and-effect diagram is also known as the Ishikawa diagram (after its founder) and the fishbone diagram because of its resemblance to the skeleton of a fish.

***c* chart** The type of control used to track attributes data that represent the number of defects or errors observed per inspection unit; the inspection unit represents the opportunity for defects to occur and will be equal to one for each sample (i.e., a predefined number of objects, predefined number of square meters of material, predefined number of lineal yards, etc.).

Census The process of observing all members of a target population.

Center gauge A device used in metalworking for checking the angles of lathe screw cutting tool profiles during grinding operations; it is commonly used for hand grinding using a bench grinder.

Central tendency A single measurement that is the most representative of the entire set for a population or process.

Chart A quality management tool used to descriptively display statistical data so that patterns and trends can be understood and analyzed; together with graphs, charts are listed as one of the seven old tools used in statistical process control applications.

Changeover The process of changing a process from the production of one product to the production of another; includes setup activities when the machine is down and runup activities during which machine controls are adjusted until the process begins to produce output of an acceptable quality level.

Changeover time A measure of how long it takes to produce acceptable product for the next run, measured from the completion time of the previous run; the sum of setup and runup times.

Check sheet One of the seven old tools of statistical process control, a check sheet is a data collection tool, normally handwritten, used to record raw data in a format that can be easily understood by everyone.

Closed structure question A question on a survey questionnaire that elicits a direct response; examples are multiple choice, multiple response, dichotomies (questions that can be answered yes or no), or rank lists (questions that ask for judgments using a Likert scale).

Coaxiality The condition in which two or more part features share a common axis.

Coefficient of determination Commonly referred to as r^2, this index represents the proportion of the total observed variation in a response variable that can be explained due to its statistical association with one or more independent variables; this index takes on values between 0 and +1.

Collective wisdom An organization's informal repository of information that includes know-how, experience, insights, communication and social skills, analytical and deductive capabilities, and the ability people have to work together.

Comment cards A method for obtaining customer feedback whereby customers are invited to return a card indicating their level of satisfaction, listing any specific complaints, and providing any suggestions or comments that can be used to improve service.

Common cause variability Random variability that is present in all samples; normally considered to be systemic. On control charts common cause variation is the only source present if there is no evidence of any nonrandom patterns in time-ordered data.

Communications The process by which information is exchanged between individuals or machines through a common system of symbols, signs, or behavior.

Company newsletters A method used to facilitate internal communications whereby periodic written publications are distributed to the workforce for the purpose of conveying information concerning corporate strategy, business decisions, customer profiles and requirements, new technology, and quality goals and initiatives.

Company-wide quality control (CWQC) A term used to describe a quality implementation strategy that involves every division of an organization and every worker at every level, requiring the integration of functions that may have traditionally been independent, such as materials management, methods and procedures, product and process development and design, and final inspection and testing.

Competence management A management process that focuses on enriching the quality of inputs to the human resource management system resulting in a more competent workforce.

Competency knowledge map A mapping of competencies to one or a group of individuals; at the individual level the map is a competency profile that documents training, skills, positions, and career path for one person; for more than one person the map provides a broad documentation of skills, tasks, and core expertise, cross-linked to the individuals who possess them.

Competitive intelligence Information concerning the external environment in which the organization operates.

Complete experiment A design of experiments in which runs are conducted for all possible treatment combinations; with a complete experiment all main effects and interaction effects can be independently estimated.

Component supplier management (CSM) A computerized knowledge base that maintains information on parts and their respective suppliers with the ability to cross-reference information.

Compound questions The structure of a survey question that within a single question asks the respondent to in fact answer multiple questions; survey objectives are better served by breaking down compound questions into multiple simpler questions with greater focus.

Computer-supported cooperative work A conventional information source that combines enabling technologies with an understanding of how humans perform work in groups. Enabling technologies include computer networks, computer hardware, computer software (sometimes referred to as groupware), and associated techniques.

Concentricity A condition in which two or more part features (cylinders, cones, spheres, etc.) in any combination share a common center.

Concept knowledge map A concept knowledge map (often referred to simply as a concept map) is a useful tool for organizing and classifying content with which relationships among concepts can be visualized; on the map concepts are connected in a downward-branching hierarchical structure with linking action–verb phrasing that articulate the nature of each relationship.

Concurrent engineering (CE) A systematic approach to the integrated concurrent design of products and their related processes, including manufacture and support. The integrated approach forces designers to consider all elements of the product life cycle from conception through disposal, including quality, cost, schedule, and user requirements. This approach is also called simultaneous manufacturing, integrated product development, and integrated product and process design.

Condition monitors Portable or in-line instruments used to predict the probability of failures of critical components. Specific condition monitoring technologies include vibration measurement and analysis, infrared thermography, oil analysis and tribology, ferrography, ultrasonics, and motor current analysis.

Confidence interval (CI) An interval estimate of some population or process parameter that defines the likelihood of containing the true value of the parameter as stipulated by a confidence coefficient. In regression models the CI represents the area within which a stipulated percentage of all regression lines will fall for any selected random sample.

Confounding In an experimental design the inability to separate out main effects from higher order interactive effects; confounding occurs when a design must be conducted by running only a fraction of all the possible treatment combinations.

Constraint management A management philosophy that emphasizes the achievement of key organizational goals by identifying and relentlessly trying to eliminate those constraints that are barriers to goals achievement; also called theory of constraints (TOC).

Consumer's risk The probability that a shipment of poor quality, defined to be the lot tolerance percent defective (LTPD), will be accepted. Also the Type II or β error that is the probability of failing to reject a shipment that is of poor quality.

Contingency theory (CT) The theory that recognizes that business decisions are based on an organization's dependencies with its environment. As opportunities arise, an organization draws on a survival instinct to make whatever changes are necessary to differentiate itself from its competitors.

Continuous distribution A distribution that describes events over a continuous range of possible outcomes, with the probability of any specific outcome equal to zero.

Continuous improvement The term used to describe any program that emphasizes the importance of identifying the impediments (within primary or supporting value streams) that limit an organization's ability to deliver sustained value to its stakeholders, and once identified facilitates the discovery of root causes and takes action to either remove them or nullify their effects.

Continuous processes Industrial processes that are characterized by standardized products produced in large volumes, typically on a 24-hour, 7-day basis, using automated specialized equipment that is typically prohibitively expensive to start, stop, or shut down.

Control The fifth and final tactic of the Six Sigma DMAIC model during which solutions are implemented and documented and sustainability of the improvement effort is ensured.

Control charts Also known as Shewhart charts, control charts are tools used in statistical process control to determine whether a production process is in a state of statistical control. A control chart is a run chart with control limits, which represent the maximum and minimum allowable values for any individual plotted point. The control chart is one of the seven old, or basic, tools of quality control.

Control limits The maximum and minimum values that would be expected in a given sample taken from a process with an assumed mean and standard deviation. The limits are typically set at plus and minus three standard deviations from the assumed mean. Control limits are one tool for identifying the presence of special (assignable) causes of variation.

Controls Devices or sets of devices that are used to manage, command, direct, or regulate the behavior of other devices or systems to achieve some desired goals or outcomes.

Control tollgate A meeting that is held at the conclusion of the Improve step in the DMAIC Six Sigma process; the purpose is a final review and to close the project.

Coordinate measuring machine (CMM) A high-tech device used for measuring geometrical dimensions and consists of a measuring probe that is guided around the geometrical profile of a work piece using a computer and measurement software. As the probe moves over the surface, the coordinates of selected points are fed into the computer and used to accurately determine key dimensions.

Correlation A state in which a relationship exists between two or more statistical variables the values of which occur together in a manner that cannot be explained on the basis of random chance alone.

Correlation coefficient An index that is computed by dividing the covariance of two variables by the product of their standard deviations. This index, which takes on values between −1 and +1, is also known as the Pearson product-moment correlation coefficient and measures the strength and direction of any linear association that exists between two variables.

Cost of quality (COQ) The cost associated with failing to produce a product or service of high quality the first time through. COQ has two components—the cost of conformance (includes all costs associated with producing quality at the source) and the cost of nonconformance (includes scrap, rework, and postproduction costs such as warranties, litigation, and customer defections).

Craft guilds An association of craftspeople in a particular trade. The earliest guilds are believed to have been formed in India circa 3800 BC.

Craft system The basis of the medieval economy under which goods and services were produced through craft guilds that controlled quality, distribution, and training.

Critical path The longest time path through a project activity network. The critical path represents the minimum completion time for the project and any delays with regard to activities on the critical path will result in delays in the project.

Critical to assembly (CTA) A specification of a product feature that is critical to all parts fitting together properly when mated in a final assembly operation.

Critical to function (CTF) A specification of a product feature that is critical to the product doing what the customer expects and what the manufacturer promises.

Critical to manufacture (CTM) A specification of a product feature that is critical to the efficient manufacture of the product in the production quantities required.

Culture The set of shared attitudes, values, goals, and practices that characterize an institution or organization.

Curtailment inspection The act of stopping the inspection process as soon as a decision can be made—that is, as soon as the number of nonconformities in a sample exceeds the stipulated acceptance number.

Customer empowerment A customer's freedom to influence the quality of a service provided; for example being able to choose the time and place for service delivery or being able to select from an array of service choices.

Customer quality awards A scheme used to facilitate internal communications for the purpose of goal setting and clarifying customer service quality expectations.

Cycle time The production rate of a process expressed as the amount of time required to produce one unit of output.

Cylindricity A condition of a surface of revolution in which all points of a surface are equidistant from a common axis.

Data Raw facts, measurements, or observations with no established relationships to each other, or to some external problem, issue, or objective.

Data mining An artificial intelligence–based information source that sorts through large data sets or databases and extracts relevant and useful information.

Data warehouse A contemporary information source that is a computer repository of historical data collected from operational systems, then queried and analyzed using techniques such as data mining.

Databases A conventional information source that consists of computerized collections of data with an underlying structure (schema) that describes each element and the relationships between them.

Decision A determination arrived at following a mental process that considers relevant factors.

Decision support system A conventional information source that consists of computer-based systems that draw from knowledge assets to provide support for decision making where problems are complex and unstructured.

Defective parts per million opportunities (DPMO) The quality measure typically used in a Six Sigma environment.

Define The first tactic of the Six Sigma DMAIC model during which the project is justified by establishing a business case, clarifying roles, and setting milestones.

Define-Measure-Analyze-Improve-Control (DMAIC) The continuous improvement model of Six Sigma that includes the five tactics: Define, Measure, Analyze, Improve, and Control.

Define tollgate A meeting that is held at the conclusion of the Define step in the DMAIC Six Sigma process. The purpose is to review progress, and the meeting is led by the team leader (usually a black belt), and is attended by team members, the project sponsor, the project champion, a master black belt, and the process owner.

Defining relation In a fractional design of experiments the set of interaction columns in the design matrix that is equal to a column of all plus signs; from this the entire aliasing structure can be determined.

Delivery lead time The period of time that elapses between the receipt of a customer's order and delivery of the final product; also called lead time.

Demand flow technology (DFT) A manufacturing philosophy that is concerned with producing a wide mix of high-quality products in the shortest time possible at a competitive cost. Every aspect of the product cycle is managed—from the time the product is ordered until it is shipped, and the emphasis is on rapidly building products to customer order.

Dendrogram A tree diagram that displays a hierarchy of objectives, which branches out to a hierarchy of ways and means for accomplishing those objectives. Also called a systematic diagram and providing more focus than an affinity or relations diagram, this is one of the seven new tools of quality management.

Deployment flowchart An output process flowchart that partitions the process by individual responsibility areas.

Design for Six Sigma (DFSS) A design methodology rooted in the Six Sigma process improvement principles that has the objective of matching customer and business needs in order to create design solutions. As an integrated product development approach, DFSS is a process generation approach in contrast to the process improvement motive of Six Sigma.

Design for the environment (DFE) A green product design process, also known as ecodesign, that is concerned with the environmental impacts of raw materials, manufacturing processes, end use of the product, distribution of the product, and end-of-life management (EOLM).

Design failure mode and effects analysis (DFMEA) A failure mode and effects analysis (FMEA) that is concerned with identifying potential failure modes associated with a product's design that would lead to product malfunctions, reduced reliability, a shortened product life, or safety hazards to customers while using the product.

Design generator (In a design of experiments) determines the alias structure—that is, which effects are confounded and cannot be separated in the analysis.

Design for installability The process by which features are included in a product design that facilitate the ease and effectiveness with which the user can get a product to function as intended and within a reasonable timeframe.

Design for maintainability/serviceability The process by which features are included in a product design that facilitate the in-service maintenance and serviceability of a product.

Design for manufacturability (DFM) The process by which ease and cost of manufacture is considered in a product's design, giving due consideration to manufacturing capabilities with a view toward optimizing the fabrication, assembly, test, procurement, shipping, delivery, service, and repair; and assuring the best cost, quality, reliability, regulatory compliance, safety, time-to-market, and customer satisfaction.

Design for reliability The process by which features are included in a product design that help improve a product's reliability over its useful service life.

Design for supportability (DFS) A concept that requires design engineers to incorporate in the product design process the support system that will be required once the product is placed into service. DFS considers factors that cover a product's entire life cycle.

Design for testability and inspectability The process by which features are included in a product design that facilitate final testing, inspection, and acceptance at the end of the manufacturing process.

Design for updateability and upgradeability The process by which features are included in a product design that take into account the future need to either update or upgrade product functionality.

Design for usability, trainability, and documentation The process by which features are included in a product design that take into account average human capabilities and the ease with which humans will be able to use the product.

Design of experiments (DOE) A structured organized method for determining the relationship between factors (independent variables) affecting a process and the output of that process (response variable).

Design to cost (DTC) The process by which a design takes into account a product's cost performance over its entire life cycle.

Detection The process of discovering the true character, the existence, the presence, or the fact of some object of an investigation; with respect to an FMEA detection is the assessment of the likelihood that the cause of the failure mode or the failure mode itself will be detected before the impact of the effect reaches the customer.

Deviation-amplifying feedback The term given to a form of positive feedback used in strategic planning where information acquired from the environment is used to justify an organization's stated purpose and the continuance with its programs and strategic direction.

Deviation-correcting feedback The term given to the various forms of negative feedback that are used in strategic planning where a controller mechanism continually measures the gap between existing and desired states, and transmits instructions for corrective action that will close the gap.

Dial caliper A vernier caliper that is equipped with a dial that can be directly read eliminating the need to consult a vernier scale. Most models permit the dial to be locked at a predetermined setting to enable simple go/no go inspections.

Dial plug gauge A device for measuring the dimensions of holes; unlike other bore gauges that require the transfer of measurements to remote measuring tools, dial plug gauges are equipped with dial indicators that enable operators to directly read measurements while they are being taken.

Digital caliper A dial caliper in which the analog dial has been replaced by an electronic digital display; this permits the measured value to be directly read from the display in either metric or imperial units. Like dial calipers, the slide of a digital caliper can usually be locked to enable simple go/no go inspections.

Dimensional tolerances The permissible limits (maximum and minimum) of variation specified on a mechanical drawing with respect to physical dimensions.

Discrete distribution A distribution that has a finite number of outcomes each with a positive probability.

Discrimination The ability of a measurement process to differentiate between units of process output.

Discussion forums A conventional information source that uses media to foster open discussions or expressions of ideas. The medium normally used is the Internet, so discussion forums are sometimes referred to as Internet, or Web, forums.

Dissatisfaction information system A formal system that a service provider uses to capture instances of customer dissatisfaction, such as complaints, claims, returns, refunds, recalls, repeat services, litigation, replacements, downgrades, warranties, and mistakes.

Distribution A graph that defines the values that a random variable can assume and the respective probabilities of these values.

Divider caliper An instrument used to measure distances between two points; two sharp points on the instrument are placed on each of the reference points, respectively; the distance scale is then read and transferred to a ruler to obtain the actual distance between the two points. Divider calipers are useful for measuring distances on maps or scale drawings.

Document management A conventional information source that utilizes the computerized tracking and storage of electronic documents and digital images of paper documents. The document management system must have capabilities in storage, retrieval, filing, security, archiving, retention, distribution, document flow, and document creation.

Double sampling An attributes acceptance sampling plan under which a shipment may be accepted, rejected, or subjected to a second sample on the basis of the results of the first sample with size n_1. If a second sample of size n_2 is required, the results of both samples are combined to determine whether the lot is accepted or rejected.

Drum The constraint production activity in a theory of constraints (TOC) implementation; the drum sets the pace for the line or value stream of which it is a part, drawing its requirements directly from customer demand; it then communicates these requirements to the originating activity in the line called the gate.

Drum-buffer-rope (DBR) A system utilized in a Theory of Constraints (TOC) production environment that is designed to get the greatest productivity out of a constraining activity (the drum); the concept involves establishing a detailed work schedule for the drum, typically pulled from customer's demand, buffering the line with in-process inventories to ensure the drum activity is never starved of material and that flow is not disrupted due to variability effects, and using a triggering mechanism to time the release of work orders into the system to ensure a continuous and efficient flow (the rope).

Ecodesign A green product design process, also known as design for the environment (DFE), which is concerned with the environmental impacts of raw materials, manufacturing processes, end use of the product, distribution of the product, and end-of-life management (EOLM).

Economic batch quantity The optimum quantity to produce during a single production run; calculated using a formula that takes into consideration the trade-off between the variable costs of holding inventories and the fixed costs of production setups. Also called economic batch size.

Economic batch size The optimum quantity to produce during a single production run; calculated using a formula that takes into consideration the trade-off between the variable costs of holding inventories and the fixed costs of production setups. Also called economic batch quantity.

Economies of repetition The value derived from reducing the variability in product, process, or method; the idea behind the cellular manufacturing approach in which some of the economies of mass production can be achieved while maintaining the ability to respond to customer needs with small batch quantities and product variety.

Electronic mail A conventional information source that comprises a method for composing, sending, storing, and receiving messages over electronic communications systems. Commonly abbreviated to *email,* the term applies to the Internet system based on the simple mail transfer protocol (SMTP) and to intranet systems where mail messages are exchanged between users internally.

Electronic publishing A conventional information source that includes the digital publishing of ebooks, electronic articles, electronic reports, and so on; also includes the development of digital libraries and catalogs.

Elevate The fourth step of the theory of constraints (TOC) cycle that involves breaking the constraint—for example, if the constraint is a production bottleneck, adding capacity at that workstation elevates the constraint. At the completion of the TOC cycle some other activity will become the constraint and the cycle repeats.

Empathy The projection of a spirit of caring and the delivery of individualized attention. Empathy includes ease of access, effectiveness of communications, and how well the customer is understood by the service provider; one of the five areas surveyed using the SERVQUAL instrument.

Empowerment The creation of opportunities for learning and new skills development to support multiskilling, participative decision making, and the use of self-managed work teams.

End-of-life management (EOLM) Management practices that demonstrate responsibility for a product's disposition at the end of its useful life; these practices include such strategies as take-back, recycle, and salvage.

Engineering method A structured approach that applies the scientific method to the solution of some previously identified problem.

Engineering tolerance (ET) The acceptable range of values for a product feature as specified on the set of engineering drawings; the difference between the highest acceptable measurement (the upper specification, or tolerance, limit [USL]) and the lowest acceptable measurement (the lower specification, or tolerance, limit [LSL]).

Entropy A measure of the unavailable energy in a closed thermodynamic system that is also usually considered to be a measure of the system's disorder. In business systems entropy is the degree of disorder or uncertainty in a system or a trend toward chaos and degradation.

Enumerative study An investigation that is concerned with some fixed quantity of material and uses sampling data to estimate the parameters of that finite population.

Environment Everything that is not contained within a system's boundaries.

Excellence The notion in a service environment that the provider is willing and able to effectively handle problems and queries and will do whatever it takes to understand and adapt to changing needs; a principal criterion for customer satisfaction.

Executive information systems A conventional information source, sometimes called business intelligence, that provides computer information, using graphical displays and user-friendly interfaces that facilitate the internal and external information needs that senior executives require in order to effectively execute an organization's strategy.

Experimental space The set of all possible treatments (combinations of factors and levels) for a particular design of experiments.

Expert system An artificial intelligence–based information system in which a computer software program captures and applies the collective knowledge and analytical skills of human subject-matter experts to make decisions or solve; also known as a knowledge-based system.

Expertise A set of special skills or knowledge derived from training or experience.

External activities The term that represents all the things that an operator can do to get ready for the next job while the current job is still running. Include activities like fetching and prepositioning the necessary tools. Moving activities from an internal to an external classification is key to changeover time reduction.

Exploit The second step of the theory of constraints (TOC) cycle during which an organization will do whatever is necessary to obtain the maximum performance from the constraining factor (e.g., production bottleneck).

Face-to-face interviews Meetings between a data collection agent and each respondent for the purpose of obtaining survey information in person.

Fact In quality management, a piece of information that is perceived to have objective reality that can form the basis for rational decision making.

Factorial design A design of experiments (DOE) that consists of two or more factors, each with discrete possible values or "levels," and whose experimental units take on all possible combinations of these levels across all such factors.

Factors In a design of experiments the variables that are thought to be key influencers of some measured response of interest; the process drivers that are independent, and generally controllable, variables.

Failure mode and effects analysis (FMEA) A procedure for analyzing a system for failure modes and classifying each mode as to its severity and effect on the system; root causes of failures are investigated, prioritized, and eliminated.

Family-friendly policies and practices Personnel policies and practices intended to help working parents balance personal and family responsibilities with work. Such practices include maternity (and paternity) leave benefits, allowances for personal and sick time, breastfeeding arrangements, on-site child care, and counseling referral services.

Feedback Information on outputs or outcomes that is used as the basis for modifying future system behavior. Also one of the five factors used in computing the motivating potential score (MPS) of a job that represents the degree to which a worker obtains information on performance, from personal observation, supervisors, and other sources. The other factors are skill variety, task identity, task significance, and autonomy.

Feeler gauge A measuring devices comprised of small blades of steel of different thicknesses used to measure small clearances.

Ferrography A predictive maintenance (PdM) technology that is concerned with the identification of particles suspended in lubricating fluids to determine the association between the size, shape, composition, and concentration of particles and the wear characteristics of machine components.

Finished goods Products that have exited a primary value chain, have been placed into temporary or long-term storage, and have not yet been distributed to the next entity down the supply chain (e.g., wholesaler, retailer, consumer).

First time yield The number of acceptable units of output from a process divided by the number of units introduced to the process as input; also called first pass yield.

Fishbone diagram One of the seven old tools of statistical process control, the fishbone diagram (that derives its name from its resemblance to the skeleton of a fish) provides a structure for exploring all the potential or real causes (or inputs) that result in a single effect (or output). Potential causes are typically arranged using four major categories—manpower, materials, methods, and machinery are common for manufacturing applications, and equipment, policies, procedures and people are common for service applications. The fishbone diagram is also known as the Ishikawa diagram (after its founder) and the cause-and-effect diagram.

Fishtail gauge A device used in metalworking for checking the angles of lathe screw cutting tool profiles during grinding operations; it is commonly used for hand grinding using a bench grinder.

5S workplace discipline A philosophy that defines a methodology for managing a workplace by instilling a discipline aimed at improved efficiency, reduced waste and better workflow.

Five whys A methodology used to explore the cause and effect relationships underlying a particular problem; the method involves stating a problem or issue followed by a sequence of five "whys" each triggering an answer aimed at getting the problem-solver closer to a true root cause.

Fixed effects model A design of experiments in which several treatment combinations are run to determine if and how the response variable is affected. This analysis leads to an estimate of the range of response variable values that would be generated by setting the process variables (factors) at a particular treatment.

Flatness The condition of a part surface where all elements lie in a single plane.

Flex-time Flexible working arrangements that are offered by the management of some high-performing organizations to accommodate the various lifestyles of its employees.

Flexible capacity A strategy employed by manufacturing organizations to improve the ability to respond by investing in backup machines and multitasking its workforce.

Flexible flow design A design that is organized around processes to accommodate the low-volume high-variety nature of the product mix.

Flexible flows Process designs that suit low-volume, high-variety production with chaotic flow patterns and where fast response is required.

Flexible job descriptions The term used to depict the lack of specificity in a job description that enables employees to spread their wings, grow their positions, and learn how to make more substantial contributions to their respective organizations.

Flow The smooth, progressive movement of materials, products, information, and people through a sequence of activities, without interruption and without the creation of waste.

Flowchart A graphical representation of a process showing inputs, outputs, and all activities—how they relate, and how material or people flow through the system; also called a process flowchart or a process map.

Flow rate A measure of the average rate at which units of product flow past a specified point in a process; also called throughput rate.

Flow shop A high-volume low-variety industrial process that uses highly specialized automated equipment to produce discrete products; assembly lines are examples of flow shop designs.

Flow strategy A management strategy that determines how the factors of production will be configured (that is, the transformation subsystem design) in order to support an organization's competitive priorities and marketing strategy.

Flow time (FT) The total time an order resides in the value stream; the difference between the time that an order reaches the first value-adding workstation and the time that the order is ready for shipment. Also called throughput time, production cycle time, and sojourn time.

Focus group A small group of individuals selected to meet periodically to discuss issues concerning requirements and service delivery.

Fractional design A design of experiments consisting of a carefully chosen subset (fraction) of the total number of treatment combinations possible. The treatments included in the experiment are carefully chosen to expose information about the most important factors while expending only a small fraction of the effort that would be required for a full experiment.

Frame A finite representation of a larger, often infinite, representation of a population from which samples are taken.

Free state condition The state of being in an unassembled condition when no machine process is operating to change shape or form; examples are rubber parts (e.g., o-rings), plastic parts (e.g., thin plastic sheets), and thin-walled parts (e.g., thin cylinders).

Fringe benefits An employment benefit (such as a 401k match or a paid holiday) granted by an employer that has a monetary value but does not affect basic wage rates or salaries.

Functional layouts A plant layout design that groups together in departments similar functions; this grouping can be by machine, task, operation, and so on. Process-oriented layouts suit conditions where production runs are of moderate size and product variety is not great. Functional layouts are also called process-oriented layouts.

Fuzzy synthetic evaluation A multiple-criteria decision technique that uses fuzzy set theory to aggregate and evaluate criteria. A process known as defuzzification is used to transform the fuzzy set of aggregate criteria to a crisp set of quality-ordered weights that guide the decision process.

Gainsharing An incentive-based compensation system that ties wage increases and/or bonuses to increased productivity in areas that are directly under the control of workers.

Gantt charts A type of bar chart used for managing project schedules, in which start and finish dates, and the status of each activity are displayed.

Gap gauge A type of go/no go gauge that can simultaneously check both upper and lower specifications of a gap dimension. The device is equipped with two pairs of jaws—the outermost pair is set at the upper specification limit and the innermost pair at the lower specification limit. An acceptable part will pass through the first pair of jaws and stop at the second. There are also versions of gap gauges designed to check the distance between threads on a threaded part.

Gate The first activity in a production line that is linked directly to the drum (constraint) workstation which provides the trigger for order releases; the gate then pushes material toward the drum.

Gemba visit A visit to the site where a product or service is used, permitting direct observation of problems that arise, workarounds that are applied, and capabilities or services that are never used.

Gender bias Wording in survey questions that is gender-specific; such wording can bias the survey responses and should be avoided.

General tolerance A stipulated product dimension or feature that is not critical to function, manufacture, or assembly.

Generating vector Used to generate the columns of the design matrix for various fractional designs of experiments, such as Plackett-Burman screening designs.

Genetic algorithm An artificial intelligence– based information source that uses computer-driven search heuristics to obtain exact or approximate solutions to optimization problems.

Geometrical notation The convention used in a design of experiments that uses a +1 to indicate the high level of a factor and a –1 to indicate the low level.

Geometrical product specifications (GPS) A set of international geometrical tolerancing standards that have been approved and published by the International Standards Organization (ISO).

Geometrical tolerancing The process of specifying the permissible limits (maximum and minimum) of variation on a mechanical drawing with respect to a part's geometry.

Global 8D An eight-step (called disciplines) approach that is used to identify and eliminate the causes of a failure in a process after the failure has occurred.

Goals Those aspirations set by an organization's leadership that define how the organization intends to achieve its purpose.

Go/no go gauge An inspection tool used to check a work piece against its allowed tolerances; it derives its name from its use—that is, checking the requirement that each work piece inspected pass one test (Go) and fail to pass another (No Go). The go/no go gauge is a fundamental tool used for the collection of attributes data for statistical process control applications.

Goniometer An instrument for measuring angles; numerous design variations are available, each specialized to a particular application.

Graph A quality management tool used to descriptively display statistical data so that patterns and trends can be understood and analyzed; together with charts, graphs are listed as one of the seven old tools used in statistical process control applications.

Green product lines The lines dedicated to the set of those individual products that account for most of the production volume (e.g., 5% of the total product mix may account for 50% of the total production volume). It is worth investing considerable effort in trying to create clean production flow lines for the products in the green product classification.

Green products Products (or services) that are environmentally and socially responsible in their production, use, and disposal. Responsibility implies that these products (or services) and the organizations that produce them are accountable to and respectful of the places and people that provide and use them.

Group process The human ability to work together and to cooperate, coordinate, and delegate.

Group technology A manufacturing philosophy in which the parts having similarities (e.g., geometry or manufacturing process) are grouped together to achieve a high level of integration for the purpose of implementing cellular manufacturing and reducing production batch sizes.

Groupware A conventional information source that consists of software that has been specifically designed to help a group of people achieve a set of common goals. Examples include helping a team create new ideas, new designs, solve quality problems, or implement change. Sometimes referred to as collaborative software.

Growth needs A measure of how well an individual will respond to jobs that have a high motivating potential as determined by the individual's responses on a growth needs strength (GNS) questionnaire.

Growth needs strength (GNS) A measure of individual preferences concerning likes and dislikes, and what constitutes the "ideal" job; the GNS and MPS can be used to match workers to jobs that best suit them.

Heijunka A Japanese term (meaning to make flat and level) used to refer to a methodology that involves breaking down the total volume of work orders into some small scheduling intervals and for each interval designing a repetitive mixed model schedule; this technique has been successful in absorbing fluctuations in demand without the need for inventory buffers; an important tool in lean production.

Hermaphrodite caliper An instrument used to scribe a line on a work piece that is a set distance from the work piece edge. A bent leg tracks along the edge of the work piece while the sharp point of the scriber makes its mark, ensuring a scribed line that is parallel to the edge. Hermaphrodite calipers are also called oddleg calipers and oddleg jennys.

High-performing work organizations (HPWOs) Organizations that are characterized by self-managed work teams; employee involvement, participation, and empowerment; quality inspired management; integrated production technologies; and continual learning.

Histogram A graphical display of tabulated frequencies that is used in statistics to provide structure and meaning to raw data; in quality management histograms are used to provide an approximate picture of the underlying distributions of process outputs. The histogram is one of the seven old tools used in statistical process control applications.

Hoshin kanri A strategic planning methodology that uses Plan-Do-Check-Act (PDCA) cycle to create goals, choose control points (measurable milestones), and link daily control activities to company strategy.

House of quality (HOQ) A graphic tool used in quality function deployment (QFD) for defining the relationship between customer desires and the firm/product capabilities; also called the house of the Toyota production system (HOTPS).

House of the Toyota production system (HOTPS) A graphic tool used in quality function deployment (QFD) for defining the relationship between customer desires and the firm/product capabilities; also called the house of quality (HOQ).

Hypermedia A conventional information source that is an extension of hypertext, in which graphics, audio, video, plain text, and hyperlinks combine to create a nonlinear medium of information. Use of the World Wide Web is a good example of a hypermedia platform.

Hypertext A conventional information source that employs a user interface paradigm to organize material in a nonlinear fashion and embed automated cross-references to other documents, called hyperlinks.

Hypotheticals Survey questions that require respondents to create fabricated or imaginary situations; such questions might begin with "What if?" or "How would you?"; hypotheticals should be avoided in good questionnaire design.

Identify The first step of the theory of constraints (TOC) cycle during which an organization will strive to identify factors (constraints) that are inhibiting success.

Improve The fourth tactic of the Six Sigma DMAIC model during which workable solutions are generated that will eliminate the confirmed causes of variability or nullify their effects.

Improve tollgate A meeting that is held at the conclusion of the Improve step in the DMAIC Six Sigma process. The purpose is to review progress, and the meeting is led by the team leader (usually a black belt), and is attended by team members, the project sponsor, the project champion, a master black belt, and the process owner. The main purpose is to propose and justify improvements.

Improvement The act of enhancing value or excellence in the delivery of a service or the production of a tangible product as perceived by those who require the service or use the product.

Inaccessible information Information or knowledge that survey respondents are not likely to possess or have ready access to; the requirement for such knowledge should be avoided in questionnaire designs.

Inappropriate assumptions An arguable premise embedded in a survey question—for example, assuming that all respondents would reply to some issue in a positive manner and disallowing any responses that might be negative; inappropriate assumptions create bias in survey data and should be avoided.

Incentives Any factors that provide the motives behind particular actions, explain the reasons for choosing between alternatives, or influence the way that humans behave, reason, or make decisions.

Independence The statistical property such that the occurrence of one event makes it neither more nor less probable that one or other events will occur. Two events, A and B, are independent if the probability of event A conditioned on event B is equal to the probability of event A, and the probability of event B conditioned on event A is equal to the probability of event B.

Individual attributes The mix of skills and human traits that contribute to making each member of a workforce unique, including such factors as education, experience, attitudes, physical capabilities, preferences, wit, and intelligence.

Information Data that have been structured by creating relationships and connectivity between the various data points.

Information engineering The discipline that is concerned with the design, maintenance, operation, and improvement of an organization's information acquisition, storage, and retrieval systems.

Information retrieval (IR) A conventional information source that can include the searching, recovery and dissemination of information in documents, the documents themselves, metadata that describe documents, or information within databases. Search engines are the most visible and widely used forms of IR.

Information workflow management A conventional information source that provides structure and control of the movement of documents through a work process. The design of management software to adequately support information workflow is essential.

Infrared thermography A predictive maintenance (PdM) technology that is concerned with the use of an infrared imaging and measurement camera to see and measure thermal energy emitted from critical process parameters, such as the detection of underperforming heat exchangers; or dysfunctional pumps, gearboxes, bearings, or motors.

Inputs All the resources acquired from the environment necessary for the effective operation of all primary and support value streams within the business system. These include personnel, information, energy, capital, and materials; and any legal, political, ethical, or bureaucratic constraints.

Inspection The act of comparing the measurement of a product feature to the stipulated requirement and deciding whether the product conforms or fails to conform to those requirements.

Inspector consistency A measure of how closely two or more inspectors will agree when they independently inspect the same item. Inspector inconsistency in variables measurement is a component of the reproducibility error and inspector inconsistency in attributes measurement is the major source of measurement error.

Integrated product and process design (IPPD) A systematic approach to the integrated concurrent design of products and their related processes, including manufacture and support; the integrated approach forces designers to consider all elements of the product life cycle from conception through disposal, including quality, cost, schedule, and user requirements. This approach is also called simultaneous manufacturing, integrated product development, and concurrent engineering.

Integrated product development (IPD) A systematic approach to the integrated concurrent design of products and their related processes, including manufacture and support. The integrated approach forces designers to consider all elements of the product life cycle from conception through disposal, including quality, cost, schedule, and user requirements. This approach is also called simultaneous manufacturing, integrated product and process design, and concurrent engineering.

Integration The process of forming, coordinating, or blending elements, concepts, components, or people into a cohesive and unified whole.

Intelligent agents An artificial intelligence-based source in which robots or logical models embedded within real- time software are capable of exhibiting some form of intelligence, such as learning, adaptation, and reasoning, that will decide if, when, and what actions are appropriate. Its intelligence comes from the manner in which the agent interacts with its environment, which simulates how a human being would behave in a similar situation.

Intelligent process control (IPC) An information technology system that has the capability of continually learning how process behavior is affected by critical process parameters and automatically adjusting those parameters to achieve optimal performance.

Interaction effect In an experimental design the average change in the response variable observed from deliberate simultaneous changes in two or more of the factors.

Internal activities The term that represents all the things that cannot be carried out while a line is up and running (e.g., changing out dies). A machine must be down and therefore not producing product while internal activities are being conducted; moving activities from an internal to an external classification is key to changeover time reduction.

Internal communications The internal system within an organization that is used for the exchange of information and transfer of knowledge. May be oral or written, face-to-face or virtual, one-on-one or in a small group; constitutes a downward, upward, and horizontal means of addressing organizational concerns.

Internet A popular medium for obtaining survey data in which respondents are typically asked to either fill out a questionnaire and return by e-mail or access a Web site and complete the questionnaire online.

Interrelationship digraph A tool that aids in the discovery of cause-and-effect relationships and in segregating major drivers from minor ones. Also called a relations diagram, the interrelationship digraph is one of the seven new tools of quality management.

Interval scale A scale of measurement where the distance between any two adjacent units of measurement (intervals) is the same but the zero point is arbitrary; scores on an interval scale can be added and subtracted but cannot be meaningfully multiplied or divided. In quality management some variables data will be measured on an interval scale.

Intranets A conventional information source that consists of private versions of the Internet that comprise internal computer networks that utilize Internet protocols, network connectivity, and may also employ the public telecommunication system to share information with employees in a secure environment.

Intrinsic motivating potential The inbuilt ability that a job's design has to excite and drive positive behaviors and attitudes of those who perform them.

Intuition The power or faculty of attaining direct knowledge or cognition without evident rational thought and inference; a decision skill that includes the ability to recognize patterns without being able to explain how.

Inventory A form of waste that results from overproduction, incurring costs related to handling, space, interest charges, people, and paperwork. Excessive inventories also cause other sources of waste such inefficiencies due to cluttered workspaces and covering a myriad of problems that would otherwise be recognized and solved. Inventory is one of the seven deadly wastes of lean production; in theory of constraints (TOC) inventory is defined as what is spent on the materials and other items the organization intends to convert to throughput.

Ishikawa diagram One of the seven old tools of statistical process control, the Ishikawa diagram (which is named for its founder) provides a structure for exploring all the potential or real causes (or inputs) that result in a single effect (or output). Potential causes are typically arranged using four major categories—manpower, materials, methods, and machinery are common for manufacturing applications, and equipment, policies, procedures and people are common for service applications. The Ishikawa diagram is also known as the fishbone diagram because of its resemblance to the skeleton of a fish and the cause-and-effect diagram.

Jidoka The Japanese word for autonomation, which roughly translated means quality at the source, is a policy that requires operators to stop a production line anytime they see a problem, and track down and eliminate the root causes.

Job enlargement The management practice of increasing the scope of a job by extending the range of skills and tasks; the opposite of a work simplification strategy; a horizontal expansion of a job.

Job enrichment The management practice of increasing the intellectual capacity of a job by providing autonomy, decision-making, and problem-solving responsibilities to employees; the vertical expansion of a job.

Job rotation The management practice that moves employees around assigning them to different tasks in order to gain insights into the processes of a company and to increase job satisfaction through job variation.

Job share The term used to represent one job position that is performed by (co-shared) two people.

Job shop A process design arrangement where general purpose equipment is located and can be used to produce customized orders; the job shop design is well suited where small volumes and a very high variety of parts must be accommodated.

Just-in-time (JIT) A philosophy that supports the lean production goals of lower costs and improved response by requiring that materials are ordered only when and in the exact quantities needed; the implementation of JIT in both internal and external material control has resulted in substantial reductions in raw materials inventories and freed up valuable floor space that could be reassigned to value-adding operations.

Kaikaku The Japanese term that refers to a sudden overpowering event that aims to tear down some process (or system, product, service, etc.) and put it back together in an improved way; aimed at achieving breakthrough levels that offer order-of-magnitude improvements; also known as kaizen blitz.

Kaizen The Japanese word for continuous improvement and when deployed as a lean production tool, kaizen projects focus on taking a method apart and rebuilding it in an improved way, then standardizing on the better method.

Kaizen blitz A sudden overpowering event that aims to tear down some process (or system, product, service, etc.) and put it back together in an improved way; aimed at achieving breakthrough levels that offer order-of-magnitude improvements; also known by the Japanese term kaikaku.

Kanban A signaling system used to trigger upstream order releases in a pull scheduling environment. Typically cards are used as the signaling medium—however, other devices such as plastic markers, balls, or empty containers have been used.

KJ method A tool used to organize ideas into categories based on the perception of natural relationships or common themes; often used in brainstorming to help a group generate a number of ideas and to reduce the list to a smaller number of manageable categories. The KJ method, sometimes called the affinity diagram, is one of the seven new tools of quality management.

Knowledge Information that has been put into a meaningful context together with a practicable interpretation that can be used to guide decisions and actions.

Knowledge base An organization's collective wisdom concerning how processes behave and why they behave the way they do, including the complex interrelationships that exist between technologies and work design.

Knowledge-based system An artificial intelligence–based information source in which a computer-based decision model utilizes the collective knowledge and analytical skills of human subject matter experts to guide the decision process; also known as knowledge-based systems; also known as an expert system.

Knowledge creation The process resulting from combining prior knowledge, conducting new contextual analyses, applying different analytical tools, looking at data from new perspectives, or socialization—the application of communication and teaming skills to learn from others.

Knowledge mapping A method that is used to break a broad issue down into increasing levels of detail to gain a better understanding of existing knowledge about the issue.

Knowledge sharing The internal transfer of ideas, learnings, and expertise between individuals, within groups, and across departments or other subsystem interfaces.

Knowledge utilization The degree to which knowledge is transferred within an organization and the individuals within the working environment apply new knowledge to improve performance.

Knowledge workers The term given to workers in the information age who are required to use factual information to design, manage, or deliver goods and services.

Latent pathogens Inherent flaws in the design of a system that make it impossible to avoid errors in decision-making.

Leadership To direct the operations, activity, or performance of others, especially having the ability to inspire, counsel, and guide in such a way that those who follow perform at their best potential.

Leading questions Survey questions that do not legitimize all possible responses.

Lead time The period of time that elapses between the receipt of a customer's order and delivery of the final product; also called delivery lead time.

Lean consumption The term used to refer to the application of lean production principles to the process of acquiring and consuming products and services.

Lean maintenance The application of lean principles to minimize waste in the maintenance function and maximize uptime of production facilities.

Lean manufacturing A manufacturing philosophy that is capable of producing more with less—less waste, less human effort, less manufacturing space, less investment in tools, and less engineering time to develop a new product; a generic term that is applied to tools and practices that have their roots in the Toyota Production System (TPS). Lean manufacturing is also called lean production, particularly when applied to non-manufacturing industries.

Lean production A production philosophy that is capable of achieving greater value with fewer resources and has as its main tenet the elimination of waste; a generic term that is applied to tools and practices that have their roots in the Toyota Production System (TPS). Lean production is also called lean manufacturing when specifically applied to manufacturing industries.

Lean service The term used to refer to lean production implementations in the service sector; includes insurance, healthcare, government, and other administrative applications.

Lean Sigma A business improvement philosophy that combines tools from Lean Manufacturing and Six Sigma that results in the simultaneous improvement in quality and response; also called Lean Six Sigma.

Lean Six Sigma (LSS) A business improvement philosophy that combines tools from Lean Manufacturing and Six Sigma that results in the simultaneous improvement in quality and response; also called Lean Sigma.

Learning The act of gaining knowledge or understanding of or skill in by study, instruction, or experience. The term can also be used to refer to the knowledge acquired.

Least material condition (LMC) The condition of a part feature or size wherein the feature contains the least (minimum) amount of material permitted by the specified tolerances; for example, the largest hole size and the smallest shaft size.

Leverage a decision skill that includes the ability to exploit a situation, capitalize on its strengths, turn problems into opportunities, and spot weaknesses.

Line balancing The process of allocating tasks to sequentially linked workstations so that the capacities of all workstations are equalized and efficiency losses along the line are minimized.

Linearity Applied to a measurement process, refers to the ability to hold its calibration across a range of measured values.

Line flows Process designs that suit high volume, low variety production with linear flow patterns.

Little's law A principle that states the long-term steady-state relationship between work in process (WIP), flow time (FT), and throughput (TP); specifically, $WIP = FT \times TP$.

Loaded questions Survey questions that use implicit endorsements or labels to deliberately influence respondent choice.

Lot tolerance percent defective The poorest quality level that a consumer is willing to accept in a single lot; also referred to as lower quality limit (LQL) or rejectable quality level (RQL).

Loyalty The act of a customer being unswerving in its allegiance to a service provider; earned when customers believe the provider will go the extra mile and provide the unexpected to satisfy their needs; a principal criterion for customer satisfaction.

Mail A popular medium for obtaining survey data whereby either the U.S. Postal Service or internal mail systems are used to distribute and collect survey instruments.

Main effect In an experimental design the average change in the response variable observed from deliberate changes in one of the factors.

Maintenance and reliability management (MRM) The system within an organization that includes the acquisition and deployment of all resources and mechanisms necessary to trigger the need for maintenance and to deliver a high quality of maintenance to all operating units.

Maintenance, repair, and operating (MRO) supplies All the supplies needed by an organizational system that are not considered to be direct material costs that can be allocated to a product; included are such items as repair parts, tools, janitorial supplies, restroom soap and paper products, electrical supplies, maintenance supplies (e.g., rags and lubricants), printer and copier paper, ink cartridges, and safety supplies (e.g., ear plugs, safety glasses, hair nets).

Maintainability In a maintenance management system the probability of successfully repairing and restoring a machine or process within a given time.

Management principles A comprehensive and fundamental set of laws, doctrines, or assumptions; or a code of conduct that is judiciously employed by those in leadership positions as the means to effectively accomplish goals and desired outcomes.

Management staff meetings A method used for facilitating internal communications between those holding management positions for the purpose of sharing insights and information and achieving consistency on quality perceptions and goals.

Manufacturing resources planning (MRPII) A computer software system that integrates a number of functional modules that are used for the effective planning and utilization of a manufacturer's resources; the principal objective is to integrate human skills, computer technologies, and data to better manage business operations.

Master production schedule (MPS) A time-phased plan of what a manufacturing organization expects to produce, in what quantities, and on what dates in the future; data on the MPS provides input to other planning tools that deal with materials control, capacity, and human resources.

Matrix data analysis A sophisticated statistical methodology, such as principal components analysis, that is useful in determining those product or service features that result in the most customer satisfaction, and those that can be eliminated with little or no customer dissatisfaction; one of the seven new tools of quality management.

Matrix diagram A tool that shows the relationship between two or more sets of factors; one of the seven new tools of quality management.

Maximum material condition (MMC) The condition of a part feature wherein the feature contains the maximum amount material within the stated limits; for example the minimum hole size or the maximum shaft diameter.

Mean The expected value of some random variable of interest; also called the population mean. For a particular data set, the mean is the arithmetic average—that is, the sum of all the observations divided by the number of observations.

Mean time between failures (MTBF) The mean (average) time between failures of a system, and is calculated as the sum of the average operating times without failure and the average repair times when a failure has occurred. The concept of MTBF is typically attributed to the "useful life" of the device and does not include premature failures or end of life when the repair is either not an option or cannot be economically justified.

Measure The second tactic in the Six Sigma DMAIC model during which the focus is on the design and implementation of a data collection strategy. The data collected provides a baseline that can be used to evaluate the merits of any future proposed changes.

Measured response In a designed experiment a variable that cannot be controlled but can be measured; typically the experimental design seeks to learn how the manipulation of variables that can be controlled affect the value of the measured response.

Measure tollgate A meeting that is held at the conclusion of the Measure step in the DMAIC Six Sigma® process; the purpose is to review progress; the meeting is led by the team leader (usually a black belt), and is attended by team members, the project sponsor, the project champion, a master black belt, and the process owner. The main focus is on the tools used, the logic employed, and the conclusions reached.

Measurement process The procedures, tools, and personnel used to collect the data that provides feedback on whether a primary or support value stream is performing as intended.

Measurements The calculation of the magnitude of some attributes of an object, such as its length or weight, relative to a unit of measurement. Measurements usually involve using a measuring instrument that is calibrated to some standard.

Mechanistic organizational structure One that is formal, nonparticipative, hierarchical, tightly controlled, and inflexible.

Median That number in a sample, population, or probability distribution that separates the upper half (50%) from the lower half (50%).

Memory recall The human ability to remember events that occurred in the past; survey questions should refrain from asking questions requiring memory recall, particularly for a time period in excess of six months prior.

Metacognition A decision skill that includes the ability to look beyond the obvious, size up a situation, and see things others have missed.

Meta-management A term used to describe the development and practice of an explicit theory of quality management that goes beyond definitions, sets of principles, rules, and so on, that draws from a coherent body of cross-disciplinary knowledge that treats quality as a strategic issue. Meta-management is similar to Quality Inspired Management.

Metaphors A decision skill that includes the ability to mentally create situations different from the decision scenario for the purpose of generating improved decision options.

Metrology The science, processes, and equipment that have to do with measurement.

Micrometer A device used for obtaining measurements of thicknesses, outer and inside diameters of shafts, and depths of slots with a precision equal to ten thousandths of an inch or one thousandth of a millimeter.

Mind reading A decision skill that includes the ability to listen and go beyond the written or spoken word, and to understand the intent behind such communications.

Mismatched response sets A set of responses on a survey question that fails to match the question; this causes confusion on the part of respondents and can bias the survey results.

Mizusumashi The Japanese term that refers to a position in lean production that is created to perform all of the tasks which do not recur each cycle, such as replenishing materials, removing filled containers, and bringing in or setting up empty cartons; the purpose of the water spider job is to maintain a smooth uninterrupted work flow; also known as the water spider.

Mode The most frequent value that a random variable assumes in a sample, population, or probability distribution.

Moment-of-truth analysis A procedure for quickly analyzing the type and quality of responses that result from instantaneous customer contacts.

Motion inefficiencies A form of waste that results from excessive motions that workers use to perform their required tasks; these include such things as searching for tools, reaching for tools and parts, and walking to tend machines and fetch materials. This is one of the seven deadly wastes of lean production.

Motivating potential score (MPS) A measure of the intrinsic motivating potential of work.

Motor current analysis A predictive maintenance (PdM) technology that is concerned with monitoring electric motors to detect faults such as broken or cracked bars; high resistance; faults to ground; and internal faults such as turn-to-turn, coil-to-coil, and phase balance.

Moving Range (*MR*) chart A control chart used to determine the stability (statistical control) of short-term process variability and monitor variability performance for processes where variables measurements are appropriate and rational subgroups cannot be formed. The *MR* chart captures the short-term variability that is present in the time period between two successive measurements from the process. The *MR* chart is used together with an *X* chart.

Muda The Japanese term for any activity outcome that is wasteful or unproductive and does not add value to the product or service being produced; the elimination of muda is the primary objective of lean production. Muda is one of three forms of waste identified in the Toyota Production Systems—the other two are mura and muri.

Multi-agent systems An artificial intelligence based information source that consists of several software agents (software that are capable of deciding when and if action is appropriate) that are collectively capable of reaching goals that are difficult to achieve by an individual agent.

Multidimensionality A form of survey question that requires respondents to simultaneously render judgments on multiple levels; such compound questions should be avoided.

Multimedia A conventional information source that refers to the computer-controlled integration of text, graphics, drawings, video, animation, audio, and any other media where every type of information can be represented, stored, transmitted, and processed digitally; passively watching a non-interactive movie is an example of a multimedia environment.

Multiobjective optimization (MOO) An approach to solving multi-criteria optimization problems that applies an algorithm that seeks to find Pareto optimal solutions to problems obtained by combining the different objectives in different 'reasonable' ways; the algorithm selects the satisficing solution which offers the best compromise between competing objectives.

Multiple sampling An attributes acceptance sampling plan that permits a specified number of samples to be inspected before a final decision can be reached; a decision to accept the lot, reject the lot, or take another sample is made as each subsequent sample is collected and inspected; the sampling continues until a clear decision to accept or reject can be reached, or until the pre-specified number of units have been inspected.

Mura The Japanese term for inconsistency or variation in a system, process, or activity; the identification and elimination of mura is a primary objective of Six Sigma. Mura is one of three forms of waste identified in the Toyota Production Systems—the other two are muda and muri.

Muri The Japanese term for overburdening or putting stress on a system, process, or activity by demanding unreasonable outcomes or timeframes; the elimination of muri can be achieved through good systems design and implementation. Muri is one of three forms of waste identified in the Toyota Production Systems – the other two are muda and mura.

Mutual exclusivity The condition in survey questionnaire design where only one response is permitted in multiple response questions.

Mystery Shoppers Persons employed who anonymously shop a service provider's facilities or its competitors to determine first-hand the quality of services delivered.

Naturalistic decision making Uses perception, intuition and judgment to make good rather than optimal decisions.

Natural language processing (NLP) A conventional information source that relies on the automated generation of normal sounding human languages from information held in computer databases; and conversely, the conversion of human language into formal representations that computer programs can process and manipulate.

Natural tolerance The range of output values that are produced by a process; for a process that is in statistical control the natural tolerance is considered to be equal to approximately six process standard deviations.

Neural networks An artificial intelligence-based information source in which a network of interconnecting artificial neurons are programmed to mimic the properties of biological neurons (i.e., the human brain); also referred to as artificial neural networks this tool is used to gain an understanding of biological neural networks or for solving artificial intelligence problems; also referred to as artificial neural networks (ANN).

Neurons The basic building blocks of an artificial neural network (ANN); neurons receive one or more inputs and sums them to produce outputs, called synapses that propagate to the next layer through nonlinear transfer functions that assign weights or exit the system as part or all of the system output; also called artificial neurons, semi-linear units, Nv neurons, binary neurons, or McCulloch-Pitts neurons.

Nominal group technique (NGT) A creative decision-making method for use among groups of many sizes, who want to make their decision quickly, as by a vote, but want everyone's opinions taken into account (as opposed to traditional voting, where only the largest group is considered).

Nominal scale The lowest order measurement scale used solely for the purpose of identifying the attributes of objects or classes of objects. Objects on a nominal scale can only be compared with respect to equivalence or non-equivalence—the measurement assigned implies no quantitative relationships or ordering among them. In quality management, attributes data is measured on a nominal scale.

Nonexhaustive response sets Choices provided on surveys that omit valid choices that respondents might wish to make.

Nonnormal data Data that does not follow a standard normal distribution.

***np* chart** The type of control used to track attributes data that represents the number of defectives observed in a sample; sample size must be constant.

Null hypothesis A problem statement that sets up a statistical analysis; the statement is presumed to be true until or unless statistical evidence in the form of a statistical hypothesis test indicates otherwise.

Occurrence Used as a measure in a failure mode and effects analysis (FMEA) to reflect the likelihood that a particular cause of a failure mode will occur during the life of a product or process.

Oddleg caliper An instrument used to scribe a line on a work piece that is a set distance from the work piece edge; a bent leg tracks along the edge of the work piece while the sharp point of the scriber makes its mark, ensuring a scribed line that is parallel to the edge. Oddleg calipers are also called hermaphrodite calipers and oddleg jennys.

Oddleg jenny An instrument used to scribe a line on a work piece that is a set distance from the work piece edge; a bent leg tracks along the edge of the work piece while the sharp point of the scriber makes its mark, ensuring a scribed line that is parallel to the edge. Oddleg jennys are also called hermaphrodite calipers and oddleg calipers.

Oil analysis and tribology A predictive maintenance (PdM) technology that is concerned with the design, friction, wear, and lubrication of interacting surfaces in relative motion (as in bearings or gears) and with monitoring the quality of oil and other lubricants to detect deterioration in critical functional characteristics in lubricants, friction build-up, or excessive wear.

One-piece flow The term used to indicate the production condition where one work piece at a time is transferred smoothly and continuously between successive workstations along a value stream.

One-sided specification limit Engineering requirement that specifies either a maximum permissible value (the upper specification limit) or a minimum permissible value (the lower specification limit) for some product feature.

Open-ended question A type of question on a survey questionnaire that requires the respondent to write an answer in his/her own words.

Operating characteristic (OC) curve A plot of the probability of accepting a shipment, P_A, versus the fraction nonconforming, p.

Operating expense In Theory of Constraints (TOC) operating expense is defined as what an organization spends in order to convert inventory into throughput.

Operational definition A clear concise statement that can be understood by everyone the same way.

Opportunity flowchart A process flowchart that keeps separate those process steps that add value from those that add cost (but no value).

Ordinal scale A measurement scale that assigns values to objects based on their ranking with respect to one another; based on the ordinal measurement assigned, each object in a set of objects can be considered equal to, greater than, or less than every other object in the set.

Organic organizational structure One that is flat, has few rules, is team-oriented and participative, and can easily adapt to change.

Outcomes The effects, either immediate or ultimate, that outputs have on subsequent systems.

Output process chart A flowchart used to map a network of subsystems using a small set of standard and easily recognizable symbols; can produce insights into how the components interrelate to produce a finished product or service.

Outputs System or process deliverables that represent the cumulative effect of all transformations that occur up to the point where a unit exits the value stream or process.

Overall equipment effectiveness (OEE) score A measure of maintenance performance expressed as a percentage; the score is the product of availability, speed (production efficiency), and yield all expressed as percentages.

Overall plant effectiveness (OPE) score A measure of overall plant maintenance which is a computed aggregate score based on all OEE scores in the plant; the OPE is computed by taking a weighted average of the availability, efficiency, and yield computations respectively for each component/process/system and multiplying these weighted averages together.

Overproduction Producing goods or services for which there is no demand or which cannot be sold within a reasonable period of time; one of the seven deadly wastes in lean production.

Parallelism A condition of a surface, line, or axis which is equidistant at all points from a datum plane or axis.

Parameter Some characteristic of a population or sampling frame that conveys important information.

Pareto chart A bar chart that displays the frequencies of items in descending order of occurrence, serving as a graphical aid to differentiate the vital few from the trivial many. The Pareto chart is one of the seven old tools of statistical process control and is used to help decision-makers focus on the most important issues.

Part coding and classification analysis (PCA) A Group Technology method under which each part is assigned a unique code that conveys information concerning product features and/or process requirements.

Participatory empowerment A management strategy that gives employees partial decision-making authority while management maintains veto powers.

***p* chart** The type of control used to track attributes data that represents the proportion of defectives observed in a sample; sample size can be constant or variable.

People finder A conventional information source that consists of on line directories that can be accessed to locate people or businesses; usually organized by demographics, such as country, region, industry, etc.; also called yellow pages.

Perception The act of becoming aware or acquiring an understanding; in quality management the term is used to refer to what someone believes to be reality which in most cases will differ in varying degrees from the truth.

Perpendicularity A condition of a surface, axis, or line which is 90 degrees from a datum plane or axis.

Personal touch The ability of a service provider to convey to a customer that the provider genuinely cares about their needs and is willing to do what it takes to satisfy them; a principal criterion for customer satisfaction.

Plackett-Burman designs Efficient screening designs in design of experiments that are used when only main effects are of interest; effective for efficiently eliminating factors that are observed to have minimal effects on the response.

Plan-Do-Check-Act (PDCA) cycle An iterative four-step problem-solving process used for problem solving and quality improvement; also known as the Deming Cycle, the Shewhart Cycle, the Deming Wheel, and the Plan-Do-Study-Act Cycle.

Planning The psychological process of thinking about the activities required to create a desired future on some scale. Planning is the first step of the PDCA improvement cycle and is therefore the fundamental building block of quality.

Poka-yoke A term borrowed from the Japanese that is applied to the concept of mistake-proofing a product or process, by incorporating safe-guards in the design that make it impossible for humans to make mistakes. Poke-yoka techniques can drive defects out of products and processes and substantially improve quality and reliability.

Policy deployment A strategic planning methodology that uses Plan-Do-Check-Act (PDCA) cycle to create goals, choose control points (measurable milestones), and link daily control activities to company strategy.

Population The set of all measurements of a characteristic of interest from a universe.

Position A location tolerance that is relative to a specified datum (point or plane) stipulated on a mechanical drawing and can refer to a point, line, plane, or surface.

Precedence diagram A tool for describing activities in a project or operation that uses a network diagram with nodes to represent activities and arrows to show dependencies and prescribed sequences.

Precision In a measurement process the term used to define how much variation exists among multiple measurements are made of the same item; precision is the standard deviation of the distributions of measurements of an item around the mean measurement.

Predictable A process condition that is referred to as statistical control; the process output of a predictable process is stable over time and both future and past output can be confidently predicted. Consequently it is possible to evaluate how well the process is doing relative to customer expectations.

Prediction A statement or claim that a particular event will occur in the future, usually accompanied by a probability statement of the calculated risks involved.

Prediction interval A range of values that can be used to predict, within a stipulated level of confidence, the distribution of individual points; provides an estimate of individual values of some variable of interest.

Predictive maintenance (PdM) A maintenance policy that uses condition monitoring devices that are designed to detect and correct failures before they occur thereby preventing equipment breakdowns.

Pretesting The practice of conducting a small-scale pilot survey to validate survey length, identify ambiguities or other problems with questions; pre-testing is desirable to improve an instrument before implementing the full-scale survey.

Preventive maintenance (PM) A maintenance policy that avoids disruptions to production by scheduling maintenance in advance; its primary goal is to prevent failures before they occur by preserving and enhancing equipment reliability.

Primary value stream All the activities and their defined sequences that are required to add value to the mix of products and services that are sold and delivered to customers; work-in-process (WIP) inventories are defined in the context of a primary value stream.

Priority vector A vector of relative importance scores reflecting the subjective and personal preferences of decision makers who are following the analytic hierarchy process (AHP) to solve a multiple criteria problem.

Proactive maintenance (PaM) A maintenance policy that focuses on root causes rather than active symptoms, faults, or machine wear conditions, with the objective of extending equipment life.

Process capability A measure of the innate ability that a process has to meet requirements; a process is capable with respect to some quality feature if its natural tolerance (*NT*) for that feature is less than or equal to the engineering tolerance (*ET*).

Process capacity The maximum flow rate for any machine, process, or workstation.

Process decision program chart (PDPC) A tool used to display the many alternative paths that can occur during the execution of a plan, including contingencies, so that strategies for dealing with them can be developed in advance; one of the seven new tools of quality management.

Process failure mode and effects analysis (PFMEA) A failure mode and effects analysis (FMEA) that is concerned with identifying potential failure modes associated with a process that could impact product quality or reliability, reduce process reliability, cause customer dissatisfaction, create safety hazards or cause injury to workers, or cause harm to the environment.

Process flowchart A graphical representation of a process showing inputs, outputs, and all activities—how they relate, and how material or people flow through the system; also called a flowchart or a process map.

Process knowledge map A graphical representation of knowledge and its sources, mapped to a business process; knowledge that either drives the process or results from its operation is shown, including tacit knowledge (expertise, experience, and intuition), explicit knowledge (codified and documented in procedural manuals), and customer knowledge.

Process map A graphical representation of a process showing inputs, outputs, and all activities—how they relate, and how material or people flow through the system; also called a flowchart or process flowchart.

Process method A form or waste that occurs due to redundant, unnecessary, or extra processing steps, lack of proper maintenance, or the failure to employ the principles of design for manufacturability (DFM) in product designs; one of the seven deadly wastes in lean production.

Process-oriented layout A plant layout design that groups together in departments similar functions; this grouping can be by machine, task, operation, etc. Process-oriented layouts suit conditions where production runs and product variety are of moderate size. Process-oriented layouts are also called functional layouts.

Process owner The single person who is given the sole responsibility for the outputs of an entire integrated value stream; an essential concept in Six Sigma® organizations.

Producer's risk The probability that a shipment of good quality, defined to be the acceptable quality level (AQL) will be rejected; also the Type I or α error which is the probability of falsely rejecting a shipment that is good.

Product family Products that can be grouped together by manufacturing process or geometry to achieve a high level of integration between design and manufacturing functions; group technology methods can be used to form rational families; the basis for cellular manufacturing.

Product life cycle The succession of stages a product goes through from the time it is produced to the time it has outlived its usefulness and must be disposed; this includes production, transportation, storage, distribution, use, and disposal.

Product life-cycle costs The costs associated with producing, distributing, servicing, and disposing of a product over its useful life; life-cycle costs typically include all direct production costs, inspections, scrap, rework, marketing distribution costs, warranty claims, returns, product replacement costs, and the costs associated with an environmentally-appropriate disposal.

Product-oriented layout A plant layout design that arranges factors of production based on product groupings. Product-oriented layouts suit conditions where production runs are large and product variety is not great.

Production batch The production quantity that is scheduled to be completed in its entirety at a workstation; once completed the entire batch is transferred to the next workstation in the routing sequence.

Production cycle time The total time an order resides in the value stream; the difference between the time that an order reaches the first value-adding workstation and the time that the order is ready for shipment; also called throughput time, flow time, and sojourn time.

Production flow analysis (PFA) A Group Technology method that groups products with similar production routings together into families.

Production loss The production that is lost during the run-up period during process changeovers due to the production of unacceptable output (scrap) and a reduction in process capacity while machines undergo operator interventions and fine-tuning.

Program evaluation and review technique (PERT) A management tool used to analyze the tasks involved in completing a project, especially the time needed to complete each task, to identify the minimum time needed to complete the total project, and to highlight and closely monitor the activities that are critical to completing the project in minimum time.

Project A unique collection of tasks undertaken in a defined sequence to create a desired product or service. Projects are usually temporary and will not likely be repeated.

Projected tolerance zone A tolerance zone that extends beyond a feature by a specified distance. Projected tolerance zones help ensure that mating parts fit during assembly.

Psychological testing A tool used in many human resource management systems to try and match personality types to job responsibilities.

Pull scheduling A system of production scheduling that pulls product through the value-stream in a continuous flow rather than pushing it through in batches; generally synonymous with Just-In-Time (JIT) inventory management, pull scheduling is a fundamental tool used in lean production.

Purpose The reason why any organization exists, usually stated in terms of the products or services offered and the customer group or sector served.

Push scheduling A system of production scheduling that produces orders in batches (which are typically the size of the entire order); these batch quantities are started into production at the first value-adding activity shown on the routing sheet and only when the entire batch is completed at that workstation, is the job "pushed" to the next activity specified in the routing.

Work begins at the first activity shown on the routing sheet and is then "pushed" from workstation to workstation until all activities are complete. Under push scheduling the common procedure is to release jobs to individual departments.

Push information technology An artificial intelligence based information content delivery system whereby information from a central server is "pushed" across the Internet to the desktop of a user's computer screen.

Quality assurance (QA) A term used to describe programs with activities that aim to provide the evidence needed to document that quality procedures are being appropriately followed on the job and that processes are appropriately designed to meet quality requirements.

Quality at the source The term used to indicate that quality is the responsibility of every employee, work group, department, or supplier; the philosophy that quality must be built rather than inspected into a product or service.

Quality circles Teams composed of workers who meet regularly to discuss workplace improvements and make presentations to management concerning ideas on how to increase productivity, reduce waste, improve quality, or address any other issues or problems.

Quality control The term used to describe systems that are dedicated to ensuring that products or services are designed, produced, and delivered in a manner that meets or exceeds customers' expectations.

Quality function deployment (QFD) An integrated product and process development (IPPD) strategic planning tool that helps an organization to transform customer needs, as defined by the voice of the customer (VOC), into engineering characteristics (and appropriate test methods) for a product or service, prioritizing each product or service characteristic while simultaneously setting development targets for product or service.

Quality inspired management (QIM) A term used to describe a management paradigm that is holistically committed to the practice of an explicit theory of quality management that goes beyond definitions, sets of principles, rules, and so on, and that draws from a coherent body of cross-disciplinary knowledge that treats quality as a strategic issue committed to business excellence and the needs of stakeholders. Similar to meta management.

Radar chart A tool that graphically displays the gaps and their relative magnitudes among five to ten performance areas; also called a spider chart.

Radius gauge A device used to check compliance with requirements for specified radii. Radius gauges are available for checking both concave and convex profiles.

Randomized order An important discipline in conducting a design of experiments; the randomization of the order of runs, particularly where replicates are part of the experiment, help to eliminate bias due to time or sequence.

Randomness A lack of order, purpose, cause, or predictability.

Random sample A group of objects selected entirely by chance so that each member of the population has the same probability of being chosen; in practice a stratified method of sampling may need to be employed to ensure uniform representation.

Random sampling The procedure of selecting objects from a production process of a sufficient size and in such a manner that the raw data collected can be statistically analyzed to satisfy pre-specified information needs about the process.

Range The difference between the highest and lowest value in a sample of data.

Range (*R*) chart A control chart used to determine the stability (statistical control) of short-term process variability and monitor variability performance for processes where variables measurements are appropriate and rational subgroups can be formed. The *R* Chart captures the variability of all sources that are active within a subgroup and are best applied when the subgroup size (n) is less than or equal to 7.

Rank listed questions Survey questions that require respondents to rank list (e.g., provide a score on a scale from 1 to 10) some response.

Ratio scale The highest scale of measurement in which numeric measurements assigned to objects in a set can be used to determine equality and ordinality among all the objects in the set; additionally the intervals between the measurements and the ratios of the measurements have meaning as all measurements are generated relative to a fixed "zero" point. In quality management most of the variables data collected will be measured using a ratio scale.

Rational analysis A decision skill that includes the ability to apply the scientific method to develop workable decision options.

Raw stocks Materials held in inventory in an unaltered state; the materials inputs to the business system. Raw stocks can be in the form of a natural state as extracted or harvested from the earth, or can be semi-finished materials in the form of components or parts that have been outsourced to other systems for pre-processing; also called raw materials.

Raw materials Materials held in inventory in an unaltered state; the materials inputs to the business system. Raw stocks can be can be in the form of a natural state as extracted or harvested from the earth, or can be semi-finished materials in the form of components or parts that have been outsourced to other systems for pre-processing; also called raw stocks.

Rectifying inspection An attributes acceptance sampling strategy under which a lot that is not accepted is subjected to a 100% screening inspection operation to separate the conforming materials from the nonconforming. Conforming material can then proceed to production while the nonconforming material is returned to the supplier to be replaced with good product. Under this strategy, even when a shipment is accepted any nonconforming units found in the sample are also replaced by the supplier with conforming product.

Relations diagram A tool that aids in the discovery of cause-effect relationships and in segregating major drivers from minor ones; also called an interrelationship digraph the relations diagram is one of the seven new tools of quality management.

Regardless of feature size (RFS) The condition where the tolerance of form, runout, or location must be met irrespective of where the feature lies within its size tolerance.

Regression A statistical technique that examines the relationship between a dependent variable (the response) and any number of independent (explanatory or process) variables.

Reliability The term that is applied in maintenance management systems to denote the probability that a piece of equipment or process will perform its intended function over its expected operation time or life; unreliability occurs when process fail; also defined as one of the five areas surveyed using the SERVQUAL instrument as the ability to deliver what was promised in a dependable manner and without errors.

Reliability-centered maintenance (RCM) A maintenance strategy that combines into an integrated comprehensive asset management system the benefits of reactive and proactive approaches, utilizing the features of run-to-failure (RTF), preventive maintenance (PM), predictive maintenance (PdM), and proactive maintenance (PaM).

Remainder cell The production cell, in a cellular manufacturing plant, that is designed to operate as a traditional job shop and perform all tasks that the Group Technology methods could not assign to any of the normal cells.

Repeatability The measurement error that results from one person not being able to achieve identical measures when using the same measuring device and taking numerous measurements of the same item.

Repeatable A process condition that is referred to as statistical control; the process output of a repeatable process is stable over time and both future and past output can be confidently predicted. Consequently it is possible to evaluate how well the process is doing relative to customer expectations.

Replicate A repeat run at a given treatment combination in a design of experiments; replicates make it possible to dampen out random variation by averaging observed effects, and to estimate the within treatment variances.

Reproducibility The measurement error that results from different people not being able to achieve identical measures when measuring the same item or one person measuring the same item using different devices.

Residual The difference between an observed value of some variable of interest and the value predicted by a model.

Resolution In a measurement process refers to the smallest quantity unit that can be measured using a particular measuring device; if the resolution is low, discrimination between units of output is likely to be poor; in a design of experiments the degree to which the estimates of main effects are aliased (confounded) with two-level, three-level, and higher interactions.

Responsiveness The willingness to help customers, be flexible to their unique requirements, and to deliver prompt service; one of the five areas surveyed using the SERVQUAL instrument.

Revenue The term that refers to the income that an organization receives from its normal business activities, usually from the sale of goods or services to customers.

Risk priority number (RPN) A metric used in a failure mode and effects analysis (FMEA) that represents the relative risk of each failure mode, computed as the product of three factors: severity, occurrence and detection.

Roadrunner mentality A term in a Theory of Constraints (TOC) environment that refers to a production line ethic that dictates making extraordinary efforts to produce when there is work to do and use the time when there is no work to relax and recover.

Rolled throughput yield (RTY) A metric that represents the probability that a part can pass through a series of processes defect free, and takes rework into consideration.

Roundness A condition on a surface of revolution (cone, sphere, cylinder) where all points of the surface intersected by any plane perpendicular to a common axis (in the case of a cone or cylinder) or passing through a common center (in the case of a sphere) are equidistant from the center.

Run A sequence of consecutive points that lie on the same side of the centerline on a control chart.

Run chart A graph that displays observed process data in time-ordered sequence.

Runout tolerances How far a surface of a part feature is permitted to deviate from the form specified on the mechanical drawing during one full rotation of the part on a datum axis; the two types of run-out tolerances are circular and total.

Run-to-failure (RTF) A maintenance policy that requires equipment to run with little or no maintenance surveillance, repairing or replacing it only when it can no longer perform its intended function.

Runup The time period early in a new production run during which the operators make adjustments until the process begins to produce a sustained stream of output that is acceptable quality.

Satisficing Making a choice that suffices to fulfill the minimum requirements to achieve an objective, without special regard for utility maximization or optimization of one's preferences.

Scatter diagram A graphical plot of a pair of variables that can be used to graphically observe any relationships that exist between the two; this is one of the seven old tools used in statistical control applications and is also called a scatterplot.

Scatterplot A graphical plot of a pair of variables that can be used to graphically observe any relationships that exist between the two; this is one of the seven old tools used in statistical control applications and is also called a scatter diagram.

Scientific method A structured method of scientific inquiry under which researchers propose hypotheses as explanations of phenomena, and design experimental studies to test these hypotheses.

Scrap, rework, and defects A form of waste caused by failing to deliver acceptable quality on the first pass through a process; this waste is not restricted to those items rejected or the cost of rework but causes other sources of waste including increase wait time for downstream operations and the need for additional capacity to accommodate the scrap yield losses. This is one of the seven deadly wastes of lean production.

Screening designs In a design of experiments a specific type of fractional factorial design, usually resolution III, that minimizes the number of runs required to eliminate factors that have minimal influence on the response variable.

Self-check inspection An inspection procedure that requires an individual worker to check his/her own performance.

Self-directed work team (SDWT) A self-managed work team that has the authority to shape its own future by making decisions on compensation, discipline, hiring and firing, and supplier certifications.

Self-managed work team (SMWT) An autonomous work group that can, at its discretion, develop and execute its own methods to achieve common goals.

Self-management A form of workplace decision-making in which the employees themselves agree on choices (for issues like customer care, general production methods, scheduling, division of labor etc.), instead of the traditional authoritative supervisor telling workers what to do, how to do it and where to do it.

Self-management empowerment A management strategy that gives front-line employees considerable autonomy through self-managed work teams to follow through on their recommendations up to and including implementing changes.

Sequential sampling A type of attributes acceptance sampling plan that permits a decision to accept, reject, or continue sampling after each unit inspected; as the total number of units sampled increases, the criteria for accepting and rejecting are changed such that, theoretically, the sampling could continue until all units in the shipment have been inspected.

Service quality gap The difference between the quality of a service that is actually delivered and the quality of service that was expected or required.

SERVQUAL A structured survey methodology for measuring service quality and relies on a customer's ability to qualitatively "score" performance in the areas of assurance, empathy, reliability, responsiveness, and tangibles.

Setup The activities required to convert a machine or process from the production of one product to the production of another; can include re-tooling, adjustments, material transfer mechanisms, etc.; setups can be minor if two consecutive products are similar or major if consecutive products are dissimilar.

Seven deadly wastes The categories of waste that Toyota identified as having a direct impact on profitability; these are overproduction, wait time, transportation, processing methods, inventories, motion inefficiencies, and defects.

Severity Used as a measure in a failure mode and effects analysis (FMEA) to reflect how serious the effect of a failure mode is on the customer.

Shape The profile of the entire set of measurements for a population or process (e.g., unimodal, symmetrical, continuous, etc.) determined by viewing a histogram plot of sample data.

Shared ownership The term used in high involvement organizations (HIO) to represent a participative style of management under which employees are empowered and are permitted to play an expanded role in planning and operating decisions and given more discretion in how they perform their assigned tasks.

Shorthand notation A convention used in which a series of lowercase letters are used to define the particular treatments in an experimental design; if a lowercase letter is present in a description the corresponding factor is set in the design at its high level for that particular treatment combination—if absent the corresponding factor is run at its low level. The symbol "(1)" is used to depict a run where all factors are set at their low levels.

Signal-to-noise ratio (SNR) A measure of the relative size of a measurement error and is computed by taking the ratio of the observed variation (as estimated by the standard deviation of process measures) and the measurement variation (as estimated by the standard deviation from measurement studies).

Simultaneous manufacturing (SM) A systematic approach to the integrated concurrent design of products and their related processes, including manufacture and support; the integrated approach forces designers to consider all elements of the product life cycle from conception through disposal, including quality, cost, schedule, and user requirements. This approach is also called integrated product and process design, integrated product development, and concurrent engineering.

Single sampling An attribute acceptance sampling plan that calls for a single sample of size n to be selected from a shipment and inspected. If the number of nonconforming units in the sample, d, is less than or equal to an acceptance number, c, the lot is accepted; if $d > c$, the lot is rejected.

Six honest servants Called the six honest serving men by Rudyard Kipling, these are *what, why, when, how, where*, and *who*. Together the six honest servants represent a questioning sequence that when applied to processes can lead to understanding and improvement.

Six Sigma (SS) A set of practices originally developed by Motorola that has widespread application in the systematic improvement of industrial processes by identifying and removing sources of variation (mura).

Six Sigma score A score that is derived by making a 1.5 standard deviation adjustment in calculating process capability; this adjustment accounts for the likelihood of a mean shift over time even in the case of a stable system.

Skill variety One of the five factors used in computing the motivating potential score (MPS) of a job and represents the degree to which a job requires different skills, abilities, or talents; the other factors are task identity, task significance, autonomy, and feedback.

Small hole gauge An instrument used to measure the diameter of small holes; the gauge is set to a size smaller than the bore, inserted, and adjusted by rotating a knurled knob until light pressure is felt against the walls of the bore. The gauge is then withdrawn and the measurement transferred to a remote measuring tool such as a micrometer or vernier caliper.

Socially favored positions Values or beliefs that are generally accepted as true, such as efficiency, honor, and integrity. Survey questions should buffer questions that deal with socially favored positions with some appropriate lead-in phrasing.

Sojourn time The total time an order resides in the value stream; the difference between the time that an order reaches the first value-adding workstation and the time that the order is ready for shipment; also called throughput time, production cycle time, and flow time.

Source inspection An inspection procedure under which individual workers are responsible for checking the quality of their work and if defects are found, to identify and correct the root causes.

Special cause variability Sporadic and non-random fluctuations in process data due to assignable causes; present in some samples, absent in others. The purpose of a control chart is to expose the presence of special cause variation.

Spider chart A tool that graphically displays the gaps and their relative magnitudes among five to ten performance areas; also called a radar chart.

Stacking tolerances The practice of building a composite tolerance (for example, a clearance) by adding standard deviations to obtain the predicted resultant part variation.

Stable A process condition that is referred to as statistical control; the process output of a stable process can be predicted forward and backward in time, and it is possible to evaluate how well the process is doing relative to customer expectations.

Standard deviation A measure of the spread of the values of a probability distribution, population, or sample; in the case of a population parameter it is usually denoted with the lower case Greek letter σ (sigma) and for samples it is depicted by a lower case English letter s; it is defined as the square root of the variance and, unlike the variance, it measures the spread of the data about the mean in the same units as the data; also called variability.

Standard deviation (*s*) chart A control chart used to determine the stability (statistical control) of short-term process variability and to monitor variability performance for processes where variables measurements are appropriate and rational subgroups can be formed. The *s* Chart captures the variability of all sources that are active within a subgroup and is preferred when the subgroup size (n) is greater than or equal to 8.

Stakeholder A party with an interest in or that is affected by an organization's operations, and in addition has the power to influence the organization's success or failure.

Statistic A numerical quantity calculated from a sample that is used to estimate the parameters of a distribution; the term is used for the function used to calculate the numerical value and for the value itself, which is computed from a sample outcome and is independent of the sample's distribution.

Statistical control A process state in which all active sources of variability are due to common causes—i.e., all sources of variability have been captured within a sample and each sample is statistically like all other samples; a process in statistical control is stable, repeatable, and predictable. Process parameters such as mean, dispersion, and shape can be estimated with confidence.

Statistical process control (SPC) The use of statistical tools to monitor, adjust, and improve industrial processes. The aim of SPC is to gain an understanding of how processes behave, why they behave the way they do, and root causes of variation and poor quality.

Statistical tolerancing The use of statistical theory to predict distributions of composite parts or assemblies where two or more components are mated in an assembly; statistical tolerancing is based on the premise that it is highly unlikely with random parts selection that parts at maximum or least material conditions will be mated.

Stockless production A variation of the Toyota Production System (TPS) that uses customer demand to pull work through a manufacturing facility, only producing work as required; this philosophy substantially reduces work-in-process inventory levels.

Stock options Incentive programs that provide employees with the right to purchase company stock at a given price during a certain period of time in the future.

Storyboard displays A method used to facilitate internal communications whereby display boards are posted in the workplace for the purpose of posting important written communications (e.g., memos, letters, reports, etc.), highlighting projects (including objectives, methodology, and progress), and displaying written customer feedback (e.g., letters of appreciation or dissatisfaction, results of customer satisfaction surveys, etc.)

Storytelling A decision skill that includes the ability to return, and vicariously take others, to a problem situation in order to analyze a decision that was made and its subsequent effects.

Straightness A condition where an element of a surface or axis is a straight line.

Stratification One of the seven old tools used in statistical control applications, stratification is a technique that sorts data into meaningful classifications so that clues as to the what, where, when, who, how, and why relative to an issue at hand can easily be found.

Strengths-weaknesses-opportunities-threats (SWOT) analysis An important step that is deployed in strategic planning to help an organization establish its purpose by evaluating its strengths and weaknesses and matching those to the opportunities and threats in the markets served.

Subordinate The third step of the theory of constraints (TOC) cycle during which an organization makes all non-constraint resources subordinate to the constraint; in a production environment this may require making significant changes to current (and generally long-established) ways of doing things at the non-constraint resources.

Successive check inspection An inspection that is performed by the next person in a value stream or by an objective evaluator such as a supervisor or group leader; if a defect is detected feedback is immediately provided to the responsible operator so that the error can be corrected and appropriate action can be taken to prevent recurrence.

Supply chain An extended enterprise that expands the primary value stream beyond the organization's system boundaries to include external supplier systems upstream and the customers' supply and distribution points downstream.

Support value stream A value stream that includes all support function and service activities required by the workstations along the primary value stream.

Surveys The most widely used instrument to gather data concerning customer opinions; can be either self-administered or administered by professional interviewers.

SWOT analysis A planning tool used to evaluate the strengths, weaknesses, opportunities, and threats involved in assessing a proposed business venture or developing corporate strategy.

Symmetry The condition in which a feature or features is symmetrically disposed about the centerline of a datum.

Synchronous manufacturing The term given to the hybrid process of pull/push scheduling that is characteristic of the Theory of Constraints (TOC) in which the drum (constraint) activity pulls its schedule from customer demand and the gate (first) activity pulls its schedule from the drum; the gate then pushes material toward the drum and the drum pushes material toward the customer.

System A set of interacting or interdependent entities or components, real or abstract, forming an integrated whole. The entities work together synergistically to achieve a set of outcomes that would not be possible through independent actions of the system components.

Systematic diagram A tree diagram that displays a hierarchy of objectives, which branches out to a hierarchy of ways and means for accomplishing those objectives; also called a dendrogram and providing more focus than an affinity or relations diagram, this is one of the seven new tools of quality management.

System boundaries The partitions that form a separation between what belongs within a business system from those things that are external to the business system.

Taguchi loss function A concept that establishes a cost associated with customer dissatisfaction relative to how much a product's performance deviates from a target value; this concept clearly establishes the importance of focusing on both average and variation in quality improvement initiatives.

Takt time The maximum time allowed to produce a product in order to meet demand; takt time is computed by taking the ratio of the time available for production to the number of units of output that must be produced in that time.

Tangibles The physical facilities, equipment, and overall appearance of service providers; one of the five areas surveyed using the SERVQUAL instrument.

Target costing (TC) A technique used in Design to Cost (DTC) strategies in which a product is designed in such a way that it meets all function requirements without exceeding a predetermined target cost.

Targeted interviews The process by which a small random (or stratified) sample (representative of customers) is selected for face-to-face interviews to determine the level of customer satisfaction with a service delivery.

Task cycle A meeting pre-planning tool that can help keep the meeting on purpose and to achieve desired outcomes within the allotted time.

Task identity One of the five factors used in computing the motivating potential score (MPS) of a job and represents the extent to which a job requires completion of a whole and identifiable unit of output; the other factors are skill variety, task significance, autonomy, and feedback.

Task significance One of the five factors used in computing the motivating potential score (MPS) of a job and represents the degree to which a job impacts the lives of other people, the organization, and the environment; the other factors are skill variety, task identity, autonomy, and feedback.

Team A group of people working together to achieve a common set of goals.

Team charter The document generated in the Define step of the DMAIC cycle of Six Sigma that makes the business case for a project; the charter explains why the project has a high priority, its expected impacts on the organization's bottom line, and a list of specific objectives.

Team meetings A method used to facilitate internal communications whereby groups of employees meet for the purpose of making announcements, sharing business news, and soliciting anecdotal feedback on experiences with customers.

Team mind A group decision skill that includes the ability to collaborate with others where a group decision is required, and to act and think as one.

Technical communications The process of conveying useable information about a specific technological subject to an intended audience for the purpose of performing an action or making a decision based on its contents.

Technical documentation The term used to describe technical communications when the delivery medium is paper.

Technologies All the plant, equipment, software, procedures, and know-how required by all of the business system components to achieve an organization's goals and purpose.

Telecommuting A work arrangement under which employees enjoy flexibility in working location and hours whereby a daily commute to a central workplace is replaced by telemcommunication links established from home, coffee shops, internet cafes, or any other appropriate locations.

Telephone A popular medium for obtaining survey data in which personnel are hired to administer survey instruments to respondents either at their place of business or homes by telephone.

Telescopic bore gauge An instrument used to transfer the internal dimension of a hole to a remote measuring tool such as a micrometer or vernier caliper.

Theory of constraints (TOC) A management philosophy that emphasizes the achievement of key organizational goals by identifying and relentlessly trying to eliminate those constraints that are barriers to goals achievement; also called constraint management.

Thread gauge a specialized go/no go plug gauge designed for checking threaded parts.

Throughput (TP) The average rate of output from a production process, expressed in the number of units per unit time (e.g., number of units per hour); in Theory of Constraints (TOC), throughput is the rate at which money is generated through sales and is considered to be the value-added contribution of a production system.

Throughput rate A measure of the average rate at which units of product flow past a specified point in a process; also called flow rate.

Throughput time The total time an order resides in the value stream; the difference between the time that an order reaches the first value-adding workstation and the time that the order is ready for shipment; also called flow time, production cycle time, and sojourn time.

TILMAG A creative tool that uses brainstorming together with word associations to help groups identify solutions by analyzing paired combinations of ideal solution elements.

Time series A sequence of plotted data points, measured typically at successive times, commonly spaced at uniform time intervals.

Tolerance frame A boxed feature on a mechanical drawing that conveys geometric tolerancing information that includes a symbol depicting the particular geometric characteristic, the allowable tolerance with any appropriate modifiers, and any relevant datum points or planes.

Tolerances of form A geometrical tolerance that states how far an actual surface or feature is permitted to vary from the desired form as specified on a mechanical drawing; form tolerances include flatness, straightness, roundness, cylindricity, surface profile, and line profile.

Tolerances of location A tolerance that states how far a part feature may vary from the location specified in a mechanical drawing as related to datums or other features; location tolerances include position, concentricity, coaxiality, and symmetry.

Tolerances of orientation A geometric tolerance that states how far an actual surface or feature is permitted to vary from the desired position as specified on a mechanical drawing; orientation tolerances include parallelism, perpendicularity, angularity.

Total productive maintenance (TPM) One of the tools of lean production, TPM tasks machine operators to perform much, sometimes all, of the routine maintenance at their respective work stations; the idea is that since the work station is wholly the domain of one person or team, the maintenance which includes housekeeping duties will be appropriate, effective, and thorough.

Total quality management (TQM) A management philosophy under which everyone in an organization strives to achieve customer satisfaction through continuous incremental improvements, radical quantum improvements, teaming, and the organization-wide application of statistical process control.

Toyota production system (TPS) The philosophy which was used to organize the manufacturing and logistics functions at Toyota that has become the basis for the lean manufacturing philosophy.

Traceability A metrology property that establishes the degree to which measurements can be compared.

Transformation A function that converts a non-normal set into a set that is normally distributed; there are many transformations available—the most common ones used in quality management are the Box-Cox and Johnson transformations.

Transformation flow process chart A process flow chart that shows the actual steps, or operations required in performing some task or sequence of tasks; standard symbols are used to capture transformations (operations that add value), transportations (those steps that add place utility), and delays, inspections, and storages (all of which add cost but not value).

Transportation and materials handling A form of waste that can arise due to the transportation and double or triple handling of raw and finished goods and is caused by a poorly conceived layout of the factory floor and storage facilities, resulting in long distance transportation and over-handling of materials; one of the seven deadly wastes in lean production.

Treatment combinations In a factorial two-level experiment the combinations of factors set at their respective high and low levels.

Trust The assured reliance on the character, ability, strength, or truth of someone or something (usually a supplier); the belief that the provider of products or services will deliver on its promises; a principal criterion for customer satisfaction.

Truth A judgment, proposition, or idea that is true or accepted as true. In quality management what is accepted as fact can differ from the truth—for example due to measurement error, sampling error, or poor communications.

Two-sided specification limits A set of engineering requirements that specifies a maximum permissible value (the upper specification limit) and a minimum permissible value (the lower specification limit).

Type I error Sometimes called the Alpha error, in statistical analyses this is the probability of rejecting as false a hypothesis that is actually true.

Type II error Sometimes called the Beta error, in statistical analyses this is the probability of failing to reject (i.e., accepting as true) a hypothesis that is actually false.

***u* chart** The type of control used to track attributes data that represents the number of defects or errors observed per inspection unit; the inspection unit represents the opportunity for defects to occur and varies from sample to sample (i.e., different number of objects, different number of square meters of material, different number of lineal yards, etc.). The sample size is the number of inspection units in each sample and, unlike with other control chart techniques the sample size need not be integral.

Ultrasonics A predictive maintenance (PdM) technology that is concerned with the use of ultrasound to detect air and gas leaks, and also to monitor the operating condition of mechanical components such as bearings and gearboxes.

Unfamiliar terms and jargon Language that is specific to a particular discipline or technology and not commonly understood by lay people; such language should be avoided in survey instruments.

Universe The collection of all entities that possess some characteristic of interest.

Unnecessary complexity The use of complex phrasing, terminology, double negatives, and the like in the wording of questions on survey instruments; such unnecessary complexity should be avoided as it can overtax the respondent's memory burden.

Unnecessary delays A form of waste that results from waiting—for example, idle workers who have completed the required amount of work and are waiting for more, or employees who spend much time watching machines but are powerless to prevent problems; one of the seven deadly wastes of lean production.

Value analysis (VA) A systematic method to improve the value of goods and services by increasing the value; this is accomplished by either improving the function or reducing the cost. Value analysis is also called value engineering.

Value engineering (VE) A systematic method to improve the value of goods and services by increasing the value; this is accomplished by either improving the function or reducing the cost. Value engineering is also called value analysis.

Value stream A system's transformation process that includes all the tasks, skills, and knowledge required to convert a system's inputs into desirable outputs.

Variable sample size When random samples selected from a process are of differing sizes; this is rare in variables measurement and occurs mostly with attributes measurement (e.g., with 100% inspection).

Variables measurement A measurement process whereby data are collected using an interval or ratio scale. The data are typically continuous random variables.

Variability A measure of how dispersed the individual data points are around the measure of central tendency (average) for a process distribution; a term commonly used in place of standard deviation.

Variance A measure of the statistical dispersion of a random variable population, probability distribution, or sample, computed by averaging the squared distance of its possible values from the expected value.

Variance-tracking matrix A matrix tool that can help an organization analyze all activities in a value stream, the associated sources of variability, and how these sources affect downstream processes.

Vernier caliper An instrument used to measure internal or external dimensions using metric and imperial scales that can provide precision to one hundredth of a millimeter or one thousandth of an inch. Some models are equipped with a depth probe that is capable of accessing deep grooves that other measuring tools cannot reach.

Vibration measurement and analysis A predictive maintenance (PdM) technology that is concerned with the monitoring of critical rotating components to detect early signs of bearing wear to prevent seizure.

Vicious cycle The spiraling destructive behavior that can result from positive feedback.

Virtuous cycle The spiraling beneficial behavior that can result from positive feedback.

Visual Displays and Controls Visual signals used to facilitate communications on the shop floor; to be effective must be instantly identifiable and understood by all observers; include information displays, layouts, material storage and handling tools, color-coding, and poka-yoke devices.

Voice of the customer (VOC) The term used to describe the process of capturing a customer's requirements; this is an essential component of the Quality Function Deployment methodology and forms the deviation-amplifying form of positive feedback in strategic planning.

Voice of the engineer (VOE) An essential component of the Quality Function Deployment, the term used to describe the process of translating customer requirements (the voice of the customer) into design and technical requirements.

Voice of the process The term used to describe the function of a control, a tool designed to provide process information, which is dependent on sampling strategies employed, including sample size determination, frequency, and specifically how individual sampling units are selected.

Voice of the stakeholder (VOS) The term used to describe the process of capturing feedback that can be used to determine the needs of all stakeholder groups. The notion of satisfying stakeholder needs, rather than restricting the feedback to customers, broadens the scope of the deviation-amplifying feedback component of strategic planning.

Waste Anything that does not add value to the customer or anything the customer is unwilling to pay for; the Japanese define waste (muda) as anything other than the minimum amount of equipment, materials, parts, space, and worker's time which are absolutely essential to add value to the product.

Water spider The term given to a position in a lean production environment that is created to perform all of the tasks which do not recur each cycle, such as replenishing materials, removing filled containers, and bringing in or setting up empty cartons; the purpose of the water spider job is to maintain a smooth uninterrupted work flow; also known by the Japanese word mizusumashi.

Weak grammatical format A survey question that is structured in such a way that respondents have difficulty linking the response alternatives to the content of the question; such awkward grammatical structures should be avoided in questionnaire designs.

Weighted mean A modification of the analytic hierarchy method (AHP) in which weights are assigned to each of the criteria in a multiple criteria decision model.

Wisdom Knowledge that has been transformed into institutional learning.

Work design The collective elements that prescribe the task and work sequence that must be followed to create a product or service.

Work-in-process (WIP) The term that applies to all unfinished jobs that reside up and down the primary value streams of a business system; WIP includes any job that has experienced some value-adding process but is not yet complete and cannot be shipped out in its current state to a customer.

World class manufacturing (WCM) A set of concepts, principles, policies, and techniques for managing and operating a manufacturing company; having its roots in the Toyota Production System (TPS) it primarily focuses on continual improvement in quality, cost, lead time, flexibility, and customer service.

***X* chart** A chart used to determine process stability (state of statistical control) and to monitor process average where variables measurements are appropriate and rational subgroups cannot be formed, and the process must be tracked using the individual (raw) data points. The *X* chart is used in conjunction with an *MR* chart. Care must be taken in evaluating the chart since the plotted data may not be normally distributed.

X, MR* charts** A pair of charts used to determine the state of control and monitor performance for processes where variables measurements are appropriate and rational subgroups cannot be formed. In this case individual measurements must be analyzed rather than the average of samples. The *MR* chart provides an estimate of the short-term variability and the ***X chart provides estimates of any long-term variability.

$\bar{X}$ Chart A chart used to determine process stability (state of statistical control) and to monitor process average where variables measurements are used and rational subgroups can be formed. The $\bar{X}$ Chart typically captures some of the variability within subgroups and any variability that is time related (occurring between subgroups), and to be properly interpreted should be used in conjunction with either an *R* Chart or *s* Chart.

$\bar{X}$, *R* charts A pair of charts used to determine the state of control and monitor performance for processes where variables measurements are appropriate and rational subgroups can be formed (typically less than *n* [=] 7). The *R* chart captures variability within subgroups and the $\bar{X}$ chart captures variability within and between subgroups.

$\bar{X}$, *s* charts A pair of charts used to determine the state of control and monitor performance for processes where variables measurements are appropriate and rational subgroups can be formed (any size *n* but especially n = 7). The *R* chart captures variability within subgroups and the $\bar{X}$ chart captures variability within and between subgroups.

Yellow pages On line directories that can be accessed to locate people or businesses; usually organized by demographics, such as country, region, industry, etc.; also called people finder.

Zero defects (ZD) Step 7 of Philip Crosby's 14-step quality improvement process, zero defects reflects the goal of meeting the genuine needs of customers every time.

INDEX

A

B

C

D

H

I

N

O

S

T